EVOLUTION OF FLIGHTLESSNESS IN RAILS (GRUIFORMES: RALLIDAE): PHYLOGENETIC, ECOMORPHOLOGICAL, AND ONTOGENETIC PERSPECTIVES

ORNITHOLOGICAL MONOGRAPHS

Edited by

DAVID A. WIEDENFELD
Sutton Avian Research Center
P.O. Box 2007
Bartlesville, OK 74005

Ornithological Monographs, published by the American Ornithologists' Union, has been established for major papers too long for inclusion in the Union's journal, *The Auk.* Publication has been made possible through the generosity of the late Mrs. Carll Tucker and the Marcia Brady Tucker Foundation, Inc.

Copies of *Ornithological Monographs* may be ordered from Buteo Books, 3130 Laurel Road, Shipman, VA 22971. Price of *Ornithological Monographs* 53: $10.00 ($9.00 for AOU members). Add $4.00 for handling and shipping charge in U.S., and $5.00 for all other countries. Make checks payable to Buteo Books.

Author of this issue, Bradley C. Livezey

Library of Congress Control Number 2003104538

Printed by Allen Press, Inc., Lawrence, Kansas 66044

Issued October 17, 2003

Ornithological Monographs, No. 53 x + 654 pp.

Copyright © by the American Ornithologists' Union, 2003

ISBN: 1-891276-35-2

Cover: Skull (left side, lateral perspective) of the flightless Takahe (*Porphyrio hochstetteri*; USNM 612797), a critically endangered swamphen of South Island, New Zealand. Halftone (pencil) illustration by author.

EVOLUTION OF FLIGHTLESSNESS IN RAILS (GRUIFORMES: RALLIDAE): PHYLOGENETIC, ECOMORPHOLOGICAL, AND ONTOGENETIC PERSPECTIVES

By:

BRADLEY C. LIVEZEY

Section of Birds
Carnegie Museum of Natural History
Pittsburgh, Pennsylvania 15213-4080 USA
E-mail: livezeyb@carnegiemuseums.org

ORNITHOLOGICAL MONOGRAPHS NO. 53
PUBLISHED BY
THE AMERICAN ORNITHOLOGISTS' UNION
WASHINGTON, D.C.
2003

DEDICATION

This work is dedicated to my family for fostering a respect for the natural world, to Philip Humphrey for stimulating my interest in avian flightlessness, to Richard Zusi for sharing his professional ethics and anatomical skills, to Jay Apt for a revitalization of the scientific enterprise at the Carnegie, and to Julie Cameron for years of loving support.

TABLE OF CONTENTS

ABSTRACT	1
INTRODUCTION	4
THE FAMILY RALLIDAE	5
Ecomorphological Diversity	5
Flightless Rails: Early Discovery and Description	8
Flightless Rails: Extinct, Endangered, and Secure	15
Flightless Rails of New Zealand: Discovery, Debate, and Destruction	23
Taxonomy and evolutionary inference	23
Hawkins' Rail: cranial homoplasy and biogeography	23
Two smaller rails from the Chathams: apparent versus genuine juvenility	25
Refugium sublimus: a tale of two takahes	29
Undeclared Pacific Campaigns: A Legacy of Tropical Devastation	30
Relevance of casualties	30
*Hawaiian Crake (*Porzana sandwichensis*)*	33
*Wake Island Rail (*Gallirallus wakensis*)*	33
*Laysan Crake (*Porzana palmeri*)*	34
MODALITIES AND LOSS OF AVIAN FLIGHT	35
Diversity of Avian Flight	35
Locomotory Modules of Birds	41
DIVERSE PERCEPTIONS OF AVIAN FLIGHTLESSNESS	42
Religious and Popular Views	42
Previous Scientific Studies of Flightless Carinates	44
OBJECTIVES OF THIS STUDY	46
MATERIALS AND METHODS	47
CLASSIFICATION AND INCLUDED TAXA	47
SPECIMENS AND RELATED DATA	53
Skin Specimens	53
Skeletal Specimens	54
Subfossil Elements	55
Fluid-Preserved Specimens	55
Miscellaneous Information	56
GENERAL DESCRIPTIVE CONVENTIONS	56
Definition of Flightlessness	56
Anatomical Nomenclature	57
DISSECTIONS OF PECTORAL MUSCULATURE	57
CLASSES OF MENSURAL DATA	58
Integument	58
Skeleton	58

Pectoral Musculature	59
NATURE OF MENSURAL DATA	59
Interspecific Comparability of Variables	59
Measurement (Pure) Error	60
PHYLOGENY-BASED COMPARISONS	60
UNIVARIATE AND BIVARIATE MORPHOMETRICS	62
Analysis of Variance	62
Estimation of Mean Body Masses, Lengths, and Areas of Wings	62
Analyses of Allometry	63
Methodological and theoretical essentials	63
Allometry of wing size with body mass	66
Other allometric relationships	66
Estimation of Missing Data	67
Ratios and Proportions	67
Statistical properties	67
Dimorphism ratios and surrogate metrics	67
Wing loadings and aspect ratios	71
Intra-appendicular skeletal proportions	72
NONPARAMETRIC STATISTICS	72
MULTIVARIATE MORPHOMETRICS	72
Analytical Commonalities and Approach	72
K-Means Cluster Analysis	74
Methodological essentials	74
Grouping subfossil elements by sex	74
Principal Component Analysis	74
Canonical Analysis and Discriminant Function Analysis	75
Correspondence Analysis	77
Methodological essentials	77
Taxonomic, morphological, and ecogeographical associations	77
Software for Morphometric Analyses	78
RESULTS	78
PHYLOGENY-BASED EXPLORATIONS	78
Analytical Considerations	78
Specific Phylogenetic Patterns and Implications	80
MORPHOMETRICS OF EXTERNAL CHARACTERS	84
Univariate Comparisons of Means and Variances	84
Interspecific differences among means	84
Differences in group variances	93
Univariate Sexual Dimorphism	93
Univariate Generalities of the Integument	95
Interspecific and Intersexual Patterns in Body Mass	96
Bivariate Correlations and Regressions	96
Relative wing lengths	96
Body mass and sexual size dimorphism	101
Hierarchical correlation structure	101
Ratios and Proportions	103
Wing loadings	103
Aspect ratios	105

- Multivariate Patterns 105
  - *Principal component analyses* 105
  - *Canonical analyses* 113
- Qualitative Characters of Plumage 136
  - *General characteristics of plumage* 136
  - *Numbers of remiges and rectrices* 138
  - *Shape and structure of remiges and rectrices* 138
  - *Microstructure of feathers* 141
  - *Miscellaneous features of the integument* 147
- MORPHOMETRICS OF SKELETAL CHARACTERS 147
  - Univariate Morphometrics 147
    - *Interspecific differences among means* 147
    - *Differences in group variances* 173
  - Univariate Sexual Dimorphism 175
  - Univariate Skeletal Generalities 177
  - Hierarchical Correlation Structure 178
  - Intra-Appendicular Skeletal Proportions and Ratios 181
    - *Proportions within pectoral limb* 181
    - *Proportions within pelvic limb* 183
    - *Interappendicular ratios: humerus versus femur* 188
  - Multivariate Comparisons 190
    - *K-means cluster analyses* 190
    - *Principal component analyses* 190
    - *Appendicular allomorphosis* 203
    - *Canonical analyses* 209
  - Qualitative Characters of the Skeleton 231
  - Interspecific Scaling Between Appendicular Cross Sections and Lengths 242
- PECTORAL MUSCULATURE 243
  - Descriptive Myology 243
    - *Scope and emphasis* 243
    - *Essential comparisons of musculature* 245
  - Allometry of Breast Musculature 275
  - Multivariate Analyses of Myological Measurements 278
    - *Analytical issues* 278
    - *Standard principal component analysis (PCA)* 278
    - *Modified principal component analysis (PCA∗)* 282
    - *Principal component analysis of residuals from body mass (PCA|mass)* 286
  - Salient Flightlessness-Related Trends in Pectoral Musculature 290
    - *Qualitative signatures* 290
    - *Quantitative commonalities* 291
- SUMMARY OF APOMORPHY IN FLIGHTLESS RAILS 292
  - Qualitative and Univariate Metrics 292
  - Multivariate Metrics 297
- CORRESPONDENCE ANALYSIS OF FORM AND CIRCUMSTANCE 306
- DISCUSSION 308
  - GROSS RAMIFICATIONS OF FLIGHTLESSNESS IN RAILS 308

  Giantism, Dwarfism, and Stasis 308
   Patterns and agents of change in size 309
   Generalities of giantism 311
   Generalities of dwarfism 311
   Insularity, changes in size, and flightlessness 312
  Pectoral Allometry and Flightlessness: Falling Off the Line 315
   Allometry of avian flight 316
   Parallels of nonavian flight 322
   Variation in flighted and flightless rails 323
FINE-SCALE MANIFESTATIONS OF FLIGHTLESSESS IN RAILS 323
  Integumentary Changes 323
   Shapes of wings and remiges 326
   Microstructural changes in remiges 327
   Shapes of tails and rectrices 327
   Variation in body plumage 328
  Cranial Morphotypes 329
   General avian patterns 329
   Patterns among the Rallidae 329
  Pectoral Skeleton 331
   Pectoral girdle 331
   Pectoral appendage 332
   Microstructural likelihoods 334
  Pelvic Skeleton 334
   Shifts in pelvic robustness and proportions 334
   Margins of safety in pelvic limb 336
  Pectoral Musculature 336
   General avian patterns 337
   Flightlessness and pectoral musculature 337
   Histological and biochemical likelihoods 339
SEXUAL DIMORPHISM: INTERPLAY OF PHYLOGENY AND FUNCTION 339
  Phylogenetic Baselines 340
   Methodological limitations 340
   Preliminary overview for the Rallidae 340
  Ecological Implications of Sexual Dimorphism 342
   Primary ecofunctional hypotheses 342
   Rensch's rule revisited 343
  Sexual Selection and Sexual Dimorphism 344
   Phylogenetic perspectives 344
   Sexual selection and body size 344
   Sexual selection and sexual dichromatism 345
  Sexual Dimorphism and Flightlessness 347
   Taxa other than the Rallidae 347
   Evidence within the Rallidae 349
MANIFOLD APOMORPHY OF FLIGHTLESSNESS IN RALLIDS 350
  Relative Apomorphy of Flightlessness in Rails 351
   Simple indices 351
   Multivariate ordinations 352
  Mediation of Change at an Evolutionary Crossroad 354

ONTOGENETIC MECHANISMS UNDERLYING AVIAN FLIGHTLESSNESS 356
   Study and Generalities of Avian Ontogeny 357
      Investigational motivations and obstacles 357
      Establishment of empirical baselines 357
   Heterochronic Perturbation, Developmental Constraint, and Evolution
   357
      Bauplan *as both limit and avenue of change* 358
      Genetic foundations of ontogenetic change 358
      Heterochrony: nomenclature and diagnosis 359
   Empirical Support for Heterochrony in Birds 361
      Generalities of avian heterochrony 361
      Evidence in other flightless neognathous birds 362
      Evidence in flightless Rallidae 365
      Pectoral growth axes and alar proportions 369
PHYLOGENETIC IMPLICATIONS OF AVIAN FLIGHTLESSNESS 372
   Evolutionary Directionality and Avian Flightlessness 373
      Unidirectionality of change, orthogenesis, and flightlessness 373
      Functional integration, evolutionary stability, and flightlessness 375
   Flightlessness as Source of Homoplasy 376
      Avian flightlessness and global homoplasy 376
      Flightlessness in rails: uniquely rampant parallelism 376
   Cladogenetic Corollaries: Is There Speciation After Flightlessness? 377
      Paedomorphs and ancestry 377
      Ratites and phylogenetic perspectives on flightlessness 378
      Speciation involving flightless neognathous birds 380
GEOGRAPHICAL CONTEXTS OF FLIGHTLESSNESS IN RAILS 382
   Biogeographical Realms: Patterns at a Global Scale 383
   Insularity: Evolution at Local Scales 389
      Historical background 389
      Colonization of islands 391
      Assembly of insular communities 392
      Islands as refugia 394
      Genetic implications: founders, bottlenecks, and drift 395
      Antiquity of isolation: differentiation and endemism 398
      Extinction on islands: stochasticity and anthropogenic inevitability
      399
   Demography of Insular Lineages: The Equilibrium Model Revisited .. 402
      Essentials of classical concepts 402
      Extensions deriving from insular rails 403
   Insularity: Isolation Both Arduous and Accommodating 404
      Dual challenge of colonization 404
      Total balances and diverse allocations 405
   Troglomorphy as Continental Parallel to Insularity 405
      Discovery and distribution 405
      Evolutionary analogies and interpretational differences 406
MIGRATORY HABIT AND FLIGHTLESSNESS 407
   Ecological Implications of Migratory Habit 407
      Generalities of pattern 408

  Physiological demands of migration 408
 Migration, Vagrancy, and Colonization 410
  Distributional and migratory patterns of rails 410
  Vagrancy and permanent migratory stopovers 411
  Rails as exceptional "weedy" colonists of islands 413
ECOPHYSIOLOGICAL CORRELATES OF FLIGHTLESSNESS IN RAILS 415
 Home Range and Territoriality 415
  General avian predictions and patterns 415
  Spatiotemporal ecology of flightless rallids 416
  Comparison with other flightless birds 417
 Metabolism, Thermodynamics, and Locomotory Modules 418
  General avian principles 418
  Metabolic implications of flightlessness in ratites 420
  Inferences for flightless neognathous birds 421
 Foraging and Diet 423
  General avian principles 423
  Generalities and exceptions of foraging by rallids 424
  Comparisons with other flightless birds 426
  Are flightless rails specialists or generalists as foragers? 427
 Flightlessness, Mobility, and Reproduction 427
  General avian principles and sectional organization 427
  Mating systems 428
  Location and structure of nests 430
  Sizes of clutches and eggs 431
  Parental care and development of hatchlings 434
FOUR CASE HISTORIES EXEMPLIFYING PRINCIPLES 437
 Selection of Exemplars 437
 Case Histories from Diverse Taxa, Latitudes, and Habitats 438
  Swamphens (Tribe Porphyriornithini) 438
  Typical rails (Subtribe Rallina) 438
  Crakes (Subtribe Crecina) 440
  Moorhens (Subtribe Fulicarina) 441
 Emergent Patterns and the *r–K* Continuum 442
 Adaptive Landscapes and Avian Flightlessness 444
BET-HEDGING: VAGRANCY AS MINORITY TACTIC 446
A METAPOPULATIONAL APPROACH TO AVIAN FLIGHTLESSNESS 447
 Metapopulations and Evolution of Life Histories 447
  General tenets 447
  Applicability to the Rallidae 448
 Polymorphism of Form and Habit 449
  Evolutionary origin and maintenance of polymorphisms 449
  Dispersal polymorphisms 450
 Phenotypic Plasticity and Evolutionary Change 453
  Threshold traits 453
  Developmental reaction norms 454
AVIAN FLIGHTLESSNESS AS EVOLUTIONARY PHENOMENON 457
 Flight as Locomotory Blessing or Energetic Burden 457
 Flightlessness as Degeneration Versus Developmental Economy 458

    Preconditions, Causes, Currencies, and Convergence of Change 460
    Modern Perspectives: Regressive, Progressive, and Neutral 464
  FLIGHTLESS BIRDS AND CONSERVATION ON ISLANDS 468
    Opportunities Lost and Investigational Priorities 468
    Anthropogenic Impacts and Conservational Imperatives 470
      Critical losses, historical and ongoing 471
      Insular futures both bleak and promising 472
ACKNOWLEDGMENTS 472
LITERATURE CITED 474
APPENDICES 622
  APPENDIX 1.   Body masses and wing lengths of rails 622
  APPENDIX 2.   Statistical details and sources of data 627
  APPENDIX 3.   Anatomical terms and abbreviations 633
  APPENDIX 4.   Symbols used in mathematical expressions 642
  APPENDIX 5.   Solution of morphological trajectory 642
  APPENDIX 6.   Generalizations of life histories of rails 644
  APPENDIX 7.   Clutch sizes and dimensions of eggs of rails 651

EVOLUTION OF FLIGHTLESSNESS IN RAILS (GRUIFORMES: RALLIDAE): PHYLOGENETIC, ECOMORPHOLOGICAL, AND ONTOGENETIC PERSPECTIVES

BRADLEY C. LIVEZEY

*Section of Birds, Carnegie Museum of Natural History,
Pittsburgh, Pennsylvania 15213-4080 USA*

ABSTRACT.—More than 50 species of flightless rails (Gruiformes: Rallidae) have been discovered on islands throughout the world, including members from most of the tribes and genera of the family. In the present study, qualitative and morphometric analysis of 3,220 study skins, more than 1,200 associated (complete and partial) skeletons, approximately 4,000 disassociated subfossil elements, and pectoral dissections of 41 fluid-preserved specimens formed the primary basis for investigation. Analyses emphasized statistical comparisons of flightless species with closest flighted relatives, augmented by analyses of data on body mass, wing areas, wing lengths, clutch sizes, egg dimensions, and ecophysiological parameters. These were integrated with a companion cladistic analysis and current evolutionary theory.

Flightless members of the Rallidae span more than two orders of magnitude in body mass. Univariate comparisons of skin specimens confirmed a repeated pattern of relatively or absolutely shortened wings and (with a few exceptions) tails in flightless taxa. Greatest reductions in relative wing size were evident in *Habroptila wallacii*, *Gallirallus australis*-group, *Tricholimnas* spp., *Habropteryx insignis*, *Amaurornis ineptus*, and *Tribonyx mortierii*. These shifts were confounded by diverse changes in body size; most flightless species were characterized by increases in body size of various magnitudes (greatest in *Porphyrio mantelli*-group, *Nesotrochis debooyi*, *Gallirallus australis*-group, *Tricholimnas lafresnayanus*, *Aphanapteryx bonasia*, *Erythromachus leguati*, *Diaphorapteryx hawkinsi*, and *Amaurornis ineptus*), whereas a minority showed substantial decreases (*Cabalus modestus*), modest decreases (*Dryolimnas aldabranus*, *Rallus recessus*, *Gallirallus wakensis*, *Atlantisia rogersi*, most flightless *Porzana*, and *Tribonyx hodgenorum*), or virtual stasis (*Porzana palmeri*) in directly measured or estimated body masses. Sexual dimorphism was significant in virtually all external dimensions, although magnitudes of these differences were substantially less than within-sex differences between congeners. Although confounded by subspecific variation in some taxa, indications were found of inflated variance in some external dimensions (e.g., tail length) of flightless species relative to flighted relatives (e.g., *Amaurornis ineptus*).

Limited data on wing loadings—ratios of body mass divided by area of wings—confirmed that two flightless species had significantly higher values than flighted relatives of similar size. Only the estimate for bulky *Porphyrio hochstetteri* exceeded the "threshold of flightlessness" of Meunier, whereas the value for tiny, flightless *Porzana atra* was roughly one fourth of the threshold value. The latter indicates the inapplicability of this criterion in taxonomic groups (e.g., Rallidae) in which reductions of the pectoral musculature are critical to flightlessness.

Principal component analyses (PCAs) and canonical analyses (CAs) of study skins provided multidimensional discrimination of species and sexes within key clades with respect to both size and shape. These not only confirmed the variably pronounced reductions in relative wing length and overall size in flightless species indicated by univariate analyses, but revealed that corresponding multivariate shifts were exceptionally great in *Porphyrio hochstetteri, Porzana sandwichensis,* and *Amaurornis ineptus,* and that sexual dimorphism was exaggerated in *P. hochstetteri, Habroptila wallacii, Gallirallus owstoni,* and *Cabalus modestus.* PCAs of lengths of extracted remiges revealed that flightless species, in addition to differences in overall size, were characterized by disproportionately short (in extreme cases, absent) distal primary remiges (i.e., had more rounded wings). Remiges displayed several important trends associated with flightlessness: reductions in length relative to body size; variably pronounced changes in shape; disproportionate shortening of the distalmost remiges, resulting in comparatively rounded wings; losses of the distalmost one or two remiges primarii and several remiges secundarii (a minority of taxa); and microanatomical reductions in the integrity of margins of vanes ("fringing").

Univariate comparisons confirmed the relative and (in some cases) absolute reductions in lengths of wing elements, and also quantified the reductions in dimensions of elements of the pectoral girdle and widths of appendicular elements. These shifts were accompanied by increased size of the cranium and pelvic apparatus in a number of flightless taxa (e.g., *Porphyrio hochstetteri, Gallirallus australis*-group, *Amaurornis ineptus,* and *Tribonyx mortierii*). Subfossil coots (*Fulica chathamensis*-group and *F. newtoni*) largely qualify as allometrically enlarged versions of typical congeners, comparable to two large Andean coots (*Fulica cornuta* and *F. gigantea*). Univariate sexual dimorphism was significant in most rallids. However, intersexual differences in bill lengths of several subfossil rails (*Diaphorapteryx hawkinsi, Aphanapteryx bonasia,* and possibly *Cabalus modestus*) were exceptionally great and suggestive of intersexual differences in feeding niche.

Bivariate correlations within flightless species differed from those for flighted species, notably in the low correlations between most sternal measurements and other osteological variables, a pattern indicative of the virtual disjunction between sternum and other skeletal elements in flightless species. Comparisons of proportions within the pectoral limb revealed that the antebrachium, carpometacarpus, and (to a generally lesser degree) the phalanges were disproportionately short and the brachium was disproportionately long in flightless species. These patterns and the disproportionately robust alulae in flightless rails are consistent with the effects of two largely perpendicular developmental axes acting on both the skeletal and muscular derivatives of the mesoderm in the avian pectoral limb: a primary, proximal–distal growth axis; and a secondary, cranial–caudal growth axis that principally affects the manus. Proportions within the pelvic limb showed a diversity of shifts associated with the loss of flight, one of the most marked being a disproportionate elongation of the pedal digits in highly aquatic *Fulica*. Ratios of humerus length divided by femur length provided a remarkably robust indicator of flight capacity of rallids (with the exception of natatorial *Fulica*), with ratios for flighted taxa averaging above 0.90, whereas those for flightless taxa averaged below 0.90. A PCA of detailed matrices of skulls displayed the diversity of size and bill manifested by species of the Rallidae, among which the most extreme bill shapes (and probably foraging modes) were those of several flightless species (e.g., *Capellirallus karamu, Diaphorapteryx hawkinsi, Aphanapteryx bonasia,* and *Erythromachus leguati*). Pectoral allomorphosis (intraspecific allometry) displayed higher slopes in many flightless species, consistent with termination of pectoral growth at an earlier stage of skeletal development through heterochrony. CAs of skeletal measurements within clades confirmed relative magnitudes of shifts related to loss of flight that were broadly consistent with those apparent in PCAs, and confirmed significant increases in sexual dimorphism in most flightless lineages. Qualitative changes associated with flightlessness were found in most pectoral elements (especially the humerus and sternum), with many extending to the extinct adzebills (Gruiformes: Aptornithidae), and corroborated homoplasy among flightless species. Tallies of these apomorphies indicated that the most-derived flightless lineages were *Porphyrio hochstetteri, Habroptila wallacii, Gallirallus australis*-group, *Cabalus modestus, Capellirallus karamu,* and *Diaphorapteryx hawkinsi.*

Comparisons of the pectoral musculature of rails revealed that reductions in bulk and cranial extents of mm. pectoralis et supracoracoideus were the most conspicuous myological changes. As a percentage of mean body mass, these underwent reductions among flightless taxa as high as 15% (*Gallirallus australis* and *Tribonyx mortierii*) and as low as 5–6% (*Dryolimnas aldabranus* and *Gallinula comeri*). Also typical of most flightless rails was an increase in the prominence of m. cucullaris capitis pars clavicularis (associated with the caudal regression of the apex carina sterni), attenuation of mm. biceps brachii et humerotriceps, greater distal extent of m. pronator superficialis relative to the underlying (foreshortened) radius, and a corresponding increase in the impressio m. brachialis relative to the ulna. A minority of flightless rails also showed variably pronounced weakening of fibrous portions of m. rhomboideus profundus, m. flexor digitorum superficialis, and m. ulnometacarpalis ventralis, and increased conformational variation in several muscles of the manus (mm. abductor alulae capita dorsale et ventrale, and m. extensor brevis alulae). PCAs of mean muscle measurements indicated greatest morphometric shifts in *Gallirallus australis, G. wakensis, Atlantisia rogersi, Porzana palmeri,* and *Gallinula comeri,* patterns not entirely congruent with reductions in breast muscles. A correspondence analysis of ecomorphological variables principally discriminated three groups: small crakes on small, extremely isolated islands (*Porzana palmeri* and *P. atra*), large terrestrial species from New Zealand (*Porphyrio hochstetteri* and *Gallirallus australis*), and robust, aquatic species from moderately large islands (*Habroptila wallacii, Amaurornis ineptus,* and *Tribonyx mortierii*).

Most flightless rails manifest variably pronounced increases in size, and in accordance with the substantial literature on giantism, these shifts appear to confer selective advantages related to thermodynamics, procurement of mates, territoriality, capacity for fasting, and interspecific competition. These gains were accompanied by negative implications, including greater total energetic requirements, diminished capacity for stealth, and vulnerability to selected environmental and predatory agents, with the latter contributing to the minority of flightless rails showing dwarfism. Changes in body size are accompanied by allometric changes in numerous, fundamental ecophysiological parameters, among which are several critical to flight capacity. Departures from familial isometry in relative wing size accompany flightlessness in most cases, but in rails reductions in pectoral musculature (and the associated skeleton) appear to be paramount, changes that were associated with variably pronounced changes in the integument, modifications in bill shape, increased sexual dimorphism, energetically efficient reductions in basal metabolic rates, and changes in reproductive and dietary parameters. The latter are consistent with r–K shifts in life histories, and most of the ecological changes are typical of insular birds. Heterochrony, combining pectoral paedomorphosis with (in most taxa) peramorphosis of the axial and pelvic complexes, appears to underlie most anatomical apomorphies related to flightlessness in rails. Morphological and ecological changes in flightless rails provide a strong qualitative analogy with those of vertebrate and invertebrate endemics of caves (troglomorphs).

Phylogenetic reconstructions of rallids are replete with morphological homoplasy, apparent irreversibility of the apomorphy associated with flightlessness, and only a few candidates for speciation following the loss of flight. Many rails show metapopulational demographic characteristics, and a number of migratory species show high vagrancy and qualify as consummate colonists of islands. These qualities suggest that a number of flighted rails, especially a core group with high fecundities and long-distance migratory patterns, may maintain dispersal polymorphisms in which a minority of progeny are predisposed to vagrancy and colonization of insular habitats. Insular colonizations occasioned thereby essentially represent "permanent migratory stopovers" followed by evolutionary refinements for year-round residency.

The highly convergent morphology of flightless rails indicates a shared, readily triggered, canalized bifurcation in ontogeny that leads to the morphological and physiological changes that result in flightlessness. Conditional advantages of resources redirected in flightless lineages (e.g., conversion of investments in musculature, pectoral skeletons, and metabolic characteristics) are substantial, as indicated by the exorbitant anatomical and physiological requirements of the primary capacity surrendered, migration. The potential for this transformation may be preserved through dispersal polymorphisms and bet-hedging against overdependence on ephemeral, variable, natal breeding locales. Alternative patterns of dispersal also may be accelerated in some rallids through selectively maintained, environmentally induced plasticity through threshold traits or developmental reaction-norms within small demes subject to founder effects, genetic drift, and population bottlenecks. Distributions of flightless rails are explainable by a complex history of colonizations by flighted ancestors, a scant number of colonizations or near-island expansions by flightless lineages, and extinctions related to small demes, marginal habitats, earthquakes, volcanoes, El Niño–La Niña events, and (especially) tsunamis of islands during recent millennia. Many flightless rails encountered ecological opportunities beyond those of continental confamilials. Selective advantages under these circumstances were accrued through decreased clutch size, increased egg size, and protracted developmental periods.

Flightlessness in rails represents the selectively advantageous, ontogenetically mediated conversion of anatomical and caloric assets of the pectoral apparatus and associated metabolic parameters related to flight toward multiple evolutionary alternatives of intensified selective importance in insular habitats and the adoption of a nonmigratory lifestyle. The evolutionary scenario can be summarized as follows: migration-imposed anatomical and physiological requirements for migration preconditioned key rallids for a conversion of resources; vagrancy (possibly enhanced by polymorphism of dispersal and accelerated cladogenetic capacity maintained among metapopulations) provided opportunities for insular colonization; one or a suite of similar alternative, heterochronic, developmental avenue(s) retained by key rallids (perhaps triggered and hastened as threshold traits or by developmental reaction norms) facilitated the anatomical and physiological transformation to the local optimal, flightless phenotype(s) after colonization; successful colonization may have been advanced by differences in preferred stopover habitats between sexes and ages, and the acceleration of kin-selected altruism among close relatives migrating in concert; concomitant changes in size carried multiple allometrically related changes in physiology and metabolism; and despite a resilience to natural disasters (notably tsunamis), anthropogenic agencies ultimately led to extinction for most flightless lineages effectively by breaching key aspects of insularity essential to their provision of refuge. Accordingly, the "ideal avian colonist" would possess a combination of a capacity to modulate metabolic and physiological parameters; manifest dispersal polymorphism that includes long-distance, gregarious vagrancy as one component tactic; comparatively high sexual dimorphism or a potential for such; and expanded ontogenetic variance and cladogenesis that facilitates evolutionary changes in size and pectoral paedomorphosis.

INTRODUCTION

I have heretofore asked the question concerning Mauritius henns [Rallidae: *Aphanapteryx bonasia*] and dodos [Raphidae: *Raphus cucullatus*], thatt seeing those could neither fly nor swymme, beeing cloven footed and withoutt wings on an island far from any other land, and none to bee seence elce where, how they shold come thither?—P. Mundy (ca. 1638), as transliterated by Temple (1919:83)

Who does not know that if we rear some bird of our own climate in a cage and it lives there for five or six years, and if we then return it to nature by setting it at liberty, it is no longer able to fly like its fellows, which have always been free? The slight change of environment for this individual has indeed only diminished its power of flight, and doubtless has worked no change in its structure; but if a long succession of generations of individuals of the same race had been kept in captivity for a considerable period, there is no doubt that even the *structure* of these individuals would gradually have undergone *notable changes*. Still more, if instead of a mere continuous captivity, this environmental factor had been further accompanied by a change to a very different climate; and if these individuals had by degrees been habituated to other kinds of food and other activities for seizing it, these factors when combined together and become *permanent* would have unquestionably given rise imperceptibly to a new race with quite special characters.—Lamarck (1809: vol. 1, 228), as translated by Elliot (1984:110, emphasis added)

As Professor Owen has remarked [cf. Owen 1848a, 1851, 1856], there is no greater anomaly in nature than a bird that cannot fly; yet there are several in this state.... As the larger ground-feeding birds seldom take flight except to escape danger, it is probable that the nearly wingless condition of several birds, now inhabiting or which lately inhabited several oceanic islands, tenanted by no beast of prey, has been caused by disuse.—Darwin (1859:106)

Since, therefore, the Struthious birds [ratites] all have perfect feathers, and all have rudimentary wings which are anatomically those of true birds, not the rudimentary forelegs of reptiles, and since we know that in many higher groups of birds—as the pigeons and rails—the wings have become more or less aborted, and the keel of the sternum greatly reduced in size by disuse, it seems probable that the very remote ancestors of the rhea, the cassowary, and the Apteryx [kiwi] were true flying birds ....—Wallace (1880:444)

Lamarck's hypothesis of the way of work of the secondary evolutionary cause of Species, by the influence, viz., of circumstances exciting or checking the exercise of parts, is more intelligible, more applicable in connexion with observed facts, to the before-cited ornithic cases [flightless *Porphyrio* and *Gallirallus*] than is Darwin's or Wallace's 'Natural Selection'—Owen (1882:695)

... the rails, as a group, may in the past have been slow to acquire the power of full flight, or even of flight in moderate perfection, while some, like the fossil forms mentioned above [*Aphanapteryx* and *Diaphorapteryx*], may never have acquired it at all. Rails seem, in fact, to have been primarily and constitutionally either bad fliers or bad 'triers'.—Lowe (1928a:111)

The Family Rallidae

Ecomorphological Diversity

Generalities of life history of a taxonomic group are critical to an understanding of the context and significance of a repeated evolutionary trend (e.g., flightlessness) among members of the taxon, especially where the trend in question is rare among birds generally and is functionally counterintuitive. Rails (family Rallidae) are a speciose and phenotypically diverse family of primarily terrestrial, typically chickenlike birds. They are members of the order Gruiformes (cranes and allies), an amalgam of families that has proven resistant to anatomical diagnosis and systematic classification (Livezey 1998). The type genus *Rallus* derives from Linnaeus (1758) and was erected for the widespread Water Rail (*R. aquaticus*) of the Palaearctic region, the specific epithet correctly conveying the habitat of the species. Other rails originally described by Linnaeus (1758)—Mediterranean Swamphen (*Porphyrio porphyrio*), Corn Crake (*Crex crex*), Sora (*Porzana carolina*), Common Moorhen (*Gallinula chloropus*), and Black Coot (*Fulica atra*)—solidified the perception of rails as furtive denizens of marshes, moors, and moist meadows.

Rails are typified by bilaterally compressed bodies, relatively short tails, short, rounded wings, strong, variably elongated bills, and, with the exception of the primitive Nkulengu Rail (*Himantornis haematopus*), a tufted glandula uropygialis (Van Tyne and Berger 1976; Ripley 1977; P. B. Taylor 1996, 1998; Livezey 1998). In the field, many species are known for their stealthy movements through dense vegetation and reluctance to take flight (even when pursued by dogs), whereas others (notably *Fulica*) forage unwarily on open water. During the 250 years following Linnaeus, the Rallidae were revealed to be diverse in form, global in modern and fossil distribution, and uniquely disposed to flightlessness, characteristics that enabled members of the family to dominate the endemic avifauna of oceanic islands.

The visual impenetrability of the habitats typical of the Rallidae undoubtedly contributed to an unfamiliarity with the group and predictably promulgated a view of the Rallidae as occult. Indeed, the calls of most Rallidae on their breeding grounds are common, but to see the birds in life is another matter altogether. This disparity between visual and acoustical detectability of rails has led to the increasingly widespread use of census techniques that rely on tallies of calls as indicators of abundance (e.g., Bart et al. 1984; Johnson and Dinsmore 1986a; Robert and Laporte 1997; Bramley and Veltman 2000). In that the acoustical properties of habitats are important determinants in the evolution of signaling systems (Endler 1992), the dense vegetative communities typically inhabited by rails probably are related to the importance of penetrating, distinctive, and variably complex vocalizations of members of the family (G. W. Kaufmann 1983; Clapperton and Jenkins 1984, 1987; Clapperton 1987). The family name, believed onomatopoetic of the rasping noises made by common members of the family in Europe (Gruson 1972; Morris 1973), affirms acoustical distinctness of member species as focal to the public perception of the group and leitmotif of a considerable literature. The appellation "rail" evidently derived from the French word *reille* (Picardy dialect), which then passed through the Old French words *raale* and *raalle* and Middle French word *rasle*. The Latin roots of the word are thought to be *rasclare* and *rasiculare*.

Variation in body mass among extant Rallidae exceeds that in any other single family in the order Gruiformes (Dunning 1993). Upland rallids include sparrow-sized crakes averaging 25 g in body mass (e.g., Yellow-breasted Crake [*Micropygia schomburgkii*]) and fowl-sized wood-rails approaching 1 kg in mass (Giant Wood-Rail [*Aramides ypecaha*]), whereas comparatively natatorial forms encompass the modest-sized Spot-breasted Crake (*Porzana porzana*; 80 g) and the ponderous swamphens (*Porphyrio*; 750–2,500 g for extant species) and coots (*Fulica*;

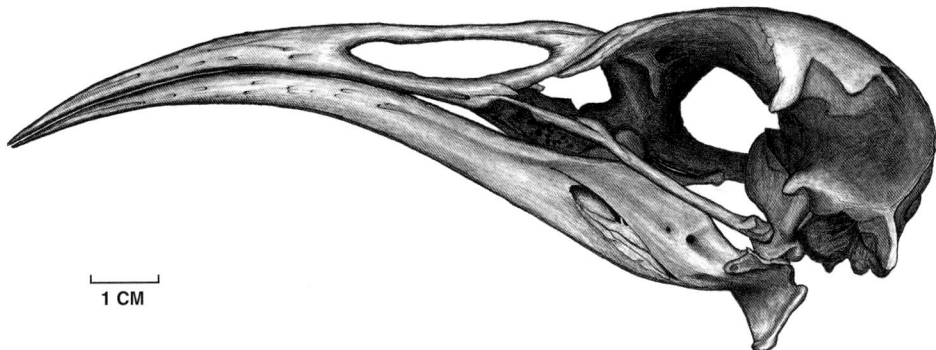

Fig. 1. Skull of *Diaphorapteryx hawkinsi* (BMNH A·1913), an extinct, flightless rail of the Chatham Islands, New Zealand. Specimen is osteologically complete with the exception of ossa lacrimales et pterygoidea. Executed by B. C. Livezey.

500–2,500 g). Dimensions of subfossil skeletal elements indicate that several extinct taxa (e.g., *Porphyrio mantelli, Diaphorapteryx hawkinsi,* and *Fulica chathamensis*) also attained substantial body sizes, including the largest for the family (estimates below).

Rails inhabit all major continental regions exclusive of the Holarctic and a multitude of oceanic islands (Ripley 1977; Taylor 1996, 1998), yielding one of the widest geographical distributions for a single family of terrestrial vertebrates (Olson 1973a). Rails can be found skulking in tussock grass or fragmented copses, along margins of ponds and swamps, or in dense, moist meadows. However, the extinction of many insular endemics has become an unfortunate legacy of the family during the last several centuries, and many species of rail (including many flightless taxa) are known only from bones moldering in caves or lava tubes where the birds were entombed centuries earlier (Fig. 1). These extinctions were caused primarily by hunting by humans, destruction of habitat, or anthropogenic introduction of predators, competitors, or disease (Hahn 1963; Greenway 1967; Olson 1989, 1991). Although commonly considered marsh-dwellers, rails occupy a broad range of habitats, including mangroves, sparsely vegetated atolls, desolate islands of southernmost New Zealand, cool-temperate woodlands, tropical forests, moors, and grasslands (Ripley 1977; Taylor 1996, 1998). Many species of Rallidae are sedentary, whereas others are migratory; some of the latter are renowned for remarkable incidents of long-distance vagrancy. A capacity for long-distance migration of many continental rails contrasts with the reluctance of most species to take flight (even when pursued) and the relative frequency of permanent flightlessness among insular members of the family (Olson 1977; Ripley 1977).

The early zoological literature is replete with speculations regarding avian flightlessness (Lamarck 1809; Darwin 1830, 1859; Owen 1848a, 1866a, 1879, 1882; Wallace 1880). Like other examples of evolutionary reduction, flightlessness initially was attributed to the heritable effects of simple disuse (Lamarck 1809), rationalizations involving retarded growth and degeneration within an orthogenetic paradigm (e.g., Cope 1868, 1872), or preservation of unrefined features in relictual taxa (Darwin 1859). The early ornithological literature understandably was dominated by generalities taken from the comparatively familiar ratites and penguins, impressions of questionable relevance to flightless carinates such as the Rallidae. Nevertheless, the implications of avian flightlessness for nascent theories of evolutionary change were profound, and the phenomenon passed from the marginalia of natural history to a metaphor for decreased utility, functional senescence, and evolutionary change (e.g., Fürbringer 1888, 1902; Wiglesworth 1900; Gadow 1902; Lowe 1926, 1928a, b, 1933, 1934, 1935; Goldschmidt 1940; Simpson 1944; de Beer 1975). However, these diverse discussions led to no consensus for the phenomenon in general or among the Rallidae (the family accounting for a majority of flightless species of carinates) for more than a century after the publication of the magnum opus by Darwin (1859).

FLIGHTLESS RAILS: EARLY DISCOVERY AND DESCRIPTION

> As, however, there are strictly aquatic forms of birds deprived, by a low development and special modification of the wings, of the power of flight, so also there are, in other natural groups of birds, aberrant forms similarly debarred from the privilege and enjoyment of the characteristic kind and field of locomotion of their class. Apart from the true *Struthionidæ*, we have an instance of this in the *Brachypteryx* or modified Rail [*Gallirallus australis*] of New Zealand; [and] the Dodo [*Raphus cucullatus*] . . . .—Owen (1848a:8)

Since the time of the first European explorers, accounts included mention of flightless rails. These early narratives included 17th-century references to the "hen(ne)s" and "fowles" of Mauritius Island (20°15′S, 57°30′E) and Rodriguez Island (19°41′S, 63°23′E) by voyagers in the Indian Ocean (quoted by Strickland and Melville 1848; Olson 1977), and notations of similar birds on Ascension Island (7°57′S, 14°22′W) in the mid-Atlantic (P. Mundy, in Temple 1919; Temple and Anstey 1936). Among the illustrations to derive from these early accounts was the striking rendition of *Aphanapteryx bonasia* executed by G. Hoefnagel (ca. 1610), a work discovered by Frauenfeld (1868) and published in a paper by Milne-Edwards (1868).

In the seminal work by Linnaeus (1758), six flightless birds (three ratites and three carinates) were described formally: Ostrich (*Struthio camelus*), Greater Rhea (*Rhea americanus*), Double-wattled Cassowary (*Casuarius casuarius*), Jackass Penguin (*Spheniscus demersus*), Great Auk (*Pinguinus impennis*), and the Dodo (*Raphus cucullatus*), with only the Dodo being extinct at the time of description. The flightless birds described during the remaining decades of the 18th century and the first decades of the 19th century perpetuated this compositional bias—the ratites *Dromaius novaehollandiae* (Latham, 1790) and *Apteryx australis* Shaw, 1813, and the penguins *Aptenodytes patagonicus* Miller, 1778, *Pygoscelis papua, P. antarctica,* and *Eudyptes chrysocome* (Forster, 1781). Most references to flightless birds in the early classics of natural history focused largely or completely on the ratites (e.g., Buffon 1770–1786; Latham 1781–1802; Cuvier 1802). The discovery of remains of the moas (Dinornithiformes) in New Zealand by Owen (1848a), an ornithological cause célèbre by any standard, solidified the focal role accorded ratites among flightless birds. By the end of the 19th century, all 11 modern species of ratites (as well as a number of extinct forms) had been described, as had 17 of 18 modern species of penguin (Peters 1931; Brodkorb 1963; Falla and Mougin 1979; Mayr 1979). The ratites and penguins came to epitomize avian flightlessness, a generalization understandable given the conspicuousness of these taxa, the timing of their description, and that these remain the only major avian groups in which flightlessness uniformly characterizes all members.

The 19th century saw the discoveries by Darwin (1830) during the voyage of the *Beagle,* including seminal observations and collections of flightless steamer-ducks (*Tachyeres* [Livezey and Humphrey 1992]), as well as the initial descriptions of the Galápagos finches (Geospizinae), with the latter ultimately considered paradigmatic of evolutionary radiations (Schluter 2000a, b). Descriptions of modern species of Hawaiian honeycreepers (Drepanididae), the great majority of which appeared during the late 19th century (Greenway 1968), revealed an equally impressive but less-studied diversity; like the Rallidae, this insular group would prove most enlightening only upon the subsequent discovery of an extinct majority of members (Amadon 1950; Bock 1970; Dobzhansky 1977; Pimm and Pimm 1982; McKinney and McNamara 1991). By contrast, only a few species of Galápagos finches are thought to have been extirpated since the first description of member taxa, local extinctions of subspecific populations aside (Grant 1986). It was during the century following the publication of Linnaeus (1758) that most families ultimately found to include flightless members were identified (Table 1), an era that also saw the transformation of ornithology from an avocation of the affluent to a natural science (Farber 1982).

During these early years of ornithological survey, flightless rallids of a variety of shapes and sizes were discovered and, with few exceptions, largely forgotten. This stealthful passage of taxa, unscrutinized by most naturalists of the day, was hastened by the rapid, widespread extinction of many taxa shortly following description (Figs. 2–4). The first flightless rail to be described formally was the tame and resourceful Weka (*Gallirallus (a.) australis* (Sparrman, 1786)) of the South Island of New Zealand, a find followed shortly by the description of the much rarer Hawaiian Crake (*Porzana sandwichensis* (Gmelin, 1789)). From a geographical perspective, these inaugural examples of flightless rallids proved prophetic, in that New Zealand and Hawaii would come to be known for an unparalled diversity and ultimate destruction of most or all of their endemic flightless rails. Evidence indicates that the greatest diversity of rallids sympatric on a single island in the Hawaiian group (three *Porzana,* one *Gallinula,* and one *Fulica*) approximates that originally endemic to Chatham Island (*Porphyrio* "*chathamensis,*" *Gallirallus dieffenbachii, Cabalus modestus, Diaphorapteryx hawkinsi,* and *Fulica chathamensis*), suggesting that five may represent an upper limit for species richness of rallids on a single, small oceanic island. However, each of the main islands of New Zealand, however, sustained a higher diversity of rallids (including four or five endemic, flightless species), exceeding many continental regions in this respect.

By the time of the publication of the *Origin of Species* (Darwin 1859), seven flightless species of rail had been described, as had flightless members of several other carinate families (e.g., two ducks, a parrot, and the solitaire). Darwin (1859) cited three examples of avian flightlessness, in both the contexts of disuse and specialization: Ostrich (*Struthio*), kiwi (*Apteryx*), and steamer-duck (*Tachyeres*). Although the description of new flighted rallids reached its apex at the middle of the 19th century, the description of flightless species peaked at the turn of the century (Fig. 5). By this time, flightless birds from a dozen carinate families had

TABLE 1. Flightless carinate birds (excluding Spheniscidae) described before 1900, listed in taxo-chronological order. Higher-order taxonomy and that of nonrallids follow Peters (1931, 1934, 1937) and Brodkorb (1963, 1964, 1967, 1971). Taxonomy of Rallidae follow Livezey (1998). Rallids are shown in boldface. Taxa preceded by daggers were diagnosed by using fossil and subfossil material.

Flightless species	Order; family	Original binomen and authorship
Pinguinus impennis	Charadriiformes: Alcidae	*Alca impennis* Linnaeus, 1758
Raphus cucullatus	Columbiformes: Raphidae	*Struthio cucullatus* Linnaeus, 1758
Gallirallus australis	Gruiformes: Rallidae	*Rallus australis* Sparrman, 1786
Tachyeres brachypterus	Anseriformes: Anatidae	*Anas cinerea* Latham, 1789
Porzana sandwichensis	Gruiformes: Rallidae	*Rallus sandwichensis* Gmelin, 1789
†*Pezophaps solitaria*	Columbiformes: Raphidae	*Didus solitarius* Gmelin, 1789
†*Tribonyx mortierii*	Gruiformes: Rallidae	*Tribonyx mortierii* Du Bus, 1840
Gallirallus dieffenbachii	Gruiformes: Rallidae	*Rallus dieffenbachii* G. R. Gray, 1843
Anas aucklandica	Anseriformes: Anatidae	*Anas aucklandica* G. R. Gray, 1844
Strigops habroptilus	Psittaciformes: Psittacidae	*Strigops habroptilus* G. R. Gray, 1845
†*Aptornis otidiformis*	Gruiformes: Aptornithidae	*Aptornis otidiformis* Owen, 1848
Porphyrio mantelli	Gruiformes: Rallidae	*Notornis mantelli* Owen, 1848
†**Aphanapteryx bonasia**	Gruiformes: Rallidae	*Apterornis bonasia* Sélys-Longchamps, 1848
Porzana monasa	Gruiformes: Rallidae	*Rallus monasa* Kittlitz, 1858
Habroptila wallacii	Gruiformes: Rallidae	*Habroptila wallacii* G. R. Gray, 1860
Tricholimnas lafresnayanus	Gruiformes: Rallidae	*Gallirallus lafresnayanus* Verreaux and Des Murs, 1860
Gallinula nesiotis	Gruiformes: Rallidae	*Gallinula nesiotis* Sclater, 1861
†*Cnemiornis calcitrans*	Anseriformes: Anatidae	*Cnemiornis calcitrans* Owen, 1865
Nesoclopeus poecilopterus	Gruiformes: Rallidae	*Rallina poeciloptera* Hartlaub, 1866
†**Fulica newtoni**	Gruiformes: Rallidae	*Fulica newtoni* Milne-Edwards, 1867
Rollandia microptera	Podicipediformes: Podicipedidae	*Podiceps micropterus* Gould, 1868
Tricholimnas sylvestris	Gruiformes: Rallidae	*Ocydromus sylvestris* Sclater, 1869
†*Aptornis defossor*	Gruiformes: Aptornithidae	*Aptornis defossor* Owen, 1871
Pareudiastes pacificus	Gruiformes: Rallidae	*Pareudiastes pacificus* Hartlaub and Finsch, 1871
Cabalus modestus	Gruiformes: Rallidae	*Rallus modestus* Hutton, 1872
†**Erythromachus leguati**	Gruiformes: Rallidae	*Erythromachus leguati* Milne-Edwards, 1874
†*Diatryma gigantea*	Diatrymiformes: Diatrymidae	*Diatryma gigantea* Cope, 1876
Dryolimnas aldabranus	Gruiformes: Rallidae	*Rallus gularis* var. *aldabrana* Günther, 1879

TABLE 1. Continued.

Flightless species	Order: family	Original binomen and authorship
Amaurornis ineptus	Gruiformes: Rallidae	*Megacrex ineptus* D'Albertis and Salvadori, 1879
Habropteryx insignis	Gruiformes: Rallidae	*Rallus insignis* Sclater, 1880
Porphyrio hochstetteri	Gruiformes: Rallidae	*Notornis hochstetteri* Meyer, 1883
Aramidopsis plateni	Gruiformes: Rallidae	*Rallus plateni* Blasius, 1886
†*Phorusrhacos longissimus*	Gruiformes: Phorusrhacidae	*Phorusrhacos longissimus* Ameghino, 1887
Gallirallus greyi	Gruiformes: Rallidae	*Ocydromus greyi* Buller, 1888
†*Tribonyx repertus*	Gruiformes: Rallidae	*Porphyrio reperta* De Vis, 1888
Nesoclopeus woodfordi	Gruiformes: Rallidae	*Rallina woodfordi* Ogilvie, 1889
†*Cnemiornis gracilis*	Anseriformes: Anatidae	*Cnemiornis gracilis* Forbes, 1891
†*Diaphorapteryx hawkinsi*	Gruiformes: Rallidae	*Aphanapteryx hawkinsi* Forbes, 1892
Porzana palmeri	Gruiformes: Rallidae	*Porzanula palmeri* Frohawk, 1892
Gallinula comeri	Gruiformes: Rallidae	*Porphyriornis comeri* Allen, 1892
†*Fulica chathamensis*	Gruiformes: Rallidae	*Fulica chathamensis* Forbes, 1892
Gallirallus sharpei	Gruiformes: Rallidae	*Stictolimnas sharpei* Büttikofer, 1893
†*Fulica prisca*	Gruiformes: Rallidae	*Fulica prisca* Hamilton, 1893
Podiceps taczanowskii	Podicipediformes: Podicipedidae	*Podiceps taczanowskii* Berlepsch and Stolzmann, 1894
Gallirallus owstoni	Gruiformes: Rallidae	*Hypotaenidia owstoni* Rothschild, 1895
Compsohalieus harrisi	Pelecaniformes: Phalacrocoracidae	*Phalacrocorax harrisi* Rothschild, 1898

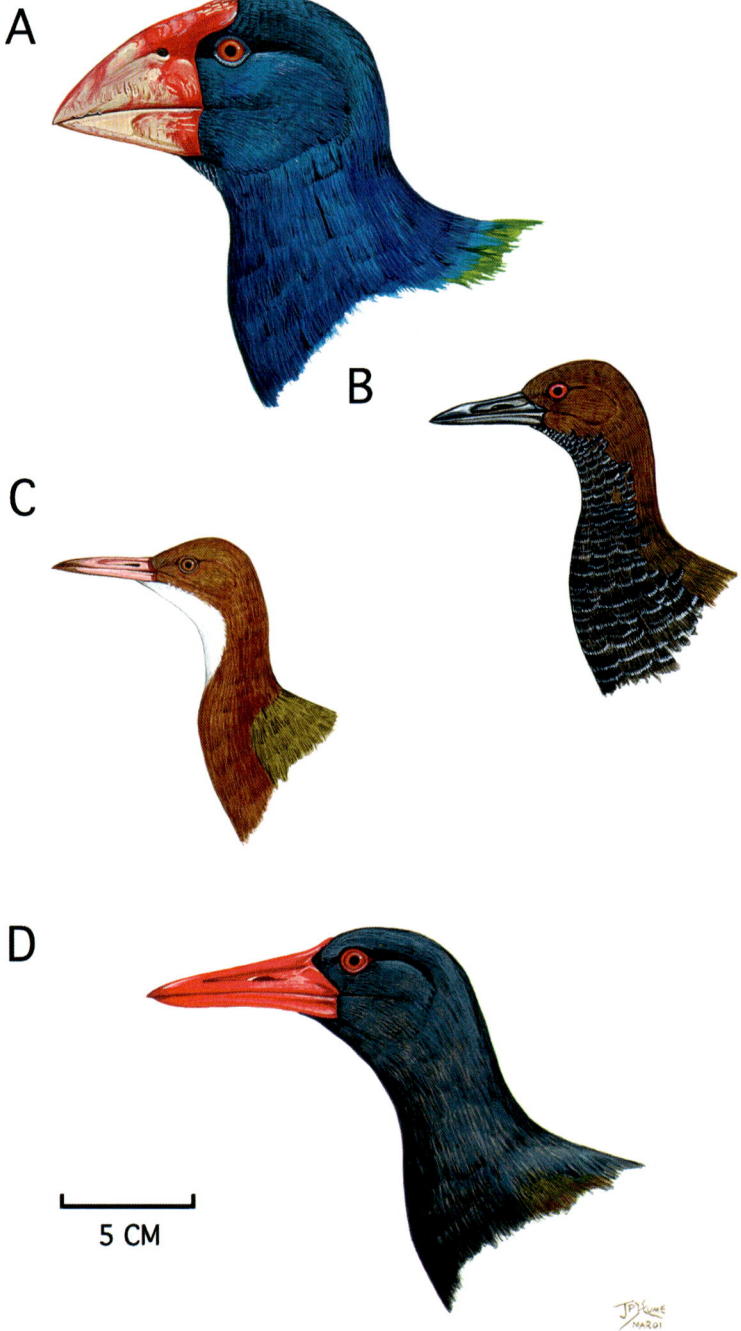

FIG. 2. Flightless Rallidae, I (cameo): **A,** South Island Takahe (*Porphyrio hochstetteri*; BMNH 1856·5·28·1, sex indeterminate); **B,** New Britain Barred-Rail (*Habropteryx insignis*; BMNH 81·3·29·90, holotype, male); **C,** Aldabra White-throated Rail (*Dryolimnas aldabranus*; BMNH 1968·43·102, male); and **D,** Wallace's Rail (*Habroptila wallacii*; BMNH 1939·12·9·3677, male). Acrylic paintings by Julian Pender Hume.

Fig. 3. Flightless Rallidae, II (cameo): **A,** South Island Weka (*Gallirallus australis*; BMNH 52·1·17·19, male); **B,** Hutton's Rail (*Cabalus modestus,* entire; BMNH 92·10·31·9, male); **C,** Dieffenbach's Rail (*G. dieffenbachii*; BMNH 42·9·29·12, holotype, sex indeterminate); **D,** Fiji Bar-winged Rail (*Nesoclopeus poecilopterus*; BMNH 89·11·3·68, male); **E,** Wake Island Rail (*G. wakensis,* entire; BMNH 1908·12·21·7, sex indeterminate); **F,** Lord Howe Island Rail (*Tricholimnas sylvestris*; BMNH 98·12·2·538, male); and **G,** New Caledonian Rail (*T. lafresnayanus*; BMNH 70·12·3·3, male). Acrylic paintings by Julian Pender Hume.

FIG. 4. Flightless Rallidae, III (cameo): **A,** Laysan Crake (*Porzana palmeri,* entire; BMNH 1939·12·9·21, male); **B,** Hawaiian Crake (*Porzana sandwichensis,* entire; BMNH 1939·12·9·553, sex indeterminate); **C,** New Guinea Flightless Bushhen (*Amaurornis ineptus*; BMNH 1911·12·20·340, male); **D,** Inaccessible Island Crake (*Atlantisia rogersi,* entire; BMNH 1953·55·105, male); **E,** Tasmanian Native-hen (*Tribonyx mortierii*; BMNH 89·11·1·403, sex indeterminate); **F,** Samoan Moorhen (*Pareudiastes pacificus*; BMNH 1923·9·7·10, sex indeterminate); **G,** Gough Island Moorhen (*Gallinula comeri*; BMNH 1922·12·6·95, male). Acrylic paintings by Julian Pender Hume.

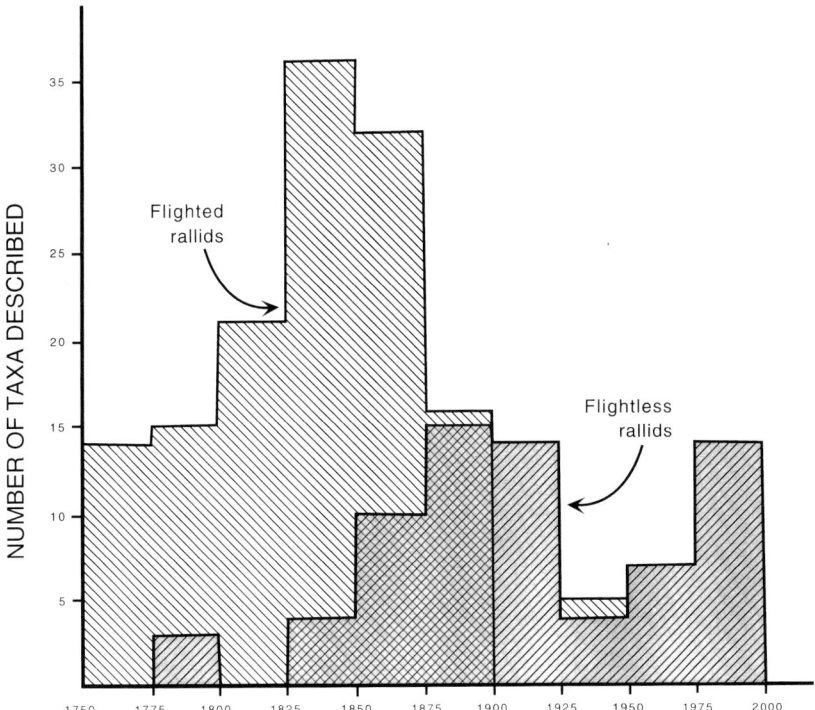

FIG. 5. Histogram of numbers of species of Rallidae described during 25-year intervals from the publication of Linnaeus (1758) to the present.

been described (Table 1) from sites around the globe, with a higher proportion of insular rails endemic to the Southern Hemisphere in broad agreement with the differential distribution of oceanic areas and islands between the hemispheres (Fig. 6). Characteristics associated with flightlessness and other unusual apomorphies (e.g., decurvature or elongation of the bill and large body size), depite convergence, impressed early taxonomists at least as much as plumage patterns or the more informative features of subfossil remains (Olson 1973a, 1977). This emphasis on general dissimilitude led to the initial allocation or prompt elevation of a substantial majority of the flightless rallids described before 1950 to monotypic genera. Subsequently, taxonomists (cf. Peters 1934; Olson 1973a, 1977; Ripley 1977) have relegated a number of these narrowly defined (often monotypic) genera (e.g., *Notornis, Habroptila, Cabalus, Tricholimnas, Aramidopsis, Nesolimnas, Habropteryx, Nesoclopeus, Porzanula, Pennula, Megacrex, Tribonyx, Porphyriornis, Pareudiastes, Edithornis,* and *Palaeolimnas*) to synonymy with more-inclusive genera.

FLIGHTLESS RAILS: EXTINCT, ENDANGERED, AND SECURE

Evidence is growing of widespread losses of insular endemics in Pacific Oceania, including many flightless rails (Olson 1989, 1991; Steadman and Kirch 1990, 1998; Steadman and Pahlavan 1992; Steadman and Rolett 1996; Kirch 1997; Steadman and Justice 1998; Steadman et al. 1999). A number of flightless rallids, especially in the New Zealand region (J. H. Fleming 1939a, b; C. A. Fleming

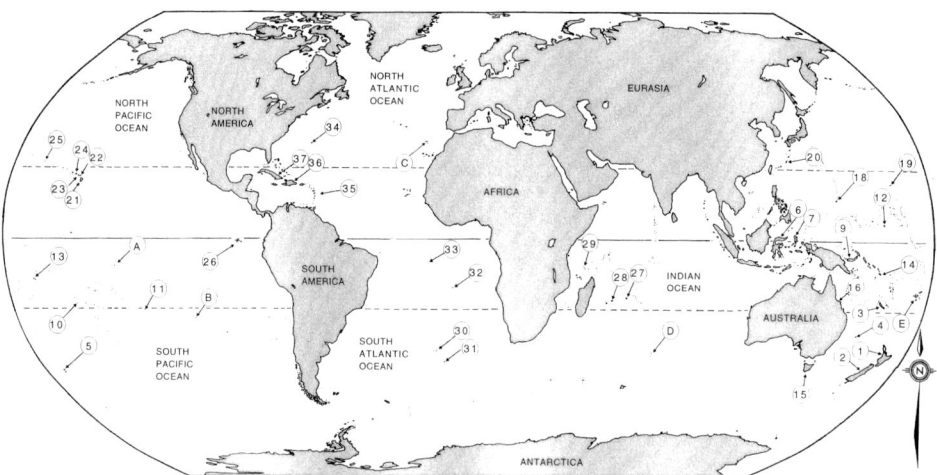

FIG. 6. Global distributions of selected flightless members of the Rallidae: 1, North Island, New Zealand; 2, South Island, New Zealand; 3, New Caledonia; 4, Lord Howe Island; 5, Chatham Islands; 6, Sulawesi; 7, Halmahera; 8, Fiji; 9, New Britain; 10, Mangaia; 11, Henderson Island; 12, Kusaie Island; 13, Savai'i, West Samoa; 14, San Cristobal, Solomon Islands; 15, Tasmania; 16, Australia (mainland); 17, New Guinea; 18, Guam; 19, Wake Island; 20, Okinawa; 21, Hawaii, Hawaiian Islands; 22, Maui, Hawaiian Islands; 23, Molokai, Hawaiian Islands; 24, Oahu, Hawaiian Islands; 25, Laysan Island, Hawaiian Islands; 26, Galápagos Islands; 27, Rodriguez, Mascarene Islands; 28, Mauritius, Mascarene Islands; 29, Aldabra, Seychelle Islands; 30, Tristan da Cunha and Inaccessible Island; 31, Gough Island, Tristan group; 32, St. Helena Island; 33, Ascension Island; 34, Bermuda; 35, Greater and Lesser Antilles; 36, Haiti; 37, Cuba. Islands hosting poorly represented taxa are lettered: A, French Polynesia; B, Easter Island; C, Madeira; D, Amsterdam Island; E, Fiji. Islands from which taxa lacking formal descriptions (rallid from Fernando de Noronha, Brazil), adequate samples, or sufficiently detailed analyses (e.g., *Hovacrex roberti, Nesotrochis picapicensis, Gallirallus ripleyi, G. rovianae,* and *Porzana rua*), or those of a controversial nature (e.g., *Tricholimnas conditicius* and *Porzana* [*Rallus*] *nigra*) are excluded.

1960; McDowall 1969; Meredith 1991; Anderson 1997), appear to have been extirpated by human immigrants before the arrival of Europeans, including (Table 2) New Caledonian Takahe (*Porphyrio kukwiedei*); St. Helena Island Swamphen (*Aphanocrex podarces*); several species of cave-rails (*Nesotrochis* spp.); Snipe-billed Rail (*Capellirallus karamu*); two endemic rallids of the Mascarenes, herein treated under separate genera (*Aphanapteryx* and *Erythromachus*); Hawkins' Rail (*Diaphorapteryx hawkinsi*); Ascension Island Crake ("*Atlantisia*" *elpenor*), referral to a monotypic genus under proposal by Bourne et al. (pers. comm.) to reflect uncertainty of generic membership (Livezey 1998); a suite of crakes (*Porzana* spp.) from the Hawaiian Islands; St. Helena Crake (*P. astrictocarpus*); two native-hens (moorhens) from Australasia (*Tribonyx* spp.); and three coots (*Fulica*), two species from Australasia and one from Mauritius. Responsibility of prehistoric humans (Rouse 1986) and European colonists for these extinctions is problematic, but evidence indicates culpability of humans in at least some cases.

Most extinct flightless rallids, and some vulnerable populations of flighted forms, were extirpated during historic times, principally as a result of European exploration during the 17th, 18th, and 19th centuries (Table 2; Lydekker 1891; Rothschild 1907a, b; Lambrecht 1933; Vestjens 1963; Greenway 1967; Halliday 1978a; Fuller 1987, 2001; Day 1989). Several species were extirpated so rapidly

TABLE 2. Geographic distributions and status of flightless Rallidae. Species known only from subfossil skeletal elements are preceded by daggers. Forms lacking formal descriptions are excluded, including an undescribed species of rail from the Madeira Islands (Pieper 1985). See also reviews by Collar and Stuart (1985), Collar and Andrew (1988), Collar et al. (1992, 1994), and P. B. Taylor (1996, 1998).

Scientific name	Distribution	Status	References
†*Porphyrio mantelli*	North Island, New Zealand	Extinct, 19th century	Owen 1948a, b, 1882; Forbes 1891, 1923; Medway 1967; Olson 1977; Andrews 1988; Trewick 1996a; Trewick and Worthy 2001
Porphyrio hochstetteri	South Island, New Zealand	Endangered	Williams 1957, 1960; Williams and Miers 1958; Reid 1967, 1974b, 1978; Mills and Lavers 1974; Reid and Stack 1974; Mills 1975, 1978, 1985; Mills and Mark 1977; Morris 1977; Lavers and Mills 1984; Beauchamp and Worthy 1988; Bunin 1995; Bunin and Jamieson 1995, 1996a, b; Bunin et al. 1997; Craig 2001; Eason et al. 2001; Maxwell 2001
†*Porphyrio kukwiedei*	New Caledonia, S. Pacific	Extinct, Holocene (?)	Balouet and Olson 1989; Balouet 1991
Porphyrio albus	Lord Howe Island, S. Pacific	Extinct, ca. 1834	Hindwood 1932, 1940, 1965; Fuller 1987, Hutton 1991; Garnett 1993
†*Porphyrio* sp.	New Ireland, New Guinea	Extinct, ca. 1,500 PB	Steadman et al. 1999
†*Aphanocrex podarces*	St. Helena Island. S. Atlantic	Extinct, ca. 1500	Ashmole 1963a; Wetmore 1963; Olson 1973b, 1975b; Rowlands et al. 1998
Habroptila wallacii	Halmahera. N. Moluccas	Vulnerable	Stresemann 1931; de Haan 1950; White and Bruce 1986
†*Nesotrochis debooyi*	Greater and Lesser Antilles	Extinct, 1600s or later	Wetmore 1918, 1922a, 1937, 1938, 1956, 1960a
†*Nesotrochis steganinos*	Haiti, West Indies	Extinct, 1600s or later	Olson 1974a, 1978
†*Nesotrochis picapicensis*	Cuba, West Indies	Extinct, 1600s or later	Fischer and Stephan 1971b
Cyanolimnas cerverai	Cuba, West Indies	Critically endangered	Garrido 1985; Garrido and Kirkconnel 2000
Dryolimnas aldabranus	Aldabra, Seychelles	Rare	Salvan 1970; Penny and Diamond 1971; Huxley and Wilkinson 1977, 1979; Wilkinson and Huxley 1978; Hambler et al. 1993
†*Rallus recessus*	Bermuda, N. Atlantic	Extinct, ca. 10,000 BP	Wetmore 1960a; Olson 1977, 1997; Olson and Wingate 2000, 2001
†*Rallus ibycus*	Bermuda, N. Atlantic	Extinct, Pleistocene	Wetmore 1960a; Olson and Wingate 2000
Gallirallus sharpei	Probably S. Pacific region	Extinct, 19th century	Olson 1986
Gallirallus australis	South Island, New Zealand	Uncommon and local	Coleman et al. 1983; Brothers and Skira 1984; Robertson and Beauchamp 1985; Beauchamp 1989

TABLE 2. Continued.

Scientific name	Distribution	Status	References
Gallirallus greyi	North Island, New Zealand	Rare and local, declining	Carroll 1963a–c; Robertson 1976; Robertson and Beauchamp 1985; Beauchamp 1986, 1987a, b, 1988, 19981a, b; MacMillan 1990; Beauchamp et al. 1993, 2000; Bramley 1996, 2001; Beauchamp and Chambers 2000; Bramley and Veltman 2000
Gallirallus dieffenbachii	Chatham Islands, New Zealand	Extinct, ca. 1842	See text; Andrews 1896c; Archey and Lindsay 1924; Olson 1985
Gallirallus owstoni	Guam, S. Pacific	Extinct in wild	Strophlet 1946; Perez 1968; Jenkins 1979; Pimm 1987; Savidge 1987; Haig et al. 1990, 1994; Haig and Ballou 1995; Fritts and Rodda 1998
Gallirallus rovianae	Central Solomon Islands	Near-threatened	Diamond 1991
†*Gallirallus* sp.	New Ireland, New Guinea	Extinct, ca. 2,000 PB	Steadman et al. 1999
Gallirallus wakensis	Wake Island, N. Pacific	Extinct, 1945	See text; Rothschild 1903; Ripley 1977; Olson 1996
Tricholimnas lafresnayanus	New Caledonia, S. Pacific	Critically endangered	Layard and Layard 1882; Vuilleumier and Gochfeld 1976; Stokes 1979
Tricholimnas sylvestris	Lord Howe Island, S. Pacific	Endangered	Hull 1909; Bassett-Hull 1910; Iredale 1910; Mathews 1928, 1936; Hindwood 1940; Mayr 1941; McKean and Hindwood 1965; Disney and Smithers 1972; Recher and Clark 1974; Fullagar and Disney 1975, 1981; Miller and Kingston 1980; Fullagar et al. 1982; Lourie-Fraser 1982; Fullagar 1985; Miller and Mullette 1985; Hutton 1991; Brook et al. 1997
Tricholimnas conditicius	Marshall Islands (?), S. Pacific	Extinct	Bangs 1930; Greenway 1952; Walters 1987; Olson 1992
Nesoclopeus poeciloperus-group	Fiji, Guadacanal, and Isabel Island	Locally common (Isabel), or endangered or extinct	Harrison and Parker 1967; Watling 1982; Blaber 1990; Webb 1992; Taylor 1996; Kratter et al. 2001; Hadden 2002
Aramidopsis plateni	Sulawesi, S. Pacific	Vulnerable	Meyer and Wiglesworth 1898; Stresemann 1941; Coomans de Ruiter 1947; White and Bruce 1986; Lambert 1989; Andrew and Holmes 1990
Cabalus modestus	Chatham Islands, New Zealand	Extinct, ca. 1900	See text; Hutton 1872a, b, 1873a; Travers 1872; Buller 1873a, b, 1888, 1905; Archey and Lindsay 1924; Olson 1975b, 1977; Tennyson and Millener 1994; Millener 1999

TABLE 2. Continued.

Scientific name	Distribution	Status	References
†*Capellirallus karamu*	North Island, New Zealand	Extinct, ca. 1600	Falla 1954; Medway 1967; Scarlet 1970; Olson 1975b
Habropteryx insignis	New Britain, S. Pacific	Locally common	Bishop 1983; Diamond 1987
Habropteryx okinawae	Okinawa, Ryuku Islands	Endangered	Yamashina and Mano 1981; Thiede 1982; Brazil 1985, 1991; Vuilleumier et al. 1992; Ikenaga 1983; Kuroda et al. 1984; Harato and Ozaki 1993; Ikenaga and Gima 1993
†*Aphanapteryx bonasia*	Mauritius Island, Mascarenes	Extinct, ca. 1700	Sélys-Longchamps 1848; Strickland and Melville 1848; Frauenfeld 1868, 1869; Milne-Edwards 1868, 1869; Newton and Gadow 1893; Meinertzhagen 1912; Piveteau 1945; Hachisuka 1953; Olson 1977; Cowles 1987
†*Erythromachus leguati*	Rodriguez Island, Mascarenes	Extinct, ca. 1700	Milne-Edwards 1874, 1875; Günther and Newton 1879; Hachisuka 1953; Olson 1977
†*Diaphorapteryx hawkinsi*	Chatham Island, New Zealand	Extinct, ca. 1600	Forbes 1892a–e, 1893a–e; Andrews 1896b; Olson 1975b; Millener 1999
Atlantisia rogersi	Inaccessible Island, S. Atlantic	Vulnerable	Lowe 1928a; Holdgate and Wace 1959; Voous 1962; Wace and Holdgate 1976; Richardson 1984; Fraser 1989; Ryan et al. 1989; Fraser et al. 1992
†*Mundia elpenor*	Ascension Island, C. Atlantic	Extinct, late 17th century	Ashmole 1963b; Olson 1973b
Porzana sandwichensis	Hawaii, Hawaiian Islands	Extinct, ca. 1844	Dole 1869, 1879; Finsch 1898; Henshaw 1902; Perkins 1903, 1913; Olson and James 1982a, b, 1991; Olson 1999
Porzana palmeri	Laysan Island, Hawaiian Islands	Extinct, 1944	See text; Rothschild 1893a, b, 1900; Fisher 1903, 1906; Bryan 1915; Bryan and Greenway 1944; Baldwin 1945, 1947; Olson 1996, 1999
Porzana monasa	Kusaie I., Caroline Islands	Extinct, ca. 1828	Ripley 1977; Steadman 1986a
Porzana atra	Henderson Island, S. Pacific	Vulnerable	Bourne and David 1983; Forsberg et al. 1983; Graves 1992; Benton and Spencer 1995; Wragg 1995
†*Porzana piercei*	Bermuda, N. Atlantic	Extinct, late Pleistocene (?)	Olson and Wingate 2000
†*Porzana astrictocarpus*	St. Helena Island, S. Atlantic	Extinct, late Pleistocene (?)	Wetmore 1963; Olson 1973b, 1975c; Rowlands et al. 1998
†*Porzana rua*	Mangaia, S. Cook Islands	Extinct, late Pleistocene (?)	Steadman 1986a

TABLE 2. Continued.

Scientific name	Distribution	Status	References
†*Porzana ziegleri*	Oahu, Hawaiian Islands	Extinct, late Pleistocene (?)	Olson and James 1982a, b, 1991; James et al. 1987; James and Olson 1991
†*Porzana menehune*	Molokai, Hawaiian Islands	Extinct, late Pleistocene (?)*	Olson and James 1982a, b, 1991; James et al. 1987; James and Olson 1991
†*Porzana keplerorum*	Maui, Hawaiian Islands	Extinct, late Pleistocene (?)	Olson and James 1982a, b, 1991; James et al. 1987; James and Olson 1991
†*Porzana ralphorum*	Oahu, Hawaiian Islands	Extinct, late Pleistocene (?)	Olson and James 1982a, b, 1991; James et al. 1987; James and Olson 1991
†*Porzana severnsi*	Maui, Hawaiian Islands	Extinct, late Pleistocene (?)	Olson and James 1982a, b, 1991; James et al. 1987; James and Olson 1991
Amaurornis ineptus	New Guinea	Uncommon and local	Rand 1942; Ripley 1964; Rand and Gilliard 1967; Coates 1985; Beehler et al. 1986; Gregory 1996
Pareudiastes pacificus	Savai'i, W. Samoa	Extinct, ca. 1874	Hartlaub and Finsch 1871; Sclater 1874; Ripley 1977
Pareudiastes silvestris	San Cristobal, Solomon Islands	Critically endangered	Mayr 1933; Cain and Galbraith 1956; Olson 1975a; Hay 1986
Tribonyx mortierii	Tasmania	Locally common	Fletcher 1909, 1912; Mathews 1911, 1913; Ridpath 1964, 1972a-c; Ridpath and Moreau 1966; Ridpath and Meldrum 1968a, b; Holyoak and Sager 1970; Maynard Smith and Ridpath 1972; Kraus et al. 1975; Goldizen et al. 1993, 2000
Tribonyx repertus	Australia (mainland)	Extinct, Pleistocene (?)	Olson 1975d
†*Tribonyx hodgenorum*	New Zealand proper	Extinct, ca. 1600	Scarlett 1955a; Olson 1975b
Gallinula nesiotis	Tristan da Cunha, S. Atlantic	Extinct, ca. 1861	Carmichael 1818; Sclater 1861; Stresemann 1953; Beintema 1972
Gallinula comeri	Gough Island, S. Atlantic	Locally common	Clarke 1905; Broekhuysen and MacNae 1949; Hagen 1952; H. F. I. Elliott 1953, 1957; Wilson and Swales 1958; Holdgate 1965; C. C. H. Elliott 1969, 1970; Clancey 1981
†*Fulica chathamensis*	Chatham Islands, New Zealand	Extinct, ca. 1600	Andrews 1896c; Scarlett 1955b; Olson 1975b, 1977; Millener 1980, 1981, 1999
†*Fulica prisca*	Main islands, New Zealand	Extinct, ca. 1600	Andrews 1896c; Scarlett 1955b; Olson 1975b, 1977; Millener 1980, 1981, 1999; Worthy 1999a
†*Fulica newtoni*	Mauritius Island, Mascarenes	Extinct, late 17th century	Milne-Edwards 1867; Olson 1977; Mourer-Chauviré et al. 1999

* This and perhaps other subfossil crakes from the Hawaiian archipelago may have persisted into historic times (Olson 1999).

after discovery that the taxa are known only from illustrations and descriptions from life (e.g., the hypothetical "*Gallirallus pacificus*" of Tahiti; Ripley 1977; Olson and Steadman 1987; Walters 1988; Taylor 1998), or are represented by a single skin specimen of a species of unknown flight status and lacking locality and date of collection (e.g., *Gallirallus sharpei* [Olson 1986]). For the majority of insular endemics extirpated during this period (e.g., *Porphyrio albus* [Lord Howe Island, 500 km east of Australia], *Gallirallus dieffenbachii* [Chatham Islands, South Pacific], *Porzana monasa* [Kusaie Island, Caroline Islands, North Pacific], *P. sandwichensis* [Island of Hawaii, North Pacific], *Pareudiastes pacificus* [Savai'i, Samoa, South Pacific], and *Gallinula nesiotis* [Tristan da Cunha, South Atlantic]), extirpation was complete before more than a few skins were preserved (Oliver 1955). Internal anatomy of these taxa is known only from skeletal elements subsequently recovered from subfossil deposits (e.g., *G. dieffenbachii*), imaged by X-ray in study skins (e.g., *P. monasa* [Steadman 1986a]), or a single fluid-preserved specimen (e.g., *Pareudiastes pacificus*).

Protracted human destruction on oceanic islands has led to losses of multiple avian endemics, many of which may have disappeared before formal description (e.g., Rabor 1959). Tallies of extinct avian endemics increasingly appear to be a function of the intensity of historical survey and preservation, especially for the Pacific islands of the Hawaiian ridge, Austral Seamount chain, and MacDonald Seamount group (Mielke 1989). In light of continuing recoveries of new, flightless species from Pacific islands (Steadman 1989a, b, 1993a, b, 1995, 1997a, b; Steadman and Intoh 1994; Worthy et al. 1999), and the likelihood that many island groups having ages of the scale of 10^7 years coalesced into larger areas or were placed in closer proximity by lowered sea levels during the Pleistocene (Watling 1982; Rodda 1994; Carson and Clague 1995), it seems implausible that substantive and remote islands did not sustain flightless rails before human occupation. Examples of islands conducive to avian endemism (Steadman 1995) include Réunion, Mascarenes, Indian Ocean; Easter Island, easternmost member of the Tuamato Ridge Archipelago (fragmentary remains of two endemic rallids recovered; Steadman et al. 1994); Tahiti, South Pacific (also in Tuamato group, purportedly supported two unsubstantiated, extinct species); Taiwan, North Pacific; Falkland Islands, South Atlantic; and the Galápagos, an isolated group of volcanic "hotspot" origin (Windley 1995), where an endemic crake (*Laterallus spilonotus*) approaches flightlessness (Franklin et al. 1979). Apparent absence of endemic rallids on the Seychelles (Gaymer et al. 1969) and Kerguelen Island (Weimerskirch et al. 1989) seems likely to be an artifact of preservation rather than a genuine reflection of former distributions throughout the Indian Ocean. Recent discovery of fragmentary remains of rails from the Madeira Islands of the North Atlantic (Pieper 1985) suggests a strong likelihood of extinct, flightless rallids having inhabited the more-isolated Azores (36–39°N, 25–31°W) as well. Before the very early immigration of humans, even comparatively nearshore island groups (e.g., Canary Islands, São Tomé, Principé, Comoro Islands, and Socotra) may have supported endemic, flightless rails, thereby extending the faunal cataclysm to other oceanic regions.

Although the evidence of extirpated rallids increasingly documents widespread destruction, the extreme estimate by Steadman (1995)—that one to four species of flightless rails may have been endemic to each island (indicating that ~2,000 species, 20% of all modern avian species, were extirpated in Oceania)—may represent an overestimate. For example, there is a likelihood that extinction-level events (e.g., tsunamis) would have devastated largely terrestrial groups such as rallids on low-elevation islands (MacArthur and Wilson 1967; see below). However, estimates by Steadman (1997c) for the comparatively vagile and arboreal Columbidae of Polynesia, a group hunted to a comparable degree as the Rallidae by human immigrants (Steadman et al. 1999), corroborate the severity of the faunal catastrophe caused by human immigrants. Of course, any estimate of taxonomic diversity is contingent on the species concept adopted (Cracraft 1983, 1988a; Frost and Kluge 1994; Zink and McKitrick 1995), a choice with significant implications for radiations such as the *Gallirallus philippensis* complex and several Pacific *Porzana* (Livezey 1998), and is doubly problematic for taxa represented only by fragmentary skeletal remains.

Still other flightless rallids survived into the 20th century (Table 2), only to be wiped out by comparatively recent destruction of habitat and introductions of predators by humans, much of which was related to the Pacific campaign of World War II (Taylor 1996, 1998; see below). Notable among these were the dubiously flightless Assumption Island Rail (*Dryolimnas abbotti* [Meade-Waldo 1908]) and two rallids extirpated during the early 1940s: the Wake Island Rail (*Gallirallus wakensis*; see below) and Laysan Crake (*Porzana palmeri*; see below). The subspecies of the White-browed Crake formerly endemic to Iwo Jima (*Porzana cinerea brevipes*), although capable of flight, suffered a similar fate; the population evidently was extirpated by introduced cats (*Felis catus*) by 1925 (Greenway 1967; Ripley 1977).

Unfortunately, some continental and other insular members of the Rallidae are in jeopardy, including a number of species capable of flight (W. B. King 1980, 1981; Williams and Given 1981; Collar and Stuart 1985; Fjeldså 1985; Ripley and Beehler 1985; Collar and Andrew 1988; Eddleman et al. 1988; Stinson et al. 1991; Collar et al. 1992; Collar 1993; Engilis and Pratt 1993). Negative impacts of humans on habitats (especially wetlands) and damage caused by commensals of humans (including predators, herbivores, and introduced diseases) have been the causes underlying most declines of flightless rails and other avian endemics of oceanic islands (Fosberg et al. 1983; Ralph and van Riper 1985; van Riper et al. 1986; Olson 1989, 1991; Baker 1991; Ritter and Savidge 1999). Flightless rails are especially vulnerable, and members of a number of genera are classified as threatened or endangered (BirdLife International 2000), some of which already may have become extinct (Table 2), including the South Island Takahe (*Porphyrio hochstetteri*; see below); Zapata Rail (*Cyanolimnas cerverai*), described as having "very weak" flight (Garrido and Kirkconnell 2000); Guam Rail (*Gallirallus owstoni*); New Caledonian Rail (*Tricholimnas lafresnayanus*); Lord Howe Rail (*T. sylvestris*); Okinawan Barred-Rail (*Habropteryx okinawae*); and San Cristobal Moorhen (*Pareudiastes silvestris*).

The Guam Rail, having survived tremendous damage to its habitat during World War II (Jenkins 1979, 1983), survives only by means of a captive-rearing program, the population in the wild having been extirpated within the span of a few decades by the accidentally introduced brown tree snake (*Boiga irregularis* [Savidge 1987]). A trial reintroduction of 16 birds was made in November 1998 (Fritts and Rodda 1998). The Okinawa Barred-Rail, also considered endangered, was known only from a few skeletal elements (Olson 1977) and was believed extinct until a remnant population was discovered in 1978, and the external anatomy, ecology, and behavior were described soon thereafter (Yamashina and Mano 1981).

Populations of a minority of flightless rallids remain secure to the present (Table 2; Taylor 1996, 1998), although a number of flightless rallids currently maintaining stable populations remain vulnerable to extirpation, in part because of restricted distributional ranges (M. Williamson 1981; Watkins and Furness 1986; Marshall 1988; Johnson and Statterfield 1990; Kuroda et al. 1984; Pimm et al. 1994a, b). These comparatively fortunate species include (BirdLife International 2000) Wallace's Rail (*Habroptila wallacii*); Aldabra Islands Rail (*Dryolimnas aldabranus*); North Island Weka (*Gallirallus greyi*) and South Island Weka (*G. australis*), although the latter is significantly more numerous than its northern sister-species; Ysabel Bar-winged Rail (*Nesoclopeus* [*woodfordi*] *immaculatus*), previously classified as threatened, was found to be locally common on Isabel, Solomon Islands (Kratter et al. 2001); Platen's Rail (*Aramidopsis plateni*); New Britain Barred-Rail (*Habropteryx insignis*); Inaccessible Island Crake (*Atlantisia rogersi*); Henderson Island Crake (*Porzana atra*); New Guinea Waterhen (*Amaurornis ineptus*); Tasmanian Native-hen (*Tribonyx mortierii*); and Gough Island Moorhen (*Gallinula comeri*). In addition to dedicated efforts of conservationists, populations of the South Island Weka owe the comparative stability to the robust constitution and opportunistic behavior of the species; Wekas are renowned for brazen foraging near human habitation and are known to be predators of the eggs of other New Zealand endemics (Atkinson and Bell 1973; St. Clair and St. Clair 1992).

Flightless Rails of New Zealand: Discovery, Debate, and Destruction

Taxonomy and evolutionary inference.—In addition to providing standardized names for species, alpha-taxonomic decisions form the primary framework for inferences concerning phylogenetic history, biogeography, and evolutionary trends. As illustrated by the intertwined controversies pertaining to species limits and flightless in steamer-ducks (Livezey and Humphrey 1992), the social and professional milieu in which taxonomic diagnoses of rails transpired profoundly influence theories of biogeography, ontogeny, insular ecology, and speciation as pertained to the study of avian flightlessness. Furthermore, assumptions concerning conspecific status of many insular endemics carried critical implications for the study and conservation of these variably divergent evolutionary lineages. In the following accounts, the importance of taxonomic confusion for the study and precarious status of flightless rails—spanning decades and debated to the present day—is exemplified by several endemics of New Zealand.

Hawkins' Rail: cranial homoplasy and biogeography.—Following a postscript to a short notice on New Zealand birds (Forbes 1891), Forbes (1892a) announced the discovery of the skeletal remains of a new, flightless rail in a telegram published in *Nature* (Fig. 1). Forbes (1892b–d) followed these announcements closely by other brief notices. The remains were found in dune deposits on Chatham Island, the largest member of an island group located 600 km southeast of the main islands of New Zealand (44°S, 176°W) and discovered by Captain Broughton on the *Chatham* in 1791 (Fig. 7). Beginning with the arrival of Polynesians ca.

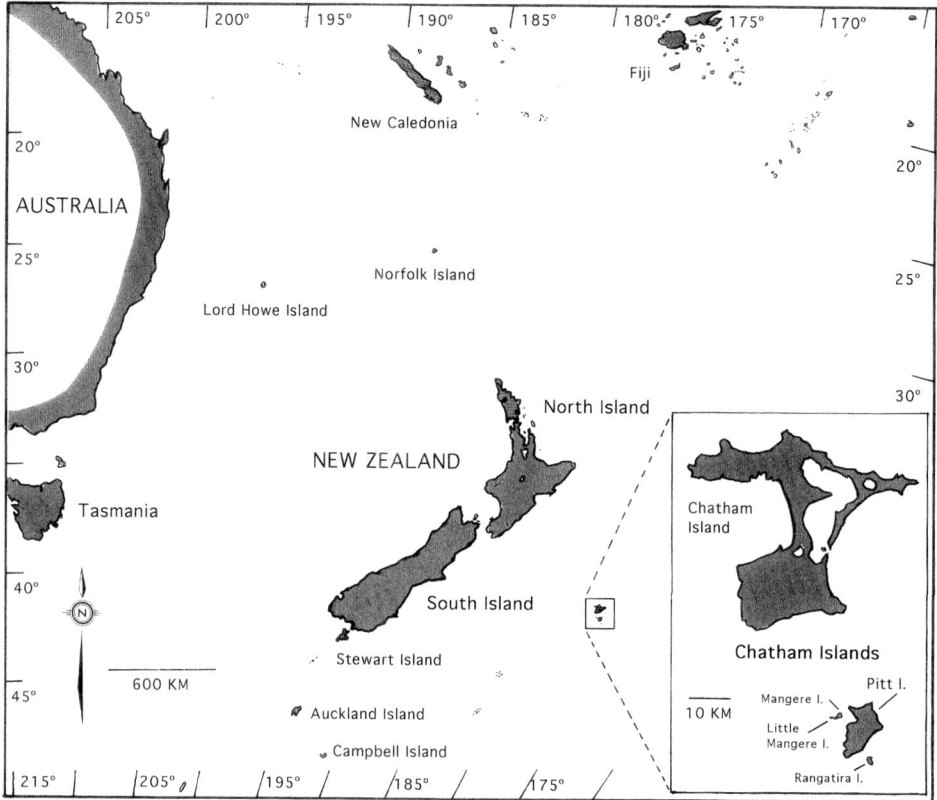

FIG. 7. Map of the New Zealand region, with detailed inset of the Chatham Islands, a small archipelago lying 865 km east of the main islands of New Zealand.

450 BP (McFadgen 1994), these islands were destined to become an abattoir for a diverse assemblage of endemic birds, including several flightless rails (Millener 1999). The new species was named *Aphanapteryx hawkinsi,* in honor of Mr. W. Hawkins, a local collector (Forbes 1892a), and henceforth the species was known in vernacular contexts as Hawkins' Rail. The first generic assignment by Forbes (1892a) was based on superficial similarities with subfossil, flightless rails known from the Mascarenes (e.g., *Aphanapteryx* and *Erythromachus* [see Olson 1977]). Subsequently, following the suggestion of A. Newton, Forbes (1892e, 1893a) erected the new genus *Diaphorapteryx* for the species, although he (Forbes 1893b–d) later reverted to his referral to the genus *Aphanapteryx* based on specimens of the latter at Cambridge. The type specimen for *Diaphorapteryx hawkinsi,* listed erroneously as a skeleton by Brodkorb (1967), may be among a series of skulls overlooked by Dawson (1958) but subsequently rediscovered at the British Museum (Olson 1975b).

A series of comparatively detailed descriptions and illustrations were published within the next decade (Andrews 1896b–d; Milne-Edwards 1896), impacts of which were both to stabilize recognition of the monotypic genus *Diaphorapteryx* for Hawkins' Rail and to lead to confusion concerning the biogeographical implications of flightlessness. Despite this generic dissassociation from the Mascarene rallids, Rothschild (1907b:196) persisted in the view that "... these extinct Chatham Island birds ... are representative species of the forms peculiar three hundred years ago, to the Mascarene Islands ..." and that "... *Diaphorapteryx* and *Palaeolimnas* [= *Fulica chathamensis*] take the place on the Chatham Islands of the Mascarene *Aphanapteryx broecki* [= *bonasia*] and *Palaeolimnas newtoni*." The taxonomic puzzlement of Forbes (1892a–c, 1893a, b) and the dubious parallels drawn by Rothschild (1907b) are understandable in light of the fact that the phylogenetic relationships of both *Diaphorapteryx* and *Aphanapteryx* remain problematic, principally because of pronounced apomorphy related to flightlessness and the absence of anatomical specimens other than subfossil remains (Livezey 1998). Henceforth, *Diaphorapteryx* was included as a distinct genus in most compendia (e.g., Lambrecht 1933) and inventories (e.g., Oliver 1930, 1955), and most authors judged similarities with *Aphanapteryx* to be convergent.

Perhaps the most apomorphic of all rallids, this sizeable species is known only from deposits in dunes and adjacent crevices on the main island of the Chatham group (Turbott 1990), and on the basis of anatomical study the taxon remains sui generis (Livezey 1998). *Diaphorapteryx* is estimated to have had a height in life of roughly 42 cm (16–17 inches [Olson 1977:368]), and its stature, remarkably reduced wings, robust pelvic limbs, and disproportionately large, strongly decurved bill were among the most distinctive aspects of its anatomy (Andrews 1896b; Olson 1975b). Early descriptions of the extinct avifauna of the Chathams by resident Polynesians indicate that *Diaphorapteryx* was reddish brown in color, slept and foraged in flocks, and sought food in rotten logs. The species probably was extirpated by humans during the late 16th or early 17th centuries (J. Cooper and C. Fisher, pers. comm.). Similarities, notably decurvature of the bill (Fig. 1), with other flightless rallids of the region (e.g., *Cabalus modestus, Gallirallus australis,* and *G. dieffenbachii*) and elsewhere (especially the Mascarene *Aphanapteryx* [Milne-Edwards 1869]), attracted speculation in early works (Andrews 1896c, d).

Two smaller rails from the Chathams: apparent versus genuine juvenility.—A rail described by G. R. Gray as *Rallus dieffenbachii* in a narrative of the travels of Dieffenbach (1843) was known for decades solely from the holotype (British Museum of Natural History [BMNH] 42·9·29·12), a study skin collected in 1840 on the islet of Wharedauri in the Chatham group (Warren and Harrison 1973; Fig. 3). The new rail was figured by J. E. Gray (1844) and included in the classification by G. R. Gray (1862), and often listed under various, more narrowly delimited genera of the period (e.g., *Ocydromus* and *Hypotaenidia*). Dieffenbach (fide Forbes 1893d:531–532) stated that the rail "... was formerly very common, but since cats and dogs have been introduced it has become very scarce ...."

Almost from its original description, a debate commenced concerning the taxonomic status and geographical distribution of *dieffenbachii* and subsequently described allies, one that primarily turned on issues of contemporaneous theories of paedomorphosis and limits on ecological similarity among sympatric relatives. In plumage pattern and size, *Gallirallus dieffenbachii* closely resembled members of the *G. philippensis* complex, but *dieffenbachii* differed most conspicuously from the latter in its evident flightlessness and distinctly decurved bill. Buller (1868, 1873a), the preeminent ornithologist residing in New Zealand during this period, argued against the proposal (e.g., Finsch 1868) that *dieffenbachii* was conspecific with *G. (p.) assimilis* or the larger *philippensis* assemblage; Buller (1874a) later reaffirmed the distinctness of *dieffenbachii* in response to concerns expressed by Hutton (1873a, b). However, Buller was to adopt an opposite view in the most enduring taxonomic controversy concerning the rallids from the Chatham Islands, one critical to the recognition of paedomorphosis in flightless rails. The debate ensued upon the description by Hutton (1972a, b) of a new, smaller flightless rail, *Rallus modestus* (Fig. 3), later referred to a new genus *Cabalus* by Hutton (1872c), the distinctness of which was supported by the accounts of the same expedition by Travers (1872).

However, Buller (1873a) averred that *C. modestus* merely represented juvenile specimens of *dieffenbachii,* relegating the former to synonymy. Hutton (1873a, b) and Travers (1872) resisted this proposal, circumstantiating the original specimens and adducing the likely maturity of at least one of the type series. Hutton (1873a) also stated that the native Morioris referred to *modestus* by the name *Matirakahu,* but to *dieffenbachii* as *Mohoriki.* Buller (1873b, 1874a) defended his position, and bristled (Buller 1874a:94): "I presume that an author who undertakes to write the history of the birds of any country is at liberty to form his own judgment as to who are 'competent' authorities . . . . It is neither usual nor necessary in such cases to 'give the names.' "

In contrast to his endorsement of the sympatry of closely related, flightless rallids—for example, the diversity of forms of weka or "*Ocydromus*" (*Gallirallus australis*-group) to which Buller (1873a, 1874b, 1877, 1882, 1888, 1892a) and others (e.g., Hutton 1873c; Sharpe 1875) accorded species status on the main islands of New Zealand (Oliver 1968)—Buller (1873a–c, 1874a) rationalized growing concerns about variation and classification of *Gallirallus,* including hybridization between rails with domestic fowl (*Gallus*) as a contributory factor (Buller 1876). Finsch (1874, 1875), however, differed with Buller both with respect to the status of *modestus* and the taxonomy of "*Ocydromus.*" The hypothesis of hybridization was refuted by Murie (1889) on anatomical grounds, and Buller (1874c) reversed his stand on the status of *C. modestus,* a change of mind repeated by Buller (1875) in response to the critiques by Finsch (1874) of the monograph by Buller (1973a). The capitulation by Buller (1874c:511) included a quotation from a letter by A. Newton dated 13 December 1874, in which Newton noted that "fresh evidence" provided by Hutton (including skeletal elements) had convinced J. Murie of the distinctness of *C. modestus.*

A qualified endorsement of the synonymy of *dieffenbachii* and *modestus* by Sharpe (1875) prompted Hutton (1879:454) and Travers (1882, 1883) to bolster support for species status for *modestus.* Buller (1878, 1892b, 1896) reaffirmed the status of *dieffenbachii* and *modestus* as separate species, including the supportive anatomical comparisons by Murie (1888), but revealed an ambivalence regarding the distinctness of the two forms (Buller 1882, 1892b) and maintained the possibility that the original collection of specimens of *modestus* were all juveniles, contrary to published documentation (Buller 1888). Ultimately, Buller (1905) acquiesced once more, retaining *modestus* separately in *Cabalus,* while assigning *dieffenbachii* to the monotypic genus *Nesolimnas* proposed by Andrews (1896c), a taxon given only tepid endorsement (e.g., Buller 1905; Oliver 1930, 1955; Lambrecht 1933; Peters 1934). Unfortunately, in England the late 19th century saw a revival of the notion that *C. modestus* merely represented juvenile specimens of *G. dieffenbachii,* a setback fostered by Sharpe (1875), a 2-year period of published indecision by Forbes (1892d, 1893d–f), and the omission of *modestus* as a member of *Cabalus* by Sharpe (1893a, b). Six months later, Forbes (1893g, h) joined with Sharpe (1893b–e, 1894) in recognizing the two species as distinct. However, in February 1894 a further misfortune of publication, in the serial *Catalogue of Birds in the British Museum* (Sharpe 1894), repeated the synonymy of *C. modestus* with *G. dieffenbachii.* Fortunately, this *lapsus* was corrected among the addenda to the volume by Sharpe (1894), and the distinctness of *modestus* was reaffirmed strongly by Newton (1896). Thus, recognition of both *modestus* and *dieffenbachii* as full species, predicted by Salvadori (1893a, b), prevailed before the turn of the century.

Taxonomy resolved, the fates of the two endemics were increasingly becoming recognized to be were sealed. Martin (1885) unsuccessfully had appealed to preserve both *modestus* and *dieffenbachii* through the purchase of island reserves. Buller (1888:122) translated a letter from a contemporary resident of Chatham Island that "... the Moeriki [*dieffenbachii*], disappeared in the third year after the occupation of this island by the Maoris." Under the auspices of Forbes and W. Rothschild, W. Hawkins searched for living specimens of *Cabalus modestus* and *G. dieffenbachii* during 1892 (Forbes 1893c). This belated effort resulted in the rediscovery of dwindling numbers of the former on the islet of Mangere, but established the apparent loss of the latter (Forbes 1893c; Buller 1896). Kirk (1895: 10) concluded that *C. modestus* was "... on the verge of extinction, if it be not already extinct ... the first act of the settler [of Mangere] having been to capture all the specimens of the *Cabalus* that he could find, in order to realize their market-value." Eradication of *modestus* was realized within the decade (Buller 1905), and Rothschild (1907b:209–210) listed both *dieffenbachii* and *modestus* as "Quite Extinct.—Externally Known."

Skin specimens of these two focal species were inadequate for resolution of the evolutionary bases for similarity of form and flightlessness. Fortunately, Forbes (1893c) and Falla (1960) reported subfossil skeletal remains referred to *G. dieffenbachii* on Pitt Island, and abundant subfossil remains of both *G. dieffenbachii* and *C. modestus* were found later on Mangere, the majority recovered from limestone crevices that entrapped and preserved skeletal elements of both rallids en masse (Tennyson and Millener 1994; Millener 1999). Evidence indicates that the two species were sympatric (Fig. 7) at least on Chatham (900 km^2), Pitt (62 km^2), and Mangere (<1 km^2) subsequent to the last glacial period (Olson 1975b; Ripley 1977; Turbott 1990; Tennyson and Millener 1994).

Limiting similarity of sympatric relatives or Gause's principle (Wattel 1973), a generality consistent with the biological species concept (BSC) and related to the concepts of ecological isolation and resource partitioning (e.g., Lack 1971a; MacArthur 1972a; Cody 1974; Schoener 1974), is of questionable explanatory power in many instances (e.g., *Porzana* of Hawaii [Table 3]). Reproductive isolation was sine qua non of species status under the BSC, the dominant perspective of ornithological systematics during the 20th century (e.g., Mayr 1963, 1969, 1976; Ripley 1977; American Ornithologists' Union 1983). This view held that limited differentiation, conterminous distributions, and evidence of hybridization were characteristic of subspecific status. Most other authorities, although recognizing

TABLE 3. Known taxa of recently extirpated and subfossil crakes (*Porzana*) of the major islands in the Hawaiian archipelago, including Laysan Island. Sequenced in order from easternmost (youngest) to westernmost (oldest) member of the westward-moving hotspot archipelago emanating from the Loihi Seamount, with reductions in area and maximal elevation resulting from erosion and subsidence following volcanic origin (Carson and Clague 1995). After Olson and James (1991), who averred that (p. 49) "... as many as five additional species are represented," which remain without formal description or names; it is probable that other, comparatively poorly known islands (e.g., Kaho'olawe, Lanai, Niihau, and Kaula) hosted *Porzana* as well.

Island	Area (km²)	Maximal elevation (m)	Taxon by body size		
			Small	Medium	Large
Hawaii	10,410	4,205	"small" *Porzana* sp.	*P. sandwichensis*	"large" *Porzana* sp.
Maui	1,885	3,055	*P. keplerorum*	"medium" *Porzana* sp.	*P. severnsi*
Molokai	671	1,515	*P. menehune* (smallest)	—	—
Oahu	1,549	1,231	*P. ziegleri*	—	*P. ralphorum*
Kauai	1,437	1,598	—	"medium" *Porzana* sp.	"large" *Porzana* sp.
Laysan	4	11	*P. palmeri*	—	—

that *G. dieffenbachii* probably was closely related to *G. philippensis,* elevated the former to species status (e.g., Scarlett 1979; Livezey 1998).

Refugium sublimus: a tale of two takahes.—Similar in some ways to the foregoing controversy, a pair of unique, flightless rails from the main islands of New Zealand suffered taxonomically from a combination of shared peculiarities and comparatively subtle differences. This couplet of congeners offers unique insights into the evolutionary processes and timescales associated with the morphological changes underlying avian flightlessness, an opportunity recognized by Buller (1894). Owen (1848b) described a large rail (*Notornis mantelli*) from the North Island of New Zealand based on subfossil elements collected by W. Mantell during the previous year (Andrews 1988). European settlers in New Zealand adopted the Maori name for the bird—*Takahe.* Although remarkable for its size and obvious pectoral reduction, preliminary comparisons of the material nevertheless permitted the early recognition of the close relationship between *Notornis* and the widespread genus *Porphyrio* (Owen 1848a, b). Additional skeletal material was recovered from a number of other sites on the North Island (Kinsky 1970). The ultimate extinction of the species was considered likely years earlier (Buller 1888), a fate attributed largely to the ravages of rats (*Rattus* sp. [Smith 1893:518]) or introduced ferrets (*Mustela putorius* [Buller 1898]). The last living specimen of the North Island Takahe appears to have been captured in 1894 (Phillipps 1959), before the preservation of a single skin specimen.

Meanwhile, on the South Island of New Zealand, a flightless relative of *Notornis* was known from a handful of specimens collected during the 19th century (Fig. 2). The first specimen procured in the south was taken on Resolution Island in 1849, and purchased for study by W. Mantell, also the discoverer of the North Island form (Andrews 1986); Mantell reportedly saw the skin of a recently killed North Island specimen hanging in a Maori village (Swinton 1958). The discovery of the first skin specimen was reported widely (e.g., Mantell 1850, 1852; Gould 1852a, b), and for some years was interpreted as evidence of the survival of *Notornis mantelli* (described from the North Island based solely on osteological material) on the South Island of New Zealand. However, the southern form was described as taxomically distinct by Meyer (1883), who included it with the northern taxon described by Owen (1848b) in the narrowly conceived genus *Notornis* as a distinct species (*N. hochstetteri*), a proposal also finding favor with Buller (1905).

Additional skin specimens of the South Island form were collected in 1851 on Secretary Island and 1879 in an area north of Mararoa River (Watson 2001), and Buller (1881:239) was persuaded that "... many yet survive to reward the future search of the Southern naturalist." Early works on the South Island Takahe outnumbered the known specimens (e.g., Park 1888a, b, 1890; Melland 1889; Benham 1898a; Henry 1899; Forbes 1901, 1923). The last specimen secured during the 19th century was taken near Lake Te Anau in 1898 (Benham 1898a; Buller 1898; Ripley 1977; Andrews 1988). Skeletal remains of the southern form also were preserved (Parker 1885a, b), and preliminary study of the internal anatomy of *P. hochstetteri* ensued (Benham 1898b, 1899).

For the next 50 years, the South Island Takahe was not reported again, and the species generally was considered to have followed its northern relative to extinction (e.g., Henry 1899). Therefore, the ornithological community celebrated the discovery by G. B. Orbell in 1948 of a population of South Island Takahes, although prior signs of the survival of species were found (Reid 1974a), and other observers subsequently came forward to report earlier encounters in the field (Reid 1978). This remnant population inhabited a few alpine valleys (750–1,200 m in altitude) in the Murchison Mountains in the extreme southwestern part of the South Island (Falla 1949, 1951; Fleming 1951; Turbott 1951a, b, 1967; Williams 1952, 1960; Wisely 1956). Despite intensive management and protection (Williams et al. 1976; Bunin 1995; Ryan 1997), this population remains critically endangered (Ballance 2001; Lee 2001). Recovery of the population increasingly rests on captive breeding (Reid 1969; Bunin and Jamieson 1996a; Bunin et al. 1997; Maxwell and Jamieson 1997; Eason and Williams 2001; Jamieson and Ryan 2001).

The nominate form of the Takahe differs from its southern counterpart principally in its substantially larger size (Williams 1960; Scarlett 1972), although subtle, largely allometric differences in proportions also are evident (Trewick 1996a; Trewick and Worthy 2001). The two populations are treated commonly as subspecies, either as the sole species in the genus *Notornis* (Peters 1934; Kinsky 1970; but see Oliver 1955) or as a species in the closely related genus *Porphyrio* (Ripley 1977). Most recently, Trewick (1996a) and Livezey (1998) advocated a return to specific rank for the two takahes and the inclusion of the species in the senior genus *Porphyrio* (Figs. 8, 9), a conservatism followed by some (e.g., Trewick and Worthy 2001) but ignored by others (e.g., Eason et al. 2001). The relationships of these forms to two other extinct, flightless swamphens in the region (*P. albus* [Lord Howe Island] and *P. kukwiedei* [New Caledonia]) and an unnamed "huge flightless" *Porphyrio* sp. from New Ireland (Steadman et al. 1999:2566) remain uncertain, although at least the latter two may belong to a clade including flightless forms (Balouet and Olson 1989; Garnett 1993; Trewick 1996a; Livezey 1998).

The differentiation of conterminous, sister populations of flightless rails on the North and South islands of New Zealand, is paralleled by parapatric populations of Weka (*Gallirallus greyi* and *G. australis*), the gruiform genus *Aptornis* (Owen 1870, 1879; Hamilton 1892; Andrews 1896e; Livezey 1994), and the anseriform genus *Cnemiornis* (Hector 1873a, b; Owen 1875, 1879; Livezey 1989a; Worthy et al. 1997). The survival of a relictual, southern population of these flightless swamphens, formerly widespread in New Zealand (Reid 1978), as well as their unique ecology (Trewick 1996b), has inspired considerable discussion of the likely causes of the demise of the northern and near-demise of the southern forms of this group. Two primary hypotheses—anthropogenic agencies and climatic effects—have won advocates, and are considered by most experts to have acted solely or in combination on "*Notornis*" and *Tribonyx* (Baird 1984, 1985, 1986, 1991a, 1992; Mills et al. 1984, 1988). The combination of climatic and human factors have been implicated in other avian extinctions (Burney et al. 1997; Worthy 1999a, b). Given the indisputable, profoundly negative impacts of humans on the endemic birds of New Zealand (King 1984) and the swamphen of Lord Howe Island (Hindwood 1940), at least partial culpability of humans seems inescapable.

UNDECLARED PACIFIC CAMPAIGNS: A LEGACY OF TROPICAL DEVASTATION

Relevance of casualties.—Many species of flightless rails were extirpated within decades of discovery. Specific examples can provide compelling insights into

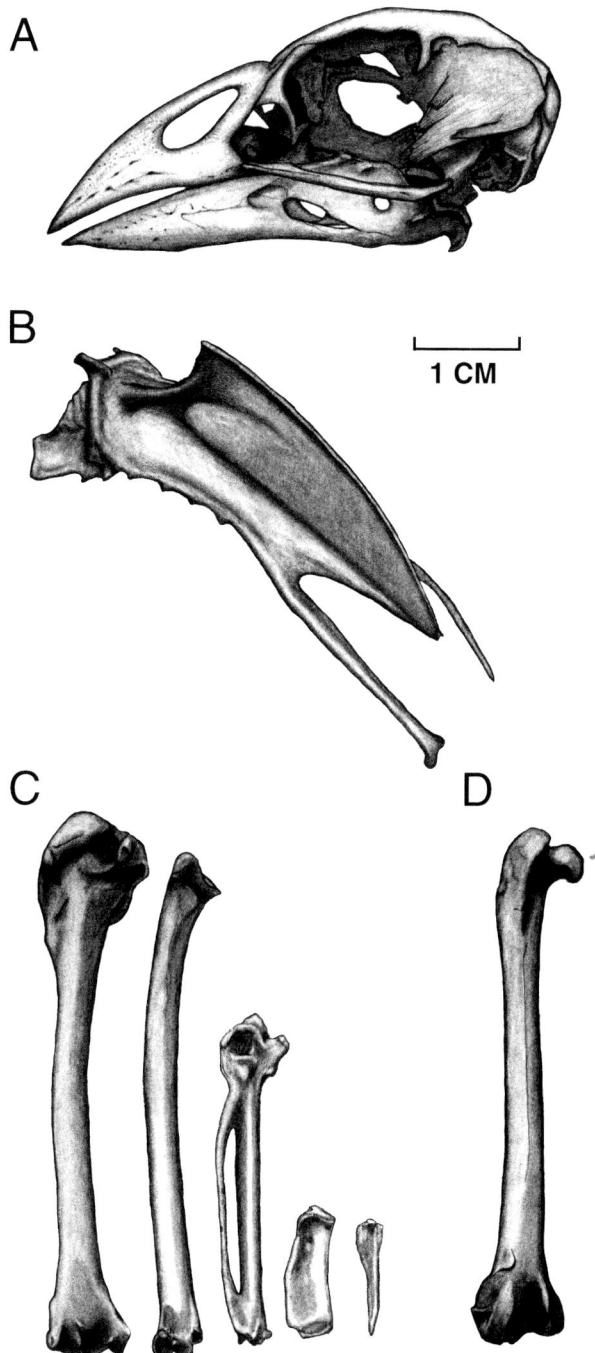

FIG. 8. Selected skeletal elements of *Porphyrio melanotus* (KUMNH 57529): **A**, skull and mandibula (lateral view); **B**, sternum (oblique, ventrolateral view); **C**, left to right (left elements, ventral views), humerus, ulna, carpometacarpus, and ossa digiti majoris; and **D**, femur (right, cranial view). Anatomical abbreviations are defined in Appendix 3. By B. C. Livezey.

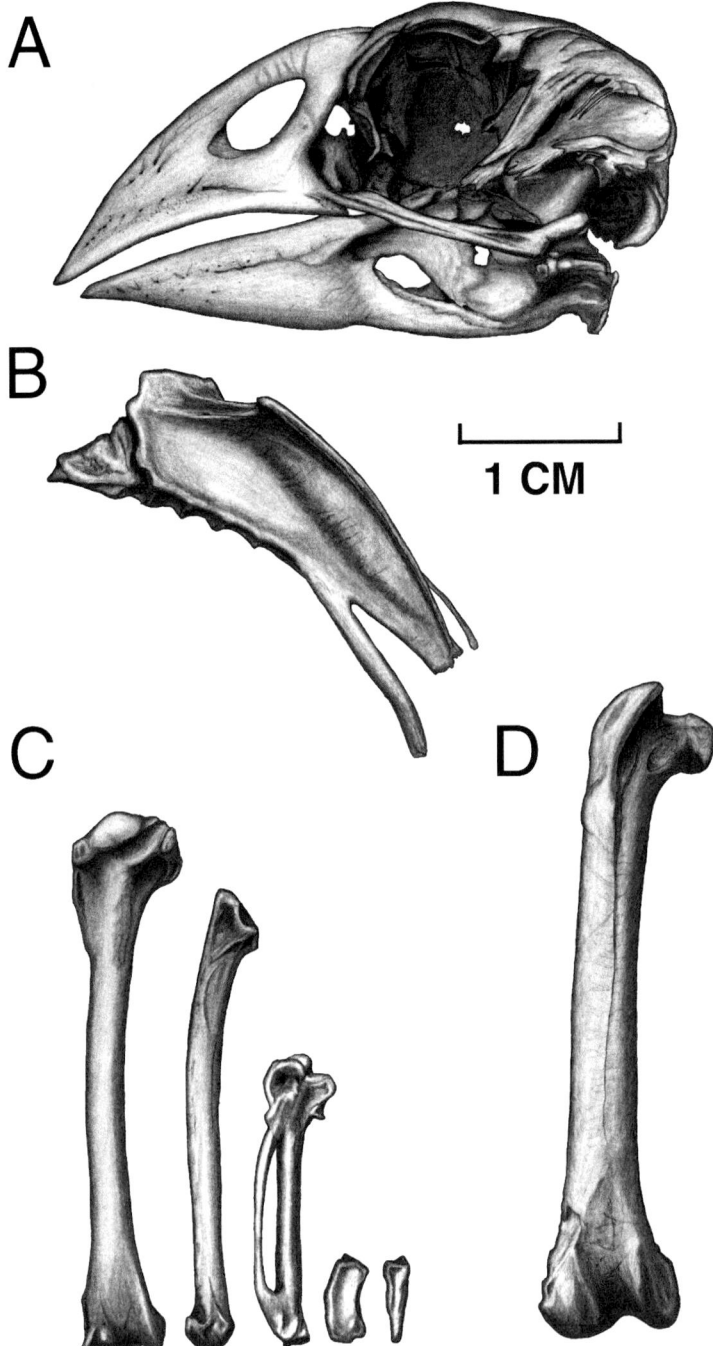

FIG. 9. Selected skeletal elements of *Porphyrio hochstetteri* (USNM 612797), illustrating megacephaly, paedomorphic reduction of the sternum (especially carina sterni), relatively underdeveloped elements of the pectoral limb, and peramorphic pelvic elements of this flightless species: **A,** skull and mandibula (lateral view); **B,** sternum (oblique, ventrolateral view); **C,** left to right (left elements, ventral views), humerus, ulna, carpometacarpus, and ossa digiti majoris; and **D,** femur (right, cranial view). Anatomical abbreviations are defined in Appendix 3. By B. C. Livezey.

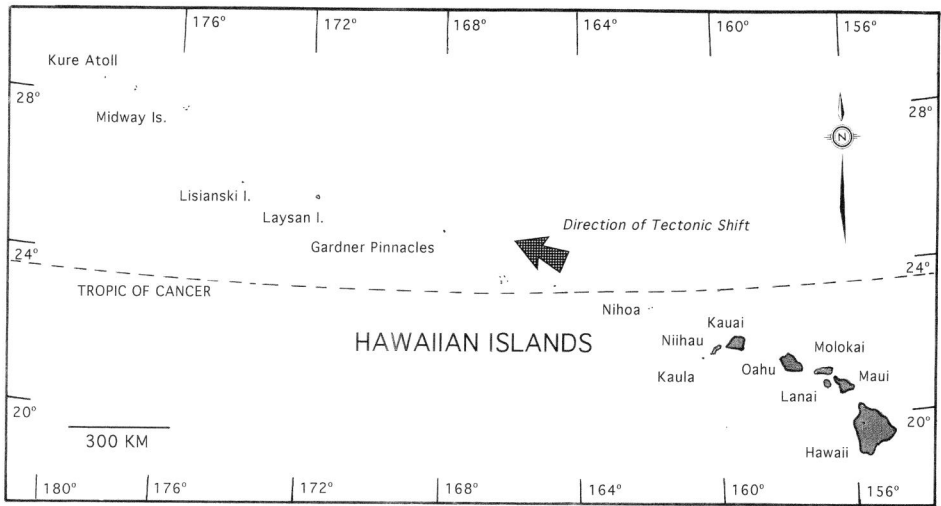

FIG. 10. Map of the Hawaiian Islands, with vectors indicating movement of underlying plates with respect to volcanic hotspot (currently coincident with large, easternmost island of archipelago, Hawaii). See Table 3 for details.

these losses, and in so doing indicate the vulnerabilities, avenues for conservation, and possible "overspecialization" of these uniquely apomorphic, evolutionary lineages. The following accounts describe three comparatively well-documented examples of the widespread devastation of flightless rails from a region of unique diversity of flightless birds, the Pacific islands.

*Hawaiian Crake (*Porzana sandwichensis*).*—Formerly endemic to the largest and geologically youngest of the Hawaiian group (Fig. 10), Hawaii proper (19°30′N, 156°30′E), this species is perhaps most comparable among New Zealand rallids to *Gallirallus dieffenbachii* in its history after contact with Europeans. Hunting, destruction of habitat, and the destruction caused by both intentionally and inadvertently introduced predators and pathogens probably all contributed to the extirpation of *Porzana sandwichensis,* some of which undoubtedly preceded the arrival of Europeans (Olson 1999) and led to extinction of many congeners (Table 3). Last recorded reliably in the mid 1860s, only seven skin specimens of the species are known (Olson and James 1994; Olson 1999). However, variation in plumage in these few skins motivated early taxonomists to recognize two species (*sandwichensis* and *millsi*), considered in recent decades as age-related plumages (Greenway 1967; Ripley 1977) or possible subspecies (Olson 1999). The meager data on life history derives primarily from secondhand accounts recorded in general faunal compendia. These limited data indicate that the species, like most insular crakes, evidently inhabited both open country and woodlands, subsisted on a generalized diet of invertebrates, nested on the ground, and probably was preyed upon by native raptors (Wilson and Evans 1890–1899; Rothschild 1900, 1907a; Henshaw 1902; Perkins 1903).

*Wake Island Rail (*Gallirallus wakensis*).*—Described by Rothschild (1903), most aspects of the life history of the Wake Island Rail (e.g., egg size) never will be known. Wake Island, to which *Gallirallus wakensis* was endemic, is one of the most isolated of Pacific islands (19°18′N, 166°35′E), one having little elevational or ecological diversity and a total area of only 23 km^2 (Fuller 1987, 2001). Early visitors to the island reported that the species was both common and tame (accounts from 1923 by A. Wetmore fide Ripley 1977), and breeding evidently was most common during the late summer (Fuller 1987, 2001).

Circumstantial evidence establishes beyond reasonable doubt that *Gallirallus wakensis* was extirpated by means of the most direct and intense human exterminations suffered by an avian species during modern times. Although the ultimate annihilation of the species presumably was not premeditated per se, the purposeful and unrestrained killing of birds cannot be contested (Greenway 1967), a chronicle that contrasts with the typically collateral extirpation of insular rails through agrarian activities. Considered common before World War II, not a single record of a living specimen was made after the occupation of the island by poorly provisioned Japanese soldiers in 1945 (Blackman 1945; Munro 1946), and it is generally accepted that the species was hunted to extinction by the starving occupational forces within a period of a few years, approximately 40 years after its formal description.

*Laysan Crake (*Porzana palmeri*).*—This former endemic of tiny Laysan Island (25°50′N, 171°50′E), in the Leeward Islands (1,200 km northwest of Niihau Island, Hawaiian Islands), was first observed upon the discovery of the island by the sailors of the *Moller* in 1828 (Kittlitz in Rothschild 1900); the species was extinct 116 years later (Fig. 4). Originally described by Frohawk (1892), this crake was placed in the new genus *Porzanula* to reflect both its small size and obvious affinities to the typical and widespread crakes of the genus *Porzana.* Its membership among the crakes never was doubted, but relationships within this complex remain poorly resolved (Livezey 1998; Olson 1999). The former abundance and late demise of the species provided opportunities for scientific collection (Greenway 1967). Olson (1999) tallied 259 specimens of *Porzana palmeri* in museums worldwide, comprising many study skins, a respectable number of skeletons, and a few fluid-preserved specimens (Wood et al. 1982; Wood and Schnell 1986).

The life history of the species is known largely from the firsthand accounts by Rothschild (1893a), Schauinsland (1899), Fisher (1906), Bryan (1915), Baldwin (1945, 1947), and Munro (1947); additional data were summarized by Ripley (1977). Although flightless, this small, inquisitive crake employed movements of the wings for balance during terrestrial locomotion, which are pectoral accommodations of broader utility among bipedal tetrapods and in some taxa accompanied by movements of the tail (Clark and Alexander 1975; Heglund et al. 1982a; Heglund and Taylor 1988; Gatesy and Biewener 1991; Gatesy and Dial 1993). Like many insular endemics (Humphrey et al. 1987), the species was tame and unsuspicious of humans and vulnerable to other introduced mammalian threats. Frohawk (1892) reported an evening "chorus" of vocalizations variously described by subsequent observers. Dill and Bryan (1912) estimated the population on Laysan to number approximately 2,000, implying densities approaching 800 birds per square kilometer of habitable land; populations introduced to Midway Atoll also attained high densities (Bryan 1915; Blackman 1945; Ely and Clapp 1973). Laysan Crakes built variably canopied terrestrial nests of dry vegetation, laid only two or three comparatively large eggs (31 × 21 mm), and the young foraged independently within a few days of hatching. The diet of young and adult crakes was varied and opportunistic, and included eggs of other birds, meat scraps, maggots, insects, and some vegetable matter. Laysan Crakes evidently relied on the availability of freshwater for survival (Grant 1947).

European hares (*Oryctolagus cuniculus*)—introduced by commercial miners of guano in 1903 and exterminated in 1923—destroyed much of the native vegetation of the island and were a principal cause of the extinction of *Porzana palmeri* on Laysan, although rails were seen commonly on Laysan as late as 1918 (Bailey 1956; Ely and Clapp 1973; Olson 1999). By April 1923, only two crakes were observed during a survey of the entire island by the *Tanager* expedition (Olson 1996). Tragically, although several populations of Laysan Crake introduced on islands in the Midway Atoll (Fisher and Baldwin 1946)—Eastern Island in 1891 and Sand Island in 1910—persisted until 1943 (when extirpated by shipborne rats), the species was not reestablished on Laysan Island after the extinction of the original population. In what was probably the first reintroduction intended to preserve an avian species, eight *P. palmeri* were released on Laysan in April 1923; by then the island was largely defoliated (A. Wetmore fide Olson 1996). At least two of these introduced birds died within a month, and the failure of the attempted triage was confirmed by subsequent collectors on the island, despite the removal of the hares and subsequent regrowth of native vegetation (Bailey 1956).

The extirpation of the Laysan Crake is especially lamentable in light of the fate of the sympatric Laysan Duck (Anatidae: *Anas laysanenesis*), formerly more widespread within the Hawaiian archipelago (Cooper et al. 1996), which preceded the crake to the brink of extinction. The suggestion by Ripley (1977:21) that interspecific competition between *A. laysanensis* and *Porzana palmeri* was important if not critical in the ultimate demise of the latter has not received empirical support. The wild population of the duck reached a nadir of six birds in 1911 (Warner 1963), but conservation efforts (deriving from the progeny of a single fertile female and subsequent captive breeding) proved successful in the preservation of the species (Moulton and Weller 1984).

The recurrent pattern among the decimated rallids of the Pacific is that a century of contact with European immigrants is likely to drive flightless taxa toward the brink, if not over the brink, of extinction. To date, restoration efforts have proven ineffective, especially if efforts are concentrated on the endangered species per se instead of land use on the islands. All efforts are undermined during periods of human warfare or economic destitution in the regions of interest.

Modalities and Loss of Avian Flight

Diversity of Avian Flight

Regardless of taxon, a number of anatomical parameters determine the feasibility and characteristics of flight, with critical dimensions ranging from body mass (Lull 1906; Brown 1953; Rayner 1988a, 1989, 1990, 1996; Ellington 1991; Marden 1994; Spedding and Lissaman 1998; Butler and Bishop 1999) to vortices and drag (Rayner 1979a, b). Modern birds vary significantly in flying abilities (Rayner 1988a; Norberg 1990; Spedding 1992), including groups characterized by highly maneuverable flight having substantial capacity for changes in direction or (de)acceleration, for example, aerial foragers and many arboreal songbirds

(Hails 1979; U. M. Norberg 1979, 1981a, 1986); energetically consumptive hovering powered by unique wing movements, for example, hummingbirds (Rayner 1979a, 1985a; Wells 1993a, b; Chai and Dudley 1999; Cotton 1996; Chai and Millard 1997; Chai et al. 1997); strong but undulating flight, for example, woodpeckers (Rayner 1985b; Tobalske 1996); flight in which critical lift is provided by rising air columns, for example, soaring birds (Taber 1932a, b; Pennycuick 1972, 1998a; Mendelsohn et al. 1989); sustained, primarily gliding flight in open, windy environments (Pennycuick 1960, 1971, 1982; Parrott 1970; Tucker and Parrott 1970; Wood 1973; Tucker 1987); and swift flight powered by virtually constant wingbeats, for example, gallinaceous birds and waterfowl (R. H. J. Brown 1948, 1951, 1961; Norberg and Norberg 1971; Van Tyne and Berger 1976; Pennycuick 1990; Kuz'mina 1992; Hedenström 1993).

To the extent that the nature of flight capacity of fossil taxa can be adduced, all flightless birds are descendants of species possessed of the last type of flight, that is, power fliers. In light of this fact, it is surprising that more flightless galliforms have not been described, although the number of flightless megapodes known to have inhabited Pacific islands continues to increase (Martin and Steadman 1999). Not surprisingly, the Rallidae fall among the power fliers, and although the family manifests substantial diversity of form, it tends to be characterized as well by low wingspans and wing areas relative to body mass. The latter generalities also apply to galliforms, waterfowl, grebes, and larger alcids, which also include flightless members (Rayner 1988a). Given the functional diversity and obvious utility of flight, it is remarkable that members of more than 30 taxonomic families of birds have lost the capability of flight secondarily (Table 4; Livezey 1995a).

The most massive members of a number of avian genera, especially those with refined diving abilities, also approach the threshold of flightlessness, including loons (Gaviidae) and some waterfowl (Anatidae: *Somateria, Mergus,* and *Biziura*), a tendency inferred both from available data on wing loading and field observations of locomotory behavior (Magnan 1922; Poole 1938; Schorger 1947; Savile 1957, pers. comm.; Meunier 1959a, b; Meinertzhagen, unpubl.; Templin 1977; Livezey 1989b, 1990, 1993a–c; Guillemette 1994). Large size may limit premigratory hyperphagia in *Somateria* (Guillemette 2001). Temporary flightlessness resulting from seasonal gains in body fat also has been reported for the Eurasian Coot (*Fulica atra*) of the Caspian Sea (M. Patrikeev fide P. B. Taylor 1998). Fjeldså (1981) suggested that age-related gains in body mass in the Giant Coot (*Fulica gigantea*) may render adults effectively flightless and restrict dispersal among lakes to immature birds. Like genuinely flightless species, these taxa poised at the threshold of flightlessness also are power fliers. For example, in most Anatini (sensu Livezey 1986a, 1991), rapid wingbeats (approximately five per second) are essential to the maintenance of air speeds at which adequate lift counters the acceleration of gravity on these heavily wing-loaded birds (Meinertzhagen 1955; Hartman 1961; Greenewalt 1962; Raikow 1973; Livezey 1990).

In some taxa, individuals experience temporary flightlessness stemming from various causes. These include atrophy of pectoral musculature, which in some groups is related to molt (Rosser and George 1985, 1987; Piersma 1988; Gaunt et al. 1990; Marks et al. 1990); and molt of remiges, which often is associated with seasonal movements and temporary occupancy of refugia (Murphy 1936;

TABLE 4. Taxonomic, geographical, and stratigraphic distributions of flightless birds, after Livezey (1995). Daggers precede taxa known only from (sub)fossil remains.

Taxonomic group(s)	Flightless members			No. of species		Key references
	Affected taxa	Period of occurrence	Distribution	Fossil	Modern	
†Alvarezsauriformes						
Alvarezsauridae	All members	Upper Cretaceous	Mongolia, South America	4+	0	Perle et al. 1993, 1994; Chiappe 1995a, b; Chiappe et al. 1998; Padian and Chiappe 1998
†Patagopterygiformes						
Patagopterygidae	*Patagopteryx*	Upper Cretaceous	South America	1	0	Alvarenga and Bonaparte 1992; Chiappe 1996
†Hesperornithiformes						
Hesperornithidae	All members	Cretaceous	Europe, N. and S. America	4+	—	Marsh 1880; L. D. Martin 1984
Baptornithidae	All members	Cretaceous	N. (possibly S.) America	2+	—	Martin and Tate 1976
Paleognathae						
†Dinornithiformes	All members	Miocene–Recent	New Zealand	11	0	Owen 1879; Archey 1941; Oliver 1949; Cracraft 1976a, b; Worthy 1988a, b, 1990; Kooyman 1991
Apterygidae	All members	Pleistocene–Recent	New Zealand	0	3	Owen 1841; Parker 1892; McGowan 1982
Casuariidae	All members	Pleistocene–Recent	Australia, New Guinea	0	3	Pycraft 1990; Raikow 1985a
Dromaiidae	All members	Pleistocene–Recent	Australia	1	2	Pycraft 1990; Raikow 1985a
†Aepyornithidae	All members	Pleistocene–Recent	Madagascar (Africa, Europe)	2–9	—	Andrews 1896a; Wiman 1935
Struthionidae	All members	Pliocene–Recent	Eurasia, Africa	5+	1–2	Pycraft 1990; Lang 1956; Webb 1957
Rheidae	All members	Eocene–Recent	South America	1	2	Pycraft 1900; Müller 1963
Galliformes						
Megapodiidae	†*Sylviornis**	Recent	New Caledonia, Fiji	1	0	Poplin and Mourer-Chauviré 1985; Worthy et al. 1999

TABLE 4. Continued.

Taxonomic group(s)	Flightless members		Distribution	No. of species		Key references
	Affected taxa	Period of occurrence		Fossil	Modern	
Anseriformes						
†Dromornithidae	All members	Miocene–Recent (?)	Australia	8	—	Rich 1979; Murray and Megirian 1998
Anatidae	*Cnemiornis*	Pleistocene–Recent	New Zealand	2	—	Livezey 1989a; Worthy et al. 1997
	Branta	Pleistocene	Hawaii	2	0	Olson and James 1991; Livezey 1996a
	Thambetochenini	Pleistocene	Hawaii	4	—	Olson and James 1991; Livezey 1996a
	Tachyeres	Recent	South America	0	3	Livezey and Humphrey 1986
	Anas	Recent	New Zealand, Amsterdam I.	1+	1+	Livezey 1990, 1993a; Olson and Jouventin 1996
	†*Chendytes*	Pleistocene	Pacific N. America	2	—	Livezey 1993c, 1995c
	Mergus	Recent	New Zealand	1	1	Livezey 1989b, 1995c; Millener 1999
Sphenisciformes						
Spheniscidae	All members	Eocene–Recent	Southern Hemisphere	18+	18	Schreiweis 1982; Livezey 1989c
Podicipediformes						
Podicipedidae	Involves three genera	Recent	Central and S. America	0	3	Livezey 1989d
Pelecaniformes						
Phalacrocoracidae	*Compsohalieus*	Recent	Galápagos Islands	0	1	Livezey 1992a
Anhingidae	†*Meganhinga*	Miocene	Chile	1	0	Alvarenga 1995
†Plotopteridae‡	—	Oligocene–Miocene	N. Pacific	3	—	Olson and Hasegawa 1979, 1996
Ciconiiformes						
Ardeidae	†*Megaphoyx*§	Recent	Rodriguez, Mascarene Islands	1	—	Hachisuka 1937
	Nycticorax sp.	Recent	Niue Island, South Pacific	1	0	Worthy et al. 1998

TABLE 4. Continued.

Taxonomic group(s)	Flightless members			No. of species		Key references
	Affected taxa	Period of occurrence	Distribution	Fossil	Modern	
Threskiornithidae	†*Apteribis*	Pleistocene	Hawaii	2	—	Olson and Wetmore 1976; Olson and James 1991
	†*Xenicibis*	Pleistocene	West Indies	1	—	Olson and Steadman 1977, 1979
	†*Borbonibis*§	Recent	Réunion, Mascarene Islands	1	1	Mourer-Chauviré et al. 1995a, b
†Gastornithiformes‖						
Gastornithidae	*Gastornis*	Paleocene	Europe, Asia	3	—	Owen 1856; Olson 1985; L. D. Martin 1992
Diatrymidae	*Diatryma*	Eocene	Europe, N. America	4+	—	Olson 1985; Witmer and Rose 1991; Andors 1992
Gruiformes						
Mesitornithidae‖	*Mesitornis, Monias*	Pleistocene–Recent	Madagascar	0	3	Evans et al. 1996; Livezey 1998
†Phorusrhacidae	—	Oligocene–Pliocene	N. and S. America, Europe	10+	—	Lydekker 1893; Andrews 1899; Sinclair and Farr 1932
†Bathornithidae	cf. *Paracrax*	Eocene–Miocene	N. America	3+	—	Cracraft 1971, 1973a; Livezey 1998
Rhynochetidae‖	*Rhynochetos*	Pleistocene–Recent	New Caledonia	1	1	Parker 1869; Hunt 1996; Livezey 1998
†Aptornithidae	*Aptornis*	Pleistocene–Recent	New Zealand	2	0	Owen 1870, 1872; Livezey 1994, 1998
Gruidae	†*Grus cubensis*	Pleistocene	Cuba	1	0	Fischer and Stephan 1971a
Rallidae	Involves 16+ genera	Pleistocene–Recent	Oceania worldwide	25+	30+	Olson 1973a; Taylor 1996; present study
Charadriiformes						
Alcidae	†Mancallinae	Miocene–Pleistocene	Pacific N. America	8	0	Livezey 1988
	†*Pinguinus*	Pliocene	Coastal N. Atlantic	1	1	Livezey 1988
Strigiformes						
Strigidae	†*Ornimegalonyx*	Pleistocene	Cuba	1+	0	Arredondo 1977
Caprimulgiformes						
Aegothelidae	†*Megaegotheles*	Recent	New Zealand	1[?]	—	Scarlett 1968; Rich and Scarlett 1977; Olson et al. 1987

TABLE 4. Continued.

Taxonomic group(s)	Flightless members			No. of species		Key references
	Affected taxa	Period of occurrence	Distribution	Fossil	Modern	
Psittaciformes						
Psittacidae	*Strigops habroptilus*	Recent	New Zealand	0	1	Livezey 1992b
Columbiformes						
†Raphidae	*Raphus, Pezophaps*	Recent	Mascarene Islands	0	2+	Owen 1866a, Livezey 1993b
Columbidae	Undescribed	Recent	Fiji	0	1	Worthy et al. 1999
	†*Dysmoropelia*‖	Pleistocene	St. Helena, S. Atlantic	0	1	Olson 1975c
Coraciiformes						
Upupidae	*Upupa antaios*#	Pleistocene	St. Helena, S. Atlantic	0	1	Olson 1975c
Passeriformes						
Acanthisittidae	†Several genera	Pleistocene–Recent	New Zealand	3	1	Millener 1988, 1989; Millener and Worthy 1991
Atrichornithidae#	*Atrichornis*	Recent	New Zealand	0	2	Raikow 1985b, Rich et al. 1985; Zusi 1985
Emberizidae	†*Emberiza alcoveri*	Pleistocene–Holocene	Canary Islands	0	1	Rando et al. 1999

* Classification tentative.
‡ Ordinal affiliation tentative.
§ Synonymy with *Threskiornis* and alleged Reunion Solitaire (Raphidae: *Raphus solitarius*), as well as flight status, remain controversial.
‖ Composition and position of this order remain controversial.
¶ Flightless condition questionable or disputed; similar tendencies toward pectoral reduction evident in an insular vireo (Olson 1994).
# Perhaps weakly flighted.

Gullion 1953a; Sullivan 1965; Woolfenden 1967; Salomonsen 1968; Haukioja 1971; Jehl 1990; Marks et al. 1990; Marks 1993). Many Rallidae also are considered to be rendered flightless by synchronous molt of remiges (Livezey 1998); the resultant vulnerability presumably is eased by the comparatively refined ability to swim and dive (Weller 1999), which is less easily accommodated in more-terrestrial groups. Aerodynamic effects of gaps in remiges during molt (Hedenström and Sunada 1999) and associated changes in pectoral musculature and ectoparasitic loads (Harper 1999) also may constrain patterns of molt in flighted birds. Other factors that can temporarily impair flight include seasonal fluctuations in body mass (Jehl 1997) and short-term weight gains during periods of intense feeding (Humphrey and Livezey 1982; Guillemette 1994).

In a case similar to that of Flying Steamer-Ducks (Anatidae: *Tachyeres patachonicus*), in which some males are at least temporarily flightless (Humphrey and Livezey 1982; Livezey and Humphrey 1986), male Musk Ducks (Anatidae: *Biziura lobata*) are substantially larger than female conspecifics and possess wing loadings sufficiently great so as to compromise flight (Hobbs 1956; Dickison 1962; Lowe 1966; McCracken et al. 2000). Still other avian taxa, including some extinct forms, approach(ed) permanent flightlessness to varying degrees (e.g., Stonor 1942; Raikow 1985b; Zusi 1985; Olson 1990). Predictably, even temporary flightlessness in continental communities can increase vulnerability to predation (Oring 1964), as can temporary gains in body mass (Gosler et al. 1995; Metcalfe and Ure 1995; Kullberg et al. 1996).

LOCOMOTORY MODULES OF BIRDS

The functional implications of the medium in which species effect movements are the principal selective agents for anatomical refinements of terrestrial, arboreal, or aquatic locomotion among flighted Aves (R. Å. Norberg 1986; Raikow 1985a; Denny 1990; Liem 1990; U. M. Norberg 1990; Bruinzeel et al. 2000). The appendicular refinements manifested among Aves are complicated and enriched by the generality that most birds possess multiple locomotor modules (Gatesy and Dial 1996a), that is, they have achieved locomotory proficiency in at least two media. Among the Rallidae, locomotory modules other than aerial include terrestrial and aquatic (Sigmund 1959). Goslow (1989) considered the evolutionary innovations permitted by such locomotory diversity, which turn in part on the bipedal terrestrial locomotion of birds, a capacity that frees the forelimb for flight or the sacrifice thereof. However, functional reductions of the forelimb in most other tetrapods would have critical implications for *all* modes of locomotion, including terrestrial.

However, the subtlety of the anatomical changes associated with the loss of flight among carinate birds, coupled with an understandable preoccupation with the evolutionary development of complex structures as opposed to their diminution (e.g., Arthur 1984, 1988, 1997; Atchley and Hall 1991; Hall 1996; Hansen and Martins 1996; Wagner and Altenberg 1996; Wagner et al. 1997; Wagner and Schwenk 2000), belie the significance of the functional and evolutionary trade-offs represented by avian flightlessness. Flightless birds are distinguished from their plesiomorphically flighted relatives by a convergent sacrifice of the locomotory option deemed quintessential of Aves. Despite considerable variation in efficiencies in alternative locomotory modes (e.g., Bruinzeel et al. 2000), pelvic anatomy provides the clearest indications of specialization for ambulatory locomotion. However, the morphological correlates of terrestrial habits generally are less pronounced than their natatorial counterparts (Stolpe 1932; Raikow 1985a), and for this reason most such comparisons are based on the relative length and robustness of appendicular elements.

Diverse Perceptions of Avian Flightlessness

The great bird will take its first flight, from the back of the great Cecero, filling the universe with wonder, filling all writings with its fame, and eternal glory to the nest where it was born.—Leonardo da Vinci (1505: inside back cover), translation by Marinoni (1976:81)

But the conditions of life were altogether too easy for it; a superabundance of food, to be had for the picking of it up, and an absence of any enemies to interfere with it, produced their usual result, and degeneration set in—a result, I may remark, by not means confined to the Dodo, but one of which, under similar circumstances, proud Man himself furnishes conspicuous examples.—Wiglesworth (1900:24)

Birds, in short, from the first were destined by Nature to possess the air . . . . In birds the fore-limb has in all cases served as an organ of flight; even where this member has been reduced to the merest vestige, it is clear that the modelling thereof is that of a wing. . . . The factor which, more than any other, has secured for birds the high place which they hold in the affections of men, is unquestionably that of flight. Thereby they ever keep themselves, as it were, before the public, and give life and beauty to the world around them.—Pycraft (1910:1–2)

. . . the strongly built New Guinea Flightless Rail, rejoicing in the inappropriate scientific species name *ineptus*—a formidable creature quite able to defend itself and hardly seeming 'inept' to me . . . we found perching solitary high up in mangrove bushes.—Ripley (1977:xv–xvi)

Religious and Popular Views

Historical interpretations and directions of science are influenced by prevailing, nonempirical views of the times. The sentiments expressed by Leonardo da Vinci (1505) in the *Codex* speak clearly to the reverance with which humans view the flight of birds. Da Vinci was motivated to study avian wings, at least in part, by an abiding interest in angelic imagery and an obsession to contrive a machine by which a human might ascend to the heavens. Admiration for avian flight is mirrored in the works of classical artists: angels and many earlier pagan deities were depicted unmistakably as possessing the wings of birds, replete with distinguishable, often realistically enumerated remiges (Rowland 1978; Janson 1991; Fellows 1995; Knapp 1999). Migratory flocks, one of the most conspicuous examples of avian flight, were considered by the ancient Greeks to be messengers from the Olympian gods (Dorst 1962). A heavenly perspective on birds among Christians accords with the first mention of the group in *Genesis* (verse 20), where creation purportedly was directed to ". . . let birds fly beneath the dome of the sky" and a command (verse 28) to humans to extend their domination to ". . . the birds of the air" (C. L. Harris 1981). In his single scientific dialogue, Plato cast the creation of birds, one of the first tasks assigned to the lesser gods by the "creator," as corresponding to the provision of the air itself (C. L. Harris 1981).

The classical view of avian flight is as clearly defined as the loss of this faculty, with avian flightlessness having been characterized as exemplifying ineptitude (e.g., specific epithet of *Amaurornis ineptus*); stupidity (Cuppy 1951); degeneration, sometimes explicitly from original perfection (Owen 1866a; Wiglesworth 1900; Lull 1935); atrophy (Owen 1866a, b); arrested development (Darwin 1859); monstrosity (Goldschmidt 1940); clumsiness (Wiglesworth 1900); indolence (Cunningham 1871); aberration or anomaly of nature (Owen 1848a; Beddard 1898; Wiglesworth 1900); grotesqueness (Halliday 1978a); anachronism (Amadon 1966b); or "terminal" phylogenetic status (Gould 1966). The Ostrich, the first flightless bird (excluding domesticated fowl) to become widely known in Europe, was used to symbolize gluttony, bloodlust, parental negligence, forgetfulness, cowardice, imbecility, hypocrisy, and sloth, among other qualities (Rowland 1978); its flightlessness was likened to "... the man weighed down with 'affecyons and othire vanytes' " (R. Rolle, 14th century, fide Rowland 1978:114).

Even where flightlessness was viewed less fancifully, the functional sacrifice of flight by avian lineages is accorded an exaggerated status among comparable evolutionary phenomena, for example, the more extensive reduction represented by limblessness in reptiles (Stokely 1947; Gans 1975; Lande 1978). By contrast, many evolutionary trade-offs in structure and function such as relative respiratory efficiency, capacity for terrestrial flight, or refinements for feeding are rarely or never cited in this context. This disparity of perception often turns on the historical extinction of many flightless birds, in which ultimate demise is taken as evidence of evolutionary folly. Ironically, in light of the fates of many flighted endemics on oceanic islands subsequent to the arrival of humans (James and Olson 1991; Freed 1999; BirdLife International 2000), the retention of flight may not have saved insular rallids from the coming anthropogenic onslaught. Subjected to virtual ignorance by the public were (are) many flightless lineages extinguished before the historical narratives by Europeans, including extinct anseriforms *Cnemiornis* of New Zealand (Owen 1875, 1879) and moa-nalos of Hawaii (Olson and James 1991).

Unlike the publicized instances of evolutionary "links" among fossil taxa (e.g., *Archaeopteryx* [C. L. Harris 1981]) or modern groups achieving semipopular notoriety (e.g., *Opisthocomus*), flightless birds (with the exception of ratites) seldom were held to be instrumental in the reconstruction of relationships among natural groups (e.g., Olson 1985; Sibley and Ahlquist 1990). Although perennial discussions of the evolutionary implications of *Archaeopteryx* are understandable, speculations in recent decades concerning the flight capacity of this fossil taxon and the origin of avian flight are secondary at best for phylogenetic purposes (Padian and Chiappe 1998), and primarily reflect the traditional equation of flight capacity with avian status. In contrast, flightless birds were viewed as somewhat paradoxical in the retention of some characters identifying them as birds (e.g., feathers) but deprived of the one capacity (flight) considered quintessentially avian.

Although mistaken (Halliday 1978a), the popular view of the Dodo as an aberrant, overly specialized, degenerate, evolutionary oddity preordained to extinction ascended to legendary status (cf. Owen 1866a; Cuppy 1941; Greenway 1967). Flightless rails may conform most closely with this popular perception in being (for the most part) small, trusting, insular, and vulnerable to the guile of the outside world. With respect to extirpation, the ultimate demise of the Great Auk, the last of which were dealt a coup de grâce by hunters informed of the rarity of the species (Livezey 1988; Gaskell 2000), certainly contributes to the fabled helplessness and naiveté of flightless birds.

PREVIOUS SCIENTIFIC STUDIES OF FLIGHTLESS CARINATES

The singular curiosity of avian flightlessness stimulated numerous commentaries in the classical ornithological literature, many of which included examples from both ratites and carinates (e.g., Strickland and Melville 1848; Owen 1879; Wiglesworth 1900; Henry 1903; Lucas 1916; Boubier 1934). In a classic series of works, Lowe (1928a, b, 1930, 1933, 1934, 1935, 1939, 1942, 1944) wrestled with the related issues of neoteny, recapitulation, and phylogeny of flightless birds. Prone to view many flightless birds as anachronistic relicts of land masses connected during ancient times, Lowe found only limited reception of his ideas (Livezey 1995a). Goldschmidt (1940) cited the flightless Galapagos Cormorant (Snow 1966; Livezey 1992a) as an example of his new evolutionary concept of the "hopeful monster," but contributed little information on this remarkable bird not presented in the classic work by Gadow (1902). A half-century later, a critical need for anatomical study of avian flightlessess was reaffirmed by George and Berger (1966).

The century-long debate concerning the taxonomy and flightlessness of steamer-ducks (Anatidae: *Tachyeres*) was sustained into the 20th century with the landmark works by Lowe (1934) and Murphy (1936). This dispute, seemingly resolved by Murphy (1936), was rejoined with the description of a new, flightless species in the genus in the early 1980s (Humphrey and Livezey 1982; Livezey and Humphrey 1986, 1992). As one of the first flightless taxa to be formally described (Table 1), the genus *Tachyeres* also is notable for the fact that three of four extant species are flightless. The steamer-ducks were the most renowned of flightless anseriforms until a spate of extinct, flightless waterfowl were described recently from the Hawaiian Islands (Olson and James 1991). The study of steamer-ducks led to parallel examinations of flightlessness in a number of other carinate groups, including alcids (Livezey 1988), grebes (Livezey 1989d), other waterfowl (Livezey 1989b, 1990, 1993c), penguins (Livezey 1989c), cormorants (Livezey 1992a), parrots (Livezey 1992b), columbiforms (Livezey 1993b), and the gruiform *Aptornis* (Livezey 1994).

Despite the number of flightless Rallidae (Table 1), flightless taxa in taxonomic groups characterized by large size or conspicuously modified pectoral limbs (e.g., ratites and penguins) have received comparatively more study. Among terrestrial neognaths, the Dodo came to epitomize the popular view of avian flightlessness, including great bulk, purportedly ponderous gait, and tameness interpreted as stupidity, the latter conducive to its ultimate extirpation (Strickland and Melville 1848; Hachisuka 1953; Livezey 1993b), but despite this notoriety the phylogeny of the raphids remains only poorly understood (Livezey 1993b; Shapiro et al. 2002). Olson (1973a) revived interest in avian flightlessness and its ontogeny, begun by the classic study of *Atlantisia rogersi* by Lowe (1928a), with a comparative study of flightlessness in several insular rallids of the South Atlantic, including two known only from subfossil remains. Subsequent studies of flightless rails include the myology of *Gallirallus australis* by McGowan (1986), further studies of *Atlantisia rogersi* by Ryan et al. (1989) and Fraser et al. (1992), and the taxonomic works by Diamond (1991) and Olson and James (1991). During this same period, ecological works on flightless birds included that by Weller (1975) on the Auckland Islands Flightless Teal (*Anas aucklandica*) and an overview of avian flightlessness by Mlíkovsky (1982).

In a survey of flightlessness in insects and birds, Roff (1994a) noted the likely evolutionary independence of flightlessness in most carinate birds, with penguins being a notable exception (Livezey 1989c), but rightly pointed out that uncertain relationships within some avian groups having multiple flightless members (notably the rails) render possible the plesiomorphy of flightlessness in some cases. McNab (1994a) summarized available physiological data for flightless birds, including several rails, and confirmed that flightlessness conferred significant benefits (where ecologically permissible) through enhanced metabolic efficiency, and subsequently predicted physiological gains for the Kakapo (Psittacidae: *Strigops habroptilus* [McNab and Salisbury 1995]). Trewick (1996a) performed a morphometric analysis of the flightless swamphens (*Porphyrio*). He concluded that three species should be recognized and that the group may be polyphyletic. Sequence data for these and a number of other Rallidae of the New Zealand region formed the basis for a companion phylogenetic analysis by Trewick (1997a) and a closely related assessment restricted to two rails endemic to the Chatham Islands (Trewick 1997b).

In a family-level survey of relative wing length and flightlessness, McCall et al. (1998) attempted to control for higher-order phylogenetic differences in body form by using the phylogenetic hypothesis proposed by Sibley and Ahlquist (1990), and confirmed that nonpasseriform groups characterized by relatively short wings (e.g., waterfowl and alcids) appear to be predisposed to flightlessness. However, despite the optimistic retrospective cited (Mooers and Cotgreave 1994), the phylogenetic hypothesis by Sibley and Ahlquist (1990) has been challenged from a number of methodological and philosophical perspectives (Cracraft 1987, 1992; Houde 1987; Sarich et al. 1989; Lanyon 1992; Harshman 1994). Cubo and Arthur (2001) tested for correlated morphological changes associated with avian flightlessness and found that pectoral reduction, pelvic enlargement, and apomorphy of the skull co-occurred in a higher-order phylogenetic context. Also, interfamilial relationships inferred in a number of subsequent phylogenetic analyses—including molecular (e.g., Helm-Bychowski and Cracraft 1993; Hackett et al. 1995; Groth and Barrowclough 1999; Mindell et al. 1997, 1999; Klicka et al. 2000), morphological (e.g., McKitrick 1991a; Griffiths 1994; Livezey 1997a), and combined approaches (Lee et al. 1997)—have differed importantly from the those depicted by Sibley and Ahlquist (1990). Disputed groupings include the Rallidae and closely related families (Houde 1994; Houde et al. 1997; Trewick 1997a, b; Livezey 1998). Phylogenetic issues aside, assessments by McCall et al. (1998) were limited to family-level patterns among selected, extant taxa in relative wing length, and therefore did not consider changes in body size per se, sexual dimorphism, feather shape, skeletal features, or pectoral musculature within groups (including the Rallidae), and made no attempt to evaluate underlying ontogenetic mechanisms, evolutionary preconditions, or likely agencies of natural selection. Despite uncertainties in some aspects of the phylogeny of the Rallidae, however, definition of subgroups including flightless taxa is possible with current reconstructions (notably Livezey 1998). Inferences deriving from this level of precision provide important insights into this important evolutionary phenomenon and are the focus of the following analyses and integration with modern theory.

OBJECTIVES OF THIS STUDY AND ORGANIZATION OF THIS PAPER

This paper describes a comprehensive examination of the morphological bases, probable ontogenetic mechanisms, phylogenetic patterns, and ecogeographical correlates of flightlessness in the rails. Throughout, an attempt was made to integrate these findings with historical and modern perspectives of popular and scientific origin. The phenomenon of avian flightlessness stands at the nexus of multiple of evolutionary, morphological, and historical issues—for example, evolutionary trends, homoplasy, adaptive compromise, insularity, ontogeny, and extinction—and the diversity of form and circumstance manifested by flightless rallids exceeds mere curiosity or aberration of nature. Simply in terms of sheer numbers, flightlessness has evolved at a frequency in the Rallidae that exceeds by an order of magnitude any other taxonomic group of neognathous birds and uniquely qualifies the Rallidae for the comparative study of this multifaceted evolutionary phenomenon.

Flightless rails came to exploit insular habitats ranging from upland forests (e.g., *Gallirallus dieffenbachii* and *Diaphorapteryx hawkinsi*) to open water (e.g., *Fulica chathamensis*) in the Chatham Islands alone (Fig. 7; Atkinson and Millener 1991), with examples at a global scale (Fig. 6) representing these extremes and virtually all intermediate habitats known from oceanic islands. This contrasts with the comparatively narrow ranges of habitats exploited by flightless members of other avian families (Table 4). Most examples are (were) restricted to upland habitats (ratites, galliformes, geese, raphids, a parrot, West Indian owls, raphids, and acanthisitids), to ecotonal zones between uplands and wetlands (ibises and some *Anas*), or to largely or wholly aquatic habitats (most anatids, a cormorant, alcids, and penguins). Flightless rails, as will be documented, underwent diverse changes in body size and cranial conformation, and even expanded to include diurnal, crespuscular, and nocturnal specialists (Ripley 1977; Atkinson and Millener 1991). This survey includes a number of extinct species, many of which are known only from subfossil remains and include some of the most apomorphic flightless members of the family (e.g., *Diaphorapteryx hawkinsi, Aphanapteryx bonasia,* and *Capellirallus karamu*).

Some of the salient characteristics of the better-known flightless rails have been the subject of commentary since the 19th century, and brief accounts of these discussions are included in the Introduction. Although the present analysis details these and other qualitative features, this study emphasizes the application of modern statistical techniques, ontogenetic information, and phylogenetic frameworks to accomplish family-wide syntheses and explicitly historical reconstructions. Both a detailed presentation of data and a thorough review of the literature are appropriate for the monographic format. The latter are detailed in the Materials and Methods, and applied to the present data in the Results. Throughout, an attempt was made to thoroughly integrate the published wealth of literature to these issues, including methodological details to enable parallel analyses of other taxonomic groups, and to provide both historical and state-of-the-art perspectives on pertinent topics. Finally, the Discussion considers newly developed evolutionary theory with respect to the morphological and ecological generalities of flightless rails and their flighted relatives, most not applied previously in this or any other ornithological context, both those inferred to be essential preconditions as well as those potentially causal with respect to the loss of flight. The dimensions examined include the importance of ontogeny in evolutionary change, recent theory bearing on developmental genetics and demographics, maintenance of multiple modes of dispersal, and the possible relevance of the concepts of innovation, adaptation, specialization, and progress with respect to avian flightlessness.

MATERIALS AND METHODS

CLASSIFICATION AND INCLUDED TAXA

The Rallidae comprise approximately 135 modern species (tally varying with authority) and a number of fossil and subfossil species (Peters 1934; Brodkorb 1967; Wolters 1975; Bock and Farrand 1980; Taylor 1996). Recent systematic assessments of the family include those by Sibley and Ahlquist (1990) and Sibley et al. (1993) based on DNA hybridization, and those by Houde et al. (1997), Trewick (1997a, b), and Slikas et al. (2002) based on DNA sequence data. However, the studies by Olson (1973a), Ripley (1977), P. B. Taylor (1996, 1998), and Livezey (1998) were the only comprehensive surveys of the Rallidae available for this study.

Accordingly, I follow the last work for purposes of inferences of closest relatives of flightless forms and for delimitation of species in the present study. As in previous works (e.g., Livezey 1997c), the latter classification was largely consistent with the phylogenetic species concept and treated many taxa as full species that were previously relegated to subspecific status, in part to communicate through taxonomy the evolutionary differentiation of insular forms (Hazevoet 1996). In this context, I point out that my use of parenthetical taxa as the second component in trinomials is to group members of superspecies, as opposed to interpolating ranks transitional between species and subspecies as advocated by Amadon and Short (1976), a compromise especially critical for widespread, polytypic forms such as the *P. porphyrio*-group, *Gallirallus philippensis*-group, and *G. pectoralis*-group (e.g., Sclater 1868; Rothschild 1893b; Ridgway and Friedmann 1941; Voous 1961, 1962; Dowsett and Dowsett-Lemaire 1980). However, two departures from the classification proposed by Livezey (1998) were adopted here. First, the original genus *Erythromachus* of Milne-Edwards (1874) was resurrected for the flightless subfossil rail *leguati* from Rodriguez, referred to *Aphanapteryx* by Günther and Newton (1879), provisionally maintained as congeneric with *A. bonasia* of Mauritius by Olson (1973a, 1977), and consonant with the treatment by Mourer-Chauviré et al. (1999).

Secondly, similar phylogenetic uncertainties necessitated a provisional substitution of the generic classification of the subfossil rail "*Atlantisia*" *elpenor* (Olson 1973a, 1977) to "*Atlantisia*" (Bourne et al., pers. comm.) to reflect the uncertainty of phylogenetic relationships of this taxon (sensu Livezey 1998). Olson (1977: 355) revealed that his concept of the enlarged *Atlantisia* met the conventional definition of polyphyly: "*Atlantisia elpenor* probably descended from a volant pro-*Rallus* stock that independently gave rise to flightless forms on Ascension [*elpenor*], St. Helena [*podarces*], and Inaccessible Island [*rogersi*] in the Tristan da Cunha group." A third case, the referral by Olson (1973a) of *Aphanocrex podarces* (Wetmore 1963) to the modern genus *Atlantisia,* also failed to be confirmed (Livezey 1998), and the original binomen is recognized herein; in this context, it is noted that the evolutionary speculations offered for this species by Rowlands et al. (1998) were made within the earlier inclusion of the taxon within *Atlantisia.* The referral of these two flightless, subfossil rallids to *Atlantisia* by Olson (1973a) is especially problematic given the persistent uncertainties regarding the phylogenetic relationship of the comparatively well-known, extant type species of the genus, *A. rogersi* (Livezey 1998).

All modern flightless species of Rallidae and their closest relatives were included for analysis. In this context, "modern" refers to taxa for which at least one study skin or fluid-preserved specimen is available. The rare, poorly known *Rallus wetmorei,* considered to possess "emerging flightlessness" by Ripley (1977:136), was not included because of an extreme paucity of specimens. Data were collected from a series of study skins of the Galapagos Crake (*Laterallus spilonotus*) because of its weak flight and relatively short wings (Gifford 1913; Franklin et al. 1979). Unfortunately, neither skeletons nor fluid-preserved specimens of this species were available.

Members of a few completely flighted genera were included to broaden the bases for comparison: *Porphyrula, Gymnocrex, Aramides, Canirallus, Rougetius, Ortygonax, Laterallus, Coturnicops, Crex,* and *Gallicrex* (Table 5). Three large rallids that have been described as showing labored or reluctant flight also were sampled (especially for assessment of allometry with increasing body size): Chestnut Rail (*Eulabeornis castaneoventris*), a large-bodied resident of mangroves of tropical Australasia considered to have weak flight (Barnes 1997; P. B. Taylor 1998); Giant Coot (*Fulica gigantea*), a sedentary resident of Andean lakes, in which adults attain weights that evidently render them flightless at least temporarily, although this inference is controversial (Ripley 1977; Fjeldså 1981; P. B. Taylor 1998); and Horned Coot (*F. cornuta*), another massive, Andean endemic described as "reluctant" to fly at times (P. B. Taylor 1998). Adequately represented subfossil rallids also were included in analyses (Table 2), especially a number from the New Zealand region (Worthy and Holdaway 1994, 1996; Worthy 1998a, b; Millener 1999) and several described since the review by Olson (1977) or within the last few years (e.g., Olson and Wingate 2000, 2001). In total, this scheme resulted in the inclusion of 82 modern species or species groups (sensu Livezey 1998; Table 5), of which about one-third are (were) flightless or nearly so (Olson 1973a, 1977; Ripley 1977).

Poor samples or inaccessibility of specimens necessitated the exclusion of several other subfossil taxa in some analyses: *Porphyrio paepae* (Steadman 1988); *Nesotrochis picapicensis* (Fischer and Stephan 1971b), although some data were taken from the description; *Dryolimnas augusti* (Mourer-Chauviré et al. 1999); *Gallirallus ripleyi* (Steadman 1986a), *G. minor* (Hamilton 1893; Olson 1975b), *G. huiatua* (Steadman et al. 2000), an undescribed rail from Fiji (Worthy et al. 1999), an undescribed rail from the Brazilian island of Fernando de Noronha (Olson 1977, 1982); *Porzana rua* (Steadman 1985, 1986a), several additional unnamed *Porzana* noted by Olson and James (1991); *Tribonyx repertus* (Olson 1975d); and *Hovacrex roberti* (Andrews 1897). A number of additional Pleistocene rallids known from very meager material, of dubious taxonomic validity, or inferred not to have been flightless (Olson 1974b, 1975b, 1977) also were ex-

TABLE 5. Sample sizes by taxon and type of specimen included in comparative groups of Rallidae. Flightless species are preceded by an asterisk. Skins primarily comprise study (round) skins, although some tallies include a minority of flat skins and mounts. Skeletons are associated and prepared from individual birds. Totals include complete skeletons and partial (trunk) skeletons; the latter are distinguished (P) if such composed entire sample. Subfossils are disassociated, unfossilized skeletal elements; total includes all individual elements measured.

Comparative groups and taxa	Numbers of specimens			Spirit specimen(s) dissected
	Skins	Skeletons	Subfossil	
Swamphens				
Porphyrio porphyrio-group*	466	108	—	AUSM 451, 452 (melanotus)
Porphyrio albus	2	—	—	—
*Porphyrio mantelli	—	—	133	—
*Porphyrio hochstetteri	14	30	—	NZNM (uncataloged)
*Porphyrio kukwiedei	—	—	11	—
Porphyrula alleni	53	8	—	—
Porphyrula martinica	34	30	—	—
Porphyrula flavirostris	54	4	—	—
*Aphanocrex podarces	—	—	28	—
Wallace's Rail and allies				
Gymnocrex rosenbergii	22	—	—	—
Gymnocrex plumbeiventris	27	1P	—	USNM 505787
*Habroptila wallacii	16	6	—	USNM 506649
Wood-rails and allies				
Eulabeornis castaneoventris	33	1P	—	—
Aramides ypecaha	30	15	—	—
Aramides cajanea-group*	30	27	—	CMNH 4457
Canirallus oculeus	45	1	—	—
Canirallus kioloides	40	3	—	—
*Nesotrochis debooyi	—	—	74	—
Nesotrochis steganinos	—	—	21	—
*Nesotrochis picapicensis	—	—	~20†	—
Rougetius rougetii	39	1P	—	—
Zapata Rail and allies				
*Cyanolimnas cerverai	9	1P	60	USNM 343159
Ortygonax sanguinolentus	20	1P	—	USNM 227903
Ortygonax nigricans	20	—	—	—

TABLE 5. Continued.

Comparative groups and taxa	Numbers of specimens			Spirit specimen(s) dissected
	Skins	Skeletons	Subfossil	
White-throated rails				
Dryolimnas cuvieri	59	6	—	BMNH 1935·10·19·1
*Dryolimnas aldabranus	28	16	—	BMNH 1979·10·1
*Dryolimnas abbotti	8	1P	—	—
Rallus sensu stricto				
Rallus aquaticus	28	39	—	BMNH 1926·3·10·1
Rallus limicola	30	39	—	CMNH 2300
Rallus ibycus	—	—	128	—
Rallus longirostris-group*	30	34	—	—
Rallus elegans-group*	30	38	—	CMNH 2119
*Rallus recessus	—	—	303	—
Gallirallus and allies				
Gallirallus pectoralis-group	42	9	—	—
Gallirallus striatus	29	6	—	—
*Gallirallus sharpei	1	—	—	—
Gallirallus australis	99	32‡	—	USNM 511701
Gallirallus scotti	10	15	—	—
*Gallirallus greyi	40	9	—	—
Gallirallus philippensis-group*	48	40	—	USNM 542344
*Gallirallus dieffenbachii	1	—	536	—
Gallirallus owstoni	29	32	—	USNM 503792
Gallirallus rovianae	1	—	—	—
*Gallirallus wakensis	58	30	—	USNM 289313
*Tricholimnas lafresnayanus	9	—	8	—
*Tricholimnas sylvestris-group	89	6	—	USNM 18407 (nominate)
*Nesoclopeus poecilopterus-group	19	2P	—	BMNH 1940·12·8·87 (nominate)
Aramidopsis plateni	14	—	—	—
Cabalus modestus	28	—	388	—
*Capellirallus karamu	—	—	231	—
*Habropteryx insignis	23	1P	—	—
Habropteryx (t.) celebensis	29	—	—	—
Habropteryx (t.) torquatus	49	14	—	USNM 318003, DEL 37754
Habropteryx (t.) sulcirostris	15	—	—	—
*Habropteryx okinawae	3	1	—	UMMZ 225360

FLIGHTLESSNESS IN RAILS 51

TABLE 5. Continued.

Comparative groups and taxa	Numbers of specimens			Spirit specimen(s) dissected
	Skins	Skeletons	Subfossil	
Problematic subfossil rails*				
*Aphanapteryx bonasia	—	—	202	—
*Erythromachus leguati	—	—	22	—
*Diaphorapteryx hawkinsi	—	—	600	—
Crakes and allies§				
*Atlantisia rogersi	45	8	—	USNM 320106, DEL 20762
*"Atlantisia" elpenor	—	—	192	—
Laterallus spilonotus	19	—	—	—
Laterallus leucopyrrhus	10	28	—	CMNH 5322
Coturnicops noveboracensis	30	35	—	CMNH 2333
Crex crex	30	24	—	USNM 539754
Crex albicollis	31	6	—	—
*Porzana sandwichensis	7	1P	—	—
*Porzana palmeri	89	14	—	BMNH 1940.12.8.13
Porzana carolina	30	31	—	CMNH 6718
*Porzana piercei	—	—	142	—
Porzana pusilla	39	24	—	—
Porzana tabuensis	36	11	—	AMNH 10024, BMNH 1971-26-9
*Porzana monasa‖	2	—	—	—
*Porzana atra	25	4	—	AMNH 9901, BMNH 1913·3·4·8
Porzana flavirostra	32	38	—	CMNH 1479
Porzana olivieri	5	—	—	—
Porzana bicolor	25	—	—	—
*Porzana astrictocarpus	—	—	27	—
*Porzana ziegleri	—	15¶	15¶	—
*Porzana menehune	—	—	19¶	—
*Porzana keplerorum	—	—	2¶	—
*Porzana ralphorum	—	—	27¶	—
*Porzana severnsi	—	—	—	—
Bushhens				
Amaurornis olivaceus	20	9	—	AMNH 10717
Amaurornis moluccanus	55	5	—	—
Amaurornis ruficrissus	11	—	—	—
Amaurornis phoenicurus	60	31	—	—

TABLE 5. Continued.

Comparative groups and taxa	Numbers of specimens			Spirit specimen(s) dissected
	Skins	Skeletons	Subfossil	
Amaurornis akool	42	1P	—	—
Amaurornis isabellinus	59	1P	—	—
**Amaurornis ineptus*	26	—	—	YPM 528
Moorhens and allies				
Gallicrex cinerea	39	10	—	—
**Pareudiastes pacificus*	8	—	—	BMNH 1874.2.20.38
**Pareudiastes silvestris*	1	—	—	—
Tribonyx ventralis	86	45	—	AMNH 7081
**Tribonyx mortierii*	47	23	—	AMNH 10079
**Tribonyx hodgenorum*	—	—	140	—
**Gallinula nesiotis*	4	5P	—	—
*Gallinula comeri	37	6	—	BMNH 1922.12.6.221
Gallinula tenebrosa	30	5	—	—
Gallinula angulata	30	1	—	—
Gallinula chloropus-group#	162	58	—	USNM 540933, BMNH 1989.19.2
Coots				
Fulica armillata	32	4	—	—
Fulica cornuta	19	2	—	—
Fulica gigantea	42	11	—	—
Fulica americana-group*	90	48	—	CMNH 6765
Fulica atra	36	38	—	—
**Fulica chathamensis*	—	—	405	—
*Fulica prisca	—	—	130	—
*Fulica newtoni	—	—	56	—

* Taxonomy follows Livezey (1998). The second and third subgroups (including *Habroptila* and *Aramides*) are collectively referred to here as "basal" rails.
† Limited mensural data taken from Fischer and Stephan (1971a).
‡ An additional 23 skeletons of undetermined species in the *Gallirallus australis*-group sampled.
§ Monophyly of genus *Porzana* unconfirmed (Livezey 1998).
|| Based on X rays of two skin specimens figured by Steadman (1986a).
¶ Supplemented in some cases by data compiled by Olson and James (1990).
# Includes series of developing juveniles.

cluded from study here and in the companion phylogenetic analysis (Livezey 1998).

The genus *Aptornis* (Owen 1848a, b), comprising two extraordinary species of extinct, flightless Gruiformes endemic to New Zealand, formerly was considered to be allied with the Rallidae (e.g., Fürbringer 1888; Oliver 1945, 1955). However, in recent years, the genus has been inferred to be more closely related to the Kagu (Rhynochetidae: *Rhynochetos jubatus*) of New Caledonia (Parker 1869; Olson 1975a, 1985; Livezey 1998; but see Weber and Hesse 1995; Houde et al. 1997). Inclusion of the uniquely derived *Aptornis* was limited to comparisons where insights pertaining to comparatively extreme changes related to flightlessness in rallids were deemed especially likely.

SPECIMENS AND RELATED DATA

SKIN SPECIMENS

A total of 3,220 skin specimens from more than 80 species or species groups of Rallidae (sensu Livezey 1998) were sampled for this analysis (Table 5). Together with supplementary information where available, this comprised approximately 20,000 measurements. Study skins, flat skins, or mounted specimens of all modern species of Rallidae in the analysis were examined, with most taxa being assessed on the basis of series of specimens. The sole exception pertained to the extinct, evidently flightless *Porzana monasa*. Two unique skin specimens were collected during 1827–1828 by Kittlitz, and are held at the Academy of Sciences, Zoological Institute, St. Petersburg, Russia. Traditional dimensions of the skin for this species were taken from the literature. Direct study of extinct *Porphyrio albus* (Hindwood 1932, 1940, 1965; Fuller 1987, 2001; Hutton 1991; Garnett 1993) included one specimen held at Merseyside Museum, Liverpool, U.K. (Sharpe 1894; Rothschild 1907a; Wagstaffe 1978). The holotype, held in Vienna (Salvin 1873; Ripley 1977; P. B. Taylor 1996, 1998), was measured for this study by E. Bauernfeind, who used instructions provided by the author. Dieffenbach's Banded-Rail (*Gallirallus dieffenbachii*), although represented by abundant subfossil skeletal elements, is known from but a single skin specimen. This skin, the holotype (Knox and Walters 1994), was studied during several visits to the BMNH. The unique specimen of *Gallirallus* (*Stictolimnas*) *sharpei* (Rijksmuseum van Natuurlijke Historie; Olson 1986), the unique holotype of *G. rovianae* (American Museum of Natural History; Diamond 1991), and a recently acquired flat skin of the newly described *Habropteryx okinawae* (Museum of Zoology, University of Michigan) also were examined directly. Additional mensural data for rare rallids held in foreign museums generously were provided on request by curatorial staff who were provided with written instructions.

The unique specimen of the so-called Gilbert Island Rail (*Tricholimnas conditicius* [Bangs 1930]), the taxonomic status of which remains in dispute (Greenway 1952; Walters 1987; Olson 1992), was included in this study as a member of the *Tricholimnas sylvestris*-group, but exemplifies the broader problem of representation and contingent delimitation of taxa. The taxonomic validity of the Tahitian Crake (*Porzana* [*Rallus*] *nigra*), based solely on a brief description by Gmelin (1789) and a painting by G. Forster made during Cook's second voyage, is disputed (Ripley 1977; Olson and Steadman 1987; Walters 1988; Taylor 1998).

Information subsequently referred to this taxon suggests that the species was flightless and may have occurred on both Tahiti and Mehetia islands, and was extirpated on these islands by introduced mammalian predators during the late 19th and early 20th centuries, respectively (Greenway 1967; Bruner 1972; W. B. King 1981; Day 1989; Taylor 1998). Subsequent illustrations attributed to this taxon include those by J. F. Miller (Shaw 1784) and Keuleman (in Rothschild 1907a: fig. 1, pl. 26). Given the absence of anatomical material and contradictory information (P. B. Taylor 1998), this taxon was excluded from further consideration here. Similar circumstances pertain to the reported presence of an endemic, flightless *Porphyrio* on the Mascarenes (Schlegel 1866).

In addition, descriptions of new taxa (Worthy 1997a, b; Lambert 1998a, b) and reports of still-undocumented forms continue (Baltzer 1990; Steadman 1992, 1999; P. B. Taylor 1998; Worthy et al. 1998), and a number of phylogenetic relationships in the Rallidae remain unresolved (Livezey 1998). For example, taxonomic divisions and number of included species (by whatever species concept) are uncertain in several genera including flightless members or that are closely related to genera having flightless members (Livezey 1998). These include the *Porphyrio porphyrio*-group (Trewick 1996a, 1997a; Sangster 1998), taxa commonly included within *Aramides cajanea* (Bangs 1907; Hellmayr and Conover 1942; Meyer de Schauensee 1966), the "superspecies" comprising *Rallus longirostris* and *R. elegans* (Oberholser 1937; Ridgway and Friedmann 1941; Ripley 1977; Avise and Zink 1988; Bledsoe 1988; Olson 1997; Eddleman and Conway 1998), and the *Gallirallus philippensis* complex (Ripley 1977; P. B. Taylor 1997, 1998) and close allies (Trewick 1997a, b). In these instances, where samples permitted, a conservative approach was adopted (sensu Livezey 1998); that is, diagnosably distinct terminal lineages were distinguished in plots and morphometric summaries.

Adequate sampling of osteological specimens necessitated use of collections worldwide through loans or visitation (see Acknowledgments). Most specimens of extant taxa were studied during visits to the American Museum of Natural History (New York), U.S. National Museum of Natural History (Washington, D.C.), and the Natural History Museum (U.K.).

Skeletal Specimens

A total of 1,200 associated (complete or partial) skeletons from more than 65 species or species groups of Rallidae (sensu Livezey 1998) were measured in this study (Table 5), approximating 45,000 measurements. Many extant species of Rallidae are poorly represented in osteological collections (Wood and Schnell 1986), whereas some extinct rails are better known osteologically than are a number of extant confamilials (Olson 1977). Complete skeletons were unavailable for a number of flightless rallids and close relatives (Wood and Schnell 1986), including *Gymnocrex rosenbergii*, *G. plumbeiventris*, *Rougetius rougetii*, *Cyanolimnas cerverai*, *Tricholimnas lafresnayanus*, *Aramidopsis plateni*, *Pareudiastes pacificus*, *Edithornis silvestris*, and several species of both *Porzana* and *Amaurornis*. Inclusion of partial skeletons retained during preparation of study skins—in which preserved elements typically were limited to certain parts of the vertebral column, both appendicular girdles, femora, and (often) proximal humerus—resulted in larger sample sizes for some elements (e.g., sterna) than for complete skeletons in some taxa.

In several cases, osteological specimens (Wood and Schnell 1986) were augmented by elements removed from study skins (*Gymnocrex plumbeiventris, Eulabeornis castaneoventris, Rougetius rougetii, Habropteryx insignis, Nesoclopeus poecilopterus,* and *Amaurornis ineptus*) and measurements of elements during dissection of fluid-preserved specimens (*Gymnocrex plumbeiventris, Cyanolimnas cerverai, N. poecilopterus, Gallirallus okinawae, A. ineptus,* and *Pareudiastes pacificus*). Mensural data for skeletal elements of *C. cerverai,* deriving largely from subfossil elements recovered from the Isle of Pines (Olson 1974b), were supplemented by data given by Fischer and Stephan (1971b; under *Rallus sumiderensis*). Lengths of most major appendicular elements from a single specimen of *Amaurornis ineptus* (UF 41507) were provided by J. K. Sailer (Florida Museum of Natural History). Finally, alar phalanges were measured by using parallax-corrected of images in X-ray photographs of skin specimens for two extinct rallids of the Chatham Islands—*Gallirallus dieffenbachii* (BMNH 1842·9·29·12, holotype) and *Cabalus modestus* (BMNH 1893·6·24·5, female).

SUBFOSSIL ELEMENTS

In total, more than 4,000 subfossil elements representing 28 species of Rallidae (sensu Livezey 1998) were included for study (Table 5), representing more than 12,000 measurements. The best-known subfossil rails (e.g., *Diaphorapteryx hawkinsi, Cabalus modestus,* and *Fulica chathamensis*) typically were represented by adequate numbers of all major skeletal elements, and permitted robust estimates of all included dimensions for both putative sex classes. Where summary statistics of samples of comparable (but not strictly identical) dimensions were available—for example, for *Porphyrio mantelli* (Trewick and Worthy 2001), *Gallirallus dieffenbachii* (Cracraft 1973a), and *Diaphorapteryx hawkinsi* (Cracraft 1973a)—estimates of first and second moments were reassuringly similar. However, even for the best-represented subfossil rallids, undamaged examples of comparatively fragile elements (e.g., maxillary rostrum, furcula, or distal alar phalanges) were rare, and most subfossil rallids (e.g., *Nesotrochis debooyi, Aphanapteryx bonasia, Capellirallus karamu,* and "*Atlantisia*" *elpenor*) lacked considerably more elements.

In some subfossil taxa, unavailability or poor condition of elements significantly limited the scope of analysis. For example, *Gallirallus ripleyi* and *Porzana rua,* both described and judged to have been flightless by Steadman (1986a), were represented too poorly for inclusion in these analyses or for confirmation of flight status. Similar considerations necessitated the exclusion of two other poorly represented subfossil rallids, *Hovacrex roberti* of Madagascar (Andrews 1897), and *Porphyrio paepae* of the Marquesas Islands (Steadman 1988). A recently discovered, unnamed, evidently flightless *Porphyrio,* and another flightless *Gallirallus* sp. from New Ireland (Steadman et al. 1999) were not available for study. Finally, comparisons of *Nesotrochis picapicensis* of Cuba were limited to figures and tabulated measurements presented by Fischer and Stephan (1971b; under the genus *Fulica*) because the specimens could not be located (O. H. Garrido, pers. comm.).

FLUID-PRESERVED SPECIMENS

Species of Rallidae included in pectoral dissections were determined largely by availability of targeted taxa (Wood et al. 1982). Two unique anatomical specimens of endangered or extinct species—*Nesoclopeus poecilopterus* and *Pareudiastes pacificus*—were made available for limited dissection at the BMNH (Blandamer and Burton 1979). Seventeen flightless species were dissected, and these necessitated the myological study of 20 species of closely related, flighted rails (Table 5). With several exceptions (detailed below), dissections encompassed the musculature of the entire wing and pectoral girdle of one side of each specimen and included the removal of the remiges intact for measurement in seriatim.

MISCELLANEOUS INFORMATION

To supplement measurements made directly from specimens, essential data on external measurements, body masses, and wing areas of Rallidae were taken primarily from 10 key published compilations (Müllenhoff 1885; Magnan 1912, 1913a, b, 1922; Poole 1938; George and Nair 1952; Meunier 1959a; Hartman 1961; Cramp and Simmons 1980) and unpublished tabulations and wing tracings by R. Meinertzhagen (Snow 1987). Additional anatomical information for rallids—including body masses, composition and sizes of eggs, dimensions of major skeletal elements, and other metric data—was taken from a critical but widely scattered literature. A summary of body masses and wing lengths of Rallidae also was prepared (Appendix 1), and published sources for other ancillary data also were compiled (Appendix 2). Volumes and masses of eggs were taken from the compilations by Schönwetter (1960a); restriction of the analysis to a single taxonomic family rendered unnecessary the more sophisticated methods of estimation by Preston (1953, 1974) and Hoyt (1979).

During dissections, a number of sources were consulted for study of the musculature, including those including both qualitative and metric comparisons (Hudson 1937; Hudson and Lanzillotti 1955, 1964; Hudson et al. 1966, 1969, 1972; Vanden Berge 1970; Livezey 1990, 1992a, b) and largely descriptive works (Berger 1956a, b, 1966; Kuroda 1961; Simic and Andrejevic 1963, 1964; George and Berger 1966; Nagamura et al. 1974, 1976; Raikow 1985a; Rosser and George 1986; McKitrick 1991b). In combination with the monograph by Ripley (1977), the timely publication of several recent compendia on rails (P. B. Taylor 1996, 1998) obviated the compilation of an exhaustive bibliography for documentation of the general habits and distribution of the Rallidae. However, inferences regarding the foraging habits of rails known only from skeletal remains, are problematic, but there is precedent for such extrapolations based on extant relatives (Hespenheide 1973; Lauder 1981, 1995; Kooyman 1991; Hertel 1995), especially where a phylogenetic hypothesis is available (Lauder 1982, 1990, 1991; Witmer and Rose 1991; Witmer 1995).

GENERAL DESCRIPTIVE CONVENTIONS

DEFINITION OF FLIGHTLESSNESS

Throughout this work, I reserved the term "flightless" to those avian taxa in which adults are characterized by the permanent inability to achieve takeoff and maintain level, powered flight for a significant distance unassisted by substantial head winds or net downward trajectory. The young of all birds pass through a period of flightlessness during early development, but all individuals of flightless species are incapable of flight throughout life, with the possible exception of some marine populations of *Tachyeres patachonicus* (Humphrey and Livezey 1982;

Livezey and Humphrey 1986, 1992). The term "flighted" is used herein in reference to those lineages capable of aerial locomotion described above, as opposed to the common term "flying," with the latter restricted herein to references to birds in the act of flight. The term "flighted" is intentionally distinctive and permits logical reference to individuals or classes thereof, uncategorized in other aspects of their life histories, but possessing the capacity for flight. The term "flighted" is adopted in lieu of the term "volant," the latter having been applied by some to tetrapods capable of extended but passive, aerial descent, for example, some primates (Dermoptera) and tree squirrels (Rodentia, Sciuridae: *Glaucomys*).

The related terms "fledging" (the process) and "fledgling" (a bird at the completion of the process) are of inconsistent usage in the ornithological literature (Middleton and Prigoda 2001). Inasmuch as there is no evidence of members of any flightless species of rail having passed through even a flighted developmental stage, however brief, in this work the term "fledging" refers to the acquistion of a complete complement of feathers (including remiges), which in flighted rails generally preceeds first flight by a week or more (P. B. Taylor 1996, 1998). "Fledglings" are individuals at the developmental stage at which the acquisition of remiges reaches completion.

Anatomical Nomenclature

The nomenclature advocated by the International Committee on Avian Anatomical Nomenclature was used in the description of anatomical characters (Baumel and Raikow 1993; Baumel and Witmer 1993; Clark 1993a, b; Vanden Berge and Zweers 1993), as opposed to perpetuating the vernacular scheme for osteology popularized by Howard (1929). Supplementary anatomical resources that were consulted included Butendieck (1980), Butendieck and Wissdorf (1982), and Schummer et al. (1992). Abbreviations used are listed in Appendix 3.

Dissections of Pectoral Musculature

Dissections of the musculature of the pectoral girdle and limb were accomplished by using standard techniques, with orientation of fibers enhanced by iodine stain (Bock and Shear 1972). Dissections included selected aspects of the overlying integument (e.g., propatagium and accessory tendons and ligaments), with comparative information taken from Brown et al. (1994, 1995). Most dissections were unilateral, with the right side typically chosen to facilitate comparisons across taxa and illustration. The microscope used was a Nikon SMZ-U binocular dissection scope. Preliminary figures were prepared by using a Nikon drawing tube; the resultant pencil drawings were refined through reference to the specimens, transferred in ink to Mylar drafting film, and these were digitally scanned for labeling and publication.

Essential references consulted during dissections that pertained to Gruiformes and (especially) the Rallidae were Fisher and Goodman (1955), Berger (1956a, b, 1966), Allen (1962), George and Berger (1966), and Rosser et al. (1982). Especially useful were the myological works by Fisher and Goodman (1955) on the Whooping Crane (*Grus americana*), Allen (1962) and Tipton (1962) on the Limpkin (*Aramus guarauna*), Rosser (1980) and Rosser et al. (1982) on the American Coot (*Fulica americana*), and McGowan (1986) on the flightless Weka (*Gallirallus australis*).

CLASSES OF MENSURAL DATA

INTEGUMENT

Most mensural data were continuous and analytical protocols depended on objective and precision (Rohlf 2001). Six measurements were made on study skins (Livezey 1989b–d, 1990, 1992a, b, 1993a, b), most of which conform with traditional methodology (Baldwin et al. 1931). These were culmen length, bill depth at gonys, wing length, tail length, tarsus length, and middle-toe length. In those taxa in which a frontal shield (crista carnosa frontalis) is characteristic but varying with age and sex (e.g., *Porphyrio, Habroptila, Amaurornis ineptus, Tribonyx, Pareudiastes, Gallinula,* and *Fulica*), culmen length was measured exclusive of this caudal extension of the integument. For *Tribonyx,* opacity and conformation of the rhamphotheca and shield made measurement of culmen length exclusive of the shield especially challenging. The most reliable comparisons of bill length in this and other genera having "shields" were based on skeletal specimens (below). Wing length (chord of flattened wing) was measured by using dividers and a ruler (small birds) or an end-stop ruler (large birds). Bill depth was measured by using dial calipers with an approximate precision of ± 0.5 mm based on informal estimates of "pure error" (repeated measurements of the same dimensions on a single individual). Other external dimensions were measured by using dividers and a ruler. Although wing length can be used as an index to body size in many applications (Rand 1961a, b; James 1970; Lack 1971a; Snyder and Wiley 1976; Payne 1984; Jehl and Murray 1986; Zink and Remsen 1986), it has proven suboptimal in others (Rising 1988), and is inappropriate a fortiori where relative wing length is itself a variable of primary interest (e.g., McCall et al. 1998). In addition, where molt or limitations of preparations permitted, counts of primary remiges were conducted on both wings. Qualitative characters of remiges also were compiled for flightless species and their close relatives. Wing areas were measured from tracings of specimens of extended wings (or those prepared by others with freshly collected birds) by using a compensating polar planimeter; these areas were doubled to estimate the total wing areas of individual birds (Raikow 1973; Blem 1975). Wing loadings were calculated as the ratio of body mass (g) divided by total wing area (cm^2) (Clark 1971).

Although body mass is variable (Amadon 1943a) and is only one of many indices to size (Piersma and Davidson 1991), mean body masses for avian taxa are critical indicators of a suite of important ecological parameters (Nice 1938; Clark 1979; Mendelsohn et al. 1989). Distributions of avian body masses are known for several spatiotemporal scales and a diversity of taxonomic groups (Van Valen 1973a; Uchmanski 1985; Maurer and Brown 1988; Bush 1993; Niklas 1994; Blackburn and Gaston 1996). Unlike some other sexually dimorphic traits (e.g., Andersson and Andersson 1994), mean body mass is a reliable measure of sexual size dimorphism (Livezey and Humphrey 1984a). Surrogate estimators of overall size can approximate body mass closely in practical applications (Amadon 1947; Senar and Pascual 1997), the most accurate of which tend to be multivariate (e.g., Willig et al. 1986; Gilliland and Ankney 1992). Some of these were employed herein.

SKELETON

Forty-one skeletal dimensions were used in this study (including one comprising three summed dimensions), most of which are self-explanatory and all of which were measured by using Helios dial calipers equipped with opposing needle-points and long, flat surfaces providing measurements at a scale of 0.1 mm. Except in those specimens for which fewer than one of a pair of elements was available or one was afflicted with deformity, measurements were taken without regard for sidedness. This practice implicitly assumes bilateral symmetry, a condition subject to stabilizing selection for a multitude of reasons (Charnov 1993), whereas significant asymmetry tends to be rare, subject to negative selection, or indicative of developmental instability or stress (Parsons 1992; Møller and Swaddle 1997; Graham et al. 1998; Møller 1998). However, several included measurements are not standard, differed from those used in previous works (Livezey and Humphrey 1984a, 1986, 1992; Livezey 1986b, 1988, 1989a–d, 1990, 1992a, b, 1993a–c), or posed special challenges for species showing strong, flightlessness-related changes. Only maximal and least widths at midpoints of appendicular elements (MWM and LWM, respectively) and shaft widths of the tarsometatarsus were not standard. In addition to these previously described measurements, a detailed suite of 21 measurements of the skull was compiled for an assessment of cranial variation in the Rallidae, generally restricted to single exemplars of taxa.

PECTORAL MUSCULATURE

In addition to qualitative description and illustration, a suite of 81 standardized measurements of most pectoral muscles or named parts thereof was compiled for dissected specimens (details of measurements given in tables of summary statistics). Measurements of muscles were made by using needle-point dial calipers; preservational variation and malleability of soft tissues magnified mensural error relative to that of dimensions of the integument and substantially more than that of the skeleton (approximately 0.5–1.0 mm). Abbreviations adopted for names of muscles and related anatomical features in tables and figures are listed in Appendix 3.

NATURE OF MENSURAL DATA

INTERSPECIFIC COMPARABILITY OF VARIABLES

Although use of dial calipers, with a mechanical precision exceeding 0.1 mm, contributed to a confidence in the accuracy of measurements compiled, there were additional sources of error which were quantified or controlled less easily. In general, skeletal variables likely were measured most precisely because of the clean, hard surfaces presented. However, among skeletal dimensions, several measurements of the sternum were confounded by interspecific variation in the conformation of this element (especially carina depth and basin widths), and several dimensions of the skull were subject to warping or variation in angles of articulation among specimens. Similarly, among external measurements, wing length proved comparatively variable because of differences among individuals in wear, and length of the middle toe (digit III) was troublesome in a substantial number of study skins because of variation in the position of the toe in preserved specimens. Measurements of muscles were least precise because of the inherent difficulty of delimiting boundaries of soft, variably distinguishable bodies of muscle tissue. Counters to such challenges consisted of limiting the vast majority of measurement to the author and to defining measurements so as to minimize error stemming from such nuisance morphological attributes.

MEASUREMENT (PURE) ERROR

"Pure" or measurement error is variance attributable to inconsistency of data collection (Yezerinac et al. 1992), as opposed to natural or "fabricational" variation (Seilacher 1975) among individuals not attributable to traditional grouping variables (e.g., species or sex). Measurement error can be treated as a separate partition of variance in morphometric data (Bailey and Byrnes 1990) and may achieve levels having practical analytical implications in fine-scale comparisons (e.g., Grant 1979a, b; Zink 1983, 1986). However, detailed quantification of measurement error (e.g., variation among repeated measurements of single specimens) was not undertaken in the present study for three reasons: the vast majority of external measurements and all skeletal and myological data were compiled by the author; pilot assessments revealed that the magnitude of such variance was low relative to intersexual (e.g., Eason et al. 2001), interspecific, and other, less interesting sources of variance (e.g., wear of feathers or subfossil elements); and inclusion of this factor in the present study was deemed prohibitive. However, regardless of the low relative magnitude of such variance, measurement error can be assumed to have increased estimates of residual variance in statistical tests, thereby rendering associated tests of main effects somewhat conservative.

The possibility that within-locality morphometric variation may magnify among-locality differences—that is, the Kluge–Kerfoot effect (Kluge and Kerfoot 1973; Johnston 1976a; Sokal 1976; Rohlf et al. 1983)—was not a problem per se in these analyses in that the finest scale of study applied was phylogenetic species sensu Livezey (1998). However, a related artifact afflicting relative variation, in which widespread taxa comprising multiple subspecific taxa (e.g., *Rallus longirostris, Porzana pusilla,* and *Fulica atra*) manifested predictably inflated variances relative to insular endemics, was apparent in some comparisons. Where these discrepancies may have affected statistical tests or other comparisons, the potential for limited comparability was noted or the widespread taxon subdivided accordingly (e.g., *Fulica atra,* nominate and *australis* groups).

PHYLOGENY-BASED COMPARISONS

Comparative studies encompassing taxa of differing degrees of phylogenetic relationship have been fundamental throughout the history of evolutionary study (Ridley 1983; Eldredge 1985; Rieppel 1988). In addition to simple mapping of attributes on phylogenetic reconstructions (e.g., Livezey 1995b–d, 1996a–c, 1997a, b), refinements for comparative assessments based on phylogeny have been detailed in recent years (Funk and Brooks 1990; Brooks and McLennan 1991; Harvey and Pagel 1991; Gittleman and Luh 1992); these can provide insights into the interplay of ecological factors and historical predisposition (Ligon 1993) and bridge within-lineage variation with phylogenetically ramified traits (Winkler 2000). The crux of these approaches is the avoidance of redundant tallies of "apomorphic" traits in descendants that are attributable, by means of phylogenetic hypotheses, to as few as one evolutionary change in an ancestral node. Roff (1992) cited a doubly relevant example in the fleas (Siphonaptera, in excess of 2,000 species), in which uniform flightlessness presumably derives from one or at most a few ancestral (synapomorphic) event(s), whereas flightlessness in several other orders of insects (Dermaptera, Diptera, Mallophaga, and Anoplura) presents more complex phylogenetic patterns.

The most widely used tests based on phylogenetic hierarchies are the univariate "concentrated-change test" and the bivariate "correlated-change test" (Maddison 1990). These are conditional on the nature of the data to be explored (e.g., morphological, behavioral, or life-historical), and assumptions imposed on rates and types of permissable evolutionary change (van Rhijn 1984, 1990; Vogl and Wagner 1990; Diniz-Filho 2001). Unfortunately, with the appreciation of the importance of phylogeny for evolutionary insights came the realization that virtually nothing was known of the phylogenetic relationships of most taxonomic groups targeted for study. The trees used for such examinations of purported evolutionary patterns are critical (Smith and Patterson 1988; Sillén-Tullberg 1993). For example, the phylogenetic assessment of selected traits in cranes by Mooers et al. (1999) based on the tree by Sibley and Ahlquist (1990) would be altered significantly if optimized instead on the phylogeny inferred by Livezey (1998). Optimizations and other phylogeny-based tests herein were based in large part on the companion study of the phylogeny of gruiforms (Livezey 1998). Although fossil taxa can be important for phylogenetic reconstruction (Donoghue et al. 1989; Nee et al. 1994), the virtual absence of information on life histories renders such forms opaque to most comparative assessments. Although some historical likelihoods can be surmised amidst substantial uncertainty regarding phylogenetic relationships (Martins 1996), the inclusion of the additional missing data characteristic of fossil taxa may worsen problems of polytomy and undermine support of retained nodes, both of which can lead to problems in performing phylogeny-based comparisons (Donoghue and Ackerly 1996). Finally, a recent examination of the effects of taxon sampling, incorporation of information on branch lengths, and statistical method showed that all of these considerations carry critical implications for power, bias, and critical values of phylogenetically based comparisons (Ackerly 2000).

Unresolved polytomies remain the bête noire of phylogenetic models of character evolution and anathema to associated statistical tests. Despite substantial advances in the resolution of relationships among the Rallidae (Livezey 1998) and higher-order groups (Livezey and Zusi 2001), many nodes remain unresolved and many terminal taxa lack data of critical attributes. Therefore, fully resolved tress (even provisionally) were required to permit at least exploratory assessments. Resolution of polytomies for these purposes were consistent with the set of shortest trees originally recovered (Livezey 1998), but were further resolved conditional on ancillary criteria, notably biogeographical plausibility and reasonable judgments based on characters of plumage, behavior, and molecular data (Olson 1973a, 1975b, 1977; Ripley 1977; Trewick 1997a, b). Unfortunately, all available data and ancillary insights do not permit fine-scale resolution of phylogenetic relationships among flightless rails, even within single archipelagos such as the Hawaiian or Chatham islands (Trewick 1997a, b; Coyne and Price 2000).

Following an assessment of limited sequence data and morphometric comparisons, Trewick (1997b: fig. 6) entertained at least three sets of phylogenetic hypotheses and associated evolutionary scenarios viable for the New Zealand members of *Gallirallus* and closely related *Cabalus* and *Capellirallus* alone. There is limited support for the latter two taxa being sister lineages and therefore congeners (substantive decurvature of the mandibula in excess of that shared by *Gallirallus dieffenbachii,* an extensive symphysis mandibulae, and extensive amphikinesis [Livezey 1998]), which favors inclusion of this arrangement among the alternative topologies considered (cf. Trewick 1997b: fig. 6*ii*). These and other hypotheses are broadly consistent with available evidence pertaining to phylogenetic relationships of flora and fauna and the geological history of the main islands of New Zealand and the Chatham Islands (Craw 1988). Quantitative assessments of traits based on fully resolved phylogenies (e.g., Thorpe and Malhotra 1996, 1998) must await a fine-scale analysis of *Gallirallus philippensis* and derivatives. Unfortunately, such molecular reconstructions may be more challenging than originally anticipated (e.g., Marshall and Baker [1999] for *Fringilla* of Atlantic islands), at least for the fine-scale determinations typically sought (e.g., probable sequence of avian colonizations on islands).

Phylogeny-contingent tests of evolution and correlation among attributes were accomplished by using MACCLADE© (Maddison and Maddison 2000).

UNIVARIATE AND BIVARIATE MORPHOMETRICS

ANALYSIS OF VARIANCE

Untransformed linear skeletal dimensions were compared by using two-way analysis of variance (ANOVA) with species and sex as fixed effects, in which levels of significance (sensu critical level α, and level of significance $1 - \alpha$) generally employed the conventional values of 0.05 and 0.95, respectively. Orthogonal contrasts between selected groups of interest also were specified to partition overall differences meaningfully. Where main effects were significant, pairwise differences were assessed a posteriori by using Bonferroni adjustments; robustness of such multiple comparisons was confirmed as well through the alternative protocols of Scheffé, Tukey, and Duncan (Miller 1981). In some tests, small sample sizes limited power significantly; therefore, the implicit confirmation of the null hypothesis should be viewed with caution (Levinton 1982a). In spite of the rigor to be gained through the clear definition of competing hypotheses (Strong et al. 1979; Strong 1980; Gotelli and Graves 1996), the use of null hypotheses in many ecoevolutionary contexts should be implemented and interpreted with care, especially in light of the probabilistic advantage conferred upon the pattern designated as the null hypothesis (Quinn and Dunham 1983).

Coefficients of variation (SD/$\bar{x}$, expressed as percentages) were used to compare the relative variability of measurements between groups (Lewontin 1966; Lande 1977), with a significance test available (Sokal and Braumann 1980). Equality of group variances was tested by using Levene's tests (Brown and Forsythe 1974; Schultz 1985), especially with respect to hypothesis of greater variability in some attributes of flightless species. Partial correlations (e.g., between variables x and y), corrected for variance attributable to a covariate (e.g., z), were symbolized as $r(x, y|z)$.

ESTIMATION OF MEAN BODY MASSES, LENGTHS, AND AREAS OF WINGS

Diverse means of estimating size of birds known only from skeletal remains have been used (e.g., Northcote 1982; Wiedenfeld 1985; Damuth and MacFadden 1990), including estimates of body masses of some subfossil rallids given (without detail) by Atkinson and Millener (1991). Mean lean body masses of provisional sex classes of subfossil rallids were estimated by using type-I regressions (Draper and Smith 1981; Weisberg 1985; McArdle 1988; Chatterjee et al. 1999) of mean body masses of modern rallids on two separate skeletal indicators of body mass, after transformation of data to natural logarithms (base e). Three predictor variables were used for estimates of body mass, all of which revealed exceptionally low errors about the regression lines and lacked outliers (Daniel and Wood 1980; Barnett and Lewis 1984). The first estimator was mean femur length, included in previous works by the author (Livezey 1988, 1989b, 1993b, c) and widely available in the literature. The second variable was an estimate of the mean cross-sectional area of the tibiotarsus at its midpoint (CAT), in which the mean MWM and LWM of the element were used as estimates of the major and minor axes of the ellipse, respectively. Symbolically, the tibiotarsal cross-sectional estimate is given as:

$$CATI = \pi \cdot (MWM/2) \cdot (LWM/2).$$

In other applications (e.g., interspecific scaling of relative robustness of limb elements), similar estimators were used for other appendicular elements having roughly elliptical cross-sectional profiles (e.g., humeri, ulnae, radii, and femora). A third estimator for approximation of body mass (as well as in other assessments of allometry) was the cross-sectional area of the tarsometatarsus at its midpoint (CATA). Conceptually similar to the estimate for the tibiotarsus, this variable was available for some subfossil taxa for which a complete specimen of the longer and more fragile tibiotarsus was not available (e.g., *Porzana rua* and *Gallinula nesiotis*), or for modern taxa for which bones extracted from study skins lacked a complete tibiotarsus (e.g., *Gymnocrex rosenbergii, Amaurornis akool,* and *A. isabellinus*). Like other surrogates for size, cross-sectional areas are not free of imprecision (Motani 2001). Because of the rectangular conformation of the corpus tarsometatarsi, the cross-sectional area of the element (CATA) was approximated as:

$$CATA = (MWM) \cdot (LWM).$$

Linear regressions (type I) based on log-transformed data were employed for estimates of mean wing lengths of selected subfossil rallids based on mean lengths of the underlying skeletal elements (Bochenski and Bochenski 1993), although this procedure is reliable only for taxa in which lengths of remiges do not deviate greatly from those of typical, flighted species. Unfortunately, estimates of wing areas could not be made based on mean wing lengths for flightless rails, because it has been documented that the wings of flightless rails tend to become more rounded distally through differential shortening of primary remiges, for example, Barbour (1928: fig. 3) for *Cyanolimnas cerverai* and Lowe (1928a) for *Atlantisia rogersi*.

ANALYSES OF ALLOMETRY

Methodological and theoretical essentials.—Allometry has been defined simply as "... the study of size and its consequences" (Gould 1966:587). Study of allometry typically is performed within bivariate contexts in which the primary focus is the quantification of scaling of one variable with another (Huxley 1932; Teissier 1948, 1955; Hayami and Matsukuma 1970; Alexander 1971; Corruccini

1972; Sweet 1980; Harvey 1982; Lande 1985a; Zeng 1988). In addition to the search for fundamental laws of scaling in biology (e.g., West et al. 2000), selection for increased body size has been interpreted to have led to the evolution of "bizarre" or "maladaptive" structures and exemplied the evolutionary constraint and "inevitability" attributed to some observed allometric trends (e.g., Trueman 1922; Simpson 1949, 1953a; Dodson 1960). However, with closer scrutiny, such cases have been debunked (e.g., Gould 1972, 1974), and allometry generally is viewed from traditional evolutionary perspectives (Bonner and Horn 2000). The significance of ecophysiological scaling with body size enjoyed a renaissance of sorts in the mid-1980s, during which four books dedicated to the subject appeared within 2 years (Peters 1983; Calder 1984; Schmidt-Nielsen 1984; Jungers 1985), and a recent volume (Brown and West 2000) affirmed a continued relevance to researchers in diverse fields. Each of these reviews emphasized total mass as the estimator of size, although fresh body mass is among the most rarely recorded data for bird specimens (Fisher 1955), despite important early works that relied on such data (e.g., Heinroth 1922; Huxley 1927).

Allometric methods are most often employed to assess changes in selected linear or two-dimensional morphometric characters with another considered indicative of size (e.g., Radinsky 1985), with primary emphasis placed on the exponent or slope (White and Gould 1965). Isometry generally is used in reference to the preservation of constant shape with varying size (see challenges of this semantic distinction, below). One perspective to be gained from allometric analysis is the quantitative elucidation of geometric or functional similarity (Günther 1975; Günther and Morgado 1982; Cubo and Casinos 1997; Berrigan and Seger 1998). Variables considered with respect to size in studies of interspecific morphological allometry include brain size (Bennett and Harvey 1985a, b; Pagel and Harvey 1992) and locomotory capacity (Fleagle 1985). In intraspecific contexts, relative growth of selected body parts with body mass for single species was the seminal focus of study (Lumer 1936, 1939; Lumer et al. 1942; Reeve and Huxley 1945; Richards and Kavanagh 1945; Laird 1965, 1966a, b; Laird et al. 1968; Shea 1985a; Riska 1986; Petrie 1992; Björklund 1994), but subsequently was extended to a diversity of ecological and physiological attributes that emphasized taxonomic scales of higher order (Zar 1969; Reynolds 1977; Platt and Silvert 1981; Smith et al. 1986; McArdle 1988; Riska 1991).

Allometry occurs at multiple scales (Cheverud 1982; Shea 1985b) and has been extended to a limited degree to multivariate applications (Jolicoeur 1963; Hopkins 1966; Klingenberg and Froese 1991). Typically, for a given group of taxa, intraspecific allometry and interspecific allometry manifest different slopes (Gould 1966), and although intraspecific slopes generally are less than those for the same variates evaluated at interspecific scales (Green et al. 2001), the evolutionary principles bearing on these trends remain a point of controversy (Kozlowski and Weiner 1997).

Identification of interspecific trends and outliers were primary foci in the present study. These were undertaken while cognizant of several methodological concerns related to interspecific analyses of allometry, including optimal estimators of essential parameters and fitting of curves (Kidwell and Chase 1967; Manaster and Manaster 1975; Prothero 1986; Martin and Barbour 1989), bias associated with estimates based on group averages (Welsh et al. 1988), and violations of the statistical assumption of independence in analyses based on related taxa (Pagel and Harvey 1988, 1989). The utility of allometric analysis for the diagnosis of heterochrony has broadened the relevance of the technique to systematists (Klingenberg 1998).

I followed standard notation for the bivariate allometric equation and associated parameters (Huxley 1932; Gould 1966; Sprent 1972; Kuhry and Marcus 1977; Schmidt-Nielsen 1984; Rayner 1985c; Jolicoeur 1990). The nonlinear (general) form relating dependent variate y to independent variable x is:

$$y = ax^b.$$

The linear form of this equation, in which both variates are log-transformed, reveals the basis for the names of the parameters as standard for linear regressions:

$$\log y = \log a + b \log x,$$

wherein $\log a$ is the intercept and b is the slope.

Although most applications employ simple linear regression of log-transformed dependent variables on independent variables primarily because of ease of calculation of standard errors of estimates of slope and intercept, this approach (like any application of standard linear regression) is unrealistic in attributing all residual variance to the dependent variate (Snedecor and Cochran 1967, 1980; Weisberg 1985; McArdle 1988; Chatterjee et al. 1999), and can lead to biased estimates (Zar 1968a; Sprugel 1983). Nonlinear regressions avoid an additional problem with the simple linear approach—that of minimizing log-transformed residuals rather than untransformed residuals—but nonlinear estimates of parameters suffer higher influence from outliers (Seim and Saether 1983), and this method suffers from the same unilateral attribution of error to the dependent variable as the more direct linear approach.

Gould (1979) pointed out that the intercepts of allometric equations can present challenges of interpretation, a perspective not as widely appreciated during the heyday of island biogeography as among contemporary morphometricians. Another symptom of interpretational excess is the attribution of almost any morphological trend or difference between taxa solely to allometry, examples of which include Dobzhansky (1977), who regarded the striking diversity of bill shape and size within the Hawaiian honeycreepers (Drepanididae) and Galápagos finches (Geospizinae) as the product of allometry, and Houck et al. (1990), who relegated the morphological variation among and within taxa of *Archaeopteryx* to simple ontogenetic allometry within a single species.

Most estimates of allometric slope (b) presented here were based on geometric-mean regressions, in which the geometric mean of estimates derived from standard and inverse (log–log) linear regressions is used (Ricker 1984). A direct means for estimation of the geometric-mean slope is the quotient of the traditional linear slope divided by the correlation coefficient for the line. Higher-order relationships among morphometric variables were displayed by way of agglomerative cluster analyses based on these coefficients (Anderberg 1973), despite the limitations of hierarchical representation of multidimensional similarities (de Queiroz and Good 1997).

Where two groups share allometric slopes (b), as indicated by the absence of significant differences between groupwise estimates thereof ($b_1 \approx b_2 \equiv b$), the group intercepts (a_1 and a_2) can be compared by using the scale ratio or criterion (s) of geometric similarity (White and Gould 1965; Gould 1968, 1971, 1972):

$$s = \left(\frac{a_1}{a_2}\right)^{1/(1-b)}, \quad b \neq 1.$$

This criterion estimates the ratio of the magnitudes of the dependent variable (y) for members of the two groups at the same value of the independent variable (x); however, if isometry pertains, the common slope (b) is one, rendering the expression mathematically undefined.

Allometry of wing size with body mass.—Allometric analyses of relative wing size are diverse both with respect to taxa and underlying data (e.g., Brower and Veinus 1981; Pennycuick 1988; Rayner 1988a). Among comparatively massive power fliers, relative wing length is an informative index to flight capacity and general locomotory habits, a relationship that is progressively more precise with more narrowly defined taxonomic groups (Hartman 1961; Greenewalt 1962; Livezey 1988, 1989b–d, 1990, 1992a, b, 1993a–c; Rakotomanana 1998). In many groups considered to date, flightless members appear as variably extreme outliers, including flightless dabbling ducks as modest outliers (Livezey 1990) and the Galapagos Cormorant as an extreme outlier (Livezey 1992a). In bivariate contexts wherein both variables are log-transformed, the slope of the line representing isometry of wing area with body mass has a slope of 2/3; this corresponds to the ratio of the physical dimensionality (Gould 1966) of the two variables involved, wing area (two-dimensional) divided by mass (three-dimensional).

Other allometric relationships.—Other applications of regression analyses in the context of allometry used in this study included a test for increasing sexual size dimorphism with increasing body mass among species (Rensch 1950, 1959). To reduce artifacts of scale (Reiss 1986, 1989), data were treated in various ways, including two in which mean body masses were log-transformed and related to different metrics of sexual dimorphism. Allometry within and among species, based on a range of methodologies, has been the subject of theoretical works and empirical syntheses (Bonner 1952; Cock 1966; Arthur 1982a, b, 1984, 1988, 1997; Cheverud et al. 1983; Goodwin 1984; Starck 1993; Wolpert et al. 1997).

Allometry of development underlies many of the differences produced through heterochrony (Klingenberg 1998), and changes in developmental allometry can produce functionally influential changes in body size (Koehl 2000). Differences in scaling of selected skeletal elements with body size within species (allomorphosis), especially between flighted and flightless species, can serve as an index to such heterochrony in the absence of known-age developmental series of specimens. Allometry among adults within species can be considered an approximation of ontogenetic allometry (Gould 1966). However, in taxa showing significant sexual size dimorphism, marked allometry of selected dimensions with body mass can result in pronounced intersexual differences in shape that are essentially artifacts of size differences (Abouheif and Fairbairn 1997). Therefore, where sample size permitted comparisons confined to flighted congeners partitioned by both species and sex, consistent differences in within-group allometry across genera were interpreted as reflections of relative ontogenetic scaling at the terminus of growth, thereby providing ancillary insights into heterochrony related to flightlessness (Livezey and Humphrey 1986; Livezey 1989d).

The disproportionate increase in the robustness of the supporting skeleton with increases in body mass was among the first allometric trends recognized (e.g., Galilei 1637), and the evolutionary implications of this relationship remain of interest (Alexander 1996a, Løvtrup and Mild 1979; Lanyon 1981; Biewener 1982, 2000; Rubin and Lanyon 1984). The scaling of robustness of limb elements with increasing mean length was explored interspecifically by the fitting of allometric curves among species means, wherein robustness was estimated by estimates of cross-sectional areas of shafts (see above) and separately performed for three major elements of both the pectoral and pelvic appendages for flighted and flightless species.

ESTIMATION OF MISSING DATA

Before multivariate analysis, data sets for external, sternal, and skeletal measurements were subjected to a procedure that provided proxies for missing data (Table 6). Estimates were based on stepwise regressions of available measurements for the same taxon (excluding juveniles), and the precision of the estimates generally was comparable to that attending the use of dial calipers, hence the surrogate figures were superior to mere ersatz interpolations. No more than one half of the measurements were estimated for specimens (i.e., 3, 2, and 20 measurements for external, sternal, and [complete] skeletal records, respectively), although the vast majority of specimens for which estimates were made lacked only one or two data points (Table 6). These procedures resulted in 507 estimates involving 414 study skins (2.7% of the data set for 3,220 skin specimens) and 2,104 estimates involving 564 complete or partial skeletons (4.8% of the data set for 1,200 skeletal specimens). Assignments and revisions of recorded data on sex were based on multivariate methods, described below (discriminant functions and K-means cluster analyses).

RATIOS AND PROPORTIONS

Statistical properties.—Numerical properties of ratios, some of which pose special problems for distributional assumptions and hypothesis tests, have been the subject of substantial study (Atchley et al. 1976; Corruccini 1977; Albrecht 1978, 1979; Atchley 1978; Atchley and Anderson 1978). Accordingly, in this work, ratios and related estimates of dispersion (e.g., standard deviations for ratios of a sample) generally are presented as heuristic aids, with P-values associated with intergroup comparisons presented with appropriate caution. Proportions constituted by skeletal elements within the wings (humerus, ulna, and carpometacarpus) and legs (femur, tibiotarsus, tarsometatarsus, and middle toe), a special class of ratios in which the numerator and denominator are not strictly independent, were compared by using two-way ANOVAs of proportions transformed to arcsines of square roots. Relative thicknesses (shaft widths divided by shaft lengths) of selected pelvic elements were compared by using two-way ANOVAs of ratios transformed to natural (base e) logarithms. Selected ratios of skeletal measurements were compared by using ANOVAs of similarly transformed data; analyses based on proportions transformed to arcsines of square roots produced inferentially identical results.

Dimorphism ratios and surrogate metrics.—Within taxa, ratios of mean measurements (e.g., body mass) of males divided by those for females were used as

TABLE 6. Numbers of specimens of modern Rallidae for which sex was determined or reassigned by mensural criteria, and those for which missing measurements were estimated, by taxon and type of specimen. Taxonomy follows Livezey (1998). Counts exclude juveniles and trunk skeletons (latter exceeded limit adopted for estimation of data).

Taxon	Study skins (counts, percentage of total)			Skeletons (counts, percentage of total)		
	Sex estimated, reassigned	Specimens lacking measurements	Entries estimated	Sex estimated, reassigned	Specimens lacking measurements	Entries estimated
Porphyrio porphyrio-group	93, 5 (20.9)	82 (17.5)	89 (3.1)	20, 0 (21.7)	57 (62.0)	167 (4.4)
Porphyrio hochstetteri	3, 0 (21.4)	4 (28.6)	5 (6.0)	24, 0 (80.0)	17 (56.7)	125 (10.2)
Porphyrula alleni	3, 2 (9.6)	4 (7.7)	5 (1.6)	5, 1 (75.0)	3 (37.5)	6 (0.2)
Porphyrula martinica	1, 0 (3.3)	3 (10.0)	4 (2.2)	5, 0 (16.7)	13 (43.3)	28 (2.3)
Porphyrula flavirostris	7, 0 (13.2)	7 (13.2)	11 (3.5)	0, 0 (0)	1 (25.0)	13 (7.9)
Gymnocrex rosenbergii	8, 1 (42.9)	3 (14.3)	3 (2.4)	—	—	—
Gymnocrex plumbeiventris	15, 0 (55.6)	7 (25.9)	9 (5.6)	—	—	—
Habroptila wallacii	11, 0 (68.8)	3 (18.8)	3 (3.1)	0, 0 (0)	1 (33.3)	1 (2.4)
Eulabeornis castaneoventris	7, 1 (24.2)	3 (9.1)	4 (2.0)	—	—	—
Aramides ypecaha	0, 1 (3.3)	4 (13.3)	4 (2.2)	1, 0 (8.3)	4 (33.3)	9 (1.8)
Aramides cajanea-group	9, 2 (37.9)	0 (0)	0 (0)	12, 5 (32.7)	26 (50.0)	90 (4.2)
Canirallus oculeus	13, 0 (29.5)	1 (2.3)	1 (0.4)	—*	1 (100.0)	2 (3.8)
Canirallus kioloides	4, 1 (12.5)	6 (15.0)	6 (2.5)	—*	1 (33.3)	6 (11.3)
Rougetius rougetii	4, 0 (10.8)	5 (13.5)	7 (3.2)	—	—	—
Cyanolimnas cerverai	0, 0 (0)	0 (0)	0 (0)	—	—	—
Ortygonax sanguinolentus	4, 1 (25.0)	3 (15.0)	5 (4.2)	—	—	—
Ortygonax nigricans	1, 1 (10.5)	4 (21.1)	4 (3.5)	—	—	—
Dryolimnas cuvieri	16, 3 (32.2)	7 (11.9)	8 (2.3)	3, 1 (80.0)	2 (40.0)	2 (0.8)
Dryolimnas aldabranus	6, 0 (21.4)	1 (3.6)	1 (0.6)	8, 0 (53.3)	10 (66.7)	50 (1.9)
Dryolimnas abbotti	0, 1 (12.5)	1 (12.5)	2 (4.2)	—	—	—
Rallus aquaticus	0, 0 (0)	3 (10.0)	3 (1.7)	4, 0 (10.3)	17 (43.6)	58 (3.6)
Rallus limicola	0, 2 (6.7)	2 (6.7)	4 (2.2)	0, 2 (5.1)	26 (66.7)	83 (5.1)
Rallus longirostris-group	0, 2 (6.7)	0 (0)	0 (0)	2, 0 (5.9)	26 (76.5)	67 (4.8)
Rallus elegans-group	0, 0 (0)	0 (0)	0 (0)	5, 0 (13.5)	27 (73.0)	118 (7.8)
Gallirallus pectoralis-group	9, 0 (21.4)	3 (7.1)	4 (1.6)	—*	7 (87.5)	26 (7.9)
Gallirallus striatus	2, 0 (6.9)	3 (10.3)	3 (1.7)	0, 0 (0)	3 (60.0)	9 (4.4)
Gallirallus sharpei	—*	0 (0)	0 (0)	—	—	—
Gallirallus australis-group	37, 1 (25.3)	25 (16.7)	35 (3.9)	18, 0 (31.0)	—	148 (5.5)
Gallirallus philippensis-group	7, 2 (19.1)	8 (17.0)	10 (3.5)	17, 0 (47.2)	22 (61.1)	67 (4.5)
Gallirallus dieffenbachii	—*	0 (0)	0 (0)	—	—	—

TABLE 6. Continued.

Taxon	Study skins (counts, percentage of total)			Skeletons (counts, percentage of total)		
	Sex estimated, reassigned	Specimens lacking measurements	Entries estimated	Sex estimated, reassigned	Specimens lacking measurements	Entries estimated
Gallirallus owstoni	4, 0 (13.8)	2 (6.9)	2 (1.2)	4, 0 (12.5)	22 (68.8)	58 (4.4)
Gallirallus wakensis	11, 0 (19.0)	11 (19.0)	16 (4.6)	3, 0†	0 (0)	0 (0)
Tricholimnas lafresnayanus	4, 0 (44.4)	2 (22.2)	3 (5.6)	—	—	—
Tricholimnas sylvestris	5, 2 (13.8)	4 (4.6)	4 (0.8)	6, 0†	2 (33.3)	10 (3.1)
Nesoclopeus poecilopterus-group	6, 1 (23.3)	3 (10.0)	4 (2.2)	—	—	—
Aramidopsis plateni	1, 3 (28.6)	3 (21.4)	3 (3.6)	—	—	—
Cabalus modestus	8, 2 (35.7)	2 (7.1)	2 (1.2)	—	—	—
Habropteryx insignis	2, 0 (10.0)	6 (30.0)	6 (5.0)	—	—	—
Habropteryx torquatus-group	9, 3 (12.9)	25 (26.9)	35 (6.3)	3, 0 (23.1)	4 (30.8)	32 (6.0)
Habropteryx okinawae	0, 0 (0)	0 (0)	0 (0)	—*	—	—
Atlantisia rogersi	20, 1 (46.7)	4 (8.9)	9 (3.3)	—*	1 (50.0)	3 (3.7)
Laterallus spilonotus	0, 0 (0)	1 (5.3)	2 (1.8)	—	—	—
Laterallus leucopyrrhus	0, 0 (0)	0 (0)	0 (0)	14, 0 (51.9)	16 (59.3)	43 (3.9)
Coturnicops noveboracensis	2, 1 (10.0)	0 (0)	0 (0)	10, 1 (31.4)	26 (74.3)	105 (7.3)
Crex crex	0, 0 (0)	1 (3.3)	1 (0.6)	0, 0 (0)	9 (37.5)	17 (1.7)
Crex albicollis	2, 2 (12.3)	4 (12.9)	4 (2.2)	2, 0 (33.3)	5 (83.3)	14 (4.4)
Porzana sandwichensis	—*	1 (14.3)	1 (2.4)	—	—	—
Porzana palmeri	5, 1 (6.7)	7 (7.9)	7 (1.3)	14, 0†	8 (66.7)	37 (7.5)
Porzana carolina	0, 0 (0)	0 (0)	0 (0)	1, 0 (3.3)	20 (66.7)	47 (3.8)
Porzana pusilla	2, 1 (7.7)	0 (0)	0 (0)	7, 0 (35.0)	13 (65.0)	76 (9.3)
Porzana tabuensis	0, 1 (2.8)	6 (16.7)	6 (2.8)	11, 0†	9 (75.0)	41 (8.3)
Porzana monasa‡	0, 0 (0)	1 (50.0)	1 (8.3)	—	—	—
Porzana atra	0, 0 (0)	4 (16.0)	6 (4.0)	1, 0 (25.0)	2 (50.0)	10 (6.1)
Porzana flavirostra	0, 0 (0)	3 (9.7)	6 (3.2)	15, 1 (43.2)	27 (73.0)	109 (7.2)
Porzana olivieri	1, 0 (20.0)	2 (40.0)	2 (6.7)	—	—	—
Porzana bicolor	8, 1 (36.0)	5 (20.0)	6 (4.0)	—	—	—
Amauromis olivaceus-group	3, 3 (7.1)	9 (10.6)	10 (2.0)	2, 1 (21.4)	4 (28.6)	7 (1.2)
Amauromis phoenicurus	0, 1 (1.7)	3 (5.0)	4 (1.1)	4, 0 (12.9)	13 (41.9)	20 (1.6)
Amauromis akool	1, 0 (2.4)	0 (0)	0 (0)	—	—	—
Amauromis isabellinus	11, 2 (22.0)	20 (33.9)	29 (8.2)	—	—	—
Amauromis ineptus	3, 1 (16.7)	5 (20.8)	7 (4.9)	—	—	—
Gallicrex cinerea	0, 0 (0)	8 (21.1)	8 (3.5)	0, 0 (0)	1 (10.0)	3 (0.7)
Pareudiastes pacificus	—*	0 (0)	0 (0)	—	—	—
Pareudiastes silvestris	0, 0 (0)	0 (0)	0 (0)	—	—	—

TABLE 6. Continued.

	Study skins (counts, percentage of total)			Skeletons (counts, percentage of total)		
Taxon	Sex estimated, reassigned	Specimens lacking measurements	Entries estimated	Sex estimated, reassigned	Specimens lacking measurements	Entries estimated
Tribonyx ventralis	30, 2 (37.2)	6 (7.0)	8 (1.6)	10, 0 (23.3)	14 (32.6)	31 (1.8)
Tribonyx mortierii	18, 1 (40.4)	11 (23.4)	16 (5.7)	11, 0 (50.0)	13 (59.1)	100 (11.1)
Gallinula nesiotis	4, 0§	1 (25.0)	1 (25.0)	0, 0 (0)	—	—
Gallinula comeri	4, 0 (10.8)	5 (13.5)	7 (3.2)	2, 0 (100.0)	1 (50.0)	2 (1.9)
Gallinula tenebrosa	2, 0 (6.7)	5 (6.7)	5 (2.8)	3, 0 (21.4)	10 (71.4)	28 (3.8)
Gallinula angulata	0, 0 (0)	0 (0)	0 (0)	0, 0 (0)	0 (0)	0 (0)
Gallinula chloropus-group	14, 3 (13.6)	15 (12.0)	18 (2.4)	12, 2 (23.7)	33 (55.9)	58 (2.4)
Fulica armillata	1, 0 (3.1)	5 (15.6)	5 (2.6)	0, 0 (0)	1 (25.0)	1 (0.6)
Fulica cornuta	1, 0 (5.3)	4 (21.1)	4 (3.5)	0, 0 (0)*	2 (100.0)	2 (1.9)
Fulica gigantea	5, 1 (14.6)	7 (17.1)	8 (3.3)	0, 0 (0)	9 (90.0)	26 (6.3)
Fulica americana-group	0, 1 (3.3)	1 (3.3)	1 (0.6)	1, 0 (3.1)	4 (12.5)	18 (1.4)
Fulica alai	10, 1 (20.4)	17 (31.5)	17 (5.2)	1, 0 (6.3)	15 (93.8)	98 (14.9)
Fulica atra	3, 2 (13.9)	3 (8.3)	3 (1.4)	1, 2 (8.1)	16 (43.2)	33 (2.2)
All taxa	460, 63 (16.7)	414 (13.2)	507 (2.7)	262, 16 (25.8)	564 (52.3)	2,104 (4.8)

* Sample treated as entirely of indeterminate sex.
† Entire sample partitioned into provisional sex groups based on *K*-means clustering.
‡ Skin measurements provided by V. Loskot (pers. comm.); skeletal data based on X rays of two skin specimens figured by Steadman (1986a).
§ Sample partitioned into provisional sex groups based on classification functions based on sister taxon *G. (Porphyriornis) comeri*.

indices to univariate sexual size dimorphism (henceforth termed "dimorphism ratios"). Compared means generally were not transformed for simplicity of interpretation among closely related measurements or for the same ratios among closely related species (e.g., Lindenfors and Tullberg 1998). Where interspecific comparisons encompass substantial size differences, scale of the incorporated means becomes important (e.g., comparison of dimorphism ratios of body mass in crakes and coots), prompting log-transformation of included means before division (e.g., Livezey and Humphrey 1984a). Such ratios approach the simple index employed by Leutenegger and Cheverud (1985), log(male mean − female mean), a precaution similar to that employed by McGillivray (1985) for *Bubo virginianus*.

Wing loadings and aspect ratios.—The aerodynamic and mechanical implications of anatomical form for avian flight have been recognized for centuries, including the insights by Leonardo da Vinci (1505; translation by Marinoni 1976) concerning distribution of mass, movement of the wings, properties of feathers (remiges, alula, and rectrices), and production of lift. Increasingly sophisticated functional and evolutionary interpretations have been undertaken since the middle of the 19th century (Gladkov 1937a; Pennycuick 1986; Lighthill 1974). Following the recommendation by Clark (1971), wing loading was defined in this study by using metric units for body mass (g) and wing area (cm^2), as follows:

$$\text{wing loading} \equiv (\text{body mass})/(\text{wing area}),$$

wherein wing area was the doubled area of a single, extended wing. Despite comparatively high variances associated with both body mass and wing area (e.g., individual variation in body mass, wear of feathers, and imprecision of measurement), wing loading provides the best single index to flight capacity in birds. Wing loadings tend to be high in power fliers, that is, taxa dependent on high air speeds to create sufficient lift), and comparatively low in soaring birds, that is, species exploiting rising air columns as an additional or primary source of lift (Hedenström and Alerstam 1992; Hedenström 1993). Among taxa in the former group, wing loadings generally provide a reliable reflection of the power required to become airborne, achieve the air speed required to generate adequate lift, and thereby maintain level (powered) flight (Rayner 1988a, 1993, 1996; Hedenström and Møller 1992; Hedenström and Alerstam 1995).

Among modern carinates, mean wing loadings of flightless species are notably greater than those of flighted congeners (Livezey and Humphrey 1986; Livezey 1988, 1989b–d, 1990, 1992a, b, 1993a–c). Among power fliers having comparable pectoral muscles (e.g., Anatidae, Podicipedidae, and Phalacrocoracidae), flightless members generally have mean wing loadings approaching or exceeding 2.0 $g \cdot cm^{-2}$, consistent with the theoretical threshold of flightlessness of 2.5 $g \cdot cm^{-2}$ proposed by Meunier (1951). Flightlessness in steamer-ducks (*Tachyeres*) conformed closely with the latter figure (Humphrey and Livezey 1982; Livezey and Humphrey 1986).

Wingspan alone can influence flight characteristics, notably parameters of gliding flight (Tucker 1987). However, aspect ratios provide additional information on flight abilities, especially maneuverability and lift, with marked differences in this parameter typically limited to comparisons among higher taxonomic groups

(Pennycuick 1975, 1978; U. M. Norberg 1981a, b; Rayner 1988a; Spear and Ainley 1997; Hertel and Ballance 1999). This ratio is defined as:

$$\text{aspect ratio} \equiv (\text{wingspan})^2 / (\text{wing area}),$$

and through cancellation of units, the resultant measure is dimensionless. Traditionally, wingspan is measured from an intact specimen with both wings completely extended and includes the width of the back between the wings (Pennycuick 1975; Warham 1977); therefore, neither published data on standard wing lengths nor study skins can be used to estimate this dimension. As a result, data on aspect ratios of avian taxa are rare. Difficulties of measurement and appression of wings to the body in typical study skins precluded the routine assessment of wing curvature, a variable of some aerodynamic consequence in birds (Taber 1932a, b). For similar reasons, assessments of other details of wing shape (e.g., pointedness, gaps between extended remiges, and relative length of the alula [Rensch 1938; Kokshaysky 1973; U. M. Norberg 1981, 1995a, b; Mulvihill and Chandler 1991; Senar et al. 1994; Mönkkönen 1995; Lockwood et al. 1998]) were limited to qualitative comparisons of closely related flighted and flightless taxa.

Intra-appendicular skeletal proportions.—Skeletal wing length (SWL) was defined as the sum of the lengths of the humerus, ulna, carpometacarpus, and the two major phalanges of the major alar digit. SLL was defined as the sum of the lengths of the femur, tibiotarsus (including crista cnemialis), tarsometatarsus, and the three proximal phalanges of the third pedal digit. In modern taxa, the proportions of the total lengths of the wing and leg composed by the constituent elements were calculated by using associated specimens; that is, the proportions of total limb length represented by each element within individual birds were averaged across individuals for comparisons of groups. Intra-appendicular proportions of subfossil rallids were based on unweighted means for available specimens of each element.

Nonparametric Statistics

For those variables for which specific distributional assumptions were not valid, routine nonparametric statistics were employed (Conover 1980; Fienberg 1980; Agresti 1984). For example, tests of similitude of multivariate distances between taxa based on different suites of variables and assessments of congruence between corresponding elements of eigenvectors derived by different means (i.e., vector correlations) were tested for bivariate relationships by using Spearman rank correlations (Snedecor and Cochran 1967, 1980), herein symbolized as r_s.

Multivariate Morphometrics

Analytical Commonalities and Approach

Many data in this study were associated (i.e., suites of measurements pertaining to the same specimen or taxon) and therefore amenable to representation by using multivariate techniques (Gnanadesikan 1977; Mardia et al. 1979; Reyment 1991). However, certain characteristics of the data precluded particular multivariate approaches or distinguished other methods as optimal. First, suites of osteological measurements were based on linear dimensions of disarticulated skeletons, hence analyses contingent on positional relationships between elements were not applicable. Similarly, variation in orientation of parts of study skins stemming from preparatory differences rendered such holistic methods of description inappropriate for external measurements. Second, although homologous landmarks were the primary means by which the linear dimensions of skins or skeletal elements were identified and compiled (G. R. Smith 1990), use of graphical deformations of Cartesian networks (Thompson 1961; Bookstein 1977; Corruccini 1988) was limited to preliminary explorations and not used to describe deformations of shape.

For similar reasons, the resultant data did not describe static, essentially two-dimensional outlines or contours effectively summarized by methods intended for analysis of closed or open curves (e.g., splines, Bezier curves, or Fourier descriptors [Waters 1977; Younker and Ehrlich 1977; Bookstein et al. 1982; Ehrlich et al. 1983; Rohlf and Archie 1984; Lestrel 1997]), procrustean reconstruction of geometrically related points (Borg and Lingoes 1987; Rohlf and Slice 1990), or other approaches optimized for landmark data (Bookstein 1978, 1980, 1984, 1991; Brower and Veinus 1978; Humphries et al. 1981; Strauss and Bookstein 1982; Bookstein et al. 1985). Although outline-based methods (MacLeod 1999) are theoretically applicable to single skeletal elements (e.g., sterna and humeri), variation in single elements was assessed by comparatively familiar multivariate methods (described below) also applied to suites of measurements descriptive of multiply jointed, three-dimensional skeletons and encompassing multiple elements and structures. Use of outline-based methods for description perimeters defined by the remiges was not possible in that naturally extended wings were not available for most taxa, and attempts to estimate these by using extracted remiges proved unsuccessful. Fortunately, pervasive multivariate patterns in bones and suites of feather measurements generally are revealed by diverse analytical methods, that is, redundant patterns (those presumably reflecting important underlying mechanisms) are robust to a substantial extent to the details of the comparative techniques employed, given adequate sample sizes (Oxnard 1969a, b, 1978). Symbols used in accompanying figures are given in Appendix 4.

Several problems invited the use of curvilinear regressions, readily implemented through polynomial regressions (Rohlf 1990). Polynomial regressions of order p for two variables x and y, having independent random error term $\epsilon \sim N(0, \sigma_e^2)$, assumed the parametric form:

$$y = \beta_0 x^0 + \beta_1 x + \beta_2 x^2 + \beta_3 x^3 + \cdots + \beta_p x^p + \epsilon.$$

Notable among these applications were the quantification of curvilinearity in log-transformed plots (i.e., where allometric relationships approach asymptotes at extreme values), including those relating body masses of adults with masses of eggs or clutches; and parsimonious fitting of heuristic curves of associated pairs of measurements from developmental series of preserved specimens for which age (time) was not available for use as independent variable and for which standard growth curves (e.g., logistic, Gompertz, and von Bertalanffy) therefore could not be used (Ricklefs 1968, 1969, 1973, 1983a; Konarzewski et al. 1998; Ricklefs and Starck 1998a, b; Starck and Ricklefs 1998a–c). Although coefficients of higher-order terms pose problems of interpretation, such curves provide a graphical alternative to simple citation of outliers.

K-MEANS CLUSTER ANALYSIS

Methodological essentials.—With the rare exception of the preservation of the elements of individual specimens in situ (e.g., some *Thambetochen* [Olson and James 1991]), subfossil elements generally are collected as aggregations of disassociated elements. This intermingling of elements not only precluded multivariate techniques for which complete (associated) skeletons are required, but such collections also can be assumed to include elements from both sexes (assuming that included taxa can be segregated by other, preferably qualitative diagnostic criteria). Hence, samples of subfossil elements represent demographic versions of mixture distributions (Titterington 1985; McLachlan and Basford 1988; Kaufmann and Rousseeuw 1990).

Grouping subfossil elements by sex.—Although sympatric species of rallids generally were separable by qualitative characters, the grouping of elements by sex could be approached only by statistical means. Therefore, before statistical comparisons, sex classes (within species) of elements with adequate sample sizes (humeri, radii, ulnae, femora, tibiotarsi, and tarsometatarsi) were estimated through K-means clustering (McLachlan and Basford 1988; Kaufmann and Rousseeuw 1990). In the case of discrimination of two sexes within a single species, $K = 2$. This method iteratively reallocates specimens into two optimally discriminatory sex groups by using variance-standardized Euclidean distances based on the dimensions available for each element. Beginning with two seed vectors derived from initial appraisals of central tendencies suggested by the univariate distributions of measurements, the procedure generally accomplished the desired partitioning of samples in a single interation; in a few cases, seed vectors were revised so as to effect partitions having within-group variances of comparable magnitudes (a condition typically substantiated in samples for which sexes were known). An application of this technique to samples of subfossil elements of raphids was described by Livezey (1993b).

As indicated by the multifactorial means employed to infer sex of an extant flightless rail for which comparatively complete data are available (Eason et al. 2001), sexual dimorphism and the allocation of specimens to sex classes is not a trivial problem for some taxa of Rallidae. Distinctness of the resultant sex groups was assessed by a subsequent one-way ANOVA. Samples comprising too few specimens for partitioning with clustering algorithms were sorted into provisional sex groups based on assessments of apparent bimodalities of univariate distributions of associated measurements. For multivariate comparisons (see below), ratios of available measurements for other species–sex groups were used to approximate those few dimensions lacking for another, related taxon–sex group.

PRINCIPAL COMPONENT ANALYSIS

Principal component analysis (PCA; Appendix 2) is a common means of multivariate comparison of morphological dimensions among taxonomic or taxon–sex groups (Jolliffe 1986; Flury 1988; Jackson 1991), a special case of a class of techniques used in factor analysis (Harman 1967; Lawley and Maxwell 1971; Catell 1978; Cooper 1983; Bartholomew 1987). Associated anatomical measurements typically show strong structure (e.g., Livezey 1986b, 1988, 1989b–d, 1990, 1992a, b, 1993b, c), and deviate substantially from eigenstructures emanating from random data (Stauffer et al. 1985). Because this technique seeks synthetic, mutually orthogonal, multivariate axes that are maximally explanatory of dispersion among included points regardless of group membership, group means frequently are analyzed to avoid differential influence on axes by differing sample sizes, although methods for multiple-group analyses have been employed (Airoldi and Flury 1988a, b; Thorpe 1988). This robustness sometimes is achieved through distortion among covariance structures (Houle et al. 2002). In the present study, applications of PCA are consistent with exploratory factor analysis, in which no assumptions about eigenstructures are assumed a priori; this contrasts with confirmatory factor analysis, in which both the number and composition of factors are predetermined (Reyment and Jöreskog 1993; Basilevsky 1994).

In this study, PCA primarily was used in the sense of R-analysis (Jackson 1991), that is, for assessing relationships among n observations in a space defined by p measurements (dimension), and having maximal rank of $min \, (n - 1, p)$. Such implementations were applied to mean vectors of suites of measurements for species (Appendix 2). The suites of measurements (dimension p) were those for study skins ($p = 6$), complete skeletons ($p = 41$), partial skeletons in which dimensions of the most fragile, often absent elements (alar and pedal phalanges and furcula) either were lacking or excluded ($p = 35$), detailed skulls ($p = 25$), and sterna ($p = 5$). Where means for groups were analyzed, estimates were based on all available data for each variable. Consequently, sample sizes for the constituent means may vary for a given group, a situation that is typical in subfossil taxa for which associated skeletons were unavailable (e.g., Radinsky 1985; Livezey 1988, 1993b, c). Suites of myological measurements were compared by using PCA because most taxa were represented for these data by single specimens; such limited samples are not suitable for analysis by several alternative means.

CANONICAL ANALYSIS AND DISCRIMINANT FUNCTION ANALYSIS

Canonical (variate) analysis (CA; Appendix 2), the two-group method first proposed by Fisher (1936), derives mutually orthogonal multivariate axes that maximally discriminate predefined (here taxon–sex) group means relative to pooled within-group variances (Pimentel 1979; Albrecht 1980; Campbell and Atchley 1981; Gittins 1985; McLachlan 1992; Huberty 1994). Analyses presented were limited to linear discriminant functions (Lachenbruch 1975; Tabachnick and Fidell 1989), as opposed to quadratic discriminant functions, which implicitly assume homogeneity of within-group covariance matrices (Reyment 1962; Johnson and Wichern 1982; James and McCulloch 1990).

Discriminatory axes, termed canonical variates (CVs), were based on subsets of m external or skeletal measurements (log-transformed to homogenize variances among variables [Bryant 1986]) that were backstep-selected from the complete set of measurements based on partial F-statistics. Magnitude of total among-group variation represented by the set of CVs for a given statistical sample is explicitly parameterized by using likelihood ratios (Wilks' λ) corresponding to the appropriate tripartite degrees of freedom: m (variables included in model); g (groups) $- 1$; and n (total specimens) $- g$. In that tables of critical values for such likelihood ratios are comparatively uncommon and the statistic itself is unfamiliar to most morphometricians, the large-sample approximation to the F-distribution is cited in the text (degrees of freedom for which are functions f_1 and f_2 of dimensions m, g, and n). Symbolically, this convergence in distributions between Wilks' λ and the more-familiar F-ratio is given as:

$$\text{Wilks' } \lambda_{(m,g-1,n-g)} \xrightarrow{D} F_{(f_1[m,g,n]), f_2[m,g,n])}, \quad \text{as } n \to \infty.$$

Multivariate differences between specific groups were tested by using associated pairwise F-statistics and Bonferroni P-values. Mahalanobis distances (D) were calculated to summarize the multivariate distances between group centroids.

Canonical analysis, like all multivariate methods for which underlying theory was based on Gaussian distributions, assumes multivariate normality of the measurements analyzed (Mardia 1975; Davis 1980), a condition generally approached or met with associated anatomical dimensions. These multivariate distances, together with related metrics (e.g., Pillai's trace and Wilks' λ), were used to summarize intersexual differences within species. Although single specimens can be plotted a posteriori on the resultant axes, only groups represented by adequate samples (i.e., informative regarding within-group covariance structure) were used to define the CVs.

Interpretation of CVs was more challenging, in that these variates are corrected for pooled within-group size and therefore typically none of the variates clearly represents the highly correlated variation in size among groups. Interpretations of variates largely were restricted to Spartan assessments of the coefficients involved in substantial contrasts (a function of algebraic signs and magnitudes of coefficients), bolstered by correlations of mean group body masses with variates as indications of the size included by each CV and direct examination of the first principal component (PC-I) of the pooled within-group covariance matrices. The size-correction represented by such standardization incorporates that effect by shearing, that is, partitioning from the variance of interest that attributable to within-group size (Humphries et al. 1981; Bookstein et al. 1985; Rohlf and Bookstein 1987). This approach includes all measurements in the interpretation of CVs (regardless of the subset of variables entered into the model), and provides measures of pattern that are independent of pooled within-group variance and comparatively stable with respect to magnitude of intergroup differences and multicolinearity of variables.

Jackknifed classifications (Lachenbruch and Mickey 1968) associated with the CAs, confidences of which are reflected in associated posterior probabilities of assignment to each of the compared groups, were used to cross-validate and determine the sexes of specimens lacking this information. In this study, classifications were of traditional or complete form, in which all individuals were classified (ultimately) into predefined groups. This is to be distinguished from partial discriminant analysis, in which cases not meeting established criteria of assignment remain of indeterminate membership (M. E. Johnson 1987); however, in the stagewise implementation that follows, the temporary retention of unclassified cases during preliminary passes accords with the intention of partial classificatory techniques. Classifications were executed in three successive passes for skin specimens (excluding juveniles): specimens of unknown sex were assigned if posterior probability of classification (P_c) was 0.900, and sex was reassigned if $P_c \geq 0.950$; specimens of unknown sex were assigned if $P_c \geq 0.750$, and sex was reassigned if $P_c \geq 0.975$; and remaining specimens of unknown sex were assigned to the sex for which P_c me 0.50, and sex was reassigned only for specimens for which sex was assigned in the second or third steps on the basis of the larger P_c. A similar three-pass procedure was implemented for skeletal specimens, except that the greater precision of mensural data for skeletons permitted the P_c for reassignment of sex to 0.950 and 0.990 for the first and second passes, respectively.

A total of 460 unsexed study skins (involving 53 modern species) and 262 skeletons (involving 36 modern species) were classified by this protocol (Table 6). Exceptions to this procedure, limited to taxa having adequate numbers of skeletal specimens reliably assigned to sex and for which K-means clustering was employed (see section above) were *Porphyrula alleni, Tricholimnas sylvestris, Porzana tabuensis, P. palmeri, Amaurornis moluccanus,* and *Gallinula* (*c.*) *pyrrhorrhoa*. Several others simply lacked adequate skeletal material for sex-specific analyses, including *Gallirallus pectoralis, Atlantisia rogersi, Gallinula sandvicensis,* and *Fulica alai*. Subfossil rails seldom were represented adequately for inclusion in CAs of skeletal measurements even as composite vectors of means (with exceptions being "*Atlantisia*" *elpenor* and *Fulica chathamensis*), in part because some of the variables most useful for distinguishing flightless rails (lengths of alar phalanges and dimensions of carina sterni) were lacking for most of these extinct taxa. Furthermore, because samples of associated skeletons were not available for any subfossil rallid, the few taxa that could be included were eligible only for plotting a posteriori and therefore did not contribute to the definition of the discriminatory axes (e.g., plotting flightless *F. chathamensis* on axes contrasting modern, flighted congeners).

CORRESPONDENCE ANALYSIS

Methodological essentials.—Like multidimensional scaling (Davison 1983), correspondence analysis graphically assesses relationships among classes of variables by means of patterns of shared observations among multiway frequency tables (Greenacre 1984; Lebart et al. 1984; Palmer 1993), the contributions of which are indicated by two criteria: distances of points vis-à-vis the origin (i.e., point having zero coordinates on both axes), indicating the magnitude of relationships among variables (contra P. J. Taylor 1998); and the relative polar position(s) of these points (approximated by the four quadrants of the plane, sequenced in clockwise fashion), reflecting the nature or sign of the relationships (Greenacre 1984; Moran and Gornbein 1988).

Taxonomic, morphological, and ecogeographical associations.—In this paper, the primary use of correspondence analysis was to assign flightless species of Rallidae (objects) to a comparatively sparse multiway contingency table of categorical variables. In this context, such variables included characteristics of the species per se (e.g., estimated anatomical changes from flighted relatives, dietary group, and status at present) and ecogeographical parameters of the islands inhabited (e.g., distance from continent, area, and latitude). The analysis was designed to reduce information on associations into a very limited number of summary axes and statistics, and to assess the magnitudes and nature of relationships among the variables with respect to these summary axes by way of one or more plots simultaneously displaying characteristics of both flightless rails and the ecogeographical circumstances under which they evolved (Appendix 2).

SOFTWARE FOR MORPHOMETRIC ANALYSES

Mensural data were imported into analytical programs by using Microsoft Excel©. Morphometric analyses, principally traditional methods (James and McCulloch 1990; Marcus 1990; Marcus et al. 1996), primarily were executed by using BMDP© statistical software (Dixon 1992), especially programs 1D, 5D, 6D, 7D, 8D, 1M, 2M, 4M, 7M, AM, KM, 1R, 3R, 5R, 6R, and 3S. In addition, customized combinations of programs were designed to accomplish novel analyses, for example, modified PCAs (programs 4M and 6R), PCAs of pooled within-species covariance matrices (programs AM and 4M), and cluster analyses of correlation coefficients for pooled within-species correlation matrices (programs AM and 1M).

Digital refinement of graphical presentations and labeling of anatomical illustrations were executed with Photoshop© (Adobe Systems 2000) and Canvas™ (Deneba Software 1998); limited plotting and mathematical applications were implemented with Mathematica© (Wolfram 1996). Several novel approaches to graphical presentation of summary findings from the foregoing analyses (e.g., dot charts) were inspired by Cleveland (1985).

RESULTS

PHYLOGENY-BASED EXPLORATIONS

ANALYTICAL CONSIDERATIONS

Despite recent efforts to reconstruct phylogenetic relationships of the Rallidae with sequence data (Trewick 1997a, b) and morphological characters (Livezey 1998), many nodes and placements of taxa remain unresolved, and these deficits curtailed a number of phylogeny-based assessments. These points of uncertainty, as well as substantial missing data for many modern and subfossil taxa for optimizations of interest, precluded empirically grounded, family-wide phylogenetic analyses of ecomorphological attributes for rails. Nonetheless, a companion analysis (Livezey 1998) established monophyly of a number of clades having flighted and flightless members (Fig. 11), thereby justifying the independent analyses of the losses of flight in these phylogenetically delimited subgroups that follow this section. Specifically, some taxa either are the only flightless members of their genus (e.g., *Dryolimnas aldabranus, Amaurornis ineptus,* and *Gallinula nesiotis*-group) or were inferred to represent the flightless sister-group of a flighted clade (e.g., *Cyanolimnas cerverai* and *Habroptila wallacii*), thereby obviating the need for wider phylogenetic resolution for tree-based assessments. A few other flightless genera can be compared provisionally to a single modern genus because of higher-order phylogenetic position and biogeographical considerations (e.g., *Nesotrochis* with *Aramides*; *Aphanapteryx* with Old World *Rallus*).

Most vexing from the phylogenetic standpoint, however, is the poor resolution among a critical set of genera of typical rails from the Australasian region, namely *Gallirallus* and allies *Tricholimnas, Nesoclopeus, Cabalus, Capellirallus,* and *Diaphorapteryx* (Livezey 1998). For the former set of genera, trees used for optimizations and tests for correlated changes were based on those presented by Livezey (1998), with unresolved portions subjected to alternative topological arrangements based variously on suggestions by Olson (1973a, 1975b), taxonomically limited inferences by Trewick (1997a, b), and an attempt to render proposed

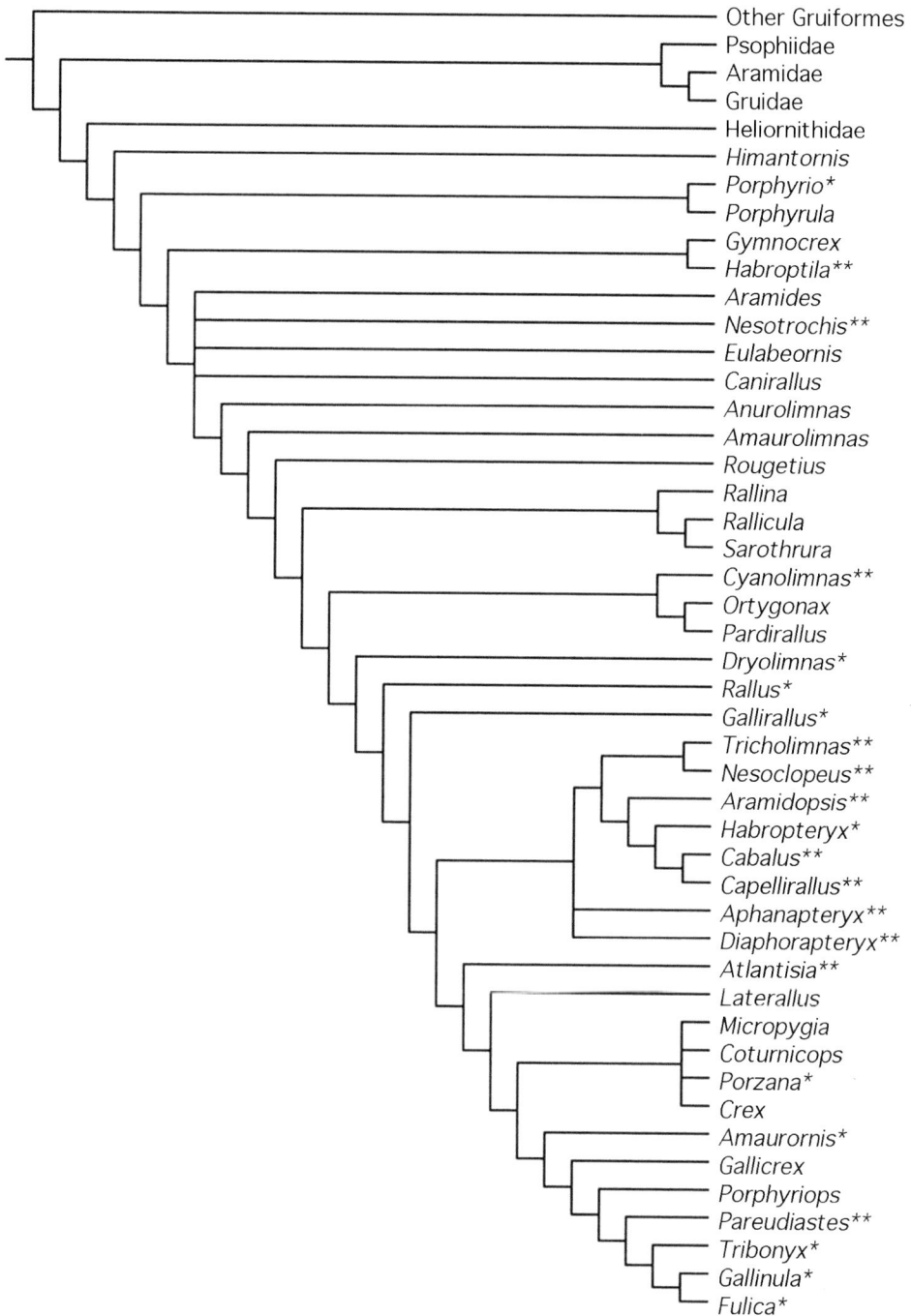

FIG. 11. Simplified phylogenetic tree for families of Grues and genera of Rallidae based on Livezey (1998). Genera including both flighted and flightless members are followed by a single asterisk; genera including only flightless members are followed by two asterisks.

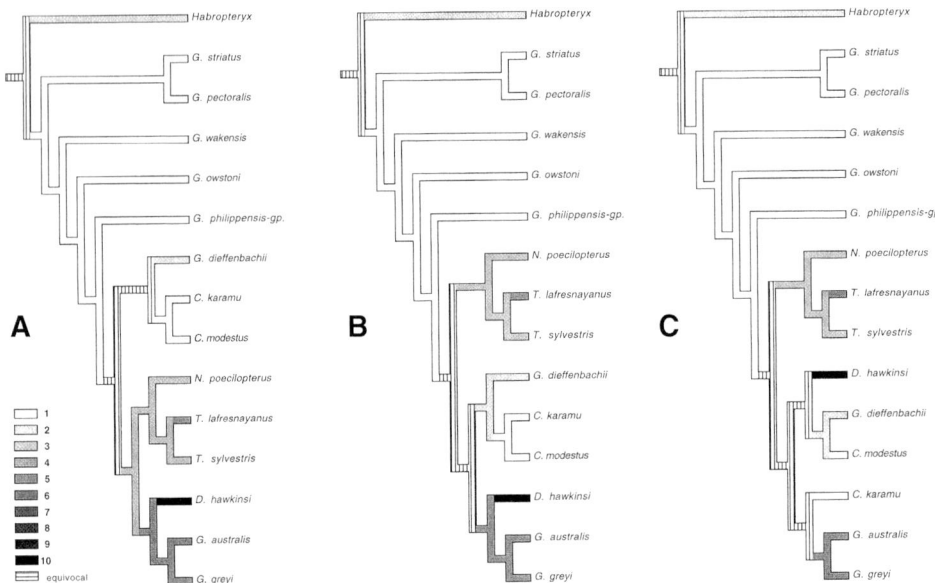

FIG. 12. Fully dichotomized variants of the phylogeny of *Gallirallus* and allied genera (Livezey 1998), in which mean body masses are symbolized by density of branch shading: **A,** topology in which monophyly of flightless rallids of Australasian region was conserved, in which *Capellirallus karamu* is shown as sister group of *Cabalus modestus*, *Diaphorapteryx hawkinsi* is shown as sister group of *Gallirallus australis*-group, and parsimony of body mass is high (length = 2,733); **B,** topology in which monophyly of flightless rallids of New Zealand was conserved, in which *Capellirallus karamu* is shown as sister group of *Cabalus modestus* and parsimony of body mass is moderately high (length = 2,880); **C,** topology in which monophyly of flightless rallids of the Chatham Islands was conserved and parsimony of body mass is moderately low (length = 3,446).

relationships consistent with the biogeographical distributions of insular taxa (Fig. 12A–D). Taxa lacking the attributes to be optimized were trimmed from the topology (e.g., *Capellirallus* and *Diaphorapteryx* were excluded for all considerations other than flight status, estimated body mass, or bill shape). Unfortunately for *Habropteryx* and probable sister-group *Aramidopsis,* exceptionally poor data for most or all flightless members rendered phylogenetic optimizations fruitless.

Several other genera including flightless members either remain of indeterminate monophyly (*Porzana* and *Crex*) or intergeneric relationship (*Atlantisia*). In the former case, critical characters were optimized on trees consistent with available phylogenetic analyses (Livezey 1998) and favored from the standpoint of biogeography (Fig. 13). Flightless moorhens and coots were placed with moderate precision within the Fulicarina (Livezey 1998), permitting phylogenetic optimizations for all flightless members (data permitting) at a tribal level (Fig. 14).

SPECIFIC PHYLOGENETIC PATTERNS AND IMPLICATIONS

Although poorly resolved portions of the only generic or species-level phylogenetic hypothesis remain (Livezey 1998) and unavailability of data afflicts a number of crucial taxa (e.g., body mass and wing lengths of subfossil species), the genus-level framework (including establishment of monophyly and composition of genera, tribes, and subfamilies) provides an essential subdivision of taxa

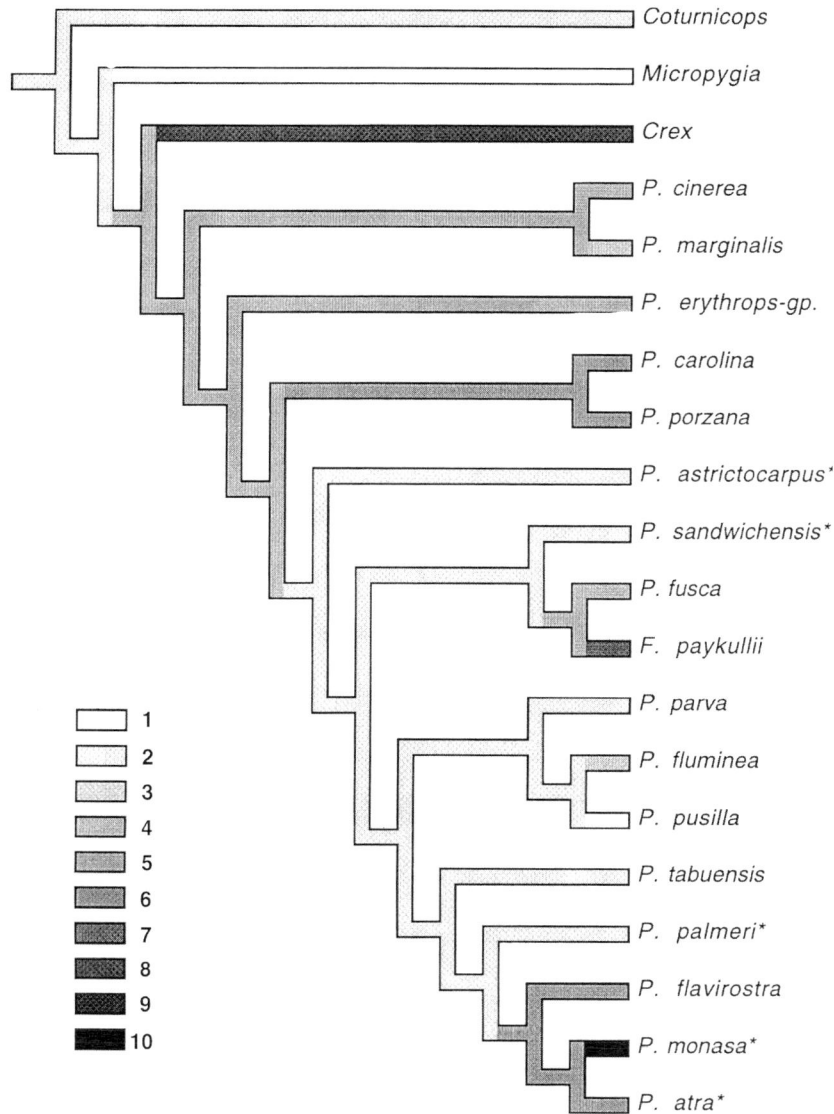

FIG. 13. Fully dichotomized variant of the phylogeny of Pacific *Porzana* (after Livezey 1998), including other genera of subtribe Crecina as outgroups, showing optimization of mean body masses (length = 353). Taxa lacking data for body masses (*P. olivieri, P. bicolor,* and Hawaiian subfossils) were omitted, and flightless taxa are indicated by asterisks.

for standard univariate and multivariate analyses. Unfortunately, implementation of quantitative tests of correlation or concentrated changes require complete resolution, no missing data, and binary states of the characters involved, a challenge exacerbated by the confounding effects of parallelism on reconstructions (Marques and Gnaspini 2001). Transformations of states and deletions of taxa that would meet these conditions rendered the analyses less informative, therefore less-formal approaches were employed. Moreover, species-level proposals made by other authors, suggested by biogeographic distributions, and consistent with the phylo-

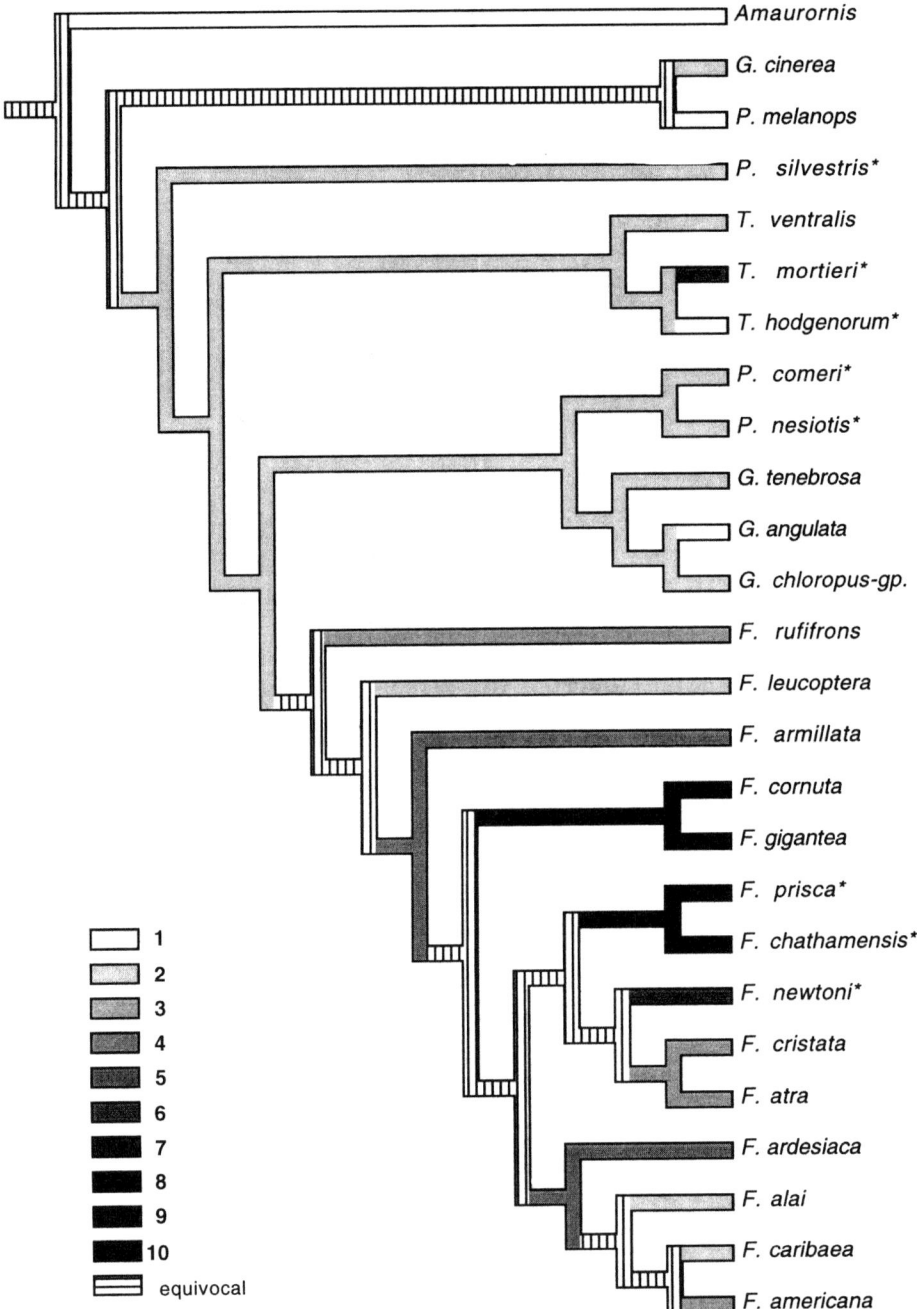

FIG. 14. Fully dichotomized variant of the phylogeny of moorhens, coots, and allies (subtribe Fulicarina), after Livezey (1998), showing optimization of mean body masses (length = 6,113) when using *Amaurornis* as outgroup. One taxon lacking data for body mass (*Pareudiastes pacificus*) was omitted, and flightless species are indicated by asterisks.

genetic information available (Livezey 1998) permitted some more-detailed explorations (Figs. 12–14).

The topological variants depicted for *Gallirallus* and allies demonstrated the poor resolution of the relationships of this group, in which several biogeographically compelling arrangements render one or more of the included "genera" para- or polyphyletic (Fig. 12). Until an empirical resolution of this complex is available, however, the constituent "genera" are retained to illustrate clearly the uncertainties of relationships among members, as opposed to merging all members into a single genus *Gallirallus* (cf. Olson 1973a) and thereby obscuring the weak evidence supporting the included subgroups or the diversity of form that engendered a conservative treatment of genera in historical and recent works (Livezey 1998). These uncertainties notwithstanding, several provisional topologies for *Gallirallus* and allies proved instructive for optimizations (Fig. 12). Of particular interest was the pattern of body size, bill shape, and sexual dimorphism in the clade, and possible evidence for a direct relationship between mean body mass and magnitude of sexual size dimorphism among species (Rensch 1950, 1959; Reiss 1989).

Two topological variants consistent with increasingly nested biogeographic patterns (Fig. 12A, B) permitted more parsimonious optimizations of body mass than the proposal perserving monophyly of Chatham endemics (Fig. 12C). All three examples presented suggested a modest, direct association between body mass and flightlessness, a trend supported largely by a number of comparatively massive taxa, most of which are not likely sister-taxa: *Gallirallus australis*-group, *Diaphorapteryx hawkinsi*, *Nesoclopeus poecilopterus*-group, and both species of *Tricholimnas*. The overall tendency for flightless *Gallirallus*-type rallids to be massive was diminished by such flightless "dwarfs" as *Gallirallus wakensis* and *Cabalus modestus*, taxa almost certainly having evolved flightlessness and small size independently of each other (Fig. 12A). This variant did not delimit an association between body mass and sexual size dimorphism or a notable phylogenetic pattern in bill shape.

Topological variant C for *Gallirallus* and allies (Fig. 12), in which monophyly of Chatham Island endemics (*Gallirallus dieffenbachii*, *Cabalus modestus*, and *Diaphorapteryx hawkinsi*) was preserved, defined a clade comprising three taxa showing uniquely high sexual dimorphism in bill lengths and other skeletal dimensions, and three of four members of the group showing strong decurvature of the bill. The latter feature was one of several morphological characters supporting a sister-group relationship between *Cabalus modestus* and *Capellirallus karamu*, the latter confined to the North Island of New Zealand (Livezey 1998). For these reasons, this specific topology is comparatively appealing (at least with respect to biogeographic criteria), but the preservation of monophyly of Chatham endemics implies a significant revision of generic taxonomy and contributed little toward an explanation of the evolutionary processes that led to the unique morphologies of *Cabalus modestus* and (especially) *Diaphorapteryx hawkinsi* (Fig. 12C).

Optimization of available data on body masses on a geographically based trimmed tree consistent with available phylogenetic information (Livezey 1998) for Pacific crakes (subtribe Crecina) revealed that little or no association occurred between body mass and flightlessness in this group (Fig. 13). Consideration of the Liliputian examples of flightless subfossil crakes from Hawaii of uncertain interrelationships (e.g., *Porzana menehune* or *P. ziegleri*) would undermine further any association between these two variables. If anything distinguishes the phylogenetic pattern of flightlessness and form among the Pacific crakes, it is the widespread occurrence of flightlessness and the absence of apparent morphological correlates beyond the pectoral apparatus (e.g., changes in body mass, increased sexual dimorphism, or size and decurvature of the bill).

Mapping of body masses on a phylogenetic tree for the moorhens and coots (subtribe Fulicarina) based on the analyses by Livezey (1998) revealed that a limited association exists between increased body mass and the evolutionary loss of flight (Fig. 14). This pattern largely reflects the comparatively large sizes of *Tribonyx mortierii* and the subfossil *Fulica* (the latter only provisionally flightless), and several exceptions are notable (*Tribonyx hodgenorum* and *Gallinula nesiotis*-group). The allometric enlargement and incipient flightlessness of the Andean *Fulica gigantea* and *F. cornuta* suggests an evolutionary pathway to flightlessness in coots that may have pertained to the subfossil *Fulica* of New Zealand and the Chatham Islands (Fig. 14).

Morphometrics of External Characters

Univariate Comparisons of Means and Variances

Interspecific differences among means.—Significant differences among species–sex groups in mean measurements of study skins were the rule in most taxonomic partitions (e.g., genera or closely related groups of genera), and were the combined outcome of generally adequate sample sizes, subtle but substantial differences among species and (to a lesser extent) sexes, and comparatively limited variation within species–sex groups (Table 7). Where significant differences were not revealed, most of the groups involved either were very closely related (e.g., flighted members of *Porphyrio porphyrio* superspecies), showed negligible sexual dimorphism, were of the minority of cases for which sample sizes were too small for reasonable statistical power, or group differences were based in large part on samples allocated to sex a posteriori or measured by third parties (Table 7). Consequently, for the sake of brevity and focus, a comprehensive summary of test statistics pertaining to differences among means will not be presented; instead, the discussion is limited to those comparisons of primary interest (i.e., those pertaining to flightlessness) and will emphasize those in which an unexpected absence of significant differences was indicated or where differences exceeded expectations. In this section, differences among taxa will be considered, whereas intersexual differences in means and variances are considered under sexual dimorphism.

Among the greater swamphens (*Porphyrio*, $n = 481$), the extraordinary dimensions of the Takahe (*P. hochstetteri*) contributed in large part to the great differences among taxa (ANOVA, $P < 0.0001$; Bonferroni *t*-tests, $P < 0.0001$) in all six external measurements (Table 7). Modest differences among flighted *Porphyrio* and within *Porphyrula* also contributed to interspecific variation, although pairwise comparisons within these groups often failed to achieve critical values. The Takahe was uniquely large in all dimensions except wing length, a dimension in which some of the larger flighted taxa approached or exceeded

TABLE 7. Univariate summary statistics ($\bar{x} \pm$ SD [n]) for selected measurements (mm) of skin specimens of modern Rallidae, by species and sex. Superspecies within *Porphyrio*, *Aramides*, *Rallus*, *Habropteryx*, *Amaurornis*, *Gallinula*, and *Fulica* follow Livezey (1998). Three recently described rallids (*Gymnocrex talaudensis* [Lambert 1998a], *Gallirallus rovianae* [Diamond 1991], and *Amaurornis magnirostris* [Lambert 1998b]) are excluded because wing lengths, tail lengths, or both were not available for these species.

Taxonomic group	Sex	Culmen length	Bill depth	Wing length	Tail length	Tarsus length	Middle-toe length
Swamphens							
Porphyrio porphyrio	M	52.5 ± 2.5 (20)	24.3 ± 0.8 (20)	264.3 ± 9.9 (20)	98.3 ± 9.4 (20)	98.7 ± 4.6 (20)	96.8 ± 4.8 (20)
	F	50.2 ± 2.7 (15)	23.1 ± 1.2 (14)	260.4 ± 8.7 (15)	93.6 ± 5.1 (15)	87.3 ± 4.7 (15)	93.7 ± 5.8 (15)
Porphyrio madagascariensis	M	46.3 ± 4.7 (11)	23.2 ± 1.5 (10)	247.7 ± 12.7 (11)	87.9 ± 6.6 (11)	89.5 ± 6.7 (11)	92.3 ± 6.3 (11)
	F	47.2 ± 2.7 (17)	22.1 ± 1.4 (16)	243.6 ± 13.8 (17)	85.4 ± 6.0 (17)	86.6 ± 5.0 (17)	87.8 ± 5.2 (17)
Porphyrio poliocephalus	M	48.2 ± 3.3 (28)	22.4 ± 1.5 (20)	253.8 ± 14.5 (27)	90.1 ± 6.9 (28)	93.9 ± 6.5 (28)	94.0 ± 5.3 (28)
	F	46.9 ± 2.4 (15)	22.1 ± 0.9 (14)	259.9 ± 7.4 (16)	93.4 ± 6.5 (16)	96.2 ± 3.9 (16)	93.7 ± 4.1 (16)
Porphyrio viridis	M	43.6 ± 2.2 (18)	20.7 ± 0.5 (14)	246.2 ± 8.0 (17)	85.1 ± 6.0 (16)	91.4 ± 4.2 (18)	90.2 ± 1.7 (18)
	F	41.2 ± 1.4 (12)	19.5 ± 0.8 (10)	239.5 ± 6.7 (8)	80.5 ± 5.8 (11)	86.2 ± 4.4 (12)	84.4 ± 3.6 (12)
Porphyrio melanopterus	M	49.1 ± 4.2 (13)	22.4 ± 1.5 (12)	245.1 ± 14.5 (13)	86.3 ± 6.7 (13)	91.8 ± 6.5 (13)	79.7 ± 5.1 (12)
	F	44.9 ± 2.9 (15)	20.0 ± 1.4 (15)	228.9 ± 12.6 (15)	80.7 ± 5.9 (14)	82.5 ± 4.9 (15)	76.4 ± 4.4 (15)
Porphyrio indicus	M	44.0 ± 3.1 (22)	19.7 ± 0.9 (18)	226.2 ± 10.5 (22)	97.0 ± 6.7 (22)	82.5 ± 6.7 (22)	79.6 ± 5.0 (22)
	F	41.3 ± 2.9 (15)	18.2 ± 1.1 (15)	223.9 ± 7.7 (15)	79.5 ± 7.3 (15)	81.1 ± 5.5 (15)	76.2 ± 6.1 (15)
Porphyrio bellus	M	55.0 ± 3.5 (8)	26.7 ± 2.4 (8)	281.3 ± 10.0 (7)	103.3 ± 8.7 (8)	99.3 ± 4.7 (8)	93.3 ± 4.8 (8)
	F	49.0 ± 2.0 (5)	24.5 ± 0.8 (5)	208.4 ± 7.7 (5)	99.0 ± 9.6 (5)	90.8 ± 2.6 (5)	84.0 ± 3.1 (5)
Porphyrio melanotus	M	52.4 ± 2.9 (50)	24.0 ± 1.1 (46)	266.8 ± 9.6 (50)	98.2 ± 9.6 (49)	99.1 ± 4.6 (49)	91.9 ± 4.1 (50)
	F	47.8 ± 2.6 (33)	21.4 ± 1.4 (29)	252.1 ± 10.7 (32)	95.4 ± 5.7 (33)	90.0 ± 4.1 (33)	84.4 ± 4.3 (33)
Porphyrio albus	—	51.0 ± 5.7 (2)	23.4 ± 4.8 (2)	229.0 ± 1.0 (2)	78.5 ± 0.7 (2)	83.5 ± 2.1 (2)	75.5 ± 4.9 (2)
Porphyrio samoensis	M	46.5 ± 3.4 (50)	21.4 ± 1.6 (48)	228.8 ± 9.0 (49)	78.2 ± 6.6 (44)	84.1 ± 4.8 (50)	77.4 ± 5.6 (50)
	F	43.6 ± 3.0 (57)	19.8 ± 1.3 (49)	221.9 ± 10.0 (57)	79.3 ± 6.9 (52)	81.6 ± 5.0 (57)	75.5 ± 4.3 (57)
Porphyrio ellioti	M	47.9 ± 3.0 (24)	23.1 ± 1.3 (23)	233.0 ± 13.6 (22)	79.4 ± 8.5 (21)	90.4 ± 4.4 (24)	79.3 ± 5.0 (24)
	F	43.9 ± 3.0 (24)	20.8 ± 1.2 (24)	222.8 ± 14.7 (24)	76.2 ± 8.1 (22)	83.5 ± 4.5 (24)	73.5 ± 5.4 (24)
Porphyrio pulverulentus	M	48.8 ± 1.5 (4)	23.3 ± 1.5 (4)	245.0 ± 5.3 (3)	84.3 ± 6.9 (4)	90.5 ± 2.6 (4)	89.5 ± 2.5 (4)
	F	46.5 ± 3.5 (6)	22.6 ± 0.9 (6)	239.3 ± 3.7 (7)	87.0 ± 5.8 (8)	88.0 ± 5.6 (10)	84.3 ± 3.8 (10)
Porphyrio hochstetteri	M	75.9 ± 1.6 (7)	48.6 ± 2.1 (6)	245.3 ± 5.4 (7)	111.7 ± 5.9 (7)	99.9 ± 3.0 (7)	75.4 ± 3.6 (7)
	F	71.0 ± 1.9 (6)	43.5 ± 1.6 (6)	225.6 ± 10.0 (5)	105.6 ± 4.5 (7)	93.7 ± 5.8 (7)	73.6 ± 2.8 (7)
Porphyrula alleni	M	28.2 ± 1.1 (23)	9.0 ± 0.5 (22)	153.7 ± 6.0 (23)	66.6 ± 3.5 (22)	54.3 ± 2.3 (23)	53.6 ± 1.8 (23)
	F	28.1 ± 1.2 (28)	8.6 ± 0.3 (27)	150.0 ± 3.9 (29)	63.5 ± 3.6 (29)	52.5 ± 3.0 (29)	50.9 ± 1.8 (29)
Porphyrula martinica	M	31.7 ± 0.7 (14)	10.6 ± 0.6 (12)	181.6 ± 5.6 (14)	69.1 ± 3.5 (13)	61.6 ± 3.3 (14)	58.9 ± 2.4 (14)
	F	30.7 ± 1.1 (16)	10.0 ± 0.4 (15)	176.9 ± 5.1 (16)	68.2 ± 4.1 (16)	61.5 ± 2.9 (16)	57.8 ± 2.2 (16)
Porphyrula flavirostris	M	25.2 ± 0.9 (25)	7.6 ± 0.4 (23)	136.5 ± 4.0 (26)	69.7 ± 4.9 (25)	44.3 ± 2.3 (26)	47.9 ± 2.6 (26)
	F	24.7 ± 1.1 (27)	7.2 ± 0.4 (26)	131.5 ± 5.8 (26)	68.9 ± 4.2 (25)	43.9 ± 1.5 (27)	47.0 ± 1.8 (17)

86   ORNITHOLOGICAL MONOGRAPHS NO. 53

TABLE 7. Continued.

Taxonomic group	Sex	Culmen length	Bill depth	Wing length	Tail length	Tarsus length	Middle-toe length
Wallace's Rail and allies							
Gymnocrex rosenbergii	M	41.6 ± 1.5 (5)	7.5 ± 0.2 (5)	191.7 ± 11.7 (6)	64.8 ± 5.2 (6)	73.0 ± 1.9 (6)	36.5 ± 1.9 (6)
	F	41.5 ± 2.1 (15)	6.9 ± 0.3 (14)	197.0 ± 4.2 (15)	67.2 ± 4.2 (15)	73.7 ± 2.6 (15)	36.5 ± 1.5 (15)
Gymnocrex plumbeiventris	M	50.3 ± 2.5 (15)	6.9 ± 0.4 (13)	188.1 ± 9.7 (15)	61.9 ± 4.7 (16)	61.9 ± 1.7 (15)	38.5 ± 2.2 (16)
	F	49.8 ± 2.6 (9)	6.8 ± 0.4 (10)	190.4 ± 6.1 (11)	68.5 ± 5.9 (11)	60.8 ± 2.8 (10)	39.2 ± 1.2 (11)
Habroptila wallacii	M	75.3 ± 5.9 (4)	13.9 ± 1.0 (4)	165.6 ± 5.8 (5)	61.8 ± 6.3 (5)	91.2 ± 3.0 (5)	63.8 ± 3.3 (5)
	F	69.9 ± 4.6 (11)	11.4 ± 0.8 (11)	169.3 ± 5.2 (10)	60.4 ± 4.7 (11)	83.0 ± 3.3 (11)	57.4 ± 4.0 (11)
Wood-rails and allies							
Eulabeornis castaneoventris	M	57.7 ± 2.7 (18)	11.9 ± 0.9 (18)	219.1 ± 7.9 (17)	118.1 ± 8.3 (18)	72.0 ± 4.0 (18)	51.4 ± 2.8 (18)
	F	54.3 ± 2.0 (15)	11.0 ± 0.7 (15)	201.9 ± 10.8 (14)	115.3 ± 5.9 (14)	69.8 ± 3.2 (14)	48.5 ± 2.9 (15)
Aramides ypecaha	M	76.7 ± 2.7 (16)	12.8 ± 0.5 (16)	225.8 ± 6.6 (16)	89.9 ± 4.5 (16)	88.1 ± 3.6 (14)	71.3 ± 2.9 (16)
	F	72.7 ± 2.8 (14)	11.7 ± 0.7 (14)	221.7 ± 5.7 (14)	88.4 ± 4.2 (14)	85.3 ± 3.7 (12)	69.4 ± 3.3 (14)
Aramides cajanea	M	54.0 ± 2.8 (11)	10.3 ± 0.7 (11)	183.0 ± 9.6 (11)	62.2 ± 6.4 (11)	74.8 ± 2.2 (11)	54.5 ± 1.8 (11)
	F	51.1 ± 1.3 (7)	9.5 ± 0.4 (7)	178.1 ± 3.3 (7)	60.7 ± 2.5 (7)	71.4 ± 3.5 (7)	51.3 ± 1.3 (7)
Aramides albiventris	M	68.0 ± 6.9 (3)	10.5 ± 0.8 (3)	196.7 ± 6.7 (3)	61.0 ± 6.6 (3)	86.7 ± 4.2 (3)	62.7 ± 2.1 (3)
	F	62.1 ± 4.1 (8)	9.5 ± 0.6 (8)	181.9 ± 7.8 (8)	55.1 ± 4.9 (8)	81.1 ± 3.3 (8)	58.5 ± 2.7 (8)
Canirallus oculeus	M	41.0 ± 1.3 (26)	8.5 ± 0.4 (26)	174.7 ± 4.8 (25)	65.0 ± 3.3 (26)	55.6 ± 2.6 (26)	41.0 ± 1.2 (26)
	F	39.6 ± 1.6 (18)	8.4 ± 0.4 (18)	173.6 ± 6.8 (18)	63.8 ± 5.3 (18)	55.3 ± 2.6 (18)	41.2 ± 1.8 (18)
Canirallus kioloides	M	30.7 ± 2.0 (22)	8.5 ± 0.4 (21)	134.0 ± 3.5 (22)	53.6 ± 4.1 (22)	47.0 ± 2.2 (21)	36.2 ± 1.9 (20)
	F	29.3 ± 2.1 (18)	7.7 ± 0.4 (17)	131.7 ± 3.4 (18)	54.4 ± 4.5 (18)	44.9 ± 1.7 (18)	34.4 ± 1.3 (17)
Rougetius rougetii	M	33.5 ± 1.7 (19)	6.1 ± 0.4 (17)	132.8 ± 3.1 (19)	47.2 ± 2.5 (17)	53.5 ± 2.3 (19)	44.6 ± 2.2 (19)
	F	32.3 ± 1.9 (18)	5.5 ± 0.4 (16)	130.9 ± 3.0 (18)	46.2 ± 1.8 (17)	52.5 ± 2.2 (18)	44.0 ± 1.9 (18)
Zapata Rail and allies							
Cyanolimnas cerverai	M	45.0 ± 5.7 (2)	8.1 ± 0.7 (2)	110.0 ± 1.4 (2)	41.0 ± 1.4 (2)	45.0 ± 4.2 (2)	40.5 ± 0.7 (2)
	F	40.6 ± 2.1 (7)	7.5 ± 0.4 (7)	102.0 ± 2.4 (7)	39.1 ± 2.4 (7)	41.0 ± 4.5 (7)	38.3 ± 1.7 (7)
Ortygonax sanguinolentus	M	52.0 ± 1.7 (9)	5.8 ± 0.3 (8)	137.2 ± 7.3 (10)	64.7 ± 5.4 (10)	47.6 ± 2.6 (10)	45.6 ± 2.1 (10)
	F	47.8 ± 3.7 (9)	5.2 ± 0.3 (9)	130.0 ± 4.1 (10)	60.6 ± 4.2 (10)	46.4 ± 2.0 (10)	41.5 ± 2.4 (10)
Ortygonax nigricans	M	52.5 ± 2.5 (12)	5.6 ± 0.5 (12)	132.8 ± 6.6 (13)	56.9 ± 4.0 (12)	50.9 ± 3.0 (13)	47.2 ± 1.8 (13)
	F	51.8 ± 4.0 (5)	5.5 ± 0.5 (6)	127.8 ± 3.2 (6)	50.2 ± 3.2 (6)	47.7 ± 2.0 (6)	46.2 ± 1.9 (6)
White-throated rails							
Dryolimnas cuvieri	M	44.9 ± 1.6 (26)	7.5 ± 0.4 (26)	154.4 ± 5.6 (25)	62.0 ± 5.1 (26)	53.1 ± 3.1 (26)	46.4 ± 1.9 (27)
	F	39.6 ± 2.0 (32)	6.7 ± 0.5 (32)	147.5 ± 3.3 (31)	57.6 ± 3.3 (31)	48.6 ± 2.4 (32)	43.7 ± 2.6 (32)
Dryolimnas aldabranus	M	46.7 ± 3.3 (15)	6.9 ± 0.7 (15)	117.3 ± 5.6 (15)	37.9 ± 4.0 (15)	46.4 ± 2.0 (15)	39.0 ± 2.3 (15)
	F	44.8 ± 2.6 (13)	6.4 ± 0.3 (13)	116.1 ± 4.8 (13)	38.4 ± 4.8 (12)	45.6 ± 2.0 (13)	38.2 ± 1.6 (13)
Dryolimnas abbotti	M	41.7 ± 2.9 (6)	7.2 ± 0.5 (6)	135.8 ± 3.1 (5)	51.8 ± 5.0 (5)	44.3 ± 0.8 (6)	41.3 ± 2.7 (6)
	F	38.5 ± 0.7 (2)	6.8 ± 0.4 (2)	135.0 ± 1.4 (2)	57.5 ± 2.1 (2)	43.0 ± 0.0 (2)	39.0 ± 1.4 (2)

TABLE 7. Continued.

Taxonomic group	Sex	Culmen length	Bill depth	Wing length	Tail length	Tarsus length	Middle-toe length
Rallus sensu stricto							
Rallus aquaticus	M	41.7 ± 2.7 (15)	4.9 ± 0.4 (15)	120.3 ± 5.4 (15)	47.1 ± 4.3 (15)	43.3 ± 2.9 (15)	42.7 ± 2.6 (15)
	F	39.2 ± 3.0 (15)	4.4 ± 0.4 (14)	117.9 ± 8.2 (15)	49.3 ± 3.6 (15)	40.6 ± 2.5 (14)	39.9 ± 1.7 (15)
Rallus limicola	M	41.2 ± 1.7 (11)	3.8 ± 0.4 (11)	106.0 ± 2.5 (13)	43.5 ± 3.5 (13)	38.0 ± 1.5 (13)	37.2 ± 1.7 (13)
	F	36.8 ± 1.9 (17)	3.6 ± 0.4 (17)	97.9 ± 2.9 (17)	40.1 ± 3.4 (17)	34.4 ± 1.0 (17)	33.2 ± 1.2 (17)
Rallus longirostris-group	M	61.2 ± 4.3 (17)	6.8 ± 0.4 (17)	145.9 ± 9.2 (17)	61.7 ± 4.7 (17)	53.9 ± 3.4 (17)	48.4 ± 3.1 (17)
	F	58.1 ± 3.5 (13)	5.6 ± 0.3 (13)	134.3 ± 6.8 (13)	55.8 ± 2.6 (13)	48.6 ± 1.5 (13)	42.2 ± 2.0 (13)
Rallus elegans-group	M	62.5 ± 3.3 (15)	7.5 ± 0.4 (15)	170.5 ± 4.1 (15)	65.0 ± 4.4 (15)	64.6 ± 3.4 (15)	54.1 ± 2.5 (15)
	F	59.0 ± 2.3 (15)	6.4 ± 0.3 (15)	156.0 ± 4.7 (15)	61.6 ± 3.1 (15)	57.5 ± 2.6 (15)	50.2 ± 1.7 (15)
Gallirallus and allies							
Gallirallus pectoralis-group	M	34.9 ± 1.6 (16)	4.1 ± 0.2 (18)	98.9 ± 2.7 (18)	44.2 ± 2.6 (17)	32.7 ± 1.5 (18)	30.5 ± 1.7 (18)
	F	31.8 ± 1.2 (23)	3.9 ± 0.2 (23)	5.7 ± 3.1 (23)	40.2 ± 2.9 (22)	30.7 ± 1.4 (23)	29.1 ± 1.5 (23)
Gallirallus striatus	M	36.4 ± 2.9 (20)	6.3 ± 0.4 (20)	116.4 ± 6.2 (19)	38.2 ± 3.6 (20)	39.1 ± 2.1 (20)	35.6 ± 2.0 (20)
	F	32.7 ± 1.7 (7)	5.9 ± 0.4 (9)	114.4 ± 5.9 (9)	37.1 ± 3.3 (9)	36.8 ± 1.9 (9)	34.2 ± 1.7 (9)
Gallirallus sharpei	—	26.0 (1)	6.3 (1)	140.0 (1)	62.0 (1)	40.0 (1)	36.0 (1)
Gallirallus australis	M	49.6 ± 3.1 (48)	11.7 ± 0.8 (47)	182.6 ± 10.8 (46)	125.0 ± 10.5 (45)	63.8 ± 3.9 (51)	55.2 ± 3.3 (51)
	F	46.1 ± 3.4 (46)	10.5 ± 0.9 (45)	170.7 ± 9.4 (43)	117.3 ± 12.4 (44)	59.3 ± 3.8 (46)	51.1 ± 3.2 (46)
Gallirallus greyi	M	45.6 ± 2.4 (20)	11.0 ± 0.6 (20)	178.9 ± 13.6 (18)	100.1 ± 11.9 (17)	63.2 ± 4.3 (20)	51.4 ± 2.0 (20)
	F	42.4 ± 2.0 (14)	9.6 ± 0.7 (14)	167.7 ± 10.7 (14)	90.8 ± 8.7 (14)	58.7 ± 2.5 (15)	46.6 ± 2.3 (15)
Gallirallus philippensis-group	M	32.8 ± 1.7 (26)	6.5 ± 0.4 (25)	137.3 ± 6.6 (24)	57.8 ± 5.3 (22)	44.0 ± 1.9 (25)	35.5 ± 1.7 (26)
	F	29.3 ± 1.8 (21)	5.6 ± 0.3 (20)	130.5 ± 8.6 (21)	55.0 ± 5.0 (20)	41.3 ± 2.1 (21)	33.5 ± 1.5 (21)
Gallirallus dieffenbachii	—	37.0 (1)	7.6 (1)	121.0 (1)	68.0 (1)	39.0 (1)	36.0 (1)
Gallirallus owstoni	M	41.9 ± 0.9 (12)	8.4 ± 0.5 (11)	123.9 ± 3.8 (11)	45.6 ± 4.4 (12)	53.2 ± 1.9 (12)	41.2 ± 1.8 (12)
	F	36.2 ± 1.6 (17)	7.5 ± 0.4 (17)	115.8 ± 4.5 (17)	40.6 ± 5.1 (17)	47.4 ± 1.9 (17)	39.0 ± 1.9 (17)
Gallirallus wakensis	M	28.8 ± 1.3 (33)	5.5 ± 0.3 (30)	90.5 ± 5.7 (29)	34.0 ± 3.2 (33)	36.3 ± 1.4 (31)	32.6 ± 1.3 (33)
	F	26.0 ± 1.1 (25)	5.1 ± 0.4 (22)	89.6 ± 4.9 (25)	35.2 ± 3.6 (25)	34.2 ± 0.8 (21)	30.7 ± 1.0 (25)
Tricholimnas lafresnayanus	M	58.3 ± 4.9 (4)	85.5 ± 4.2 (4)	182.3 ± 12.0 (3)	107.0 ± 2.8 (2)	62.3 ± 4.0 (4)	50.3 ± 2.2 (4)
	F	53.6 ± 1.8 (5)	78.6 ± 7.2 (5)	180.4 ± 9.0 (5)	100.8 ± 5.3 (5)	62.4 ± 1.9 (5)	46.2 ± 1.3 (5)
Tricholimnas sylvestris	M	53.2 ± 2.6 (39)	9.5 ± 0.3 (39)	135.5 ± 5.2 (46)	58.7 ± 5.7 (39)	50.9 ± 1.9 (39)	48.2 ± 1.7 (39)
	F	48.0 ± 3.0 (45)	8.4 ± 0.6 (47)	140.4 ± 4.9 (38)	58.5 ± 3.5 (47)	48.8 ± 1.6 (47)	45.3 ± 1.5 (47)
Nesoclopeus poeciloptierus-group	M	45.7 ± 2.7 (17)	10.4 ± 0.7 (17)	161.9 ± 13.3 (17)	70.8 ± 6.3 (17)	64.0 ± 3.2 (17)	47.3 ± 3.2 (16)
	F	42.8 ± 3.8 (13)	9.6 ± 0.7 (13)	160.0 ± 9.0 (14)	69.8 ± 6.7 (13)	61.2 ± 3.9 (14)	43.9 ± 2.4 (14)
Aramidopsis plateni	M	55.8 ± 2.5 (5)	8.1 ± 0.6 (6)	150.0 ± 4.3 (6)	31.5 ± 4.6 (6)	60.5 ± 2.3 (6)	50.5 ± 1.4 (6)
	F	51.6 ± 3.9 (7)	7.7 ± 0.6 (8)	144.8 ± 4.4 (8)	31.4 ± 3.4 (7)	59.9 ± 5.4 (8)	46.9 ± 2.6 (8)
Cabalus modestus	M	36.8 ± 1.7 (15)	5.1 ± 0.3 (15)	87.6 ± 3.9 (15)	37.3 ± 2.7 (15)	30.6 ± 0.9 (15)	29.3 ± 1.7 (15)
	F	30.8 ± 1.5 (13)	4.7 ± 0.2 (13)	78.8 ± 3.1 (13)	34.3 ± 3.6 (11)	27.4 ± 2.3 (13)	25.9 ± 1.5 (13)

TABLE 7. Continued.

Taxonomic group	Sex	Culmen length	Bill depth	Wing length	Tail length	Tarsus length	Middle-toe length
Habropteryx insignis	M	50.1 ± 4.8 (9)	9.8 ± 0.8 (9)	149.0 ± 3.5 (6)	39.0 ± 2.6 (6)	66.7 ± 3.8 (9)	48.3 ± 3.1 (9)
	F	47.1 ± 2.1 (11)	8.5 ± 0.5 (11)	139.2 ± 3.7 (11)	38.4 ± 5.3 (11)	63.3 ± 2.5 (11)	45.9 ± 2.0 (11)
Habropteryx torquatus	M	41.7 ± 2.6 (21)	7.7 ± 0.5 (19)	150.9 ± 6.1 (20)	57.4 ± 5.2 (20)	53.3 ± 2.7 (22)	43.8 ± 1.6 (22)
	F	37.6 ± 2.5 (27)	6.8 ± 0.4 (23)	143.5 ± 6.8 (23)	54.2 ± 5.1 (25)	49.9 ± 2.1 (27)	39.7 ± 1.8 (27)
Habropteryx sulcirostris	M	44.0 ± 2.8 (8)	8.8 ± 0.7 (7)	156.9 ± 8.4 (7)	57.9 ± 5.3 (7)	57.3 ± 2.4 (8)	44.1 ± 2.2 (8)
	F	40.9 ± 3.2 (7)	8.1 ± 0.9 (7)	141.3 ± 4.4 (6)	52.6 ± 3.2 (5)	53.1 ± 3.8 (7)	41.0 ± 3.7 (7)
Habropteryx celebensis	M	39.1 ± 1.7 (16)	8.2 ± 0.3 (16)	151.4 ± 4.8 (16)	57.1 ± 4.5 (15)	53.5 ± 1.5 (13)	43.2 ± 2.5 (15)
	F	36.9 ± 1.1 (13)	7.2 ± 0.3 (12)	146.2 ± 5.4 (12)	52.4 ± 4.5 (13)	50.8 ± 2.3 (11)	40.3 ± 2.1 (13)
Habropteryx okinawae	M	52.5 ± 0.7 (2)	9.5 (1)	145.0 (1)	58.0 (1)	62.0 ± 4.2 (2)	45.0 (1)
	F	45.0 (1)	—	140.0 (1)	53.0 (1)	59.0 (1)	—
Atlantisia and allies							
Atlantisia rogersi	M	21.4 ± 1.2 (24)	3.7 ± 0.2 (24)	56.2 ± 2.0 (24)	30.7 ± 2.4 (24)	25.3 ± 1.3 (24)	26.1 ± 1.5 (24)
	F	17.9 ± 1.2 (19)	3.5 ± 0.2 (17)	53.5 ± 1.4 (20)	29.6 ± 2.4 (21)	24.3 ± 1.7 (21)	25.7 ± 1.2 (19)
Laterallus leucopyrrhus	M	17.8 ± 0.8 (5)	4.7 ± 0.2 (5)	82.6 ± 1.8 (5)	46.0 ± 3.1 (5)	34.2 ± 1.4 (5)	30.8 ± 1.3 (5)
	F	16.4 ± 0.9 (5)	4.4 ± 0.2 (5)	84.0 ± 1.4 (5)	46.0 ± 1.9 (5)	33.0 ± 1.0 (5)	29.4 ± 0.5 (5)
Laterallus spilonotus	M	16.6 ± 0.5 (7)	3.7 ± 0.2 (7)	68.3 ± 1.8 (8)	23.5 ± 1.6 (8)	24.4 ± 1.2 (8)	24.5 ± 0.8 (8)
	F	15.3 ± 0.6 (11)	3.3 ± 0.2 (11)	66.2 ± 1.8 (11)	22.1 ± 2.3 (11)	22.6 ± 1.9 (11)	23.9 ± 1.4 (11)
Coturnicops noveboracensis	M	13.6 ± 0.7 (14)	4.2 ± 0.3 (14)	88.9 ± 1.5 (14)	34.0 ± 3.1 (14)	24.9 ± 1.2 (14)	24.4 ± 1.3 (14)
	F	13.1 ± 0.7 (16)	3.9 ± 0.2 (16)	82.4 ± 1.2 (16)	31.9 ± 1.8 (16)	23.4 ± 1.3 (16)	22.8 ± 0.9 (16)
Crex crex	M	21.3 ± 1.2 (15)	5.7 ± 0.3 (15)	138.9 ± 3.6 (15)	44.8 ± 2.6 (15)	40.2 ± 1.7 (14)	32.4 ± 1.4 (15)
	F	21.1 ± 1.5 (15)	5.5 ± 0.4 (15)	136.1 ± 6.5 (15)	44.3 ± 6.5 (15)	39.0 ± 1.3 (15)	31.0 ± 2.1 (15)
Crex albicollis	M	28.5 ± 1.5 (18)	7.2 ± 0.3 (18)	113.3 ± 5.3 (18)	43.7 ± 2.8 (18)	38.5 ± 1.8 (18)	35.9 ± 1.9 (18)
	F	25.8 ± 1.4 (13)	6.8 ± 0.2 (11)	104.0 ± 5.4 (13)	41.3 ± 2.6 (11)	36.1 ± 2.3 (13)	34.1 ± 1.6 (13)
Gray crakes and allies							
Porzana sandwichensis	—	19.1 ± 1.3 (7)	5.3 ± 0.5 (7)	67.6 ± 5.2 (7)	15.6 ± 1.7 (7)	29.0 ± 2.2 (7)	28.8 ± 1.7 (7)
Porzana palmeri	M	19.4 ± 1.0 (58)	4.8 ± 0.2 (56)	58.8 ± 2.4 (57)	22.6 ± 2.7 (57)	25.3 ± 1.4 (58)	25.8 ± 1.2 (58)
	F	17.8 ± 1.2 (31)	4.6 ± 0.3 (29)	58.1 ± 2.8 (30)	22.9 ± 2.7 (31)	24.2 ± 1.8 (31)	24.4 ± 1.3 (31)
Porzana carolina	M	21.5 ± 1.2 (15)	5.4 ± 0.4 (15)	109.9 ± 4.2 (15)	47.4 ± 3.4 (15)	35.4 ± 1.5 (15)	35.8 ± 1.4 (15)
	F	19.5 ± 0.8 (15)	5.0 ± 0.2 (15)	102.5 ± 2.5 (15)	47.2 ± 2.9 (15)	32.7 ± 1.1 (15)	33.1 ± 1.1 (15)
Porzana pusilla	M	17.1 ± 1.1 (22)	4.5 ± 0.4 (21)	89.8 ± 5.2 (22)	42.2 ± 4.6 (22)	29.1 ± 1.1 (22)	30.9 ± 1.6 (22)
	F	16.0 ± 0.8 (17)	4.3 ± 0.3 (17)	86.8 ± 3.2 (17)	40.5 ± 2.9 (17)	27.9 ± 1.0 (17)	30.2 ± 1.9 (17)
Porzana tabuensis	M	19.2 ± 1.7 (21)	4.2 ± 0.4 (17)	79.7 ± 4.0 (21)	40.9 ± 5.5 (20)	29.0 ± 2.2 (21)	27.6 ± 2.0 (21)
	F	17.9 ± 1.5 (15)	3.9 ± 0.3 (14)	78.3 ± 5.0 (15)	39.6 ± 4.7 (15)	28.4 ± 2.0 (15)	26.7 ± 2.3 (15)
Porzana monasa	—	20.5 ± 0.7 (2)	5.1 ± 0.2 (2)	78.0 ± 1.4 (2)	38.0 ± 0.0 (2)	33.0 ± 1.4 (2)	32.5 ± 0.7 (2)
Porzana atra	M	24.1 ± 1.5 (11)	5.7 ± 0.4 (11)	82.9 ± 3.1 (11)	40.1 ± 3.1 (12)	36.3 ± 1.3 (12)	32.1 ± 1.2 (12)
	F	21.9 ± 0.8 (12)	5.4 ± 0.2 (12)	83.2 ± 1.8 (12)	40.0 ± 2.7 (13)	35.3 ± 1.4 (13)	31.2 ± 1.4 (13)

TABLE 7. Continued.

Taxonomic group	Sex	Culmen length	Bill depth	Wing length	Tail length	Tarsus length	Middle-toe length
Porzana flavirostra	M	23.9 ± 1.6 (15)	6.2 ± 0.6 (15)	104.1 ± 3.4 (14)	40.0 ± 3.4 (15)	41.9 ± 1.6 (16)	40.9 ± 1.4 (16)
	F	22.1 ± 1.5 (15)	5.7 ± 0.4 (15)	100.4 ± 2.6 (15)	38.7 ± 2.5 (14)	40.0 ± 2.8 (15)	38.9 ± 1.8 (15)
Porzana olivieri	M	27.2 ± 1.4 (2)	5.9 ± 0.1 (2)	105.5 ± 2.1 (2)	51.5 ± 2.1 (2)	41.5 ± 2.1 (2)	40.5 ± 0.7 (2)
	F	23.7 ± 2.1 (3)	5.1 ± 0.2 (3)	102.0 (1)	51.7 ± 3.5 (3)	36.3 ± 1.2 (3)	37.0 ± 1.0 (3)
Porzana bicolor	M	25.4 ± 1.6 (14)	5.6 ± 0.4 (13)	115.6 ± 3.2 (14)	53.9 ± 1.4 (14)	40.9 ± 1.4 (14)	39.1 ± 1.4 (14)
	F	23.8 ± 1.5 (11)	5.3 ± 0.3 (10)	111.4 ± 3.8 (10)	54.3 ± 4.3 (10)	38.6 ± 1.8 (11)	36.2 ± 1.2 (11)
Bushhens							
Amaurornis olivaceus	M	37.8 ± 1.3 (5)	10.5 ± 0.4 (5)	168.4 ± 4.0 (5)	57.4 ± 4.1 (5)	61.6 ± 5.5 (5)	53.8 ± 2.2 (5)
	F	35.0 ± 2.3 (15)	9.4 ± 0.6 (14)	157.5 ± 7.0 (15)	56.2 ± 5.7 (15)	60.3 ± 4.0 (15)	52.7 ± 3.0 (15)
Amaurornis moluccanus	M	34.3 ± 1.6 (29)	9.4 ± 0.4 (29)	138.9 ± 8.2 (27)	52.0 ± 7.5 (27)	56.5 ± 3.2 (28)	49.9 ± 2.7 (29)
	F	31.4 ± 1.5 (25)	8.5 ± 0.4 (25)	133.3 ± 7.3 (25)	48.9 ± 8.1 (22)	53.0 ± 2.0 (25)	46.9 ± 2.8 (25)
Amaurornis ruficrissus	M	32.7 ± 1.2 (6)	9.1 ± 0.8 (5)	141.0 ± 3.7 (6)	54.8 ± 4.0 (6)	52.5 ± 2.3 (6)	46.5 ± 2.4 (6)
	F	27.8 ± 0.8 (5)	8.0 ± 0.4 (5)	139.6 ± 9.6 (5)	56.2 ± 8.8 (5)	50.4 ± 1.5 (5)	44.2 ± 2.3 (5)
Amaurornis phoenicurus	M	39.2 ± 2.5 (29)	9.1 ± 0.5 (28)	166.9 ± 8.1 (28)	66.3 ± 6.6 (28)	58.7 ± 2.9 (29)	54.9 ± 2.2 (29)
	F	36.0 ± 2.4 (31)	8.1 ± 0.5 (31)	157.4 ± 5.5 (31)	64.0 ± 4.3 (30)	55.5 ± 2.0 (31)	52.0 ± 2.4 (31)
Amaurornis akool	M	31.6 ± 1.9 (21)	7.8 ± 0.4 (21)	128.5 ± 4.4 (21)	58.1 ± 4.0 (21)	52.7 ± 3.0 (21)	45.7 ± 2.4 (21)
	F	30.3 ± 1.3 (21)	7.2 ± 0.5 (21)	125.4 ± 4.4 (21)	57.3 ± 3.8 (21)	50.6 ± 2.8 (21)	44.2 ± 2.3 (21)
Amaurornis isabellinus	M	36.0 ± 1.6 (27)	9.6 ± 0.4 (24)	170.6 ± 6.0 (27)	68.0 ± 6.1 (27)	64.5 ± 3.2 (27)	54.1 ± 2.8 (23)
	F	33.5 ± 1.8 (32)	8.7 ± 0.4 (29)	160.1 ± 5.9 (30)	61.8 ± 7.9 (24)	61.7 ± 2.8 (32)	53.3 ± 2.5 (28)
Amaurornis ineptus	M	68.9 ± 5.4 (7)	15.8 ± 1.3 (6)	182.8 ± 7.2 (6)	33.0 ± 5.4 (7)	97.3 ± 4.2 (7)	65.6 ± 3.6 (7)
	F	65.1 ± 4.1 (16)	14.2 ± 1.5 (16)	176.6 ± 7.4 (17)	34.9 ± 2.9 (15)	91.8 ± 2.8 (17)	64.6 ± 2.8 (16)
Moorhens and allies							
Gallicrex cinerea	M	37.6 ± 1.4 (20)	9.4 ± 0.5 (15)	204.3 ± 4.5 (20)	75.4 ± 3.9 (20)	72.7 ± 2.6 (19)	70.0 ± 3.7 (20)
	F	32.7 ± 1.5 (18)	8.2 ± 0.4 (17)	176.1 ± 7.9 (18)	67.2 ± 4.7 (18)	65.2 ± 5.1 (17)	64.2 ± 3.5 (18)
Pareudiastes pacificus	—	31.0 ± 2.9 (8)	7.4 ± 0.4 (8)	123.3 ± 7.8 (8)	32.6 ± 5.3 (8)	43.8 ± 2.0 (8)	35.1 ± 2.0 (8)
Pareudiastes silvestris	—	41.0 (1)	12.1 (1)	149.0 (1)	39.0 (1)	58.0 (1)	48.0 (1)
Tribonyx ventralis	M	32.1 ± 2.0 (51)	9.7 ± 0.4 (48)	215.1 ± 8.4 (52)	78.0 ± 5.0 (51)	63.0 ± 2.4 (52)	47.5 ± 2.4 (52)
	F	29.1 ± 2.0 (34)	8.7 ± 0.3 (32)	201.2 ± 8.5 (34)	75.0 ± 4.5 (34)	58.0 ± 2.8 (34)	44.3 ± 2.2 (34)
Tribonyx mortierii	M	38.5 ± 2.0 (30)	13.4 ± 0.6 (26)	194.8 ± 8.6 (31)	92.8 ± 3.2 (28)	86.0 ± 4.4 (32)	63.8 ± 3.0 (32)
	F	38.6 ± 1.6 (14)	12.4 ± 0.4 (13)	192.9 ± 9.2 (15)	73.7 ± 4.4 (15)	83.4 ± 3.7 (15)	62.2 ± 3.2 (15)
Gallinula nesiotis	M	32.0 (1)	11.1 (1)	153.0 (1)	67.0 (1)	52.0 (1)	55.0 (1)
	F	31.7 ± 1.5 (3)	10.0 ± 0.9 (2)	134.0 ± 5.3 (3)	57.7 ± 7.4 (3)	52.3 ± 1.2 (3)	55.3 ± 2.9 (3)
Gallinula comeri	M	30.8 ± 3.0 (16)	11.1 ± 0.5 (16)	148.0 ± 4.7 (16)	64.3 ± 3.4 (14)	53.9 ± 2.1 (17)	55.1 ± 2.2 (17)
	F	29.4 ± 3.0 (19)	10.2 ± 0.5 (19)	140.2 ± 2.6 (19)	61.7 ± 4.0 (18)	50.8 ± 2.2 (19)	52.2 ± 2.2 (19)

TABLE 7. Continued.

Taxonomic group	Sex	Culmen length	Bill depth	Wing length	Tail length	Tarsus length	Middle-toe length
Gallinula tenebrosa	M	34.8 ± 2.9 (16)	8.6 ± 0.7 (15)	188.6 ± 10.3 (16)	64.5 ± 4.7 (16)	60.4 ± 4.6 (16)	67.7 ± 6.2 (16)
	F	33.5 ± 2.5 (14)	8.3 ± 0.7 (11)	187.1 ± 12.5 (14)	62.1 ± 4.2 (14)	59.7 ± 5.4 (14)	65.5 ± 4.7 (13)
Gallinula angulata	M	22.4 ± 1.2 (14)	6.9 ± 0.6 (14)	139.1 ± 4.0 (14)	57.4 ± 2.2 (14)	39.8 ± 2.8 (14)	46.8 ± 3.0 (14)
	F	21.3 ± 1.1 (15)	6.7 ± 0.3 (15)	133.7 ± 4.2 (15)	55.2 ± 3.5 (15)	37.5 ± 1.8 (15)	44.1 ± 1.8 (15)
Gallinula chloropus	M	31.6 ± 2.9 (14)	7.9 ± 0.5 (12)	168.9 ± 5.9 (14)	65.2 ± 2.2 (14)	52.6 ± 2.5 (14)	57.4 ± 3.7 (14)
	F	29.0 ± 1.7 (19)	7.3 ± 0.3 (15)	165.4 ± 3.6 (16)	61.8 ± 1.8 (19)	50.8 ± 1.7 (19)	54.8 ± 2.7 (19)
Gallinula pyrrhorrhoa	M	31.8 ± 2.3 (9)	7.9 ± 0.3 (8)	160.1 ± 6.8 (9)	67.7 ± 3.4 (9)	52.0 ± 2.7 (9)	57.9 ± 4.6 (9)
	F	31.1 ± 1.5 (11)	7.6 ± 0.4 (11)	158.6 ± 7.3 (11)	65.8 ± 3.8 (11)	51.8 ± 2.1 (11)	57.8 ± 3.7 (11)
Gallinula cachinnans	M	33.5 ± 1.8 (14)	8.0 ± 0.4 (14)	175.8 ± 5.8 (14)	68.7 ± 4.4 (14)	58.2 ± 2.0 (14)	65.4 ± 3.2 (14)
	F	32.2 ± 2.0 (15)	7.5 ± 0.5 (15)	166.6 ± 5.5 (16)	65.4 ± 5.3 (16)	53.9 ± 2.9 (16)	61.8 ± 3.8 (16)
Gallinula galeata	M	34.0 ± 2.2 (6)	9.0 ± 0.5 (6)	218.3 ± 7.7 (6)	86.3 ± 3.6 (6)	65.2 ± 3.9 (6)	70.2 ± 5.3 (6)
	F	31.8 ± 1.9 (5)	8.6 ± 0.6 (5)	209.2 ± 9.1 (5)	83.4 ± 4.6 (5)	58.4 ± 4.4 (5)	64.4 ± 2.4 (5)
Gallinula sandvicensis	M	32.7 ± 2.1 (22)	8.3 ± 0.4 (21)	180.3 ± 4.4 (22)	66.6 ± 6.0 (23)	59.9 ± 5.0 (23)	63.2 ± 5.1 (23)
	F	30.8 ± 2.8 (8)	7.6 ± 0.8 (7)	170.1 ± 4.7 (7)	65.3 ± 4.6 (8)	55.1 ± 2.9 (8)	58.4 ± 4.9 (8)
Coots							
Fulica armillata	M	36.7 ± 3.0 (15)	10.3 ± 0.7 (13)	197.8 ± 9.4 (15)	48.0 ± 4.2 (14)	66.1 ± 5.2 (15)	81.7 ± 7.5 (15)
	F	36.9 ± 2.1 (17)	9.8 ± 0.6 (15)	193.2 ± 6.6 (17)	45.9 ± 4.2 (17)	64.2 ± 2.3 (17)	81.1 ± 2.9 (17)
Fulica cornuta	M	52.1 ± 3.3 (12)	13.6 ± 0.8 (11)	288.6 ± 9.6 (12)	69.1 ± 7.5 (12)	82.6 ± 5.3 (10)	98.9 ± 4.6 (12)
	F	49.7 ± 3.4 (7)	13.3 ± 0.7 (6)	280.1 ± 10.0 (7)	65.9 ± 7.2 (7)	80.4 ± 4.4 (7)	93.6 ± 5.6 (7)
Fulica gigantea	M	52.3 ± 6.8 (21)	12.7 ± 0.9 (19)	262.5 ± 18.8 (20)	61.5 ± 6.0 (21)	99.2 ± 5.6 (21)	112.1 ± 8.7 (21)
	F	49.2 ± 6.1 (18)	11.9 ± 0.7 (17)	269.7 ± 13.2 (20)	64.1 ± 5.3 (20)	95.1 ± 5.5 (20)	111.8 ± 5.1 (20)
Fulica americana	M	34.8 ± 1.9 (16)	10.0 ± 0.6 (16)	193.9 ± 7.6 (16)	53.0 ± 4.3 (16)	60.2 ± 2.1 (16)	73.2 ± 4.2 (16)
	F	34.3 ± 2.6 (14)	9.1 ± 0.6 (13)	180.5 ± 6.4 (14)	46.9 ± 4.1 (14)	55.1 ± 2.7 (14)	69.4 ± 3.5 (14)
Fulica alai	M	35.3 ± 2.0 (28)	10.3 ± 0.5 (23)	184.8 ± 7.9 (28)	47.1 ± 5.1 (29)	59.2 ± 2.8 (29)	72.5 ± 3.1 (29)
	F	33.0 ± 1.8 (25)	9.5 ± 0.6 (18)	174.4 ± 4.6 (23)	43.8 ± 2.8 (25)	54.2 ± 1.4 (25)	66.6 ± 3.1 (25)
Fulica atra (*atra*-group)	M	38.6 ± 1.9 (15)	10.8 ± 0.7 (13)	208.8 ± 6.0 (15)	58.7 ± 4.2 (15)	62.4 ± 2.9 (15)	78.9 ± 4.9 (15)
	F	36.0 ± 2.0 (21)	9.6 ± 0.6 (20)	197.4 ± 5.8 (21)	52.1 ± 3.0 (21)	58.4 ± 2.6 (21)	73.8 ± 3.5 (21)

flightless *P. hochstetteri* (Table 7). The reputedly flightless, extinct *Porphyrio albus* was not adequately represented for formal inclusion in the statistical analyses, but summary statistics suggest that it possessed comparatively short wings for a swamphen of such robustness (Table 7). Interspecific differences within *Porphyrula* ($n = 135$) generally were less pronounced than those among species of *Porphyrio* because of the absence of a uniquely large, flightless member; the primary contributor to interspecific differences was the comparatively diminutive, paedomorphic *Porphyrula flavirostris* (Table 7).

Comparisons of external measurements among the "basal rails" ($n = 277$; *Gymnocrex, Habroptila, Aramides, Eulabeornis, Canirallus,* and *Rougetius*) demonstrated highly significant interspecific differences in all six external dimensions (ANOVA, $P < 0.0001$), an outcome predictable given the generic diversity compared (Table 7). Flightless *Habroptila wallacii* is possessed of a substantial, falchion-shaped bill, and was the largest in both bill measurements. *Habroptila* also was approximately equal to the more massive *Aramides ypecaha* in tarsus length, second only to *Aramides* in middle-toe length, but among the smallest in tail length, and the smallest of all the basal rails in wing length (Table 7).

Ortygonax and *Cyanolimnas* ($n = 48$) manifested significant interspecific variation among means (ANOVA; $P < 0.0005$), wherein the flightless Zapata Rail (*C. cerverai*) was smaller than either species of *Ortygonax* in all dimensions but bill height (Table 7). However, the shortness of the wings of the latter exceeded other dimensions in magnitude of differences, and wing lengths of *Cyanolimnas* did not overlap with those of *Ortygonax* and averaged only two-thirds those of *Ortygonax* (Table 7). A similar pattern in mean tail length characterized these taxa (Table 7), a finding confirming the description by Bond (1980:70) of the tail of *Cyanolimnas* as "short and decomposed." Within the white-throated rails (*Dryolimnas*; $n = 95$), the six measurements of study skins (Table 7) showed significant interspecific differences in means (ANOVA; $P < 0.0005$); subsequent pairwise comparisons revealed that most of the interspecific effects were attributable to the comparatively small, flightless *D. aldabranus* (Bonferroni *t*-tests). Univariate comparisons, as well as other analyses and comparisons presented below, were consistent with the prevailing view that the extinct *D. abbotti* was not flightless.

Four species of the genus *Rallus* ($n = 122$) showed significant differences in all external measurements (ANOVA; $P < 0.0001$), with the greater part of this variation (Table 7) stemming from differences between the two smaller species (*R. aquaticus* and *R. limicola*) and the two larger species (*R. longirostris* and *R. elegans*). In comparisons of 16 species comprising samples of both sexes in the Pacific genera *Gallirallus, Cabalus, Tricholimnas, Nesoclopeus, Aramidopsis,* and *Habropteryx* ($n = 634$), interspecific differences were significant in all six external dimensions (ANOVA; $P < 0.0001$). Lengths of the wing and tail were reduced in most flightless taxa relative to other dimensions. However, shifts in tail length varied markedly, with some taxa increasing tail length (e.g., *Gallirallus australis* and *Tricholimnas lafresnayanus*), whereas other taxa (e.g., *Aramidopsis plateni* and *Habropteryx insignis*) approached external taillessness (Table 7). In two extinct species represented by single specimens, *Gallirallus sharpei* showed only a suggestion of the reductions in wing and tail indicative of flightlessness, whereas *G. dieffenbachii* clearly showed reduction of the wing (Table 7).

Fourteen species of crakes (*Coturnicops, Atlantisia, Crex, Laterallus,* and *Porzana*) were sampled in sufficient quantity ($n = 454$) to permit two-way comparisons of species–sex groups (Table 7). For these taxa, interspecific differences were significant (ANOVA; $P < 0.0001$) in all six external dimensions compared (Table 7), with the largest differences being associated with the relatively short wings and tails of *A. rogersi* and *P. palmeri* (with *P. atra* being intermediate), and the relatively short toes of *Coturnicops noveboracensis*. *P. palmeri* had short toes as well, but in concert with other dimensions. Reduction in tail length among flightless crakes was typical, although the relative truncation evident in *A. rogersi, P. sandwichensis,* and *P. palmeri* was exceptionally great. In two species of crakes lacking adequate samples of specimens of known sex—*Porzana sandwichensis* ($n = 7$) and *P. monasa* ($n = 2$)—the former showed truncation of both wing and tail, whereas the latter only manifested definitive shortening of the wing (Table 7).

Seven taxa of bushhens (*Amaurornis*, including flightless "*Megacrex*" *ineptus*) were subjected to two-way comparisons ($n = 270$), and interspecific differences were highly significant in all six external measurements (ANOVA; $P < 0.0001$). Comparisons of means (Table 7) and examination of pairwise comparisons (Bonferroni *t*-tests) substantiated that the great majority of the species effects stemmed from the generally large dimensions of flightless *A. ineptus,* the means for which (exclusive of length of wing and tail) were at least twice those of all flighted congeners (Table 7). The two aerofunctional dimensions—lengths of wing and tail—departed markedly from this pattern of general enlargement (Table 7), with the former being only marginally longer (especially closely approached by likely sister-species *A. isabellinus*) and the latter being distinctly shorter in flightless *A. ineptus* than in its generally smaller, flighted congeners (Table 7).

Twelve taxa of moorhens and allies permitting two-way comparisons (*Gallicrex, Gallinula* including *Porphyriornis,* and *Tribonyx*; $n = 404$), confirmed interspecific differences (ANOVA; $P < 0.0001$) in all six external variables (Table 7). One-way comparisons of species (sexes pooled), which permitted the inclusion of flightless *Pareudiastes pacificus* ($n = 8$) and *P. silvestris* ($n = 1$), confirmed the highly significant ($P < 0.0001$) species effects in all six dimensions. Both analyses, pairwise comparisons of means (Bonferroni *t*-tests), and examination of summary statistics (Table 7) revealed that much of the variation in dimensions unrelated to flight derived from differences among three broad groups: small species (*Gallicrex* and *Gallinula* exclusive of *Porphyriornis*), medium-sized species (*Porphyriornis, Pareudiastes,* and *Tribonyx ventralis*), and the large *Tribonyx mortierii* (Table 7). One departure from this general pattern pertained to the relatively short toes of *Pareudiastes pacificus,* with which no other taxon overlapped (Table 7). Although clearly related to flight capacity, wing length showed significantly different patterns across genera. Wing lengths of *Pareudiastes* and the subgenus *Porphyriornis* were the shortest among all taxa of moorhens compared, with the shortest absolute wing lengths occurring in *Pareudiastes pacificus* (Table 7). In the comparatively massive *Tribonyx mortierii,* wing lengths were only slightly shorter than in its much smaller, flighted congener *T. ventralis* (Table 7). However, wing lengths in all flightless moorhens were substantially smaller relative to dimensions indicative of general body size than in flighted relatives (Table 7). Tail lengths varied differently with respect to both genus and flight capacity; the rectrices were of uniform length in *Gallicrex* and *Gallinula,* exceptionally foreshortened in both species of *Pareudiastes,* and comparatively long in *Tribonyx.* Within the last genus, the flightless *T. mortierii* had only slightly longer rectrices than its much smaller congener, *T. ventralis* (Table 7).

Six species of coots (*Fulica*) were subjected to two-way comparisons of external measurements ($n = 212$), and all of them showed highly significant interspecific differences in all six dimensions analyzed (ANOVA; $P < 0.0001$). Examination of pairwise comparisons (Bonferroni *t*-tests) and examination of summary statistics indentified the primary source of these highly significant effects as the differences in size between the two large coots of the Andes (*F. gigantea* and *F. cornuta*) and all other congeners. Generally, the former pair averaged 50% larger than the four more-typical species in most external measurements (Table 7). Moreover, within these two partitions, included species were comparatively uniform in overall conformation; differences between species within partitions generally did not differ in one or more measurements (same-sex comparisons), including several dimensions between *F. gigantea* and *F. cornuta* (Table 7).

Differences in group variances.—In most of the genera and supergeneric groups examined, significant differences in variances (Levene's test; $P < 0.05$) were detected in most external variables among species, between sexes, and (less frequently) species–sex interactions (i.e., intersexual differences in variances occurred in some species in a group but not in others). This generality characterized most variables among species of *Porphyrio* ($P < 0.05$), six genera of basal rallids, four species of *Rallus,* 16 species of Pacific rails (*Gallirallus* and allies), and *Fulica gigantea* and *F. cornuta* (middle-toe lengths), but comparisons revealed that those groups showing comparatively high variability were those for which geographical delimitation of taxa, subspecific variation, or source of data remained problematic (Table 7). In several other groups, differences in variances among species–sex groups were significant but remained unexplained and evidently unrelated to flightlessness, including 14 taxa of crakes and moorhens sensu lato (Fulicarina exclusive of *Fulica* [Table 7]). Virtual homogeneity of variances among species and sexes (Levene's test; $P < 0.05$) was indicated in *Dryolimnas.* Heterogeneity of variances among species–sex groups (Levene's test; $P < 0.05$) that appear to pertain to the loss of constraint related to flight (even if effects of size were minimized by comparisons of coefficients of variation) was found in *Amaurornis* (bill height and tail length), indicating that flightless *A. ineptus* is not only giant but variable in at least these external dimensions.

UNIVARIATE SEXUAL DIMORPHISM

In the greater swamphens (*Porphyrio*), sexual dimorphism was marked in five of the six external measurements compared (Table 7; ANOVA; $P < 0.0001$), with tail lengths showing only marginal intersexual differences (ANOVA; $P < 0.05$). In addition, species–sex interaction effects were pronounced in the former five dimensions (ANOVA; $P < 0.005$) and marginal in tail lengths (ANOVA; $P < 0.05$), indicating that magnitude of sexual dimorphism differed among species in most or all dimensions. Examination of means for sexes within species and the results of pairwise comparisons (Bonferroni *t*-tests) indicated that *P. melanotus* was most dimorphic of the taxa compared, but that the closely related, flightless *P. hochstetteri* also showed substantial dimorphism, despite the small sample size for the latter (Table 7). Lesser swamphens (*Porphyrula*) showed sexual dimorphism comparable to that inferred for *Porphyrio*, but unlike the latter, no differences were found among species in the magnitude of sexual differences.

The five genera of basal rallids including flightless *Habroptila wallacii* showed significant sexual dimorphism in all external measurements (ANOVA; $P < 0.0001$). However, any specific inferences to be made concerning *H. wallacii* were compromised by the dubious identification of sexes in a majority of the skins of this taxon (Table 7). Allocation to sex a posteriori and variable preparation of skins may account for the variable sexual differences in *Habroptila*, accounting for the significant species–sex interaction effects (ANOVA; $P < 0.05$).

Intersexual differences of meso-rallids (*Ortygonax* and *Cyanolimnas*) were significant in external measurements (ANOVA; $0.001 < P < 0.01$), but the absence of significant species–sex interaction effects ($P > 0.05$) indicated that magnitudes of sexual dimorphism were comparable across taxa. Intersexual differences in external measurements in *Dryolimnas* were significant (ANOVA; $P < 0.01$), except in wing length (ANOVA; $P > 0.65$). Species–sex interaction effects generally were weak (ANOVA; $0.05 < P < 0.20$), indicating that sexual dimorphism was of comparable magnitude in all three species in the genus.

Substantial intersexual differences were detected in the four species of *Rallus* compared (ANOVA; $P < 0.0005$) in all six external measurements (Table 7). Species–sex interaction effects in bill height ($P < 0.0001$), wing length ($P < 0.005$), tail length ($P < 0.001$), tarsus length ($P < 0.01$), and middle-toe length ($P < 0.05$) apparently stemmed principally from subspecific variation in the two larger species (Table 7). Among the 16 species of *Gallirallus* and allies compared (Table 7), intersexual differences were pronounced in all six external dimensions of skin specimens (ANOVA; $P < 0.0001$). Species–sex interaction effects in *Gallirallus* and allies were significant in bill height ($P < 0.0001$), wing length ($P < 0.0005$), tail length ($P < 0.01$), tarsus length ($P < 0.01$), and middle-toe length ($P < 0.0005$). Exaggerated sexual dimorphism was indicated in some of the dimensions compared in a minority of species (Table 7), notably several flightless taxa (*Tricholimnas sylvestris*, *G. australis*, *G. owstoni*, and *Cabalus modestus*).

In the 14 species of crakes (Crecina) for which samples permitted comparisons by species and sex (Table 7), intersexual differences were highly significant (ANOVA; $P < 0.0001$) in all but tail length; overall sexual differences in the latter were only marginally significant (ANOVA; $P < 0.05$). Species–sex interaction effects were highly significant only in wing length (ANOVA; $P < 0.0001$); examination of pairwise comparisons (Bonferroni t-tests) indicated that sexual dimorphism in this dimension (Table 7) was comparatively great in three sizable, migratory species (*Coturnicops noveboracensis*, *Porzana carolina*, and *P. albicollis*) and comparatively small in two flightless species (*P. atra* and *P. palmeri*). Sexual differences in flightless *Atlantisia rogersi* were of a magnitude comparable to the majority of other crakes (Table 7).

The seven taxa of bushhens (*Amaurornis*) permitting two-way comparisons manifested significant intersexual differences (ANOVA) in culmen length ($P < 0.0001$), bill height ($P < 0.0001$), wing length ($P < 0.0001$), tarsus length ($P < 0.0001$), and middle-toe length ($P < 0.0001$). Magnitudes of these differences were exceptionally high throughout the genus (Table 7). Species–sex interaction effects, indicating taxonomic differences in magnitude of sexual dimorphism, were significant in wing length ($P < 0.05$), and there only marginally so. Examination of group means and a posteriori pairwise comparisons (Bonferroni *t*-tests) indicated that this heterogeneity reflected the comparatively high sexual dimorphism in wing lengths of species other than *A. akool* and the allospecies *A. moluccanus* and *A. ruficrissus* (Table 7).

The 12 taxa of moorhens sensu lato permitting comparisons of sexes confirmed significant intersexual differences in all six external dimensions (ANOVA; $P <$ 0.0001). Species–sex interaction effects were found for culmen length ($P <$ 0.0005), bill height ($P < 0.0005$), wing length ($P < 0.0001$), tail length ($P <$ 0.01), and tarsus length ($P < 0.0005$). Only middle-toe length lacked significant variation in sexual dimorphism among taxa ($P > 0.10$). Examination of pairwise comparisons (Bonferroni *t*-tests) indicated that the interspecific differences in magnitudes of sexual dimorphism in five of the six external dimensions largely reflected the exceptionally large intersexual differences in *Gallicrex cinerea* (Table 7), one of the few species of Rallidae also characterized by conspicuous sexual dichromatism of plumage (Livezey 1998). Mean intersexual differences for this species were at least two to three times as great as those for other moorhens.

The six species of coots (*Fulica*) subjected to two-way comparisons were characterized by significant intersexual differences in means of all six external measurements: culmen length ($P < 0.001$), bill height ($P < 0.0001$), wing length ($P < 0.0001$), tail length ($P \approx 0.07$), tarsus length ($P < 0.0001$), and middle-toe length ($P < 0.0001$). Species–sex interaction effects were significant in wing length ($P < 0.0005$), tail length ($P < 0.05$), and middle-toe length ($P < 0.05$). Pairwise comparisons (Bonferroni *t*-tests) and summary statistics revealed the nature of this heterogeneity in sexual dimorphism (Table 7). Wing lengths showed weak female-larger dimorphism in *F. gigantea*, whereas this measurement showed disproportionate male-larger dimorphism in *F. americana*, *F. alai*, and *F. atra*; tail lengths showed a similar, but more tenuous female-larger dimorphism in *F. gigantea*; and sexual dimorphism in middle-toe lengths ranged from essentially zero in *F. gigantea* to pronounced in *F. cornuta*, *F. alai*, and *F. atra*. In light of the modest magnitude of the reversal in dimorphism in *F. gigantea* and variation deriving from third-party provision of data for this species, the heterogeneity in sexual dimorphism may prove to be artifactual.

UNIVARIATE GENERALITIES OF THE INTEGUMENT

Comparisons of means of external measurements revealed that statistical significance was the rule in interspecific comparisons, with both classificatory and flightlessness-related partitions being contributory. In aggregate, most flightless taxa tended toward increased body size, although instances of approximate stasis (e.g., *Pareudiastes*, *Porphyriornis*, and *Porzana atra*) or apparent dwarfism (e.g., *Dryolimnas aldabranus*, *Gallirallus wakensis*, *Atlantisia rogersi*, and the smallest *Porzana*) were evident. Of the six external variables compared by univariate means, lengths of the wing and tail were the external dimensions most predictably related to flight capacity, but these departures from the flighted norm varied across taxonomic groups (Table 7). Variances within species–sex groups differed in many cases, but the majority of these appeared to derive largely from sources other than flight capacity. Uncertain allocation of specimens to sex classes (primarily *Habroptila*), merging of variably differentiated subspecific taxa (e.g., some members of the *Porphyrio porphyrio* complex), heterogeneity in measurement error (notably third-party data for *Fulica gigantea*), and the confounding, broad association between mean and variance (e.g., *Amaurornis ineptus*) often were implicated in variation among second moments for groups. Univariate sexual dimorphism was typical in most dimensions of most taxa regardless of flight capacity. Exceptionally great intersexual differences were associated either with loss of flight capacity (*Porphyrio hochstetteri*, with predisposition indicated in likely sister-species *P. melanotus*), comparatively intense migratory habit (e.g., *Coturnicops noveboracensis* and *Porzana carolina*), or extreme and anatomically pervasive, sexually selected dimorphism including marked dichromatism (e.g., *Gallicrex cinerea*), which is rare among the Rallidae.

INTERSPECIFIC AND INTERSEXUAL PATTERNS IN BODY MASS

Mean body masses of flighted rallids ranged from 25 g for *Micropygia flaviventer* to approximately 2,400 g for *Fulica gigantea* (Appendix 1; average for sexes). Modern flightless rallids, many of which lack recorded data on body mass, range at least from a minimum of 40 g for *Atlantisia rogersi* to a maximum of more than 2,700 g for *Porphyrio hochstetteri* (Appendix 1). Osteologically based estimates of body masses of subfossil rallids indicate that extinct, flightless rallids encompassed body masses as low as 25 g (small *Porzana*) and exceeded 4 kg for *Porphyrio mantelli* (Table 8). Although body masses of captive-reared males of the smaller *P. hochstetteri* can exceed 4 kg (P. B. Taylor 1998), a similar estimate for the mean body mass for *P. mantelli* is descriptive of a rail of ponderous body size. The estimated body masses of the endemic, terrestrial rallids of the Chatham Islands are ~60 g for *Cabalus modestus*, ~340 g for *Gallirallus dieffenbachii*, and ~1,900 g for *Diaphorapteryx hawkinsi*. The three species, when ordered by size, comprise a graded series in which body mass increases interspecifically by a factor of six. Also notable was that *Aphanapteryx bonasia* exceeded 1 kg in body mass, and the two extinct *Fulica* from the New Zealand region were only slightly smaller than the extant, weakly flighted *F. gigantea* and *F. cornuta* (Appendix 1). Where separate estimates for provisional sex groups were feasible (see *K*-means clustering), no strong evidence was found for unusually large sexual size dimorphism in the subfossil taxa, including those in which evidence for exceptional dimorphism of the bill was detected (e.g., *Aphanapteryx bonasia, Erythromachus leguati,* and *Diaphorapteryx hawkinsi*).

Where the same taxa were considered, the estimated body masses presented here compare well with those compiled by Atkinson and Millener (1991), but less favorably with those presented by Holdaway (1999). The latter estimates were based on more-general models derived from a wide range of taxonomic groups by Campbell and Tonni (1983) and Anderson et al. (1985). Log-scale regression models among taxa predictably shift toward higher slopes and lower intercepts as the diversity of taxonomic groups included in the estimators is increased (Gould 1966; Cheverud 1982; Shea 1985b; Rayner 1985c) and therefore tend to provide biased estimates relative to those based on intensive sampling of the taxonomic group of interest.

BIVARIATE CORRELATIONS AND REGRESSIONS

Relative wing lengths.—As is typical of most avian families (Hartman 1961; Greenewalt 1962), rails (even including flightless species) show a direct association between wing size and body mass (Fig. 15). A regression of mean wing length on mean body mass for all flighted species of Rallidae for which data for these two variables were available in the literature or in label data (Appendix 1; pooling *G. philippensis*-group) provided a quantitative estimate of this relationship for the Rallidae and taxonomic and functional subgroups thereof. For 107 flighted taxa, a virtually isometric relationship ($\hat{b} \pm$ SE $= 0.34 \pm 0.01$) was inferred with high statistical significance ($F = 1,647.1$; d.f. $= 1, 105$; $P < 0.0001$; $r = 0.97$). The linear allometric regression for 14 flightless species (Table 7; Appendix 1) also was highly significant ($F = 385.3$; d.f. $= 1, 12$; $P < 0.0001$; $r = 0.98$), with an estimated slope ($\hat{b} \pm$ SE) of 0.33 ± 0.02. The models for flighted and flightless species differed significantly ($F = 52.0$; d.f. $= 2, 117$; $P < 0.00001$). The model for flighted rallids was slightly improved (partial $F = 5.52$; d.f. $= 1, 104$; $P \leq 0.01$; multiple $r = 0.97$) by the addition of a negative quadratic term, which was an order of magnitude smaller than the linear term. This nonlinear component (Fig. 15) was largely attributable to the disproportionately short wings of *Fulica gigantea* and *F. cornuta*, an inference confirmed by the insignificance of the quadratic term if these two taxa were excluded from the analysis (partial $F = 1.06$; d.f. $= 1, 102$; $P > 0.25$). Hence the simple linear model was considered sufficiently explanatory for finer comparisons of effects of taxon and flight capacity. The high correlation coefficients rendered negligible the difference between the slopes, differing by a factor of $1/r$, of the standard and geometric-mean regressions.

Flighted and flightless taxa differed in intercept terms ($\hat{a} = 3.13$ and 2.95, respectively). When using the criterion (s) of geometric similarity (White and Gould 1965; Gould 1971, 1972), this transposition is summarized (by using an average $\hat{b}$ of 0.33) by a slightly lower criterion of geometric similarity (s) of 1.09. This index means that the mean wing length of a flighted rallid would be 9% longer on average than that of a flightless species at the same (log-transformed) body mass. To the extent that slopes were of comparable magnitude, the intercept term for flighted rallids (3.13) was similar to those for other comparatively heavy power fliers (Hartman 1961; Greenewalt 1962) (e.g., waterfowl [Livezey and Humphrey 1986; Livezey 1990] and grebes [Livezey 1989d]), and significantly lower than several others (e.g., cormorants [Phalacrocoracidae; Livezey 1992a]).

Separate regressions for flighted species (Appendix 1; excluding unique *Himantornis*) in nine taxonomic groups (swamphens [10 species], basal rallids [14 species], meso-rallids [four species], typical rallids [11 species], flufftails and allies [15 species], crakes [28 species], bushhens [five species], moorhens [10 species], and coots [nine species]) documented significantly different models among groups ($F = 4.84$; d.f. $= 16, 88$; $P < 0.00001$). Seven of the nine groups manifested highly significant linear relationships between mean wing lengths and body mass ($P < 0.005$). Only the poorly sampled meso-rallids (*Ortygonax*, *Pardirallus*, and *Dryolimnas*) and bushhens (*Amaurornis*) had insignificant slopes ($P > 0.05$). Of the remaining groups, allometric models ranged from those with high intercepts and low slopes (swamphens [$\hat{a} = 3.60$, $\hat{b} = 0.29 \pm 0.02$; $P < 0.0001$], moorhens [$\hat{a} = 3.53$, $\hat{b} = 0.28 \pm 0.06$; $P < 0.005$], and coots [$\hat{a} = 3.43$, $\hat{b} = 0.29 \pm 0.03$; $P < 0.0001$]) to groups described by low intercepts and high slopes (basal genera [$\hat{a} = 3.29$, $\hat{b} = 0.32 \pm 0.04$; $P < 0.0001$], flufftails and allies [$\hat{a} = 3.25$, $\hat{b} = 0.31 \pm 0.02$; $P < 0.0001$], and the typical rails [$\hat{a} = 3.28$, $\hat{b} = 0.30$

TABLE 8. Estimates of mean body masses (g) of subfossil and other selected species of Rallidae, by provisional sex group, based on linear regressions of selected osteological dimensions of the pelvic limb. Stepwise linear regression of $\log_e$-transformed variables significantly entered into model ($P < 0.05$). Estimates are presented for all specimens pooled, followed (where partitioning feasible) by separate estimates for provisional (bracketed) or actual sex groups. Final model for $\log_e$-transformed data ($n = 86$): mass = 2.11443 + 2.3220 (tibiotarus maximal width at midpoint); $R^2_{adj} = 0.9597$, SE (estimate) = 0.2212. Single elements of partial skeletons were limited to linear regressions based on femoral measurements (*Gymnocrex rosenbergii* and *Tricholimnas lafresnayanus*) or tibiotarsal least width at midpoint (*Porzana rua* and *P. monasa*). Former model ($n = 85$) included three estimator variables, $R^2_{adj} = 0.9286$, SE (estimate) = 0.2968. Latter, single-predictor model ($n = 86$) was comparably precise, $R^2_{adj} = 0.9399$, SE (estimate) = 0.2700.

		Measurements available	
Taxon	Sex	Femoral and tibiotarsal	Single elements or part
Porphyrio mantelli	All	4,141	—
	[M]	4,565	—
	[F]	3,899	—
Porphyrio kukwiedei	All	2,316	—
Aphanocrex podarces	All	702	—
Gymnocrex rosenbergii	All	—	385
Habroptila wallacii	All	886	—
Nesotrochis debooyi	All	1,338	—
Nesotrochis steganinos	All	385	—
Cyanolimnas cerverai	All	241	—
Dryolimnas abbotti	All	192	—
Rallus ibycus	All	102	—
	[M]	108	—
	[F]	91	—
Rallus recessus	All	224	—
	[M]	244	—
	[F]	210	—
Gallirallus dieffenbachii	All	340	—
	[M]	360	—
	[F]	310	—
Gallirallus wakensis	All	113	—
Tricholimnas lafresnayanus	All	—	868
Nesoclopeus poecilopterus-group*	All	487	—
Cabalus modestus	All	58	—
	[M]	62	—
	[F]	55	—
Capellirallus karamu	All	240	—
	[M]	235	—
	[F]	196	—
Habropteryx insignis†	All	515	—
Aphanapteryx bonasia	All	1,143	—
	[M]	1,326	—
	[F]	1,075	—
Erythromachus leguati	All	573	—
Diaphorapteryx hawkinsi	All	1,910	—
	[M]	2,095	—
	[F]	1,742	—
"*Atlantisia*" *elpenor*	All	83	—
	[M]	91	—
	[F]	80	—

TABLE 8. Continued.

Taxon	Sex	Measurements available	
		Femoral and tibiotarsal	Single elements or part
Porzana sandwichensis*	All	49	—
Porzana palmeri	All	45	—
Porzana rua†	All	—	72
Porzana monasa†	All	—	150
Porzana piercei	All	33	—
	[M]	37	—
	[F]	32	—
Porzana astrictocarpus	All	45	—
Porzana ziegleri	All	28	—
Porzana menehune	All	24	—
Porzana keplerorum	All	33	—
Porzana severnsi	All	92	—
Amaurornis isabellinus*	All	318	—
Tribonyx hodgenorum	All	278	—
	[M]	313	—
	[F]	263	—
Fulica chathamensis	All	1,847	—
	[M]	1,910	—
	[F]	1,725	—
Fulica prisca	All	1,910	—
	[M]	2,035	—
	[F]	1,707	—
Fulica newtoni	All	1,287	—

* Measurements were taken from elements removed from skin specimens (see text).
† Measurements were based on tabulations or figures presented by Steadman (1986a).

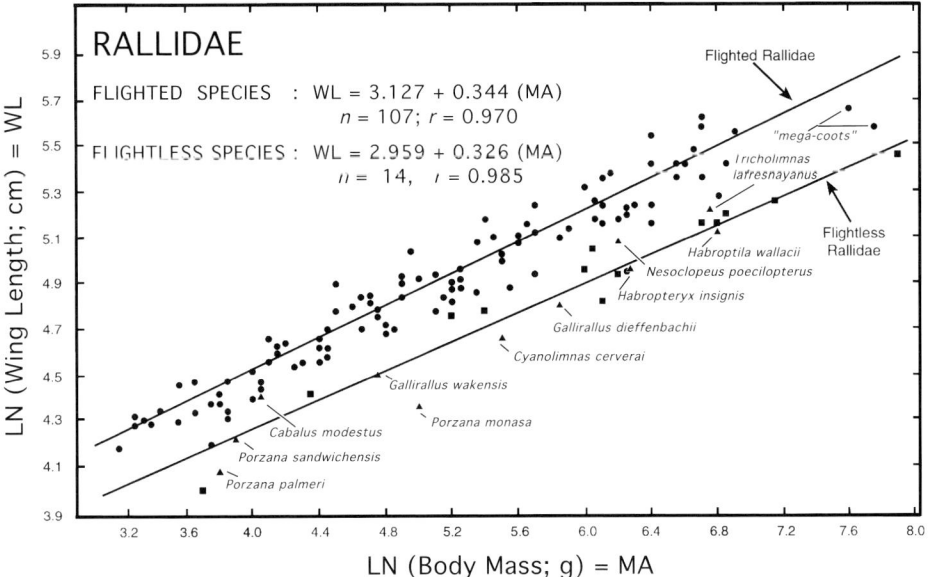

FIG. 15. Bivariate plot of log-transformed, mean body masses and wing lengths of species of Rallidae (Appendix 1). Separate regression lines were fitted for flighted (circles) and flightless (squares) species; triangles signify taxa for which body mass was estimated.

± 0.03; $P < 0.0001$]). The crakes were described by a model that combined an extraordinarily low intercept with a slope indicative of significant positive allometry ($\hat{a} = 2.77$, $\hat{b} = 0.43 \pm 0.04$; $P < 0.0001$). Plotting species for which body mass was estimated in association with these allometric regressions confirmed the tendency for flightless species of all sizes to possess variably low relative wing lengths (e.g., *Habroptila wallacii, Cyanolimnas cerverai, Gallirallus dieffenbachii, G. wakensis, Cabalus modestus, Tricholimnas lafresnayanus, Nesoclopeus poecilopterus, Habropteryx insignis, Porzana sandwichensis,* and *P. palmeri*), including some extreme outliers relative to flighted rallids (Fig. 15).

A surrogate for body mass in morphometric studies of birds (especially intraspecific) is mean wing length (e.g., Hamilton 1961; Rand 1961a; Visser 1976; Fjeldså 1977; Bochenski and Bochenski 1993). However, in light of the obvious implications of flightlessness for the interspecific relationship between wing length and body size, this alternative is not appropriate in the present context. However, tarsus length is a standard measurement of study skins that generally provides flight-independent information of size within reasonably constrained taxonomic groups (Cock 1966; Grant 1966b, 1971), and this dimension is available for a number of flightless rallids lacking data on body mass (Table 7). Furthermore, the utility of the ratio of mean wing length to mean tarsus length as an index of flight capacity has been shown for flighted and flightless congeners in the Anatidae (e.g., Livezey and Humphrey 1992: fig. 24). A linear regression of mean wing lengths on mean tarsus lengths (log-transformed data) for 69 flighted species of Rallidae revealed a strong interspecific relationship between these two external dimensions ($F = 791.0$; d.f. = 1, 67; $P < 0.0001$; $r = 0.96$). The estimated slope ($\hat{b} \pm$ SE) for the standard regression (0.92 ± 0.03) indicated slight negative allometry (i.e., isometry for two unidimensional variables would have unit slope), but statistical significance of this difference in slopes was only marginal ($P < 0.10$). A regression of wing length on tarsus length for 29 flightless rallids was of comparable significance ($F = 213.5$; d.f. = 1, 27; $P < 0.0001$; $r = 0.94$), for which the estimated slope (0.93 ± 0.06) did not differ significantly from that for isometry. In light of the similarity of slopes, the significant overall difference between the two regressions ($F = 16.3$; d.f. = 2, 94; $P < 0.00001$) indicates that the two virtually isometric relationships differ by a transposition of intercepts ($\hat{a}$), wherein that for flighted rallids (1.39) was predictably higher than that for flightless confamilials (1.18).

These same primary study species of Rallidae also confirmed the similitude of correlation between wing length and body mass (Table 7). The linear allometric equation relating wing length to body mass for 54 flighted species confirmed that tarsus length ($F = 541.7$; d.f. = 1, 53; $P < 0.0001$; $r = 0.96$), and the estimate of slope ($\hat{b} \pm$ SE = 0.34 ± 0.01) indicated virtual isometry of wing length (one-dimensional variable) with respect to body mass (three-dimensional variable). A slight negative curvilinearity was confirmed for the flighted taxa by a marginally significant reduction in residual variance (partial $F = 7.29$; d.f. = 1, 51; $P \leq 0.01$; multiple $r = 0.91$), which was attributable to the disproportionately short wings of the two large Andean species of coots (*Fulica gigantea* and *F. cornuta*), as shown by the insignificance of the quadratic term if these two species were excluded from analysis (partial $F = 1.39$; d.f. = 1, 50; $P > 0.25$). The latter curvilinearity was not indicated in the analysis based on tarsus lengths. The wing–

tarsus relationship among flightless rallids revealed a pattern comparable to the general case. However, a highly significant difference between the separate linear models based on body masses for flighted and flightless rallids ($F = 49.9$; d.f. = 2, 65; $P < 0.00001$) reaffirmed the difference in intercepts between the groups (flightless taxa, $â = 2.95$; flighted taxa, $â = 3.18$) inferred from the more-inclusive data set, and one comparable to those models based on tarsus lengths.

Body mass and sexual size dimorphism.—Rensch (1950, 1959) inferred that magnitude of sexual size dimorphism generally increased with mean body size among species, a generality disputed as artifactual by Reiss (1986, 1989). To test for such a pattern among the species of the Rallidae, the mean differences in body masses of male and female conspecifics were regressed against the mean mass of the species (Appendix 1), and this regression was performed for all species ($n = 100$) and was restricted to flighted species ($n = 91$). Also, three specific metrics of sexual mass dimorphism were used as dependent variable in these regressions. Model a used the difference between the raw (untransformed) mean masses of the sexes (i.e., mean for males − mean for females). Model b used the difference between the log-transformed mean masses of the sexes (i.e., ln(mean for males) − ln(mean for females). Model c used the log-transformed difference between the mean masses of the sexes (i.e., ln(mean for males − mean for females)). Where differences in mean masses of sexes included negative numbers, an equal constant was added to all data before log-transformations to avoid mathematical singularities. In the analysis of raw differences between means for the sexes (model a), untransformed mean masses for the species were used as the independent variable; the independent variable (mean body mass) was log-transformed in models a and b.

Regressions based on raw differences (model a) resulted in strongly positive correlations for all species ($r = 0.87$; $P < 0.0001$) and analysis restricted to flighted rails ($r = 0.86$; $P < 0.0001$). Regressions based on differences between log-transformed masses (model b) revealed substantially lower correlations, which, nevertheless, remained significant for all species ($r = 0.32$; $P < 0.005$) and flighted species ($r = 0.32$; $P < 0.005$). This reduction in proportion of variance due to sexual size dimorphism evidently resulted at least as much from log-transformation of the independent variable as that of the dependent variable. Finally, log-transformed differences (model c) produced correlations of intermediate magnitude, both for all Rallidae ($r = 0.75$; $P < 0.0001$) and for flighted species alone ($r = 0.72$; $P < 0.0001$). Taken together, these exercises indicate that a significant and positive correlation was found between mean body masses of species and magnitude of sexual differences among the Rallidae. Although there was some inflation of the correlation related to simple scaling effects, the relationship remained significant under the most stringent transformation regimes. Flightlessness was not an important confounding factor in any of the models.

Hierarchical correlation structure.—A hierarchy of correlations among six external measurements for 69 flighted species of Rallidae (total $n = 2,283$) emerged by means of a cluster analysis of the pooled within-species correlation matrices (Figs. 16, 17). All correlation coefficients between pairs of the six external measurements within flighted species of Rallidae fell within the interval $0.23 < r < 0.67$. Two couplets of anatomically related variables showed predictably high correlations within species ($r \geq 0.74$): culmen length and bill height, and lengths

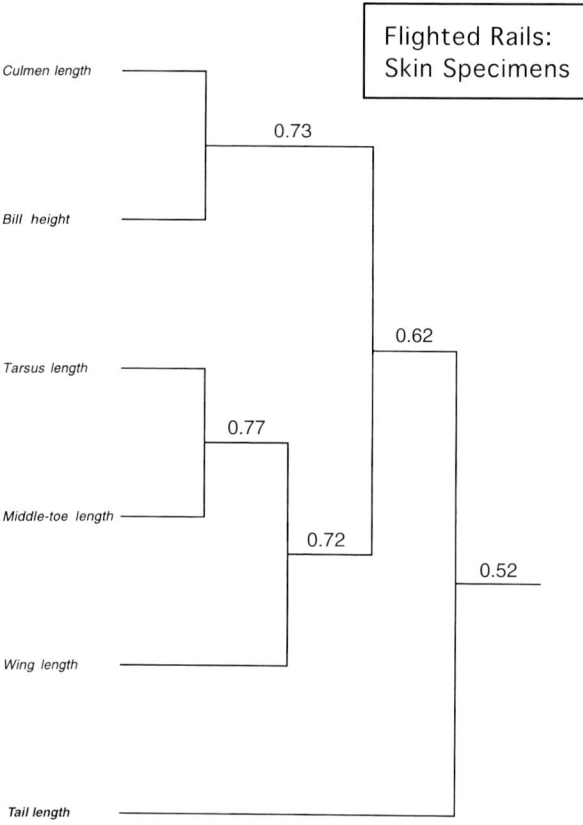

FIG. 16. Dendrogram depicting hierarchy of pairwise associations among six external measurements for 69 flighted species of Rallidae based on a cluster analysis of the pooled within-species correlation matrix (pooled $n = 2,283$).

of the tarsus and middle toe. These two couplets were then united at slightly higher mean correlation to each other than this foursome was to wing length. The last to be joined for flighted species (i.e., the least correlated with other dimensions within species) was tail length (Fig. 16).

A similar exercise for 22 flightless species of rallids (total $n = 1,245$) indicated generally higher correlation coefficients between pairs of the six external measurements within flightless species ($0.47 < r < 0.77$), and revealed a similar pair of fundamental couplets of bill and pelvic dimensions (Fig. 17). However, wing length was correlated more tightly with the pelvic couplet in flightless taxa than the latter was to the pair of bill measurements. Together these five dimensions were joined last by tail length, but at a slightly higher correlation than for flighted confamilials (r at last linkage approximating 0.30 and 0.45 in flighted and flightless species, respectively). Despite the limited dimensionality of the suite of external measurements, weak evidence was found that external measurements of flightless taxa are generally more highly correlated with each other than their pairwise counterparts in flighted taxa, and that wing length followed pelvic dimensions more closely in flightless species than in flighted relatives.

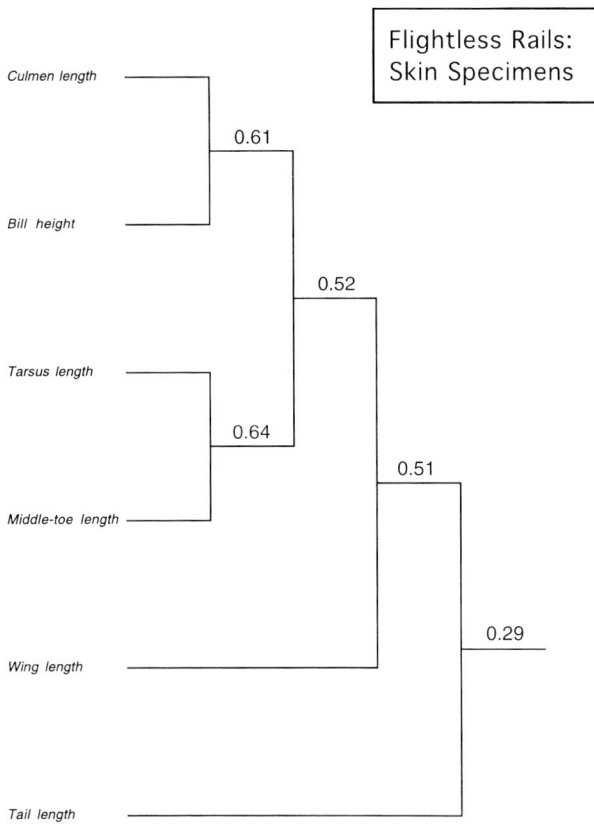

FIG. 17. Dendrogram depicting hierarchy of pairwise associations among six external measurements for 22 flightless species of Rallidae based on a cluster analysis of the pooled within-species correlation matrix (pooled $n = 1,245$).

RATIOS AND PROPORTIONS

Wing loadings.—Estimates of wing loadings for 33 species of Rallidae (g/cm^2) were available from the literature and various combinations of direct measurement (Table 9), and a first examination confirmed that wing loadings were associated directly with body mass among flighted species within avian families (Hartman 1961; Greenewalt 1962). Although these figures reveal that wing loadings of rallids averaged lower than members of most other nonpasseriform families having a minority of flightless members (e.g., Anatidae, Podicipedidae, Phalacrocoracidae, and Alcidae [Hartman 1961; Livezey and Humphrey 1986; Livezey 1988, 1989d, 1990, 1992a, 1993c]), rails were characterized by high wing loadings relative to mean body masses (Hartman 1961: fig. 6). Of the species of Rallidae for which estimates of wing loading were feasible (Table 9), only two were flightless (*Porphyrio hochstetteri* and *Porzana atra*). Both of the flightless examples, especially massive *P. hochstetteri*, affirm the predicted increase in wing loadings with loss of flight. The estimate for *P. hochstetteri* (2.76 g/cm^2) also exceeded the estimated threshold of flightlessness of 2.5 g/cm^2 of Meunier (1951), a critical value that did not incorporate information on relative masses of the pectoral mus-

TABLE 9. Wing loadings (g·cm^{-2}) of selected species of Rallidae. Taxonomy and sequence follow Livezey (1998). Summary statistics are mean ± standard deviation and range, with sample size for wing areas in parentheses. Where associated body masses were not available, numerator of estimates were means compiled for corresponding species and sex (Appendix 1).

Taxon	Wing loading	Basis for estimate
Porphyrio poliocephalus	0.79 — (1)	Spread wing with associated body mass (sex undetermined)
Porphyrio madagascariensis	0.63 ± 0.05 0.60–0.66 (2)	Wing areas with associated body masses (1 male, 1 female)
Porphyrio melanotus	0.93 ± 0.04 0.90–0.95 (2)	Spread wing (female) and area of undetermined sex, both without associated body mass
Porphyrio hochstetteri	2.76 — (1)	Area of undetermined sex and without associated body mass
Porphyrula martinica	0.50 — (12)	Associated data from Hartman (1961)
Aramides ypecaha	0.66 — (1)	Spread wing with associated body mass (male)
Aramides cajanea	0.63 ± 0.10 0.55–0.77 (4)	Spread wing with body mass (1 male); spread wings without masses (1 male, 2 females)
Sarothrura boehmi	0.30 — (1)	Spread wing with associated body mass (male)
Rallus limicola	0.39 ± 0.02 0.37–0.41 (3)	Spread wings without associated body masses (1 male, 2 females)
Rallus aquaticus	0.96 ± 1.93 0.40–0.67 (24)	Wing areas with associated body masses (13 males, 11 females); Meinertzhagen (unpubl.)
Rallus longirostris	0.73 ± 0.06 0.66–0.77 (5)	Spread wings without associated body masses (4 males, 1 female)
Rallus elegans	0.63 ± 0.29 0.42–0.84 (2)	Spread wing without associated body mass (male), and one ratio from Poole (1938)
Gallirallus pectoralis	0.46 ± 0.05 0.43–0.52 (5)	Spread wings without associated body masses (2 males, 2 females), and one ratio from Müllenhoff (1885)
Gallirallus striatus	0.39 — (1)	Spread wings without associated body mass (female)
Gallirallus philippensis-group	0.60 ± 0.09 0.51–0.76 (8)	Spread wings without associated body masses (2 males, 5 females, 1 of undetermined sex)
Laterallus leucopyrrhus	0.36 — (1)	Spread wing without associated body mass (female)
Laterallus albigularis	0.36 — (16)	Hartman (1961)
Crex crex	0.49 — (1)	Associated data from Magnan (1922)
Porzana porzana	0.40 ± 0.01 0.40–0.41 (2)	Spread wings with associated body masses (1 male, 1 female); Meinertzhagen (unpubl.)
Porzana carolina	0.53 ± 0.04 0.48–0.57 (4)	Spread wings without associated body masses (22 males, 1 female, 1 of undetermined sex)
Porzana pusilla	0.30 ± 0.07 0.25–0.35 (2)	Spread wing with (male) and without (female) associated body masses
Porzana fluminea	0.42 0.42–0.43 (2)	Spread wings with (sex undetermined) and without (male) associated body masses
Porzana tabuensis	0.31 ± 0.01 0.30–0.32 (2)	Spread wings with and without associated body masses (1 female each)
Porzana atra	0.58 ± 0.04 0.54–0.68 (10)	Tracings of spread wings with associated body masses (3 males, 6 females, 1 juvenile)
Porzana flavirostra	0.41 ± 0.02 0.40–0.43 (2)	Spread wings without associated body masses (sexes undetermined)
Amaurornis olivaceus	0.53 ± 0.03 0.51–0.56 (2)	Spread wings without associated body masses (1 male, 1 female)
Amaurornis ruficrissus	0.54 ± 0.04 0.51–0.57 (2)	Spread wings without associated body masses (1 male, 1 female)

TABLE 9. Continued.

Taxon	Wing loading	Basis for estimate
Tribonyx ventralis	0.75 ± 0.08 0.64–0.83 (4)	Spread wings without associated body masses (3 males, 1 of undetermined sex)
Gallinula tenebrosa	0.92 ± 0.10 0.81–1.08 (7)	Spread wings without associated body masses (2 males, 4 females, 1 of undetermined sex)
Gallinula chloropus-group	0.82 ± 0.11 0.55–1.07 (72)	Spread wings with associated body masses (40 males, 32 females); Meinertzhagen (unpubl.)
Fulica americana	1.13 ± 0.18 0.84–1.35 (8)	Spread wings without associated body masses (3 males, 3 females, 2 of undetermined sex)
Fulica atra	1.20 ± 0.17 1.63–0.84 (36)	Spread wings with associated body masses (21 males, 15 females); Meinertzhagen (unpubl.)
Fulica cristata	1.42 ± 0.00 1.42–1.42 (2)	Spread wings with associated body masses (2 males); Meinertzhagen (unpubl.)

culature. The importance of the latter in the flightlessness of many rallids is demonstrated by the estimated wing loading for small, flightless *Porzana atra* (0.58 ± 0.04 g/cm^2), roughly one fourth of the threshold inferred by Meunier (1951).

Aspect ratios.—For the reasons given above, direct measurements of aspect ratios are comparatively rare for most taxonomic families of birds, including the Rallidae. Hartman (1961) compiled aspect ratios for the following five flighted species of rail ($\bar{x}$ ± SD, n): *Porphyrula martinica* (2.29 ± 0.04, 14), *Aramides cajanea* (1.67 ± 0.04, 9), *Laterallus albigularis* (1.65 ± 0.04, 16), *Gallinula cachinaans* (2.09 ± 0.03, 11), and *Fulica americana* (2.32 ± 0.10, 9). These estimates compared favorably with figures given by Hartman (1961) for representatives of closely related gruiform families: *Aramus guarauna* (2.55 ± 0.50, 2) and *Heliornis fulica* (2.07 ± 0.03, 5). Given that this dimensionless ratio tends to decrease with a tendency toward short, broad wings, the rounding of wings evident in flightless rallids (see below) suggests that aspect ratios of rallids would average lower in flightless members than in their flighted relatives. Because aspect ratio is a reflection of maneuverability in flight, changes following the loss of flight capacity reflect the indirect effects of selection against nonfunctional structures as mediated by ontogeny and not functional adaptations in themselves.

MULTIVARIATE PATTERNS

Principal component analyses.—PCA served as a natural launching point for multivariate analyses in that it was used to compare differences among means for taxa, with within-taxon variation being relegated to the more powerful but assumption-laden CAs to follow. In comparisons of the externum, PCA was used in direct assessments of relationships of variables (Q-mode PCA), relationships among taxa based on measurements of study skins (R-mode PCA of skins) and remiges (R-mode PCA of remiges), and within-species multivariate dispersion common to flighted and flightless taxa (R-mode PCA of pooled within-species covariance matrices).

The general characteristics of the relationships among skin measurements were assessed by using separate Q-mode PCAs for flighted and flightless rallids (Table 10). In a morphometric context, this method tends to ordinate variables by overall scale (mean size) on the first component and display relative multivariate com-

TABLE 10. Mean correlation coefficients ($\bar{r}$) and summary statistics for first two principal components of separate Q-mode analyses of correlation matrices for six external measurements of 69 flighted and 29 flightless species of Rallidae, by taxonomic group. Numbers of flighted (n_f) and flightless species (n_g), respectively, are given in parentheses after genus names.

Genera (n_f, n_g)	Flighted species (69)		Flightless species (29)	
	PC-I	PC-II	PC-I	PC-II
Porphyrio and *Porphyrula* (14, 2)	0.984	−0.158	0.949	−0.080
Gymnocrex and *Habroptila* (2, 1)	0.980	0.138	0.982	0.137
Eulabeornis, Aramides, Canirallus, and *Rougetius* (7, 0)	0.985	0.106	—	—
Cyanolimnas, Ortygonax, and *Dryolimnas* (4, 2)	0.980	0.186	0.988	0.099
Rallus, Gallirallus, Cabalus, Tricholimnas, Habropteryx, and allies (10, 13)	0.982	0.166	0.983	−0.041
Atlantisia, Laterallus, Crex, Coturnicops, and *Porzana* (11, 5)	0.994	−0.008	0.983	0.015
Amaurornis, Gallicrex, Pareudiastes, Gallinula, and *Tribonyx* (15, 6)	0.996	−0.035	0.970	0.050
Fulica (6, 0)	0.979	−0.095	—	—
Eigenvalue (λ_i)	67.241	1.090	27.751	0.729
Percentage of variance explained	97.5	1.6	95.7	2.5

monalities on the second component (with outliers being relatively independent of other variables). In both flighted (69 species) and flightless (29 species) rallids, the first component separated three broad scales of skin measurements (Table 10; Figs. 18, 19): a comparatively minute measurement (bill height), dimensions of intermediate magnitude (lengths of culmen, tarsus, middle toe, and tail), and a variable having uniquely great mensural scale (wing length). In both groups, all taxa showed very high correlations with the first component (Table 10). However, the second components had much lower correlations in taxon space, and the multivariate relationships among variables differed between the two groups. Among flighted rallids, culmen length was singled out as comparatively independent of other skin measurements (Fig. 18), whereas in flightless species tail length was uniquely disassociated from other external dimensions (Fig. 19).

Standard R-mode PCA of the six external measurements for 98 species of Rallidae identified three components of particular interest in the present context, which together summarized 95% of the total dispersion among species (Table 11). Predictably, the first component (PC-I) represented a general size axis having strong, positive correlations with all six variables. Scores on this axis were highly correlated with mean body masses (Table 11). Accordingly, *Porphyrio hochstetteri* had the highest score on this axis and a collection of tiny crakes were the smallest taxa sampled (Fig. 20). Changes in size on PC-I associated with flightlessness (Fig. 20) included increases (e.g., *P. hochstetteri, Habroptila wallacii, Gallirallus australis*-group, *Tricholimnas sylvestris, T. lafresnayanus, Habropteryx okinawae, Porzana atra, Amaurornis ineptus, Tribonyx mortierii,* and *Pareudiastes* spp.), decreases (e.g., *Dryolimnas aldabranus, Gallirallus wakensis, Cabalus modestus, Atlantisia rogersi,* and *Porzana palmeri*), and cases of virtual stasis (e.g., *Porphyrio albus, Gallirallus owstoni,* and *Gallinula nesiotis*-group).

Principal component II and PC-III had very similar eigenvalues, and correla-

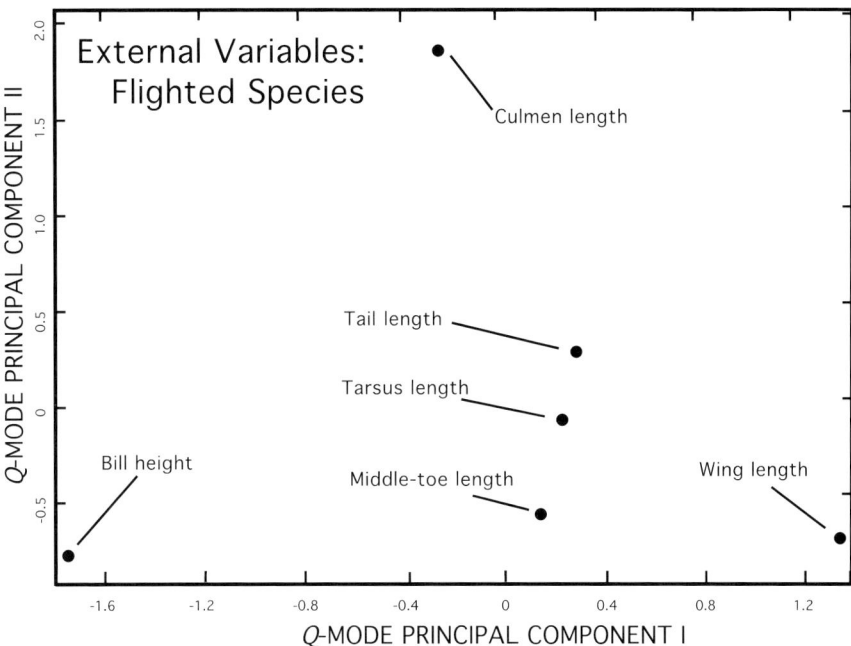

FIG. 18. Plot of six external measurements on first two Q-mode principal components for flighted species of Rallidae.

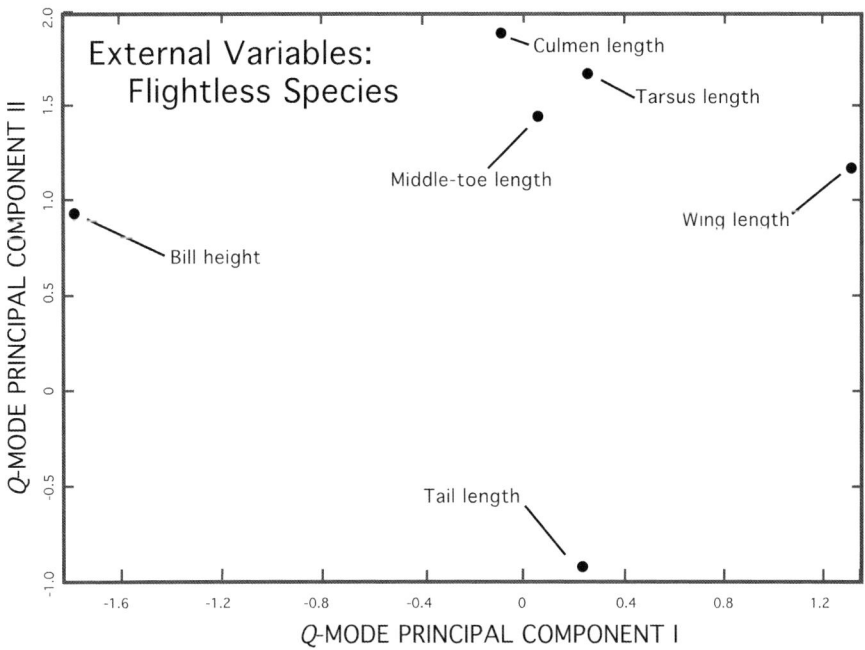

FIG. 19. Plot of six external measurements on first two Q-mode principal components for flightless species of Rallidae.

TABLE 11. Correlation coefficients (r) and summary statistics for first four principal components of six external skin measurements (mm) for 98 species of Rallidae.

Variable	Correlation coefficient (r)			
	PC-I	PC-II	PC-III	PC-IV
Culmen length	0.779	0.614	−0.071	−0.085
Bill height	0.937	−0.201	−0.204	−0.198
Wing length	0.958	−0.016	0.113	0.187
Tail length	0.812	−0.073	0.562	−0.125
Tarsus length	0.974	0.088	−0.061	0.086
Middle-toe length	0.926	−0.135	−0.099	0.294
Correlation (r) with body mass ($n = 69$)	0.910	0.083	−0.198	0.060
Eigenvalue (λ_i)	0.746	0.064	0.057	0.030
Percentage of variance explained	82.0	7.0	6.3	3.3

tions between these two components and the original variables suggest that the comparatively modest proportions of total variance summarized by PC-II and PC-III together display changes in body proportions orthogonal to size (not depicted). PC-II contrasted culmen length with bill height and middle-toe length, whereas PC-III contrasted bill height with tail length. With respect to these shifts in external shape, changes were diverse in direction, largely as a result of the taxonomically divergent shifts in lengths of bill and tail, changes that are only marginally related to capacity for flight (Table 11).

A fourth axis (PC-IV) accounted for a significant portion of the variation among

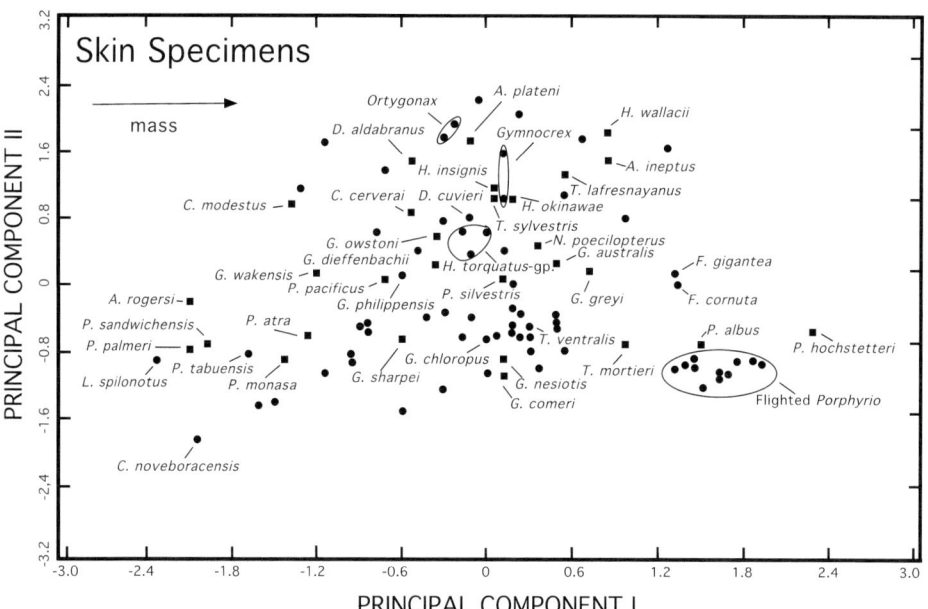

FIG. 20. Plot of mean scores for 98 species of Rallidae on first two standard (R-mode) principal components of six external measurements. Flighted species are signified by circles, and flightless species are signified by squares.

taxa, and principally contrasted lengths of the wing and middle toe with bill height and tail length (Table 11). Scores revealed that this dimension augmented the discrimination of comparatively natatorial taxa (notably subtribe Fulicarina, especially large *Fulica*) from other taxa, especially those characterized by deep bills or elongated tails (e.g., *Gallirallus dieffenbachii* and *Porphyrio hochstetteri*).

An *R*-mode PCA of pooled within-species covariance matrices of external variables partitioned intraspecific, multivariate variation common to species included in the analysis. This portion of total sample variance accounts for all variance other than that attributed to taxon, including both sexual differences and variation with sexes for each species. Because CA employs a standardization of data that is based on multivariate dispersion described by a pooled within-group covariance matrix, examination of the qualities of the variance described is useful. Regardless of whether the species included in the estimate of within-group dispersion were flighted, flightless, or combined both groups, the first three PCs of the pooled within-group covariance matrices were extremely similar (Table 12). In all three partitions, PC-I was interpretable as a general size axis and accounted for more than 10% more of total variance in flightless species than in flighted species (Table 12). Furthermore, correlations between corresponding elements of PC-I of the standard (*R*-mode) PCA and those of pooled within-species covariance matrices for the six external measurements were similar in eigenstructure whether the latter pertained to flighted ($r_s = 0.77$; $P < 0.05$), flightless ($r_s = 0.60$), or all species ($r_s = 0.60$). In all three analyses, PC-II contrasted tail length with all others. PC-III was similar for the three analyses and contrasted wing length with all others (Table 12).

Lengths of the calami and rachises of primary remiges for 41 species of Rallidae were recorded, either during dissection of pectoral musculature or from a small series of critical specimens for which skeletal elements were removed from one side of a single skin specimen, with a template for extended wings of rallids based on that for *Porzana carolina* (Fig. 21). A standard PCA of these data (with *R*-matrix) identified three readily interpretable multivariate axes that together summarized more than 98% of the total variance in these data. PC-I was highly correlated with all variables ($r_i > 0.91$, $\forall\ 1 = 1, \ldots, 20$), represented a general size axis for the 20 dimensions of primary remiges, and was highly correlated with mean body masses of these taxa (Table 13; Fig. 22). PC-II was positively correlated with lengths of the three outermost primary remiges and negatively so with lengths of the five innermost primary remiges, essentially representing relative pointedness of the wings (Table 13). PC-III represented a contrast between lengths of calami and rachises of all but the outermost primary remex.

A plot of species on the first two components, in which the implied vectors of change are shown between flightless taxa and their flighted relatives, revealed that modest reductions in overall wing size and a marked shift toward comparatively rounded wings (i.e., smaller wings showing truncation or loss of several outermost remiges) characterized changes associated with flightlessness in most taxa (Fig. 22). Notable exceptions to this general pattern were the dwarfed *Dryolimnas aldabranus* (substantial reduction in size with no change in shape), and the three giant lineages, *Gallirallus australis*, *Nesoclopeus poecilopterus*, and *Amaurornis ineptus* (modest increase in size with significant increase in rounding). PC-IV (not shown) represented only a modicum of the total variance (Table 13), but this axis

TABLE 12. Correlation coefficients (r) and summary statistics for first three principal components of pooled within-species covariance matrices for six external measurements for 69 flighted, 22 flightless, and 91 species of Rallidae.

Variable	Flighted species ($n = 2,283$)			Flightless species ($n = 1,245$)			All species ($n = 3,528$)		
	PC-I	PC-II	PC-III	PC-I	PC-II	PC-III	PC-I	PC-II	PC-III
Culmen length	0.571	−0.138	0.342	0.647	−0.042	0.279	0.603	−0.068	0.318
Bill height	0.582	−0.175	0.279	0.698	−0.117	0.234	0.660	−0.102	0.256
Wing length	0.975	−0.094	−0.202	0.974	−0.161	−0.158	0.973	−0.159	−0.168
Tail length	0.592	0.784	0.184	0.780	0.626	0.018	0.720	0.693	0.031
Tarsus length	0.625	−0.317	0.576	0.787	−0.194	0.459	0.728	−0.204	0.523
Middle-toe length	0.608	−0.299	0.596	0.794	−0.155	0.481	0.732	−0.159	0.547
Eigenvalue (λ_i)	94.144	20.553	14.538	266.039	37.049	19.970	153.045	25.976	15.269
Percentage of variance explained	67.8	14.8	10.5	78.5	10.9	5.9	74.5	12.6	7.4

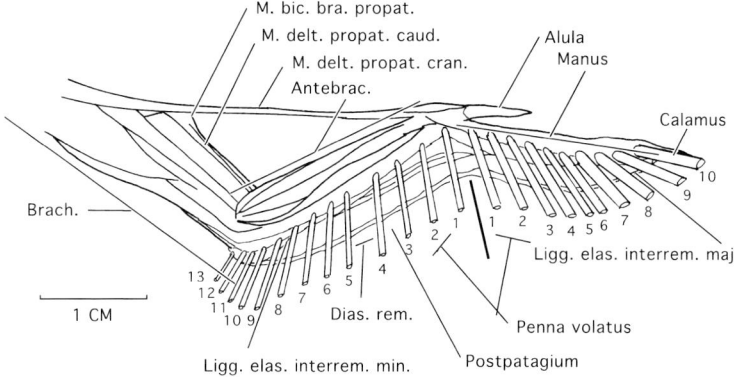

FIG. 21. Diagrammatic illustration of the relative positions of remiges in the extended right wing of *Porzana carolina* (CMNH 2007), dorsal view.

TABLE 13. Correlation coefficients (r) and summary statistics for first four principal components of lengths of extracted primary remiges for 41 species of Rallidae. Analysis was based on measurements from single specimens (dissections or bone-salvaged study skins) except for five species for which data from two specimens were averaged (*Porphyrio melanotus, P. hochstetteri, Gymnocrex plumbeiventris, Porzana tabuensis,* and *P. atra*). In *P. hochstetteri*, an apparent 11th remex primari was treated as the first remex secundari for comparison of outlines (see text).

	Correlation coefficient (r)			
Variable	PC-I	PC-II	PC-III	PC-IV
Remex primari 10, calamus	0.915	0.281	0.153	0.224
rhachis	0.924	0.227	0.260	0.088
Remex primari 9, calamus	0.951	0.247	−0.043	−0.053
rhachis	0.974	0.085	0.153	−0.113
Remex primari 8, calamus	0.967	0.185	−0.076	−0.076
rhachis	0.975	0.020	0.125	−0.155
Remex primari 7, calamus	0.976	0.142	−0.106	−0.048
rhachis	0.986	−0.071	0.101	−0.088
Remex primari 6, calamus	0.981	0.118	−0.113	−0.034
rhachis	0.987	−0.105	0.078	−0.069
Remex primari 5, calamus	0.984	0.022	−0.145	−0.011
rhachis	0.982	−0.168	0.064	−0.030
Remex primari 4, calamus	0.987	−0.003	−0.103	−0.014
rhachis	0.976	−0.205	0.060	0.005
Remex primari 3, calamus	0.982	−0.008	−0.161	0.041
rhachis	0.971	−0.224	0.057	0.043
Remex primari 2, calamus	0.978	−0.009	−0.185	0.063
rhachis	0.965	−0.242	0.054	0.058
Remex primari 1, calamus	0.974	−0.029	−0.155	0.091
rhachis	0.961	−0.229	0.014	0.098
Correlation (r) with body mass ($n = 23$)	0.873	−0.196	0.054	0.082
Eigenvalue (λ_i)	18.815	0.516	0.309	0.149
Percentage of variance explained	94.1	2.6	1.5	0.7

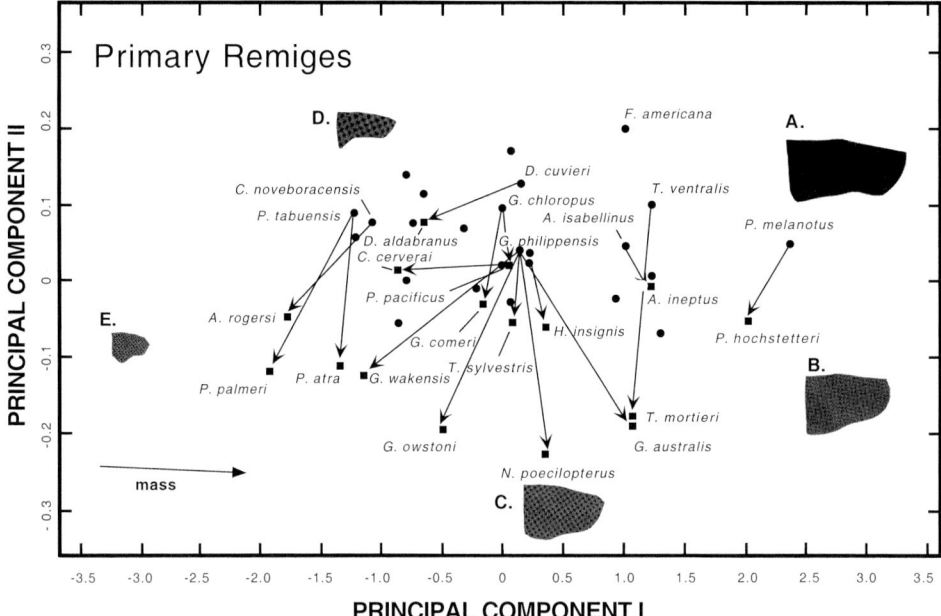

FIG. 22. Plot of mean scores for 41 species of Rallidae on first two standard (R-mode) principal components of measurements of extracted primary remiges. Flighted species are signified by circles, flightless species are signified by squares, and inferred vectors of change connect flightless species and their flighted relatives. Approximate orientation of vector for mean body mass of taxa is indicated. Reconstructed outlines of wings (based on template from *Porzana carolina*; Fig. 21): **A**, *Porphyrio melanotus* (Australian Museum 452); **B**, *Porphyrio hochstetteri* (NZNM A-2); **C**, *Nesoclopeus poecilopterus* (BMNH 1940·12·8·8); **D**, *Porzana carolina* (CM 6718); and **E**, *Atlantisia rogersi* (AMNH 320106).

usefully extracted residual variance in relative length of the 10th primary remex, singling out those flightless species having lost this (and only this) quill (*Gallirallus owstoni, G. wakensis,* and *Porzana atra*). The next axis (PC-V; not shown or tabulated) primarily singled out the only species (*Porzana palmeri*) analyzed that had lost one or more additional primary remiges (see below).

Inclusion of measurements of extracted secondary remiges added a substantial amount of additional information to the comparisons, but because many of the taxa lacked a substantial number of the secondary remiges through original preparation of the specimens, molt, damage, or disease, this extended analysis was limited to only 27 taxa. As in the foregoing analysis, PC-I for primary and secondary remiges was a well-defined general size axis (eigenvalue = 41.6, ~85% of total variance) that largely reflected intergeneric differences in body size. However, PC-II was a comparatively complex contrast that combined a measure of rounding of the wing (i.e., relative lengths of the outer and inner remiges) with a measure of lengths of calami relative to associated rachises (the latter reflected by PC-III for primary remiges).

A plot of the smaller series of taxa (not figured) confirmed the general trend in flightless lineages toward decreased size of remiges, accompanied by an increased rounding of the wing combined with a reduction in the relative lengths of calami. However, patterns in this analysis of higher dimension but lower tax-

TABLE 14. Standardized coefficients and summary statistics for first three canonical variates and two flighted–flightless contrasts based on external dimensions stepwise-selected ($P < 0.05$) from six skin measurements for 30 taxon–sex groups of Porphyriornithini. *Porphyrio albus,* for which only two specimens of undetermined sex exist, also was included a posteriori in plots.

Variable	Canonical variate*			Flightless contrast	
	I	II	III	Both genera†	*Porphyrio* only‡
Culmen length	−0.119	−0.327	−0.083	0.150	—
Bill height	0.889	−0.654	0.118	1.085	1.198
Wing length	0.296	0.799	−0.510	−0.392	−0.478
Tail length	−0.168	−0.269	0.929	0.156	0.229
Tarsus length	0.288	0.020	−0.506	0.169	—
Middle-toe length	−0.181	0.733	0.756	−0.587	−0.579
Eigenvalue	65.5	6.54	1.36	9.72	9.27
Cumulative variance (%)	88.0	96.9	98.7	—	—
Canonical R	0.992	0.931	0.759	0.952	0.950

* Wilks' λ (d.f. = 6, 30, 587) = 0.00039; $\hat{F}_{180, 3,441}$ = 53.7 ($P \ll 0.0001$).
† Wilks' λ (d.f. = 6, 1, 586) = 0.093; $\hat{F}_{6, 581}$ = 941.2 ($P \ll 0.0001$).
‡ Wilks' λ (d.f. = 4, 1, 457) = 0.097; $\hat{F}_{4, 454}$ = 1,051.7 ($P \ll 0.0001$); excluded six species–sex groups of *Porphyrula*.

onomic diversity differed from that based only on primary remiges, in that *Dryolimnas aldabranus* manifested a change in wing shape (as well as a decrease in size), and the giant flightless *Tribonyx mortierii* was shown, in addition to *Gallirallus australis* and *Nesoclopeus poecilopterus,* as having undergone a modest general increase in lengths of the remiges. However, the latter three species also have undergone increases in body size (not shown in the analyses of remiges), which together resulted in a reduction in remex size relative to body size. As in the analysis of primary remiges, immediately subsequent components (PC-IV and PC-V) emphasized residual variance associated with the loss of one or more of the outermost primaries.

Canonical analyses.—Where samples of multiple groups are represented by associated suites of attributes or measurements, CA provides a powerful means for multivariate discrimination of groups while conserving information on dispersion within groups. Phylogenetic subdivision of taxonomic groups permitted the analyses to focus on multivariate differences among comparatively closely related taxa, both overall differences and contrasts distilling differences related to flightlessness (Tables 14–22; Figs. 23–31). The technique was used as well to derive family-wide contrasts of flighted and flightless species (Table 23; Fig. 32). Furthermore, the technique has the advantage of incorporating information on both within-group and among-group variation, especially where augmented by ANOVA of scores on the resultant canonical axes (see below), and permitting formal, parametric tests of interspecific and intersexual effects through specification of appropriate multivariate analyses of variance (MANOVAs) (Tables 24–26). However, the canonical axes generally are less readily interpretable in terms of size and shape than those based on PCAs of morphological measurements, although this can be overcome to some extent through examination of correlations of mean scores on the axes with mean body masses (see below). One affirmation of the discriminatory power of CA in the present context was the number of CVs in each taxonomic subgroup for which significant interspecific, intersexual, or

TABLE 15. Standardized coefficients and summary statistics for first three canonical variates and two flighted–flightless contrasts based on external dimensions stepwise-selected ($P < 0.05$) from six skin measurements for 20 taxon–sex groups of basal Rallidae. The group comprises *Gymnocrex, Habroptila, Aramides, Eulabeornis, Canirallus,* and *Rougetius.*

Variable	Canonical variate*			Flighted–flightless contrast	
	I	II	III	All genera†	Sister genera‡
Culmen length	0.499	−0.088	0.446	−0.502	—
Bill height	0.118	0.552	−0.688	−0.384	−0.763
Wing length	0.353	−0.700	−0.167	−0.543	0.482
Tail length	0.268	0.004	−0.590	—	—
Tarsus length	0.242	−0.257	0.439	−0.371	—
Middle-toe length	0.141	0.814	0.118	−0.249	−0.462
Eigenvalue	60.66	18.58	14.67	6.63	34.53
Cumulative variance (%)	59.5	77.7	92.0	—	—
Canonical R	0.992	0.974	0.968	0.932	0.986

* Wilks' λ (d.f. = 6, 19, 257) = 0.0000012; $\hat{F}_{114, 1458}$ = 123.2 ($P \ll 0.0001$).
† Wilks' λ (d.f. = 5, 1, 257) = 0.1311; $\hat{F}_{5, 253}$ = 335.4 ($P \ll 0.0001$).
‡ Wilks' λ (d.f. = 3, 2, 58) = 0.0281; $\hat{F}_{3, 56}$ = 644.5 ($P \ll 0.0001$); contrasted flightless *Habroptila* with two species of *Gymnocrex.*

species–sex interaction effects in scores were found through two-way ANOVAs. In most instances, significant differences among group means (species, sexes, or both) were found on all CVs derived (with the maximum number being six for all but the taxonomically depauperate *Dryolimnas,* the meso-rallids *Cyanolimnas* and *Ortygonax,* and sampled species of *Fulica*), including many axes accounting for 1% or less of the total variation among groups.

Interspecific differences in the swamphens (*Porphyrio,* including flightless "*Notornis*" *hochstetteri* and *P. albus,* and *Porphyrula*) were found on all six CVs (ANOVA of scores; $P < 0.05$). Flightless *P. hochstetteri* was discriminated from flighted congeners on both the first canonical variate (CV-I) and (especially) the second canonical variate (CV-II; Table 14; Fig. 23). CV-I was dominated by bill

TABLE 16. Standardized coefficients and summary statistics for first three canonical variates and flighted–flightless contrast based on external dimensions stepwise-selected ($P < 0.05$) from six skin measurements for six species–sex groups of pardiralline rallids. Comparisons included two species of *Ortygonax* and monotypic, flightless *Cyanolimnas.*

Variable	Canonical variate*			Flighted–flightless contrast†
	I	II	III	
Culmen length	—	—	—	0.379
Bill height	0.609	0.166	0.786	−0.723
Wing length	−0.545	−0.007	0.120	0.488
Tail length	−0.594	−0.347	0.510	0.549
Tarsus length	—	—	—	—
Middle-toe length	−0.287	0.889	−0.022	—
Eigenvalue	19.68	1.56	0.79	13.41
Cumulative variance (%)	89.2	96.3	99.8	—
Canonical R	0.976	0.780	0.664	0.965

* Wilks' λ (d.f. = 4, 5, 42) = 0.01019; $\hat{F}_{20, 130}$ = 19.5 ($P < 0.0001$).
† Wilks' λ (d.f. = 4, 1, 42) = 0.0694; $\hat{F}_{4, 39}$ = 130.7 ($P < 0.0001$).

TABLE 17. Standardized coefficients and summary statistics for first three canonical variates and flighted–flightless contrast based on external dimensions stepwise-selected ($P < 0.05$) from six skin measurements for six species–sex groups of *Dryolimnas*.

Variable	Canonical variate*			Flighted–flightless contrast†
	I	II	III	
Culmen length	−0.859	0.605	−0.140	−0.914
Bill height	0.206	−0.063	−0.923	0.318
Wing length	0.908	0.096	0.262	0.834
Tail length	0.282	−0.150	−0.497	0.369
Tarsus length	0.132	0.605	0.689	—
Middle-toe length	—	—	—	—
Eigenvalue	20.74	1.44	0.51	10.34
Cumulative variance (%)	91.2	97.5	99.8	—
Canonical R	0.977	0.768	0.582	0.955

* Wilks' λ (d.f. = 5, 5, 89) = 0.0118; $\hat{F}_{25, 317}$ = 29.2 ($P < 0.0001$).
† Wilks' λ (d.f. = 4, 1, 89) = 0.0882; $\hat{F}_{4, 868}$ = 222.4 ($P < 0.0001$).

height and lengths of the wing and tarsus, and strongly correlated with body mass ($r = 0.98$; $P \ll 0.01$). CV-I primarily discriminated the two genera, although interspecific differences within genera (especially between *P. hochstetteri* and flighted congeners) and sexual differences within species also contributed to the variance on this axis (ANOVA of scores; $P < 0.05$). CV-II primarily distinguished flightless *P. hochstetteri* from all other taxa (Fig. 23). Coefficients of variables indicates that this delineation resulted from the relatively short wings and toes of the flightless species (Table 14); mean scores on CV-II showed a moderate, negative correlation with mean body masses ($r = -0.42$; $P < 0.05$). Also notable

TABLE 18. Standardized coefficients and summary statistics for first four canonical variates and two flighted–flightless contrasts based on external dimensions stepwise-selected ($P < 0.05$) from six skin measurements for 40 taxon–sex groups of "typical" rails. "Typical" rails here comprise Rallini exclusive of *Dryolimnas* (treated separately). General analysis also included samples of three species lacking adequate numbers of specimens of determined sex (*Gallirallus dieffenbachii*, *G. sharpei*, and *Habropteryx okinawae*).

Variable	Canonical variate*				Flighted–flightless contrast	
	I	II	III	IV	All genera†	Without *Rallus*‡
Culmen length	0.338	1.047	−0.069	0.385	0.445	−0.230
Bill height	−0.562	−0.557	0.186	0.761	−0.964	−0.578
Wing length	−0.341	0.053	0.086	−0.623	—	0.271
Tail length	−0.311	−0.001	−0.945	0.007	—	−0.153
Tarsus length	−0.400	−0.287	0.494	−0.654	−0.407	−0.321
Middle-toe length	−0.104	0.399	0.053	0.288	—	−0.274
Eigenvalue	47.64	13.45	7.61	5.08	5.46	3.51
Cumulative variance (%)	62.9	80.7	90.8	97.5	—	—
Canonical R	0.990	0.965	0.940	0.914	0.919	0.882

* Wilks' λ (d.f. = 6, 41, 707) = 0.00001; $\hat{F}_{246, 4,182}$ = 107.7 ($P \ll 0.0001$).
† Wilks' λ (d.f. = 3, 1, 707) = 0.1547; $\hat{F}_{3, 705}$ = 1,283.9 ($P \ll 0.0001$).
‡ Wilks' λ (d.f. = 6, 1, 595) = 0.2219; $\hat{F}_{6, 5,908}$ = 344.8 ($P < 0.0001$); excluded eight species–sex groups of *Rallus* sensu stricto.

TABLE 19. Standardized coefficients and summary statistics for first three canonical variates and two flighted–flightless contrasts based on external dimensions stepwise-selected ($P < 0.05$) from six skin measurements for 28 taxon–sex groups of crakes (Crecini). General analysis also included samples of two species lacking adequate numbers of specimens of determined sex (*Porzana monasa* and *P. sandwichensis*).

Variable	Canonical variate*			Flighted–flightless contrast	
	I	II	III	All genera†	*Crex* and *Porzana*‡
Culmen length	−0.201	−0.639	0.087	−0.340	−0.241
Bill height	0.208	−0.228	−0.639	0.109	—
Wing length	0.992	0.507	−0.210	1.101	0.998
Tail length	−0.163	−0.264	0.871	−0.209	—
Tarsus length	0.252	−0.263	−0.018	—	0.139
Middle-toe length	−0.064	−0.337	0.240	—	—
Eigenvalue	59.09	11.61	6.21	21.05	14.50
Cumulative variance (%)	72.7	87.0	94.7	—	—
Canonical R	0.992	0.960	0.928	0.977	0.967

* Wilks' λ (d.f. = 6, 29, 424) = 0.0001; $\hat{F}_{174, 2447}$ = 81.3 ($P \ll 0.0001$).
† Wilks' λ (d.f. = 4, 1, 417) = 0.0453; $\hat{F}_{4, 414}$ = 2,179.0 ($P \ll 0.0001$); excluded two species for which sex was undetermined.
‡ Wilks' λ (d.f. = 3, 2, 321) = 0.0645; $\hat{F}_{3, 319}$ = 1,542.0 ($P < 0.0001$); excluded *Laterallus*, *Atlantisia*, and *Coturnicops*.

was the modest shift shown by *P. albus* toward the uniquely derived position of *P. hochstetteri* on the first two CVs (Fig. 23). The essential differences in proportions associated with flightlessness indicated by CV-II for swamphens were confirmed by canonical contrasts between *P. hochstetteri* and other taxa (Table 14). None of the remaining axes augmented the discrimination of flightless *P. hochstetteri* appreciably, although all four showed significant interspecific differences in scores, and CV-III and CV-V included significant intersexual differences (ANOVA of scores; $P < 0.05$).

Ten species of basal rallids representing six genera manifested substantial diversity of form (Table 15). CV-I represented variation among groups of compa-

TABLE 20. Standardized coefficients and summary statistics for first three canonical variates and flighted–flightless contrast based on external dimensions stepwise-selected ($P < 0.05$) from six skin measurements for 14 taxon–sex groups of *Amaurornis*.

Variable	Canonical variate*			Flighted–flightless contrast†
	I	II	III	
Culmen length	−0.479	−0.292	0.896	0.538
Bill height	−0.353	−0.074	−0.682	0.319
Wing length	−0.396	0.915	−0.231	0.250
Tail length	0.452	0.118	0.339	−0.449
Tarsus length	−0.368	−0.190	−0.325	0.390
Middle-toe length	0.023	0.151	0.213	—
Eigenvalue	29.40	3.89	1.71	21.26
Cumulative variance (%)	81.6	92.4	97.2	—
Canonical R	0.983	0.892	0.794	0.977

* Wilks' λ (d.f. = 6, 13, 256) = 0.0011; $\hat{F}_{78, 1390}$ = 43.7 ($P < 0.0001$).
† Wilks' λ (d.f. = 5, 1, 256) = 0.0449; $\hat{F}_{5, 256}$ = 1,071.3 ($P < 0.0001$).

TABLE 21. Standardized coefficients and summary statistics for first three canonical variates and two flighted–flightless contrasts based on external dimensions stepwise-selected ($P < 0.05$) from six skin measurements for 24 taxon–sex groups of moorhens. The group comprised *Gallicrex*, *Tribonyx*, *Gallinula*, and *Porphyriornis*; general analysis also included samples of two species lacking adequate numbers of specimens of determined sex (*Pareudiastes pacificus* and *P. silvestris*).

Variable	Canonical variate*			Flighted–flightless contrast	
	I	II	III	All genera†	*Tribonyx* only‡
Culmen length	−0.047	0.174	0.341	−0.251	—
Bill height	−0.394	−0.810	−0.223	−0.641	−0.594
Wing length	−0.333	0.966	−0.188	0.285	0.426
Tail length	−0.215	−0.227	0.085	0.316	−0.343
Tarsus length	−0.790	−0.127	0.213	−0.773	−0.466
Middle-toe length	0.834	0.064	0.814	0.927	−0.253
Eigenvalue	26.36	13.30	8.37	2.77	31.72
Cumulative variance (%)	52.6	79.2	95.9	—	—
Canonical R	0.982	0.964	0.945	0.857	0.985

* Wilks' λ (d.f. = 6, 23, 371) = 0.00006; $\hat{F}_{138, 2142}$ = 66.4 ($P < 0.0001$).
† Wilks' λ (d.f. = 6, 1, 378) = 0.0449; $\hat{F}_{6, 373}$ = 172.2 ($P < 0.0001$); included *Pareudiastes*.
‡ Wilks' λ (d.f. = 5, 1, 129) = 0.0306; $\hat{F}_{5, 125}$ = 792.9 ($P \ll 0.0001$).

rable magnitude to that for swamphens (eigenvalues of 60.7 and 65.5, respectively), although this axis accounted for a substantially smaller proportion of the total variation among groups in the basal rallids (Table 15). Coefficients of variables (Table 15) and a strong correlation with mean body masses ($r = 0.93$; $P \ll 0.01$) indicated that this variate essentially reflected general body size (Fig. 24). In light of the standardization to which CA imposes variances, this partition of size would correspond to that in addition to variation among specimens within groups. CV-II for basal rallids essentially contrasted lengths of the wing and middle toe with bill height (Table 15), an axis uncorrelated with body mass ($r = 0.13$; $P > 0.05$). This proportionality placed flightless *Habroptila wallacii* at one

TABLE 22. Standardized coefficients and summary statistics for first three canonical variates and contrast of "mega" taxa based on external dimensions stepwise-selected ($P < 0.05$) from six skin measurements for 12 species–sex groups of coots (*Fulica*).

Variable	Canonical variate*			Flighted–flightless contrast†
	I	II	III	
Culmen length	—	—	—	—
Bill height	−0.354	−0.623	−1.00	0.236
Wing length	0.597	−0.633	0.562	−0.666
Tail length	0.216	−0.119	0.109	−0.226
Tarsus length	0.692	0.500	−0.509	−0.661
Middle-toe length	0.143	0.571	0.513	—
Eigenvalue	30.02	2.29	0.65	21.29
Cumulative variance (%)	90.6	97.5	99.5	—
Canonical R	0.984	0.834	0.628	0.977

* Wilks' λ (d.f. = 5, 11, 200) = 0.0051; $\hat{F}_{55, 911}$ = 35.3 ($P < 0.0001$).
† Wilks' λ (d.f. = 4, 1, 200) = 0.0449; $\hat{F}_{4, 197}$ = 1,048.4 ($P < 0.0001$).

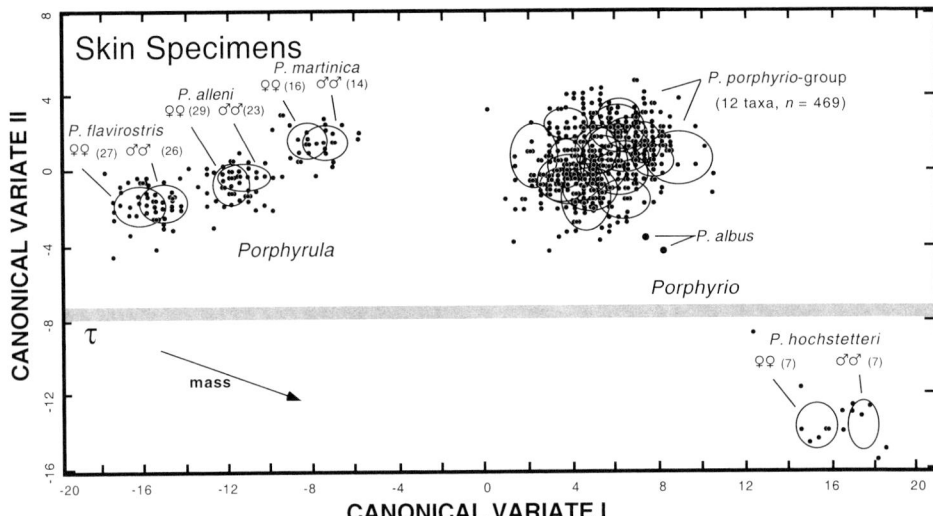

Fig. 23. Plot of 30 species–sex groups of swamphens (*Porphyrio* and *Porphyrula*) on the first two canonical variates for six external measurements. Except for *Porphyrio albus*, ellipses for groups delimit summary statistics for scores ($\bar{x} \pm$ SD). Approximate orientation of vectors for mean body masses of taxa ("mass") and apparent threshold of flightlessness (τ, positioned on the flightless side of boundary) are indicated.

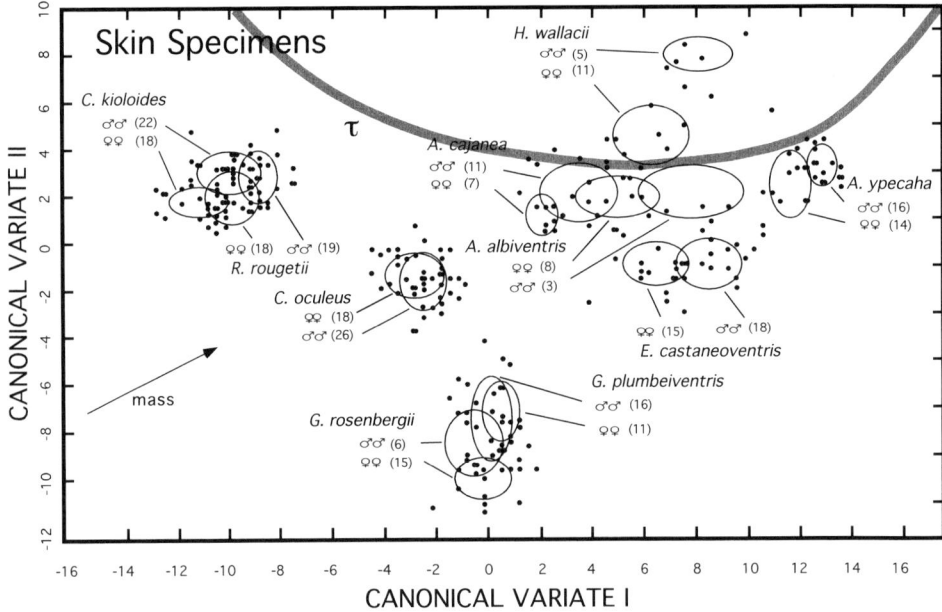

Fig. 24. Plot of 20 species–sex groups of basal rallids (*Gymnocrex, Habroptila, Aramides, Eulabeornis, Canirallus,* and *Rougetius*) on the first two canonical variates for six external measurements. Ellipses for groups delimit summary statistics for scores ($\bar{x} \pm$ SD). Approximate orientation of vector for mean body masses of taxa ("mass") and apparent threshold of flightlessness (τ, positioned on the flightless side of boundary) are indicated.

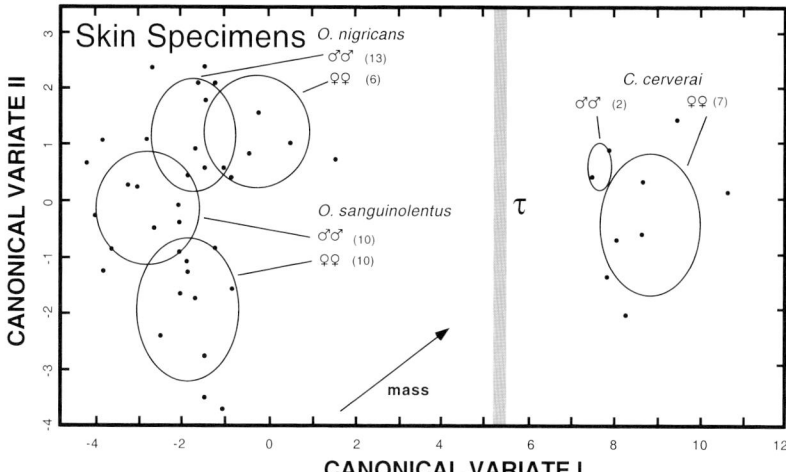

FIG. 25. Plot of six species–sex groups of meso-rallids (*Ortygonax* and *Cyanolimnas*) on the first two canonical variates for six external measurements. Ellipses for groups delimit summary statistics for scores ($\bar{x} \pm$ SD). Approximate orientation of vector for mean body masses of taxa ("mass") and apparent threshold of flightlessness (τ, positioned on the flightless side of boundary) are indicated.

extreme and its sister genus (*Gymnocrex*) at the other, with all other included genera (*Canirallus, Rougetius, Eulabeornis,* and *Aramides*) being intermediate in this respect (Fig. 24). Contrasts between *H. wallacii* and flighted relatives were comparatively simple contrasts between wing length and all other included variables, axes that were oblique to both CV-I and CV-II, but coefficients indicate that the contrast between flighted and flightless species was most similar to CV-II discriminating all species–sex groups (Table 15). Of the remaining four canonical axes, all of which included significant interspecific and three of which in-

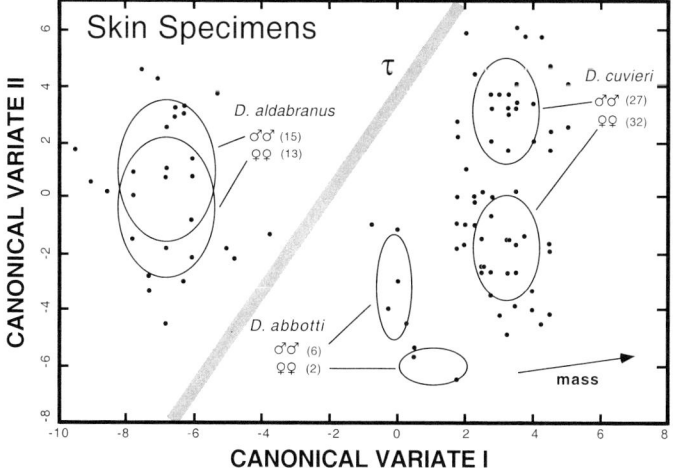

FIG. 26. Plot of six species–sex groups of white-throated rails (*Dryolimnas*) on the first two canonical variates for six external measurements. Ellipses for groups delimit summary statistics for scores ($\bar{x} \pm$ SD). Approximate orientation of vector for mean body masses of taxa ("mass") and apparent threshold of flightlessness (τ, positioned on the flightless side of boundary) are indicated.

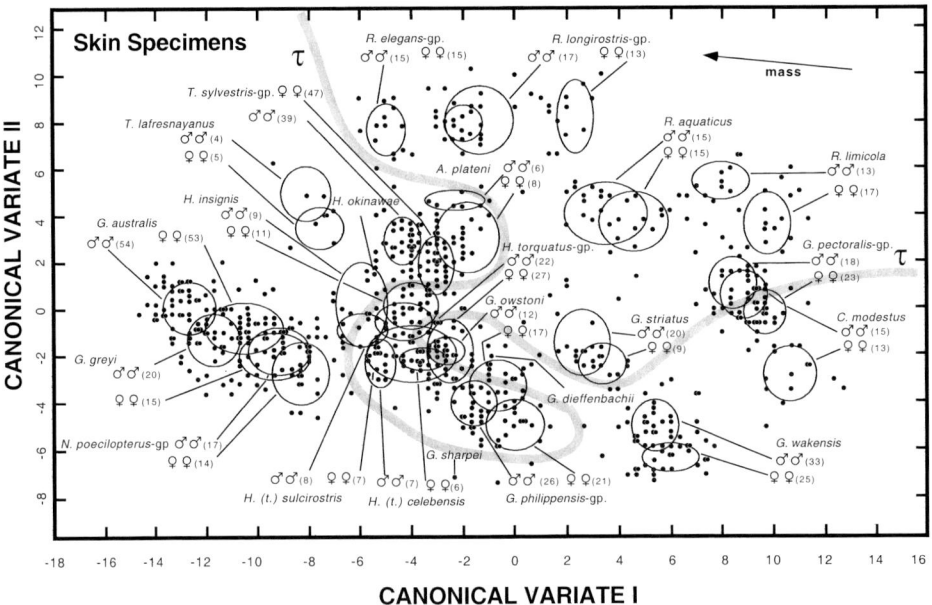

FIG. 27. Plot of 40 species–sex groups of typical rails (tribe Rallini exclusive of *Dryolimnas*) on the first two canonical variates for six external measurements. Except for taxa represented by inadequate samples (*Gallirallus dieffenbachii, G. sharpei,* and *Habropteryx okinawae*), ellipses for groups delimit summary statistics for scores ($\bar{x} \pm$ SD). Approximate orientation of vector for mean body masses of taxa ("mass") and apparent threshold of flightlessness (τ, positioned on the flightless side of boundary) are indicated.

cluded significant intersexual differences (ANOVA of scores; $P < 0.05$), only CV-IV contributed substantial separation of flightless *H. wallacii* from other taxa. In large part, this axis reflected the relatively long tarsi and short toes of *H. wallacii*, with secondary discrimination based on the relatively deep bills of this flightless species.

Despite the limited morphometric dimensionality afforded by the two species of meso-rallids (*Ortygonax* and flightless *Cyanolimnas cerverai*), three CVs were extracted that described significant differences among species–sex groups (ANOVA of scores; $P < 0.05$). CV-I for these meso-rallids accounted for almost 90% of the total variance among groups, principally discriminated flightless *C. cerverai* from *Ortygonax* (Fig. 25), and essentially contrasted wing length with the three other variables significantly entered into the model (Table 16). CV-I was not significantly correlated with body mass ($r = 0.18$; $P > 0.05$). This first axis was similar in composition (but opposite in sign) to a contrast between flightless *C. cerverai* and flighted *Ortygonax* (Table 16), confirming that CV-I summarized the variance associated with flightlessness (and perhaps other intergeneric differences). CV-II for meso-rallids accounted for differences between species of *Ortygonax* and (uniquely for this set of taxa) sexual dimorphism common to all three species (ANOVA of scores; $P < 0.05$), a discrimination largely effected by a contrast between lengths of the tail and middle toe (Table 16; Fig. 25) and significantly correlated with mean body mass ($r = 0.98$; $P < 0.01$). CV-III for meso-rallids represented residual differences between *Ortygonax nigricans* and the other

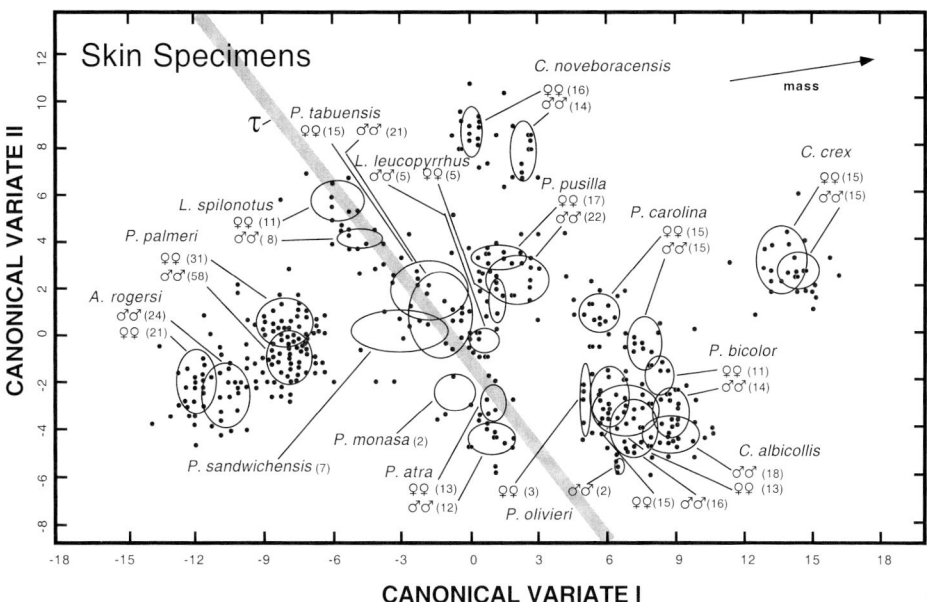

FIG. 28. Plot of 28 species–sex groups of crakes (subtribe Crecina) on the first two canonical variates for six external measurements. Except for taxa represented by inadequate samples (*Porzana monasa* and *P. sandwichensis*), ellipses for groups delimit summary statistics for scores ($\bar{x} \pm$ SD). Approximate orientation of vector for mean body masses of taxa ("mass") and apparent threshold of flightlessness (τ, positioned on the flightless side of boundary) are indicated.

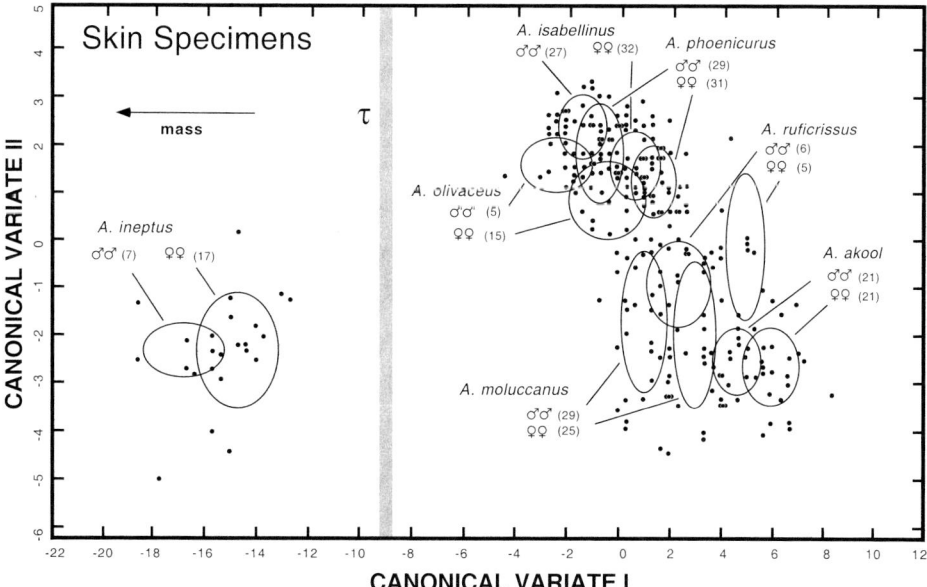

FIG. 29. Plot of 14 species–sex groups of bushhens (*Amaurornis*) on the first two canonical variates for six external measurements. Ellipses for groups delimit summary statistics for scores ($\bar{x} \pm$ SD). Approximate orientation of vector for mean body masses of taxa ("mass") and apparent threshold of flightlessness (τ, positioned on the flightless side of boundary) are indicated.

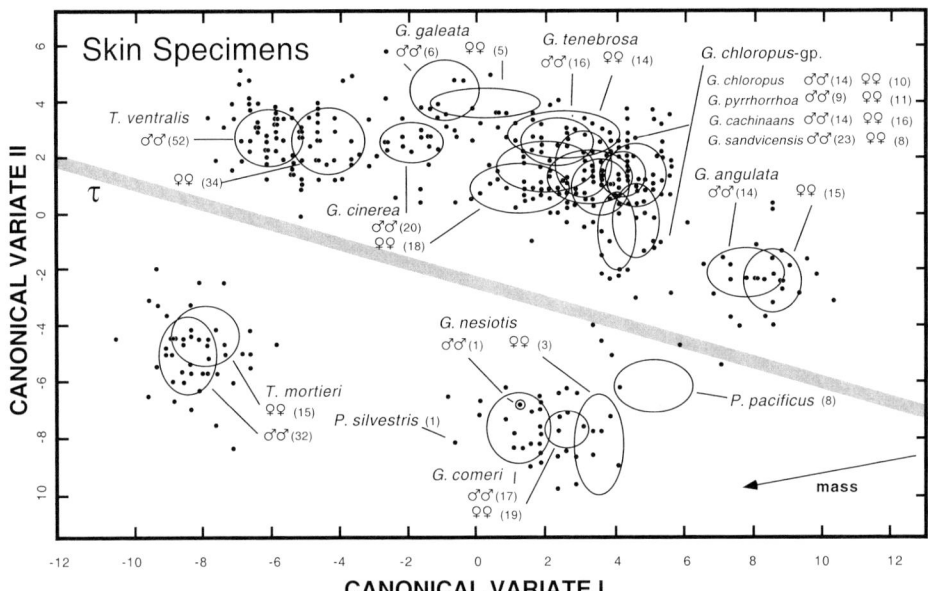

Fig. 30. Plot of 24 species–sex groups of moorhens (*Gallicrex, Tribonyx, Gallinula,* and *Porphyriornis*) on the first two canonical variates for six external measurements. Except for taxa lacking adequate samples (*Pareudiastes pacificus* and *P. silvestris*), ellipses for groups delimit summary statistics for scores ($\bar{x} \pm$ SD). Approximate orientation of vector for mean body masses of taxa ("mass") and apparent threshold of flightlessness (τ, positioned on the flightless side of boundary) are indicated.

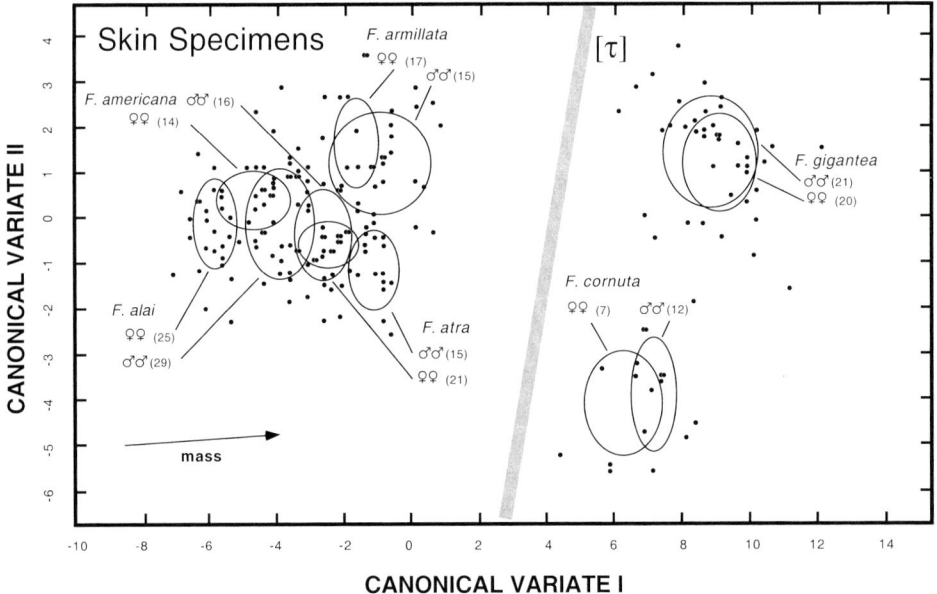

Fig. 31. Plot of 12 species–sex groups of coots (*Fulica*) on the first two canonical variates for six external measurements. Ellipses for groups delimit summary statistics for scores ($\bar{x} \pm$ SD). Approximate orientation of vector for mean body masses of taxa ("mass") and provisional threshold of flightlessness ([τ], positioned on the flightless side of boundary) are indicated.

TABLE 23. Standardized coefficients and summary statistics for flighted–flightless contrast based on external measurements stepwise-selected ($P < 0.05$) from six variables for 96 species groups of Rallidae. Size of problem dictated that sexes be pooled within taxa for analysis.

Variable	Coefficient
Culmen length	−0.642
Bill height	−0.271
Wing length	0.897
Tail length	0.064
Tarsus length	−0.081
Middle-toe length	0.548
Eigenvalue*	1.65
Cumulative variance (%)	—
Canonical R	0.789
Correlation with body mass ($n = 70$)	0.745

* Wilks' λ (d.f. = 6, 1, 3,033) = 0.3781; $\hat{F}_{6, 3028}$ = 829.9 ($P \ll 0.0001$).

two included taxa (ANOVA of scores; $P < 0.005$) as well as supplemental sexual dimorphism (ANOVA of scores; $P < 0.0001$), an axis that crossed delimitations of genus or flight capacity and emphasized relative bill height and tail length (Table 16).

Comparable in its modest dimensionality to the analysis of meso-rallids (Table 17), the CA of external measurements of *Dryolimnas* identified three variates possessing significant differences among means of species–sex groups (ANOVA of scores; $P < 0.05$). CV-I for *Dryolimnas* contrasted culmen length with the four other variables entered significantly. Of the latter, wing length was the primary contributor (Table 17). CV-I was strongly correlated with mean body mass ($r = 0.92$; $P < 0.01$), and placed flightless, dwarfed *Dryolimnas aldabranus* at the low extreme, extinct *D. abbotti* at an intermediate position, and flighted *D. cuvieri* at the high extreme (Fig. 26). The strong association between reduced body size and flightlessness in *Dryolimnas* (Fig. 26) resulted in a decided similarity between CV-I and a contrast between *D. aldabranus* and its flighted congeners, with the latter largely contrasting lengths of the culmen and wing (Table 17). Additional interspecific differences, largely distinguishing *D. abbotti* from its two congeners and uncorrelated with mean body mass ($r = 0.47$; $P > 0.05$), were incorporated into CV-II for *Dryolimnas* (ANOVA of scores; $P < 0.0001$), as were essentially all intersexual differences for members of the genus (ANOVA of scores; $P < 0.0001$). CV-III for this genus extracted residual differences between *D. abbotti* and its congeners (ANOVA of scores; $P < 0.0001$), and stressed magnitudes of bill height and tail length relative to tarsus length (Table 17).

The sheer diversity of included taxa, incomplete resolution of phylogeny, and the multiple instances of flightlessness in the typical rails (subtribe Rallina, exclusive of *Dryolimnas* [Livezey 1998]) rendered simultaneous inference of morphometric patterns in this group challenging at best, despite the exclusion of several flightless allies (e.g., *Diaphorapteryx hawkinsi, Aphanapteryx bonasia, Erythromachus leguati,* and *Capellirallus karamu*) lacking skin specimens. Interspecific differences in mean scores were highly significant on all six CVs (ANOVA of scores; $P < 0.0001$), and intersexual differences were significant on the

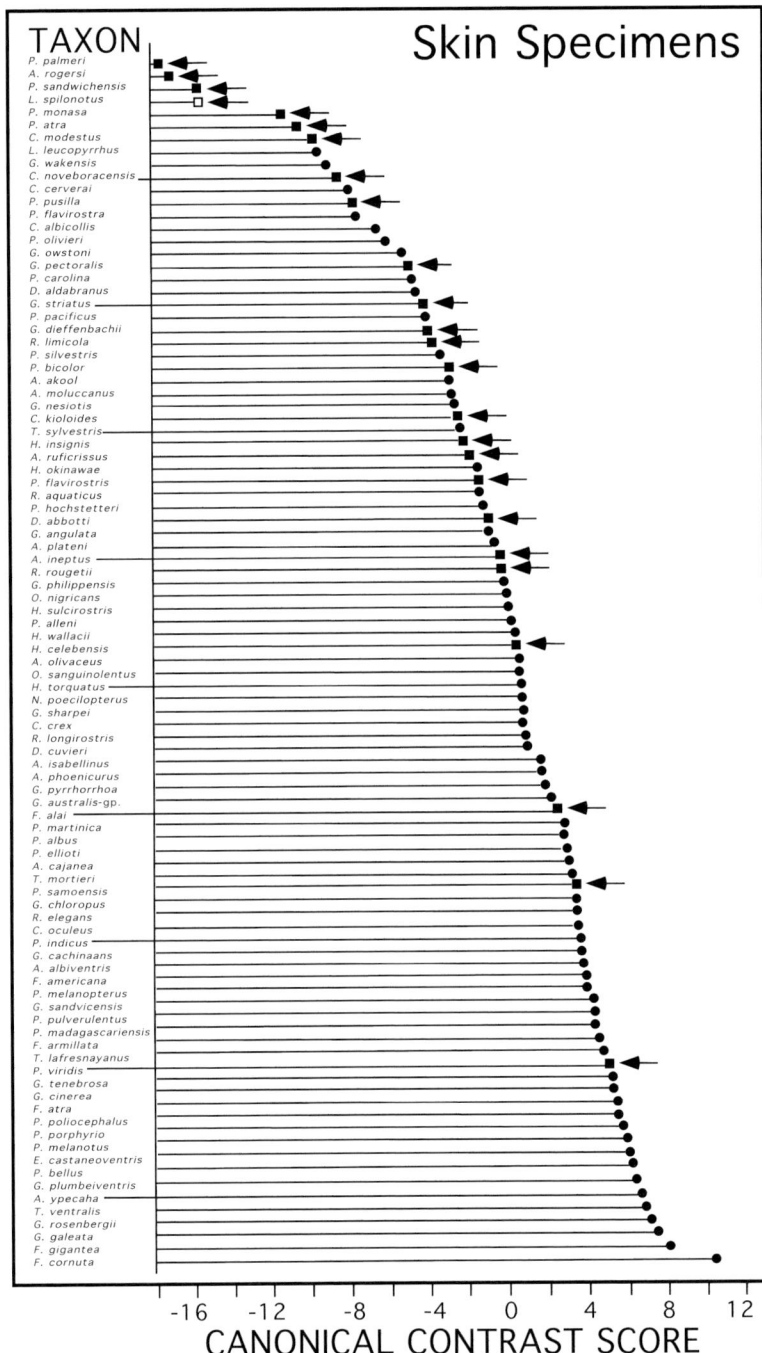

FIG. 32. Dot chart of mean scores for 96 species of Rallidae on a global contrast of flighted and flightless taxa for six external measurements. Flighted species are symbolized by circles, flightless species are symbolized by squares and emphasized by arrows, and uncertain flight status of *Laterallus spilonotus* is indicated by a hollow square.

TABLE 24. Summary statistics for selected main effects for selected flightless species of Rallidae and flighted relatives based on stepwise multivariate analyses of variance ($P < 0.05$) of six external measurements. Approximate F-statistics correspond to Wilks' λ (see text).

Taxonomic group	Interspecific effect		Intersexual effect		Flight effect	
	$\hat{F}_{v,\delta}$	P-value	$\hat{F}_{v,\delta}$	P-value	$\hat{F}_{v,\delta}$	P-value
Porphyrio and *Porphyrula*	$\hat{F}_{6,581} = 1{,}639.9$	<0.0001	$\hat{F}_{2,585} = 86.1$	<0.0001	$\hat{F}_{6,581} = 941.5$	<0.0001
Porphyrio	$\hat{F}_{6,42} = 367.9$	<0.0001	$\hat{F}_{2,456} = 70.0$	<0.0001	$\hat{F}_{4,581} = 1{,}051.6$	<0.0001
Basal Rallidae*	$\hat{F}_{5,253} = 804.2$	<0.0001	$\hat{F}_{2,256} = 60.0$	<0.0001	$\hat{F}_{3,454} = 335.4$	<0.0001
Habroptila and *Gymnocrex*	$\hat{F}_{3,56} = 602.6$	<0.0001	$\hat{F}_{2,57} = 22.8$	<0.0001	—†	
Cyanolimnas and *Ortygonax*	$\hat{F}_{4,39} = 97.6$	<0.0001	$\hat{F}_{3,40} = 10.5$	<0.0001	$\hat{F}_{4,39} = 130.7$	<0.0001
Dryolimnas	$\hat{F}_{3,87} = 29.8$	<0.0001	$\hat{F}_{1,89} = 24.3$	<0.0001	$\hat{F}_{4,86} = 222.4$	<0.0001
Typical Rallidae‡	$\hat{F}_{4,704} = 375.9$	<0.0001	$\hat{F}_{3,705} = 187.2$	<0.0001	$\hat{F}_{3,705} = 1{,}283.9$	<0.0001
Gallirallus and close relatives§	$\hat{F}_{5,591} = 57.6$	<0.0001	$\hat{F}_{3,593} = 137.4$	<0.0001	$\hat{F}_{6,590} = 344.8$	<0.0001
Crakes (Crecina)§	$\hat{F}_{6,412} = 220.8$	<0.0001	$\hat{F}_{5,413} = 39.3$	<0.0001	$\hat{F}_{4,414} = 2{,}179.0$	<0.0001
Crex and *Porzana*	$\hat{F}_{6,316} = 1{,}291.3$	<0.0001	$\hat{F}_{4,318} = 34.7$	<0.0001	$\hat{F}_{5,319} = 1{,}542.0$	<0.0001
Amaurornis	$\hat{F}_{4,253} = 503.2$	<0.0001	$\hat{F}_{3,254} = 68.9$	<0.0001	$\hat{F}_{5,252} = 1{,}071.3$	<0.0001
Moorhens sensu lato‖	$\hat{F}_{6,330} = 526.2$	<0.0001	$\hat{F}_{2,334} = 44.7$	<0.0001	$\hat{F}_{4,332} = 1{,}252.7$	<0.0001
Gallinula	$\hat{F}_{3,204} = 282.3$	<0.0001	$\hat{F}_{2,205} = 27.8$	<0.0001	$\hat{F}_{3,204} = 282.4$	<0.0001
Tribonyx	$\hat{F}_{5,125} = 792.9$	<0.0001	$\hat{F}_{3,127} = 53.0$	<0.0001	—¶	
Fulica	$\hat{F}_{3,198} = 354.7$	<0.0001	$\hat{F}_{2,199} = 39.3$	<0.0001	$\hat{F}_{4,197} = 1{,}048.4$¶#	<0.0001

* Paraphyletic series of genera (*Gymnocrex*, *Habroptila*, *Aramides*, *Eulabeornis*, *Canirallus*, and *Rougetius*).
† Equivalent to intergeneric effect in this context.
‡ *Rallus*, *Gallirallus*, and close allies, following Livezey (1998).
§ Following Livezey (1998).
‖ *Gallicrex*, *Pareudiastes*, *Gallinula* (including *Porphyriornis*), and *Tribonyx*.
¶ Equivalent to interspecific effect in this context.
# *Fulica gigantea* and *F. cornuta* treated as flightless for this comparison.

TABLE 25. Summary statistics for sex-related interaction effects for selected flightless species of Rallidae and flighted relatives based on stepwise multivariate analyses of variance ($P < 0.05$) of six external measurements. Approximate F-statistics correspond to Wilks' λ (see text).

Taxonomic group	Species–sex interaction		Flight–sex interaction	
	$\hat{F}_{v,b}$	P-value	$\hat{F}_{v,b}$	P-value
Porphyrio and *Porphyrula*	$\hat{F}_{1,586} = 4.4$	<0.05	$\hat{F}_{2,585} = 4.8$	<0.01
Porphyrio	$\hat{F}_{1,457} = 7.2$	<0.01	$\hat{F}_{2,456} = 4.9$	<0.01
Basal Rallidae*	$\hat{F}_{2,256} = 5.3$	<0.01	$\hat{F}_{3,255} = 11.2$	<0.0001
Habroptila and *Gymnocrex*	$\hat{F}_{3,56} = 13.1$	<0.0001	$\hat{F}_{3,56} = 11.0$	<0.0001
Cyanolimnas and *Ortygonax*	$\hat{F}_{1,42} = 2.1$	>0.15	$\hat{F}_{1,42} = 0.8$	>0.35
Dryolimnas	$\hat{F}_{1,89} = 5.6$	<0.05	$\hat{F}_{1,89} = 4.1$	<0.05
Typical Rallidae†	$\hat{F}_{3,705} = 4.8$	<0.005	$\hat{F}_{2,706} = 7.4$	<0.001
Gallirallus and close relatives	$\hat{F}_{3,593} = 5.8$	<0.001	$\hat{F}_{1,595} = 4.1$	<0.05
Crakes (Crecina)	$\hat{F}_{2,416} = 6.8$	<0.005	$\hat{F}_{1,417} = 3.1$	<0.10
Crex and *Porzana*	$\hat{F}_{2,320} = 8.3$	<0.005	$\hat{F}_{1,321} = 7.0$	<0.01
Amaurornis	$\hat{F}_{1,256} = 2.5$	>0.10	$\hat{F}_{1,256} = 2.3$	>0.10
Moorhens sensu lato‡	$\hat{F}_{1,335} = 2.7$	>0.10	$\hat{F}_{2,370} = 8.3$	<0.0005
Gallinula	$\hat{F}_{1,206} = 6.1$	<0.05	$\hat{F}_{2,205} = 5.7$	<0.01
Tribonyx§	$\hat{F}_{3,127} = 12.8$	<0.0001	$\hat{F}_{3,127} = 12.8$	<0.0001
Fulica	$\hat{F}_{1,200} = 9.7$	<0.005	$\hat{F}_{2,199} = 8.7$‖	<0.0001

* Paraphyletic series (*Gymnocrex, Habroptila, Aramides, Eulabeornis, Canirallus,* and *Rougetius*).
† *Rallus, Gallirallus,* and close allies, following Livezey (1998).
‡ *Gallicrex, Pareudiastes, Gallinula* (including *Porphyriornis*), and *Tribonyx*.
§ Tests identical for this taxon.
‖ *Fulica gigantea* and *F. cornuta* treated as flightless for this comparison.

first four CVs ($P < 0.0001$). CV-I for typical rails conformed with the general pattern for other groups in being uniquely highly correlated with mean body masses ($r = -0.96$; $P \ll 0.01$), on which massive *Gallirallus australis*-group had the lowest scores and comparatively petite *Cabalus modestus* had the highest scores (Fig. 27). Coefficients of variables for CV-I indicated that all but culmen length varied directly with this index to size, reflecting a variation in bill length among included taxa that is unrelated to other external dimensions (Table 18).

The range of scores shown by flightless taxa on CV-I foreshadowed the difficulties of delineating species by flight capacity in this group, a complexity exacerbated by the range of values assumed by flightless species on CV-II and subsequent axes. CV-II for typical rails was uncorrelated with body mass ($r = -0.03$; $P \gg 0.05$); essentially contrasted lengths of the culmen and middle toe with bill height and tarsus length (Table 18); and amplified the separation of groups by flight capacity, sex, and (to a lesser extent) discriminated *Rallus* from *Gallirallus* in general (Fig. 27). A bivariate plot of taxa on the first two variates tended to position flightless taxa in the lower left part of the plane and flighted taxa in the upper right, with a sinuous, diagonally oriented partition between the two groups. This nonlinear threshold between flighted and flightless taxa indicated an ambiguity of flight status for the flighted *Gallirallus philippensis*-group and *Habropteryx torquatus*-group by intrusions into the flightless domain (Fig. 27).

The third CV further separated flightless taxa in general and *Aramidopsis plateni* and *Habropteryx insignis* in particular (not shown), principally displaying a contrast between tail length and tarsus length (Table 18); species with high scores were typified by relatively short to obsolescent tails. CV-IV largely contrasted bill dimensions with lengths of the wing and tarsus (Table 18), and especially singled

TABLE 26. Multivariate sexual dimorphism in selected species of Rallidae, documented by Mahalanobis' distances (D) and statistics from associated multivariate analyses of variance, based on m variables stepwise-selected ($P < 0.05$) from six external measurements in single-taxon analyses. Flightless taxa are marked by asterisks. Pillai's trace is numerically equal to the average squared canonical correlation in the two-group (unidimensional) case. Wilks' λ is numerically equal to the Hotelling–Lawley trace and Roy's maximum root (Morrison 1976) in the two-group (unidimensional) case. Approximate F-statistics correspond to Wilks' λ (U-statistic), with m (variables included in model), g (groups) $- 1$, and n (total specimens) $- g$ degrees of freedom. Accuracy of allocation is the percentage of specimens assigned to correct sex in jackknifed classifications.†

Taxon	r Males	r Females	m	D	Pillai's trace	Eigenvalue	R	Wilks' λ	$\hat{F}$	Accuracy of classification
Porphyrio (p.) bellus	8	5	2	3.3	0.75	2.96	0.864	0.25	14.8****	85
Porphyrio (p.) indicus	22	15	4	2.6	0.63	1.72	0.795	0.37	13.8*******	87
Porphyrio (p.) samoensis	50	57	3	1.4	0.33	0.49	0.574	0.67	16.9*******	74
Porphyrio (p.) pulverulentus	4	10	3	4.6	0.83	4.96	0.912	0.17	16.5*****	100
Porphyrio (p.) porphyrio	20	15	1	1.3	0.29	0.41	0.540	0.71	13.6*****	77
Porphyrio (p.) poliocephalus	28	16	2	1.2	0.26	0.35	0.511	0.74	7.3****	66
Porphyrio (p.) madagascariensis	11	17	2	1.3	0.30	0.42	0.545	0.70	5.3*	79
Porphyrio (p.) viridis	18	12	2	3.4	0.75	3.01	0.866	0.25	40.7*******	93
Porphyrio (p.) ellioti	24	24	2	2.1	0.54	1.17	0.734	0.46	26.2*******	81
Porphyrio (p.) melanotus	50	33	3	3.1	0.70	2.29	0.835	0.30	60.4*******	93
Porphyrio (p.) melanopterus	13	15	2	1.9	0.50	1.00	0.708	0.50	12.6*******	75
*Porphyrio hochstetteri**	7	7	2	4.4	0.85	5.67	0.922	0.15	31.2*******	100
Porphyrula alleni	23	29	2	1.9	0.47	0.90	0.689	0.53	22.2*******	83
Porphyrula martinica	14	16	3	1.5	0.37	0.58	0.605	0.63	5.0**	70
Porphyrula flavirostris	26	27	2	1.2	0.28	0.39	0.531	0.72	9.8*****	72
Gymnocrex rosenbergii	6	15	1	2.2	0.51	1.05	0.715	0.49	19.9******	91
Gymnocrex plumbeiventris	16	11	2	1.4	0.42	0.73	0.648	0.58	8.7****	74
*Habroptila wallacii**	5	11	2	5.4	0.88	7.08	0.936	0.12	46.0*******	100
Eulabeornis castaneoventris	18	15	3	2.4	0.61	1.55	0.779	0.39	15.0*******	88
Aramides (c.) cajanea	11	7	2	3.1	0.72	2.50	0.845	0.29	18.8*******	89
Aramides (c.) albiventris	3	8	1	1.6	0.38	0.61	0.616	0.62	5.5*	82
Aramides ypecaha	16	14	2	2.1	0.55	1.22	0.741	0.45	16.5*******	90
Canirallus oculeus	26	18	1	1.0	0.19	0.24	0.440	0.81	10.1****	73
Canirallus kioloides	22	18	2	1.7	0.63	1.72	0.795	0.37	31.8*******	85
Rougetius rougetii	19	18	1	1.6	0.39	0.64	0.623	0.61	22.2*******	76
[*Cyanolimnas cerverai**]	2	7	2	5.7	0.88	7.12	0.937	1.12	21.5*****	100
Ortygonax sanguinolentus	10	10	3	4.2	0.83	4.88	0.911	0.17	26.0*******	95

TABLE 26. Continued.

Taxon	n Males	n Females	m	D	Pillai's trace	Eigenvalue	R	Wilks' λ	F	Accuracy of classification
Ortygonax nigricans	13	6	2	2.6	0.62	1.66	0.790	0.38	13.3*****	84
Dryolimnas (c.) cuvieri	27	32	2	3.0	0.70	2.32	0.836	0.30	65.0********	90
*Dryolimnas (c.) aldabranus**	15	13	1	1.0	0.21	0.26	0.454	0.79	6.8*	68
[*Dryolimnas (c.) abbotti*]	6	2	2	8.8	0.95	19.19	0.975	0.05	48.0*****	100
Rallus aquaticus	15	15	3	2.8	0.68	2.12	0.824	0.32	18.4******	87
Rallus limicola	13	17	2	4.5	0.84	5.36	0.918	0.16	72.4*******	97
Rallus longirostris-group	17	13	1	2.4	0.75	2.97	0.865	0.25	83.2********	97
Rallus elegans	15	15	2	4.4	0.84	5.13	0.915	0.16	69.3********	97
Gallirallus pectoralis	18	23	2	2.8	0.67	2.07	0.821	0.33	39.4*******	90
Gallirallus striatus	20	9	1	1.1	0.23	0.30	0.481	0.77	8.1**	76
*Gallirallus (a.) australis**	55	53	2	1.6	0.40	0.67	0.633	0.60	35.1*******	78
*Gallirallus (a.) greyi**	20	15	2	3.0	0.70	2.33	0.836	0.30	37.2******	94
Gallirallus philippensis-group	26	21	2	2.7	0.65	1.87	0.807	0.35	41.1*******	94
*Gallirallus owstoni**	12	17	4	5.3	0.88	7.34	0.938	0.12	44.0*******	100
*Gallirallus wakensis**	33	25	2	2.9	0.68	2.16	0.827	0.32	59.5********	91
Tricholimnas lafresnayanus	4	5	1	2.3	0.64	1.74	0.797	0.36	12.2**	78
*Tricholimnas sylvestris**	39	47	4	3.3	0.74	2.77	0.857	0.27	56.1*******	92
Nesoclopeus poecilopterus-group*	18	13	2	2.6	0.64	1.79	0.800	0.36	25.0*******	84
*Aramidopsis plateni**	6	8	2	3.6	0.79	3.66	0.886	0.22	20.1******	93
*Cabalus modestus**	14	14	2	4.2	0.82	4.67	0.908	0.18	58.4********	96
*Habropteryx insignis**	9	11	3	4.2	0.83	4.91	0.911	0.17	26.2*******	100
Habropteryx (t.) torquatus	22	27	2	3.0	0.70	2.29	0.834	0.30	52.7********	92
Habropteryx (t.) celebensis	16	13	2	3.7	0.78	3.63	0.885	0.22	4.7*******	97
Habropteryx (t.) sulcirostris	8	7	1	1.7	0.46	0.84	0.675	0.54	10.9**	87
*Atlantisia rogersi**	24	21	3	2.6	0.64	1.78	0.800	0.36	24.3*******	93
[*Laterallus leucopyrrhus*]	5	5	2	3.1	0.75	3.02	0.867	0.25	10.6******	90
Laterallus spilonotus	8	11	2	3.9	0.81	4.20	0.899	0.19	33.6******	100
Coturnicops noveboracensis	14	16	1	4.9	0.86	6.36	0.930	0.14	178.0*******	100
Crex crex	15	15	1	0.8	0.15	0.17	0.381	0.86	4.7*	70
Crex albicollis	18	13	2	3.1	0.71	2.44	0.842	0.29	34.1*******	94
*Porzana palmeri**	58	31	2	1.7	0.40	0.67	0.632	0.60	28.6*******	79
Porzana carolina	15	15	2	2.8	0.68	2.16	0.827	0.32	29.1*******	87
Porzana pusilla	22	17	2	1.6	0.39	0.63	0.622	0.61	11.4******	62
Porzana tabuensis	21	15	1	0.8	0.14	0.17	0.377	0.86	5.6*	67
*Porzana atra**	12	13	2	2.4	0.60	1.52	0.777	0.40	16.8******	88
Porzana flavirostra	16	15	1	1.2	0.29	0.41	0.539	0.71	11.9***	65

TABLE 26. Continued.

Taxon	n		m	D	Pillai's trace	Eigenvalue	R	Wilks' λ	$\hat{F}$	Accuracy of classification
	Males	Females								
[*Porzana olivieri*]	2	3	2	7.9	0.96	24.89	0.981	0.04	24.9*	100
Porzana bicolor	14	11	1	2.2	0.57	1.33	0.755	0.43	30.6******	88
Amaurornis (o.) olivaceus	5	15	1	2.0	0.47	0.87	0.682	0.53	15.7****	85
Amaurornis (o.) moluccanus	29	25	3	3.2	0.73	2.63	0.851	0.28	43.9******	94
[*Amaurornis (o.) ruficrissus*]	6	5	1	4.7	0.87	6.75	0.933	0.13	60.7******	100
Amaurornis phoenicurus	29	31	3	2.9	0.68	2.12	0.824	0.32	39.5******	92
Amaurornis akool	21	21	1	1.3	0.30	0.43	0.547	0.70	17.1*****	71
Amaurornis isabellinus	27	32	2	2.8	0.67	1.99	0.816	0.34	55.6******	90
Amaurornis ineptus*	7	17	3	2.5	0.59	1.42	0.766	0.41	9.5******	79
Gallicrex cinerea	20	18	2	4.7	0.86	5.91	0.925	0.15	103.4******	97
Tribonyx ventralis	52	34	4	3.4	0.74	2.87	0.861	0.26	58.2******	98
Tribonyx mortierii*	32	15	1	1.7	0.39	0.64	0.624	0.61	28.7******	83
Gallinula nesiotis-group*	18	22	2	3.1	0.71	2.44	0.842	0.29	45.2******	93
Gallinula tenebrosa	16	14	1	0.5	0.07	0.07	0.261	0.93	2.1	57
Gallinula angulata	14	15	1	1.3	0.32	0.46	0.562	0.68	12.5****	69
Gallinula (c.) chloropus	14	19	2	2.1	0.53	1.14	0.730	0.47	17.2******	88
Gallinula (c.) pyrrhorrhoa	9	11	1	0.8	0.14	0.17	0.380	0.86	3.0	60
Gallinula (c.) cachinnans	14	16	2	2.2	0.56	1.27	0.749	0.44	17.2******	80
Gallinula (c.) galeata	6	5	1	1.6	0.44	0.77	0.660	0.56	7.0*	73
Gallinula (c.) sandvicensis	23	8	2	2.8	0.62	1.63	0.789	0.38	22.9******	97
Fulica armillata	15	17	1	0.5	0.05	0.05	0.227	0.95	1.6	66
Fulica cornuta	12	17	1	1.1	0.24	0.31	0.486	0.76	5.3*	68
Fulica gigantea	21	20	2	1.5	0.38	0.60	0.614	0.62	11.5******	78
Fulica (a.) americana	16	14	3	3.0	0.70	2.36	0.838	0.30	20.5******	90
Fulica (a.) alai	29	25	3	3.0	0.70	2.30	0.835	0.30	38.3******	94
Fulica atra, atra-group	15	21	3	2.7	0.65	1.84	0.805	0.35	19.6******	92

† Asterisks reflect levels of significance: none, $P > 0.05$; *, $P \leq 0.05$ **, $P \leq 0.01$; ***, $P \leq 0.005$; ****, $P \leq 0.001$; *****, $P \leq 0.0005$; ******, $P \leq 0.0001$.

out two flightless species (*Tricholimnas sylvestris* and *Cabalus modestus*) having long bills relative to their limbs.

No combination of these four axes unequivocably circumscribed the flightless species of typical rails in a bivariate space, indicating that despite conformance with the generality that flightless rails have relatively short wings, standard external measurements fail to concisely discriminate flightless members of the typical rails (comprising taxa disparate in size and relative lengths of bills, tarsi, and tails) from their flighted relatives. This conclusion was supported by explicit unidimensional contrasts of species in these genera by flight capacity. The contrast for all genera of typical rails emphasized culmen length relative to two other dimensions (with wing length not being entered significantly) and accounting for little more of the among-group variance than that incorporated by CV-IV. A contrast excluding *Rallus* proved to be an unremarkable index to relative wing length accounting for a comparable proportion of the dispersion among group means (Table 18). Neither canonical contrast of flighted and flightless typical rails with external measurements resolved an ordination in which the two groups were dichotomized. Instead, these attempts educed the inadequacy of such data for diagnosis of flight capacity in these taxa (especially for dwarfed flightless species and flighted *Habropteryx*). The contrast including all genera resulted in the following sequence of taxa, from lowest to highest mean scores (with males having lower means than females in all taxa for which sex was specified): *Gallirallus australis, G. greyi, Nesoclopeus poecilopterus*-group, *Habropteryx insignis, H. okinawae, H. sulcirostris, H. celebensis, G. owstoni, Tricholimnas sylvestris, T. lafresnayanus, Aramidopsis plateni, H. torquatus, G. sharpei, G. dieffenbachii, Rallus elegans, G. philippensis*-group, *G. striatus, G. wakensis, R. longirostris, R. aquaticus, Cabalus modestus, G. pectoralis,* and *R. limicola*.

Second only to the typical rails in taxonomic diversity of flightless forms, the external measurements of crakes (including *Atlantisia, Laterallus, Coturnicops, Crex,* and *Porzana*) spanned six CVs showing signficant differences among groups. Two extinct, flightless crakes (*Porzana sandwichensis* and *P. monasa*) were included in the analyses, but because the small samples of specimens available for these taxa were of unknown sex (Table 7), these groups were represented as single taxonomic groups. All six CVs included highly significant interspecific differences in mean scores (ANOVA of scores; $P < 0.0001$), and the first three of these also included significant intersexual differences ($P < 0.005$). CV-I for external measurements of crakes accounted for almost three fourths of the total dispersion among groups. Scores on CV-I were highly correlated with mean body masses ($r = 0.85$; $P < 0.01$), but coefficients of variables indicated that the size reflected by CV-I also was influenced by relative wing length (Table 19). Positions of taxa on CV-I affirmed this interpretation: small, brevipennate *Atlantisia rogersi* and *Porzana palmeri* had the lowest scores; intermediate-sized species (including *P. sandwichensis, P. monasa, P. atra,* and *P. tabuensis*) assumed a range of somewhat larger scores; and large, strongly flighted *Crex crex* had the highest scores (Fig. 28).

The second CV essentially accounted for residual differences among groups in relative wing length (Table 19) that was uncorrelated with mean body masses ($r = -0.30$; $P > 0.05$). Three flightless species (*Atlantisia rogersi, Porzana atra,* and *P. monasa*) had the lowest scores on this axis, followed closely by two other flightless crakes (*P. palmeri* and *P. sandwichensis*). The plane spanned by the first two variates displayed flighted and flightless crakes neatly partitioned by a diagonal line (Fig. 28). The position of *Laterallus spilonotus* as approaching the threshold of flightlessness for crakes was noteworthy, in that observations indicate that this species is weakly flighted if not flightless.

The third CV for crakes largely contrasted bill height with tail length (Table 19). In addition to amplifying differences among flighted taxa (not shown), this axis demarcated flightless species having relatively thick bills and short tails (*Porzana sandwichensis* and *P. palmeri*) from others having intermediate (*P. atra*) or opposite proportions (*Atlantisia rogersi* and *P. monasa*). Interspecific differences (ANOVA of scores; $P < 0.0001$), unconfounded by intersexual differences ($P > 0.25$), also were detected on CV-IV, CV-V, and CV-VI for crakes (not shown). Mean scores indicated that none of these last three CVs amended the morphological correlates of flightlessness in crakes: CV-IV primarily separated *Crex crex* and *Atlantisia rogersi* from *Porzana flavirostra* and *P. pusilla*, with other species being intermediate on this axis; CV-V distinguished *Laterallus spilonotus* and *P. flavirostra* from other species; and CV-VI principally separated *L. leucopyrrhus* from other crakes.

In canonical contrasts resolving the essential differences between species–sex groups of crakes solely on the basis of flight capacity (one contrast included all genera and one was limited to *Crex* and *Porzana*), wing length was contrasted with variable combinations of other, significantly entered dimensions (Table 19). The flight contrast for all genera of crakes, unlike those for typical rails in which manifold changes in size were associated with loss flight, displayed an intuitive ordination of species. The sequence of species on this variate (Table 19), in order of increasing mean score and in which females had lower average scores than male conspecifics, was *Atlantisia rogersi, Porzana palmeri, Laterallus spilonotus, P. sandwichensis, P. monasa, P. tabuensis, P. atra, L. leucopyrrhus, P. pusilla, Coturnicops noveboracensis, P. olivieri, P. flavirostra, Crex albicollis, P. carolina*, and *C. crex*. The contrast limited to *Crex* and *Porzana* preserved the same sequence for included members of these two genera. The logical rankings recovered for the crakes (excepting the transposition of *P. tabuensis* and *P. atra*), in stark contrast to those among typical rails, evidently stems from the absence of both marked dwarfism and giantism among flightless species, a situation that compromised a linear ordination of taxa with respect to capacity for flight.

The first CV for external measurements of species–sex groups of the bushhens (*Amaurornis*) integrated much of the variation of interest in the present context, accounting for 82% of the differences among groups (Table 20), including highly signficiant differences both among species and between sexes in mean scores (ANOVA of scores; $P < 0.0001$), being strongly (negatively) correlated with mean body masses ($r = -0.99$; $P \ll 0.01$), and reflecting the vast majority of the divergence of flightless *A. ineptus* from its flighted congeners (Fig. 29). However, coefficients of variables on CV-I unexpectedly emphasized relative tail length as the primary discriminator of flight capacity among groups, with middle-toe length contributing negligibly to this axis (Table 20). Predictably from the strong correlation between CV-I and body mass, males on average were plotted as comparatively more flightless than female conspecifics, a proportionality that would be expected to have practical implications for *A. ineptus* in that no other species of *Amaurornis* approached the apparent threshold of flightlessness suggested by external dimensions (Fig. 29). The virtually complete summary of variation among *Amaurornis* related to flight capacity by CV-I was confirmed by the close similarity of coefficients (irrespective of a reversal of sign) for this contrast for the genus (Table 20).

The second CV for external measurements of *Amaurornis* conveyed an additional 10% of the total variation among groups (Table 20). CV-II essentially represented relative wing length not subsumed by CV-I, residual variation that was not significantly correlated with mean body mass ($r = -0.15$; $P > 0.05$) but included significant interspecific differences in scores (ANOVA of scores; $P < 0.0001$), and which separated flightless *A. ineptus* and three flighted species (*A. moluccanus*, *A. ruficrissus*, and *A. akool*) from three other flighted congeners (*A. olivaceus*, *A. isabellinus*, and *A. phoenicurus*), with the former group having relatively shorter wings (Fig. 29). CV-III contrasted lengths of the culmen, tail, and middle toe with bill height, wing length, and tarsus length (Table 20); included significant interspecific dispersion of means (ANOVA of scores; $P < 0.0001$); and primarily amplified the discrimination of *A. phoenicurus* and *A. akool* from other *Amaurornis*. Although CV-IV and CV-VI for *Amaurornis* provided modest but significant discrimination of species (ANOVA of scores; $P < 0.01$), and CV-IV and CV-V contributed additional discrimination of sexes within species (ANOVA of scores; $P < 0.01$), none of these last three variates highlighted morphometric corollaries of flightlessness in *Amaurornis*.

Another group having multiple flightless members is the moorhens (*Gallicrex*, *Gallinula*, *Pareudiastes*, and *Tribonyx*). All six CVs for external measurements of these taxa revealed highly significant interspecific differences (ANOVA of scores; $P < 0.0001$). The first, third, fourth, and fifth variates commingled these with significant intersexual differences (ANOVA of scores; $P < 0.0001$). CV-I for external measurements of these taxa accounted for roughly one half of the total variation among groups and essentially contrasted middle-toe length with all other dimensions (Table 21); mean scores on this axis were highly (negatively) correlated with mean body mass ($r = -0.78$; $P < 0.01$). The gamut of body sizes manifested by flightless moorhens rendered CV-I relatively uninformative concerning the morphometric correlates of flightlessness, with only the comparatively small size and score of flighted *Gallinula angulata* imposing a suggestion of flight-related discrimination based on these data (Fig. 30).

The second CV for external measurements of moorhens extracted much of the interspecific differences associated with flightlessness (Fig. 30). Coefficients for this axis revealed that it contrasted wing length and (to a lesser extent) culmen length with other dimensions exclusive of the generally contrarian middle-toe length (Table 21); mean scores on CV-II were uncorrelated with mean body mass ($r = -0.04$; $P \gg 0.05$). CV-III for these taxa revealed additional differences among species (not shown), in which the comparatively diminutive *Gallinula angulata* had uniquely low scores, most flightless species (*Gallinula nesiotis*-group and *Pareudiastes*) and *Tribonyx ventralis* had intermediate scores, and other species (including the large, flightless *Tribonyx mortierii*) had high scores.

The remaining three CVs each provided additional discrimination of the two species of the flightless Pacific genus *Pareudiastes* from all other moorhens (adding 2%, 1%, and 0.5% of total dispersion among groups), regardless of flight capacity of the latter (not shown). These differences did not consider the possible contributions of sexual dimorphism in either species of *Pareudiastes,* because neither species in the genus permitted partitioning by sex (Table 7). These last three axes particularized slight differences among groups in size-standardized proportionalities: CV-IV contrasted the lengths of tarsus and tail with the four other variables; CV-V contrasted lengths of the culmen and tarsus with other variables; and CV-VI contrasted lengths of the culmen and tail with the remaining external dimensions. Together these last three variates accentuated the unusually short pelvic appendages and tails of *Pareudiastes* relative to other dimensions, subtle residual proportionalities not accounted for by the first three CVs.

A canonical contrast of flighted and flightless moorhens of all genera successfully divided the two groups on a single multivariate axis that contrasted culmen length, bill height, and tarsus length with lengths of the wing, tail, and middle toe (Table 21). This contrast ranked the species, in order of increasing mean scores (with males having lower mean scores than female conspecifics, where determinable), as follows: *Pareudiastes silvestris, P. pacificus, Tribonyx mortierii, Gallinula comeri, G. nesiotis, T. ventralis, Gallicrex cinerea, Gallinula sandvicensis, G. pyrrhorrhoa, G. tenebrosa, G. chloropus, G. cachinaans, G. galeata,* and *G. angulata.* The sequence of taxa on this ordination, inter alia, indicates that among the flighted species, *Tribonyx ventralis* most closely approached the proportions characteristic of flightlessness in moorhens; among the members of the *Gallinula chloropus* superspecies, insular *G. sandvicensis* most closely approached flightless moorhens in form; and across all taxa, proportions of males tended to approach those of flightless taxa more than female conspecifics. A contrast of *Tribonyx ventralis* and flightless *T. mortierii* differed in composition from that for moorhens generally (Table 21), and was a comparatively simple contrast between wing length and other variables (exclusive of culmen length, not entered significantly in this restricted analysis).

Although no extant coot (*Fulica*) is truly flightless, the pronounced negative pectoral allometry (Fig. 31) shown by the Andean coots *F. gigantea* and *F. cornuta,* and the similar tendencies inferred for several subfossil congeners (*F. newtoni, F. chathamensis,* and *F. prisca*) generally considered to have been flightless or weakly flighted, justify comparable morphometric assessments of this group. The 12 species–sex groups of *Fulica* manifested ample diversity of form (Table 22), minimally comprising three dimensions for depiction (Fig. 31). The first three CVs for external variables of coots included highly significant interspecific differences in scores (ANOVA of scores; $P \ll 0.0001$), and marginal differences among species were detected ($P < 0.05$) on CV-IV as well (Table 22). Sexual differences within species also were highly significant on CV-I (ANOVA of scores; $P < 0.0001$), with supplemental sexual dimorphism of substantial magnitude ($P < 0.0001$) but varying among species (species–sex interaction effects; $P < 0.005$) revealed on CV-III.

The first CV for *Fulica* accounted for more than 90% of the total dispersion among groups (Table 22), was strongly correlated with mean body mass ($r = 0.98$; $P \ll 0.01$), and discriminated groups in canonical space through a contrast of bill height with other external dimensions exclusive of culmen length. The latter variable was not entered significantly in the model for this genus because variation in this feature was redundant with interspecific differences in one or more of the variables retained. CV-I primarily discriminated *F. gigantea* and *F. cornuta* from other, more-typical congeners (Fig. 31). CV-II for coots amplified differences between *F. armillata* and *F. gigantea* and the four other species (Fig. 31). This axis was not significantly correlated with body mass ($r = -0.23$; $P > 0.05$) and contrasted lengths of the tarsus and middle toe with the other three variables entered (Table 22). CV-III, which contrasted bill height and tarsus length with other dimensions (Table 22), augmented the discrimination between sexes to varying degrees among species and supplemented the separation of *F. alai* from its congeners (not shown). CV-IV contributed marginal ancillary separation of two of the smaller species (*F. armillata* and *F. atra*) from other included taxa. A canonical contrast between the two large Andean species *Fulica* and their smaller congeners was only marginally enlightening, in that this unidimensional abstraction significantly incorporated only four of the six dimensions analyzed, and accomplished this strongly size-related discrimination through a comparatively unintuitive contrast between bill height and three other variables (Table 22).

Given the general power of CA and the demonstrated efficacy of such analysis to distinguish flightless and flighted relatives in most groups of rallids (Tables 14–22), the feasibility of a canonical contrast between all flighted and flightless rails based on the six external measurements was attempted for all sampled taxa from the entire family. The resultant analysis, in which sexes were pooled within species to accommodate limitations of problem size, identified a single numerical abstraction that maximally discriminated the two groups (relative to pooled within-group variation) across the 96 species of Rallidae sampled (Table 23). The canonical axis essentially contrasted culmen length and bill height with lengths of the wing and middle toe, in that the standardized coefficients for two of the variables were of sufficiently small magnitude to be discounted for purposes of interpretation (Table 23). Mean scores on the canonical contrast for all Rallidae were moderately correlated with mean body mass ($r = 0.74$; $P \ll 0.01$), that is, 55% of the variation among taxa (R^2) on this axis was explainable simply by body mass. In light of the divergent trends in size associated with flightlessness in several of the subgroups analyzed (e.g., *Dryolimnas*, typical rails, and crakes), as well as the probably substantial violation of the statistical assumption of homogeneity of within-group covariance matrices across such diverse taxa, it was not unexpected that the the discrimination of flightless species throughout the family when using the paltry suite of external measurements would be but an heuristic exercise (Fig. 32).

The importance of body size in separating rallids generally and the extraordinary pectoral reduction in several of the smallest flightless species (e.g., *Porzana palmeri, P. sandwichensis, Atlantisia rogersi,* and *Cabalus modestus*) resulted in the correct grouping of small flightless taxa among other predominantly flightless taxa; that is, these taxa had low scores on the canonical contrast (Fig. 32). However, although most larger flightless species were shifted toward lower scores than would have been predicted on the basis of size alone (e.g., *Pareudiastes silvestris, Tricholimnas sylvestris, Habropteryx insignis,* and *Porphyrio hochstetteri*), ambiguous placements of a number of small, flighted rallids (e.g., *Porzana tabuensis, Laterallus leucophyrrus, Coturnicops noveboracensis, Gallirallus pectoralis,* and *Amaurornis akool*) and large flightless species (e.g., *Habroptila wallacii, Nesoclopeus poecilopterus*-group, *Gallirallus australis*-group, and *Tricholimnas la-*

fresnayanus) were evident (Fig. 32). The generally massive stature and deceptively long (but subtly modified) wings and tails of the last group of taxa apparently concealed the flightlessness more readily diagnosable in more narrow taxonomic contexts or on the basis of other, comparatively trenchant, anatomical traits. The extremely low score of *Laterallus spilonotus,* a species of uncertain flight status, as well as the relatively high scores of two species of impeachable flight status (*Gallirallus sharpei* and *Porphyrio albus*) are noteworthy, shortcomings of the global discrimination notwithstanding (Fig. 32).

Interpretative complexities of CVs aside, the stepwise procedure that underlies stepwise discriminant analysis permits the specification of selected MANOVAs, and the subsets of variables significantly entered in these models can be used to derive Mahalanobis distances between pairs of groups. Given the statistical power of this methodology, documentation of significant differences between naturally delimited groups (species or sexes) is readily accomplished in most applications.

Comparisons of species and sexes in the subgroups of the Rallidae discussed above revealed similarly strong structure, in which multivariate differences among species and between sexes were highly significant ($P < 0.0001$) in all cases (Table 24). Although not strictly comparable, exceptionally high F-statistics tended to pertain to interspecific comparisons involving more than one genus. Exceptions to this general pattern included the relatively small values associated with two groups comprising comparatively similar genera (meso-rallids and crakes), and the extraordinarily high F-value deriving from the contrast of the two species of *Tribonyx* (Table 24). The high interspecific F-statistic for *Tribonyx* is consistent with other multivariate comparisons of external measurements within genera or those involving several closely related genera where one or more included species is flightless (Table 24). Throughout the family, especially where comparisons were limited to congeners, the multivariate F-statistic associated with flight capacity (or the loss thereof) was generally substantially greater than those deriving from overall differences between species irrespective of flight capacity, and substantially greater than those associated with intersexual differences in the same taxonomic groups (Table 24). That is, the effect of flightlessness on multivariate differences in external measurements, regardless of the details of the changes in proportions involved, was larger in magnitude than interspecific or intersexual differences, complications of intergeneric differences aside.

Multivariate interaction effects in external measurements within taxa quantified differences in magnitudes of effects attributable to one grouping factor (e.g., sex) among different values of another grouping factor (e.g., species). In most taxonomic groups of rallids, species–sex interaction effects were significant ($P < 0.05$) in MANOVAs of external measurements (Table 25). Even those taxa in which such effects were not substantiated (*Cyanolimnas* and *Ortygonax, Amaurornis,* and moorhens), these differences approached significance ($0.10 < P < 0.15$), with marginalities probably reflecting inadequate samples for critical taxa (Tables 7, 25). This generality of interspecific heterogeneity of sexual dimorphism was consistent with the spectrum of magnitudes of sexual dimorphism in external dimensions estimated for the Rallidae sampled, although the pattern of relative magnitudes was complex (Table 26). A similar heterogeneity in magnitude of sexual dimorphism pertained to flight capacity. Multivariate interaction effects between flight status and sex were significant ($P < 0.05$) in external measurements (Table

25) for all but two of the same poorly sampled taxonomic subsets identified above (*Cyanolimnas* and *Ortygonax*, and *Amaurornis*), with crakes closely approaching significance ($P < 0.10$). These findings suggest that the fabric of interspecific patterns in magnitudes of sexual dimorphism were related, at least in part, to the confounding interaction between sexual dimorphism and flightlessness (Table 26).

Summaries of summary statistics pertaining to multivariate sexual dimorphism in external measurements (Table 26) revealed flightless species characterized by significantly greater sexual dimorphism than flighted relatives, summarized by Mahalanobis distances (D) between species: *Porphyrio hochstetteri* with D roughly twice that for flighted congeners, *Habroptila wallacii* with D at least twice that for *Gymnocrex* spp., *Gallirallus owstoni* with D roughly twice that of flighted congeners, *Cabalus modestus* with D roughly 1.5 times that of flighted allies, and *Gallinula nesiotis*-group with D roughly 1.5 times that of flighted congeners. Notable in this context was the exceptional sexual dimorphism of a flighted moorhen (Table 26), *Gallicrex cinerea*, one of a small minority of species of the Rallidae showing sexual dichromatism (Livezey 1998). One additional flightless species showed suggestively larger dimorphism, *Cyanolimnas cerverai* with D approximately 1.5 times that of *Ortygonax* spp., but sample sizes undermined significance (Tables 25, 26). One flightless species, one inferred to have undergone a reduction in overall size, *Dryolimnas aldabranus* with D no more than one third that of flighted congeners, showed multivariate dimorphism in external dimensions of a smaller magnitude than that estimated for flighted relatives. However, most other flightless rallids, including *Gallirallus australis*-group, *G. wakensis*, *Tricholimnas* spp., *Aramidopsis plateni*, *Habropteryx insignis*, *Atlantisia rogersi*, *Porzana palmeri*, *P. atra*, *Amaurornis ineptus,* and *Tribonyx mortierii*, exhibited sexual dimorphism in external measurements of magnitudes comparable to those shown by flighted relatives (Table 26).

QUALITATIVE CHARACTERS OF PLUMAGE

General characteristics of plumage.—As exemplified by the storied *Cabalus modestus,* some flightless rallids exhibit a generalized aspect of juvenility in definitive body plumage. However, this appearance can involve at least two distinct classes of characters: retention of plumage patterns similar to those found in juveniles of close relatives; and degeneration in gross and microscopic structure of the plumage, in which contour feathers (regardless of color pattern) assume a weakened or poorly developed aspect. The latter was found in a number of flightless rallids to variable degrees, being significant in flightless members of most genera (Table 27). Notably, several extinct rallids of questionable flight status (*Porphyrio albus, Gallirallus sharpei,* and *Gallinula comeri*) manifested negligible changes in general plumage structure, a condition consistent either with a retention of (probably weak) flight or comparatively recent loss of flight in these taxa (Table 27), and *Dryolimnas abbotti* showed no apomorphies of the plumage, consistent with its exclusion from consideration as flightless. The most substantial structural changes in body plumage were found in several profoundly derived dwarf lineages (e.g., *Cabalus modestus, Atlantisia rogersi,* and *Porzana sandwichensis*) and a number of the larger flightless taxa (e.g., *Porphyrio hochstetteri, Habroptila wallacii, Tricholimnas* spp., and *Amaurornis ineptus*). Unfortunately, the condition of the plumage of several extremely derived, subfossil rallids (e.g.,

TABLE 27. Qualitative changes in structure and shapes of primary and secondary remiges of flightless species of Rallidae. Based on subjective comparisons with flighted, close relatives. Conditions either were treated as binary or were partitioned into several classes (unchanged vs. minor, moderate or major shortening). Minor—reduced, but distinctly longer than tectrices primariae dorsales majores; moderate—reduced, approximating tectrices primariae dorsales in length; and major—reduced, distinctly shorter than tectrices primariae dorsales. Based on growth patterns—in which remiges primarii precede remiges secundarii, and tectrices primariae dorsales majores precede remiges in ontogeny—these conditions represent progressively paedomorphic states. Remiges alulae typically parallel the tectrices majores in relative lengths. Brackets indicate that inferred changes are marginal.

Taxon	Relative lengths of primary remiges*	Relative lengths of secondary remiges*	Rounding of wing tip defined by remiges	Integrity of vexillae of primary remiges	Asymmetry of vexillae of primary remiges	Rounding of apices of primary remiges	Rigidity of rachises of primary remiges	Craniocaudal curvature of primary remiges
Porphyrio albus	Unchanged	Unchanged	Unchanged	Unchanged	Unchanged	Unchanged	Unchanged	Unchanged
Porphyrio hochstetteri	Moderate	Moderate	Increased	Decreased	Decreased	Increased	Decreased	Decreased
Habroptila wallacii	Minor	Major	Increased	Decreased	Decreased	Increased	Decreased	Decreased
Cyanolimnas cerverai	Minor	Minor	Increased	Decreased	Decreased	Increased	Decreased	Decreased
Dryolimnas aldabranus	Unchanged	Unchanged	Unchanged	Unchanged	Unchanged	Unchanged	[Decreased]	Unchanged
Gallirallus sharpei	Unchanged	Unchanged	Unchanged	Unchanged	Unchanged	Unchanged	Unchanged	Unchanged
Gallirallus australis	Moderate	Moderate	Increased	Decreased	Decreased	Increased	Decreased	Decreased
Gallirallus greyi	Moderate	Moderate	Increased	Decreased	Decreased	Increased	Decreased	Decreased
Gallirallus dieffenbachii	Unchanged	Unchanged	Increased	Decreased	Decreased	Increased	Decreased	Unchanged
Gallirallus owstoni	Minor	Major	Increased	Decreased	Decreased	Increased	Decreased	Decreased
Gallirallus wakensis	Minor	Major	Increased	Decreased	Decreased	Increased	Decreased	Decreased
Tricholimnas lafresnayanus	Moderate	Moderate	Increased	Decreased	Decreased	Increased	Decreased	Unchanged
Tricholimnas sylvestris	Moderate	Major	Increased	Decreased	Decreased	Increased	Decreased	Decreased
Nesoclopeus poecilopterus	Minor	Moderate	Increased	Decreased	Decreased	Increased	Decreased	Decreased
Nesoclopeus woodfordi	Minor	Major	Increased	Decreased	Decreased	Increased	Decreased	Unchanged
Aramidopsis plateni	Moderate	Major	Increased	Decreased	Decreased	Increased	Decreased	Decreased
Cabalus modestus	Minor	Major	Increased	Decreased	Decreased	Increased	Decreased	Decreased
Habropteryx insignis	Minor	Moderate	Increased	Decreased	Decreased	Increased	Decreased	Decreased
Habropteryx okinawae	Unchanged	Unchanged	Increased	Decreased	Decreased	[Increased]	Unchanged	Decreased
Atlantisia rogersi	Major	Major	Increased	Decreased	Decreased	Increased	Decreased	Decreased
Porzana sandwichensis	Moderate	Major	Increased	Decreased	Decreased	Increased	Decreased	Decreased
Porzana palmeri	Moderate	Moderate	Increased	Decreased	Decreased	Increased	Decreased	Decreased
Porzana atra	Minor	Moderate	Increased	Decreased	Decreased	Increased	Decreased	Decreased
Amaurornis ineptus	Moderate	Moderate	Increased	Decreased	Decreased	Increased	Decreased	Unchanged
Pareudiastes pacificus	Minor	Minor	Increased	Decreased	Decreased	Increased	Decreased	Unchanged
Pareudiastes silvestris	Minor	Minor	Increased	Decreased	Decreased	Increased	Decreased	Unchanged
Tribonyx mortierii	Minor	Minor	Increased	Decreased	Decreased	Increased	Decreased	Unchanged
Gallinula nesiotis	Minor	Minor	Increased	Unchanged	Unchanged	Increased	Unchanged	Unchanged
Gallinula comeri	Minor	Minor	Increased	Unchanged	Unchanged	Increased	Unchanged	Unchanged

Diaphorapteryx hawkinsi, Aphanapteryx bonasia, and *Capellirallus karamu*) must remain the subject of conjecture because of the unavailability of even single feathers for these taxa.

An objective assessment of the paedomorphic retention of juvenile plumage patterns is more problematic. Comparatively obvious examples of paedomorphic plumages among flightless rallids include those of *Gallirallus dieffenbachii, G. australis*-group, *Tricholimnas sylvestris, Cabalus modestus,* and possibly *Atlantisia rogersi.* In a minority of cases (e.g., *G. dieffenbachii* and *C. modestus* as paedomorphs of *G. philippensis*), retention of juvenile plumages substantiated earlier conjectures (Olson 1973a), although in many cases such definitive plumages were as easily described as nondescript or cryptic. As noted previously, at least one flighted rallid (*Porphyrula flavirostris*) shows marked paedomorphism of plumage pattern (cf. Ripley 1977; Taylor 1998). Equivocations of assessment notwithstanding, clear cases of paedomorphic plumage patterns in flightless rallids remain few relative to other changes of the plumage associated with flightlessness, and can be considered among the rarer signatures of paedomorphosis in flightless rails (Tables 27–29), and the occurrence in at least one flighted species renders this only ambiguously indicative of flightlessness in the rails. The early extirpation of a number of the more-derived flightless taxa (e.g., *Capellirallus karamu, Diaphorapteryx hawkinsi, Aphanapteryx bonasia, Erythromachus leguati,* and Hawaiian *Porzana*) precludes a knowledge of their plumages. However, the relative osteological apomorphy shown by these taxa (below) suggests that at least some of these taxa would show comparatively extreme changes in the integument as well.

Numbers of remiges and rectrices.—Numbers of rectrices and (especially) remiges vary little within taxonomic families of birds (Gadow 1888; Verheyen 1958; Stresemann 1963), and only rarely are reduced in flightless or flight-impaired carinates (Livezey 1990, 1993b). Likewise, the vast majority of rallids conserve the plesiomorphic numbers of primary remiges (10) and show substantial conservatism with respect to changes in the numbers of rectrices (Livezey 1998), including a distinguishable diastema (Pycraft 1899; Steiner 1956). Therefore, the general pterylography of most rallids (including a number of flightless species) can be approximated by using any flighted rallid (Fig. 22) and would conform with that described for *Fulica atra* by Jeikowski (1971). However, such uniformity, would not extend precisely to numbers of secondary remiges and associated coverts (Table 28). A reduction in the number of primary remiges in *Porzana palmeri* was noted by Greenway (1973) and Van Tyne and Berger (1976), and additional cases of losses of primary remiges were documented here among other flightless rallids (Table 28). Losses of primary remiges in rallids were restricted to the distalmost members, as shown by vestigial remiges at the end of the wing in *P. palmeri* and *Atlantisia rogersi,* and characterized only a minority of flightless rallids (Figs. 33–37), a pattern consistent with relative rates of development (Gadow 1888). Apparent augmentation of the primary remiges in *Porphyrio hochstetteri,* suggested by initial appraisals of dissected wings during this study, ultimately was attributed to a relative shift of the distalmost secondary remex immediately distal to the wrist, caused by the shortening of the underlying skeletal elements and the absence of losses among the distalmost primary remiges.

Shape and structure of remiges and rectrices.—A series of gross structural changes in the flight feathers (remiges and rectrices) was observed to varying

TABLE 28. Qualitative changes in general structure of body plumage and modal numbers of remiges (excluding remicle) of flightless species of Rallidae. Conclusions are based on comparisons with flighted, close relatives. "Degeneration" of plumage refers to loss of rigidity and integrity of margins or contour feathers, resulting in increased looseness and semiplumaceous aspect of body plumage. Degeneration was classified as negligible, minor, or moderate. Brackets indicate weakly substantiated inferences, and dashes indicate that requisite specimens were not available (data for *Porzana monasa* and all subfossil taxa were unavailable).

Taxon	Structural degeneration of body plumage	Modal number of primary remiges (change)	Modal number of secondary remiges (change)
Porphyrio albus	Negligible	10 (0)	—
Porphyrio hochstetteri	Moderate	10 (0)†	~15 (0)
Habroptila wallacii	Moderate	10 (0)	—
Cyanolimnas cerverai	Negligible	10 (0)	~13 (0)
Dryolimnas aldabranus	Negligible	10 (0)	—
Gallirallus sharpei	Negligible	10 (0)	—
Gallirallus australis	Minor	10 (0)	13 (0)
Gallirallus greyi	Minor	10 (0)	—
Gallirallus dieffenbachii	Minor	10 (0)	—
Gallirallus owstoni	Minor	9 (−1)	13 (0)
Gallirallus wakensis	Minor	9 (−1)	~11 (−1 or −2)
Tricholimnas lafresnayanus	Moderate	10 (0)	—
Tricholimnas sylvestris	Moderate	10 (0)	13 (0)
Tricholimnas conditicius	Moderate	10 (0)	—
Nesoclopeus poecilopterus	Minor	10 (0)	~13 (0)
Nesoclopeus woodfordi	Minor	10 (0)	—
Nesoclopeus immaculatus	Minor	10 (0)	—
Aramidopsis plateni	Moderate	10 (0)	—
Cabalus modestus	Moderate	9 (−1)	—
Habropteryx insignis	Minor	10 (0)	—
Habropteryx okinawae	Negligible	10 (0)	—
Atlantisia rogersi	Moderate	9 (−1)	11 (−1)†
Porzana sandwichensis	Moderate	9 (−1)	—
Porzana palmeri	Minor	7–8 (−2 or −3)	~11 (−1)
Porzana atra	Minor	9 (−1)	~12 (0)
Amaurornis ineptus	Moderate	10 (0)	—
Pareudiastes pacificus	Negligible	10 (0)	13 (0)
Pareudiastes silvestris	Minor	10 (0)	—
Tribonyx mortierii	Minor	10 (0)	~13 (0)
Gallinula nesiotis	Negligible	10 (0)	—
Gallinula comeri	Negligible	10 (0)	~12 (0)

* Count reflects interpretation of apparent augmentation of remiges primarii by one as resulting from truncation of manus relative to overlying integument (see text).
† Based on comparisons with *Crex crex, Coturnicops noveboracensis, Laterallus leucopyrrhus*, flighted *Porzana*, and small, flighted *Rallus*.

degrees among flightless rallids (Tables 27–29; Fig. 38), a diverse pattern of apomorphy that corresponds only to a limited degree with other indicators of flightlessness (e.g., relative wing length, and osteological changes). Unlike the generality applicable to most birds, in which lengths and diameters of primary remiges are isometric with body mass whereas flexural stiffness of primary remiges is negatively allometric with body mass (Worcester 1996), flightless rails show considerable variation in such parameters that bear little obvious relationship with body mass (Tables 27–29; Fig. 38).

Generally, flightless rallids more frequently showed changes in ridgity, profile,

TABLE 29. Qualitative changes in form and number of rectrices of flightless species of Rallidae. Conditions either were treated as binary or were partitioned into several categories based on subjective comparisons with flighted, close relatives: unchanged versus minor, moderate, or major. Magnitude of reduction: minor—reduced, but not markedly; moderate—reduced markedly, but tail not obsolete; major—greatly reduced, tail obsolete. Brackets indicate marginal changes, and dashes indicate that specimens with developed, undamaged rectrices were not available (data for *Porzana monasa* and all subfossil taxa were unavailable).

Taxon	Lengths of rectrices	Rigidity of rectrices	Integrity of vexillae of rectrices	Modal number of rectrices
Porphyrio albus	—	Reduced	Reduced	
Porphyrio hochstetteri	Minor shortening	Reduced	Reduced	Unchanged
Habroptila wallacii	Minor shortening	Reduced	Reduced	Reduced
Cyanolimnas cerverai	Minor shortening	Unchanged	Reduced	Unchanged
Dryolimnas aldabranus	Unchanged	Unchanged	Unchanged	Unchanged
Gallirallus sharpei	Unchanged	Unchanged	Unchanged	Unchanged
Gallirallus australis	Unchanged	Reduced	Reduced	[Unchanged]
Gallirallus greyi	Unchanged	Reduced	Reduced	[Unchanged]
Gallirallus dieffenbachii	Unchanged	Reduced	Reduced	Unchanged
Gallirallus owstoni	Minor shortening	Unchanged	Reduced	[Reduced]
Gallirallus wakensis	Minor shortening	Reduced	Reduced	Unchanged
Tricholimnas lafresnayanus	Unchanged	Reduced	Reduced	Reduced
Tricholimnas sylvestris	Minor shortening	Reduced	Reduced	Unchanged
Tricholimnas conditicius	Minor shortening	Reduced	Reduced	[Reduced]
Nesoclopeus poecilopterus	Unchanged	Reduced	Reduced	[Reduced]
Nesoclopeus woodfordi	Unchanged	Reduced	Reduced	[Reduced]
Nesoclopeus immaculatus	Unchanged	Reduced	Reduced	[Reduced]
Aramidopsis plateni	Major shortening	Reduced	Reduced	[Reduced]
Cabalus modestus	Major shortening	Reduced	Reduced	[Reduced]
Habropteryx insignis	Moderate shortening	Reduced	Reduced	[Unchanged]
Habropteryx okinawae	Unchanged	Unchanged	Unchanged	—
Atlantisia rogersi	Moderate shortening	Reduced	Reduced	Unchanged
Porzana sandwichensis	Major shortening	Reduced	Reduced	Reduced
Porzana palmeri	Moderate shortening	Reduced	Reduced	Unchanged
Porzana atra	Minor shortening	Unchanged	Reduced	Unchanged
Amaurornis ineptus	Major shortening	Reduced	Reduced	Reduced
Pareudiastes pacificus	Major shortening	Reduced	Reduced	Reduced
Pareudiastes silvestris	Major shortening	Reduced	Reduced	Reduced
Tribonyx mortierii	Minor shortening	Reduced	Reduced	Unchanged
Gallinula nesiotis	Unchanged	Unchanged	Unchanged	Unchanged
Gallinula comeri	Unchanged	Unchanged	Unchanged	Unchanged

and curvature of the remiges and rectrices than in relative lengths (Tables 28, 29). As with the structural changes manifested by the body plumage, several purportedly flightless species showed negligible structural changes in the flight feathers (e.g., *Porphyrio albus* and *Gallirallus sharpei*), and several other species known to be flightless also showed minimal changes (e.g., *Dryolimnas aldabranus* and *Gallinula nesiotis*-group). The most extreme structural changes in flight feathers were characteristic of *Porphyrio hochstetteri, Gallirallus australis*-group, *Tricholimnas sylvestris, Aramidopsis plateni, Atlantisia rogersi,* and *Porzana sandwichensis.*

Relative shortening of the remiges, extreme cases of which show tectrices (coverts) that almost completely conceal the underlying remiges, was made more conspicuous in some species having disproportionately elongate remiges alulae (e.g., the latter were 70 mm long in dissected *Nesoclopeus poecilopterus*). In those

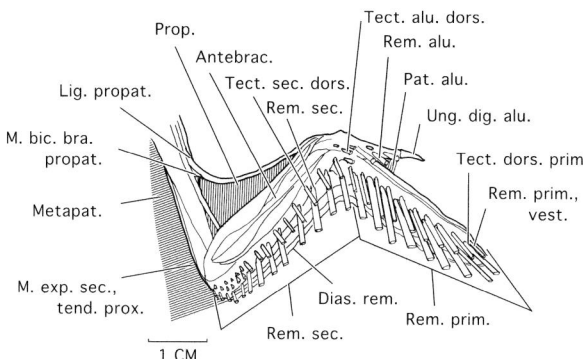

FIG. 33. Diagrammatic illustration of the remiges and tectrices majores of *Gallirallus wakensis* (USNM 289313), dorsal view.

taxa having remiges truncated relative to the overlying coverts, care is advised in measuring wing lengths, in that uncritical handling of the specimen may result in the resultant wing lengths being a variable reflection of lengths of remiges or coverts. The distinctive increase in relative lengths of the remiges alulae and tectrices majores of the remiges primarii et secundarii was accompanied by a disproportionately large unguis alularis (alular or wing spurs) in most flightless species, with the latter being most notable in the least massive species (Table 27). In combination with the disproportionate reduction in distal skeletal elements in the pectoral limb (see below), these integumentary signatures of flightlessness produce the peculiar aspect of a generally diminutive, distally truncated wing bearing loosely structured feathers, in which coverts approach remiges in lengths and a disproportionately robust unguis alularis is retained.

Microstructure of feathers.—The superficiality of the plumage of a bird insured that modifications thereof figured in the descriptions of flightless rails from the

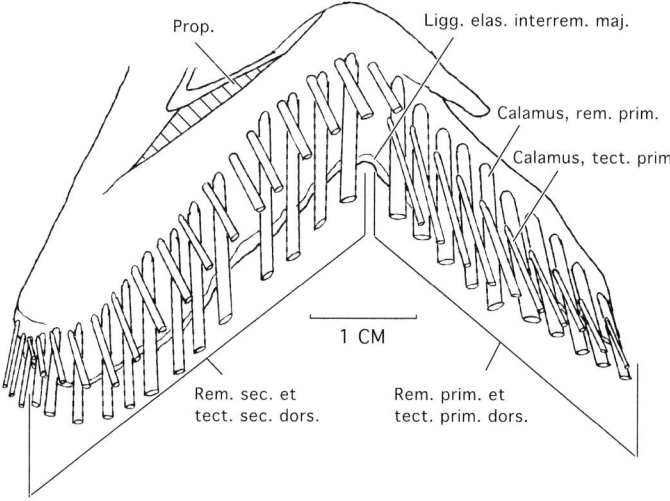

FIG. 34. Diagrammatic illustration of the remiges and tectrices majores of *Nesoclopeus poecilopterus* (BMNH 1940·12·8·87), dorsal view.

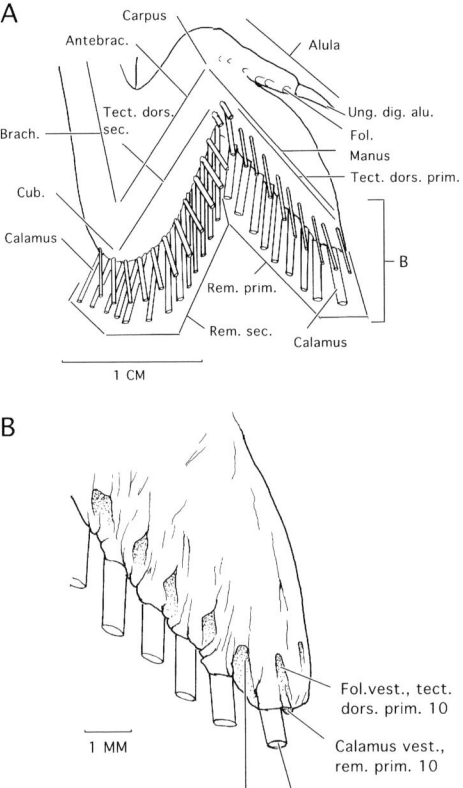

Fig. 35. Diagrammatic illustration of the remiges and tectrices majores of *Atlantisia rogersi* (USNM 320106), dorsal views: **A,** remiges et tectrices primariorum et secundariorum; and **B,** detail depicting aborted, outermost (10th) remex primari.

early ornithological literature to the present day. The loose, hairlike structure of the body feathers of flightless species, as well as reduced rigidity of the remiges, were noted frequently (e.g., Lowe 1928a; Ripley 1977; P. B. Taylor 1998), whereas McGowan (1989:543) performed a comparative survey (including flighted and flightless rails) and reported that few structural differences were found between flighted and flightless carinates. Moreover, it is known that the functional properties of remiges extend beyond aerodynamics to storage of elastic energy in calami (Pennycuick and Lock 1976).

Possible structural corollaries between structure of feathers and terrestrial and aquatic habit (Chandler 1916; Rijke 1967, 1968, 1970a, b) were not of obvious importance, evidently because degree of such specialization among rallids varied little. Nevertheless, variation in the microstructure of remiges existed, variation that produced some of the gross differences detectable in superficial examination. Images of ventral and dorsal views of barbae from the vexillum internum of pars pennacea (at approximate midpoint of the rachis) of the distal remiges secundarii of selected taxa were made with a scanning electron microscope (SEM). Taxa imaged by SEM were *Gallirallus philippensis, G. australis, Cabalus modestus, Nesoclopeus poecilopterus, Tricholimnas sylivestris, Habropteryx insignis, Later-*

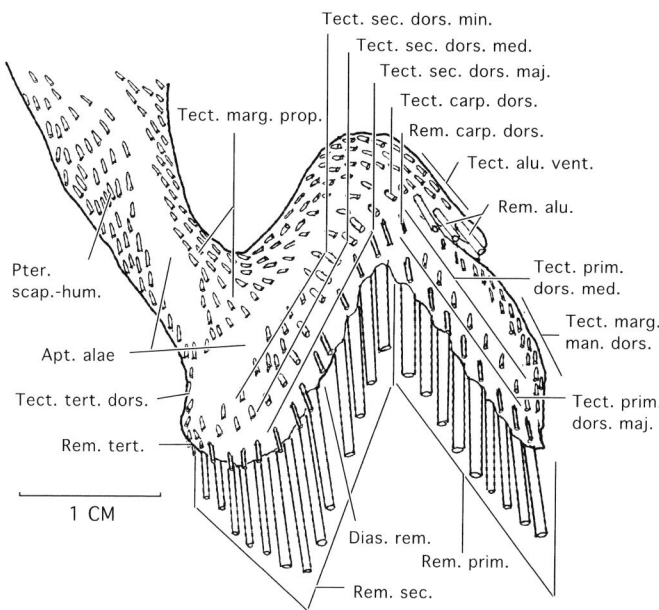

FIG. 36. Diagrammatic illustration of the remiges and tectrices majores of *Porzana atra* (AMNH 9901), dorsal view.

allus jamaicensis, Atlantisia rogersi, Porzana palmeri, Amaurornis olivaceus, A. isabellinus, A. ineptus, Tribonyx ventralis, and *T. mortierii.*

At lower magnifications (Fig. 39), landmarks include crita ventralis (including tegmen and villi) and crista dorsalis of barba, with comparatively simple barbulae proximales (with rugae proximales having dorsal, gutterlike arcus dorsalis) and the comparatively foliate barbula distalis (comprising basis [including dens ventralis] and pennula [comprising cilia dorsales, cilia ventrales, and ventrally directed hamuli]). Most flightless rallids tended to have comparatively short bases

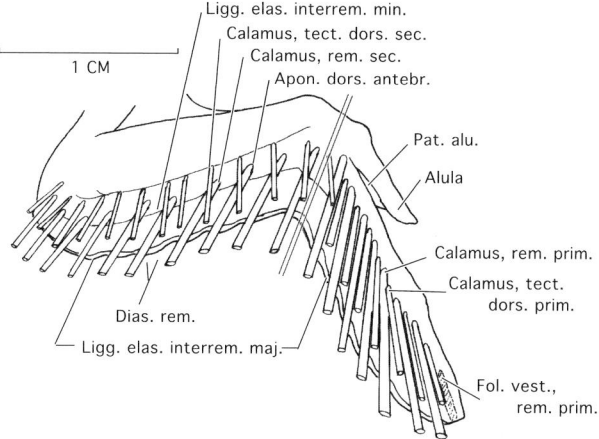

FIG. 37. Diagrammatic illustration of the remiges and tectrices of *Porzana palmeri* (BMNH 1940·12·8·13), including aborted, outermost (10th) remex primari, dorsal view.

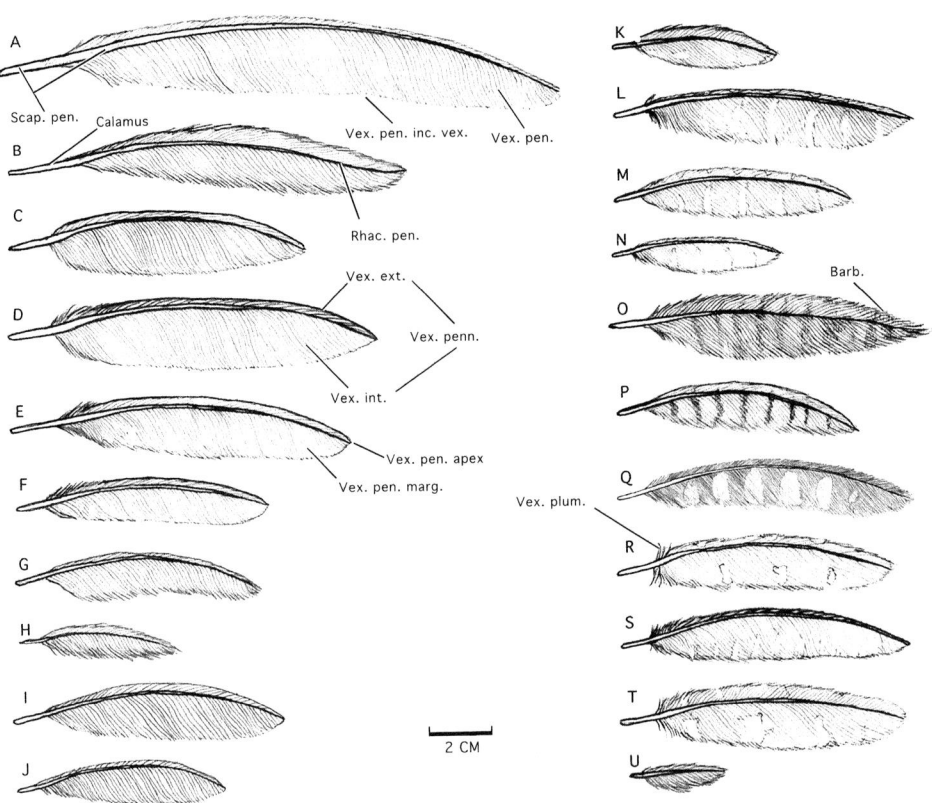

FIG. 38. Illustrations (natural size) of outermost primary remiges (dorsal surfaces) of selected members of the Rallidae (right side, unless otherwise indicated): **A,** *Porphyrio melanotus* (SY-452); **B,** *Porphyrio hochstetteri* (NZNM-1); **C,** *Gymnocrex plumbeiventris* (AMNH 545554, left); **D,** *Eulabeornis castaneoventris* (AMNH 545692, left); **E,** *Aramides cajanea* (CM 4457); **F,** *Rougetius rougetii* (AMNH 545411); **G,** *Ortygonax sanguinolentus* (USNM 227903); **H,** *Cyanolimnas cerverai* (USNM 343159); **I,** *Dryolimnas cuvieri* (BMNH 1935·10·19·1); **J,** *Dryolimnas aldabranus* (BMNH 1979·10·1); **K,** *Rallicula rubra* (AMNH 338586, left); **L,** *Gallirallus philippensis goodsoni* (USNM 542344); **M,** *Gallirallus owstoni* (USNM 503792); **N,** *Gallirallus wakensis* (USNM 289313); **O,** *Gallirallus australis* (USNM 511701, left); **P,** *Tricholimnas sylvestris* (USNM 18407); **Q,** *Nesoclopeus (p.) poecilopterus* (BMNH 1940·12·8·87); **R,** *Nesoclopeus (p.) woodfordi* (AMNH 545539, left); **S,** *Habropteryx torquatus* (USNM 318003); **T,** *Habropteryx insignis* (AMNH 333796, left); **U,** *Atlantisia rogersi* (USNM 320106); **V,** *Laterallus leucopyrrhus* (CM 5322); **W,** *Coturnicops noveboracensis* (CM 2333); **X,** *Crex crex* (USNM 539754); **Y,** *Micropygia schomburgkii* (AMNH 323254, left); **Z,** *Porzana carolina* (CM 6718); **AA,** *Porzana tabuensis* (AMNH 10224); **BB,** *Porzana atra* (AMNH 9901, left); **CC,** *Porzana palmeri* (BMNH 1940·12·8·13); **DD,** *Neocrex erythrops* (AMNH 525544, left); **EE,** *Amaurornis akool* (AMNH 142922); **FF,** *Amaurornis isabellinus* (AMNH 546502); **GG,** *Amaurornis ineptus* (AMNH 265851); **HH,** *Pareudiastes pacificus* (BMNH 1874·2·20·38); **II,** *Gallinula (c.) cachinaans* (USNM 122197); **JJ,** *Gallinula comeri* (BMNH 1922·12·6·221); **KK,** *Tribonyx ventralis* (AMNH 7081); **LL,** *Tribonyx mortierii* (AMNH 10079); **MM,** *Fulica americana* (CM 6765). Abbreviations for anatomical structures are given in Appendix 3.

of the barbulae distales and few, only incompletely formed hamuli. These conditions result in a larger portion of the pars pennacea of respective remiges (and other pennae contornae) being pars pennacea sine barbulis (i.e., a broader, loosely organized fringe), sections that reduce the zona impendens (regions of adjacent

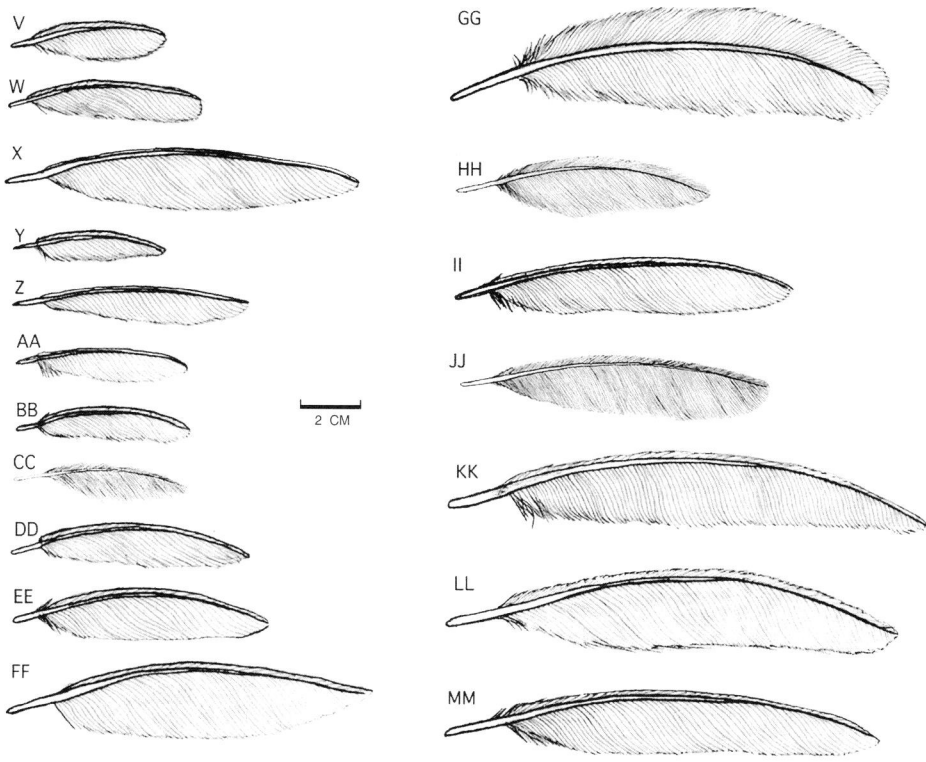

FIG. 38. Continued.

remiges that overlap and engage in limited interlocking). That is, the essential microstructural difference giving rise to the looseness of remiges and pennae contornae in general among flightless rallids is, as observed by Lowe (1928a) in *Atlantisia rogersi,* that completely formed and functional hamuli were comparatively restricted to the barbulae proximales (especially distally in the vexillum), with the barbulae distales bearing imperfectly formed hamuli and therefore effecting at most weak interlocking with opposing rugae distales (Fig. 39). Topographical variation in the rudimentation of hamulae in remiges (and other pennae contornae), both among feathers and tracts and within feathers, appears to account for much of the structural "degeneration" manifested to varying degrees in flightless rallids.

Like other apomorphies of the integument (Tables 27–29), the distribution and relative degree of these modifications varied among the flightless rallids, and these variable interspecific differences were obscured further by variation among feathers (with remiges being more strongly structured than other feathers), within feathers (wear and frictional degradation of microstructure increasing toward the apices of feathers), age of individual birds, and the interval since the feather examined was produced through molt. Some species (e.g., *Tricholimnas sylvestris, Cabalus modestus, Atlantisia rogersi,* and *Porzana palmeri*) were characterized by pronounced looseness of the body feathers and remiges, a general difference that was confounded by the structural effects of wear, age, and preservation of study skins

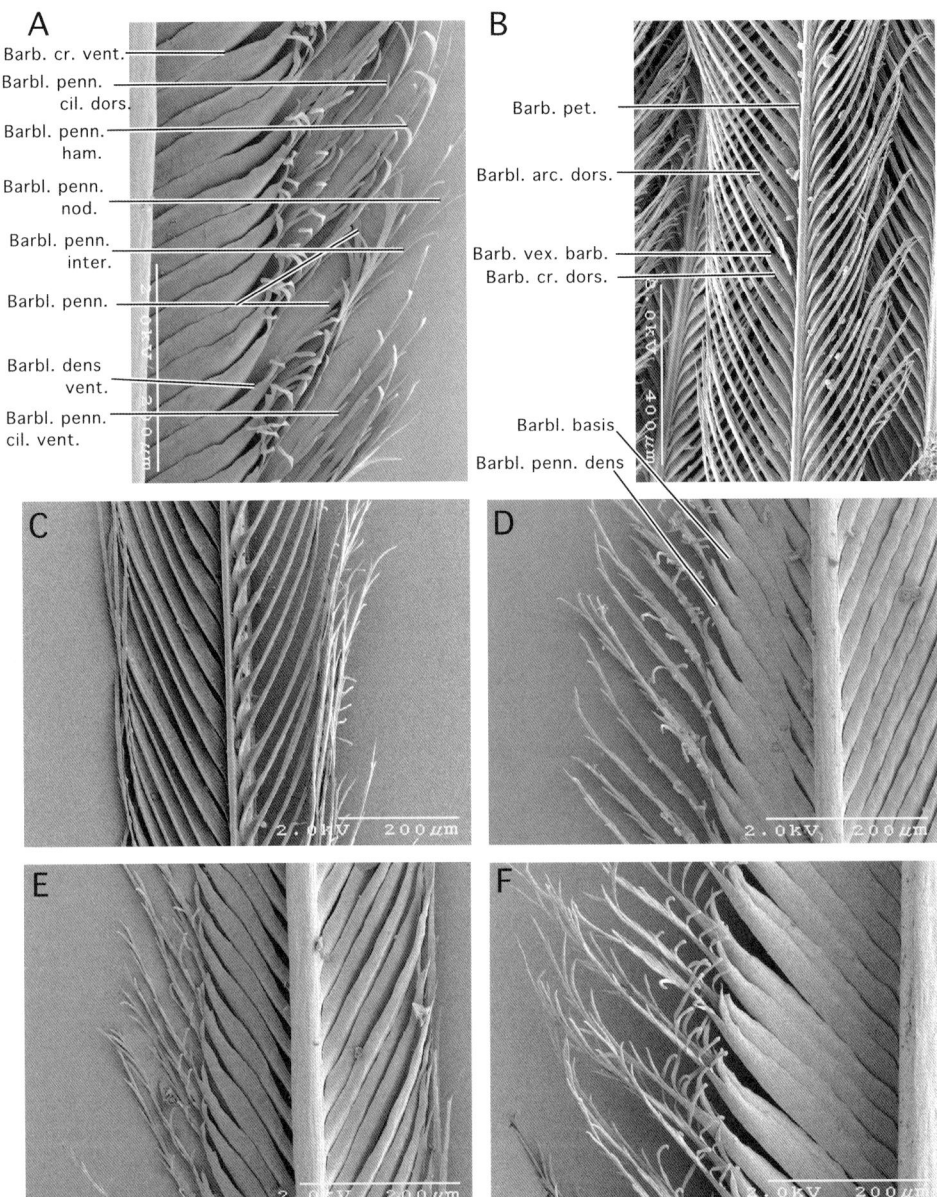

FIG. 39. Anatomical and microanatomical features of pennae contornae, followed by scanning electron microphotographs of sections of barbae from the vexillum internum of remiges secundarii (~sixth) of selected species of Rallidae (dorsal views unless otherwise indicated): **A**, *Gallirallus philippensis goodsoni* (CM 11606), **B**, *G. australis* (CM 24250), **C**, *Cabalus modestus* (CM 24270), **D**, *Tricholimnas sylvestris* (AMNH 545306), **E**, *Atlantisia rogersi* (AMNH 300590), and **F**, *Tribonyx mortierii* (AMNH 546453). Abbreviations for anatomical structures are given in Appendix 3, individual scales (μ) given.

(Fig. 39). In addition, segments of some remiges of flighted rallids such as *Gallirallus philippensis* or *Laterallus jamaicensis* can generate aspects at least as "loose" or "worn" as those of flightless species such as *Habropteryx insignis* and *Amaurornis ineptus,* a finding indicative that reliance on single specimens for interspecific generalizations is fraught with difficulties. In addition, the qualitative comparisons undertaken here were conservative in that close relatives of flightless species were used as "controls," for example, *Gallirallus philippensis* was used to represent a typical flighted rail even though this species itself is insular and may have undergone limited apomorphic changes convergent with those observed in flightless relatives.

Accordingly, a conclusive comparison would have to be based on stratified samples of specimens and feathers, and samples would have to be taken at parallel suites of micropositioned excisions in each, with the differences made quantitative through counts of key structures per unit length of barbule (e.g, number of complete hamulae). Quantification and detailed comparisons of structural variation in the feathers of rails must await the systematic sampling of series of skins and standardized preparations of such specialized surveys (e.g., Rutschke 1960; Brom 1986; Müller and Patone 1998; Dove 2000).

Comparably modest modifications of the microstructure of contour feathers were documented in *Strigops* by Livezey (1992b). The structural apomorphies found in flightless members of the Rallidae were qualitatively similar but much less extensive than those documented for ratites; the latter show additional structural changes and extraordinary degrees of modification of feathers throughout the plumage (McGowan 1989). The extreme apomorphies of the body feathers of ratites were among the most readily diagnosed sources of taxonomic variation and functional corollaries in surveys of the avian integument (Chandler 1916; Stresemann 1932; Sick 1937), and provide a standard for diagnoses of apomorphy in flightless members of other avian groups (Brom and Prins 1989).

Miscellaneous features of the integument.—Comparatively prominent ungues alulares and use of them by nestlings have been noted in several flightless rallids, for example, *Gallirallus australis* (Beauchamp 1998b) and *Atlantisia rogersi* (Lowe 1928a). Although relatively large on the generally diminutive wings of flightless rails (Figs. 33–37), ungues alulares are present in all rallids (Livezey 1998) and are known from a number of avian families (Wetmore 1922b; Tordoff 1952; Rand 1954; Stephan 1992), and use of the claw in locomotion has been documented in flighted rails as well (e.g., *Coturnicops noveboracensis* and *Gallinula chloropus* [P. B. Taylor 1998]). The relevance of the disproportionately robust ungues alulares and tectrices majores of the remiges primarii and (especially) secundarii in the wings of highly derived, flightless rallids to avian developmental patterns and paedomorphosis is considered below.

Morphometrics of Skeletal Characters

Univariate Morphometrics

Interspecific differences among means.—Summary statistics of skeletal measurements for modern and subfossil rallids (Tables 30–33), directly partitioned by sex directly or (for subfossil species) indirectly by using K-means clustering (see details under multivariate statistics), revealed a diversity of variational scales per-

TABLE 30. Univariate summary statistics ($\bar{x} \pm$ SD (n)) for four measurements (mm) of skulls of modern and subfossil species of Rallidae, by species and sex. Provisional sex groups or approximations are given in brackets. Superspecies within *Porphyrio*, *Aramides*, *Rallus*, *Habropteryx*, *Amaurornis*, *Gallinula*, and *Fulica* follow Livezey (1998).

Taxonomic group	Sex	Maxilla length from zona flexoria	Cranial Length	Cranial Depth	Cranial Width
Swamphens					
Porphyrio madagascariensis	M	42.0 ± 2.1 (9)	39.1 ± 1.2 (9)	25.8 ± 1.0 (9)	24.4 ± 0.9 (9)
	F	41.8 ± 1.4 (8)	38.4 ± 0.8 (8)	24.5 ± 0.5 (7)	23.8 ± 0.8 (8)
Porphyrio melanotus	M	47.6 ± 2.3 (30)	41.5 ± 1.0 (32)	27.6 ± 1.4 (31)	25.0 ± 0.5 (32)
	F	43.5 ± 2.3 (20)	38.9 ± 0.8 (20)	25.9 ± 1.1 (19)	23.9 ± 0.6 (20)
Porphyrio mantelli	[M]	65.0 ± 3.6 (3)	54.4 ± 1.0 (9)	32.0 ± 0.9 (9)	35.1 ± 0.9 (9)
	[F]	56.1 ± 3.3 (7)	51.5 ± 0.3 (3)	30.4 ± 0.3 (3)	33.3 ± 0.8 (3)
Porphyrio hochstetteri	M	64.0 ± 2.5 (10)	51.2 ± 2.1 (11)	31.9 ± 1.4 (11)	32.8 ± 1.8 (11)
	F	62.2 ± 2.7 (18)	49.8 ± 1.8 (17)	31.0 ± 1.3 (18)	32.4 ± 1.0 (18)
Aphanocrex podarces	—	—	40.6 ± 1	24.6 (1)	24.3 (1)
Porphyrula alleni	M	26.2 ± 1.4 (3)	29.4 ± 1.4 (3)	18.3 ± 0.8 (3)	18.5 ± 1.5 (3)
	F	23.7 ± 1.9 (5)	27.4 ± 0.8 (5)	17.6 ± 0.7 (5)	17.8 ± 0.7 (5)
Porphyrula martinica	M	29.2 ± 1.5 (15)	31.7 ± 0.6 (14)	19.0 ± 0.4 (14)	19.7 ± 0.4 (15)
	F	27.4 ± 0.6 (15)	31.0 ± 0.6 (15)	18.7 ± 0.4 (15)	19.4 ± 0.5 (15)
Porphyrula flavirostris	M	23.6 ± 2.1 (3)	25.8 ± 0.9 (3)	15.6 ± 0.5 (3)	15.8 ± 0.3 (3)
	F	23.9 (1)	25.6 (1)	15.4 (1)	15.7 (1)
Wallace's Rail and allies					
Gymnocrex plumbeiventris	—	52.7 (1)	—	25.6 (1)	22.7 (1)
Habroptila wallacii	M	67.2 ± 1.9 (3)	47.1 ± 1.3 (3)	26.0 ± 0.6 (3)	28.0 ± 0.9 (3)
	F	61.6 (1)	44.4 (1)	25.8 (1)	27.8 (1)
Cave-rails and allies					
Eulabeornis castaneoventris	F	49.0 (1)	—	—	—
Aramides ypecaha	M	68.3 ± 3.1 (4)	43.2 ± 1.3 (4)	27.2 ± 1.7 (4)	26.5 ± 0.9 (4)
	F	64.6 ± 2.6 (8)	42.1 ± 0.9 (8)	26.0 ± 0.3 (7)	25.4 ± 0.5 (7)
Aramides cajanea-group	M	52.2 ± 4.0 (27)	39.9 ± 1.0 (26)	23.7 ± 0.6 (26)	23.4 ± 0.6 (26)
	F	48.7 ± 4.5 (23)	38.8 ± 1.2 (23)	23.3 ± 0.8 (23)	23.1 ± 0.6 (23)
Canirallus oculeus	—	38.0 (1)	33.5 (1)	21.4 (1)	21.1 (1)
Canirallus kioloides	—	29.2 ± 0.9 (3)	31.5 ± 0.9 (3)	19.1 ± 0.3 (3)	19.7 ± 0.1 (3)
Nesotrochis debooyi	—	—	—	27.5 (1)	29.1 ± 0.8 (2)
Rougetius rougetii	M	36.0 (1)	32.5 (1)	—	18.3 (1)

TABLE 30. Continued.

Taxonomic group	Sex	Maxilla length from zona flexoria	Length	Cranial Depth	Width
White-throated rails					
Dryolimnas cuvieri	M	44.1 ± 2.5 (3)	34.1 ± 1.5 (3)	20.2 ± 0.2 (3)	19.4 ± 0.4 (2)
	F	40.6 ± 0.6 (2)	32.3 ± 0.1 (2)	19.6 ± 0.3 (2)	19.0 ± 0.6 (2)
Dryolimnas aldabranus	M	46.8 ± 2.1 (9)	31.3 ± 0.7 (9)	18.5 ± 0.4 (8)	18.0 ± 0.3 (7)
	F	40.3 ± 1.5 (5)	29.9 ± 0.3 (5)	17.7 ± 0.2 (5)	17.2 ± 0.3 (5)
Dryolimnas abbotti	—	43.0 (1)	31.6 (1)	18.8 (1)	18.3 (1)
Rallus sensu stricto					
Rallus aquaticus	M	44.5 ± 2.2 (17)	28.4 ± 0.6 (17)	16.7 ± 0.4 (16)	16.1 ± 0.3 (16)
	F	40.6 ± 2.2 (21)	26.8 ± 1.0 (21)	16.2 ± 0.2 (20)	15.5 ± 0.3 (20)
Rallus limicola	M	41.1 ± 1.6 (21)	25.8 ± 0.8 (19)	15.3 ± 0.5 (19)	14.7 ± 0.3 (20)
	F	37.0 ± 0.9 (13)	24.2 ± 0.8 (17)	14.9 ± 0.4 (16)	14.3 ± 0.5 (17)
Rallus ibycus	—	[40.0 ± —] (3)	28.4 (1)	15.9 ± 0.4 (3)	15.6 ± 0.2 (3)
Rallus longirostris-group	M	61.3 ± 3.8 (15)	33.6 ± 1.4 (17)	18.5 ± 0.4 (16)	18.0 ± 0.5 (15)
	F	59.0 ± 3.9 (17)	32.7 ± 1.1 (15)	17.9 ± 0.5 (13)	17.6 ± 0.4 (15)
Rallus elegans-group	M	61.5 ± 2.9 (21)	35.4 ± 0.9 (24)	19.6 ± 0.4 (23)	18.8 ± 0.3 (23)
	F	57.9 ± 1.6 (8)	34.0 ± 0.6 (11)	19.0 ± 0.6 (10)	18.4 ± 0.5 (10)
Rallus recessus	[M]	63.7 ± 1.9 (4)	32.7 ± 0.6 (11)	18.0 ± 0.6 (11)	16.6 ± 0.2 (11)
	[F]		30.1 ± 1.3 (2)	17.2 ± 0.1 (2)	15.9 ± 0.8 (2)
Gallirallus and allies					
Gallirallus pectoralis-group	—	34.0 ± 2.7 (8)	25.7 ± 0.8 (8)	15.6 ± 0.5 (7)	15.1 ± 0.4 (8)
Gallirallus striatus	M	34.7 ± 2.4 (3)	29.0 ± 0.4 (3)	16.5 (1)	16.0 ± 0.3 (2)
	F	34.0 ± 0.3 (2)	27.6 ± 1.3 (2)	16.3 ± 0.2 (2)	16.3 ± 0.6 (2)
Gallirallus australis	M	51.3 ± 2.6 (18)	43.7 ± 1.0 (17)	25.2 ± 0.9 (17)	25.2 ± 0.7 (17)
	F	45.3 ± 4.4 (4)	42.5 ± 1.6 (4)	24.3 ± 0.6 (4)	24.4 ± 0.5 (4)
Gallirallus greyi	M	47.0 ± 2.3 (13)	40.9 ± 1.3 (11)	23.7 ± 0.9 (10)	23.6 ± 0.4 (13)
	F	44.1 ± 2.6 (4)	40.0 ± 0.8 (4)	23.1 ± 1.2 (4)	23.5 ± 0.9 (4)
Gallirallus philippensis-group	M	32.2 ± 2.5 (15)	31.2 ± 1.1 (15)	18.4 ± 0.7 (14)	17.9 ± 0.4 (14)
	F	30.4 ± 2.0 (18)	30.5 ± 1.1 (18)	17.8 ± 0.3 (17)	17.6 ± 0.6 (17)
Gallirallus dieffenbachii	[M]	42.0 ± 1.1 (27)	35.2 ± 0.7 (27)	21.6 ± 0.6 (27)	21.6 ± 0.3 (27)
	[F]	39.0 ± 1.4 (16)	33.4 ± 0.9 (16)	21.1 ± 0.7 (16)	21.0 ± 0.5 (16)
Gallirallus owstoni	M	40.3 ± 2.1 (14)	34.0 ± 1.0 (10)	18.9 ± 0.6 (10)	18.7 ± 0.5 (13)
	F	36.3 ± 1.9 (17)	32.2 ± 0.9 (11)	18.7 ± 0.5 (8)	18.0 ± 0.3 (15)
Gallirallus wakensis	M	29.3 ± 0.4 (2)	26.3 ± 0.1 (2)	15.6 ± 0.1 (2)	17.1 ± 0.4 (2)
	F	24.8 (1)	24.8 (1)	15.5 (1)	16.7 (1)

TABLE 30. Continued.

Taxonomic group	Sex	Maxilla length from zona flexoria	Length	Cranial Depth	Width
Tricholimnas sylvestris	M	50.9 ± 0.7 (4)	37.8 ± 1.8 (4)	22.2 ± 1.2 (4)	22.3 ± 1.4 (4)
	F	44.7 ± 2.5 (2)	35.4 ± 0.4 (2)	21.2 ± 0.9 (2)	20.2 ± 0.8 (2)
Nesoclopeus poecilopterus-group	M	43.5 (1)	—	—	22.8 (1)
Cabalus modestus	[M]	39.2 ± 1.2 (17)	25.5 ± 0.7 (17)	15.2 ± 0.4 (17)	14.5 ± 0.3 (17)
	[F]	35.2 ± 1.1 (11)	24.6 ± 0.5 (11)	15.2 ± 0.5 (11)	14.1 ± 0.3 (11)
Capellirallus karamu	—	68.5 ± 2.4 (3)	32.2 ± 1.0 (8)	19.0 ± 5.0 (6)	20.1 ± 0.2 (5)
Habropteryx insignis	—	45.9 (1)	—	—	22.5 (1)
Habropteryx torquatus-group	M	41.8 ± 1.6 (6)	34.2 ± 1.7 (6)	19.6 ± 0.5 (6)	20.0 ± 0.4 (6)
	F	37.4 ± 1.4 (7)	31.1 ± 0.7 (7)	18.8 ± 0.3 (7)	19.1 ± 0.4 (7)
Habropteryx okinawae	—	44.5 (1)	38.5 (1)	22.0 (1)	20.4 (1)
Problematic subfossil rails					
Aphanapteryx bonasia	[M]	97.0 (1)	48.0 (1)	26.6 (1)	29.7 (1)
	[F]	84.7 ± 0.5 (4)			
Erythromachus leguati	—	[78.0] (1)	38.2 ± 1.8 (2)	21.8 ± 1.1 (2)	20.8 ± 0.2 (3)
Diaphorapteryx hawkinsi	[M]	95.5 ± 3.4 (33)	51.8 ± 2.5 (33)	31.4 ± 1.1 (33)	36.2 ± 1.2 (33)
	[F]	86.6 ± 2.9 (20)	48.2 ± 2.1 (20)	29.4 ± 1.0 (20)	34.5 ± 1.0 (20)
Atlantisia and allies					
Atlantisia rogersi	—	22.3 ± 0.8 (2)	20.6 ± 0.4 (2)	12.3 ± 0.3 (2)	13.5 ± 0.1 (2)
Mudia elpenor	—	26.2 (1)	27.4 ± 0.6 (2)	14.5 ± 0.1 (6)	15.9 ± 0.2 (3)
Laterallus leucopyrrhus	M	18.2 ± 1.1 (7)	22.6 ± 0.9 (7)	14.1 ± 0.4 (7)	15.2 ± 0.6 (7)
	F	18.6 ± 1.2 (19)	22.2 ± 0.6 (19)	13.9 ± 0.4 (18)	14.6 ± 0.4 (18)
Coturnicops noveboracensis	M	16.3 ± 0.9 (14)	20.7 ± 0.5 (13)	12.6 ± 0.4 (13)	13.1 ± 0.3 (14)
	F	15.0 ± 1.2 (18)	20.2 ± 0.6 (17)	12.4 ± 0.3 (16)	12.8 ± 0.3 (17)
Crex crex	M	22.4 ± 1.0 (13)	28.3 ± 1.2 (13)	16.5 ± 0.5 (13)	16.4 ± 0.6 (13)
	F	21.3 ± 0.9 (11)	27.8 ± 1.0 (11)	16.3 ± 0.4 (11)	16.4 ± 0.6 (11)
Crex albicollis	M	26.3 ± 1.1 (2)	28.2 ± 0.8 (2)	17.0 ± 0.1 (2)	17.1 ± 0.4 (2)
	F	24.0 ± 0.3 (2)	28.1 ± 0.9 (2)	16.3 ± 0.4 (2)	17.1 ± 0.4 (2)
Gray crakes and allies					
Porzana palmeri	M	20.3 ± 0.4 (6)	20.3 ± 0.6 (6)	12.8 ± 0.4 (6)	13.3 ± 0.4 (6)
	F	19.1 ± 0.5 (6)	20.2 ± 0.6 (6)	12.8 ± 0.5 (5)	12.6 ± 0.9 (5)
Porzana carolina	M	19.0 ± 0.7 (15)	23.0 ± 0.6 (15)	13.8 ± 0.5 (14)	14.1 ± 0.3 (15)
	F	19.0 ± 0.7 (15)	23.0 ± 0.6 (15)	13.8 ± 0.5 (14)	14.1 ± 0.3 (15)

TABLE 30. Continued.

Taxonomic group	Sex	Maxilla length from zona flexoria	Length	Cranial Depth	Width
Porzana pusilla	M	18.3 ± 0.8 (8)	21.0 ± 0.9 (9)	12.8 ± 0.5 (9)	12.7 ± 0.5 (9)
	F	17.4 ± 1.2 (12)	20.3 ± 1.1 (12)	12.4 ± 0.5 (10)	12.5 ± 0.6 (12)
Porzana tabuensis	M	18.7 ± 0.5 (3)	22.8 ± 0.1 (4)	13.7 ± 0.3 (3)	14.2 ± 0.3 (4)
	F	18.8 ± 0.7 (6)	22.0 ± 0.5 (6)	13.6 ± 0.5 (6)	13.9 ± 0.4 (6)
*Porzana monasa**	—	[22.0] (1)	—	—	—
Porzana atra	M	24.3 (1)	24.1 (1)	14.6 (1)	15.5 (1)
	F	22.4 ± 0.1 (2)	23.2 ± 0.3 (3)	14.5 ± 0.1 (2)	15.2 ± 0.2 (3)
Porzana flavirostra	M	25.2 ± 0.8 (16)	25.2 ± 0.7 (15)	15.6 ± 0.2 (15)	16.3 ± 0.3 (15)
	F	23.5 ± 1.1 (15)	24.6 ± 0.6 (17)	15.4 ± 0.5 (16)	16.0 ± 0.4 (18)
Porzana piercei	—	16.8 ± 0.8 (6)	22.0 (1)	13.6 ± 0.7 (4)	13.7 ± 0.5 (3)
Porzana ziegleri†	—	—	21.3 (1)	11.4 (1)	13.1 (1)
Porzana keplerorum†	—	18.5 (1)	21.0 (1)	11.6 (1)	13.6 (1)
Porzana severnsi†	—	26.4 (1)	26.5 (1)	16.0 (1)	17.6 (1)
Waterhens					
Amaurornis olivaceus	M	37.8 ± 1.2 (5)	33.5 ± 0.6 (5)	20.6 ± 0.2 (5)	21.8 ± 0.5 (5)
	F	34.1 ± 2.7 (4)	32.5 ± 0.9 (4)	19.3 ± 1.3 (3)	20.2 ± 0.8 (4)
Amaurornis moluccanus	M	30.8 ± 2.6 (3)	31.5 ± 0.5 (3)	18.3 ± 0.6 (3)	19.7 ± 0.8 (3)
	F	28.9 ± 0.3 (2)	30.0 ± 0.4 (2)	17.4 ± 0.1 (2)	19.5 ± 0.1 (2)
Amaurornis phoenicurus	M	36.0 ± 1.9 (16)	31.5 ± 0.8 (16)	18.8 ± 0.5 (15)	19.9 ± 0.6 (15)
	F	32.9 ± 1.9 (15)	30.5 ± 1.1 (15)	18.1 ± 0.6 (15)	19.3 ± 0.5 (15)
Amaurornis akool	—	29.1 (1)	—	—	15.2 (1)
Amaurornis isabellinus	—	33.9 (1)	—	—	20.3 (1)
Amaurornis ineptus	—	58.2 (1)	—	—	27.8 (1)
Moorhens and allies					
Gallicrex cinerea	M	36.1 ± 2.1 (8)	33.7 ± 0.6 (8)	20.5 ± 0.5 (8)	20.3 ± 0.3 (8)
	F	30.4 ± 1.6 (2)	30.9 ± 0.2 (2)	18.6 ± 0.1 (2)	18.9 ± 0.8 (2)
Tribonyx ventralis	M	28.5 ± 1.2 (22)	35.1 ± 0.7 (24)	20.8 ± 0.6 (23)	20.8 ± 0.3 (24)
	F	26.8 ± 1.4 (19)	33.5 ± 1.0 (19)	20.1 ± 0.6 (19)	20.1 ± 0.5 (19)
Tribonyx mortierii	M	35.5 ± 1.2 (10)	43.5 ± 0.7 (8)	25.0 ± 0.6 (10)	24.1 ± 0.5 (10)
	F	35.3 ± 1.2 (11)	42.0 ± 0.7 (9)	24.5 ± 0.7 (9)	23.7 ± 0.5 (11)
Tribonyx hodgenorum	—	23.2 ± 2.3 (5)	31.9 ± 0.9 (3)	18.4 ± 1.6 (5)	19.0 ± 1.1 (6)
Gallinula comeri	—	31.6 ± 1.1 (2)	33.7 ± 1.1 (2)	22.2 ± 0.6 (2)	20.5 ± 0.2 (2)
Gallinula tenebrosa	M	35.0 ± 1.0 (6)	34.6 ± 0.8 (6)	21.5 ± 0.6 (5)	20.2 ± 0.5 (6)
	F	36.0 ± 1.7 (8)	34.1 ± 0.9 (8)	21.6 ± 0.9 (8)	19.7 ± 0.4 (8)

TABLE 30. Continued.

Taxonomic group	Sex	Maxilla length from zona flexoria	Length	Cranial Depth	Width
Gallinula chloropus	M	29.2 ± 1.5 (9)	31.7 ± 1.4 (10)	19.4 ± 0.9 (10)	18.3 ± 0.6 (10)
	F	27.9 ± 1.8 (17)	30.8 ± 1.6 (19)	18.7 ± 0.9 (19)	17.9 ± 0.6 (19)
Gallinula cachinnans	M	31.8 ± 1.8 (11)	32.2 ± 0.9 (11)	19.9 ± 0.6 (10)	19.0 ± 0.3 (9)
	F	29.6 ± 0.8 (8)	31.0 ± 1.1 (8)	19.0 ± 0.4 (7)	18.2 ± 0.2 (8)
Gallinula sandvicensis	—	29.5 ± 1.3 (5)	31.0 ± 1.5 (5)	19.2 ± 0.7 (5)	18.5 ± 0.1 (5)
Coots					
Fulica armillata	—	38.4 ± 1.3 (4)	37.3 ± 1.0 (4)	23.2 ± 0.8 (4)	22.4 ± 0.5 (4)
Fulica cornuta	—	55.8 ± 0.3 (2)	42.6 ± 1.4 (2)	24.4 ± 0.2 (2)	25.0 ± 0.7 (2)
Fulica gigantea	M	53.2 ± 1.1 (5)	42.4 ± 0.9 (5)	26.1 ± 0.7 (5)	26.3 ± 0.3 (5)
	F	50.8 ± 2.5 (4)	42.3 ± 0.9 (4)	25.0 ± 0.9 (4)	26.1 ± 0.5 (4)
Fulica americana	M	35.7 ± 1.3 (16)	33.6 ± 0.8 (16)	20.7 ± 0.3 (16)	20.2 ± 0.4 (16)
	F	33.3 ± 1.0 (12)	32.0 ± 0.7 (15)	20.0 ± 0.4 (13)	19.3 ± 0.4 (13)
Fulica alai	—	33.2 ± 1.9 (7)	32.3 ± 1.0 (4)	20.4 ± 0.8 (5)	20.2 ± 0.5 (7)
Fulica antra (*atra*-group)	M	36.8 ± 2.3 (14)	35.3 ± 1.3 (15)	21.8 ± 0.6 (14)	21.1 ± 0.8 (15)
	F	34.9 ± 2.1 (21)	34.8 ± 1.2 (20)	21.6 ± 0.8 (20)	20.9 ± 0.9 (20)
Fulica chathamensis	[M]	50.4 ± 0.8 (9)	41.9 ± 0.7 (23)	25.2 ± 1.1 (23)	24.6 ± 0.5 (23)
	[F]	47.1 ± 2.1 (3)	39.9 ± 1.0 (12)	23.5 ± 0.8 (12)	24.2 ± 0.8 (12)
Fulica prisca	—	—	42.3 (1)	26.5 ± 0.5 (3)	23.3 ± 0.3 (3)

* Measurements based on X ray photographs in Steadman (1986a).
† Samples based in part on measurements provided by Olson and James (1991).

taining to taxonomic group, sex, and flight capacity. Salient differences in means and variances were subjected to formal statistical testing (two-way ANOVA) where samples allowed; pairwise comparisons (Bonferroni t-tests) and informal comparisons with other groups followed. As a rule, statistical significance of differences was high for skeletal comparisons, a reflection of both manifold diversity of form and the precision of skeletal measurements (the latter were conducive to low within-group variances and increased statistical power). Unfortunately, skeletal data were less widely available than measurements of study skins, precluding statistical comparisons for a number of taxa (e.g., *Gymnocrex* spp., *Ortygonax* spp., *Cyanolimnas cerverai, Tricholimnas lafresnayanus, Amaurornis ineptus, Fulica chathamensis,* and *F. newtoni*). However, subfossil remains permitted informal consideration of a number of taxa not available as study skins. As in the comparisons of external measurements, the focus here remains on those comparisons related to loss of flight or those in which an unexpected absence or magnitude of differences was indicated. Differences among taxa will be considered first, whereas intersexual differences in means and variances are considered under sexual dimorphism.

Three species of greater swamphens (*Porphyrio*; $n = 105$) were sampled in sufficient quantity to permit formal statistical comparisons of skeletons (*P. madagascariensis, P. melanotus,* and flightless *P. hochstetteri*), with unassociated subfossil remains for one other flightless congener (*P. mantelli*) available for informal consideration (Tables 30–33). Interspecific differences were highly significant (ANOVA; $P \leq 0.0001$) in all 41 dimensions analyzed (only a subset of which are tabulated). A posteriori pairwise comparisons (within sexes) revealed that these interspecific effects were due largely to the distinct characteristics of flightless *P. hochstetteri,* although most variables differed less dramatically but significantly (Bonferroni t-tests, $P < 0.05$) between the two flighted taxa as well (*P. melanotus* was generally larger than *P. madagascariensis,* especially males). In comparison with its more-typical congeners, flightless *P. hochstetteri* was characterized by being larger in dimensions of the skull, scapula, coracoid, furcula height, sternal widths, two dimensions of the pelvic girdle, and proximal segments of the pelvic limb, but absolutely smaller in least depth of the carina sterni and lengths of the alar and pedal phalanges (Tables 30–33). These differences were shared and generally of greater magnitude in the extinct, generally larger *P. mantelli.* Available elements indicate that the poorly represented *P. kukwiedei* had undergone comparable or greater pectoral reduction (Tables 30–33).

Comparisons of lesser swamphens (*Porphyrula*; $n = 42$) confirmed significant interspecific differences (ANOVA; $P < 0.01$) in all 41 of the dimensions analyzed, with *P. flavirostris* averaging the smallest of the three species (Tables 30–33), a pattern consistent with mean body masses (Appendix 1) and notable relative to the paedomorphic plumage pattern of this species (Livezey 1998). Summary statistics of extinct *Aphanocrex podarces,* qualitatively allied with *Porphyrula* (Livezey 1998), indicate that this species was distinguished by a comparatively large cranium and only weakly reduced pectoral elements (Tables 30–33).

Of the basal genera of rallids (Livezey 1998), samples of skeletons permitted the formal analysis of only three taxa ($n = 69$)—*Habroptila wallacii, Aramides cajanea*-group, and *A. ypecaha*)—of which only the first is flightless. Limited mensural data for additional taxa permitted limited informal comparisons (Tables

TABLE 31. Univariate summary statistics ($\bar{x} \pm$ SD (n)) for six measurements (mm) of the pectoral girdle of modern and subfossil species of Rallidae, by species and sex. Provisional sex groups or approximations are enclosed by brackets. Superspecies within *Porphyrio, Aramides, Rallus, Habropteryx, Amaurornis, Gallinula,* and *Fulica* follow Livezey (1998).

Taxonomic group	Sex	Scapula length	Coracoid length
Swamphens			
Porphyrio madagascariensis	M	62.2 ± 1.4 (8)	36.0 ± 1.8 (8)
	F	59.2 ± 2.7 (9)	35.5 ± 1.6 (9)
Porphyrio melanotus	M	67.6 ± 2.5 (32)	40.5 ± 1.7 (33)
	F	62.2 ± 2.3 (19)	37.6 ± 1.1 (20)
Porphyrio mantelli	[M]	73.3 ± 4.1 (5)	46.0 ± 2.0 (5)
	[F]		41.2 (1)
Porphyrio hochstetteri	M	75.6 ± 3.4 (13)	43.0 ± 1.7 (14)
	F	72.2 ± 2.1 (16)	41.3 ± 1.2 (18)
Porphyrula alleni	M	42.9 ± 2.7 (3)	24.4 ± 1.1 (3)
	F	38.5 ± 1.6 (5)	23.0 ± 2.4 (5)
Porphyrula martinica	M	46.4 ± 1.3 (15)	26.6 ± 0.7 (15)
	F	44.2 ± 1.7 (14)	25.6 ± 0.9 (14)
Porphyrula flavirostris	M	33.8 ± 0.8 (3)	20.1 ± 0.6 (3)
	F	32.5 (1)	—
Wallace's Rail and allies			
Habroptila wallacii	M	51.6 ± 0.7 (3)	32.7 ± 0.6 (3)
	F	48.7 ± 2.3 (3)	31.1 ± 0.4 (3)
Cave-rails and allies			
Eulabeornis castaneoventris	—	56.5 (1)	34.3 (1)
Aramides ypecaha	M	60.1 ± 2.9 (4)	40.2 ± 1.0 (4)
	F	56.9 ± 1.7 (8)	38.4 ± 1.1 (8)
Aramides cajanea-group	M	49.3 ± 1.5 (24)	31.2 ± 1.1 (27)
	F	47.6 ± 1.8 (24)	30.1 ± 1.5 (23)
Canirallus oculeus	—	41.0 (1)	26.1 (1)
Canirallus kioloides	—	36.5 ± 1.4 (3)	23.1 ± 0.4 (3)
Nesotrochis debooyi	—	57.5 (1)	33.1 (1)
Nesotrochis steganinos	—	—	—
Zapata Rail and allies			
Cyanolimnas cerverai	—	31.8 (1)	18.9 ± 1.2 (3)
Ortygonax sanguinolentus	—	41.5 (1)	23.1 (1)
White-throated rails			
Dryolimnas cuvieri	M	43.0 ± 1.6 (3)	25.8 ± 0.3 (3)
	F	43.0 ± 2.3 (3)	26.3 ± 1.4 (3)
Dryolimnas aldabranus	M	35.0 ± 1.0 (10)	22.0 ± 0.5 (10)
	F	33.3 ± 0.6 (6)	20.5 ± 0.4 (6)
Dryolimnas abbotti	—	—	23.7 (1)
Rallus sensu stricto			
Rallus aquaticus	M	36.9 ± 1.2 (17)	21.0 ± 0.7 (17)
	F	34.9 ± 1.7 (21)	19.9 ± 0.7 (21)
Rallus limicola	M	33.1 ± 1.1 (20)	18.6 ± 0.5 (21)
	F	30.6 ± 1.4 (16)	17.2 ± 0.8 (18)
Rallus ibycus	[M]	—	18.4 ± 7.7 (8)
	[F]	—	16.2 ± 4.5 (5)
Rallus longirostris-group	M	46.3 ± 1.4 (17)	27.8 ± 0.9 (17)
	F	44.2 ± 2.1 (15)	26.6 ± 1.0 (16)
Rallus elegans-group	M	51.0 ± 1.9 (22)	30.6 ± 1.0 (25)
	F	48.1 ± 1.3 (12)	28.7 ± 0.5 (11)
Rallus recessus	[M]	40.6 ± 2.3 (6)	23.9 ± 0.5 (14)
	[F]		21.7 ± 0.7 (6)

FLIGHTLESSNESS IN RAILS

TABLE 31. Extended.

	Sternum				
		Corpus		Carina depth	
Carina length	Length	Least width	Oblique	Perpendicular
---	---	---	---	---
55.6 ± 3.0 (9)	62.2 ± 3.0 (9)	13.3 ± 0.8 (9)	20.7 ± 1.0 (9)	15.7 ± 0.9 (9)
53.4 ± 3.4 (9)	57.4 ± 4.5 (9)	12.9 ± 1.0 (9)	20.4 ± 1.0 (9)	15.3 ± 0.8 (9)
64.3 ± 4.0 (31)	70.0 ± 4.0 (29)	14.0 ± 1.1 (32)	22.8 ± 1.3 (32)	17.0 ± 1.1 (31)
58.1 ± 2.7 (20)	63.2 ± 2.9 (20)	13.8 ± 1.1 (20)	21.1 ± 1.3 (20)	16.3 ± 1.3 (20)
50.7 ± 4.1 (5)	66.6 ± 6.0 (5)	22.2 ± 1.6 (7)	21.1 ± 2.0 (12)	6.5 ± 0.9 (10)
51.4 ± 2.7 (10)	65.8 ± 3.4 (10)	24.6 ± 2.6 (11)	20.7 ± 1.6 (11)	7.6 ± 1.1 (10)
49.0 ± 2.0 (17)	62.9 ± 2.3 (17)	23.5 ± 1.3 (18)	20.5 ± 0.8 (17)	7.1 ± 0.8 (17)
36.2 ± 4.3 (2)	38.4 ± 4.3 (2)	8.9 ± 0.6 (3)	13.1 ± 0.6 (3)	11.4 ± 1.0 (3)
30.3 ± 2.3 (5)	34.0 ± 1.9 (5)	9.3 ± 2.0 (5)	12.9 ± 0.7 (5)	10.7 ± 0.5 (5)
39.6 ± 2.3 (15)	41.3 ± 1.7 (15)	9.1 ± 0.4 (15)	14.5 ± 0.6 (15)	12.7 ± 0.7 (15)
38.2 ± 2.4 (14)	40.4 ± 2.2 (15)	8.6 ± 0.6 (15)	14.1 ± 0.8 (15)	12.3 ± 0.8 (15)
28.4 ± 0.2 (3)	30.9 ± 0.8 (3)	6.8 ± 0.8 (3)	10.7 ± 1.1 (3)	90.1 ± 0.7 (3)
29.0 (1)	32.2 (1)	7.6 (1)	12.0 (1)	10.0 (1)
49.6 ± 2.9 (3)	56.3 ± 2.3 (3)	10.6 ± 0.4 (3)	12.6 ± 1.2 (3)	6.3 ± 0.7 (3)
46.0 ± 0.8 (3)	52.0 ± 0.5 (3)	9.9 ± 0.9 (3)	12.5 ± 0.5 (3)	6.1 ± 0.5 (3)
50.5 (1)	52.8 (1)	11.7 (1)	14.8 (1)	12.3 (1)
76.1 ± 2.4 (4)	73.8 ± 1.7 (4)	15.1 ± 2.0 (4)	19.1 ± 1.1 (4)	18.6 ± 0.9 (4)
76.6 ± 4.6 (8)	73.6 ± 3.0 (8)	14.5 ± 0.6 (8)	17.4 ± 1.5 (8)	16.7 ± 1.7 (8)
58.4 ± 2.9 (23)	57.7 ± 2.9 (24)	10.3 ± 0.9 (26)	14.0 ± 0.8 (25)	13.1 ± 0.7 (24)
57.6 ± 2.9 (23)	57.1 ± 2.8 (23)	10.2 ± 0.8 (23)	14.0 ± 1.0 (24)	13.0 ± 1.0 (23)
48.3 (1)	46.3 (1)	10.5 (1)	12.3 (1)	12.3 (1)
43.5 ± 1.8 (3)	45.5 ± 2.2 (3)	8.2 ± 0.1 (2)	12.5 ± 0.6 (3)	10.7 ± 0.9 (3)
—	—	—	—	—
—	—	11.6 (1)	10.7 (1)	5.2 (1)
25.4 (1)	30.0 (1)	8.8 (1)	9.1 (1)	4.2 (1)
40.1 (1)	41.6 (1)	7.4 (1)	13.8 (1)	12.3 (1)
45.7 ± 0.9 (3)	46.3 ± 2.6 (3)	7.8 ± 0.7 (3)	13.3 ± 1.7 (3)	12.5 ± 1.3 (3)
43.8 ± 4.5 (3)	44.1 ± 3.8 (3)	7.9 ± 0.6 (3)	13.1 ± 1.7 (3)	12.3 ± 1.5 (3)
31.0 ± 1.7 (7)	35.2 ± 0.7 (10)	7.5 ± 0.2 (10)	10.7 ± 0.5 (7)	8.0 ± 0.3 (7)
30.1 ± 1.0 (5)	33.9 ± 0.6 (6)	6.6 ± 0.2 (5)	10.0 ± 0.4 (5)	7.7 ± 0.4 (5)
41.4 (1)	41.3 (1)	7.5 (1)	11.3 (1)	11.1 (1)
36.2 ± 1.2 (17)	37.8 ± 1.1 (17)	5.6 ± 5.2 (17)	12.8 ± 0.6 (17)	11.6 ± 0.5 (17)
33.8 ± 1.6 (20)	35.7 ± 1.4 (19)	5.2 ± 0.3 (19)	12.3 ± 0.6 (19)	11.2 ± 0.7 (19)
31.5 ± 1.1 (20)	33.1 ± 1.1 (19)	5.3 ± 0.4 (20)	11.9 ± 0.6 (19)	10.9 ± 0.7 (19)
28.8 ± 1.6 (12)	30.9 ± 0.6 (15)	4.9 ± 0.3 (17)	11.7 ± 0.6 (13)	10.5 ± 0.7 (13)
18.3 ± 1.2 (3)	23.4 ± 0.5 (3)	7.6 ± 0.9 (4)	8.1 ± 0.6 (7)	4.1 ± 0.4 (7)
48.3 ± 4.2 (16)	49.9 ± 3.1 (16)	6.3 ± 1.1 (16)	16.2 ± 1.1 (17)	15.4 ± 1.1 (17)
45.5 ± 2.7 (16)	47.2 ± 2.2 (17)	5.9 ± 0.6 (17)	15.5 ± 1.0 (16)	14.5 ± 0.9 (16)
55.1 ± 2.7 (25)	54.9 ± 2.2 (24)	7.0 ± 0.6 (22)	17.1 ± 0.8 (24)	16.4 ± 0.6 (24)
50.6 ± 2.2 (10)	51.0 ± 1.8 (11)	6.5 ± 0.6 (11)	16.3 ± 0.9 (10)	15.6 ± 0.8 (9)
30.5 ± 1.4 (6)	35.0 ± 1.1 (6)	7.5 ± 0.5 (6)	11.6 ± 1.0 (6)	8.4 ± 0.6 (6)
27.7 ± 0.7 (15)	32.2 ± 1.0 (15)	7.2 ± 0.6 (15)	10.6 ± 0.2 (15)	7.5 ± 0.5 (15)

TABLE 31. Continued.

Taxonomic group	Sex	Scapula length	Coracoid length
Gallirallus and allies			
Gallirallus pectoralis-group	—	30.7 ± 1.0 (7)	17.3 ± 0.6 (8)
Gallirallus striatus	M	37.4 ± 1.3 (3)	23.3 ± 2.2 (4)
	F	37.7 ± 0.3 (2)	21.8 ± 0.2 (2)
Gallirallus australis	M	51.4 ± 1.9 (17)	32.4 ± 1.0 (18)
	F	45.4 ± 1.3 (13)	28.7 ± 1.1 (12)
Gallirallus greyi	M	49.9 ± 3.8 (4)	31.2 ± 2.5 (4)
	F	44.8 ± 1.6 (4)	28.5 ± 1.1 (4)
Gallirallus philippensis-group	M	39.8 ± 1.8 (15)	22.8 ± 1.1 (16)
	F	39.3 ± 1.9 (20)	22.4 ± 0.9 (21)
Gallirallus dieffenbachii	[M]	36.9 ± 1.1 (19)	23.1 ± 0.6 (23)
	[F]	34.3 ± 1.1 (9)	21.3 ± 0.5 (12)
Gallirallus owstoni	M	40.8 ± 1.9 (14)	22.7 ± 1.2 (13)
	F	38.8 ± 1.6 (18)	21.3 ± 0.8 (16)
Gallirallus wakensis	M	29.1 ± 0.8 (20)	16.4 ± 0.6 (19)
	F	28.2 ± 0.9 (10)	15.6 ± 0.3 (10)
Tricholimnas lafresnayanus	—	—	23.5 (1)
Tricholimnas sylvestris	M	45.3 ± 2.6 (4)	26.1 ± 0.5 (4)
	F	41.8 ± 0.6 (2)	24.3 ± 0.8 (2)
*Nesoclopeus poecilopterus**	—	49.0 (1)	28.0 (1)
Cabalus modestus	[M]	22.1 ± 0.8 (14)	12.6 ± 0.4 (14)
	[F]	20.0 ± 0.5 (14)	11.6 ± 0.3 (14)
Capellirallus karamu	[M]	28.0 (1)	15.3 ± 0.4 (7)
	[F]	24.1 ± 0.9 (3)	13.7 ± 0.1 (2)
Habropteryx torquatus-group	M	42.9 ± 0.3 (4)	26.0 ± 0.6 (4)
	F	40.2 ± 1.5 (8)	23.7 ± 0.9 (8)
Habropteryx okinawae	—	42.0 (1)	26.8 (1)
Problematic subfossil rails			
Aphanapteryx bonasia	[M]	50.5 ± 2.9 (3)	38.6 ± 0.2 (2)
	[F]		32.1 ± 1.2 (5)
Erythromachus leguati	—	—	
Diaphorapteryx hawkinsi	[M]	64.6 ± 2.0 (21)	40.6 ± 1.8 (21)
	[F]	58.9 ± 1.6 (12)	37.0 ± 1.6 (24)
Atlantisia and allies			
Atlantisia rogersi	—	18.9 ± 1.2 (6)	10.6 ± 0.5 (7)
Mundia elpenor	[M]	25.6 ± 0.7 (4)	15.6 ± 0.2 (5)
	[F]		14.8 ± 0.4 (11)
Laterallus leucopyrrhus	M	24.8 ± 0.6 (7)	14.7 ± 0.5 (7)
	F	23.6 ± 1.1 (21)	14.2 ± 0.8 (20)
Coturnicops noveboracensis	M	27.5 ± 1.4 (15)	15.8 ± 0.7 (115)
	F	25.6 ± 1.8 (19)	14.8 ± 0.4 (20)
Crex crex	M	43.5 ± 2.1 (13)	22.1 ± 1.0 (13)
	F	41.5 ± 1.8 (11)	21.5 ± 0.6 (11)
Crex albicollis	M	33.1 ± 2.7 (2)	20.1 ± 0.8 (2)
	F	34.2 ± 0.8 (2)	19.3 ± 0.8 (2)
Gray crakes and allies			
Porzana palmeri	M	19.8 ± 1.2 (6)	11.0 ± 0.2 (6)
	F	19.3 ± 0.7 (6)	10.6 ± 0.3 (6)
Porzana carolina	M	32.3 ± 1.3 (13)	17.2 ± 0.5 (15)
	F	30.6 ± 1.2 (15)	16.4 ± 0.3 (15)
Porzana pusilla	M	26.2 ± 1.8 (9)	14.4 ± 0.9 (10)
	F	25.0 ± 1.5 (14)	13.8 ± 0.8 (15)
Porzana tabuensis	M	24.5 ± 1.1 (5)	14.4 ± 0.2 (5)

TABLE 31. Continued, Extended.

	Sternum			
	Corpus		Carina depth	
Carina length	Length	Least width	Oblique	Perpendicular
29.5 ± 2.1 (8)	31.4 ± 1.6 (8)	4.6 ± 0.4 (8)	10.2 ± 0.6 (8)	9.7 ± 0.6 (8)
39.3 ± 4.0 (3)	40.0 ± 3.0 (3)	5.1 ± 0.4 (4)	13.0 ± 1.6 (3)	12.0 ± 0.9 (3)
38.5 ± 2.1 (2)	39.0 ± 1.6 (2)	5.0 ± 0.2 (2)	12.6 ± 1.1 (2)	12.2 ± 1.1 (2)
34.0 ± 1.9 (19)	41.0 ± 1.9 (19)	11.3 ± 0.9 (19)	12.7 ± 1.0 (18)	7.3 ± 0.6 (18)
29.9 ± 1.4 (13)	36.0 ± 1.8 (13)	10.1 ± 1.0 (13)	10.6 ± 1.0 (13)	6.1 ± 0.8 (13)
32.9 ± 1.1 (4)	40.3 ± 1.4 (4)	11.6 ± 0.9 (4)	10.8 ± 2.2 (4)	5.7 ± 0.5 (4)
29.8 ± 0.9 (4)	35.7 ± 1.7 (4)	10.1 ± 0.6 (3)	10.5 ± 2.0 (4)	5.6 ± 0.2 (4)
40.9 ± 2.5 (14)	40.8 ± 1.8 (15)	6.9 ± 0.9 (15)	13.1 ± 1.1 (15)	12.6 ± 1.0 (15)
41.9 ± 3.1 (21)	41.3 ± 2.6 (20)	6.3 ± 0.8 (21)	13.3 ± 1.0 (20)	12.5 ± 1.0 (20)
30.3 ± 1.5 (20)	36.2 ± 1.4 (20)	9.5 ± 0.5 (20)	12.6 ± 0.7 (20)	8.5 ± 0.6 (20)
27.8 ± 1.2 (9)	33.0 ± 0.9 (9)	9.0 ± 0.6 (9)	11.3 ± 0.6 (9)	7.7 ± 0.6 (9)
32.2 ± 2.0 (14)	36.5 ± 1.3 (14)	7.5 ± 0.7 (14)	11.0 ± 0.8 (14)	8.5 ± 0.7 (14)
29.9 ± 1.8 (18)	34.8 ± 1.6 (18)	7.5 ± 0.8 (17)	10.8 ± 0.65 (18)	8.0 ± 0.8 (18)
21.5 ± 1.1 (20)	23.9 ± 1.4 (20)	6.8 ± 0.5 (20)	7.9 ± 0.6 (20)	5.3 ± 0.8 (20)
21.1 ± 1.9 (10)	23.8 ± 1.3 (10)	6.6 ± 0.3 (10)	7.5 ± 0.4 (10)	5.4 ± 0.5 (10)
—	—	10.5 (1)	13.0 (1)	7.0 (1)
34.0 ± 1.8 (4)	38.3 ± 1.6 (4)	10.3 ± 0.9 (4)	11.1 ± 0.7 (4)	8.1 ± 1.0 (4)
30.5 ± 0.8 (2)	34.9 ± 0.2 (2)	9.3 ± 0.8 (2)	11.3 ± 0.4 (2)	8.5 ± 0.1 (2)
44.0 (1)	47.0 (1)	16.0 (1)	12.0 (1)	8.5 (1)
12.0 ± 0.7 (18)	16.3 ± 0.7 (18)	7.0 ± 0.5 (18)	6.7 ± 0.6 (18)	2.9 ± 0.9 (18)
10.5 ± 0.5 (10)	15.0 ± 0.7 (10)	6.7 ± 0.5 (10)	6.2 ± 0.6 (10)	2.4 ± 0.3 (10)
17.1 ± 0.5 (2)	23.3 ± 0.5 (3)	9.0 ± 0.7 (6)	8.1 ± 1.0 (8)	1.2 ± 0.4 (8)
40.9 ± 2.1 (5)	43.0 ± 1.8 (5)	8.2 ± 0.5 (5)	13.3 ± 0.8 (5)	11.6 ± 1.1 (5)
37.1 ± 1.8 (8)	39.6 ± 1.7 (8)	7.5 ± 0.6 (8)	12.2 ± 0.8 (8)	10.5 ± 0.8 (8)
34.1 (1)	40.0 (1)	10.3 (1)	12.7 (1)	8.5 (1)
[37.0] (1)†	[52.0] (1)†	15.9 ± 3.6 (4)	15.9 ± 2.9 (5)	7.5 ± 2.9 (6)
32.9 ± 4.3 (3)	41.5 ± 2.1 (3)	10.7 ± 0.3 (3)	12.4 ± 0.8 (4)	8.1 ± 1.1 (4)
52.2 ± 2.4 (13)	60.0 ± 2.0 (13)	16.9 ± 1.2 (13)	13.2 ± 1.4 (13)	5.8 ± 0.7 (13)
42.2 ± 3.9 (12)	55.0 ± 3.3 (12)	15.2 ± 1.1 (12)	12.8 ± 1.2 (12)	5.5 ± 0.9 (12)
7.7 ± 0.9 (8)	12.5 ± 0.8 (8)	5.7 ± 0.3 (8)	6.4 ± 0.6 (8)	2.4 ± 0.1 (7)
16.4 ± 1.0 (8)	20.2 ± 1.3 (8)	6.7 ± 0.6 (12)	5.2 ± 0.4 (13)	2.4 ± 0.4 (12)
22.0 ± 1.2 (7)	24.7 ± 1.5 (7)	6.1 ± 0.4 (6)	8.6 ± 0.5 (7)	6.8 ± 0.4 (7)
20.6 ± 1.5 (17)	23.9 ± 1.4 (17)	5.8 ± 0.4 (18)	8.4 ± 0.5 (18)	6.4 ± 0.5 (18)
24.3 ± 1.4 (12)	26.6 ± 1.1 (12)	5.8 ± 0.3 (11)	10.8 ± 0.3 (12)	9.5 ± 0.5 (12)
23.3 ± 1.0 (14)	25.6 ± 1.1 (13)	5.5 ± 0.3 (16)	10.3 ± 0.5 (16)	9.2 ± 0.6 (16)
44.0 ± 0.4 (12)	40.7 ± 1.6 (13)	6.3 ± 0.5 (13)	14.4 ± 0.9 (13)	14.4 ± 0.9 (13)
42.6 ± 1.9 (11)	39.4 ± 1.7 (11)	6.4 ± 0.8 (11)	13.8 ± 0.8 (10)	13.8 ± 0.8 (10)
34.5 ± 0.8 (2)	35.9 ± 2.2 (2)	5.6 ± 0.1 (2)	11.5 ± 0.1 (2)	10.2 ± 0.5 (2)
32.0 ± 1.3 (2)	34.3 ± 1.3 (2)	6.7 ± 0.4 (2)	11.2 ± 0.1 (2)	10.2 ± 0.3 (2)
11.1 ± 0.7 (6)	14.3 ± 0.6 (6)	5.9 ± 0.5 (6)	5.7 ± 0.7 (6)	3.9 ± 0.4 (6)
10.9 ± 0.7 (5)	14.5 ± 0.9 (5)	5.5 ± 0.5 (5)	5.9 ± 0.9 (5)	3.8 ± 0.4 (5)
27.5 ± 1.3 (12)	29.0 ± 1.1 (12)	7.6 ± 0.6 (14)	11.2 ± 0.5 (12)	10.4 ± 0.4 (12)
25.7 ± 1.6 (14)	27.3 ± 1.2 (14)	7.0 ± 0.5 (15)	11.0 ± 0.8 (15)	9.9 ± 0.6 (15)
23.3 ± 2.6 (7)	25.0 ± 1.9 (7)	5.0 ± 0.5 (8)	9.6 ± 0.9 (9)	8.8 ± 0.7 (9)
21.0 ± 2.2 (15)	23.4 ± 1.5 (15)	5.0 ± 0.6 (14)	9.4 ± 0.6 (15)	8.4 ± 1.0 (15)
19.4 ± 1.1 (4)	23.0 ± 1.0 (5)	5.2 ± 0.5 (5)	8.3 ± 0.4 (5)	6.5 ± 0.8 (5)

TABLE 31. Continued.

Taxonomic group	Sex	Scapula length	Coracoid length
	F	24.1 ± 1.0 (6)	14.0 ± 0.4 (6)
Porzana atra	M	27.1 (1)	15.5 (1)
	F	25.5 ± 1.9 (3)	14.5 ± 0.3 (2)
Porzana flavirostra	M	30.6 ± 1.5 (16)	17.9 ± 0.7 (17)
	F	29.9 ± 1.0 (16)	17.5 ± 0.7 (18)
Porzana piercei	[M]	17.5 (1)	11.3 ± 0.2 (9)
	[F]		10.6 ± 0.1 (5)
Porzana keplerorum‡	—	—	—
Porzana severnsi‡	—	22.6 (1)	12.2 (1)
Waterhens			
Amaurornis olivaceus	M	43.0 ± 1.1 (5)	25.1 ± 0.8 (5)
	F	40.9 ± 0.8 (4)	24.2 ± 1.3 (4)
Amaurornis moluccanus	M	41.3 ± 1.6 (3)	23.5 ± 1.0 (3)
	F	37.9 ± 0.5 (2)	21.8 ± 0.5 (2)
Amaurornis phoenicurus	M	41.9 ± 1.4 (15)	24.9 ± 0.8 (16)
	F	39.9 ± 1.9 (15)	23.6 ± 0.9 (15)
*Amaurornis ineptus**	—	57.0 (1)	36.0 (1)
Moorhens and allies			
Gallicrex cinerea	M	55.5 ± 2.3 (8)	30.8 ± 1.6 (8)
	F	45.7 ± 0.1 (2)	25.3 ± 1.3 (2)
*Pareudiastes pacificus**	—	33.0 (1)	20.0 (1)
Tribonyx ventralis	M	55.7 ± 1.5 (25)	29.8 ± 1.0 (24)
	F	52.0 ± 2.3 (19)	27.9 ± 1.3 (19)
Tribonyx mortierii	M	64.7 ± 1.2 (8)	35.6 ± 1.2 (8)
	F	63.9 ± 3.0 (9)	35.0 ± 1.6 (9)
Tribonyx hodgenorum	[M]	37.2 ± 2.2 (3)	20.2 ± 0.7 (12)
	[F]		18.5 ± 0.7 (8)
Gallinula nesiotis	—	46.1 ± 1.6 (5)	24.9 ± 1.3 (5)
Gallinula comeri	—	49.1 ± 1.0 (4)	26.5 ± 1.0 (4)
Gallinula tenebrosa	M	60.3 ± 1.9 (6)	33.4 ± 0.8 (6)
	F	58.8 ± 2.6 (8)	32.1 ± 1.1 (8)
Gallinula chloropus	M	49.8 ± 3.6 (8)	27.0 ± 2.1 (10)
	F	48.0 ± 3.3 (16)	26.5 ± 1.9 (19)
Gallinula cachinaans	M	52.6 ± 2.0 (12)	29.4 ± 1.1 (12)
	F	49.3 ± 2.5 (8)	27.5 ± 0.9 (9)
Gallinula sandvicensis	—	47.2 ± 0.9 (4)	26.9 ± 0.5 (5)
Coots			
Fulica armillata	—	66.1 ± 2.8 (4)	36.1 ± 1.9 (4)
Fulica cornuta	—	88.1 ± 2.3 (2)	48.6 ± 1.3 (2)
Fulica gigantea	M	87.1 ± 2.9 (6)	48.4 ± 1.3 (6)
	F	79.6 ± 6.2 (5)	45.4 ± 2.2 (5)
Fulica americana	M	60.3 ± 2.5 (17)	32.7 ± 1.4 (17)
	F	55.9 ± 2.3 (15)	30.5 ± 1.2 (15)
Fulica alai	M	60.8 ± 2.8 (7)	32.1 ± 1.7 (7)
	F	55.5 ± 3.3 (7)	29.4 ± 0.9 (8)
Fulica atra (*atra*-group)	M	62.6 ± 2.2 (15)	33.7 ± 1.7 (15)
	F	61.1 ± 2.5 (21)	32.5 ± 1.6 (22)
Fulica chathamensis	[M]	76.0 ± 2.2 (19)	44.0 ± 1.4 (18)
	[F]	71.0 ± 1.5 (7)	41.5 ± 0.9 (14)
Fulica prisca	[M]	76.0 ± 2.8 (2)	42.4 ± 1.3 (13)
	[F]		39.6 ± 1.6 (4)
Fulica newtoni	—	—	[41.0 (1)]§

* Measurements taken on elements during myological dissection.
† Measurement estimated from photographs of mounted specimen at Mauritius Institute taken by J. Hume.
‡ Samples based in part on measurements tabulated by Olson and James (1991).
§ Measurement based on Mourer-Chauviré et al. (1999: fig. 131).

TABLE 31. Continued, Extended.

	Sternum			
	Corpus		Carina depth	
Carina length	Length	Least width	Oblique	Perpendicular
18.4 ± 1.1 (6)	22.0 ± 0.8 (6)	5.0 ± 0.5 (6)	8.6 ± 0.7 (6)	6.9 ± 0.7 (6)
15.5 (1)	21.8 (1)	6.4 (1)	10.1 (1)	5.8 (1)
15.2 ± 0.3 (3)	21.8 ± 0.5 (3)	6.2 ± 0.6 (3)	9.4 ± 0.1 (3)	5.7 ± 0.2 (3)
27.1 ± 1.0 (13)	30.5 ± 0.7 (13)	6.9 ± 0.6 (12)	10.1 ± 0.5 (15)	8.2 ± 0.6 (15)
27.1 ± 1.3 (17)	30.4 ± 1.3 (16)	6.6 ± 0.5 (16)	10.0 ± 0.4 (16)	8.0 ± 0.4 (16)
—	—	—	—	—
—	—	—	—	—
—	—	—	5.2 (1)	1.6 (1)
12.0 (1)	18.3 (1)	13.1 (1)	7.0 (1)	1.0 (1)
39.2 ± 0.8 (4)	41.9 ± 0.9 (5)	9.4 ± 0.6 (5)	11.1 ± 0.8 (4)	9.1 ± 0.4 (4)
37.7 ± 0.9 (4)	40.9 ± 0.6 (4)	8.9 ± 1.0 (4)	10.6 ± 0.7 (3)	8.6 ± 0.4 (3)
38.7 ± 2.7 (3)	40.6 ± 2.8 (3)	8.3 ± 0.4 (3)	13.2 ± 0.6 (3)	11.9 ± 0.7 (3)
36.5 ± 0.7 (2)	39.3 ± 0.1 (2)	8.4 ± 0.1 (2)	12.7 ± 0.5 (2)	10.8 ± 0.4 (2)
42.8 ± 2.4 (16)	43.6 ± 2.0 (16)	8.5 ± 0.6 (16)	12.5 ± 0.9 (16)	11.5 ± 0.8 (16)
39.4 ± 2.0 (15)	40.3 ± 1.8 (15)	7.7 ± 0.6 (15)	11.8 ± 0.7 (15)	11.1 ± 0.5 (15)
50.0 (1)	54.0 (1)	23.0 (1)	12.5 (1)	8.0 (1)
59.7 ± 3.4 (8)	57.4 ± 3.1 (8)	7.5 ± 0.7 (8)	16.1 ± 0.8 (8)	15.7 ± 1.0 (8)
48.3 ± 5.7 (2)	47.8 ± 4.7 (2)	5.4 ± 0.9 (2)	14.2 ± 0.9 (2)	13.4 ± 0.8 (2)
24.0 (1)	32.0 (1)	10.0 (1)	12.0 (1)	7.0 (1)
54.8 ± 2.5 (26)	51.5 ± 1.8 (25)	14.4 ± 1.0 (25)	17.1 ± 0.8 (25)	16.2 ± 0.7 (25)
50.1 ± 3.0 (19)	47.5 ± 2.6 (18)	13.2 ± 0.8 (19)	16.0 ± 0.9 (19)	15.0 ± 0.8 (19)
54.0 ± 2.0 (9)	55.8 ± 1.7 (9)	20.8 ± 1.4 (9)	13.2 ± 1.2 (9)	9.6 ± 1.2 (9)
51.6 ± 2.4 (9)	54.2 ± 2.4 (9)	19.4 ± 0.6 (9)	13.2 ± 0.7 (9)	10.4 ± 1.3 (9)
{ 25.0 (1)	{ 27.6 (1)	13.1 (1)	{ 7.5 (1)	—
34.6 ± 2.4 (5)	38.9 ± 2.4 (5)	14.1 ± 1.5 (5)	11.8 ± 0.7 (5)	8.3 ± 0.7 (5)
38.6 ± 1.2 (4)	42.3 ± 2.1 (4)	15.9 ± 0.6 (4)	12.0 ± 0.8 (4)	8.5 ± 0.8 (4)
60.1 ± 1.6 (6)	57.8 ± 1.9 (6)	15.6 ± 0.9 (6)	17.9 ± 1.2 (6)	16.9 ± 0.4 (5)
59.3 ± 1.6 (8)	55.5 ± 3.4 (8)	15.5 ± 1.1 (8)	17.1 ± 1.0 (8)	15.8 ± 1.0 (6)
46.6 ± 3.0 (10)	45.4 ± 3.0 (10)	12.7 ± 1.3 (10)	15.7 ± 1.2 (10)	14.6 ± 1.1 (10)
44.5 ± 3.4 (18)	43.4 ± 3.1 (18)	112.1 ± 1.0 (19)	14.7 ± 1.3 (19)	13.7 ± 1.4 (19)
50.5 ± 2.6 (12)	48.8 ± 0.8 (12)	14.5 ± 0.8 (12)	15.9 ± 0.6 (12)	15.1 ± 0.7 (12)
46.7 ± 2.6 (8)	45.5 ± 2.3 (8)	14.0 ± 1.3 (8)	15.3 ± 1.1 (8)	13.9 ± 0.8 (8)
42.1 ± 1.3 (4)	41.1 ± 1.1 (4)	14.0 ± 0.7 (3)	14.0 ± 0.6 (5)	12.8 ± 0.8 (5)
64.6 ± 3.7 (4)	62.7 ± 3.7 (4)	22.6 ± 1.3 (4)	19.0 ± 2.2 (4)	16.7 ± 1.9 (3)
84.8 ± 2.9 (2)	82.7 ± 1.6 (2)	30.7 ± 2.1 (2)	27.8 ± 1.2 (2)	22.5 ± 1.4 (2)
84.7 ± 2.8 (6)	85.6 ± 2.4 (6)	30.7 ± 1.1 (6)	23.9 ± 1.4 (6)	20.1 ± 0.8 (3)
80.0 ± 4.8 (5)	78.7 ± 5.6 (5)	27.7 ± 2.9 (5)	23.0 ± 1.3 (5)	20.0 ± 1.3 (4)
56.1 ± 3.0 (17)	54.6 ± 2.3 (17)	19.1 ± 0.7 (17)	18.3 ± 0.7 (16)	16.6 ± 0.8 (16)
51.5 ± 2.3 (12)	50.0 ± 1.8 (12)	17.1 ± 1.2 (15)	17.5 ± 0.7 (15)	15.4 ± 0.8 (14)
55.5 ± 2.1 (7)	54.2 ± 1.8 (6)	18.7 ± 1.2 (7)	18.2 ± 0.7 (6)	15.7 ± 1.0 (7)
49.9 ± 3.3 (7)	49.7 ± 3.2 (7)	17.5 ± 0.8 (7)	16.8 ± 1.2 (7)	14.8 ± 0.8 (7)
60.2 ± 2.4 (15)	57.0 ± 2.2 (15)	18.8 ± 1.0 (15)	18.5 ± 1.0 (15)	16.8 ± 0.9 (15)
58.0 ± 3.1 (22)	55.0 ± 2.5 (22)	18.2 ± 1.2 (21)	18.4 ± 1.1 (22)	16.5 ± 1.1 (21)
72.8 ± 1.8 (19)	72.9 ± 1.7 (19)	27.2 ± 1.1 (19)	20.0 ± 1.0 (19)	15.6 ± 1.3 (19)
68.7 ± 2.5 (5)	70.1 ± 1.6 (5)	26.0 ± 0.7 (5)	18.7 ± 0.4 (5)	15.2 ± 0.6 (5)
{ 64.9 ± 2.9 (4)	{ 66.1 ± 2.9 (4)	27.7 ± 1.4 (5)	{ 19.8 ± 1.4 (7)	14.3 ± 1.8 (7)
66.5 (1)	65.3 (1)	20.0 (1)	19.9 (1)	17.0 (1)

TABLE 32. Univariate summary statistics ($\bar{x} \pm$ SD (n)) for six measurements (mm) of the pectoral limb of modern and subfossil species of Rallidae, by species and sex. Provisional sex groups or approximations are enclosed by brackets. Superspecies within *Porphyrio*, *Aramides*, *Rallus*, *Habropteryx*, *Amaurornis*, *Gallinula*, and *Fulica* follow Livezey (1998).

Taxonomic group	Sex	Humerus Corpus length	Humerus Caput width	Ulna corpus length	Carpometacarpus corpus length	Digitalis majoris, length Phalanx proximalis	Digitalis majoris, length Phalanx distalis
Swamphens							
Porphyrio madagas-	M	75.9 ± 3.6 (9)	16.1 ± 0.8 (9)	68.3 ± 3.1 (8)	46.4 ± 1.9 (9)	18.2 ± 0.9 (9)	15.9 ± 0.8 (8)
cariensis	F	74.3 ± 3.0 (9)	15.9 ± 0.6 (9)	67.4 ± 3.0 (8)	45.9 ± 1.9 (9)	18.2 ± 0.7 (9)	15.5 ± 0.8 (9)
Porphyrio melanotus	M	86.2 ± 3.0 (33)	18.4 ± 0.5 (33)	80.7 ± 2.8 (33)	53.9 ± 1.9 (32)	21.0 ± 0.9 (32)	18.3 ± 0.7 (22)
	F	80.3 ± 2.1 (19)	16.9 ± 0.4 (20)	75.3 ± 2.0 (20)	50.6 ± 1.3 (20)	20.0 ± 0.7 (20)	17.4 ± 0.7 (17)
Porphyrio mantelli	[M]	90.2 ± 2.6 (14)	21.3 ± 1.2 (14)	79.5 ± 0.5 (3)	66.0 (1)		
	[F]	84.9 ± 2.0 (8)	19.3 ± 0.6 (8)	74.1 ± 1.5 (4)			
Porphyrio hochstetteri	M	84.6 ± 2.4 (13)	19.6 ± 0.7 (13)	71.1 ± 2.3 (11)	45.2 ± 1.6 (11)	13.8 ± 0.6 (8)	12.1 ± 0.7 (8)
	F	82.3 ± 2.0 (18)	19.0 ± 0.7 (18)	70.0 ± 2.0 (18)	44.4 ± 1.3 (17)	13.6 ± 0.6 (15)	12.1 ± 1.0 (13)
Porphyrio kukwiedei	[M]	80.5 (1)	17.1 (1)				
	[F]	74.1 ± 0.1 (3)	15.9 ± 0.5 (3)	59.2 ± 0.1 (2)			
Aphanocrex podarces	—		13.0 ± 0.1 (2)	48.5 (1)	35.1 ± 1.5 (3)	10.5 (1)	
Porphyrula alleni	M	48.1 ± 1.8 (3)	9.9 ± 0.7 (3)	42.6 ± 2.0 (3)	30.4 ± 1.8 (3)	12.6 ± 0.8 (3)	11.0 ± 1.3 (3)
	F	42.7 ± 1.6 (5)	9.1 ± 0.4 (5)	38.0 ± 1.5 (5)	26.8 ± 0.9 (5)	10.7 ± 0.3 (5)	9.3 ± 0.6 (5)
Porphyrula martinica	M	52.6 ± 1.1 (15)	10.8 ± 0.4 (15)	46.0 ± 1.2 (15)	32.2 ± 0.7 (15)	13.3 ± 0.4 (15)	12.0 ± 0.5 (14)
	F	50.0 ± 1.4 (15)	10.2 ± 0.4 (15)	43.7 ± 1.3 (15)	30.8 ± 0.7 (15)	12.8 ± 0.4 (15)	11.4 ± 0.5 (14)
Porphyrula flavirostris	M	37.0 ± 0.4 (3)	7.7 ± 0.1 (3)	31.1 ± 1.2 (3)	22.6 ± 1.0 (3)	9.8 ± 0.2 (3)	8.9 ± 0.4 (3)
	F				22.3 (1)	9.9 (1)	8.4 (1)
Wallace's Rail and allies							
Gymnocrex plumbeiventris	—			51.7 (1)			10.3 (1)
Habroptila wallacii	M	61.0 ± 1.1 (3)	12.1 ± 0.3 (3)	50.0 ± 0.2 (3)	31.5 ± 0.6 (3)	10.6 ± 0.2 (3)	9.7 ± 0.7 (3)
	F	59.4 (1)	11.2 ± 0.1 (3)	48.9 (1)	30.6 (1)	9.6 (1)	8.7 (1)
Cave-rails and allies							
Eulabeornis							
castaneoventris	—		14.1 (1)	51.8 (1)	36.2 (1)	13.0 (1)	12.1 (1)
Aramides ypecaha	M	73.6 ± 1.3 (4)	17.5 ± 0.3 (4)	65.3 ± 1.7 (4)	45.1 ± 0.6 (4)	18.1 ± 0.4 (4)	15.2 ± 0.7 (4)
	F	71.2 ± 1.5 (8)	16.7 ± 0.4 (8)	63.8 ± 2.4 (8)	43.0 ± 1.1 (8)	17.0 ± 0.7 (8)	14.0 ± 0.6 (8)
Aramides cajanea-group	M	56.9 ± 1.9 (27)	12.5 ± 0.4 (27)	49.5 ± 1.9 (26)	53.8 ± 1.3 (27)	12.6 ± 0.6 (27)	11.0 ± 0.7 (24)
	F	55.2 ± 2.4 (24)	12.3 ± 0.5 (24)	48.4 ± 1.9 (23)	33.2 ± 1.6 (23)	12.7 ± 0.7 (24)	11.0 ± 0.8 (19)
Canirallus oculeus	—	49.6 (1)	11.5 (1)	50.4 (1)	29.5 (1)	10.8 (1)	—

TABLE 32. Continued.

Taxonomic group	Sex	Humerus Corpus length	Caput width	Ulna corpus length	Carpometacarpus corpus length	Digitalis majoris, length Phalanx proximalis	Phalanx distalis
Canirallus kioloides	—	37.2 ± 0.9 (3)	9.2 ± 0.7 (3)	34.8 ± 1.1 (2)	22.7 ± 0.3 (3)	9.5 ± 0.2 (3)	5.9 ± 2.1 (3)
Nesotrochis debooyi	[M]	60.8 ± 3.1 (3)	13.0 ± 0.4 (3)	45.2 ± 0.2 (4)	26.6 ± 1.4 (4)	—	—
	[F]			41.6 ± 2.1 (2)		—	—
Nesotrochis steganinos	[M]	48.2 ± 1.5 (3)	11.7 ± 0.3 (3)	33.0 ± 1.2 (2)	21.4 ± 1.5 (3)	—	—
	[F]	43.3 ± 0.2 (2)	10.5 ± 0.0 (2)			—	—
*Nesotrochis picapicensis**	—	[49.5] (4)	—	—	—	—	—
Rougetius rougetii	—	47.0 (1)	9.0 (1)	41.8 (1)	27.4 (1)	9.5 (1)	8.3 (1)
Zapata Rail and allies							
Cyanolimnas cerverai	[M]	36.1 ± 1.0 (9)	7.2 ± 0.2 (9)	27.5 ± 1.1 (8)	18.2 ± 0.1 (2)	—	—
	[F]	34.5 ± 0.7 (5)	6.8 ± 0.3 (5)	24.1 ± 0.7 (6)		—	—
White-throated rails							
Dryolimnas cuvieri	M	50.4 ± 0.9 (3)	10.4 ± 0.3 (3)	43.5 ± 1.5 (3)	29.3 ± 0.8 (3)	10.2 ± 0.4 (3)	9.7 ± 0.2 (3)
	F	48.2 ± 0.3 (2)	10.2 ± 0.5 (3)	41.5 ± 1.3 (2)	28.5 ± 0.3 (2)	10.3 ± 0.6 (2)	9.1 ± 0.0 (2)
Dryolimnas aldabranus	M	40.5 ± 1.0 (10)	8.7 ± 0.2 (10)	33.6 ± 0.9 (10)	22.2 ± 0.6 (10)	8.2 ± 0.3 (10)	7.4 ± 0.3 (6)
	F	37.7 ± 0.4 (5)	8.2 ± 0.2 (6)	31.1 ± 0.4 (3)	20.7 ± 0.6 (5)	7.6 ± 0.2 (5)	7.1 ± 0.1 (4)
Dryolimnas abbotti	—	44.6 (1)	9.8 (1)	37.9 (1)	25.3 (1)	—	—
Rallus sensu stricto							
Rallus aquaticus	M	40.3 ± 1.3 (7)	80.3 ± 0.3 (17)	32.7 ± 1.3 (17)	21.5 ± 0.8 (1)	8.1 ± 0.4 (16)	7.6 ± 0.4 (16)
	F	38.2 ± 1.3 (21)	7.8 ± 0.3 (21)	30.8 ± 1.1 (21)	21.1 ± 0.8 (21)	7.6 ± 0.4 (21)	7.1 ± 0.4 (19)
Rallus limicola	M	37.3 ± 1.2 (20)	7.3 ± 0.2 (21)	30.9 ± 1.2 (21)	20.7 ± 0.8 (21)	7.4 ± 0.3 (21)	6.9 ± 0.3 (20)
	F	35.0 ± 1.5 (17)	6.8 ± 0.2 (18)	28.7 ± 1.5 (18)	19.3 ± 1.1 (17)	6.9 ± 0.4 (17)	6.5 ± 0.4 (16)
Rallus ibycus	[M]	33.8 ± 0.6 (8)	6.9 ± 0.2 (8)	25.6 ± 0.8 (8)	16.5 ± 3.0 (6)	—	—
	[F]	30.6 ± 1.3 (5)	6.2 ± 0.2 (5)	23.2 ± 0.5 (5)	15.3 ± 2.9 (6)	—	—
Rallus longirostris-group	M	56.6 ± 1.7 (17)	11.3 ± 0.5 (17)	45.8 ± 2.2 (16)	30.0 ± 1.2 (17)	11.0 ± 0.4 (17)	9.9 ± 0.4 (15)
	F	54.1 ± 1.7 (15)	10.5 ± 0.6 (17)	45.8 ± 2.2 (15)	28.9 ± 0.8 (16)	10.5 ± 0.6 (17)	9.8 ± 0.5 (15)
Rallus elegans-group	M	61.7 ± 2.1 (25)	12.6 ± 0.4 (25)	51.6 ± 1.9 (24)	31.3 ± 0.6 (25)	12.2 ± 0.6 (25)	11.2 ± 0.7 (22)
	F	57.8 ± 1.4 (11)	11.5 ± 0.3 (12)	47.8 ± 1.5 (9)	32.1 ± 1.0 (12)	11.7 ± 0.5 (11)	10.4 ± 0.5 (10)
Rallus recessus	[M]	45.5 ± 0.4 (10)	9.7 ± 0.2 (10)	35.0 ± 0.9 (15)	22.8 ± 0.4 (12)	8.1 (1)	—
	[F]	43.9 ± 0.8 (18)	9.5 ± 0.2 (18)	32.0 ± 0.7 (6)	22.1 ± 0.9 (15)		—
Gallirallus and allies							
Gallirallus pectoralis-group	—	35.5 ± 1.1 (8)	6.8 ± 0.3 (8)	29.1 ± 0.7 (8)	19.2 ± 0.7 (8)	7.0 ± 0.2 (8)	6.2 ± 0.2 (6)
Gallirallus striatus	M	41.7 ± 2.6 (3)	8.7 ± 0.6 (3)	35.1 ± 2.4 (3)	23.9 ± 1.5 (3)	8.6 ± 0.5 (3)	8.0 ± 1.1 (2)
	F	41.6 ± 1.3 (2)	8.5 ± 0.4 (2)	36.0 ± 1.3 (2)	24.0 ± 1.3 (2)	8.5 ± 0.4 (2)	8.0 ± 0.8 (2)
Gallirallus australis	M	56.6 ± 2.3 (15)	13.2 ± 0.7 (19)	42.7 ± 2.1 (18)	28.8 ± 1.4 (18)	9.9 ± 0.7 (17)	9.8 ± 0.7 (15)
	F	50.7 ± 1.6 (13)	11.6 ± 0.5 (13)	38.4 ± 1.2 (12)	25.9 ± 1.0 (12)	8.9 ± 0.6 (10)	8.7 ± 0.7 (8)
Gallirallus greyi	M	54.8 ± 3.8 (4)	12.9 ± 0.5 (4)	41.2 ± 2.7 (4)	27.6 ± 2.4 (4)	9.3 ± 0.8 (3)	8.3 (1)

TABLE 32. Continued.

Taxonomic group	Sex	Humerus			Ulna corpus length	Carpometacarpus corpus length	Digitalis majoris, length	
		Corpus length	Caput width				Phalanx proximalis	Phalanx distalis
Gallirallus philippensis-group	F	50.0 ± 1.5 (4)	11.7 ± 0.6 (4)		37.2 ± 1.4 (4)	25.0 ± 1.1 (4)	8.5 ± 0.2 (4)	8.7 ± 0.6 (4)
	M	47.3 ± 1.8 (16)	9.5 ± 0.4 (16)		42.4 ± 2.0 (16)	26.9 ± 1.3 (16)	10.0 ± 0.8 (15)	9.4 ± 0.8 (13)
	F	47.0 ± 1.6 (19)	9.2 ± 0.6 (21)		40.1 ± 1.8 (18)	26.6 ± 1.5 (19)	9.8 ± 0.6 (18)	9.2 ± 0.6 (13)
Gallirallus dieffenbachii	[M]	44.9 ± 0.8 (33)	10.0 ± 0.3 (33)		33.6 ± 0.7 (15)	23.2 ± 3.5 (23)	8.4 ± 0.3 (5)	7.8 (1)‡
	[F]	42.4 ± 0.8 (24)	9.5 ± 0.4 (24)		31.3 ± 1.0 (14)	20.9 ± 0.7 (17)		
Gallirallus owstoni	M	46.0 ± 2.8 (14)	9.4 ± 0.6 (14)		37.9 ± 2.3 (14)	24.6 ± 1.5 (14)	8.5 ± 0.6 (14)	9.0 ± 0.6 (11)
	F	44.1 ± 1.9 (18)	8.8 ± 0.3 (18)		36.2 ± 1.7 (18)	23.2 ± 1.6 (18)	8.0 ± 0.6 (18)	8.3 ± 0.6 (17)
Gallirallus wakensis	M	33.7 ± 0.4 (2)	7.3 ± 0.2 (12)		28.1 ± 0.6 (2)	18.9 ± 0.5 (2)	7.3 ± 0.3 (2)	6.7 ± 0.1 (2)
	F		6.9 ± 0.4 (4)					
Tricholimnas lafresnayanus	—	31.3 (1)			25.0 (1)	16.6 (1)	6.1 (1)	5.9 (1)
Tricholimnas sylvestris	M	51.0 ± 2.5 (4)	11.2 ± 1.1 (4)		41.3 ± 1.0 (4)	26.9 ± 0.7 (4)	10.3 ± 0.7 (4)	9.9 ± 0.2 (4)
	F	47.5 ± 1.1 (2)	10.3 ± 0.1 (2)		39.3 ± 0.9 (2)	26.5 ± 0.5 (2)	10.0 ± 0.6 (2)	8.9 ± 0.2 (2)
Nesoclopeus poeciloperus-group	—	54.0 (1)	—		45.6 ± 0.6 (2)	31.2 ± 1.2 (2)	11.1 ± 1.3 (2)	10.1 ± 0.1 (2)
Cabalus modestus	[M]	25.7 ± 0.8 (14)	5.0 ± 0.2 (14)		19.9 ± 0.5 (9)	13.4 ± 0.3 (11)	5.0 (1)‡	4.5 (1)‡
	[F]	23.8 ± 0.4 (14)	4.8 ± 0.1 (14)		18.1 ± 0.4 (19)	12.2 ± 0.4 (17)		
Capellirallus karamu	[M]	27.5 ± 0.5 (9)	5.6 ± 0.2 (9)		16.2 ± 0.5 (5)	10.4 ± 0.8 (4)	—	—
	[F]	25.8 ± 0.7 (12)	5.2 ± 0.4 (12)		14.8 ± 0.3 (4)			
Habropteryx insignis	—				41.1 (1)	25.8 (1)	9.0 (1)	9.0 (1)
Habropteryx torquatus-group	M	48.6 ± 1.3 (5)	10.5 ± 0.3 (5)		43.2 ± 1.1 (5)	28.9 ± 0.7 (6)	10.2 ± 0.7 (5)	9.4 ± 0.5 (6)
	F	46.6 ± 1.6 (7)	9.7 ± 0.4 (8)		40.7 ± 1.5 (7)	27.1 ± 0.6 (7)	9.5 ± 0.4 (7)	8.9 ± 0.3 (7)
Habropteryx okinawae	—	50.5 (1)	10.3 (1)		43.9 (1)	28.3 (1)	10.1 (1)	—
Problematic subfossil rails								
Aphanapteryx bonasia	[M]	65.0 ± 0.8 (7)	14.4 ± 0.5 (7)		50.6 ± 0.8 (94)	28.8 ± 2.5 (6)	13.7 ± 3.4 (4)	8.6 (1)
	[F]	59.0 ± 2.2 (7)	13.4 ± 0.7 (7)		43.3 ± 2.1 (4)			
Erythromachus leguati	—	48.1 ± 3.3 (2)	11.7 ± 1.3 (2)		40.1 ± 0.5 (4)	—	—	—
Diaphorapteryx hawkinsi	[M]	65.8 ± 1.8 (34)	16.3 ± 1.2 (34)		44.3 ± 1.2 (13)	27.7 ± 0.5 (13)	13.0 ± 3.4 (5)	—
	[F]	60.3 ± 1.7 (21)	15.1 ± 1.2 (21)		40.5 ± 1.6 (18)	25.3 ± 0.7 (26)		
Atlantisia and allies								
Atlantisia rogersi	—	20.2 ± 0.5 (2)	3.9 ± 0.2 (5)		13.4 ± 0.1 (2)	9.4 ± 0.1 (2)	4.0 ± 0.1 (2)	3.4 ± 0.2 (2)
Mundia elpenor	[M]	26.6 ± 0.5 (12)	6.0 ± 0.2 (12)		20.5 ± 0.5 (13)	14.3 ± 0.5 (5)	5.3 ± 0.2 (5)	4.5 (1)
	[F]	24.4 ± 0.5 (3)	5.6 ± 0.3 (3)		19.3 ± 0.6 (4)	13.4 ± 0.3 (8)		
Laterallus leucopyrrhus	M	27.2 ± 0.9 (7)	5.5 ± 0.2 (7)		22.2 ± 0.6 (7)	15.7 ± 0.5 (7)	6.2 ± 0.3 (7)	5.0 ± 0.3 (7)
	F	26.0 ± 1.2 (20)	5.3 ± 0.3 (21)		21.1 ± 0.3 (20)	14.9 ± 0.7 (19)	5.8 ± 0.3 (20)	4.4 ± 0.3 (19)
Coturnicops noveboracensis	M	29.9 ± 1.0 (15)	6.0 ± 0.2 (15)		25.6 ± 0.9 (15)	17.7 ± 0.6 (15)	6.2 ± 0.2 (15)	5.5 ± 0.3 (15)
	F	28.4 ± 0.8 (20)	5.7 ± 0.2 (19)		24.3 ± 0.8 (20)	16.8 ± 0.4 (20)	5.8 ± 0.2 (20)	5.3 ± 0.3 (18)
Crex crex	M	45.0 ± 1.6 (13)	9.1 ± 0.5 (13)		40.3 ± 1.3 (13)	26.9 ± 1.0 (13)	9.4 ± 0.5 (13)	9.2 ± 0.5 (12)

TABLE 32. Continued.

Taxonomic group	Sex	Humerus Corpus length	Humerus Caput width	Ulna corpus length	Carpometacarpus corpus length	Digitalis majoris, length Phalanx proximalis	Digitalis majoris, length Phalanx distalis
Crex albicollis	F	44.4 ± 1.3 (11)	9.1 ± 0.3 (11)	39.7 ± 1.4 (11)	26.2 ± 1.5 (11)	9.1 ± 0.4 (11)	8.8 ± 0.7 (11)
	M	39.0 ± 1.9 (2)	8.1 ± 0.1 (2)	36.0 ± 4.5 (2)	21.9 ± 1.0 (2)	8.0 ± 0.3 (2)	7.8 (1)
	F	38.2 ± 0.5 (2)	7.9 ± 0.0 (2)	33.0 ± 0.4 (2)	21.9 ± 0.7 (2)	7.9 ± 0.6 (2)	7.3 ± 0.7 (2)
Gray crakes and allies							
Porzana sandwichensis	—	—	—	14.9 (1)	11.0 (1)	4.0 (1)	3.4 (1)
Porzana palmeri	M	20.4 ± 0.4 (5)	4.2 ± 0.2 (5)	15.1 ± 0.3 (7)	10.4 ± 0.4 (7)	3.9 ± 0.2 (7)	3.6 ± 0.2 (6)
	F	19.5 ± 0.2 (5)	4.1 ± 0.2 (6)	14.5 ± 0.5 (6)	10.1 ± 0.5 (6)	3.7 ± 0.5 (6)	3.3 ± 0.3 (5)
Porzana carolina	M	33.7 ± 0.7 (15)	6.8 ± 0.2 (15)	28.5 ± 0.5 (15)	20.4 ± 0.4 (15)	7.5 ± 0.4 (15)	7.3 ± 0.3 (14)
	F	32.4 ± 0.9 (15)	6.3 ± 0.2 (15)	27.1 ± 0.9 (15)	19.4 ± 0.5 (15)	7.1 ± 0.3 (15)	6.6 ± 0.3 (13)
Porzana pusilla	M	27.1 ± 1.2 (9)	5.4 ± 0.4 (9)	22.1 ± 0.9 (8)	15.6 ± 0.9 (9)	6.0 ± 0.4 (9)	5.3 ± 0.4 (7)
	F	25.7 ± 1.8 (10)	5.0 ± 0.4 (12)	20.7 ± 1.4 (12)	14.9 ± 0.9 (11)	5.8 ± 0.3 (9)	5.0 ± 0.3 (7)
Porzana tabuensis	M	28.0 ± 0.7 (5)	5.2 ± 0.2 (5)	21.5 ± 0.5 (4)	15.1 ± 0.6 (4)	5.9 ± 0.2 (4)	5.3 ± 0.4 (2)
	F	27.0 ± 0.7 (6)	5.1 ± 0.2 (6)	20.8 ± 0.5 (6)	14.7 ± 0.8 (6)	5.6 ± 0.4 (5)	5.0 ± 0.3 (3)
Porzana monasa†	—	—	—	—	[18.0] (1)	[7.0] (1)	[5.0] (1)
Porzana atra	M	28.2 (1)	5.9 (1)	21.5 (1)	14.7 (1)	6.0 (1)	4.5 (1)
	F	28.0 ± 0.7 (3)	5.7 ± 0.0 (3)	21.1 ± 0.5 (3)	14.3 ± 0.3 (3)	5.4 ± 0.2 (3)	4.8 ± 0.3 (3)
Porzana flavirostra	M	33.4 ± 1.3 (17)	6.9 ± 0.4 (17)	27.8 ± 1.3 (16)	20.0 ± 0.9 (17)	7.3 ± 0.3 (16)	6.0 ± 0.3 (15)
	F	32.5 ± 0.9 (17)	6.8 ± 0.2 (18)	27.4 ± 1.2 (18)	19.8 ± 0.9 (18)	7.3 ± 0.3 (16)	6.0 ± 0.4 (14)
Porzana piercei	[M]	22.3 ± 0.6 (7)	4.4 ± 0.2 (7)	16.4 ± 0.5 (8)	11.5 ± 0.2 (10)	—	—
	[F]	20.6 ± 0.7 (10)	4.3 ± 0.1 (10)	15.6 ± 0.2 (7)	10.6 ± 0.8 (5)	—	—
Porzana rua†	—	25.2 (1)	4.9 (1)	19.4 (1)	—	—	—
Porzana astrictocarpus	—	20.3 ± 0.5 (4)	4.2 ± 0.1 (4)	—	11.2 (1)	—	—
Porzana ziegleri§	—	18.2 ± 0.5 (4)	3.5 ± 0.1 (3)	12.9 ± 0.3 (4)	8.4 ± 0.1 (3)	—	—
Porzana menehune§	—	13.5 ± 0.4 (5)	3.0 ± 0.1 (3)	9.0 ± 0.3 (10)	6.3 ± 0.1 (5)	—	—
Porzana keplerorum§	—	15.0 ± 0.7 (8)	2.9 ± 0.2 (4)	12.5 (1)	6.7 (1)	—	—
Porzana ralphorum§	—	26.6 (1)	5.1 (1)	—	—	—	—
Porzana severnsi§	—	24.5 ± 0.6 (5)	5.7 ± 0.8 (5)	17.8 (1)	11.6 ± 0.0 (2)	—	—
Waterhens							
Amaurornis olivaceus	M	50.1 ± 1.6 (5)	10.5 ± 0.2 (5)	44.3 ± 1.4 (5)	29.7 ± 0.4 (5)	10.7 ± 0.8 (4)	9.4 ± 0.3 (5)
	F	48.2 ± 1.8 (4)	9.8 ± 0.4 (4)	42.0 ± 1.9 (4)	28.1 ± 1.4 (4)	9.9 ± 0.5 (3)	9.1 ± 0.1 (4)
Amaurornis moluccanus	M	46.2 ± 2.0 (3)	9.5 ± 0.5 (3)	40.6 ± 2.1 (3)	26.6 ± 1.1 (3)	9.9 ± 0.5 (3)	8.7 ± 0.2 (3)
	F	43.8 ± 0.0 (2)	8.8 ± 0.3 (2)	37.8 ± 0.4 (2)	24.4 ± 0.7 (2)	9.7 ± 0.7 (2)	7.7 ± 0.5 (2)
Amaurornis phoenicurus	M	48.3 ± 1.6 (16)	10.3 ± 0.3 (16)	42.8 ± 1.5 (16)	29.4 ± 1.3 (16)	11.1 ± 0.4 (14)	9.7 ± 0.6 (15)
	F	46.4 ± 1.6 (15)	9.7 ± 0.3 (15)	40.7 ± 1.8 (15)	27.9 ± 1.3 (15)	10.3 ± 0.4 (13)	8.7 ± 0.3 (14)
Amaurornis akool	—	—	—	33.1 (1)	—	8.2 (1)	6.9 (1)
Amaurornis isabellinus	—	—	—	46.7 (1)	30.3 (1)	11.8 (1)	10.4 (1)
Amaurornis ineptus	—	—	—	45.3 (1)	29.6 (1)	11.0 (1)	10.0 (1)

TABLE 32. Continued.

Taxonomic group	Sex	Humerus		Ulna corpus length	Carpometacarpus corpus length	Digitalis majoris, length	
		Corpus length	Caput width			Phalanx proximalis	Phalanx distalis
Moorhens and allies							
Gallicrex cinerea	M	63.6 ± 2.9 (8)	14.0 ± 0.6 (8)	59.9 ± 3.0 (8)	38.0 ± 1.8 (8)	14.7 ± 0.8 (8)	12.5 ± 0.7 (8)
	F	53.7 ± 1.2 (2)	11.4 ± 0.4 (2)	50.3 ± 1.8 (2)	32.0 ± 1.2 (2)	12.6 ± 0.1 (2)	10.6 ± 0.1 (2)
Pareudiastes pacificus‖	—	41.0 (1)	—	29.0 (1)	21.0 (1)	7.0 (1)	6.0 (1)
Tribonyx ventralis	M	62.9 ± 1.9 (24)	13.6 ± 0.5 (24)	55.4 ± 2.2 (24)	42.1 ± 1.5 (24)	15.8 ± 0.6 (24)	15.0 ± 0.7 (21)
	F	58.5 ± 2.2 (19)	12.6 ± 0.5 (19)	50.6 ± 2.2 (19)	38.4 ± 1.4 (19)	14.4 ± 0.6 (18)	13.7 ± 0.6 (18)
Tribonyx mortierii	M	66.1 ± 2.0 (10)	14.8 ± 0.3 (10)	49.2 ± 1.7 (10)	36.7 ± 1.2 (10)	14.2 ± 0.7 (10)	13.1 ± 0.8 (7)
	F	65.2 ± 2.4 (11)	14.3 ± 0.4 (11)	48.3 ± 2.1 (11)	36.0 ± 1.8 (11)	13.8 ± 0.8 (8)	12.8 ± 0.7 (6)
Tribonyx hodgenorum	[M]	41.7 ± 0.8 (14)	8.2 ± 0.4 (14)	31.3 ± 1.8 (7)	21.6 ± 0.7 (6)	—	—
	[F]	39.0 ± 1.0 (11)	7.6 ± 0.3 (11)	27.9 ± 0.7 (11)	19.6 ± 0.4 (8)	—	—
Gallinula nesiotis	—	—	11.1 (1)	—	—	—	—
Gallinula comeri	—	53.6 ± 3.2 (2)	11.7 ± 0.4 (3)	42.0 ± 2.2 (2)	29.8 ± 1.8 (2)	10.6 (1)	10.2 (1)
Gallinula tenebrosa	M	66.6 ± 2.3 (6)	14.6 ± 0.4 (6)	54.8 ± 1.7 (5)	40.5 ± 1.3 (6)	15.5 ± 0.7 (6)	14.4 ± 0.6 (6)
	F	63.4 ± 2.5 (8)	13.9 ± 0.8 (8)	52.5 ± 2.1 (8)	39.1 ± 2.0 (8)	14.9 ± 0.9 (8)	13.9 ± 1.1 (8)
Gallinula chloropus-group	M	51.8 ± 3.1 (10)	11.4 ± 0.9 (10)	42.2 ± 3.2 (10)	31.8 ± 2.2 (10)	11.8 ± 0.7 (10)	10.9 ± 0.6 (7)
	F	50.8 ± 3.6 (19)	10.9 ± 0.8 (19)	41.4 ± 2.6 (19)	31.3 ± 1.8 (19)	11.5 ± 0.7 (18)	10.8 ± 0.8 (15)
Gallinula cachinaans	M	56.7 ± 1.6 (12)	12.3 ± 0.4 (12)	46.2 ± 2.0 (12)	34.8 ± 1.2 (12)	13.0 ± 0.5 (12)	12.3 ± 0.7 (12)
	F	53.4 ± 1.7 (8)	11.4 ± 0.5 (8)	44.1 ± 1.3 (8)	32.9 ± 0.9 (8)	12.0 ± 0.5 (8)	11.0 ± 0.4 (8)
Gallinula sandvicensis	—	51.3 ± 1.0 (5)	11.1 ± 0.2 (5)	40.8 ± 1.2 (5)	30.8 ± 0.6 (5)	11.0 ± 0.2 (5)	10.5 ± 0.3 (5)
Coots							
Fulica armillata	—	81.4 ± 3.3 (4)	16.8 ± 0.5 (4)	68.5 ± 2.8 (4)	44.0 ± 0.8 (4)	15.8 ± 0.5 (4)	14.8 ± 0.5 (4)
Fulica cornuta	—	111.4 ± 0.5 (2)	23.4 ± 0.4 (2)	101.9 ± 0.7 (2)	63.4 ± 0.6 (2)	22.4 ± 0.4 (2)	20.6 (1)
Fulica gigantea	M	108.3 ± 5.8 (6)	23.1 ± 0.9 (6)	95.5 ± 5.4 (6)	59.0 ± 3.5 (6)	11.2 ± 1.3 (4)	19.3 ± 1.5 (4)
	F	101.3 ± 4.2 (4)	22.1 ± 1.1 (4)	89.7 ± 3.1 (4)	56.2 ± 1.9 (4)	20.8 ± 0.9 (2)	18.5 ± 0.7 (2)
Fulica americana-group	M	72.8 ± 2.1 (17)	14.5 ± 0.5 (17)	64.2 ± 2.1 (17)	41.1 ± 1.5 (17)	14.9 ± 0.7 (17)	13.1 ± 0.5 (17)
	F	67.0 ± 2.4 (14)	13.2 ± 0.4 (15)	58.9 ± 2.0 (15)	37.8 ± 1.3 (15)	13.6 ± 0.4 (15)	12.0 ± 0.6 (14)
Fulica alai	M	71.2 ± 1.8 (7)	14.7 ± 0.3 (7)	62.1 ± 1.5 (7)	39.2 ± 1.2 (7)	14.7 ± 0.5 (6)	14.7 (1)
	F	65.7 ± 2.4 (9)	13.5 ± 0.5 (9)	57.0 ± 2.1 (9)	37.2 ± 1.5 (9)	13.7 ± 0.6 (8)	11.7 ± 0.9 (4)
Fulica atra (*atra*-group)	M	75.1 ± 3.4 (15)	15.1 ± 0.6 (15)	66.4 ± 3.5 (15)	41.5 ± 2.0 (15)	15.7 ± 0.7 (15)	13.5 ± 0.8 (13)
	F	73.3 ± 3.9 (22)	14.6 ± 0.7 (22)	64.6 ± 0.7 (22)	40.6 ± 2.1 (22)	15.2 ± 0.9 (22)	13.2 ± 0.9 (19)
Fulica chathamensis	[M]	95.9 ± 2.4 (33)	20.4 ± 0.6 (33)	798.2 ± 1.5 (22)	49.6 ± 1.0 (25)	17.7 ± 0.4 (6)	16.4 ± 0.6 (5)
	[F]	90.6 ± 1.6 (22)	19.4 ± 0.6 (22)	74.6 ± 1.3 (14)	46.6 ± 1.2 (12)		
Fulica prisca	[M]	96.0 ± 2.2 (10)	20.5 ± 0.7 (10)	77.4 ± 2.7 (5)		16.5 (1)	—
	[F]	89.8 ± 1.5 (16)	19.1 ± 0.4 (16)				—
Fulica newtoni	—	87.5 ± 3.2 (2)	18.3 ± 0.5 (4)	73.3 ± 0.0 (2)	49.1 (1)¶	—	—

* Data taken from Fischer and Stephan (1971b). Entries enclosed by brackets are midpoints of reported ranges in the absence of means.
† Data taken from Steadman (1986a).
‡ Measurements made from X ray images of study skins taken by J. Cooper (BMNH).
§ Samples based in part on measurements tabulated by Olson and James (1991).
‖ Measurements taken from elements during dissection of pectoral apparatus.
¶ Measurement based on Mourer-Chauviré et al. (1999: table 10).

TABLE 33. Univariate summary statistics ($\bar{x} \pm$ SD (n)) for six measurements (mm) of the pelvic girdle and limb of modern and subfossil species of Rallidae, by species and sex. Provisional sex groups or approximations are enclosed by brackets. Superspecies within *Porphyrio*, *Aramides*, *Rallus*, *Habropteryx*, *Amaurornis*, *Gallinula*, and *Fulica* follow Livezey (1998). Entries enclosed by brackets are sums of average lengths of three proximal phalanges (as opposed to averages of sums for associated elements), and are presented only for taxa having samples for each of the three included phalanges. Parentheses enclose the range of sample sizes for each phalanx.

Taxonomic group	Sex	Interacetabular width of pelvis	Femur Corpus length	Femur Caput width	Tibiotarsus corpus length	Tarsometatarsus corpus length	Digit III total length
Swamphens							
Porphyrio madagascariensis	M	22.1 ± 1.1 (9)	76.1 ± 3.3 (9)	12.6 ± 0.7 (9)	135.0 ± 7.1 (8)	93.3 ± 6.3 (7)	87.2 ± 2.9 (3)
	F	22.1 ± 1.7 (8)	74.2 ± 3.6 (9)	12.4 ± 0.8 (9)	133.5 ± 5.4 (9)	92.2 ± 4.1 (8)	84.7 ± 2.8 (2)
Porphyrio melanotus	M	25.5 ± 1.0 (29)	79.1 ± 2.5 (33)	14.0 ± 0.6 (33)	144.0 ± 4.8 (30)	99.9 ± 3.5 (32)	88.1 ± 3.3 (31)
	F	23.8 ± 1.3 (16)	73.3 ± 1.9 (19)	12.7 ± 0.4 (20)	134.7 ± 3.6 (18)	94.1 ± 2.3 (20)	83.8 ± 2.8 (18)
Porphyrio mantelli	[M]	—	110.5 ± 2.5 (14)	25.9 ± 0.4 (20)	134.7 ± 3.6 (4)	120.7 ± 2.7 (14)	—
	[F]	42.6 ± 3.7 (2)	102.4 ± 2.7 (11)	12.5 ± 1.8 (11)	130.0 ± 3.9 (5)	114.9 ± 2.3 (8)	—
Porphyrio hochstetteri	M	38.3 ± 2.4 (11)	103.7 ± 2.3 (14)	21.7 ± 1.1 (13)	158.8 ± 3.7 (13)	97.2 ± 2.2 (13)	72.9 ± 2.2 (9)
	F	35.9 ± 1.6 (20)	97.8 ± 2.6 (20)	20.8 ± 1.0 (20)	152.1 ± 3.8 (19)	93.6 ± 2.6 (19)	71.0 ± 2.7 (18)
Porphyrio kukwiedei	[M]	—	66.3 ± 0.2 (4)	12.7 ± 0.0 (4)	—	109.0 ± 0.7 (2)	—
	[F]	—	63.5 ± 0.3 (3)	11.9 ± 0.4 (3)			
Aphanocrex podarces		—	48.7 ± 0.6 (3)	7.7 ± 0.4 (3)	107.7 ± 1.3 (2)	71.7 ± 1.9 (4)	[55.0] (1–2)
			44.6 ± 2.2 (5)	7.1 ± 0.4 (5)			
Porphyrula alleni	M	14.6 ± 0.3 (3)	48.7 ± 0.6 (3)	7.7 ± 0.4 (3)	88.3 ± 4.4 (3)	60.6 ± 3.8 (3)	57.2 ± 0.6 (2)
	F	13.6 ± 0.8 (5)	44.6 ± 2.2 (5)	7.1 ± 0.4 (5)	78.7 ± 3.8 (5)	52.4 ± 1.8 (5)	51.0 ± 2.5 (5)
Porphyrula martinica	M	15.8 ± 0.7 (15)	52.1 ± 1.5 (14)	8.6 ± 0.4 (15)	96.4 ± 2.8 (14)	64.0 ± 1.4 (15)	60.4 ± 1.8 (12)
	F	15.0 ± 0.8 (15)	94.8 ± 1.6 (14)	8.2 ± 0.3 (15)	92.7 ± 3.0 (14)	61.6 ± 2.1 (1)	58.7 ± 2.1 (13)
Porphyrula flavirostris	M	13.2 ± 0.6 (3)	38.0 ± 0.5 (3)	6.2 ± 0.1 (3)	70.5 ± 2.0 (3)	45.8 ± 1.5 (3)	50.3 ± 0.8 (3)
	F	13.6 (1)	—	6.0 (1)	72.9 (1)	48.3 (1)	52.6 (1)
Wallace's Rail and allies							
Gymnocrex plumbeiventris	—	—	—	—	—	61.9 (1)	42.1 (1)
Habroptila wallacii	M	23.4 ± 0.7 (3)	81.9 ± 0.6 (3)	15.0 ± 0.4 (3)	125.6 ± 1.5 (3)	90.0 ± 1.7 (3)	61.8 ± 1.8 (3)
	F	22.9 ± 1.8 (3)	77.1 ± 1.2 (3)	13.6 ± 0.5 (3)	118.0 (1)	83.4 (1)	57.9 (1)
Cave-rails and allies							
Eulabeornis castaneoventris	—	23.4 (1)	72.1 (1)	13.9 (1)	—	70.4 (1)	46.5 (1)
Aramides ypecaha	M	23.4 ± 0.9 (4)	79.0 ± 1.8 (4)	15.2 ± 0.3 (4)	132.2 ± 4.5 (4)	89.9 ± 2.2 (8)	72.4 ± 2.7 (4)
	F	22.1 ± 1.5 (8)	75.7 ± 1.2 (8)	19.0 ± 0.6 (8)	128.8 ± 2.9 (8)	88.6 ± 2.2 (8)	69.3 ± 1.6 (8)
Aramides cajanea-group	M	17.8 ± 1.2 (26)	62.8 ± 2.7 (26)	11.6 ± 0.5 (26)	104.5 ± 4.8 (25)	75.2 ± 4.2 (26)	51.1 ± 3.6 (27)
	F	17.4 ± 1.3 (23)	61.7 ± 3.6 (24)	11.3 ± 0.7 (24)	102.2 ± 6.6 (23)	74.3 ± 5.0 (23)	50.5 ± 3.9 (23)

TABLE 33. Continued.

Taxonomic group	Sex	Interacetabular width of pelvis	Femur Corpus length	Femur Caput width	Tibiotarsus corpus length	Tarsometatarsus corpus length	Digit III total length
Canirallus oculeus	—	18.7 (1)	53.0 (1)	9.7 (1)	84.4 (1)	56.5 (1)	39.2 (1)
Canirallus kioloides	—	17.9 ± 0.2 (3)	44.3 ± 1.2 (3)	8.2 ± 0.6 (3)	67.3 ± 1.3 (3)	41.8 ± 2.6 (3)	35.4 ± 0.3 (3)
Nesotrochis debooyi	[M]	—	77.4 ± 1.9 (12)	16.4 ± 0.5 (12)	125.5 ± 3.9 (6)	75.9 ± 4.7 (8)	—
	[F]	—	71.5 ± 1.9 (5)	15.3 ± 1.2 (5)	115.1 ± 3.9 (12)		—
Nesotrochis steganinos	—	24.1 (1)	57.9 ± 0.1 (2)	12.5 ± 1.5 (2)	89.0 (1)	56.5 ± 2.1 (2)	—
*Nesotrochis picapicensis**	—	—	61.2 ± 3.7 (2)	—	92.1 ± 5.8 (2)	62.6 ± 0.8 (2)	—
Rougetius rougetii	—	—	—	—	78.3 (1)	54.5 (1)	—
Zapata Rail and allies							
Cyanolimnas cerverai	[M]	16.6 (1)	48.2 ± 0.9 (5)	9.0 ± 0.4 (5)	[60.8] (82)†	40.7 ± 9.1 (8)	—
	[F]		45.6 ± 1.2 (5)	8.7 ± 0.3 (5)		39.8 ± 9.2 (7)	—
Ortygonax sanguinolentus	—	14.3 (1)	48.0 (1)	7.5 (1)	—	—	—
White-throated rails							
Dryolimnas cuvieri	M	15.6 ± 0.4 (3)	51.8 ± 0.7 (3)	9.0 ± 0.3 (3)	79.9 ± 1.4 (3)	51.2 ± 1.4 (3)	44.4 ± 1.5 (3)
	F	15.6 ± 0.2 (3)	50.8 ± 2.3 (3)	8.8 ± 0.6 (3)	75.5 ± 1.0 (2)	48.5 ± 1.5 (2)	42.0 ± 1.3 (2)
Dryolimnas aldabranus	M	14.1 ± 0.9 (9)	45.0 ± 2.0 (10)	8.1 ± 0.3 (10)	69.5 ± 1.5 (10)	44.8 ± 1.2 (10)	38.6 ± 1.3 (9)
	F	13.6 ± 0.5 (6)	42.2 ± 0.7 (6)	7.5 ± 0.1 (6)	63.7 ± 1.1 (5)	41.3 ± 0.9 (5)	35.5 ± 0.5 (3)
Dryolimnas abbotti	—	15.2 (1)	47.1 (1)	8.1 (1)	70.9 (1)	44.9 (1)	—
Rallus sensu stricto							
Rallus aquaticus	M	11.5 ± 0.2 (16)	42.2 ± 1.3 (15)	6.9 ± 0.3 (16)	67.6 ± 2.1 (16)	43.8 ± 1.7 (16)	43.0 ± 1.9 (17)
	F	11.0 ± 0.5 (18)	39.7 ± 1.6 (20)	6.4 ± 0.3 (20)	63.5 ± 2.3 (20)	40.5 ± 1.9 (21)	39.7 ± 1.9 (20)
Rallus limicola	M	10.6 ± 0.5 (21)	37.7 ± 1.3 (21)	6.0 ± 0.2 (21)	57.6 ± 2.2 (19)	37.6 ± 2.0 (21)	35.3 ± 1.2 (21)
	F	10.0 ± 0.5 (18)	35.0 ± 1.6 (17)	5.4 ± 0.3 (15)	52.9 ± 2.8 (18)	33.3 ± 1.8 (18)	31.6 ± 1.6 (18)
Rallus ibycus	[M]	13.1 ± 0.1 (5)	40.3 ± 1.0 (9)	6.8 ± 0.3 (9)	53.1 ± 2.4 (5)	33.1 ± 0.8 (10)	[34.9] (1–2)
	[F]	12.0 ± 0.5 (3)	35.8 ± 1.7 (4)	6.3 ± 0.2 (4)	48.9 ± 1.1 (8)	28.4 ± 1.0 (6)	
Rallus longirostris-group	M	13.8 ± 0.7 (16)	55.6 ± 2.2 (16)	9.2 ± 0.5 (17)	82.9 ± 4.4 (17)	52.8 ± 3.1 (17)	44.7 ± 2.8 (16)
	F	13.4 ± 0.9 (14)	52.3 ± 1.8 (16)	8.5 ± 0.5 (16)	79.6 ± 3.4 (15)	49.8 ± 2.6 (17)	43.3 ± 2.4 (16)
Rallus elegans-group	M	14.9 ± 0.7 (21)	60.3 ± 1.6 (25)	10.2 ± 0.5 (25)	94.0 ± 3.1 (21)	61.7 ± 3.1 (24)	52.3 ± 2.5 (25)
	F	14.0 ± 0.5 (10)	55.7 ± 1.4 (11)	9.2 ± 0.3 (12)	86.7 ± 2.7 (11)	56.6 ± 2.4 (11)	47.9 ± 2.0 (11)
Rallus recessus	[M]	15.5 ± 0.1 (12)	49.8 ± 0.4 (10)	8.9 ± 0.3 (10)	70.4 ± 1.3 (15)	42.8 ± 0.9 (17)	[40.5] (3–16)
	[F]	15.5 ± 0.8 (14)	47.9 ± 0.6 (18)	8.5 ± 0.4 (18)	65.1 ± 2.6 (12)	39.3 ± 1.5 (15)	
Gallirallus and allies							
Gallirallus pectoralis-group	—	10.8 ± 0.5 (8)	36.7 ± 1.2 (7)	5.7 ± 0.3 (8)	51.5 ± 1.2 (7)	31.7 ± 0.8 (6)	29.7 ± 0.9 (6)

TABLE 33. Continued.

Taxonomic group	Sex	Interacetabular width of pelvis	Femur Corpus length	Femur Caput width	Tibiotarsus corpus length	Tarsometatarsus corpus length	Digit III total length
Gallirallus striatus	M	12.1 ± 0.7 (3)	43.7 ± 2.2 (3)	7.0 ± 0.5 (3)	61.1 ± 2.9 (3)	38.4 ± 1.9 (3)	35.5 ± 2.0 (3)
	F	11.6 ± 0.1 (2)	43.1 ± 1.6 (2)	6.9 ± 0.4 (2)	60.5 ± 1.3 (2)	37.7 ± 1.0 (2)	34.0 ± 0.3 (2)
Gallirallus australis	M	22.6 ± 1.4 (16)	77.9 ± 3.6 (19)	15.3 ± 0.9 (19)	114.8 ± 4.2 (18)	65.8 ± 2.5 (18)	54.2 ± 2.1 (19)
	F	20.1 ± 0.8 (11)	69.6 ± 2.6 (13)	13.2 ± 0.7 (13)	102.5 ± 4.2 (13)	58.5 ± 3.0 (13)	49.5 ± 1.8 (11)
Gallirallus greyi	M	22.1 ± 1.6 (4)	76.0 ± 5.1 (4)	15.2 ± 0.9 (4)	113.1 ± 7.5 (4)	6.5 ± 5.9 (4)	52.8 ± 4.3 (4)
	F	20.2 ± 1.6 (4)	67.3 ± 2.5 (4)	13.7 ± 0.5 (4)	98.9 ± 4.4 (4)	57.7 ± 1.2 (4)	48.3 ± 1.1 (4)
*Gallirallus philippensis-*group	M	13.9 ± 1.1 (15)	47.3 ± 2.1 (16)	8.0 ± 0.6 (16)	67.4 ± 3.8 (16)	42.2 ± 2.8 (15)	34.4 ± 1.9 (15)
	F	13.7 ± 0.6 (20)	46.6 ± 1.9 (21)	7.8 ± 0.4 (21)	66.4 ± 2.6 (19)	41.4 ± 1.7 (19)	33.7 ± 2.0 (17)
Gallirallus dieffenbachii	[M]	17.6 ± 0.6 (29)	56.4 ± 0.9 (22)	10.5 ± 0.3 (22)	78.6 ± 1.4 (27)	45.9 ± 0.8 (22)	[41.6] (6–18)
	[F]	16.6 ± 0.7 (28)	52.8 ± 1.4 (36)	9.9 ± 0.5 (36)	73.5 ± 2.0 (14)	42.8 ± 1.5 (24)	
Gallirallus owstoni	M	15.7 ± 0.6 (14)	53.6 ± 2.3 (14)	9.1 ± 0.6 (1)	78.4 ± 3.8 (14)	51.2 ± 2.7 (14)	40.4 ± 2.2 (13)
	F	15.3 ± 0.6 (18)	50.9 ± 1.8 (18)	8.6 ± 0.4 (18)	74.7 ± 2.7 (18)	48.1 ± 2.2 (18)	38.3 ± 1.9 (18)
Gallirallus wakensis	M	13.5 ± 0.4 (20)	36.8 ± 0.6 (20)	6.9 ± 0.2 (20)	59.0 ± 0.1 (2)	360 ± 0.3 (2)	32.0 ± 0.3 (2)
	F	12.9 ± 0.3 (10)	35.2 ± 0.4 (10)	6.4 ± 0.2 (10)			
Tricholimnas lafresnayanus	—	23.2 (1)	68.2 (1)	14.1 (1)	55.2 (1)	33.3 (1)	30.2 (1)
Tricholimnas sylvestris	M	19.2 ± 1.6 (3)	61.1 ± 0.8 (4)	11.6 ± 1.1 (4)	86.3 ± 1.1 (4)	51.4 ± 1.2 (4)	47.1 ± 2.9 (4)
	F	17.7 ± 0.6 (2)	57.6 ± 2.2 (2)	11.2 ± 1.7 (2)	83.5 ± 0.1 (2)	47.7 ± 0.2 (2)	43.5 ± 3.5 (2)
Nesoclopeus poeciloptererus-group							
Cabalus modestus	[M]	9.8 ± 0.3 (15)	31.0 ± 0.7 (14)	5.2 ± 0.2 (14)	49.9 ± 1.2 (12)	30.2 ± 1.3 (15)	44.6 (1)
	[F]	9.4 ± 0.3 (13)	29.1 ± 0.6 (14)	4.9 ± 0.2 (14)	45.5 ± 0.9 (16)	27.6 ± 0.6 (13)	[26.9] (5–25)
Capellirallus karamu	[M]	14.4 ± 0.3 (9)	46.9 ± 1.2 (24)	8.8 ± 0.2 (24)	70.0 ± 2.2 (18)	42.0 ± 1.6 (17)	[38.2] (6–22)
	[F]	13.5 ± 0.5 (12)	43.9 ± 1.0 (17)	8.2 ± 0.2 (17)	65.2 ± 1.2 (11)	38.7 ± 0.7 (19)	
Habropteryx insignis	—					63.3 (1)	42.4 (1)
Habropteryx torquatus-group	M	16.5 ± 0.3 (5)	53.8 ± 1.7 (5)	9.8 ± 0.4 (5)	83.4 ± 2.6 (5)	53.7 ± 1.9 (6)	42.5 ± 1.4 (6)
	F	15.2 ± 0.2 (8)	50.1 ± 1.5 (8)	8.8 ± 0.2 (8)	76.3 ± 2.2 (7)	49.1 ± 1.6 (7)	38.4 ± 1.1 (7)
Habropteryx okinawae	—	18.7 (1)	61.2 (1)	11.7 (1)		58.3 (1)	
Problematic subfossil rails							
Aphanapteryx bonasia	[M]	27.6 ± 2.8 (4)	74.6 ± 1.5 (7)	16.3 ± 0.6 (7)	120.9 ± 3.0 (13)	78.6 ± 1.8 (16)	[58.5] (7–8)
	[F]		69.1 ± 1.8 (4)	15.7 ± 0.6 (4)	108.9 ± 3.7 (26)	70.9 ± 2.1 (19)	[53.0] (7–12)
Erythromachus leguati	—	19.7 ± 0.5 (3)	58.0 ± 5.5 (2)	13.0 ± 0.8 (2)	90.7 ± 7.1 (2)	58.1 ± 0.6 (2)	
Diaphorapteryx hawkinsi	[M]	31.2 ± 1.3 (29)	96.6 ± 2.3 (35)	22.1 ± 0.8 (35)	136.4 ± 3.0 (29)	72.2 ± 2.0 (31)	[79.5] (9–23)
	[F]	28.7 ± 1.4 (23)	87.3 ± 3.1 (20)	20.1 ± 1.1 (20)	126.2 ± 3.7 (25)	65.3 ± 1.6 (21)	[70.1] (11–17)

TABLE 33. Continued.

Taxonomic group	Sex	Interacetabular width of pelvis	Femur Corpus length	Femur Caput width	Tibiotarsus corpus length	Tarsometatarsus corpus length	Digit III total length
Atlantisia and allies							
Atlantisia rogersi	—	9.6 ± 0.5 (6)	24.3 ± 0.8 (8)	4.1 ± 0.2 (8)	41.9 ± 1.7 (2)	24.8 ± 0.3 (2)	26.0 ± 1.1 (2)
Mundia elpenor	[M]	10.6 ± 0.3 (6)	34.8 ± 0.6 (8)	59.3 ± 0.2 (8)	56.9 ± 1.2 (8)	36.8 ± 0.6 (8)	[29.3] (3–9)
	[F]		32.9 ± 0.6 (8)	57.6 ± 0.1 (8)	54.1 ± 0.9 (8)	34.5 ± 0.8 (8)	
Laterallus leucopyrhus	M	9.4 ± 0.5 (7)	31.2 ± 1.2 (7)	4.9 ± 0.3 (7)	50.3 ± 1.9 (7)	33.5 ± 1.4 (7)	30.5 ± 1.2 (7)
	F	9.1 ± 0.4 (20)	29.8 ± 1.6 (20)	4.7 ± 0.4 (21)	48.7 ± 2.2 (20)	32.4 ± 1.7 (20)	29.7 ± 1.7 (16)
Coturnicops noveboracensis	M	9.7 ± 0.5 (12)	29.5 ± 1.1 (15)	4.6 ± 0.4 (15)	41.5 ± 1.7 (14)	25.0 ± 1.0 (15)	24.6 ± 1.3 (15)
	F	9.3 ± 0.4 (18)	27.9 ± 0.8 (20)	4.2 ± 0.3 (20)	38.5 ± 1.2 (20)	53.0 ± 0.6 (20)	23.1 ± 0.8 (16)
Crex crex	M	12.4 ± 0.9 (13)	45.9 ± 1.6 (13)	7.2 ± 0.4 (13)	62.0 ± 2.2 (12)	39.5 ± 1.6 (13)	31.8 ± 2.0 (13)
	F	11.9 ± 0.6 (11)	45.3 ± 1.5 (11)	6.8 ± 0.4 (11)	61.4 ± 2.3 (10)	38.8 ± 2.3 (11)	31.5 ± 1.2 (10)
Crex albicollis	M	11.4 (1)	43.3 (1)	6.9 (1)	59.6 ± 2.8 (2)	37.2 ± 3.0 (2)	35.2 ± 2.3 (2)
	F	11.6 ± 0.8 (2)	40.7 ± 1.4 (2)	6.5 ± 0.3 (2)	58.5 ± 1.4 (2)	37.2 ± 0.1 (2)	34.3 ± 1.3 (2)
Gray crakes and allies							
Porzana sandwichensis	—	—	—	—	—	27.0 (1)	
Porzana palmeri	M	9.9 ± 0.3 (6)	25.4 ± 0.5 (6)	4.2 ± 0.2 (6)	39.3 ± 0.9 (6)	24.3 ± 0.7 (7)	24.9 ± 1.0 (7)
	F	9.6 ± 0.2 (5)	24.4 ± 0.4 (6)	4.2 ± 0.1 (6)	37.2 ± 1.1 (6)	22.7 ± 0.5 (6)	23.3 ± 0.6 (5)
Porzana carolina	M	10.8 ± 0.4 (13)	34.4 ± 0.9 (14)	5.5 ± 0.2 (15)	55.5 ± 1.4 (15)	33.8 ± 1.4 (15)	35.4 ± 1.8 (13)
	F	10.4 ± 0.4 (14)	32.6 ± 1.0 (14)	5.0 ± 0.2 (15)	51.2 ± 1.8 (15)	31.6 ± 1.4 (15)	32.5 ± 1.5 (13)
Porzana pusilla	M	8.8 ± 0.3 (9)	28.7 ± 1.3 (9)	4.2 ± 0.3 (9)	45.7 ± 1.5 (8)	28.7 ± 1.3 (9)	31.0 ± 1.5 (6)
	F	8.7 ± 0.6 (1)	27.2 ± 1.4 (14)	4.1 ± 0.3 (14)	43.5 ± 2.1 (11)	27.3 ± 1.2 (12)	28.7 ± 1.7 (9)
Porzana tabuensis	M	9.3 ± 0.4 (4)	30.0 ± 0.9 (5)	4.7 ± 0.2 (5)	48.3 ± 0.8 (4)	31.8 ± 1.0 (4)	29.7 ± 0.8 (4)
	F	9.3 ± 0.3 (6)	28.8 ± 1.2 (6)	4.4 ± 0.4 (6)	44.7 ± 2.3 (4)	28.9 ± 1.9 (6)	27.7 ± 1.5 (5)
Porzana monasa‡	M	—	—	—	[61.0] (1)	[35.6] (2)	[33.0] (2)
Porzana atra	F	11.5 (1)	34.6 (1)	5.8 (1)	54.2 (1)	35.6 (1)	32.7 (1)
Porzana flavirostra	M	11.2 ± 0.2 (3)	33.4 ± 0.7 (3)	5.3 ± 0.2 (3)	53.0 ± 0.8 (2)	34.3 ± 0.1 (2)	31.4 ± 0.8 (3)
	F	10.7 ± 0.5 (16)	36.8 ± 1.4 (15)	6.1 ± 0.3 (17)	61.8 ± 2.4 (16)	42.4 ± 1.1 (15)	41.2 ± 1.6 (17)
Porzana piercei	[M]	10.6 ± 0.7 (14)	36.8 ± 1.4 (15)	5.7 ± 0.3 (18)	60.9 ± 2.8 (18)	40.8 ± 1.6 (18)	39.7 ± 1.5 (17)
	[F]	9.7 ± 0.3 (5)	25.0 ± 0.6 (5)	4.1 ± 0.2 (5)	38.2 ± 0.9 (6)	22.6 ± 0.5 (10)	[25.2] (1–5)
Porzana rua‡	—		23.7 ± 0.4 (10)	3.9 ± 0.2 (10)	35.8 ± 0.8 (8)	21.2 ± 0.6 (7)	
Porzana astrictocarpus†	—	—	34.7 (1)	5.2 (1)	—	34.7 (1)	—
			26.2 ± 1.2 (8)	4.5 ± 0.2 (8)	43.3 ± 0.9 (4)	26.7 ± 0.6 (5)	—

TABLE 33. Continued.

Taxonomic group	Sex	Interacetabular width of pelvis	Femur Corpus length	Femur Caput width	Tibiotarsus corpus length	Tarsometatarsus corpus length	Digit III total length
Porzana ziegleri‡	—	8.6 (1)	24.0 ± 1.0 (7)	3.9 ± 0.2 (4)	33.3 (1)	21.7 ± 0.9 (12)	—
Porzana menehune†	—	—	20.9 ± 0.7 (3)	3.6 ± 0.1 (3)	30.0 ± 1.4 (2)	18.0 ± 0.8 (9)	—
Porzana keplerorum‡	—	—	23.0 ± 1.8 (6)	3.7 ± 0.3 (6)	35.2 ± 2.5 (7)	20.7 ± 2.0 (9)	—
Porzana ralphorum‡	—	—	—	—	—	35.7 (1)	—
Porzana severnsi‡	—	11.7 (1)	34.3 ± 1.2 (5)	5.9 ± 2.5 (5)	54.7 ± 1.9 (6)	36.0 ± 1.5 (6)	—
Waterhens							
Amaurornis olivaceus	M	17.7 ± 0.5 (5)	56.5 ± 1.6 (5)	9.9 ± 0.5 (5)	97.3 ± 2.6 (5)	65.9 ± 2.2 (5)	54.0 ± 1.7 (4)
	F	15.1 ± 1.3 (4)	53.6 ± 3.1 (4)	9.0 ± 0.8 (4)	92.7 ± 7.6 (4)	61.6 ± 5.2 (4)	51.9 ± 3.0 (4)
Amaurornis moluccanus	M	14.4 ± 0.6 (3)	48.8 ± 1.9 (3)	8.1 ± 0.2 (3)	83.6 ± 4.7 (3)	54.7 ± 3.1 (3)	46.8 ± 2.2 (3)
	F	13.1 ± 1.3 (2)	46.7 ± 0.4 (2)	7.6 ± 0.1 (2)	77.7 ± 0.4 (2)	52.1 ± 0.6 (2)	45.3 ± 0.4 (2)
Amaurornis phoenicurus	M	14.6 ± 0.6 (—)	48.9 ± 1.6 (16)	8.6 ± 0.3 (16)	87.8 ± 2.6 (15)	58.3 ± 2.2 (16)	54.0 ± 2.1 (16)
	F	13.5 ± 0.7 (5)	46.8 ± 1.5 (15)	8.1 ± 0.4 (15)	83.6 ± 2.5 (1)	54.9 ± 1.8 (15)	51.9 ± 2.4 (15)
Amaurornis akool	—	—	—	—	—	49.2 (1)	42.8 (1)
Amaurornis isabellinus	—	—	—	—	—	66.0 (1)	53.0 (1)
Amaurornis ineptus	—	—	—	—	124.3 (1)	87.5 (1)	60.1 (1)
Moorhens and allies							
Gallicrex cinerea	M	15.8 ± 0.7 (8)	59.7 ± 2.3 (8)	10.3 ± 0.7 (8)	109.9 ± 4.7 (8)	72.0 ± 4.3 (7)	68.7 ± 2.5 (8)
	F	13.6 ± 0.1 (2)	51.6 ± 2.8 (2)	8.5 ± 0.6 (2)	95.4 ± 4.7 (2)	63.6 ± 2.1 (2)	63.3 ± 1.1 (2)
Tribonyx ventralis	M	20.1 ± 1.0 (25)	57.1 ± 1.6 (24)	11.3 ± 0.5 (24)	98.9 ± 3.2 (24)	63.2 ± 2.7 (24)	46.5 ± 1.6 (22)
	F	19.0 ± 1.0 (18)	53.0 ± 1.9 (18)	10.6 ± 0.5 (19)	90.8 ± 4.0 (18)	58.2 ± 2.3 (19)	42.4 ± 1.9 (18)
Tribonyx mortierii	M	26.0 ± 0.4 (6)	84.4 ± 3.4 (8)	16.4 ± 0.7 (8)	139.6 ± 5.6 (9)	87.4 ± 4.3 (10)	61.2 ± 1.2 (9)
	F	25.2 ± 1.0 (8)	80.5 ± 3.1 (9)	16.3 ± 0.7 (9)	132.5 ± 5.6 (11)	82.1 ± 3.9 (11)	59.4 ± 1.9 (11)
Tribonyx hodgenorum	[M]	19.2 ± 1.1 (5)	57.9 ± 1.7 (20)	10.5 ± 0.5 (20)	76.2 ± 1.5 (11)	45.3 ± 1.6 (13)	—
	[F]		53.8 ± 1.6 (17)	9.5 ± 0.3 (17)	72.2 ± 1.4 (16)	41.1 ± 1.3 (26)	—
Gallinula nesiotis	—	16.8 (1)	53.8 ± 1.2 (2)	10.3 ± 0.4 (2)			
Gallinula comeri	—	20.7 ± 1.1 (5)	56.9 ± 3.5 (4)	11.0 ± 0.7 (5)	89.7 ± 8.1 (2)	52.4 ± 4.2 (2)	54.8 ± 3.8 (2)
Gallinula tenebrosa	M	19.0 ± 0.4 (5)	63.5 ± 2.5 (6)	12.1 ± 0.3 (6)	107.1 ± 3.0 (6)	66.5 ± 2.5 (6)	73.9 ± 3.4 (6)
	F	18.9 ± 1.1 (8)	59.9 ± 2.3 (8)	11.4 ± 0.7 (8)	102.7 ± 4.5 (8)	62.9 ± 2.2 (7)	70.8 ± 1.7 (6)
Gallinula chloropus	M	16.1 ± 1.3 (10)	49.7 ± 4.1 (9)	9.1 ± 0.6 (9)	82.4 ± 6.6 (10)	50.1 ± 4.0 (10)	56.2 ± 3.8 (9)
	F	15.2 ± 0.9 (17)	48.7 ± 3.1 (18)	8.8 ± 0.7 (19)	82.3 ± 4.6 (17)	49.7 ± 2.9 (19)	55.8 ± 3.4 (19)
Gallinula cachinnans	M	16.8 ± 0.7 (12)	54.2 ± 1.3 (12)	9.7 ± 0.5 (12)	92.6 ± 3.0 (12)	57.7 ± 1.6 (12)	65.7 ± 1.6 (12)
	F	16.0 ± 0.6 (8)	51.2 ± 1.9 (8)	9.1 ± 0.3 (8)	88.0 ± 2.8 (8)	55.0 ± 1.6 (8)	62.4 ± 2.6 (8)
Gallinula sandvicensis	—	16.6 ± 0.3 (3)	51.1 ± 1.3 (5)	9.5 ± 0.2 (5)	88.1 ± 1.9 (5)	55.2 ± 1.6 (5)	58.3 ± 1.7 (5)

TABLE 33. Continued.

Taxonomic group	Sex	Interacetabular width of pelvis	Femur Corpus length	Femur Caput width	Tibiotarsus corpus length	Tarsometatarsus corpus length	Digit III total length
Coots							
Fulica armillata	—	17.8 ± 1.1 (4)	62.7 ± 1.4 (4)	13.1 ± 0.4 (4)	122.7 ± 4.0 (4)	69.9 ± 1.9 (4)	83.0 ± 0.9 (4)
Fulica cornuta	—	26.0 ± 0.7 (2)	78.9 ± 0.8 (2)	16.2 ± 0.5 (2)	148.0 ± 4.1 (2)	88.7 ± 0.7 (2)	101.0 ± 2.3 (2)
Fulica gigantea	M	25.0 ± 1.2 (5)	84.4 ± 4.7 (6)	18.2 ± 1.1 (6)	165.2 ± 10.7 (6)	101.0 ± 6.4 (6)	110.6 ± 3.0 (4)
	F	25.6 ± 2.1 (3)	81.9 ± 3.0 (4)	17.9 ± 0.4 (4)	161.9 ± 4.2 (4)	100.3 ± 3.3 (4)	109.7 ± 5.1 (2)
Fulica americana-group	M	15.3 ± 0.8 (17)	54.6 ± 1.8 (17)	10.4 ± 0.4 (17)	100.2 ± 3.6 (15)	58.4 ± 3.0 (17)	70.6 ± 3.4 (17)
	F	14.2 ± 0.6 (15)	50.1 ± 2.0 (15)	9.6 ± 0.4 (15)	91.8 ± 4.0 (15)	53.6 ± 2.3 (15)	64.8 ± 2.5 (15)
Fulica alai	M	15.4 ± 0.7 (6)	55.0 ± 1.7 (7)	10.4 ± 0.4 (7)	102.2 ± 3.8 (6)	59.4 ± 2.8 (7)	71.9 ± 4.2 (7)
	F	15.1 ± 0.9 (9)	50.4 ± 2.1 (9)	9.4 ± 0.5 (9)	92.9 ± 4.5 (9)	54.1 ± 2.4 (9)	66.4 ± 2.2 (9)
Fulica atra (*atra*-group)	M	15.7 ± 0.8 (15)	55.2 ± 2.3 (15)	11.2 ± 0.5 (15)	102.8 ± 4.9 (14)	60.4 ± 2.5 (15)	73.6 ± 3.7 (14)
	F	15.3 ± 0.7 (20)	53.7 ± 2.7 (20)	10.6 ± 0.7 (22)	100.3 ± 5.1 (21)	57.7 ± 3.1 (22)	71.7 ± 4.9 (21)
Fulica chathamensis	[M]	23.2 ± 1.1 (32)	85.1 ± 2.3 (23)	18.7 ± 1.1 (23)	161.7 ± 3.1 (33)	94.7 ± 1.6 (26)	[106.7] (14–19)
	[F]	22.3 ± 1.0 (20)	80.4 ± 1.4 (29)	17.3 ± 0.5 (29)	154.5 ± 2.2 (23)	89.0 ± 1.6 (28)	[101.3] (9–17)
Fulica prisca	[M]	24.1 ± 0.9 (6)	85.3 ± 1.4 (12)	19.0 ± 0.8 (12)	157.5 ± 5.1 (15)	91.4 ± 3.8 (16)	
	[F]	21.8 ± 0.6 (5)	76.6 ± 4.5 (13)	16.7 ± 1.3 (13)	145.6 ± 4.2 (13)	81.9 ± 3.5 (12)	
Fulica newtoni	[M]	} 20.7 ± 0.6 (2)	} 80.9 ± 5.8 (3)	} 17.1 ± 0.8 (3)	141.0 ± 2.5 (6)	85.9 ± 1.1 (13)	} [102.4] (2–4)
	[F]				128.8 ± 2.5 (9)	81.0 ± 1.8 (10)	

* Data taken from Fischer and Stephan (1971b). Entries enclosed by brackets are midpoints of reported ranges.
† Samples based in part on measurements tabulated by Olson and James (1991).
‡ Data taken from Steadman (1986a).

30–33). Nonetheless, interspecific differences in these three species were highly significant ($P < 0.0001$) in all skeletal dimensions compared. Summary statistics revealed that *H. wallacii* was comparable to or slightly smaller than medium-sized *A. cajanea* in pectoral dimensions, but comparable to or slightly smaller than large *A. ypecaha* in other measurements, with this combination of differences accounting for the interspecific effects and together reflecting the pectoral reduction of *H. wallacii* (Tables 30–33). Other less well-represented taxa of "basal" rallids (Tables 30–33) either were comparable to *Aramides* in general body form (e.g., *Eulabeornis*), were distinctly smaller but displayed the proportions typical of flighted rallids (e.g., *Canirallus* and *Rougetius*), or showed variably strong indications of pectoral reduction at least as great as that evident in *Habroptila* (e.g., *Nesotrochis* spp.).

Two species of white-throated rails (*Dryolimnas cuvieri* and *D. aldabranus*; n = 22) were amenable to formal comparisons of skeletons (two-way ANOVA), with only limited, informal comparisons of summary statistics feasible for extinct congener *D. abbotti* (osteologically limited to subfossil remains) and two poorly represented meso-rallids (*Ortygonax sanguinolentus* and flightless *Cyanolimnas cerverai*). Formal comparisons of skeletal variables of flighted *Dryolimnas cuvieri* and flightless *D. aldabranus* revealed a pattern of divergence distinct from those indicated for flightless swamphens or basal rallids (Tables 30–33). Interspecific differences were significant ($P < 0.05$) in all but 3 of the 41 variables compared (bill length and the two furcular widths), averaging larger in respective sex-groups in *D. cuvieri* in the other 38 dimensions. Summary statistics for *D. abbotti* indicated that *D. abbotti* was intermediate to its two extant congeners in size, with relative sizes indicating that *D. abbotti* did not share the disproportionate reduction in selected pectoral dimensions indicated for *D. aldabranus* (Tables 30–33). Similar assessments of summary statistics revealed that *Cyanolimnas cerverai* was larger than *Ortygonax sanguinolentus* in dimensions of the pelvic apparatus but substantially smaller than its flighted relative in available pectoral dimensions (Tables 30–33).

Typical rails (*Rallus*, *Gallirallus*, *Tricholimnas*, and *Habropteryx*) having adequate samples for two-way analysis of skeletal dimensions comprised 11 species (n = 327), of which 4 (*T. sylvestris*, *G. australis*-group, *G. owstoni*, and *G. wakensis*) were flightless. An additional nine taxa were included for informal comparisons because they were represented by samples too limited for formal analysis or were represented only by unassociated subfossil elements (Tables 30–33). These were congeners of the foregoing taxa, members of allied genera (*Nesoclopeus*, *Cabalus*, and *Capellirallus*), or specimens of three problematic, monotypic genera most comparable morphologically with typical rails (*Diaphorapteryx*, *Aphanapteryx*, and *Erythromachus*). Among those amenable to two-way ANOVA, taxon effects were highly significant in all 41 skeletal variables ($P < 0.0001$). Consideration of summary statistics of other, less well-represented taxa added to the interspecific differences in osteological dimensions. Most notable were the modest pectoral changes of *Rallus recessus*; pronounced relative pectoral reductions indicated for flightless *R. ibycus* (cf. *G. owstoni*), *Nesoclopeus poecilopterus* (cf. *T. sylvestris*), *G. dieffenbachii* (cf. *G. wakensis*), *Habropteryx insignis*, and *H. okinawae*; and the extreme pectoral reductions of *Cabalus modestus* and *Capellirallus karamu* (Tables 30–33). Summary statistics compiled here for *C. mo-*

destus and *G. dieffenbachii* (Tables 30–33) compared very closely to those for samples (not partitioned to sex) presented by Trewick (1997b).

Crakes (*Laterallus, Coturnicops, Crex,* and *Porzana*) having adequate samples for two-way analysis of skeletal dimensions comprised 10 species ($n = 213$), of which 2 (*P. atra* and *P. palmeri*) were flightless. A number of additional species, either congeneric with the aforementioned taxa or members of allied taxa, were available for comparisons of summary statistics but were represented by unsexed specimens or disassociated subfossil elements not suitable for two-way statistical comparisons (Tables 30–33). These secondary species included a number of flightless forms: *Atlantisia rogersi,* "*Atlantisia*" *elpenor, P. sandwichensis, P. monasa, P. piercei, P. rua, P. astrictocarpus,* and five other *Porzana* from the Hawaiian Islands. Those taxa permitting two-way comparisons manifested significant interspecific differences in all 41 skeletal dimensions (ANOVA; $P < 0.05$). Comparisons of summary statistics for these taxa and those not subjected to formal analyses revealed that a considerable range in general body size is encompassed by crakes, ranging from the smaller subfossil endemics of the Hawaiian Islands (*P. menehune* and *P. ziegleri*) to the comparatively large species of *Crex* (Tables 30–33). Flightless species, regardless of whether samples permitted two-way comparisons, showed variably pronounced but significant pectoral reduction. The most pronounced diminutions of pectoral elements (relative to nonpectoral dimension) were present in *A. rogersi* and most of the subfossil Hawaiian *Porzana*, and notably less substantial reductions were found in *P. atra* and *P. piercei* (Tables 30–33).

Only two taxa of bushhens (*Amaurornis*) were represented adequately for two-way analysis (*A. phoenicurus* and *A. olivaceus*-group). Three others (*A. akool, A. isabellinus,* and flightless *A. ineptus*) were limited to data permitting only informal comparison against the summary statistics of these benchmark, flighted groups ($n = 45$). Despite the limited disparity between the two species subject to two-way analyses (both flighted and of medium size), 19 of 41 skeletal dimensions showed significant interspecific differences (ANOVA; $P < 0.05$). Summary statistics for congeners represented by very limited samples indicated that *A. akool* was smaller and *A. isabellinus* of approximately equal overall size to the better-sampled species (Tables 30–33). However, to the extent permitted by the very limited skeletal data available, flightless *A. ineptus* departed substantially from the proportions of its flighted congeners, being much larger in cranial and pelvic dimensions but approximately equal to or slightly smaller in pectoral dimensions.

Six taxa of moorhens (sensu lato) were represented by samples ($n = 153$) suitable for two-way analysis (*Gallicrex cinerea, Gallinula chloropus*-group, *G. tenebrosa, G. comeri, Tribonyx ventralis,* and *T. mortierii*), with four others having limited data permitting limited, informal comparisons (*Pareudiastes pacificus, T. hodgenorum, G. nesiotis,* and *G. sandvicensis* [Tables 30–33]). This diverse assemblage manifested highly significant intersexual differences in all 41 of the skeletal dimensions compared (ANOVA; $P < 0.05$). Summary statistics for these taxa and those not amenable to formal analyses revealed that flightless species tended to be larger than flighted relatives in all but pectoral dimensions (Tables 30–33), a generality epitomized by the contrast between the comparatively well-sampled flighted *T. ventralis* and its flightless congener *T. mortierii*.

Only two taxa of coots (*Fulica*) were amenable to two-way ANOVA (*F. amer-*

icana and *F. atra* [*n* = 69]). Four other species were sampled for which available specimens permitted one-way ANOVA among taxa (*F. armillata, F. gigantea, F. cornuta,* and *F. alai* [*n* = 32]), and three more (comprising the only flightless species in the genus) were represented by samples of subfossil elements of marginal, subanalytical quality (*F. chathamensis, F. prisca,* and *F. newtoni* [Tables 30–33]). Despite a predictable affinity between the two well-represented species of *Fulica,* interspecific differences were significant ($P < 0.05$) in all but 2 of the 41 skeletal dimensions compared, with the Eurasian *F. atra* averaging slightly but significantly larger (sex for sex) than *F. americana* of the New World (Tables 30–33). One-way comparisons of the enlarged series of six species confirmed highly significant interspecific differences ($P < 0.0001$) in all 41 skeletal dimensions, reflecting in large part the differences in overall size between *F. gigantea* and *F. cornuta* and other species of coots. Summary statistics for the three subfossil species indicated that these species were comparable in overall size to the two large Andean species (e.g., dimensions of the skull and pelvic limb), but showed relative reductions in most pectoral measurements (Tables 30–33).

Differences in group variances.—In the greater swamphens (*Porphyrio*), differences in group variances (two-way Levene's tests) of skeletal dimensions largely were restricted to interspecific variation. Significant differences among species were found in 14 of 41 measurements compared ($P < 0.05$), whereas intersexual effects and species–sex interaction effects were significant ($P < 0.05$) in only 6 and 3 variables, respectively (Tables 30–33). The interspecific differences derived principally from the comparatively large variances of some measurements in *P. hochstetteri*; large size and the proportion of skeletons assigned to sex a posteriori render this pattern less than compelling. Among the lesser swamphens (*Porphyrula*), differences in group variances achieved significance ($P < 0.05$) among species and between sexes in six and one measurements, respectively. Some variables (notably depth of the carina sterni) showed greater relative variation than others (e.g., lengths of limb elements), notably in some dimensions of comparatively late-ossifying elements in flightless species. For example, coefficients of variation ($SD/\bar{x}$) for perpendicular depth of the carina sterni approximated 5% in flighted species, but was 10% or more in flightless *Porphyrio* (Table 31).

Three taxa of basal rallids were formally compared (*Aramides cajanea*-group, *A. ypecaha* and *Habroptila wallacii*). Very few significant differences in within-group variances (two-way Levene's test; $P < 0.05$) were found in the 41 skeletal dimensions, with only interspecific effects in 3 variables, intersexual effects in 1, and species–sex interactions in 2. Neither these nor the summary statistics for other, poorly represented basal taxa indicated meaningful patterns in variances. However, as in the swamphens, some variables (notably carina depths of the sternum) showed higher relative variances than other dimensions, especially in flightless species (Tables 30–33).

The two species of white-throated rails amenable to formal analyses (*Dryolimnas cuvieri* and *D. aldabranus*) manifested significant interspecific differences ($P < 0.05$) in within-group variances in 10 skeletal dimensions, intersexual differences in 4, and species–sex interaction effects in 5 (4 of which also showed significant main effects). These effects were comparatively frequent in dimensions of the cranium, pectoral girdle, and proximal pectoral limb (Tables 30–33). Informal comparisons of summary statistics for extinct *D. abbotti* and the two poorly represented meso-rallids (*Ortygonax sanguinolentus* and *Cyanolimnas cerverai*) revealed no obvious patterns in within-group variances. Comparisons of coefficients of variation ($SD/\bar{x}$) indicated no obvious heterogeneity among variables in relative variances within *Dryolimnas,* whereas data were too few for assessments of *Cyanolimnas* or *Ortygonax.*

For the typical rails permitting two-way analysis of within-group variances, significant interspecific effects (Levene's tests, $P < 0.05$) were found in all but 4 of the 41 skeletal dimensions. Exceptions were ulna length, least depth of the carina sterni, maximal width of the furcula, and length of pedal digit III, all of which tended to high variance in all rails and were prone to somewhat inflated measurement error. Intersexual differences in group variances were significant in only nine dimensions, with species–sex interactions limited to one of these. Examination of summary statistics indicated that many of the interspecific effects were due to the anomalously high variances for the small sample of *Tricholimnas sylvestris* (in part a reflection of increased measurement error attending the inclusion of mounted skeletons), the variably inflated variances of the polytypic *Rallus longirostris* and polymorphic *Gallirallus australis,* and the probably artifactually low variances for the minimal sample available for *G. wakensis* (Tables 30–33). The additional species not included in formal comparisons revealed no informative departures from the commonalities pertaining to the foregoing species, bearing in mind the limitations and conditions on these estimates. Notable once again in all typical rails was the higher relative variation of a subset of the variables, especially the two depths of the carina sterni; coefficients of variation ($SD/\bar{x}$) for the latter generally approximated (where samples were partitioned by sex) twice those of other dimensions, for example, lengths of appendicular elements (Tables 30–33).

The 10 species of crakes (*Laterallus, Coturnicops, Crex,* and *Porzana*) having adequate samples for two-way analysis of within-group variances in skeletal dimensions (Levene's tests; $P < 0.05$), of which 2 (*P. atra* and *P. palmeri*) were flightless, showed interspecific differences in variances in 29 of 41 variables. No dimensions showed significant ($P > 0.05$) intersexual heterogeneity in group variances, and only two (both of which also showed interspecific differences) manifested significant species–sex interactions in group variances. Those skeletal variables showing significant heterogeneity of variances included dimensions of all parts of the skeleton, and did not indicate any meaningful functional patterns (Tables 30–33), evidently resulting in large part from variances in a few comparatively large, widespread species (e.g., *Crex crex* and *C. albicollis*). Summary statistics for the taxa of crakes not amenable to formal two-way comparisons did not suggest marked, meaningful heterogeneity in group variances.

The two adequately represented taxa of bushhens (*Amaurornis olivaceus*-group and *A. phoenicurus*) manifested significant interspecific effects ($P < 0.05$) in 24 of 41 skeletal dimensions in two-way Levene's comparisons of within-group variances. Summary statistics indicated that much of this heterogeneity reflected the modest mensural differences evident among the two taxa included within the *A. olivaceus*-group (*A. olivaceus* proper and *A. moluccanus*), and hence were not of relevance to flightlessness in the genus (Tables 30–33). Intersexual differences (five instances) and species–sex interaction effects (four instances) in group variances were limited to dimensions scattered throughout the skeleton.

The six taxa of moorhens and allies that were represented by skeletal samples permitting two-way analysis of group variances (Levene's tests; *Gallicrex cinerea, Gallinula chloropus*-group, *G. tenebrosa, G. comeri, Tribonyx ventralis,* and *T. mortierii*) showed significant interspecific effects for 33 of 41 skeletal variables compared. Two of these variables also manifested significant intersexual and species–sex interaction effects (both shaft widths of appendicular elements). With respect to the intersexual differences, the only grouping apparent in lacking interspecific effects were several dimensions of the sternum, an element notorious for comparatively high components of variance attributable to individual developmental anomalies and vagaries of measurement. Summary statistics for four other species in this group for which samples precluded formal analyses indicated no additional noteworthy patterns (Tables 30–33).

The two species of coots permitting two-way analyses of group variances (*Fulica atra* and *F. americana*) showed significant ($P < 0.05$) interspecific effects in 12 of 41 skeletal dimensions compared; intersexual and species–sex interaction effects were negligible, and were significant ($P < 0.05$) in only 2 and 1 of 41 variables, respectively. One-way comparisons of the larger set of six species of *Fulica* were not informative indicators of intraspecific variances in that there was no assurance of comparability of taxa with respect to the relative proportions of sexes within each.

Univariate Sexual Dimorphism

Univariate differences between the sexes in skeletal measurements, as indicated by corresponding effects in two-way ANOVA, revealed that sexual dimorphism of the skeleton is characteristic of rails. Subfossil species were not subjected to these formal two-way comparisons, but summary statistics (Tables 30–33) permitted informal assessments of dimorphism in those taxa provisionally partitioned into sexes based on K-means clustering (detailed below).

The greater swamphens (*Porphyrio*) manifested highly significant dimorphism of all skeletal measurements that were compared ($P < 0.05$), with species–sex interaction effects significant ($P < 0.05$) in 22 of the 41 dimensions as well. Pairwise comparisons (Bonferroni t-tests) and summary statistics (Tables 30–33) indicated that *P. melanotus* was characterized by exceptionally high sexual dimorphism relative to the other two species tested. Where samples permitted confident estimates, sexual dimorphism in subfossil *P. mantelli* appeared to exceed that of *P. hochstetteri* and to approach the magnitude found for *P. melanotus* (Tables 30–33). Lesser swamphens (*Porphyrula*) showed significant intersexual differences ($P < 0.05$) in 14 of 41 dimensions compared, with 4 of these (and 2 others) showing significant species–sex interaction effects ($P < 0.05$). Summary statistics and pairwise comparisons indicated that these effects largely reflected the comparatively low sexual dimorphism evident in the limited sample of *P. flavirostris* (Tables 30–33).

The three taxa of basal rallids amenable to formal two-way analyses (*Aramides cajanea*-group, *A. ypecaha,* and *Habroptila wallacii*) showed significant ($P < 0.05$) intersexual differences in 15 of 41 skeletal dimensions, with species–sex interaction effects also significant ($P < 0.05$) in 4 of these (Tables 30–33). Examination of summary statistics for these taxa and other, less well-represented taxa indicated that sexual size dimorphism was generally evident in both the large

A. ypecaha and flightless *H. wallacii,* and suggested that sexual dimorphism was substantial in flightless *Nesotrochis* (Tables 30–33).

The two species of white-throated rails subjected to two-way ANOVAs of skeletal dimensions (*Dryolimnas cuvieri* and *D. aldabranus*) showed significant ($P < 0.05$) intersexual differences in 26 of 41 variables. Two dimensions of the coracoid showed both interspecific and species–sex interaction effects, in this case reflecting sexual dimorphism only in *D. aldabranus*. Informal comparisons of summary statistics for *Cyanolimnas cerverai* indicated that sexual dimorphism was comparable in magnitude to that evident in most other rallids; limited samples available for extinct *D. abbotti* and *Ortygonax sanguinolentus* precluded assessments of sexual dimorphism (Tables 30–33).

For typical rails for which samples permitted two-way analysis, intersexual differences were detected ($P < 0.05$) in all but 1 of the 41 skeletal dimensions compared; species–sex interaction effects also were significant in all but 1 of these and were detected in 1 additional dimension as well. Accordingly, it can be inferred that skeletal dimensions show strong sexual dimorphism in these taxa, and that the magnitude of this dimorphism varies among taxa. Examination of summary statistics (Tables 30–33) and a posteriori pairwise comparisons indicated that the heterogeneity of sexual effects principally reflected the comparatively great dimorphism of a minority of species (e.g., *Gallirallus australis*-group and *G. wakensis*) and the comparatively low dimorphism (relative to within-group variances) evident in several others (e.g., *Tricholimnas sylvestris, G. philippensis, G. striatus,* and *G. owstoni*). Among the species for which formal analysis was not possible, *K*-means clustering of subfossil elements into provisional sex groups revealed exceptional dimorphism throughout the skeleton in *Diaphorapteryx hawkinsi,* previously misconstrued as anomalously high intraspecific variation unrelated to gender (Andrews 1896b; Cracraft 1973a), and comparably extreme bimodality in bill lengths of *Aphanapteryx bonasia* (Tables 30–33).

The 10 species of crakes (*Laterallus, Coturnicops, Crex,* and *Porzana*) having adequate samples for two-way analysis of skeletal dimensions (ANOVA), 2 of which (*P. atra* and *P. palmeri*) were flightless, only 28 of 41 skeletal dimensions showed significant intersexual effects ($P < 0.05$). One of these showed significant species–sex interaction effects ($P < 0.05$). The only obvious pattern in these differences was the absence of sexual dimorphism in measurements of the furcula and pelvis; both of the latter elements were prone to comparatively high individual variation (e.g., variation in numbers of vertebrae synsacrales), and this may have undermined the tests. None of the other crakes not included in formal analyses showed notable departures from the general pattern of modest but significant sexual dimorphism of skeletal dimensions (Tables 30–33).

Samples permitted assessments of skeletal sexual dimorphism in only two taxonomic groups having comparable body sizes (*Amaurornis olivaceus*-group and *A. phoenicurus*), but significant intersexual differences (ANOVA; $P < 0.05$) were confirmed in 33 of 41 skeletal dimensions analyzed. Three additional variables showed significant species–sex interaction effects, indicating that one of the two species showed sexual dimorphism whereas the other did not. Summary statistics (Tables 30–33) revealed that *A. phoenicurus* was characterized by more substantial sexual dimorphism than *A. olivaceus*-group in the latter cases, in part because of the higher within-group variances that resulted from the pooling of two allospecies in the latter taxon.

The six taxa of moorhens and allies represented by samples suitable for two-way ANOVA (*Gallicrex cinerea*, *Gallinula chloropus*-group, *G. tenebrosa*, *G. comeri*, *Tribonyx ventralis*, and *T. mortierii*) showed significant intersexual effects in all 41 skeletal dimensions that were compared ($P < 0.05$), with 17 variables also showing significant species–sex interaction effects (indicative of differences in magnitude of sexual dimorphism among species). A posteriori pairwise comparisons (Bonferroni t-tests) revealed that the latter reflected the extraordinarily high sexual skeletal dimorphism of *G. cinerea* (one of the few rallids showing sexual dichromatism of the plumage), substantial but intermediate sexual dimorphism of *Tribonyx ventralis* and (to a lesser extent) *T. mortierii*, and comparatively modest sexual differences in other taxa of moorhens (Tables 30–33). In *G. cinerea*, skeletal dimensions of males generally were at least 10% larger than those of female conspecifics (Tables 30–33). Data were too limited for even qualitative assessments of sexual dimorphism (Tables 30–33) in four other moorhens represented (*Pareudiastes pacificus*, *Tribonyx hodgenorum*, *Gallinula nesiotis*, and *G. sandvicensis*).

The two species of coots (*Fulica*) for which samples permitted two-way analyses showed significant intersexual differences in 40 of 41 skeletal dimensions that were compared, of which 16 also manifested significant species–sex interaction effects (indicative of interspecific differences in magnitude of sexual dimorphism). Summary statistics and a posteriori pairwise comparisons documented that where sexual dimorphism varied in magnitude between species, mean sexual differences generally were larger in the small *F. americana* than in *F. atra* (Tables 30–33). Summary statistics for two subfossil species amenable to K-means clustering (*F. chathamensis* and *F. prisca*) indicated that sexual dimorphism was at least as great in these two flightless species as in the smaller, flighted congeners (Tables 30–33).

UNIVARIATE SKELETAL GENERALITIES

Comparisons of skeletal dimensions by way of univariate protocols revealed that flightlessness is associated with shortening of wing elements and comparable or greater diminution of the sternum relative to other osteological portions of the body (e.g., skull and pelvic limb). Most instances of flightlessness and concomitant reductions in relative size of the pectoral apparatus were accompanied by increases in overall body size, which in combination with the relative pectoral truncation or approximate stasis created in many flightless taxa an aspect of enlargement of the body in which the pectoral apparatus appeared retarded. Exceptions to this generalization were apparent dwarfed lineages (e.g., *Gallirallus wakensis*, *Atlantisia rogersi*, and most flightless *Porzana*), in which lineages evidently underwent overall decreases in body size accompanied by even more pronounced pectoral reductions. Still other flightless taxa were distinguished by other exceptional apomorphies unrelated to the pectoral apparatus, including decurvature of the bill (e.g., *Gallirallus dieffenbachii*, *Aphanapteryx bonasia*, *Erythromachus leguati*, and *Diaphorapteryx hawkinsi*), combined in some with extraordinarily pronounced sexual dimorphism (e.g., *Aphanapteryx* and *Diaphorapteryx*). Some indications also were found that some dimensions of skeletal elements directly related to flight and subject to comparatively late ossification (e.g., carina sterni) showed higher relative variation (coefficients of variation, $SD/\bar{x}$), especially in flightless species, than the majority of skeletal dimensions compared (e.g., lengths of appendicular elements).

The major subgroups of flightless birds can be delimited by the combined differences in lengths of the pectoral and pelvic limbs or relative appendicular size associated with the loss of flight, although these more-inclusive metrics will exclude a number of taxa that remain imperfectly represented for one or more critical limb elements (e.g., *Porphyrio mantelli, Cyanolinmas cerverai, Habropteryx okinawae, Amaurornis ineptus,* and *Fulica newtoni*). Mean SWL and SLL revealed three main partitions among flightless rallids, based on approximations compared among congeners or closely related genera: flightless species showing absolute increases or virtual stasis in SWL with increased SLL, that is, species showing allometric giantism (e.g., *Porphyrio hochstetteri, Habroptila wallacii, Gallirallus australis*-group, *Tricholimnas sylvestris, Porzana atra,* and *Gallinula comeri*); flightless species showing absolute decreases in both SWL and SLL, that is, species having undergone generalized dwarfism (e.g., *Dryolimnas aldabranus, Gallirallus owstoni, G. wakensis, Cabalus modestus, Atlantisia rogersi, Porzana palmeri,* and *P. menehune*); and flightless species showing evolutionary disjunction between pectoral and pelvic limbs, in which absolute decreases in SWL are combined with virtual stasis or increases of SLL, that is, those species having undergone pectorally reductive giantism (e.g., *Gallirallus dieffenbachii, Capellirallus karamu, Diaphorapteryx hawkinsi, Tribonyx mortierii,* and *Fulica chathamensis*).

Although a useful preliminary summary, the complexity of changes within and between appendages and other parts of the skeleton, and the potential for such associated changes for obscuring evolutionary changes in form, point to the need for explicitly bivariate and multivariate comparisons to summarize and ordinate changes associated with flightlessness. The shortcomings of such a simplistic, univariate classification are indicated as well by the frequency of species falling ambiguously close to two of the three groups (e.g., *Porphyrio hochstetteri, Gallirallus australis, Tricholimnas sylvestris,* "*Atlantisia*" *elpenor,* and *Gallinula comeri*). Still others preclude confident assignment because of uncertainties concerning the most closely related, flighted taxon against which to make comparisons (e.g., *Aphanapteryx bonasia*). This univariate approach fails to discern the nuances of possibly changing proportions that accompanied overall increases in limb lengths, let alone the composite diversity of such that derive from a consideration of both limbs and other anatomical regions across lineages. More-sophisticated methods are required to elucidate the underlying correlative complexes that generate these disparate flightless phenotypes, and to quantify nonredundantly the magnitudes and directions of the associated interspecific and intersexual shifts.

HIERARCHICAL CORRELATION STRUCTURE

Bivariate correlation coefficients (r) among variables within species typically reflect developmentally constrained subsets of morphometric space (e.g, Power 1971; Cracraft 1976a; Livezey 1989c; Johnston 1992a), and are among the most familiar indices to intimacy of relationships between pairs of variables. Cluster analyses of variables based on pooled within-species correlation matrices for

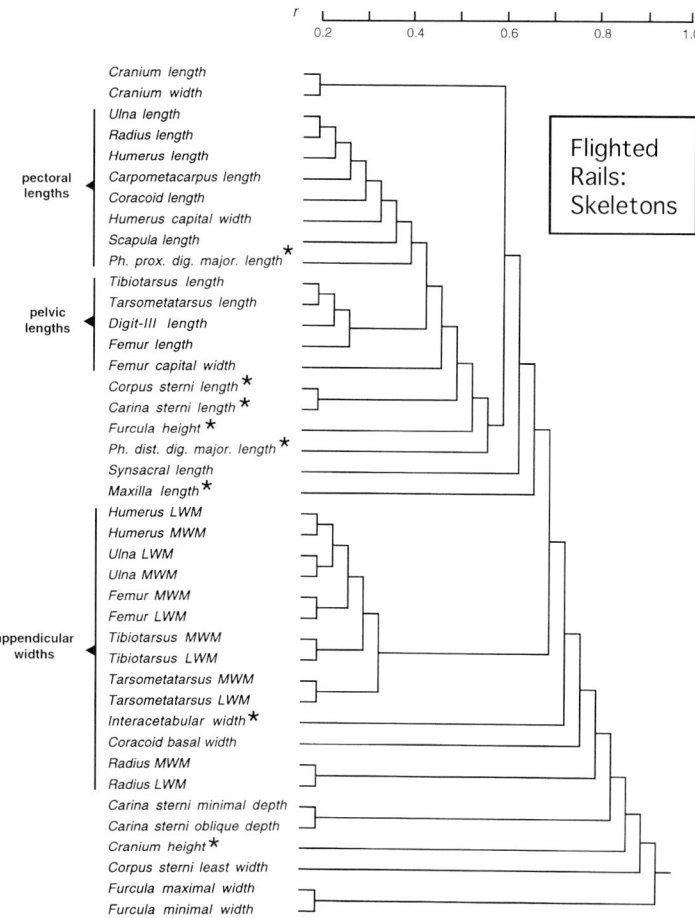

FIG. 40. Dendrogram depicting hierarchy of pairwise associations among 41 skeletal measurements for 43 flighted species of Rallidae based on a cluster analysis of pooled within-species correlation matrix (pooled $n = 817$).

flighted and flightless species (Figs. 40, 41) summarize intraspecific, pairwise correlations between variables or groups of variables across taxa, here partitioned by flight capacity. The phenogram for skeletal variables of flighted taxa (Fig. 40) summarized average correlations within 43 species (total $n = 817$); that for flightless taxa (Fig. 41) depicted the hierarchy of correlations between the same skeletal variables for 8 species (total $n = 179$). These figures revealed a number of closely correlated groups of variables that were common to both flighted and flightless taxa, including lengths of major elements of the pectoral and pelvic limbs; widths of the major limb elements, most closely linked within elements and limbs; and width and length of the cranium (Figs. 40, 41). Also similar between the two phenograms were the greater correlations among lengths of appendicular elements, regardless of which limb was involved or proximity of the element to the trunk within the appendage (Figs. 40, 41). Commonalities between the two groups also included variables characterized by comparatively low correlations with other skeletal dimensions. These loosely associated variables principally involved

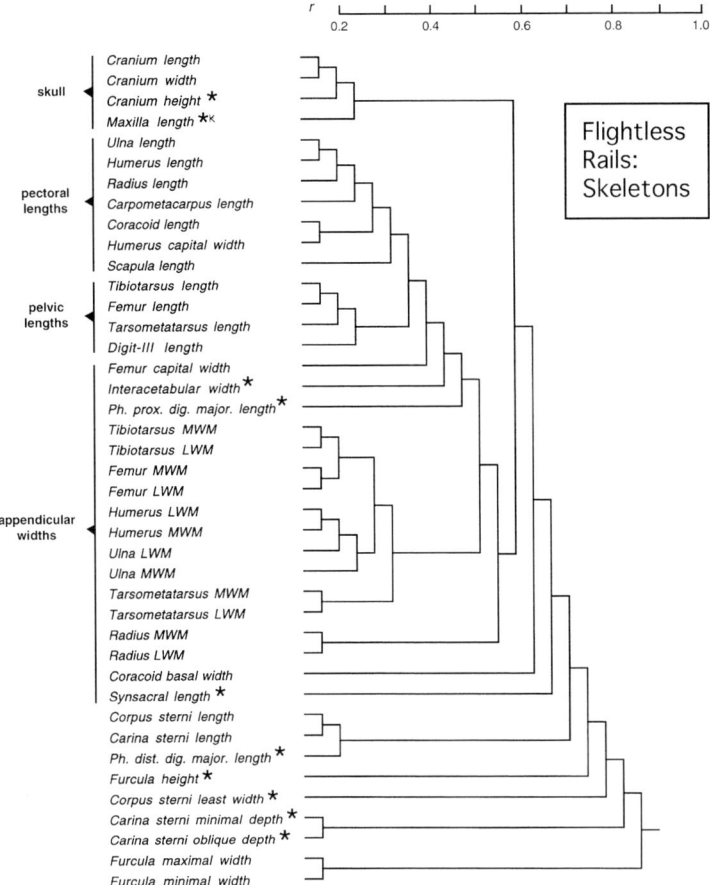

FIG. 41. Dendrogram depicting hierarchy of pairwise associations among 41 skeletal measurements for eight flightless species of Rallidae based on a cluster analysis of pooled within-species correlation matrix (pooled $n = 179$).

widths of the elements of the pectoral girdle (sternum, coracoid, and especially furcula).

Generalities of intraspecific correlations that differed with flight capacity were of special interest. Of the four measurements of the skull, all four were tightly intercorrelated in the flightless taxa (Fig. 41), whereas only width and length of the cranium were highly correlated in the flighted taxa (Fig. 40). More intuitive, however, was the comparatively loose correlation between sternal lengths and the depth of the carina sterni in flightless rallids. Together with widths within the pectoral girdle, these variables were among the last to be clustered with all other skeletal dimensions in the flightless taxa (Fig. 41), whereas in flighted rallids the former were closely correlated with appendicular lengths (Fig. 40). Flightless rallids also were distinguished from flighted species in the comparatively low correlations between lengths of the two phalanges of the major alar digit and other appendicular lengths; length of the distal phalanx in flightless rallids was especially poorly correlated with other skeletal dimensions (Fig. 41). Both of the latter generalities established a low intraspecific association between the lengths of the sternum and distalmost alar elements and other skeletal dimensions common to all eight flightless rallids analyzed.

INTRA-APPENDICULAR SKELETAL PROPORTIONS AND RATIOS

Proportions within pectoral limb.—Univariate comparisons demonstrate that flightless rallids tend to be characterized by truncated pectoral elements (Table 34), although the magnitudes of these reductions vary among taxa and can be obscured by concomitant changes in overall body size and sexual dimorphism to a limited degree (Tables 30–33). However, with all such potentially confounding factors held constant, the possibility remains that such flightlessness-related reductions may not occur to the same degree in each element, thereby effecting changes in skeletal proportions, and these alterations of proportions may vary taxonomically or with changes in overall body size. Two-way ANOVA of proportions within the pectoral limb and ratios of elements between appendages revealed that, regardless of taxon or element in question, interspecific differences were highly significant for most ($P < 0.0001$), although often numerically small (with most differences being less than 3% of total appendicular length). Despite the comparatively precise differences such bivariate metrics revealed between species, intersexual or species–sex interaction effects in these ratios and proportions were not significant ($P \gg 0.05$). Moreover, in only a minority of instances were interspecific differences in variances among taxonomic or sex groups significant (Levene's tests; $P < 0.05$), further simplifying generalizations. This inferential clarity permitted the following summaries to focus narrowly on taxonomic differences in means, with special emphasis on patterns of differences between flighted and flightless relatives.

Rails capable of flight tend to possess pectoral limbs in which the primary skeletal segments decrease monotonically in contributory proportions proximodistally. That is, within the alar skeleton, proportions of total SWL decrease from a maximum in the brachium (humerus), through the parallel elements of ulna and radius (antebrachium), the proximal manus (carpometacarpus), and with the least segments being the distalmost elements or phalanges of the major digit (Tables 30–33). This array of proportions (Pr) of the humerus, ulna, carpometacarpus, and proximal and distal phalanges of the major digit can be summarized algebraically for flighted members of the Rallidae as (Fig. 42):

$$Pr \text{ (humerus)} \geq Pr \text{ (ulna)} \gg Pr \text{ (carpometacarpus)} \gg Pr \text{ (proximal phalanx)}$$
$$\geq Pr \text{ (distal phalanx)},$$

wherein magnitudes of the inequalities are symbolized as $\gg \approx 0.1$, $> \approx 0.05$, and $\geq \approx 0$–0.03. In addition to those taxa for which samples permitted formal two-way analyses, proportions for exemplary specimens for most other osteologically known rallida agreed with these general patterns, and the combined tallies of taxa examined were 67 flighted and 20 flightless species. Exceptions to this familial generality exist. For example, *Rallus longirostris* exhibits alar proportions in which the first proportion is unusually high (37%), and the first proportion is unusually small (0.34%) in all members of the subtribe Rallina (*Rallina, Rallicula,* and *Sarothrura*). The segment manifesting the greatest uniformity among flighted rallids is that for the carpometacarpus (Table 34), the proportion for which was

TABLE 34. Ranges for mean intra-appendicular proportions and skeletal wing lengths for major elements of the pectoral appendage for selected flighted and flightless species or species groups of Rallidae.

Taxonomic group (k species, n)	Mean skeletal wing lengths	Range in proportion (%) of skeletal wing length by element				
		Humerus	Ulna	Carpometacarpus	Dig. maj. phal. I	Dig. maj. phal. II
Porphyrio porphyrio-group (2, 53)	223.0, 252.2	33.0, 33.7	30.5, 31.0	20.7	8.1, 8.2	7.0, 7.1
Porphyrio hochstetteri (20)	224.9	37.1	31.5	19.9	6.1	5.4
Porphyrula spp. (3, 53)	133.9–152.1	33.4–33.7	28.4–29.7	20.6–21.0	8.5–9.0	7.4–8.2
Aramides spp. (2, 55)	162.2, 211.8	34.0, 34.6	30.2, 30.4	20.6, 20.7	7.8, 8.2	6.8
Canirallus kioloides (2)	109.4	34.0	31.8	20.8	8.6	6.6
Habroptila wallacii (4)	161.4	37.6	30.8	29.4	6.4	5.9
Dryolimnas cuvieri (5)	140.9	35.2	30.3	20.6	7.3	6.7
Dryolimnas aldabranus (10)	108.4	36.2	29.9	19.9	7.3	6.7
Flighted *Rallus, Gallirallus,* and *Habropteryx* (8, 170)	96.1–168.1	34.7–36.8	29.4–30.8	19.6–20.5	7.1–7.3	6.5–6.8
Aphanapteryx bonasia (–)	159	39	29	18	9	5
Gallirallus wakensis (3)	91.3	36.0	29.6	19.8	7.5	7.0
Gallirallus owstoni (28)	123.4	36.7	30.1	19.4	6.7	7.0
Gallirallus australis (31)	139.6	38.4	29.0	19.4	6.6	6.6
Gallirallus dieffenbachii (–)	115	38	28	19	7	7
Cabalus modestus (–)	66	38	28	19	7	7
Nesoclopeus poecilopterus (1)	153.0	35.3	29.4	20.9	7.8	6.5
Tricholimnas sylvestris (6)	136.9	36.4	29.7	19.5	7.4	7.0
Coturnicops, Crex, Laterallus, and flighted *Porzana* (8, 132)	74.1–129.5	34.5–36.8	28.8–30.9	20.1–21.1	7.1–8.0	6.4–7.3
Porzana atra (4)	73.9	37.9	28.7	19.5	7.5	6.4
Porzana palmeri (9)	52.6	37.9	28.3	19.7	7.4	6.7
Atlantisia rogersi (2)	50.3	40.1	26.6	18.6	7.9	6.9
"Atlantisia" elpenor (–)	70	37	29	20	8	6
Flighted *Amaurornis* (2, 40)	136.9–137.5	34.4–35.0	30.3–30.6	20.3–20.8	7.6–7.8	6.5–6.7
Amaurornis ineptus (–)	165.7	36.5	30.4	19.3	7.7	6.2
Gallicrex, flighted *Gallinula* and *Tribonyx* (4, 104)	149.2–187.0	33.1–34.9	28.5–31.7	20.1–22.0	7.8–8.2	6.6–7.8
Tribonyx mortierii (13)	179.0	36.9	27.4	20.5	7.9	7.2
"Typical" *Fulica* (4, 72)	301.6–321.2	35.3–36.3	30.5–31.3	19.6–20.0	7.0–7.4	6.4–6.6
Fulica gigantea and *F. cornuta* (2, 7)	301.6–321.2	34.8–35.7	31.5–31.9	19.6–19.9	7.0–7.1	6.3–6.4
Fulica chathamensis (–)	254	37	30	19	7	6

0.20–0.21 for 65 of 67 flighted rallids (with exceptions being 0.25 for *Gallinula tenebrosa* and 0.22 for *Tribonyx ventralis*).

The proportions of skeletal elements within the wing underwent variably pronounced modifications associated with the loss of flight ($P < 0.0001$ for interspecific effects, unless otherwise noted, in two-way ANOVAs of proportions), but several general patterns emerged that were shared by a number of the flightless taxa having adequate osteological samples (Table 34; Fig. 42). Estimates of alar proportions based on single specimens of *Aramides calopterus* and *A. saracura* closely conformed to those for the two well-sampled species of *Aramides* (Table 34). Interspecific, flightless-associated differences in alar proportions were extremely slight in *Dryolimnas* (Table 34). The flighted profile of proportions among eight flighted species of *Rallus* and *Gallirallus* (Table 34) was mirrored by single exemplars of three species of *Rallina,* one species of *Rallicula,* and three species of *Sarothrura.* Unfortunately, incomplete resolution of phylogenetic relationships within these taxa precluded pairwise assessments of these shifts. Nevertheless, broad comparisons indicated increases in brachial proportions and decrease in the carpometacarpus and digits proportions in the wings of flightless members of the tribe (Table 34).

Eight flighted crakes possessed alar proportions closely similar to that for other flighted rallids (Table 34), with all but one of these segments encompassed by the four flighted species of *Porzana* alone. In addition, the proportions of another five flighted species of crakes represented by single exemplars (*Crex egregia, Micropygia flaviventer, Porzana porzana, P. fluminea,* and *P. fusca*) conformed with these intervals as well. Proportions in the wings of two flighted bushhens (*Amaurornis phoenicurus* and *A. olivaceus*-group) agreed with those of other flighted rallids, but differed from those for one skeleton of the flightless congener *A. ineptus* (Table 34). Similar patterns pertained to four flighted moorhens and allies, despite substantial differences in overall size. Four species of coots (*Fulica*) of typical size were consistent with moorhens in alar proportions (Table 34). Proportions of three additional species of coots (*F. rufifrons, F. leucoptera,* and *F. cristata*) represented by single exemplars also fell within these ranges. Proportions for *F. gigantea* and *F. cornuta* suggested slight shifts ($P < 0.0005$), most notably a relative increase in the antebrachium (Table 34). Estimates for unassociated subfossil elements of *F. chathamensis* indicated that the species had undergone only modest shifts toward those typical of flightless rallids, that is, a slight relative elongation of the brachium and a foreshortening of the antebrachium and manus.

Proportions within pelvic limb.—Although less obviously pertinent to comparisons related to flight capacity, changes in pelvic proportions associated with the loss of flight were sought among those taxa having sufficient samples. Fortunately, the same combination of favorable statistical properties found for the pectoral limb characterized the pelvic appendage. Unless otherwise indicated, interspecific differences were highly significant (two-way ANOVA; $P < 0.0001$) and heterogeneity of variances was not significant (Levene's test; $P > 0.05$). As for the assessments of lengths and proportions within the pectoral limb, unavailability of requisite elements precluded estimates for a number of taxa, both modern and subfossil forms (Table 35).

Flighted rails showed modest differences in the proportions within the pelvic

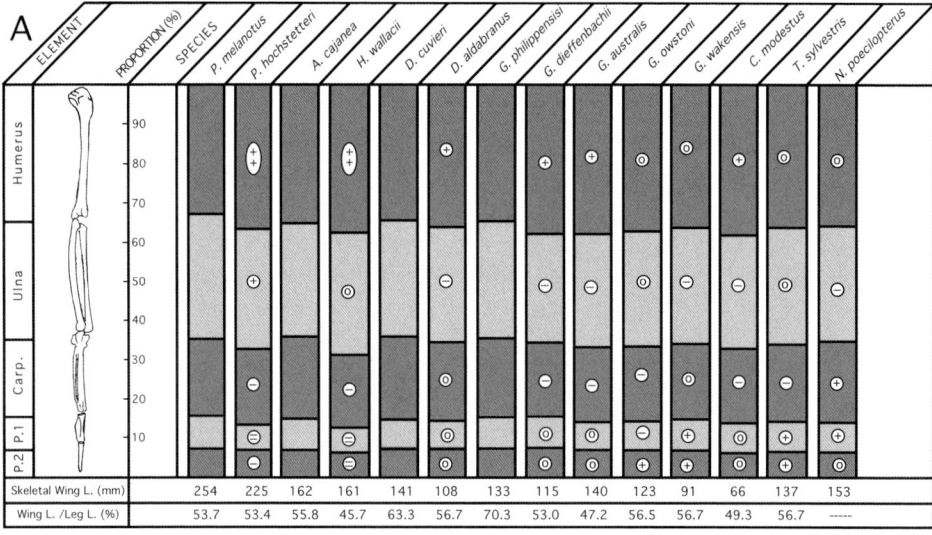

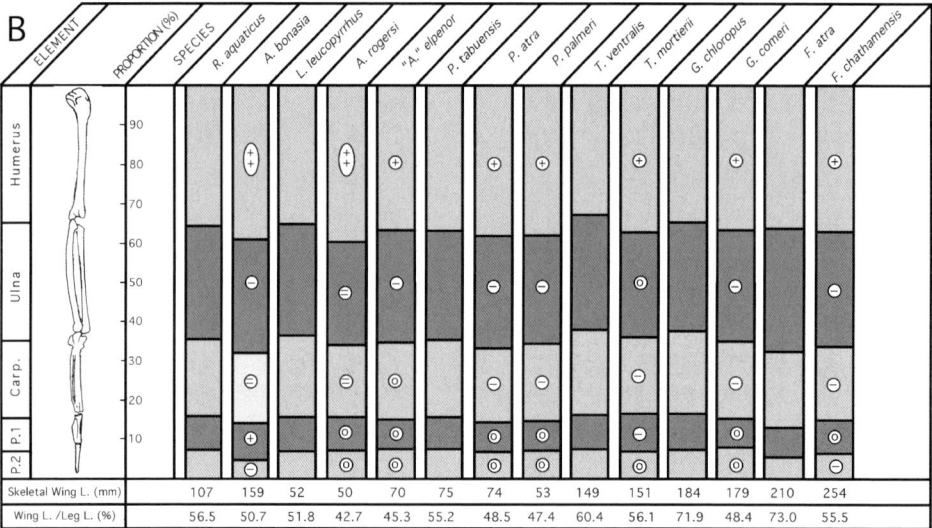

FIG. 42. Two series of bar charts depicting mean appendicular proportions within the major skeletal segments of the pectoral appendage of selected flighted and flightless species of Rallidae. **A,** *Porphyrio melanotus, P. hochstetteri, Aramides cajanea, Habroptila wallacii, Dryolimnas cuvieri, D. aldabranus, Gallirallus philippensis, G. dieffenbachii, G. australis, G. owstoni, G. wakensis, Cabalus modestus, Tricholimnas sylvestris,* and *Nesoclopeus poecilopterus.* **B,** *Rallus aquaticus, Aphanapteryx bonasia, Laterallus leucopyrrhus, "Atlantisia" rogersi, "Atlantisia" elpenor, Porzana atra, P. palmeri, Tribonyx ventralis, T. mortierii, Gallinula chloropus, G. comeri, Fulica atra,* and *F. chathamensis.* Directions of inferred shifts in flightless rails are indicated by enclosed signs; + = increase, ++ = great increase, 0 = no change, − = decrease, and − − = great decrease.

limb, which although small, were critical for assessments of shifts related to the loss of flight. SLL, on the other hand, varied greatly at all taxonomic levels and between the sexes (two-way ANOVA; $P < 0.0001$), and showed moderately pronounced interspecific heterogeneity in within-group variances as well (Levene's tests; $P < 0.001$). Nevertheless, most flighted rallids possessed a suite of proportions within the pelvic limb that are summarized as follows (Table 35; Fig. 43):

$$Pr \text{ (femur)} \ll Pr \text{ (tibiotarsus)} \gg Pr \text{ (tarsometatarsus)}$$
$$\geq Pr \text{ (digit III, exclusive of phalanx ungualis)},$$

wherein magnitudes of the inequalities are symbolized as $\ll$ and $\gg \approx 0.15$, $> \approx 0.10$, and $\geq \approx 0$–0.05. The most notable exception to this generality pertained to the long toes of *Porphyrula* and *Fulica*. The middle toes of other flighted rallids averaged only 17.4–23.0% (Table 35).

Within *Porphyrio*, the large, flightless *Porphyrio hochstetteri* was characterized by proportions differing markedly in all segments of the pelvic limb exclusive of the tarsometatarsus (Fig. 43): elongated femur, elongated tibiotarsus, unchanged tarsometatarsus, and truncated middle toe. Among the basal rallines, flightless *Habroptila wallacii* had proportions that did not differ interspecifically for the tibiotarsus and presented no clear parallels in comparison with two flighted species of *Aramides* and two flighted species of *Canirallus* (Table 35).

Among typical rails, *Dryolimnas* showed virtually identical proportions within the pelvic limb regardless of flight capacity (Table 35). Among the typical rails that were analyzed, eight flighted species of *Rallus*, *Gallirallus*, and *Habropteryx* manifested comparatively diverse pelvic proportions. Of the four flightless members of this group that were formally compared, only *G. owstoni* had proportions that fell within all four intervals established for flighted species. *Tricholimnas sylvestris* deviated from flighted relatives only in tarsometatarsal proportions, whereas tiny *G. wakensis* departed from the pelvic proportions of flighted relatives in two or more of the (nonfemoral) segments (Table 35; Fig. 43). Disassociated subfossil elements provided estimates of proportions for seven additional allied rallids (*Rallus ibycus*, *R. recessus*, *G. dieffenbachii*, *Cabalus modestus*, *Capellirallus karamu*, *Aphanapteryx bonasia*, and *Diaphorapteryx hawkinsi*). Except for *R. recessus*, all of the latter were characterized by pelvic proportions similar to comparatively derived, flightless rails, despite a remarkable range in overall size (Table 35; Fig. 43).

Eight flighted species of crakes (*Coturnicops*, *Crex*, *Laterallus*, and *Porzana* of diverse size) spanned a comparatively broad spectrum of pelvic proportions (Table 35). Although there was a tendency for the three flightless crakes thtat were analyzed (*Atlantisia rogersi*, *Porzana atra*, and *P. palmeri*) to possess relatively short femora and long tibiotarsi, the pelvic proportions of the three flightless taxa manifested sufficient heterogeneity so as to obscure patterns simply attributable to flight capacity (Table 35; Fig. 43). Subfossil elements of "*Atlantisia*" *elpenor* provided estimates of pelvic proportions that (at least for the tibiotarsus) were consistent with flightlessness. However, similar estimates for subfossil *Porzana piercei* departed from proportions encompassed by flighted congeners only in its slightly truncated tarsometatarsi and elongated digits (Table 35).

Two flighted species of *Amaurornis* manifested close similitude of pelvic proportions in which interspecific differences in tibiotarsal and tarsometatarsal segments were atypically low ($P < 0.01$ and $P < 0.005$, respectively). These proportions were very similar to those exhibited by the aberrant moorhen *Gallicrex cinerea* (Table 35). *Gallinula* revealed a numerically small but distinct shift in

TABLE 35. Ranges for mean skeletal leg lengths and intra-appendicular proportions and for major elements of the pelvic appendage, and humerus : femur ratios for selected flighted and flightless species or species groups of Rallidae. Sample sizes (n) given as dashes indicate estimates were based on unassociated elements.

Taxonomic group (k species, n)	Mean skeletal leg lengths	Range in proportion (%) of skeletal leg length by element				Ratio between length of humerus and femur
		Femur	Tibiotarsus	Tarsometatarsus	Digit III	
Porphyrio porphyrio-group (2, 47)	370.0, 401.8	19.2	34.3, 35.0	23.4, 24.3	21.5, 23.1	1.00, 1.09
Porphyrio mantelli	—	—	—	—	—	0.83
Porphyrio hochstetteri (27)	419.9	23.7	36.7	22.6	17.1	0.83
Porphyrula spp. (3, 33)	204.5–267.3	18.6–19.5	34.4–35.3	22.4–23.4	22.3–24.6	0.97–1.01
Aramides spp. (2, 58)	291.2, 366.1	21.0, 21.4	35.5, 35.6	24.3, 25.6	17.4, 19.2	0.90–0.94
Nesotrochis spp. (2, —)	—	—	—	—	—	0.08, 0.81
Canirallus spp. (2, 4)	188.6, 233.1	22.7, 23.5	35.6, 36.2	22.2, 24.2	16.8, 18.7	0.84–0.94
Habroptila wallacii (4)	353.3	22.8	35.0	25.0	17.2	0.75
Dryolimnas cuvieri (5)	222.5	22.9	35.1	22.5	19.5	0.97
Dryolimnas aldabranus (12)	194.7	22.8	35.1	22.7	19.5	0.90
Cyanolimnas cerverai (—)	—	—	—	—	—	0.76
Flighted *Rallus*, *Gallirallus*, and *Habropteryx* (8, 173)	149.6–261.6	21.5–24.6	34.3–35.6	22.0–23.1	18.0–21.8	0.91–1.03
Rallus ibycus (—)	154	25	33	20	23	0.84
Rallus recessus (—)	199	25	34	21	21	0.92
Aphanapteryx bonasia (—)	318	23	36	24	18	0.86
Erythromachus leguati (—)	—	—	—	—	—	0.83
Gallirallus wakensis (3)	160.9	22.8	35.9	21.8	19.5	0.90
Gallirallus owstoni (31)	217.0	24.0	35.2	22.8	18.0	0.86
Gallirallus australis (46)	296.4	24.9	36.6	21.1	17.5	0.73
Gallirallus dieffenbachii (—)	215	25	35	21	20	0.81
Cabalus modestus (—)	134	22	36	22	20	0.83
Capellirallus karamu (—)	—	—	—	—	—	0.58
Diaphorapteryx hawkinsi (—)	—	—	—	—	—	0.68
Tricholimnas sylvestris (6)	241.4	24.8	35.4	20.8	19.0	0.83
Habropteryx okinawae (—)	—	—	—	—	—	0.83

TABLE 35. Continued.

Taxonomic group (k species, n)	Mean skeletal leg lengths	Range in proportion (%) of skeletal leg length by element				Ratio between length of humerus and femur
		Femur	Tibiotarsus	Tarsometatarsus	Digit III	
Coturnicops, Crex, Laterallus, and flighted *Porzana* (8, 148)	116.3–178.2	21.5–24.8	34.3–35.6	22.0–23.1	18.0–21.8	0.87–1.02
Porzana piercei (–)	108	22	34	20	23	0.88
Porzana atra (2)	154.3	21.8	35.0	22.7	20.5	0.83
Porzana palmeri (11)	110.8	22.5	34.6	21.2	21.7	0.80
Atlantisia rogersi (2)	117.9	21.4	35.5	21.1	22.0	0.80
"*Atlantisia*" *elpenor* (–)	154	22	36	23	19	0.77
Flighted *Amaurornis* (2, 40)	243.4, 252.7	19.7, 20.8	35.2, 35.6	23.3, 23.7	19.9, 21.8	0.91–0.99
Gallicrex, flighted *Gallinula* and *Tribonyx* (4, 104)	247.2–304.9	19.2–21.6	34.5–37.2	21.3–23.7	17.4–23.7	1.04–1.10
Tribonyx hodgenorum	—	—	—	—	—	0.72
Tribonyx mortierii (15)	359.7	22.7	37.4	23.2	16.7	0.79
"Typical" *Fulica* (4, 72)	272.2–338.2	18.5–19.2	35.2–36.3	20.5–20.7	24.0–24.2	1.30–1.37
Fulica gigantea and *F. cornuta* (8)	416.5, 458.7	18.1, 18.9	35.5, 35.8	21.3, 22.0	24.0, 24.2	1.27, 1.41
Fulica chathamensis and *F. prisca* (–)	436, 422	19	36	21	24	1.13, 1.15

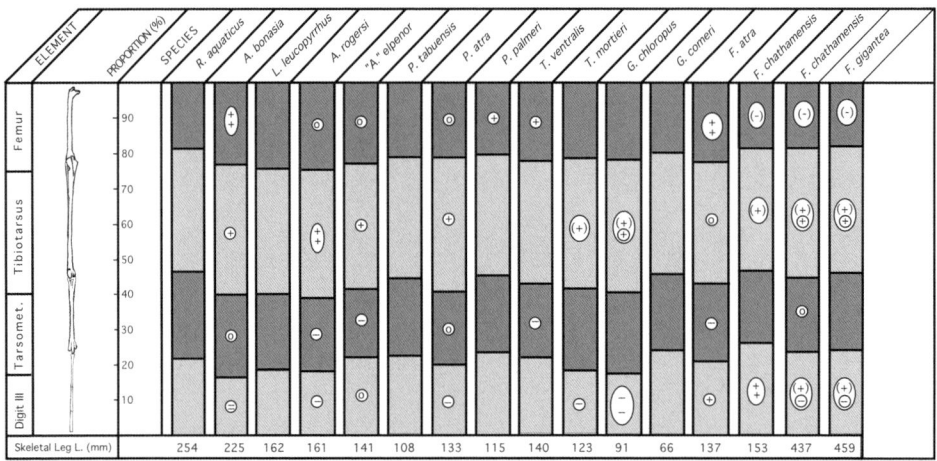

FIG. 43. Bar charts depicting mean appendicular proportions within the major skeletal segments of the pelvic appendage of selected flighted and flightless species of Rallidae: *Porphyrio melanotus, P. hochstetteri, Gallirallus philippensis, G. australis, G. wakensis, Laterallus leucopyrrhus, Atlantisia rogersi, "Atlantisia" elpenor, Porzana tabuensis, P. palmeri, Gallinula chloropus, G. comeri, Tribonyx ventralis, T. mortierii, Fulica atra, F. chathamensis,* and *F. gigantea.* Directions of inferred shifts in flightless rails as in Figure 42.

proportions, in which two flighted species shared virtually identical pelvic proportions, shifts extended further still by flightless *G. comeri.* A parallel suite of pelvic proportions characterized *Tribonyx*; flighted *T. ventralis* showed relatively truncated femoral segments and elongated toes in comparison with corresponding dimensions of flightless *T. mortierii,* and both species of *Tribonyx* differed from *Gallinula* by similar differences in proportions (Table 35; Fig. 43).

Among the highly aquatic coots (*Fulica*), four species of typical size shared a well-defined suite of pelvic proportions that diverged from those of all other rallids, with the most marked features being the relatively short femora and elongate toes (Fig. 43). *F. gigantea* and *F. cornuta* differed only slightly from their smaller congeners. Estimates based on disassociated subfossil elements of *F. chathamensis* and *F. prisca* revealed that these extinct taxa possessed pelvic limbs of comparable size and similar skeletal proportions to those of the two large Andean species (Table 35).

Interappendicular ratios: humerus versus femur.—Given the functional relationships between major wing elements and flight capacity (Table 34; Fig. 42) and between major skeletal elements of the leg and overall size (Table 35; Fig. 43), ratios involving elements from both appendages hold a potential for a comparatively simple index to flight capacity that would be robust to a reasonable range of body size. A further advantage of such ratios is that these effectively correct for size within taxa, rendering intersexual differences or species–sex interactions insignificant (two-way ANOVA; $P > 0.05$), thereby increasing power of interspecific comparisons. The comparative commonness of the humerus and femur (either is the most likely to be retained in trunk skeletons of modern specimens or as subfossil remains [available for 112 species of Rallidae; Fig. 44]) makes these proximal elements logical candidates to derive such a ratio. Only

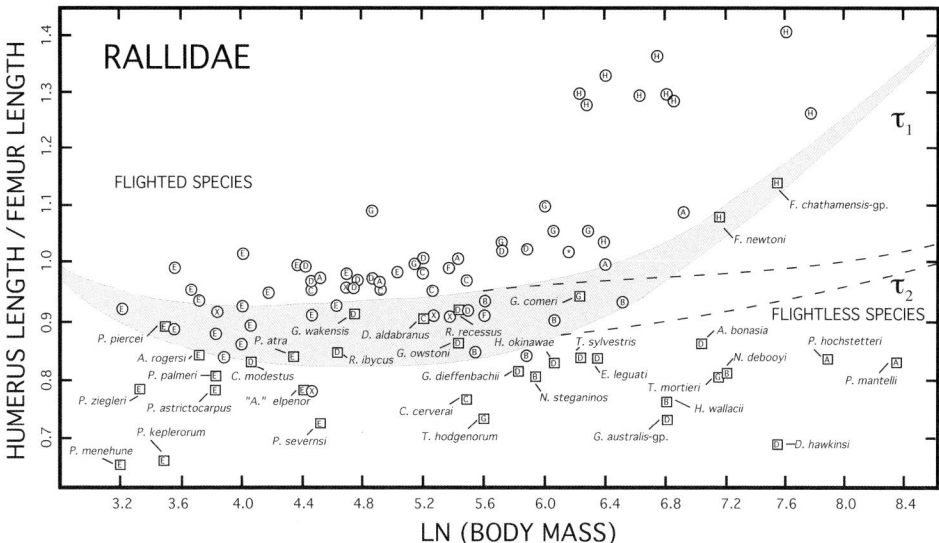

FIG. 44. Ratios of humerus length divided by femur length for 106 species of Rallidae, plotted against $\log_e$-transformed mean body masses or estimates thereof (Table 8). Flighted species are depicted as circles and flightless species are depicted as squares. Stippled zone delimited by smoothed boundaries approximates the threshold of flightlessness (τ_1, positioned on the flightless side of boundary). Alternative threshold (τ_2) admits subfossil *Fulica* for inclusion among flighted taxa.

where fundamental functional differences characterize compared taxa can such ratios become significantly misleading.

For flighted species of Rallidae represented by reasonable samples, mean ratios of the lengths of humeri and femora (within individuals) manifested significant diagnostic potential for most genera (Table 35). Although doubt remains as to the flight capacity of the extinct coots (*F. chathamensis, F. prisca,* and *F. newtoni*), the exceedingly high ratios for these species (Table 35) reveal this ratio to be uninformative for classification of flight capacity for species of large size and natatorial specialization (Fig. 43). Excluding several estimates for species represented by single skeletal specimens, only a few large, terrestrial species (e.g., *Aramides*) had mean values for this ratio coincident with the threshold value of 0.9. The estimate for *Rallus recessus* of Bermuda (0.92) suggests that this species was approaching flightlessness, but a determination of genuine flightlessness is not possible on the basis of this single index (Table 35).

Accordingly, excluding *Fulica* and noting the caveat regarding *Aramides,* the humerus to femur ratio averages 0.90–1.41 for flighted members of the Rallidae, and 0.58–0.90 for flightless confamilials (with the maximum for flightless taxa pertaining to nanitic *D. aldabranus*), and evidently provides a useful index for discriminating flighted from flightless species of rails. That is, flighted species are characterized by humerus to femur ratios in excess of 0.9, and flightless species generally have a ratio less than 0.9 (some considerably less). However, caution is advisable for flighted species that are large and cursorial (e.g., *Aramides*) or possess pelvic appendages adapted for diving (*Fulica*).

MULTIVARIATE COMPARISONS

K-means cluster analyses.—This method effected the provisional partitioning of sufficiently represented samples of subfossil elements into provisional sex-groups for comparison with modern species. For most elements and taxa, such partitions were made with high statistical significance (Tables 36, 37). In most taxa, intersexual differences in the vast majority of skeletal measurements approximated 5% (Tables 30–33), a finding in agreement with intersexual differences in skeletal measurements in a variety of other carinate birds (Livezey and Humphrey 1984a; Livezey 1989b–d, 1992a, b, 1993b), but greater than that of the virtually monomorphic Alcidae (Livezey 1988).

Although statistics for the clustering were not tabulated, a number of subfossil rallids permitted the provisional allocation of bill lengths into gender classes. In most of these taxa, sexual dimorphism of bill lengths was inferred a posteriori to be of unremarkable magnitude (cf. *Porphyrio mantelli, Rallus ibycus,* and *Fulica chathamensis*). Despite the recognition of substantial dimorphism generally in *Diaphorapteryx hawkinsi,* previously cited simply as intraspecific variation (Andrews 1896b; Cracraft 1973a), the apparent dimorphism in bill lengths of this species is remarkable (Table 30; Fig. 45). *K*-means clustering of skulls of *D. hawkinsi* resulted in highly significant differences in bill lengths (skeletal analogue of culmen length) between provisional sex groups, regardless of whether partitions were based on complete skulls (four measurements; $F = 107.6$; d.f. $= 1, 51$; $P \ll 0.001$) or solely on bill lengths ($F = 126.7$; d.f. $= 1, 51$; $P \ll 0.001$). Both of these F-statistics were at least four times those for intersexual differences in other cranial dimensions compared for this species (Table 30). This disparity in F-values (statistics that are squared standardized differences between groups) concurs with differences in group means: bill lengths differed by a factor of 10% (~9 mm), whereas the average difference between three other skull measurements for the same sample and partitions was 5% (Table 30). Based on general trends in ramphothecas of rallids, this magnitude in skeletal bill lengths corresponds to at least an intersexual mean difference of 2 cm in life.

A comparable degree of bill dimorphism was indicated in the much smaller sample of elements for *Aphanapteryx bonasia* ($F = 480.2$; d.f. $= 1, 3$; $P < 0.001$; Table 30). Also, Günther and Newton (1879) alluded to exceptionally great variation in bill length among specimens of *Erythromachus leguati* and raised the possibility of sexual dimorphism, but samples available in the present survey were not adequate for confirmation of the probable dimorphism that this variation probably reflects (Table 30). Mean bill lengths for similarly sorted subsamples of *Cabalus modestus* suggested slightly increased sexual dimorphism in this taxon as well, especially if mean difference in bill length (4 mm) is considered relative to the generally small skeletal dimensions of this taxon (Tables 30–33), a pattern also indicated by the moderate bimodality in lengths of premaxillae of *C. modestus* presented by Trewick (1997b: fig. 5).

Principal component analyses.—As a means of direct examination of relationships among variables, separate Q-mode PCAs of skeletal data were conducted for partial skeletons of flighted and flightless rallids (Table 38). As with similar analyses of anatomical measurements (e.g., Johnston 1992a, b), this method ordinated variables by overall scale (mean size) on the first component and dis-

TABLE 36. Summary statistics (F-value, [d.f.]) for K-means clustering of selected postcranial measurements to provisional gender-groups of subfossil elements of flightless species of Rallidae (part I).

Variable	Porphyrio mantelli	Nesotrochis debooyi	Cyanolimnas cerverai	Rallus ibycus	Rallus recessus	Gallirallus dieffenbachii	Cabalus modestus	Capellirallus karamu
Humerus length	25.6 (1, 20)	—	10.6 (1, 12)	34.4 (1, 11)	30.7 (1, 26)	130.8 (1, 55)	56.9 (1, 26)	38.9 (1, 19)
Caput width	20.8 (1, 20)	—	7.9 (1, 11)	35.4 (1, 11)	15.2 (1, 6)	28.3 (1, 55)	18.1 (1, 26)	7.0 (1, 16)
Radius length	—	15.1 (1, 4)	—	10.6 (1, 6)	28.1 (1, 17)	58.4 (1, 28)	77.0 (1, 20)	—
Ulna length	34.0 (1, 5)	—	45.6 (1, 12)	40.6 (1, 11)	56.6 (1, 19)	51.7 (1, 27)	90.6 (1, 26)	20.4 (1, 7)
Carpometacarpus length	—	—	—	45.0 (1, 10)	37.5 (1, 25)	7.5 (1, 38)	91.3 (1, 26)	—
Coracoid length	19.6 (1, 3)	—	—	30.7 (1, 11)	58.1 (1, 18)	71.2 (1, 33)	50.9 (1, 26)	40.0 (1, 7)
Sternum carina length	—	—	—	—	27.1 (1, 14)	10.1 (1, 13)	34.0 (1, 26)	—
Carina oblique depth	—	—	—	—	13.8 (1, 19)	24.9 (1, 27)	4.1 (1, 26)	—
Synsacral length	—	—	—	17.9 (1, 24)	41.7 (1, 24)	80.6 (1, 55)	39.8 (1, 26)	32.0 (1, 16)
Interacetabular width	—	—	—	9.0 (1, 3)	20.1 (1, 7)	23.9 (1, 49)	11.8 (1, 26)	18.5 (1, 12)
Femur length	61.3 (1, 23)	28.5 (1, 10)	15.7 (1, 8)	38.6 (1, 11)	69.0 (1, 26)	119.2 (1, 56)	59.1 (1, 26)	65.1 (1, 35)
Caput width	32.8 (1, 23)	6.2 (1, 12)	1.4 (1, 8)	14.7 (1, 11)	12.0 (1, 11)	27.2 (1, 56)	13.5 (1, 26)	52.0 (1, 38)
Tibiotarsus length	20.5 (1, 7)	16.0 (1, 10)	—	18.9 (1, 11)	49.6 (1, 25)	94.0 (1, 39)	125.9 (1, 26)	32.1 (1, 23)
Tarsometatarsus length	25.9 (1, 20)	—	37.4 (1, 30)	108.9 (1, 14)	65.5 (1, 30)	73.9 (1, 44)	42.7 (1, 26)	70.0 (1, 34)

TABLE 37. Summary statistics (F-value, [d.f.]) for K-means clustering of selected postcranial measurements to provisional gender-groups of subfossil elements of flightless species of Rallidae (part II).

Variable	Aphanapteryx bonasia	Diaphorapteryx hawkinsi	Mundia elpenor	Porzana piercei	Tribonyx hodgenorum	Fulica prisca	Fulica chathamensis
Humerus length	36.8 (1, 9)	123.4 (1, 53)	43.7 (1, 13)	27.8 (1, 15)	51.2 (1, 23)	72.1 (1, 24)	83.9 (1, 53)
Caput width	9.5 (1, 10)	137 (1, 53)	5.7 (1, 12)	3.0 (1, 15)	17.7 (1, 23)	42.0 (1, 24)	38.8 (1, 51)
Radius length	18.5 (1, 5)	22.2 (1, 4)	10.3 (1, 4)	—	—	—	42.7 (1, 33)
Ulna length	19.6 (1, 4)	52.1 (1, 29)	10.7 (1, 12)	16.2 (1, 13)	32.4 (1, 16)	—	92.7 (1, 34)
Carpometacarpus length	—	114.9 (1, 37)	28.6 (1, 13)	12.7 (1,13)	44.9 (1, 12)	—	62.6 (1, 35)
Coracoid length	48.8 (1, 5)	52.5 (1, 43)	14.1 (1, 14)	45.9 (1, 12)	9.1 (1, 18)	13.0 (1, 15)	32.9 (1, 30)
Sternum carina length	—	49.9 (1, 18)	—	—	—	—	15.5 (1, 20)
Synsacral length	—	109.0 (1, 50)	—	—	—	19.9 (1, 9)	52.8 (1, 50)
Femur length	23.9 (1, 7)	156.8 (1, 50)	39.3 (1, 14)	23.6 (1, 13)	56.9 (1, 35)	40.5 (1, 23)	83.6 (1, 50)
Caput width	2.3 (1, 8)	60.7 (1, 53)	4.2 (1, 14)	1.7 (1, 13)	42.4 (1, 34)	25.4 (1, 23)	34.8 (1, 50)
Tibiotarsus length	59.8 (1, 29)	125.0 (1, 52)	29.7 (1, 14)	25.7 (1, 12)	49.0 (1, 25)	43.2 (1, 26)	92.2 (1, 54)
Tarsometatarsus length	99.2 (1, 27)	1716 (1, 50)	35.0 (1, 13)	29.0 (1, 15)	77.6 (1, 37)	46.8 (1, 26)	150.1 (1, 51)

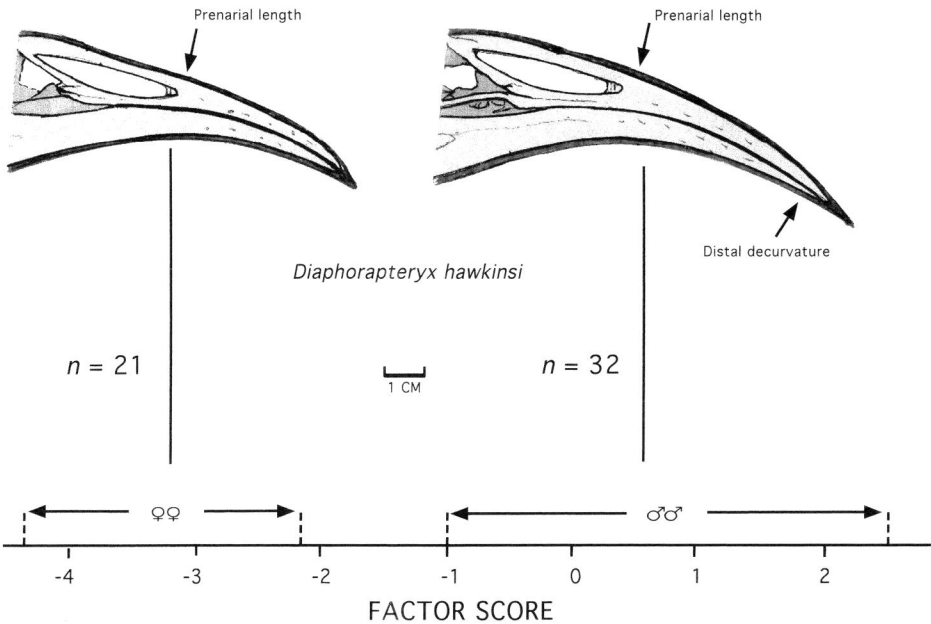

FIG. 45. Histogram of skeletal bill lengths of *Diaphorapteryx hawkinsi*, as partitioned into provisional sex groups by using *K*-means cluster analysis (males 95.7 ± 3.2, females 86.7 ± 2.9). Skulls of presumptive male and female specimens are depicted above (BMNH 3278 and 3293·1, respectively). Key points of difference are indicated.

TABLE 38. Mean correlation coefficients ($\bar{r}$) and summary statistics for first two principal components of separate *Q*-mode analyses of correlation matrices for 35 measurements from "reduced" skeletons of 39 flighted and 23 flightless species of Rallidae, by taxonomic group. Numbers of flighted (n_f) and flightless species (n_g), respectively, for each taxonomic group are given in parentheses.

	Flighted species (39)		Flightless species (23)	
Genera (n_f, n_g)	PC-I	PC-II	PC-I	PC-II
Porphyrio and *Porphyrula* (5, 2)	0.993	0.008	0.992	0.013
Habroptila (0, 1)	—	—	0.997	0.006
Aramides and *Canirallus* (4, 0)	0.997	−0.020	—	—
Dryolimnas (1, 1)	0.999	−0.042	0.993	−0.087
Rallus, Gallirallus, Cabalus, Tricholimnas, Habropteryx, and allies (8, 10)	0.996	−0.061	0.993	0.013
Atlantisia, Laterallus, Crex, Coturnicops, and *Porzana* (8, 5)	0.997	−0.029	0.991	0.035
Amaurornis, Gallicrex, Pareudiastes, Gallinula, and *Tribonyx* (7, 3)	0.997	0.023	0.993	−0.048
Fulica (6, 1)	0.993	0.108	0.983	−0.141
Eigenvalue (λ_i)	38.672	0.131	22.638	0.172
Percentage of total variance explained	99.2	0.3	98.4	0.7

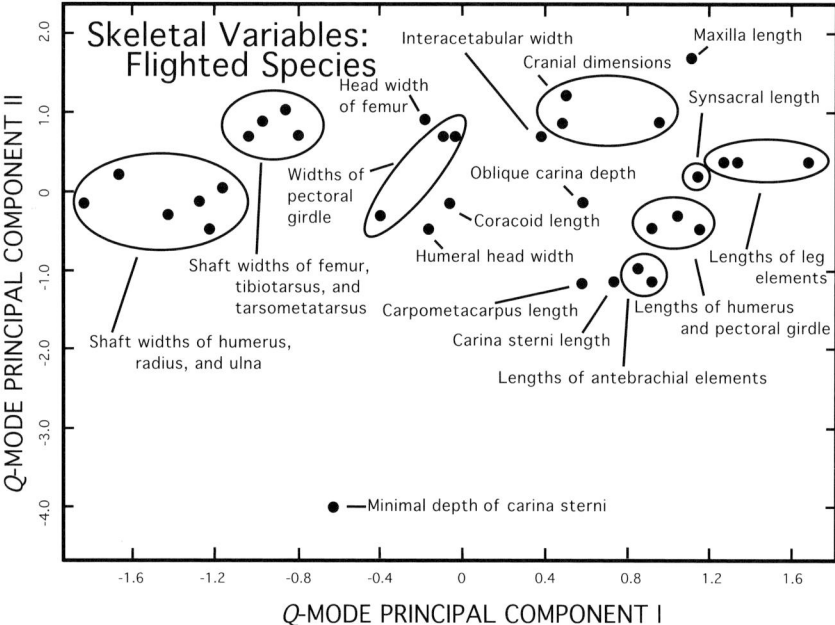

FIG. 46. Plot of 35 skeletal measurements on first two Q-mode principal components for flighted species of Rallidae.

played relative multivariate commonalities on the second (with outliers being relatively independent of other variables). In both flighted (39 species) and flightless (23 species) rallids, the first component separated three broad partitions of skeletal measurements on the basis of overall scale (Table 38; Figs. 46, 47): a series of comparatively minute measurements (shaft widths of appendicular elements), measurements of intermediate magnitude (capital widths of appendicular elements and lateromedial widths and depths of axial elements), and a collection of measurements having variably large means across species (lengths of appendicular and axial elements). In both groups, all taxa showed very high correlations with the first component (Table 38), a finding consistent with the general importance of mensural scale in comparisons of variables.

However, the second Q-mode components derived for the two groups were characterized by much lower correlations with taxonomic vectors, and the multivariate relationships among variables differed between the two groups. In flighted rallids, least width of the corpus sterni was identified as comparatively independent of other skeletal measurements (Fig. 46). Reconsideration of sternal shape in rallids confirmed that the relative width of this element varied considerably among genera of Rallidae, regardless of flight status (Livezey 1998; Table 31). However, in flightless species, minimal (perpendicular) depth of the carina sterni was uniquely disassociated from other skeletal dimensions (Fig. 47), which is predictive of a relatively prominent role for depths of the carina sterni in the separation of flighted and flightless taxa in standard multivariate analyses.

Primarily for purposes of providing a comparatively detailed overview of skeletal anatomy most intimately associated with foraging and diet, an R-mode PCA was performed based on an enriched suite of 21 measurements of skulls (most of

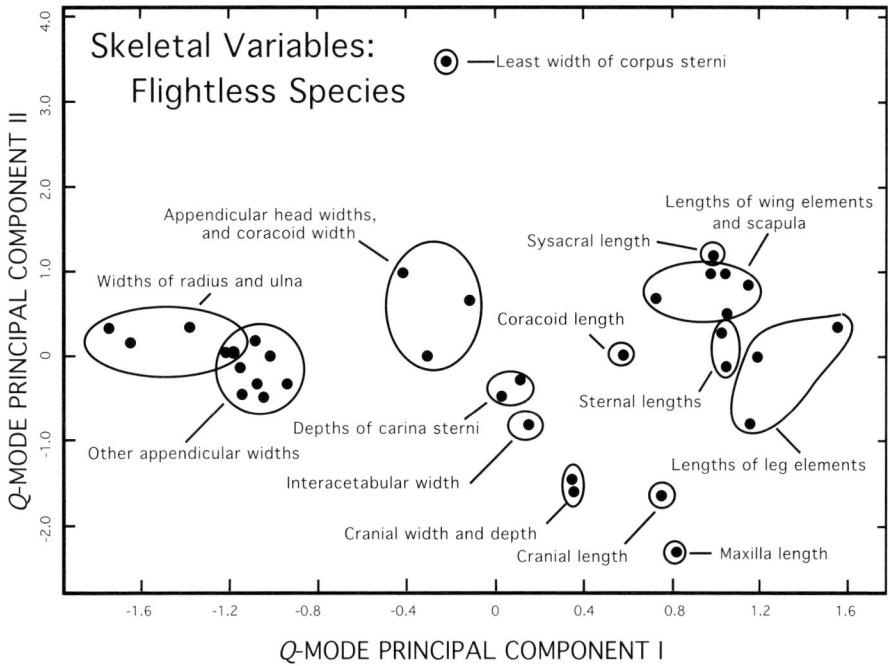

FIG. 47. Plot of 35 skeletal measurements on first two Q-mode principal components for flightless species of Rallidae.

which were not approximated in the PCAs of whole skeletons) and an enlarged series of 95 species representing most modern genera of Rallidae. As for most morphometric applications involving anatomical dimensions, PC-I for detailed skulls was indicative of overall size, here of the skull, and strongly correlated with mean body mass (Table 39). Accordingly (Fig. 48), the smallest species of rail (both least massive and having small skulls) had the lowest scores on PC-I (e.g., *Sarothrura* and crakes) and a majority of taxa were of intermediate size. The several taxa plotted with highest scores either were known to be both massive and to possess skulls of comparatively great size (e.g., *Porphyrio hochstetteri*), or represented extinct taxa having large skulls but for which body masses are not known (*Porphyrio mantelli* and *Diaphorapteryx hawkinsi*). Plots of putative male and female specimens of hyperdimorphic *Diaphorapteryx hawkinsi* revealed these means to be as far apart as those of different species in many other genera (Fig. 49).

The second PC was typical of strongly defined axes of shape (Table 39). PC-II comprised those variables that reflected length of the bill (e.g., lengths of the mandibula, maxilla, and parts thereof). These were contrasted with length of the cranium proper and most widths of the skull. Simply stated, PC-II reflected relative bill length, a dimension uncorrelated with body mass (Table 39). Taxa with relatively long, typically decurved bills (e.g., *Capellirallus karamu, Aphanapteryx bonasia,* and *Erythromachus leguati*) scored highly on this component, whereas taxa with short, deep bills (e.g., *Porphyrio* and members of the subtribe Fulicarina) were assigned low scores (Fig. 48). Basal genera and typical rallids were

TABLE 39. Correlation coefficients (r) and summary statistics for first three principal components of 21 detailed measurements of skulls for 95 species of Rallidae.

Variable	Correlation coefficient (r)		
	PC-I	PC-II	PC-III
Maxilla, total length	0.835	0.532	0.054
Height at zona flexoria craniofacialis	0.963	−0.037	0.129
Prenarial length	0.885	0.369	−0.146
Postnarial length	0.848	−0.283	0.111
Prenarial height	0.933	−0.306	−0.077
Cranium, length	0.943	0.137	0.196
Width	0.952	0.072	0.103
Height	0.955	0.066	0.167
Zona flexoria craniofacialis, width	0.955	−0.089	0.143
Apertura nasale, length	0.590	0.750	0.261
Maximal width	0.845	−0.126	0.278
Os nasale, processus maxillaris, rostrocaudal width	0.713	−0.493	0.163
Ramus maxillaris, minimal height	0.883	−0.356	−0.094
Pila supranasalis, width at midpoint	0.951	−0.060	−0.122
Width at rostral terminus	0.958	−0.072	−0.046
Width at caudal terminus	0.930	0.005	0.150
Mandibula, total length	0.851	0.508	0.090
Length of symphysis	0.818	0.437	−0.313
Maximal height at angulus	0.969	0.033	0.071
Height at margo caudalis of symphysis	0.945	−0.162	−0.124
Width at margo caudalis of symphysis	0.918	−0.278	−0.084
Correlation (r) with body mass ($n = 83$)	0.917	−0.066	0.373
Eigenvalue (λ_i)	3.058	0.489	0.095
Percentage of variance explained	77.2	12.3	2.4

clustered at appealingly intermediate scores on both PC-I and PC-II (Fig. 48). Although a number of exceptional positions in the bivariate plot were occupied by flightless rallids possessed of either large skull size or relatively long bills, or both (e.g., *C. karamu, E. leguati, A. bonasia, Diaphorapteryx hawkinsi,* and flightless *Porphyrio*), most flightless rallids (e.g., flightless *Gallirallus,* crakes, and moorhens) were unremarkable in the proportions of their skulls and fell among flighted congeners (Fig. 48).

An R-mode PCA of means of 41 cranial and postcranial measurements for 55 species was maximally effective in dimensional reduction, encapsulating approximately 95% of the multivariate variation in the first two components alone (Table 40). As in most analyses, PC-I for complete skeletons conveyed general skeletal size (Table 40), and scores on the axis were directly related to mean body masses for species ($r = 0.97$; $P \ll 0.01$). Only two variables were distinctly lower in correlation with PC-I, length of the rostrum maxillae (bill length) and minimal (perpendicular) depth of the carina sterni, both of which had been singled out by Q-mode PCA as incorporating comparatively large components of independent variances. PC-II for complete skeletons contributed another 5% of the total dispersion among taxa, and correlations indicated this axis to be a contrast between dimensions of the pectoral apparatus (excluding only sternal width and several variable shaft widths) and all other skeletal dimensions (Table 40). Scores on PC-II were weakly negatively correlated with mean body masses ($r = -0.31$; $P < 0.05$). Although PC-II and PC-III contributed small but informative interspecific

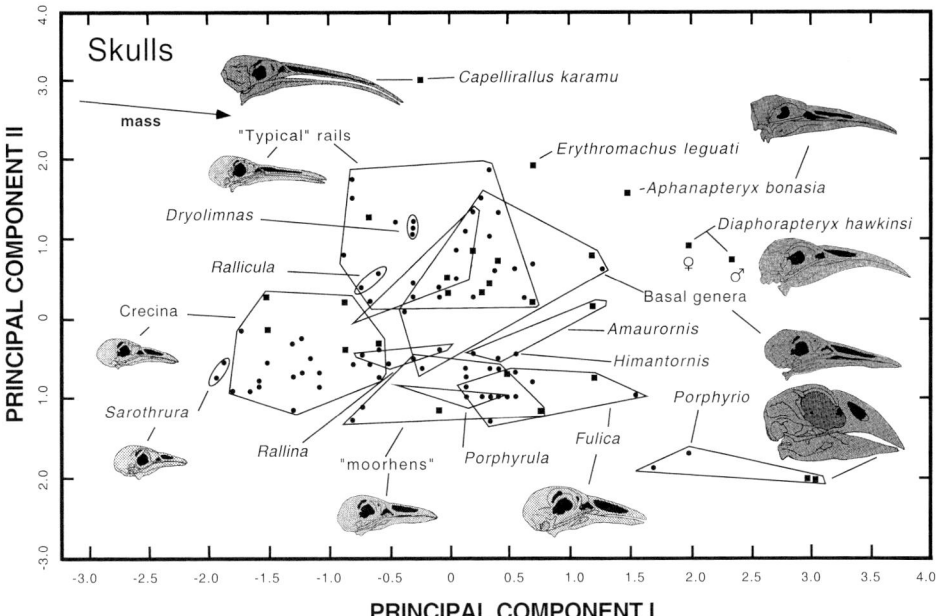

FIG. 48. Plot of mean scores for 95 species of Rallidae on first two standard (*R*-mode) principal components of 21 measurements of the skull. Approximate orientation of vector for mean body mass of taxa is indicated. Flighted species are symbolized by circles, flightless by squares, and threshold-flightless *Fulica* by triangles. Inset line drawings (not to scale): *Porphyrio hochstetteri* (USNM 612797); *Aramides cajanea* (CM 10049); *Sarothrura rufa* (CM 10452); *Rallus limicola* (CM 11027); *Capellirallus karamu* (Canterbury Museum 20615); *Diaphorapteryx hawkinsi* (BMNH A·3078); *Aphanapteryx bonasia* (Mauritius Institute, mounted skeleton on exhibit lacking specimen label); *Porzana carolina* (CM 13777); *Gallinula chloropus* (CM 13041); and *Fulica americana* (CM 13870).

dispersion among skeletons, neither was germane to morphological changes associated with flightlessness (Table 40).

A bivariate plot of PC-I and PC-II for complete skeletons (Fig. 49), respectively, provided a clear and intuitive partitioning of general skeletal size and relative pectoral size. Moreover, scores on PC-II alone permitted a linear cutoff point between flighted and flightless rallids (Fig. 49). This value for PC-II that serves as a provisional threshold of flightlessness (τ), approximately -0.20, is especially interesting in that flight status was not included in the definition of axes (as it was in canonical contrasts targeting flight capacity), and therefore the clarity of separation of flightless rallids is an emergent property of the interspecific variation in skeletal data as opposed to specified a priori. Although the distribution of scores on PC-I was not surprising, essentially reflecting the skeletal size, the positions of taxa on PC-II, the inferred changes associated with flightlessness, and the relationship between the latter and body size provided a succinct summary heretofore not attained (Fig. 49). Flightless taxa spanned a continuum of apomorphy, as reflected by scores on PC-II, in which taxa showing comparatively slight apomorphy (e.g., *Dryolimnas aldabranus, Porzana atra,* and *Fulica chathamensis*) fell immediately below the emergent threshold of -0.20, extremely derived species fell at or near scores of -2.0 or more (*Porphyrio hochstetteri, Habroptila wallacii, Gallirallus australis, Cabalus modestus, Aphanapteryx bon-*

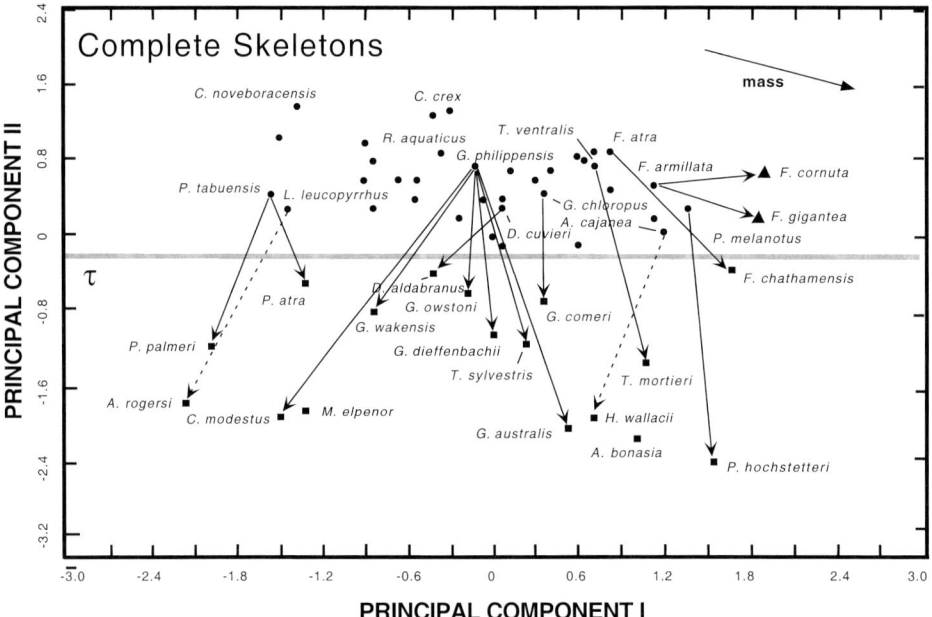

FIG. 49. Plot of mean scores for 55 species of Rallidae on first two standard (*R*-mode) principal components of 41 measurements of complete skeletons. Flighted species are symbolized by circles, flightless by squares, and threshold-flightless *Fulica* by triangles. Inferred vectors of change connect flightless species and their flighted relatives. Approximate orientation of vector for mean body masses of taxa ("mass") and apparent threshold of flightlessness (τ, positioned on the flightless side of boundary) are indicated.

asia, Atlantisia rogersi, and *"Atlantisia" elpenor*), and comparatively mainstream flightless taxa (*Gallirallus dieffenbachii, G. owstoni, G. wakensis, Tricholimnas sylvestris, Porzana palmeri, Gallinula comeri,* and *Tribonyx mortierii*) fell in between these extremes (Fig. 49).

Although only a comparatively small number of taxa qualified for this analysis, plotting of inferred "vectors of change" between flightless taxa and their closest flighted relatives was feasible (Fig. 49). The magnitudes of these inferred changes varied to some degree from the rankings of relative apomorphy indicated by simple scores on PC-II in that the starting points represented by flighted taxa also varied to a limited degree (Fig. 49). Comparatively minor shifts were indicated for *Porzana atra* and *Dryolimnas aldabranus,* whereas the shift suggested for *Fulica chathamensis* was moderately large despite its marginal position with respect to the threshold of flightlessness (τ; Fig. 49). Correspondences between relative apomorphy (i.e., score on PC-II) and inferred magnitudes of total change became increasingly strong with increases in one or both of PC-I and PC-II, a predictable outcome given that points corresponding to extreme apomorphy only could be attained by substantial shifts beyond the inferred threshold (τ) of flightlessness (Fig. 49).

An *R*-mode PCA of reduced suites of skeletal measurements deleted only six measurements but permitted the inclusion of seven additional, critical taxa. These new taxa included several additional flightless species (*Porphyrio mantelli, Rallus*

TABLE 40. Correlation coefficients (r) and summary statistics for first four principal components of 41 skeletal measurements for 55 species of Rallidae.

Variable	Correlation coefficient (r)			
	PC-I	PC-II	PC-III	PC-IV
Rostrum maxillae, length	0.716	−0.367	0.473	−0.113
Cranium, length	0.925	−0.307	0.181	−0.047
Maximal width	0.910	−0.348	0.091	−0.003
Height perpendicular to palate	0.943	−0.266	0.098	−0.022
Sternum, length of carina	0.916	0.340	0.131	−0.074
Length of corpus	0.960	0.205	0.136	−0.066
Minimal width of corpus	0.853	−0.132	−0.468	−0.157
Oblique depth of carina	0.909	0.334	0.011	−0.031
Minimal depth of carina	0.706	0.687	0.091	−0.032
Furcula, height	0.982	0.010	0.033	−0.032
Minimal width at midpoint	0.935	0.002	−0.101	0.081
Maximal width at midpoint	0.910	0.074	−0.078	−0.205
Scapula scapus length	0.987	0.099	−0.045	−0.029
Coracoid, length	0.995	0.041	0.044	−0.028
Crista articularis width	0.983	−0.057	−0.100	−0.081
Humerus, length	0.990	0.093	−0.007	−0.022
Intertubercular width of caput	0.996	0.041	−0.005	−0.027
Minimal width at midpoint	0.996	0.040	0.024	0.003
Maximal midpoint width	0.993	0.076	0.017	−0.007
Radius, length	0.970	0.191	0.009	0.010
Minimal width at midpoint	0.901	0.014	−0.058	0.347
Maximal midpoint width	0.996	−0.015	−0.001	−0.011
Ulna, length	0.977	0.179	0.020	0.007
Minimal width at midpoint	0.995	0.007	−0.016	0.017
Maximal midpoint width	0.994	0.029	0.037	−0.006
Carpometacarpus length	0.974	0.207	−0.012	0.022
Digitalis majoris				
Phalanx proximalis length	0.968	0.197	−0.035	0.074
Phalanx distalis length	0.954	0.224	−0.042	0.084
Synsacral length	0.978	−0.111	−0.063	−0.095
Pelvis, interacetabular width	0.908	−0.335	−0.010	−0.013
Femur, length	0.963	−0.190	0.142	0.002
Caput width	0.961	−0.251	0.051	−0.054
Minimal width at midpoint	0.958	−0.255	0.069	−0.038
Maximal width at midpoint	0.962	−0.251	0.047	−0.028
Tibiotarsus, length	0.978	−0.131	−0.008	0.076
Minimal width at midpoint	0.960	−0.259	0.057	0.011
Maximal width at midpoint	0.968	−0.230	0.025	−0.017
Tarsometatarsus, length	0.958	−0.124	0.060	0.121
Craniocaudal width at midpoint	0.971	−0.216	−0.012	0.034
Lateromedial width at midpoint	0.935	−0.291	−0.007	0.126
Digit III length (excluding unguis)	0.937	0.023	−0.152	0.136
Correlation (r) with body mass ($n = 47$)	0.968	−0.311	−0.121	−0.063
Eigenvalue (λ_i)	6.137	0.352	0.121	0.063
Percentage of variance explained	89.7	5.1	1.8	0.9

recessus, Capellirallus karamu, Diaphorapteryx hawkinsi, Porzana severnsi, and *Tribonyx hodgenorum*). As for PC-I for complete skeletons, here PC-I depicted general skeletal size, and summarized 86% of interspecific dispersion (Table 41). Scores on this axis were strongly correlated with mean body mass ($r = 0.77$; $P \ll 0.01$), and were virtually identical to relative positions on PC-I for complete skeletons (Fig. 50).

TABLE 41. Correlation coefficients (r) and summary statistics for first four principal components of 35 measurements from "reduced" skeletons for 62 species of Rallidae.

Variable	Correlation coefficient (r)			
	PC-I	PC-II	PC-III	PC-IV
Rostrum maxillae, length	0.671	−0.409	0.535	−0.189
Cranium, length	0.922	−0.327	0.158	−0.016
Maximal width	0.889	−0.394	0.068	0.026
Height perpendicular to palate	0.938	−0.288	0.079	−0.010
Humerus, length	0.984	0.131	−0.123	−0.048
Intertubercular width of caput	0.993	0.087	−0.009	−0.027
Minimal width at midpoint	0.994	0.069	0.011	0.004
Maximal midpoint width	0.992	0.085	0.012	−0.008
Radius, length	0.950	0.266	−0.029	−0.027
Minimal width at midpoint	0.910	0.045	−0.030	0.350
Maximal midpoint width	0.994	−0.009	−0.016	−0.011
Ulna, length	0.955	0.259	−0.018	−0.022
Minimal width at midpoint	0.992	0.063	−0.035	0.023
Maximal midpoint width	0.989	0.104	0.008	0.022
Carpometacarpus length	0.945	0.302	−0.050	0.003
Scapula scapus length	0.984	0.117	−0.055	−0.029
Coracoid, length	0.993	0.077	0.038	−0.018
Crista articularis width	0.985	−0.027	−0.103	−0.038
Sternum, length of carina	0.902	0.375	0.089	−0.064
Length of corpus	0.956	0.227	0.102	−0.066
Minimal width of corpus	0.830	−0.206	−0.471	−0.159
Oblique depth of carina	0.889	0.344	0.006	−0.083
Minimal depth of carina	0.597	0.787	0.064	0.020
Synsacral length	0.970	−0.150	−0.075	0.101
Pelvis, interacetabular width	0.902	−0.346	−0.017	−0.066
Femur, length	0.961	−0.219	0.101	0.031
Caput width	0.953	−0.286	0.036	−0.015
Minimal width at midpoint	0.951	−0.285	0.043	0.003
Maximal width at midpoint	0.955	−0.279	0.024	0.005
Tibiotarsus, length	0.976	−0.142	−0.022	0.040
Minimal width at midpoint	0.950	−0.296	0.042	0.033
Maximal width at midpoint	0.962	−0.258	0.011	−0.003
Tarsometatarsus, length	0.955	−0.110	0.028	0.068
Craniocaudal width at midpoint	0.957	−0.266	−0.003	0.036
Lateromedial width at midpoint	0.913	−0.356	0.001	0.125
Correlation (r) with body mass ($n = 48$)	0.774	−0.410	−0.291	0.182
Eigenvalue (λ_i)	5.361	0.560	0.120	0.053
Percentage of variance explained	86.0	9.0	1.9	0.8

The second PC for reduced skeletons summarized an additional 9% of interspecific variation. Scores on this axis were weakly (negatively) correlated with body mass ($r = -0.41$; $P < 0.05$). Coefficients for variables on PC-II were extremely similar, respectively, to those for PC-II for complete skeletons, and indicated that the axis represented a contrast between the pectoral skeleton and other dimensions (Table 41). PC-III and PC-IV for reduced skeletons, based on coefficients of variables (Table 41) and plots of scores, were not informative concerning flightlessess. Of the taxa newly added for multivariate consideration, most striking in position were the extreme apomorphy of *Capellirallus karamu*, *Porzana severnsi*, and (to a lesser extent) *Diaphorapteryx hawkinsi* (Fig. 50). The

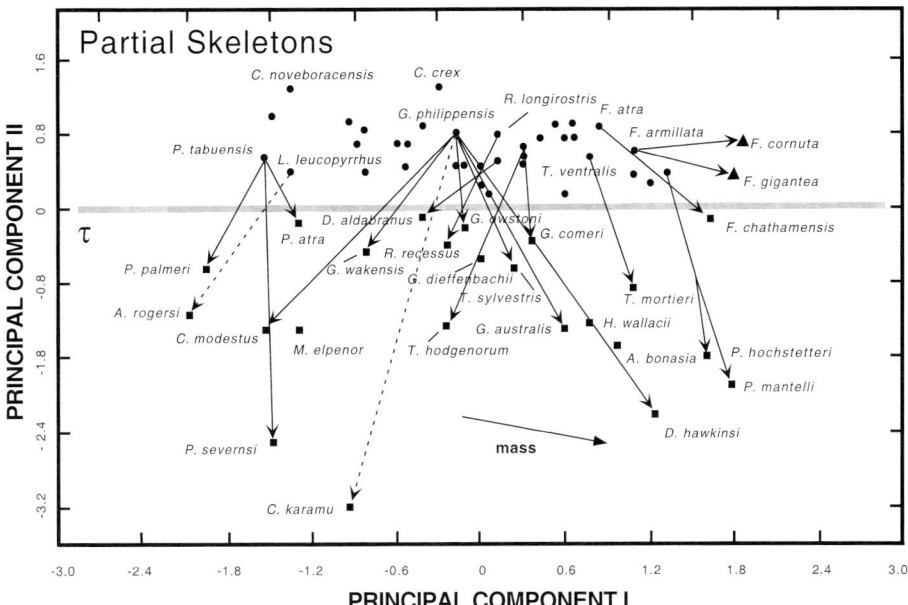

FIG. 50. Plot of mean scores for 62 species of Rallidae on first two standard (*R*-mode) principal components of 35 measurements of partial ("reduced") skeletons. Inferred vectors of change connect flightless species and their flighted relatives. Approximate orientation of vector for mean body masses of taxa ("mass") and apparent threshold of flightlessness (τ, positioned on the flightless side of boundary) are indicated.

inferred shifts for *C. karamu* and *D. hawkinsi* were augmented by a significant apparent decrease and increase, respectively, in overall size (Fig. 50). The moderately greater shifts on both PC-I (general skeletal size) and PC-II (relative size of pectoral skeleton) inferred for the extinct *Porphyrio mantelli* in comparison with extant *P. hochstetteri* (Fig. 50) are interpretable as largely a shift attributable to the common ancestor of these sister taxa before the vicariant separation of the North and South islands of New Zealand (Livezey 1998). The large *Rallus recessus* from Bermuda manifested a small reduction in overall size (PC-I) in conjunction with the modest shift toward reduction of the pectoral apparatus (PC-II), and although limited in magnitude, the latter was sufficient to place this taxon confidently among flightless rallids (Fig. 50).

An *R*-mode PCA of sterna alone permitted a narrowly focused examination of interspecific trends associated with flightlessness in this justifiably focal element, and also admitted a considerable number of additional species to the comparisons. Accordingly, a PCA of mean sternal measurements for 109 species of Rallidae identified two PCs that summarized essentially all variation in this single element (Table 42). PC-Is for sterna, which accounted for more than 99% of the total variance of the data set, reflected size sensu lato and were strongly positively correlated ($P < 0.01$) with mean body masses (Table 42). Magnitudes of the correlations indicated that one variable (least sternal width) was substantially less closely correlated with this axis than were other measurements analyzed (Table 42). Residual variance in least sternal width, melded with a contrast between oblique depth of the carina sterni and essentially all other variables (with perpen-

TABLE 42. Correlation coefficients (r) and summary statistics for first two principal components of five sternal measurements (mm) for 109 species of Rallidae.

Sternal variable	Correlation coefficient (r)	
	PC-I	PC-II
Carina length	0.844	0.426
Corpus length	0.772	0.547
Corpus least width	0.496	0.828
Carina oblique depth	0.853	0.414
Carina perpendicular depth	1.000	−0.000
Correlation (r) with body mass ($n = 83$)	0.673	0.804
Eigenvalue (λ_i)	1,707.03	0.29
Percentage of variance explained	>99.9+	<0.1

dicular width of the carina sterni being rendered redundant), characterized PC-II for sterni (Table 42). Alternatively interpreted, PC-II represented relative carina depth, reflecting information on position and perpendicular depth of the carina.

A bivariate plot of scores for taxa on PC-I and PC-II for sterna was highly informative (Fig. 51). First, flightless and flighted taxa were demarcated by a gently curving line, one oriented roughly diagonally and having a positive first derivative (Fig. 51). Second, the large species of coots (*Fulica*) occupied unique positions to the middle and upper right of the plot. Two subgroups of this genus, comprising the three subfossil taxa (*F. chathamensis, F. prisca,* and *F. newtoni*)

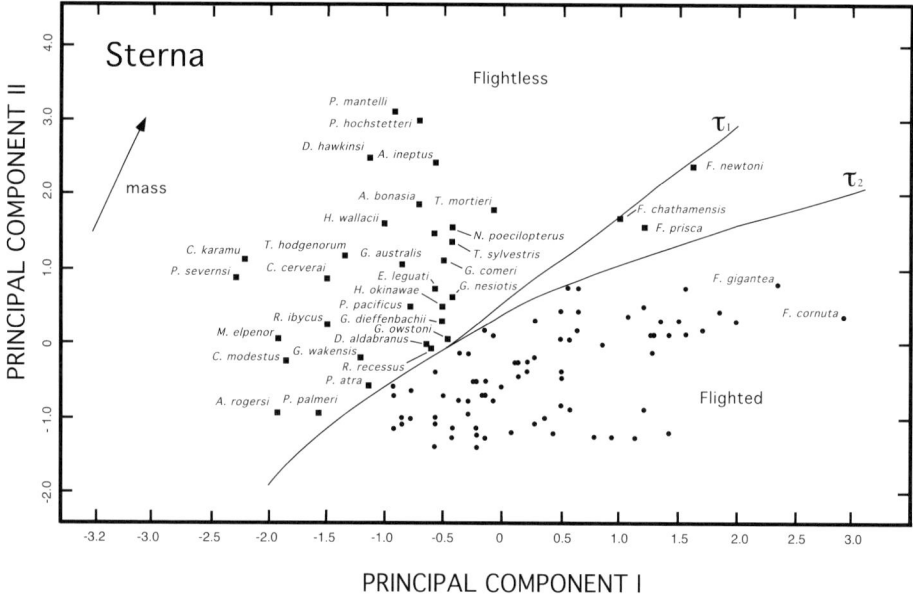

FIG. 51. Plot of mean scores for 109 species of Rallidae on first two standard (R-mode) principal components of five measurements of sterna. Approximate orientation of vector for mean body masses of taxa ("mass"), and apparent thresholds of flightlessness (τ_1 and τ_1, including and excluding subfossil *Fulica* among flightless taxa, respectively) are indicated, with the symbol τ positioned on the flightless side of boundary.

and the two extant forms (*F. gigantea* and *F. cornuta*), suggest the further subdivision of the plane described by the two sternal components, in which the subfossil *Fulica* occupy a zone between the two large Andean coots and all manifestly flightless rails (Fig. 51). Third, within the flightless species, taxa range from forms having sterna only slightly deviating from those of flighted confamilials to those characterized by comparatively extreme apomorphy (i.e., taxa to the upper left of the plot). The latter include several comparatively small species (e.g., *Porzana severnsi* and *Capellirallus karamu*) and a few moderately large taxa (*Diaphorapteryx hawkinsi* and flightless *Porphyrio*).

A suite of R-mode PCAs of the pooled within-species covariance matrices for flighted, flightless, and all species of Rallidae was performed primarily as a means of ascertaining the strong congruence between the first component (within-species size, used in standardizing distances in CAs, below) and those derived in interspecific contexts (among-species size, above); and to gain a multivariate insight into the structure of bivariate correlations suggested by a cluster analysis of variables with two of the same matrices (Figs. 40, 41). Although PC-I in the within-species context generally is specified by a size axis having scores strongly positively correlated with all of the original variables, more detailed examination of magnitudes of correlations generally reveals that these PC-Is are far from isometric (Table 43). This outcome indicates that correction for within-species size will not remove all size-related information in many comparisons (e.g., CAs of species–sex groups). Nonetheless, correlations between the PC-Is of the standard (R-mode) analysis and those of pooled within-species covariance matrices for the 41 skeletal measurements were similar in eigenstructure, whether the latter pertained to flighted ($r_s = 0.42$; $P = 0.01$), flightless ($r_s = 0.39$; $P < 0.05$), or all species ($r_s = 0.42$; $P = 0.01$).

The first PCs for the within-species covariance matrices revealed that correlations among skeletal variables exclusive of sternal dimensions are slightly stronger with respect to size in flightless species than in flighted species (Table 43). Although fewer taxa were incorporated into the matrix for the former, this general pattern suggests that a greater proportion of skeletal variation in flightless rails (including dimensions of wing elements) is attibutable to simple (intraspecific) size, with the notable exception of the comparatively independent sternal dimensions (Table 43). However, in both groups, PC-I conformed with eigenvectors consistent with general size axes, with only a minority of measurements having low enough correlation coefficients with this component to indicate unusually high measurement error (e.g, widths of the furcula). The components for the matrix in which all species were pooled converged closely to those for flighted species, with a slight increase in some correlations deriving largely from the comparatively strongly defined PC-I for flightless taxa (Table 43). Subsequent components varied little among the partitions, although flightless rails were characterized by a PC-II in which a more strongly negative contrast in most pelvic dimensions was manifested (Table 43).

Appendicular allomorphosis.—Allometric slopes ($\hat{b}$) relating lengths of pectoral elements with overall body size were higher in flightless species than in their flighted relatives in several previous ornithological applications (Livezey and Humphrey 1986; Livezey 1989d) and studies of mammals (Christiansen 1999). This pattern was interpreted to reflect relative ontogenetic rates at terminus of

TABLE 43. Correlation coefficients (r) and summary statistics for first two principal components of pooled within-species covariance matrices for 41 skeletal measurements for 43 flighted, 8 flightless, and all 51 species of Rallidae.

Variable	Flighted species (n = 817)			Flightless species (n = 179)			All species (n = 996)		
	PC-I	PC-II	PC-III	PC-I	PC-II	PC-III	PC-I	PC-II	PC-III
Rostrum maxillae, length	0.677	−0.075	−0.171	0.702	0.430	0.042	0.686	−0.002	0.055
Cranium, length	0.739	0.031	−0.129	0.776	0.034	0.182	0.747	0.018	0.133
Maximal width	0.639	−0.011	−0.077	0.666	0.174	0.186	0.641	0.004	0.103
Height perpendicular to palate	0.531	0.038	−0.150	0.657	0.172	0.142	0.564	0.039	0.120
Sternum, length of carina	0.761	0.525	−0.216	0.629	0.515	0.020	0.727	0.565	0.097
Length of corpus	0.773	0.503	−0.204	0.707	0.503	0.038	0.754	0.533	0.096
Minimal width of corpus	0.423	0.171	−0.066	0.412	0.161	−0.002	0.413	0.142	0.050
Oblique depth of carina	0.598	0.300	0.106	0.552	0.327	−0.032	0.586	0.299	−0.081
Minimal depth of carina	0.545	0.272	0.053	0.330	0.388	−0.063	0.500	0.310	−0.054
Furcula, height	0.709	0.212	0.122	0.635	0.241	0.044	0.692	0.224	−0.046
Minimal width at midpoint	0.340	0.171	0.063	0.421	0.187	−0.067	0.360	0.170	−0.035
Maximal width at midpoint	0.334	0.063	0.000	0.263	0.178	−0.048	0.320	0.085	0.008
Scapula scapus length	0.808	0.293	0.099	0.895	0.253	0.008	0.829	0.272	−0.057
Coracoid, length	0.871	0.123	0.091	0.923	0.101	0.028	0.883	0.110	−0.033
Crista articularis width	0.576	0.237	−0.002	0.626	0.120	0.118	0.587	0.196	0.049
Humerus, length	0.906	0.138	0.291	0.940	0.119	−0.058	0.913	0.132	−0.178
Intertubercular width of caput	0.819	0.221	0.053	0.863	0.126	0.043	0.829	0.194	−0.019
Minimal width at midpoint	0.685	0.152	0.038	0.715	0.174	0.036	0.692	0.144	−0.022
Maximal midpoint width	0.653	0.193	−0.035	0.707	0.116	0.109	0.665	0.159	0.053
Radius, length	0.879	0.134	0.357	0.843	0.187	−0.140	0.869	0.153	−0.258
Minimal width at midpoint	0.549	0.105	−0.035	0.628	0.230	0.086	0.570	0.114	0.035
Maximal midpoint width	0.513	0.028	−0.015	0.555	0.093	−0.047	0.523	0.040	−0.006
Ulna, length	0.901	0.135	0.328	0.913	0.162	−0.088	0.900	0.149	−0.218
Minimal width at midpoint	0.633	0.067	−0.032	0.666	0.129	0.094	0.641	0.077	0.039
Maximal midpoint width	0.645	0.103	−0.003	0.678	0.096	0.005	0.651	0.084	−0.007
Carpometacarpus length	0.871	0.154	0.299	0.876	0.118	−0.115	0.870	0.155	−0.211
Digitalis majoris									
Phalanx proximalis length	0.758	0.083	0.188	0.743	0.094	−0.103	0.751	0.103	−0.145
Phalanx distalis length	0.665	0.069	0.194	0.697	0.119	−0.079	0.670	0.093	−0.144
Synsacral length	0.699	−0.013	−0.405	0.715	−0.155	0.676	0.702	−0.039	0.640
Pelvis, interacetabular width	0.596	0.052	−0.124	0.781	−0.040	0.051	0.644	−0.002	0.083
Femur, length	0.807	−0.239	0.030	0.956	−0.082	−0.069	0.846	−0.216	−0.063
Caput width	0.767	−0.005	−0.158	0.779	−0.041	0.091	0.760	−0.029	0.122
Minimal width at midpoint	0.621	0.032	−0.144	0.703	−0.088	0.082	0.634	−0.026	0.110

TABLE 43. Continued.

Variable	Flighted species (n = 817)			Flightless species (n = 179)			All species (n = 996)		
	PC-I	PC-II	PC-III	PC-I	PC-II	PC-III	PC-I	PC-II	PC-III
Maximal width at midpoint	0.602	0.025	−0.090	0.707	−0.102	0.056	0.624	−0.041	0.080
Tibiotarsus, length	0.937	−0.230	−0.013	0.959	−0.213	−0.079	0.942	−0.231	−0.007
Minimal width at midpoint	0.588	0.030	−0.144	0.743	0.020	0.146	0.626	0.001	0.114
Maximal width at midpoint	0.640	0.046	−0.145	0.751	−0.011	0.093	0.662	−0.003	0.118
Tarsometatarsus, length	0.897	−0.283	−0.062	0.906	−0.244	−0.149	0.900	−0.266	−0.025
Craniocaudal width at midpoint	0.607	0.076	−0.122	0.756	−0.079	0.029	0.644	0.015	0.086
Lateromedial width at midpoint	0.555	0.152	−0.110	0.735	0.057	0.048	0.601	0.105	0.074
Digit III length (excluding unguis)	0.837	−0.305	−0.088	0.898	0.047	−0.009	0.844	−0.222	0.006
Eigenvalue (λ_i)	7,918.8	865.3	455.4	11,574.8	905.5	688.2	8,507.2	867.1	476.4
Percentage of total variance	68.5	7.5	3.9	73.5	5.8	4.4	69.1	7.0	3.9

TABLE 44. Intraspecific allometric coefficients ($\hat{b} \pm$ SE) or allomorphosis of three skeletal pectoral elements with corresponding first principal component (PC-I) of nonpectoral dimensions, by species and sex, for three flighted species and flightless relatives in three taxonomic tribes of Rallidae.

Species	Sex (n)	Pectoral element (proximal → distal)		
		Humerus	Ulna	Carpometacarpus
Porphyrio melanotus	Males (33)	0.05 ± 0.01	0.05 ± 0.01	0.05 ± 0.01
	Females (20)	0.02 ± 0.01	0.01 ± 0.01	0.02 ± 0.01
Porphyrio hochstetteri	Males (12)	0.08 ± 0.02	0.10 ± 0.02	0.11 ± 0.04
	Females (18)	0.10 ± 0.01	0.11 ± 0.01	0.11 ± 0.02
$F(6, 75)$, P-value		31.2 (<0.0001)	61.6 (<0.00001)	74.5 (<0.00001)
Gallirallus philippensis	Males (16)	0.03 ± 0.01	0.03 ± 0.02	0.03 ± 0.02
	Females (19)	0.03 ± 0.01	0.05 ± 0.01	0.05 ± 0.02
Gallirallus owstoni	Males (14)	0.12 ± 0.02	0.12 ± 0.02	0.13 ± 0.02
	Females (18)	0.08 ± 0.02	0.09 ± 0.02	0.13 ± 0.04
Gallirallus australis-group	Males (28)	0.15 ± 0.03	0.16 ± 0.03	0.24 ± 0.04
	Females (22)	0.12 ± 0.02	0.12 ± 0.02	0.17 ± 0.03
$F(10, 105)$, P-value		51.3 (<0.00001)	38.8 (<0.00001)	37.2 (<0.00001)
Porzana tabuensis	Both (11)	0.04 ± 0.01	0.04 ± 0.01	0.06 ± 0.02
Porzana palmeri	Both (12)	0.03 ± 0.01	0.05 ± 0.01	0.08 ± 0.02
$F(2, 19)$, P-value		44.0 (<0.00001)	56.9 (<0.00001)	15.9 (<0.0001)
Tribonyx ventralis	Males (24)	0.04 ± 0.02	0.05 ± 0.02	0.06 ± 0.02
	Females (19)	0.03 ± 0.01	0.04 ± 0.01	0.03 ± 0.01
Tribonyx mortierii	Males (10)	0.10 ± 0.04	0.12 ± 0.05	0.11 ± 0.06
	Females (11)	0.11 ± 0.08	0.11 ± 0.10	0.19 ± 0.11
$F(6, 56)$, P-value		5.2 (<0.005)	23.6 (<0.0001)	31.7 (<0.0001)

growth (i.e., sampled by skeleton at fledging); flightless birds were interrupted at an earlier point in ontogeny of the pectoral apparatus, an inference that goes beyond mere documentation of the relatively reduced size of these components. In each set of assessments, limb elements were analyzed relative to positions of specimens on PC-I of skeletal dimensions exclusive of the appendage under scrutiny. Differences in allometric relationships were tested by using standard F-statistics for two-parameter linear regressions and comparisons of estimates of allometric slopes in light of respective standard errors of these estimates. The standard transformation of all metrics to natural logarithms for allometric analyses precluded a confounding of simple interspecific differences in overall size with the trends of interest. Samples of complete skeletons for each included species–sex group were only adequate for three groups comprising both flighted and flightless species: *Porphyrio melanotus* and *P. hochstetteri*; *Gallirallus philippensis*, *G. owstoni*, and *G. wakensis*; and *Tribonyx ventralis* and *T. mortierii* (Table 44). A fourth comparison also was performed for an additional pair of taxa for which samples precluded subdivision of the sexes (*Porzana tabuensis* and *P. palmeri*), but these small crakes manifested comparatively small intersexual differences in skeletal measurements (Tables 30–33).

Allometric patterns of pectoral elements within species of *Porphyrio* (Table 44), regardless of sex, consistently revealed that the major skeletal segments of the wing were undergoing more rapid growth (i.e., had higher slopes) in flightless *P. hochstetteri* than they were in smaller, flighted *P. melanotus* (F-statistics comparing regressions; $P < 0.0001$). Direct pairwise comparisons of allometric slopes

($\hat{b}$) revealed, within elements and sexes, that values for flightless *P. hochstetteri* were roughly 1.5–2.0 times those for *P. melanotus,* but sexes within species were described by similar allometric slopes (Table 44). Examination of the limited data suggested a trend for increasingly large differences in allometric slopes between respective species–sex groups as one passes from proximal to more distal elements (i.e., humeri to carpometacarpi). This pattern results solely from increases in slopes in flightless *P. hochstetteri* (Table 44), and is consonant with the general avian signatures in which appendicular growth is initiated earliest proximally and last distally (e.g., Levinton 1988).

Identical comparisons made for typical rallids (*Gallirallus*) documented qualitatively identical, but quantitatively more pronounced patterns (F-tests; $P < 0.00001$). Within elements, slopes for intraspecific scaling were two to five times as great in the two flightless taxa (*G. owstoni* and *G. wakensis*) as in flighted *G. philippensis.* Intersexual differences within species were not significant. The magnitude of increase in allometric slope in flightless species increased as comparisons of elements proceeded from proximal to distal elements (Table 44).

Flighted and flightless *Porzana,* samples of which were too meager for sex-specific analyses, revealed little concerning relative appendicular allomorphosis (Table 44). Interspecific differences (sexes pooled) were significant for the entire model (i.e., the paired estimates of intercepts and slopes; $P < 0.0001$). Comparisons and slopes and errors thereof revealed that the interspecific differences pertained to the intercepts, and species differed in mean lengths of pectoral elements (Tables 30–34) but not in the slopes at terminus of growth. One pattern suggested by the comparisons within *Porzana* shared by those for other genera was a modest increase in slopes as one moves distally through the pectoral limb, a shift more pronounced in flightless *P. palmeri* (Table 44).

Pectoral allomorphosis for *Tribonyx* confirmed most of the the patterns revealed in those for *Porphyrio* and *Gallirallus* (Table 44). As in those other genera, allometric slopes ($\hat{b}$) for flightless *T. mortieri* exceeded those for respective groups of flighted *T. ventralis* by more than two times (Table 44), and accounted for a substantial proportion of the overall differences between allometric models for species (F-statistics; $P < 0.005$). Also in common with the other two sex-specific comparisons, intersexual differences were negligible, but unlike comparisons for *Porphyrio* and *Gallirallus,* allometric slopes remained virtually constant across elements and did not manifest any indication of proximal-to-distal gradients in ontogenetic schedules (Table 44).

Parallel comparisons of allomorphosis for pelvic elements (Table 45) revealed pronounced differences in allometric slopes, but unlike trends in pectoral allomorphosis, these manifested a distinctly different and less redundant pattern in slopes. In *Porphyrio,* flightless *P. hochstetteri* was characterized by nonsignificant allomorphosis in lengths of pelvic elements ($\hat{b} \approx 0$), whereas flighted *P. melanotus* maintained significant allomorphosis in lengths of leg elements (Table 45). Sexes within species were alike in all comparisons. An ontogenetic interpretation of these marked differences in static allomorphosis would suggest that although individuals of the smaller, flighted *P. melanotus* maintained modest but significant elongation of pelvic elements at the terminus of skeletal growth, specimens of the larger, flightless *P. hochstetteri* had reached asymptotic, essential stasis of pelvic development at the terminus of growth (Table 45), despite its overall per-

TABLE 45. Intraspecific allometric coefficients ($\hat{b} \pm$ SE) or allomorphosis of three skeletal pelvic elements with corresponding first principal component (PC-I) of nonpelvic dimensions, by species and sex, for three flighted species and flightless relatives in three taxonomic tribes of Rallidae.

Species	Sex (n)	Pelvic element (proximal → distal)		
		Femur	Tibiotarsus	Tarsometatarsus
Porphyrio melanotus	Males (33)	0.082 ± 0.029	0.078 ± 0.026	0.077 ± 0.030
	Females (20)	0.060 ± 0.035	0.069 ± 0.027	0.069 ± 0.027
Porphyrio hochstetteri	Males (12)	−0.001 ± 0.005	0.007 ± 0.007	0.008 ± 0.008
	Females (18)	−0.005 ± 0.007	−0.000 ± 0.006	0.009 ± 0.007
$F(6, 75)$, P-value		39.1 (<0.00001)	21.7 (<0.00001)	13.0 (<0.00001)
Gallirallus philippensis	Males (16)	0.036 ± 0.011	0.041 ± 0.013	0.062 ± 0.014
	Females (19)	0.023 ± 0.008	0.021 ± 0.006	0.024 ± 0.008
Gallirallus owstoni	Males (14)	0.032 ± 0.009	0.032 ± 0.009	0.040 ± 0.010
	Females (18)	0.019 ± 0.008	0.018 ± 0.007	0.022 ± 0.010
Gallirallus australis-group	Males (28)	0.116 ± 0.020	0.094 ± 0.016	0.106 ± 0.027
	Females (22)	0.084 ± 0.011	0.083 ± 0.012	0.093 ± 0.018
$F(10, 105)$, P-value		39.9 (<0.00001)	46.3 (<0.00001)	33.9 (<0.00001)
Porzana tabuensis	Both (11)	0.071 ± 0.062	0.039 ± 0.065	0.026 ± 0.111
Porzana palmeri	Both (12)	0.022 ± 0.008	0.030 ± 0.010	0.027 ± 0.012
$F(2, 19)$, P-value		2.1 (>0.10)	1.2 (>0.30)	2.4 (>0.10)
Tribonyx ventralis	Males (24)	0.013 ± 0.014	0.007 ± 0.015	−0.006 ± 0.022
	Females (19)	0.007 ± 0.015	0.025 ± 0.016	0.018 ± 0.015
Tribonyx mortierii	Males (10)	0.069 ± 0.034	0.066 ± 0.032	0.079 ± 0.048
	Females (11)	−0.055 ± 0.035	−0.057 ± 0.039	−0.085 ± 0.044
$F(6, 56)$, P-value		16.8 (<0.00001)	7.7 (<0.00001)	8.3 (<0.00001)

amorphic pelvic skeleton (Tables 30–33; Figs. 8, 9). No evidence was found of a proximal-to-distal increase in allometric slope late in skeletal development in either species of *Porphyrio*.

In *Gallirallus*, a very different picture emerged in pelvic allomorphosis (Table 45). Across all elements and in both sexes, flighted *G. philippensis* and *G. owstoni* showed virtually identical patterns in slopes of allomorphosis (Table 45), indicating that both taxa extended limited elongation of pelvic elements until the terminus of skeletal growth. Flightless *G. australis* revealed a very different pattern in pelvic allomorphosis (Table 45): estimates of slopes ($\hat{b}$) were approximately three to four times as large as those for respective estimates in *G. philippensis* or *G. owstoni*, and within *G. australis*, slopes for males averaged roughly 25% higher than those for female conspecifics. These figures suggest that the peramorphic pelvic elements of large, flightless *G. australis* were undergoing rapid elongation at the terminus of skeletal ontogeny. *G. australis* was similar to its two congeners in the uniformity of slopes across all three pelvic elements (Table 45). By contrast, the marginally useful samples available for *Porzana tabuensis* and *P. palmeri* were uniform regarding pelvic allomorphosis, indicating no signficant differences among models of allomorphosis for any of the three pelvic elements (Table 45). Moreover, where slopes indicated significant regressions, the models did not deviate meaningfully from those for other groups.

Pelvic allomorphosis in flighted *Tribonyx ventralis* and flightless *T. mortierii* revealed yet another pattern of ontogeny at the terminus of skeletal growth (Table 45), in which both species showed extremely limited to insignificant allometric slopes ($\hat{b}$) in all three major elements of the pelvic appendage. Slightly greater slopes tended to characterize males relative to females of both species (Table 45). In summary, however, it seems that elongation of the principal skeletal elements of the pelvic appendage of *Tribonyx,* regardless of sex or flight status, had reached their respective, different asymptotes before or approximately coincident with terminus of development, and that the pronounced peramorphosis of the pelvic limb characteristic of flightless *T. mortierii* (Tables 30–33) is effected comparatively early in ontogeny.

Taken together, these examples substantiate a generality in the variation among adults that is consistent with the relatively early attainment of asymptotic size in the pelvic limb that contrasts to variable degrees the earlier stage of development of the pectoral limb in flightless rallids. This disparity of relative development of the two locomotory units mirrors the earlier, rapid growth of the pelvic apparatus and delayed, generally slower growth of the pectoral apparatus in birds generally (Butler and Bishop 1999). This pattern also contributes further evidence of paedomorphosis in flightless rails by extending support for this hypothesis to intraspecific scaling in skeletons of flighted and flightless relatives.

Canonical analyses.—As for the parallel applications for external dimensions, CAs provided the most powerful separation of associated suites of skeletal measurements for samples partitioned by species and sex, and comparisons that conformed most closely with distributional assumptions were limited to closely related taxa. Interpretation of the canonical axes themselves (descriptors maximizing differences among group means relative to within-group variances) is of secondary priority to the quantification of differences. The latter was exacerbated in the case of skeletal measurements in that different subsets of the variables were retained through the stepwise-selection procedures in the various taxonomic groups, an unpredictability worsened by small sample sizes for a number of the taxa compared. This complication overlays a common condition in which such variates meld information on size and shape.

Nevertheless, the general differences among groups reflected by these axes were interpretable through a two-stage procedure: first, determination of general changes in proportions *among* anatomical regions through an examination of the general pattern of signs of the coefficients for corresponding variables (Tables 46–52); and second, comparisons of signs and magnitudes of coefficients for measurements *within* regions or individual skeletal elements retained by the stepwise procedure in light of relative differences among univariate group means of skeletal measurements (Tables 30–33). These interpretations were limited largely to those axes summarizing discrimination of flightless taxa from flighted relatives. Availability of associated samples of skeletal specimens limited these comparisons to only seven major taxonomic groups, and reduced the numbers of genera and species represented within most of these partitions. For example, among basal rallids, *Habroptila* was represented by only four skeletal specimens, and the most closely related genus against which it could be compared skeletally was *Aramides.*

A CA of skeletal measurements of *Porphyrio* (four flighted and one flightless species; $n = 108$) and *Porphyrula* (three flighted species; $n = 42$) extracted seven CVs incorporating significant interspecific variation (ANOVA of scores; $P < 0.0001$). Five of the first eight CVs included significant ($P < 0.05$) intersexual variation, and nine of the total of 15 variates revealed significant species–sex

TABLE 46. Standardized coefficients and summary statistics for first two canonical variates and two flighted–flightless contrasts for skeletal measurements stepwise-selected ($P < 0.05$) from 41 variables for 16 species–sex groups of Porphyrionithini.

Variable	Canonical variate* I	Canonical variate* II	Flighted–flightless contrast All taxa†	Flighted–flightless contrast Porphyrio‡
Maxilla length	−0.451	−0.054	0.390	0.577
Cranium, length	−0.054	−0.017	—	—
Maximal width	−0.439	−0.323	0.422	0.450
Height perpendicular to palate	0.095	0.404	—	—
Sternum, length of carina	0.579	−0.166	−0.459	—
Length of corpus	−0.471	0.007	0.394	—
Minimal width of corpus	−0.276	−0.076	—	0.468
Minimal depth of carina	0.360	0.157	—	−0.389
Furcula, height	—	—	−0.390	—
Minimal width at midpoint	0.008	0.126	—	—
Maximal width at midpoint	—	—	—	−0.451
Scapula scapus length	−0.100	−0.191	—	—
Coracoid, length	0.393	−0.309	—	—
Crista articularis width	−0.097	−0.047	—	—
Humerus, intertubercular width of caput	0.211	0.702	—	−0.556
Maximal width at midpoint	0.787	0.041	−0.751	−0.960
Radius, length	−0.267	0.505	0.813	—
Ulna, length	—	—	−0.871	—
Minimal width at midpoint	0.043	0.067	—	—
Digitalis majoris, phalanx proximalis length	0.458	0.449	−0.466	−0.833
Synsacral length	−0.294	0.256	0.279	0.447
Pelvis, interacetabular width	−0.363	−0.149	—	—
Femur, length	−0.961	−0.031	1.105	1.093
Minimal width at midpoint	−0.729	−0.164	0.573	0.820
Tibiotarsus, length	0.315	−0.301	−0.546	−0.731
Maximal width at midpoint	—	—	0.423	0.868
Tarsometatarsus, length	0.354	0.243	—	—
Lateromedial width at midpoint	−0.285	−0.054	0.237	—
Digit III length	0.208	0.165	−0.455	—
Eigenvalue	352.18	98.60	226.55	242.31
Cumulative variance (%)	75.8	97.1	—	—
Canonical R	0.999	0.995	0.998	0.998

* Wilks' λ (d.f. = 25, 15, 134) < 10^{-7}; $\hat{F}_{375, 1,471}$ = 10.0 ($P \ll 0.0001$).
† Wilks' λ (d.f. = 16, 1, 134) = 0.00439; $\hat{F}_{16, 119}$ = 1,685.0 ($P < 0.0001$).
‡ Wilks' λ (d.f. = 14, 1, 98) = 0.00411; $\hat{F}_{14, 85}$ = 1,471.1 ($P < 0.0001$).

interaction effects ($P < 0.05$), indicating that sexual differences were evident in a subset of the species on the corresponding axes. The CVs were based on a subset of 25 variables selected stepwise from the total of 41 skeletal measurements analyzed (Table 46). Despite the inference of variation among mean scores of species–sex groups in more than one half of the total suite of variates extracted, the great majority of total variation among groups and the entirety of that pertaining to flightlessness was summarized by the first two CVs (Table 46; Fig. 52).

The first CV for skeletons of swamphens alone accounted for threefourths of the dispersion among groups (Table 46). CV-I primarily discriminated flightless *Porphyrio hochstetteri* from all other taxa (Fig. 52); scores on this axis were strongly (negatively) correlated with mean body mass ($r = -0.85$; $P < 0.01$). Interpretation of CV-I based on the coefficients of variables retained (Table 46)

TABLE 47. Standardized coefficients and summary statistics for first two canonical variates and flighted–flightless contrast for skeletal measurements stepwise-selected ($P < 0.05$) from 41 variables for six species–sex groups of basal rails (*Habroptila* and *Aramides*).

Variable	Canonical variate* I	Canonical variate* II	Flighted–flightless contrast†
Cranium, length	0.370	0.713	−1.108
Maximal width	0.519	−0.221	—
Sternum, length of carina	—	—	−1.046
Length of corpus	—	—	1.040
Oblique depth of carina	—	—	−0.740
Minimal depth of carina	−1.470	0.019	1.814
Furcula, height	0.686	−0.303	—
Minimal width at midpoint	0.106	−0.639	—
Maximal width at midpoint	−0.728	0.161	0.558
Coracoid, length	−0.575	−0.230	—
Humerus, intertubercular width of caput	−0.573	−0.663	1.215
Minimal width at midpoint	−0.220	−0.342	—
Radius, length	—	—	−0.868
Minimal width at midpoint	0.093	0.181	—
Maximal width at midpoint	−0.635	−0.004	0.807
Ulna, minimal width at midpoint	1.013	−0.022	−0.683
Carpometacarpus length	1.927	0.132	−1.131
Digitalis majoris, phalanx proximalis length	−0.540	−0.219	0.906
Femur, length	1.583	0.931	−1.680
Width of caput	1.037	0.399	−1.032
Maximal width at midpoint	—	—	−0.698
Tibiotarsus, length	−1.239	−1.205	1.336
Tarsometatarsus, length	−1.509	0.811	1.655
Lateromedial width at midpoint	0.867	−0.216	—
Digit III length	−0.077	−1.157	—
Eigenvalue	54.49	30.94	46.04
Cumulative variance (%)	61.7	96.7	—
Canonical R	0.991	0.984	0.989

* Wilks' λ (d.f. = 20, 5, 60) = 0.00008; $\hat{F}_{100, 205}$ = 12.2 ($P \ll 0.0001$).
† Wilks' λ (d.f. = 17, 1, 60) = 0.0213; $\hat{F}_{17, 44}$ = 119.2 ($P < 0.0001$).

revealed that this axis distinguished *P. hochstetteri* from other swamphens on the basis of its larger skull (especially the more massive bill and broader cranium), relatively smaller pectoral girdle (emphasizing the disproportionately reduced carina sterni and relatively broad coracoid), a diminutive pectoral limb (including the relatively short, narrow proximal elements and the absolutely reduced elements of the manus), a substantially enlarged pelvis and leg (weighting the robust femur especially heavily), and absolutely shorter toes. These shifts in proportions not only reflected the diminution of the pectoral apparatus in flightless *P. hochstetteri*, but also highlighted the shifts within the pelvic limb associated with increased terrestrial habitus (i.e., relative reductions in the tibiotarsus, tarsometatarsus, and toes) and megacephaly associated with herbivory (Fig. 9). In this context, it is noted that flighted species of *Porphyrio* were slightly more terrestrial in general body form than the species of *Porphyrula*, which are smaller (Fig. 52). Canonical contrasts were broadly consistent with CV-I in the signs of the variables retained, although the narrowed scope of these comparisons even further limited the number of measurements that could be included without redundancy and incurring the statistical problems associated with multicolinearity (Table 46).

TABLE 48. Standardized coefficients and summary statistics for two canonical variates and flighted–flightless contrast based on skeletal measurements stepwise-selected ($P < 0.05$) from 41 variables for four species–sex groups of *Dryolimnas*.

Variable	Canonical variate*		Flighted–flightless contrast†
	I	II	
Maxilla length	3.149	−1.160	—
Cranium, length	—	—	3.154
Maximal width	−8.157	4.011	—
Sternum, length of carina	—	—	8.909
Length of corpus	8.562	1.165	—
Minimal width of corpus	3.992	−1.230	—
Furcula, minimal width at midpoint	−6.714	0.110	−8.547
Maximal width at midpoint	0.450	−2.357	—
Coracoid, crista articularis width	2.555	1.132	—
Humerus, length	3.961	1.133	9.373
Femur, length	—	—	9.493
Width of caput	−3.946	−0.259	—
Maximal width at midpoint	—	—	−7.365
Tibiotarsus, length	0.188	−4.206	—
Maximal width at midpoint	—	—	3.292
Tarsometatarsus, length	—	—	−6.654
Craniocaudal width at midpoint	1.768	−1.698	—
Lateromedial width at midpoint	—	—	−9.672
Eigenvalue	1,932.72	61.13	5,599.06
Cumulative variance (%)	96.9	99.9	—
Canonical R	0.999	0.992	1.000

* Wilks' λ (d.f. = 11, 3, 16) < 0.00001; $\hat{F}_{33,18}$ = 38.2 ($P \ll 0.0001$).
† Wilks' λ (d.f. = 9, 1, 16) = 0.00018; $\hat{F}_{9,8}$ = 4,976.9 ($P < 0.0001$).

The second CV for swamphens added another 21% of the dispersion among groups in skeletal measurements (Table 46) and largely separated the lesser swamphens (*Porphyrula*) from the greater swamphens (*Porphyrio*), with lesser discrimination of *P. hochstetteri* from its congeners, or of sexes within species (Fig. 52). The primary contrast between genera resulted in a moderate correlation with mean body mass ($r = 0.74$; $P < 0.01$). The remaining 13 variates extracted together accounted for less than 3% of the dispersion among groups, and served mainly to discriminate among the flighted species of *Porphyrio*. Mean scores indicated that none of these remaining variates revealed important discrimination of flightless *Porphyrio hochstetteri* or substantive sexual dimorphism peculiar to this species, hence these lesser axes will not be considered further.

Analysis of skeletons of basal rallids, including flightless *Habroptila wallacii* ($n = 4$) and two flighted species of *Aramides* ($n = 62$), extracted five CVs based on 20 of the 41 skeletal measurements compiled (Table 47). Of these five axes, three (CV-I, CV-II, and CV-IV) included significant differences ($P < 0.001$) among species in mean scores, two (CV-I and CV-IV) incorporated significant intersexual differences ($P < 0.0005$) across species, and three (CV-III, CV-IV, and CV-V) included significant species–sex interaction effects ($P < 0.0001$), indicating sexual dimorphism in a subset of the included taxa. CV-I for basal rallids summarized more than one half of the total dispersion among groups (Table 47), was directly correlated with mean body mass ($r = 0.99$; $P < 0.01$), and explained most of the skeletal distinctness of flightless *Habroptila* relative to *Aramides* (Fig. 53). Coefficients for CV-I indicated that the axis was a subtle commingling of

TABLE 49. Standardized coefficients and summary statistics for first three canonical variates and two flighted–flightless contrasts based on skeletal measurements stepwise-selected ($P < 0.05$) from 41 variables for 23 species–sex groups of "typical" rails (subtribe Rallina, excluding *Dryolimnas*). A single taxon lacking sexual partitioning was included (*Gallirallus pectoralis*-group).

Variable	Canonical variate* I	Canonical variate* II	Canonical variate* III	Flighted–flightless contrast All taxa†	Flighted–flightless contrast No *Rallus*‡
Maxilla length	0.429	0.215	0.772	0.249	−0.224
Cranium, length	−0.139	−0.124	−0.282	—	—
Maximal width	−0.580	−0.006	−0.212	−0.670	−0.536
Height perpendicular to palate	−0.265	−0.100	−0.044	—	—
Sternum, length of carina	0.492	−0.368	0.082	0.480	0.435
Length of corpus	−0.358	−0.343	0.167	—	—
Oblique depth of carina	0.282	0.236	0.075	0.279	—
Minimal depth of carina	—	—	—	0.144	0.362
Furcula, height	0.056	0.137	0.085	—	—
Minimal width at midpoint	−0.031	−0.157	−0.262	—	—
Scapula scapus length	−0.279	0.289	0.069	—	—
Coracoid, length	−0.329	−0.161	0.033	—	0.404
Crista articularis width	0.023	0.113	−0.110	—	—
Humerus, length	0.209	−0.650	0.416	—	—
Intertubercular width of caput	0.254	0.154	0.316	0.281	—
Minimal width at midpoint	−0.052	0.087	0.232	—	—
Maximal width at midpoint	−0.042	0.027	−0.073	—	—
Radius, length	−0.251	−0.247	−0.167	—	—
Minimal width at midpoint	−0.084	0.087	−0.014	—	—
Maximal width at midpoint	0.115	−0.160	0.064	—	—
Ulna, length	0.840	−0.592	−0.904	0.557	0.940
Minimal width at midpoint	0.191	0.148	0.023	—	—
Maximal width at midpoint	−0.060	−0.122	0.149	—	—
Carpometacarpus length	0.036	0.183	−0.047	—	—
Digitalis major, phalanx proximalis length	−0.015	−0.338	−0.285	—	—
Phalanx distalis length	0.230	0.256	−0.090	−0.408	−0.542
Synsacral length	−0.141	−0.210	0.188	—	—
Pelvis, interacetabular width	−0.045	−0.083	−0.300	−0.292	—
Femur, width of caput	−0.422	−0.050	−0.251	−0.394	—
Minimal width at midpoint	0.153	0.153	0.093	—	—
Tibiotarsus, length	−1.065	0.678	0.269	−0.981	−1.028
Minimal width at midpoint	0.221	0.079	0.050	—	—
Maximal width at midpoint	0.148	−0.408	−0.175	0.229	0.344
Tarsometatarsus, length	0.646	−0.552	0.241	0.809	0.702
Craniocaudal width at midpoint	−0.091	0.056	0.200	—	—
Lateromedial width at midpoint	−0.276	0.086	−0.107	−0.353	−0.473
Digit III length	0.301	0.954	0.220	—	−0.554
Eigenvalue	154.89	30.93	28.87	17.93	14.81
Cumulative variance (%)	67.2	80.6	93.1	—	—
Canonical R	0.996	0.984	0.983	0.973	0.968

* Wilks' λ (d.f. = 6, 22, 293) < 0.00001; $\hat{F}_{792,4,972}$ = 10.5 ($P \ll 0.0001$).
† Wilks' λ (d.f. = 14, 1, 286) = 0.0528; $\hat{F}_{14,273}$ = 273.0 ($P < 0.0001$).
‡ Wilks' λ (d.f. = 12, 1, 146) = 0.0588; $\hat{F}_{12,135}$ = 166.6 ($P < 0.0001$).

proportions that confounded discrimination of genera with significant discrimination between species of *Aramides* (Fig. 53). Characteristics distinguishing *Habroptila* were its relatively robust cranium (especially relative to its bill), markedly small pectoral girdle (especially the depth of the carina sterni), distal reduction of the pectoral limb (principally in the major digit), and a disproportionately

TABLE 50. Standardized coefficients and summary statistics for first four canonical variates and two flighted–flightless contrasts based on skeletal measurements stepwise-selected ($P < 0.05$) from 41 variables for 20 species–sex groups of crakes (Crecina). A single additional taxon (*Atlantisia rogersi*) lacking sexual partitioning was included in the general analysis.

	Canonical variate*				Flighted–flightless contrast	
Variable	I	II	III	IV	All genera[†]	*Crex* and *Porzana*[‡]
Maxilla length	−0.141	0.034	−0.030	−0.429	—	—
Cranium, length	0.019	−0.126	−0.157	−0.021	—	−0.297
Maximal width	−0.076	−0.200	−0.053	0.352	—	—
Height perpendicular to palate	—	—	—	—	—	−0.445
Sternum, length of carina	0.173	−0.209	0.022	0.130	0.521	0.457
Length of corpus	0.097	−0.194	0.314	0.346	—	−0.492
Minimal width of corpus	−0.103	0.021	0.275	−0.275	—	−0.260
Oblique depth of carina	−0.019	0.002	−0.322	−0.026	—	—
Minimal depth of carina	0.222	0.262	0.813	−0.180	0.442	—
Furcula, minimal width at midpoint	0.003	−0.035	−0.293	−0.290	—	—
Scapula scapus length	1.130	0.227	−0.259	−0.403	—	—
Coracoid, length	0.113	0.158	0.252	0.215	—	—
Crista articularis width	0.214	−0.211	−0.115	0.016	0.191	0.325
Humerus, length	0.110	−0.077	−0.040	−0.595	—	−0.577
Minimal width at midpoint	—	—	—	—	—	0.486
Maximal width at midpoint	−0.045	−0.239	0.140	0.130	—	—
Radius, length	0.093	0.475	−0.579	−0.230	—	—
Ulna, length	0.817	−0.174	0.636	0.628	1.050	0.591
Maximal width at midpoint	0.340	0.188	−0.123	0.210	—	—
Carpometacarpus length	−0.009	−0.374	0.343	−0.041	—	—
Digitalis majoris, phalanx proximalis						
Length	0.035	−0.000	0.095	0.246	—	0.436
Phalanx distalis length	0.137	0.029	−0.168	−0.570	—	—
Synsacral length	−0.086	0.065	−0.058	−0.290	—	—
Pelvis, interacetabular width	0.228	0.332	−0.034	0.128	−0.767	—
Femur, length	0.253	0.439	−1.542	0.300	−0.285	—
Width of caput	—	—	—	—	—	0.418
Tibiotarsus, length	−0.664	−0.726	0.470	−0.135	—	−0.729
Minimal width at midpoint	−0.184	−0.012	0.137	−0.403	—	—
Maximal width at midpoint	−0.127	−0.078	−0.208	−0.065	—	—
Tarsometatarsus, length	0.013	−0.549	−0.363	0.798	—	0.421
Craniocaudal width at midpoint	−0.200	0.196	−0.186	−0.099	−0.255	—
Lateromedial width at midpoint	−0.363	−0.331	−0.296	0.122	−0.403	−0.323
Digit III length	−0.704	0.022	1.042	−0.697	—	−0.719
Eigenvalue	94.81	44.41	23.53	9.35	15.54	151.95
Cumulative variance (%)	51.0	74.9	87.5	92.5	—	—
Canonical R	0.995	0.989	0.979	0.950	0.969	0.997

* Wilks' λ (d.f. = 30, 20, 184) < 10^{-6}; $\hat{F}_{600, 2677}$ = 11.6 ($P \ll 0.0001$).
[†] Wilks' λ (d.f. = 8, 1, 183) = 0.06045; $F_{8, 176}$ = 341.9 ($P < 0.0001$).
[‡] Wilks' λ (d.f. = 5, 1, 103) = 0.05039; $F_{15, 111}$ = 1,124.4 ($P < 0.0001$).

TABLE 51. Standardized coefficients and summary statistics for first four canonical variates and two flighted–flightless contrasts based on skeletal measurements stepwise-selected ($P < 0.05$) from 41 variables for 15 species-sex groups of moorhens. The group comprises *Gallicrex*, *Gallinula* (including single specimens of each sex of *G. comeri*), and *Tribonyx*. A single additional taxon (*Gallinula sandvicensis*) lacking sexual partitioning was included in the general analysis.

Variable	Canonical variate*				Flighted–flightless contrast	
	I	II	III	IV	All genera†	*Tribonyx* only‡
Cranium, maximal width	−0.467	−0.247	0.214	−0.026	−0.487	−0.447
Sternum, length of corpus	0.021	0.073	0.613	0.112	—	—
Minimal width of corpus	−0.027	0.089	−0.931	−0.298	—	−0.537
Minimal depth of carina	0.178	−0.261	−0.471	−0.293	−0.460	—
Furcula, minimal width at midpoint	0.254	0.018	−0.091	0.122	—	—
Maximal width at midpoint	−0.044	−0.138	0.023	0.246	—	—
Coracoid, length	−0.292	0.153	−0.076	0.822	—	—
Humerus, intertubercular width of caput	0.031	−0.158	0.748	−0.264	—	—
Minimal width at midpoint	—	—	—	—	−0.298	—
Maximal width at midpoint	0.409	−0.057	−0.138	0.246	—	—
Radius, length	1.244	−0.012	1.397	0.034	−0.900	—
Maximal width at midpoint	—	—	—	—	−0.298	0.377
Carpometacarpus length	0.253	−0.570	−1.712	−0.632	−0.601	1.314
Digitalis majoris, phalanx proximalis						
Length	−0.315	−0.662	0.004	−0.435	−0.389	0.679
Phalanx distalis length	—	—	—	—	0.514	—
Synsacral length	−0.231	−0.023	−0.331	−0.418	—	—
Femur, length	−1.049	0.879	−0.644	0.264	1.742	−1.894
Minimal width at midpoint	−0.292	−0.344	0.215	0.121	—	—
Maximal width at midpoint	−0.255	0.259	−0.325	−0.154	0.277	—
Tarsometatarsus, length	−0.524	−0.730	1.148	0.008	—	—
Lateromedial width at midpoint	−0.106	0.307	0.155	−0.105	0.344	—
Digit III length	0.855	1.301	0.003	−0.512	−0.373	—
Eigenvalue	109.5	71.25	32.59	4.91	23.71	276.15
Cumulative variance (%)	49.0	80.9	95.5	97.7	—	—
Canonical R	0.995	0.993	0.985	0.911	0.980	0.998

* Wilks' λ (d.f. = 19, 14, 129) < 10⁻⁶; $\bar{F}_{266,1295} = 16.8$ ($P \ll 0.0001$).
† Wilks' λ (d.f. = 12, 1, 125) = 0.04047; $F_{12,114} = 225.2$ ($P < 0.0001$).
‡ Wilks' λ (d.f. = 6, 1, 60) = 0.00361; $\bar{F}_{6,55} = 2,531.3$ ($P < 0.0001$).

TABLE 52. Standardized coefficients and summary statistics for three canonical variates and contrast between "typical" and "mega" taxa based on skeletal measurements stepwise-selected ($P < 0.05$) from 41 variables for nine species or species–sex groups of coots. Three species also were included in the general analysis for which partitioning by sex was not possible (including *Fulica gigantea* and *F. cornuta*).

Variable	Canonical variate*			Flight contrast†
	I	II	III	
Maxilla length	−0.321	0.634	0.461	−0.711
Maximal width	0.057	0.067	−0.671	—
Height perpendicular to palate	0.131	−0.731	−0.728	0.548
Sternum, length of carina	−0.008	−0.857	0.217	—
Furcula, maximal width at midpoint	0.098	0.648	0.302	−0.393
Coracoid, crista articularis width	0.100	0.537	−0.175	—
Humerus, length	−0.167	−1.043	−1.202	—
Minimal width at midpoint	—	—	—	0.440
Radius, minimal width at midpoint	0.161	−0.177	0.514	—
Ulna, length	−0.365	−0.871	3.471	—
Minimal width at midpoint	—	—	—	−0.553
Carpometacarpus length	0.483	1.187	−1.271	—
Pelvis, interacetabular width	−0.629	0.239	0.060	−0.707
Femur, length	−0.171	1.568	0.703	—
Width of caput	−0.186	−0.830	0.064	—
Minimal width at midpoint	−0.369	−0.092	−0.055	−0.390
Maximal width at midpoint	0.017	0.526	−0.303	—
Tibiotarsus, length	0.065	−0.949	−1.110	—
Minimal width at midpoint	−0.271	0.623	−0.206	—
Tarsometatarsus, length	−0.852	0.053	0.136	—
Craniocaudal width at midpoint	0.007	0.482	0.786	—
Lateromedial width at midpoint	—	—	—	—
Digit III length	0.923	0.030	−0.446	0.403
Eigenvalue	34.68	5.54	3.73	9.93
Cumulative variance (%)	73.1	84.8	92.7	—
Canonical R	0.986	0.920	0.888	0.953

* Wilks' λ (d.f. = 20, 8, 92) < 0.00008; $\hat{F}_{160, 559} = 8.8$ ($P < 0.0001$).
† Wilks' λ (d.f. = 8, 1, 95) = 0.09152; $\hat{F}_{8,88} = 109.2$ ($P < 0.0001$).

elongated pelvic limb in which the femur is enlarged relatively more than distal elements (Table 47). A canonical contrast limited to differences pertaining to faculty of flight in basal rallids, although reversed in sign, indicated a similar suite of proportional shifts (Table 47).

The second CV for basal rallids contributed an additional 35% of the total dispersion among groups (Table 47). This axis primarily discriminated *Aramides ypecaha* from the other two species, with lesser discrimination between *A. cajanea* and *Habroptila wallacii* (Fig. 53). The extreme position of massive *A. ypecaha* resulted in a strong, negative correlation between mean scores on CV-II and body mass ($r = -0.97$; $P < 0.01$). However, the comparatively small component of CV-II attributable to the discrimination of *H. wallacii* does not justify a detailed examination of this axis (Table 47). The three remaining CVs accounted for vanishingly small fractions of dispersion among groups and no vital insights into skeletal changes associated with flightlessness in *Habroptila* (Table 47): CV-III summarized sexual differences within *Aramides*; and CV-IV and CV-V principally depicted minuscule, residual sexual dimorphism confined to *Habroptila*.

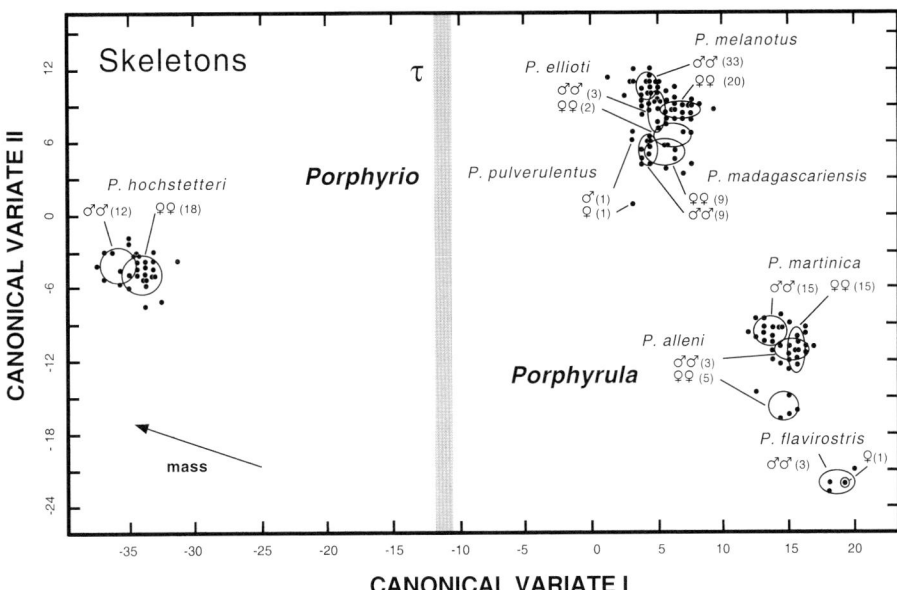

FIG. 52. Plot of 16 species–sex groups of swamphens (*Porphyrio* and *Porphyrula*) on first two canonical variates for stepwise-selected subset of 41 skeletal measurements. Ellipses for groups delimit summary statistics for scores ($\bar{x} \pm$ SD). Approximate orientation of vector for mean body masses of taxa ("mass") and apparent threshold of flightlessness (τ, positioned on the flightless side of boundary) are indicated.

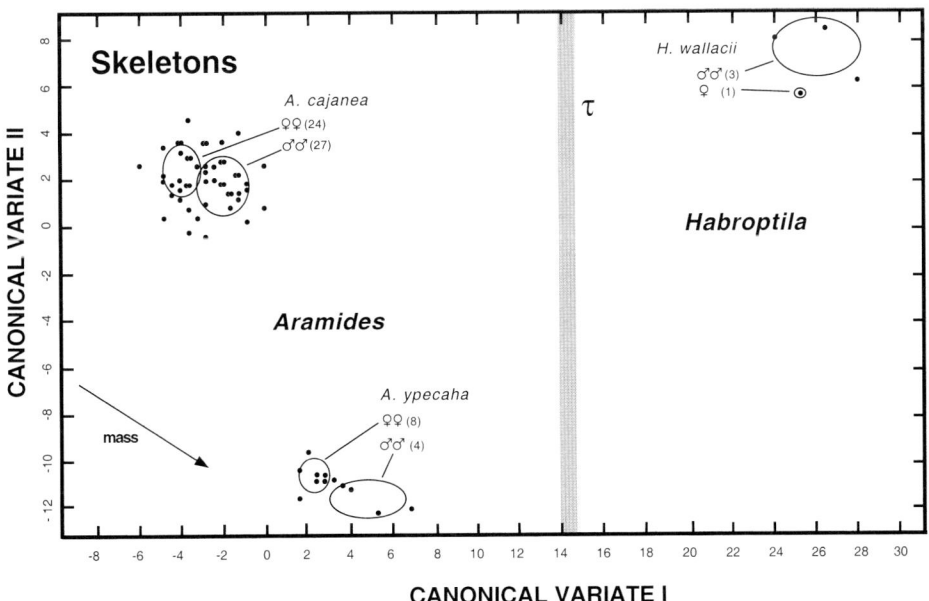

FIG. 53. Plot of six species–sex groups of basal rallids (*Habroptila* and *Aramides*) on first two canonical variates for stepwise-selected subset of 41 skeletal measurements. Ellipses for groups delimit summary statistics for scores ($\bar{x} \pm$ SD). Approximate orientation of vector for mean body masses of taxa ("mass") and apparent threshold of flightlessness (τ, positioned on the flightless side of boundary) are indicated.

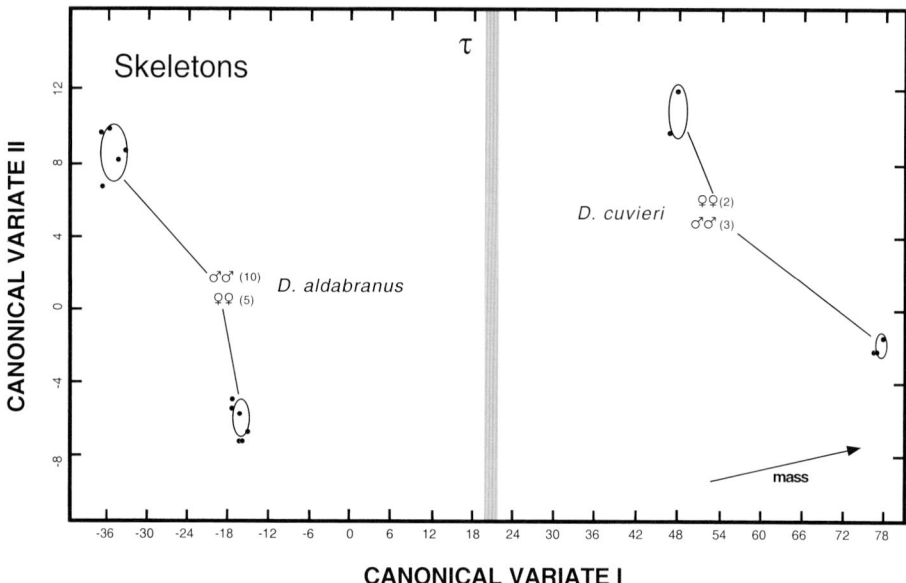

FIG. 54. Plot of six species–sex groups of white-throated rails (*Dryolimnas*) on first two canonical variates for stepwise-selected subset of 41 skeletal measurements. Ellipses for groups delimit summary statistics for scores ($\bar{x} \pm$ SD). Approximate orientation of vector for mean body masses of taxa ("mass") and apparent threshold of flightlessness (τ, positioned on the flightless side of boundary) are indicated.

Given the lack of associated skeletal material for *Dryolimnas abbotti*, only four species–sex groups of white-throated rails (*Dryolimnas*; $n = 20$) were available for analysis, limiting the dimensionality of the canonical space to three (Table 48). On these axes, interspecific and intersexual effects were highly significant (ANOVA of scores; $P < 0.0001$) on CV-I and CV-II, with CV-III including only intersexual differences ($P < 0.05$).

The first CV for white-throated rails alone accounted for more than 96% of the total dispersion among the four species–sex groups (Table 48) and included the great majority of the discrimination between flighted *Dryolimnas cuvieri* and flightless *D. aldabranus* (Fig. 54). Scores on this axis were directly correlated with mean body masses of the four groups ($r = 0.99$; $P < 0.01$). The limited samples and few groups in this comparison limited the number of skeletal variables retained nonredundantly in the discrimination of these two similar congeners to only 11 of 41 skeletal variables, rendering direct interpretation of coefficients for CV-I of *Dryolimnas* problematic (Table 48). What can be inferred regarding *D. aldabranus* concerns relative lengths of the bill and cranium, a diminutive pectoral girdle (summarized primarily through dimensions of the sternum and coracoid), a relatively short pectoral limb (with humerus length most efficiently reflecting a reduction approaching uniformity throughout the appendage), and the retention of robust pelvic limb (especially well summarized by widths of femur and tarsometatarsus) in the face of an overall reduction in size that accompanied the loss of flight (Tables 30–33).

A canonical contrast confined to interspecific differences affirmed these inter-

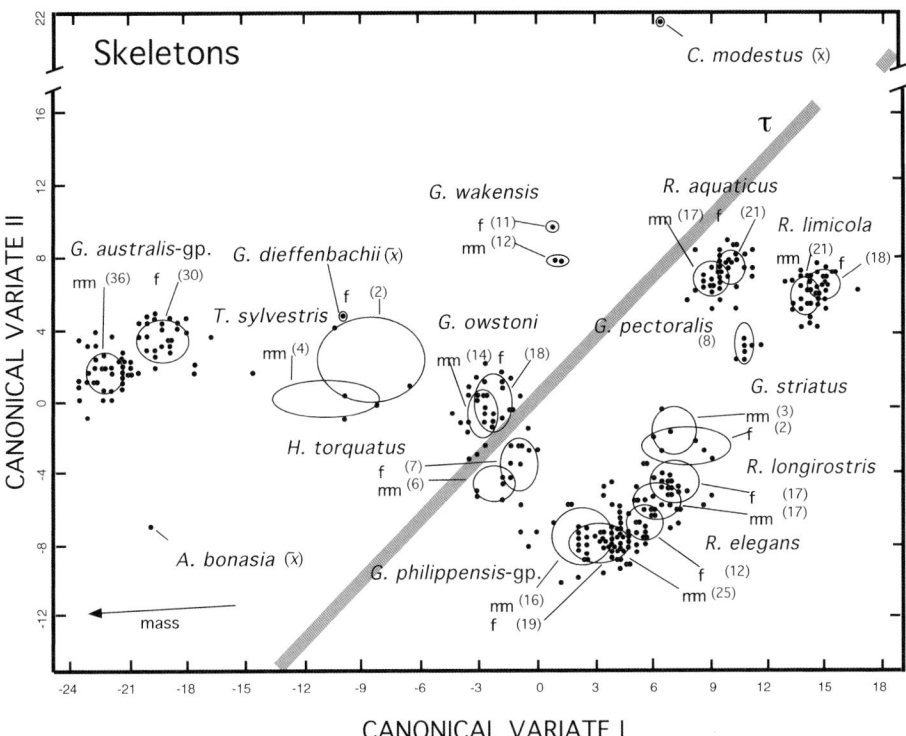

FIG. 55. Plot of 23 species–sex groups of "typical" rails (*Rallus, Gallirallus,* and allied genera) on first two canonical variates for stepwise-selected subset of 41 skeletal measurements. Except for taxa represented by composite mean vectors plotted a posteriori (*Gallirallus dieffenbachii* and *Cabalus modestus*), ellipses for groups delimit summary statistics for scores ($\bar{x} \pm$ SD). Approximate orientation of vector for mean body masses of taxa ("mass") and apparent threshold of flightlessness (τ, positioned on the flightless side of boundary) are indicated.

pretations, although the inclusion of only nine variables in the axis minimized its interpretability (Table 48). Consonant with patterns in most other taxonomic groups, sexual dimorphism on CV-I was such that males, in comparison with their female conspecifics, were shifted in proportions toward those characteristic of flightless taxa (Fig. 54). The CV-II for skeletons of *Dryolimnas* integrated residual interspecific and intersexual differences uncorrelated with mean body masses ($r = -0.15$; $P \gg 0.05$), and contributed only 3% of the total dispersion among groups (Table 48). The modicum of intersexual discrimination afforded by skeletal CV-III for *Dryolimnas* did not merit detailed consideration (Table 48).

The myriad forms of typical rails (subtribe Rallina [Livezey 1998]), even after the separate consideration of one member genus (*Dryolimnas*), span a rich spectrum of skeletal size and shape. CAs were performed to discriminate 11 species that permitted sex-specific discrimination (i.e., accounting for 22 groups) and 1 species (*G. pectoralis*) for which sexes were not determinable, bringing the groups defined a priori ($n = 316$) to a total of 23 (Table 49). Also, mean measurements for three extinct taxa lacking associated skeletons (*Gallirallus dieffenbachii, Cabalus modestus,* and *Aphanapteryx bonasia*) were evaluated a posteriori for plotting (Fig. 55). Fifteen CVs for skeletal measurements of typical rails were extracted, based on 36 of the 41 measurements submitted for analysis, of which 13 displayed significant interspecific differences in mean scores ($P < 0.05$). Intersexual differences were found on 8 of the 15 CVs (including the first 7), with significant species–sex interaction effects on 12 (including 6 of the first 7), 4 of which were accompanied by significant interspecific effects. These findings indicate that at least a subset of the taxa showed sexual dimorphism on all but CV-VIII. As in other taxonomic groups, the preponderance of variation among groups was summarized by the first few CVs; for typical rails, the first three axes alone summarized more than 93% of the total dispersion among groups (Table 49).

The first CV for skeletons of typical rails accounted for roughly two thirds of the total dispersion among groups (Table 49), revealed highly significant interspecific and intersexual differences in scores ($P < 0.0001$), included substantial separation of flighted and flightless species (Fig. 55), and its scores were highly (negatively) correlated with mean body masses ($r = -0.90$; $P < 0.01$). Coefficients for variables on CV-I (Table 49) revealed that the axis distinguished groups having low scores (predominately flightless) by a contrast in proportions that emphasized robust skulls (especially cranium), sterna having carinae relatively reduced, pectoral appendages characterized by disproportionately truncated antebrachial elements (with secondary weight placed on relative shortening of the manus), and pelvic limbs having disproportionately elongated tibiotarsi and truncated tarsometatarsi and digits. On CV-I, *Gallirallus australis*-group, *Aphanapteryx bonasia*, *Tricholimnas sylvestris,* and *G. dieffenbachii* had the lowest scores (i.e., the most extreme proportions), whereas several other flightless species characterized by less extreme shifts in form and (to a lesser extent) small body size (e.g., *Gallirallus wakensis,* and especially *Cabalus modestus*) occupied positions on CV-I comparable to those of many flighted species (Fig. 55). As noted in many other contexts, scores on CV-I tended to place males, relative to their female conspecifics, more in the direction of flightless taxa (Fig. 55).

The second CV for skeletons of typical rails contributed an additional 13% of total dispersion among groups, included highly significant interspecific and intersexual differences in scores ($P < 0.0001$), and scores on CV-II were not correlated with mean body masses ($r = -0.04$; $P > 0.05$). CV-II amplified discrimination among both flighted and flightless members, and (a posteriori) proved critical in the discrimination of *Cabalus modestus,* the high score for which proved to be maximal (Fig. 55). Coefficients for variables on CV-II indicated that higher scores were associated with relative elongation of the bill, especially pronounced truncation throughout the pectoral girdle and appendage (especially marked in antebrachial elements), unremarkably proportioned pelvic limbs, and (to a lesser extent) small size generally (Table 49).

The third CV for typical rails added almost as much separation among groups as CV-II for these taxa (detailed above), and scores on CV-III included both interspecific and intersexual effects (ANOVA of scores; $P < 0.0001$). However, examination of positions of groups on CV-III revealed that this axis primarily discriminated *Rallus* (especially *R. elegans* and *R. longirostris*) from *Gallirallus, Tricholimnas,* and *Habropteryx,* regardless of the flight status of the latter taxa. Scores for the remaining variates likewise affirmed that the skeletal proportions critical to the discrimination of flightless members of the typical rails largely were subsumed by the first two CVs (Fig. 55), an interpretation consistent with the eigenvalues and coefficients of canonical contrasts of typical rallids grouped strictly by flight capacity (Table 49).

Ten species of crakes in four genera (*Crex, Laterallus, Coturnicops,* and *Porzana*) were represented by samples of skeletons partitioned into sex groups for CAs, to which was added a single species (*Atlantisia rogersi*) lacking information on sex. In addition to these groups defined a priori, a single extinct taxon ("*Atlantisia*" *elpenor*) was plotted a posteriori by using a composite of mean measurements. The 21 groups formally analyzed ($n = 205$) sustained the definition of 15 CVs, based on 30 of 41 skeletal variables analyzed, of which 10 included significant differences among species (ANOVA of scores; $P < 0.05$). Intersexual differences were found as significant main effects ($P < 0.05$) on only 4 of the 15 CVs, although species–sex interaction effects were significant ($P < 0.05$) in scores on 11 of the 15 axes. The latter combination of findings indicated that multivariate sexual dimorphism in skeletons of crakes was generally less pronounced than in most other subgroups of the Rallidae, and that (where present) it tended to be peculiar to individual genera or species of the tribe.

The first CV for skeletal measurements of crakes accounted for slightly more than one half of the total dispersion among groups (Table 50) and included highly significant differences in mean scores among species ($P < 0.0001$) but not between sexes ($P > 0.35$). Scores on this axis were significantly correlated with mean body masses of groups ($r = 0.64$; $P < 0.01$). Positions of taxa on CV-I revealed that this axis integrated much of the variation attributable to flight capacity; mean scores for all flightless taxa were less than -11, whereas those for flighted crakes where greater than -8 (Fig. 56). Coefficients for CV-I indicated that this axis distinguished flightless crakes from their flighted relatives by a combination of the following features: relatively long bills, shortened pectoral appendages (disproportionately reduced in distal elements), generally diminutive pectoral girdles having disproportionately shallow carinae, pelvic limbs relatively large but unremarkable in proportions, and (in extreme cases) small overall body size (Table 50).

The second CV of skeletal measurements of crakes added another 25% of the total dispersion among groups (Table 50), and included highly significant interspecific and intersexual differences (ANOVA of scores; $P < 0.0001$). Scores were significantly (negatively) correlated with mean body masses of groups ($r = -0.68$; $P < 0.01$). Positions of taxa on CV-II indicated that this axis substantially augmented the separation of flightless species (Fig. 56), in which flightless species tended to have higher mean scores than flighted taxa (with *Coturnicops* being exceptionally high among the latter). Coefficients of variables for CV-II further separated flightless taxa from most flighted relatives through differences in relative proportions within the distal segments of the wing, comparatively subtle proportions within the pectoral girdle (including the relationship between length and depth of the carina sterni), and modest differences in proportions within the pelvic limb in which proximal elements tended to be slightly longer relative to more-distal elements (Table 50).

Together, the first two CVs for skeletons of crakes encompassed the dimensions critical to the discrimination of flightless species (Fig. 56). Subsequent axes (e.g., CV-III and CV-IV) summarized significant taxonomic (principally residual intergeneric) differences (ANOVA of scores; $P < 0.0001$), but these did not bear on

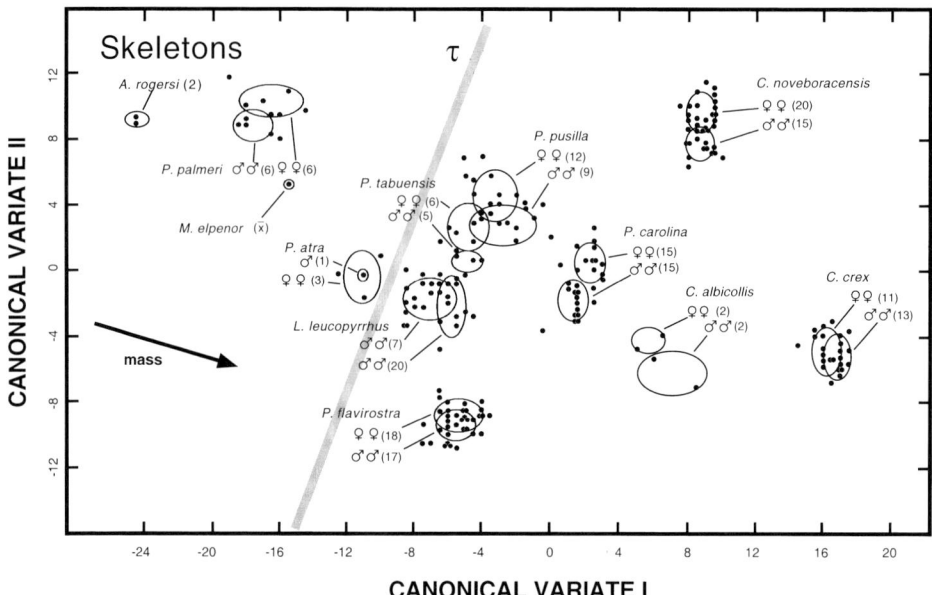

Fig. 56. Plot of 20 species–sex groups of crakes (Subtribe Crecina) on first two canonical variates for stepwise-selected subset of 41 skeletal measurements. Except for a taxon represented by a composite mean vector plotted a posteriori ("*Atlantisia*" *elpenor*), ellipses for groups delimit summary statistics for scores ($\bar{x} \pm$ SD). Approximate orientation of vector for mean body masses of taxa ("mass") and apparent threshold of flightlessness (τ, positioned on the flightless side of boundary) are indicated.

the discrimination of taxa by flight capacity. This inference is consistent with the variance incorporated by canonical contrasts targeting the essential differences associated with the faculty of flight, as were the broad interpretations permitted by the coefficients of variables in these lower-dimensional analyses (Table 50). Consequently, the CVs for crakes subsequent to CV-I and CV-II will not be detailed.

Samples of skeletons of moorhens of known sex were available for *Gallicrex cinerea,* four species of *Gallinula* sensu stricto (*G. chloropus, G. cachinaans, G. tenebrosa,* and *G. comeri*), and both modern species of *Tribonyx.* A small series of *Gallinula sandvicensis* of undetermined sex ($n = 5$) also was sampled (Tables 30–33). To maximize the information derived, CAs were performed with all 15 groups ($n = 144$) comprising 14 species–sex groups amenable to both one-way (interspecific) and two-way (species–sex) ANOVA of scores, plus one additional species group suitable only for one-way (interspecific) comparisons of scores.

The first 7 of the 14 CVs for the skeletons of moorhens included significant interspecific differences in mean scores ($P < 0.0001$). Significant intersexual differences ($P < 0.05$) were detected on the first, fourth, fifth, eighth, and 10th CVs. CV-I for skeletons of moorhens accounted for approximately 50% of the total dispersion among groups (Table 51) and largely served to distinguish flightless *Tribonyx mortierii* from other species (Fig. 57). The comparatively large size of this flightless species resulted in a strong (negative) correlation between scores on CV-I and mean body masses of the groups ($r = -0.89; P < 0.01$). The

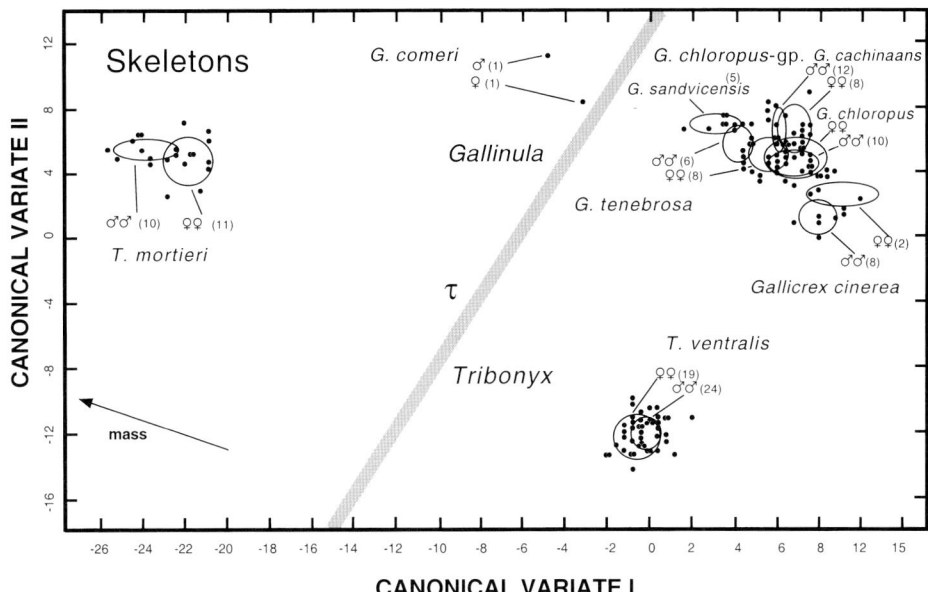

FIG. 57. Plot of 15 species–sex groups of moorhens (*Gallicrex, Tribonyx,* and *Gallinula*) on first two canonical variates for stepwise-selected subset of 41 skeletal measurements. Ellipses for groups delimit summary statistics for scores ($\bar{x} \pm$ SD). Approximate orientation of vector for mean body masses of taxa ("mass") and apparent threshold of flightlessness (τ, positioned on the flightless side of boundary) are indicated.

pertinence of CV-I to flight capacity also is consistent with the intermediate position of weakly flighted *Gallinula (Porphyriornis) comeri* (Fig. 57). Coefficients of variables on the axis, which retained 19 of 41 measurements analyzed, indicated that this dimension principally distinguished *T. mortierii* by a combination (Table 51) of its relatively massive skull, disproportionately short antebrachium and carpometacarpus, relatively shallow carina sterni and diminutive furcula, and large pelvic limb especially characterized by an elongated femur and truncated toes. Although fewer variables were retained in the canonical contrasts of flight for moorhens, which were performed both for all species and for *Tribonyx* alone, the general patterns of signs and magnitudes of coefficients conformed generally with those of CV-I (with slight, oblique influence by those reflected by CV-II) for all taxa (Table 51).

The second CV for skeletons of moorhens summarized an additional 30% of the dispersion among groups and largely discriminated flighted *Tribonyx ventralis* from all other species regardless of general size or flight capacity (Fig. 57). Accordingly, scores on CV-II were not correlated with mean body masses ($r = 0.19$; $P > 0.05$) and the proportions reflected by this axis were not informative about the morphological corollaries of flightlessness (Table 51). The third CV-III contributed another 15% of the intergroup dispersion for moorhens (Table 51), and principally discriminated *Gallicrex cinerea* from all other taxa (not shown). Of the remaining CVs for moorhens, the CV-V was notable for its discrimination of *Gallinula comeri* from other taxa. Coefficients for this minor axis (not tabulated) indicated that weakly flighted or flightless *G. comeri* was distinguished from other

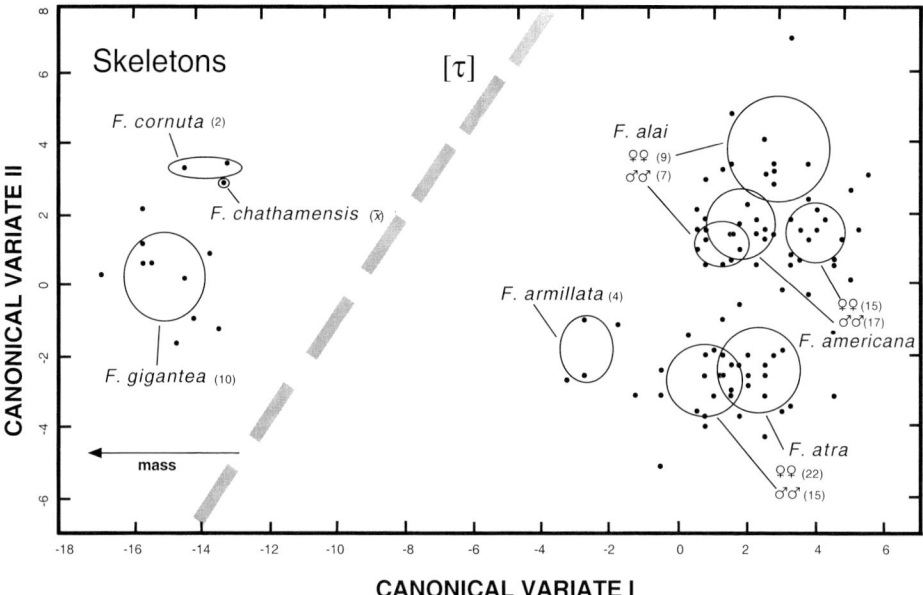

FIG. 58. Plot of nine species–sex groups of coots (*Fulica*) on first two canonical variates for stepwise-selected subset of 41 skeletal measurements. Except for a taxon represented by a composite mean vector plotted a posteriori (*Fulica chathamensis*), ellipses for groups delimit summary statistics for scores ($\bar{x} \pm$ SD). Approximate orientation of vector for mean body masses of taxa ("mass") and provisional threshold of flightlessness ([τ], positioned on the flightless side of boundary) are indicated.

moorhens (including *T. mortierii*) by a unique proportion dominated by relative lengths of the radius and tarsometatarsus.

Skeletal samples of coots (*Fulica*) permitted only three species to be compared within a two-way (species–sex) context (Tables 30–33). Restriction of analyses to one-way comparisons of species permitted the inclusion of *F. armillata* ($n = 4$) and the two large Andean species *F. gigantea* and *F. cornuta* ($n = 10$ and 2, respectively). Therefore, CAs of skeletons of *Fulica* were performed that discriminated nine groups, comprising all six species ($n = 101$); for those taxa adequately represented (*F. americana, F. alai,* and *F. atra*; $n = 85$), groups were partitioned with respect to both species and sex. Comparisons of scores a posteriori were based on corresponding one-way (interspecific) and two-way (species–sex) ANOVAs of six species and six species–sex groups, respectively. Six of the eight CVs for skeletons of coots included significant ($P < 0.01$) interspecific differences in scores; five of these also manifested significant ($P < 0.05$) differences between the sexes of the three species permitting such comparisons.

The first CV for skeletons of *Fulica* accounted for almost three fourths of the total dispersion among groups (Table 52), was strongly (negatively) correlated with mean body masses ($r = -0.97$; $P < 0.01$), and provided the vast majority of the discrimination between *F. gigantea* and *F. cornuta* and other species (Fig. 58). Coefficients for the 20 of 41 skeletal variables significantly retained in the discrimination of coots for CV-I (Table 52), in conjunction with univariate comparisons of these measurements (Tables 30–33), revealed that the large coots were distinguished from other coots by a combination of their relatively large bills, pectoral appendages showing slight relative truncation, and pelvic limbs having disproportionately longer tarsometarsi and short toes (Table 52).

The second CV for skeletons of coots contributed an additional 11% of the total intergroup dispersion (Table 52), and principally ordinated *Fulica atra* and *F. armillata* at the low extreme; *F. gigantea* at an intermediate position; and *F. cornuta, F. americana,* and *F. alai* among the highest scores (Fig. 58). Scores on CV-II were not significantly correlated with mean body masses ($r = 0.27$; $P > 0.05$). Coefficients for included variables indicated that this axis largely integrated relative bill length and proportions among elements within each limb (Table 52).

The remaining six CVs for skeletons of *Fulica* together accounted for only 15% of the total multivariate dispersion among group means (Table 52). Summary statistics for scores on the four of these lesser axes including significant interspecific differences in scores effected the following discriminations: CV-III primarily separated *F. cornuta* from *F. armillata,* with other species being intermediate; CV-IV distinguished *F. alai* and *F. atra* from *F. armillata,* with other taxa assuming intermediate scores; and CV-V and CV-VI primarily amplified the separation of *F. cornuta* from its congeners. As such, none of these variates merit detailed consideration in the context of skeletal changes in *Fulica* associated with the generalities of giantism and reduced flight capacity.

Mean measurements for extinct, purportedly flightless *Fulica chathamensis* indicated that this species closely resembled *F. cornuta* in the skeletal proportions displayed by CV-I and CV-II (Fig. 58). However, *F. chathamensis* on CV-III showed greater similitude with *F. gigantea* and (especially) *F. armillata* than with *F. cornuta* (Table 52). Examination of the available data for the inadequately represented *F. prisca* and the poorly known *F. newtoni* suggests that these forms also would be positioned in relative proximity to *F. cornuta* and *F. gigantea* on these canonical axes (Tables 30–33).

Informative dimensions recorded from skeletal specimens and patterns of differences associated with flightlessness across taxonomic groups permitted a canonical contrast of of complete skeletons of all 52 species of Rallidae sampled with respect to flight capacity, resolved a single discriminatory axis that incorporated 25 of 41 variables, and provided highly significant discrimination of species by flight class (Table 53; Fig. 59). Scores on this contrast, with several additional subfossil species plotted a posteriori), were only weakly correlated with body mass ($r = 0.29$; $P = 0.05$), that is, only 8% (R^2) of the variance in scores is shared by mean body masses (Table 53).

Relative positions of species on this contrast successfully placed radically derived, flightless taxa of diverse size (e.g., *Atlantisia rogersi, Cabalus modestus,* and *Aphanapteryx bonasia*) at one extreme and flighted, generally large species (flighted *Porphyrio, Fulica,* and *Gallinula*) at the other. Intervening taxa followed a largely intuitive order with respect to apomorphy, but only loosely suggestive of body size (Fig. 59). Coefficients for included variables were consonant with the contrasts commonplace within taxonomic subgroups, in which dimensions of the pectoral apparatus (especially antebrachium and sternum) were of opposite sign to other skeletal dimensions (Table 53). Noteworthy is that, with the exception of *Porzana tabuensis,* all taxa having mean scores greater than -3.5 were flightless, and that all of the large coots (*Fulica*) fell among other large, flighted taxa, with *F. chathamensis* having a mean score of 3.8 (Fig. 59). A canonical

TABLE 53. Standardized coefficients and summary statistics for flighted–flightless contrasts based on skeletal measurements stepwise-selected ($P < 0.05$) from 41 variables for 52 species and 97 species–sex groups of Rallidae. Taxa lacking sexual partitioning were included as single samples in contrast of species–sex groups (*Gallirallus pectoralis*-group, *Atlantisia rogersi*, *Gallinula comeri*, *G. sandvicensis*, *Fulica armillata*, *F. cornuta*, and *F. gigantea*), and adequately represented subfossil taxa were scored a posteriori (*Gallirallus dieffenbachii*, *Cabalus modestus*, *Aphanapteryx bonasia*, "*Atlantisia*" *elpenor*, and *Fulica chathamensis*).

Variable	Coefficients by grouping	
	Species*	Species–sex†
Maxilla length	−0.322	−0.298
Cranium, length	−0.315	−0.313
Maximal width	−0.214	−0.219
Height perpendicular to palate	0.111	—
Sternum, length of carina	0.557	0.393
Minimal depth of carina	0.242	0.268
Scapula scapus length	−0.121	—
Humerus, length	−0.242	−0.239
Intertubercular width of caput	0.224	0.216
Minimal width at midpoint	0.159	0.187
Maximal width at midpoint	0.137	—
Ulna, length	0.497	0.368
Minimal width at midpoint	—	0.120
Maximal width at midpoint	0.149	—
Carpometacarpus length	0.389	0.257
Digitalis majoris, phalanx proximalis length	0.152	0.192
Phalanx distalis length	−0.117	−0.150
Pelvis, interacetabular width	−0.210	−0.208
Femur, length	−0.137	−0.132
Width of caput	−0.366	−0.324
Minimal width at midpoint	−0.191	−0.245
Tibiotarsus, length	−0.239	−0.248
Minimal width at midpoint	−0.207	−0.205
Tarsometatarsus, length	—	0.161
Craniocaudal width at midpoint	−0.139	−0.095
Lateromedial width at midpoint	−0.172	−0.150
Digit III length	0.497	0.467
Eigenvalue	13.11†	14.80†
Canonical R	0.964	0.968
Correlation with body mass‡	0.287	−0.068

* Wilks' λ (d.f. = 25, 1, 997) < 0.0708; $\hat{F}_{25, 973} = 510.4$ ($P \ll 0.0001$).
† Wilks' λ (d.f. = 23, 1, 950) < 0.0633; $\hat{F}_{23, 928} = 597.1$ ($P \ll 0.0001$).
‡ $n = 48$ for species group; $n = 90$ for species–sex groups.

contrast of complete skeletons in which groups were partitioned both by species and sex (Table 53) resolved a similar ordination of taxa. A notable difference was the finding that males of most species tended to have lower scores (i.e., were closer to flightless taxa) than their female conspecifics, with the sexes for some taxa (e.g., *Gallirallus australis*-group and *Tricholimnas sylvestris*) being separated from each other by one or more groups from other taxa.

Canonical contrasts of all members of the family Rallidae by flight capacity, based solely on the basis of sternal dimensions, provided similar insights. Whether for species groups or species–sex groups, these comparisons emphasized the prominence of the carina sterni relative to other dimensions of the element (Table

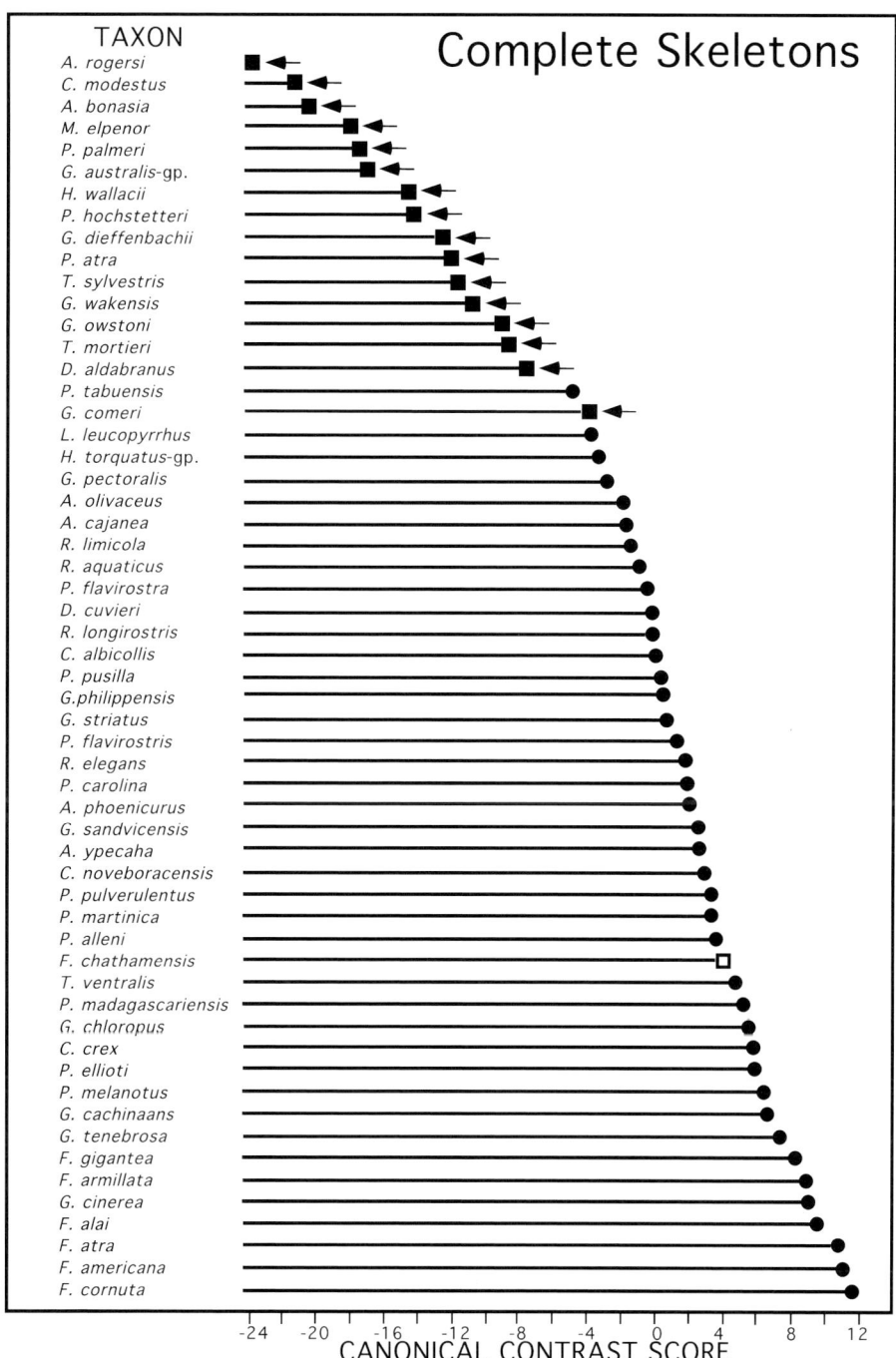

Fig. 59. Dot chart of mean scores for 52 species of Rallidae on a global canonical contrast of flighted and flightless taxa for a stepwise-selected subset of 41 skeletal measurements (five additional species plotted a posteriori). Flighted species are symbolized by circles, flightless species are symbolized by squares and emphasized by arrows, and taxon of uncertain flight status is shown as hollow squares.

TABLE 54. Standardized coefficients and summary statistics for flighted–flightless contrast based on sternal measurements stepwise-selected ($P < 0.05$) from five variables for 52 species groups and 97 species–sex groups of Rallidae. Taxa lacking sexual partitioning were included as pooled samples (*Gallirallus pectoralis*-group, *Atlantisia rogersi*, *Gallinula comeri*, *Fulica armillata*, *F. cornuta*, and *F. gigantea*), and adequately represented subfossil taxa were scored a posteriori (*Porphyrio mantelli*, *Eulabeornis castaneoventris*, *Rallus ibycus*, *R. recessus*, *Gallirallus dieffenbachii*, *Habropteryx okinawae*, *Cabalus modestus*, *Capellirallus karamu*, *Aphanapteryx bonasia*, *Erythromachus leguati*, *Diaphorapteryx hawkinsi*, "*Atlantisia*" *elpenor*, *Porzana severnsi*, *Amaurornis ineptus*, *Gallinula nesiotis*, *Tribonyx hodgenorum*, *Fulica chathamensis*, *F. prisca*, and *F. newtoni*).

	Coefficient by grouping	
Sternal variable	Species*	Species–sex‡
Length of carina	1.138	0.905
Length of corpus	−0.533	−0.412
Minimal width of corpus	−0.253	−0.272
Oblique depth of carina	0.145	0.222
Minimal depth of carina	0.431	0.495
Eigenvalue	6.01	5.38
Canonical R	0.926	0.918
Correlation with body mass‡	0.612	0.548

* Wilks' λ (d.f. = 5, 1, 1,057) < 0.1426; $\hat{F}_{5,1,053}$ = 1,266.1 ($P \ll 0.0001$).
† Wilks' λ (d.f. = 5, 1, 1,008) < 0.00001; $\hat{F}_{5,1,004}$ = 1,080.5 ($P \ll 0.0001$).
‡ n = 48 for species groups; n = 94 for species–sex groups.

54). Reduced dimensionality permitted the inclusion of a number of other extinct taxa in the plots, notably the extremely derived taxa *Porzana severnsi*, *Capellirallus karamu*, and *Rallus ibycus* (Fig. 60). Positions of taxa on this intuitively appealing ordination were consistent with other ordinations. Reduced informativeness of these anatomically restricted contrasts, in part, led to an increased confounding with mean body mass ($r = 0.61$; $P < 0.01$), in which 38% (R^2) of the variance in sternal scores was explainable simply by body size (with flightless species tending to be smaller than flighted species in the sampled taxa for which body masses were available).

As for skin specimens, examination of effects of interest in stepwise MANOVAs of skeletal variables confirmed highly significant differences among species, between sexes, or associated with loss of flight (Table 55). Not unexpectedly, differences among species tended to exceed in magnitude those between sexes (Table 55). Although the number of genera assessed for flight effects was reduced for skeletons, substantial effects attributable to the loss of flight were confirmed for *Porphyrio*, crakes, and *Tribonyx*. An exceptionally large estimate for interspecific or flight-related effects in *Dryolimnas* may be in part an inflation related to small sample sizes (Table 55).

Interaction effects in MANOVAs of skeletal measurements indicated that intersexual differences varied among species or between flight classes in all taxonomic subgroups, with especially notable species–sex interaction effects documented for *Gallirallus* and allies, *Tribonyx*, and (cognizant of the caveat regarding sample sizes) *Dryolimnas* (Table 56). Substantial interactions between flight class and sex in skeletal measurements were detected for the typical rails and *Dryolim-*

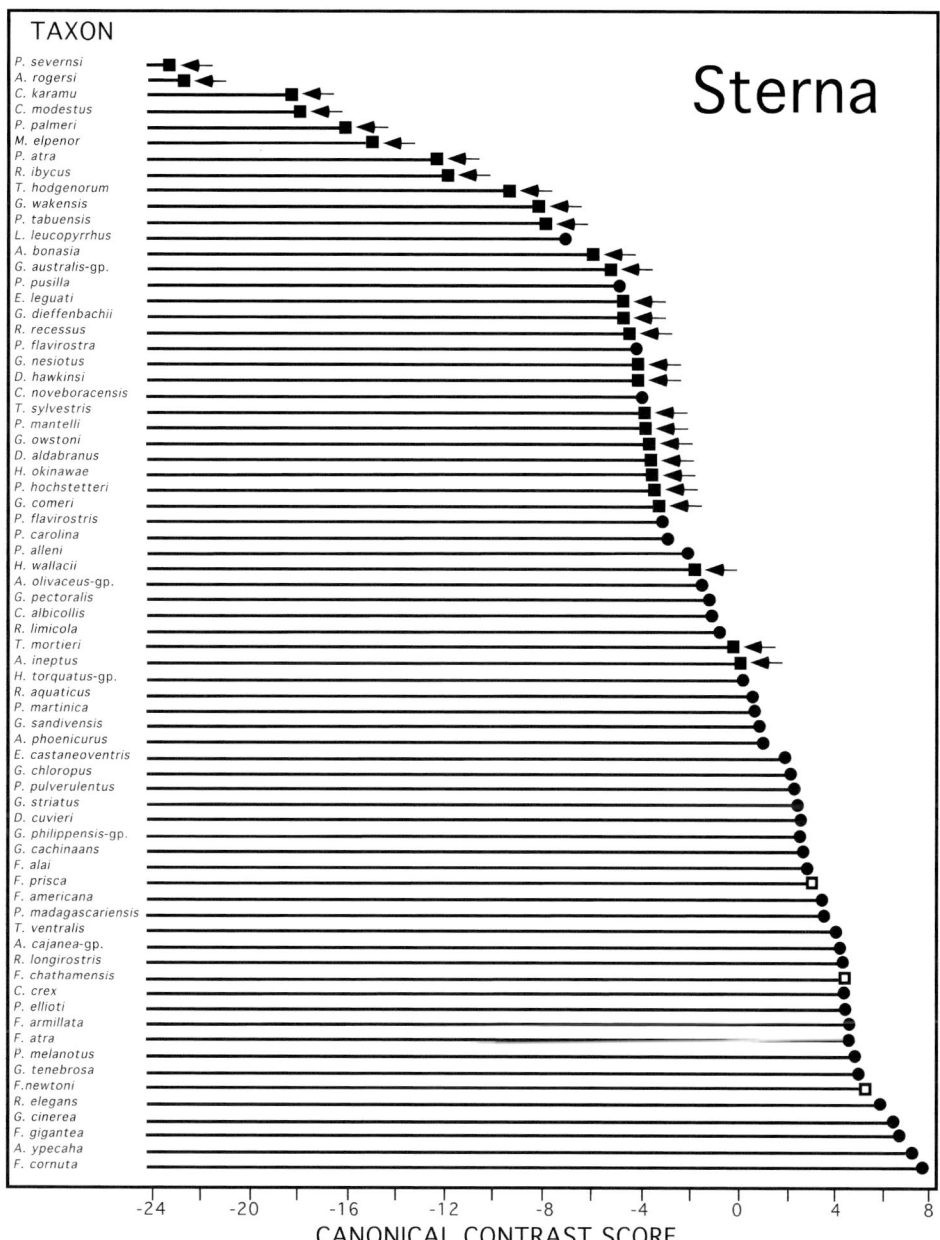

FIG. 60. Dot chart of mean scores for 52 species of Rallidae on a global canonical contrast of flighted and flightless taxa for a stepwise-selected subset of five sternal measurements (19 additional species plotted a posteriori). Flighted species are symbolized by circles, flightless species are symbolized by squares and emphasized by arrows, and taxa of uncertain flight status are shown as hollow squares.

TABLE 55. Summary statistics* for selected main effects for selected flightless species of Rallidae and flighted relatives based on stepwise multivariate analyses of variance ($P < 0.05$) of 41 osteological measurements. Approximate F-statistics correspond to Wilks' λ (see text).

Taxonomic group	Interspecific effect $\hat{F}_{v,\delta}$	P-value	Intersexual effect $\hat{F}_{v,\delta}$	P-value	Flight effect $\hat{F}_{v,\delta}$	P-value
Porphyrio and *Porphyrula*	$\hat{F}_{12, 123} = 556.1$	<0.0001	$\hat{F}_{7, 128} = 10.7$	<0.0001	$\hat{F}_{26, 119} = 1,685.0$	<0.0001
Porphyrio	$\hat{F}_{15, 84} = 541.3$	<0.0001	$\hat{F}_{2, 97} = 17.3$	<0.0001	$\hat{F}_{14, 85} = 1,471.1$	<0.0001
Habroptila and *Aramides*	$\hat{F}_{20, 41} = 115.4$	<0.0001	$\hat{F}_{6, 55} = 7.7$	<0.0001	$\hat{F}_{17, 44} = 119.2$	<0.0001
Dryolimnas	$\hat{F}_{9, 8} = 4,977.0$	<0.0001	$\hat{F}_{10, 7} = 97.5$	<0.0001	—*	—
Typical Rallidae†	$\hat{F}_{17, 270} = 113.2$	<0.0001	$\hat{F}_{9, 278} = 20.3$	<0.0001	$\hat{F}_{14, 273} = 349.7$	<0.0001
Gallirallus and close relatives	$\hat{F}_{16, 131} = 22.7$	<0.0001	$\hat{F}_{10, 137} = 10.0$	<0.0001	$\hat{F}_{12, 135} = 166.6$	<0.0001
Crakes (*Crecina*)	$\hat{F}_{10, 174} = 677.7$	<0.0001	$\hat{F}_{4, 180} = 13.2$	<0.0001	$\hat{F}_{8, 176} = 341.9$	<0.0001
Crex and *Porzana*	$\hat{F}_{8, 118} = 865.8$	<0.0001	$\hat{F}_{3, 123} = 16.0$	<0.0001	$\hat{F}_{15, 111} = 1,124.4$	<0.0001
Gallinula and *Tribonyx*	$\hat{F}_{10, 108} = 922.9$	<0.0001	$\hat{F}_{3, 115} = 10.7$	<0.0001	$\hat{F}_{10, 108} = 255.8$	<0.0001
Gallinula‡	$\hat{F}_{10, 48} = 67.2$	<0.0001	$\hat{F}_{3, 55} = 6.3$	<0.0001	$\hat{F}_{8, 50} = 64.7$	<0.0001
Tribonyx	$\hat{F}_{6, 55} = 2,531.3$	<0.0001	$\hat{F}_{4, 57} = 25.0$	<0.0001	—*	—
Fulica	$\hat{F}_{11, 69} = 29.4$	<0.0001	—§		$\hat{F}_{8, 88} = 109.2$	<0.0001

* Equivalent to interspecific effect in this context.
† *Rallus*, *Gallirallus*, and close allies.
‡ *Gallinula* includes single representatives of male and female "*Porphyriornis*" *comeri*.
§ Not possible because the "mega" species of *Fulica* treated as flightless were not adequately partitioned as to sex.

TABLE 56. Summary statistics for sex-related interaction effects of selected flightless species of Rallidae and flighted relatives based on stepwise multivariate analyses of variance ($P < 0.05$) of 41 osteological measurements. Approximate F-statistics correspond to Wilks' λ (see text).

Taxonomic group	Species–sex interaction $\hat{F}_{v,\delta}$	P-value		Flight–sex interaction $\hat{F}_{v,\delta}$	P-value
Porphyrio and *Porphyrula*	$\hat{F}_{2, 133} = 8.2$	<0.0005		$\hat{F}_{6, 129} = 5.5$	<0.0001
Porphyrio	$\hat{F}_{4, 95} = 4.3$	<0.005		$\hat{F}_{1, 98} = 2.2$	>0.10
Habroptila and *Aramides*	$\hat{F}_{4, 57} = 5.2$	<0.001		$\hat{F}_{22, 39} = 1{,}490.2$	<0.0001
*Dryolimnas**	$\hat{F}_{10, 7} = 126.9$	<0.0001	=	$\hat{F}_{10, 7} = 126.9$	<0.0001
Typical Rallidae†	$\hat{F}_{4, 283} = 9.2$	<0.0001		$\hat{F}_{4, 283} = 6.9$	<0.0001
Gallirallus and close relatives	$\hat{F}_{11, 136} = 15.2$	<0.0001		$\hat{F}_{4, 143} = 5.8$	<0.0005
Crakes (Crecina)	$\hat{F}_{1, 183} = 4.3$	<0.05		$\hat{F}_{1, 183} = 2.6$	>0.10
Crex and *Porzana*	$\hat{F}_{5, 121} = 4.5$	<0.001		$\hat{F}_{2, 124} = 3.1$	<0.05
Gallinula and *Tribonyx*	$\hat{F}_{3, 115} = 11.7$	<0.0001		$\hat{F}_{4, 114} = 7.4$	<0.0001
Gallinula‡	$\hat{F}_{8, 50} = 4.5$	<0.0005		$\hat{F}_{4, 54} = 5.7$	<0.001
Tribonyx	$\hat{F}_{10, 1} = 10.5$	<0.0001	=	$\hat{F}_{10, 1} = 10.5$	<0.0001
Fulica	$\hat{F}_{3, 77} = 5.7$	<0.005		—§	—

* Flight–sex interaction effect equivalent to species–sex interaction in this two-species context.
† *Rallus*, *Gallirallus*, and close allies.
‡ Samples of flightless *Gallinula* (*Porphyriornis*) *comeri* limited to single represents of each sex.
§ Not possible because the "mega" species of *Fulica* treated as flightless were not partitioned to sex.

nas, with the latter being suspiciously large (Table 56). Estimators of multivariate sexual dimorphism in species having minimally adequate samples (e.g., *Habroptila* not included) indicated a pattern of variation in dimorphism (Table 57) that differed in several important ways from that indicated by skin measurements (Table 26). Notable among the taxa permitting assessments by using both skin and skeletal data were *Porphyrio hochstetteri*, which showed comparatively high dimorphism in skin measurements not demonstrated in skeletal variables, *Dryolimnas aldabranus*, which showed suspiciously high sexual dimorphism in osteological dimensions but comparatively low dimorphism in external measurements, and *Porzana palmeri*, which had high sexual dimorphism in skeletal variables but only modest dimorphism in skin specimens.

Flightless species for which the two data sets indicated similar patterns in relative dimorphism were *Gallirallus owstoni* (dimorphism was comparatively high in both data sets), and *Gallirallus australis*-group and *Tribonyx mortierii* (with dimorphism in both species being unremarkable regardless of data set). Also notable was the exceptionally great sexual dimorphism shown by *Gallicrex cinerea* (Table 57), one of the few (flighted) species of Rallidae also possessing sexual dichromatism, a finding consistent with the estimate based on skin measurements (Table 26). In a few cases (e.g., *T. mortierii*), MANOVA documented highly significant flight–sex interaction effects (Table 56) but intersexual distances were of unremarkable magnitude (Table 57). These apparent discrepancies reflect the fact that the former incorporates differences in the direction of multivariate dimorphism (i.e., relative contributions of variables), whereas intersexual distances measure simple magnitude of dimorphism within each species.

QUALITATIVE CHARACTERS OF THE SKELETON

A number of skeletal apomorphies tended to be associated with flightlessness in rallids (Figs. 61–66), although flightless taxa manifested great variability in

TABLE 57. Multivariate sexual dimorphism in selected species of Rallidae, documented by Mahalanobis' distances (D) and statistics from associated multivariate analyses of variance, based on m variables stepwise-selected ($P < 0.05$) from 41 skeletal measurements in single-taxon analyses. Flightless taxa are marked by asterisks. Pillai's trace is numerically equal to the average squared canonical correlation in the two-group (unidimensional) case. Wilks' λ is numerically equal to the Hotelling–Lawley trace and Roy's maximum root (Morrison 1976) in the two-group (unidimensional) case. Approximate F-statistics correspond to Wilks' λ (U-statistic), with m (variables included in model), g (groups) $-$ 1, and n (total specimens) $-$ g degrees of freedom. Accuracy of allocation is the percentage of specimens of correct sex in jackknifed classifications.

Taxon	n Males	n Females	m	D	Pillai's trace	Eigenvalue	R	Wilks' λ	$\hat{F}$	Accuracy of classification
Porphyrio (p.) madagascariensis	9	9	9	16.2	0.99	73.84	0.993	0.013	65.64***	100
Porphyrio (p.) melanotus	33	20	16	7.3	0.93	12.95	0.963	0.072	29.13***	100
Porphyrio hochstetteri*	12	18	13	13.0	0.98	43.18	0.989	0.023	53.16***	100
Porphyrula martinica	15	15	18	19.2	0.99	99.05	0.995	0.010	60.53***	100
Aramides cajanea-group	27	23	23	8.1	0.94	16.86	0.972	0.056	19.06***	96
Aramides ypecaha	4	8	4	7.5	0.94	14.92	0.968	0.063	26.11**	100
Dryolimnas (c.) cuvieri	3	2	2	17.9	0.99	128.58	0.996	0.077	128.58*	100
Dryolimnas (c.) aldabranus*	10	5	7	33.2	0.99	282.40	0.998	0.004	282.40***	100
Rallus aquaticus	17	21	11	7.8	0.94	15.92	0.970	0.059	37.63***	100
Rallus limicola	21	18	10	8.2	0.95	17.52	0.973	0.054	49.05***	100
Rallus longirostris-group	17	17	15	11.0	0.97	32.19	0.985	0.030	38.63***	100
Rallus elegans-group	25	12	20	28.0	0.99	181.19	0.997	0.005	144.95***	100
Gallirallus australis-group*	36	30	12	4.8	0.86	5.88	0.924	0.145	25.97***	97
Gallirallus philippensis-group	16	19	9	5.2	0.88	7.07	0.936	0.124	19.64***	94
Gallirallus owstoni*	14	18	18	16.1	0.98	67.90	0.993	0.015	49.04***	100
Habropteryx torquatus	6	7	7	18.2	0.99	233.33	0.998	0.004	166.66***	100
Laterallus leucopyrrhus	7	20	16	22.8	0.99	108.19	0.995	0.009	67.62***	100
Coturnicops noveboracensis	15	20	21	23.7	0.99	145.77	0.997	0.007	90.24***	100
Crex crex	13	11	9	9.1	0.96	22.24	0.978	0.043	34.60***	100
Porzana palmeri	6	6	7	26.3	0.99	207.79	0.998	0.005	118.74***	100
Porzana carolina	15	15	12	15.9	0.99	67.37	0.993	0.015	95.44***	100
Porzana pusilla	9	12	7	4.5	0.84	5.38	0.918	0.157	9.99**	91
Porzana tabuensis	5	6	5	18.8	0.99	107.16	0.995	0.009	107.16***	100

TABLE 57. Continued.

Taxon	n		m	D	Pillai's trace	Eigenvalue	R	Wilks' λ	F̂	Accuracy of classification
	Males	Females								
Porzana flavirostra	17	18	20	19.9	0.99	104.40	0.995	0.009	73.08***	100
Amaurornis olivaceus-group	8	6	2	4.2	0.83	5.00	0.913	0.167	27.51***	100
Amaurornis phoenicurus	16	15	18	12.1	0.98	39.34	0.988	0.025	26.22***	100
Gallicrex cinerea	8	2	4	34.1	0.99	232.07	0.998	0.004	290.08***	100
Tribonyx ventralis	24	19	23	14.3	0.98	52.97	0.991	0.019	43.76***	100
Tribonyx mortierii*	10	11	11	13.4	0.98	49.73	0.990	0.020	40.69***	100
Gallinula tenebrosa	6	8	6	12.0	0.98	41.04	0.988	0.024	47.88***	100
Gallinula chloropus-group	22	27	4	2.8	0.53	1.11	0.726	0.473	12.25***	87
Fulica gigantea	6	4	4	7.8	0.95	18.19	0.974	0.052	22.74**	100
Fulica (a.) americana	17	15	13	8.6	0.95	19.54	0.975	0.049	27.05**	100
Fulica (a.) alai	7	9	6	19.3	0.99	104.48	0.995	0.009	156.71***	100
Fulica atra, atra-group	15	22	18	12.0	0.97	36.99	0.987	0.026	36.99***	100

† Asterisks reflect levels of significance: *, $P \leq 0.01$; **, $P \leq 0.0005$; ***, $P \leq 0.0001$.

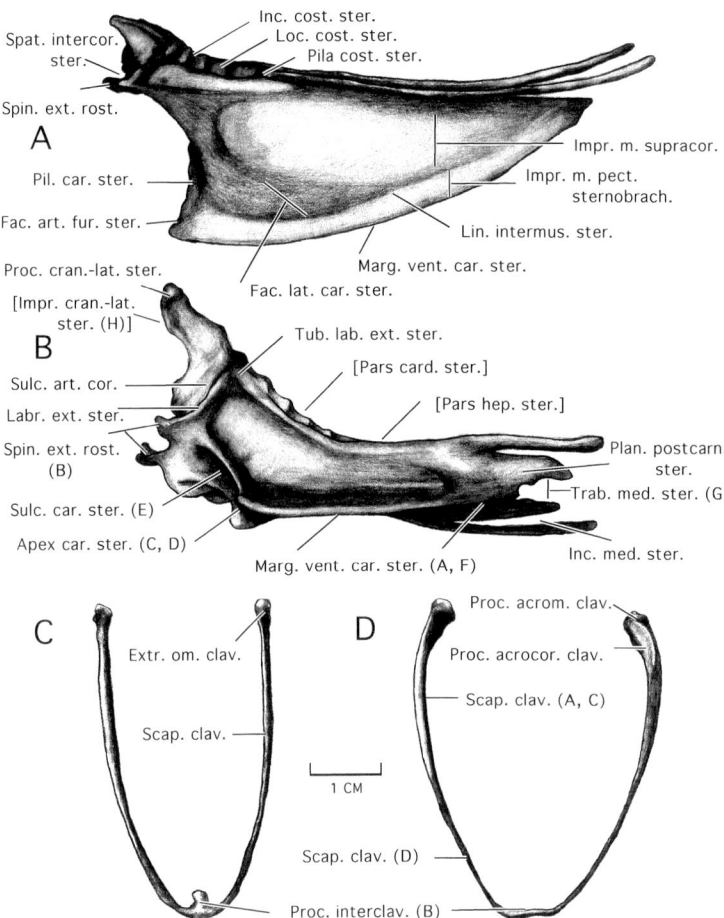

FIG. 61. Illustrations of sterna and furculae of *Rallus longirostris* (CM 11353) and *Gallirallus australis* (USNM 19021), with flightlessness-related apomorphies indicated (letters corresponding to Tables 56 and 57): **A**, sternum of *R. longirostris* (lateral aspect of facies muscularis; facies visceralis concealed above); **B**, sternum of *G. australis* (lateroventral oblique aspect of facies muscularis); aspect furcula (cranial aspect); **C**, furcula of *R. longirostris* (cranial aspect); and **D**, furcula of *G. australis* (cranial aspect). See Appendix 3 for explanation of abbreviations for anatomical features (those in brackets are concealed from view). By S. Townsend.

these features (Tables 58–64). A number were shared as well by the flightless gruiform *Aptornis,* although problems of comparability attended coding a number of features for this perhaps uniquely apomorphic genus, a member of the sister-group of the suborder Grues (Livezey 1998).

Nine qualitative characters of the sternum were found in at least one species of flightless rallid (Figs. 61, 62), ranging from several features of the carina sterni that were found in almost all flightless species (notably excluding subfossil *Fulica*) to a feature of the corpus sterni unique to *Porzana severnsi* (Table 58). One of these, a deepening of impressio m. sternocoracoidei on the processus craniolateralis of the corpus sterni (Table 58: character H), is confounded in some flightless rallids by a disproportionate elongation of the processus craniolateralis itself.

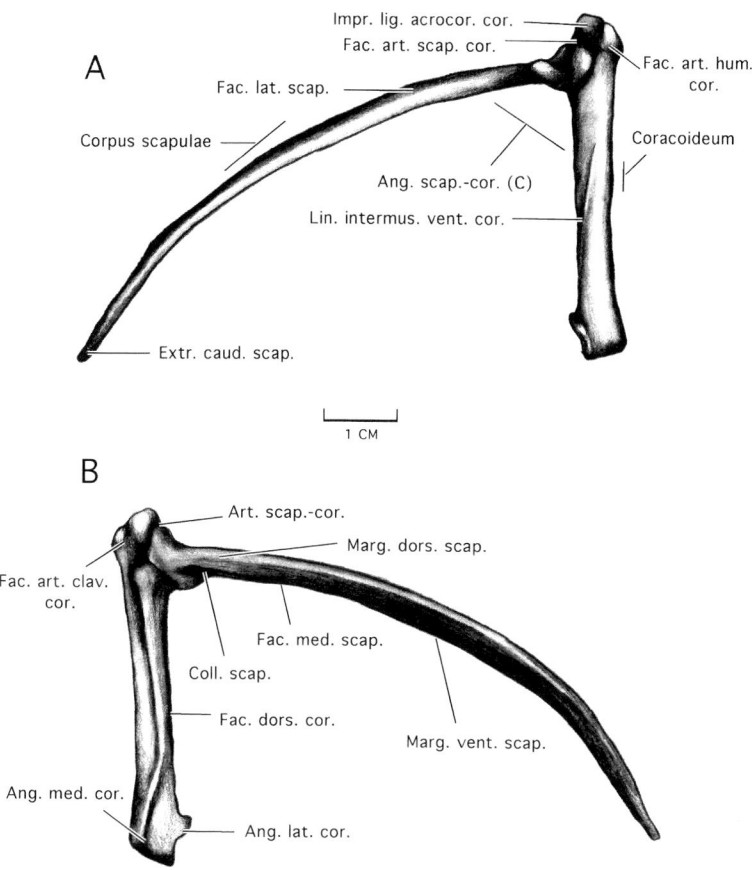

FIG. 62. Right coracoid and scapula of *Rallus longirostris* (CM 11353) in articulation: **A**, lateral aspect; and **B**, medial aspect. Acute angulus scapulocoracoidei (Table 60) of flighted rallids shown (letter corresponding to Table 58). See Appendix 3 for explanation of abbreviations for anatomical features. By S. Townsend.

Neither of these apomorphies were reflected in the modest suite of measurements analyzed in the morphometric comparisons (e.g., Tables 31, 40–42, 46–54; Figs. 49–60). General aspects of the sterna of comparatively derived flightless rails are distinguished by a combination of three primary conformational shifts: extension of the processus craniolateralis; a retreat of the carina sterni (indicated by both a reduction in depth and a caudal displacement of the apex carina); and a truncation of the caudal margin of the element, acting especially on the central margo caudalis of corpus sterni and disproportionately less on both processes caudolaterales (Fig. 67).

None of these characters were found to be unequivocally diagnostic of flightlessness in rallids, that is, none were shared by all flightless rallids and found in no flighted confamilial (Livezey 1998). To the extent that the sternum of comparatively distantly related *Aptornis* could be compared with those of members of the Rallidae, the former showed substantial but not extreme apomorphy of these sternal features (Table 58). This interspecific range of sternal apomorphy identified several taxa showing extreme degrees of osteological changes that

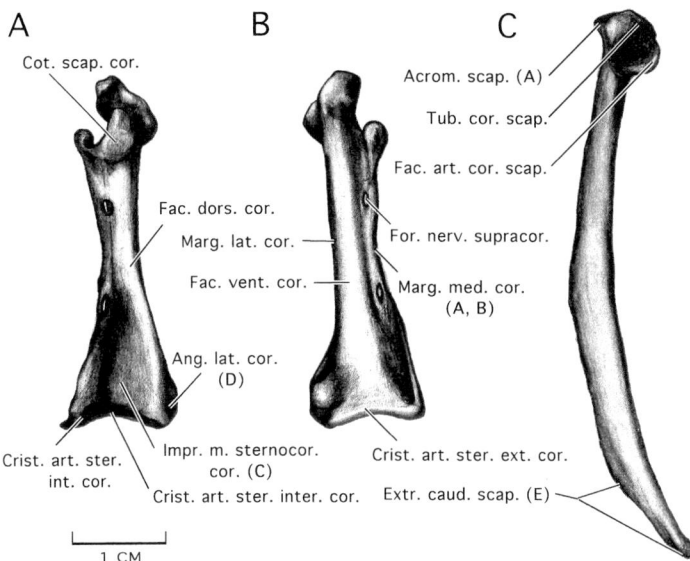

FIG. 63. Right coracoid and scapula of *Gallirallus australis* (USNM 19021) with flightlessness-related apomorphies indicated (letters corresponding to Tables 58 and 59): **A,** coracoideum (dorsal aspect); **B,** coracoideum (ventral aspect); and **C,** scapula (lateral aspect). See Appendix 3 for explanation of abbreviations for anatomical features. The tuberculum tendo proximalis m. expansor secundariorum (Table 58, character B) is not evident. By S. Townsend.

would be mirrored by other pectoral elements (e.g., *Cabalus modestus, Capellirallus karamu,* and *Atlantisia rogersi*), as well as several showing minimal change (e.g., subfossil *Fulica*). The three endemics from Bermuda (*Rallus ibycus, R. recessus,* and *Porzana piercei*) showed moderate to strong apomorphy of the sternum (Table 58). However, with the exception of apomorphies of the humerus, these three extinct rallids show only minor qualitative changes in other pectoral elements (below). Alternative comparative methods applied to sterna include a canonical contrast of measurements (Table 54; Fig. 60) and a broader suite of diagrams of the element for rallids (Fig. 67).

Five such features of the furcula (Fig. 61) were tallied for flightless rallids, all of which were possessed by a substantial majority of the species as well as *Aptornis* (Table 59). Three of five apomorphies of the scapula (Figs. 62, 63) were shared by the majority of flightless rallids and *Aptornis* (Table 60), with two being restricted to flightless *Porphyrio* and *Diaphorapteryx* and noncomparable in *Aptornis*. Flightlessness-related apomorphies of the coracoid (Figs. 62, 63) were distributed much more sparsely among rallids; only *Capellirallus karamu* was typified by all four. Few of these features were comparable in the uniquely conformed coracoid of *Aptornis* (Table 61).

Enlargement of the angulus coracoscapularis (Table 62; Fig. 62), long recognized in the ratites and many flightless carinates (e.g., Owen 1879; Fürbringer 1888, 1902; Knöpfli 1918; Nauck 1930a; Kaelin 1941; Livezey 1989d, 1992a) and inferred by some to be paedomorphic in rails (e.g., Olson 1973a), was confirmed as indicative of flightlessness in a number of rallids, especially those showing profound anatomical changes in general (e.g., *Porphyrio hochstetteri, Gallir-*

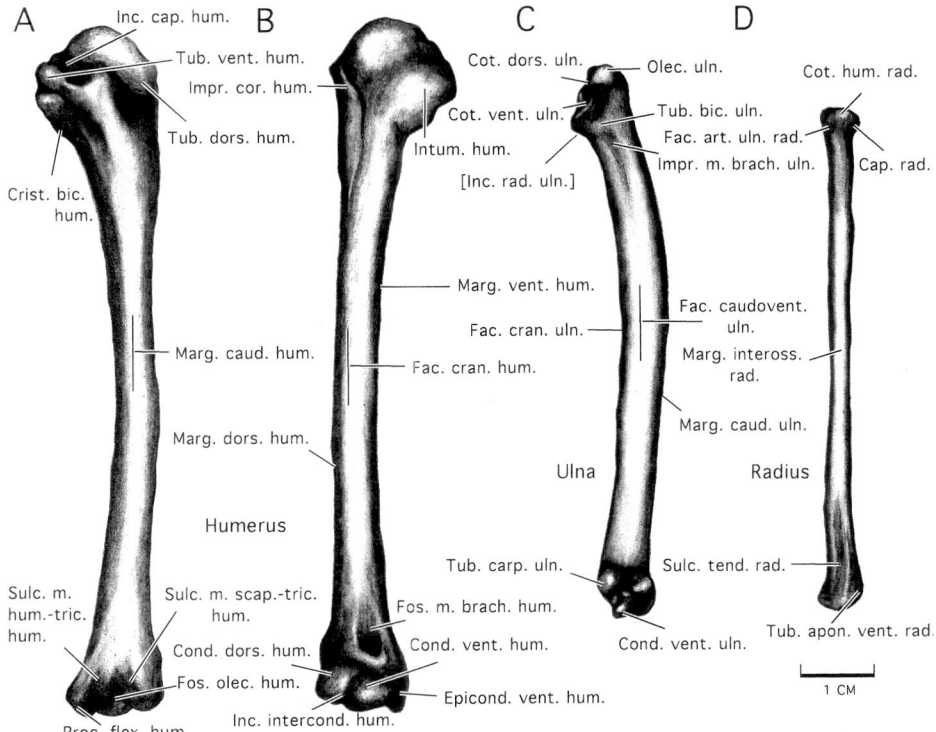

FIG. 64. Right humerus, ulna, and radius of *Rallus longirostris* (CM 11353) with flightlessness-related apomorphies indicated (letters corresponding to Tables 61 and 62): **A,** humerus (caudal aspect); **B,** humerus (cranial aspect); **C,** ulna (ventral aspect); and **D,** radius (dorsal aspect). See Appendix 3 for explanation of abbreviations for anatomical features. By S. Townsend.

allus australis-group, *Atlantisia rogersi*, and *Porzana palmeri*). Angles exhibited by flightless rallids formed a virtual continuum of states from acute to obtuse. However, a number of other flightless rallids showed such modest changes as to prevent confident assignment to one of the two broad apomorphic classes of angles (right and obtuse). The pectoral girdles of many more taxa were not preserved in articulation and precise reconstructions were not feasible (e.g, most subfossil species and many modern species for which articulated skeletons were not available). Nonetheless, the tendency toward enlargement of this angle in flightless lineages was sustained, especially among those taxa showing both pectoral reduction and increased body size (Table 62).

Qualitative apomorphies of the humerus observed among flightless rallids (Figs. 64, 65) were comparatively numerous (15 features; Table 63), and included characters found in almost all species assessed (e.g., reduced depth of fossa pneumotricipitalis, and increased depth of fossa m. brachialis) to several that were limited to relatively few taxa (e.g., obsolesence of cristae pectoralis et bicipitalis, and marked angularity of proximal surface of facies caudalis of corpus). Where comparable, most of these apomorphies also characterized *Aptornis* (Table 63). As in most other apomorphies of skeletal elements, several taxa were exceptional in showing almost all of the features (some to an extreme degree, e.g., *Cabalus*

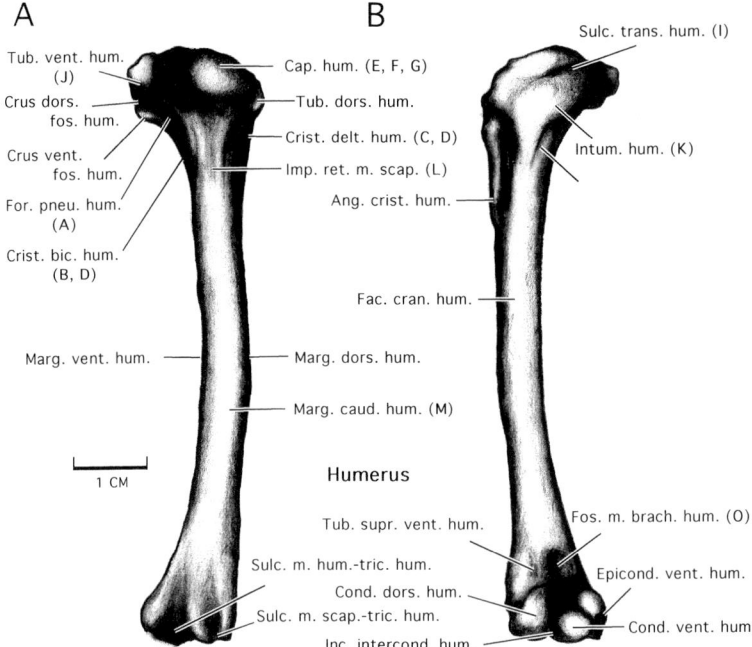

FIG. 65. Right humerus of *Gallirallus australis* (USNM 19021) with flightlessness-related apomorphies indicated (letters corresponding to Table 61): **A,** humerus (caudal aspect); and **B,** humerus (cranial aspect). See Appendix 3 for explanation of abbreviations for anatomical features. By S. Townsend.

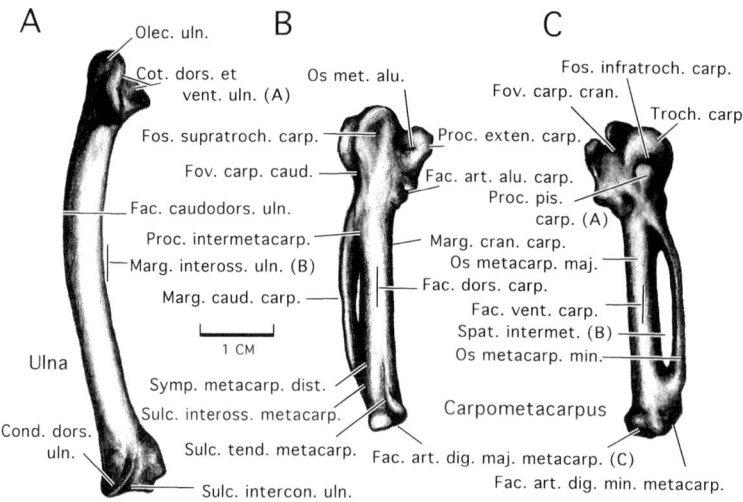

FIG. 66. Right ulna and carpometacarpus of *Gallirallus australis* (USNM 19021) with flightlessness-related apomorphies indicated (letters corresponding to Table 62): **A,** ulna (dorsal aspect); **B,** carpometacarpus (dorsal aspect); and **C,** carpometacarpus (ventral aspect). See Appendix 3 for explanation of abbreviations for anatomical features. By S. Townsend.

TABLE 58. Apomorphies of sterna of flightless species of Rallidae and Aptornithidae, based on subjective comparisons with close, flighted relatives. Codes are absent (○), present (×), extreme (××), or not comparable (NC). Apomorphies are identified by corresponding letters.* Brackets indicate marginal conditions and dashes that specimens were not available. Sternal data were completely lacking for *Porphyrio kukwiedei, Nesotrochis debooyi, N. picapicensis, Porzana astrictocarpus, P. ziegleri, P. menehune, P. ralphorum, P. sandwichensis,* and *Tribonyx repertus.*

Species	Flightlessness-related apomorphy of sternum*								
	A	B	C	D	E	F	G	H	I
Porphyrio mantelli	×	×	×	○	×	○	×	○	○
Porphyrio hochstetteri	×	×	×	○	×	○	×	○	○
Habroptila wallacii	×	×	×	×	×	×	×	×	○
Nesotrochis steganinos	×	×	×	—	○	—	—	×	○
Cyanolimnas cerverai	[×]	×	×	○	○	○	[×]	○	○
Dryolimnas aldabranus	×	○	[×]	○	×	○	○	○	○
Rallus recessus	×	×	×	○	×	—	○	—	○
Rallus ibycus	×	×	×	○	×	—	○	—	○
Gallirallus australis	×	×	×	×	×	×	×	○	○
Gallirallus greyi	×	×	×	×	×	×	×	○	○
Gallirallus dieffenbachii	×	×	×	○	×	○	×	○	○
Gallirallus owstonii	[×]	×	○	○	○	○	○	○	○
Gallirallus wakensis	×	×	×	○	○	○	○	○	○
Tricholimnas lafresnayanus	×	×	×	[×]	×	—	—	○	○
Tricholimnas sylvestris	×	×	×	○	×	○	×	○	○
Nesoclopeus poecilopterus	[×]	×	○	○	[×]	○	○	○	○
Cabalus modestus	×	××	××	NC	×	○	×	○	○
Capellirallus karamu	××	××	××	NC	×	×	×	○	○
Habropteryx okinawae	×	×	×	○	×	○	×	—	○
Aphanapteryx bonasia	×	×	×	○	×	—	—	○	○
Erythromachus leguati	×	×	×	○	×	○	×	○	○
Diaphorapteryx hawkinsi	×	×	×	×	×	×	×	×	○
Atlantisia rogersi	××	××	××	○	○	○	×	○	○
"Atlantisia" elpenor	×	×	×	×	×	×	×	×	○
Porzana piercei	×	×	×	×	×	×	×	×	○
Porzana palmeri	×	×	×	○	○	○	○	○	○
Porzana atra	×	×	×	○	○	○	○	○	○
Porzana keplerorum	×	×	×	×	NC	—	—	×	○
Porzana severnsi	×	×	×	×	[×]	×	×	○	×
Amaurornis ineptus	×	×	×	○	○	○	×	—	○
Pareudiastes pacifica	×	×	×	○	○	○	○	○	○
Gallinula nesiotis	×	×	×	○	○	○	○	○	○
Gallinula comeri	×	×	×	○	○	○	○	○	○
Tribonyx mortierii	[×]	○	[×]	○	[×]	×	[×]	○	○
Tribonyx hodgenorum	×	×	×	○	○	—	—	○	○
Fulica chathamensis	×	○	○	○	○	○	○	○	○
Fulica prisca	○	○	○	○	○	○	○	○	○
Fulica newtoni	○	×	○	○	○	○	○	○	○
Aptornithidae	×	×	×	×	○	×	×	×	○

* A, carina sterni reduced ventrally, margo ventralis straight; B, spina externa bifurcated to obsolete, associated with medial separation of sulci articulares coracoidei; C, carina sterni, apex carina, and margo cranialis dorsocaudally shifted; D, carina sterni, apex carina bifurcate; E, carina sterni, margo cranialis, sulcus carinae deepened; F, carina sterni, caudal terminus on facies muscularis relatively cranial, producing abbreviated planum postcarinale; G, corpus sterni, margo caudalis, and trabecula mediana poorly ossified, shortened; H, corpus sterni, processus craniolateralis, impressio m. sternocoracoidei extraordinarily deep, suggesting relatively large m. sternocoracoideus reducing kinesis of articulatio sternocoracoidale; I, corpus sterni, irregular fenestrae perforating basin, reflecting incomplete ossification.

TABLE 59. Apomorphies of claviculae of flightless species of Rallidae and Aptornithidae, based on subjective comparisons with close, flighted relatives. Codes are absent (○) or present (×). Apomorphies are identified by corresponding letters.* Brackets indicate marginal conditions and dashes that specimens were not available. Clavicular data were completely lacking for *Porphyrio kukwiedei, Nesotrochis* spp., *Tricholimnas lafresnayanus, Rallus ibycus, Habropteryx okinawae, Erythromachus leguati, Porzana astrictocarpus, P. piercei,* subfossil *Porzana* from Hawaii, *P. sandwichensis, Tribonyx repertus, T. hodgenorum,* and *Fulica newtoni.*

Species	Flightlessness-related apomorphies of claviculae*				
	A	B	C	D	E
Porphyrio mantelli	×	○	×	○	×
Porphyrio hochstetteri	×	○	×	○	×
Habroptila wallacii	×	×	×	×	×
Cyanolimnas cerverai	○	[×]	○	×	—
Dryolimnas aldabranus	○	○	○	○	○
Rallus recessus	○	○	○	○	○
Gallirallus australis	×	×	×	×	×
Gallirallus greyi	×	×	×	×	×
Gallirallus dieffenbachii	×	×	○	○	×
Gallirallus owstonii	○	[×]	○	○	○
Gallirallus wakensis	○	×	○	○	×
Tricholimnas sylvestris	×	×	×	×	×
Nesoclopeus poecilopterus	×	×	×	×	×
Cabalus modestus	×	×	×	×	×
Capellirallus karamu	—	×	×	×	×
Habropteryx okinawae	×	—	○	×	×
Aphanapteryx bonasia	×	×	○	○	×
Diaphorapteryx hawkinsi	×	×	×	×	×
Atlantisia rogersi	×	×	×	×	×
"Atlantisia" elpenor	×	×	×	×	[×]
Porzana palmeri	×	×	×	×	○
Porzana atra	[×]	○	○	○	○
Amaurornis ineptus	×	×	×	×	×
Pareudiastes pacifica	×	—	×	×	×
Gallinula nesiotis	○	○	×	○	×
Gallinula comeri	○	○	×	○	×
Tribonyx mortierii	○	×	○	[×]	○
Fulica chathamensis	○	×	○	×	×
Fulica prisca	○	×	○	×	×
Aptornithidae	×	—	×	×	×

* A, scapus claviculae, torsion or craniocaudal compression, especially dorsally; B, extremitas sternalis claviculae, reduction or loss of processus (dorsalis) interclavicularis; C, scapus claviculae, increase in lateral divergence, especially dorsally; D, scapus claviculae, reduction of width ventrally; E, scapus claviculae, reduction or loss of craniocaudal curvature.

modestus, Capellirallus karamu, Diaphorapteryx hawkinsi, Aphanapteryx bonasia, and *Atlantisia rogersi*), whereas other taxa showed comparatively few (e.g., *Dryolimnas aldabranus, Gallirallus owstoni,* and *Fulica newtoni*).

Flightlessness-related features of the more-distal alar elements comprised two features of the ulna and three of the carpometacarpus (Table 64; Fig. 66). The two ulnar characters were shown by roughly one third of the flightless rallids for which this element was available. Those taxa showing this element were among the taxa consistently showing qualitative apomorphies throughout the pectoral apparatus (e.g., *Porphyrio hochstetteri, Gallirallus australis*-group, *Cabalus mo-*

TABLE 60. Apomorphies of scapulae of flightless species of Rallidae and Aptornithidae, based on subjective comparisons with close, flighted relatives. Codes are absent (○), present (×), extreme (××), or noncomparable (NC). Apomorphies are identified by corresponding letters.* Brackets indicate marginal conditions and dashes that specimens were not available. Scapular data were completely lacking for *Porphyrio kukwiedei, Nesotrochis picapicensis, Tricholimnas lafresnayanus, Habropteryx okinawae, Aphanapteryx leguati, Porzana astrictocarpus,* subfossil *Porzana* from Bermuda and from Hawaii (exclusive of *P. severnsi*), *P. sandwichensis, Amaurornis ineptus, Tribonyx repertus,* and *Fulica newtoni.*

Species	Flightlessness-related apomorphies of scapula*				
	A	B	C	D	E
Porphyrio mantelli	○	×	×	○	×
Porphyrio hochstetteri	○	×	×	○	×
Habroptila wallacii	×	×	×	×	×
Nesotrochis debooyi	○	×	—	○	○
Nesotrochis steganinos	—	×	—	—	—
Cyanolimnas cerverai	○	×	×	○	○
Dryolimnas aldabranus	○	○	×	○	○
Rallus recessus	○	×	—	○	○
Rallus ibycus	○	×	—	○	○
Gallirallus australis	×	×	×	○	○
Gallirallus greyi	×	×	×	○	○
Gallirallus dieffenbachii	○	×	[×]	○	○
Gallirallus owstoni	○	×	×	○	○
Gallirallus wakensis	[×]	×	×	○	○
Tricholimnas sylvestris	[×]	×	×	○	○
Nesoclopeus poecilopterus	—	—	[×]	○	—
Cabalus modestus	×	×	×	○	○
Capellirallus karamu	×	×	×	○	○
Habropteryx okinawae	○	×	—	○	○
Aphanapteryx bonasia	×	×	×	○	○
Diaphorapteryx hawkinsi	[×]	×	×	×	×
Atlantisia rogersi	×	×	×	○	○
"*Atlantisia*" *elpenor*	○	×	×	○	○
Porzana piercei	○	×	—	○	○
Porzana atra	[×]	×	[×]	○	○
Porzana palmeri	○	×	×	○	○
Porzana severnsi	×	×	×	○	○
Pareudiastes pacifica	—	—	×	○	—
Tribonyx hodgenorum	[○]	×	—	○	○
Tribonyx mortierii	×	×	×	○	○
Gallinula nesiotis	○	○	×	○	○
Gallinula comeri	○	○	×	○	○
Fulica chathamensis	×	×	×	○	○
Fulica prisca	×	×	×	○	○
Aptornithidae	×	NC	××	NC	NC

* A, extremitas cranialis scapulae, acromion reduced, rounded; B, collum scapulae, modal reduction or loss of tuberculum m. expansor secundariorum; C, angulus coracoscapularis, enlargement from acute to obtuse; D, synostosis coracoscapularis (infrequent); E, corpus scapulae, modal "flanging" caudally.

destus, Capellirallus karamu, Aphanapteryx bonasia, Erythromachus leguati, and *Diaphorapteryx hawkinsi,* but not *Atlantisia rogersi*). Two of the three carpometacarpal apomorphies showed similar distributions, whereas one (partial occlusion of spatium intermetacarpale) was only typical of the sister-species *Porphyrio mantelli* and *P. hochstetteri* (Table 64).

TABLE 61. Apomorphies of coracoidea of flightless species of Rallidae and Aptornithidae, based on subjective comparisons with close, flighted relatives. Codes are absent (○), present (×), extreme (××), or noncomparable (NC). Apomorphies are identified by corresponding letters.* Brackets indicate marginal conditions and dashes that specimens were not available. Coracoidal data were completely lacking for *Porphyrio kukwiedei*, *Nesotrochis steganinos*, *N. picapicensis*, *Erythromachus leguati*, *Porzana astrictocarpus*, *P. piercei*, subfossil *Porzana* from Hawaii (exclusive of *P. severnsi*), *P. sandwichensis*, *Amaurornis ineptus*, and *Tribonyx repertus*.

Species	Flightlessness-related apomorphies of coracoidea*			
	A	B	C	D
Porphyrio mantelli	×	○	○	○
Porphyrio hochstetteri	×	○	○	○
Habroptila wallacii	○	○	○	○
Nesotrochis debooyi	○	○	×	○
Cyanolimnas cerverai	×	○	○	○
Dryolimnas aldabranus	○	○	×	○
Rallus recessus	○	○	○	○
Rallus ibycus	×	○	○	○
Gallirallus australis	×	○	○	○
Gallirallus greyi	×	○	○	○
Gallirallus dieffenbachii	○	○	○	○
Gallirallus owstoni	○	○	○	○
Gallirallus wakensis	○	○	×	○
Tricholimnas lafresnayanus	×	○	○	○
Tricholimnas sylvestris	×	○	○	○
Nesoclopeus poecilopterus	[×]	○	—	—
Cabalus modestus	×	○	×	○
Capellirallus karamu	×	×	××	×
Habropteryx okinawae	×	○	○	○
Aphanapteryx bonasia	×	○	○	○
Diaphorapteryx hawkinsi	×	○	○	○
Atlantisia rogersi	×	○	○	○
"*Atlantisia*" *elpenor*	○	○	○	○
Porzana piercei	○	○	○	○
Porzana palmeri	○	○	○	○
Porzana atra	○	○	○	○
Porzana severnsi	○	○	○	○
Pareudiastes pacifica	×	○	○	○
Tribonyx hodgenorum	×	○	○	○
Tribonyx mortierii	○	○	○	○
Gallinula nesiotis	○	○	○	○
Gallinula comeri	○	○	○	○
Fulica chathamensis	○	○	○	○
Fulica prisca	○	○	○	○
Fulica newtoni	○	—	○	—
Aptornithidae	NC	NC	NC	×

* A, corpus coracoidei, margo medialis bowed; B, extremitas sternalis coracoidei, margo medialis, "crista medialis" obsolete; C, extremitas sternalis coracoidei, facies dorsalis, impressio m. sternocoracoidei shallow (×) or virtually indistinguishable (××); D. extremitas sternalis coracoidei, margo lateralis, processus lateralis lacking.

INTERSPECIFIC SCALING BETWEEN APPENDICULAR CROSS SECTIONS AND LENGTHS

Whichever of the competing principles posited to explain interspecific trends in relative robustness of skeletal appendicular elements (e.g., symmorphosis, elasticity, or margins of safety), the profound reduction in stress experienced by pectoral elements of flightless birds predictably should be associated with reductions in strength of the structural constituents of the wing. General shortening and reductions in width of most flightless rallids, although diverse in magnitude (Tables 30–33), was verified directly, but interspecific trends in widths (or preferably estimates of cross section or circumference) relative to lengths required special techniques. Accordingly, a suite of traditionally familiar and widespread allometric analyses was employed, in which interspecific scaling of estimated cross-sectional areas with corresponding lengths of major appendicular elements were compared between flighted and flightless rails (Table 65).

However, patterns indicated by these assessments were not predictable from a superficial estimate of likely functional stresses and other parameters, which would lead to the comparatively intuitive expectation that flighted species should show scaling of appendicular robustness of the pectoral limb that is greater than that of flightless confamilials, whereas the opposite expectation would pertain to pelvic elements. Confounded in part by the impossibility of strict pairwise composition of the taxonomic samples in each class (e.g., the existence of a number of genera containing only flighted or flightless rails, several quite distinctive morphologically), as well as proportionately greater pure error, several elements did not conform with the simple predictions. Among the pectoral elements compared, scaling of robustness of humeri did not differ significantly between groups, and ulnae showed slightly higher scaling of robustness in flightless species than in flighted species (Table 65). Among pelvic elements subjected to this approach, the femur did not conform with expectations based on predicted stress, and showed slightly higher scaling of robustness in flighted species than in flightless (Table 65). If representative, these departures from logical expectation may indicate that the shortening of antebrachial elements undergone by flightless lineages led by ontogenetic canalization to disproportionately stout elements (at least in comparatively small and large species), and that the stress to which femora are subjected in takeoff and landing by flighted rails supersede ambulatory stresses in flightless rallids.

Pectoral Musculature

Descriptive Myology

Scope and emphasis.—Traditional descriptions of the origins, corpora, and insertions of each muscle of the pectoral girdle and limb of one or more exemplars of the rallids that were dissected, the vast majority of which would duplicate the accounts provided by Rosser (1980) for *Fulica,* were waived for this analysis. Instead, descriptive accounts of the pectoral musculature of rails dissected in this study are limited to differences apparently associated with the loss of flight, including comparisons with myological study of other flightless birds (e.g., Livezey 1990, 1992a, b); general differences among taxa that were not reported by Rosser (1980) for *Fulica americana* or McGowan (1986) for *Gallirallus australis,* with secondary reference to the works by Fisher and Goodman (1955) and Berger (1956a) on *Grus* and to Vanden Berge (1970) on the Ciconiiformes; anomalous variation within taxa; and differences between findings made from the present work and the descriptions by Lowe (1928a), Rosser (1980), and McGowan (1986). The accounts that follow generally treat proximal muscles before distal muscles,

TABLE 62. Angulae coracoscapulares of adequately represented species of Rallidae, allocated into one of three ranges: acute (<90°), approximately right (~90°), or obtuse (>90°). Brackets indicate estimate based on disarticulated elements; asterisks mark flightless species.

Taxa	Angulus coracoscapularis
Swamphens	
Porphyrio porphyrio-group	Acute
*Porphyrio mantelli**	Obtuse (slight)
*Porphyrio hochstetteri**	Obtuse (slight)
Porphyrula alleni	Acute
Porphyrula martinica	Acute
Porphyrula flavirostris	Acute
Wallace's Rail and allies	
Gymnocrex plumbeiventris	Acute
*Habroptila wallacii**	Right
Cave-rails and allies	
Eulabeornis castaneoventris	Acute
Aramides ypecaha	Acute
Aramides cajanea-group	Acute
Canirallus oculeus	Acute
Canirallus kioloides	Acute
Zapata Rail and allies	
Ortygonax sanguinolentus	Acute
Ortygonax nigricans	Acute
Pardirallus maculatus	Acute
White-throated rails	
Dryolimnas cuvieri	Acute
*Dryolimnas aldabranus**	Acute (approaching right)
Rallus sensu stricto	
Rallus aquaticus-group	Acute
Rallus limicola	Acute
Rallus longirostris-group	Acute
Rallus elegans-group	Acute
Gallirallus and allies	
Gallirallus pectoralis	Acute
Gallirallus muelleri	Acute
Gallirallus striatus	Acute
*Gallirallus australis**	Right (approaching obtuse)
*Gallirallus greyi**	Right (approaching obtuse)
Gallirallus philippensis-group	Acute
*Gallirallus owstoni**	Right (slight)
*Gallirallus wakensis**	Right
Habropteryx torquatus-group	Acute
Problematic subfossil rails	
*Aphanapteryx bonasia**	[Right or obtuse]
*Erythromachus leguati**	—
*Diaphorapteryx hawkinsi**	[Obtuse]
Atlantisia and allies	
*Atlantisia rogersi**	Right (approaching obtuse)
Laterallus jamaicensis	Acute
Laterallus leucopyrrhus	Acute
Coturnicops noveboracensis	Acute
Crex crex	Acute

TABLE 62. Continued.

Taxa	Angulus coracoscapularis
Gray crakes and allies	
Porzana palmeri*	Right (approaching obtuse)
Porzana carolina	Acute
Porzana albicollis	Acute
Porzana pusilla	Acute
Porzana tabuensis	Acute
Porzana monasa*	—
Porzana atra*	Approximately right
Porzana flavirostra	Acute
Waterhens	
Amaurornis olivaceus-group	Acute
Amaurornis phoenicurus	Acute
Amaurornis ineptus*	Approximately right
Moorhens and allies	
Gallicrex cinerea	Acute
Tribonyx ventralis	Acute
Tribonyx mortierii*	Right
Gallinula nesiotis*	Right
Gallinula comeri*	Right
Gallinula tenebrosa	Acute
Gallinula chloropus-group	Acute
Coots	
Fulica armillata	Acute
Fulica cornuta	Acute (slightly enlarged)
Fulica gigantea	Acute (slightly enlarged)
Fulica americana-group	Acute
Fulica atra	Acute

and consider superficial elements of the musculature before deeper elements. Throughout an effort was made to relate myological features to associated arthrological and osteological characters.

Essential comparisons of musculature.—Initial preparation for dissection underscored the disparity of gross appendicular robustnessness in extremely derived examples of flightless rails) e.g., *Atlantisia rogersi* [Fig. 68]) and foreshadowed corresponding diminution of the underlying pectoral musculature. The largest of the pectoral muscles (m. pectoralis pars sternobrachialis [thoracica]) manifested the most pronounced differences in robustness between flighted and flightless relatives on superficial examination (Figs. 69–72). Prime examples of well-differentiated muscles throughout the pectoral girdle and limb were those of migratory *Coturnicops noveboracensis* and (to lesser degree) *Crex crex,* especially relative to body size. In all rallids, origiones m. pectoralis were extensive, with the primary origo being from the facies sterni, despite the generality that rallids have comparatively small mm. pectoralis et supracoracoideus (Hartman 1961; Rayner 1988a).

Origo m. pectoralis also extended to the ventral portions of the costae sternales (Fig. 72) (pars costobrachialis of Simic and Andrejevic [1963, 1964] and adopted by Vanden Berge and Zweers [1993]) and the lateroventral margins of the claviculae (pars sternobrachialis [in part] of Simic and Andrejevic [1963, 1964]). The belly of m. pectoralis sensu lato also enveloped the ventral surfaces of the

TABLE 63. Apomorphies of humeri of flightless species of Rallidae and Aptornithidae, based on subjective comparisons with close, flighted relatives. Codes are absent (O), present (X), extreme (XX), or not comparable (NC). Apomorphies are identified by corresponding letters.* Brackets indicate deviations were marginal and dashes that specimens were not available. Qualitative humeral characters were completely lacking for *Nesotrochis picapicensis*, *Nesoclopeus poecilopterus*-group, *Habropteryx insignis*, *Porzana sandwichensis*, *Amaurornis ineptus*, *Pareudiastes pacificus*, *P. silvestris*, and *Tribonyx repertus*.

Species	A	B	C	D	E	F	G	H	I	J	K	L	M	N	O
Porphyrio mantelli	X	X	X	O	X	X	X	O	[X]	X	X	X	O	O	X
Porphyrio hochstetteri	X	X	X	O	X	X	X	O	[X]	X	X	X	O	O	X
Porphyrio kukwiedei	O	X	X	O	X	O	X	X	X	O	O	X	O	O	X
Habroptila wallacii	X	X	X	O	X	X	O	O	X	O	X	X	O	O	X
Nesotrochis debooyi	X	X	X	O	X	X	X	X	X	X	X	X	X	O	X
Nesotrochis steganinos	O	X	X	O	O	X	O	O	X	O	O	X	O	O	O
Cyanolimnas cerverai	X	X	X	O	X	X	O	X	[X]	O	X	X	X	O	[X]
Dryolimnas aldabranus	X	O	X	X	X	X	O	X	O	[X]	O	X	O	O	X
Rallus recessus	X	X	X	X	X	X	O	X	X	X	O	X	X	O	X
Rallus ibycus	X	X	X	X	X	X	X	X	X	X	X	X	O	O	X
Gallirallus australis	X	X	X	X	X	X	X	X	X	X	X	X	[X]	O	X
Gallirallus greyi	X	X	XX	X	X	X	X	X	X	X	X	X	[X]	O	X
Gallirallus dieffenbachii	X	X	X	X	X	X	X	X	X	O	O	X	X	O	X
Gallirallus owstoni	X	O	X	X	X	O	O	X	X	O	O	X	O	O	O
Gallirallus wakensis	X	X	O	X	X	[X]	O	X	X	O	O	X	O	O	X
Tricholimnas lafresnayanus	X	X	X	X	X	X	O	X	X	O	O	X	O	O	—
Tricholimnas sylvestris	X	X	X	X	X	X	O	X	X	O	X	X	O	X	X
Cabalus modestus	X	X	X	X	X	X	X	X	X	O	X	X	O	O	X
Capellirallus karamu	X	X	X	X	X	X	X	X	X	X	X	—	O	O	O
Habropteryx okinawae	X	X	X	X	[X]	X	O	X	[X]	X	X	—	O	O	X
Aphanapteryx bonasia	X	X	X	X	X	X	X	X	X	X	X	X	O	O	X
Erythromachus leguati	X	X	X	O	X	X	X	X	X	O	O	X	X	O	X
Diaphorapteryx hawkinsi	X	X	X	O	X	X	X	X	X	O	O	X	O	O	O
Atlantisia rogersi	X	X	X	O	O	O	X	X	O	X	O	X	X	O	X
"Atlantisia" elpenor	X	O	O	O	X	O	O	X	X	O	O	X	X	O	X
Porzana piercei	X	O	O	O	O	O	X	X	O	X	O	X	O	O	X
Porzana astrictocarpus	X	X	O	O	X	O	X	X	O	X	O	X	O	O	X
Porzana palmeri	X	X	X	O	X	O	X	O	O	O	O	X	O	O	X
Porzana atra	O	O	X	O	X	X	O	X	O	O	O	X	O	X	X
Porzana ziegleri	X	X	O	O	X	X	O	X	O	X	O	X	O	X	X

TABLE 63. Continued.

Species	Flightlessness-related apomorphies of humerus*														
	A	B	C	D	E	F	G	H	I	J	K	L	M	N	O
Porzana menehune	X	O	O	O	O	O	O	O	O	O	O	X	O	[X]	X
Porzana keplerorum	X	X	O	O	X	O	O	O	O	O	O	X	O	X	X
Porzana ralphorum	X	X	X	O	X	X	X	X	X	X	O	X	O	X	X
Porzana severnsi	X	X	O	O	X	O	X	O	O	O	O	O	O	O	O
Gallinula nesiotis	X	O	X	O	X	O	O	O	O	O	O	O	O	O	O
Gallinula comeri	X	O	X	O	X	X	O	O	O	O	O	X	O	O	O
Tribonyx hodgenorum	O	O	X	O	X	X	[X]	X	X	X	X	X	O	O	[X]
Tribonyx mortierii	X	O	X	O	X	X	O	O	X	O	X	X	O	O	O
Fulica chathamensis	X	O	O	O	X	X	X	O	X	O	X	X	O	O	[X]
Fulica prisca	X	O	O	O	X	O	X	O	X	O	[X]	X	O	O	[X]
Fulica newtoni	O	O	O	O	O	O	O	O	O	O	X	X	O	O	X
Aptornithidae	X	X	X	XX	XX	NC	X	NC	X	XX	X	X	XX	[O]	[O]

* A, extremitas proximalis humeri, fossa pneumotricipitalis, reduction in depth; B, extremitas proximalis humeri, crista bicipitalis, distal reduction; C, extremitas proximalis humeri, crista pectoralis, dorsal prominence reduced, but more prominent, notched, and thickened cranially; D, extremitas proximalis humeri, cristae pectoralis et bicipitalis obsolete; E, extremitas proximalis humeri, caput, foreshortened, flattened; F, extremitas proximalis humeri, caput, dorsally rotated relative to corpus humeri; G, extremitas proximalis humeri, caput, caudally deflected relative to corpus humeri; H, extremitas proximalis humeri, sulcus ligamentosus transversus, deep and extends farther dorsally; I, extremitas proximalis humeri, tuberculum ventrale, disproportionately large; J, extremitas proximalis humeri, tuberculum ventrale, proximally shifted, approximately equal to caput in proximal extent; K, extremitas proximalis humeri, crista bicipitalis and intumescentia, cranially bowed, comparatively concave; L, extremitas proximalis humeri, facies caudalis, impressio retinaculum m. scapulotricipitis deep, distinctly margined; M, corpus humeri, sharp, tending to subangular, dorsoventral curvature; N, corpus humeri, facies caudalis, marked angularity and dorsoventral compression proximally; O, extremitas distalis humeri, fossa m. brachialis, exceptionally deep.

TABLE 64. Qualitative changes in ulnae and carpometacarpi of flightless species of Rallidae and Aptornithidae, based on subjective comparisons with flighted, close relatives. Codes are absent (○), present (×), extreme (××), or noncomparable (NC), with direction indicated by associated sign. Apomorphies are identified by corresponding letters. Brackets indicate marginal conditions and dashes that specimens were not available.

| | Flightlessness-related apomorphies* | | | | |
| | Ulna | | Carpometacarpus | | |
Species	A	B	A	B	C
Porphyrio mantelli	×	×+	×	×	×
Porphyrio hochstetteri	×	×+	×	×	×
Porphyrio kukwiedei	×	—	—	—	—
Habroptila wallacii	×	×+	×	○	×
Nesotrochis debooyi	×	×+	×	○	×
Nesotrochis steganinos	×	—	[○]	○	○
Cyanolimnas cerverai	○	○	×	○	○
Dryolimnas aldabranus	○	○	○	○	×
Rallus recessus	○	○	○	○	○
Rallus ibycus	○	○	○	○	○
Gallirallus australis	×	×+	○	○	×
Gallirallus greyi	×	×+	○	○	×
Gallirallus dieffenbachii	○	○	○	○	×
Gallirallus owstoni	○	○	○	○	○
Gallirallus wakensis	○	○	○	○	○
Tricholimnas lafresnayanus	○	×+	—	—	—
Tricholimnas sylvestris	○	×+	×	○	×
Nesoclopeus woodfordi	○	○	[×]	○	○
Cabalus modestus	×	×	×	○	×
Capellirallus karamu	×	××−	×	○	×
Habropteryx insignis	○	×	○	○	○
Habropteryx okinawae	○	×−	○	○	○
Aphanapteryx bonasia	[×]	○	×	○	×
Erythromachus leguati	[×]	×	[○]	○	[○]
Diaphorapteryx hawkinsi	×	×−	×	○	×
Atlantisia rogersi	○	○	×	○	○
"Atlantisia" elpenor	○	○	×	○	×
Porzana piercei	○	○	○	○	○
Porzana astrictocarpus	—	—	×	[×]	×
Porzana palmeri	○	○	○	○	[×]
Porzana atra	○	○	○	○	[×]
Porzana ziegleri	○	○	○	○	○
Porzana menehune	○	×−	○	[×]	○
Porzana keplerorum	○	○	○	○	○
Porzana severnsi	○	○	○	○	○
Amaurornis ineptus	×	○	×	○	×
Tribonyx hodgenorum	×	×+	×	○	○
Tribonyx mortierii	×	×+	×	○	×
Gallinula nesiotis	○	○	○	○	○
Gallinula comeri	○	○	○	○	○
Fulica chathamensis	○	○	○	○	[×]
Fulica prisca	○	○	○	○	[×]
Fulica newtoni	○	○	—	—	—
Aptornithidae	××	××−	××	NC	NC

* Ulna: A, extremitas proximalis ulnae, processus cotylaris dorsalis and cotylaris ventralis disproportionately large; B, corpus ulnae, relative craniocaudal curvature great (+) or lessened (−). Carpometacarpus: A, extremitas proximalis carpometacarpi, processus pisiformis reduced; B, corpus carpometacarpi, spatium intermetacarpale, partial occlusion by osseus lamina (*Porphyrio (m.) hochstetteri*, NZNM 24057); C, extremitas distalis carpometacarpi, cranially directed curvature.

coracoid and mm. supracoracoideus et coracobrachialis caudalis (Figs. 71–72). However, in a minority of flightless species, reduction of m. pectoralis pars sternobrachialis (thoracica) extended to the skeletal elements from which origo m. pectoralis arose; for example, in *Nesoclopeus poecilopterus*, pars thoracica was confined largely to the sternum, barely extending onto the lateral surfaces of the ventral ends of the adjacent costae sternales. In all flightless species, m. pectoralis pars sternobrachialis (thoracica) was reduced in all dimensions to varying degrees, notably length, width, and depth of the belly, changes that were roughly proportional to reductions of the underlying sternum. M. pectoralis pars sternobrachialis (thoracica) was extremely thin in *Atlantisia rogersi*, but nonetheless extended laterally to include the ventral extremities of the costae sternales.

Masses of the two major breast muscles (mm. pectoralis et supracoracoideus) displayed a predictable general pattern of relative reduction with the loss of flight in members of the Rallidae (Figs. 69–73). Whether based on the few published data for fresh specimens or estimates based on birds dissected in the present study, the proportion of mean body mass composed by mm. pectoralis et supracoracoideus in flighted rallids was approximately two to eight times (on average three to four) greater than that in flightless relatives (Tables 66, 67). The decrease in relative mass of breast muscles necessarily was most marked in those flightless species in which breast muscles were absolutely smaller, but mean body mass was larger, than in flighted relatives (e.g., *Porphyrio hochstetteri*, *Habroptila wallacii*, *Cyanolimnas cerverai*, *Gallirallus owstoni*, *Tricholimnas sylvestris*, *Nesoclopeus poecilopterus*, *Porzana atra*, *Tribonyx mortierii*, and *Gallinula comeri*). Smaller decreases in relative mass of breast muscles were indicated in flightless lineages characterized by larger body size and absolutely larger (but relatively smaller) breast muscle masses, that is, those with pronounced negative allometry, but absolute increases in mass of breast muscles pertains (e.g., *Gallirallus australis*-group, *Habropteryx okinawae*, and *Amaurornis ineptus*). Perhaps most notable were the decreases in relative masses of breast muscles indicated for several flightless taxa having small overall body masses than their flighted relatives (i.e., those in which dwarfism is inferred), in that these taxa showed substantial decreases in the former despite an overall reduction in body size that obscured to some extent the diminution of breast musculature (e.g., *Dryolimnas aldabranus*, *Gallirallus wakensis*, and probably *Atlantisia rogersi*).

The pectoralis complex also included a distinct pars subcutanea thoracica and variably differentiated, generally indistinct pars propatagialis in all rallids examined (Figs. 69–71). For example, pars subcutanea thoracica was well developed and extended 35 mm obliquely lateroventrad in flightless *Nesoclopeus poecilopterus*. The fibers that act primarily on tendo propatagialis, nominally constituting pars propatagialis, formed a vaguely distinguishable portion of the laterocranial extremity of m. pectoralis, confined to a region lateral to the furcula. As McGowan (1986) observed in *Gallirallus australis*, m. pectoralis pars propatagialis was only weakly discernible in most rallids, regardless of taxon or flight capacity (Fig. 74). For example, pars propatagialis was at least as well differentiated in *N. poecilopterus* and *Atlantisia rogersi* as in most flighted species dissected, although in *A. rogersi* the propatagial slip lacked a tendon entirely, with the insertio propatagialis comprising only fleshy fibers. By contrast, pars propatagialis was stout in *Pareudiastes pacificus* but small and distinctly tendinous in *Porzana palmeri*.

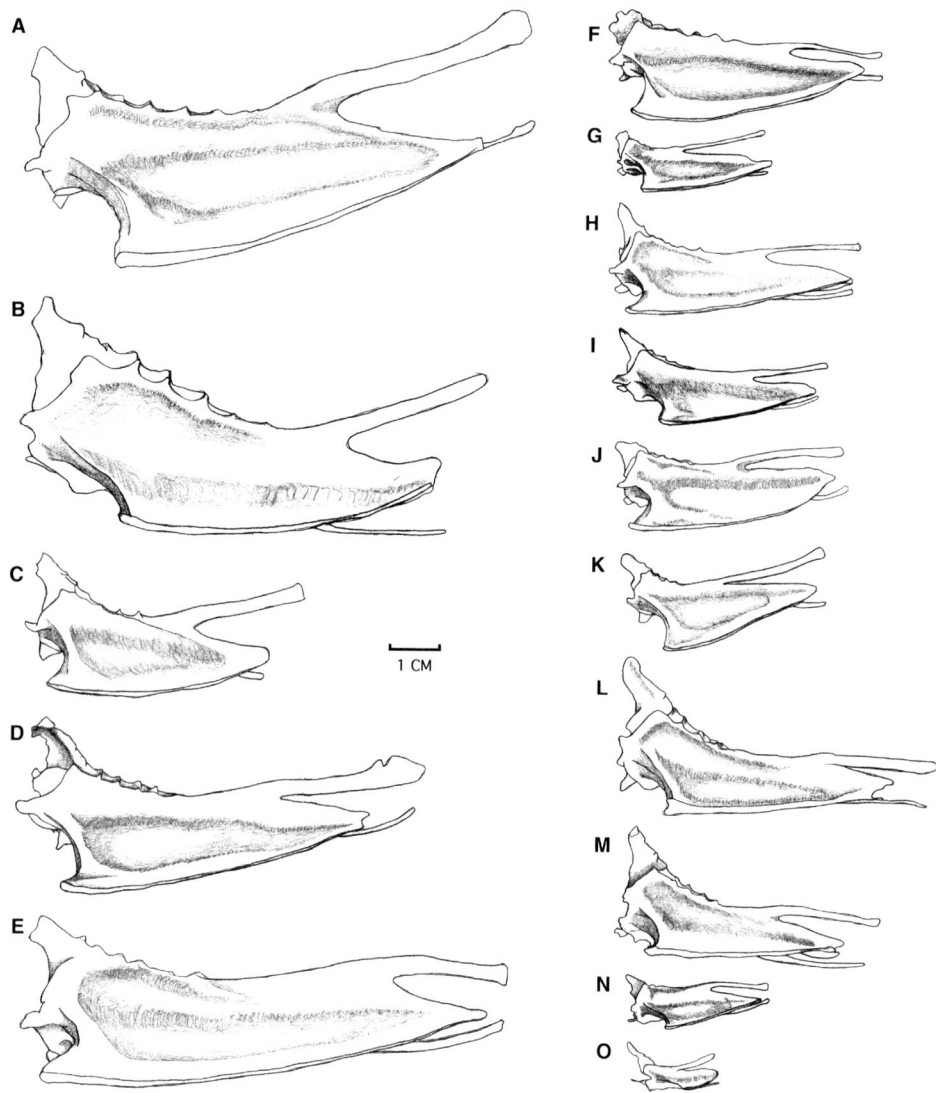

Fig. 67. Diagrammatic illustrations of sterna (corpus sterni) of selected species of Rallidae, lateroventral views: **A,** *Porphyrio melanotus* (BMNH 1872·10·25·25); **B,** *P. hochstetteri* (BMNH 1874·9·3); **C,** *Porphyrula alleni* (BMNH 1961·11·1); **D,** *Eulabeornis castaneoventris* (CSIRO 30); **E,** *Aramides ypecaha* (BMNH 1898·12·3·95); **F,** *Canirallus kioloides* (BMNH 1929·8·4·11); **G,** *Sarothrura elegans* (BMNH 1997·34·1); **H,** *Dryolimnas abbotti* (BMNH 1910·4·8·1); **I,** *D. aldabranus* (BMNH 1989·38·4); **J,** *Amaurolimnas concolor* (BMNH 1847·6·16·86); **K,** *Rallus aquaticus* (BMNH 1986·36·4); **L,** *Gallirallus australis* (BMNH 1896·2·16·40); **M,** *Tricholimnas sylvestris* (BMNH 1939·12·9·3705); **N,** *Laterallus leucopyrrhus* (BMNH 1967·1·3); **O,** *Atlantisia rogersi* (BMNH 1924·1·2·2); **P,** *Porzana pusilla* (BMNH 1897·5·10·2); **Q,** *P. palmeri* (BMNH 1972·1·63); **R,** *Amaurornis phoenicurus* (BMNH 1993·5·1); **S,** *Gallinula chloropus* (BMNH 1975·93·1); **T,** *G. (n.) nesiotis* (BMNH 1952·3·8); **U,** *G. (n.) comeri* (BMNH 1974·19·2); **V,** *Tribonyx ventralis* (BMNH 1966·50·7); **W,** *T. mortierii* (BMNH 1970·4·2); **X,** *Fulica americana* (BMNH 1977·81·3).

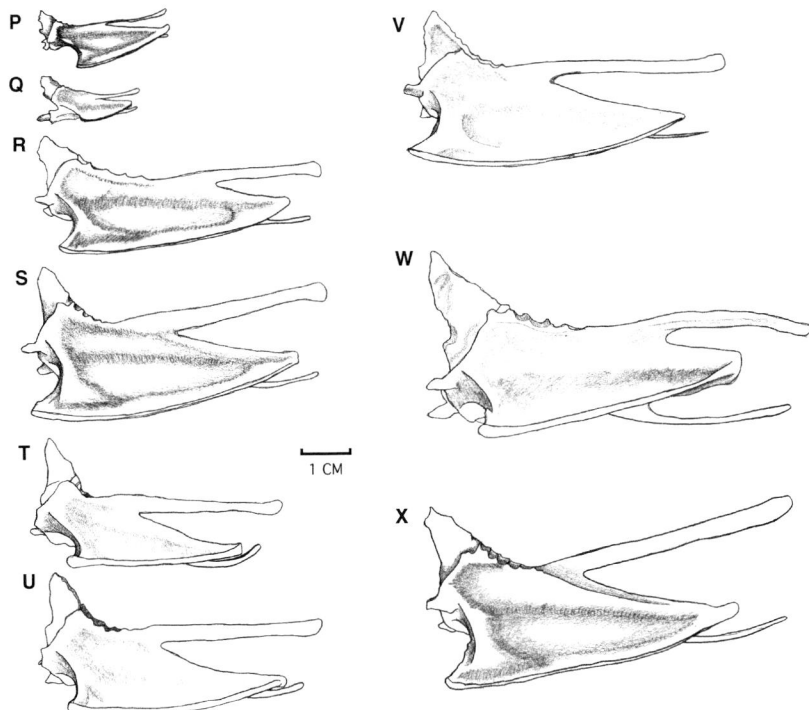

FIG. 67. Continued.

The second-largest pectoral muscle (m. supracoracoideus) showed negligible interspecific variation in the extent of the origin, with the substantial differences in bulk of the muscle instead corresponding to the extent of the membrana sternocoracoclavicularis craniomedially and the relative depth and length of the carina sterni (specifically impressio m. supracoracoideus) caudomedially (Figs. 72–74). M. supracoracoideus was weakly developed in *Gallirallus wakensis* and *Atlantisia rogersi*, with the condition in the former being especially striking in light of the comparatively larger body size of the species. However, in flightless and diminutive *Porzana palmeri* (Fig. 74), m. supracoracoideus was disproportionately large relative to the overlying m. pectoralis pars sternobrachialis (thoracica).

TABLE 65. Statistics of interspecific allometry ($Y = aX^b$) of estimated cross-sectional areas with lengths of major appendicular elements for flighted and flightless species of Rallidae.

Element	Flighted species ($n = 40$)		Flightless species ($n = 25$)		Intergroup difference	
	Intercept ($\hat{a}$)	Slope ($\hat{b} \pm$ SE)	Intercept ($\hat{a}$)	Slope ($\hat{b} \pm$ SE)	F-value	Significance (P)
Humerus	−5.95	2.02 ± 0.07	−6.95	2.19 ± 0.07	1.94	>0.15
Radius	−6.55	1.97 ± 0.12	−5.93	1.92 ± 0.13	11.52	<0.0001
Ulna	−5.14	1.88 ± 0.04	−5.08	1.92 ± 0.09	21.57	<0.00001
Femur	−8.14	2.43 ± 0.06	−7.60	2.37 ± 0.07	7.55	<0.005
Tibiotarsus	−7.18	2.08 ± 0.06	−8.25	2.28 ± 0.08	26.24	<0.00001
Tarsometatarsus	−5.52	1.98 ± 0.08	−6.42	2.19 ± 0.14	24.21	<0.00001

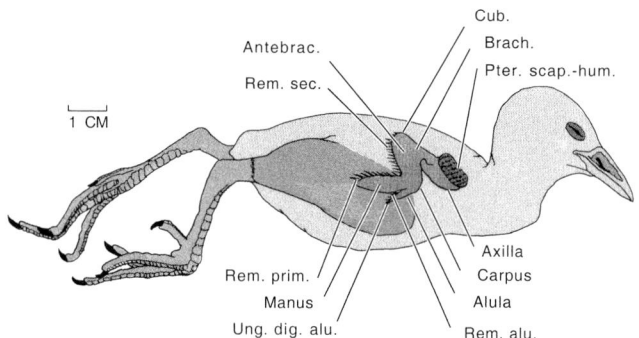

FIG. 68. Corpus of *Atlantisia rogersi* (DMNH 20762), lateral view with feathers removed, illustrating gross relative size of pectoral appendage. Anatomical abbreviations are defined in Appendix 3.

Perhaps the most striking aspect of the superficial breast musculature associated with flightlessness in rallids was the relative position of the articulatio sternoclavicularis (i.e., of the apophysis furculae and apex carina sterni) and the associated caudal extension and orientation of insertio m. cucullaris capitis pars clavicularis (Figs. 69–72, 73). In most flightless species having undergone significant reduction of the carina sterni, the apex carina was positioned caudad relative to the corpus sterni, with the pila carinae sloping caudomediad to the overlying musculature. In those taxa manifesting this condition, the articulatio sternoclavicularis was shifted correspondingly caudad, as were the bilaterally paired bellies of m. cucullaris capitis pars clavicularis (Figs. 69–72). This repositioning of skeletal and muscular structures in flightless rallids resulted in a variably striking ventral aspect in which the angulus defined by the furcula is comparatively acute and accentuated by the identically oriented m. cucullaris capitis pars clavicularis, with the latter producing a chevron-shaped aspect in ventral view (Figs. 69–72).

Dorsally, the superficial layers of muscles showed modest, but taxonomically

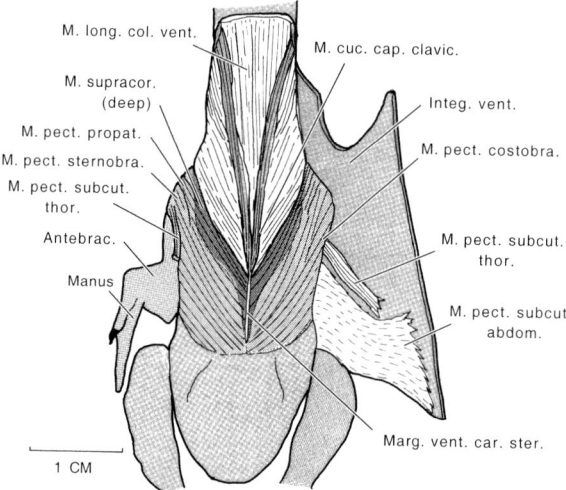

FIG. 69. Pectoral musculature of *Atlantisia rogersi* (DMNH 20762), breast, superficial layer, ventral view. Anatomical abbreviations are defined in Appendix 3.

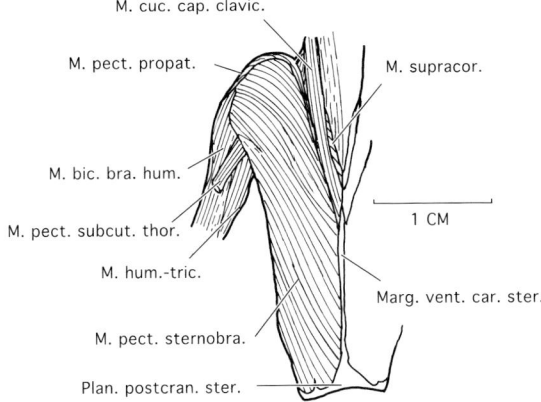

FIG. 70. Pectoral musculature of *Porzana palmeri* (BMNH 1940·12·8·13), breast, superficial layer, ventral view. Anatomical abbreviations are defined in Appendix 3.

varying alterations associated with the loss of flight (Figs. 75–87). Mm. latissimus dorsi partes cranialis et caudalis were thinner in flightless species, with less obvious narrowing of sheets as well. However, the specimen of *Gallirallus australis* dissected here was unusual, in that origiones mm. latissimus dorsi cranialis et caudalis were essentially continuous, lacking the distinct, variably broad separation between partes cranialis et caudalis found in all other rallids. The two bellies became distinct distally, with the insertions being of typical conformation. Other variations in this muscle complex included the apparently bipartite conformation of m. latissimus dorsi cranialis in *Habropteryx okinawae* (each having origins of 7 mm width). M. latissimus dorsi pars propatagialis was well developed in *Tricholimnas sylvestris* (origo 6 mm, insertio 9 mm). This specimen of *T. sylvestris* also included several dermal components craniad to m. latissimus dorsi propatagialis, which originated in the dermis of the laterocervical region and were interpreted to represent m. cucullaris capitis pars propatagialis. Similar proximal dermal components of the propatagium were detected in *Nesoclopeus poecilopterus*

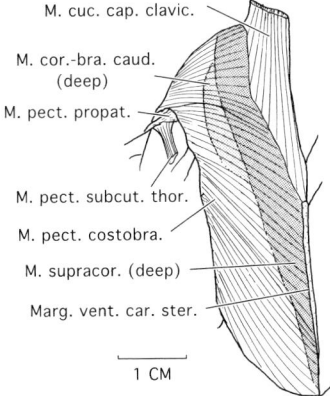

FIG. 71. Pectoral musculature of *Pareudiastes pacificus* (BMNH 1874·2·20·38), breast, superficial layer, dorsolateral view. Anatomical abbreviations are defined in Appendix 3.

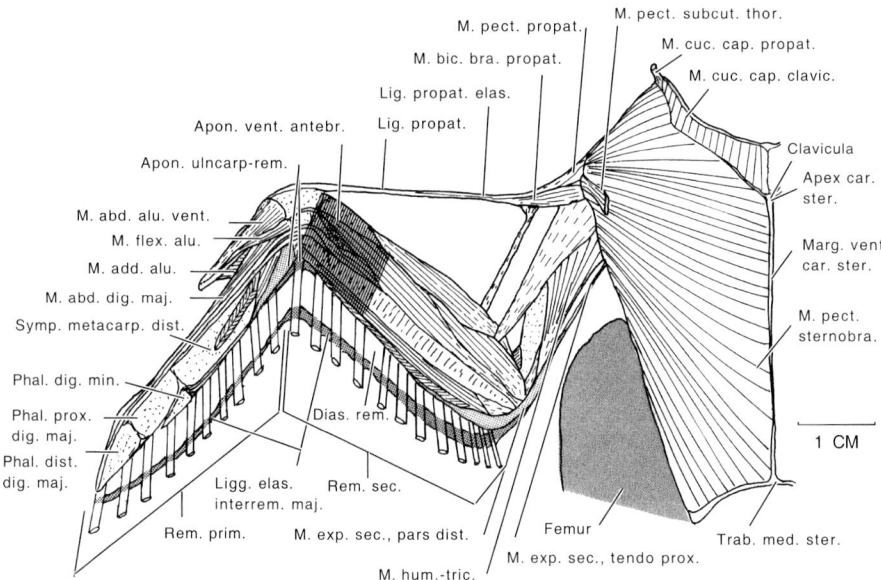

FIG. 72. Pectoral musculature of *Gallinula comeri* (BMNH 1922·12·6·221), breast and extended wing, superficial layer (remiges attached), ventral view. Anatomical abbreviations are defined in Appendix 3.

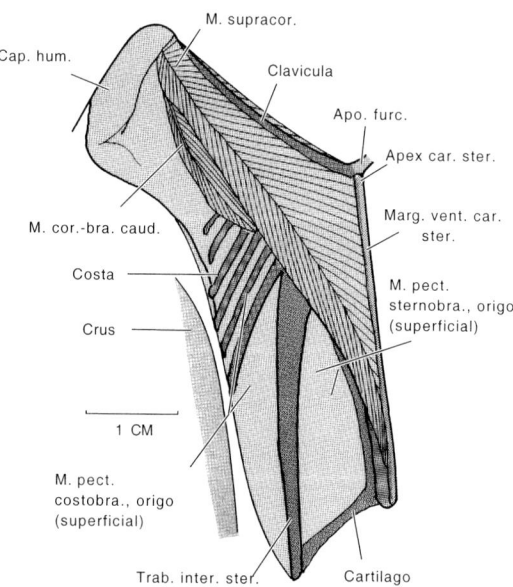

FIG. 73. Pectoral musculature of *Gallinula comeri* (BMNH 1922·12·6·221), breast, deep layer (with extent of overlying m. pectoralis indicated by stippled overlay), ventral view. Anatomical abbreviations are defined in Appendix 3.

TABLE 66. Masses (g) of fresh breast muscles of species of Rallidae. Estimates were based on single specimens unless sample size (*n*) follows mean. Percentages of body mass were based on species means (Appendix 1) if figures for dissected specimens not given. Flightless species are marked by asterisks.

Taxon	Mass ($n > 1$), % of mean body mass		Source
	M. pectoralis	M. supra-coracoideus	
Porphyrio melanotus	93.5 (33), 9.8%	12.9 (33), 1.4%	B. Reid fide McNab (1994a)
*Porphyrio hochstetteri**	34.9 (3), 1.4%	9.8 (3), 0.4%	B. Reid fide McNab (1994a)
Porphyrula martinica	— (3), 8.9%	— (3), 1.1%	Hartman (1961)
	— (6), 8.6%	— (6), 1.3%	Hartman (1961)
Aramides cajanea-group	— (6), 8.6%	— (6), 1.3%	Hartman (1961)
Gallirallus philippensis-group	[11.4], 5.8%	[2.2], 1.1%	Beauchamp (1987a)
*Gallirallus owstoni**	15.8 (5), 6.9%	1.8 (0.8%)	S. L. Olson fide McNab (1994a), amended by single specimen
Laterallus albigularis	— (6), 7.0%	— (6), 1.3%	Hartman (1961)
Crex crex	[24.3], 15.7%	[3.4], 2.2%	Beauchamp (1987a)
Amaurornis ineptus	~22, 2.4%	—	B. McNab (pers. comm.)
Gallinula chloropus-group	— (3), 10.2%	— (3), 1.5%	Hartman (1961)
Fulica (a.) americana	— (3), 8.2%	— (3), 1.2%	Hartman (1961)

(Fig. 85). These relatively small dermal components may occur in all rallids to variably limited degrees, but these tiny patagial slips were detected only in some of the largest specimens dissected, which in this study coincidentally were flightless.

However, contrary to the description by McGowan (1986), m. latissimus dorsi pars metapatagialis was present and broad in *Gallirallus australis* and all other rallids, and typically was a single-layered sheet of narrower width arising from a portion of the underlying origo m. latissimus dorsi caudalis (Figs. 75–77, 79, 84, 86–87). In *Porzana palmeri,* m. latissimus dorsi metapatagialis was represented only by a thin, fiberless fascia having a 5-mm origo and extending obliquely 1–5 mm laterad to the spinae dorsales vertebrales, inserting in the proximal margin of the metapatagium (Figs. 76, 79). However, this muscle was reasonably robust in other flightless rallids; for example, it was broad, well developed, and fibrous throughout in *Nesoclopeus poecilopterus.*

Musculus rhomboideus superficialis extended significantly craniad to m. latissimus dorsalis pars cranialis, whereas margo cranialis of m. rhomboideus profundus approximated that of m. latissimus dorsi pars cranialis in all rallids examined (Figs. 76–80). The fibrous portions of m. rhomboideus superficialis and (especially) m. rhomboideus profundus began comparatively caudad in some flightless species (e.g., one of two specimens of *Porphyrio hochstetteri, Habroptila wallacii, Gallirallus australis, Habropteryx okinawae, Atlantisia rogersi, Porzana palmeri,* and *Tribonyx mortierii*). However, m. rhomboideus profundus was well developed in the flightless taxa *Dryolimnas aldabranus, Nesoclopeus poecilopterus, Atlantisia rogersi,* and *Pareudiastes pacificus,* as in the flighted taxa *Coturnicops noveboracensis, Laterallus leucopyrrhus, Amaurornis olivaceus, Tribonyx ventralis,* and *Fulica americana* (Figs. 76, 77, 79, 88–91). Several other variants were noted in m. rhomboideus superficialis: fibers were limited to the lateral half of the fascia in the cranial half of the belly in *Gallirallus owstoni,* and fibers were lacking in

TABLE 67. Masses (g) of dehydration-corrected, fluid-preserved breast muscles of selected species of Rallidae. Data were derived during dissections performed in present study, unless otherwise noted. Flightless species are preceded by asterisks. Estimates were based on single specimens except for *Porphyrio melanotus*, *P. hochstetteri*, and *Atlantisia rogersi* ($n = 2$ each). Masses of muscles removed from one side were doubled to estimate total raw masses. Intraspecific comparisons (five flighted and two flightless species) of fresh and fluid-preserved masses of the two breast muscles revealed that the former averaged 1.64 ± 0.36 and 1.66 ± 0.57 times as large as the latter, respectively (approximate factor of 5/3 for each). This factor was used to correct the figures presented. Percentages of body mass were based on unweighted averages of sex-class means presented in Appendix 1. Those based on estimated body masses (Table 8) are enclosed by square brackets.

	Corrected mass (percentage of body mass)	
Taxon	M. pectoralis	M. supracoracoideus
Porphyrio melanotus	109.9 (11.2)	13.5 (1.4)
*Porphyrio hochstetteri	37.8 (1.4)	10.1 (0.4)
Gymnocrex plumbeiventris	29.7 (10.2)	5.0 (1.7)
*Habroptila wallacii	17.2 [1.9]	4.6 [0.5]
Aramides cajanea	46.9 (11.2)	9.2 (2.2)
Ortygonax sanguinolentus	20.8 (15.2)	3.3 (2.4)
*Cyanolimnas cerverai	5.3 [2.2]	1.0 [0.4]
Dryolimnas cuvieri	33.0 (13.7)	6.6 (2.7)
*Dryolimnas aldabranus	14.9 (8.1)	3.3 (1.8)
Rallus aquaticus	19.5 (17.0)	4.3 (3.7)
Rallus limicola	13.9 (16.7)	2.0 (2.4)
Rallus elegans	46.5 (13.2)	6.6 (1.9)
Gallirallus philippensis	25.4 (12.8)	5.3 (2.7)
*Gallirallus australis	29.7 (3.3)	7.9 (0.9)
*Gallirallus owstoni	12.2 (5.4)	2.3 (1.0)
*Gallirallus wakensis	2.1 [1.9]	0.5 [0.4]
*Tricholimnas sylvestris	10.9 (2.2)	3.3 (0.7)
*Nesoclopeus poecilopterus	16.5 [3.4]	6.6 [1.4]
*Habropteryx torquatus**	18.2 (7.4)	3.3 (1.3)
*Habropteryx okinawae	21.1 (4.9)	5.6 (1.3)
*Atlantisia rogersi†	0.8 (2.1)	0.2 (0.5)
Laterallus leucopyrrhus†	3.6 (8.1)	1.0 (2.2)
Coturnicops novaboracensis	7.3 (13.0)	1.3 (2.4)
Crex crex	20.8 (13.7)	3.6 (2.4)
*Porzana palmeri	1.0 (2.2)	0.3 (0.7)
Porzana carolina	8.3 (10.2)	1.3 (1.6)
*Porzana flavirostra**	6.9 (8.3)	2.0 (2.4)
Porzana tabuensis	4.3 (10.2)	0.7 (1.6)
*Porzana atra	3.3 (4.3)	0.7 (0.9)
*Amaurornis olivaceus**	12.9 (6.7)	2.6 (1.4)
*Amaurornis ineptus	36.0 (3.7)	9.6 (1.0)
*Pareudiastes pacificus	7.6 (—)	1.7 (—)
Tribonyx ventralis	57.4 (14.3)	8.9 (2.2)
*Tribonyx mortierii	21.1 (1.6)	5.0 (0.4)
*Gallinula comeri	17.5 (3.4)	3.6 (0.7)
Gallinula chloropus-group*	26.4 (7.9)	5.0 (1.5)
Fulica (a.) americana	68.3 (11.2)	8.3 (1.4)
Mean; range percentage (n)		
Flighted species (15)	12.9; 10.2–17.0	2.2; 1.4–3.7
Flightless species (16)	3.3; 1.4–8.1	0.8; 0.4–1.8

* Muscle masses were suspiciously low because specimens were captive, emaciated, or immature at time of death. These data were excluded from summary tallies.
† A single pair of data (0.32 and 0.02, respectively) reported by T. Cassidy and A. C. Kemp fide McNab (1994a), were excluded because both were substantially lower than my own estimates and details of dissection were not available.

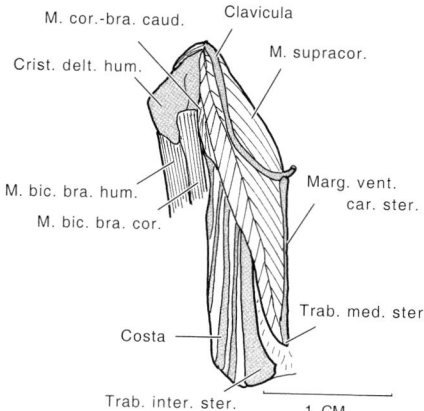

Fig. 74. Pectoral musculature of *Porzana palmeri* (BMNH 1940·12·8·13), breast, deep layer, ventral view. Anatomical abbreviations are defined in Appendix 3.

a 20-mm segment in the middle of the belly in *Amaurornis ineptus*. In *G. australis*, the cranialmost 9 mm of the 45-mm origin of m. rhomboideus superficialis was entirely without fibers, comprising a simple, translucent fascia between the processes dorsales vertebrales and the scapus scapulae.

In all rallids, mm. serratus superficialis pars cranialis et caudalis were embedded in a common fascia, the density of which varied significantly with flight capacity and body size (Figs. 76, 77, 79, 88–91). Indicative of the variation inherent in this muscle complex was the uniquely two-layered, four-parted m. serratus superficialis caudalis in *Aramides cajanea*. M. serratus superficialis metapatagialis was present (but readily overlooked without careful removal of fat and use of muscle stain) in all rallids; for example, in *Porzana palmeri* the tiny dermal slip was fibered but had a maximal width of 2 mm at origo costalis.

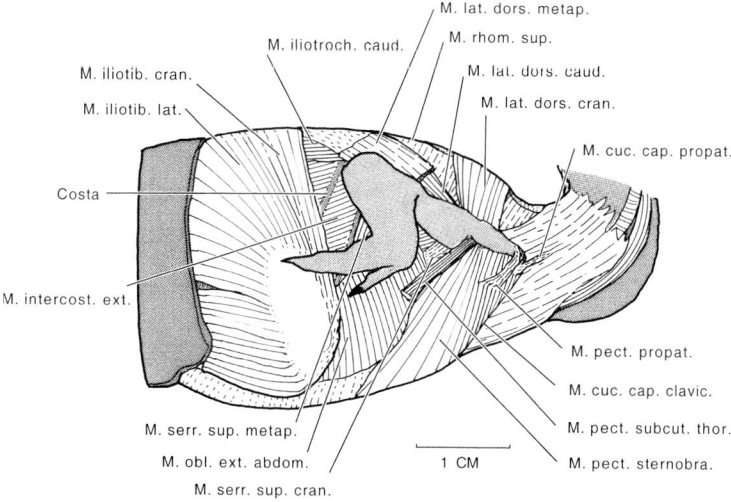

Fig. 75. Pectoral musculature of *Atlantisia rogersi* (DMNH 20762), breast and back, superficial layer, lateral view. Anatomical abbreviations are defined in Appendix 3.

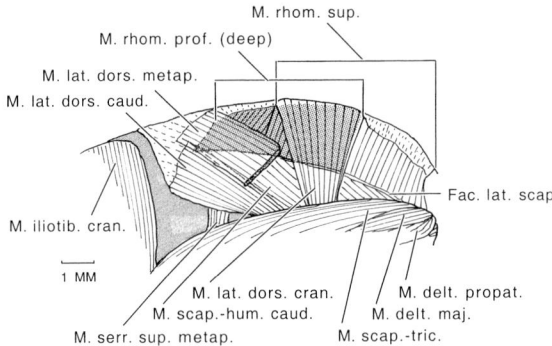

FIG. 76. Pectoral musculature of *Porzana palmeri* (BMNH 1940·12·8·13), articulatio omalis, middle layer, dorsolateral view. Some details of dorsum, damaged in primary specimen, were based on partial dissection of supplementary specimen USNM 290223. Anatomical abbreviations are defined in Appendix 3.

Numbers of fasciculi composing m. serratus profundus showed reduction in some taxa regardless of flight status (*Dryolimnas aldabranus, Tricholimnas sylvestris, Nesoclopeus poecilopterus, Rallus aquaticus, Gallirallus philippensis, G. wakensis, Habropteryx torquatus, H. okinawae, Coturnicops noveboracensis, Crex crex, Laterallus leucopyrrhus, Porzana* spp., *Amaurornis ineptus*, and *Pareudiastes pacificus*) from the count of six typical of the rallids examined, in part varying with body size. Lesser variation was detected in the number of fasciculi composing m. serratus superficialis caudalis (with the typical condition being three), but no association with flight capacity was evident in this pattern (Figs. 88–91). Moreover, the functional impact of this variation in either m. serratus

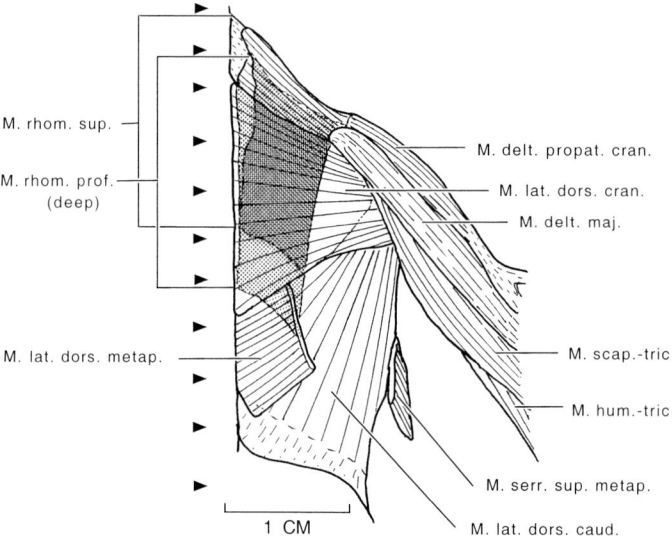

FIG. 77. Pectoral musculature of *Pareudiastes pacificus* (BMNH 1874·2·20·38), dorsum, regio omalis, and proximal brachium, dorsolateral view, superficial layer, with selected components of deep layers indicated by overlay. Anatomical abbreviations are defined in Appendix 3. Arrowheads mark positions of vertebrae thoracicae.

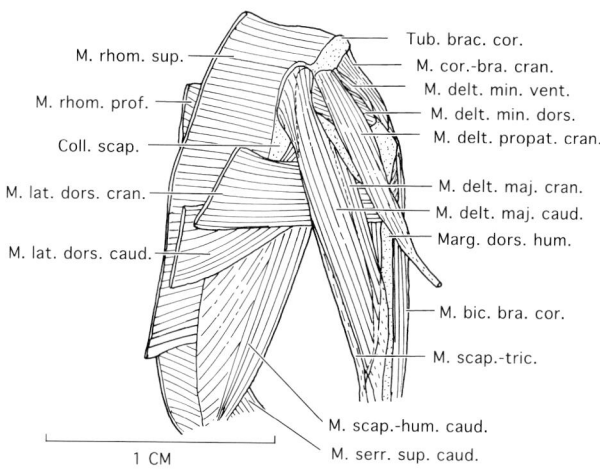

FIG. 78. Pectoral musculature of *Atlantisia rogersi* (DMNH 20762), articulatio omalis, dorsolateral view. Anatomical abbreviations are defined in Appendix 3.

superficialis or m. serratus profundus is unclear, in that although a high number might distribute the adductive action over a greater portion of the scapus scapulae, the total power of the adduction could be offset by substantial, possibly compensatory, variation in the sizes of the remaining fasciculi.

Contrary to the inference by McGowan (1986), both caput dorsale and caput ventrale of m. subcoracoideus were present in *Gallirallus australis,* as in other

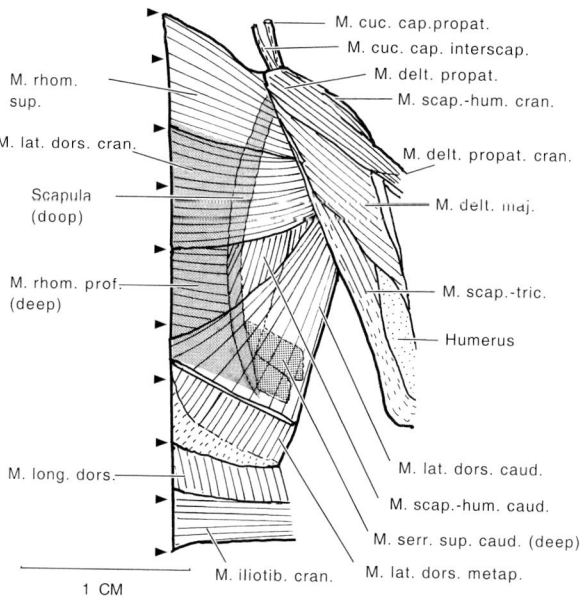

FIG. 79. Pectoral musculature of *Porzana palmeri* (BMNH 1940·12·8·13), back and proximal brachium, dorsal view, superficial layer, with selected components of deep layers shown in stippled overlay. Anatomical abbreviations are defined in Appendix 3. Arrowheads mark positions of vertebrae thoracicae.

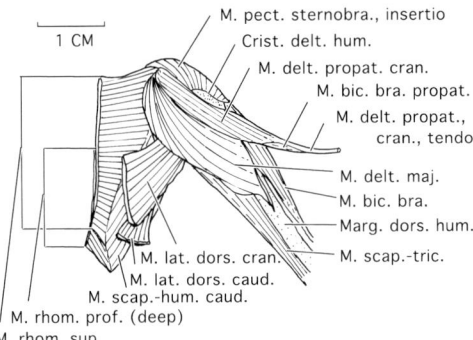

FIG. 80. Pectoral musculature of *Pareudiastes pacificus* (BMNH 1874·2·20·38), articulatio omalis, superficial layer, dorsolateral view. Anatomical abbreviations are defined in Appendix 3.

rallids examined (Figs. 88–91). However, in one specimen of *Porphyrio hochstetteri*, m. subcoracoideus caput mediale was extremely thin and poorly distinguished, and in this species m. sternocoracoideus was reduced to an afibrous, tendinous fascia. In all of the species compared, the adductive action of m. sternocoracoideus was stabilized by a closely adjacent complex of ligamenta sterocoracoidea. Other comparatively occult muscles of the pectoral girdle (e.g., mm. subscapularis caput laterale and scapulohumeralis cranialis) also were present in all rallids studied, although of these two bellies, measurement of only the former was possible and this required abduction of the humerus from the scapula for access (Figs. 83, 88–91). Manipulation of the shoulder girdle was difficult in all specimens dissected, in that the articulatio omalis (shoulder) was reinforced strongly by the combined effects of the ligamenta acrocoracoclaviculares, coracoscapulares, procoracoclaviculares, acromioclaviculares, scapuloclaviculares, acrocoracohumerale, coracohumerale dorsale, et scapulohumerales (Vanden Berge and Zweers 1993). This complexity of ligamenta can be envisioned more simply by noting that it essentially comprises dorsoventral pairs of ligamenta (further strengthened by a few caudal or cranial constituents) uniting each pair of osseous elements that enclose the cavitas glenoidalis (scapula, coracoideum, and humerus).

Muscles aligned along the brachium proper, m. biceps brachii, m. biceps propatagialis, and m. triceps brachii (both partes scapulotriceps et humerotriceps), as

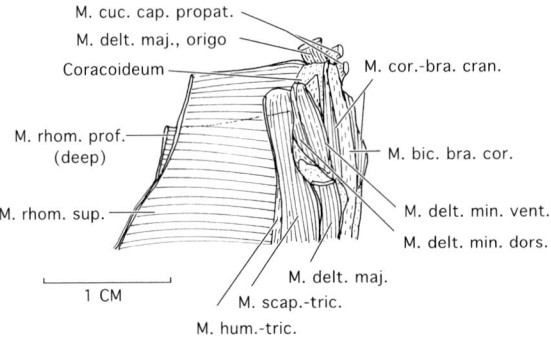

FIG. 81. Pectoral musculature of *Gallirallus wakensis* (USNM 289313), articulatio omalis, middle and deep layers, dorsolateral view. Anatomical abbreviations are defined in Appendix 3.

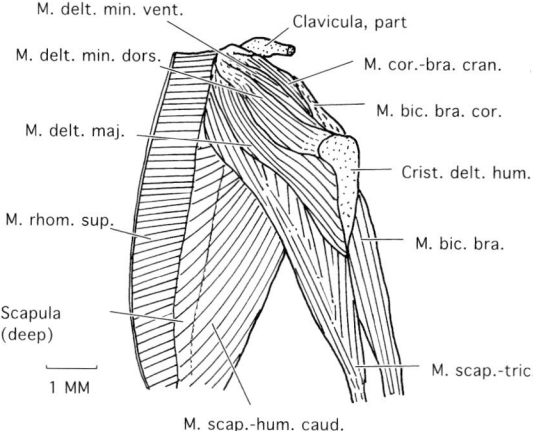

FIG. 82. Pectoral musculature of *Porzana palmeri* (BMNH 1940·12·8·13), back and proximal brachium and regio omalis, superficial layer, dorsal view. Some anatomical details damaged in primary specimen were based on partial dissection of supplementary specimen USNM 290223. Anatomical abbreviations are defined in Appendix 3.

well as obliquely oriented m. deltoideus major, tended to be disproportionately slender in the most severely derived flightless species (Figs. 72, 77–82, 85–87), with at least the first three tending toward strips of uniform width as opposed to fusiform bellies in flightless taxa (e.g., *Gallirallus wakensis* and *Gallinula comeri*). However, no reductions were obvious in relative sizes of m. biceps brachii et propatagialis in some flightless rallids (e.g., *Pareudiastes pacificus*). Contrary to the report by Beddard (1898:323), *Gallirallus australis* (his *Ocydromus earlei*) possessed a typically conformed m. biceps pars propatagialis. No significant subdivisions of origo m. humerotriceps were detected in this study, whereas such have been described by other investigators (McGowan 1986; Vanden Berge and Zweers 1993). As affirmed by previous myological descriptions of members of

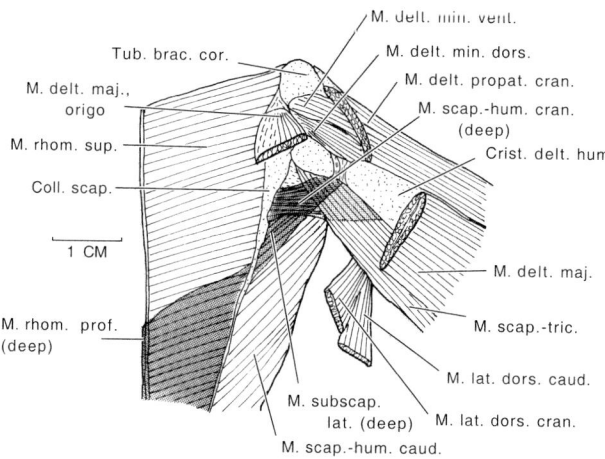

FIG. 83. Pectoral musculature of *Tricholimnas sylvestris* (USNM 18407), articulatio omalis, superficial and deep layers, dorsolateral view. Anatomical abbreviations are defined in Appendix 3.

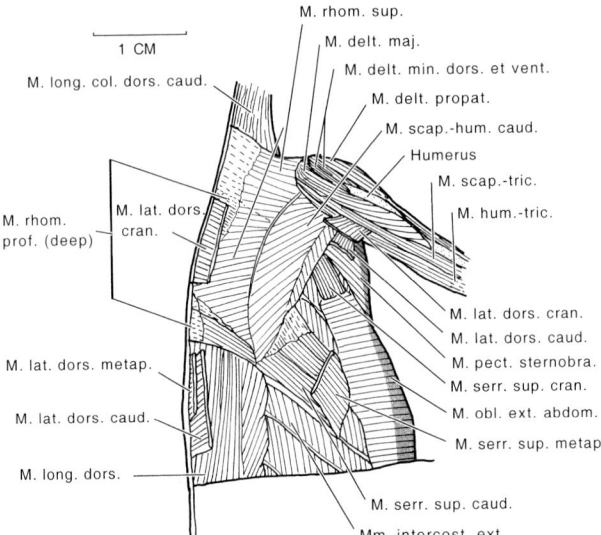

FIG. 84. Pectoral musculature of *Atlantisia rogersi* (DMNH 20762), back and side, deep layer, dorsolateral view. Anatomical abbreviations are defined in Appendix 3.

the Gruiformes (e.g., Fisher and Goodman 1955; George and Berger 1966; Rosser 1980; McGowan 1986), m. coracotriceps was not detected in any of the rallids examined here. Other aspects of the brachial and cubital musculature for rallids in the present study were consistent with those of previous works (e.g., Rosser 1980; McGowan 1986), and included involvement of retinaculum m. scapulotricipitis with origio m. scapulotriceps, absence of os sesamoideum m. scapulotricipitis, insertions of m. scapulotricipitis on the apex dorsoproximalis of the ulna (impressio m. scapulotricipitis) and m. humerotricipitis on the olecranon, and

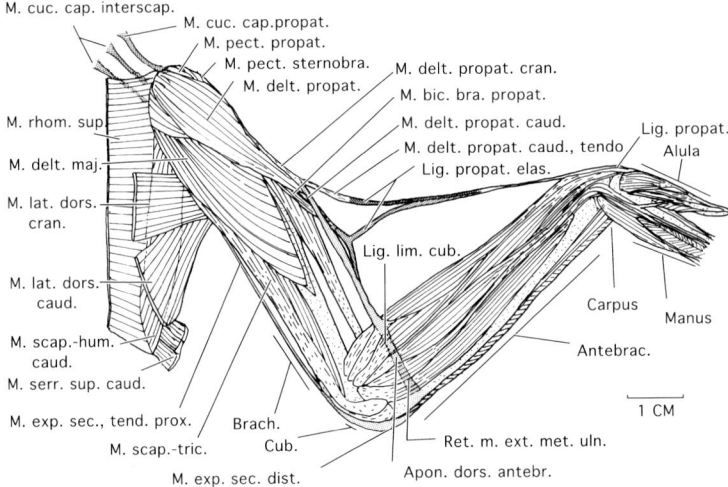

FIG. 85. Pectoral musculature of *Nesoclopeus poecilopterus* (BMNH 1940·12·8·87), back and extended wing (remiges removed), superficial layer, dorsal view. Anatomical abbreviations are defined in Appendix 3.

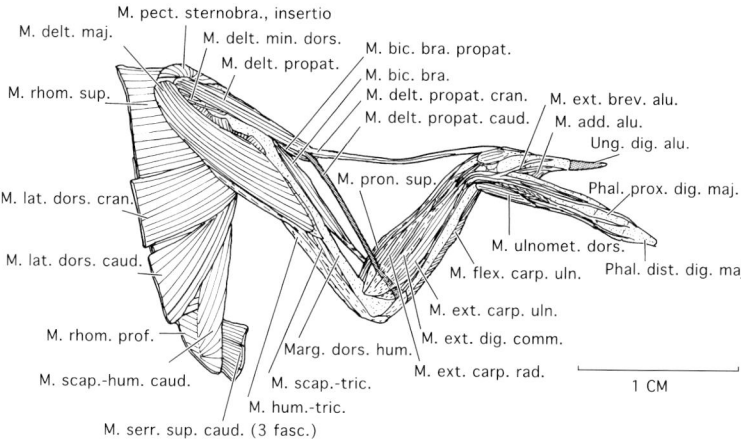

FIG. 86. Pectoral musculature of *Atlantisia rogersi* (DMNH 20762), back and extended wing, superficial layer (remiges removed), dorsal view. Anatomical abbreviations are defined in Appendix 3.

united by the ligamentum tricipitale, and presence of two differentiable capita coracoideum et humerale for m. biceps brachii (Fig. 72).

Musculus deltoideus pars major was present and conspicuously developed in all rallids examined, but in no specimen were capita cranialis et caudalis clearly distinguishable; the muscle inserted proximally by fleshy fibers directly attaching to the humerus and distally by a variably extensive tendinous aponeurosis (Figs. 78, 81–83). Mm. deltoideus minor capita dorsalis et ventralis were poorly differ-

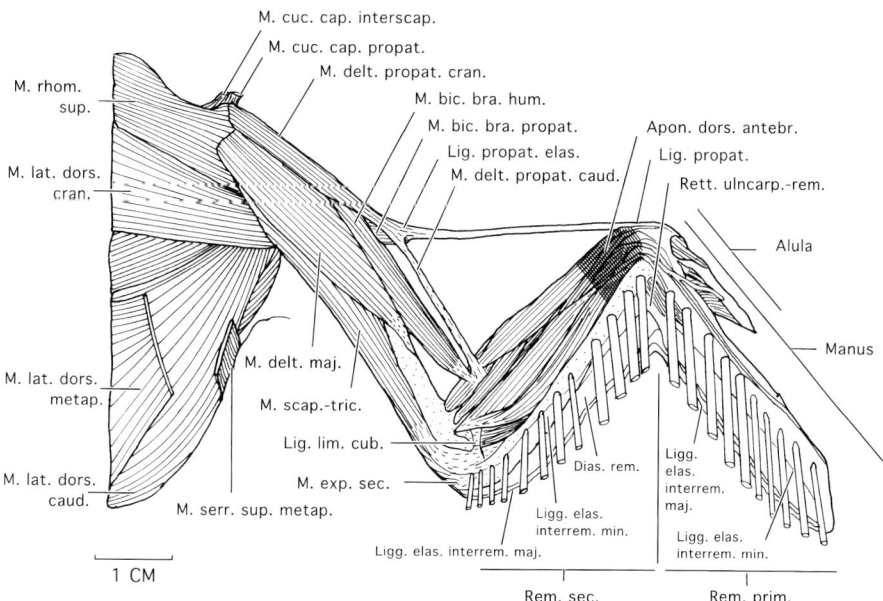

FIG. 87. Pectoral musculature of *Gallinula comeri* (BMNH 1922·12·6·221), back and extended wing, superficial layer (remiges attached), dorsal view. Anatomical abbreviations are defined in Appendix 3.

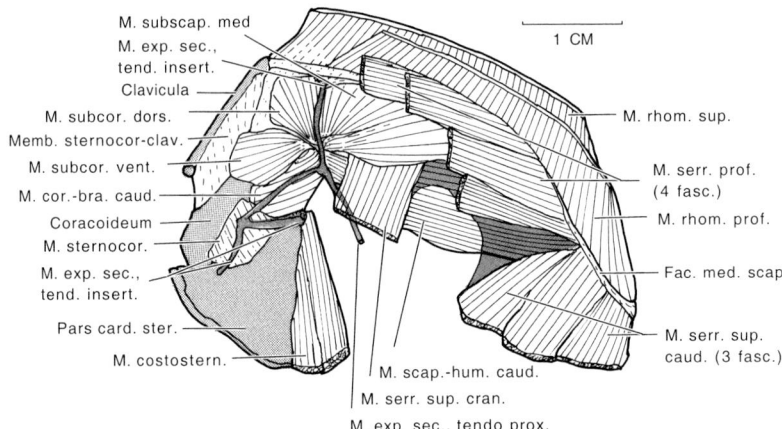

FIG. 88. Pectoral musculature of *Nesoclopeus poecilopterus* (BMNH 1940·12·8·87), regio omalis, superficial layer, medial view. Anatomical abbreviations are defined in Appendix 3.

entiated in some taxa, notably several flightless species (e.g., *Habropteryx okinawae*), a condition also observed in some flighted taxa (e.g., *H. torquatus*). Capita dorsalis et ventralis of m. deltoideus minor were virtually indistinguishable in *Gallinula comeri* and the juvenile specimen of *G. (c.) cachinaans,* whereas the two capita were distinct in the adult specimen of the *G. chloropus* complex; whether two capita were distinguishable or not, the muscle inserted on tuberculum dorsale humeri (Figs. 78, 81–83).

As noted by McGowan (1986) for *Gallirallus australis,* all parts of the fine musculature of the propatagium were present in flightless rallids, and provided only quantitative changes in muscle bellies or extents as evidence of reduced function (Figs. 72, 85–87). Partes radialis et ulnaris of tendo brevis of m. deltoideus propatagialis, caput caudale were variably distinguishable craniad to the

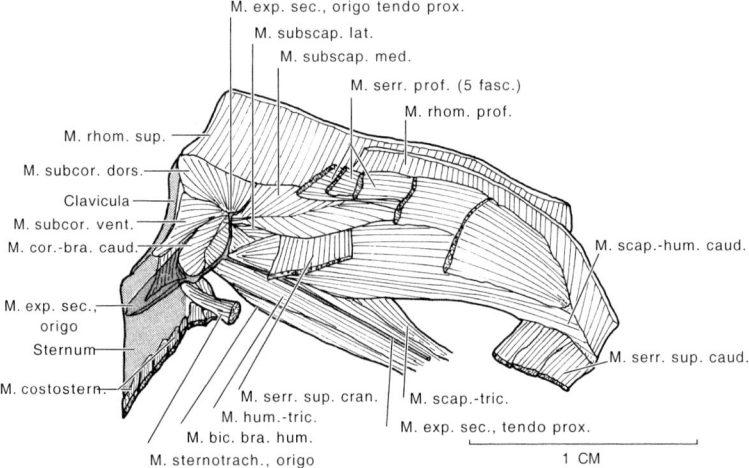

FIG. 89. Pectoral musculature of *Atlantisia rogersi* (DMNH 20762), regio omalis, superficial layer, medial view. Anatomical abbreviations are defined in Appendix 3.

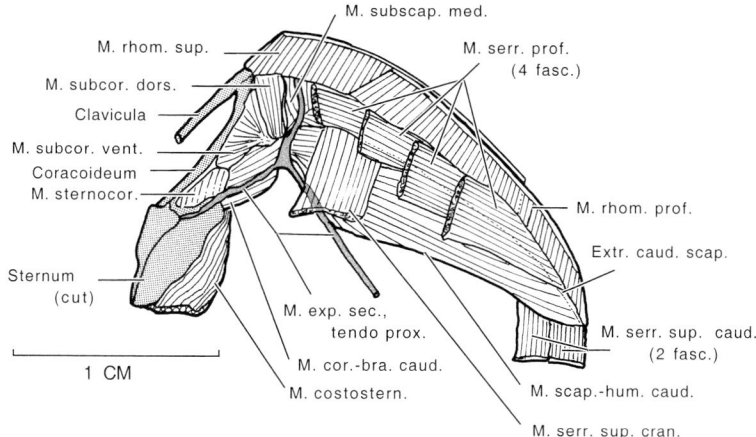

FIG. 90. Pectoral musculature of *Porzana palmeri* (BMNH 1940·12·8·13), regio omalis, superficial layer, medial view. Anatomical abbreviations are defined in Appendix 3.

antebrachial musculature in all rallids, being clearly separable in some (e.g., one *Porphyrio hochstetteri*). In all species, caput caudale (formerly pars brevis of m. tensor propatagialis) showed distinct but tightly coherent tendons of insertion for the radius and ulna. M. deltoideus propatagialis tended to be narrow and thin in flightless species, especially relative to body size. In *Porzana palmeri*, the fibers of the proximal belly were restricted to the proximal portion.

As was especially clear in the century-old, but well-preserved specimen of *Nesoclopeus poecilopterus*, a cranialmost complex of tendons of the propatagium comprising separate components was typical of all rallids examined (Fig. 85). Proximally, the fleshy belly of m. deltoideus propatagialis (formerly m. tensor propatagialis) gave rise to ligamentum propatagiale (formerly comprising tendo longus and tendo brevis), which was joined by tendo m. biceps propatagialis near the divergence of ligamentum limitans cubiti (formerly tendo brevis) of m. deltoideus propatagialis (Fig. 85). Tendo m. biceps propatagialis remained distinct

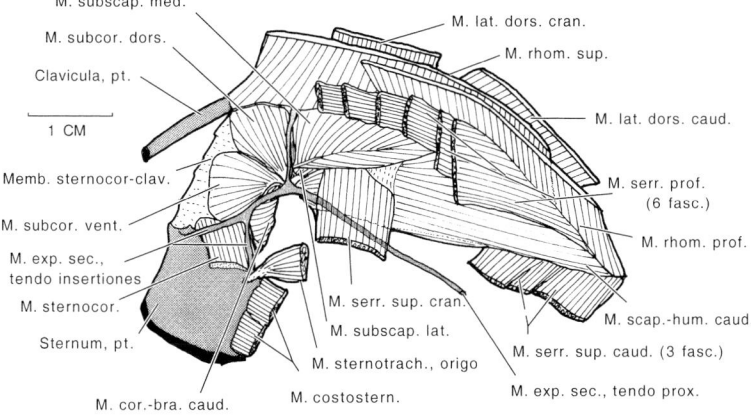

FIG. 91. Pectoral musculature of *Gallinula comeri* (BMNH 1922·12·6·221), regio omalis, superficial layer, medial view. Anatomical abbreviations are defined in Appendix 3.

from, but tightly ensleeved with, ligamentum propatagiale throughout much of the length of the propatagium, uniting immediately proximal to their common insertion by broad tendinous fascia on the manus. The distinctness of tendo m. brachialis propatagialis distal to the bifurcation of both caput craniale and caput caudale of m. deltoideus pars propatagialis was considered by Rosser (1980) to be unique to the Rallidae among the Gruiformes. A significant expanse of pars elasticum (elasticity of which was conserved for more than a century in spirit) also was confirmed within the distal portion of ligamentum propatagiale of *Nesoclopeus poecilopterus* (Fig. 85).

Musculus expansor secundariorum is largely dermal in nature but comprising origins on two elements of the pectoral girdle, passing along the brachium, superficial and deep bellies on the proximocaudal portion of the antebrachium (composed of smooth muscle), and inserting on the remiges secundarii via the ligamenta interremigales (Berger 1956b; Vanden Berge and Zweers 1993). This muscle (including proximal tendon, and deep and superficial distal bellies), which acts primarily to maneuver the secondary remiges, was present in all rallids including flightless species (Figs. 85, 87). The origin of m. expansor secundariorum on the scapula, inferred to be unique among gruiform families by Rosser (1980), was present in all rallids studied (Figs. 88–91). The tuberculum on the scapula that marked this origin varied considerably (detectable in skeletal specimens), and showed a slight tendency toward reduction associated with flightlessness (Olson 1973b). In *Atlantisia rogersi*, the distal tendon (inserting serially on the calami of the remiges secundarii) was extremely fine and delicate, and the tendon of origin on facies dorsalis of the coracoid was not bifurcated as in most other rallids (Figs. 88, 89, 91), including such flightless species as *Habropteryx okinawae* and *Pareudiastes pacificus*. Where the tendon was single, it arose with ligamentum sternoprocoracoideum from the pila coracoidea. The second point of origin for the tendon (where present) arose from processus craniolateralis sterni within a tendinous fascia shared by both tendons of origin.

The dorsal muscles of the antebrachium revealed similarly suggestive but non-diagnostic corollaries of flightlessness confounded by interspecific differences in body size (Figs. 85–87, 92–94). The flexor and extensor musculature of the dorsal antebrachium was covered by dense, translucent fascial sheets, the aponeurosis dorsalis antebrachii, which was tightly adherent proximally to the ligamentum limitans cubiti. Cranially, capita dorsale et ventrale of m. extensor carpi radialis were tightly united in most rallids studied, especially the flightless taxa (e.g., *Porphyrio hochstetteri*). Typically, the complex comprised a comparatively truncated, thin caput dorsale and a longer caput ventrale that was fused proximally with its ventral counterpart and tightly bound but differentiable as the capita narrowed distally to tendons of insertion (Fig. 95). Exceptional again among flightless rallids was *Pareudiastes pacificus*, in which caput dorsale was comparable in the length of its fibrous portion to that of caput ventrale. Deep to this complex was m. supinator, which revealed a similar pattern in which flightless taxa tended to have relatively thin, distally truncated bellies not confidently defined without the use of muscle stain (Figs. 92–95).

Between the radius and ulna in dorsal perspective, a subtly challenging group of muscles was found to be similar in the rallids studied here and elsewhere (Rosser 1980; McGowan 1986), and in other neognathus birds generally (Berger

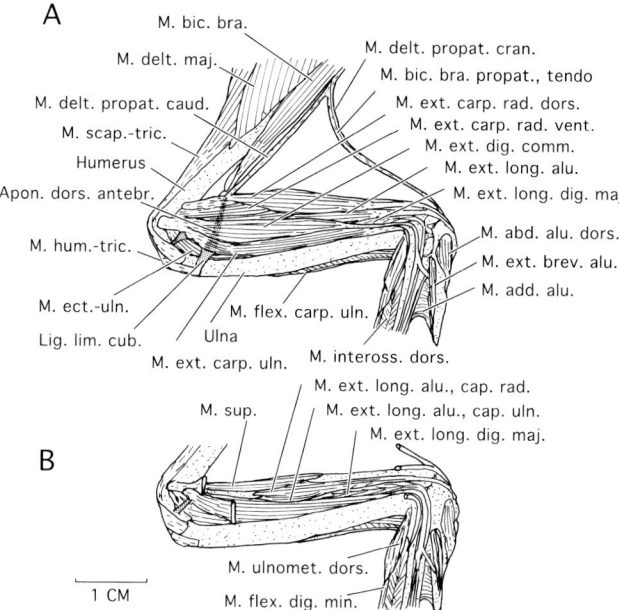

FIG. 94. Pectoral musculature of *Gallinula comeri* (BMNH 1922·12·6·221), distal brachium, antebrachium, and proximal manus, dorsal views: **A,** superficial layer; and **B,** deep layer. Anatomical abbreviations are defined in Appendix 3.

As for the dorsal antebrachial musculature, the flexor and extensor musculature of the ventral antebrachium is covered by dense, translucent fascial sheets, the aponeurosis ventralis antebrachii, which is adherent to the prominent ligamentum humerocarpale throughout. Distally this aponeurosis blended with the dense, fanlike covering of the ventral surface of the carpus and base of os metacarpale minus, aponeurosis ulnocarporemigalis and the anchoring retinaculum flexorium (Fig. 72). Of the cranial musculature on the ventral surface of the antebrachium (Figs. 96–

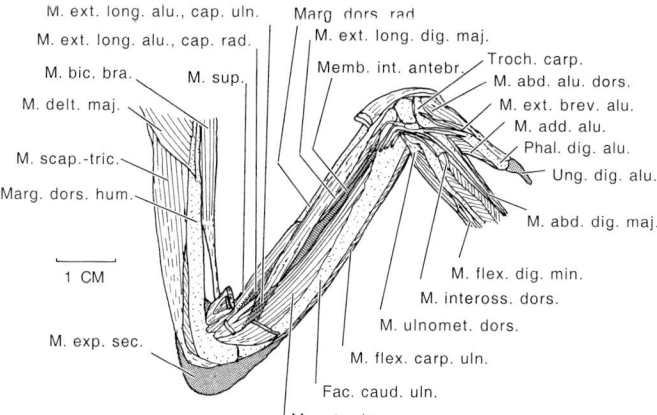

FIG. 95. Pectoral musculature of *Habroptila wallacii* (USNM 506649), distal brachium, antebrachium, and proximal manus, deep layer, ventral view. Anatomical abbreviations are defined in Appendix 3.

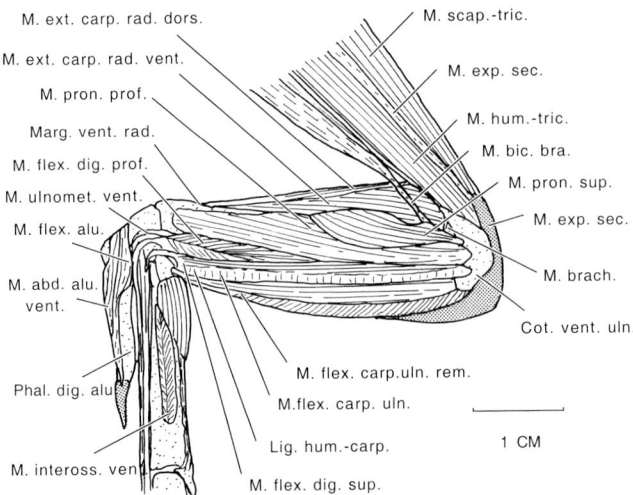

FIG. 96. Pectoral musculature of *Cyanolimnas cerverai* (USNM 343159), distal brachium, antebrachium, and proximal manus, superficial layer, ventral view. Anatomical abbreviations are defined in Appendix 3.

99), m. pronator superficialis appeared relatively large in some flightless rallids, extending essentially to the extremitas distalis of the foreshortened radius (e.g., *Gallirallus australis* and *Pareudiastes pacificus*). Parallel conditions were evident in m. brachialis (ventral view) and m. ectepicondylo-ulnaris (dorsal view) relative to the underlying ulna (Figs. 98B, 99B). Both conditions were suggestive of an evolutionary truncation of underlying skeletal elements without a proportionate shortening of muscles having extensive origins along these elements. The relatively large m. brachialis of flightless rallids also was reflected in the conspicuously deep impressio m. brachialis noted in skeletal specimens (Figs. 64, 65).

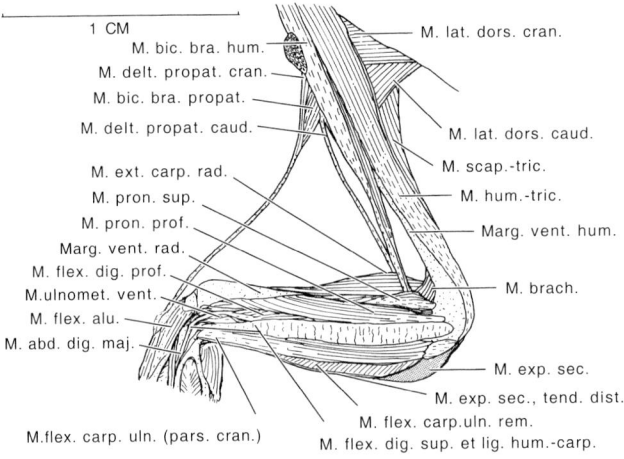

FIG. 97. Pectoral musculature of *Atlantisia rogersi* (DMNH 20762), distal brachium, antebrachium, and proximal manus, superficial layer, ventral view. Anatomical abbreviations are defined in Appendix 3.

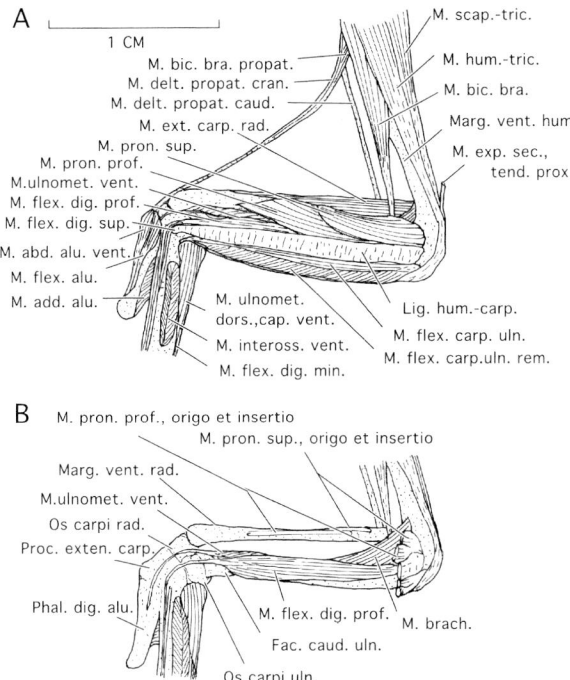

FIG. 98. Pectoral musculature of *Porzana palmeri* (BMNH 1940·12·8·13), distal brachium, antebrachium, and proximal manus, ventral views: **A,** superficial layer; and **B,** deep layer. Anatomical abbreviations are defined in Appendix 3.

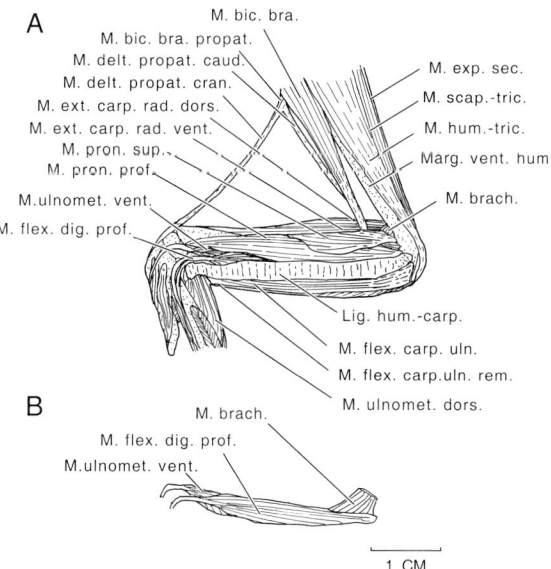

FIG. 99. Pectoral musculature of *Gallinula comeri* (BMNH 1922·12·6·221), distal brachium, antebrachium, and proximal manus, ventral views: **A,** superficial layer; and **B,** deep layer. Anatomical abbreviations are defined in Appendix 3.

Among the caudal components of the musculature of the ventral antebrachium, variation in m. flexor digitorum superficialis merits special attention. Tendo m. flexor digitorum superficialis was very weakly developed in *Cyanolimnas cerverai,* such that the width of the overlying ligamentum humerocarpale was less than one half of that of muscle belly proper, resembling the superficial tendinous fascia giving rise to the tendon of insertion of m. pronator superficialis (Fig. 96). This condition was in marked contrast to the normal condition in neognathous birds in which the ligamentum humerocarpale completely envelopes the muscle belly ventrally (Figs. 97–99), with the tendon of insertion of m. flexor digiti superficialis diverging from the enclosing ligamentum only at its craniodistal extremity where it passes beneath (with tendo insertii m. flexor digitorum profundus) the retinaculum flexorium on facies ventralis of os carpi ulnare (e.g., Fisher and Goodman 1955; Berger 1966; George and Berger 1966; Vanden Berge 1970; Livezey 1990, 1992a, b). Throughout much of the length of the ligamentum humerocarpale, a variably dense tendinous fascia (septum humerocarpale) extended caudally from margo caudalis of the ligamentum to the ventral bases of the remiges secundariorum (Fig. 72). Distally, the two tendines mm. flexores digitorum passed through a tendinous sheath (retinaculum flexorium) proximal to their respective insertions. The tendon of m. flexor digiti superficialis inserts on apex cranioventralis of phalanx proximalis digiti majoris (Rosser 1980), and that of m. flexor digiti profundus, passes around processus pisiformis of the extremitas proximalis carpometacarpi to insert on phalanx distalis digiti majoris (Figs. 96, 98, 99; Vanden Berge and Zweers 1993).

Caudal and parallel to m. flexor digitorum superficialis and the closely associated ligamentum humerocarpale was m. flexor carpi ulnaris (Figs. 96–99). In all of the rallids compared here, and in previous studies of rallids (Rosser 1980; McGowan 1986) and other neognathous birds (e.g., Fisher and Goodman 1955; Berger 1966; George and Berger 1966), this muscle comprised partes cranialis et caudalis, with the latter being renamed pars remigalis by Vanden Berge and Zweers (1993). The approximately parallel arrangement of fibers of pars cranialis and the oblique orientation of fibers in pars caudalis (remigalis), relative to the dividing septum humerocarpale, was evident in all species studied, as were the insertions of pars cranialis on processus muscularis of os carpi ulnare and that of pars caudalis on ligamentum elasticum interremigale minor (Figs. 72, 96–99).

The smallest component of the deep, caudal musculature of the antebrachium, m. ulnometacarpalis ventralis, generally occupied less than one quarter of the length of the ulna, lying almost completely obscured by the immediately overlying and adherent m. flexor digitorum profundus (Figs. 96–99). Although small in all rallids studied, this muscle was particularly thin and proximally restricted in flightless species (Fig. 100). Tendo m. ulnometacarpalis ventralis passed through sulcus tendineus on facies ventralis of os carpi radiale, beneath tendo m. extensor metacarpi radialis, and angled dorsad to insert on facies ventralis of extremitas proximalis carpometacarpi.

Muscles of the manus (including the digits) manifested only limited variation that tended to be associated with flightlessness, including components positioned on the dorsal (Figs. 101–107) and ventral aspects (Figs. 108–110) of the distal wing. Most muscles of this distalmost segment of the wing differed interspecifically only in size, and this was confounded by differences in body size among

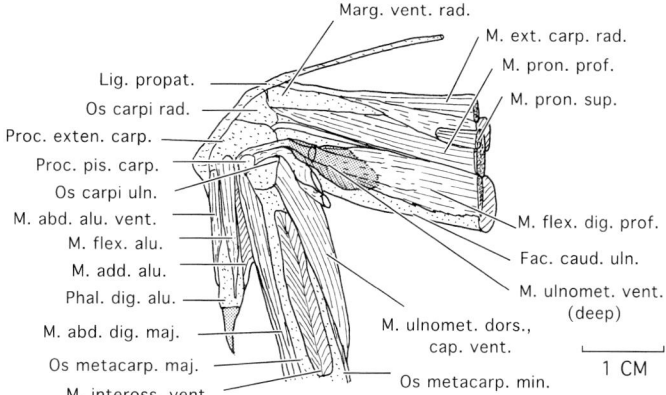

FIG. 100. Pectoral musculature of *Tribonyx mortierii* (AMNH 10079), detail of distal antebrachium, carpus, and proximal manus, ventral view, superficial layer, with deep m. ulnometacarpalis ventralis indicated in stippled overlay. Anatomical abbreviations are defined in Appendix 3.

species. These qualitatively invariant muscles included mm. interosseus dorsalis et ventralis, m. adductor alulae, m. flexor alulae, m. abductor digiti majoris, and m. flexor digiti minoris (Figs. 101–110). Qualitative variation was most notable in two muscles inserting on phalanx proximalis digiti alulae—mm. abductor alulae (comprising capita dorsale et ventrale) and m. extensor brevis alulae. M. abductor alulae caput dorsale showed considerable variation, being bi- or trilobate and extending distocaudally to partly cover m. extensor brevis alulae in one specimen of *Porphyrio hochstetteri* and in *Gallirallus australis* (Figs. 101, 102). M. extensor brevis alulae manifested additional variation, most notably a tendency toward reduction in some flightless species (e.g., this muscle was vestigial in *Gallirallus wakensis* and *Atlantisia rogersi*). However, other details of the distalmost musculature of the wing were uniform in all rallids examined, regardless of flight capacity. Such structural details that showed minimal to no variation within or among species included the tightly associated bellies of muscles on the ventral aspect of the alula (Figs. 108–110), and a bifurcated tendon of insertion of m. interosseus dorsalis on phalanx proximalis digiti majoris (Figs. 105, 106).

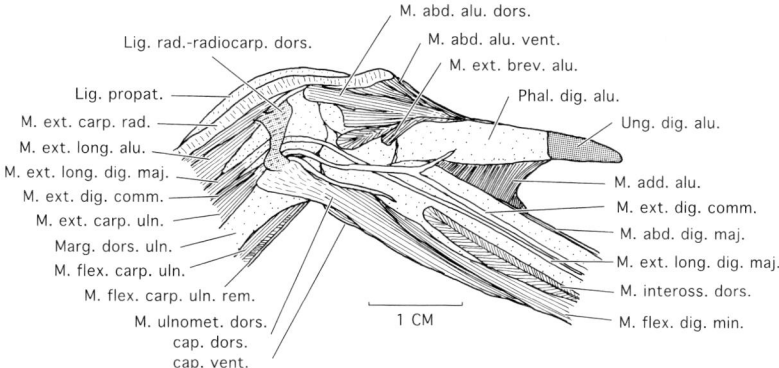

FIG. 101. Pectoral musculature of *Porphyrio hochstetteri* (NZNM-2), distal antebrachium, carpus, and proximal manus, superficial layer, dorsal view. Anatomical abbreviations are defined in Appendix 3.

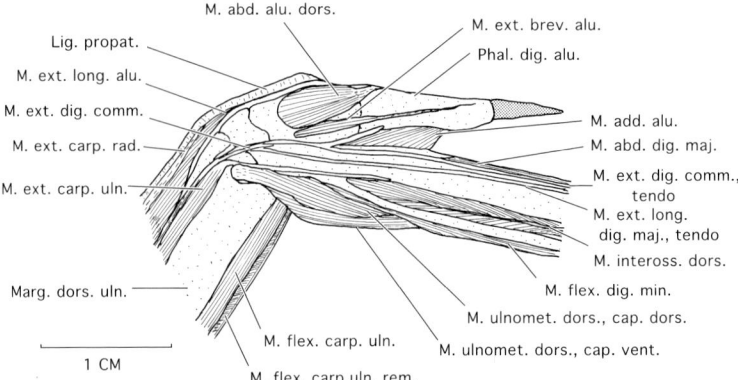

FIG. 102. Pectoral musculature of *Habropteryx okinawae* (UMMZ 225360), detail of distal antebrachium, carpus, and proximal manus, superficial layer, dorsal view. Anatomical abbreviations are defined in Appendix 3.

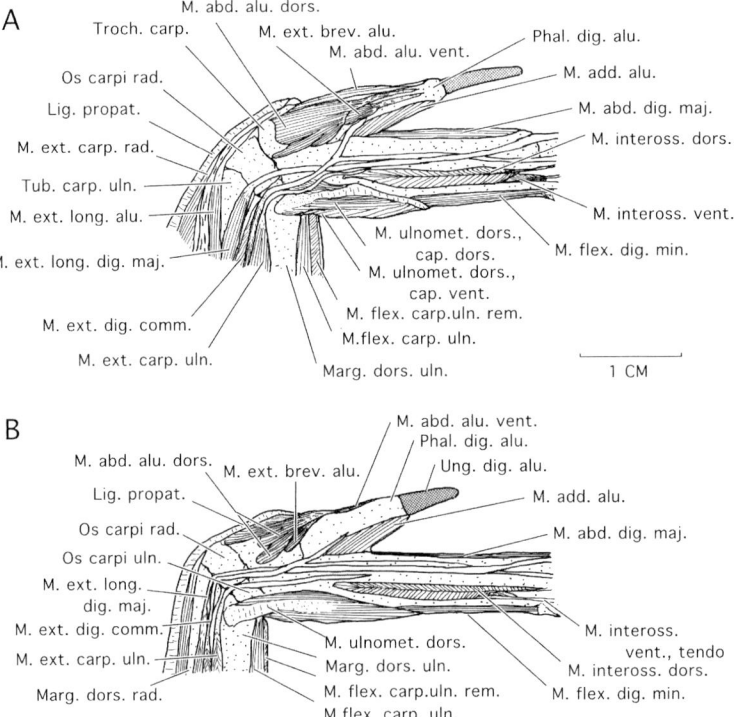

FIG. 103. Pectoral musculature of **A,** *Gallirallus australis* (USNM 511701), and **B,** *Porphyrio hochstetteri* (NZNM-1), distal antebrachium, carpus, and proximal manus, superficial layer (dorsal view) showing conformational variation in mm. extensor brevis alulae et abductor alulae caput dorsale. See Appendix 3 for explanation of abbreviations for anatomical features.

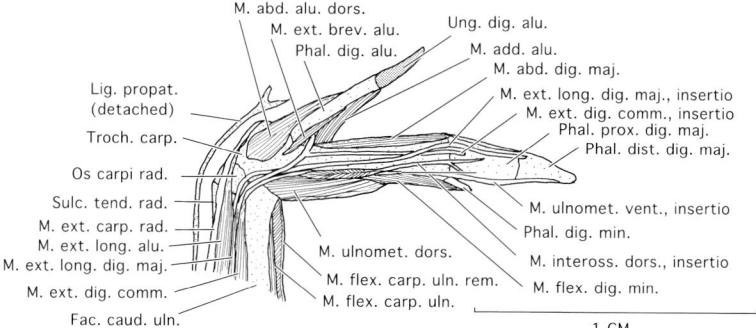

FIG. 104. Pectoral musculature of *Atlantisia rogersi* (DMNH 20762), distal antebrachium, carpus, and manus, superficial layer, dorsal view, tendo longus detached at insertion. Anatomical abbreviations are defined in Appendix 3.

ALLOMETRY OF BREAST MUSCULATURE

Reductions in the relative or (in extreme cases) absolute reductions in the major breast muscles (mm. pectoralis sternobrachialis et supracoracoideus) of flightless rails were responsible for the most visually conspicuous and functionally obvious of the changes related to the loss of flight evinced by myological dissections (Tables 66, 67). At an interfamilial scale, the members of the Rallidae (even in studies confined to flighted species) appear as moderately distant outliers from the allometric relationship for mass of breast muscles with body mass, falling distinctly below points for most other avian families (Hartman 1961; Greenewalt 1962; Rayner 1988a).

The present analysis provided an opportunity for an examination of such allometry at a finer scale, at least with respect to flightlessness across a number of genera (Table 66). Fitting of allometric relationships for flighted and flightless species of Rallidae revealed virtual isometry between mean masses for the breast

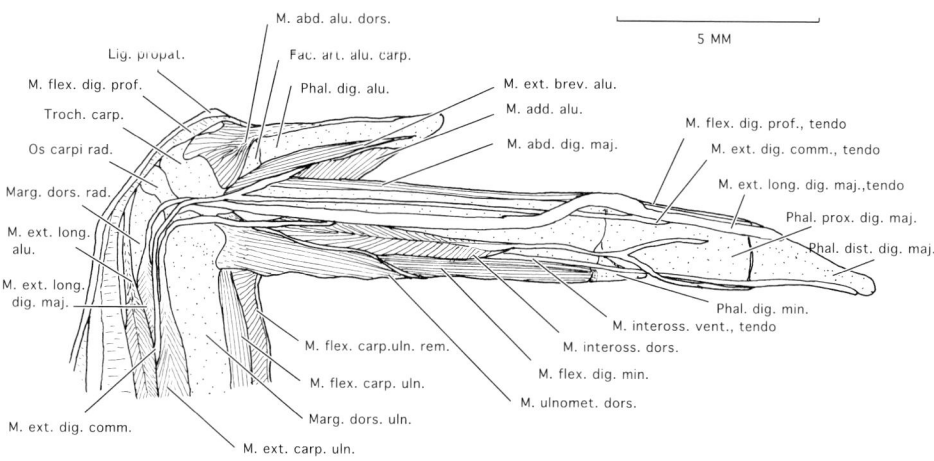

FIG. 105. Pectoral musculature of *Porzana palmeri* (BMNH 1940·12·8·13), detail of distal antebrachium, carpus, and proximal manus, superficial layer, dorsal view. Anatomical abbreviations are defined in Appendix 3.

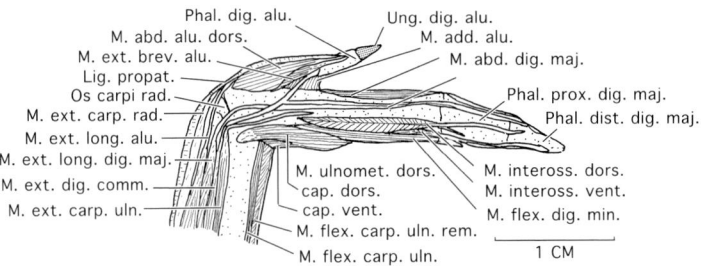

FIG. 106. Pectoral musculature of *Pareudiastes pacificus* (BMNH 1874·2·20·38), distal antebrachium, carpus, and manus, superficial layer, dorsal view. Anatomical abbreviations are defined in Appendix 3.

muscles with the entire body, and each model manifested high statistical precision ($P < 0.0001$ for all). The standard linear models for masses of breast muscles relative to mean body mass (M_B) for flighted rallids ($n = 15$) were:

mass (m. pectoralis)
$= -1.88 + (0.97 \pm 0.05)M_B$ ($r = 0.98$; $F_{1,13} = 351.0$),

mass (m. supracoracoideus)
$= -3.30 + (0.89 \pm 0.08)M_B$ ($r = 0.95$; $F_{1,13} = 125.4$), and

mass (mm. pectoralis et supracoracoideus)
$= -1.67 + (0.96 \pm 0.05)M_B$ ($r = 0.98$; $F_{1,13} = 317.4$).

The corresponding models for flightless rallids ($n = 16$) were:

mass (m. pectoralis)
$= -3.10 + (0.92 \pm 0.05)M_B$ ($r = 0.92$; $F_{1,14} = 76.9$),

mass (m. supracoracoideus)
$= -4.66 + (0.95 \pm 0.10)M_B$ ($r = 0.92$; $F_{1,14} = 81.6$), and

mass (mm. pectoralis et supracoracoideus)
$= -2.91 + (0.96 \pm 0.10)M_B$ ($r = 0.92$; $F_{1,14} = 79.5$).

Despite the obvious similarity in allometric slopes ($\hat{b}$)—the mean for two separate muscles was 0.93 for both flighted and flightless species—the allometric

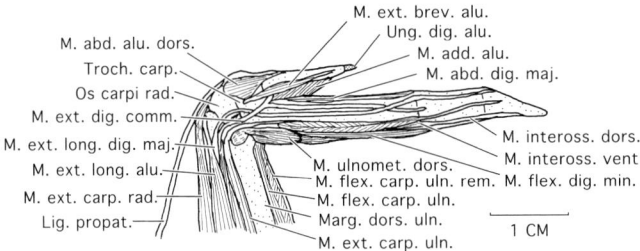

FIG. 107. Pectoral musculature of *Gallinula comeri* (BMNH 1922·12·6·221), distal antebrachium, carpus, and manus, superficial layer, dorsal view. Anatomical abbreviations are defined in Appendix 3.

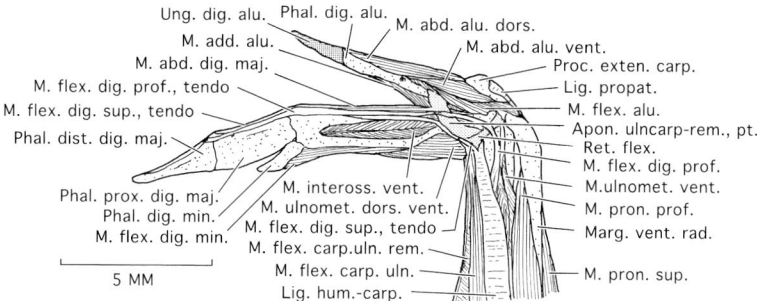

FIG. 108. Pectoral musculature of *Atlantisia rogersi* (DMNH 20762), distal antebrachium, carpus, and manus, superficial layer, ventral view. Anatomical abbreviations are defined in Appendix 3.

relationships for corresponding muscles differed markedly between flighted and flightless rails: m. pectoralis ($F = 53.4$; d.f. $= 2, 27$; $P < 0.00001$), m. supracoracoideus ($F = 23.6$; d.f. $= 2, 27$; $P < 0.00001$), and combined ($F = 48.7$; d.f. $= 2, 27$; $P < 0.00001$). As with the allometry between with length and body mass, these findings are consistent with pairs of allometric relationships showing virtually geometrical isometry (when using geometric-mean estimates for combined muscles, $\hat{b} = 0.99$) and common slope, but in which flighted and flightless taxa differ appreciably in intercepts. The resultant transposition of allometric intercepts ($\hat{a}$) in these relationships for which the common slope (b) is essentially equal to 1.0 (isometry) precludes the evaluation of the scale ratio (s) of White and Gould (1965), in that the exponent of the function under these circumstances is undefined (i.e., the exponent includes a divisor of $1 - b$, here being zero). In lieu of this formal parameterization, comparison of the mean masses of breast muscles reveals that mm. pectoralis et supracoracoideus of flightless rallids averaged about one fourth and one third, respectively, of those of flighted rallids (Table 67), and the foregoing models confirm that these differences are maintained isometrically.

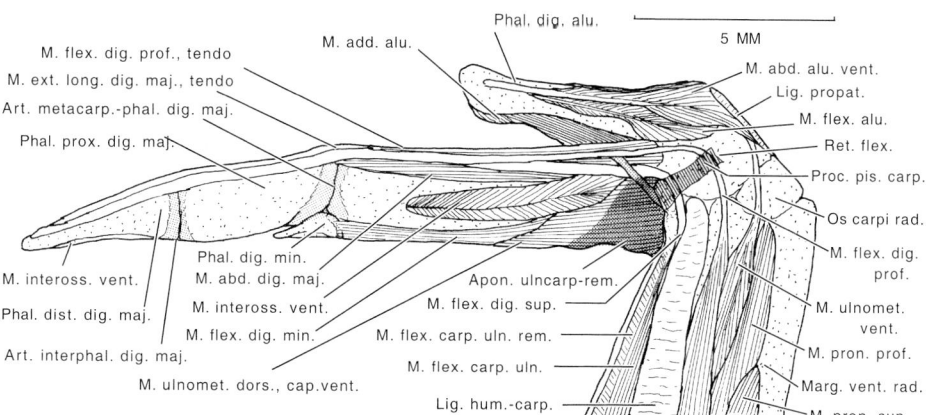

FIG. 109. Pectoral musculature of *Porzana palmeri* (BMNH 1940·12·8·13), detail of distal antebrachium, carpus, and proximal manus, superficial layer, ventral view. Anatomical abbreviations are defined in Appendix 3.

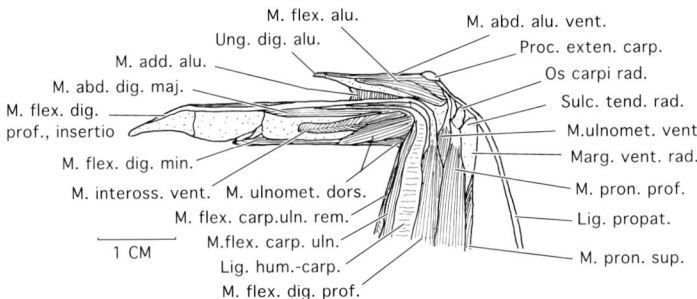

FIG. 110. Pectoral musculature of *Gallinula comeri* (BMNH 1922·12·6·221), distal antebrachium, carpus, and manus, superficial layer, ventral view. Anatomical abbreviations are defined in Appendix 3.

MULTIVARIATE ANALYSES OF MYOLOGICAL MEASUREMENTS

Analytical issues.—With few exceptions, data for each species were derived from single dissected specimens. This reduced the precision of positions of taxa in analyses derived from these data, for example, as compared to PCA of mean skeletal measurements. Also, limitations of dissections to the pectoral limb resulted in components indicative of general size being limited to that of the pectoral musculature, as opposed to a more inclusive assessment based on the entire musculature (including axial and pelvic components). The only exception to the problem of generality of size in comparisons of muscle measurements was analysis of residuals of measurements from mean body masses of the taxa concerned (see below). Third, the myological work generated a comparatively high dimensionality of measurements (81 variables), which, although informative and balanced against a reasonable representation of taxa, made for complexity of interpretation and comparison of multivariate axes. For these reasons, as well as emergent patterns described below, three different variants of PCA were applied to the myological data. The nontraditional analyses were chosen in particular to rectify the confounding of size and shape caused by outlying flightless taxa (PCA*) or the restriction of muscle measurements to those of the pectoral apparatus (PCA or residuals from mean body mass). Findings confirmed through all three methods were considered the best-substantiated of the myometric inferences.

Standard principal component analysis (PCA).—The first standard PC for muscle measurements was characteristic of axes descriptive of general size, having high, positive correlations with all measurements (Table 68) and a very high correlation with mean body masses of the taxa included ($r_i = 0.89$, $\forall\ i = 1, \ldots, p$; $P \ll 0.01$). PC-I alone accounted for more than 80% of the total dispersion among points (Table 68). Taxa were ordinated on this axis in a manner largely consonant with mean body mass, with tiny *Atlantisia rogersi* having the smallest score and *Porphyrio* spp. having the largest scores (Fig. 111). However, a modest obliquity of scores on PC-I with respect to body mass was evident at the largest extreme, however, in that *Porphyrio melanotus* had a slightly higher score on PC-I than its larger, flightless congener *P. hochstetteri*; a similar, modest transposition also pertained to flighted and flightless species of *Tribonyx* (Fig. 111).

The second PC-II for muscle measurements was defined by lesser correlations of both positive and negative sign (Table 68), and scores on this axis were significantly but less strongly correlated with mean body masses ($r = 0.33$; $P =$

0.05). However, the latter association explained the aforementioned discrepancies between scores on PC-I and body mass, in that the vector sum of correlations between the first two PCs and body mass revealed that body mass was oriented diagonally within this plane, and that taxa were ordinated along this line in a manner consistent with body mass (Fig. 111). Positions of flightless rails on PC-II mirrored to a significant extent the foreshortening of underlying skeletal elements that characterized these taxa, but a portion of the variance expressed by PC-II involved facets of the pectoral musculature not directly related to those of the associated skeleton (Table 68).

Correlation coefficients revealed that PC-II principally contrasted one dimension of the serratus complex and dimensions of the major muscles of the breast (m. pectoralis) and propatagium (m. biceps pars propatagialis) with the widths of origins of two components of the serratus complex (mm. serratus superficialis cranialis et profundus), one muscle acting on the brachium (m. latissimus dorsi pars caudalis), widths of the origins or bellies of three muscles acting on the antebrachium (mm. biceps brachii caput coracoideum, humertriceps, et pronator profundus), two dimensions of muscles acting on the manus and digits (mm. extensor metacarpi radialis pars ventralis et abductor alulae pars dorsalis), and the width of the origin of the major dermal slip acting on the metapatagium. These correlations indicate that flightless taxa had disproportionately reduced breast and biceps muscles (relative to a sample of muscles scattered throughout the pectoral apparatus) in comparison with their flighted relatives. The positions of juveniles relative to adult conspecifics (two species) implied comparatively minor shifts in parallel to those normative of flightless species, and suggested that a portion of the myological changes characteristic of flightless species reflected immaturity or were interpretable as paedomorphic (Fig. 111).

Residual myological shifts not included in PC-II, but showing consistent directionality with flightlessness, were relegated to PC-III (Table 68). Correlation coefficents for the third component primarily contrasted a muscle acting on the proximal antebrachium (m. brachialis) and two alular muscles (mm. extensor brevis alulae et adductor alulae) with a minority of muscles acting principally to restrain the scapula, brachium, and propatagium (Table 68). As on PC-II, flightless species tended to have higher scores on PC-III than their flighted relatives, indicating that loss of flight is accompanied by a disproportionate reduction in the bulk of m. brachialis and much of the alular musculature. The myological commonalities of flightless species on the two shape axes (PC-II and PC-III) were obfuscated by the diverse initial conditions approximated by their respective flighted relatives and the range of body sizes represented, and were only obvious when the inferred vectors of change between pairs of flighted and flightless relatives were superimposed on plots (Fig. 112). The latter vectors of approximate shift displayed a strong similarity of change associated with flightlessness (signified by the virtually uniform directionality of left-to-right shifts shown by flightless taxa on this shape plot), within which a minority of taxa (*Cyanolimnas cerverai, Gallirallus wakensis, Atlantisia rogersi, Porzana palmeri,* and *Pareudiastes pacificus*) were inferred to have undergone variably pronounced reductions in the overall bulk of the pectoral musculature in conjunction with the changes in shape (Fig. 112). An additional insight provided by this bivariate depiction of shape of the pectoral musculature is that the differences in position between two juvenile

TABLE 68. Correlation (r) and summary statistics for first three standard principal components of 81 measurements of pectoral muscles for 35 species of Rallidae, grouped following Raikow (1985a). Taxa are represented by single dissections, with the exception of *Porphyrio melanotus*, *P. hochstetteri*, *Atlantisia rogersi*, and *Porzana palmeri* (points based on averaged from two specimens). Also plotted (Figs. 111, 112) but not used in derivation of components were juvenile specimens of *Habropteryx torquatus* and *Gallinula chloropus*. Abbreviations for muscles are given in Appendix 3.

Variable by primary element*	Correlation coefficient (r)		
	PC-I	PC-II	PC-III
Patagia and respiration			
M. lat. dors. metap. MOW	0.801	0.255	0.075
M. serr. sup. metap. MOW†	0.709	−0.012	0.171
M. pect. subcut. thor. IW	0.803	0.026	−0.041
M. sternocor. MOW	0.880	0.088	−0.093
M. delt. propat. MOW	0.790	−0.168	0.158
FBL	0.944	−0.017	0.118
MBW	0.865	−0.152	0.039
M. delt. propat. tend. brev. IW	0.756	0.017	0.326
M. bic. propat. FBL	0.867	−0.004	0.101
Scapula			
M. rhom. sup. MOW	0.930	0.078	0.064
M. rhom. prof. MOW	0.936	−0.036	0.002
M. serr. sup. cran. MOW	0.796	0.408	−0.002
M. serr. sup. caud. MOW†	0.700	−0.285	0.271
M. serr. prof. MOW†	0.856	0.217	0.314
Humerus			
M. lat. dors. cran. MOW	0.888	0.144	0.134
M. lat. dors. caud. MOW	0.823	0.322	0.233
M. scap.-hum. cran. MOW	0.834	0.094	0.079
M. scap.-hum. caud. TBL	0.929	0.067	0.104
M. subscap. med. MOW	0.899	0.164	−0.043
M. subcor. vent. MOW	0.800	0.176	−0.012
M. subcor. dors. MOW	0.854	0.184	0.132
M. pect. sternobra. TBL	0.987	0.020	−0.006
MBW	0.932	−0.225	−0.071
Mass	0.935	−0.312	−0.038
M. supracor. TBL	0.975	0.026	−0.032
MBW	0.897	−0.304	−0.103
Mass	0.959	−0.159	−0.061
M. cor.-brach. cran. FBL	0.909	0.141	0.075
MBW	0.854	0.133	0.117
M. cor.-brach. caud. FBL	0.936	0.114	0.077
M. delt. maj. cran. et caud. MOW	0.657	0.029	0.442
IW	0.943	−0.075	0.071
TBL	0.976	−0.030	0.003
MBW	0.930	−0.053	0.073
M. delt. min. dors. TBL	0.917	0.089	−0.091
M. delt. min. vent. TBL	0.965	−0.008	−0.043
Antebrachium			
M. bic. bra. cor. MOW	0.801	0.283	−0.030
TBL	0.984	0.013	−0.145
M. biceps brach. hum. TBL	0.960	0.047	−0.008
MBW	0.798	−0.278	0.312
M. scap.-tric. TBL‡	0.983	0.058	−0.070
MBW	0.892	0.042	−0.014
M. hum.-tric. TBL‡	0.986	0.021	−0.006
MBW	0.886	0.202	0.008
M. brach. IW	0.859	0.089	−0.324

TABLE 68. Continued.

Variable by primary element*	Correlation coefficient (r)		
	PC-1	PC-II	PC-III
M. pron. sup. FBL	0.936	−0.036	−0.041
MBW	0.908	0.027	0.055
M. pron. prof. FBL	0.969	0.004	−0.034
MBW	0.893	0.215	0.007
M. ect.-uln. FBL	0.958	−0.025	−0.010
M. sup. FBL	0.956	0.040	−0.077
Proximal manus			
M. flex. carp. uln. cran. FBL	0.971	−0.028	−0.083
MBW	0.830	0.166	0.099
M. flex. carp. uln. caud. FBL	0.943	−0.017	−0.172
M. ulnomet. dors. FBL	0.939	−0.014	0.044
M. ulnomet. vent. FBL	0.891	0.023	−0.124
M. ext. metacarp. rad. dors. FBL	0.946	0.043	−0.082
MBW	0.893	0.027	0.178
M. ext. metacarp. rad. vent. FBL	0.977	−0.000	−0.029
MBW	0.878	0.233	0.113
M. ext. metacarp. uln. FBL	0.956	−0.002	−0.149
MBW	0.920	−0.002	0.052
Digiti			
M. flex. dig. sup. FBL	0.970	−0.086	−0.079
MBW§	0.919	−0.013	−0.038
M. flex. dig. prof. FBL	0.961	−0.083	−0.129
MBW	0.934	−0.102	0.003
M. ext. dig. comm. FBL	0.967	−0.016	−0.025
MBW	0.891	0.171	−0.074
M. ext. long. dig. maj. prox. FBL	0.943	−0.033	−0.130
M. ext. long. alu. uln. FBL	0.930	0.037	−0.128
M. ext. long. alu. rad. FBL	0.910	0.191	−0.060
M. ext. brev. alu. FBL	0.816	0.147	−0.422
M. abd. alu. dors. FBL	0.853	0.281	−0.055
M. abd. alu. vent. FBL	0.882	0.068	−0.036
M. add. alu. OW	0.872	0.088	−0.277
M. flex. alu. FBL	0.854	0.081	0.091
M. abd. dig. maj. FBL	0.949	−0.064	−0.007
M. inteross. dors. FBL	0.934	−0.101	−0.167
M. inteross. vent. FBL	0.947	−0.092	−0.112
MBW	0.876	0.192	−0.190
M. flex. dig. min. FBL	0.944	−0.148	−0.006
Correlation (r) with body mass‖	0.891	0.328	0.059
Eigenvalue (λ_i)	13.546	0.445	0.355
Percentage of variance explained	80.3	2.6	2.1

* Dimensions abbreviated as follows: MOW, maximal width at origo; IW, width at insertio; FBL, length of fibrous belly, or belly exclusive of tendines; TBL, total length of belly and tendines; MBW, maximal width of belly; Mass, total mass of damp, detached muscle.
† Summed over all fasciculi.
‡ Distal limit taken to be apex of angulus of elbow.
§ Defined as width at midpoint of the enclosing ligamentum humerocarpale.
‖ $n = 35$.

specimens of flighted rails (*Habropteryx torquatus* and *Gallinula chloropus*) and their adult counterparts were of similar direction as the shifts shown by adult specimens of flightless relatives, confirming that paedomorphosis of the musculature includes both size and shape components (Fig. 112).

Modified principal component analysis (PCA*).—An alternative method to the preceeding is the modified version of PCA (PCA*), in which the tendency for

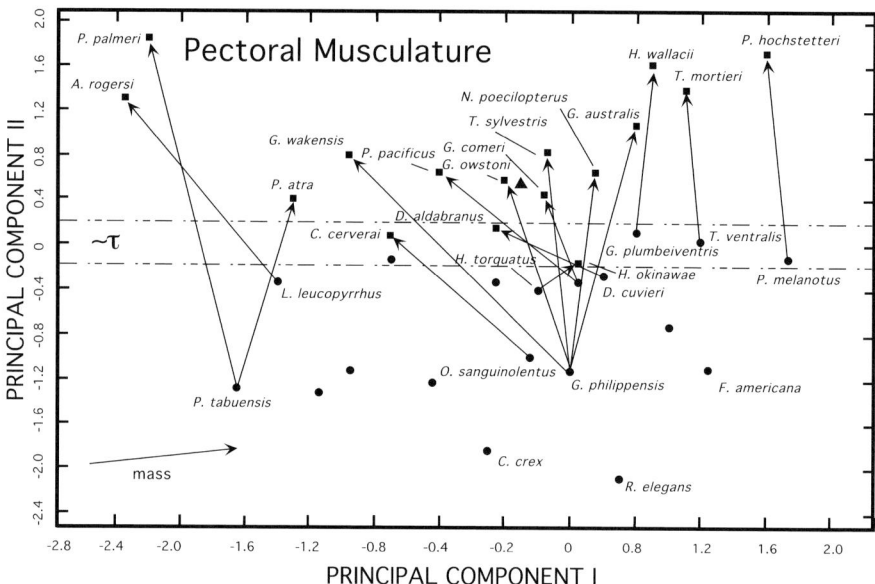

Fig. 111. Plot of mean scores for 35 species of Rallidae on first two standard (R-mode) principal components of 81 measurements of the pectoral musculature. Flighted species are symbolized by circles, flightless species by squares, and a single juvenile specimen of *Habropteryx torquatus* by a triangle. Inferred vectors of change connect flightless species and their flighted relatives. Approximate orientation of vector for mean body masses of taxa ("mass") and approximate threshold of flightlessness ($\sim\tau$) are indicated.

outlying, flightless species to confound the partitioning of size from flightlessness-related shape was countered (see Appendix 2). For the myolometric data, this method defined a first modified component (PC-I*) that corresponded to general size, and did so more clearly than the first component for the standard PCA (Table 69) in that both the correlation coefficients between original variables and PC-I* frequently were higher and the correlation between scores on PC-I* and mean body mass was decreased only negligibly ($r = 0.87$, $P \ll 0.01$; a reduction of only 0.02). Positions of taxa on PC-I*, although broadly congruent with rankings by mean body mass, are best understood when it is recalled that this size metric reflects general size of pectoral muscles, which may have changed variously with overall body size among lineages (Fig. 113).

The second modified component (PC-II*) conveyed the multiplicity of variables manifesting changes in the loss of flight, contributed another 5% of the total variance in the data set, and was not significantly correlated with body mass (Table 69). Correlation coefficients for the axis were mostly positive but of widely variable magnitudes, with dimensions of only two dermal slips (mm. pectoralis subcutaneous thoracicus et biceps propatagialis) having negative correlations of considerable magnitude. The combined positions of taxa on PC-I* and PC-II* revealed five broad classes of trajectories related to the loss of flight (Fig. 113): taxa having undergone decreases in overall size of pectoral musculature (PC-I*) and increases in the proportions indicated by PC-II* (e.g., *Tricholimnas sylvestris, Porzana palmeri, Pareudiastes pacificus,* and *Gallinula comeri*); taxa having undergone decreases in overall size of pectoral musculature (PC-I*) and decreases

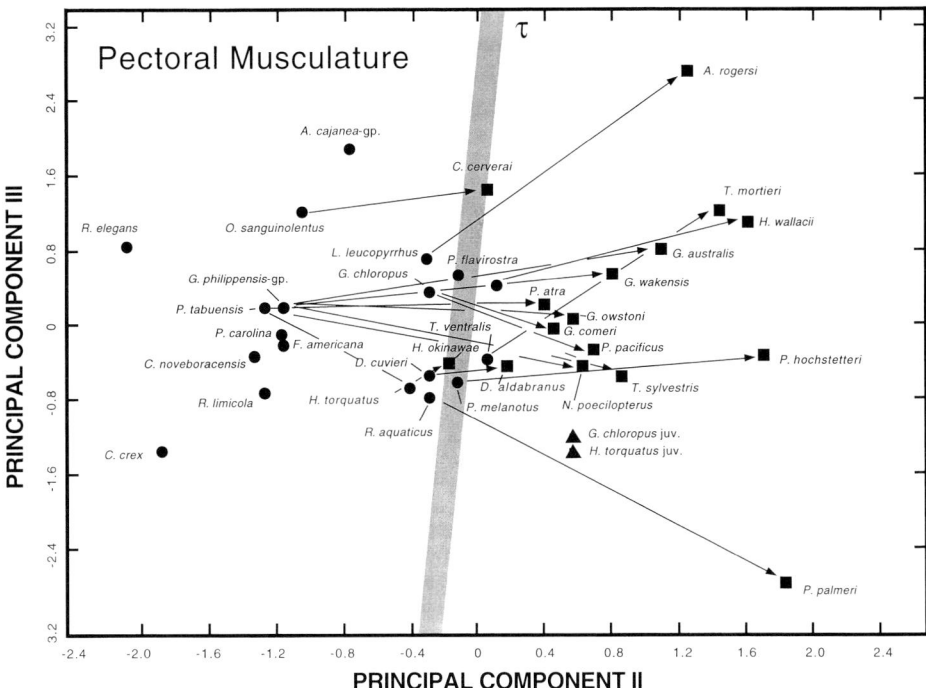

FIG. 112. Plot of mean scores for 35 species of Rallidae on the second and third standard (R-mode) principal components of 81 measurements of the pectoral musculature (i.e., the two primary axes of "shape"). Flighted species are symbolized by circles, flightless species by squares, and juvenile specimens of *Habropteryx torquatus* and *Gallinula chloropus* by triangles. Approximate orientation of vector for mean body mass of taxa and approximate threshold of flightlessness (~τ) are indicated.

in the proportions indicated by PC-II* (e.g., *Cyanolimnas cerverai, Dryolimnas aldabranus, Gallirallus wakensis,* and *Atlantisia rogersi*); taxa having undergone increases in overall size of pectoral musculature (PC-I*) and increases in the proportions indicated by PC-II* (e.g., *Habropteryx okinawae*); taxa having undergone increases in overall size of pectoral musculature (PC-I*) and decreases in the proportions indicated by PC-II* (e.g., *Habroptila wallacii* and *Nesoclopeus poecilopterus*); and a taxon that underwent a modest decrease in overall size of the pectoral musculature and virtually no change in proportions depicted by PC-II* (*Porphyrio hochstetteri*).

The third modified component (PC-III*) accounted for another 3% of the total variance and, like PC-II*, was uncorrelated with body mass, and scores indicated that the axis included important information concerning the transition to flightlessness (Table 69). Perusal of the correlations indicated that this axis essentially contrasted the dimensions of a suite of muscles including the major breast muscles (mm. pectoralis sternobrachialis et supracoracoideus), several major flexors and extensors of the digitus majoris (mm. flexor digiti superficialis, flexor digiti profundus, et interossea), and two dermal slips (mm. latissimus dorsi metapatagialis et biceps propatagialis) with a diverse assemblage of muscles from throughout the pectoral apparatus (most notably mm. serratus superficialis pars cranialis, latissimus dorsi pars caudalis, extensor metacarpi radialis pars ventralis, and most

TABLE 69. Correlation coefficients (r) and summary statistics for first three custom principal components (PC*) of 81 measurements of pectoral muscles for 35 species of Rallidae, grouped following Raikow (1985a). Taxa are represented by single dissections, with the exception of *Porphyrio melanotus, P. hochstetteri, Atlantisia rogersi,* and *Porzana palmeri* (points based on averages from two specimens). Also plotted (Fig. 114) but not used in derivation of components were juvenile specimens of *Habropteryx torquatus* and *Gallinula chloropus.* Abbreviations for muscles are given in Appendix 3.

Variable by primary element*	Correlation coefficient (r)		
	PC-I*	PC-II*	PC-III*
Patagia and respiration			
M. lat. dors. metap. MOW	0.748	0.496	−0.309
M. serr. sup. metap. MOW†	0.719	0.189	−0.075
M. pect. subcut. thor. IW	0.846	−0.155	−0.052
M. sternocor. MOW	0.859	0.449	−0.079
M. delt. propat. MOW	0.780	0.291	0.185
FBL	0.931	0.351	0.076
MBW	0.847	0.453	0.245
M. delt. propat. tend. brev. IW	0.813	−0.443	−0.102
M. bics. propat. FBL	0.885	0.033	0.019
Scapula			
M. rhom. sup. MOW	0.900	0.593	−0.149
M. rhom. prof. MOW	0.898	0.630	0.148
M. serr. sup. cran. MOW	0.788	0.371	−0.631
M. serr. sup. caud. MOW†	0.702	0.020	0.329
M. serr. prof. MOW†	0.864	0.136	−0.467
Humerus			
M. lat. dors. cran. MOW	0.871	0.386	−0.268
M. lat. dors. caud. MOW	0.828	0.191	−0.535
M. scap.-hum. cran. MOW	0.804	0.395	−0.118
M. scap.-hum. caud. TBL	0.913	0.381	−0.134
M. subscap. med. MOW	0.866	0.586	−0.245
M. subcor. vent. MOW	0.792	0.341	−0.277
M. subcor. dors. MOW	0.843	0.393	−0.308
M. pect. sternobra. TBL	0.974	0.715	0.009
MBW	0.902	0.572	0.548
Mass	0.909	0.489	0.798
M. supracor. TBL	0.959	0.546	0.045
MBW	0.880	0.292	0.717
Mass	0.933	0.537	0.574
M. cor.-brach. cran. FBL	0.897	0.455	−0.310
MBW	0.861	0.224	−0.308
M. cor.-brach. caud. FBL	0.926	0.344	−0.240
M. delt. maj. cran. et caud. MOW	0.687	−0.032	−0.111
IW	0.940	0.227	0.266
TBL	0.953	0.681	0.184
MBW	0.921	0.509	0.167
M. delt. min. dors. TBL	0.886	0.678	−0.135
M. delt. min. vent. TBL	0.933	0.736	0.069
Antebrachium			
M. bic. bra. cor. MOW	0.764	0.535	−0.363
TBL	0.964	0.745	0.056
M. biceps brach. hum. TBL	0.941	0.541	−0.061
MBW	0.834	−0.284	0.460
M. scap.-tric. TBL‡	0.955	0.843	−0.099
MBW	0.863	0.624	−0.015
M. hum.-tric. TBL‡	0.968	0.756	0.000
MBW	0.877	0.488	−0.374
M. brach. IW	0.809	0.745	−0.086

TABLE 69. Continued.

Variable by primary element*	Correlation coefficient (r)		
	PC-I*	PC-II*	PC-III*
M. pron. sup. FBL	0.927	0.419	0.167
MBW	0.909	0.276	−0.060
M. pron. prof. FBL	0.961	0.544	0.074
MBW	0.869	0.572	−0.387
M. ect.-uln. FBL	0.956	0.358	0.145
M. sup. FBL	0.930	0.698	−0.052
Proximal manus			
M. flex. carp. uln. cran. FBL	0.951	0.648	0.165
MBW	0.838	0.370	−0.277
M. flex. carp. uln. caud. FBL	0.917	0.662	0.122
M. ulnomet. dors. FBL	0.920	0.362	0.129
M. ulnomet. vent. FBL	0.866	0.562	−0.031
M. ext. metacarp. rad. dors. FBL	0.917	0.684	−0.075
MBW	0.904	0.260	−0.049
M. ext. metacarp. rad. vent. FBL	0.960	0.649	0.075
MBW	0.882	0.312	−0.474
M. ext. metacarp. uln. FBL	0.939	0.634	0.077
MBW	0.920	0.219	0.038
Digiti			
M. flex. dig. sup. FBL	0.951	0.655	0.318
MBW§	0.889	0.650	−0.020
M. flex. dig. prof. FBL	0.939	0.573	0.351
MBW	0.936	0.296	0.339
M. ext. dig. comm. FBL	0.954	0.625	0.079
MBW	0.872	0.499	−0.276
M. ext. long. dig. maj. prox. FBL	0.926	0.508	0.184
M. ext. long. alu. uln. FBL	0.900	0.655	−0.014
M. ext. long. alu. rad. FBL	0.905	0.349	−0.384
M. ext. brev. alu. FBL	0.774	0.614	−0.072
M. abd. alu. dors. FBL	0.831	0.280	−0.342
M. abd. alu. vent. FBL	0.878	0.353	−0.078
M. add. alu. OW	0.860	0.587	−0.107
M. flex. alu. FBL	0.885	0.076	−0.196
M. abd. dig. maj. FBL	0.937	0.483	0.185
M. inteross. dors. FBL	0.904	0.663	0.310
M. inteross. vent. FBL	0.922	0.618	0.314
MBW	0.860	0.465	−0.267
M. flex. dig. min. FBL	0.946	0.330	0.165
Correlation (r) with body mass‖	0.870	−0.171	−0.137
Correlation (r) with PC-I*	—	−0.410	0.173
Eigenvalue (λ_i)	11.020	0.838	0.460
Percentage of variance explained	73.1	5.6	3.5

* Dimensions abbreviated as follows: MOW, maximal width at origo: IW, width at insertio; FBL, length of fibrous belly, or belly exclusive of tendines; TBL, total length of belly and tendines; MBW, maximal width of belly; Mass, total mass of damp, detached muscle.
† Summed over all fasciculi.
‡ Distal limit taken to be apex of angulus of elbow.
§ Defined as width at midpoint of the enclosing ligamentum humerocarpale.
‖ $n = 35$.

muscles acting on the alula). Examination of relative scores revealed that flightless species tended to have lower scores on PC-III* than did their flighted relatives (Fig. 113).

Scores on PC-III*, taken together with those on the other shape-axis PC-II*, revealed that transitions to flightlessness were accompanied by a two-dimensional

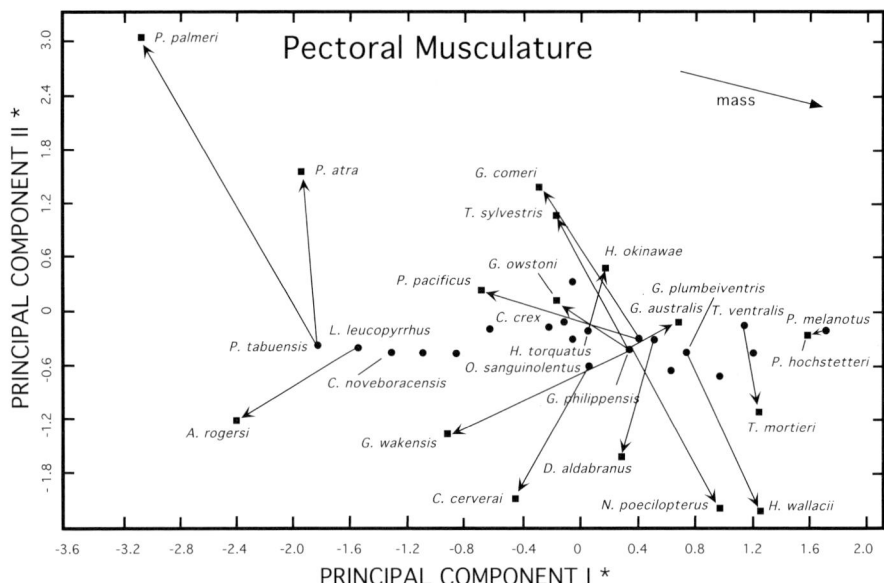

Fig. 113. Plot of mean scores for 35 species of Rallidae on the first two modified (*R*-mode) principal components (PCA∗) of 81 measurements of the pectoral musculature. Inferred vectors of change connect flightless species and their flighted relatives. Approximate orientation of vector for mean body mass of taxa is indicated.

space descriptive of relative proportions of pectoral musculature, and that encapsulated a semicircular radiation of shifts in shape, including (Fig. 114): species that underwent variable decreases on PC-II∗, and modest (e.g., *Cyanolimnas cerverai*, *Dryolimnas aldabranus*, and *Nesoclopeus poecilopterus*), moderate (*Habroptila wallacii*, *Gallirallus wakensis*, and *Tribonyx mortierii*), or pronounced decreases (*Atlantisia rogersi*) on PC-III∗; species showing limited change on PC-II∗ and moderately large decreases on PC-III∗ (e.g., *Porphyrio hochstetteri*, *Gallirallus australis*, and *G. owstoni*), shifts most similar to that differentiating juvenile *Gallinula* and *Habropteryx* from adult conspecifics; and species showing variably pronounced increases on PC-II∗, and either no change (*Habropteryx okinawae*), moderate decreases (e.g., *Tricholimnas sylvestris*, *Porzana atra*, and *Gallinula comeri*), or a large decrease (*Porzana palmeri*) on PC-III∗.

Principal component analysis of residuals from body mass (PCA|mass).—A final refinement was made to remove size as intuitively understood from subsequent dimensions interpretable as shape, whereby all variation attributable to mean body masses was defined as size and residuals therefrom were subjected to standard PCA for definition of shape (Table 70). The variance in muscle measurements attributable to mean body mass accounted for almost two thirds of the total myolometric dispersion, was highly significant for all dimensions ($16.9 < F_i < 221.1$, $P < 0.0005$), and provided a predictable ordination of species by overall size (Fig. 115).

The first PC for the residuals of body mass (PC-I|mass) axis accounted for more than one half of the residual variance (Table 70). Correlation coefficients for variables were all positive but ranged in magnitude from 0.11 to 0.95, a

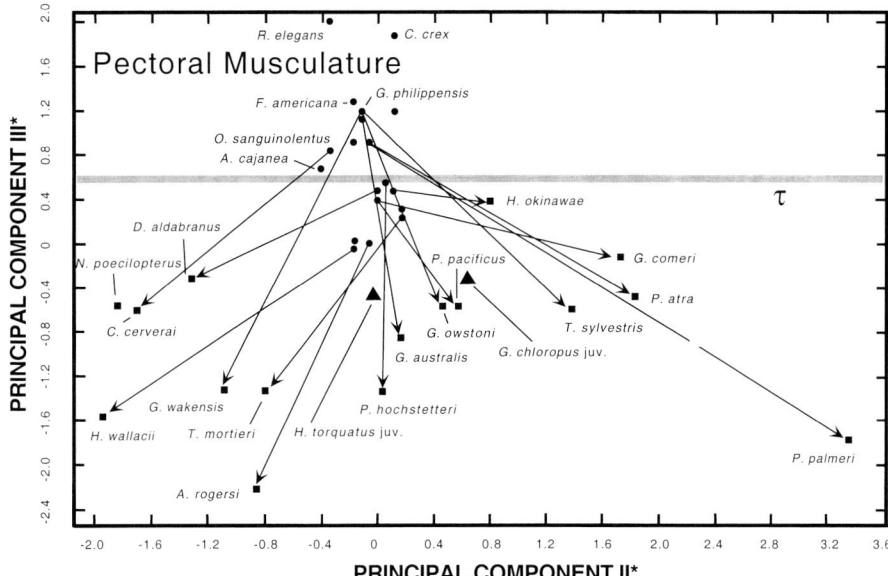

FIG. 114. Plot of mean scores for 35 species of Rallidae on the second and third modified (R-mode) principal components (PCA*) of 81 measurements of the pectoral musculature (i.e., the two primary axes of "shape"). Flighted species are symbolized by circles, flightless species by squares, and single juvenile specimens of *Habropteryx torquatus* and *Gallinula chloropus* are indicated by triangles. Approximate orientation of vector for mean body masses of taxa ("mass") and approximate threshold of flightlessness (τ, positioned on the flightless side of boundary) are indicated.

condition characteristic of first shape components that are not strictly orthogonal to the associated size axis (Table 70). Furthermore, positions of flightless taxa in comparison with their flighted relatives indicated that this first component for residuals largely served as a distillation of shifts associated with the loss of flight, in which flightless taxa tended to have markedly lower scores (Fig. 115). In combination with the changes in overall size inferred previously, these shifts produced a downward radiation of flightless taxa in the bivariate plot, the magnitudes of which reflected both the starting position represented by the flighted taxon and the ending position signified by the flightless relative (Fig. 115). Magnitudes of correlation coefficients indicated that the essential contrast achieved by this axis was a contrast between the comparatively massive muscles of the breast, pectoral girdle, and proximal wing elements and those of the manus, digits, and associated dermal components (Table 70).

The second PC for residuals from body mass (PC-II|mass) accounted for one tenth of the variation summarized by the first residual axis (described above). Although both it and the third component for residuals (PC-III|mass) provided additional discrimination of flightless taxa from their flighted relatives, these minor components of the transitions provide no significant insights and are relegated to supplements to the overall distances compared below.

SALIENT FLIGHTLESSNESS-RELATED TRENDS IN PECTORAL MUSCULATURE

Qualitative signatures.—Despite the structural complexity of the musculature of the avian pectoral girdle and wing (more than 55 muscles or parts thereof),

TABLE 70. Correlation coefficients (r) and summary statistics for first three principal components of residuals of 81 measurements of pectoral muscles from $\log_e$-transformed mean body masses (PCA|mass) for 35 species of Rallidae, grouped following Raikow (1985a). Taxa are represented by single dissections, with the exception of *Porphyrio melanotus*, *P. hochstetteri*, *Atlantisia rogersi*, and *Porzana palmeri* (points based on averages from two specimens). Also plotted (Fig. 115) but not used in derivation of components were juvenile specimens of *Habropteryx torquatus* and *Gallinula chloropus*. Abbreviations for muscles are given in Appendix 3.

	Correlation coefficient (r)		
Variable by primary element*	PC-I (residual)	PC-II (residual)	PC-III (residual)
Patagium and respiration			
M. lat. dors. metap. MOW	0.307	0.107	0.341
M. serr. sup. metap. MOW†	0.504	−0.262	0.194
M. pect. subcut. thor. IW	0.507	0.214	0.108
M. sternocor. MOW	0.601	0.163	−0.022
M. delt. propat. MOW	0.632	−0.139	0.195
FBL	0.781	−0.322	0.053
MBW	0.773	0.046	0.178
M. delt. propat. tend. brev. IW	0.460	−0.417	0.258
M. bic. propat. FBL	0.539	−0.427	−0.164
Scapula			
M. rhom. sup. MOW	0.654	−0.188	−0.152
M. rhom. prof. MOW	0.757	−0.056	−0.124
M. serr. sup. cran. MOW	0.115	0.207	0.178
M. serr. sup. caud. MOW†	0.507	−0.579	−0.320
M. serr. prof. MOW†	0.364	−0.423	0.457
Humerus			
M. lat. dors. cran. MOW	0.430	−0.163	0.167
M. lat. dors. caud. MOW	0.313	−0.105	0.525
M. scap.-hum. cran. MOW	0.479	−0.142	0.047
M. scap.-hum. caud. TBL	0.656	−0.336	−0.198
M. subscap. med. MOW	0.600	0.010	−0.009
M. subcor. vent. MOW	0.367	0.121	−0.089
M. subcor. dors. MOW	0.500	0.009	0.379
M. pect. sternobra. TBL	0.935	−0.000	−0.012
MBW	0.922	−0.020	−0.103
Mass	0.953	0.017	−0.092
M. supracor. TBL	0.884	0.011	−0.077
MBW	0.862	0.024	−0.201
Mass	0.932	0.119	−0.034
M. cor.-brach. cran. FBL	0.602	0.043	0.223
MBW	0.527	−0.256	−0.024
M. cor.-brach. caud. FBL	0.659	−0.250	−0.026
M. delt. maj. cran. et caud. MOW	0.366	0.229	0.607
IW	0.812	−0.261	−0.119
TBL	0.907	−0.101	−0.078
MBW	0.843	0.077	0.207
M. delt. min. dors. TBL	0.679	0.220	−0.079
M. delt. min. vent. TBL	0.855	0.075	−0.084
Antebrachium			
M. bic. bra. cor. MOW	0.304	0.088	−0.061
TBL	0.925	−0.007	−0.164
M. biceps brach. hum. TBL	0.803	−0.126	−0.236
MBW	0.692	−0.404	0.107

TABLE 70. Continued.

	Correlation coefficient (r)		
Variable by primary element*	PC-I (residual)	PC-II (residual)	PC-III (residual)
M. scap.-tric. TBL‡	0.916	0.097	−0.170
MBW	0.668	0.121	0.235
M. hum.-tric. TBL‡	0.940	−0.038	−0.099
MBW	0.628	0.092	0.401
M. brach. IW	0.590	0.462	−0.175
M. pron. sup. FBL	0.820	−0.085	−0.098
MBW	0.718	0.084	0.351
M. pron. prof. FBL	0.879	−0.012	−0.033
MBW	0.611	0.294	0.546
M. ect.-uln. FBL	0.843	−0.127	−0.101
M. sup. FBL	0.815	0.054	−0.206
Proximal manus			
M. flex. carp. uln. cran. FBL	0.896	0.030	−0.227
MBW	0.413	0.123	0.366
M. flex. carp. uln. caud. FBL	0.837	0.191	−0.177
M. ulnomet. dors. FBL	0.798	−0.088	0.038
M. ulnomet. vent. FBL	0.717	0.180	−0.002
M. ext. metacarp. rad. dors. FBL	0.762	0.184	−0.028
MBW	0.678	0.012	0.453
M. ext. metacarp. rad. vent. FBL	0.894	−0.008	−0.111
MBW	0.521	0.127	0.498
M. ext. metacarp. uln. FBL	0.837	0.135	−0.238
MBW	0.731	−0.076	−0.029
Digiti			
M. flex. dig. sup. FBL	0.915	−0.007	−0.214
MBW§	0.716	−0.022	−0.177
M. flex. dig. prof. FBL	0.892	0.052	−0.263
MBW	0.804	−0.052	−0.243
M. ext. dig. comm. FBL	0.896	0.035	0.090
MBW	0.586	0.351	0.275
M. ext. long. dig. maj. prox. FBL	0.801	0.106	−0.243
M. ext. long. alu. uln. FBL	0.763	0.282	0.078
M. ext. long. alu. rad. RBL	0.613	0.188	0.129
M. ext. brev. alu. FBL	0.489	0.658	−0.089
M. abd. alu. dors. FBL	0.293	0.358	0.242
M. abd. alu. vent. FBL	0.596	0.016	0.006
M. add. alu. OW	0.676	0.293	−0.164
M. flex. alu. FBL	0.571	0.010	0.150
M. abd. dig. maj. FBL	0.847	−0.032	−0.071
M. inteross. dors. FBL	0.848	0.096	−0.219
M. inteross. vent. FBL	0.876	0.049	−0.134
MBW	0.521	0.340	−0.101
M. flex. dig. min. FBL	0.818	−0.069	−0.053
Eigenvalue (λ_j)	3.197	0.362	0.302
Percentage of variance explained‖	51.5	5.8	4.9

* Dimensions abbreviated as follows: MOW, maximal width at origo: IW, width at insertio; FBL, length of fibrous belly, or belly exclusive of tendines; TBL, total length of belly and tendines; MBW, maximal width of belly; Mass, total mass of damp, detached muscle.
† Summed over all fasciculi.
‡ Distal limit taken to be apex of angulus of elbow.
§ Defined as width at midpoint of the enclosing ligamentum humerocarpale.
‖ Percentages of residual variance for regression on log-transformed mean body masses of species ($n = 35$), with the regression accounting for 63% of the total variance.

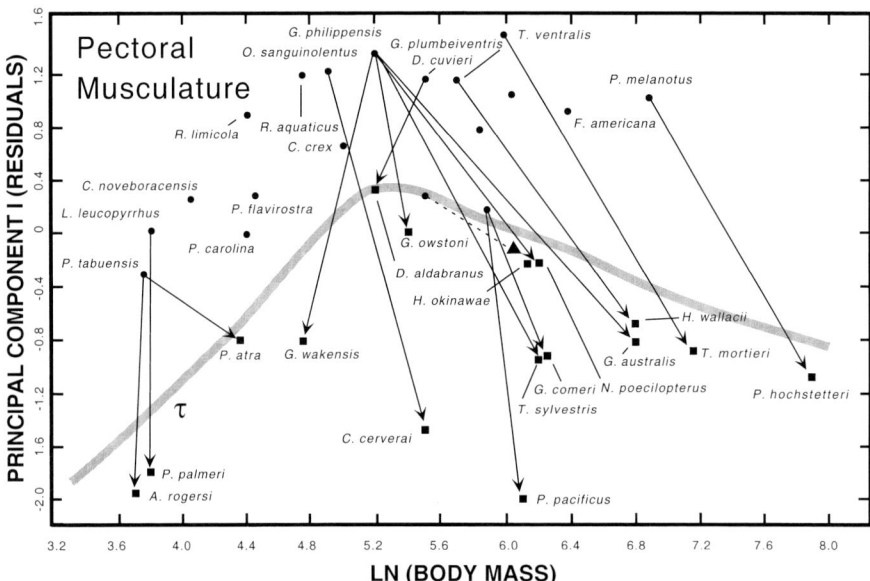

FIG. 115. Plot of mean scores for 35 species of Rallidae on the first two (*R*-mode) principal components of residuals of 81 measurements of the pectoral musculature corrected for mean body masses. Flighted species are symbolized by circles, flightless species by squares, and a single juvenile specimen of *Habropteryx torquatus* is indicated by a triangle. Approximate orientation of vector for mean body masses of taxa ("mass") and approximate threshold of flightlessness (τ, positioned on the flightless side of boundary) are indicated.

qualitative distinctions between 17 flightless rails and 20 flighted confamilials were few. Undoubtedly, this uniformity stemmed in part from obfuscation of subtle details by the vagaries of sex, age, provenance, and preservation of the specimens examined. Some aspects of the musculature suggestive of reduction (e.g., robustness of the muscle bellies, extent of the fibrous portion of the muscles relative to adjacent skeletal elements, or development of tendons) characterized a number of flightless rails but defied precise description or comparison among species of different body sizes. Also, even flighted rails are notable for comparatively small pectoral musculature and disproportionately large pelvic muscles (Hartman 1961; Greenewalt 1962, 1975). For example, out of more than 80 (flighted) avian taxa compiled by Hartman (1961: table 3), the single rallid sampled (*Laterallus albigularis*) had the smallest relative mass of pectoral muscles and the second-highest relative mass of pelvic muscles. Accordingly, loss of flight left few qualitative signatures on the pectoral musculature of rails to be revealed through intrafamilial comparisons (cf. Lowe 1928a; McGowan 1986), a finding consistent with comparable myological studies of other carinate birds (e.g., Livezey 1989d, 1990, 1992a, b), penguins excepted (Lowe 1933; Schreiweis 1982). Also profoundly different in the magnitudes of flightlessness-related muscular changes were those noted for ratites, which include losses of entire muscles and parts thereof (Fürbringer 1888; Lowe 1928b; McGowan 1982).

The most conspicuous qualitative features of flightless rails concern the substantial reduction of m. pectoralis pars sternobrachialis (thoracica), a feature perhaps most readily revealed by a dorsoventral shallowness of the carina sterni, caudal shift of the apex carina, articulatio sternoclavicularis, and associated conformational changes in m. pectoralis pars sternobrachialis (thoracica) et m. cucullaris capitis pars clavicularis (Figs. 69–72). Parallel reductions in robustness and extent also characterized m. supracoracoideus (Figs. 73, 74), conformational apomorphies reflected only to a limited extent by the impressio m. supracoracoideus on the underlying sternum (Figs. 61, 67). Shortening and reductions in the fibrous portions of several patagial slips (mm. latissimus dorsi pars metapatagialis et serratus superficialis pars metapatagialis), as well as reductions in fibrous portions of mm. rhomboideus profundus were secondary indicators of flightlessness in rallids. Other, less uniformly distributed muscular characters indicative of flightlessness (Figs. 69–110) included distinctly feeble or truncated bellies of the brachium (e.g., mm. biceps brachii, scapulotriceps, et deltoideus major), antebrachium (mm. extensor digiti communis, extensor carpi ulnaris, et flexor ulnometacarpalis ventralis), and manus (e.g., m. extensor brevis alulae). Conversely, several muscles of the antebrachium (mm. pronator superficialis, brachialis, et ectepicondylo-ulnaris) appeared in situ to be hyperdeveloped in some flightless rallids, evidently an aspect stemming from the reduction of the underlying radius and ulna in these species. This condition of one of these antebrachial muscles (m. brachialis) represents the myological counterpart of a notable osteological feature of most flightless rails, the relatively deep and extensive impressio m. brachialis (Table 64).

Quantitative commonalities.—Perhaps the most striking myological finding was that flightless rallids have breast muscles that comprised an average of 2.2% of their body masses, an estimate 9–13% less than that inferred for flighted confamilials (Table 67). This average difference compares favorably with published figures for fresh muscles weights, being intermediate between the extremely small relative muscle masses for *Porphyrio hochstetteri* and the only slightly reduced muscle masses reported for *Gallirallus owstoni* (Table 66). Both sets of data are doubly remarkable in light of the fact that even flighted rallids as a group have small breast muscles in comparison with other families of neognathous birds (Hartman 1961), a generality that would tend to diminish the clarity of differences between subgroups of the family in this regard.

Comparatively subtle but nonetheless informative patterns emerged from the multivariate analyses of linear measurements of pectoral muscles (Tables 68–70; Figs. 111–115). In all multivariate assessments, most flightless species were inferred to have undergone variably large increases in general size, with a minority of rails (*Dryolimnas aldabranus, Gallirallus wakensis,* and *Cabalus modestus*) and crakes (e.g., *Atlantisia rogersi* and *Porzana palmeri*) manifesting modest dwarfism (Figs. 111–115). Despite modest interpretational challenges posed by varying partitions of shape by different algorithms (PCA, PCA∗, and PCA|residuals from mass) and a tendency for the first two approaches to confound size of some of the pectoral muscles with general size, flightless taxa generally were characterized by disproportionate reductions in the major breast muscles (mm. pectoralis sternobrachialis et supracoracoideus), dermal slips associated with the propatagium (m. biceps propatagialis) or metapatagium (m. serratus superficialis metapatagialis), and a suite of flexors and extensors acting on the digiti majoris et alularis. These shared patterns reveal that flightlessness tends to affect the major, calorically expensive breast muscles most intensively, followed by the (much smaller) pectoral muscles that have little or no function in the absence of the demands of flight, for example, patagial slips and muscles acting on the manus and digits that are critical for aerial maneuverability (Tables 68–70). Across all myological analyses, these overall shifts in muscular shape were comparatively great in *Porzana palmeri* and *Atlantisia rogersi,* and comparatively slight in *Dryolimnas aldabranus, Habropteryx okinawae,* and *Gallinula comeri,* with other flightless taxa having undergone moderately large but variable shifts in proportions (Figs. 111–115).

Summary of Apomorphy in Flightless Rails

Qualitative and Univariate Metrics

A number of variably interrelated indices to the changes undergone by flightless rails have been presented. Such indices to changes in body mass and measurements of the externum provide robust, synthetic insights into apomorphy (Table 71), with the most intuitively obvious being absolute and relative changes in body mass. The latter fell into three broad groups of inferred change: dwarfed species having undergone variably pronounced decreases of approximately one tenth to three fourths of the mass of respective flighted relatives (e.g., *Dryolimnas aldabranus, Rallus recessus, Gallirallus wakensis, Cabalus modestus, Atlantisia rogersi,* most subfossil Hawaiian *Porzana,* and *Tribonyx hodgenorum*); species showing only limited size changes, having undergone increases up to a doubling of the mass of respective flighted relatives (e.g., *Cyanolimnas cerverai, Rallus ibycus, Gallirallus dieffenbachii, G. owstoni, Capellirallus karamu, Habropteryx okinawae,* "*Atlantisia*" *elpenor,* four *Porzana* spp., *Pareudiastes silvestris, Gallinula nesiotis*-group, and *Fulica newtoni*); and (c) truly giant species having attained masses 2–10 times those of their putative flighted relatives (e.g., *Porphyrio mantelli*-group, *Habroptila wallacii, Nesotrochis debooyi, Gallirallus australis*-group, *Tricholimnas lafresnayanus, Aphanapteryx bonasia, Erythromachus leguati, Diaphorapteryx hawkinsi, Porzana monasa, Amaurornis ineptus, Tribonyx mortieri,* and *Fulica chathamensis*-group).

However, estimated changes in body mass (Table 71) were conditioned on the designation of flighted relatives for each flightless rail, and alternative choices can significantly modify the inferred shifts. Some alternatives are plausible on phylogenetic and biogeographic grounds. For example, proposal of either *Rallus limicola* or a member of the *R. longirostris-elegans* complex as sister-group to either of the subfossil *Rallus* from Bermuda is reasonable on biogeographical grounds; the taxon used as primary species for comparison was made from the standpoint of parsimony of size differences (i.e., small *R. ibycus* with *R. limicola,* and large *R. recessus* with comparatively maritime *R. longirostris*). However, comparison of both subfossil species of *Rallus* from Bermuda with small *R. limicola* implies that *R. recessus* underwent an increase of 170% (as opposed to a 31% decrease), whereas use of large *R. longirostris* for both comparisons leads to the inference that *R. ibycus* underwent a decrease of 65% (instead of a 23% increase). Substitution of *R. elegans* for *R. longirostris,* evident sister-taxa of similar size (Livezey 1998; Appendix 1), resulted in negligible changes in these estimates.

Another external character having obvious implications for flight capacity is relative wing length (Fig. 15), which for most flightless taxa (for which these data were available) also showed absolute reductions in comparisons with flighted relatives despite substantial increases in body size in some (Table 71). Exceptions were among the taxa showing giantism (e.g., *Gallirallus australis*-group, *Tricholimnas lafresnayanus,* and *Amaurornis ineptus*). More reliable indices to flight capacity were departures of wing lengths of flightless species from those predicted for flighted rails of equal body masses (Fig. 15) or comparisons of ratios of wing lengths to cube roots of respective body masses (Table 71). The latter measures confirmed variable reductions in relative wing lengths of flightless rails, with residuals from the predicted length for a flighted rail of equal mass generally indicating greater reductions in larger species than the ratio with body mass because the former took into account the allometry of these two variables (Fig. 15), and the latter tended to minimize inferred shifts through comparison of closely related taxa (Table 71).

Among the array of structural changes in body feathers of flightless rails (Tables 27–29), the simplest indicator of integumentary apomorphy is number of primary remiges, an index that classifies *Gallirallus owstoni, G. wakensis, Cabalus modestus, Atlantisia rogersi, Porzana sandwichensis, P. palmeri* (extreme), and *P. atra* as having undergone notable changes among taxa for which study skins are available (Table 28). Most other apomorphies of the integument concerned magnitudes of qualitative changes and shortening of primary remiges (Table 71). These indices provided mutually consistent estimates of apomorphy of remiges in flightless rallids, defining a continuum with weakly modified species (e.g., *Dryolimnas aldabranus, Amaurornis ineptus,* and *Gallinula comeri*) at one extreme, and species showing highly derived remiges (e.g., *Porphyrio hochstetteri, Cyanolimnas cerverai, Gallirallus australis, G. wakensis, Atlantisia rogersi,* and *Tribonyx mortierii*) at the opposite pole (Table 71).

Although truncation of the tail typified most flightless rails, size of the glandulae uropygiales does not decline uniformly with flightlessness (Johnston 1988), and there was greater variation in changes in lenths of the tail than of the wing (Table 71). A substantial portion of this variation was accounted for by the confounding effects of changing body size, whereby reductions corrected for relative body masses confirmed reductions in the vast majority of flightless rallids (Table 71). Exceptions to this generality, in which flightless taxa had longer tails relative to their body mass than their flighted relatives, were *Gallirallus australis* (large), *Gallirallus dieffenbachii* (marginal), and *Tricholimnas lafresnayanus* (intermediate). Perhaps more informative were those taxa showing remarkably pronounced shortening of the tail: *Habroptila wallacii, Cyanolimnas cerverai, Aramidopsis plateni, Habropteryx insignis, Porzana sandwichensis, P. palmeri, P. monasa, Amaurornis ineptus,* and *Pareudiastes pacificus* (Table 71).

Straightforward skeletal and myological changes having obvious import for flight capacity generally confirmed the patterns of relative apomorphy evinced by the foregoing external differences, but cast light on some shifts not obvious in the externum (Table 72). Although most species showing larger changes in external wing lengths (Table 71) also were characterized by commensurate shifts in SWLs (e.g., *Gallirallus dieffenbachii, G. owstoni, G. wakensis, Cabalus modestus,* and *Porzana palmeri* [Table 72]), there were notable departures from this con-

TABLE 71. Inferred changes in body mass (g) and lengths (mm) of wing, primary remiges, and tail associated with flightlessness in the Rallidae. Mean difference in body mass (g) between flightless species (mass estimated for subfossil species) and closest flighted relative, both as raw difference and percentage of mass for flighted species (Table 8; Appendix 1). Where no single flighted species was designated for comparison (Table 5), flightless species were compared with mean for closest flighted relatives for which data were available. Changes in wing lengths are antilogarithms of residuals from allometric relationship (Fig. 9) between wing length and body mass for flighted rallids, followed by simple difference in length between compared taxa (Table 7). Standardized wing lengths are simple ratios of wing length divided by cube-root of mean body mass for flighted and flightless relatives, respectively (Tables 7, 8; Appendix 3) showing presumed direction of change. Qualitative comparisons of remiges (Table 28) are defined as 2(sum of integer-ranked categorical assessments). Reduction in lengths of remiges are simple differences (flightless species – flighted species) in total length (mm) of remex (calamus and rachis), with percentage of total length in flighted species given in parentheses. Mass-relativized differences (flightless species – flighted species) were divided by cube-roots of body masses ($g^{-1/3}$) of respective species. Changes in length of tail are simple difference (flightless species – flighted species) in total length (mm) of tail, with percentage of total length of remex in flighted species given in parentheses. Mass-relativized differences are (flightless species – flighted species) in total length (mm) of tail, divided by cube-roots of body masses ($g^{-1/3}$) of respective species.

Flightless species	Change in mean body mass (%)	Change in wing length relative to		Mass-standardized wing lengths	Qualitative changes in remiges	Reduction in length of primary remex 9		Reduction in tail length	
		Allo-metric curve	Flighted relative			Simple (%)	Relative to mass	Simple (%)	Relative to mass
Porphyrio mantelli	+3,156 (+320%)	−27	−32	27.7 → 17.5	2.50	−36 (−15%)	−13.7	+12 (+12%)	−1.9
Porphyrio hochstetteri	+1,733 (+175%)	—	−47	—	0.00	—	—	—	—
Porphyrio albus	—	—	—	—	2.50	—	—	—	—
Habroptila wallacii	+548 (+162%)	−67	−25	27.5 → 17.5	—	—	—	−4 (−7%)	−3.1
Nesotrochis debooyi	+904 (+208%)	—	—	—	—	—	—	—	—
Nesotrochis steganinos	−49 (−11%)	—	—	—	—	—	—	—	—
Cyanolimnas cerverai	+83 (+52%)	−46	−28	24.6 → 17.0	2.00	−33 (−36%)	−14.4	−18 (−31%)	−4.4
Dryolimnas aldabranus	−58 (−24%)	−20	−34	24.3 → 20.6	0.00	−17 (−16%)	−3.0	−22 (−36%)	−2.9
Rallus recessus	−100 (−31%)	—	—	—	—	—	—	—	—
Rallus ibycus	+19 (+23%)	—	—	—	—	—	—	—	—
Gallirallus australis-group	+649 (+326%)	−57	+35	24.2 → 18.4	2.50	+19 (+15%)	−12.5	+52 (+92%)	+2.7
Gallirallus dieffenbachii	+141 (+71%)	−48	−17	24.2 → 17.5	1.25	—	—	+12 (+21%)	+0.1
Gallirallus owstoni	+28 (+14%)	−28	−19	24.2 → 19.7	2.50	−4 (−31%)	−7.5	−13 (−24%)	−2.6
Gallirallus wakensis	−86 (−43%)	−26	−49	24.2 → 18.6	2.50	−67 (−52%)	−9.3	−22 (−39%)	−2.5
Tricholimnas lafresnayanus	+669 (+336%)	−53	+46	24.2 → 19.4	2.50	—	—	+48 (+84%)	+1.3
Tricholimnas sylvestris	+296 (+149%)	−55	0	24.2 → 17.6	2.75	−6 (−5%)	−6.6	+2 (+4%)	−2.2

TABLE 71. Continued.

Flightless species	Change in mean body mass (%)	Change in wing length relative to		Mass-standardized wing lengths	Qualitative changes in remiges	Reduction in length of primary remex 9		Reduction in tail length	
		Allometric curve	Flighted relative			Simple (%)	Relative to mass	Simple (%)	Relative to mass
Nesoclopeus poecilopterus	+288 (+145%)	−31	−7	24.2 → 20.5	2.25	+15 (+12%)	−3.8	+14 (+25%)	−0.7
Cabalus modestus	−141 (−71%)	−9	−55	24.2 → 21.7	2.50	—	—	−21 (−37%)	−0.4
Capellirallus karamu	+41 (+21%)	—	—	—	—	—	—	—	—
Aramidopsis plateni	—	—	+3	—	2.50	—	—	—	—
Habropteryx insignis	+269 (+109%)	−52	−4	—	2.25	+1 (+1%)	−4.5	−24 (−44%)	−4.1
Habropteryx okinawae	+187 (+43%)	−42	−6	23.6 → 18.8	1.25	—	—	−17 (−31%)	−1.6
Aphanapteryx bonasia	+1,028 (+894%)	—	—	—	—	—	—	−0.3 (−1%)	—
Erythromachus leguati	+458 (+398%)	—	—	—	—	—	—	—	—
Diaphorapteryx hawkinsi	+1,711 (+860%)	—	—	—	—	—	—	—	—
Mundia elpenor	+38 (+84%)	—	—	—	—	—	—	—	—
Atlantisia rogersi	−5 (−13%)	−26	−23	23.1 → 15.8	3.00	−49 (−60%)	−11.8	−16 (−34%)	−4.1
Porzana sandwichensis	+7 (+17%)	−20	−17	24.5 → 18.6	2.75	Remex absent		−25 (−61%)	−7.3
Porzana palmeri	+3 (+7%)	−25	−27	24.5 → 16.3	2.50	Remex absent		−17 (−43%)	−5.1
Porzana monasa	+108 (+257%)	−50	−7	24.5 → 14.7	—	—	—	−2 (−6%)	−4.4
Porzana atra	+35 (+83%)	−19	−2	24.5 → 19.5	2.00	−15 (−21%)	−7.2	−0 (−1%)	−2.2
Porzana piercei	−48 (−59%)	—	—	—	—	—	—	—	—
Porzana astrictocarpus	−5 (−10%)	—	—	—	—	—	—	—	—
Porzana ziegleri	−18 (−43%)	—	—	—	—	—	—	—	—
Porzana menehune	−18 (−43%)	—	—	—	—	—	—	—	—
Porzana keplerorum	−9 (−21%)	—	—	—	—	—	—	—	—
Porzana severnsi	+50 (+119%)	—	—	—	—	—	—	—	—
Amaurornis ineptus	+751 (+348%)	−64	+30	25.0 → 18.2	2.25	+5 (+3%)	−6.6	−31 (−48%)	−6.1
Pareudiastes pacificus	—	—	−45	—	1.75	−20 (−15%)	—	−33 (−50%)	—
Pareudiastes silvestris	+147 (+49%)	−37	−19	25.0 → 19.4	1.75	—	—	−26 (−40%)	−4.3
Tribonyx mortierii	+892 (+222%)	−74	−11	27.8 → 17.8	1.75	−23 (−12%)	−10.5	+7 (+9%)	−2.7
Tribonyx hodgenorum	−123 (−31%)	—	—	—	—	—	—	—	—
Gallinula nesiotis	+97 (+32%)	−40	−26	25.0 → 19.3	1.00	—	—	−3 (−4%)	−0.9
Gallinula comeri	+210 (+69%)	−51	−24	25.0 → 18.0	1.00	−7 (−5%)	−4.0	−2 (−4%)	−1.5
Fulica chathamensis	+1,168 (+172%)	—	—	—	—	—	—	—	—
Fulica prisca	+1,231 (+181%)	—	—	—	—	—	—	—	—
Fulica newtoni	+608 (+89%)	—	—	—	—	—	—	—	—

TABLE 72. Summary of inferred mensural and qualitative changes in pectoral skeleton and musculature of adequately represented flightless species of Rallidae. Simple differences are given in lengths of pectoral and pelvic limbs between flightless taxon and its closest flighted relative (Tables 32, 33), with percentage difference relative to length for flighted relative. Relative wing lengths are simple ratios of skeletal wing length divided by cube-root of mean body mass for flighted and flightless relatives, respectively (Tables 7, 8, 32), with arrow of presumed change. Pectoral proportions are simple differences (flightless − flighted) in raw proportions for five major skeletal segments (Fig. 36). Pelvic proportions are simple differences (flightless − flighted) in raw proportions for four major skeletal segments (Fig. 37). Scores of qualitative apomorphy are simple counts (Tables 58–64), in which a plus sign indicates uncertainty and noncomparable states were treated as apomorphies. Scores of qualitative apomorphy are simple counts (Tables 58–64), in which a plus sign indicates uncertainty and noncomparable states were treated as apomorphies. Muscle masses are differences in sums of mm. pectoralis et supracoracoideus (Table 66), both raw and as percentage of mass of breast muscles for flighted species. Differences (flightless species − flighted species) in ratios of breast-muscle masses were divided by mean body masses of respective species (Table 66; Appendix 1).

Flightless species	Change in length (mm) of appendicular skeleton (%)		Relative wing length, interspecific shift	Change in proportions (%) of major segments within appendages		Qualitative pectoral apomorphies		Change in mass (g) of breast muscles relative to	
	Pectoral	Pelvic		Pectoral	Pelvic	Girdle	Limb	Flighted relative	Body mass
Porphyrio hochstetteri	−27 (−11%)	+18 (+5%)	25.3 → 16.1	4.1, 0.5, −0.8, −2.1, −1.7	4.5, 1.7, −1.7, −4.4	12	16	−76 (−61%)	−11%
Habroptila wallacii	−1 (−1%)	+62 (+21%)	21.7 → 16.8	3.0, 0.6, −1.3, −1.4, −0.9	1.4, −0.6, −0.6, −0.2	17	11	−13 (−37%)	−9%
Dryolimnas aldabranus	−32 (−23%)	−28 (−13%)	22.6 → 19.1	1.0, −0.4, −0.7, 0.0, 0.0	−0.1, 0.0, 0.1, 0.0	5	5	−22 (−55%)	−6%
Rallus recessus	—	−32 (−14%)	—	—	1.6, −1.1, −1.2, 1.8	5+	9	—	—
Rallus ibycus	—	−7 (−4%)	—	—	2.4, −1.5, −2.1, 2.2	6+	9	—	—
Gallirallus australis-group	+7 (+5%)	+107 (+57%)	22.8 → 14.5	3.0, −1.2, −0.6, −0.8, −0.4	0.1, 1.3, −0.9, −0.5	16	16	+7 (+22%)	−15%
Gallirallus dieffenbachii	−18 (−14%)	+26 (+14%)	22.8 → 16.5	2.6, −1.7, −0.8, −0.3, 0.0	−0.8, −0.3, −1.0, 2.0	10	12	—	—
Gallirallus owstoni	−10 (−7%)	+28 (+15%)	22.8 → 20.2	1.3, −0.1, −0.6, −0.7, 0.0	−0.8, −0.1, 0.8, 0.0	5	3	−16 (−53%)	−9%
Gallirallus wakensis	−42 (−31%)	−28 (−15%)	22.8 → 18.9	0.6, −0.6, −0.2, 0.1, 0.0	−2.0, 0.6, −0.2, 1.5	8	6	−28 (−92%)	−13%
Tricholimnas sylvestris	+4 (+3%)	+52 (+28%)	22.8 → 17.3	1.0, −0.5, −0.5, 0.0, 0.0	0.0, 0.1, −1.2, 1.0	13	14	—	−13%

TABLE 72. Continued.

Flightless species	Change in length (mm) of appendicular skeleton (%)		Relative wing length, interspecific shift	Change in proportions (%) of major segments within appendages		Qualitative pectoral apomorphies		Change in mass (g) of breast muscles relative to	
	Pectoral	Pelvic		Pectoral	Pelvic	Girdle	Limb	Flighted relative	Body mass
Nesoclopeus poecilopterus	+20 (+15%)	—	22.8 → 19.5	−0.1, −0.8, 0.9, 0.4, −0.5	—	10+	—	−17 (−54%)	−11%
Cabalus modestus	−67 (−50%)	−55 (−29%)	22.8 → 17.1	2.6, −1.8, −0.8, −0.4, 0.0	−2.8, 0.7, 0.0, 2.0	16	16	−8 (−25%)	—
Capellirallus karamu	—	+2 (+1%)	—	—	−0.8, 0.5, −1.0, 1.5	18+	16	—	—
Aphanapteryx bonasia	+51 (+48%)	−128 (−32%)	22.1 → 15.2	2.6, −0.4, −2.2, 1.7, −1.8	1.5, 0.5, 1.8, −3.8	11+	15	—	—
Diaphorapteryx hawkinsi	—	+177 (+93%)	—	—	0.2, 0.7, −4.0, 3.0	19	16	—	—
Mundia elpenor	−4 (−5%)	+12 (+8%)	40.0 → 16.0	1.1, −0.2, −0.6, 0.0, −0.3	0.7, 1.4, −0.1, −2.0	15	10	—	—
Atlantisia rogersi	−23 (−32%)	−25 (−17%)	40.0 → 34.5	4.2, −2.6, −2.0, −0.1, 0.6	0.1, 0.9, −2.0, 1.0	13	12	−4 (−78%)	−8%
Porzana palmeri	−21 (−29%)	−24 (−18%)	21.5 → 14.8	1.1, −0.2, −0.4, −0.4, −0.2	0.8, 0.1, −1.3, 0.4	9	8	−4 (−74%)	−9%
Porzana atra	0 (0%)	+19 (+14%)	2.15 → 17.4	1.1, 0.2, −0.6, −0.3, −0.5	0.1, 0.5, 0.2, −0.8	7	6	−1 (−20%)	−7%
Tribonyx mortierii	−5 (−3%)	+104 (+41%)	25.0 → 16.4	3.8, −1.5, −1.5, −0.3, −0.6	1.1, 0.2, −0.5, −0.7	10	12	−40 (−61%)	−15%
Gallinula comeri	−2 (−1%)	+8 (+3%)	21.5 → 18.9	2.0, 0.3, −0.8, −0.9, 0.7	2.3, 0.5, −0.8, −2.2	6	3	−10 (−33%)	−5%
Fulica chathamensis	+44 (+21%)	+149 (+52%)	23.8 → 20.7	1.7, −0.9, −0.4, −0.3, −0.3	0.1, 0.6, 0.5, −1.2	7	9	—	—

gruence of metrics, including species having large reductions in external wing lengths but only modest truncation in skeletal wing elements (e.g., *Habroptila wallacii, Tribonyx mortierii,* and *Gallinula comeri*), and species characterized by only limited shortening of the external wing but having undergone substantial reductions in the underlying skeleton (e.g., *Dryolimnas aldabranus*). However, these varying discrepancies between external and SWLs belied a distinct modality in the shifts in skeletal proportions within the wing that accompanied flightlessness, in which relative lengths of the brachium (humerus) increased, relative lengths of the antebrachium (ulna and radius) or proximal manus (carpometacarpus) significantly decreased, and the proportions of the remaining elements tended to show marginal increases or lesser reductions (Table 72). Also, *Porphyrio hochstetteri* and *Aphanapteryx bonasia* were characterized by aberrant pectoral proportionalities in which the distalmost elements (phalanges of digiti majoris) manifested the greatest reductions in relative contributions to SWL (Table 72).

By contrast, reductions in skeletal lengths of the pelvic appendage were not obscured by overlying, specialized feathers, and corresponded rather closely with relative changes in body mass. A notable deviation from this correspondence was the disproportionate elongation of the leg in *Fulica chathamensis,* a diving species in which pelvic propulsion would be at a premium (Tables 71, 72). However, proportions of the four major skeletal segments within the pelvic limb displayed diverse changes that were partitionable into three broad classes: species in which proportions of the proximal segments increased and the distal segments decreased (most taxa compared), species in which the reverse pattern was manifested (e.g., most *Gallirallus* spp., *Tricholimnas sylvestris,* and *Cabalus modestus*), and species in which the contributions of the proximal and distal extremities increased while the intermediate two segments decreased in relative proportions (e.g., *Rallus ibycus* and *R. recessus*). Four exceptional sets of pelvic proportionalities also were found (Table 72): *Dryolimnas aldabranus* (essentially no change detected), *Diaphorapteryx hawkinsi* (dominated by decreased contribution by tarsometatarsus and increased contribution by the middle toe), *Capellirallus karamu* (four segments alternated between increased and decreased proportionalities), and *Fulica chathamensis* (decreased relative length of middle toe while all other segments accounted for larger proportions of the pelvic limb).

Elements of the pectoral skeleton revealed a subset of flightless taxa as showing comparatively more qualitative apomorphies in most of those elements available: *Habroptila wallacii, Gallirallus australis*-group, *G. dieffenbachii, Tricholimnas sylvestris, Cabalus modestus, Capellirallus karamu, Aphanapteryx bonasia, Erythromachus leguati, Atlantisia rogersi,* and *Porzana severnsi* (Tables 58–64). Tallies of these across elements confirmed this ranking of relative change related to flightlessness, one that was only loosely correlated with change in body size (Tables 71, 72).

Reductions in relative masses of breast muscles in flightless species showed substantial concordance with the metrics of apomorphy based on skins and (especially) skeletons (several taxa were not available for dissection), singling out *Porphyrio hochstetteri* (extreme), *Habroptila wallacii* (extreme), *Cyanolimnas cerverai, Gallirallus australis, G. wakensis, Atlantisia rogersi, Amaurornis ineptus,* and *Tribonyx mortierii* as having undergone the greatest reductions (Table 67). Slightly more sophisticated indices that incorporated information concerning the masses of breast muscles or body masses of close flighted relatives affirmed most of these initial assessments (Table 72), identifying species showing the greatest reductions in breast muscles (e.g., *P. hochstetteri, G. australis, G. wakensis,* and *T. mortierii*), and revealing the small breast muscles of several others as the result, in part, of small initial or derived body sizes (e.g., *A. rogersi* and *Porzana palmeri*).

MULTIVARIATE METRICS

A summary of multivariate distances, estimated separately for study skins, remiges, skeletons (complete, partial, skulls, and sterna), and pectoral musculature by using PCA (Table 73),z permitted an assessment of general extremity of flightless taxa (distances from the centroid for all rallids analyzed) and apomorphy related to flightlessness (distances from respective flighted relatives). Estimates of multivariate apomorphy also were possible for some taxa based on CA (Tables 74, 75; Figs. 32, 59, 60). Moreover, rank correlations provided a means for comparing these indices against each other across species. Finally, these metric estimators of extremity and apomorphy can be related to patterns of qualitative and univariate apomorphy in the integument (Tables 27–39, 71) and skeletal elements (Tables 58–64, 72) to provide a synthesis of the diverse manifestations of flightlessness that occur among the members of the Rallidae. Inasmuch as the distances deriving from PCAs permitted the inclusion of a substantially greater number of flightless taxa than those from CAs (the former required only a vector of mean dimensions), the former are considered as a primary means of comparison; the sample-based statistics associated with the CA-based Mahalanobis distances were treated largely in a confirmatory capacity (Tables 74, 75).

Euclidean distances based on PCAs revealed largely consonant indications of flightlessness-related apomorphy. That is, certain flightless species tended to have a comparatively large D_T regardless of data set, whereas others tended to show a comparatively small D_T in most or all data sets (Table 73). Sample sizes permitting, this general correspondence resulted in significant rank correlations among distances between flightless species and their flighted relatives (D_T) regardless of the data upon which the distances were based: skins versus skeletons ($r_s = 0.36$; $P < 0.10$), skins versus partial skeletons ($r_s = 0.44$; $P < 0.05$), skins versus sterna ($r_s = 0.35$; $P < 0.10$), remiges versus partial skeletons ($r_s = 0.55$; $P < 0.05$), skeletons versus partial skeletons ($r_s = 0.83$; $P < 0.001$), partial skeletons versus skulls ($r_s = 0.46$; $P < 0.05$), and partial skeletons versus sterna ($r_s = 0.79$; $P < 0.001$). Only distances between related taxa (D_T) based on the pectoral musculature showed no significant correlations with those based on all other data sets, although this partly reflected the small sample sizes available for such comparisons ($n = 12–16$). These metrics identified a subset of flightless species as having undergone both the greatest apomorphic changes and coming to occupy extreme morphometric positions within the Rallidae for multiple data sets: *Porphyrio hochstetteri* (extreme), *Habroptila wallacii, Tricholimnas lafresnayanus, Cabalus modestus, Aramidopsis plateni, Aphanapteryx bonasia* (extreme), *Diaphorapteryx hawkinsi* (extreme), *Porzana sandwichensis, P. severnsi,* and *Amaurornis ineptus* (Table 73). These patterns agreed broadly with Mahalanobis distances based on skins and skeletons, with the correspondence being lessened by the reduced role of size (by correction for pooled within-species covariances) in the latter (Tables

TABLE 73. Euclidean distances between flightless species of Rallidae and flighted relatives (D_T) and Mahalanobis distances between flightless taxa and centroids for all taxa (D_C) in principal component analyses of measurements of study skins, primary remiges, skeletons (complete and reduced), detailed skulls, sterna, and pectoral musculature. Number of principal components on which distances were based (k) and original dimensionalities of data sets (p) were as follows (k, p): skins (4, 6), primary remiges (4, 20), complete skeletons (4, 35), skulls (3, 21), sterna (2, 5), and pectoral musculature (5, 81). Euclidean distances between contrasted pairs (D_T) were for standardized distances in space spanned by k components. Mahalanobis distances to centroids of all taxa in space spanned by k components (D_C) ere given as χ^2/k. Where multiple alternatives for flighted species were listed, the first adequately represented taxon was used.

Flightless and flighted taxa contrasted	Skin specimens		Primary remiges		Complete skeletons		Reduced skeletons		Detailed skulls		Sterna		Musculature	
	D_T	D_C	D_T	D_C	D_T	D_C	D_T	D_C	D_T	D_C	D_T	D_C	D_T	D_C
Porphyrio mantelli vs. *P. melanotus*	3.91	6.07	—	—	—	—	2.42	2.03	1.15	5.54	3.87	5.29	—	—
Porphyrio hochstetteri vs. *P. melanotus*	1.10	1.24	2.59	1.82	3.09	2.12	2.32	1.39	1.10	5.08	3.65	4.73	1.99	1.50
Porphyrio albus vs. *P. melanotus*	2.14	1.29	—	—	—	—	—	—	—	—	—	—	—	—
Habroptila wallacii vs. *Gymnocrex* sp./*Aramides cajanea*	2.75	0.90	0.86	0.17	1.83	1.60	1.46	1.00	1.54	1.02	1.82	1.79	1.80	0.95
Cyanolimnas cerverai vs. *Ortygonax sanguinolentus*	1.83	0.92	0.93	0.35	0.96	0.36	0.86	0.43	0.82	0.84	2.08	1.46	1.46	0.60
Dryolimnas aldabranus vs. *D. cuvieri*	—	—	—	—	—	—	1.41	0.72	—	—	1.19	0.19	1.05	0.39
Rallus recessus vs. *R. longirostris*	—	—	—	—	—	—	—	—	—	—	2.13	0.20	—	—
Rallus ibycus vs. *R. limicola*	0.89	0.63	—	—	—	—	—	—	—	—	2.15	1.16	—	—
Gallirallus sharpei vs. *G. philippensis*-group	1.92	1.57	2.75	2.12	2.93	1.25	2.36	0.72	1.00	0.39	2.39	0.95	3.42	1.28
Gallirallus australis-group vs. *G. philippensis*-group	1.42	1.10	—	—	1.86	0.33	1.45	0.17	0.90	0.18	1.57	0.17	—	—
Gallirallus dieffenbachii vs. *G. philippensis*-group	1.86	0.40	3.85	2.90	1.50	0.27	1.11	0.18	1.23	0.15	0.14	0.13	2.06	0.09
Gallirallus owstoni vs. *G. philippensis*-group	1.75	0.55	2.78	1.30	2.15	0.39	1.79	0.34	1.28	0.15	1.87	0.72	3.32	2.26
Gallirallus wakensis vs. *G. philippensis*-group	2.07	1.81	—	—	—	—	—	—	—	—	—	—	—	—
Tricholimnas lafresnayanus vs. *G. philippensis*-group	1.59	0.50	0.98	0.21	1.95	0.46	1.52	0.22	1.52	0.27	1.74	0.27	2.15	0.32
Tricholimnas sylvestris vs. *G. philippensis*-group	—	—	—	—	—	—	—	—	—	—	—	—	—	—
Nesoclopeus poecilopterus-group vs. *G. philippensis*-group	1.22	0.36	2.89	2.73	2.00	1.55	—	—	0.88	0.30	2.60	1.30	2.27	0.42
Cabalus modestus vs. *G. philippensis*-group	0.99	1.20	—	—	—	—	2.63	1.02	2.53	1.00	2.46	1.74	—	—
Capellirallus karamu vs. *G. philippensis*-group	—	—	—	—	—	—	4.26	3.38	1.00	3.23	3.40	3.07	—	—
Aramidopsis plateni vs. *Habropteryx torquatus*-group	2.91	2.16	—	—	—	—	—	—	—	—	—	—	—	—
Habropteryx insignis vs. *H. torquatus*-group	2.11	1.07	2.91	0.43	—	—	—	—	0.67	0.47	—	—	—	—
Habropteryx okinawae vs. *H. torquatus*-group	1.22	0.54	—	—	—	—	—	—	—	—	1.07	0.30	0.76	0.40
Aphanapteryx bonasia vs. *Rallus aquaticus*	—	—	—	—	3.43	1.79	2.70	1.09	3.02	2.18	3.10	1.97	—	—
Erythromachus leguati vs. *Rallus aquaticus*	—	—	—	—	—	—	—	—	1.84	1.35	2.03	0.45	—	—
Diaphorapteryx hawkinsi vs. *G. philippensis*-group	—	—	—	—	—	—	3.43	2.07	3.19	1.76	3.75	3.77	—	—
Mundia elpenor vs. *Laterallus leucopyrrhus*	—	—	—	—	2.20	1.47	1.72	0.94	1.91	0.20	1.19	1.84	—	—

TABLE 73. Continued.

Flightless and flighted taxa contrasted	Skin specimens		Primary remiges		Complete skeletons		Reduced skeletons		Detailed skulls		Sterna		Musculature	
	D_r	D_c	D_r	D_c	D_r	D_c	D_r	D_c	D_r	D_c	D_r	D_c	D_r	D_c
Atlantisia rogersi vs. *L. leucopyrrhus*/*Coturnicops noveboracensis*	1.98	1.39	1.48	0.70	2.55	2.43	0.88	1.69	1.41	0.95	1.05	2.28	3.06	3.14
Porzana sandwichensis vs. *P. tabuensis**	4.02	4.01	—	—	—	—	—	—	—	—	—	—	—	—
Porzana palmeri vs. *P. tabuensis**	2.56	2.40	2.61	5.24	1.87	1.59	1.57	1.26	1.94	1.30	0.70	1.71	4.93	3.54
Porzana monasa vs. *P. tabuensis**	0.87	0.76	—	—	—	—	—	—	—	—	—	—	—	—
Porzana atra vs. *P. tabuensis**	1.04	0.63	2.32	0.96	1.45	0.62	0.87	0.47	1.80	0.55	0.49	0.78	1.91	0.47
Porzana severnsi vs. *P. tabuensis**	—	—	—	—	—	—	3.51	3.43	1.53	0.27	2.31	2.92	—	—
Amaurornis ineptus vs. *A. isabellinus*/*olivaceus*	4.13	3.47	1.51	0.88	—	—	—	—	1.30	0.61	2.35	3.08	—	—
Pareudiastes pacificus vs. *Gallinula chloropus*-group	2.83	0.70	2.42	0.77	—	—	—	—	—	—	1.77	0.46	1.20	0.33
Pareudiastes silvestris vs. *G. chloropus*-group	3.38	1.13	—	—	—	—	—	—	—	—	—	—	—	—
Tribonyx mortierii vs. *T. ventralis*	1.17	0.54	3.28	1.26	2.17	0.94	1.70	0.86	1.13	1.28	2.35	1.64	1.65	1.22
Tribonyx hodgenorum vs. *T. ventralis*	—	—	—	—	—	—	2.51	1.49	0.69	0.82	2.99	1.70	—	—
Gallinula nesiotis vs. *G. chloropus*-group	2.09	0.22	—	—	—	—	—	—	—	—	1.75	0.73	—	—
Gallinula comeri vs. *G. chloropus*-group	2.20	0.35	1.47	0.59	1.36	0.61	1.11	0.46	1.07	0.47	1.90	1.04	4.03	0.30
Fulica cornuta vs. *F. armillata*	1.21	1.70	—	—	0.83	1.56	0.81	1.65	1.17	1.39	1.46	4.42	—	—
Fulica gigantea vs. *F. armillata*	1.36	3.13	—	—	0.85	1.22	0.80	1.38	—	—	0.80	3.08	—	—
Fulica chathamensis vs. *F. atralamericana*	—	—	—	—	1.49	1.06	1.30	1.02	1.00	1.19	1.24	1.94	—	—
Fulica prisca vs. *F. atralamericana*	—	—	—	—	—	—	—	—	—	—	1.48	1.92	—	—
Fulica newtoni vs. *F. atralamericana*	—	—	—	—	—	—	—	—	—	—	2.04	4.13	—	—

* Distances based on comparisons with the less well-represented *Porzana pusilla* were very similar, but tending to be somewhat larger.

TABLE 74. Multivariate differences between selected flightless rallids and a flighted relative, documented by mean within-sex Mahalanobis distances (D) and associated statistics from multivariate analyses of variance, based on m variables stepwise-selected ($P < 0.05$) from six external measurements. Pillai's trace is numerically equal to the average squared canonical correlation in the two-group (unidimensional) case. Wilks' λ is numerically equal to the Hotelling–Lawley trace and Roy's maximum root (Morrison 1976) in the two-group (unidimensional) case. Approximate F-statistic corresponds to Wilks' λ (U-statistic) with m (variables included in model), g (groups) $- 1$, and n (total specimens) $- g$ degrees of freedom. Levels of significance for all comparisons were $P \leq 0.0001$.

Flightless and flighted taxa contrasted (n_g)	D	m	Pillai's trace	Eigenvalue	R	Wilks' λ	F
Porphyrio hochstetteri (14) vs. *P. melanotus* (83)	17.4	3	0.975	38.910	0.987	0.0251	1,180.3
Habroptila wallacii (16) vs. *Gymnocrex plumbeiventris* (27)	12.9	3	0.975	38.511	0.987	0.0253	475.0
Cyanolimnas cerverai (9) vs. *Ortygonax sanguinolentus* (20)	11.6	3	0.962	25.661	0.981	0.0375	196.7
Dryolimnas aldabranus (28) vs. *D. cuvieri* (59)	9.3	3	0.952	19.775	0.976	0.0481	533.9
Gallirallus dieffenbachii (1) vs. *G. philippensis*-group (47)	7.7	5	0.562	1.283	0.750	0.4380	10.5
Gallirallus australis-group (142) vs. *G. philippensis*-group (47)	8.6	3	0.934	14.064	0.966	0.0664	857.9
Gallirallus owstoni (29) vs. *G. philippensis*-group (47)	7.7	5	0.935	14.492	0.967	0.0646	197.1
Gallirallus wakensis (58) vs. *G. philippensis*-group (47)	8.6	3	0.950	18.914	0.975	0.0502	624.2
Tricholimnas lafresnayanus (9) vs. *G. philippensis*-group (47)	12.8	2	0.959	23.347	0.979	0.0411	595.4
Tricholimnas sylvestris (86) vs. *G. philippensis*-group (47)	12.2	4	0.972	34.722	0.986	0.0280	1,093.7
Nesoclopeus poecilopterus-group (31) vs. *G. philippensis*-group (47)	11.2	3	0.969	31.425	0.985	0.0308	754.2
Cabalus modestus (18) vs. *G. philippensis*-group (47)	13.6	6	0.978	45.304	0.989	0.0216	498.3
Aramidopsis plateni (14) vs. *Habropteryx torquatus*-group (93)	9.1	4	0.905	9.546	0.951	0.0948	238.6
Habropteryx insignis (20) vs. *H. torquatus*-group (93)	6.4	2	0.859	6.079	0.927	0.1413	328.3
Habropteryx okinawae (1) vs. *H. torquatus*-group (93)	4.1	1	0.152	0.179	0.389	0.8485	16.3
Atlantisia rogersi (45) vs. *Laterallus leucopyrrhus* (10)	15.4	2	0.974	38.129	0.987	0.0256	953.2
Atlantisia rogersi (45) vs. *Coturnicops noveboracensis* (30)	18.2	3	0.988	83.805	0.994	0.0118	1,927.5
Atlantisia rogersi (45) vs. *Crex crex* (30)	27.9	2	0.995	196.469	0.997	0.0051	6,876.4
Laterallus spilonotus (19) vs. *L. leucophyrrhus* (10)	9.2	2	0.968	29.813	0.984	0.0325	357.8
Porzana sandwichensis (7) vs. *P. tabuensis* (36)	10.6	3	0.942	16.363	0.971	0.0576	207.3
Porzana monasa (2) vs. *P. tabuensis* (36)	4.7	3	0.539	1.169	0.734	0.4611	12.9
Porzana atra (25) vs. *P. tabuensis* (36)	6.5	3	0.916	10.897	0.957	0.0841	199.8
Porzana palmeri (89) vs. *P. tabuensis* (36)	7.6	4	0.921	11.638	0.960	0.0791	343.3
Amaurornis ineptus (24) vs. *A. isabellinus* (59)	15.9	4	0.979	47.661	0.990	0.0206	905.6
Pareudiastes pacificus (8) vs. *Gallinula chloropus* (33)	13.4	3	0.968	30.205	0.984	0.0321	362.5
Tribonyx mortierii (47) vs. *T. ventralis* (86)	12.2	5	0.969	31.715	0.985	0.0306	792.9
Gallinula nesiotis (4) vs. *G. cachinaans* (30)	9.9	4	0.918	11.217	0.958	0.0818	78.5
Gallinula comeri (36) vs. *G. cachinaans* (30)	11.4	5	0.971	33.867	0.986	0.0287	399.6
Fulica cornuta (19) vs. *F. armillata* (32)	9.7	1	0.958	22.746	0.979	0.0421	1,069.1
Fulica gigantea (4) vs. *F. armillata* (32)	8.3	3	0.947	17.703	0.973	0.0535	395.4

TABLE 75. Multivariate differences between selected flightless rallids and a flighted congener, documented by mean within-sex Mahalanobis distances (D) and associated statistics from multivariate analyses of variance, based on m variables stepwise-selected ($P < 0.05$) from 41 skeletal measurements. Pillai's trace is numerically equal to the average squared canonical correlation in the two-group (unidimensional) case. Wilks' λ is numerically equal to the Hotelling–Lawley trace and Roy's maximum root (Morrison 1976) in the two-group (unidimensional) case. Approximate F-statistic corresponds to Wilks' λ (U-statistic) with m (variables included in model), g (groups) $-$ 1, and n (total specimens) $-$ g degrees of freedom. Levels of significance for all comparisons were $P \leq 0.0001$.

Flightless and flighted taxa contrasted (n_s)	D	m	Pillai's trace	Eigenvalue	R	Wilks' λ	F
Porphyrio hochstetteri (30) vs. *P. melanotus* (53)	47.3	14	0.998	517.4	0.999	0.0019	2,439.1
Dryolimnas aldabranus (15) vs. *D. cuvieri* (5)	159.3	9	0.999	5,599.1	0.999	0.0002	4,976.9
Gallirallus australis-group (66) vs. *G. philippensis*-group (35)	25.6	12	0.994	153.6	0.997	0.0065	1,100.7
Gallirallus owstoni (32) vs. *G. philippensis*-group (35)	19.1	18	0.990	95.4	0.995	0.0104	243.8
Gallirallus wakensis (3) vs. *G. philippensis*-group (35)	49.4	17	0.994	179.3	0.997	0.0056	189.8
Tricholimnas sylvestris (6) vs. *G. philippensis*-group (35)	24.3	16	0.987	73.7	0.993	0.0134	101.3
Porzana atra (4) vs. *P. tabuensis* (11)	63.4	7	0.999	859.6	0.999	0.0012	614.0
Porzana palmeri (12) vs. *P. tabuensis* (11)	100.0	12	0.999	3,007.6	0.999	0.0003	2,005.1
Tribonyx mortierii (21) vs. *T. ventralis* (43)	34.4	6	0.996	276.1	0.998	0.0036	2,531.3
Gallinula comeri (2) vs. *G. chloropus*-group (49)	34.6	18	0.980	48.8	0.990	0.0201	81.3

TABLE 76. Summary statistics for total and first two (largest) axes for correspondence analysis of associations among four morphological metrics, five ecogeographical characteristics, and 19 flightless species of Rallidae. Mass is total frequency of category over sample (i.e., $n/n_..$). Quality is proportion of χ^2 distance from origin in two dimensions over that for all (19) dimensions (χ^2_2/χ^2_{19}). Inertia is proportion of eigenvalue (λ_i) for row to total of eigenvalues for all (19) dimensions ($\lambda_i/\Sigma \lambda_i$). Score is the factor score for category. R^2 axis is the squared correlation coefficient of category with axis, and the sum of these for all (19) axes equals quality. Contribution is the relative contribution of category to relative inertia of the axis.

Variable	Category	Mass	Total (19 axes)			Axis I			Axis II		
			Quality	Inertia	Score	R^2 axis	Contribution	Score	R^2 axis	Contribution
Change in body mass	Modest decrease	0.026	0.460	0.074	−0.004	0.000	0.000	0.165	0.010	0.002
	Modest increase	0.042	0.474	0.058	0.610	0.271	0.032	−0.225	0.037	0.006
	Large increase	0.032	0.356	0.068	−0.809	0.302	0.043	0.163	0.012	0.002
Decrease in wing length	Small	0.021	0.517	0.079	1.187	0.376	0.061	−0.315	0.026	0.006
	Medium	0.042	0.501	0.058	0.066	0.003	0.000	0.683	0.340	0.053
	Large	0.037	0.556	0.063	−0.753	0.331	0.043	−0.601	0.211	0.036
Interspecific D (skins)	Small	0.058	0.343	0.042	0.226	0.070	0.006	0.421	0.243	0.028
	Large	0.042	0.343	0.058	−0.311	0.070	0.008	−0.578	0.243	0.038
Interspecific D (skeletons)	Small	0.011	0.632	0.089	2.026	0.483	0.089	−1.122	0.148	0.036
	Medium	0.053	0.029	0.047	0.158	0.028	0.003	−0.015	0.000	0.000
	Large	0.037	0.446	0.063	−0.805	0.378	0.049	0.341	0.068	0.012
Population status	Stable	0.032	0.555	0.068	−0.862	0.343	0.048	−0.245	0.028	0.632
	Endangered	0.037	0.716	0.063	0.217	0.027	0.004	0.329	0.063	0.011
	Extinct	0.032	0.333	0.068	0.160	0.172	0.024	−0.139	0.009	0.002
Typical habitus	Terrestrial	0.068	0.774	0.032	0.403	0.352	0.023	−0.438	0.415	0.036
	Aquatic	0.032	0.774	0.068	−0.874	0.352	0.050	−0.948	0.415	0.077
Island area	Tiny	0.021	0.211	0.079	0.741	0.147	0.024	−0.189	0.010	0.002
	Small	0.016	0.228	0.084	0.741	0.103	0.018	−0.632	0.075	0.017
	Medium	0.021	0.508	0.079	0.560	0.084	0.014	1.196	0.381	0.082
	Large	0.016	0.240	0.084	−0.602	0.068	0.012	−0.796	0.119	0.027
	Very large	0.011	0.332	0.089	−1.184	0.165	0.030	−0.553	0.036	0.009
	Huge	0.016	0.339	0.084	−1.085	0.221	0.038	0.454	0.039	0.009
Latitudinal interval	Tropical	0.042	0.213	0.058	−0.318	0.074	0.009	−0.377	0.103	0.016
	Warm temperate	0.021	0.575	0.079	1.263	0.425	0.069	−0.406	0.044	0.009
	Cool temperate	0.037	0.551	0.063	−0.358	0.075	0.010	0.662	0.256	0.044
Distance to continent	Small	0.037	0.405	0.063	−0.548	0.175	0.023	−0.445	0.116	0.020
	Medium	0.042	0.640	0.058	−0.118	0.010	0.001	0.887	0.573	0.090
	Large	0.021	0.657	0.079	1.194	0.380	0.062	−0.996	0.264	0.057

TABLE 76. Continued.

Variable	Category	Mass	Total (19 axes)		Inertia	Axis I			Axis II		
			Quality			Score	R^2 axis	Contribution	Score	R^2 axis	Contribution
Flightless species	Porphyrio hochstetteri	0.005	0.117		0.095	−0.728	0.029	0.006	1.255	0.088	0.022
	Habroptila wallacii	0.005	0.210		0.095	−1.278	0.091	0.018	−1.427	0.113	0.029
	Cyanolimnas cerverai	0.005	0.239		0.095	−0.818	0.037	0.007	−0.659	0.024	0.006
	Dryolimnas aldabranus	0.005	0.074		0.095	0.255	0.004	0.001	−0.373	0.008	0.002
	Gallirallus australis	0.005	0.166		0.095	−1.081	0.065	0.013	0.936	0.049	0.012
	G. dieffenbachii	0.005	0.135		0.095	0.560	0.017	0.003	1.452	0.117	0.030
	G. owstoni	0.005	0.179		0.095	0.469	0.012	0.002	1.240	0.085	0.022
	G. wakensis	0.005	0.069		0.095	0.799	0.035	0.007	−0.003	0.000	0.000
	Tricholimnas sylvestris	0.005	0.054		0.095	0.233	0.003	0.001	−0.114	0.001	0.000
	Nesoclopeus poecilopterus	0.005	0.121		0.095	−0.452	0.011	0.002	0.777	0.034	0.009
	Cabalus modestus	0.005	0.218		0.095	0.442	0.011	0.002	1.358	0.103	0.026
	Habropteryx okinawae	0.005	0.236		0.095	0.768	0.033	0.006	0.734	0.030	0.008
	Atlantisia rogersi	0.005	0.138		0.095	−0.074	0.000	0.000	0.672	0.025	0.006
	Porzana palmeri	0.005	0.337		0.095	2.007	0.224	0.044	−1.312	0.096	0.025
	P. atra	0.005	0.288		0.095	2.044	0.232	0.045	−0.932	0.048	0.012
	Amaurornis ineptus	0.005	0.208		0.095	−1.445	0.116	0.023	−0.830	0.038	0.010
	Pareudiastes pacificus	0.005	0.169		0.095	−0.074	0.000	0.000	−1.736	0.167	0.043
	Tribonyx mortierii	0.005	0.151		0.095	−1.551	0.134	0.026	−0.448	0.011	0.003
	Gallinula comeri	0.005	0.060		0.095	−0.077	0.000	0.000	−0.590	0.019	0.005
Eigenvalue (% inertia)	Total (47 categories)		3.700 (100%)			0.485 (13.1%)			0.369 (10.0%)		

74, 75). This correspondence accounted for significant rank correlations between apomorphy (D_T) and extremity (D_C) for most several data sets sharing adequate numbers of taxa for comparison: skins ($r_s = 0.43$; $P < 0.01$), remiges ($r_s = 0.72$; $P < 0.005$), skeletons ($r_s = 0.44$; $P < 0.05$), partial skeletons ($r_s = 0.44$; $P < 0.05$), sterna ($r_s = 0.35$; $P < 0.05$), and pectoral musculature ($r_s = 0.34$; $P < 0.10$).

Accordingly, it is clear that loss of flight in the Rallidae is not identified with any single sine qua non of anatomy or evident from any single metric abstraction. Instead, flightlessness in the Rallidae is manifested by a somewhat variable constellation of morphological features, many of which are cryptic and not discernible in cursory examination of skin specimens, but which are collectively conducive to limited similarity in gross appendicular proportions and obscured to some degree by differing initial conditions (i.e., plesiomorphic states) and inferred changes in body size (i.e., apomorphies) associated with the loss of flight.

CORRESPONDENCE ANALYSIS OF FORM AND CIRCUMSTANCE

An analytical segue between the shifts in form associated with the loss of flight and generalizations of the environmental circumstances under which these evolved was provided by a correspondence analysis of summary metrics for flightless rallids and parameters of the islands that these species inhabit(ed). This method reduces the multivariate associations among categorical variables to comparatively few axes (Appendix 2), thereby permitting a simultaneous display of apomorphy of flightless rallids and characteristics of habitats. However, unlike conventional multivariate techniques such as PCA and CA estimates of missing data were impossible, and therefore a variate-rich correspondence analysis was necessarily limited to 19 species for which information was comparatively complete (Table 76). The resultant bivariate plot of the correspondence analysis (displaying the associations summarized in the first 2 of 19 axes) represented roughly one fourth of the total associations among the 19 species and nine categorical descriptors of islands (Table 76). This finding indicated that the dimensionality of associations among apomorphy of species and characteristics of islands was understandably complex and not readily distilled into a subspace spanned by only a few synthetic axes. Nevertheless, display of one fourth of this multitude of associations in a single plot revealed several noteworthy generalizations (Fig. 116).

Interpretation of plots of multiple correspondence analyses merits special commentary, in that the disposition of points with respect to axes (and the quadrants defined thereby), the relative distances among points, and the distances of points from the origin (intersection of the two axes shown in the plot) all bear on the nature and intensity of associations depicted (Greenacre 1984). To begin with, constellations of points at comparatively great distances from the origin represent associations of relatively high significance; similarly, points clustered near the origin reveal little if anything about associations among the variables (and their values) involved. Accordingly, confining attention to outlying groups of points indicated that a group of four points at the high end of axis I (*Porzana palmeri, P. atra*, small skeletal changes, and large distances between islands and nearest continent) contributed much to the information conveyed by this dimension. Several other variables approached significance in this direction (small change in wing and warm temperate islands). Two other less-compact groups of values were po-

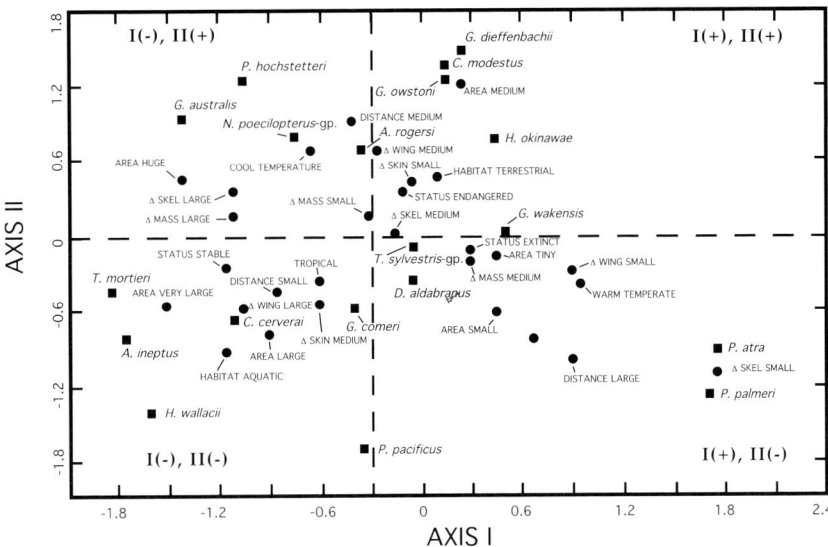

FIG. 116. Bivariate plot of four morphological (triangles) and five ecogeographical (circles) variables for 19 flightless species of Rallidae on the first two axes from a correspondence analysis. Flighted species are symbolized by circles and flightless species are symbolized by squares.

sitioned at low values of axis I (Fig. 116), one in which points scored highly on axis II (*Porphyrio hochstetteri, Gallirallus australis,* and huge island areas) and another characterized by low scores on axis II (*Habroptila wallacii, Amaurornis ineptus, Tribonyx mortieri,* very large island areas, and aquatic habitats). Two sets of characteristics had positions bordering on membership in the latter two constellations: near the former (*Nesoclopeus poecilopterus,* medium isolation of island, and large changes in skeletons and masses), and near the latter (*Cyanolimnas cerverai,* large changes in wings, and large island areas). Taken together, these comparatively disparate clusters of associated taxa and variables indicate that axis I displayed a comparatively large partition of total association attributable to a contrast between small, only moderately apomorphic crakes of well isolated, warm islands and moderately massive, largely aquatic rallids of diverse relationships inhabiting very large islands (Fig. 116).

A slightly less important partition of total association was summarized on axis II (Table 76), which was dominated by two clusters of points having high values on this axis (Fig. 116). One was in the upper left quadrant (*Porphyrio hochstetteri, Gallirallus australis,* and islands having immense areas), and one was in the upper right quadrant (*Gallirallus dieffenbachii, G. owstoni, Cabalus modestus,* and medium areas of islands). These two groups of characteristics were countered graphically principally by two essentially isolated taxonomic points having moderately low scores on axis II (*Habroptila wallacii* and *Pareudiastes pacificus*), with several other parameters or taxa having scores on axis II bordering on membership (*Porzana atra, P. palmeri,* small skeletal changes, and large isolation distances of islands). These opposing groups reveal axis II to be essentially a contrast between four New Zealand endemics and several, widely scattered rallids of other areas of the Pacific region (Fig. 116), groups that appear to be distinguished largely by differences in areas and relative isolation of the respective islands inhabited.

The approximate sphericity of associations among the axes derived from correspondence analysis (Table 76), the generality of the inferences possible from the two most explanatory axes, and the finding that the majority of taxa and parameters of inhabited islands fell so close to the origin as to be of negligible, shared association (Fig. 116), confirmed that flightless rails encompass a remarkable diversity of form and have succeeded in a wide variety of ecological circumstances. With the exception of a few comparatively intuitive, salient gradients among these features, flightless rallids and the islands upon which they evolved are shown to defy succinct numerical summary or multivariate interpretation. Further insights and generalities synthetic of taxa and circumstances will require more focused and novel means of interpretation that incorporate supplementary ontogenetic and ecological information specific to the problem at hand.

DISCUSSION
GROSS RAMIFICATIONS OF FLIGHTLESSNESS IN RAILS
GIANTISM, DWARFISM, AND STASIS

... variety in size ... would, in a corresponding degree, render the act of flight more difficult and laborious. Consequently, if that act were not needed for the acquisition of food, it might seldom or never be exercised in the absence of any enemy from which it would offer a way of escape. By long disuse of the wings, continued through successive generations, those organs, agreeably with Lamarck's theory of the 'Origin of Species,' would become enfeebled, and ultimately atrophied to the degree exemplified in *Apteryx* and *Dinornis*. The legs, then monopolizing the functions of locomotion, would attain, through the concomitant force and frequency of exercise, proportional increase of power and size. Under these conditions may be comprehended, by *veræ causæ*, the origin of the great flightless Anserine ... *Cnemiornis calcitrans*. It has become such through no choice or selection, but by a combination of circumstances enforced, with operative conditions of organic vitality, first taught us by the immortal author of the 'Philosophie Zoologique.' ... The same course of cognition, so guided, leads to the same conclusion as to the origin of *Notornis,* of *Aptornis,* [and] of *Dinornis*. The tendency to variation in size and proportions, after the reduction and loss of wings, leads to minor modifications of such flightless genera. ... The genus *Notornis* is now known to be represented by species, living in the present generation of New-Zealand colonists, in localities nearly one hundred miles apart, and which have belonged to a once gregarious family.—Owen (1882:693–694)

In the first instance [raphids and flightless rallids] the absence of enemies on the insular areas inhabited by them leads to the loss of the power of flight through the reduction of the wing by disuse, or by the natural selection of the less strongly flying individuals; by this profound change of habit the organism is thrown out of equilibrium with its environment, and variations, eliminated under ordinary circumstances, survive. The reason why these variations tend towards an increase of bulk may be that the expenditure of energy is much less.—Andrews (1896b:77)

In the broadest sense, environment is so complex and subtle that the optimum may shift in what appears at first sight to be a stable environment. Thus, animals capable of self-directed locomotion the optimum for size shows a tendency to shift slowly in one direction, toward larger size, a tendency that may be counteracted by other influences and abruptly reversed, but is nevertheless usual in the evolution of such groups. Even though individual animals may be perfectly adapted at a particular size level, in the population as a whole there is a constant tendency to favor a size slightly above the mean. The slightly larger animals have a very small but in the long run, in large populations, decisive advantage in competition for food and for reproductive opportunities and in escaping enemies.—Simpson (1944:85–86)

Adaptations for flight carry certain disadvantages, particularly those that go along with comparatively small size.—Colbert (1955:171)

Body size has often been portrayed as one of the most direct links between microevolution and macroevolution.—Jablonski (1996:256)

Patterns and agents of change in size.—The ecological importance and evolutionary implications of body size have been obvious to biologists as long as reasonably precise estimates of size have been available, and there exist a number of reviews of the general patterns and implications of both giantism (Haldane 1928; Guthrie 1984) and dwarfism (Pregill 1986; Roth 1990, 1992; Hanken and Wake 1993; Lister 1996; Miller 1996; Anderson and Handley 2002), the latter notably among Amphibia (Hanken 1984, 1985; Trueb and Alberch 1985; Yeh 2002). Comfortably consistent with the capitalistic view of more being better, Cope's law (macroevolutionary increase in size within a lineage) received exceptional attention from the evolutionarily inclined and was based principally on trends toward greater body size evident in many fossil assemblages of vertebrates (Cope 1885, 1896; Newell 1949; Kramer 1960; Forster 1964; Gould 1970a; Stanley 1973; Bonner 1974; Pennycuick 1975; Hayami 1978; Brown and Maurer 1986; LaBarbera 1986; Benton 1990; Carroll 1997; Lindenfors and Tullberg 1998). Simpson (1944) noted that evolutionary increases in body size have been inferred from the fossil record of vertebrates so commonly that relative size came to be used as a criterion of ancestry, whereby large fossil taxa were disqualified as plausible ancestors of smaller taxa. Unfortunately, much of the reasoning applied by Simpson (1944:85–86) to this generality was characterized by group-selectionist arguments, and focused on marginal advantages of large size for a minority of individuals (this contributing to long-term increases in size), while maintaining mean body size of the population at a lower adaptive optimum (thereby preserving present fitness of the group). An idiosyncratic explanation of giantism proposed by Edinger (1942) involved a pathology of the pituitary gland (i.e., an acromegaly afflicting entire lineages).

With few exceptions, the primacy of body size with respect to the ecological parameters of organisms has been substantiated by diverse empirical studies (LaBarbera 1986, 1989; Jablonski 1996), despite diverse trends in size among anatomical elements obscured by patterns of body mass (e.g., McGillivray 1985). These studies include those bearing on the coexistence of closely related species of birds (Nudds and Wickett 1994) and flightless rails (Olson 1975b, 1986), thereby providing a comparatively sound basis for arguments requiring size-related selection in birds as a premise. Evolutionary increases in size are not only comparatively common, but the apparent magnitudes of such changes were interpreted by Løvtrup et al. (1974) as contradicting the assumption that mutations effecting large changes are necessarily deleterious (Fisher's axiom).

The intensity of natural selection on body size among adult birds was documented convincingly through a report by Bumpus (1899) of weather-related mortality in *Passer,* data subsequently were subjected to a series of increasingly refined analyses (Harris 1911; Calhoun 1947; Grant 1972a; Johnston et al. 1972; O'Donald 1973; Lowther 1977; Pugesek and Tomer 1996). The evolutionary impact of such a punctuated episode has been balanced by a number of direct quantifications of natural selection on body size and allometric correlates over longer periods (Fleischer and Johnston 1982; Grant and Grant 1987, 2000), works that were bolstered by the confirmation of heritable variation in body size (Grant 1983b; Noordwijk 1984; Price et al. 1984a; B. R. Grant and P. R. Grant 1989), but conditional on a comparative paucity of data bearing directly on the relationship between ontogeny and adult size and fitness (Gebhardt-Henrich and Richner 1998; Schew and Ricklefs 1998) possibly confounding effects of introgression (Faivre et al. 1999). However, it is noteworthy in this regard that no convincing evidence has been marshaled to validate Cope's rule for avian taxa (Damuth 1993), despite the fact marginal evidence for a positive relationship between mean body size and caloric dominance has been marshaled for birds (Brown and Maurer 1986; Juanes 1986; Cotgreave and Harvey 1992). Empirical evidence for convergent changes in both increased and decreased size in *Cerion* was presented by Gould (1984).

McKinney and McNamara (1991) discussed both giantism and dwarfism as targets of selection in heterochrony, and the empirical evidence for the greater frequency with which large body size is favored over small size. Large body size is generally favored where competition among consumers is important relative to efficiency of resource use (Maynard Smith and Brown 1986) and on islands (Case 1978c). Important exceptions include the vulnerability of large lineages under conditions of major environmental perturbation or where large size magnifies exposure to predation, including that by humans (McKinney and McNamara 1991). The evolution of body size can be viewed in both micro- and macroevolutionary scales, although most attention has been paid to the latter (Maurer et al. 1992; Sheldon 1993). Changes in body size inferred for most, if not all, insular flightless birds have been interpreted to represent anagenetic (as opposed to cladogenetic) trends (McKinney 1990a, b; Whittaker 1998). These general arguments become more diffuse under the scrutiny of finer-scale changes and the capacity for specific developmental fields to undergo independent evolutionary change (McKinney and McNamara 1991).

A preoccupation with evolutionary change and related preconceptions regarding trends in body size were undermined further by substantive consideration of the evolutionary bases of macroevolutionary stasis (Grassé 1977; Stanley 1979, 1990; P. G. Williamson 1981; Charlesworth and Lande 1982; Charlesworth et al. 1982; Levinton 1982a, 1983; Wake et al. 1983; Barnard 1984; Wilkens 1993). These were paralleled by study of the mechanisms underlying evolutionary invariance (Vermeij 1974; Williamson 1987; Stearns 1993), and the bradytelic lineages were envisioned as examples of retarded evolutionary change (Simpson 1953a; Cracraft 1984).

Generalities of giantism.—McKinney (1988b:333) cited environmental correlates fostering larger body size in animals, including regularly abundant food (all taxa), cooler temperatures (many vertebrates), large prey size (predators), seasonality (all), and postboreality in vertebrates (see also Fogden 1972; Levinton 1982b; Martin 1996). The latter, which implies a relaxation of selectively maintained size limitations related to locomotion, parallels several corollaries of avian flightlessness. To these McKinney and McNamara (1991:376) added several more factors of possible relevance to birds: low-nutrient food (after abundant food), and sexual selection (after seasonality). Increased body mass has obvious implications for centers of mass, ambulatory balance, posture, arthrological accommodation within limbs, and tendency toward graviportality (Alexander 1971, 1977a, 1980, 1983a, b, 1984, 1985a, b).

These implications entail a downside for large body size in multiple parameters of life history. For example, most modes of locomotion (with an exception being diving) are more expensive in absolute terms for individuals of large body mass, and therefore flightlessness ameliorates a major component of these disadvantages for birds. A general relationship between brain size and body mass, although taxonomically variable, exemplifies the potential for overly simplistic inferences regarding evolutionary trends in body size (Bennet and Harvey 1985a, b). Increased size can affect factors of safety both with respect to avian flight (Vogel 1988) or its loss (Alexander 1984). Moreover, selection for change in size may lead to changes in shape as a by-product of constraint (Gould 1966; Clutton-Brock and Harvey 1979; Pagel and Harvey 1989; Riska 1989; McKinney and McNamara 1991; Kozlowski 1996). Finally, even where selective advantages of large size are established, genetically determined covariance structures can constrain evolutionary change (Schluter 1996a), and environmentally imposed variation can retard or negate the realization of significant changes (Larsson et al. 1998). However, there is overwhelming evidence that insularity per se is associated with changes in body size (either increased or decreased), with shifts in part attributed to the implications of body size for community structure and an accommodation by colonists to other residents of the island. Attribution to insularity of changes in body size is difficult, in part because typical shifts vary with taxonomic group and initial body sizes, and are fostered by diverse ecological circumstances (McNab 2001). For example, Forster (1964) and Lomolino (1985) summarized data for insular mammals and found that small species tended to undergo increased body size, whereas large species tended to become smaller after colonization.

Generalities of dwarfism —Increasingly, evolutionary decreases in body size, or dwarfism' have been the subject of study (Marshall and Corruccini 1978; Price and Grant 1984; Griffith 1990; Bush 1993; Alexander 1996a; Rieppel 1996). That profound decrease in size can accompany colonization of islands is beyond reasonable doubt, perhaps best substantiated by the dwarfed elephant of Malta (*Elephas falconeri*), which underwent a reduction of 75% in height to approximately 1 m at adulthood (Lister 1993, 1996). Despite bringing needed balance to the theory of body size, some investigations involving dwarfisms (notably those of insular mammals) have been characterized by interpretational excesses regarding postcolonizational selection, an underemphasis concerning possible plesiomorphy or founder effects on differentiation and speciation, and a prevalence of ad hoc explanations (Berry 1998; Grant 1998c). Nevertheless, phylogenetically sound examples of miniaturization among birds are known. For example, small size of the hummingbirds (Trochilidae) poses thermodynamic challenges while conferring locomotory advantages for foraging through size-contingent apomorphy of the flight mechanism (e.g., Cotton 1996). For the minority of rallids manifesting dwarfism in association with flightlessness, there is a possibility that body size is inversely related to rate of speciation (Dial and Marzluff 1988; Searle 1996), that is, dwarfism might hasten other apomorphic changes traditionally accorded taxonomic recognition. However, the relationship between ontogeny, size, and evolutionary rates is a complex one (Arthur 1982a; Atchley and Hall 1991), and most perceived evolutionary trends derive from higher-order comparisons.

Insularity, changes in size, and flightlessness.—Owen (1866a:70) compared the large size of *Raphus cucullatus* and *Pezophaps solitaria* with numerous flightless representatives of six taxonomic families of ratites, penguins (*Aptenodytes*), waterfowl (*Cnemiornis*), alcids (*Pinguinus impennis*), and several gruiforms (*Aptornis* and *Porphyrio mantelli*), and elevated the importance of flightlessness to avian body size to an unprecedented level, concluding that these examples ". . . point to the disuse of wings in flight as the main condition of increase of size in species of birds . . . ." In that many of the species cited and most flightless birds are both insular and appear to have undergone increases in body size (Livezey and Humphrey 1986; Livezey 1988, 1989a–d, 1990, 1992a, b, 1993a–c), one might expect that clear generalities would emerge regarding changes in size in flightless rails on islands. Based on mammals, an island rule unified the advantages of both giantism and dwarfism in the absence of continental competitors and predators to effect a shift in mean masses of taxa in insular mammaliam communities toward an intermediate optimum (~1 kg), a selection regime that tends to make small species larger and large species smaller and acts most intensively on taxa of either extreme in size (Van Valen 1973b; Damuth 1993). The approximate optimum for mammals of 1 kg exceeds the body masses of all but the most massive of flighted rails (e.g., *Porphyrio, Aramides,* and *Fulica*), and if such an intermediate optimum existed for insular avian communities the value probably would be substantially lower.

Even if global patterns of change in body size are substantiated, the underlying causes of insular giantism and dwarfism (nanism) remain in dispute, with proposed candidates for insular tetrapods including intraspecific competition, interspecific competition (including character displacement), relaxed predation (e.g., absence of all but aerial predators) or absence of predation (Brown and Wilson 1956; Schoener and Gorman 1968; Williams 1972; Cox and Moore 1993; Reyment 1983; Schüle 1993; Adler and Levins 1994). Recently, Polo and Carrascal (1999) inferred that body size tended to be larger in birds characteristic of open habitats than those in woodlands, especially among ground-foraging species, implying that increases in body size may be especially favored on arid islands supporting habitats at early successional stages. These apparently contradictory explanations at least share the view that shifts are interpreted as a selective response to the body sizes of other colonists, available prey, and predators, with a largely untested role for directional selection toward local optima (e.g., Schoener 1975; Schwaner and Sarre 1988; Diamond et al. 1989; Forsman 1991; Cox and Moore 1993). However, most assessments of such shifts unfortunately may be confounded by other, variably controversial but competition-neutral generalities concerning body size, including Bergmann's and Allen's rules (Calhoun 1947; Snow 1954, 1958; Williamson 1958; James 1968, 1970; Kendeigh 1969, 1970, 1972, 1976; Power 1969, 1970a, b; McNab 1971; Niles 1973; Murphy 1985; Zink and Remsen 1986; Geist 1987; Rising 1988; Johansson et al. 1998), and tend to lead to ad hoc scenarios that resort to special pleading where particular data fail to fit general expectations (Whittaker 1998), and many were conceived without consideration of extinct, potentially critical components of insular communities (Pregill 1986).

However, no specific selective trade-off has been demonstrated for insular birds (Damuth 1993), and if any pattern is evident among flightless rails it would be an opposite one (Table 71): lineages characterized by large body size tend to become exceptionally so (e.g., *Porphyrio* and *Fulica*). Lineages characterized by intermediate initial body sizes vary in direction of change (e.g., *Dryolimnas, Cyanolimnas, Gallirallus,* and *Gallinula*). However, lineages characterized by small initial body sizes vary from modest giantism to dwarfism (e.g., *Atlantisia* and *Porzana*). Tree-based explorations (Figs. 12–14), multivariate analyses (Tables 11, 13–22, 40–42, 46–52; Figs. 15, 20, 22–31, 49–58), and pairwise comparisons with most plausible sister-species (Tables 71, 72) indicate that most flightless species of the Rallidae, regardless of geographical distribution, underwent similar changes in body size, with the majority of taxa assuming greater body sizes with the loss of flight (Table 8; Appendix 1). Several groups indicated approximate stasis in mean body masses or dwarfism with flightlessess—*Dryolimnas, Rallus, Aphanapteryx,* and *Erythromachus,* and *Porzana* (*P. monasa* being an exceptional case of size increase). In summary, for most Rallidae, large species tend to get larger with the loss of flight (with two notable exceptions being stasis in *Rallus* and slight dwarfism in *Dryolimnas*), and small crakes (especially *Porzana*) tend to modest dwarfism with flightlessness.

This counterindication of rails and birds generally is at least consistent with two alternative proposals: higher vagility of birds may increase the likelihood of the eventual immigration of avian competitors (i.e., a reduction in effective isolation); and the multitude of size-related parameters in birds is complicated further by the impact of increasing size on flight capacity (see below). Whatever the underlying cause of the diversity of change evident of flightless rails, the process(es) is (are) manifested in the extraordinary range of size and inferred changes thereof shown by flightless rails in many taxonomic groups (Tables 8, 9, 11–23, 38–54, 65, 68–71; Figs. 15, 20, 22–32, 44, 48–60). Of the three alternative classes of change in body size that attend the loss of flight in birds (decrease, stasis, or increase [Fig. 117]), flightlessness in association with increased body size (Fig. 117A) is most common in many taxonomic groups (including the Rallidae). This finding suggests that loss of flight in concert with increased size is either of greater selective advantage, ontogenetically parsimonious, or both. In some groups, flightlessness evidently derives from the enlargement of body size accompanied by negatively allometric increases in the pectoral apparatus (e.g., some *Tachyeres*), with actual reduction in wing size (if any) occurring subsequent to attaining the threshold of flightlessness (τ). Flightlessness by such allometric giantism (Fig. 117A), which essentially requires only general size increase with the pectoral apparatus undergoing increase not commensurate with the complement of the corpus, can be derived dynamically from first principles by using relatively straightforward methods (Appendix 5), and can be depicted by a phase plane illustrative of a two-dimensional system of differential equations in which trajectories of approach to the threshold τ are readily specified by ratios of allometric coefficients pertaining to change in size of the body and the pectoral limb (Fig. 118).

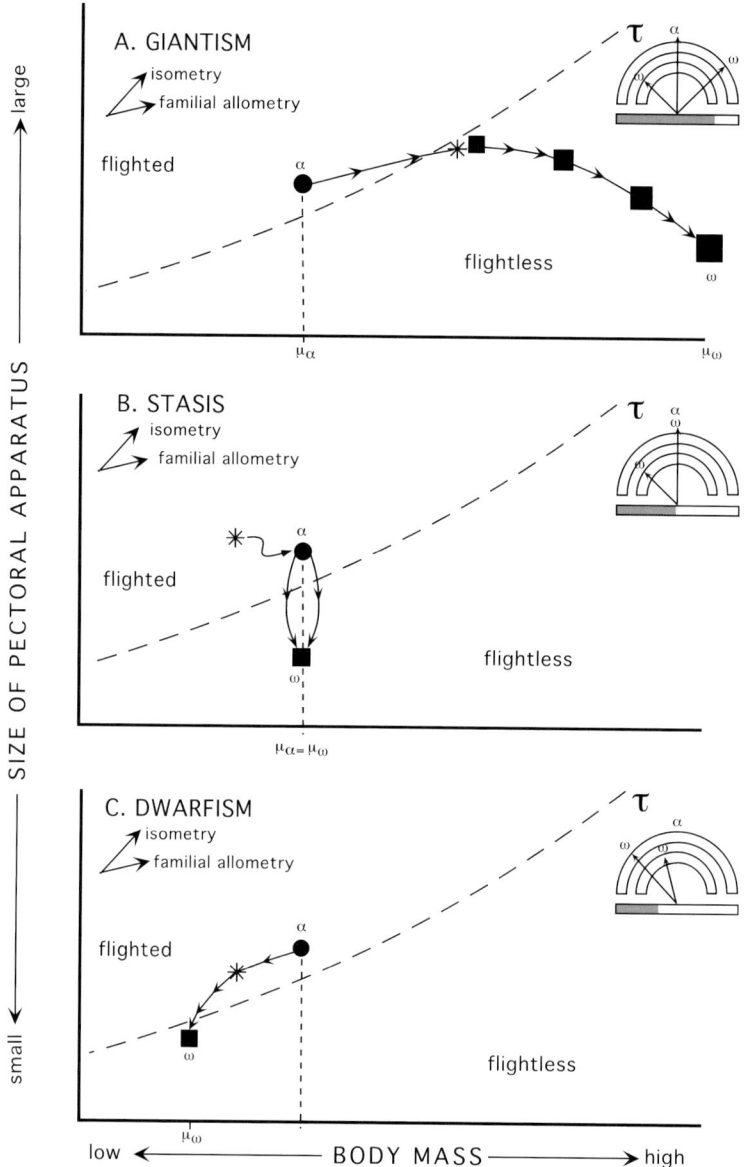

FIG. 117. Hypothetical evolutionary pathways leading to flightlessness in bivariate space defined by body size (initial values symbolized by μ_α) and size of the pectoral apparatus, in which selection intensity varies directly with $\mu_\alpha - \mu_\omega$: **A,** giantism, initially characterized by allometric increase in size but undergoing a key evolutionary innovation or event (κ) from flighted *Bauplan* upon crossing the threshold of flightlessness τ to attain optimal, flightless body size μ_ω; **B,** stasis, characterized by initial departure (allometric reorganization) from flighted *Bauplan* through size-neutral reduction in pectoral apparatus to cross the threshold of flightlessness τ while maintaining the same body size $\mu_\omega = \mu_\alpha$; and **C,** dwarfism, initially characterized by allometric decrease in size but undergoing a departure (allometric reorganization) from flighted *Bauplan* to permit crossing the threshold of flightlessness τ to attain smaller, optimal flightless body size μ_ω. Each evolutionary scenario is accompanied by an "clock diagram" corresponding to the shifts in size, shape, and age through heterochrony in the flightless species (see text), after Gould (1977).

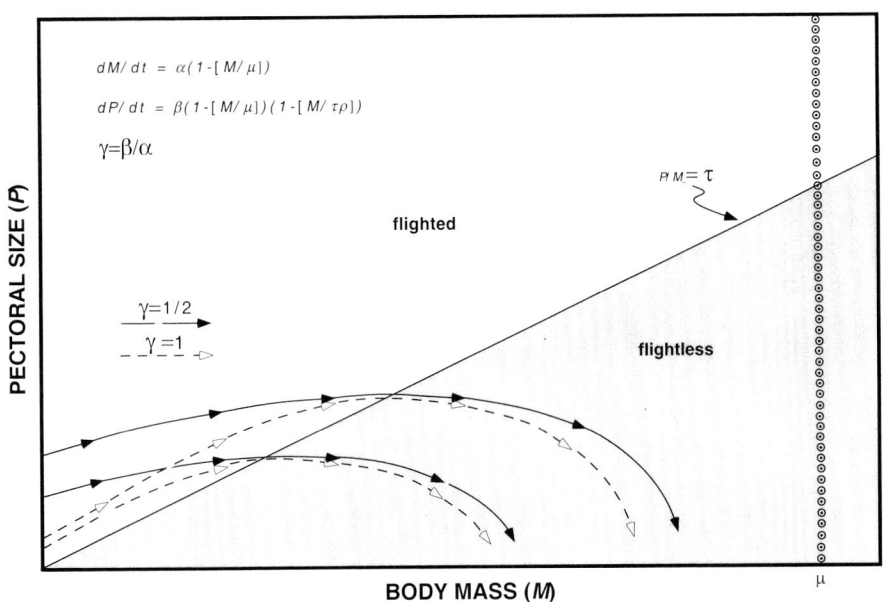

FIG. 118. Phase-plane for system of differential equations depicting evolutionary increase in body mass (M), that is logistic with respect to optimal body mass (μ) and interrelated change in size of the pectoral apparatus (P), the latter multiplicatively logistic both with respect to change in M and the threshold of flightlessness (τ). Trajectories for two initial positions and two values of ratios of scaling parameters are approximated. See Appendix 5 for explicit development of system and the derivation of trajectories under selected scaling parameters and initial conditions.

PECTORAL ALLOMETRY AND FLIGHTLESSNESS: FALLING OFF THE LINE

The bird's case is of peculiar interest. In running, walking or swimming, we consider the speed which an animal *can attain,* and the increase of speed which increasing size permits of. But in flight there is a certain necessary speed—a speed (relative to the air) which the bird *must attain* in order to maintain itself aloft, and which *must* increase as its size increases. . . . The principle of *necessary speed* . . . accounts for . . . why larger birds have a marked difficulty in rising from the ground, that is to say, in acquiring to begin with the horizontal velocity necessary for their support. . . . For the fact is that the heavy birds must fly quickly, or not at all.—Thompson (1961:30–31; emphasis in original)

There has been, in my opinion, mistaken emphasis on the non-adaptive nature of simple allometric trends in phylogeny. Proportions produced by constant α values [allometric intercepts] need not be viewed as by-products of size increase brought to expression without selective modification of genetic shape factors; constant α may, rather, reflect an ordered set of proportions specifically selected to accommodate absolute magnitude at each step of phyletic size increase. . . . allometric trends are as subject to evolutionary alteration as are morphological features. If increasing size would lead to inadaptive proportions, allometric parameters may be modified to allow further phyletic growth.—Gould (1966:618–619)

Allometry of avian flight.—Although flight is a comparatively speedy and generally economical means of transport (Tucker 1970, 1973a; Schmidt-Nielsen 1972;

Rayner 1981, 1988a, b, 1990), and can provide an important vantage point for display, search, and orientation, flight is enormously consumptive of energy in birds per unit time (Berger and Hart 1974; Bishop 1997). High energetic demand is especially true of those groups in which flight is maintained by comparatively constant flapping of the wings or those having larger body sizes or induced drag (Raveling and LeFebvre 1967; Ellington 1984; Rayner 1985a, 1988a, b, 1993, 1995). These groups of power fliers include the great majority of flightless carinates. In the latter group, the threshold of flightlessness reduces to a failure to achieve takeoff, because this entails overcoming the acceleration of gravity as well as generation of forward thrust (Marden 1987; Tobalske and Dial 2000). Challenges of takeoff and an essential role of head winds for achieving necessary air (flight) speed also are shared by the large members of the Procellariiformes (Warham 1977, 1996; Pennycuick 1982, 1989), a group employing slope-soaring flight, possessed of high aspect ratios, and among which flightlessness is unknown (Table 4). The capacity of the pelvic limb to contribute upward thrust and forward velocity assists in overcoming the challenge of takeoff in many groups, including the tubenoses (Heppner and Anderson 1985; Bonser and Rayner 1996; Warham 1996; Earls 2000). Power required to remain airborne drops precipitously after takeoff, and only begins to increase again at velocities significantly in excess of the energetic optimum for a species (Blem 1999). Once airborne, level flight can be sustained as long as velocity is sufficient to generate necessary lift, a relationship central to the determination of optimal flight speed for a given taxon (Hedenström and Alerstam 1995; Thomas and Hedenström 1998) and that underlies the typically high velocities maintained by heavy-bodied power fliers (Cottam et al. 1942; Pennycuick 1968, 1996, 1997, 1998b).

Classic syntheses of essential aerodynamic parameters in birds (Hamilton 1961; Hartman 1961; Greenewalt 1962, 1975) established virtual isometry of wing length and wing area with body mass in modern taxa (Fig. 119), that is, the slopes closely approximated those expected on geometric grounds ($\hat{b}$ = 1/3 and 2/3, respectively), with the hummingbirds (Trochilidae) excepted on the basis of their unique flight in which lift is generated on both downward and recovery strokes of the wings (Epting and Casey 1973; Feinsinger and Chaplin 1975; Feinsinger et al. 1979; Epting 1980). Although differing significantly in allometric relations between wing length and body mass (F = 51.2; d.f. = 16, 599; P < 0.00001), diverse taxa of power fliers manifest marked similarity in a close approach to isometry (Fig. 119). Likewise, flighted birds collectively manifest isometry between masses of the major breast muscles (mm. pectoralis et supracoracoideus) with body mass (Hartman 1961; Greenewalt 1962). This degree of conformity in gross body proportions among birds, together with the demonstrable implications of these for flight, strongly suggests intensive underlying stabilizing selection in Aves (Rayner 1996). Nonetheless, variation in such vital parameters is evident among taxa within taxonomic families and among larger taxonomic groups (Fig. 119). Some taxonomic groups of birds are characterized by transpositional shifts in allometry of wing size, by which increases in body size among flighted members are accompanied by comparatively profound changes in conformation or intercept (Meunier 1959a, b; Gould 1966).

Relative robustness of the pectoral musculature has functional implications of comparable magnitude to the relationship between body mass and wing size.

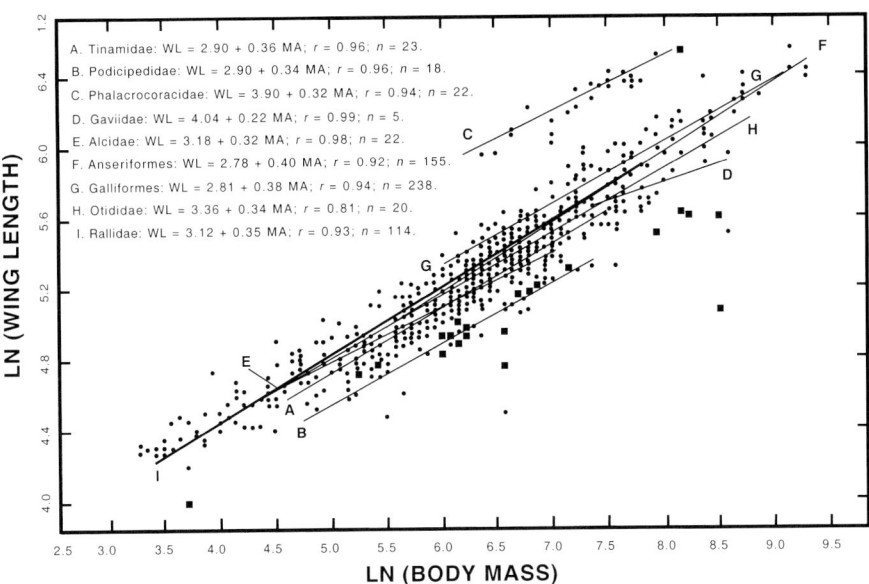

FIG. 119. Allometric relationship between mean wing length and mean body mass for selected taxonomic families of modern Aves, depicting the relative position of the Rallidae and several related families. Flightless members are indicated by squares.

Rails, even flighted species, are characterized by relatively small breast muscles (Hartman 1961; Rayner 1988a). The major breast muscles (mm. pectoralis et supracoracoideus) compose only 7–12% of total body mass in rallids, whereas in birds generally these account for at least 15% of body mass. A commensurate diminution of the other muscles of the pectoral apparatus—the "rest" muscles of Hartman (1961)—also typifies the Rallidae, a generality accompanied by comparatively heavy wing loadings and small tail areas (Hartman 1961; Rayner 1988a). Therefore, despite the general validity of relatively short wings as prima facie evidence of avian flightlessness, diagnoses of flight capacity based on this criterion among the Rallidae require supplementary information on other influential anatomical parameters (e.g., relative mass of pectoral muscles and structure of remiges), in that some rallids unexceptional in relative wing length are flightless (Table 9; Fig. 15).

Accordingly, flight capacity (ϕ) can be symbolized as a function of three general parameters:

$$\phi = g(M_B, L_w, P_f),$$

where M_B is body mass, L_w is the lift provided by the wing surfaces, and P_f is power provided by pectoral musculature (in turn a function of bulk, histology, and distribution of composite muscles), with the latter being affected by cardiovascular capacity (in turn a function of heart rate, basal metabolic rate [BMR], and aerobic efficiency). The limiting requirements placed on lift for takeoff affect the values for lift (L_w) and power (P_f) most severely. Of these essential components, the lift (L_w) provided by the wings is most difficult about which to generalize, because it is a complex function of several variables of the bird and its environment. Based on the pioneering work by von Helmholtz (1874), lift has been determined to be a function of the area of the wings (A_w), density of the air (ρ), velocity of the bird through the air (U), and a lift coefficient (C_L), and has been approximated as (Pennycuick 1969; Tucker 1973a; Schmidt-Nielsen 1984):

$$L_w = \rho/2 A_w U^2 C_L.$$

This estimate is further complicated by the inverse effect of flight speed on induced drag (proportional to U^{-2}), and the fact that both lift and drag vary with the shape of the wing (the latter varying substantially among taxonomic groups of birds).

It has been predicted theoretically and confirmed empirically that the morphometric space bounding those avian body forms that should be capable of flight can be defined broadly in terms of a suite of functional parameters that vary allometrically with body mass (Table 77; Fig. 120): isometric scaling of wing length ($M_B^{1/3}$) and wing area ($M_B^{2/3}$), closely similar to that realized for rails and other birds (Table 77; Hartman 1961; Greenewalt 1962), in comparison with the upper bound permitting powered flight (Meunier 1951); a lower bound for frequency of wingbeats ($M_B^{-1/6}$; von Helmholtz 1874; Rayner 1988a), with the corresponding upper bound for frequency of wingbeats ($M_B^{-1/3}$; Pennycuick 1968, 1975; Lighthill 1977); and the power available ($M_B^{0.72}$) and required for flight ($M_B^{1.0}$) as a function of body mass (Pennycuick 1969; Calder 1974; Rayner 1988a). The maximal body mass for a bird capable of flight (roughly 12 kg) evidently coincides with the intersection of the differentially allometric increases in power required and available for flight (Tucker 1974, 1975, 1977; Rayner 1988a; Norberg 1996; Butler and Bishop 1999; Brown et al. 2000), a size approximated by the largest extant flying birds, for example, the Trumpeter Swan (Anseriformes, Anatidae: *Cygnus buccinator*) and Kori Bustard (Gruiformes, Otididae: *Ardeotis kori*).

These fundamental allometric relations define a morphological window or morphospace of avian body forms that should be capable of flight (Fig. 120). Nevertheless, many flightless species fall within these limits, underscoring the reality that excluded factors are influential and that care must be used in the interpretation of multiple allometric relationships for organisms (e.g., Kozlowski 2000). Morphological parameters that vary with body size and impinge on flight capacity include wing loadings, minimum air speeds to overcome drag and maintain lift, relative thickening of appendicular bones, and rigidity of remiges (Table 77). Ecophysiological correlates of body size (notably thermodynamic, metabolic, and ecological) are of comparable consequence and considered below (Peters 1983; Calder 1984; Schmidt-Nielsen 1984; Tributsch 1984).

Larger species exhibit much narrower ranges of flight patterns in normal and migratory contexts than smaller species, with the comparatively massive species tending to closely approximate speeds of maximal energetic efficiency in normal locomotion (Bruderer and Boldt 2001). This suggests that functional tolerances bearing on flight capacity become increasingly stringent at higher body masses, both within and among taxa, being more permissive regarding pectoral reduction in small arboreal insectivores (Olson 1994). This boundary phenomenon is exemplified by the temporary flightlessness experienced by large individuals of spe-

TABLE 77. Allometric relationships ($Y = aM_B^b$) of selected characteristics (Y, SI units) with adult body mass (M_B, kg) for birds (where usefully partitionable, nonpasserines), after Peters (1983), Calder (1984), and Schmidt-Nielsen (1984).

Dependent variable (Y), units (symbol)	$Y = f(M_B)$	Key references
Morphological: external		
Wingspan (excluding Trochilidae), m (L_s)	$L_s = 0.75 M_B^{3.00}$	Tucker 1973a
Wing area (shorebirds), m² (A_w)	$A_w = 0.13 M_B^{0.71}$	Rayner 1979c
Wing loading (shorebirds), kg m⁻² (B_w)	$B_w = 7.69 M_B^{0.29}$	Rayner 1979c
Skin area, m² (A_s)	$A_s = 0.10 M_B^{0.67}$	Drent and Stonehouse 1971
External plumage area, m² (A_p)	$A_p = 0.08 M_B^{0.67}$	Walsberg and King 1978
Plumage mass, kg (M_f)	$M_f = 8 \times 10^{-5} M_B^{0.95}$	Turcek 1966
Morphological: internal		
Total muscle mass, kg (M_m)	$M_m = 0.48 M_B^{1.08}$	Hiebert and Calder (unpubl.), in Calder 1984
M. pectoralis mass, kg (M_p)	$M_p = 0.155 M_B^{1.00}$	Greenewalt 1962
M. supracoracoideus mass, kg (M_s)	$M_s = 0.016 M_B^{1.00}$	Greenewalt 1962
Total skeleton mass, kg (M_B)	$M_B = 0.065 M_B^{1.07}$	Prange et al. 1979
Heart mass, kg (M_h)	$M_h = 0.008 M_B^{0.91}$	Parrot 1894; Hartman 1955
Lung volume, L (V_l)	$V_l = 0.030 M_B^{1.05}$	Maina and Settle 1982
Tracheal volume, L (V_{tr})	$V_{tr} = 0.004 M_B^{1.06}$	Hinds and Calder 1971
Tidal volume, l (V_t)	$V_t = 0.013 M_B^{1.08}$	Lasiewski and Calder 1971
Gut mass, kg (M_g)	$M_g = 0.98 M_B^{1.00}$	Quiring 1950; Calder 1984
Physiological		
Basal metabolic rate, W (P_m)	$P_m = 3.76 M_B^{0.74}$	Zar 1969
Basal metabolic rate (nonpasserines), W (P_m)	$P_m = 3.64 M_B^{0.72}$	Lasiewski and Dawson 1967
Basal metabolic rate (nonpasserines), W (P_m)	$P_m = 4.47 M_B^{0.73}$	Calder 1974
Existence metabolic rate (nonpasserines), W (P_m {ϵ})	$P_m(\epsilon) = 8.17 M_B^{0.53}$	Calder 1974
Flight metabolic rate (all taxa), W (P_m (ϕ))	$P_m(\phi) = 108 M_B^{1.02}$	Berger and Hart 1974
Resting metabolic rate (nonpasserines, day), W (P_n (ρ))	$P_m(\alpha) = 4.41 M_B^{0.73}$	Aschoff and Pohl 1970a,b
Specific metabolic rate (endotherms), W/kg ($_mR_s$)	$_mR_s = 4.1 M_B^{-0.25}$	Hemmingsen 1960
Ventilation rate, L/s (V_r)	$V_r = 0.005 M_B^{0.77}$	Lasiewski and Calder 1971; Vleck and Bucher 1998
General pulse rate, s⁻¹ (F_p)	$F_p = 2.60 M_B^{-0.23}$	Calder 1968; Lasiewski and Calder 1971
Flight pulse rate, s⁻¹ (F_f)	$F_f = 8.79 M_B^{-0.15}$	Berger and Hart 1974
Maximal pulse rate, s⁻¹ (F_m)	$F_m = 8.48 M_B^{-0.16}$	Berger and Hart 1974

TABLE 77. Continued.

Dependent variable (Y, units (symbol)	$Y = f(M_B)$	Key references
Thermodynamic		
Body temperature, °C (T_B)	$T_B = 40.1 M_B^{0.04}$	Calder and King 1974
Thermal conductance, W/°C (C)	$C = 0.11 M_B^{0.49}$	Calder and King 1974
Thermal conductance, resting (nonpasserines), W/°C ($C(\alpha)$)	$C(\alpha) = 0.13 M_B^{0.54}$	Aschoff 1981
Thermal conductance, active (nonpasserines), W/°C ($C(\rho)$)	$C(\rho) = 0.19 M_B^{0.52}$	Aschoff 1981
Heat stress coefficient, W/s (S_h)	$S_h = 0.14 M_B^{0.35}$	Weathers 1981
Locomotory		
Cost of flight*, J/m (C_f)	$C_f = \hat{\alpha} M_B^{1.00}$	Pennycuick 1969; Tucker 1973a, b
Power available for flight, J/m (P_f)	$P_f = 4.19 M_B^{0.73}$	Berger and Hart 1974
Cost of biped running, J/m (C_r)	$C_r = 11.1 M_B^{0.76}$	Fedak et al. 1974
Period of wingstroke (shorebirds), s (T_w)	$T_w = 0.25 M_B^{0.19}$	Rayner 1979c
Frequency of wingbeats (shorebirds), s^{-1} (f_w)	$f_w = 4.00 M_B^{-0.19}$	Rayner 1979c
Characteristic flight speed, m/s (U_c)	$U_c = 14.6 M_B^{0.20}$	Tucker and Schmidt-Koenig 1971
Flight speed for maximal endurance, m/s (U_e)	$U_e = 5.70 M_B^{0.16}$	Rayner 1979c; Schnell and Hellack 1979
Flight speed for maximal endurance (shorebirds), m/s (U_{es})	$U_{es} = 16.1 M_B^{0.17}$	Greenewalt 1975
Flight speed for maximal range, m/s (U_r)	$U_r = 10.9 M_B^{19}$	Rayner 1979c
Maximal migratory flight range, m (R_m)	$R_m = 1.9 \times 10^6 M_B^{0.23}$	Peters 1983
Reproductive		
Clutch size† (N_c)	$N_c = 4.90 M_B^{-0.00}$	Blueweiss et al. 1978
	$N_c = 3.60 M_B^{-0.05}$	Western and Ssemakula 1982
Egg mass (all taxa)‡, kg (M_e)	$M_e = 0.40 M_B^{0.73}$	Brody 1945
(all taxa)	$M_e = 0.05 M_B^{0.77}$	Blueweiss et al. 1978
(nonpasserines)	$M_e = 0.08 M_B^{0.68}$	Cabana et al. 1982
(shorebirds)	$M_e = 0.09 M_B^{0.71}$	Ross 1979
Clutch mass, kg (M_c)	$M_c = 0.21 M_B^{0.74}$	Blueweiss et al. 1978
Egg shell thickness, m (D_s)	$D_s = 0.0001 M_B^{0.46}$	Ar et al. 1979
Incubation period§, s (T_i)	$T_i = 2.5 \times 10^6 M_B^{0.17}$	Drent 1975; Rahn et al. 1975
	$T_i = 2.4 \times 10^6 M_B^{0.16}$	Blueweiss et al. 1978
Hatchling mass, kg (M_h)	$M_h = 0.03 M_B^{0.69}$	Blueweiss et al. 1978
Postembryonic doubling time, s (T_d)	$T_d = 1.1 \times 10^6 M_B^{0.34}$	Calder 1984; after Ricklefs 1979
Energy expended during egg development, J (E_e)	$E_e = 1.6 \times 10^6 M_B^{0.95}$	Ar and Rahn 1980; Vleck et al. 1980a,b
Age at fledging (precocial), s (T_f)	$T_f = 3.8 \times 10^6 M_B^{0.20}$	Calder 1984; after Ar and Yom-Tov 1978
	$T_f = 3.9 \times 10^6 M_B^{0.37}$	Ricklefs 1973
Age at first breeding, s (T_r)	$T_r = 9.1 \times 10^7 M_B^{0.23}$	Western and Ssemakula 1982

TABLE 77. Continued.

Dependent variable (Y), units (symbol)	$Y = f(M_B)$	Key references
Demographic		
Longevity in the wild (nonpasserines)‖, s (T_l)	$T_l = 5.3 \times 10^8 M_B^{0.18}$	Lindstedt and Calder 1976
Survival time under thermoneutral fasting, s (T_s)	$T_s = 9.2 \times 10^5 M_B^{0.26}$	Peters 1983
Survival time (−13 to −18°C), s (T_c)	$T_c = 7.8 \times 10^5 M_B^{0.59}$	Calder 1974
Survival time (−1 to −9°C), s (T_m)	$T_m = 1.1 \times 10^6 M_B^{0.58}$	Calder 1974
Survival time (2 to 6°C), s (T_h)	$T_h = 1.3 \times 10^6 M_B^{0.39}$	Calder 1974
Ecological		
Total energy budget (flight foragers), W (E_t)	$E_t = 8.3 \times 10^4 M_B^{0.61}$	Walsberg 1983
Total energy budget (nonflight foragers), W (E_t)	$E_t = 7.5 \times 10^4 M_B^{0.61}$	Walsberg 1983
Home range (all taxa), m² (A_h)	$A_h = 1.6 \times 10^3 M_B^{1.23}$	Armstrong 1965
	$A_h = 1.0 \times 10^3 M_B^{1.14}$	Schoener 1968

* Approximation of intercept â complex and involves multiple parameters of bird, environment, and conditions of flight (Tucker 1973b).
† Substantial interfamilial differences exist. For example, Anatidae, $V_c = 7.0 M_B^{-0.16}$; Phasianidae, $N_c = 5.8 M_B^{-0.14}$ (Rhan et al. 1974, 1975).
‡ Estimate for intercept (a) by Rahn et al. (1975) confounded volumetric estimates with masses (Grant 1983a; Scott and Ankney 1983).
§ Incubation period is more precisely a function of egg size: $T_i = 12.03 M_e^{0.217}$ (Rahn and Ar 1974).
‖ Under assumption of Chossat's rule (Kleiber 1961).

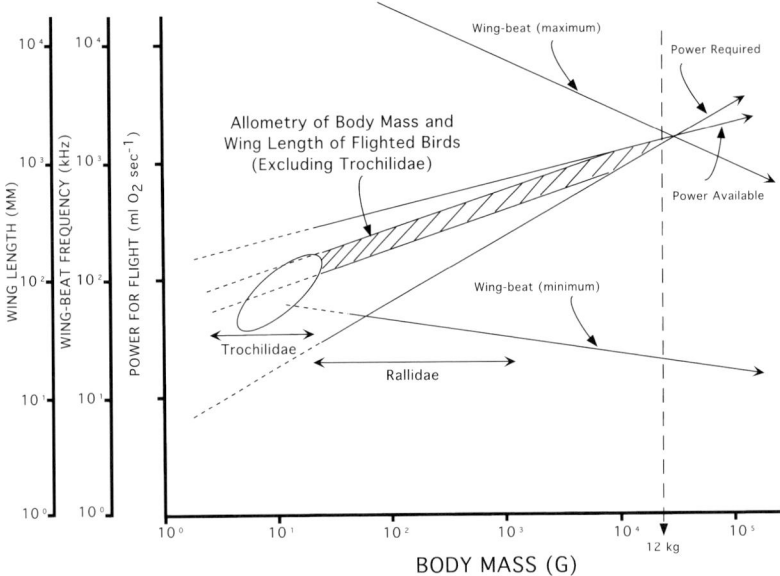

FIG. 120. Bivariate plot of allometric relationships between observed wing lengths (excluding Trochilidae, approximated separately as ellipse), power for flight (available and required), and wingbeat frequency (maximal and minimal) with mean body mass for flighted Aves. Plot is intended to be a heuristic illustration of three parameters bearing on flight capacity across a range of body masses, and largely reflecting relative scaling of parameters and the documented impossibility of avian flight at body masses greater than ~12 kg (see text). Approximate range of body masses of species of Rallidae is shown.

cies near the threshold (τ) following periods of intensive feeding (e.g., *Tachyeres patachonicus* [Humphrey and Livezey 1982; Livezey and Humphrey 1986]). Other power fliers that approach the threshold of flightlessness include eiders (*Somateria* [Guillemette 1994]) and loons (Gaviidae), notably large *Gavia adamsii* (Preston 1951; Savile 1957, 1958, 1962, pers. comm.; Norberg and Norberg 1971).

Parallels of nonavian flight.—The complex of interrelationships that enable powered flight undoubtedly contributed to the evolutionary rarity of this mode of locomotion, which is known to have arisen only three times among tetrapods, in the Pterosauria, Aves, and Chiroptera (Vermeij 1987). However, this delicate balance of relationships has been disrupted secondarily so as to render multiple avian groups apomorphically flightless (Table 4). Foraging niche and the morphological prerequisites for aerial maneuverability have imposed some convergences between insectivorous bats (Chiroptera) and aerial-feeding birds (e.g., Apodidae and Hirundinidae), as opposed to sit-and-wait predators (McNab and Bonaccorso 1995), notably comparatively low body masses, light wing loadings, and absence of flightlessness in the latter avian taxa (U. M. Norberg 1986). Body masses of bats (Mammalia: Chiroptera) are no greater than 1.5 kg, an upper bound evidently imposed both by aerial maneuverability and (in many Chiroptera, but few of the massive *Pteropus*) size-related selective constraints on echolocative frequency (U. M. Norberg 1981a, b, 1987, 1994; Norberg and Rayner 1987; Jones 1996). Reductions in pectoral limbs in other tetrapods related to changes in locomotion or transport strategies, although they probably share similar ontogenetic underpinnings (Rayner 1996), manifest distinctly different morphological patterns (Table 78).

Variation in flighted and flightless rails.—No modern rallid approaches the upper limit of body mass permitting avian flight (12 kg), and only a few species (e.g., *Fulica gigantea* and *F. cornuta*) approach the threshold of flightlessness so closely as to vary in performance. Also, relative wing lengths and theoretically attainable frequencies of wingbeats in modern members of the family Rallidae evidently are not limiting with respect to flight capacity (Greenewalt 1962). Rails are unremarkable among other terrestrially specialized power fliers in the interspecific relationship between wing length and body mass (Figs. 119, 120), and on this basis seem an unlikely source for a multitude of flightless members. Despite reports emphasizing the brevipennate condition of some flightless rallids (e.g., *Habropteryx okinawae* [Kuroda 1993]) and *Porphyriornis comeri* [Eber 1961]), few taxa qualify as marked outliers from the allometric relation characteristic of flighted confamilials (Figs. 119–121). Notable deviations from the norm for the Rallidae on the basis of external dimensions include (Table 71; Fig. 15) *Habroptila wallacii, Gallirallus australis, Tricholimnas sylvestris, T. lafresnayanus, Habropteryx insignis, Porzana monasa,* and *Tribonyx mortierii.*

As detailed below, most flightless rails appear to have lost the capacity for flight by other means, notably reductions in the pectoral musculature (source of power behind air speed) in combination with maintenance or increases of total body mass. Members of the family Rallidae as a whole tend to have comparatively weak breast musculature (Hartman 1961; Greenewalt 1962), among which flightless rails fall well below the isometry between masses of mm. pectoralis et supracoracoideus and body mass typical of flighted members (Table 77). This shift in distribution of body mass has important implications for allocation of resources by flightless rails (McNab 1994a, 2001), representing an evolutionary strategy incompletely conveyed by the morphological parameters normally descriptive of flight capacity (Figs. 117, 119, 120). This explanatory shortcoming primarily stems from the omission of other critical physiological and ecological correlates of body mass, notably thermodynamic efficiency and ecological implications of body size (Calder 1974; Lindstedt and Calder 1981; Peters 1983; Schmidt-Nielsen 1984).

FINE-SCALE MANIFESTATIONS OF FLIGHTLESSESS IN RAILS

INTEGUMENTARY CHANGES

> In the most adult example [of *Atlantisia rogersi*] . . . there is still a very evident air of immaturity; and this impression is borne out by the discovery that the remiges or wing-feathers, as regards the development of the rami and radii, exhibit microscopically a stage of evolution which has not advanced beyond that seen in the body or contour-feathers of such a volant form as *Rallus aquaticus*. . . . the rami and radii of the wing-feathers of other truly flightless rails present a similar want of development or juvenility, a condition of things which seems to be intimately connected with the subject of flightlessness . . . .—Lowe (1928a:106)

TABLE 78. Anatomical changes associated with reduction and loss of pectoral appendages in selected Tetrapoda. Characteristics listed for most mammalian examples pertain to both pectoral and pelvic appendages; in some the latter are more profound and better documented. Reduction limited to pelvic appendage in Cetacea was not included.

Taxonomic group	Pectoral apomorphies	Functional context(s)	Likely ontogenetic basis	Selected references
Amphibia				
Gymnophiona (caecilians)	Modern taxa limbless; fossils indicate sequential loss of digits and truncation of long bones initially	Fossorial locomotion	Heterochrony in ancestral node likely, but available evidence inconclusive	Lande 1978 Jenkins and Walsh 1993
Anura (frogs and toads)	Phylogenetically complex variation in girdle, synostosis of carpals, and reduction in number of phalanges	Repeated, convergent firmisterny, function of which unknown; locomotory refinements distally	Hypothesized to be neoteny, principally effected by altered chondogeny and ossification	Trueb 1973; Emerson 1978, 1988; Alberch and Gale 1983
Caudata (salamanders)	Truncation or loss of phalanges, especially in digits I and V	Streamlining of body via dwarfism and shortening of appendages	Reduction of mesenchymal cells by paedomorphosis and dwarfism	Alberch and Gale 1985
Reptilia (nonavian)				
Amphisbaenia	Complete limblessness	Fossorial locomotion	Ancestral heterochrony, but available evidence inconclusive	Essex 1927; Raynaud 1972; Lande 1978
Squamata (lizards and snakes)	Variable reduction or loss of digits (e.g., Scincidae, Teiidae) or complete limblessness (e.g., Ophidia)	Diverse, undulatory locomotion of elongated body	Heterochrony, perhaps early-acting, but available evidence inconclusive	Miller 1944; Stokely 1947; Gans 1975; Presch 1975; Greer 1987, 1990
Aves				
Palaeognathae: Dinornithidae	Complete loss of appendage	Graviportal folivory on large, predator-free islands	Fixation of early-acting "wingless" regulatory gene(s) likely	Livezey 1995a
Palaeognathae: other ratites	Loss of carina sterni; claviculae vestigial or lost; variably profound reduction or synostosis of limb elements, especially phalanges	Cursorial ambulation in absence of predators (Apterygidae) or on continents (others)	Probably comparatively ancient fixation of early-acting regulatory genes effecting paedomorphosis or pectoral apparatus	See Table 4; Owen 1848a, b, 1870, 1872, 1875, 1879; Cracraft 1974a See Table 4; Owen 1841; Fürbringer 1888, 1902; Parker 1892; Cracraft 1974a

TABLE 78. Continued.

Taxonomic group	Pectoral apomorphies	Functional context(s)	Likely ontogenetic basis	Selected references
Neognathae: Aptornithidae	Reduction of carina sterni; reduction of long bones; loss or synostosis of alar phalanges	Graviportal locomotion for strictly terrestrial omnivory in absence of terrestrial predators	Paedomorphosis of unknown subtype, effected moderately early in development	See Table 4; Livezey 1994
Neognathae: other members	Reductions of carina sterni and conformation change and truncation of appendicular elements (none lost)	Various, including aquatic and terrestrial examples; mostly insular, rarely continental habitats	Negative allometric increase in body size or variably pronounced, late-acting paedomorphosis	See Table 4; Gadow 1902; Livezey and Humphrey 1986; Livezey 1988, 1989a–d, 1990, 1992a, b, 1993a–c
Mammalia*				
Edentata	Loss of single digit	Modification for arboreal brachiation	Early heterochronic abortion likely but evidence inconclusive	Flower 1885; Grassé 1955; Weber 1928; Hofstetter 1958
Artiodactyla	Loss of 1–4 lateral digits	Modification for quadrupedal ambulatory locomotion	Heterochrony likely but not shown, despite rich paleontological record	Hildebrand 1995; Carroll 1997
Sirenia	External digits coalesced, with reduction or limited loss of osteological elements	Modification exclusively for submarine propulsion; incremental reduction of pelvic limb well known	Heterochrony effected through early-acting regulatory genes likely, but available evidence inconclusive	Rienhart 1959; Kaiser 1974; Savage 1976
Lande 1978				

*External coalescence of digits, with retention or augmentation of osteological elements in "Pinnipedia" (Carnivora: Otarioidea, Phocoidea) and Cetacea (Kellogg 1936; Slijper 1961; Thewissen and Fish 1997) not included.

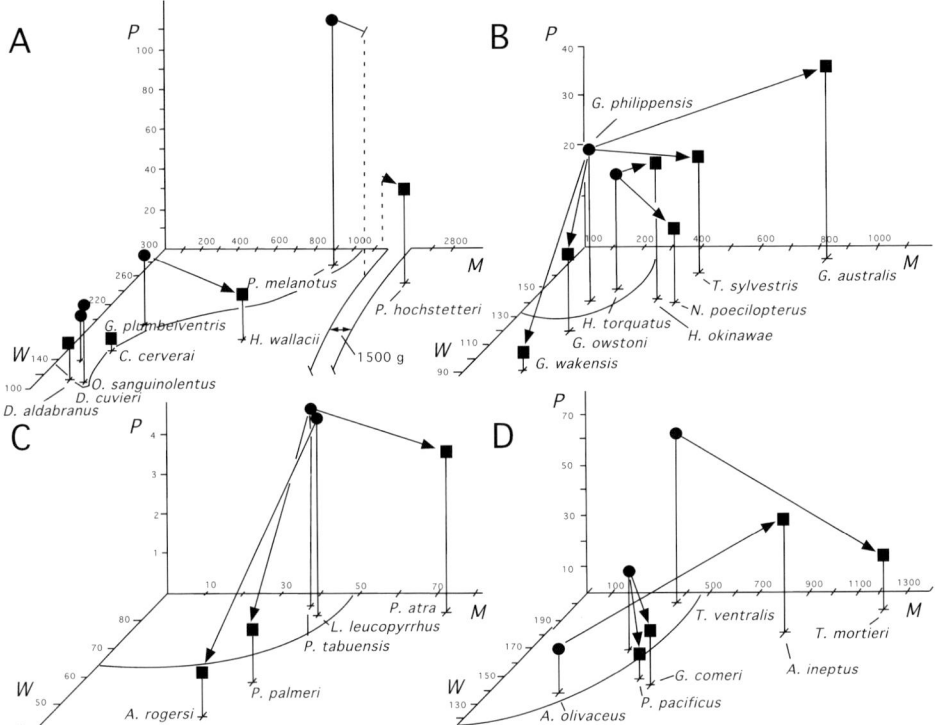

FIG. 121. Heuristic plots of selected pairs of flighted and flightless rails (connected by vectors of inferred change) on axes of body mass, wing length, and pectoral muscle size, with two-dimensional threshold of flightlessness (τ) inferred a posteriori: **A,** *Porphyrio, Gymnocrex, Habroptila, Ortygonax, Cyanolimnas,* and *Dryolimnas*; **B,** *Gallirallus, Tricholimnas, Nesoclopeus,* and *Cabalus*; **C,** *Laterallus, Atlantisia,* and *Porzana*; and **D,** *Amaurornis, Tribonyx, Gallinula,* and *Pareudiastes*.

The possession of embryonic or larval characters in the adult does not necessarily prove that the possessor is primitive, for it is equally if not more likely to be due to neoteny, and therefore phylogenetically secondary. Lowe has revealed the most interesting fact that the plumage of the ostrich and of the penguin remains throughout life in the condition of the down plumage of the chicks of flying birds. Since the wings of the ostrich and other so-called flightless birds and of penguins must have been derived by degeneration from those of flying birds, it is to be concluded that these embryonic and larval features have been secondarily prolonged and retarded by neoteny in the evolution of flightless birds and penguins from flying birds.—de Beer (1951:99–100)

Shapes of wings and remiges.—Several refinements in the shape of wings are known to have aerodynamic implications in birds (Sundevall 1886; Sy 1936; Stolpe and Zimmer 1939b), including turbulence-reducing notches in the caudal margin of the extended wing (Drovetski 1996), marked asymmetry of vexillae of remiges (Feduccia and Tordoff 1979), and other structural details (R. Å. Norberg 1985; U. M. Norberg 1990). In light of the nonfunctional variation in remiges and wing shape that characterize at least some flightless rails (e.g., Lowe 1928a; McGowan 1986), it is not surprising that some of the most conspicuous aerodynamic features of the wing have undergone variably pronounced degradation among more-derived taxa (Tables 13, 27, 28; Figs. 22, 38, 39). Several of the most conspicuous apomorphies of flightless rails (e.g., rounding of the extended wing [Fig. 22] and differential reductions in lengths and number of remiges [Tables 27, 28; Figs. 38, 39]) are interrelated. Based on the pterylographical comparisons by Lowe (1928a) for *Atlantisia rogersi* and that by Jeikowski (1971) for *Fulica*, the preliminary comparisons feasible herein (Figs. 33–37) confirm that modest losses of tectrices throughout the wing accompanied the general shortening of the appendage and the losses of one or more remiges primarii in some flightless species.

The disproportionate reduction or outright losses of the distalmost remiges primarii is one of the most widespread convergences of the integument among flightless carinates, including the Laysan Duck (*Anas laysanensis* [Moulton and Weller 1984; Livezey 1993a]) and Auckland Islands Teal (*Anas aucklandica* [Livezey 1990]). Based on the length of the vestigial carpometacarpus and apparent absence of the ossa digiti majoris of the gruiform *Aptornis*, a maximum of one to three (probably diminutive) remiges primarii could have been retained (Livezey 1994). This contrasts with the 10 remiges primarii (excluding the remicle) found in *Rhynochetos* and evidently plesiomorphic for the Gruiformes (Livezey 1998). Whether these changes in the numbers of remiges and tectrices in flightless gruiforms were (are) associated with shifts or increased variation in patterns or schedules of molt typical for the order (Stresemann and Stresemann 1966) is not known. A number of the qualitative changes in relative lengths of remiges and the unguis alularis have heterochronic underpinnings, and are considered below (see Ontogeny).

Microstructural changes in remiges.—A preliminary survey of samples of the vexilla of remiges secundarii of selected flightless rails with SEM not only confirmed the descriptions by Lowe (1928a) for *Atlantisia rogersi*, but these comparisons augmented the microapomorphies of remiges of flightless rails and extended the record of these features to many additional taxa (Fig. 39). Notable among these was the general reduction (especially in the distal portions of the vexillum) in number of functional hamulae (i.e., curvature fully developed and effecting secure locking of opposing rugae of adjacent feathers), with the homologues assuming the appearance of cilia in the remiges. These findings differ from the conclusions drawn by McGowan (1989) from a microscopic comparison of remiges of seven flightless carinates (*Atlantisia rogersi, Gallirallus australis, Pygoscelis adeliae, Tachyeres leucocephalus, Strigops habroptilus, Compsohalieus harrisi*, and *Rollandia microptera*) with those of flighted relatives (*Gallirallus philippensis, Tachyeres patachonicus, Nestor notabilis, Hypoleucos olivaceus,* and *Rollandia rolland*) and several palaeognathous taxa. McGowan (1989:543) came to a conclusion at odds with the present study: "... flight loss in carinates is accompanied by little or no change in the structure of the primary feathers, and this parallels the situation seen in the wing musculature, flightless species having pectoral muscles that are almost identical with those of their flying relatives . . . ." The latter inference may have resulted, at least in part, from inclusion of flightless species that use their wings for aquatic propulsion (e.g., *Tachyeres* and *Pygoscelis*), while undersampling the Rallidae.

Shapes of tails and rectrices.—In addition to the involvement of the tail in flight (Balmford et al. 1993; Thomas 1993, 1996a, b), this structure has been co-opted in a number of avian lineages for secondary purposes, notably sexually selected ornamentation (Alatalo et al. 1988; Barnard 1991; Thomas and Balmford 1995; Fitzpatrick 1997). Accordingly, the conformation and maneuverability of the tail often represent a selective compromise among aerodynamic and ethological functions (Fitzpatrick 1999; Møller and Hedenström 1999). This evolutionary trade-off is analogous to other traits serving multiple roles, for example, structures serving both as ornaments and armaments (Berglund et al. 1996), compromises most conspicuous in taxa (e.g., Trochilidae and Hirundinidae) combining strong sexual dimorphism with highly refined aerodynamics (Evans et al. 1994; R. Å. Norberg 1994; Møller et al. 1995, 1998; Evans and Thomas 1997; Barbosa and Møller 1999; Buchanan and Evans 2000; Park et al. 2000).

The dual roles of aerodynamic utility and ornamentation vary in importance throughout the Rallidae (Taylor 1996), and in flightless taxa the gross structure of the tail is consistent with relaxation of the former and variably pronounced intensification of the latter sources of selection. Absolute and relative changes in tails of flightless rails exceeded those in wing lengths, in large part because some flightless rails appear to have undergone elongation of the tail (Table 71). However, structural changes in the rectrices of flightless rails tended to parallel those of remiges to a greater extent (Table 29). Evidently, sexual selection acting on the tail as ornament primarily targets length and overall conformation, whereas the relaxation of selection for aerodynamic function tended to proceed unimpeded to effect a reduced investment in microstructural integrity.

In other flightless birds, reduction in the size and microstructural integrity of the tail varies substantially, but typically conforms with the general trend evident in flightless rails toward diminished length and stiffness (Livezey 1990, 1993a, b). However, the tails of the aquatic Galapagos Cormorant (*Compsohalieus harrisi*) are quite robust and those of flightless steamer-ducks (*Tachyeres* spp.) have undergone only modest shortening, evidently in both cases related to hydrodynamic function of the tail (Livezey and Humphrey 1986, 1992; Livezey 1992a). The tail of the Kakapo (*Strigops habroptilus*) is elongate and stiff, and apparently is a sexually selected ornament (Livezey 1992b). Predictably, the ratites present diverse but generally more substantive structural changes in the tail, in which rectrices vary greatly in length (primarily related to ornamentation) but retain little structural integrity, and in most taxa a reduction in the underlying pygostylus is evident (Rüppell 1977). Although the latter may represent a reversal related to the secondary loss of flight, the basal position of the ratites among Neornithes, the evolutionary transformation of the pygostylus (Gatesy and Dial 1996b), and other structural tolerances pertaining to the origin of flight (Peters and Gutmann 1985) make the determination of this polarity problematic. The variably substantive reductions in tails found in flightless rails are consistent with relaxation of structural aspects having aerodynamic implications, notably those pertaining to the provision of ancillary lift and reduction of drag (Maybury et al. 2001; Maybury and Rayner 2001; Evans et al. 2002), conditioned by those aspects bearing on the utilization of the tail for display.

Variation in body plumage.—In comparison to the structural changes substantiated by the remiges of flightless rails (Tables 27, 28; Figs. 38, 39), the microstructure of the contour feathers of the rest of the body was conspicuously compromised in only the most derived, strongly paedomorphic taxa (e.g., *Cabalus modestus* and *Atlantisia rogersi*). These most extreme cases among rallids did not exhibit changes as profound and widespread as were indicated for the Dodo (*Raphus cucullatus* [Brom and Prins 1989]) or the ratites (McGowan 1989). Other possible but unassessed changes in integumentary anatomy (Rawles 1960; Stettenheim 1972; Spearman and Hardy 1985), including conservative changes in ptilosis (Miller 1924) or pterylography (Nitzsch 1840), would appear to be have been likely in those taxa showing the most profound anatomical changes (e.g., *Diaphorapteryx* and *Aphanapteryx*).

CRANIAL MORPHOTYPES

General avian patterns.—Allometry evident in the vertebrate (especially avian) skull (Erdmann 1940; Van der Klaauw 1948, 1951, 1952; Barnikol 1952; Simonetta 1960; Bock 1964, 1966; Emerson and Bramble 1993; Hanken and Hall 1993) indicates that some cranial proportionalities are explainable as by-products of variation in size. Moreover, a number of flightless members of other avian families (Table 4) manifest variably similar changes in morphology with the loss of flight (Table 79) (e.g., change in body size [increases in most taxa], elongation and curvature of bill [*Aptornis*], and relative elongation and robustness of the pelvic limb largely correlated with increased body size [most taxa]) that are similar to those inferred for flightless rallids (Tables 71, 72). In that convergences (exclusive of those in the pectoral apparatus) largely stem from a tendency toward increased size (Figs. 119–121), the implications of which are discussed elsewhere, few of these merit special consideration. One exception is the variable tendency toward elongation and decurvature of the bill in some flightless rails, examples of which in other taxonomic families include the gruiform *Aptornis*, perhaps symplesiomorphic in part with *Rhynochetos* (Livezey 1998), and flightless ibises (Threskiornithidae), for which inferences of changes in bill length and decurvature have been made for various flightless taxa and for flighted, insular representatives (Olson and Wetmore 1976; Olson and James 1991; Mourer-Chauviré et al. 1995a, b).

Patterns among the Rallidae.—Some aspects of cranial shape were correlated with morphometric size among rails (Fig. 48). For example, a great majority of the smallest skulls (many of the crakes) also were characterized by short bills (Table 30), and with few exceptions (e.g., *Cabalus modestus* and *Capellirallus karamu*), taxa possessing relatively long, decurved bills also were of intermediate or large body size (Fig. 48). Despite the availability of remarkably detailed information on the early ontogeny of most qualitative cranial features in at least one rallid (*Fulica atra* [Macke 1969]), no quantitative data are available on development of the skull in the Rallidae; therefore the possible role of intraspecific scaling in patterns of variation among species cannot be explored. However, much of the variation in relative bill length was orthogonal to overall size of skulls (Table 41; Fig. 48), and models explanatory of cranial diversity in rails exceed mere corollaries of allometry. Notable also were the sizes of the minority of taxa manifesting exceptional sexual dimorphism in bills (*Cabalus modestus* [small body size], *Erythromachus leguati* [moderately small], *Aphanapteryx bonasia* [moderately large], and *Diaphorapteryx hawkinsi* [large]), a pattern also suggestive of independence of bill and body sizes and perhaps more closely associated with certain insular conditions (detailed below). The generally low diversity of cranial form among insular endemics in the Rallidae compared to such radiations as the drepanidids (Amadon 1950; Baldwin 1953; Richards and Bock 1973; Rai-

TABLE 79. Ecomorphological analogues of flightless species of Rallidae in other families of carinate birds.

Flightless Rallidae	Anatomical generalities	Ecological generalities	Nonrallid analogue(s)
Porphyrio hochstetteri, P. mantelli	Giantism	Intensified terrestrial herbivory, acclimatization to high altitudes	*Sylviornis, Cnemiornis, Thambetochenini, Strigops,* Raphidae
Habroptila wallacii	Giantism	Increased aquatic habits	*Anas aucklandica* (habitus only)
Nesotrochis spp.	Giantism	Forest-dwelling, graviportality	*Aptornis*
Gallirallus australis, Tricholimnas spp., *Nesoclopeus* spp.	Giantism	Generalized habitat, niche	Thambetochenini, *Aptornis*
Dryolimnas aldabranus, Gallirallus wakensis	Dwarfism	Increased utilization of upland habitats	*Apteryx, Apteribis*
Gallirallus dieffenbachii, G. owstoni, Capellirallus karamu	Variable changes in body size	Forest-dwelling, diversified diet	*Apteribis, Aptornis*
Habropteryx okinawae	Giantism	Increased arboreality, nocturnality	*Strigops*
Cabalus modestus, Atlantisia rogersi, most flightless *Porzana*	Dwarfism, paedomorphosis	Increased utilization of upland habitats, diversified diet	Acanthisittidae
Diaphorapteryx hawkinsi	Giantism	Forest-dwelling	*Aptornis*
Amaurornis ineptus	Giantism	Specialization for mangroves, arboreality, nocturnality	*Rhynochetos, Strigops*
Pareudiastes spp.	Modest increase in body size	Shift to upland habitus, possibly nocturnal activity pattern	*Thambetochen*
Gallinula nesiotis-group	Modest increase in body size	Shift to upland habitus	*Thambetochen*
Tribonyx mortierii, T. hodgenorum	Giantism	Shift to upland habitus	Thambetochenini
Fulica chathamensis, F. prisca, F. newtoni	Giantism	Shift to larger, more-alkaline bodies of water	*Tachyeres* spp., *Phalacrocorax harrisi, Rollandia micropterum*

kow 1977; James and Olson 1991), geospizine finches (Lack 1945, 1947a; Bowman 1961, 1963; Boag and Grant 1984; Grant 1986), and meliphagid honeyeaters (Keast 1968; Diamond et al. 1989) presumably reflects the distribution of insular rallids, in which divergence of form in sympatry (e.g., Chatham and Hawaiian islands) is the exception rather than the rule.

One cranial feature typical of a minority of rails, supraorbital depressions for the salt or nasal glands (sulcus supraorbitalis glandula nasalis), was not encoded by the measurements employed herein. This osteological feature reflects the relative size of a (generally bilobed) bilaterally paired gland related to maintenance of electrolytic equilibrium, and in some species contributes to direct voiding of excess salts in alkaline or marine environments (Schmidt-Nielsen and Kim 1964; Carpenter and Stafford 1970; Hughes 1970; Peaker and Linzell 1975; Holmes and Phillips 1985). These depressions were observed in the mangrove-dwelling *Eulabeornis castaneoventris,* estuarine populations of *Rallus longirostris* (Olson 1997), and the formerly lagoon-dwelling *Fulica chathamensis* (Beddard 1898; Millener 1980). Although many aspects of the skull can be phylogenetically informative (Müller 1964; Livezey 1998), ecologically informative cranial variation that derives from changes in aqueous salinity can contribute to exceptionally high intraspecific variation in the sulcus supraorbitalis (Marples 1932; Siegel-Causey 1990). This environmentally induced variation might undermine the reliability of the character for purposes of phylogenetic inference at lower levels (e.g., among partitions of a widely distributed, variably migratory rallid; but see Olson 1997), although gross aspects of the feature may be useful at higher phylogenetic scales (Livezey 1998) and indicative of the environments in which sedentary, flightless rails lived (e.g., *Fulica chathamensis* vs. *F. prisca* [Millener 1980, 1981]).

PECTORAL SKELETON

> ... with further atrophy of the pectoral muscles, the keel should disappear from the sternum and leave no trace, as it has done is several genera of birds otherwise structurally distinct, as, for example, in *Struthio, Rhea, Dromaius, Casuarius, Apteryx,* and *Dinornis,* are *Aptornis* and therewith *Notornis* [*Porphyrio hochstetteri*] to be likewise included in the subclass '*Ratitæ*' of a binary ornithological system?— Owen (1882:692–693; numbers for footnotes excluded)

Pectoral girdle.—The avian sternum displays remarkable diversity of form (Parker 1864; Fürbringer 1888), and the functional and taxonomic implications of this morphological variation have attracted considerable study by anatomists from both descriptive and ontogenetic standpoints since the middle of the 19th century (e.g., Parker 1868; Lindsay 1885; Hommes 1924; Fell 1939; Klima 1962; Lidauer et al. 1985). The exceptional attention given the developmental patterns of the avian sternum was engendered, in large part, by the comparatively late ossification of the element in modern birds, a distinction that not only facilitated limited ontogenetic inferences, but that qualified the element as among the most likely skeletal structures to reveal signatures of heterochrony (Owen 1882; Parker 1894; Andrews 1896a; Glutz von Blotzheim 1958; Livezey 1995a). Variation in sterna among flightless rallids was significant (Table 60; Figs. 51, 67), with notable changes associated with reduction of loss of flight capacity numerous and discernible without precise quantification (Figs. 8, 9, 61); many of these changes were convergent with those typical of flightless members of other carinate families (Livezey 1989a–d, 1990, 1992a, b, 1993a–c).

In contrast, the furcula showed only subtle changes in shape in flightless rallids, although overall robustness underwent a marked decrease in some of the more derived taxa (Table 61; Fig. 61). Similarly, the furcula remained relatively typical in conformation in *Raphus cucullatus* and *Pezophaps solitaria* (Livezey 1993b), whereas this element varies considerably in some taxonomic groups in both flighted and flightless members and at both interspecific and intraspecific levels (Glenny and Friedmann 1954; Glenny 1959). This plasticity of the claviculae suggests a comparatively great potential for evolutionary change and relaxation of selection acting on this element (Witte et al. 1990; Travis 1994), although the functional importance of the furcula for flight documented in *Sturnus* by Jenkins et al. (1988) suggests that the comparative uniformity in other groups such as the Passeriformes is maintained selectively.

Exclusive of reduction in the carina sterni (Table 60), an enlargement of the angulus coracoscapularis (Table 64) is the most frequently cited character shared by most flightless carinates (Owen 1879; Fürbringer 1888, 1902; Knöpfli 1918; Nauck 1930a; Kaelin 1941; Livezey 1989d, 1992a). In light of the morphogenesis of the avian pectoral girdle (Nauck 1930b; Howell 1937), the divergence of this articulation is probably paedomorphic (Livezey 1995a), at least among carinate groups. Systematic comparison of elements revealed convergent changes in the scapula and coracoideum of flightless rails (Tables 62, 63). Lambrecht (1931) and Fisher (1945) considered the relevance of the "anterior intermuscular line" of the coracoideum for flight capacity; this feature is synonymous with linea intermuscularis ventralis on facies ventralis of the corpus coracoidei, and corresponds to the variably conspicuous margin between the impressiones of mm. supracoracoideus et coracobrachialis caudalis. Although a trend in the position of this line may be discernible at the highest taxonomic levels, it is of no use at or below familial level in modern birds, and showed no consistent differences between flighted and flightless rails.

Pectoral appendage.—The most extreme reductions of the pectoral limb are found among the ratites (flightless Palaeognathae). In the manus, the carpometacarpi of adult ostriches (Struthionidae), rheas (Rheidae), casuaries (Casuariidae), and Dromornithidae retain complete, distinguishable ossa metacarpalia (although variably synostotic for some of their lengths), whereas those of emus (Dromaiidae), kiwis (Apterygidae), and elephant birds (Aepyornithidae) are variably synostotic with the distal antebrachium and lack distinct ossa metacarpalia and the spatium intermetacarpale (Owen 1841; Parker 1892; Pycraft 1900; Rich 1979; McGowan 1982). The only birds in which it is beyond reasonable doubt that complete loss of the pectoral limb occurred were the extinct moas (Dinornithiformes [Oliver 1949; Anderson 1989; but see Forbes 1892f]). However, the early description of *Aphanapteryx bonasia* by Milne-Edwards (1869), at which time there were no pectoral elements or skins for study, suggests that he incorrectly entertained the possibility of genuine winglessness in this rail. For example, Milne-Edwards (1869:260) stated of the included painting of *Aphanapteryx* by Hoefnagel (ca. 1610) that "No indication of wings is to be seen," although he later stated (1869:270) "The feathers of the wings are too slight and offer far too little resistance to have been of use in flight; and, besides this, the wings themselves are rudimentary."

Outright loss of skeletal elements aside, multiple changes in dimensions (Tables 32), relative proportions (Tables 34, 35), and qualitative features of the pectoral skeleton (Tables 58–64) characterized flightless rails. Several of the features found in virtually all flightless rallids—for example, truncation of the skeletal elements of the wing relative to the rest of the body (especially in distal elements) and several qualitative features of major elements (e.g., a comparatively distinct impressio m. brachialis ulnaris and moderately conspicuous conformational changes in the carpometacarpus)—are characteristic of flightless or flight-impaired waterfowl, including steamer-ducks (*Tachyeres* [Livezey and Humphrey 1986, 1992]), Auckland Islands Teal (*Anas aucklandica* [Livezey 1990]), Auckland Islands Merganser (*Mergus australis* [Livezey 1989b]), Chatham Islands Merganser (*Mergus* sp. [Millener 1999]), and the extinct seaducks of the genus *Chendytes* (Livezey 1993c). Whether any of the quantitative or qualitative changes associated with reduced flight capacity occur in the "flight-impaired" rail *Laterallus spilonotus* (Ripley 1977; Fjeldså 1981; P. B. Taylor 1998), or if the suggestive behavior of this species is entirely behavioral and related to tameness shared with a number of insular species (Humphrey et al. 1987), cannot be determined currently because of an inexplicable lack of anatomical specimens.

A subset of these features also appear in incipient stages in several taxa nearing the threshold of flightlessness (in order of decreasing apomorphy): *Mareca marecula* of Amsterdam Island (Martinez 1987; Olson and Jouventin 1996), *Euryanas finschi* of New Zealand (Worthy 1988a, 1997a), Laysan Duck (*Anas laysanensis*), and Hawaiian Duck (*Anas wyvilliana*). Several additional recently extinct or subfossil insular dabbling ducks, including Coues' Gadwall (*Mareca couesi* [Livezey 1993a]) and *Anas theodori* of Mauritius Island (Newton and Gadow 1893), also show indications of pectoral reduction based on ratios of mean wing lengths divided by tarsus lengths (Livezey 1993a). However, this trend of pectoral shifts is far from universal among insular anatids (Lack 1970b; Weller 1980). Eaton's Pintail (*Anas eatoni*), Marianas Duck (*A. oustaleti*), Andaman Teal (*A. albogularis*), Galapagos Pintail (*Anas bahamensis galapagensis*), and South Georgia Pintail (*A. (g.) georgica*) show no obvious changes in relative wing length or sternal size (Weller 1975, 1980; Stahl et al. 1984; Livezey 1993a).

Not all insular anseriforms exhibit only subtle or incipient skeletal apomorphies indicative of flightlessness. For example, the the extinct, flightless geese of Hawaii (*Thambetochen, Ptaiochen,* and *Chelychelynechen*) possessed carpometacarpi in which the narrowed spatium intermetcarpale was occluded (Olson and Wetmore 1976; Olson and James 1991), evidently through ossification of the membrana between mm. intersossea dorsalis et ventralis (Olson and James 1991; Livezey 1996a). Of comparable scale in pectoral apomorphy is the extinct *Cnemiornis* of New Zealand, a goose possessed of massive size, truly vestigial carina sterni, and extraordinarily reduced elements of the pectoral limb (Hector 1873a, b; Owen 1875; Livezey 1989a; Worthy et al. 1997). Pectoral reduction of a magnitude at least as great as shown by the foregoing genera characterized the foot-propelled diving duck *Chendytes* (Livezey 1993c).

Aside from the Rallidae, flightlessness was associated with the most profound morphological changes in the extinct Adzebill (Aptornithidae: *Aptornis*) of New

Zealand, the sister-group of the weakly flighted *Rhynochetos* (Livezey 1998). The unique truncation of the distal segments of the ossa metacarpalia majus et minus of *Aptornis*, as well as the reduction or loss of most of the features typical of the proximal portion of the carpometacarpus (Livezey 1994), represent structural reduction of a greater magnitude than that described for flightless rails or for members of any other carinate order (Owen 1875; Sinclair and Farr 1932; Olson and Wetmore 1976; Livezey and Humphrey 1986; Millener 1988, 1989; Millener and Worthy 1991; Olson and James 1991). There is no indication that the variably synostotic phalanx digiti alulae of *Aptornis* functioned as a combative spur (Livezey 1994), because the element shows none of the distal rugosity or thickening typical of the spurs derived from os metacarpale alulare or os carpi radiale in some other taxa (Rand 1954). An alternative interpretation of this feature in *Aptornis* based on conservative aspects of avian ontogeny is presented below (see Ontogeny).

Retention of complete, but variably derived elements of the manus also characterizes flightless carinates that are (or were) obligate wing-propelled diving birds (the penguins [Sphenisciformes], plotopterids [cf. Pelecaniformes], and flightless alcids [Charadriiformes]), flightless birds in which the loss of the capacity for aerial flight was accompanied by employment of the forelimb as a propulsive organ in aquatic locomotion (Kelso 1922; Olson and Hasegawa 1979, 1996; Bannasch 1986a, b, 1987; Livezey 1988, 1989c; Raikow et al. 1988).

Microstructural likelihoods.—The impact of size and locomotive strain on microstructural characteristics (e.g., histology) has been recognized for decades (Rensch 1948), but a paucity of quantitative data for the pectoral appendage of birds persists. This deficiency of data is critical, in that if structures undergoing reductive evolution are subject to the optimization of relative investments or symmorphosis (Taylor and Weibel 1981; Alexander 1984, 1998a; Ricklefs 1998; Weibel et al. 1998; Weibel 2000), then parallel shifts in modal robustness of pectoral elements should be associated with the loss of flight. That is, if the reasoning generally applied to the pelvic limb concerning apparent margins of safety are substantiated empirically (cf. Gans 1979), then the economics of reductions in these margins in such critically important structural elements in birds may be critical. Given the evidence for the role of heterochrony in pectoral reduction in flightless birds, such economics in margins of safety would correspond well with the inclusion of symmorphosis as one component at the histological level of constraint in avian development proposed by Ricklefs et al. (1998).

PELVIC SKELETON

> That mere shortness of wing and length of leg are no true characters of a natural group, even when the legs are adapted for swift course on dry land, is evinced by the singular short-winged Rail (*Brachypteryx*) at present existing in New Zealand, and which probably was manifested on a larger scale by the allied genus *Notornis*, which unquestionably belongs to the *Rallidæ* rather than to the *Struthionidæ*.—Owen (1848b:374; comparing *Gallirallus* with the ratite *Dinornis*)

Shifts in pelvic robustness and proportions.—Pelvic characteristics typical of large, variably cursorial birds adapted to walking on firm substrates represent the predicted endpoints for the hindlimb in terrestrial birds after the loss of flight.

These expectations apply to most rallids, the comparatively natatorial genus *Fulica* excepted. Generalities typical of avian taxa reliant on ambulatory movements include heavy skeletal elements in the pelvic appendage and relatively short tarsometatarsi and digits (Storer 1971a), with increased tibiotarsal proportions as lineages undergo cursorial specialization (Blechschmidt 1929; Engels 1938a, b; Maloiy et al. 1979; Bochenski 1989). In general, flightless rallids having predominately terrestrial habits conformed with the predicted shift (Fig. 43). The pelvic proportions of *Raphus cucullatus* and *Pezophaps solitaria* showed distinctly different shifts in pelvic proportions characterized by modest shifts toward graviportality (Prange et al. 1979; Alexander 1984, 1985b; Currey and Alexander 1985). *R. cucullatus* had disproportionately short tarsometatarsi, whereas *P. solitaria* had disproportionately short digits, but both possessed substantially more robust pelvic elements (Livezey 1993b). In neither species is disproportionate shortening of the femur characteristic of more-specialized terrestrial birds (Gatesy 1991).

Consistent with expectations for aquatic forms (Raikow 1970, 1973; Livezey 1986a, b; Faith 1989), flightlessness in *Fulica* was accompanied with disproportionate elongation of the tibiotarsus and digits, with these augmentations being superimposed on the specialized proportions already characteristic of flighted members of the genus (Fig. 43). This shift predictably opposes those typical of waterbirds in which flightlessness is associated with increased dependence on terrestrial locomotion (e.g., insular *Anas*), in which hindlimbs assumed a compromise between natatorial efficiency (Stolpe 1932; Raikow 1970, 1973, 1985a; Storer 1971a; Bruinzeel et al. 2000) and modest ambulatory refinements (Livezey 1990, 1993a). The differences in shifts shown by insular dabbling ducks and the flightless coots (Fig. 43) strongly indicate that the latter had moved toward increased specialization of aquatic movements in their lacustrine and estuarine habitats as opposed to a shift toward increased exploitation of uplands. The large body size, elongation of the pelvic limb (particularly the tibiotarsus), and long, narrow pelvis of *Fulica* in general, and congeners approaching flightlessness in particular, are shared by a number of flightless taxa (e.g., Hesperornithiformes, grebes, cormorants, *Tachyeres,* and *Chendytes*) in which members also are (or were) proficient foot-propelled divers (Townsend 1909, 1924; Dabelow 1925; Livezey and Humphrey 1986; Livezey 1989d, 1992b, 1993c).

Although proportions within the pelvic limb and allometry in ratites (Cracraft 1976a, b; Alexander 1983a, b; Worthy 1988b; Cubo and Casinos 1997; Abourachid and Renous 2000) can be considered extreme examples of appendicular specialization for terrestrial locomotion, these diverse, comparatively enormous taxa encompass a diverse continuum from graviportal to cursorial forms. A few flightless rallids tend toward graviportal pelvic limbs and shortening of toes (e.g., *Porphyrio mantelli*-group, *Diaphorapteryx hawkinsi,* and *Tribonyx mortierii*), but none approaches the apomorphic extremity of the ratites or raphids (Raikow 1985a; Livezey 1993b). Penguins (Sphenisciformes) represent an extreme in pelvic proportions associated with primary reliance on wing-propulsion for locomotion and relegation of the hindlimb for infrequent ambulation and protracted, erect stance during nesting (Livezey 1989c), specializations that produced convergent pelvic proportions consistent with those characteristic of cursorial landbirds (Storer 1971a).

Margins of safety in pelvic limb.—An emerging body of evidence confirms a strong correlation structure among skeletal elements within species or among individuals of closely related species (Selander and Johnston 1967; Schnell 1970a, b; Baker 1975; Johnston 1976b; Zink and Winkler 1983; Termaat and Ryder 1984; Jehl et al. 1990), an intraspecific generality that was confirmed for rails. In birds and other terrestrial tetrapods, widths of appendicular elements (and bulk of associated musculature) and the microstructural correlates of bone strength correspond generally with body mass and the loads that these elements routinely bear (Pennycuick 1967; Alexander et al. 1984; Currey 1984; Currey and Alexander 1985; Campbell and Marcus 1992; Bennett 1996). These correlations are interpretable in terms of the functional requirements and similarities of dynamic strain, elasticity, and margins of safety (McMahon 1975a, b, 1984; Alexander 1977a, b, 1980, 1984, 1985a, 1990a, b, 1996b, 1998a; Anderson et al. 1979, 1985; Løvtrup and Mild 1979; Lanyon 1981; Biewener 1982, 2000; Rubin and Lanyon 1984), although details are disputed (Garland 1983).

Although correlation structures of flightless rails differed from that for flighted confamilials to a limited extent, notably in the degree of intercorrelations among dimensions of the pectoral apparatus (Figs. 40, 41), a strong covariance structure was evident among lengths of appendicular elements that was significantly independent from a similarly pronounced covariance among widths of appendicular elements (Tables 44, 45; Figs. 46, 47). Also, multivariate comparisons of rallids confirmed strong, covarying contributions of pelvic dimensions to general skeletal size applicable to both flighted and flightless species, whereas pectoral dimension predictably included variation related to flightlessness that was largely orthogonal to that in the pelvic limb (Tables 40, 41; Figs. 49, 50). Similarly, pelvic dimensions contributed consistently to group-wise CVs most closely correlated with mean body masses (Tables 14–23, 46–53; Figs. 23–32, 52–59). In short, despite a conservative, primary complex of correlations common to the Rallidae, significant independence of covariance structures of pectoral and pelvic dimensions was evident, a decoupling that was magnified in the flightless species.

PECTORAL MUSCULATURE

> . . . all the muscles and flesh of the breast are made for the benefit and increase of the movement of the wings, with that bone of the breast all in once piece, which provides the bird with great power, with the wings woven of thick nerves and other very strong ligaments of cartilage and skin very strong with various muscles . . . , so much strength is provided for power beyond the ordinary support of the wings, it needs in its place to double and triple the movement in order to flee from its predators or pursue its prey.—Leonardo da Vinci (1505: folio 16, recto; translation by Marinoni 1976:73)

> . . . during the earlier stages of the process of reduction [of the pectoral apparatus of insular species], such birds might be expected to resemble in the state of their organs of flight our domesticated ducks.—Darwin (1868:286–287)

Indeed, one of the most surprising results obtained in comparing the flying mechanism of rails known to be truly flightless, such as *Atlantisia* and *Ocydromus* [*Gallirallus australis*], with rails which are known to make long migratory journeys, . . . is the discovery that the patagial differences as regards perfection of mechanism were very slight or vague whether the subject dissected was flightless or the reverse.

But one would have been naturally inclined to think that if ... the gradual loss of flight by default really lead[s] to such radical change in the structural details of the barbules described ..., so much the greater would have been the effect upon ... the patagium of *Atlantisia* with its associated muscles .... Nothing of the sort appears to have happened, except that ... the great pectoral [m. pectoralis sternobrachialis] is obviously thin and feeble ....—Lowe (1928a:121)

General avian patterns.—Among the first naturalists to consider the functional anatomy of birds, Leonardo da Vinci (1505) recognized that the musculature of birds must meet both routine and extreme demands, that is, proficiency merely to become airborne is not enough, but birds in typical environments must be able to execute maneuvers for evasion of predators, foraging, display, and access to nesting sites. Despite advances in evolutionary inferences for diverse anatomical systems in birds (e.g., Bock 1979, 1990; Zweers 1979; Dullemeijer 1980), much remains to be learned of the mechanics of avian flight, a state of affairs clearly demonstrated by recent discoveries concerning the functional importance of the pectoral muscles, tendons, and skeletal elements involved (Goslow et al. 1989; Dial et al. 1991; Dial 1992; Vazquez 1992, 1994). In recent decades, significant insights have been gained into the power generated by the breast muscles in (typical flapping) flight (Simpson 1983; Dial and Biewener 1993; Tobalske and Dial 1996; Dial et al. 1997) and the role of elastic-strain energy stored in muscles and other tissues (Alexander and Bennett-Clark 1977). It is in this investigational context, bolstered by the descriptive myological examinations of the flightless rallids *Atlantisia rogersi* (Lowe 1928a) and *Gallirallus australis* (McGowan 1986), that the effects of flightlessness on pectoral musculature are to be assessed.

Flightlessness and pectoral musculature.—Apomorphy of the pectoral musculature of flightless rails is substantially less than the profound changes evident in the ratites (Owen 1841; Fürbringer 1888; Lowe 1928b; McGowan 1982, 1986), with changes in rails associated with flightlessness being limited largely to reductions in size (Figs. 69–115). For example, the smallest relative masses of pectoral muscles of flightless rails (Tables 66, 67) are an order of magnitude greater than that for the Brown Kiwi (*Apteryx australis*), which is estimated to compose only 0.13% of total body mass in the latter (P. R. Millener fide McNab 1996).

However, findings in the present study indicate that McGowan (1986) was mistaken in his conclusion that *Gallirallus australis* lacks one of the two capita of m. subcoracoideus typical of nonpasseriform birds (cf. Figs. 88–91), and lacks m. serratus superficialis pars metapatagialis (e.g., Figs. 77, 87). However, perhaps more importantly, morphometric analyses of the externum (Tables 7, 11, 14–23, 71; Figs. 20–32), skeleton (Tables 30–33, 40–42, 46–54, 72; Figs. 42–54), and pectoral musculature (Tables 66–70; Figs. 111–115) of *G. australis* and other flightless rallids presented here demonstrate that the reductions in the pectoral apparatus of flightless rails are generally substantial, and in most taxa are accompanied by notable qualitative differences in the integument (Tables 27–29; Figs. 33–39), skeleton (Tables 58–64; Figs. 61–67) and major pectoral muscles (Figs. 70–110). Although the apomorphies of flightless rails are dwarfed by those found in ratites, and cognizant of variation in pectoral musculature among flighted avian taxa (Gladkov 1937b; Nair 1954a, b; Berger 1966), the findings of the present study decisively contradict the statement by McGowan (1986:343) that "... there is nothing in the skeleton [of *G. australis*] today to suggest that they are flightless."

Relative sizes of the major breast muscles (mm. pectoralis and supracoracoideus) are reflected to a large extent by the depth of the carina sterni. This relationship reliably indicates approximate reductions in the relative bulk of breast muscles in flightless rails (Figs. 51, 60, 70–74). Similar patterns of parallel structural changes in the sternum and the muscles that originate on the sternum parallel the capacity for flight in a number of flightless carinates, including grebes (Livezey 1989d), a cormorant (Livezey 1992a), a parrot (Livezey 1992b), steamerducks (Livezey and Humphrey 1986), and several other anseriforms approaching the loss of flight capacity (Livezey 1989b, 1993a). A relationship of this kind evidently can be extended to several extinct, flightless carinates for which direct measurements of muscles are not feasible (Livezey 1988, 1989a, 1993b, c). For flighted birds, m. pectoralis tends to be correlated with body mass (e.g., Marsh and Storer 1981) and generally constitutes approximately 15.5% of total body mass (Hartman 1961), although seasonal atrophy of pectoral muscles, like that documented in grebes (Piersma 1988; Gaunt et al. 1990), may complicate interspecific comparisons. As noted by Darwin (1868) and Wiglesworth (1900), a continuum of pectoral reduction occurs among domestic varieties of the Red Junglefowl (*Gallus gallus* [Hartman 1961]), Muscovy Duck (*Cairina moschata* [Hartman 1961]), and Common Mallard (*Anas platyrhynchos* [Livezey 1993a]), lineages that demonstrate the efficacy of artificial selection on traits of diminished utility. Despite substantial differences in ontogenetic patterns in altricial and precocial birds, suspected differences in the relative development of pectoral and pelvic muscles have not been demonstrated (Konarzewski et al. 1998).

Changes in the size and conformation of muscles other than the major breast components are manifest in the more derived of flightless carinates. Several such were found in the flightless rails examined, including m. cucularis capitis pars clavicularis and several other comparatively small muscles, with several species including a suite of notable changes in the mean size or apparent variation in the musculature of the wing (e.g., *Porzana palmeri* and *Atlantisia rogersi*; Figs. 70, 74–76, 78, 79, 84, 86, 93, 97). Substantial muscular apomorphy likely characterized the extinct and exceptionally derived *Cabalus modestus*, *Capellirallus karamu*, *Diaphorapteryx hawkinsi*, *Aphanapteryx bonasia*, *Erythromachus leguati*, and dwarfed *Porzana* from Hawaii. Although data on nonskeletal anatomy of members of the extinct Raphidae are few or lacking, available information indicates that a diversity of other characteristics of flighted species of the family Columbidae (Hartman 1961; Greenewalt 1962; Ricklefs 1973; Aulie 1983), including details of the pelvic musculature (Cracraft 1971), functional morphology of the tail (Baumel 1988; Baumel et al. 1990), rectrices (Balmford et al. 1993), and neuroanatomy of the pectoral musculature (Dial et al. 1987), may have undergone significant changes in *Raphus cucullatus* and *Pezophaps solitaria* as well (Livezey 1993b). The extreme reduction of the pectoral skeleton of the extinct gruiform *Aptornis* indicates that muscles that typically insert (at least in part) on the manus (e.g., m. deltoideus pars propatagialis and m. flexor carpi ulnaris [Raikow 1985a; McGowan 1986]) would have undergone profound truncation, modification of insertiones, or complete loss. Such radical changes are known among modern taxa only in the ratites (McGowan 1982). Unfortunately, absence of soft tissues for *Aptornis* precludes critical myological comparisons and direct study of the functional implications of such anatomical changes (cf. Emerson 1978, 1984, 1988), about which the remaining skeletal elements can only provide indirect, variably misleading clues (McGowan 1986).

Histological and biochemical likelihoods.—Anatomical differences define broad limits on performance capacities in living organisms. However, an understanding of the functional changes imposed by evolutionary apomorphy in the pectoral musculature, however, could be improved through comparisons of interspecific differences in histology (George and Nair 1957, 1959a–c; Nene and George 1965; Maier 1983). An inkling of the importance of such fine-scale changes can be gained by taxonomic differences in biochemical composition of muscles (Nair 1952) and the effects of seasonal variation in ultrastructure of breast muscles documented in some migratory birds (Rosser and George 1986, 1987; George et al. 1987; Gaunt et al. 1990; Evans et al. 1992). Optimally, this would be carried out in combination with consideration of architectural differences (Gans and Bock 1965) and empirical assessment of ontogeny of the muscles involved (Ricklefs and Webb 1985; Gaunt and Gans 1990; Choi et al. 1993; Ricklefs et al. 1994; Choi and Ricklefs 1997; Dietz and Ricklefs 1997). In addition, biochemical changes may pertain to pectoral reductions. Kaplan (1964) found that the concentration of lactase dehydrogenase in the breast muscles of the Laysan Duck (*Anas laysanensis*) was substantially less than that of a wild Common Mallard (*A. platyrhynchos*). Study of specimens at comparable stages in the annual cycle is critical in that molt entails substantial expenditures of energy, during which individuals must lose and regenerate approximately 25% of lean body mass over a short period of time (J. R. King 1981; Blem 1999).

Anecdotal evidence provides a general idea of the magnitude of the implications of histological change on activity patterns and individual flight performance. Langston and Rohwer (1996) inferred a relationship between timing of molt and breeding in albatrosses (Diomedeidae). Both of these processes have been shown to have critical impacts on composition of muscles and energy balances, and these are interrelated with the reattainment of the capacity of flight and the timing of breeding and migration in many flighted species (Woolfenden 1967; Ankney 1984; Rosser and George 1985, 1987; Panek and Majewski 1990; Holmgren and Hedenström 1995; Murphy 1996; Swaddle and Witter 1997; Brown and Saunders 1998). Whether such contingencies underlie, at least in part, variation in the chronology of molt in *Fulica americana* (McKnight and Hepp 1999), is not known. Nevertheless, the obvious implications of these considerations for the liberation represented by the loss of flight in insular birds are substantial (see below).

Sexual Dimorphism: Interplay of Phylogeny and Function

> ... the bill [of *Erythromachus leguati*] ... is curved in exactly the same manner and to the same degree as in *A. broecki* [*Aphanapteryx bonasia*] ... and varies extraordinarily in length, being in some specimens one third shorter than in others. The extremity of the longer beaks ... is more conspicuously curved than that of the shorter ones ... throughout the series of bones at our disposal a marked difference in size may be traced; but whether the short bills belong to smaller individuals, or whether the difference in length of beak and in size generally is attributable to sex, we are unable to say.—Günther and Newton (1879:432)

As already mentioned, there is considerable individual variation in size, both in the skull and other portions of the skeleton of this species [*Diaphorapteryx hawkinsi*]. Indeed, Mr. Forbes has suggested that there may really be several distinct species . . . there is, in fact, only one . . . . These considerable individual differences in the size of the skull and mandible are accompanied by equal differences in the dimensions of the other portions of the skeleton . . . . Most flightless birds appear to be subject to this great variability; for instance, it is well marked in *Didus, Pezophaps, Erythromachus* . . . . In different skulls of *Diaphorapteryx* the beaks are commonly dissimilar, not only in size but also in form (degree of curvature, etc.) . . . .
—Andrews (1896b:76–77)

One may conjecture that this is due to sexual selection in favour of more powerful males [in flightless *Tachyeres pteneres* relative to flighted *T. patachonicus*] being no longer counteracted by natural selection in relation to efficiency of flight.—Huxley (1943:286; based on data presented by Murphy 1936)

In all but two measurements *Diaphorapteryx hawkinsi* exhibits the greatest variability of the four rail species, ... The great variability of *D. hawkinsi* may be correlated, at least in part, with the advance [sic] state of wing reduction (relative to leg length) in this flightless species . . . . It is a generally accepted assumption that when there is a tendency to lose certain structures through a lineage, the intensity of selection on that structure is reduced and an increase of variability follows.
—Cracraft (1973a:99)

PHYLOGENETIC BASELINES

Methodological limitations.—In light of many explanations proposed for avian sexual dimorphism (phylogenetic inertia, baseline body mass, mating systems, intersexual partitioning of foraging niche, and territoriality), the possibility that flightlessness may interact with these or other influential factors is very real. However, before a consideration of evidence for an effect of flightlessness on sexual dimorphism in rails, the more widespread and better-documented effects of phylogeny, sexual selection, and body size in birds must be assessed. Absent a widely accepted, well-supported, and taxonomically inclusive phylogenetic hypothesis for modern birds, rigorous assessments of historical limits and potentialities of general morphological parameters within Aves (including sexual dimorphism) remain unattainable. Fortunately, in some cases the distribution of the attributes of interest and the phylogenetic precision required are such that patterns are comparatively obvious and robust with respect to choice of formal analytical protocols (see below for methodological challenges). Furthermore, in some of the following discussion, issues of sexual size dimorphism and sexual dichromatism are treated separately, a necessity given the sometimes divergent correlations between these components of intersexual differentiation and other parameters of interest in some taxonomic groups (Sigurjónsdóttir 1981; Reynolds and Székely 1997).

Preliminary overview for the Rallidae.—The phylogenetic analysis by Livezey (1998) provided at least a basic framework for broad patterns of sexual dimorphism in higher taxa of the Rallidae and other closely related gruiform families. Despite the substantially larger body sizes of the cranes, limpkins, and trumpeters (Gruoidea [Archibald and Meine 1996; Bryan 1996; Sherman 1996]) and the paucity of data on body masses of the sister-family of the rails (Fig. 11), the

Heliornithidae (Alvarez del Toro 1971; Bertram 1996), the modal magnitudes of intersexual differences in mean body mass among members of the Rallidae (Appendix 1) indicate no substantial family-wide departure from the magnitude of sexual size dimorphism typical of most neognathous birds, in which the mean mass of males is roughly 1.15–1.25 that of conspecific females (Livezey and Humphrey 1984a). Accordingly, the majority of rallids can be considered either to require no causal or evolutionary assessment in this respect because these taxa simply retain the plesiomorphic condition; or whatever the selection regime(s) that is (are) responsible for this widespread ratio of sex-specific size in neognathous birds, it extends to the majority of species of the Rallidae regardless of flight capacity.

Two basal members of the moorhens and coots (subtribe Fulicarina) are exceptions to this general pattern, both monotypic and flighted: *Gallicrex* and *Porphyriops,* in which the mean mass of males is approximately 1.5 times that of females (Appendix 1). Although well above average for neognathous birds (Ralls 1976; Amadon 1977), substantially greater dimorphism characterizes a small minority of taxa scattered among several families, notably the Galliformes and Anseriformes (Sigurjónsdóttir 1981; Lindén and Väisänen 1986; McCraken et al. 2000). Several extinct flightless species, including several possibly close relatives of *Gallirallus* (Fig. 12) were characterized by exceptionally great skeletal dimorphism indicative of substantial intersexual differences in body mass (Table 8), and include *Diaphorapteryx, Aphanapteryx,* and *Erythromachus.* Excluding three species in two sister-genera (*Gymnocrex* and *Habroptila*) that are suggestive but compromised by samples of specimens (Tables 5, 7, 30–33), there are no substantiated examples of reversed or female-larger sexual (size) dimorphism among the Rallidae, a minority phenomenon that characterizes most Falconiformes (Snyder and Wiley 1976; Andersson and Norberg 1981; Mueller et al. 1981; Safina 1984; Paton et al. 1994) and some Charadriiformes (Jehl and Murray 1986; Catry et al. 1999).

Intersexual differences in linear external measurements of rallids demonstrated herein (Tables 7–8, 14–22, 24–26; Figs. 52–55), whether univariate or of greater dimensionality, are broadly comparable in magnitude and form to those reported for many other neognathous taxa (excluding traditional taxonomic revisions), despite a preoccupation among the latter regarding geographic variation and climate (Gould and Johnston 1972), a predilection for passerine taxa, or a perspective of sexual dimorphism as a factor confounding interspecific diagnoses (e.g., Packard 1967; Power 1969, 1970a, b, 1979; James 1970, 1983; Johnston 1969, 1973; Johnston and Selander 1971, 1973; Grant 1979a, b). Similarly, intersexual differences in skeletal dimensions of rails in the present study (e.g., Tables 30–35, 46–52, 57; Figs. 23–31, 45, 48) were consistent with the comparatively rarer set of counterparts for other avian taxa with respect to magnitudes and orientation of multivariate sexual dimorphism documented for other avian taxa (Niles 1973; Campbell and Saunders 1976; Hamilton and Johnston 1978; Baker and Moeed 1979, 1980; Ross and Baker 1982; Handford 1983; Gibson et al. 1984; Schluter and Grant 1984a, b; McGillivray 1985; Rising 1987; Johnston 1992a, b; Engelmoer and Roselaar 1998). However, unlike a small minority of avian taxa, rails lack clear cases of single dimensions being larger in the otherwise smaller sex, as has been reported for bill lengths (externally measured) of Huias (Callaeidae:

Heteralocha acutirostris) by Burton (1974) and cranial widths of Great Horned Owls (Strigidae: *Bubo virginianus*) by McGillivray (1985).

A modest but intriguing body of evidence documents ecological segregation of the sexes within species on the wintering grounds, which in some species includes differential preferences for habitat (Lynch et al. 1985; Morton 1990; Ornat and Greenwood 1990). A perspective on insular colonization by rallids as comparable to vagrancy resulting in geographically novel migratory stopovers or wintering grounds (see below) exposes the potential of such intersexual differentiation of niches for the facilitation of sexual dimorphism of insular endemics, including those of the Rallidae (e.g., *Diaphorapteryx hawkinsi*). Furthermore, such intersexual partitioning of feeding niches might increase the likelihood of successful colonization of islands by small vagrant flocks by broadening the utilization of available habitats on the island. Clearly, the plausibility of this mechanism awaits the demonstration of such intraspecific partitioning of habitats among members of the Rallidae, because studies heretofore have involved arboreal passerines.

Also of interest is sexual dichromatism, intersexual differences in plumage pattern and coloration of soft parts, aspects of appearance other than simple size that serve to distinguish the sexes (Badyaev et al. 2001a, b). Once again, only a minority of the species of the Rallidae deviate from sexual monochromatism, the norm for the Gruiformes (Livezey 1998). Conspicuous sexual differences in plumage or soft parts are limited among the rallids to two instances: a clade comprising two flighted genera (*Rallicula* and *Sarothrura*) in the tribe Sarothrurini, in which sexual differences in size are not noteworthy; and the monotypic *Gallicrex,* characterized as well by substantial sexual size dimorphism (Livezey 1998; Appendix 1). Unlike intersexual differences in size, sexual dichromatism is associated largely if not completely with sexual selection and signaling of status within and (less importantly) between species (Green 2000). A positive correlation between the phenotypes favored by sexual and natural (sensu stricto) selection is intuitive, and balance or mutual complementarity has been documented in some natural populations (e.g., Price 1984a, b; Székely et al. 2000). The potential for traits favored by sexual selection in opposition to those directly advantageous is underscored by a framework of hypotheses premised on a negative fitness of sexually selected traits, that is, the handicap principle (Zahavi 1975; Zahavi and Zahavi 1997).

ECOLOGICAL IMPLICATIONS OF SEXUAL DIMORPHISM

Primary ecofunctional hypotheses.—Ligon (1999: table 2.1) listed seven direct benefits associated with mate choice that accrue to the fitness of females: defense of territories and included resources by males, provision of care and protection of females by males, provision of parental care to offspring, efficiency with respect to the costs of searches for mates, avoidance of interspecific hybridization and associated reduction in fitness of offspring, favorable selection pertaining to differential viability or quantity of sperm of males, and avoidance of sexually transmitted diseases or parasites. Of these proposals, those related to intersexual partitioning of foraging niche (Selander 1966; Shine 1989a, b), including terrestrial (Kilham 1965; Ligon 1968; Holyoak 1970; Wallace 1974; Hogstad 1976, 1977, 1978; Williams 1980; Ross and Baker 1982; Holmes 1986) and aquatic habitats (Jehl 1970; Oksanen et al. 1979; Nudds et al. 1981; Nudds and Kaminski 1984), and effects of body mass on seasonal mobilization of energy for reproduction (Downhower 1976), sexual differences in parental investment (Trivers 1972; Cabana et al. 1982; Hughes and Hughes 1986; Wesolowski 1994), and paternal territoriality (Price 1984b; Livezey and Humphrey 1985a, b; Livezey 1987) appear to be plausible candidates for the Rallidae. Where of sufficient magnitude, sexual size dimorphism can carry implications for metabolic rates (Wijnandts 1984; Masman et al. 1986; Maloney and Dawson 1993), but there is no evidence as yet of this among rallids. Even where ecological differences are inferred to be involved intersexual differences in morphology (but see Mueller et al. 1981), the latter instead can reflect competition between the sexes or independent shifts by the sexes toward distinct phenotypic optima that are devoid of intraspecific competition (Shine 1989a).

Although sexual dimorphism is significant in external and skeletal dimensions and exhibits a complex array of magnitudes within and between genera (Tables 24–26, 55–57), some functional or ecological interpretations emerged. The relative importance of territoriality to increased body size in males appears to be comparatively great for aquatic rallids (e.g., swamphens, moorhens, and coots) nesting and feeding in open, lacustrine habitats (Gullion 1953b; Fjeldså 1973; Craig 1976; Searcy 1982; Petrie 1984, 1986, 1988; Hunter 1987a; Cave et al. 1989; Badyaev and Qvarnström 2002), a component that in turn is intensified through sexual selection in "arms races" within and between species (Dawkins and Krebs 1979). Defense of feeding and nesting territories is primarily a responsibility of males in many taxa, and has been implicated as a primary factor (Nice 1941; Tinbergen 1957; Brown 1964, 1969; J. H. Kaufmann 1983; Price 1984a, b). However, territoriality may not be simply a matter of body size and combative prowess (cf. Dawkins and Krebs 1979), in that more subtle patterns of deterrence can be effective (Stamps and Krishnan 2001) and interspecific relationships can be ameliorated through cooperation with other species (Dugatkin 1997). The role of thermodynamic advantages in sexual dimorphism of rallids—which are implicated in patterns of dimorphism among higher-order groups of vertebrates (Greenwood and Wheeler 1985) and inferred (Johnston and Selander 1973; Johnston and Fleischer 1981) for *Passer*—also remains wholly untested.

Rensch's rule revisited.—As detailed in the foregoing analyses, a significant tendency exists among members of the Rallidae for intersexual differences in body mass to vary directly with body mass; that is, rails appear to conform with Rensch's rule. The proposed generality by Rensch (1950, 1959, 1960), although apparent in a number of taxonomic groups (Clutton-Brock 1985; Leutenegger and Cheverud 1985; Fairbairn 1997), is not without controversy (Reiss 1986, 1989; Abouheif and Fairbairn 1997). For example, Sigurjónsdóttir (1981) found that mean body size and sexual size dimorphism in Falconiformes were inversely correlated, a finding consistent with frequent departure from Rensch's rule in groups showing female-larger dimorphism (Fairbairn 1997).

Also inferred in the present study was the inconsequence of flight status to the general, family-wide relationship between intersexual differences in body mass and mean masses of species. Therefore, it appears that sexual mass dimorphism of rails, whatever the ecological or selective basis of the phenomenon or flight status of the species compared, increases allometrically with body size.

SEXUAL SELECTION AND SEXUAL DIMORPHISM

Phylogenetic perspectives.—Explorations of several morphological parameters of interest here have been attempted by using the phenetic networks by Sibley and Ahlquist (1990). Notable among these was a comparative study of the Charadriiformes, the likely sister-order of Gruiformes (Livezey 1998), by Székely et al. (2000) that indicated that sexual selection was associated more closely with sexual dimorphism than feeding ecology. Höglund (1989) inferred that lek mating systems did not promote male-biased sexual dimorphism in birds generally (i.e., dimorphism in which males are larger than females, as in the Rallidae and most Aves), a finding Nylin and Wedell (1994) considered overly negative in light of the heterogeneity among different taxonomic groups, a needless strictness in critical values adopted, and the likely obfuscation of pattern caused by recoding a continuum of magnitudes of sexual dimorphism into a binary character for use of the test of Ridley (1983). Similarly, in otherwise comparable tests for a correlation between mating system and sexual size dimorphism in birds, Höglund (1989) concluded that the correlation was not supported, whereas Oakes (1992) subjected the same data to a different statistical criterion and considered similar significance values to be confirmatory of such a correlation.

Also of interest are generalities drawn from phylogenetically oriented assessments for more-limited avian taxa, for example, those of single taxonomic orders or families. Björklund (1991) found equivocal evidence of a correlation between sexual dimorphism and mating system in grackles (*Quiscalus*) by using a method conditional on ancestral reconstructions, with one discrete character of interest. Webster (1992) assessed the same correlation in an enlarged data set for grackles and used a method that employed nondirectional metrics and continuous coding of polygyny. For the simple comparisons he found strong correlations, but, with standardized body size he found no support for the correlation and concluded that allometry accounted for the apparent relationship. Clearly, even if a phylogenetic tree is taken to be accurate, distributions and methodological details can render robust inferences problematic (Sillén-Tullberg and Temrin 1994).

Sexual selection and body size.—A substantial body of theory rests on models and studies of single species or small numbers of closely related taxa. Key among these ideas are the fundamental principles by which choices of mates between the sexes (i.e., intersexual selection) and competition among members of the same sex for mates (i.e., intrasexual selection) influence patterns of mating and effect changes in genotypic and phenotypic frequencies through time (O'Donald 1962, 1980, 1983; Selander 1965, 1972; Emlen and Oring 1977; Payne 1984; Bradbury and Davies 1987). The essential principle relevant for typical avian groups is that males compete to varying degrees and under different mating systems to win sexual access to females, with the latter often effecting a selective influence through choice of mate(s), typically via female choice. The latter tends to be of increased potency where sex ratios strongly favor females (O'Donald et al. 1974; Yom-Tov 1976; Lande 1980a; Kirkpatrick 1982a; Arnold 1983; Burley 1985; Eberhard 1996). Sexual selection often acts in combination with typical agencies of natural selection that act differentially on the sexes (Parker 1983; Greenwood and Wheeler 1985; Price and Grant 1985; but see Weatherhead and Clark 1994). Blanckenhorn et al. (1995) discussed an extreme case in a heteropteran insect in which sexual dimorphism reflects a balance between selective optima related to foraging and mating. In light of the subtle complexity of the means and covariates of sexual selection, Ghiselin (1974) warned against ad hoc explanations of the evolutionary bases for parameters of reproduction, including sexual dimorphism.

The influence of sexual selection on sexual dimorphism is related directly or indirectly to mating systems, reproductive investment, selection regimes, and genetic covariance, and an important role of sexual selection in the evolution of parental care and sexual dimorphism in birds is well established (Cabana et al. 1982; Cooke and Davies 1983; Hughes and Hughes 1986; Bradbury and Davies 1987; Halliday 1987; Lande 1987a; Partridge and Endler 1987; Davies 1991; Arnold and Duvall 1994; Badyaev et al. 2001c). However, some predictions related to sexual selection (e.g., increased intensity in colonial species leading to accelerated speciation [Marzluff and Dial 1991; Birkhead and Møller 1992]) have not been supported by phylogenetically controlled assessments (Mooers and Møller 1996). Clearly, sexual dimorphism is but one part of the milieu of reproductive parameters that interact in diverse ways to delimit the reproductive strategies of birds, and there may be important differences in the hormonal underpinnings of sexual dimorphism among avian taxa (Emerson 2000). For example, sexual size dimorphism can be correlated positively with sexual dichromatism, and correlated either positively or negatively with parental investment in Galliformes, Anseriformes, and Falconiformes (Sigurjónsdóttir 1981).

Among the Rallidae, one of the traditionally recognized variables implicated in differential sexual selection and concomitant variation in sexual dimorphism that appears to be important is mating system (Appendix 6). Of the life-historical variants represented, infrequent polygamy reported for several genera (e.g., *Porphyrio, Coturnicops, Crex, Porzana, Tribonyx,* and *Gallinula*) appears not to have effected exaggerated sexual size dimorphism (Tables 7, 26; Appendix 1). Moreover, sexual size dimorphism of the greatest magnitude among extant members of the Rallidae (e.g., that found in *Gallicrex cinerea*) is associated with monogamy (Appendix 6); the marked differentiation between the sexes in this taxon appears to enhance capacities for territorial interactions between males (P. B. Taylor 1996, 1998). Although flight capacity does not appear to be associated with departures from normal size dimorphism among extant rallids, there are examples of extinct, flightless rails that appear to have been characterized by sexual size dimorphism of exceptionally great (e.g., *Aphanapteryx, Erythromachus,* and *Diaphorapteryx*) and typical (*Porzana* spp., *Tribonyx hodgenorum,* and *Fulica chathamensis*) magnitudes (Table 8). Although those showing exaggerated dimorphism show evidence of intraspecific partitioning of niche, the potential influence of flightlessness and mating system on dimorphism must remain the subject of speculation.

Sexual selection and sexual dichromatism.—Despite the significant functional importance and energetic implications of sexual size dimorphism, intersexual differences in plumage pattern and ornamentation—sexual dichromatism—traditionally have attracted at least as much attention from early naturalists (e.g., Darwin 1859, 1871; Huxley 1914, 1938a–c, 1943) and modern ornithologists (e.g., Andersson 1982, 1994; Barnard 1991; Oakes 1992; Petrie 1992; Andersson and Andersson 1994; Møller and Birkhead 1994; Berglund et al. 1996; Møller and Hedenström 1999). The conspicuousness of sex-specific ornamentation (and associated displays) in turn engendered substantial interest in the evolutionary implications of the absence of such features in certain taxa or demographic groups within taxa, especially with respect to diminished detectability or deception of conspecifics (Rohwer 1978, 1983a, 1986; Rohwer et al. 1980; Burley 1981; Rohwer and Ewald 1981; Flood 1984; Studd and Robertson 1985; Lyon and Montgomerie 1986; Montgomerie and Lyon 1986; Hill 1988, 1996; Rohwer and Butcher 1988; Butcher and Rohwer 1989; Thompson 1991; Zack and Stutchbury 1992; Chu 1994; Piper 1997; Conover et al. 2000).

The essential generalities of the concept as applied to Aves is that many or all external intersexually differentiated phenotypic traits, especially those of an ornamental nature and conferring negligible advantage or functionality detriment, are likely candidates for sexually selected features. These structural signals of the phenotype, in combination with those serving both as sexually selected "badges" and one or more functional requirements (Berglund et al. 1996), are the targets of directional selection by potential mates, and therefore serve primarily or wholly to gain access to mates and thereby enhance reproductive fitness (Andersson 1994). Examples of "honest" signals (i.e., reliably indicative of fitness) include asymmetry of wing-patches in Common Chaffinches (*Fringilla coelebs* [Jablonski and Matyjasiak 2002]) and bill color in Common Moorhens (Fenoglio et al. 2002). Taxonomic differences in sexual dichromatism among avian taxa stem in part from variation in the proximate mechanism(s) that produce plumage color (estrogen, testosterone, luteinizing hormone, and nonhormonal factors), of which the latter three are candidates for the Rallidae (Kimball and Ligon 1999).

Conceived by Darwin (1871) to be fundamentally distinct from traits modeled by natural selection (those pertaining to survival), these features constitute special refinements for the proximate concern of mating. Subsequently, the complexity of roles of natural and sexual selection acting on traits increasingly has been recognized (Andersson 1994). Nevertheless, the distinct dimensions of sexual selection remain critical, a status perhaps best revealed by the evidence and theory that sexually selected traits can serve as strongly attractive ornaments in opposition to functional considerations, at times possibly leading to otherwise maladaptive extremes (Fisher 1930, 1958; Andersson 1994; Ligon 1999). The possibility that such signals of fitness can be otherwise disadvantageous led to the handicap principle (Zahavi 1975; Zahavi and Zahavi 1997), in which perseverance despite such impediments is a significant component of the signal of interest. Intense sexual selection and associated ornamentation are characteristic of species in which males display in traditional arenas or leks for the sole purpose of attracting mates and typically show pronounced sexual dimorphism, reduced or obsolete parental care, and (in some species) intense aggressive behavior (Bradbury 1981; Bradbury and Gibson 1983; Payne 1984; Wiley 1991; but see Höglund 1989). Evidently, structural requirements of such ornamentation can be so stringent as to select strongly against asymmetric growth; that is, frequency of (typically stress-related) asymmetry is suppressed in costly ornaments (Ligon 1999). Despite a broad empirical base for the concept, exceptions and counter-intuitive findings are known. For example, Ligon (1999: table 4.1) cited examples of plumage ornamentation of males that experimental data indicated were not of importance to selection of birds as mates by female conspecifics, some of which may pertain to some male rallids (coloration and elongated tails).

The possibility that polyandrous mating may be related to the unique, pronounced sexual dichromatism synapomorphic for the clade comprising *Rallicula* and *Sarothrura* and autapomorphic in *Gallicrex cinerea* (Livezey 1998) merits study (Appendix 6). Comparatively subtle sexual dichromatism or sex-specific differences in aspect that may serve as signals of status in social contexts are more common among rallids than in other groups of birds. To these rare distinctions might be added simple differences in dimensions that may be evident in the field but confounded by age-related variation, for example, differences in bulk, wing lengths, depth of brightly colored bills, or ungues of wings and feet (Carroll 1963c; Visser 1976; Fjeldså 1977; Craig et al. 1980; Fullagar and Disney 1981). Sexual selection favoring small but fat individuals was documented in *Gallinula chloropus* by Petrie (1983), a trait that covaries significantly with ability to acquire and defend territories, which is in turn related to reproductive success (Petrie 1984, 1986, 1988). Comparatively modest sexual dichromatism (most frequently involving differences in bill color), possibly confounded by protracted retention of femalelike subdefinitive plumages in some first-year individuals, is evident in some crakes (*Porzana porzana, P. parva,* and *P. carolina*). Whether this represents the delayed acquisition of adult plumage before the attainment of competitive condition or inadequately understood geographic variation is not clear. The latter condition is accorded added significance in light of the permanent paedomorphosis in plumage pattern in both sexes of several rallids, including a flighted swamphen (*Porphyrula flavirostris*) and several flightless species (*Cabalus modestus* and *Atlantisia rogersi*). Clearly, only the latter cases can be attributed to selectively neutral covariates of paedomorphosis targeting the pectoral apparatus (see below). Whether the several extinct, flightless rallids that showed substantial sexual size dimorphism also were characterized by sexual dichromatism (at least in bill color), unfortunately, never will be known, precluding potentially important insights into the social ramifications of flightlessness.

Sexual Dimorphism and Flightlessness

Taxa other than the Rallidae.—Upon first analysis, the secondary loss of flight may seem only remotely related to mating systems, sexual selection, and sexual dimorphism. A number of notable exceptions are known among insects, in which some taxa comprise castes or sexual groups that differ (among other characteristics) in the capacity for flight (Roff 1986a–c, 1994b, 1995; Roff and Fairbairn 1991). Although no comparable instances of intraspecific or intersexual partitions of avian taxa with respect to flight are known, several avian species approach this condition. Among steamer-ducks (Anatidae: *Tachyeres*), a genus comprising three flightless species and one flighted species, the largest males of the Flying Steamer-Ducks (*T. patachonicus*) are flightless for considerable periods of time (Humphrey and Livezey 1982; Livezey and Humphrey 1986), evidently a result of selection for large body size for maintenance of territories, in part effected by sexual means (Livezey and Humphrey 1984a, 1985a, b). The most extreme sexual size dimorphism of any anseriform bird characterizes the lek-breeding Musk Duck (Anatidae, Oxyurini: *Biziura lobata*), in which the mean body mass of males is approximately three times that of females (McCracken et al. 2000). The resultant wing loadings render flight difficult if not impossible in the heaviest males (Hobbs 1956; Dickison 1962; Lowe 1966). Finally, the flightless Auckland Islands Teal (*Anas aucklandica*), despite smaller body size and vestigial sexual dichromatism, is characterized by greater sexual size dimorphism than its flighted congeners (Livezey 1990).

Although sexual dimorphism can be functionally decisive for males in species of waterfowl near the threshold of flightlessness, insularity per se is not known to be influential in this regard (Weller 1980). For example, magnitude of sexual dimorphism in insular *Anas* did not differ from that inferred for continental relatives (Livezey 1993a), in contrast to the comparatively great size dimorphism documented in some insular species (Selander 1966; Wallace 1978) and the increased sexual size dimorphism that sometimes accompanies the evolution of flightlessness (Livezey 1989b, 1990, 1992a). Estimates of mean body masses of male and female *Chendytes lawi* (2,820 and 2,360 g, respectively) indicated that mass dimorphism in *Chendytes* was substantial but closely approximated that of extant members of the Mergini (Livezey 1993c).

Among flightless species exclusive of waterfowl, sexual dimorphism in body size in the Galapagos Cormorant is the greatest of all modern members of the family (Livezey 1992a). Likewise, the single flightless species of parrot (Psittacidae: *Strigops habroptilus*) is characterized by uniquely great sexual size dimorphism (Livezey 1992b). Intersexual differences in size were inferred to have been substantial in the Dodo (Raphidae: *Raphus cucullatus*) by Livezey (1993b). Sexual dimorphism in the related Solitaire (*Pezophaps solitaria*) equaled or exceeded that of any other carinate species of bird (Amadon 1959, 1977; Ralls 1976; Livezey and Humphrey 1984a) and led Strickland (1859) to conclude that the sexes of *P. solitaria* represented two distinct species (Livezey 1993b). Pronounced sexual dimorphism and evidently protracted developmental periods of the extinct raphids suggest that sexual and ontogenetic differences in niche characterized both species (Selander 1966; Werner and Gilliam 1984; Shine 1989a, b). The sexually dimorphic metacarpal spurs of male *P. solitaria* (Rand 1954), which frequently showed evidence of mended fractures, strongly indicated that combat occurred among males (Livezey 1993b). Accordingly, it seems likely that at least *P. solitaria* was polygynous and perhaps lek-breeding, and that the extraordinary sexual dimorphism in this species was functionally permissible in part because of flightlessness and its insular habitus (Livezey 1993b). Tarsal spurs of male galliforms are functionally analogous to the sexually dimorphic carpal spurs of *Pezophaps* and some anseriforms (e.g., *Tachyeres*), being sexually dimorphic, age-related, and larger in polygynous species (Davison 1985; Sullivan and Hillgarth 1993). The aerodynamic implications of some sexual ornamentation involving the rectrices of birds (Andersson and Andersson 1994; Balmford et al. 1994; Møller and Birkhead 1994; Fitzpatrick 1999; Møller and Hedenström 1999) lend support to the proposal that flightlessness granted additional latitude for evolutionary changes in shape in response to sexual selection. The latter interpretation accorded well with the widespread view (e.g., Colbert 1955; Yapp 1962a; Sibley and Ahlquist 1990; Feduccia 1995, 1996) so well articulated by King and King (1979:2, emphasis added): "So restrictive are the anatomical requirements of flight that, *even when the flightless birds are included,* the entire class of Aves presents greater uniformity of general structure than many single orders of fishes, amphibians and reptiles."

Sexual differences in appearance (exclusive of size) also show variable patterns with respect to insular endemism and flightlessness. One hypothesis held that the reduced sexual dichromatism of insular ducks represents the adaptive loss of isolating mechanisms rendered superfluous in the absence of sympatric, closely related species that might pose difficulties for selection of conspecific mates (Sibley 1957; Johnsgard 1963; Lack 1970b; Weller 1975, 1980). However, at the very least, the frequency of interspecific hybridization in waterfowl suggests that these mechanisms, if real, are remarkably ineffectual (Williams 1983; Livezey 1991). For example, insular reductions in dichromatism instead may be a correlate of increased involvement by males of insular endemics in brood-rearing (West-Eberhard 1983; Ketterson and Nolan 1994). This interpretation is consistent with the biparental attendance of broods typical of insular *Anas* (Weller 1980) and generally associated among anatines with reduced sexual dichromatism (Kear 1970; Lack 1974).

Evidence within the Rallidae.—In contrast to the general pattern of sexual dimorphism in the Rallidae, several flightless species suggest parallels to the exaggerated sexual dimorphism of flightless endemics in other nonpasseriform families (Table 8; Appendix 1). Although significant interactions between the statistical effects of sex and flight capacity were inferred in most genera (Tables 25, 56), consistent with the hypothesis that flightlessness permits greater intersexual differentiation under many circumstances, these translated into comparatively modest mean differences in specimens (Tables 7, 30–33). Badyaev and Martin (2000) found evidence of strong directional selection for different sexual optima in selected features in recently colonized populations of House Finches (*Carpodacus mexicanus*). This dynamic evidently varies among conspecific populations, extends to covariance structures (Badyaev and Hill 2000; Badyaev et al. 2001a), and works in concert with sexually divergent patterns of growth (Badyaev et al. 2001a, b). Given the greater antiquity of divergences giving rise to flightless rallids, it is reasonable to expect that scenarios of this kind may have occurred in multiple dimensions of flightless rails, a likelihood furthered by the associations among paedomorphosis, great sexual dimorphism, and flightlessness (McNamara 1995; see below).

More dramatic and functionally interpretable were the cases of exceptional dimorphism inferred for several extinct flightless rallids, notably *Aphanapteryx bonasia, Erythromachus leguati,* and (especially) *Diaphorapteryx hawkinsi* (Tables 8, 30–35; Fig. 45). References to extraordinary intraspecific variation in bill lengths of *D. hawkinsi* were noted by Andrews (1896b) and Cracraft (1973a). The latter also inferred that *Gallirallus dieffenbachii* and *Fulica chathamensis* exhibited unusually high variances, but interpreted this as increased variance associated with reductive evolution as opposed to flightlessness-related increase in sexual dimorphism. Furthermore, sexual dimorphism of the bill indicated in these subfossil rails may have been increased substantially by differences in the overlying integument, as in the extinct, extraordinarily dimorphic Huia (*Heteralocha acutirostris* [Burton 1974]). Taking the integument into consideration indicates that the intersexual difference in culmen lengths of *D. hawkinsi* was at least 2 cm (~20% of bill length), a difference unique for the family and undoubtedly having functional implications.

Manifold Apomorphy of Flightlessness in Rallids

There are birds which never fly. These are *ostriches,* among the terrestrial birds; and the *penguins,* and the *manchots,* among the aquatic; their wings are so small that they appear only to possess them, that they may not form too marked an exception to the rules of resemblance in the different classes of animals.—Cuvier (1802:540, emphasis in original)

The New Zealand birds afford instructive examples of the progressive loss of the volant faculty, with concomitant modifications of the parts of the skeleton giving origin to the pectoral muscles. The keel progressively shrinks from *Porphyrio* to *Tribonyx*, thence to *Notornis, Aptornis, Stringops* [sic], *Apteryx,* [and] *Dinornis.*— Owen (1882:696)

... in *Nesolimnas* [*G. dieffenbachii*] we have an annectant form linking the flying to the flightless rails. In its plumage, in the condition of its sternum, and in many other points, it reminds us of *Hypotaenidia* [flighted *Gallirallus*], while, on the other hand, in the reduction of its wings and the consequent modification of its hind-limb it approaches *Ocydromus* [*G. australis*-group]. The existence of such an intermediate type seems to give strong support to the opinion ... that the Ocydromine rails have originated from forms capable of flight at a comparatively recent date and in the islands they now inhabit.—Andrews (1896c:271)

Yet these rails [*Porzana palmeri, P. sandwichensis, P. atra, Dryolimnas aldabranus,* and *Porphyriornis nesiotis*-group] ..., generally regarded as 'flightless,' ... would appear to be 'flightless,' not from any physical or anatomical handicap, as seems evident from the dissection, but because they have lost the faculty of flight or the will to fly from simple disuse of the wings. ... On the other hand, there are, or were, certain Rails [*Gallirallus australis, Aphanapteryx,* and *Diaphorapteryx*] ... which I am much inclined to think do not or did not fly because they have not or had not acquired the structural details necessary for that function.—Lowe (1928a: 110)

Loss of limbs is one of the most extreme morphological changes in the history of tetrapods, yet it has evolved repeatedly in amphibians, reptiles, birds, and mammals.—Lande (1978:73; presumably in reference to the Dinornithiformes, the only avian group in which wings evidently were lost entirely)

Even if we had a complete fossil record for the species, we still could not determine the point in time when Wekas [*G. australis*] became flightless because there is nothing in the skeleton today to suggest that they are flightless. Indeed, the rails as a group tend to be reluctant fliers, and this has more to do with their behaviour than with any shortcomings in their anatomy.—McGowan (1986:343)

Apart from some of the rails of New Zealand and the Chatham Is., insular rails usually exhibit little morphological diversity. Other than the adaptations associated with flightlessness and the generally more robust hindlimbs associated with being more terrestrial, rails appear to require little in the way of adaptations to insular environments, so depending on the ancestral genus, flightless rails tend to differ among themselves mainly in plumage, size, and amount of reduction in the flight apparatus. Morphologically and probably behaviorally, one flightless rail is much like another and the Hawaiian species are unexceptional.—Olson (1999:2)

RELATIVE APOMORPHY OF FLIGHTLESSNESS IN RAILS

Simple indices.—Although students of flightless rails differed in the relative magnitudes or phylogenetic implications of morphological changes that they perceived, each recognized a range in the magnitudes of changes manifested by the flightless members of the Rallidae. These shared perceptions of a continuum of change were based on impressions drawn from variably limited subsets of known flightless rails, and none of these authors attempted an explicit ranking of the taxa considered. Would such a ranking of flightless rails by relative apomorphy be feasible and informative? At least for those taxa represented by adequate specimens, a number of comparative measures can be derived that serve to order taxa by inferred magnitude of change (Tables 71–75). Some of these offered substantial intuitive simplicity (e.g., changes in relative wing length, modifications in structure and numbers of remiges, or qualitative changes in skeleton), others were confounded by the effects of sample size on estimates of "distance" (Tables 71–75), and most were contingent on the reliability of the taxa inferred to be most closely related to each flightless species. Although pairwise comparison of flightless rails with their closest flighted relatives or phylogeny-based comparisons offer the advantage of explicit historicism, such approaches offer no explict ranking of all flightless species relative to a single metric or criterion.

Simple metrics for such comparisons (e.g., the ratio of wing length divided by mean body mass, that is, wing loading) at least are simple to calculate (where data are available), have an intuitive appeal, and are applicable across flightless and flighted taxa in multiple taxonomic families. Comparability across a wide range of taxonomic groups not only permits wide comparative surveys but seems appropriate in that the vast majority of the morphological changes that are attributable to flightlessness in rails (e.g., multiple expressions of pectoral reduction) are convergent with those of at least a minority of flightless members in other avian families (Tables 4, 79). Several apomorphies possessed by some flightless rallids in which the relationship to flightlessness is more ambiguous (e.g., change in body size) or characteristic of only a minority of avian lineages in which the loss of flight was comparatively ancient (e.g., pronounced hypertrophy of the pelvic limb, or truncation of pedal digits) also have parallels in other avian groups (Raikow 1985a; Table 79), although precedent for changes in body mass with loss of flight exists (Livezey and Humphrey 1986; Livezey 1988, 1989a, d, 1990, 1992a, b, 1993b, c). In fact, changes in body size at all phylogenetic scales in birds is so widespread as to defy a single explanation, and examples of substantial change in body size among flighted members of other taxonomic families are known, including *Biziura lobata* (Anatidae: Oxyurini), the Gruidae, and some passerines (Livezey 1995b, 1998; Zink 1986). Despite a sentiment expressed by some that size is a nuisance variate to be corrected before interspecific comparisons (e.g., Burnaby 1966; Mosimann and James 1979; Darroch and Mosimann 1985; Sampson and Siegel 1985; James and McCulloch 1990), herein both size and shape are considered to be relevant to evolutionary changes associated with flightlessness.

Multivariate ordinations.—In addition to the comparatively simple univariate and bivariate indices to change between species-pairs referenced above (Tables 71–75), several multivariate, taxonomically inclusive ordinations of flighted and flightless species by relative apomorphy were performed. Specifically, these were the canonical contrasts of study skins, complete skeletons, and sterna that included all taxa for which requisite data were available. Canonical contrasts by design derive a single axis that maximally discriminated rallids by presumptive flight status, but inherently summarized the total variance attributable to flight capacity (standardized by pooled within-species covariances) without a partitioning of components of that total variance attributable to size or shape (Tables 23, 53, 54; Figs. 32, 59, 60).

A canonical contrast of species by flight capacity with six standard measurements of study skins incorporated almost two thirds of the total standardized variation among species (Table 23), and effectively contrasted wing length and middle-toe length with three of four of the remaining measurements (with tail length proving of marginal utility). Not surprisingly, wing length was significantly more influential in this contrast than the covarying middle-toe length (Table 23); the latter evidently served as an index to body size that is comparatively independent of flight capacity. A dot-chart of 96 rallid taxa by their scores on this canonical contrast placed several small flightless crakes (*Porzana palmeri, Atlantisia rogersi,* and *P. sandwichensis*) and one crake commonly considered flighted but inferred to be at least flight-impaired in this study (*Laterallus spilonotus*) at one extreme, and a series of large, flighted rallids of terrestrial and aquatic habit (e.g., *Gymnocrex rosenbergii, Gallinula (c.) galeata, Fulica gigantea,* and *F. cornuta*) at the other (Fig. 32). The pattern of symbols indicative of flight capacity in the dot-chart clearly reveals that no single value or position on the axis based on skin measurements unerringly divided flightless from flighted taxa. The confounding of increasing body size with flight capacity on this axis is evident by an examination of the positions of taxa in the plot (Fig. 32), an impression corroborated by a significant correlation between mean scores and mean body masses on the canonical contrast ($r = 0.75$, $P \ll 0.01$). Consequently, although the contrast portrayed a general trend in flight capacity (Fig. 32), the comparatively limited information afforded by six measurements of study skins tended to de-emphasize the apomorphy of the larger flightless species (e.g., *Porphyrio hochstetteri*) relative to smaller species showing only subtle external changes (e.g., *Dryolimnas aldabranus*). However, in some respects, this bias is welcome, in that many measures of flight capacity (e.g., wing loading) are influenced strongly by body mass, rendering the discrimination of dwarfed flightless species (e.g., *Dryolimnas aldabranus* and *Gallirallus wakensis*) problematic.

Canonical contrasts of complete skeletons (i.e., those providing 41 measurements), whether or not sexes were distinguished within species, ranked species virtually identically (Table 53), and therefore only the simpler analysis in which sexes were pooled was plotted (Fig. 59). Although fewer taxa were represented by sufficient material to be included in the analysis, the richer suite of measurements provided substantially more power with which to discern flight capacity, and the resultant contrast of 52 taxa emphasized the magnitudes of several sternal and most alar dimensions relative to those entered from the skull and pelvic apparatus (Table 53). Scores of species on the skeletal contrast divided flightless species from their flighted confamilials with only two exceptions (Fig. 59): *Porzana tabuensis* was placed marginally within the range of scores for flightless species, and extinct *Fulica chathamensis* was placed well within the flighted species analyzed. The latter finding represents a less compelling departure from expectations given that the flight status of this giant, subfossil coot is not known with certainty (in fact it may have been capable of labored flight), and its position on the contrast reflected in part the modest but significant correlation between scores on the contrast and body mass (Table 53). Of the species for which complete skeletal data were available, *Atlantisia rogersi, Cabalus modestus,* and

Aphanapteryx bonasia were resolved to be the most apomorphic with respect to flightlessness (Fig. 59).

A canonical contrast of five sternal measurements (Table 54) provided a morphometric synthesis of osteological changes associated with flight status in the single element most profoundly indicative of flightlessness, affording a dimensionality approximating that for study skins, and that emphasized the relative robustness of the carina sterni (Table 58). Although empirically impoverished relative to comparisons based on complete skeletons, restriction of focus to sternal measurements permitted the inclusion of 20 additional species known only from very limited, often subfossil remains, many of which were flightless (Table 54; Fig. 60). Although considerably more blurring of the division of flight classes was evident, misplacements primarily were limited to the scoring of flighted crakes (e.g., *Porzana tabuensis, P. pusilla, P. flavirostra,* and *Coturnicops noveboracensis*) among flightless taxa and the purportedly flightless, subfossil coots (*Fulica prisca, F. chathamensis,* and *F. newtoni*) among the flighted taxa (Fig. 60). Once again, a significant correlation between body masses and scores on the sternal contrast explained, in part, the positions of the giant, extinct coots (Table 54), species that may have been capable of heavy flight. Despite these exceptions, an informative ordination of rallids by sternal apomorphy was achieved in which small, flightless *Porzana severnsi, Atlantisia rogersi, Capellirallus karamu,* and *Cabalus modestus* occupied one extreme and the ponderous but flighted *Fulica gigantea, Aramides ypecaha,* and *F. cornuta* defined the opposite pole (Fig. 60).

Those taxa approaching the threshold of flightlessness (τ) based on one or more of these canonical contrasts were not completely unanticipated in light of other taxonomically global assessments of bivariate or multivariate design (Figs. 15, 20, 44, 49–51). Some of the taxa traditionally assumed to be flighted (e.g., *Laterallus spilonotus*) may be described more appropriately as flight-impaired or verging on flightlessness (cf. Figs. 20, 28), and functionally comparable to other flightless rails showing limited apomorphy (e.g., *Gallirallus owstoni*). Similarly, the three subfossil coots justify only a provisional consideration as flightless, and would be classified conservatively as capable of weak flight, an allometrically compromised capacity comparable to that of the heaviest Flying Steamer-Ducks (Humphrey and Livezey 1982; Livezey and Humphrey 1986). This status is consistent with the inference by Mourer-Chauviré et al. (1999) that *Fulica newtoni* was probably capable of flight and that this mobility explained its apparent presence on both Mauritius and Réunion. The discriminatory power of the analysis of study skins would be increased substantially if some means of quantifying changes in the shape and microstructure of flight feathers was included. Similarly, even limited augmentation of the osteological contrast with measurements of the breast musculature would improve the separation of flightless and flighted taxa, but availability of data would limit prohibitively the taxa so analyzed.

MEDIATION OF CHANGE AT AN EVOLUTIONARY CROSSROAD

As demonstrated above, flightlessness in the Rallidae is manifested anatomically in large part by broadly similar shifts in proportions, but that these shifts vary substantially in magnitude (Figs. 15, 20, 32, 44, 49–51, 59, 60, 111–115). That is, the morphometric apomorphy shown by flightless rallids represents a family-wide case of rampant parallelism in manifold, but subtle qualitative changes (primarily in the pectoral apparatus) that encompass a wide range of magnitudes. One approach to the apomorphy that characterizes flightless rails (and other birds) is premised on a partitioning of morphological changes in relative size into two primary components (Fig. 117): size of the pectoral apparatus (P), and size of its complement or that of the rest of the avian corpus ($P^c = M*$). Note that, for the general case, each of these two components is symbolized as a vector of possible subcomponents, permitting the subspecification of anatomical subparts that are subjected to different selection regimes or possess different developmental parameters. However, for the preliminary explorations below, the two components will be considered to be unidimensional entries in vectors of phenotypes (symbolized by P and $M*$, respectively).

In addition to providing a simplified conceptual framework for the essential morphological changes that are associated with flightlessness in rails, this bipartitioning is consistent with current theory in which evolutionary trends are considered as departures from typical developmental constraints that turn on poised bifurcations in morphogenetic trajectories (e.g., Balon 1989; McKinney 1990a, b; Brock 2000). Such ontogenetic departures from normal or plesiomorphic developmental pathways represent decanalizations—variations from typical ontogenetic interrelations—that may be critical for "saltational changes" in body form (Salthe 1975; Scharloo 1991; Wagner 1997; Wagner et al. 1997; Rice 2000). Experimental evidence for reptiles and amphibians (Alberch and Gale 1983, 1985; Raynaud 1985; Cohn and Tickle 1999) and birds (Coelho et al. 1991, 1992; Mackem and Mahon 1991; Nelson et al. 1996; Goff and Tabin 1997) substantiates that such ontogenetic reorganizations involve regulatory genes (e.g., *Hox* genes). A concept of bifurcation of potential developmental pathways holds promise as an ontogenetic basis for the production of two selective optima through disruptive selection (Barton 1998).

Lande (1979) provided a quantitative genetic model applicable to this two-dimensional case, in which the evolutionary response of a vector of mean phenotypes (**z**)—in terms of the additive genetic covariance matrix (**G**), environmental covariance matrix (**E**), the phenotypic covariance matrix (**P** = **G** ⊕ **E**), and the vector of selection differentials (**S**)—was shown to be:

$$\mathbf{z} = \mathbf{G} \cdot \mathbf{P}^{-1} \cdot \mathbf{S}.$$

This model summarized a plausible means of evolutionary change under the constraints of genetic and environmental covariances (Roff 1992), and clearly implies that both intensity of selection and the interrelated constraints of genetic and environmental effects contribute to evolutionary response. In the context of avian flightlessness in particular, it is intuitively appealing to consider this model in terms of evolutionary changes in the two key parameters of interest (P and $M*$), that is, where:

$$\mathbf{z} = \begin{pmatrix} P \\ M* \end{pmatrix}.$$

Upon colonization of an island, the founding population of flighted rails having mean phenotype (α) is predicted to undergo total changes to an ultimate mean phenotype (ω). This transition between selective optima generally is envisioned to be effected through a number of generations, symbolized by iterations of the process of selection upon each, with possible changes in the selection differentials through time (Fig. 117). This process imposes expected shifts in the phenotypes of successive generations that reflect simple changes in selection gradients (**S**) acting on the two key parameters of interest; changes in the interrelationships specified by the genetic and environmental covariances (**G** and **E**, respectively, summing to **P**), that is, the effects of a key ontogenetic reorganization (κ) that alter the allometry imposed upon selected changes in the pectoral apparatus (*P*) and the rest of the body (*M*∗); or both of the preceding sources of evolutionary modification. As depicted in the simple differential system descriptive of mean phenotypes subjected to one variant of flightlessness through allometric giantism (Fig. 118), simple selection for increased size (i.e., parameterized by **S**) may drive the change in phenotypes until the threshold of flightlessness (τ) is reached, after which a qualitative alternation in the relationships between components (κ, parameterized by **G**) is initiated in subsequent generations through selection or genetic drift (Roff 2000).

In the simple two-dimensional case, these various pathways to change ultimately can produce in combination any of nine major categories of bidimensional apomorphy (i.e., for the simple case in which the two components are considered to be unidimensional, there are nine combinations of reduction, stasis, or increase in each of the two components). Of these nine broad categories of apomorphy (δ), the three corresponding to evolutionary increases in relative investment in pectoral size (*P*) are candidates not relevant to the question of avian flightlessness. Similarly, on both intuitive and empirical grounds, the category of evolutionary change that results from decreased nonpectoral body size (*M*∗) and static pectoral size (*P*) also is implicated in flightlessness of rails. On intuitive grounds, the remaining five phenotypic classes (δ∗) that remain (decreased, static, or increased nonpectoral size in combination with decreased pectoral size, and increased nonpectoral size in combination with static pectoral size) can be associated with flightlessness, and most if not all are represented among flightless members of the Rallidae (Tables 71, 72).

Flightlessness in rails largely represents variation in size combined with variable, evidently paedomorphic pectoral reduction, and the five categories germane to the loss of flight correspond closely with the major categories of heterochrony (see below). Ideally, with improved knowledge of the environmental circumstances that fostered the various morphotypes of flightlessness in rails, it will be possible to predict the most likely evolutionary changes associated with flightlessness in rails by insular conditions in advance of excavation and recovery of subfossil remains. Such an exercise would constitute a genuine test of the classificatory model, but in many instances this will not be possible because with the continuing passage of time, the quality of subfossil remains has declined. For example, most rallids recently described from recovered skeletal elements have not warranted inclusion in the present morphometric study (Steadman 1986a, 1988; Steadman et al. 1994), and others permit only limited univariate reconstructions (Olson and James 1991).

Ontogenetic Mechanisms Underlying Avian Flightlessness

> We cannot form a better idea of it than by imagining a Duck or Gosling enlarged to the dimensions of a Swan. It affords one of those cases ... where a species, or

a part of the organs in a species, remains permanently in an underdeveloped or infantine state. . . . the Dodo is (or rather was) a *permanent nestling,* clothed with down instead of feathers, and with the wings and tail so short and feeble, as to be utterly unsubservient to flight.—Strickland and Melville (1848:33–34; emphasis in original)

. . . it was a singular fact that this little Rail [*Cabalus modestus*] should possess in its adult plumage the exact dress which might have been expected to characterize the young of *C. dieffenbachii*; even with the evidence now before them it was difficult to believe that the birds were fully adult.—Sharpe (1893d:46)

Finally, I might add that not only do there not appear to be any signs of 'degeneration' in the second category [those considered by Lowe to be secondarily flightless, e.g., *Atlantisia rogersi*], but there seem to be none in the first [those thought primitively flightless, e.g., *Diaphorapteryx hawkinsi*]. All that seems obvious in the latter is a failure to have developed beyond a certain phase, and this seems to be especially evident in *Atlantisia,* where there is an actual advance in the evolution of the structure of the barbules in the remiges as we proceed from the juvenile to the adult stage.—Lowe (1928a:113–114)

. . . it is clear that the rate of growth of the wing of the non-flying embryo has either been already relatively retarded *before hatching* or the rate of general body-growth relatively accelerated. . . . but my reason for thinking that the flightless condition of *T. brachypterus* [Falkland Flightless Steamer-Duck] is primarily due to some physiological factor which *permanently* retards the rate of growth of the wing to a degree which is not normal is that the same sort of process exercises its restraining influence *temporarily* in the chicks of Anatidæ in general, as also in other groups like the Gallinules.—Lowe (1934:482–483, emphasis in original)

As we have seen, neoteny need not affect all structures of the body, and may be restricted to quite few. A case of this kind is provided by the plumage of the ostrich and other flightless birds and penguins, which resembles the nestling down of young flying birds. Far from meaning that the ostrich is primitive and that its plumage is 'recapitulated' in the 'chick' stage of the ontogeny of a bird like a fowl, this series must be read exactly the other way. . . . the ostrich, in retaining in the adult a type of plumage characteristic of the young of other birds, is neotenous.—de Beer (1951: 63–64)

I also believe that an understanding of regulation must lie at the center of any rapprochement between molecular and evolutionary biology; for a synthesis of the two biologies will surely take place, if it occurs at all, on the common field of development.—Gould (1977:408)

It is mainly through a net of developmental constraints that natural selection works by filtering actual phenotypes out of all possible genotypes. . . . Unfortunately, to this day very little is known about embryonic development.—Jacob (1982:43)

It's not all heterochrony.—Raff (1996:255)

STUDY AND GENERALITIES OF AVIAN ONTOGENY

Investigational motivations and obstacles.—Despite the fundamental role accorded embryology in the evolutionary synthesis—as a crucial link between genetics and morphology—a coordinated and strategic pursuit of developmental biology in an evolutionary context is conspicuously lacking (Thomson 1985; Maynard Smith 1986; Hull 1988). Accordingly, McKinney and McNamara (1991:

87) relegated all developmental processes leading from the zygote to adult phenotype to an ontogenetic black box. Among ornithologists, the traditional view that the anatomical prerequisites for flight severely restricted the limits of permissible anatomical change did little to foment inquiry (King and King 1979). Furthermore, the recapitulationist perspective (Garstang 1922; Fink 1982; Løvtrup 1989a; Mayr 1994) was adopted by Lowe (1928a, b, 1930, 1933, 1934, 1935, 1939, 1942) to argue that avian flightlessness represents a recapitulation of traits of primitively flightless ancestors. This proposal proved unappealing to ornithologists (McDowell 1948; Livezey 1995a) and fundamentally flawed (Gould 1977). Despite this professional reticence toward some ontogenetic explanations, landmark works on tetrapods (including Aves) by W. K. Parker (1866, 1868, 1888), T. J. Parker (1894), Gadow (1888), Hommes (1924), Goodrich (1930), Marples (1930, 1932), de Beer (1937, 1940), Fell (1939), Broman (1941), Lillie (1942, 1952), Montagna (1945), and Gaertner (1949), a trend ornithologically codified by Romanoff (1960) and treated quantitatively by Ricklefs (1968, 1969, 1973, 1979, 1983a, 1984) and Calder (1982b), fostered a renaissance in the study of avian ontogeny (Goddard et al. 1993).

Establishment of empirical baselines.—Considerable progress has been made in several areas of avian ontogeny, notably those aspects of development that are amenable to comparatively simple descriptive models or revealed through broad comparative surveys (e.g., Kirkwood et al. 1989). Basic study of avian development progressed, and among the fundamental ontogenetic generalities of relevance to avian flightlessness that emerged was that avian taxa spanned a range of developmental patterns in both anatomical and behavioral parameters, all operative within a context of environmental variability (Houston and McNamara 1990). This diversity subsequently was conceptualized as the altricial–precocial continuum (Nice 1962; Starck 1989; Konarzewski et al. 1998; Ricklefs and Starck 1998a, b; Ricklefs et al. 1998; Starck and Ricklefs 1998a–c).

Fortuitously for the ontogenetic interpretation of avian flightlessness, a systematically motivated tradition of ontogenetic study of ratites persisted throughout the 20th century (e.g., Broom 1907; Lutz 1942; Glutz von Blotzheim 1958; Reece and Butler 1984; Beale 1985, 1991). Also fortunate in this context was the attention paid to the ontogeny of the avian manus (e.g., Stark and Searls 1973; Hinchliffe 1977, 1989; Kieny 1977; Hinchliffe and Griffiths 1983; Hinchliffe and Gumpel-Pinot 1983; Padian and Chiappe 1998) and the order in which skeletal elements ossify (Starck 1998). Another advantage is the recognition that nonheterochronic mechanisms may assume important roles, at least in part, in some evolutionary changes (Shapiro and Carl 2001).

HETEROCHRONIC PERTURBATION, DEVELOPMENTAL CONSTRAINT, AND EVOLUTION

> ... this implies that, if you have a cascade of regulatory genes, with *A* switching on (or off) genes *B, C* and *D,* which then switch on *E, F, G* and *H,* and so on, the one gene you don't muck about with is gene *A.*—Maynard Smith (1998:14)

Bauplan *as both limit and avenue of change.*—The conservatism of early development, both as a *Bauplan* on which subsequent ontogenetic variations are imposed and as a fundamental limit on the evolutionary diversity of body form (Scharloo 1991; Amundson 1994), is central to the theory of commitment and the stabilization of form described by Levinton (1988). Accordingly, the evolutionary importance of development is twofold: as a mechanism for rapid evolution of novel phenotypes (Müller 1990; Raff et al. 1990); and paradoxically, a conservative *Bauplan* that constrains the possible directions of evolutionary change (Alberch 1980; Holder 1983; S. A. Kauffman 1983; Maynard Smith et al. 1985; McKitrick 1993). The intimate relationship between the two conceptual frameworks is exemplified by the interpretation of vestigial structures as by-products of evolutionary shifts along conservative ontogenetic trajectories in morphospace (Stone 1997, 1998). The latter is attributed, at least in part, to the limits imposed on selected traits by genetic covariances (Cheverud 1984; Wagner 1988a, b; Houle 1991), a matrix-based parameterization of the limiting role of ontogenetic constraints (i.e., interrelationships of developmental pathways) on evolutionary plasticity and change (Stearns 1986; Scharloo 1990; Streicher and Müller 1992; Arnold 1994; Stone 1996), including the acceleration of speciation (Pfennig and Murphy 2002).

Since the review by Gould (1977), the evolutionary importance of constraints and changes in developmental schedules has been the focus of renewed theoretical and empirical effort (Atchley 1987; McKinney 1988a, b; McNamara 1990; Raff 1996). Appreciation of the possibility of escape from a developmental constraint and attainment of new, macroevolutionary forms exposed a great potential for interspecific divergence mediated by ontogenetic trajectories (Bonner 1968; Jacob 1977, 1982). The scope of this realization was magnified with the recognition that selection for change in size may lead to developmentally canalized, selectively neutral or even suboptimal changes in shape (or less frequently, vice versa). The issue of by-products of selection for changes in size may be of special relevance to the increases in wing loading that accompany increased body mass in many avian lineages. Many of the specific evolutionary potentials listed by Maderson et al. (1982) and subsequently by McKinney and McNamara (1991) were proposed in the context of apparently rapid, juvenilized transitions in the fossil record, observations that were seminal for the hypothesis of punctuated equilibria (Eldredge and Gould 1972; Gould and Eldredge 1977), but that had broader implications related to rapid changes in size and shape under different environmental regimes. Balance in this regard was provided by the proposal by Soulé (1982a, b) concerning morphological variation (termed by him "allomery"), that included a reduction in variability with size or number of parts; and that extreme phenotypes are expected to show higher degrees of morphological asymmetry through developmental instability.

Genetic foundations of ontogenetic change.—Although a role for mutation has been documented (Ede 1991), the genetic bases for heterochrony remain largely unknown, and the inferred changes in ontogenetic timing have been attributed to the action of regulatory genes (Gould 1977; Slatkin 1987). Heterochrony also has been implicated in ecological apomorphy (Lawton and Lawton 1986), and conversely, ecological conditions have been implicated in the occurrence of ontogenetic patterns (McKinney 1986; Stearns 1989a). The prevalent perspective regarding ontogenetic effects on evolutionary change draws from a broader genetic context that includes the timing of genetic actions on evolutionary change (Haldane 1932), the role of regulatory genes (MacIntyre 1982), actions of genetic covariances (Atchley 1984, 1990), and the problem of embryonic "clocks" (Hall and Miyake 1995). Relatively rapid evolutionary changes in form are thought to be possible through minor changes in regulatory genes and resultant developmental ramifications (Goodwin 1982; Raff and Kaufman 1983; Arthur 1984, 1988; Anderson 1987; Hafner and Hafner 1988; McKinney 1988a, b; Wray and Wray 1989; Wray 1992).

Study of the effects of certain genes in the appendicular primordia of limbless and other varieties of domestic fowl, notably *Hox* genes (e.g., *Ghox*-4.6, *Ghox*-4.7, *Ghox*-8, *Hoxd*-11, and *Hoxd*-13), renders it likely that heterochronic changes of the kinds inferred in the pectoral apparatus of rails are mediated, at least in part, through the actions of such mechanisms (Ambros 1988; Coelho et al. 1991, 1992; Mackem and Mahon 1991; Nelson et al. 1996; Goff and Tabin 1997). A recent review by Capdevila and Izpisúa Belmonte (2000) stressed that the ultimate evolvability of the tetrapod limb may derive from the variations imposed by *Hox* genes, a potential that presumably underlies phylogenetic diversification in birds (Nemeschkal 1999).

Heterochrony: nomenclature and diagnosis.—Shifts in ontogenetic schedules over evolutionary time—heterochrony—as a source of anatomical novelties increasingly has become recognized (Gould 1982, 1988a; McNamara 1982, 1997; Shea 1983; Hall 1984a, b, 1990a, b, 1998; Levinton 1986; Raff and Wray 1989; McKinney and McNamara 1991; McKinney and Gittleman 1995; but see Raff 1996; Rice 1997), from both intraspecific and interspecific perspectives (Reilly 1997). In recent years these ontogenetic perturbations have been conceived most frequently as changes in timing, whereas von Baer (1866) originally partitioned such variation with respect to both time and space, the latter being termed "heterotopy" (Zelditch et al. 1993; Zelditch and Fink 1996). Buss (1987) emphasized that the moderately late action of heterochrony characteristic of major evolutionary change reflects the conservation of critical, early developmental patterns, while permitting comparatively rapid modification of form through variations in later stages of ontogeny.

Heterochrony often can be described by allometry of affected parts (Bonner and Horn 1982; Klingenberg 1998); diagnosis of the phenomenon has been hampered by terminological imprecision (McKinney 1999) and the uncritical extension of the concept to any evolutionary change having demonstrable ontogenetic bases (which likely would include virtually all evolutionary change in form). In the avian context, for example, uncritical acceptance of a superficial aspect of "juvenility" as diagnostic of paedomorphosis would label as paedomorphic any avian group with relatively long legs (e.g., any wading bird). Accordingly, rigor of diagnostic criteria is central. For example, some evolutionary changes in birds averred to be heterochronic have led to parallels between taxa as distantly related as ratites and passerines (Dawson et al. 1994). An extreme example of such terminological extension pertains to the Mesozoic fossil *Archaeopteryx*, for which some of the many profoundly plesiomorphic characters were interpreted as neotenous by Thulborn (1984). By contrast, heterochrony has been cited as the probable underlying cause of changes in several flighted birds (Fry 1983; Foster 1987; McDonald and Smith 1994; Andersson 1999).

Heterochrony currently is conceived as comprising six major manifestations produced by one or more developmental processes acting singly or in combination (McNamara 1986; McKinney and McNamara 1991), reformulated in large part

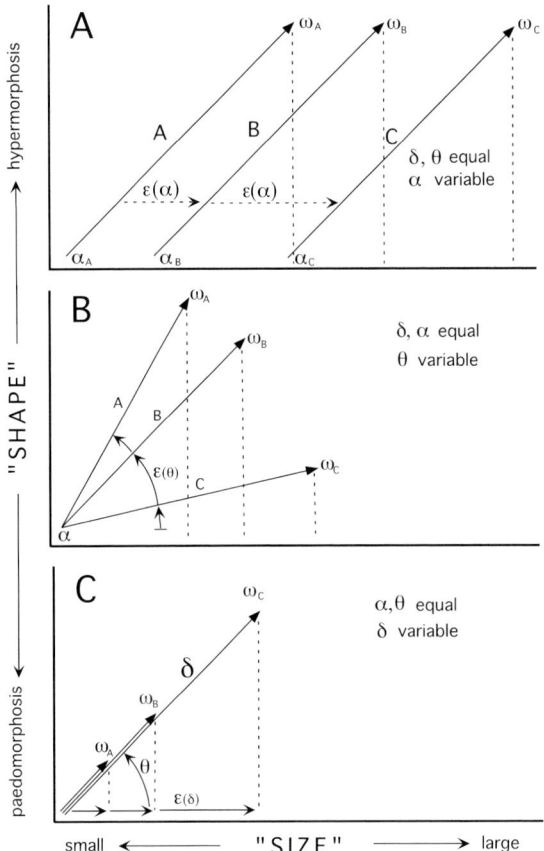

FIG. 122. Hypothetical developmental trajectories for three species (*a, b,* and *c*) differing in body size, shape, or both, as produced by three broad suites of ontogenetic mechanisms, after Alberch et al. (1979): **A,** three trajectories sharing identical orientation and duration but having different initial body sizes; **B,** three trajectories of equal duration and initial conditions but differing in orientation; and **C,** three trajectories sharing initial conditions and orientation but differing in duration. Initial morphostates are symbolized by α and terminal morphostates are symbolized by ω.

from the clock models of Gould (1977). Three classes of developmental processes give rise to peramorphosis or derived overdevelopment: predisplacement or early initiation of growth, acceleration or increased rate of growth, and hypermorphosis or prolonged growth (Fig. 122). A parallel suite of three mechanisms are credited for paedomorphosis or derived underdevelopment: postdisplacement or late initiation of growth, neoteny or reduced rate of growth, and progenesis or early terminus of growth. Acceleration and neoteny differ from the other mechanisms by entailing a dissociation or uncoupling of size and shape in ontogeny (Shea 1985a).

The recognition of variable morphological scales in the effects of heterochrony has led to a formal distinction between global scales (i.e., affecting whole organisms) and local scales (affecting subparts thereof) of heterochronic processes (Jablonski 2000). An alternative dichotomy of terms (systemic and specific, respectively) was coined by Hall (1998). The dissociability of heterochrony is a logical reflection of the substantive ontogenetic semiautonomy of different local growth fields (and resultant anatomical regions) in developing individuals (McKinney and McNamara 1991), a generality substantiated and graphically displayed in an avian context by Maunder and Threlfall (1972) and O'Connor (1984). This refinement is critical for the diagnosis of heterochrony in flightless birds, in that there are examples (detailed below) in which the pectoral apparatus manifests paedomorphosis (underdevelopment), whereas the axial components and pelvic apparatus show peramorphic (overdeveloped) aspects. However, substitutes for true age (e.g., size or other developmental stages) can be misleading with respect to ontogenetic changes that involve the surrogate variable (Shea 1988; Jones and Gould 1999). With few notable exceptions among tetrapods (e.g., Lowe 1928a; Goldschmidt 1940; de Beer 1951), amphibians, notably salamanders (Bruce 1979; Semlitsch 1985, 1987, 1990; Semlitsch and Wilbur 1989), include lineages showing facultative expression and environmental induction of paedomorphosis (Harris 1987; Jackson and Semlitsch 1993; Whiteman 1994; Whiteman et al. 1996; Ryan and Semlitsch 1998).

EMPIRICAL SUPPORT FOR HETEROCHRONY IN BIRDS

> Now the same sort of temporary retardation of the growth of the wing and its feathers occurs normally in the chicks of the Rails and Gallinules (as also in other bird groups), and if this *retardation is permanently prolonged beyond the normal time,* as it seems to have been in the Flightless Steamer Duck [cf. *Tachyeres ptenereres*], we might get, I suggest, a condition of fightlessness comparable with that seen in the ocydromine Rails of New Zealand [e.g., *Gallirallus australis*], and in *Tribonyx, Cabalus,* and *Diaphorapteryx* of the outlying islands, or in the 'Island Hens,' *Porphyriornis* and *Atlantisia,* of the South Atlantic.—Lowe (1935:426, emphasis added)

Generalities of avian heterochrony.—Heterochrony offers an alternative perspective on evolutionary change leading to reduction in size, traditionally envisioned as "degeneration" or genetically stochastic "wasting" of structures of diminished functional utility (Brace 1963; Holdgate 1965; Peters and Peters 1968; Peters 1988). The early emphasis on the ontogeny of the avian flight apparatus (including feathers) provided a reasonably sound basis for inferences concerning processes leading to flightlessness. However, other developmental events also identify ontogenetic events of possible diagnostic utility, including patterns of ossification within limbs (Müller and Streicher 1989; Müller 1991) and cranial elements (Parker 1866; de Beer 1937; Jollie 1957; Murray 1963; Murray and Drachman 1969; Lindén and Väisänen 1986).

Broad ontogenetic parameters were established for a number of taxonomic orders of birds pertinent to the study of avian flightlessness, including Galliformes (e.g., Lansdown 1970; Hinchliffe and Ede 1973; Hogg 1980, 1982; Druschel and Caplan 1991), Anseriformes (e.g., Cain 1970; Würdinger 1975; Sedinger 1986), Charadriiformes (e.g., Maillard 1948; Weber 1990; Cane 1993), and Passeriformes (e.g., Grant 1981a; Boag 1984; Tatner 1984), as well as comparative studies with other tetrapods (Kramer 1953; Oudhof 1975; Johnson 1986; Starck 1989). Based on general growth patterns, traditional expectations of paedomorphosis acting on the pectoral skeleton of birds would include small general size of the pectoral elements, which develop relatively late in avian embryos; disproportionately short distal wing elements, a condition shared by embryos of flighted species, and dubbed for evolutionary extrapolation as the "law of distal–terminal transformation" by Levinton (1988:269); vestigial carina sterni, one of the last skeletal structures to ossify in carinates; and obtuse angulus coracoscapularis, a characteristic of embryos of flighted carinates (Parker 1888; Kulczycki 1901; Knöpfli 1918; Marples 1930; Nauck 1930a, b; Schinz and Zangerl 1937a, b; Portmann 1938; Saunders 1948; Fujioka 1955; Klima 1962; Kieny 1977; Lande 1978; Abs 1983; Rofstad 1986; Levinton 1988; Milonoff and Linden 1989; Carrier and Leon 1990). The few data pertinent to timing of development of the pectoral musculature are consistent with relatively late growth of the breast muscles in avian lineages (e.g., *Cochlearius cochlearius* [Hartman 1961]).

With respect to the pectoral skeleton, the avian carpometacarpus represents the ankylosis of certain fetal ossa carpi centralia and ossa carpi distalia with the proximal termina of the ossa metacarpalia during early postnatal ontogeny (Shubin and Alberch 1986). During this process the carpometacarpus passes through a stage in which a poorly defined proximal complex and three distal cartilaginous elements are discernible (Sieglbauer 1911; Holmgren 1955), events of potential diagnostic utility of pectoral heterochrony in birds. Typical avian growth patterns include developmental stages at which distal portions of ossa metacarpale majus and metacarpale minus and distal ossa digitorum manus are not yet formed (Hamburger and Hamilton 1951); and the developing carpometacarpus comprises a proximal complex of the cartilaginous distal carpals and pisiform (corresponding to the extremitas proximalis carpometacarpi), a cranial cartilaginous primordium of phalanx digiti alulae, and distal cartilaginous precursors of ossa metacarpales majus et minus (Hinchliffe 1977; Hinchliffe and Griffiths 1983). With respect to the pectoral musculature, the developmental stages of Hamburger and Hamilton (1951) provide critical diagnostic events, junctures that are distinguishable in all avian species studied in the reviews by Starck and Ricklefs (1998a) and Ricklefs and Starck (1998a) regardless of position on the altricial–precocial spectrum of ontogeny. Redfern (1989) provided ontogenetic insights regarding primary remiges that are promising with respect to the recognition of paedomorphosis in birds.

Evidence in other flightless neognathous birds.—The recognition of "gigantic immaturity" in the Dodo by Strickland and Melville (1848) was the earliest recognition of ontogenetic change in the evolutionary loss of avian flight (Livezey 1993b). A number of early references to evolutionary change in ontogeny or pectoral agenesis (e.g., the "arrested development" of Owen [1866b:12]) fell short of inspiring mechanistic insights and essentially remained simply descriptive of relative size. For a diversity of flightless birds, the pectoral appendage was described as underdeveloped, a condition currently interpreted as paedomorphosis (Lowe 1928a; de Beer 1951; Olson 1973a, b; Feduccia 1980; James and Olson 1983; Worthy 1988a).

In the flightless raphids and most other flightless carinates (Livezey and Humphrey 1986; Livezey 1988, 1989b, d, 1990, 1992a, b, 1993a–c, 1995a), the most conspicuous and frequent osteological evidence of paedomorphosis includes a reduced or aborted carina sterni and margo caudalis of the corpus sterni, an enlarged (but rarely obtuse) angulus coracoscapularis, and the comparatively diminutive (underdeveloped) pectoral apparatus, especially the major elements of the antebrachium and manus. These generalities also are true of the apomorphic pectoral limbs of the ratites (Cracraft 1974a), in which comparatively pronounced phylogenetic differences confound comparisons. Anatomical variation superimposed on features typical of flightlessess include conformation and composition of the pectoral girdle (Fürbringer 1888) and differences in pelvic limb (Storer 1945; McGowan 1979, 1982). Several myological characteristics found in flightless carinates in other families (Livezey 1990, 1992a) were consistent with the truncated pectoral development documented in birds (Gadow 1902; Kirkham and Haggard 1915; Kipp 1959).

Waterfowl (Anseriformes: Anatidae), second only to the Rallidae in number of flightless members, run the gamut in degree of morphological changes and conspicuousness of the effects of heterochrony. The extinct moa-nalos of Hawaii (Thambetochenini [Livezey 1996a]) represent excellent examples of profound pectoral paedomorphosis (including uniquely derived carpometacarpi and remarkably reduced carina sterni) combined with pelvic peramorphosis approaching graviportality (James and Olson 1983; Olson and James 1991; Livezey, unpublished data). Unfortunately, material for study of moa-nalos is limited to subfossil elements, precluding detailed assessments of the specific heterochronic processes involved. The extant, endemic Nene (Anserini: *Branta sandvicensis*) of Hawaii, although flighted, shares less pronounced morphological changes related to intensified reliance on terrestrial locomotion (Miller 1937). Flightlessness in three flightless species of steamer-duck (*Tachyeres brachypterus, T. leucocephalus,* and especially *T. pteneres*), although largely attributable to an increase in body size without commensurate increase in wing area, also is associated with modest indications of pectoral paedomorphosis, for example, relatively shallow carina sterni and truncated remiges (Livezey 1986c; Livezey and Humphrey 1986). Flightlessness occurs (at least temporarily) in the heaviest male Flying Steamer-Ducks (*T. patachonicus*) in marine populations (Humphrey and Livezey 1982), thereby providing limited validation of the hypothesis of developmental loss of flight in *Tachyeres* proposed by Cunningham (1871). Some large anatid waterfowl exemplify dissociability of parts with respect to the actions of heterochrony (Gould 1977; McKinney and McNamara 1991), in that they manifest pectoral paedomorphosis in conjunction with cranio-pelvic peramorphosis (Livezey and Humphrey 1986; Livezey 1989a, 1993c, unpublished data).

Morphometric positions of immature specimens of insular dabbling ducks (Anatini: *Anas*) indicate that the disproportionately shortened wing elements and shallow carina sterni of *A. laysanensis* are paedomorphic (Livezey 1993a), features that are more pronounced in flightless *A. aucklandica* (Livezey 1990). Moreover, the small body sizes of insular dabbling ducks suggest that progenesis (somatically early sexual maturation) is a likely candidate for the underlying heterochronic mechanism (Livezey 1990). Progenesis, in which descendant lineages undergo generalized termination of growth relative to ancestral patterns, is likely exemplified by the drab alternate plumages and small overall size of insular endemics of *Anas* and other waterfowl (Lack 1970b; Weller 1980; Livezey 1990). In the latter, drab plumages qualitatively resemble the juvenal plumages of continental relatives (Livezey 1993a). Crypsis of incubating ducks in the presence of aerial predators of eggs and ducklings may be a confounding source of selection acting on male plumage coloration (Götmark 1993). Similar patterns are found in some insular hawks (Accipitridae), in which adults of several insular endemics (e.g., *Aviceda madagascariensis* and *Buteo brachypterus*, both of Madagascar) retain definitive plumages similar in aspect to those of immature continental congeners (Brown and Amadon 1968; Thiollay 1994).

Among seaducks (tribe Mergini), mensural data from skeletal elements for flightless, fossil *Chendytes* are consistent with the hypothesis of paedomorphosis (Livezey 1993c), although the wing elements of *Chendytes* do not show qualitative characters indicative of radical truncation of growth (cf. Parker 1888). Although most probably caused by neoteny in conjunction with significant hypermorphosis of the rest of the corpus, a conclusive diagnosis of the underlying developmental modifications in *Chendytes* is not possible with available data (Gould 1977; McNamara 1986; Fink 1988; McKinney 1988a, b). As with insular *Anas* showing initial signs of pectoral reduction and paedomorphosis (Livezey 1993a), the Auckland Islands Merganser (Mergini: *Mergus australis*) and extinct congener on the Chatham Islands show remarkably similar, comparatively subtle pectoral reductions and indications of paedomorphosis of plumage patterns (Livezey 1989b; Millener 1999).

Among nonrallid gruiforms, the reduced pectoral skeleton of *Aptornis*, especially the general truncation of the distal wing elements, is consistent with paedomorphosis of the pectoral limb (Steiner 1922; Böker 1927; Marples 1930; Montagna 1945; Sullivan 1962). The specific mechanism(s) that produced the pectoral paedomorphosis of *Aptornis* is (are) not determinable with available material (McKinney and McNamara 1991), although the comparatively radical underdevelopment of the carpometacarpus of *Aptornis* indicates that the ontogenetic shift involved was profound (Livezey 1994). The vestigial carpometacarpus of *Aptornis* resembles that of the domestic fowl (*Gallus gallus*) during stages 27–29 (Sullivan 1962), which in turn corresponds to embryonic development at 5–6 days of incubation (Hamburger and Hamilton 1951).

The specific heterochronic mechanism(s) responsible for the pectoral paedomorphosis cannot be determined without detailed developmental information for both flighted and flightless relatives (Gould 1977; Alberch 1985; McKinney and McNamara 1991), but the degree of effects and diversity of anatomical regions affected among extremely derived nonpasseriform neognaths (e.g., *Raphus* and *Aptornis*) suggest that two or more changes in rates or onset of ontogeny may have been involved. The gruiform *Aptornis* resembled the raphids (Livezey 1993b) and the Galapagos Cormorant (Livezey 1992a) in showing pronounced peramorphosis of the pelvic appendage and cranium in combination with significant paedomorphosis of the pectoral limb (Livezey 1994). This pattern of opposite apomorphy in relative size resulting from heterochrony exemplifies the developmental dissociability among different growth fields (Gould 1977; McKinney and McNamara 1991), which in turn underlies the loose correlation or independence of evolutionary changes in different morphological complexes within lineages (de Beer 1951; Levinton 1988).

Apparent peramorphosis of body size in the moa-nalos (Olson and James 1991), Galapagos Cormorant (Livezey 1992a), *Aptornis* (Livezey 1994), and raphids (Livezey 1993b) presumably is (was) associated with a protracted developmental period. This pattern may have been adopted comparatively easily in the raphids, given late attainment of locomotory capability by typical columbids (Ricklefs

1973); comparatively slow growth rates of tropical species (Ricklefs 1984); reliance on low-protein food resources (Ricklefs 1979); physiological constraints related to interspecific differences in body form (Ricklefs 1969, 1973; Case 1978b); insular relaxation of predation-related shortening of developmental periods (Lack 1968b); and parallel trends toward giantism among insular columbids (Worthy 2001). The Galapagos Cormorant and many flightless anseriforms combine pectoral paedomorphosis and axio-pelvic peramorphosis (Livezey 1992a).

In flightless carinates, especially the most extremely apomorphic lineages, characters of the integument and musculature are preserved more rarely than the underlying skeletal elements. An extreme form of pectoral truncation in the wingless variant of domestic fowl included developmentally aborted pectoral musculature (Kirkham and Haggard 1915; Waters and Bywaters 1943; Pease 1961; Lancaster 1968; Zwilling 1974; Stevens 1991), and some flightless (nondomesticated) birds share some of these morphological changes to varying degrees (Darwin 1868; Murphy 1938; Hartman 1961). Qualitative paedomorphosis of the pectoral musculature was documented in the flightless Galapagos Cormorant (*Compsohalieus harrisi*) by Livezey (1992a; see below), and was inferred indirectly for the extinct Raphidae and *Aptornis* based on comparative osteology (Livezey 1993b, 1994). Descriptions by firsthand observers and illustrations of live *Raphus* (Hachisuka 1953) suggest that adults were covered largely or completely in down but possessed remiges and rectrices, an improbable combination of states that has been cited as evidence of paedomorphosis (Strickland and Melville 1848; Hachisuka 1953; Stresemann 1958; Halliday 1978a; Livezey 1993b; but see Brom and Prins 1989). At the very least, these accounts indicate that the definitive plumage of raphids was characterized by extreme structural apomorphy in which the integrity of vanes of contour feathers was severely compromised or obsolete.

Evidence in flightless Rallidae.—The limited literature on the growth of rallids (e.g., *Porphyrio, Rallus, Porzana, Gallinula,* and *Fulica*) principally established that developmental periods vary in part as a function of adult body mass and, more importantly, that fledging or acquisition of flight capacity occurs significantly later than other hallmarks of maturation (Steinmetz 1930; Boyd and Alley 1948; Meanley and Meanley 1958; Sigmund 1958; Meanley 1969; Visser 1974; Wilkinson and Huxley 1978; Cordonnier 1985; Kaufmann 1987, 1988, 1989; Becker 1990). These data and previous studies confirm the generality of a terminal burst of alar growth that is especially pronounced in the Rallidae, in which maturation of the wing and its feathers occurs long after ambulatory capacity is attained and delayed further under nutritional stress (Karhu 1973). Karhu (1973: 6) concluded for *Gallinula chloropus* that: "Apparently the growth of wing feathers occurs independently of the weight increase, but if the chick is undernourished the growth of feathers is obviously retarded." Such extrinsic influences on pectoral growth are consistent with the variation in wing lengths reported by Visser (1976) for *Fulica atra,* variation that not only affects young birds before first acquisition of remiges, but also induces variation between annual molts among adults. The relatively late maturation of the pectoral apparatus in rails essentially provides a wide developmental window within which simple truncation or slowing of growth can effect a relative reduction in anatomical structures related to flight.

Available data on the ontogeny of rails (Table 80) indicate that the musculature of the pelvic limb in the Rallidae and other Galliformes has development that is

TABLE 80. Published data on growth parameters for species of Rallidae. Parameters are K, asymptotic size; K_L, adjusted rate constant; and $t_{10} - t_{90}$, difference between times at 10% and 90% of adult size. Compiled by Starck and Ricklefs (1998c).

Species	Sex	n	Body mass (g)	K	K_L	$t_{10} - t_{90}$	Original reference
Rallus limicola	—	1	39	0.199	—	—	Kaufmann 1987
Rallus aquaticus	—	1	136	0.078	0.115	—	Heinroth and Heinroth 1928
	Male	5–8	118	0.109	—	18.7	Salzer 1996
	Female	5–8	94	0.119	—	15.9	Salzer 1996
Rallus elegans	—	3	330	0.077	—	75.2	Meanley, in Ricklefs 1968, 1973
Crex crex	—	8–15	135	0.153	—	16.1	Salzer 1996
	Male	3	149	0.143	—	—	Schäffer, pers. comm.
	Female	3	118	0.150	—	—	Schäffer, pers. comm.
Porzana porzana	—	1	45	0.099	0.145	—	Heinroth and Heinroth 1928
	Male	10–22	76	0.210	—	17.3	Salzer 1996
	Female	10–22	76	0.209	—	18.1	Salzer 1996
Porzana carolina	—	1	67	0.149	—	—	Kaufmann 1987
	—	1	79	0.123	—	—	Kaufmann 1987
Fulica atra	—	1	500	0.174	—	—	Heinroth and Heinroth 1928
Fulica americana	—	—	515	0.041	—	—	Gullion 1954, in Ricklefs 1973

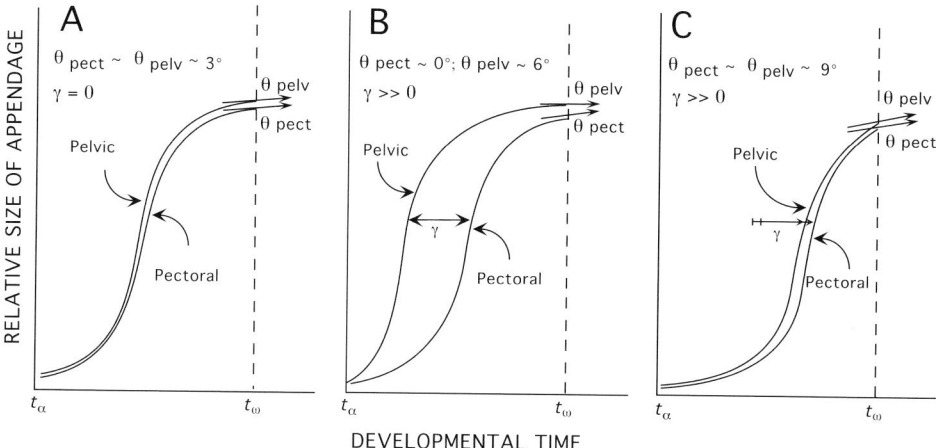

Fig. 123. Bivariate plot of relative pelvic and pectoral development against developmental time for three classes of phenotype: **A**, typical flighted rails; **B**, most flightless rails (e.g., *Porphyrio hochstetteri, Gallirallus owstoni,* and *Tribonyx mortierii*), in which variable evolutionary increases in body size are associated with comparatively protracted developmental periods; and **C**, minority of flightless rails (e.g., *Gallirallus australis*), characterized by large evolutionary increases in body size and only limited protraction of developmental periods. Slopes indicated represent estimates of the first partial derivative of the rate of growth of the apparatus in question (Tables 42, 43) at the terminus of growth, that is, $(\partial x/\partial t)(t_\omega)$. Symbols: t_α, time at hatching; t_ω, time at terminus of skeletal growth; θ_{Pect}, angle of pectoral allometry at t_ω; θ_{Pelv}, angle of pelvic allometry at t_ω; γ, offset parameter between components.

accelerated relative to that in other avian groups; constitutes a comparatively large portion of body mass at an early age; has proportions that decline with age in most species (O'Connor 1984; Starck and Ricklefs 1998c); and remains relatively robust in adults, this asymptotic pattern notwithstanding (Hartman 1961). The greater allometric slopes of pectoral elements (pectoral allomorphosis) and lower slopes for pelvic elements (pelvic allomorphosis) relative to general skeletal size within flightless rails (Tables 44, 45) are consistent with the interpretation of adults of flightless species as representative of a sample deriving from an earlier portion of the pectoral growth curve (Fig. 123). P.B. Taylor (1998) reinforced the perception that some flightless rails combined diminutive pectoral limbs with oversized pelvic musculature, an observation reaffirmed for some species in the present study (Fig. 68). The allegation that flightless rails tend to have disproportionately robust pelvic musculature is undocumented, at least compared to like-sized congeners, may simply be an artifact of comparison with a diminutive pectoral limb, and in any case may be confounded by differences in exercise (Weinstein et al. 1984). The combined effects of these varying contributions are manifested in a continuum of paedomorphic change in flightless rallids, attainable by diverse heterochronic means (Fig. 122) but usefully divisible a posteriori into three categories (Fig. 123).

A great majority of the osteological features of the pectoral apparatus that distinguish flightless rails from their flighted relatives are consistent with the hypothesis of underdevelopment or paedomorphosis (Tables 58–62; Figs. 61–67). One of the most widespread and conspicuous of anatomical characters indicative of flightlessness and characteristic of embryos of flighted carinates—obtuse angulae coracoscapulares (e.g., Olson 1973a; Cracraft 1974b)—were cited as early as Buller (1873a) for *Gallirallus australis*.

Despite substantial variation in the microstructure and pigmentation of feathers in modern birds that is at most partly related to flight capacity (Chandler 1916; Auber 1957), some qualitative differences in plumage pattern among rallids justify consideration with respect to avian ontogeny (Lillie 1942, 1952; Watterson 1942). The loose plumage characteristic of *Cabalus modestus* was noted by many of the earlier describers of the taxon, and misled some into concluding that the sample of *C. modestus* comprised a collection of juveniles of *Gallirallus dieffenbachii* (see Introduction). The definitive plumage pattern of *Porphyrula flavirostris* closely resembles that of the juvenal of *P. martinica* (Ripley 1977; Meyer de Schauensee and Phelps 1978; Taylor 1996), and is consistent with paedomorphosis of the plumage without pectoral consequences (Livezey 1998). Other, less obvious examples of paedomorphic plumages in rallids may include *Rallus wetmorei*; a thorough assessment of these possibilities must await substantially improved information on predefinitive plumages of most genera (e.g., Dickerman and Parkes 1969; Dickerman and Haverschmidt 1971). There also are inconclusive indications of delayed acquisition of definitive adult plumage (perhaps confounded by intersexual differences) in a few flighted species of *Porzana* (Livezey 1998).

Inferences of heterochrony in birds have not been limited to anatomical characters. Provine (1983, 1984, 1986) inferred a relationship between retention of flapping behavior with flight capacity and the antiquity of its evolutionary loss. Such behavioral changes can effect limited caloric economy in the absence of morphological changes, and are an example of prospective adaptation (Brock 2000). Herremans (1990:250) judged *Erythromachus leguati* to be neotenic with respect to its underdeveloped pectoral apparatus. He also concluded that flightless *Dryolimnas aldabranus* was neotenic as compared to flighted *D. cuvieri* in its "astonishing, 'infantile', inquisitive tameness." These and several other forms of nonanatomical heterochrony, for example, generalized hypermorphosis and early hatching of megapodes (Starck and Sutter 2000), require very different diagnostic indicators than required for relative underdevelopment of the pectoral apparatus. Rallids differ from raphids at least in the late attainment of locomotory capability and reliance on low-protein food resources (Livezey 1993b).

Available specimens and related data on body mass and development permit a particularly detailed comparative analysis of heterochrony in a flightlessness rail, *Porphyrio hochstetteri* (Fig. 124), one for which available data permit calibration of ontogenetic events with respect to time and thereby avoid the pitfalls of size-based estimates of developmental stage (Godfrey and Sutherland 1995). In common with other flightless rallids showing increased body size (e.g., *Gallirallus australis*), the most striking departures in growth shown by flightless *P. hochstetteri* from that for flighted *P. melanotus* are the substantially retarded growth of the pectoral apparatus (indexed by masses of mm. pectoralis et supracoracoideus) and a substantial protraction of the growth period (Fig. 124). The resultant augmentation of sexual size dimorphism (Tables 26, 57) in *P. hochstetteri* also is shown (Fig. 124). Multivariate sexual dimorphism in birds typically is initiated in early development (Lindén and Väisänen 1986; Rofstad 1986; McGillivray and Johnston 1987), and therefore can be altered through heterochrony (McNamara 1995). The steeper slope of pectoral allomorphosis in the flightless species (Tables

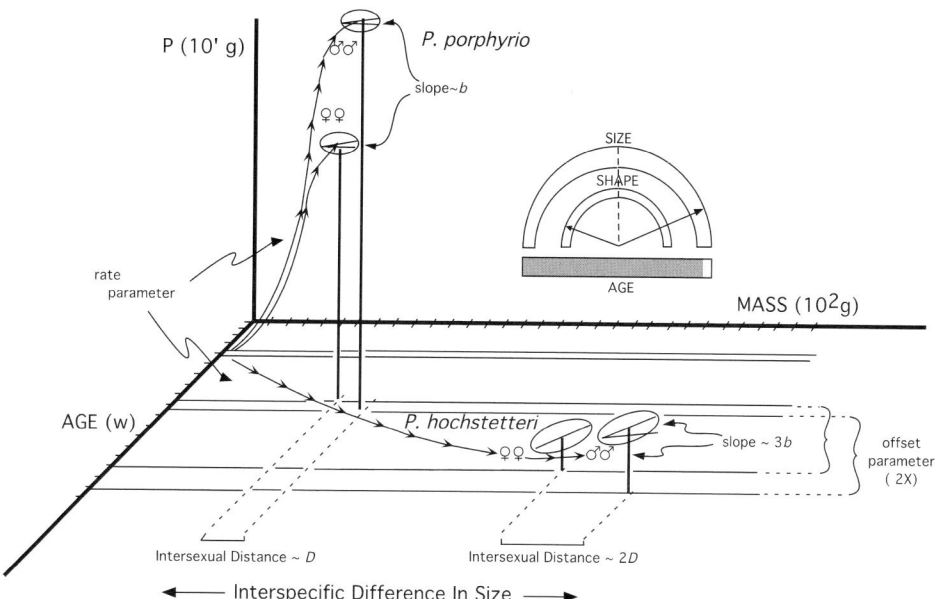

FIG. 124. Developmental trajectories for flighted and flightless swamphens (*Porphyrio*), sexes distinguished (intersexual dimorphism symbolized by *D*), spanning interval (weeks) from deposited egg to "fledging" in flighted *P. melanotus* and flightless *P. hochstetteri*. Three ontogenetic dimensions of "size" (body mass [g]), "shape" (mass of mm. pectoralis et supracoracoideus [g]), and developmental interval (weeks). Display after Alberch et al. (1979). Slopes of ellipsoids of terminal individuals based on allometry of pectoral elements relative to separate first principal components for nonpectoral elements for species–sex groups, after Livezey and Humphrey (1986: fig. 4) and Livezey (1989d: table 6), with additional parameters of "offset" and "rate" after Atchley (1987). Differences between the ontogenetic trajectories of *P. hochstetteri* to *P. melanotus* are reflected by a "clock diagram" corresponding to the apomorphic shifts in size, shape, and age (see text and Fig. 117), after Gould (1977).

44, 45) also is depicted, indicative of a comparatively juvenile position of *P. hochstetteri* with respect to pectoral development at the terminus of growth (Fig. 124). In toto, these data permit the implementation of the "clock model" of Gould (1977), in which age (at terminus of growth), shape (relative pectoral size), and overall size (body mass) of flightless *P. hochstetteri* is shown relative to the baseline position of the model calibrated for flighted *P. melanotus* (Fig. 124).

Pectoral growth axes and alar proportions.—Among flightless rallids, several tendencies pertaining to the pectoral integument, in addition changes in remiges, merit special consideration with respect to heterochrony. In most examples, several other unusual features emerged as typical: tectrices majores dorsales approaching in size that of the underlying remiges (Figs. 33–37), largely covering the latter (especially remiges secundarii) in dorsal view (e.g., *Tricholimnas sylvestris*); disproportionate elongation of the remiges alulae, in some species assuming the appearance of normal, cranially displaced remiges (e.g., *Nesoclopeus poecilopterus*); and retention of a disproportionately robust unguis alularis, especially conspicuous in flightless taxa having small body sizes (e.g., *Atlantisia rogersi* and *Porzana palmeri*). In light of the predominant pattern of pectoral truncation in flightless rallids (and other avian groups) in which distal skeletal elements undergo greater relative reductions than proximal segments (Fig. 42; Livezey 1989d, 1990, 1992a, b, 1993b, c, 1995a), these apomorphies of the overlying integument, on initial inspection, were not predictable or consistent with publicized ontogenetic generalities. Further study reveled that the first anomaly—that concerning relative lengths of the remiges and associated tectrices—provides evidence of paedomorphosis in that the development of secondary remiges and associated tectrices thereof is shifted later than that of primaries (Stresemann and Stresemann 1966; Panek and Majewski 1990), exposing the former to the actions of heterochrony for a longer period.

Further investigation revealed that all three of these characteristics, especially the relative robustness of the unguis and remiges of the alula, also are consistent with the early termination of developmental fields of birds and other tetrapods (Alberch and Alberch 1981; Hinchliffe and Gumpel-Pinot 1983; Alberch and Gale 1983, 1985; Hall 1998). Branching and segmentation of prechondrogenic condensations produce ontogenetic bifurcations in the pectoral limb, in which a primary proximodistal axis of development (humerus → ulna + radius → digital arch) undergoes a shift in orientation to a craniocaudal axis in the distal segment (i.e., the digital arch or the carpometacarpus and digits). This favors the initiation of earliest growth in the first (cranialmost) digit and sequentially through the remaining digits. That is, in the avian case, the axis is oriented as digitus alularis (I) → digitus majoris (II) → digitus minoris (III), somewhat complicated by the relatively early synostosis of elements within the carpometacarpus in modern birds (Prein 1914; Steiner 1922; Montagna 1945; Holmgren 1955; Hinchliffe 1977, 1989). This axial interruption of growth in the avian wing produces a disproportionate truncation of distal elements and a superimposed, secondary trend (progressively obvious distally); these lead to unusually well-developed cranial structures of the appendage (e.g., features of the alula vs. those of more-caudal digits, and tectrices vs. remiges), the very pattern observed among flightless rallids.

The graphical method of Cartesian transformation described by Thompson (1961) offers a means for intuitive display of ontogenetic and evolutionary change, and the method reflects uncannily the actual role of morphogenetic fields and included "catastrophes" in evolutionary change (Thom 1975, 1983). Use of the method of distorted coordinates described by Thompson (1961) in the present context would display a flightless rail (e.g., *Atlantisia rogersi* relative to *Laterallus jamaicensis*) on proximally displaced vertical coordinates (describing lateromedial contraction of the wing along the proximodistal axis) in the antebrachium and (to a slightly lesser degree) the manus, with an additional upwards (cranial) curvature of horizontal coordinates (pertaining to the craniocaudal distortion of the manus). Implications of such quantitative shifts along ontogenetic axes for the complete loss of fundamental mesodermal elements in the avian pectoral limb are not clear, although the truncations of distalmost skeletal elements in at least some ratites (Owen 1841, 1879; Cracraft 1974a) and the flightless gruiform *Aptornis* (Livezey 1994) are broadly consistent with the primary (proximodistal) axis of development. Extension of such congruence to assertions of universality of distal-to-proximal appendicular reductions in birds, analogous to Morse's law for digital reduction and its application to early avian evolution (Padian and Chiappe 1998), may not be warranted in light of the interrelated effects of multiple ontogenetic axes within the pectoral appendage of birds.

The relative prominence of ungues alulares in *Gallirallus australis* (Beauchamp 1998b) and *Atlantisia rogersi* (Fraser et al. 1992), detectable from early developmental stages, may have attracted attention out of proportion to their locomotory utility for nestlings, whereas the possible implications of these proportions for the relative effects of pectoral heterochrony on the integument and underlying skeletomuscular apparatus largely were ignored. A similar relative robustness of the digitus alularis characterized the flightless gruiform *Aptornis* (Livezey 1994) and may have contributed to the metacarpal armature of the raphid *Pezophaps solitaria* (Livezey 1993b). Such elaborations of digitus alularis were obsolete in flightless members of a number of other carinate families (Livezey and Humphrey 1986, 1992; Livezey 1989b, d, 1990, 1992a, b, 1993c).

Throughout the pectoral apparatus of flightless rails and other flightless neognathous birds is evidence for a strong ontogenetic integration of skeletal and overlying muscular elements. In many respects, the relative sizes of muscles (Figs. 68–110) exhibit magnitudes that varied in concert with the bones with which they were structurally and functionally associated. This musculoskeletal correspondence was especially marked in muscles with expansive areas of origio(nes)—for example, mm. pectoralis et supracoracoideus with respect to the corpus et carina sterni (Figs. 67, 69–74), and mm. interosseus dorsalis et ventralis with respect to the carpometacarpus (Figs. 106–110)—in which the bulks of muscle bellies appear to be simply a function of the associated bony elements. The probably artifactual nature of this correspondence, as opposed to the alternative of functional interpretation, is especially conspicuous in muscles having essentially no functional utility in flightless lineages—for example, mm. interossea, the sole action of which is to elevate and retract the digitus majoris relative to the carpometacarpus—but that fully occupy the spatium intermetacarpale that accommodates the muscle in all neognathous birds. Accordingly, the growth axes and other generalized ontogenetic parameters affecting size and proportions of the pectoral limb appear to affect both skeletal and muscular derivatives of the mesoderm constrained by strong covariances.

The importance of these developmental axes in relative robustness of the alula and its constituents in flightless carinates explains, in part, an error of myological diagnosis in an earlier description of the pectoral musculature of the flightless and pectorally paedomorphic Galapagos Cormorant (*Compsohalieus harrisi* [Livezey 1992a]). In this earlier work, two muscles of the digitus alularis were misidentified, in part because of their extraordinary robustness and relative positions on the digit of this flightless species. Specifically, the muscle labeled as m. extensor brevis alulae by Livezey (1992a: figs. 17, 20, appendix) corresponded instead to m. abductor alulae pars dorsalis. Pars ventralis of the latter muscle was labeled simply as m. abductor alulae in depictions of the ventral aspect of the manus (Livezey 1992a: figs. 18, 21), and pars dorsalis of this muscle was considered to be absent in this species. Although the position and subdivision of m. abductor alulae vary substantially even among flighted, nonpasseriform carinates (e.g., Fisher and Goodman 1955; Berger 1956a, 1966; Hudson and Lanzillotti 1964; George and Berger 1966; Hudson et al. 1969, 1972; Vanden Berge 1970; Zusi and Bentz 1978, 1984; Rosser 1980; McKitrick 1991b), the primary source of confusion arose from the exceptionally robust, uniquely fusiform and bitendinous m. extensor brevis alulae in flightless *C. harrisi*. In this flightless, foot-propelled diving species, this muscle presented an extraordinarily hypertrophied aspect, prompting it to be deemed unique to this species and of special functional utility in submarine locomotion (Livezey 1992a). However, ontogenetic considerations reveal that the disproportionate size of this otherwise minute muscle is likely attributable to the profound pectoral paedomorphosis substantiated by other characters in this species. Therefore, the name proposed for this muscle by Livezey (1992a)—m. levator alulae—is synonymous with m. extensor brevis alulae, as foreshadowed by the nomenclature by Vanden Berge and Zweers (1993:217).

Apparently, the distal reorientation of the developmental axis of the avian pectoral limb accounts for the generality that the antebrachium is disproportionately shorter than the more-distal manus in rallids and most flightless carinates (Fig. 42; Livezey 1989b, d, 1990, 1992a, b, 1993c), in that the proximodistal gradient affecting the proximal elements is retarded in the most-distal elements (carpometacarpus and digits) by this approximately 90° rotation of ontogenetic axes. The pectoral limb of the ratites evidently manifests much more pronounced effects of paedomorphosis (Livezey 1995a), in which growth of the pectoral skeleton trails greatly that of other anatomical regions and is terminated altogether immediately before (e.g., *Dromaius*) or shortly after (e.g., other modern ratites) synostosis of the carpometacarpal elements (Owen 1841; Parker 1892; Pycraft 1900; Steiner 1922; Lutz 1942; Lang 1956; Glutz von Blotzheim 1958).

PHYLOGENETIC IMPLICATIONS OF AVIAN FLIGHTLESSNESS

The form and structure of these characteristic parts [cranium and leg elements] in one of the genera (*Dinornis*) are so peculiar, that the author does not refer the genus to any known natural family of birds. Its location in the order *Struthionidæ* implies little more than an arrested development of wings, and an exaggerated development of legs, organized for progression on dry land.—Owen (1848a:8)

... As is well known, there are many types in all orders of Vertebrata which present us with rudimentary organs, as rudimental digits, feet or limbs, rudimental fins, teeth and wings. There is scarcely an organ or part which is not somewhere in a rudimental and more or less useless condition. The difficulty these cases present is, simply, whether they be persistent primitive conditions, to be regarded as ancestral types which have survived to the present time, or whether, on the other hand, they be results of a process of degeneration, and therefore of comparatively modern origin. The question, in brief, is, whether these creatures presenting these features be primitive ancestors or degenerate descendants.—Cope (1885:140–141)

Dr. Gadow [1893] has, I think, given the true explanation of the likeness of *Diaphorapteryx* and *Aphanapteryx* to one another, namely, that it is the result of parallelism of evolution .... The ancestors in the two cases, generalised rails capable of flight, were probably of different genera, or, at least, species. In the case of *Diaphorapteryx* this ancestor was most likely some widely spread form, such as *Hypotaenidia* [*Gallirallus*] *philippensis* is at the present day, individuals of which from time to time reached New Zealand, Lord Howe Island, and the Chatham Islands, the channels between which may formerly have been narrower than at present. The modified descendants of these birds are now referred to the genera *Diaphorapteryx, Cabalus,* and *Ocydromus,* the most highly modified forms being the outcome of earlier, the less altered of later colonisations.—Andrews (1896b:84)

... a sharp distinction should be drawn between birds which, except for the reduction of their wings and sternum, are true *Carinatæ* ... such as *Didus* and *Cnemiornis*, and those flightless forms included in the *Ratitæ*, which, leaving entirely out of account their wing-reduction, present numerous primitive characters which render it, at least, highly probable that they reached their present degenerate condition a very long while ago.—Andrews (1896c:260, in dissension with Milne-Edwards 1896)

In a similar way other evidence could be adduced tending to show that different groups or orders of birds present what is probably a primary constitutional facility, or the reverse, for the acquisition of flight in its full powers. And not only this, but one seems more and more drawn to the conclusion that such forms as *Atlantisia*, *Ocydromus* [*Gallirallus australis*], and *Cabalus* may be surviving examples—in fact, blind-alley representatives of a phyletic line which is distinct from the line I may call the modern present-day rail.

In a word, although none of us may believe in the polyphyletic origin of either birds or flight, it seems permissible to believe in polyphyletic evolution and to think that although the basal avian groups derived an equal potentiality to acquire flight out of the factors inherited from the primitive ancestor, all were not equally successful in its application.—Lowe (1928a:113)

... there are many instances of flightless birds, which have reverted to their original environment. There were probably ground or sea birds, however, before flight was lost. There is thus a reversal of the *direction* of evolution, but this does not actually contradict the law of irreversibility of evolution, which says that an organ once lost can never be regained and that a specialized form can never again become generalized.—Lull (1935:249)

... a number of flightless birds [have been] extinguished by colonization of a new environment by mammalian carnivores. Re-evolution of flight (contrary to Dollo's 'law') or a development of great body size and/or extreme leg length would appear to have been the paths of escape, but time was insufficient.—Mayo (1983:97)

EVOLUTIONARY DIRECTIONALITY AND AVIAN FLIGHTLESSNESS

Unidirectionality of change, orthogenesis, and flightlessness.—The paleontological literature of the 19th and early 20th centuries is replete with references to intrinsic tendencies of variously defined phylogenetic lineages to progress unidirectionally toward one or more macroevolutionary optima or goals, a finalistic march toward perfection rooted in the teleological perspectives of Lamarck (1809), Agassiz (1857), Nägeli (1865), and Eimer (1888), a concept widely referred to as orthogenesis (Simpson 1944, 1953a; Ghiselin 1980; Gould 1980c; Mayr 1980a, b; Grehan and Ainsworth 1985). Although the evolutionary loss of flight was not an attractive candidate for orthogenetic interpretation, apparent phyletic trends in body size (especially increase) frequently were the subject of such speculation, as were examples in which morphological features (e.g., size of body, antlers, or teeth) were averred to have orthogenetically exceeded selective advantage. In some instances, the concepts of orthogenesis and ultimate overspecialization were applied to several such trends or characteristics simultaneously (e.g., Schindewolf 1950), neither perspective having a legitimate basis in the realities of natural selection or related optimizations.

Dollo's law—the improbability of the reevolution of a complex feature (Gould 1970b; Brooks et al. 1988; Sanderson 1993)—is distinct from the related concept of irreversible evolution (Bull and Charnov 1985). Irreversibility is for the most part a methodological assumption, implemented a priori, that absolutely precludes reversal from an apomorphic state to one diagnosably indistinguishable from the antecedent or plesiomorphic condition, one not necessarily conditional on the inherent complexity of the trait involved. In essence, Dollo's law is an ad hoc rationalization for a perceived paucity of evidence for such evolutionary reacquisitions. This generalization was made in the absence of requisite phylogenetic reconstructions and was conditional on the vague concept of complexity, but persisted despite early insights into the genetics of reversals (e.g., Müller 1939). An example of the confusion between impossible and improbable evolutionary reversal and the attribution of exceptional explanatory power to such legalism was provided by Schindewolf (1950), as translated by Schaefer (1993:295, emphasis in original): "Unquestionable examples of a once-attained body size being secondarily reduced are almost unknown . . . . Accordingly, the evolution of size is, in general, *irreversible*." Such interpretations are not limited to morphological attributes; van Rhijn (1984, 1990) inferred a phylogenetically constrained unidirectionality of parental care and social organization in birds.

However, given the emphasis of Dollo's law on the improbability of the "reinvention" of complex structures, the apparent absence of a bona fide example of a return to the capacity for flight by a formerly flightless lineage is noteworthy (Mayo 1983). With the advent of a phylogenetic perspective on evolutionary phenomena (MacBeth 1980), Dollo's law can be formalized in terms of the practical impossibility of reevolving a complex structure (a special case of homoplasy). In the event that a well-supported, completely resolved, species-level phylogeny for the Rallidae is realized, the distribution of flight capacity among rails may offer a definitive test of Dollo's generalization in the face of newly recognized mediators (e.g., *Hox* genes).

Limitations of phylogenetic resolution and evidence of evolutionary rates of change aside, if it were to be inferred that no avian example of reversal of flightlessness had been found, it would remain to determine whether this were a reflection of ontogenetic unfeasibility, selective impossibility, or ecogeographical circumstance. Would such a revealed pattern imply that to regain the capacity of flight was simply a historical artifact of primarily human truncation of flightless lineages (i.e., a case of the "actual") or that such reversals were evolutionarily precluded (i.e., a reflection of the "possible"), sensu Jacob (1982)? Unfortunately, to distinguish between these two alternatives would require insights into ontogenetic and epigenetic programs and the evolutionary latitudes of such formational processes, a level of inferential sophistication heretofore not attained.

The notion of lineage sorting is especially appropriate in the case of flightlessness in the Rallidae, in which an independently acquired trait (flightlessness) is strongly associated with extinction in modern times (Mayo 1983:87, above). Such differential extirpation may qualify flightless rails as the casualties of "lineage sorting" (Vrba 1989; Gould 1990) effected by humans and their commensals, thereby denying such lineages the minimal number of generations required to recover the capacity for flight even if this were (onto)genetically possible. Nevertheless, it remains that there is no indisputable evidence of the reevolution of flight capacity by a flightless lineage among rails (Livezey 1998) or any other avian lineage (Livezey 1986a, c, 1991; Livezey and Humphrey 1986), thereby provisionally removing "reversals" from the factors likely to confound phylogenetic reconstructions in this group.

Functional integration, evolutionary stability, and flightlessness.—An early emphasis on the external factors conducive to evolutionary change subsequently was balanced by an appreciation of the potentially countering effects of the intrinsic complexity of organisms. The importance of architectural constraints acting on potential evolutionary change was formalized in recent decades, in part spawned by a reaction to the Panglossian view of adaptation (e.g., Gould and Lewontin 1979; Gould 1980a, b, 1997; but see Dennett 1995). In other words, the functional and selective integration of parts within structural complexes was recognized increasingly as a dynamic component of evolutionary change (Van der Klaauw 1948, 1951, 1952; Bock and von Wahlert 1965; Whyte 1965; Dullemeijer 1974, 1980, 1989; Riedl 1978; Wake et al. 1983; Hall 1996; Hansen and Martins 1996; Wagner and Altenberg 1996; Arthur 1997; Wagner et al. 1997).

This growing sense of evolutionary integration led to the discrimination of both occupied (Thomas and Reif 1993) and unoccupied morphometric space (Bookstein et al. 1985; Schindle 1990) and the notion that the number of realized body forms is limited (Gans 1979; Mayo 1983). This perception is a pragmatic parallel to the duality of the "actual" being a subset of the diversity conceived to be "possible" in the natural world (Jacob 1982; Arthur 1997). A more pragmatic and phylogenetic perspective identifies the "actual" as the subset of functionally possible, organismal solutions that were realized by a genetically integrated hierarchy of history (Brooks and Wiley 1988; Lauder 1990). That is, the asymptotic distributions of character states observed in natural lineages delimit the variation realized within a larger but finite range of evolutionary possibilities (Wake 1991; Foote 1994; Wagner 2000). In the morphometric context, the complement of the total morphospace defined by components having significant eigenvalues corresponds to the morphometric space that proved inaccessible to the known diversity of taxa, and descriptive of morphotypes structurally unfeasible, ontogenetically unattainable, or fatally disadvantaged under actual circumstances (Reyment 1980; Reyment et al. 1984). In the context of avian flightlessness, this perspective underscored the limited combinations of evolutionary circumstances and complex integration of parts that can lead to flight, improbable preconditions that potentially expand the set of successful evolutionary pathways to the surrender of this capacity.

This realization has led to the notion that functionally interrelated features can impose mutual selective constraints and evolutionary conservatism; such a complex of characters was termed an "evolutionarily stable configuration" (ESC) by Wagner and Schwenk (2000). This concept bears similarities to a model by Arthur (1988), in which ensembles of states that share high ontogenetic interattainability require a macromutational change to open a "morphological window" that permits transformations of states in another such morphogenetic set. The concept of an ESC may have special relevance to avian flightlessness, at least in the sense that the phenomenon qualifies as the *dis*-integration of the highly integrated pectoral ESC that makes avian flight possible. This perspective may explain in large part the conservative rate of reduction of pectoral structures in many flightless rallids, in that selection favoring the retention of any of the components of the pectoral ESC (e.g., pectoral muscles for steaming and combat in *Tachyeres,* and remiges for brooding and rectrices for display in rallids) may retard reductive evolution in other parts of this ESC. A view of the avian pectoral apparatus as an ESC also is consistent with the nonindependence of phylogenetic changes in components of flightless lineages (Livezey 1998). The small number of evolutionary pathways evidently leading to avian flightlessness is indicative of the functional and selective integration of the affected parts, one aspect of design limitations that contributes to the likelihood of flightlessness-related homoplasy (Wake 1991; Emerson and Hastings 1998).

FLIGHTLESSNESS AS SOURCE OF HOMOPLASY

Avian flightlessness and global homoplasy.—The confusion of plesiomorphy with paedomorphic apomorphy, in part a legacy of the biogenetic aphorism attributed to Haeckel (1866) and uncritically extended to others (e.g., von Baer 1828; Agassiz 1849), evidently induced Lowe (1926, 1928a, b, 1930, 1933, 1934, 1935, 1939, 1942, 1944) to interpret flightlessness-related characters of many avian groups as "primary" or evolutionary recapitulations, with the distinction between the two classes not always being clear (Livezey 1995a). Early interpretations of the flightlessness and biogeography of modern ratites had a broader, if less obvious, impact on the ornithological literature pertaining to neoteny and convergence (see below), as well as the utility of functional characters in phylogenetic inference (e.g., Mayr 1969, 1976).

The endeavors of ornithological systematists have been judged to have been plagued by pervasive anatomical homoplasy since decades before the advent of modern (cladistic) systematic techniques, an allegation stemming both from the morphological limitations purportedly imposed on flighted species by the functional requirements of flight (e.g., Colbert 1955; King and King 1979; Sibley and Ahlquist 1990, Feduccia 1995), and by the morphological similarities averred to have accompanied independent evolutionary losses of flight (e.g., Olson 1973a, 1999; Ripley 1977; Feduccia 1980, 1995). That is, birds were held to be especially vulnerable to misleading anatomical similarities related to flight, whether the faculty was retained or lost. Unfortunately, the coincidence of unique plesiomorphies with (variably convergent) flightlessness-related apomorphies can pose fundamental problems of diagnosis and phylogenetic reconstruction among comparatively closely related species (e.g., Livezey 1986c, 1998). Confirmation of the influential potential of characters associated with flightlessness for the phylogenetic reconstruction of the Rallidae (Livezey 1998) accorded with earlier expressions of concern (e.g., Olson 1973a) and the generality of such deceptive similarity for "syn-vestigiational traits" (Brock 2000).

Flightlessness in rails: uniquely rampant parallelism.—With the exception of the purported monophyly of three, flightless species in the genus "*Atlantisia*," proposed by Olson (1973b) but rejected by Livezey (1998), monophyly of any group of flightless rails has not been established unequivocally to date, polytypic, flightless genera such as *Nesotrochis* and *Pareudiastes* notwithstanding (but see below). The comparatively high frequency of dwarfism in the plesiomorphically small flightless *Porzana* is unlikely to represent mere serendipity, but convergence among these congeners, fostered by shared ecological circumstances and phylogenetic commonalities of ontogeny, is as plausible as genuine synapomorphic loss of flight and subsequent dispersal among islands by means other than flight. Given these likelihoods and precluding that unassailable support for monophyly of a flightless rallid clade is forthcoming, the multiple instances of flightlessness in the Rallidae qualify as a uniquely rich instance of parallelism involving a critical functionality, born of a common, if bifurcated, ontogenetic pathway.

This phenomenon of "rampant parallelism" of flightlessness is distinct in frequency and ontogenetic details from the convergent losses of flight in of most or all other neognathous taxa. Despite demonstration of flightlessness-related homoplasy in the Rallidae (Livezey 1998), even a cursory examination of the anatomy of flightlessness in other avian families reveals that the anatomical ramifications differ substantially among higher taxonomic groups (Livezey 1988, 1989a–d, 1990, 1992a, b, 1993a, b, 1994), a generality especially obvious in comparisons of the anatomy of flightless paleognathous and neognathous birds (Lutz 1942; Lang 1956; Cracraft 1974a; Livezey 1995a; Lee et al. 1997; Abourachid and Renous 2000). Accordingly, a nomenclature that reflects these realities is advisable because it is useful to refer to the anatomically similar, ontogenetically comparable flightlessness within the Rallidae as *parallelism*; whereas *convergence* should be reserved for the diagnosibly distinct shifts responsible for flightlessness in other avian families sharing broadly comparable pectoral function (e.g., ratites, anatids, grebes, ibises, cormorants, raphids, and parrots; Table 4); and *analogy* is used for those losing flight through unique pectoral specialization (e.g., penguins and alcids; Table 4).

This phenomenon of "rampant parallelism" pertaining to rallids is doubly remarkable in light of the diversity of forms and inferred changes of flightless taxa (Figs. 15, 20, 22–31, 42–44, 49–58, 111–115), the diversity of insular habitats occupied (Figs. 6, 116; see below), and the relative simplicity of the fundamental and malleable ontogenetic mechanism that generates this spectrum of functionally similar, morphological diverse changes (Figs. 117, 121–124). Accomplishment of a well-resolved, rigorously supported, species-level phylogeny will allow an assessment of the imprint of history on present-day distributional patterns and "structural" aspects of sympatric populations, and permit a conclusive test of the assumption that all instances of flightlessness in the Rallidae, otherwise rare for Aves, were truly independent events.

CLADOGENETIC COROLLARIES: IS THERE SPECIATION AFTER FLIGHTLESSNESS?

> As will be seen . . . , the characters of this group [ratites] are to a considerable extent negative characters. They are for the most part such characters as are correlated with the loss of the power of flight. We need not, therefore, lay too much stress upon them as indicative of the naturalness of the group.—Beddard (1898:493)

> But flightlessness seems to confer no special favors upon the possessors, and flightless birds seem to have traveled into an evolutionary box canyon compared to their flying relatives. . . . Flightlessness may arise in almost any group where the inability to fly is not a current disadvantage.—Wing (1956:162)

Paedomorphs and ancestry.—In a section intriguingly entitled "Paedomorphs: Not Necessarily the Best Ancestors," McKinney and McNamara (1991:345) considered the relative likelihood of the persistence and cladogenesis of paedomorphic lineages in nature. In other words, are the short-term gains realized by heterochrony necessarily in the long-term evolutionary interest (sensu subsequent phylogenetic diversification) of the lineage so modified? In extreme cases of paedomorphosis, the reductions in structural faculties seem likely to limit the capacity of the lineage to accommodate substantial environmental changes (e.g., blind, apigmentary, cavernicolous salamanders), but evolutionary arguments turning on hypothetical changes and potentials are problematic (McKinney and McNamara 1991). The characterization of flightless avian lineages as "terminal" by Gould (1966) echoed the general view of flightlessness as limiting in the evolutionary sense, and that species so afflicted, and confronted with predaceous immigrants or the catastrophes likely at geological time-scales, probably would not undergo speciation. Alternatively, the propensity for cladogenesis appears to be a heritable character of some taxonomic groups (Savolainen et al. 2002), perhaps enhanced by differentiation among niches within demes (Bolnick et al. 2003), and the former diversity of flightless Rallidae (*Gallirallus* and *Porzana*) suggests that the most widespread genera may have been endowed accordingly.

Ratites and phylogenetic perspectives on flightlessness.—The prevailing view that each instance of avian flightlessness was acquired independently and therefore convergent had its genesis in the first century of study of the modern ratites—the large flightless, paleognathous birds of the Southern Hemisphere (e.g., ostrich, rheas, emu, cassowaries, kiwis, and moas). Beginning with Merrem (1813), early systematists separated the ratites from other birds on the basis, at least in part, of the possession of acarinate sterna (e.g., Lesson 1831; Huxley 1867). Fürbringer (1888) was the first to challenge this traditional classification, and he argued explicitly that the similarities shared by ratites were the products of convergence and that each ratite lineage evolved independently from ancestors as yet unknown. This proposal of polyphyly was questioned by Gadow (1893), Newton (1896), and Beddard (1898), whereas other workers found the hypothesis persuasive (e.g., Pycraft 1900; Wetmore 1930, 1951, 1960b; McDowell 1948; Mayr and Amadon 1951; Storer 1960b, 1971b; de Beer 1964; Falla 1964; Moreau 1966).

Lowe (1926, 1928b, 1930, 1935, 1942) drew attention to the possible importance of ontogeny and phylogeny in the evolution of ratites, but his often recapitulationist views received a cool reception (Livezey 1995a). As noted by McDowell (1948:522) in a review of the palate of ratites: "Percy Lowe has advanced a theory that ... the ratites diverged from the avian stem at a time when the power of flight had not yet been attained but the fore-limbs had become rather wing-like. This theory would necessitate believing that birds had at some stage of evolution sacrificed the use of their anterior limbs in order to acquire flight at a considerably later date, a teleology not acceptable to post-Lamarckian students of evolution." McDowell (1948) linked the hypothesis of polyphyly of ratites supported by most of his contemporaries with an inference, credited (p. 543) to E. Mayr, that the pectoral changes of ratites were the result of "neoteny." The hypothesis of polyphyly was considered by McDowell (1948) to be consistent with the palatal diversity he noted in the ratites, the latter reflected by the heading (p. 536) "The Impossibility of Defining the Palaeognathous Palate." McDowell (1948) concluded that each major ratite lineage (i.e., tinamiform, casuariiform, apterygiform, and struthioniform lines) evolved both their palatal anatomies and those changes related to flightlessness independently. Although a secondary role of "adaptation" in these convergent characters was entertained by McDowell (1948), he envisioned "neoteny" as playing a critical role in avian flightlessness, not only in the pectoral reduction(s) of ratites, but also in those of flightless carinates (e.g., Spheniscidae, *Raphus cucullatus,* and *Strigops habroptilus*).

In addition to virtually equating pectoral reduction in flightless birds with paedomorphosis, McDowell (1948:540–541) extended the role of "neoteny" to other organ systems and taxa:

> There are numerous examples in zoology of neoteny having worked morphological reversals. Examples are the Sphenisci [penguins], where neoteny has produced a less fused, and therefore more reptilian, tarsometatarsus (*see* Simpson, 1946), . . . neoteny is usually not limited to a specific organ. Thus, neotenic persistence of gills in the 'perennibranchs' is accompanied by neotenic skull-characters and muscular characters, and neoteny in the tarsometatarsus of the Sphenisci is accompanied by late obliteration of the cranial sutures. We may, therefore, expect that the neoteny of the wings of ratites might affect the palate. . . . Neoteny might easily explain such a palatal characteristic as the imperfect fusion of the halves of the prevomer. . . . neoteny could also explain the pterygoid–prevomer contact, the heart of Pycraft's definition of the 'Palaeognathae.'

Romer (1968), after considering other recent discussions of the matter (Hofer 1955; Simonetta 1960; Müller 1961, 1963; Bock 1963; Meise 1963), deemed the dismissal of the similarities among ratites by McDowell (1948) an unjustified defense of the prevalent hypothesis of polyphyly. Decades later, the origins of moas from flightless anseriforms and of kiwis from flightless ibises were suggested (Olson 1983; but see Livezey 1995a), thereby providing yet another variant on the polyphyletic phylogenesis of ratites from among neognathous birds through neoteny. Similarly, anatomically naive appeals to heterochrony were made in support of preliminary molecular evidence presented by Härlid and Arnason (1998), which suggested the evolutionary derivation of the Ostrich (exclusive of other ratites) from within neognathous birds, a radical inference contradicted by contemporary analyses of DNA sequence data (Härlid et al. 1997, 1998; Tuinen et al. 1998, 2000) but indicative of a residual predisposition for the traditional view of ratites as polyphyletic.

Much support of the polyphyly of ratites originated before the acceptance of continental drift (Vuilleumier 1975), considered by Serventy (1960:101–102) to be hypothetical and ". . . unnecessary to account for recent and extinct bird and mammal distributions." Lingering resistance to the naturalness of the ratites turned on a constellation of arguments involving anatomical characters not related to flightlessness (e.g., the palate), confusion regarding phylogenetic methodology (e.g., Houde and Olson 1981; Feduccia 1985; Houde 1988), or premature inferences based on preliminary sequence data (Härlid and Arnason 1998). Mayr (1976:465) joined those advocating the likely homoplasy of characters averred by some (e.g., Cracraft 1972, 1974a) to be supportive of the monophyly of ratites, stating: "If several groups of running birds lose their power of flight independently, acquire large size, and specialize entirely in running, one would expect them to acquire the ratite complex of characters even if the stem species of this assemblage did not have these characters." The course of this debate was paralleled by the odyssey of an individual systematist, Walter Bock, who passed from support for monophyly (Bock 1963, 1976), to advocacy of polyphyly (Bock and Bühler 1988), and finally to an admission of indecision (Bock 1992).

A growing consensus among avian phylogeneticists places the ratites as monophyletic and the sister group of other modern (neognathous) birds (Cracraft 1974a, 1986, 1988b; Cracraft and Mindell 1989; Lee et al. 1997; Tuinen et al. 1998, 2000; Groth and Barrowclough 1999; Haddrath and Baker 2001), obviating speculations concerning the misdirection of reconstructions of higher-order avian phylogeny by homoplasy generated by convergent ontogenetic changes. Also, the minority view that the flightlessness of ratites is primitive (Chandler 1916; Lowe 1928b, 1935; McGowan 1979, 1982, 1986) has been countered by phylogenetic reconstructions including nodes basal to the Neornithes (e.g., Cracraft 1986, 1988b; Chiappe 1995a; Padian and Chiappe 1998). Nonetheless, the hypothesis that flightlessness among the ratites was acquired convergently would color for decades subsequent perspectives on the evolutionary ease with which flightlessness can arise in avian lineages and thereby obfuscate phylogenetic inferences (Livezey 1995a), especially where such assessments were made in the absence of explicit phylogenetic hypotheses and primitive calibrations of absolute time (e.g., Feduccia 1980, 1996; Olson 1983).

Speciation involving flightless neognathous birds.—A counterexample to the traditional view of avian flightlessness as a phylogenetic dead end is provided by the flightless steamer-ducks (Anatidae: *Tachyeres*), a genus comprising four species having a predominately continental distribution for which there is considerable support for monophyly of the three flightless members (Livezey 1986c; Corbin et al. 1988). If this phylogenetic reconstruction is correct, the hypothesis confirms the intuitive plausibility of avian cladogenesis following flightlessness, that is, that flightlessness can be plesiomorphic at least for clades of neognathous species having largely continental distributions (Livezey and Humphrey 1992). However, some examples among other flightless neognaths can be inferred to have arisen independently from flighted ancestors even in the absence of a robust phylogenetic hypothesis for the groups, because each is the sole flightless member of its taxonomic genus or family (e.g., *Rollandia microptera, Podilymbus gigas, Podiceps taczanowskii, Anas aucklandica*-group, *Mergus australis*-group, *Compsohalieus harrisi,* and *Strigops habroptilus* [Livezey 1989b, d, 1990, 1992a, b]), and each evidently evolved either on oceanic islands (most species) or under quasi-insular conditions of high-altitude lakes (Podicipedidae). A small number of neognath families include more than one flightless member for which phylogenetic hypotheses are essential for confident reconstructions of speciation patterns, including the Alcidae (Livezey 1988) and Raphidae (Livezey 1993b). The remaining instances of flightlessness among neognathous birds are those in which entire taxonomic families (or orders) are (were) flightless (e.g., Diatrymidae, Aptornithidae, Phorhusrhacidae [Table 4]). Also among the latter are the penguins (Spheniscidae), in which all known fossil and modern species are (were) flightless members and that approach the flightless Rallidae in number of species (Livezey 1989c). The problem in the latter cases of clade-wide flightlessness is reduced to the reconstruction of events leading to flightlessness in the flighted ancestors of each group (Livezey 1994, 1995a, 1998).

Despite the comparatively great frequency of flightlessness in the Rallidae, reconstructions of this kind remain speculative without knowing the exact sequence of the loss of flight and speciation events (Livezey 1998). Speciation in the Rallidae appears to conform with the broad profile for the traditional allopatric mode (Bush 1975), although likely scenarios in some instances seem to contradict traditional assumptions of the geographical scales typical for birds (Diamond 1977a). Likely cases of vicariance events initiating speciation of flightless rails include apparent sister taxa in regions having undergone coalescence or substantially reduced separation during glacial periods of low sea levels (Cooper 1989; Craw 1988; Livezey 1998; Whittaker 1998; Craw et al. 1999): *Porphyrio mantelli* and *P. hochstetteri*; *Gallirallus australis* and *G. greyi*; and *Gallinula nesiotis* and *G. comeri*. Of these, the first two couplets (*P. mantelli*-group and *G. australis*-group) may qualify as avian examples of continental speciation (sensu Diamond 1977a) on the mega-island of New Zealand. Geological events of greater antiquity (e.g., movements and coalescence of tectonic terranes in New Zealand [Craw 1988; Cooper 1989; Humphries and Parenti 1999]) probably are germane only to deeper nodes in the Rallidae, for example, relationships between endemics of the Chatham Islands (Andrews 1896b, c; Craw 1988; Livezey 1998).

Speciation or radiations of flightless rails within single archipelagoes—"intra-archipelagal speciation" of Diamond (1977a)—is likely for at least some of the *Porzana* of Hawaii (Table 3) or the *Gallinula nesiotis*-group of the Tristan da Cunha Islands. However, as with all taxonomic groups possessed of low dispersal capacities, cladogenesis requiring dispersal among islands would be limited to comparatively short distances for flightless landbirds (Selmi and Boulinier 2001). Speciation of flightless rails within the Chatham Islands remains a distant possibility, and the relegation of rails endemic to the Chathams to subspecific status (e.g., *Gallirallus philippensis dieffenbachii* [Ripley 1977]), in fact, may reflect a reluctance by many taxonomists to recognize diversification of flightless avian clades within limited geographical areas. Assumptions of this kind seem especially tenuous in light of the variation within single islands shown by some flighted passerines (e.g., *Zosterops borbonicus* complex) in the Indian Ocean (Gill 1973).

The majority of speciation events among the Rallidae that ultimately evolved flightlessness likely involved colonization of islands by continental species, with a few being attributable to the "insular speciation" of Diamond (1977a); the latter is speciation via dispersal and allopatric speciation between two archipelagoes. Given the importance of flotsam transport (Heatwole and Levins 1972) and the successful colonization of oceanic islands by tortoises, lizards, and rodents (Carlquist 1965, 1966, 1974, 1980; M. Williamson 1981; Grant 1998a–c; Whittaker 1998), it is difficult to dismiss out of hand the dispersal among islands, followed by differentiation, of flightless rallids. Moreover, asynchronous temporal variation in resources may select for dispersal dimorphism, in which a single species may comprise two or more dispersal phenotypes conducive to speciation (Mathias et al. 2001). Although not documented in birds, such polymorphism in dispersal would appear especially plausible for rallids having undergone widespread insular radiations.

Given the precarious or extinct status of many flightless rails, data bearing on colonization by rafting in flightless rails may be based optimally using other ecologically similar tetrapods. For example, shrews (Mammalia: Insectivora, Soricidae) may offer some insights into the role of such colonizations in island groups (Peltonen and Hanski 1991). Diversification within flightless lineages of rails after dispersal by rafting would be especially likely among elements in isolated island groups (e.g., the Hawaiian Islands), outlying satellites of major island groups (e.g., Chatham Islands, Norfolk Island, Lord Howe Island, and New Caledonia), and between members of innumerable island groups of the South Pacific (M. Williamson 1981; Grant 1998a–c; Whittaker 1998), perhaps by stepping-stone dispersal. Candidates for dispersal and speciation after flightlessness include *Porphyrio albus* (cf. *P. mantelli*-group) and possibly the unnamed, flightless *Porphyrio* of New Ireland (Steadman et al. 1999); a complex of endemic taxa of New Zealand and the Chatham Islands (*Gallirallus australis, G. dieffenbachii, Cabalus modestus, Capellirallus karamu,* and *Diaphorapteryx hawkinsi*), perhaps confounded by glacial coalescence among islands followed by differential, anthropogenic extinctions; the complex of *Porzana* that formerly inhabited the Hawaiian Islands, in which rafting of flightless forms would be at least as likely as multiple invasions by flighted congeners from distant sources; and the *Pareudiastes* moorhens, endemic to islands in the Solomon and Samoan groups (Olson 1975a; Ripley 1977), between which are interposed a number of potential stepping-stone islands (Gilpin 1980).

GEOGRAPHICAL CONTEXTS OF FLIGHTLESSNESS IN RAILS

This curious form [in reference to *Cabalus modestus*] must not, however, be regarded as a change produced by long isolation, but rather as an old form preserved from destruction by isolation.—Hutton (1872c:238)

Considering the restricted powers of locomotion of the several genera above cited [*Notornis* = *Porphyrio, Ocydromus* = *Gallirallus* and *Tricholimnas*], it may be inferred that the lands yielding examples of such flightless birds were not, in their primitive days, separated by such breadths of ocean as that which divides the South Island from Lord Howe's Island, or as that known as 'Cook's Straits'.—Owen (1882: 695)

Thus as regards flightless rails there would appear to be two categories, a primarily flightless and a secondarily flightless category. The first would appear to have been preserved in such insular extensions of continental land-masses as 'Lemuria' and 'Antipodea,' because there were no carnivorous foes to extinguish them, such as subsequently extinguished their flightless progenitors on the main continental landmasses; the second category we suggest was derived from volant ancestors, and found their way to such oceanic islands as are represented in the Polynesian groups. Here also they found no carnivorous foes—or, at any rate, no carnivorous mammals. Hence, they, too, have been preserved; but being originally volant they are still potentially volant, and have nearly lost the power to fly by default, having retained all their original volant armature. It may also be that they belong to a more recent geological period.—Lowe (1928a:113)

... flightlessness, comparable in skeletal and feather structure to that exhibited by many breeds of domestic fowl, has arisen among island birds of several orders where the absence of enemies lifts the penalty from such a trend ....—Murphy (1938: 537)

... the occurrence of an animal on an isolated island presupposes a lack of enemies and a reduction of the danger from epidemics. Such a habitat would therefore guarantee undisturbed evolution and favor the development of large size ... the gigantic birds of New Zealand (Dinornithidae, Moas) and Madagascar (Aepyornithidae) .... The absence of hostile predators, of course, is not the cause of size increase or any other kind of overspecialization and degeneration ... lack of predators on oceanic islands would only explain why large types have been preserved and could grow larger there and not elsewhere, which is not consistently the case anyway.—Schindewolf (1950; translated by Schaefer 1993:305–306)

On oceanic islands, evolution is faster than immigration. On continental islands . . . immigration is faster than evolution.—M. Williamson (1981:167)

. . . the evolution of flightlessness would be most likely to occur in species that face the greatest restriction in resources, so it would be most common on smaller islands (i.e., with the small resources) and in those species with the greatest resource requirements (i.e., in larger endotherms).—McNab (1994a:640)

BIOGEOGRAPHICAL REALMS: PATTERNS AT A GLOBAL SCALE

Regions of special biogeographical relevance for flightless rails include New Zealand (Williams 1962, 1973; Cracraft 1975, 1980; Salmon 1975; Beauchamp 1989), other South Pacific islands (Smithers and Disney 1969; Rich 1981; Rich et al. 1983; Schodde et al. 1983; Richardson 1984; Pauley 1990; Adler 1992; Goodman et al. 1995), Australia (Ridpath and Moreau 1966; Horton 1972; Rich 1975), Hawaii (Perkins 1913; Mayr 1943; Pratt 1979), islands of the (especially southern) Atlantic (Rand 1955; Pieper 1985; Ashmole and Ashmole 1997), and islands in the Indian Ocean, including the Mascarenes (Strickland and Melville 1848; Adler 1994). The family Rallidae possesses one of the widest global distributions of any terrestrial group of tetrapods (Olson 1973a; Livezey 1998), and although this suggests that the group is amenable to large-scale analysis (e.g., Craw et al. 1999) or more traditional assessments by geographic realms (e.g., Wallace's line [Mayr 1944]), advances will require a fully resolved phylogeny.

Nevertheless, with respect to distributions of clades within the Rallidae and allies, several patterns are evident (Fig. 11; Livezey 1998): Southern-Hemispheric distributions characterize a preponderance of the gruiform families related to the Rallidae (e.g., Cariamidae, Eurypygidae, Rhynochetidae, and Aptornithidae), and largely Neotropical or equatorial ranges characterize others (e.g., Psophiidae, Aramidae, basal Gruidae, and Heliornithidae); Southern-Hemispheric origins extend also to most basal flighted lineages within the Rallidae (e.g., *Himantornis, Gymnocrex, Eulabeornis, Canirallus, Anurolimnas, Rallicula,* and *Sarothrura*); and predominately southern affinities are shown by a majority of genera, in which only a minority of members (typically of included crown-groups) are distributed in the Northern Hemisphere (at least seasonally), a number of which include one or more flightless members (e.g., *Porphyrio, Porphyrula, Rougetius, Rallina, Rallus, Gallirallus, Laterallus, Crex, Coturnicops, Porzana, Amaurornis, Gallircrex, Gallinula,* and *Fulica*). These broad, largely phylogenetically congruent patterns effectively define limits within which distributional variation of a local scale (essentially within-clade or error variance) is expressed, much of which is not rigorously testable against current phylogenetic reconstructions, a prerequisite for alternatives for the intuitive proposals of past decades (e.g., Larson 1957; Vuilleumier 1970, 1973, 1975; Beauchamp 1989). Only provisional biogeographic explorations are possible for such focal groups as *Gallirallus* and *Porzana* (Figs. 12, 13). Several clades of generic or higher rank retain obvious geographical conguence and include one or more flightless members (Livezey 1998)—*Gymnocrex* + *Habropteryx, Aramides* + *Nesotrochis, Ortygonax* + *Cyanolimnas, Amaurornis,* and *Tribonyx*—and merely document the diversity of patterns.

The biogeographic patterns of flightless rails are principally a phenomenon of insular immigration and endemism (Fig. 6), with larger-scale inferences limited to the comparatively high incidence of flightlessness (perhaps indicative of ontogenetic propensity) in some groups (e.g., *Gallirallus* and *Porzana*). The essential importance of insular biogeography to the evolution of flightlessness among the species of Rallidae is magnified further by former presence of long-extinct flightless rails on additional islands (e.g., Steadman 1989a, b, 1993a, b, 1995, 1997a, b; Worthy et al. 1999), the limiting factor of which appears to be the preservational qualities of sand dunes, caves, or lava tubes in these localities as opposed to those of the rails themselves (e.g., Olson 1978, 1991; Carr 1980; Steadman 1986b, Schubel and Steadman 1989). Emergent patterns are that most, if not essentially all, habitable oceanic islands at one time supported population(s) of flightless rails, most of which became extinct before ornithological inventory; and no species of Rallidae would have evolved flightlessness under circumstances exclusive of oceanic islands, the inadequately known distribution and status of *Tribonyx repertus* notwithstanding.

Of the primary patterns and predictions of insularity (Table 81), most are applicable to a substantial degree to the evolutionary shifts inferred for flightless rails (Tables 7–9, 11, 13–26, 30–35, 39–42, 46–57, 68–70; Figs. 15, 20, 22–32, 42–44, 49–60, 111–115; Appendices 1, 6, 7). However, some changes were not predictable even broadly with respect to quantitative or qualitative changes, for example, those evident in bills both interspecifically and intersexually (Figs. 45, 48). Still other comparisons revealed a diversity of changes (Tables 69–74; Figs. 121–123) that defied simple application of previously formulated rules of insular change (Table 81).

Simple biogeographic opportunity leads predictably to the preponderance of flightless species being from Pacific Oceania, because the number of islands representing the number of potential evolutionary "incubators" for flightless rallids is maximal (Fig. 6) and coincides with the distributional pool of several genera characterized by great vagility (*Gallirallus* and *Porzana*). Also, there is a possibility that tropical islands host(ed) more flightless species by means of an amelioration of population bottlenecks than occurred at higher latitudes (cf. Elton 1946; Janzen 1967; Miller 1967, 1969; Vermeij 1974; Arthur 1982b; DuBowy 1988; but see Nudds 1982), thereby fostering more founder-effect speciation events (Templeton 1980; Charlesworth and Smith 1982; Nei et al. 1983; Barton and Charlesworth 1984; Barton 1989, 1998; Provine 1989; Willis and Orr 1993; Gavrilets and Hastings 1996; Slatkin 1996; Charlesworth 1997; Grant 1998a), even in the absence of concomitant genetic drift (Whitlock 1995, 1997). On the other hand, in tectonically active regions, oceanic islands may undergo frequent episodes of low population size resulting from tsunamis and volcanism (MacArthur and Wilson 1967; Nunn 1994), moderate bottlenecks that may hasten genetic drift through reduced genetic diversity (Barton 1998).

Among the few higher-order biogeographical patterns involving flightless rallids are the unsurpassed diversity of sympatric, flightless *Porzana* (Olson and James 1991) in the Hawaiian Islands (Tables 2, 3; Fig. 10), and an almost complementary diversity of flightless rallids in several clades other than *Porzana* in the New Zealand region (Table 2). The latter, in which the Chatham Islands formerly supported a remarkable diversity of endemic flightless rails relative to

TABLE 81. General predictions for insular avifaunas (with selected references) and applicability to flightless Rallidae.

Characteristic	General predictions for islands and birds	Predictions confirmed for insular rails	Selected references
Colonization and distance to source populations	Rates decrease with distance, increasing risk of extinction	Broad agreements obscured by vagility, human impacts	Gressitt 1954, 1956, 1961, 1964; Hamilton and Rubinoff 1963, 1964, 1967; Hamilton et al. 1963; MacArthur and Wilson 1967; Thornton 1967; Haila et al. 1979; Gilpin 1980; Hnatiuk 1980; Kadmon and Pulliam 1993; Martin et al. 1995; Wiggens and Møller 1997
Colonization and area of islands	Rates increase with area, decreasing risk of extinction	Moderate congruence in New Zealand region, but some areas (e.g., Hawaii) outliers in diversity	Hamilton and Rubinoff 1963, 1964, 1967; Hamilton et al. 1963; Mayr 1965; MacArthur and Wilson 1967; Grant 1969; Williams 1969; Darlington 1970; Terborgh 1973; Ball and Glucksman 1975; Gilroy 1975; Gilpin and Diamond 1976, 1982; Hnatiuk 1980; Cole 1981; Zann et al. 1990; McNab 1994a; Martin et al. 1995; Weisler 1995; Wiggens and Møller 1997
Colonization and geological age and ecological maturity of islands	Asymptotic increase in expected colonization with time and decrease in extinction with ecological maturity	New Zealand examples supportive, but role of Hawaii and isolated islands obscured by uncertain histories	Hamilton et al. 1964; Hamilton and Armstrong 1965; Schoener 1965, 1984; Grant 1966c, d, 1967; MacArthur and Wilson 1967; Lack 1969b; Harris 1973; Abbott 1974a, b, 1976a, b; Ford and Paton 1975; Strong et al. 1979; Roth 1981; Haila et al. 1983; Grant and Schluter 1984; Reed 1984; Carrascal et al. 1994; Martin and Thibault 1996
Propensity and ability of taxa for insular colonization	Favors vagility, flocking habits, ecoefficiency, generalized needs; bills large and unspecialized	Descriptive of Rallidae, some not gregarious or having bills consistent with prediction	Baker and Stebbins 1965; Mayr 1965; Lewontin 1966; MacArthur and Wilson 1967; Ouellet 1967; Grant 1969; Williams 1969; Darlington 1970; Terborgh 1973; Ball and Glucksman 1975; Terborgh et al. 1978; Cole 1981; Gilpin and Diamond 1981; Parsons 1982; Zann et al. 1990; Weisler 1995; Veltman et al. 1996; Selmi and Boulinier 2001
Change in overall size in taxa following colonization	Size of small taxa (herbivores) tends to increase; that of large taxa (herbivores and carnivores) decreases	Increases in size most typical, but variable in magnitude and cases of dwarfism known	Carlquist 1965, 1966, 1974; Grant 1965a, b, 1966a, b, 1968, 1971, 1979a, b, 1983b; Edwards 1967; MacArthur and Wilson 1967; Lack 1970a, 1971a, b, 1976; Sondaar 1976; Case 1978a, c; Dunham et al. 1978; Power 1979; Abbott 1980; Feduccia 1980; Weller 1980; Simberloff and Boecklen 1981; M. Williamson 1981; Ebenman and Nilsson 1982; Case et al. 1983; Simberloff 1983a, b, 1984; Lomolino 1985; Livezey 1989a, b, 1990, 1992a, b, 1993a, c; Kozlowski 1996
Change in body form exclusive of size in taxa following colonization	Flightlessness increases with decreased area of island and body size of colonists; decreased sexual dichromatism	Not uncommon; e.g., decurvature of bill, shifts in pelvic proportions, reduction of pectoral apparatus	Darlington 1943, 1957; Baldwin 1953; Bowman 1961, 1963; Carlquist 1965, 1966, 1974; Dobzhansky 1965; Grant 1965a, 1981b, 1983c; MacArthur and Wilson 1967; Keast 1968; Cody 1969, 1973; Lack 1970a, 1971a, 1976; Abbott 1974c; Grant et al. 1976; Sondaar 1976; Templin 1976; Diamond 1977a, 1991; Baker and Moeed 1979, 1980; Grant and Grant 1979, 1983; Baker 1980; Feduccia 1980; Weller 1980; M. Williamson 1981; Travis and Ricklefs 1983; Boag and Grant 1984; Gibson et al. 1984; Schluter and Grant 1984a, b; Guyer and Slowinski 1993; McNab 1994a

TABLE 81. Continued.

Characteristic	General predictions for islands and birds	Predictions confirmed for insular rails	Selected references
Reduction in genetic variation of founding and descendent population	Founder effect, genetic drift tends to reduce genetic variation and fixation of alleles; aided by polygyny	Evident in frequency of leucistic and other morphs, limited genetic support; probably vital to differentiation	Mathews and Iredale 1914; Mayr 1941; Cunningham 1959; MacArthur and Wilson 1967; Dobzhansky 1970; Endler 1973, 1977, 1982; Ripley 1977; Templeton 1980, 1981; Barton and Charlesworth 1984; Carson and Templeton 1984; Lande 1985b; Levinton 1986; Franklin 1987; Goodnight 1988; Brakefield 1989; Bryant 1989; de Jong 1989; Haig et al. 1990, 1994; Grant and Grant 1995; Haig and Ballou 1995; Kaneshiro 1995; Thorpe and Malhotra 1996, 1998; Hanski 1997; Le Corre and Kremer 1998
Changes in spatial behavior, sexual dimorphism, or reproductive ecology following colonization	Reduced vagility, clutch size; increased egg size and biparental care, sexual dimorphism; competitive ability, aggression variable	Reduced mobility typical; increased egg size and decreased fecundity modal; shifts unclear in dimorphism, parental care	Marshall and Harrison 1941; Darlington 1943; Carlquist 1965, 1966, 1974; Grant 1966a, 1986, 1994a, 1998a, b; Miller 1966; MacArthur and Wilson 1967; Diamond 1970, 1981; Lack 1970a, 1971a, b, 1975, 1976; Keast 1971; MacArthur et al. 1972; Abbott 1975, 1976a, b, 1977, 1978a, b, 1980, 1981; Emlen 1977; Wallace 1978; Weller 1980; Nilsson and Ebenman 1981; M. Williamson 1981; Nilsson et al. 1985a, b; Stamps and Buechner 1985; Baker and Jenkins 1987; Humphrey et al. 1987; Vitousek et al. 1995; Whittaker 1998
Change in foraging methods and diet after colonization	Flightlessness and lower basal metabolic rate under arduous conditions; broadened ecological niches, e.g., new foraging methods and dietary items	Omnivory conserved or enhanced; some taxa adding predation on eggs and sanguinvory of nestling seabirds, increased scavenging	Marshall and Harrison 1941; Crowell 1962; Hatch 1965; Amadon 1966a; Grant 1966a, 1986, 1994a, 1998a, b; MacArthur and Wilson 1967; Millikan and Bowman 1967; Harris 1968; Diamond 1970a, b, 1974; Bowman and Carter 1971; Lack 1971a, b, 1975, 1976; Disney and Smithers 1972; Abbott et al. 1975, 1977; Gorman 1975; Johnston 1975; Salomonsen 1976; Trimble 1976; Emlen 1977; Wallace 1978; Faaborg 1980, 1982; Weller 1980; M. Williamson 1981; Simberloff 1983b; Blondel 1985a, b; Stamps and Buechner 1985; Blondel et al. 1988; Atkinson and Greenwood 1989; Carrascal et al. 1994; McNab 1994a, b; James 1997
Extinction of taxa pertaining to factors exclusive of anthropogenic	Drought and storms important, buffered by ecological and elevational diversity	None documented, but storms and drought probable causes of extinctions	Lack 1942; Preston 1962a, b; MacArthur and Wilson 1967; P. S. Martin 1967, 1984; Terborgh 1974; Boucot 1975; Brown and Kodric-Brown 1977; Faaborg 1979; Karr 1982; Diamond 1984b, c; Flessa 1986; Simberloff 1986; Shaffer 1987; Marshall 1988; Unwin 1988; Donovan 1989; Maynard Smith 1989; M. Williamson 1989; Benton 1991; Simberloff and Boecklen 1991; Tracy and George 1992; Lomolino 1996; Pagel and Payne 1996

TABLE 81. Continued.

Characteristic	General predictions for islands and birds	Predictions confirmed for insular rails	Selected references
Anthropogenic impacts and vulnerability of taxa	Endemics specialized to mature communities, of large size, or having low reproductive rates or mobility are at risk	Countless examples, including well documented and informed annihilation, indirect extirpations	Williams 1962, 1964; Greenway 1967; MacArthur and Wilson 1967; Guilday 1967; McDowall 1969; Diamond and Veitch 1981; Fordyce 1982; Olson and James 1982b, 1984; Cassels 1984; Dewar 1984; Meredith et al. 1985; Temple 1985, 1986; Pimm et al. 1988, 1993, 1994a, b; Anderson 1989; Barnosky 1989; Diamond 1989a; Olson 1989; Owen-Smith 1989; Atkinson and Millener 1991; Bell 1991; Millener 1991; Alcover et al. 1992; Milberg and Tyrberg 1993; Bibby 1994a, b; James 1995; Chown et al. 1998b; MacPhee 1999
Predictions extended to quasi-insular contexts	Increased immigration, decreased extinction; increased nesting densities in some cases leading to increased predation on eggs and young	"Mega"-coots endemic to Andean lakes almost flightless; sympatric with flightless grebes on some larger lacustrine "islands"; terrestrial examples lacking	Habitat "islands" (MacArthur and Wilson 1967; Vuilleumier 1970; Galli et al. 1976; Helliwell 1976; Leck 1979; Vuilleumier and Simberloff 1980; Higgs 1981; Temple 1981; McCoy 1982; Simberloff and Abele 1982; Reed 1983; East and Williams 1984; Howe 1984; Opdam et al. 1984; Williams 1984; Raff et al. 1985; Szaro 1985; Gilpin and Soulé 1986; Martin 1988a, 1995; Lockwood and Moulton 1994; Bellamy et al. 1996; Jullien and Thiollay 1996; Grant 1998b; Mallet and Turner 1998; Prance 1998; Telleria and Santos 1999); *peninsulas* (Taylor and Regal 1978); *mountaintops* (Riebesell 1982; Nores 1995; Brühl 1997; Fjeldså et al. 1999); "*harlequin*" *habitats* (Horn and MacArthur 1972; Keast 1976); *caves* (Eigenmann 1909; Culver 1970, 1982; Vuilleumier 1973; Howarth 1991; Jones et al. 1992); *wetlands* (Maguire 1963; Sepkoski and Rex 1974; Sillen and Solbreck 1977; Holland and Jain 1981; Elmberg et al. 1994; Schulter 1998); *host–parasite systems* (Tallamy 1983)

land area (Millener 1999) and diversity of habitats (Fig. 7), involved at least six genera other than *Porzana* having endemic flightless representatives (Table 2). Perhaps the most surprising aspect of the avifauna of the Chathams is the apparent absence of a flightless swamphen (*Porphyrio*) or crake (*Porzana*), with the former being especially notable in light of the presence of the genus throughout the rest of the region (Ripley 1977; P. B. Taylor 1996, 1998). Intermittent connection of habitable islands clearly influenced insular faunas (Mielke 1989; Nunn 1994; Whittaker 1998: fig. 2.3), and may be among the key factors leading to sympatry of closely related, flightless rails within archipelagos.

At one extreme, processes germane to flightless rallids are ancient isolations among large land masses, for example, the tectonic separation of New Zealand from Australia dating to approximately 60 million years ago; and subdivision of much lesser age within these major masses, for example, both those between Tasmania and the Australian mainland, and division of the three main constituent islands of New Zealand, dating to approximately 20,000 BP (Craw et al. 1999). The latter geological subdivisions, in turn, had clear implications for flightless rallids: isolation of Tasmania almost certainly initiated the differentiation of flightless *Tribonyx mortierii* from *T. ventralis*; and separation of the North and South islands within New Zealand undoubtedly initiated the differentiation of the flightless sister-taxa *Porphyrio mantelli* and *P. hochstetteri,* and the parallel diversification of flightless *Gallirallus greyi* and *G. australis*. More complex correspondences among endemic rallids and greater New Zealand, although of perennial interest and study (Fleming 1962, 1975, 1979; Brown et al. 1968; Atkinson and Bell 1973; Bull and Whittaker 1975; Craw 1988; Cooper 1989; Southey 1989; Atkinson and Millener 1991; Millener 1991, 1999; McFadgen 1994; Craw et al. 1999), remain largely conjectural (e.g., Figs. 12, 13).

"Hot-spot" archipelagos present a special case of spatiotemporal dynamics in oceanic islands, in which member islands are displaced by movements of underlying tectonic plates, with the youngest (and typically the largest) islands in each being under volcanic augmentation while positioned over the "hot spot" of the island chain. Both the Hawaiian Islands (Carson and Clague 1995; Windley 1995) and Galápagos Islands (Simkin 1984; Christie et al. 1992; Rasmann 1997) are examples of hot-spot archipelagos, in which member islands are moving east to west or west to east, respectively, and decrease in area over time through subsidence and erosion as they move away from the hot spot. Both island groups include members having ages of $\sim 5 \times 10^6$ years, although volcanic activity has occurred in both over longer periods ($\sim 20 \times 10^6$ years), suggesting that extant populations had origins on now-submerged (extinct) islands (Carson and Clague 1995; Rasmann 1997).

Remarkably, this comparative wealth of geochronological data has provided few insights into the evolution of flightless rallids or other birds endemic to the Hawaiian or Galápagos islands. At best there appears to be a poor correspondence between the relative apomorphy of the flightless anatids and rallids and geological ages of the Hawaiian Islands (Table 3; Olson and James 1991; Livezey 1993a, 1996a, 1998), a finding plausible given the likelihood of multiple colonizations of one or more islands by flighted ancestors and subsequent exchanges of flighted and flightless descendants among islands (Fig. 10). A similar state of affairs attends the distribution of the endemic flightless cormorant of the Galápagos, which corresponds to that of prey and associated oceanic currents as opposed to antiquity of islands upon which this profoundly apomorphic species nests (Livezey 1992a). The degree of apomorphy of flightless rallids of the Mascarenes (Fig. 6; Cheke 1987; Cowles 1987; Mourer-Chauviré et al. 1999), as with other geologically intractable complexes of islands (e.g., outlying islands of New Zealand region [Figs. 7, 12]), are amenable to little more than speculation. Estimates drawn from DNA sequences have shown geochronologically consistent patterns of colonization, differentiation, and dispersal from older to newer islands (with variable amounts of reverse colonizations) for a number of radiations of Hawaiian plants (e.g., Funk and Wagner 1995; Givnish et al. 1995; but see Lowrey 1995; Wagner et al. 1995), insects (Asquith 1995; DeSalle 1995; Shaw 1995), and several birds (Tarr and Fleischer 1995).

INSULARITY: EVOLUTION AT LOCAL SCALES

> In any case there can be little doubt that these rails became flightless in the islands they now inhabit, and cannot therefore be regarded as evidence of the former extension of land; in other words, they are of no value in determining former geographical conditions, since they are themselves the outcome of the present one.—Andrews (1896b:84)

> If . . . [*Falco novaeseelandiae*] were plentiful it must have been a great inducement for the old swamp-hens [ancestral *Porphyrio mantelli*] to stay in the scrub, until at last they were *too lazy to fly,* especially those that had escaped a knock or two from the hawks. Then, with mates of like experience, there is no mystery about the founding of this notable family—a branch of which exists in the white swamp-hen of Norfolk Island [*P. albus*].—Henry (1899:54, emphasis added)

> Isolation from potential enemies or rivals may permit unusual specialization, as in flightless island birds, or encourage variability and degree of adaptive radiation, as in the fish of certain African lakes or the marsupials of Australia.—Huxley (1943: 129)

> Insularity is moreover a universal feature of biogeography.—MacArthur and Wilson (1967:3)

Historical background.—Unlike the ratites, penguins, and at least some flightless members of several other neognathous families of aquatic birds (e.g., *Tachyeres* and Podicipedidae), flightlessness in the Rallidae is interrelated inextricably with insularity. Although this obvious association has led to simplistic explanations bordering on cliché, islands increasingly have been recognized as providing unique opportunities into evolutionary insights (Murphy 1938; Mayr 1967; Herremans 1990). Islands represent natural experiments involving the interactions of colonization, divergence, and extinction, permutations of tractable size, varying complexity, estimable age, and comparatively reduced faunal and floral diversity, and barriers to dispersal that may both prevent invasion but also effect isolation and therefore hasten the processes of divergence and speciation (Grinnell 1914, 1917; Mayr 1947; Williams 1953; Lomolino 1984; Hengeveld 1997; Manne et al. 1998; Whittaker 1998). Insular biogeography has been imbued with a reverence for islands as isolated and pristine relicts of nature, qualities ironic in light of a history of human degradation and destruction.

Both naturalistic reverence and empirical interest in islands have deep historical roots. Important works include the attentions of Darwin (1859) and Wallace (1880), and were sustained by regular revisitations to the topic to the present day (Buller 1894; Gulick 1932; Lack 1949, 1975, 1976; Carlquist 1965, 1974, 1980; Diamond 1976). A modern analytical study—that of MacArthur and Wilson (1967)—was to mark a major rejuvenation of the study of things insular. Not without controversy, the model has been criticized on several points, especially regarding naiveté in assumptions regarding estimation of historical patterns of extinction and immigration based solely on current faunal richness (Gilbert 1980; Grant and Abbott 1980; Pregill and Olson 1981; Gotelli and Abele 1982; Brown and Gibson 1983; Graves and Gotelli 1983; Gotelli and Graves 1996), estimates of turnover (Nilsson and Nilsson 1983), extension of the framework to paleontological contexts (Hoffman 1985; Jablonski 1991; Hallam and Wignall 1997), covariance among predictor variables (Johnson 1975), or interpretational excesses of coefficients (Gould 1979; Rey et al. 1982). Perhaps most troubling were instances where advocates of the paradigm, faced with inconsistencies from predictions, conveniently attributed discrepancies to "nonequilibrial" conditions (Abbott and Grant 1976; Connor and Simberloff 1978, 1979, 1984; Strong et al. 1979; Connell 1980; Wiens and Rotenberry 1981; Wiens 1982, 1984, 1989; Simberloff 1983a, b; Gilpin and Diamond 1984; Strong et al. 1984; Masters and Rayner 1993). Despite these shortcomings, the theory formalized by MacArthur and Wilson (1967) provided a natural conceptual segue between population-level processes and evolutionary timescales, distilling a myriad of accidents and processes to a dynamic balance between immigration and extinction and resultant turnover of taxa (Lynch and Johnson 1974; Brown and Kodric-Brown 1977; Lomolino 1996). The elegance of the model spawned an extraordinary literature of descriptive and hypothetico-deductive studies, many of which confirmed generalities at least broadly applicable to flightless rails (Table 81).

From a phylogenetic perspective, the equilibrium paradigm was fundamentally inadequate for issues of deep time, in which the roles of colonization and extinction rates were minor in comparison with those of vicariance events for clades of potential membership, relationships within clades, and biogeographical distributions of members. The original conceptualization by MacArthur and Wilson (1967) emphasized the dynamic process acting on islands, in which islands functioned as urns among which species were moved with probabilities indicative of colonizing abilities and withdrawn in accordance with estimated extinction rates. However, hints at uses of the model for deeper histories and broader geographic scales were included by MacArthur and Wilson (1967), and these led to controversial and largely ad hoc extensions to modern faunal distributions, among which was the proposal of the "taxon cycle" of islands and archipelagos (Wilson 1961; Greenslade 1968; Ricklefs 1970, 1972a, b; Ricklefs and Cox 1972, 1978; Simberloff 1974; Mayr and Diamond 1976; Cox and Ricklefs 1977; Roughgarden and Pacala 1989; Ricklefs and Bermingham 1999).

Does the concept of taxon cycles hold promise for flightless birds in general and flightless rails in particular? This paradigm places highly apomorphic species endemic to single islands (e.g., flightless birds) in the twilight of the term of survival of a lineage in an archipelago (i.e., stage IV of the cycle), and flightless rallids are consistent at least with respect to longevity of species. A review by Grant (1998a) of the foremost points of concern regarding taxa colonizing islands also was of relevance to flightlessness in rails; these included relationships between insular characteristics and emergent evolutionary trends; bases and implications of reduced dispersal (e.g., insular insects, birds, and propagules of plants [Darlington 1943; Brinck 1948; Hesse et al. 1951; Gressitt 1954, 1956, 1961, 1964; Carlquist 1966, 1974; M. Williamson 1981; Table 81]); patterns and selective implications of changes in body size; changes in parameters of reproduction; shifts or broadening of ecological niches; and tameness toward recently arrived organisms, notably humans.

Colonization of islands.—Avian dispersal appears to be selectively driven to avoid inbreeding and the philopatry and territoriality of established adults motivates unestablished, typically younger individuals to undertake exploration for unoccupied habitat (Greenwood 1980; Greenwood and Harvey 1982; Swingland and Greenwood 1983; Johnson and Gaines 1990; Paradis et al. 1998; Colbert et al. 2001). Faunal invasions on continents, including those by humans, can be viewed as large-scale, short-term analogues of insular colonization (Elton 1958; Holdgate and Wace 1959; Cumberland 1965; Wodzicki 1965; Warner 1968; Salmon 1975; Heaney 1978; King 1984; Fitzgerald and Veitch 1985; Meredith et al. 1985; Cheke 1987; Caughley 1988; Brown 1989; Loope and Mueller-Dombois 1989; Shigesada 1997). In both contexts, the probability of successful colonization by a given species of bird involves both the colonizing abilities of the taxon (e.g., maximal flight radius) and the favorability of the island or region involved (Table 81; Baker and Stebbins 1965; Lewontin 1965; Mayr 1965; Carlquist 1966; Williams 1969; van Noordwijk and Scharloo 1981; Parsons 1982). Other potentially influential factors are areas of source populations and those of stepping-stone islands (MacArthur and Wilson 1967; Gilpin 1980; Le Corre and Kremer 1998).

Heterogeneity of family-level taxonomic composition of the avifaunas of oceanic archipelagos evinces the disparate colonizing abilities of avian taxa and (to a lesser extent) the stochasticity of the underlying processes (MacArthur and Wilson 1967; Whittaker 1998). The Hawaiian Islands were a center of extraordinary endemism (notably for Anatidae, Threskiornithidae, Rallidae, Corvidae, Turdidae, Meliphagidae, and Drepanididae) and (in nonpasseriform taxa) flightlessness, but there is no evidence of any Hawaiian representatives of several families having wide distributions and endemism in other archipelagos (e.g., Psittacidae, Columbidae, Alcedinidae, Picidae, Apodidae, or Caprimulgidae), and several families of shorebirds (Charadriiformes) of which members occur regularly as migrants but are not resident (James and Olson 1991; Olson and James 1991). Such faunal disharmony can be explained in part by the differential penetration manifested by different (sub)continental taxa to increasingly isolated island groups (Carlquist 1965, 1966, 1974; MacArthur and Wilson 1967; M. Williamson 1981; Whittaker 1998). This gradient emulates a "filter" acting on taxonomic groups. For example, rails extend through all seven zones of differential penetration from New Guinea eastward through to Easter Island (M. Williamson 1981; Steadman et al. 1994), a faculty undoubtedly enhanced by the ability of most rallids to rest on water in transit (P. B. Taylor 1998). The extremely isolated Hawaiian Islands hosted (by natural means) only 6 of more than 80 extant families of passerines and a few bats, and the former exceeded 12 species of endemic rallids (Olson and James 1991). The virtual ubiquity of rallids on oceanic islands contrasts with the regional kaleidoscope of taxa in the same regions (Olson 1973a; Ripley 1977, M. Williamson 1981; P. B. Taylor 1996, 1998), and a predilection toward peregrination by juvenile and migratory rallids seems inescapable as contributory factor.

Assembly of insular communities.—Insular endemics share several ecobehavioral traits, including reduced vigilance and inclination to escape (tameness or approachability), which sometimes is intruder-specific (Curio 1969; Humphrey et al. 1987; McNab 2001); relatively rapid evolutionary divergence, sometimes leading to "adaptive radiation" (Price et al. 1984b; Grant 1998b; Schluter 2000a, b); decreased competitive ability or broadened ecological niche related to "disharmonious" faunal communities on islands (Feinsinger and Swarm 1982; Alatalo and Gustafsson 1988; Martin 1991), thought to be conducive to speciation (Diehl and Bush 1989), critical to community structure (Moulton and Pimm 1987), but of uncertain causal mechanism (Carrascal et al. 1992); and increased sexual dimorphism and elevated interspecific aggression (e.g., Carlquist 1965, 1966, 1974; Diamond 1970a, 1981; Lack 1971a, b, 1976; Wallace 1978; Abbott 1980; Feduccia 1980; Weller 1980; M. Williamson 1981; Schluter 1996b, 2000a, b). Abbott (1980: fig. 4) graphically conceptualized the roles of resources, habitat, competitors, and predators in the general impoverishment of insular avifaunas, emphasizing the dearth of successful colonists.

Detailed comparisons of the species composition of islands within the equilibrium paradigm (e.g., Diamond 1972; Power 1972, 1975, 1976; Abbott 1974a, b; Abbott and Grant 1976; Connor and Simberloff 1978; Juvik and Austring 1979; Terborgh and Faaborg 1980; Diamond and Gilpin 1982; Gilpin and Diamond 1982, 1984; Jarvinen and Haila 1984; Simberloff and Martin 1991) led to attempts to explain recurrent patterns in terms of accepted evolutionary principles. These explanations emphasize character displacement in both morphological and ecological aspects (Brown and Wilson 1956; Grant 1972a, b, 1975, 1976, 1978, 1994a; Bulmer 1974; Eldredge 1974; Grant and Grant 1982; Fjeldså 1983b; Diamond et al. 1989; Whitehead and Walde 1993; Schluter 1994, 2000a, b) as a result of disruptive selection (Thoday 1972; Smith and Temple 1982). Other special points of interest include directional selection on size and shape of the bill (P. R. Grant 1972c; B. R. Grant 1985; Schluter and Smith 1986), magnified ecomorphological differences among species or "ecological release" (Fjeldså 1983b), sympatric speciation (Grant et al. 1976; Grant and Grant 1979, 1983), counter-rationalizations of character convergence (Cody 1969, 1973), relationships between guild structure and interspecific competition (Schoener 1965, 1984; Grant and Schluter 1984; Schluter 1988a; Carrascal et al. 1994; Martin and Thibault 1996), and consumer–resource dynamics (Morris and Knight 1996). All of these mechanisms fall within a broad concept of a coevolutionary fabric among coexisting species (Thompson 1999), but failed to provide parsimonious explanations for the surrender of flight.

Data from study of the flora and fauna of islands have led to a number of predictions regarding changes in ecology and morphology under insular regimes, many of which proved applicable to flightless rails and other birds (Table 81). Much of the literature on the "assembly" of ecological communities on islands assumed an important interplay among member taxa (especially within guilds), foremost among which has been been interspecific competition, a premise that subsequently became the focus of considerable debate (reviewed by Strong et al. 1984). The classical view holds that avian communities generally are not merely accidental conglomerates of taxa, but instead communities are posited to be sequential assemblages in which each successful colonization is viewed as consonant with or competitively superior to and former occupants of a niche or guild within the insular community. Perhaps most importantly, this perspective emphasized importance of qualities of colonists beyond mere dispersal (MacArthur 1958, 1968, 1969, 1971, 1972a, b; MacArthur and Levins 1967; Lack 1971a; Cody 1974, 1985; Cody and Diamond 1975; Pianka 1976; Diamond and Gilpin 1982), a shift of particular relevance to flightless birds.

With the possible exceptions of the Hawaiian and Chatham islands, unlike the cases of West Indian *Anolis,* honeycreepers of the Hawaiian Islands, and the geospizine finches of the Galápagos, however, flightless rails seldom qualify as having radiated to fill multiple niches on a single island or island group. However, the comparatively great endemism of rails of the Chatham Islands has inspired four essentially ad hoc assessments of competition-based community structure and differentiation in situ. Andrews (1896b) proposed the first and simplest theory, in which the three flightless typical rails of the Chathams were inferred to have arrived in order of relative, present-day apomorphy (i.e., *Diaphorapteryx hawkinsi* first, *Cabalus modestus* second, and *Gallirallus dieffenbachii* third), predicated on the assumption that relative divergence from ancestral conditions can be assumed to be indicative of the time that each lineage remained in isolation. Likewise, Olson (1975b) speculated that the comparatively modest morphological changes evident in *G. dieffenbachii* relative to its *philippensis*-like ancestor indicated that it was a more recent colonist of the Chathams than *C. modestus*. In a broader study of the avian communities of New Zealand and the Chatham Islands, Atkinson and Millener (1991) interpreted, under the classical assumption that coexistence implied the maintenance of minimal interspecific differences in niche, that the endemic rails of the Chatham Islands and their potential competitors differed in one or more essential factors such as body mass, bill shape, or general dietary group; however, these authors did not hazard a guess as to which of the three taxa of typical rails endemic to the islands arrived first.

More recently, Trewick (1997b) performed a morphological comparison of two flightless rails (*Gallirallus dieffenbachii* and *Cabalus modestus*) formerly endemic to the Chatham Islands evidently closely related to the *Gallirallus philippensis*-group (sensu lato). Trewick (1997b) agreed that these species probably had generalized diets (Falla 1954; Olson 1975b), with moderate specialization of *C. modestus* for insectivory suggested by Atkinson and Millener (1991) excepted. However, Trewick (1997b) judged that the comparatively great sexual dimorphism in bill length and inferred ecological specialization of *C. modestus* were likely the result of directional selection in the presence of a competitor (*G. dieffenbachii*) and therefore speculated that the specialized body form of *C. modestus* was evidence of a more-recent colonization.

Growing, variably controversial evidence supports the importance of interspecific interactions as contributory factors in the selection regimes acting on morphological and behavioral divergence (e.g., MacArthur and Levins 1967; Thoday 1972; Benkman 1989a, 1991, 1993; Schluter 1994; Benkman et al. 2001). Nevertheless, given the speculative and contradictory nature of the ad hoc interpretations by Olson (1975b), Atkinson and Millener (1991), and Trewick (1997b), one is faced with proposals that selection for divergence could favor either the earlier colonist (wherein a larger deme size might include more potential for change) or the later colonist (wherein small deme size might enhance change via drift, but founder effect would limit possible avenues of change). Absent a resolved phylogeny (Trewick 1997a; Livezey 1998), the relationship of morphological differences to sequence of colonization and interspecific competition (if any) remains beyond empirical test.

Islands as refugia.—Human-related introductions aside, oceanic islands generally are thought of as being predator-free (Carlquist 1965, 1974; Snow 1966; Weller 1980; M. Williamson 1981). In reality, most insular birds are vulnerable, especially as eggs or hatchlings, to both submarine and aerial predators. Published reports include observations of predation on birds by submarine fishes, submarine mammals, or other birds (Murphy 1936; Franklin et al. 1979; Fraser 1984; Nilsson et al. 1985b; Livezey and Humphrey 1986; Livezey 1988, 1989b–d, 1990, 1992a; Taborsky 1988; Baird 1991b; James and Olson 1991; Alcover and McMinn 1994; Bunin and Jamieson 1996b), with other observations substantiating the importance of these and other subaquatic predators of young and adult birds in most maritime habitats, whether insular or continental (Legendre 1941; Scheffer 1942; Leach 1943; Lowe 1943; Taverner 1943; Glegg 1945, 1947; Hamilton 1946; Hewitt 1950; Harrison 1955; Hindwood 1964; Spellerberg 1975; Brooke and Wallett 1976; Kinnear 1977; Endler 1978; Davenport 1979; French 1981; Straneck et al. 1983; Gilpin 1987; Randall et al. 1988).

Insular landbirds, as well as the young of seabirds that nest seasonally in the Galápagos, are vulnerable to aerial predators, including both diurnal hawks and nocturnal owls; similar threats pertained to landbirds endemic to the West Indies, New Zealand, and Hawaii (Wetmore 1956; M. Williamson 1981; Holdaway 1989, 1990; Livezey 1990, 1992a, Whittaker 1998). Continental rails are preyed upon by diverse taxa (Evens and Page 1986; Motta Junior 1991; P. B. Taylor 1996, 1998), of which only the mammalian and (to a lesser extent) reptilian components were reduced or absent on most oceanic islands. The nearly flightless *Laterallus spilonotus* of the Galápagos is preyed upon by the endemic subspecies of Short-eared Owl (*Asio flammeus*) where sympatric (Franklin et al. 1979). James and Olson (1991) identified the extinct owls (*Grallistrix* spp.) of the Hawaiian Islands as important predators of the extinct crake *Porzana severnsi*. Williams (1960) reported that the eggs of one flightless rallid of New Zealand, *Porphyrio hochstetteri,* have been preyed upon by another flightless rallid, *Gallirallus australis.* Moreover, fluctuations in water levels may pose the most serious threat to nesting success of continental rallids nesting in wetlands (Legare and Eddleman 2001), a factor to which many insular rallids would also be vulnerable, especially those species (e.g., *Fulica*) nesting in tidal zones of coastal habitats.

Galápagos finches respond to calls or silhouettes of raptors (Curio 1969, 1976), an indication of the persistence of such risks for these and other insular landbirds of this archipelago. Vigilance and escape behavior have been described in a number of continental rallids (Hobbs 1967, 1973; Alvarez 1993; Evens and Page 1986; McRae 1997) efficacy of which is conditional on perceptions of risk (Lima and Dill 1990). In light of the importance of antipredator strategies for survivorship of relatives (Harvey and Greenwood 1978; Woodland et al. 1980; Lima 1993, 1998) and the rapidity with which new predators can be behaviorally accommodated by prey (O'Steele et al. 2002), foraging posture, formation of foraging flocks, distribution of nests, and temporary gains in body mass may alter significantly vulnerability of adults and nests to predation (Orians 1971; T. E. Martin 1981, 1988a–c, 1992, 1995, 1998; Krause and Godin 1996; Martin and Badyaev 1996; Veasey et al. 1998). A quantitative assessment of such ethological dimensions may serve as an indirect indication of the intensity of predation that impinges on insular rallids, both flighted and flightless, and in conjunction with estimates of genetic differentiation from likely source populations and surveys of other critical resources (Martin 1987), may provide an insight into the retention of flight in the many populations of flighted rails on oceanic islands.

Given that cryptic coloration can be important in the avoidance of detection by predators (Endler 1978; Baker and Parker 1979), the obsolete sexual dichromatism and subdued plumage coloration of insular birds (Grand 1965e; Lack 1970b; Weller 1980; Livezey 1990, 1993a; Figuerola and Green 2000) probably improves concealment from aerial predators, especially during nesting and biparental attendance of broods. Given the possibility that cryptic plumages of some flightless rails may be obligatory developmental by-products of paedomorphosis, these benefits may constitute selective gains post facto. At least one alternative explanation for an absence or apomorphic reduction in bright, often sexually dichromatic plumages among insular birds has been advanced: reduced sexual selection for species identification in depauperate faunas (Sibley 1957; Johnsgard 1963; Lack 1970b; Weller 1975, 1980). Grant (1965b) suggested that the drab plumages of insular passerines may be selectively neutral by-products of reduced genetic diversity of founding populations. However, in light of the apparent importance of aerial predators on insular avifaunas, avoidance of aerial predators may be the most likely selective advantage for insular rails so characterized (e.g., *Gallirallus australis*-group and *Cabalus modestus*).

Genetic implications: founders, bottlenecks, and drift.—Two phenomena tending to reduce genetic variation in the small contingents of immigrants that colonize oceanic islands—founder effect (limited genetic variation in colonists) and drift (stochastic loss of genetic variation within small demes)—are among the most widely cited genetic mechanisms responsible for the characteristics of insular populations (Table 81), the importances of which are inversely related to deme size (Mayr 1954, 1960, 1963; Carlquist 1966, 1974; Dobzhansky 1970, Endler 1973, 1977, 1982; Barrowclough 1980; Lande 1980b, 1992; Templeton 1980, 1981; Barton and Charlesworth 1984; Carson and Templeton 1984; Boag 1988; see also Keller and Arcese 1998). Whereas the importance of founder effects in insular populations is widely recognized, notably with respect to the increased likelihood of rapid, innovative evolutionary change in isolated populations (Carson and Templeton 1984; Provine 1989; Willis and Orr 1993; but see Fisher 1930; Barton and Charlesworth 1984), the role of genetic drift has been the subject of significant debate on theoretical grounds (e.g., Fisher 1949, 1958; Wright 1969; Crow and Kimura 1970). Gillespie (2001) inferred that weakly selected alleles at closely linked loci in small populations are subject to "genetic draft" and more likely to become fixed than the same alleles in large demes.

These factors include compounding constraints on demes where immigration occurs through sequential colonization of stepping-stone islands (Gilpin 1980; Le Corre and Kremer 1998), notably in geographically complex networks such as Melanesia and Polynesia (Gressitt 1954, 1956, 1961; MacArthur and Wilson

1967); the relationship between founder effect and evolution of altruism (Eshel 1977); inbreeding (Rowley et al. 1993) and its effect on probability of extinction (Hartt and Haefner 1995) and genetic impoverishment in isolated populations (Keller et al. 2002); the "founder-flush" theory of speciation (Slatkin 1996); possibly countering effects of periodic hybridization among closely related taxa (Grant 1994b); and the differentiation of isolates despite the countering effects of gene flow (Endler 1974, 1977, 1982; Porter and Johnson 2002). In cases involving small islands and relatively mobile organisms, the issue of inbreeding sensu stricto is confounded with the genetic implications of small insular demes; for example, Miller (1947) inferred that the geographically confined population of *Porzana palmeri* had probably achieved panmixia, and Noordwijk and Scharloo (1981) described an analogous situation in *Parus* on a continental island.

The broad concept of epistasis—in which the effect of one genetic unit is contextual with respect to one or more other such units, resulting in a nonlinear phenotypic response involving more than simple, independent, additive effects—offers substantial potential both for preservation of genetic options and realization of such options during or after populational bottlenecks (Brodie 2000; Phillips et al. 2000; Wolf et al. 2000; Whitlock et al. 2002). With respect to the scenario of colonization, evolutionary changes in the normally canalizing effects of epistasis among genes (Phillips et al. 2000; Rice 2000) can be hastened by founder effects (Goodnight 1987, 1988, 2000; Meffert 2000). Such genetic reorganizations would enhance options for capitalizing on heritable diversity preserved through polymorphism and the stability conferred through epistasis (Lewontin and Kojima 1960; Kelly 2000; Wade 2000), while still being selectively constrained by the modified array of genetic interrelationships (Schluter 1996a).

The phenotypic effects of small deme sizes and the fixation of alleles uncommon or recessive in source (e.g., continental) populations may account, at least in part, for leucistic plumage variants found in insular populations of some waterfowl (Delacour 1956; Livezey 1992b). Similarly, the comparatively high incidence of variably expressed leucisticism in rallids (notably *Porphyrio melanotus* and *Gallirallus australis*-group) of New Zealand (Potts 1870; Buller 1874b, 1875, 1891, 1896, 1905; Kirk 1880; Handly 1895; Mathews and Iredale 1914; Marchant and Higgins 1993) and other Pacific islands (Mayr 1941; Ripley 1977) and the reports of buff-colored populations of *P. melanotus* (Cunningham 1959) are consistent with reduced genetic diversity leading to expression of rare or recessive alleles. With respect to insular variants in *Porphyrio,* the leucistic plumage of the extinct *P. albus* may reflect fixation of recessive alleles made possible through the contributions of founder effects and subsequent genetic drift. Wide variation in plumages within the *G. australis*-group led to early taxonomic confusion, a situation complicated further by biogeographic partitioning of this variation within the region made possible by substantial sizes of the main islands of New Zealand (e.g., Buller 1905; Mathews and Iredale 1914). Deme sizes, genetic drift, and founder effects presumably contributed to the variation in plumages of Wekas, which includes moderately pronounced melanistic morphs (e.g., most *Gallirallus "troglodytes"*) in some regions (Ripley 1977; P. B. Taylor 1996, 1998). Selective reasons for melanism (e.g., protection from ultraviolet radiation or crypsis in regions dominated by dark substrates) in this and many other avian examples have proven elusive (Majerus 1998).

An additional, comparatively less-studied potential of founder effects is per capita amplification of the impoverished genetic diversity of colonists by the composition of avian migratory flocks by especially closely related individuals (e.g., extended family groups). A scenario of insular colonization by a small founding flock of comparatively closely related individuals implies an intensification of founder effects through inbreeding depression (Fisher 1949; Willis 1996). Also, this eventuality predictably would enhance, as explicitly modeled by Eshel (1977), the potential of kin selection and related social organizations and reciprocal altruism to contribute to successful establishment (Trivers 1971; Hamilton 1972; Levin and Kilmer 1974; West-Eberhard 1975; Hull 1980), redirections of social evolution that might foster subsequent differentiation of mating systems, participation by reproductive helpers, territoriality, and the correlations among them under novel environmental circumstances (Maynard Smith and Ridpath 1972; Koenig and Pitelka 1981; Service and Rose 1985; Koenig et al. 1992).

Although great attention has been paid to the genetic implications of initial population size, with the exception of bottlenecks having conservation implications (e.g., that of Laysan Duck [Warner 1963, Moulton and Weller 1984]), reductions in genetic diversity deriving from bottlenecks significantly subsequent to colonization have received substantially less theoretical or empirical consideration. Unfortunately for such simplistic approaches, the genetics of insular populations seldom if ever can be approximated accurately by a scenario in which a founding populations of effective size N_e undergoes logistic growth with respect to some spatiotemporally stable carrying capacity K. Also, not only are mean genetic diversities affected by bottlenecks, but likewise genetic variances as well (Brakefield 1989). Islands, more so than many continental regions, are subjected to destructive or catastrophic events that affect indigenous populations negatively to varying degrees, the most severe of which undoubtedly impose pronounced population bottlenecks and associated impacts on genetic diversity.

Of special importance to most oceanic regions of the world, especially the tectonically active Pacific Rim, are several classes of natural catastrophes that evidently regularly affect oceanic islands—volcanoes, tsunamis (tidal waves), earthquakes, and cyclonic storms (Myles 1985; Prager 2000). Of these, tropical storms probably produce damage to island ecosystems at comparatively high frequency but low intensity (Bourrouilh-Le and Taladier 2000). Volcanism, the process responsible for the original emergence of islands in hot-spot archipelagos (Christie et al. 1992; Carson and Clague 1995; Windley 1995; Rassmann 1997), can cause infrequent but potentially catastrophic events on many oceanic islands, and in some cases (e.g., Krakatau, Indonesia) can effectively destroy all terrestrial life for a period of time (Diamond 1984c). The important aspect of volcanic eruptions is that they, unlike tsunamis and many storms, pose a special threat to biotic communities of uplands. Variably extensive landslides occur regularly on volcanic islands—a result of unstable edifices, steep slopes, seismicity, and rainfall (Keefer 1984; Moore et al. 1989; Urgeles et al. 1997; Glade et al. 2000; Keating and McGuire 2000; Smith and Wessel 2000)—some of which are extensive and would be fatal to all flightless landbirds throughout the affected altitudinal zones. Evidence of high-magnitude earthquakes during prehistoric times in the Pacific region suggest that these may pose additional threats to oceanic islands (Atwater 1987, 1992). Volcanic islands, especially hot-spot archipelagos such as the Hawaiian Islands, are subject to regular, variably severe local tsunamis generated by submarine landslides (Moore et al. 1994a, b; Bryant 2001); endemics of such insular refugia would be at risk from the combination of volcanic eruptions from above, landslides at mid-elevational habitats, and tsunamis in coastal communities.

At greater timescales, all islands in the Pacific region would be expected to be affected by tsunamis of varying magnitudes and frequencies (Johnson and Mader 1994; Bryant et al. 1997; Dawson and Shi 2000; Keating and McGuire 2000; Bryant 2001). The ecological impacts of tsunamis generally would be limited to coastal environments where islands are large and include high-altitude regions (e.g., Island of Hawaii). Other sites of importance to flightless rails are sufficiently low in elevation (e.g., Chatham Islands) that large-scale devastation of the flightless avifauna would be expected to occur at least rarely. Recently it was hypothesized that giant mega-tsunamis, with run-up distances on island surfaces to elevations in excess of 300 m, occur at comparatively low frequencies (at least) in the Pacific region (Moore and Moore 1984; Paskoff 1991; Young and Bryant 1992; Moore et al. 1994a, b). Although the site originally interpreted to have been affected by a megatsunami (Lanai, Hawaiian Islands, ca. 10^5 PB) remains a subject of considerable debate (Nott 1997; Bourrouilh-Le and Taladier 2000; Dawson and Shi 2000; Felton et al. 2000), this hypothesis raises the possibility of an additional source of rare, extinction-level events for low-elevational islands and bordering continental coastlines (Bryant 2001).

In addition to the possibility of tectonically produced megatsunamis, extremely rare but enormously destructive, comet-produced archaeotsunamis have been documented (e.g., Alvarez 1997). Anecdotal evidence that tsunamis generated by the Deluge Comet (ca. 8,200 BP) affected multiple areas, including the eastern tropical Pacific, northern Atlantic, and the northern and eastern Indian Ocean (Bryant 2001). The elevation-conditional impact of tsunamis on oceanic islands logically would lead to an expectation that the antiquity of flightless lineages on oceanic islands would be directly related to the maximum habitable elevation provided by the islands, virtually precluding the long-term persistence of endemic, flightless birds on otherwise accommodating, low-lying islands and atolls in tsunami-prone regions. Where long-term survival was possible, a major tsunami may lead to an opposite evolutionary likelihood, in which the significant population bottleneck and ecological changes imposed on the island would be conducive to accelerated evolutionary change, including loss of flight in birds (but see López-Fanjul et al. 2002).

Antiquity of isolation: differentiation and endemism.—Simpson (1944:110) alleged that the loss of flight in birds typically is characterized by discontinuous change in the sense that the transition is marked by no clearly intermediate forms. Recognizing the signs of paedomorphosis in some flightless rallids and assuming that heterochrony offered a potential for rapid evolutionary change, Olson (1973b: 34) suggested that the time "... needed to evolve flightlessness in rails can probably be measured in generations rather than in millennia." However, changes in ontogenetic pathways need not lead to saltatory evolutionary changes (Greenwood 1989). Worthy (1988a) reported evidence from subfossil remains that revealed modest changes in dimensions of pectoral elements in *Euryanas finschi* (Anatidae), a taxon later judged to have been flightless by Worthy and Olson (2002), and that required approximately 10,000 years. Geological ages of islands place upper bounds on the antiquity of residence of endemic taxa, and some studies have revealed predictable associations between degree of isolation, size of islands, and the degree of observed genetic divergence of faunal novelties (e.g., Johnson et al. 2000).

However, high mobility of birds before the loss of flight, possible complications of multiple invasions, phylogenetic uncertainties, successive extinctions of insular populations of a widespread taxonomic group, errors in calibration of absolute time based on sequence data (Hillis et al. 1996; Swofford et al. 1996), and (to a lesser extent) rafting and swimming after the loss of flight likely will prevent accurate reconstructions of colonization events even to the precision of 10^6 years, let alone millennia (e.g., Marshall and Baker 1999). Attempts to correlate relative apomorphy of extant flightless lineages are confounded by different antiquities of isolation and phylogenetic uncertainties deriving from extinction of intermediate lineages, as well as the possibility that changes leading to flightlessness may occur by gradual or saltatory progressions (Greenwood 1989). Furthermore, the hypothesis that the loss of flight results from gradualistic change is consistent with the intermediate conditions of several semiflightless insular rallids (e.g., *Laterallus spilonotus*), as well as those of some grebes and waterfowl (Humphrey and Livezey 1982; Livezey and Humphrey 1986; Livezey 1989b, d, 1990, 1993a). Perhaps most discouraging for such historical reconstructions, however, is that even if reasonably precise estimates of interpopulational divergences and information on ages and movements of islands are available, as in hot-spot archipelagos (Fig. 10; Christie et al. 1992; Carson and Clague 1995; Windley 1995; Rassmann 1997), multiple ad hoc scenarios of dispersal and backcrossing will remain plausible in many cases.

Uncertainties of taxonomic rank and evolutionary relationships remain even where both phylogenetic hypotheses and geochronologies are available (e.g., Cox 1990; Stern and Grant 1996), in part a result of the obfuscation of history effected by microgeographic variation within single islands and populations (e.g., Conant 1988; P. R. Grant and B. R. Grant 1989; Dennison and Baker 1991; Ryan et al. 1994; Thorpe et al. 1996; Thorpe and Malhotra 1996, 1998; Grant 1998a), gene flow from nearby continental populations (e.g., Edwards 1993), varying regimes of natural selection in the short term (Price et al. 1984b), and interspecific hybridization (Grant and Grant 1996, 1998). Isolation, area, ecological diversity, environmental changes, and altitudinal range of host islands enhance endemism (MacArthur and Wilson 1967; M. Williamson 1981). These issues take on a special nature for breeding members of the Procellariiformes (Chown et al. 1998a, b), the seasonal endemism of which raises the fundamental distinction between the selective regimes to which terrestrial, year-round avian residents of oceanic islands are subjected, probably none more so than flightless endemics. Finally, simply on the grounds of spatial scale, small body size of organisms may favor speciation (Bush 1993), an effect perhaps contributory to the diversity of flightless *Porzana* endemic to Hawaii (Olson and James 1991). However, it seems improbable that the ranges of body size found in insular birds would prove crucial as a factor in comparison to the ecological conditions and frequencies of natural catastrophes of colonized islands.

Extinction on islands: stochasticity and anthropogenic inevitability.—A romantic view of islands as pristine refugia contrasts vividly with the reality of uncounted losses of avian taxa associated with the exposure of naive species to introduced threats, often within devastatingly brief periods of time, and the finality of extinction (Merton 1978; Terborgh and Winter 1980; Steadman 1997b). Extirpation of insular endemics by humans and their commensals merits comparisons with those epidemics that strike all ages and both sexes of many species and for which neither ecocultural "remedies" nor "cures" are evident. Flightless rails may have persisted for longer periods after human immigration to oceanic islands than some larger species, the latter proving more desirable to human immigrants for food (Holdaway and Jacomb 2000).

Anthropogenic extinctions summarized by Steadman (1995), and documented by a number of key paleornithological surveys of islands in the Pacific (Olson and James 1982a, b, 1984, 1991; Steadman and Olson 1985; Steadman 1988, 1993a–c; Steadman et al. 1990; James and Olson 1991; Kirch 2000), South Atlantic (Olson 1973b, 1975c), and the Caribbean region (Olson and Wetmore 1976; Olson and Steadman 1977, 1979; Olson and Hilgartner 1982; Olson and Wingate 2000, 2001), not only reveal the magnitude of such destruction, but also suggest additional characteristics of species that are predictive of vulnerability. Critical findings for Pacific Oceania include (e.g., Steadman 1995; Steadman et al. 1999) a preponderance of extirpated species from the Rallidae (extinctions especially severe), Columbidae, and Psittacidae, with losses also notable among the Megapodiidae (Poplin and Mourer-Chauviré 1985; Steadman 1989b, 1993b; Jones et al. 1995); flightless endemics prevalent among extinct Rallidae, with examples also from Columbidae and Megapodiidae (and a flightless parrot known only from New Zealand); Hawaiian Islands supported a uniquely high diversity of finches (carduelines) and flightless Anatidae, a diversity in the former group approached only by the Galápagos (although members remained flighted in the latter); and exclusive of flightless landbirds and seabirds, the avian taxa mostly vulnerable to anthropogenic extinction were frugivorous species foraging in the middle to high canopy of forests, a dependence rendering them especially sensitive to deforestation. The comparably great diversity of megapodes endemic to Pacific islands, some of which were flightless, is explainable in part by the strong flight capacity of smaller members of the family to colonize oceanic islands and a notable facility to traverse expanses of open water between islands (Jones et al. 1995).

Ongoing declines of insular rallids (Benson and Penny 1971; Jenkins 1979, 1983; Forberg et al. 1983; Mills et al. 1984, 1988; Engbring and Pratt 1985; Holdaway 1989, 1990; Lloyd and Powlesland 1994; Miller and Pierce 1995) must be compared against those documented for some continental confamilials (e.g., Corn Crake [Stowe et al. 1993]), most of which in both categories appear to be related to loss of habitat. Any assessment of decline also should include documented mortality of flightless, insular rails related to human activities that have not (yet) placed the species in jeopardy (Taylor and Mooney 1991). As with imperiled continental forms, measures to preserve insular endemics (notably rails and the Kakapo) emphasize protection of critical habitat, although several critically endangered insular forms also require(d) captive breeding, reduction of introduced predators, supplementary feeding, or mitigation of interspecific competition (Pracy 1969; Robertson 1976; Mills et al. 1978, 1980, 1989, 1991; Lourie-Fraser 1982; Moors 1982; Moulton and Pimm 1983; Miller and Mullette 1985;

Mountainspring and Scott 1985; Powlesland et al. 1992, 1995; Powlesland and Lloyd 1994; Clout and Craig 1995; Young 1995; Bramley 1996). Fortunately, insular rails do not seem to have suffered demographic acceleration of extinction related to highly skewed sex ratios or other parameters that can lower thresholds of extinction (Lande 1987b, 1993; Pimm et al. 1988), characteristics that have proven critical for a number of introduced and natural avian colonists of islands (McLain et al. 1995; Veltman et al. 1996; Trewick 1997c; Legendre et al. 1999).

The "*Blitzkrieg*" model of the extirpation of moas in New Zealand, in which extirpation required no longer than a century after contact (Holdaway and Jacomb 2000), suggests that ecomorphological factors, associated by some with "evolutionary age" of faunas (Gaston and Blackburn 1997a, b; Sadler 1999), merely postpone the inevitable in the long term. Although climatic changes and other agencies have been implicated in devastation of populations, species, or faunas throughout evolutionary time (Emiliani 1982; Donovan 1989; Martin and Steadman 1999), and many basic ecological parameters (including population size, migratory habit and body size) remain critical to patterns of commoness and vulnerability of species (Lack 1942; Preston 1962a, b; Terborgh 1974; Root 1988a, b; Pagel and Payne 1996; Gaston and Chown 1999), the impacts of humans on insular endemics are numerous and almost uniformly negative (Table 81). Grassé (1977:129) observed that: ". . . man has massacred so many insular species that it is hard to know what share of this destruction was caused by other predators."

Comparatively recent human intervention aside, it is useful to consider flightlessness in a larger, strictly evolutionary context. Is flightlessness (on islands) an example of a local evolutionary optimum subject to global demise? It seems clear that flightlessness contributes to the propensity of proximate (local) extinctions through increased vulnerability to predation related to impaired capacity for escape, as well as reducing access to nest sites, roosting sites, or routes of travel. At an ultimate (global) scale, flightlessness may lead toward extinction by limiting the dispersion of populations through reduced dispersal capacities and rates, thereby reducing the likelihood of colonizing refugia and diminishing the capacity for movement among areas in response to changes in habitat quality among locales in the short or middle term. If the eventual extinction of flightless lineages of rails is comparatively likely, whatever the short-term success of such lineages, the apomorphy of flightlessness may serve as a predictor for the "pruning" of phylogenetic terminals (lineage sorting), serving as a negative "evolutionarily singular strategy" that is phylogenetically influential by way of truncation (Geritz et al. 1998).

Vermeij (1987:40) concluded: "If disallowed variants are often eliminated by ecological agencies, environments in which such agencies play only a minor role should allow the existence of phenotypes that in most other situations are forbidden." In the context of the present study, the "forbidden phenotype" in question is avian flightlessness, and the permissive environment is the isolation afforded by oceanic islands. Such conditions of tolerance do not persist indefinitely in a changing world, however, and the long-term jeopardy of flightless birds exemplifies the difference between evolutionary changes leading to short-term success while hastening extinction over the long-term. Gould (1977) emphasized this distinction in scale, and argued that ephemeral environmental opportunities often can be best exploited reproductively by paedomorphic lineages, with the latter clearly applicable to many flightless members of the Rallidae. However, it is arguable that the vulnerability of insular endemics to exogenic change is the forbidden characteristic, not flightlessness per se, given the numerous extinctions of flighted birds that have occurred during the same period.

DEMOGRAPHY OF INSULAR LINEAGES: THE EQUILIBRIUM MODEL REVISITED

Essentials of classical concepts.—The seminal monograph by MacArthur and Wilson (1967) principally forged a conceptual framework within which to relate the processes impinging on the flora and fauna of islands, especially the (comparatively short-term) roles of immigration, colonization, evolutionary changes, and extinction. Three primary phases of insular evolution were articulated by MacArthur and Wilson (1967), although the great majority of their treatise emphasized only the first phase (these were sequenced by Roman numerals in the original work): demographic characteristics of species that bear on their aptitude as potential colonists, preliminary adaptation to insular environments, and speciation and adaptive radiation. These can be recast as a graded series of organismic scales of relevance to insular evolution: repeated patterns regarding individuals within populations (e.g., Preston 1962a, b), divergent histories of conspecific populations to initial evolutionary changes (e.g., MacArthur and Wilson 1967), and long-term, large-scale changes in groups of evolutionary lineages in biogeographical regions (e.g., Darlington 1965; Cracraft 1973b, 1980; Craw 1988; Cooper 1989; Craw et al. 1999). On a faunistic scale, the implications are most obvious (MacArthur and Wilson 1967: fig. 22): rates of immigration are inversely related to distance from source populations, immigration is accelerated by the existence of potential stepping-stone islands, and rates of extinction are inversely related to areas and ecological diversities of islands. Accordingly, large continental islands are expected to support larger avifaunas than small, isolated islands, but predictions concerning endemicity and apomorphy such as flightlessness, a trait antithetical to dispersal, are made less confidently.

In sum, the models presented by MacArthur and Wilson (1967) primarily targeted the compositional changes in island communities—notably the countering effects of multispecific immigration (I), extinction (E), and survival (S), which is the net difference of these—as functions of distances between islands and continental source populations (D) and areas of islands (A). Ancillary effects of stepping-stone islands, ecological succession on the islands, and the biogeographic region of the source continents also were considered, with only secondary mention of the individual taxa that these comprise. Nonetheless, several of the principles set forth are fundamental to an understanding of the likelihood of a single colonizing population surviving to undergo significant evolutionary change, and these have formed the basis for a new, growing paradigm of metapopulations (see below; Hanski and Simberloff 1997).

Rudimentary explorations of such single-species scenarios were presented by MacArthur and Wilson (1967:68), but unfortunately these led to recursive sets of equations to estimate expected time to extinction for a minimal originating propagule (T_{K+1}), an exercise that required an estimate of K (carrying capacity for the species on the island), per capita rates of birth (λ_x) and death (μ_x) at population size x. Despite these mathematical hurdles, qualitative generalities emerged, including expected time to reach carrying capacity (T_K) increases rapidly with K and decreases with ratio λ/μ; and if K is not very small (a condition anathema to establishment), T_K is approximated by a function of the time to extinction of a single founding propagule (T_1), that is, $T_K \sim \lambda(\lambda - \mu)T_1 = \lambda/r \cdot T_1$, where r is the intrinsic rate of natural increase (Pielou 1977).

Extensions deriving from insular rails.—Except to the extent that vagility of a species covaries with other parameters of interest, distance to the source population is of little relevance to a formal model of the expected colonization, survival, and magnitude of evolutionary change of a species on an island. In many contexts the probability of a given species reaching a given island can be approximated by a Poisson or negative binomial distribution in which scaling parameters are functions of distance and area. Following the time at which the founding population (e.g., a gravid female or a pair of potential mates) reaches the island (defined as t_C), survivorship can be assumed to follow an exponential decline as a function of time. While the colonizing population persists (evolutionary change assumed as being zero at time of extinction, t_E), it would be subject to selection. Expected evolutionary change under an approximately constant selection regime might be hypothesized to follow uniform or exponential distributions; more complex or dynamic ramifications of selection, not considered here, could enhance realism. Putting aside the issue of immigration, a first approximation would derive from the combined outcome of survival and evolutionary change, a conditional probability at a given time t_i ($t_i - t_C$, $t_i < t_E$).

In addition to a prerequisite capacity for long-distance dispersal, species endowed with facile rates of reproduction and plasticity of ontogenetic parameters are predisposed for success as colonists, that is, successful colonization turns largely on r-selected traits (MacArthur and Wilson 1967:81, 93). These characteristics are associated in many taxa with a propensity to some heterochronically mediated changes (e.g., Gould 1977; McKinney and McNamara 1991), which in turn are implicated in most instances of avian flightlessness (see above). Once successfully established, colonizing species appear to shift increasingly toward K-selected attributes (e.g., protraction of reproductive life span) that are selectively advantageous in comparatively complex, competitive communities, and the mathematical, selection-neutral inevitability that the probability of extinction of any population increases asymptotically with time (MacArthur and Wilson 1967:149–151, fig. 20A). Although MacArthur and Wilson (1967:151) commented that the absence of predators on islands contributes to a freedom to allocate "evolutionary attention" to properties which in continental environments would conflict with capacity for escape, and went on to outline the three phases through which lineages undergoing significant evolutionary changes must pass, these authors stopped short of modeling the likelihood of evolutionary changes on islands (e.g., preconditions favoring flightlessness).

An elementary formulation of this kind, premised on the assumption that the total evolutionary change attained by a lineage before extinction (υ) is a function of the rate of evolutionary change in the lineage ($\partial \upsilon/\partial t$) and the period of time over which that the lineage persisted on the island ($\Delta t = \iota$), can be written:

$$\Delta(\upsilon) = \upsilon_\alpha - \upsilon_\omega = f\left(\frac{\partial \upsilon}{\partial t}, \iota\right).$$

Secondly, the rate of evolutionary change in a lineage is a function both of the genetic, epigenetic, morphological, and demographic parameters of the lineage in question (e.g., genetic diversity, ontogenetic propensity toward heterochrony, body size, specialization for flight, primary mode of foraging, deme size, longevity, age at reproductive maturity), as well as the interval of time over which this rate of change is permitted to operate. The latter interval of survivorship is largely a function of the ecogeological characteristics of the island colonized (e.g., geological age, distance to source population, total area through time, elevational range, susceptibility to catastrophic events, structural diversity and successional stage of ecological communities, diversity of competitors and predators, and presence of human immigrants and antiquity thereof). Symbolically:

$$\frac{\partial v}{\partial t} = g \text{ (qualities of lineage, characteristics of island)}.$$

Clearly, the capacity for change possessed by different species varies greatly; therefore, a simple examination of these relationships is achieved by the consideration of a single colonizing taxon. Expected survival of a given avian colonist (η)—the inverse of expected extinction (MacArthur and Wilson 1967: fig. 20A)—can be plotted against absolute (geological) time (t) for islands A, B, C, and D that differ in selected, influential parameters. Given the inescapable, asymptotic decline in expected survival of lineages over time, the curves for colonist η on islands A–D share a general form, but such curves differ in the time of initiation, rapidity of decline, and the asymptote approached. Accordingly, the relative opportunity or expectation for evolutionary change of lineage η on these islands is reflected by the integral of the corresponding curves over the time that the lineage persisted. That is, expected total change for species η on island A for the entire interval of time $\iota = t_\omega - t_\alpha$ can be summarized:

$$E\,[\Delta(v)] = \int_{t_\omega}^{t_\alpha} g_\eta(A).$$

INSULARITY: ISOLATION BOTH ARDUOUS AND ACCOMMODATING

Dual challenge of colonization.—The unfortunate opposition of Owen to the theory of Darwin, both in general and with respect to the evolution of avian flightlessness, was more than a little tainted by personal and religious differences (Desmond 1982). In light of the prime importance of the theory espoused by Darwin (1859) and the anatomical descriptions and discoveries of flightless birds by Owen (1870, 1872, 1875, 1879, 1882), this conflict impeded the reciprocal illumination requisite for the resolution of evolutionary issues related to avian flightlessness, and undoubtedly diminished the importance of flightless birds in the maturation of evolutionary biology during the first century of the Darwinian revolution.

As illustrated by the historical view of avian flightlessness as but a single, uniform phenomenon of natural history, it can be argued that an overly simplistic perspective on the nature of insularity for birds can obscure vital diversity of evolutionary mechanism and resultant forms. The selective environments presented by islands, permuted across myriad latitudinal gradients and ecological communities, presents at the very least dual prospects for avian colonists, barring whole taxonomic groups from long-term membership in these communities while nurturing intricate evolutionary diversity in others. Examination of the limited data indicates that insular populations of birds are typified by extended survivorship, a demographic generality associated with resource-dependent depressions in reproductive rates (Ashmole 1963c; Ricklefs 1980, 1983b; Faaborg and Arendt 1992, 1995). Insular conditions in many cases present mutually contradictory selection regimes that pose widely variable productivity and sources of sustenance on the one hand (e.g., the challenge of desert islands), and provide (at least partial) refugia from terrestrial predators on the other hand. In light of this duality of insular conditions, one might consider avian flightlessness alternatively to be indicative of either fragments of desolate habitat (i.e., evolution in extremis) or refinements under ecological conditions combining relief from predation pressure with a surfeit of unexploited resources (i.e., evolution in refugium). In other words, flightlessness may be innovation born of necessity, opportunity, or both.

Total balances and diverse allocations.—Unfortunately, no simple arithmetic or accounting can be applied to the economics of evolutionary changes giving rise to avian flightlessness, in that there is no basis upon which to assume equality, increase, or decrease in the net balance sheet between resources available to lineages before and after insular colonization. It is possible to envision specific circumstances that would favor one of these alternatives strongly. For example, a lineage that underwent a substantial increase in size, a shortening of developmental period, and an increase in net reproductive investments by adults and only a minor reduction (savings) in the pectoral apparatus after colonization could be considered well positioned for prosperity under favorable environmental regimes. Similarly, counterexamples in which body size declines slightly, the pectoral apparatus is substantially reduced, and reproductive output is diminished might indicate that the sacrifice of flight in this species was not sufficient to compensate for the total decline in resources faced by the colonists. In short, where the investment represented by the pectoral apparatus is postulated as having been sacrificed for other selected parameters to maintain fitness assumes that total energy balances remained invariant before and after the loss of flight, a simplistic frame of reference that may apply only rarely to actual circumstances.

TROGLOMORPHY AS CONTINENTAL PARALLEL TO INSULARITY

> The *Proteus,* an aquatic reptile allied to the salamanders, and living in deep dark caves under the water, has, like *Spalax* [blind mole-rats, specialized fossorial rodents], only vestiges of the organ of sight, vestiges which are covered up and hidden in the same way .... Light does not penetrate everywhere; consequently animals which habitually live in places where it does not penetrate have no opportunity of exercising their organ of sight, if nature has endowed them with one. Now animals belonging to a plan of organization of which eyes were a necessary part, must have originally had them. Since, however, there are found among them some which have lost the use of this organ and which show nothing more than hidden and covered-up vestiges of them, it becomes clear that the shrinkage and even disappearance of the organ in question are the results of permanent disuse of that organ.—Lamarck (1809, vol. 1; translated by Elliot 1984:116)

Discovery and distribution.—Beginning with the description of a cave-dwelling

TABLE 82. Evolutionary parallels and morphological–ecological convergences between avian endemics of oceanic (epigean) islands and endemics of caverns (hypogean islands) or troglomorphs, following Cope (1868), Packard (1888), Christiansen (1961, 1985, 1992), Poulson (1963, 1985), Barr (1968), Thines (1969), Culver (1982), Fong and Culver (1985), Hüppop (1985), Kane and Richardson (1985), Romero (1985), Sket (1985), Wilkens (1985, 1986, 1988, 1993), Juberthrie (1989), Jones et al. (1992), Howarth (1993), and Culver et al. (1995).

Character or circumstance	Comparable evolutionary changes on islands or in caves	
	Insular endemics (epigean)	Troglomorphs (hypogean)
"Atrophied" or "regressive" anatomical features	Wings, pectoral apparatus Sexual dichromatism	Eyes, wings Optic lobe Pigmentation Cuticle, scales
"Hypertrophied" anatomical features	Body mass, height (most) Pelvic limb Bill (some)	Body mass, length, flattening of corpus Elongation of appendages Olfactory lobe Antennae and tactile organs
Metabolic and physiological changes	Relaxed, conditioned schedules Lowered basal metabolic rate Enhanced capacity for fasting	Irregular activity and reproductive patterns Lowered basal metabolic rate Enhanced capacity for fasting
Behavioral and ecological changes	Lower fecundity (some) Increased egg size (some) Change in diel activity mode (some) Increased longevity (most?) Broader diet and foraging niche	Lower fecundity Reduced egg size Relaxation of rhythmicity Increased longevity Decreased aggregation (some) Reduced territoriality Broader diet and foraging niche
Circumstances accelerating changes	Harsh environment Small size of founding population Protracted period of isolation Intense directional selection	Harsh environment Small size of founding population Protracted period of isolation Intense directional selection

salamander (*Proteus anguinus*) from the Slovenian karst of Europe (Laurenti 1768), natural historians discovered thousands of cavernicolous species, many of which displayed to varying degrees a suite of apomorphies, including reduction or loss of eyes and pigmentation, hypertrophy of tactile sense organs, and elongation of the body (Culver 1982; Culver and Holsinger 1992; Culver et al. 1995). Organisms showing any or all of these convergent changes—comprising a minority of amphibians (e.g., *Proteus* salamanders), insects (e.g., Collembola), crustaceans (e.g., Isopoda), and fishes (e.g., Characidae and Amblyopsidae)—came to be known as "troglodytes"—later subdivided into obligate "troglobites" and facultative "troglophiles" (Culver et al. 1995)—and these shared some "troglomorphic" similarities with the more-specialized fossorial mammals (Nevo 1999). Early references to troglomorphs in the literature were influenced profoundly by the evolutionary views of the various authors, and provided an additional phenomenological chronicle of the evolution of theory (e.g., Lamarck 1809; Cope 1868; Packard 1888) in parallel with the treatment of avian flightlessness. In both contexts, early explanations focused on Lamarckian effects of disuse, with subsequent interpretations turning progressively toward arguments of selective advantages of economy (e.g., reductions or loss of any structure requiring substantial investment ontogenetically or for maintenance) or reduced vulnerability to injury (e.g., loss of eyes).

Evolutionary analogies and interpretational differences.—Continued study provided numerous additional examples of troglomorphy and more detailed characterization of the salient features thereof (Christiansen 1961, 1985, 1992; Poulson 1963, 1985; Barr 1968; Thines 1969; Culver 1982; Hüppop 1985; Romero 1985; Sket 1985; Wilkens 1985, 1986, 1988, 1993; Juberthrie 1989; Jones et al. 1992; Howarth 1993). This extended collection of examples revealed the extent of the convergent morphological and ecological changes in subterranean isolation (Table 82), with the most conspicuous change in troglomorphs—reduction or loss of sight—also being the one most obviously analogous to the reduction or loss of flight in insular birds. Troglomorphic trends, like the more taxonomically inclusive trends evident on oceanic islands (e.g., Carlquist 1965, 1966, 1974, 1980), included other, convergent evolutionary shifts (Table 82). Accordingly, caves and oceanic islands both came to be viewed as natural evolutionary laboratories characterized by comparative simplicity of environmental variables occurring in diverse combinations, but both presented interpretational problems (Culver et al. 1995). As with the study of avian flightlessness, the conspicuousness of the "regressive" trends among troglomorphs tended to preoccupy early investigators (Brace 1963; Peters and Peters 1968; Fong and Culver 1985; Kane and Richardson 1985; Peters 1988; Christiansen 1992; Wake 1992). However, recent studies of troglomorphs differ from those of flightless birds in the significant role accorded by some to essentially random genetic variation and neutral mutations in the directional changes manifested within small demes of cavernicolous organisms (Culver et al. 1995).

Migratory Habit and Flightlessness

Ecological Implications of Migratory Habit

What a paradox; to fly poorly, to occur so widely, and to evolve flightlessness so easily!—Ripley (1977:15, in reference to the Rallidae)

Given the omnipresent risk of local extinction, what can a species do about it? Briefly there are two types of strategies: either to concentrate on preventing it from happening, or else to let it happen and concentrate on reversing it by recolonization. The former is the strategy of old colonists, which have lost much of their dispersal ability, can rarely reverse local extinctions by recolonization and become restricted to stable or extensive habitats where extinction is rare. The latter is the strategy of expanding colonists, which have high dispersal rates and can therefore occupy small and unstable habitats, recolonizing as extinctions occur.—Diamond (1977b:257)

That birds can fly is their most important characteristic, and also one of their most misleading ones. It is often inferred that the power of flight guarantees birds the ability to cross water gaps or expanses of alien habitat that act as barriers to flightless animals. In fact, the realization has been growing in recent years that many flying birds *choose* not to fly across barriers of water or alien habitat: they are psychologically flightless.—Diamond (1985:18)

It requires special conditions for the birds to be able to get by without the capacity for flight during the moulting period. Rails and crakes live under cover of tall, dense vegetation in marshes and on lakeshores, and cope very well with catching food, even though they are flightless for several weeks.—Alerstam (1990:69)

Generalities of pattern.—Much of the theory surrounding the evolutionary origins and geographical patterns of avian migration is grounded in terms of interspecific competition, variability of environments, and the relative costs of long-range movements (MacArthur 1959; Ulfstrand 1963; Castro 1968, 1989; Haartman 1968; Lack 1968c; Fretwell and Lucas 1970; Fretwell 1972, 1980; Wilson 1976; Alerstam and Högstedt 1980; Cox 1985; Pienkowski et al. 1985; Piersma 1987, 1997; Holmgren and Lundberg 1993; Meltofte 1996; Newton and Dale 1996). Migration and its evolutionary origins must be considered in terms of selection acting on individuals and ecological contexts; interspecific comparisons of modes of dispersal (e.g., Diamond 1977b, quoted above) can lead, perhaps unintentionally, to implicitly group-selectionist arguments.

The primary selective advantage of migratory habit is considered to be access to spatiotemporally disjoined resources; consideration of increased mortality during migration remains largely anecdotal (e.g., Wiedenfeld and Wiedenfeld 1993) and the evolutionary origins remain a point of debate (Alerstam 1990; Berthold 1993, 1996). Evolution of the migratory habit in birds is favored selectively in species inhabiting environmentally unstable and unpredictable ecological communities, conditions that do not select for high fidelity to natal site and can favor prospecting for new, favorable habitats. Where such movements provide access to seasonally abundant resources, regular movements (i.e., migration) represent a behavioral attribute of substantial selective advantage (Alerstam and Enckell 1979). Migratory habit represents an extreme among possible modes of dispersal and potentially predispositional conditions (Johnson and Gaines 1990), and evidently evolves only under the most stringent of selection regimes and ecological circumstances (Kokko and Lundberg 2001; Mathias et al. 2001). Accordingly, a binary classification of migratory status is artificial and obscures a continuum within the trait, the latter illustrated by "partial" migration (Lundberg and Alerstam 1986; Lundberg 1987, 1988; Chan 2001). Such artificial treatments render problematic the assessment of human-induced changes in habitat on migratory species (Sutherland and Dolman 1994; Sutherland 1996), some dimensions of which have been confirmed by intraspecific variation in expression (Swingland 1983; Sutherland 1998) and genetically based, short-term changes in migratory behavior (Berthold et al. 1992). Any explanation of avian flightlessness must account for the sacrifice of migratory habit as a strategy of potentially critical magnitude, with exceptions being limited to penguins (Livezey 1989).

Physiological demands of migration.—Migration places extreme ecophysiological stresses on birds, and migrational patterns in time and space reflect a complex of trade-offs in critical parameters (Salomonsen 1955; Odum et al. 1961; Gauthreaux 1982; Sinclair 1983; Alerstam and Lindstrom 1990; Welham 1992; Alerstam and Hedenström 1998; Alexander 1998b; Farmer and Wiens 1998; Berthold 1999). The critical nature of these demands is indicated by the aerodynamic accommodation of changes in physical environment by migrating birds, for example, direction and speed of travel with respect to prevailing winds (Alerstam 1979a, b, 1991; Bloch and Bruderer 1982; Wege and Raveling 1984; Liechti 1995; Butler et al. 1997; Liechti and Bruderer 1998), mode of flight (Hedenström and Alerstam 1992; Hedenström 1993; Weber et al. 1998a, b), and formation of aerodynamically efficient flocks (Heppner 1974). Although the life histories of avian taxa can be viewed as a complex of trade-offs among critical resources, each of which is subject to natural selection (Perrins and Birkhead 1983; Feder et al. 1987; Blem 1990, 1999; Perrin 1992; Schmidt-Nielsen 1997), quantification of the net gains realized through the termination of migration or (more extremely) flight remains a direct if elusive problem.

One such economic approximation is afforded by the relationship between body mass and energetic demands of migration (Masman and Klaasen 1987), anatomical differences between migratory and nonmigratory populations that are related to flight (Fry et al. 1972; Lundgren and Kiessling 1988; Bishop et al. 1996, 1998; Rising 2001; Tellería et al. 2002), simulations of caloric consumption during migratory flight (Pennycuick 1998c), and seasonal changes in organs within populations (Nisbet et al. 1963; Marsh 1981, 1983, 1984; Lundgren and Kiessling 1985; Evans et al. 1992; Driedzic et al. 1993; Piersma et al. 1993; Piersma and Lindström 1997; Piersma 1998; Piersma and Gill 1998; Kullberg et al. 2002). There also is strong direct evidence for the magnitude of the physiological demands of migration, notably adjustments in metabolic rates (Dawson et al. 1983; Kersten and Piersma 1987; Klassen 1996; Weber and Piersma 1996), changes in intake of water (Carmi et al. 1992; Klassen 1995), rates of respiration (Crnokrak and Roff 2002), and associations with schedules of molt, another seasonal event with substantial energetic and aerodynamic implications (Yapp 1962b; Walker 1970; Owen and Ogilvie 1979; Zwarts et al. 1990; Holmgren et al. 1993; Holmgren and Hedenström 1995). The caloric demands of molt are significant (Murphy 1996), a finding mirrored by the allometric relation by Turcek (1966) that indicates that roughly 6% of body mass in birds is comprised by feathers, with smaller songbirds reaching proportions twice as high (Table 77). Furthermore, morphological subtleties such as the shape of the wings and tail (Kipp 1959; Rensch 1959; Norberg 1995a, b; Fitzpatrick 1998; Calmaestra and Moreno 2001) also appear to have been favored selectively to meet the rigors of migratory habit, with mechanical constraints among most aerodynamic parameters imposing intercorrelations among most such changes (Pennycuick 1978, 1986).

Perhaps the most direct validation of the energetic investment necessitated by flight is the increased storage of fat, proteins, and erythrocytes during migration, especially gains before departure and during stopovers and the corresponding losses during transit, generalities that are substantiated by an overwhelming body of literature (see reviews by Piersma 1990; Zwarts et al. 1990; Lindström 1991; Lindström and Alerstam 1992; Lindström and Piersma 1993; Marks and Redmond 1994; Biebach 1996; Weber and Houston 1997; Jenni and Jenni-Eiermann 1998; Weber et al. 1998b). A substantial portion of the investment in pectoral musculature for migration involves reinforcement of myofibrils and increased density of mitochondria (Evans and Davidson 1990; Evans et al. 1992), parameters known to be of functional import to migration from first principles (Pennycuick 1978). Increases in body mass before migration typically exceed 30%, and some estimates range as high as 70% (Lindström and Piersma 1993). Conversion of such stores toward other investments (e.g., breeding, territoriality, or fasting) would have obvious implications for waifs in new environments.

However, energy required for migration is not simply optimized as fuel for flight per se, but has implications for general life histories (O'Connor 1981, 1985, 1990; Pienkowski and Evans 1985; Levey and Karasov 1989; Rogers et al. 1994; Karasov 1996; Parrish 1997), aerodynamics (e.g., takeoff, flight speed, maneuverability, and rates of ascent; Tucker and Schmidt-Koenig 1971), and vulnerability to predation (Metcalfe and Furness 1984; Lima 1986, 1998; McNamara and Houston 1990, 1994; Hedenström 1992; Hedenström and Alerstam 1992, 1997; Houston et al. 1993, 1997; Witter and Cuthill 1993; Klassen and Lindström 1996; Kullberg et al. 1996; Houston 1998; Lind et al. 1999). Moreover, many of the relationships between energy stores and mortality during migration also pertain to other intervals in the annual cycles of birds, notably mortality during periods of thermal extremes, deprivation of moisture, or fasting (Blem 1976, 1990; Cherel et al. 1988, 1992, 1993; Houston and McNamara 1993; Levey and Martínez del Rio 2001).

MIGRATION, VAGRANCY, AND COLONIZATION

> Few more versatile or successfully diverse bird families could be imagined than these omnivorous, generalized marsh birds—loathe to fly, difficult to stop once in the air, and with a curious evolutionary predilection to disregard normal avian rules of migration.—Ripley (1977:21–22)
>
> These marsh birds have an uncanny ability to migrate and apparently 'pop up' somewhere else, and they seem to have hardy and persistent patterns for survival.—Ripley (1977:31)
>
> In summary, because rails and gallinules account for some spectacular instances of vagrancy, with some records thousands of kilometers from their known breeding range, and because their global distribution pattern indicates that they are champion dispersers and colonizers, we the think that extralimital records of these birds should be regarded in general as representing wild vagrants unless there is some specific reason to think otherwise.—Remsen and Parker (1990:397)

Distributional and migratory patterns of rails.—Flighted species of Rallidae rarely fly for considerable distances except during migration; during much of the year, swift and stealthful ambulatory transit evidently suffices for foraging and escape for most rallids (Ripley 1977; P. B. Taylor 1996, 1998). Given the extraordinary implications of flight for anatomy and physiology of birds (e.g., Pennycuick et al. 1988; Blem 1990; Butler and Woakes 1990; Butler 1991; Rayner 1993, 1996) and the generality that migration (where practiced) represents the most protracted, intensive use of flight normally encountered by most avian species (Dorst 1962; Pennycuick 1969; Alerstam 1990; Rayner 1990), an assessment of the surrender of migration by most flightless birds (including flightless rallids) holds substantial promise for gaining insights into the evolutionary implications of this "reductive" evolutionary phenomenon. Potential colonists face the dual hurdles of high vagility required to reach islands, and qualities of survivorship necessary to maintain a foothold, which in some lineages (e.g., many members of the Rallidae) includes the sacrifice of the apparatus by which the colonists reached the islands. Differences among taxonomic groups in efficiency of terrestrial locomotion (Bruinzeel et al. 2000) suggest that there is substantial potential for ambulatory refinement in lineages freed from the anatomical constraints imposed by flight, and the terrestrial agility of rallids (regardless of circumstance) suitably equips the group for refinements of this kind after insular colonization.

It can be assumed that flightless rallids derive principally or entirely from colonists having at least some, probably considerable, migratory abilities. Moreover, migratory habit of founding populations was crucial to the evolution of flightlessness of insular rallids in at least two ways: as enhancement of the opportunities for waifs straying to oceanic islands of sufficient isolation to engender flightless lineages, and intensification of the selective advantages of surrendering flight on islands through the increased economic advantages of reductions or practical loss of the robust flight apparatus characteristic of migratory founding populations. That is, the large pectoral investment of the "ideal progenitor" of a flightless rail has a dual nature in that it both permits the original dispersal and may constitute resources after colonization in a manner similar to economics bridging breeding and nonbreeding seasons (Doherty et al. 2001). The implication that the plesiomorphy of migratory capacity sets the stage for intensified selection in an insular context may explain the apparent rapidity with which flightlessness evolves in rails, despite the generally weak selection documented in natural settings (Kingsolver et al. 2001). This short-term change within individuals would be rendered permanent and more profound through selection for alternative allocations of resources over subsequent generations, ultimately leading to the complete and permanent loss of flight in those taxa having habits permitting this extreme.

Vagrancy and permanent migratory stopovers.—Qualification of rails as consummate vagrants is supported by extralimital records for most tribes, including *Porphyrula martinica* (Olson 1972; Siegfried and Frost 1973), *P. flavirostris* (Remsen and Parker 1990), *Pardirallus maculatus* (Parkes et al. 1978; Dod 1980), *Rallus caerulescens* (P. B. Taylor 1996, 1997, 1998), *R. longirostris* (Crawford et al. 1983), *Gallirallus philippensis* (Schodde and de Naurois 1982), *Coturnicops notatus* (P. B. Taylor 1998), *Porzana marginalis* (Taylor 1996, 1997, 1998), *Neocrex erythrops* (Lévéque et al. 1966; Osborne and Beissinger 1979; Taylor 1998; Watson and Benz 1999), *Tribonyx ventralis* (Braithwaite 1963), *Gallinula chloropus* (Ritter and Sweet 1993; Worthington 1998), and *Fulica americana* (Pratt 1987). In combination with a tendency to migrate in flocks (Tordoff and Mengel 1956; Pulich 1961; Stoddard 1962; Thompson and Ely 1989) and generalized feeding habits (see below), this renders them uniquely qualified as potential colonists.

The potential for migratory movements to foster the discovery and occupancy of new habitats holds dual, essentially opposite implications with respect to the evolution of flightlessness (Alerstam and Enckell 1979): *colonizational,* enabling the founding population to reach habitable, isolated islands; and *postcolonizational,* being strongly disadvantageous on an oceanic island distant from other habitable areas, and hastening selectively favored shifts toward sedentary habit and (in some groups) flightlessness. What genetic mechanism(s) would enable the retention of such diverse options within the genome of the colonizing species? In species showing dispersal (including migration), demographic cohorts frequently differ in the timing and destinations of movements and the means by which these habits are refined through prospecting (Reed et al. 1999). At least some of the more pugnacious rallids are known to maintain territories away from the breeding grounds (e.g., *Gallinula* [Wood 1974]), and such sequestration of resources during the first year of life may render prospecting for nontraditional wintering grounds advantageous for immature birds (Matthysen 1993). A distillation of the literature on differential migration by Cristol et al. (1999)—in which body size, social dominance, and time of arrival may act individually or in combination to determine relative migratory radii within species—raises the likelihood that comparatively small, inexperienced, and subordinate birds, especially males (Norris and Stutchbury 2002), would be induced to migrate farthest and thereby reach new, isolated of habitats more frequently than older conspecifics. An alternative view by Greenberg and Mettke-Hofmann (2001) holds that the neophilia of juveniles and ecotonal specialists qualifies them as potentially optimal colonists of islands (Morse 1971, 1977).

If the newly attained parcel of habitat is a sufficiently isolated island that permits year-round habitation, the waifs may not attempt or succeed in returning to the traditional breeding grounds. This scenario would be especially likely where the colonists are true vagrants that reached the island through a substantial departure from traditional migratory routes through inexperience, unfavorable conditions for navigation, strong winds in opposition to the normal direction, or a combination of these variables. The influence of prevailing winds on likelihood of vagrancy during migratory periods is substantial (Pennycuick 1978), and genetic or congenital misorientational behavior also may play a critical role (Alerstam 1990; Berthold 1993, 1996). Elkins (1988) proposed that some purported examples of vagrancy instead may represent pioneering that is not entirely passive, but instead the exploitation of weather-related opportunities to explore beyond normal migratory routes. Failure to depart the island would be more likely for individuals lacking well-developed philopatry or navigational skills, a condition more typical of immature birds (Reed et al. 1999). The comparatively exploratory nature of first migratory passages by young birds also increases the probability of vagrancy by immature birds. In addition, females generally show greater variation and lower philopatry than males in first migrations (Gauthreaux 1982; Alerstam 1990). Accordingly, the most likely demographic cohort to serve as colonists among rails would be young females.

Viewed in this way, islands colonized by ostensibly wayward rallids essentially amount to permanent migratory stopovers, that is, locations reached during adverse weather conditions that possessed critical combinations of isolation and habitability that favored residency over egress. Moreover, given the likelihood that even the short-term atrophy of flight musculature and energy stores would preclude a successful departure from a remote island, areas colonized in this way can be conceived as well as insular traps for migrational strays. Although some avian species show increased dietary plasticity during migration (Parrish 2000; Levey and Martínez del Rio 2001), which might persist indefinitely in waifs ultimately founding new resident populations, there is evidence of preferences and selection by migratory landbirds for specific habitats during stopovers (Moore and Aborn 2000; Petit 2000), extending in some taxa to specificities differing between sexes and age groups (Woodrey and Chandler 1997; Woodrey and Moore 1997; Yong et al. 1998; Ornat and Greenberg 1990). The latter suggests that a substantial number of oceanic islands reached by potential colonists would not sustain permanent populations for extended periods of time, a likelihood increased by a prevalence of vagrancy by juveniles. This limitation almost certainly would pertain to the numerous, sparsely vegetated atolls in the South Pacific (Nunn 1994), and represents an upper limit on estimates of pristine endemicity of rallids (cf. Steadman 1995).

Colonists of remote islands, no longer subjected to selection for maintenance of migratory ability (and therefore flight capacity), likely would be subjected to directional selection on the grounds of somatic economy toward the surrender of flight. That is, absence of predators is not the sole agent of selection promoting flightlessness in insular rallids. Instead an absence of terrestrial predators is but one factor contributing to the relaxation of selection for maintenance of flight in those lineages in which aerial foraging and migration can sacrificed. This then fosters opportunities for selectively advantageous changes in anatomical investments and gains in physiological strategies (see below). The potency of selection on body form for purposes of eluding predators can be seen, among other manifestations, in the morphological "arms race" represented by the wing forms of avian predators and their avian prey (e.g., Pennycuick et al. 1994). In this sense, anatomical and physiological refinements made possible by flightlessness are the adaptive sequelae of economic gains realized by a non-migratory lifestyle in the largely predatorfree, but often arduous environments of oceanic islands, the original footholds on which were facilitated by the "irregular or irruptive patterns of migration" (Ripley 1977:12) and wide ecological tolerances of many migratory, continental rails.

The perspective of insular colonizations as permanent stopovers raises the possibility that documented demographic and ecological characteristics of migratory flocks may hold implications for the likelihood of successful establishment of founding populations. Evidence of segregation of migratory schedules and routes by sex and age (Woodrey and Chandler 1997; Woodrey and Moore 1997; Yong et al. 1998; Bensch and Nielsen 1999), as well as habitat (T. E. Martin 1980) may suggest avenues for expansion of niches utilized by colonizing species, opportunities countered by habitat and topographic preferences shared by entire species during periods of transit (Farmer and Parent 1997; Kelly et al. 1999).

Rails as exceptional "weedy" colonists of islands.—Many species of landbird have occurred at least rarely as wind-driven waifs at sea (Alerstam 1990; Berthold 1993, 1996), but only the members of the Rallidae consistently converted these long-odds dispersal events to achieve a global distribution rivaling that of any other family of terrestrial tetrapods (Olson 1973a). The far-flung distribution of rails contrasts with those of several ecologically similar families of shorebirds (Charadriiformes), the likely sister-order of the Gruiformes (Livezey 1998). Numerous genera of semiaquatic forms in the Charadriidae (e.g., *Vanellus, Pluvialis,* and *Charadrius*) and Scolopacidae (e.g., *Scolopax, Lymnocryptes, Gallinago, Limosa, Numenius, Bartramia, Tringa, Actitis, Heteroscelus, Arenaria,* and *Calidris*) include members characterized by strongly migratory habits, terrestrial feeding methods, high vagrancy, and a tendency to make migratory stopovers on oceanic islands. The absence of a single flightless member in these groups is doubly striking in light of the occurrence of species endemic to single islands or archipelagos—for example, charadriids *Charadrius sanctaehaelenae* and *C. novaeseelandiae* of St. Helena and Chatham islands, respectively; scolopacids *Prosobonia cancellata* and *P. leucoptera* (extinct) of Tuamotu and Tahiti, respectively; scolopacids *Coenocorypha pusilla* and *C. aucklandica* of the Chatham and Auckland islands, respectively; and haematopodids *Haematopus meadewaldoi* (extinct) and *H. chathamensis* of the Canary and Chatham islands, respectively—all of which retain(ed) the power of flight.

Similarities between the Rallidae and many taxonomic families of Charadriiformes also extend, at least in some taxa, to the semiaquatic foraging habitats and specialized exploitation of comparatively challenging invertebrate prey (Graves 2001). Some migrant charadriiforms may be constrained stringently to follow traditional migratory routes (and thereby avoid vagrancy) in order to exploit local, seasonal food resources (Castro and Myers 1993; Clark and Niles 1993; Botton et al. 1994; González et al. 1996; Engilis et al. 1998), a condition that does not pertain to the generally opportunistic foraging habits and transient habitats of most migrant rails (Remsen and Parker 1990). However, in spite of this potential difference and the possibly greater predatory pressures suffered by charadriiforms that typically forage in more open habitats than most rallids, many migrant taxa of Charadriiformes presumably had numerous biogeographical and general ecological opportunities to follow the Rallidae on the evolutionary pathway to flightlessess, but evidently lacked the selection pressure(s) or ontogenetic means to do so.

Can the greater success of insular colonization by rallids be explained, at least in part, by a prevalence of r-selected, "weedy" traits by the colonists? Although members of the Scolopacidae have a moderate frequency of lek-mating and polyandry, departures both from typical monogamy and biparental care are known in some rallids as well (Appendix 6; e.g., *Tribonyx* and *Gallinula* [Garnett 1980; Gibbs et al. 1994]). Reviews of reproductive parameters (Starck and Ricklefs 1998d; Ligon 1999) indicated that shorebirds and rails show a comparable range of mating systems, nest sites, egg sizes, and developmental parameters. Clutch sizes also show broad similarities between the Charadriiformes and Rallidae, with the strong modality in the former being four with a minority having smaller mean clutch sizes (Piersma 1996a, b).

However, an analogy of rails as "weeds" among birds gains support from the relatively larger clutch sizes of flighted rails most closely related to a number of flightless species (e.g., flighted *Rallus, Gallirallus,* many *Porzana, Amaurornis,* and most Fulicarina). These flighted species average two or more in excess of the four-egg clutch typical of other rails and many charadriiforms (Appendix 7), whereas the flightless derivates of this fecund minority show reductions in clutch sizes to the familial mode of four or fewer eggs. In those rallid genera including only a very small number of flightless members (e.g., *Porphyrio, Ortygonax,* and *Dryolimnas*), flighted members have more typical clutch sizes. Also consistent with the "weed" theory are distributional patterns in these genera indicating that comparatively short dispersal events gave rise to flightlessness (Fig. 6). These early stages of dispersal and initial colonization, closely consistent with the first and second phases defined by MacArthur and Wilson (1967), are followed by a variant of the third phase described by these authors in which increasingly K-selected attributes for long-term competitiveness, including a sacrifice of flight for other attributes, are typical. Finally, the widespread, long-distance vagrancy of young rails is related, in part, to an ability to rest on water during transit, a capability (shared by anatids) that improves the likelihood of successful traversal of extensive water barriers (P. B. Taylor 1998).

It may be that only a comparatively small subset of taxa are the flighted sister-taxa of the vast majority of flightless species, with the most likely candidates being the transcontinental migrants (most from the Northern Hemisphere) that are characterized by unusually high fecundities; the latter allow for the strategy (by lessened relative penalties) of reproductive investments in vagrancy-prone offspring or (if optimization with respect to reproductive variance pertains) bet-hedging. Specific taxa qualifying for flighted members of clades giving rise to flightless species by way of long-distance, migratory peregrinations include *Rallus aquaticus, R. limicola, R. elegans, R. longirostris*-group, *Gallirallus philippensis*-group, *G. striatus, Laterallus jamaicensis, Coturnicops noveboracensis, Crex crex, Porzana carolina, P. porzana, P. parva, P. pusilla, Gallinula tenebrosa, G. chloropus*-group, *Fulica atra,* and *F. americana*). Although these taxa share several critical qualifications for disseminating insular colonists, these rallids presumably would possess different likelihoods of successful colonization of islands possessed of different habitat(s); for example, *Rallus* would be optimal for marshlands, Crecina for meadows, and *Fulica* for open wetlands. All of these high-fecundity, long-distance migratory species are consistent with a general correlation between latitude and prevalence of migratory habit in birds (Diamond 1985).

Several Pacific rallids have distributions consistent with frequent inter-island dispersal and survivorship on islands of diverse habitats (e.g., *Gallirallus philippensis*-group and *Porzana tabuensis*) that qualify as "tramp" species (sensu Diamond 1974; see also McNab 2000), and on empirical grounds appear to include or be closely related to insular populations ultimately destined to flightlessness (Olson 1973a, 1977; Ripley 1977; Livezey 1998). Although most flightless rails were restricted to single small islands or groups thereof (P. B. Taylor 1996, 1997, 1998), the inclusion of many geologically mature islands and obsolete vagility precludes any of these species being considered "supertramps" (sensu Diamond 1974). By contrast, columbids (e.g., *Ducula pacifica*) that are typical of a number of small, often geologically young islands qualify for the status of "supertramp," some of which may be specialized for these habitats through depressed mass-specific BMRs and apomorphically small body size (McNab 2000).

ECOPHYSIOLOGICAL CORRELATES OF FLIGHTLESSNESS IN RAILS

HOME RANGE AND TERRITORIALITY

> The preadaptation of Rallidae to dispersal to islands, survival on islands, and evolution into flightlessness there becomes obvious . . . . Because they are somewhat migratory in their feeding habits and territory-bound as passerine birds typically are, Rallidae are likely to occur as stragglers to islands . . . . On islands they will survive because of their broad food tolerances. Their feeding at ground level or in shallow water continues. . . . If an island can support a rail population at all (and evidently small islands can), the food supply can be exploited without resort to flight. The only further requisite for evolution into flightlessness is then the absence of predators. Even flight for evasive purposes may not be necessary.—Carlquist (1974:491)

General avian predictions and patterns.—Clearly, body size is of fundamental importance to basic ecological parameters of terrestrial organisms, both continental and insular, and figures prominently throughout the evolutionary theory and prime currencies of life-history strategies (size and age at maturity, size and number of propagules, and longevity), and the trade-offs among them (Roff 1992). Large body size in birds has been shown to be positively correlated with diet (Case 1978a), stability of population densities (Brown 1964, 1969; Verner 1977; Peters and Wassenberg 1983; Martin 1996), longevity (Botkin and Miller 1974; Lindstedt and Calder 1976, 1981; Calder 1982a, 1983a, b; Saether 1989), age at which adult body mass is attained (Taylor 1968), size of home range (Harested and Bunnell 1979), and mean height of nests (Burger and Trout 1979). Body size also tends to be correlated with prey size, especially maximal size of prey (Levinton 1982b; McKinney 1988b; McKinney and McNamara 1991), dominance-mediated foraging efficiency (Brown et al. 1978; Alatalo and Moreno 1987), and capacity for fasting (Brown et al. 1978; Cherel et al. 1988). Not surprisingly, interactions of these size-related ecological allometries can affect predator–prey dynamics (McNamara and Houston 1987; Emerson et al. 1994).

Large body size is correlated with sizes of territory and home range (McNab 1963; Schoener 1968; Mace and Harvey 1983). Avian territoriality can serve several important ecological functions and may be intra- or interspecific in nature (Nice 1941; Tinbergen 1957; Orians and Wilson 1964; Davies 1982; J. H. Kaufmann 1983), and in turn is a major determinant of the distribution of breeding birds (Fretwell and Lucas 1970). Where interference competition occurs, whether intraspecific or interspecific, large body size generally imparts special advantage (Persson 1985; Brown and Maurer 1986). Differences in body size among species generally are considered to be critical to the mitigation of interspecific competition (Wilson 1975; Roff 1981; Pöysä 1983; Simberloff 1983a, b).

Spatiotemporal ecology of flightless rallids.—Large body size and flightlessness have important implications for foraging and territoriality in flighted rallids, especially those feeding primarily in comparatively open, aquatic habitats, for example, *Porphyrio* (Craig 1976, 1977, 1979, 1980b, 1982; Hunter 1987a), *Gallinula* (Petrie 1984, 1988), and *Fulica* (Gullion 1953b; Fjeldså 1973, Cave et al. 1989). In some of these same open-habitat species, aggression between parents and offspring also is heightened (Leonard et al. 1988). Accordingly, body size and associated territoriality almost certainly characterized some extinct flightless rallids, at least those in which habitat permitted visibility and the associated elaboration of visual displays and areal defense (e.g., *Fulica chathamensis*-group); of extant examples, territoriality certainly characterizes *Tribonyx mortierii* and its flighted relative *T. ventralis* (Ridpath 1972a–c).

Current populations of flightless endemics for which human-related effects remain negligible (predator-free populations of *Gallirallus greyi* and *Tribonyx mortierii*) substantiate that natural populations of other flightless rails generally may have reached high densities. Estimates of population densities of several flightless rallids (e.g., *Atlantisia rogersi* and *Porzana palmeri* [Baldwin 1947; Fraser et al. 1992]) indicate that under pristine conditions, flightless rails often constitute one of the dominant components of their insular communities. Although the demographics of many flightless rallids never will be known, relatively high, stable population densities probably characterized large, sedentary species during pristine times, a speculation consistent with the abundance of preserved subfossil elements of extinct endemics of the Chatham Islands (e.g., *Diaphorapteryx hawkinsi, Cabalus modestus,* and *Gallirallus dieffenbachii*). High population densities characterize many vertebrates endemic to islands (e.g., Kramer and Mertens 1938; Kramer 1946; Case 1983), although this parameter is interrelated with other demographic and ecological conditions (Grant 1998a). Based on the "wild nonpasseriform" model of Lindstedt and Calder (1976) and actual and estimated body masses of selected rallids (Table 8; Appendix 1), mean longevities of flightless rails, while respectable, do not approach the life spans estimated for such ponderous, sedentary forms as the raphids (Livezey 1993b).

Evidence for increased levels of territorial activity and (in some taxa) physical combat with increased population densities or ecomorphological similarity (Cody 1969, 1974; Pianka 1976; Abbott 1977; Abbott et al. 1977; Oksanen et al. 1979; Wiens and Rotenberry 1981; Arthur 1982b; James and Boecklen 1984; Wiggins and Møller 1997) makes it doubly likely that those characteristics of advantage in the acquisition and defense of territories—e.g., body size, armature, phenotypic "badges" of adulthood, and agonsitic displays (Crawford 1978; Miskelly 1981; Rohwer 1982; Eddleman and Knopf 1985; Persson 1985; Alisauskas 1987)—would have been especially well developed in flightless rails. In addition to those cases known for extant taxa (Gullion 1951; Burger 1973; Ryan and Dinsmore 1980; Alisauskas 1987; Cave et al. 1989), the cranial indications of a crista frontalis carnosa (a frontal shield) in *Diaphorapteryx hawkinsi* are consistent with an intensification of territorial signals, perhaps seasonally enhanced in coloration (Visser 1988), in an insular rail.

The likelihood of interspecific competition being especially influential on structures of insular communities notwithstanding, an alternative means having comparable effects may apply, at least to a limited extent, in the context of birds on oceanic islands. This structuring effect is that of apparent competition or competition for predator-free space (Holt 1977; Jeffries and Lawton 1984; Holt and Kotler 1987; Schmitt 1987; Holt et al. 1994; Holt and Lawton 1994; Schmidt and Whelan 1998) that emulates at least the spatial ramifications of interspecific competition in the traditional sense. On oceanic islands, competition for space among birds subjected to reduced or negligible predation is obvious and virtually universal among pelagic seabirds (including flightless alcids and penguins), in which defense of a nest site within a dense aggregation of nesting birds is critical to nesting success (Brown 1987; Koenig and Stacey 1990; Ligon 1999). Among avian species that are permanent residents of oceanic islands, including flightless landbirds (e.g., rails), where predation may be limited to aerial and submarine threats, similar competition for favored, concealed nest sites and areas for rearing of precocious young also may be important and resemble traditional competition within and between species (e.g., Holt 1977; Schmitt 1987; Holt et al. 1994), the resemblance may extend to mortality from starvation and predation associated with such spatial structuring (McNamara and Houston 1987). Among the flightless rails known only from subfossil remains, the coots (e.g., *Fulica chathamensis*-group) are perhaps the most likely candidates for such spatial competition related to avoidance of predators of young.

Comparison with other flightless birds.—Body size has an important role in the dispersion and territoriality of other flightless carinates (Livezey 1988, 1989a, 1990, 1992a, 1993b, c). In the flightless steamer-ducks (*Tachyeres*), large body size, great sexual dimorphism, year-round residency at a single site, habitats conducive to great visability, and predictability of resources are associated with uniquely intense, evidently multifunctional territoriality (Livezey and Humphrey 1984a, 1985a, b, 1986; Livezey 1987, 1989e). Whether the extinct, flightless raphids defended feeding territories sensu stricto is not clear; although combative behavior was conspicuously developed in at least *Pezophaps solitaria* (Livezey 1993b), as in other tropical frugivores, seasonality of food supply alone would be conducive to large body size (Fogden 1972).

METABOLISM, THERMODYNAMICS, AND LOCOMOTORY MODULES

... in the carinate or flying birds loss of flight, while it has occurred (flightless rail, penguin, dodo, etc.), is relatively extremely rare. Flightless pterosaurs and bats, on the other hand, are inconceivable, as their flight mechanism involves the hind limbs which have, as a consequence, largely lost their terrestrial locomotion function; whereas birds, being double adapted, can lose their flying powers and still progress easily on the ground or in the water, as their legs are not thus involved.—Lull (1935: 327–328)

On the other hand, flight frees the tetrapod from the trammels of earth-bound locomotion; the flying vertebrate is free to move about over areas of considerable size and to cross many barriers that limit the movements of land-living animals. These advantages of flight need no elaboration; the distribution and success of the modern flying birds are indications of the benefits of being able to fly.—Colbert (1955:171–172)

One cannot stress too strongly that of all natural methods of locomotion, flight is the most highly expensive of energy; therefore, in any condition under which flight is not a necessity for a volant organism, there will be a positive selective pressure for flightlessness.—Carlquist (1974:491)

In short, we would expect that, on islands, metabolically conservative species should do well, because this reduced expense per bird aids their survival through stress situations and allows them to exist for long periods of time in this restricted situation.... On islands, especially small ones, resistance to extinction becomes all important and the nonpasserine characteristics dominate.—Faaborg (1977:911)

Thus, rails of all sizes and types seem almost equally able to colonize island archipelagos.... Multiple invasions of similar stock have occurred, spaced out in time as evidenced circumstantially by the resulting divergences in size or other special modifications such as flightlessness.... In individual cases, members of all these types or subgroups of rails have developed larger or smaller, longer-billed or shorter-billed, or flightless forms to adapt to competitive interspecific situations, a variety of food resources, a range from littoral to forest or mountains, and flightlessness for an island, predator-free environment.—Ripley (1977:19–21)

General avian principles.—Once again, physical size of organisms exerts fundamental influence over a variety of characteristics and functions of life. Among those of a thermodynamic or physiological nature are the "surface law" or the geometric relationship between surface area to volume of three dimensional bodies with varying body size, that is, surface area of a body increases as the square of an increase in linear dimension(s), whereas volume and mass (three-dimensional corollaries of linear change) vary as the cube (Calder 1984; Schmidt-Nielsen 1984). The resultant ratio of two-dimensional and three-dimensional corollaries of linear changes—the most frequently cited being the surface:volume ratio—is imputed to underlie thermodynamic efficiency of large endotherms, including those of the class Aves (Leighton et al. 1966; Walsberg and King 1978). This includes influences on metabolic rate (Gray 1981), tolerance of thermal stress (Tanner 1949; Scholander et al. 1950a–c; Weathers 1979; Walsberg 1983), relationship between energetic efficiency and mortality (Priede 1977), and association between BMR and climatic regime (Hails 1983). Similar relationships pertain to a wider array of tetrapods (e.g., Herreid and Kessel 1967; Kleiber 1972; Elgar and Harvey 1987), although, as with birds, the traditionally cited generalities fall far short of total explanatory power (Speakman 1996). Naturally occurring trends in these dimensional aspects of organisms typically are attributed to selective advantages of thermodynamics related to climatic patterns (Rensch 1948; Schindewolf 1950; James 1970, 1983; Yarbrough 1971; McMahon 1973; McKinney 1990a, b).

In concert with these comparatively direct thermodynamic corollaries of change in gross body size, a diversity of other ecophysiological parameters are known to vary allometrically with body size (Table 77; Calow 1977; Blueweiss et al. 1978; Donhoffer 1986; Sibly and Calow 1986; Barbault 1988), including vulnerability to evaporative water loss (Williams 1996a), tolerance of thermal stress, including torpor (Scholander et al. 1950a–c; McNab 1966, 1983, 1989; Dawson and Hudson 1970; Calder and King 1972; Calder 1974; Weathers 1981; Reinertsen 1983; Dawson and Marsh 1989; Marsh and Dawson 1989; Brown et al. 1993; West et al. 1997), and cardiovascular efficiency (Zeuthen 1953; Lasiewski and Calder 1971; Grubb 1983; Maina et al. 1989). Effects of ambient temperature on locomotion and activity levels were documented in birds (Paladino 1985), with torpor being an extreme case (McNab and Bonaccorso 1995), but the thermodynamics during development differ from those for adults. During ontogeny, the thermoregulatory effects of increasing body size (with body size being a better predictor of metabolic rates and thermal conductance than developmental mode) are complicated by intrinsic changes in physiology, with the rapidity with which thermoregulation is achieved being one of the fundamental differences between avian developmental modes (Visser 1998).

Metabolic rates vary with body size, and although less intuitive than such corollaries as the surface:volume ratio, are no less important from functional or evolutionary perspectives. Mass-specific BMR—the average rate at which energy is consumed by a resting organism at thermoneutrality per unit body mass (Blem 1999)—is negatively allometric with respect to mean body mass in birds, although total BMR may be more critical from an evolutionary standpoint (McNab 1999). Therefore, larger species tend to expend less energy per unit mass than smaller species under comparable conditions (Kleiber 1932, 1975; Zeuthen 1947; Altman and Dittmer 1968; Blum 1977; Gordon 1977; Wilkie 1977; McNab 1983, 1988; Donhoffer 1986; Bennett and Harvey 1987; Nagy 1987; Trevelyan et al. 1990).

In addition to the comparatively obvious implications of such rates for consumption of energy, other relationships between BMR and avian body form include stress resistance, most notably a prediction that reductions in BMR may serve as adaptations for populations under environmental stress (Parsons 1987, 1990, 1993a–c; Djawdan et al. 1997); optimal morphological structure (Economos

1979, 1982); dependence of passerine diversity on food resources (Faaborg 1977); practical relevance of ¾ exponent of allometry in BMR with body mass (Feldman and McMahon 1983); selection in tropical environments (Hails 1983); relative brain size (Martin 1981); and response to artificial selection (Jackson and Diamond 1996). In addition, examination of available data on cardiopulmonary capacities of terrestrial birds (e.g., King et al. 1992), cardiovascular rates in birds (Calder 1968), skeletal length (Frasier 1984; Daan et al. 1990), and mass-specific utilization of energy (Maurer and Brown 1988) suggests fundamental, but less well-established relationships with BMRs. Suarez (1996) estimated an upper limit for mass-specific metabolic rate, underscoring that high metabolic rates carry attendant penalties, a reality that is consistent with the tendency for some flightless birds to lower BMR if ecologically permissible.

The period required to attain adult (asymptotic) body mass is the single most variable developmental parameter in birds (Schumacher and Wolff 1967a, b; Ricklefs 1968, 1969, 1973, 1979, 1984; McNab 1970; Case 1978b; Rahn 1980). This variation in the interval required to attain asymptotic mass is in part a function of the allometric decrease in growth rate (K) with increasing adult body mass (M_B). For avian taxa similar in other respects, this relationship approximates (Ricklefs 1979):

$$K = aM_B^b,$$

where $\hat{b} = -0.34 \pm 0.03$. The similarity of the allometric coefficient (b) in this context, and that relating BMR to body mass (M_B) in birds (Schmidt-Nielsen 1977), is probably an indirect reflection of conformance of both relationships to similar, interrelated functions of energetics (Ricklefs 1979).

Relationship of body size to locomotory efficiency is less clear (Greenewalt 1977), and is strongly dependent on means of transport and medium. Energetics of terrestrial locomotion (Taylor et al. 1971, 1982; Taylor 1973, 1977, 1980; Fedak and Seeherman 1979), especially bipedal locomotion (Fedak et al. 1982; Heglund et al. 1982a, b; Gatesy and Biewener 1991; Gatesy 1995), involve scale-dependent effects (Kokshaysky 1977) and differences in efficiency among taxonomic groups (Bruinzeel et al. 2000). However, the preponderance of evidence and arguments from first principles suggest a taxonomically varying relationship between maximal metabolic rates (typically manifested in flight) and those at rest (Pennycuick 1978), a coupling with important implications for the potential of flightlessness for BMRs in birds.

Metabolic implications of flightlessness in ratites.—Flight is among the most energetically demanding activities regularly undertaken by birds (Berger et al. 1970; Gold 1973; Taylor 1973; Tucker 1973a, b; Goldspink 1977a, b, 1981; Schmidt-Nielsen 1977; Torre-Bueno and Larochelle 1978; Rayner 1982; Brackenbury 1984; Castro and Myers 1988), and one that is capable of impressive power output, cardiovascular performance, and muscular forces (Bernstein et al. 1973; Biewener et al. 1992; Bishop and Butler 1995; Norberg 1996; Bishop 1997; Dial et al. 1997). As a result, flightlessness entails the foregoing of a regularly engaged, burst-consumptive activity, and therefore may permit a heritable reduction in BMR and increased allocation of energy to alternate activities, including storage of fat for periods of fasting (Calder and Dawson 1978; Ryan et al. 1989; McNab 1994a; Grant 1998a). The caloric implications of flightlessness are underscored in the dual reductions of BMR and mean body temperature of kiwis (McNab 1996).

Darlington (1943) was one of the first to suggest that conversion of energy formerly allocated to flight and its supportive apparatus to other functions is a prime selective advantage accruing to flightless lineages in accommodating environments. With increased access to living ratites and technology for quantification of physiological activities, data were compiled for ratites that confirmed the predicted reduction in metabolic rates (Lasiewski and Dawson 1967; Crawford and Lasiewski 1968; Zar 1968b, 1969; Aschoff and Pohl 1970a), somewhat anomalous in *Apteryx* (Calder 1978; McNab 1996), as well as sexual differences within species (Maloney and Dawson 1993) and thermodynamic corollaries of large body size (Maloney and Dawson 1994). Although these exceptional examples of avian flightlessness were naturally attractive for study, the applicability of findings to flightless neognathous birds (including rails) remained in question.

Inferences for flightless neognathous birds.—Accordingly, species having undergone both the permanent loss of flight (and associated energetic and metabolic demands) and increased body size are doubly positioned to achieve greater metabolic efficiency. Giantism in combination with flightlessness characterizes a flightless megapode (Poplin and Mourer-Chauviré 1985), flightless steamer-ducks (Livezey and Humphrey 1986, 1992), flightless grebes (Livezey 1989d), the Galapagos Cormorant (Livezey 1992a), the Kakapo (Livezey 1992b), and the raphids (Livezey 1993b). However, efficiency is distinct from total demand, and giantism typically entails an increase in the total consumption of energy. Therefore, the upper limits on total energy consumption to support baseline activities and size-related reductions in efficiency of heat loss (i.e., the downside of the surface: volume ratio for large-bodied endotherms) may place an upper evolutionary limit on body size in birds (Kendeigh 1972; Kirkwood 1983). Taken in combination, flight and its narrowed range of permissible body sizes may represent in some ecological contexts both a locomotory blessing and an energetic curse. Therefore, broad-scale changes in life history may carry more important implications for metabolic parameters than simple phylogeny within the Neognathae. For example, Reynolds and Lee (1996) found no significant difference between BMRs of passerine and nonpasserine birds, a phylogenetic division having mean differences in body mass that probably exceed those of other subdivisions. An attempt to assess metabolic trends for members of the class Aves within a phylogenetic context found, unlike for mammals, no significant relationship for birds (Ricklefs et al. 1996). However, this attempt was limited by a small taxonomic sample, bias of the method of phylogenetic contrasts, and reliance on the phylogenetic hypothesis of Sibley and Ahlquist (1990).

Seasonal variations in fat deposition and body mass evidently characterized the extinct raphids (Livezey 1993b) and possibly enhanced capacity for fasting (Baldwin and Kendeigh 1938; King 1972; Pond 1978). This relationship between body size and fasting raises the possibility that such advantages are (were) gained by the largest flightless rallids (e.g., *Porphyrio hochstetteri, P. mantelli, Gallirallus australis, Diaphorapteryx hawkinsi, Aphanapteryx bonasia, Fulica chathamensis,* and *F. newtoni*). An increase in absolute caloric demands imposed by increased body mass was confirmed intraspecifically in *Fulica atra* (Hurter 1979), and clearly holds special relevance for lineages undergoing evolutionary increase in mean

TABLE 83. Absolute basal metabolic rates (BMRs; cm^3 [O$_2$]/hour) and associated body masses of species of Rallidae (fide B. McNab, pers. comm.). Taxonomy and sequence follows Livezey (1998). Asterisks precede flightless species.

Taxon	Mass (g)	BMR	Source
Porphyrio porphyrio*	973	691	McNab 1994a
*Porphyrio hochstetteri	2,764	1,227	McNab 1994a
Aramides cajanea	374	276	McNab 1994a
*Gallirallus australis†	812	331	McNab 1994a
Gallirallus philippensis‡	172	177	McNab 1994a
*Gallirallus owstoni	201	165	McNab 1994a
*Atlantisia rogersi	39	41	Ryan et al. 1989
Crex crex	96	141	Kendeigh et al. 1977
*Amaurornis ineptus	896	247	McNab (unpubl. data)
*Tribonyx mortierii	974	624	H. Ellis fide McNab 1994a
Gallinula tenebrosa	519	454	H. Ellis fide McNab 1994a
Fulica atra§	412	367	Kendeigh et al. 1977

* Assumed to be *Porphyrio (p.) melanotus* of New Zealand.
† Assumed to be *Gallirallus (a.) hectori* of the South Island, New Zealand.
‡ Assumed to be *Gallirallus (p.) assimilis* of New Zealand.
§ Assumed to be *Fulica (a.) atra* of Eurasia.

body size. Similarly, dwarfed species of flightless rallid may achieve a balance between the reduced demands on BMR made possible by fightlessness with the lessened demands for total energy. Such trade-offs would be of special advantage in arduous environments such as those inhabited by the nanitic *Gallirallus wakensis* and *Porzana laysanensis*. Unfortunately, with the exception of the demonstration of modest lowering of BMR in *Atlantisia rogersi* reported by Ryan et al. (1989) and various reductions in BMR (relative to body mass) confirmed or predicted for several other flightless neognaths (McNab 1994a; McNab and Salisbury 1995), the combined investigational demands of vivaria and sophisticated technology have limited severely the metabolic data available for flightless rallids and other neognaths, including studies of the ontogeny of BMR (Vleck et al. 1980a, b; Sutter and MacArthur 1992). For rallids, at least, comparative data for several flighted species are available (e.g., Visser 1978; Sutter and MacArthur 1992).

Ongoing work by B. K. McNab (pers. comm.) indicates that significant caloric savings may be realized by relatively modest reductions in BMR. Available data for rails (Table 83), in addition to the finding that *Amaurornis ineptus* has a BMR that is roughly 42% of that expected for a flighted rail of equal body mass (McNab, pers. comm.), indicate that adjustments of metabolic rate in flightless rails may be an important credit in the energetic balance achieved through the sacrifice of flight. Given the physiological demands imposed by diving (Woakes and Butler 1986), the potential for metabolic economy after achieving flightlessness may be comparatively limited; for example, flightless coots (*Fulica*) may have accrued smaller metabolic benefits of insular existence and associated anatomical apomorphies than their terrestrial, flightless confamilials (e.g., *Cabalus modestus* and *Diaphorapteryx hawkinsi*). Also suggestive are slight reductions in metabolic rates and masses of pectoral muscles in flighted pigeons inhabiting islands without raptors (*Accipiter*) and especially where resources appear to be limited (McNab 2000, 2001). Predictably, in light of the relatively modest anatomical changes shown by many flightless rails, these savings are (were) less than those evidently realized by some ratites

(McNab 1996), but nonetheless are substantial and almost certainly subject to strong selection where ecologically permissible.

FORAGING AND DIET

> I have often speculated on antiquity of islands, but not with your precision, or at all under the point of view of Natural Selection *not* having done what might have been anticipated.... With respect to bats at New Zealand ... not having given rise to a group of non-volant bats, it is, now you put the case, surprising; more especially as the genus of bats in New Zealand is very peculiar, and therefore has probably been long introduced.... But the first necessary step has to be shown, namely, of a bat taking to feed on the ground, or anyhow, and anywhere, except in the air. I am bound to confess I do know one single such fact, viz. of an Indian species killing frogs.—C. Darwin in letter to C. Lyell, 1 September 1860 (F. Darwin 1887, vol. 2: 128–129)

> The Galapagos cormorant is flightless. But we know that mutants producing rudimentation and other monstrosities may have a partial effect in one case or a maximum effect in another. The interpretation of this case is clear. A single mutant may produce any degree of wing rudimentation. If such a mutation occurs in a hawk, for example, the resulting monster will not survive. But if it occurs in a such a bird as cormorant, which is already organized for catching its food while swimming under water, the monstrosity will not be deleterious and might even be of the 'hopeful' type if it enhances simultaneously the swimming and diving capacity (by lessening friction). Whether a complete or a partial reduction of the wing can take place depends upon the habitat.... an island cormorant finds enough fish with very little flying; and a Galapagos cormorant can do well without flying at all.—Goldschmidt (1940:393)

General avian principles.—The vast majority of the members of the class Aves retain the capacity for flight, and most of these employ flight as an important component of the arsenal of foraging tools, the energetic requirements of that figure importantly in total energy balances (King and Farner 1961; Kendeigh et al. 1977; Wolf and Hainsworth 1978; Morse 1980; Stephens and Krebs 1986; Bryant 1988, 1991, 1997; Maurer 1996; Ricklefs 1996) and that increase directly with body mass (Klassing 1998). Movement among foraging sites or pursuing prey on the wing is only one of a suite of avian solutions for the procurement of food. Feeding niche is influenced significantly by the size of an organism, either directly in terms of the size of the furnace to be built and stoked, or indirectly through the thermodynamic and metabolic correlates of body size (above). In addition, size of the body is related broadly to other structures pertaining to foraging, particularly the size of the mouth parts and digestive tract (Stanley 1973; Zweers 1979; Peters 1983; Bednekoff and Houston 1994). Body mass also is positively correlated with efficiency of water use in landbirds (Bartholomew and Cade 1963; Walter and Hughes 1978; Nagy and Peterson 1988).

Intensity of competition among avian taxa within guilds typically is reflected by similarity of prey and periods of activity. The classical modes of partitioning niches by sympatric insectivorous birds described by MacArthur (1958, 1959), and which came to idealize resource partitioning by birds generally, unfortunately holds few insights for many terrestrial avian taxa. Regardless, relationships between sympatric competitors and predators and the foraging niches and behavioral patterns of insular birds are documented (Diamond 1970a, b; MacArthur et al. 1972; Atwood 1980; Baker-Gabb 1986; Löfgren 1995; Mettke 1995; Blazquez et al. 1997; Mettke-Hofmann 1999). Niche differences among sympatric shorebirds may be more relevant for terrestrial groups such as the Rallidae, especially with respect to the role of bill shape and leg length in foraging (Baker and Baker 1973; Baker 1979), but the seasonality and greater diversity that typify sympatric charadriiforms exceed those of most sympatric rallids (Ripley 1977; P. B. Taylor 1996, 1998). Differentiation of bill morphology is critical in other groups of variably sympatric confamilials such as crossbills (*Loxia* [Benkman 1987a, b, 1988a, b, 1989a–c, 1991, 1993; Benkman et al. 2001; Summers et al. 2002]) and syntopic combinations of geospizine finches and other landbirds (e.g., Grant 1967, 1981b; Grant et al. 1976; B. R. Grant and P. R. Grant 1982, 1989; Schluter 1988a). Such radiations involving bill specializations among comparatively closely related, resident landbirds represent important parallels for relatively short-term diversification of form among phylogenetically constrained, sympatric birds such as rails (Björklund 1996).

Generalities and exceptions of foraging by rallids.—The applicability of the term "adaptive radiation" to the flightless members of the Rallidae is debatable, but the interrelationship between loss of mobility, genetic isolation, and speciation on islands compares favorably with the utility of the concept in other contexts, for example, cichlid fishes in historical complexes of lacustrine "islands" (Turner 1999) and truly insular radiations in other birds (Amadon 1966a; Morse 1975; Grant 1981b, 1983c, 1998a, b; Schluter 1994, 1996b, 2000a, b). Extreme dietary versatility and plasticity of foraging tactics are the most important aspects of the feeding ecology of most rallids (Appendix 6). Even the comparatively vegetarian coots (*Fulica*) include eggs of other aquatic birds as an important component of their diet during breeding in some areas (Burger 1973). Opportunism of foraging and diet of rallids in general are indicated by a sampler of reported feeding behaviors, including predation on a water snake (Hoff 1975), scavenging (Madden and Schmitt 1976), subaquatic capture of mollusks while wading (Kilham 1979), consumption of feces of waterfowl (Kear et al. 1980; Phillips 1991; Starks and Peter 1991), predation on passeriform birds (Jorgensen and Ferguson 1982), predation associated with forays by army ants (Willis 1983), cannibalism (Paullin 1987), and capture of insects exposed by trampling floating plants (Möller 1992).

This diversity of feeding habit and the facility with which rallids colonize diverse, isolated patches of habitat certainly disqualifies the rallids as candidates for "neophobia" in foraging tactics (Greenberg 1983, 1984a, b, 1990). Also, use of unusual habitat (sandstone flats and tidal zones) and anvils for exploiting mollusks as food by Chestnut Rails (*Eulabeornis castaneoventris*) is consistent with the apparent expansion of foraging niche shown by some flightless endemics (below), although at most this species could be considered flight-compromised (Woinarski et al. 1998). Variable or incipient shifts toward nocturnal foraging are indicated in flightless *Cabalus* (see below; Travers 1872), and may be suggested by the evident origin of intraspecific brood parasitism in flighted *Gallinula* with nocturnal visits to the nests of hosts (McRae 1996a, b).

Increased dietary breadth in flightless rails occurs both with and without changes in bill shape. Limited information available for extinct flightless rails indicates broad, opportunistic feeding habits. The extinct *Porzana palmeri* consumed eggs of terns (*Sterna*) opened by the Laysan Finch (*Telespiza cantans*), adult and larval dipterans, coleopterans, and lepidopterans found on or near avian carcasses (Baldwin 1947), and at times also carrion (Blackman 1945). Scavenging associated with carcasses may have been critical to the success of the crakes and to have (indirectly) fostered flightlessness and pugnacity (Baldwin 1947). Whether or not the Laysan Crake was capable of opening eggs unassisted remains a point of dispute; some considered the birds capable (e.g., Bailey 1956), whereas others demurred (Schauinsland 1899; Fisher 1903; Baldwin 1947).

More typical of the Rallidae, *Cabalus modestus* was reported to prey upon emerging coleopterans (Hutton 1873a, b). Predation by Wekas on chicks can be important sources of mortality for other flightless birds (St. Clair and St. Clair 1992). Similarly varied and opportunistic feeding habits seem likely for other small, flightless rails on tropical islands, including *Gallirallus wakensis, G. owstoni,* and Hawaiian *Porzana*. The powerful structure of the bill of *Diaphorapteryx hawkinsi* (Fig. 1) is consistent with the suggestion by Atkinson and Millener (1991:174) that "... it would be surprising if neither the Giant Rail [*Diaphorapteryx hawkinsi*] nor Dieffenbach's Rail [*Gallirallus dieffenbachii*] preyed on petrel chicks or even adult birds." Not all flightless rallids show(ed) broad, diversified feeding habits. The specialized feeding habits of the Takahe—unique for *Porphyrio* in its reliance on alpine grasslands—clearly are related, directly or indirectly, to the decline of the species (Reid 1978; Baird 1984, 1985, 1986, 1991a, 1992; Mills et al. 1984, 1988; Beauchamp and Worthy 1988; Bunin and Jamieson 1995). The specialized bill morphologies of *Aphanapteryx bonasia, Erythromachus leguati,* and *Capellirallus karamu* suggest parallel specializations of diet, especially given the sexual dimorphism of the former (Table 30), but virtually nothing is known of these extinct rails.

Several apomorphies of the bill shown by a few flightless rails seem only indirectly related to the loss of flight, but are notable for having arisen several times in flightless rallids and convergently in a number of other avian families (Hofer 1945; Barnikol 1952; Zusi 1993): pronounced elongation (e.g., *Capellirallus karamu*; cf. *Apteryx, Ibidorhyncha, Nycticryphes,* and multiple scolopacids); massive deepening (*Porphyrio mantelli*-group; cf. moa-nalos [Anatidae] and *Dromas*); and strong decurvature (*Gallirallus dieffenbachii, Cabalus modestus, Capellirallus karamu, Diaphorapteryx hawkinsi, Aphanapteryx bonasia,* and *Erythromachus leguati*; cf. threskiornithids, multiple charadriiforms, *Mesitornis,* as well as variable numbers of cuculids, coraciiforms, dendrocolaptids, formicariids, drepanidids, meliphagids, and mimids). It is noteworthy that most of the flightless rallids that possessed strongly decurved bills (e.g., *G. dieffenbachii, D. hawkinsi, A. bonasia,* and *E. leguati*) evolved on islands lacking other taxa, flightless or otherwise, that possessed such specializations (e.g., kiwis, ibises, and curlews). The sole exception to this complementarity is *Capellirallus karamu,* which coexisted with apterygids on the North Island of New Zealand. This pattern is consistent with the conspicuous absence of a rallid with an elongated bill (e.g., a derivative of *Rallus*) on the Hawaiian archipelago, where at least two species of flightless ibises (Threskiornithidae) occurred (Olson and James 1991). Birds equipped with elongated and strongly decurved bills evidently are able to probe and search soft earth and forest litter or shallow water more efficiently for soft-bodied prey than those with shorter, straight bills (e.g., *Porzana*).

The remaining specialization of the bill shown by flightless rails—the extraordinary deepening of the bill of the takahes (*Porphyrio mantelli* and *P. hochstetteri*)—is interpretable as an apomorphy that facilitates the efficient use of high-elevational grasses and foraging under syntopy with potential competitors. This specialization, presumably advantageous under past regimes, is related at least in part to the precarious current status of the species, both through climatic reduction in such habitats and competition with introduced foragers for remaining resources (Turbott 1951a, 1967; Williams 1952, 1960; Reid 1974b, 1978; Mills 1975, 1978, 1985; Williams et al. 1976; Lavers and Mills 1984; Mills et al. 1984, 1988; Beauchamp and Worthy 1988; Bunin and Jamieson 1995). A notable exception among flightless rails, and one showing the greatest success in the face of changing circumstances, is the Weka (*Gallirallus australis*-group), which has augmented the already broad omnivorous diet of its flighted congeners (e.g., *G. philippensis*-group) to include comparatively greater numbers of eggs, small birds (e.g., nestling procellariiforms), lizards, and carrion, by employing diversified foraging methods (e.g., hammering of wood, entering of nesting burrows and tree hollows, and following wild pigs), and ingestion facilitated by a 40-mm gape (Marchant and Higgins 1993).

Comparisons with other flightless birds.—In addition to the much-studied ecological implications of depauperate faunas and vacant niches on islands, ecological extremity also is characteristic of many tropical islands (Carlquist 1965, 1974, 1980), and adaptation to live on such "desert islands" or other environmental challenges by insular birds represents a special case of subsistence under extreme conditions (Andreev 1999). Many insular birds show exceptionally broad diets. Particularly compelling are instances of unique feeding methods in insular birds, including comparatively high reliance on egg-eating (Hatch 1965; Harris 1968; Bowman and Carter 1971; Grant 1986), sanguinivory (Bowman and Billeb 1965; Harris 1968), tool-using (Millikan and Bowman 1967), and parasitism of sea lions (Trimble 1976). Among flightless birds, highly specialized herbivory evidently characterized the extinct moa-nalos of Hawaii (James 1997).

The Auckland Islands Teal consumes more animal prey and forages on land and at night more frequently than its continental relatives (Livezey 1990). Weller (1980) reasoned that the finer bill lamellae of insular waterfowl improved the capture of invertebrates. Whether these convergent dietary shifts result from competitive release in island communities (Lack 1970a, b; Wallace 1978; Weller 1980) or are correlates of decreased body size or simply reflect similar food resources available on oceanic islands is not clear. A similar pattern characterizes the steamer-ducks (*Tachyeres*), in which the three, large flightless species have relatively or (in some cases) absolutely fewer bill lamellae than their smaller, flighted congener (Livezey 1989e). Unlike the Anatidae, flightless members of the Rallidae do not possess elaborate lamellar or other refinements for filtration of aquatic prey; the closest approach to these specializations occurs in coots (*Fulica*). The Kakapo is a highly specialized herbivore, a comparatively rare dietary mode in birds (Morse 1975; Grajal et al. 1989) that requires a substantial digestive tract (Karasov 1990), and encumbers birds with massive ingesta (Dudley and Vermeij 1992), a characteristic generally associated with large body mass (Morton 1978; McNab 1988) and which may have hastened flightlessness in this unique parrot (James et al. 1991; Livezey 1992b; Powlesland and Lloyd 1994).

Unfortunately, most flightless neognaths are either extinct or poorly known ecologically (e.g., extinct anatids and raphids), and remain the subject of speculation. Whether flightlessness and concomitant anatomical changes were associated with a dietary shift in the extinct Auckland Islands Merganser never will be known (Livezey 1989b). Based on the meager observational evidence available, *Raphus cucullatus* and *Pezophaps solitaria* can be inferred to have been "obligate," somewhat specialized terrestrial foragers (Sherry 1990), probably relying most heavily on fallen fruit. Ingestion of bulky foods by them was facilitated by a large crop, and digestion evidently was aided by a powerful gizzard enclosing gastroliths (Strickland and Melville 1848; Owen 1866a; Hachisuka 1953; Livezey 1993b). Diets of the extinct raphids may have been similar to those of the massive, probably flightless megapode *Sylviornis neocaledoniae* of New Caledonia (Poplin and Mourer-Chauviré 1985). Acquisition of gastroliths also was typical of the ponderous, herbivorous, flightless moas (Dinornithiformes) of New Zealand (Archey 1941; Oliver 1949; A. Anderson 1989). Here again, potential morphological specializations that might have permitted more efficient exploitation of available foods by rallids (e.g., expansive crops, powerful gizzards, utilization of gastroliths, or capacity for digestion of crude vegetable matter) were not an inheritance of modern Rallidae.

Are flightless rails specialists or generalists as foragers?—The preceeding survey of foraging habits and diets of flighted and flightless rails indicates that the majority of species are (were) generalists having unusually broad behavioral repertoires for the capture of food. These versatile species include most genera in the Rallidae, including the speciose *Aramides, Rallus, Gallirallus, Laterallus, Porzana,* and *Gallinula* (Appendix 6). Several genera (notably *Porphyrio*) comprise a majority of species having comparatively broad diets, and a minority of flightless members having specialized feeding niches. Although a few more flightless rallids possessed specialized bills suggestive of dietary specialization (e.g., *Habroptila wallacii, Diaphorapteryx hawkinsi, Aphanapteryx bonasia, Erythromachus leguati,* and *Capellirallus karamu*), rarity or early extinction preclude detailed tests. These exceptional species aside, however, it seems clear from available data on diet, foraging habits, and morphology that most flightless rails, like their flighted relatives, can be characterized as generalists, differing perhaps in tendencies toward opportunism.

FLIGHTLESSNESS, MOBILITY, AND REPRODUCTION

General avian principles and sectional organization.—Natural selection hinges on differential reproduction, a criterion critical in theoretical explorations of evolutionary trade-offs (MacArthur 1962; Sanfriel 1975; Alerstam and Högstedt 1983; Nur 1988). Despite a comparative wealth of basic data and organizational schemes (e.g., classifications of avian mating systems [Ligon 1999]), the derivation of an inclusive explanation for the fundamental aspects of avian life histories has proven elusive, primarily because of the interrelationships among the parameters of reproduction (e.g., mating system, number of clutches per year, number of eggs per clutch, and egg size) and the diversity of selective agents that act upon them (Ricklefs 1974, 2000; Bryant and Tatner 1988; Partridge 1989; Bennett 1997). Reduction in body mass (and associated nutrients) of females during egg-laying and incubation has been debated as a proximate index to the costs of reproduction, including for rails (R. Å. Norberg 1981; Reznick 1985; Bryant 1988; Moreno 1989; Stearns 1992; Arnold and Ankney 1997; Thomson et al. 1998). Toward emphasis on information that serves the primary focus of this monograph, reproductive biology will be considered with respect to general avian principles, patterns in rallids (flighted and flightless), and comparisons with shifts apparently related to flightlessness in other taxonomic families.

Mating systems.—Although preliminary analyses based on the phylogenetic hypothesis of Sibley and Ahlquist (1990) have been attempted (Sillén-Tullberg and Temrin 1994; Temrin and Sillén-Tullberg 1994, 1995), much of the current knowledge derives from broad generalizations drawn from the body of monospecific studies. Reviews of avian mating systems generally point to the plesiomorphic condition of most taxonomic families as being one of monogamy, despite the inherent ambiguity of the concept (Wickler and Seibt 1983), with a growing body of evidence that polyandry (Faaborg 1981; Ward 2000) and cooperative breeding (Emlen 1978; Koenig and Pitelka 1981; Gibbons 1986, 1987; Brown 1987; Koenig and Stacy 1990; Koenig et al. 1992; Cockburn 1998) are not uncommon and in some taxa are intraspecifically variable or coincident (Orians 1969; Verner and Wilson 1969; Jenni and Collier 1972; Kendeigh 1972; Jenni 1974; Emlen and Oring 1977; Murton and Westwood 1977; Oring 1982; Knoppien 1985; Davies 1991).

In addition to the long-recognized importance of mate choice for sexual selection and signals of reproductive status in the acquisition of viable primary (and possibly extra-pair) mates in many avian taxa (Crawford 1978; Miskelly 1981; Rohwer 1982; Eddleman and Knopf 1985; Møller and Birkhead 1994), the critical energetic burden of reproduction on adults (King 1973; Calow 1979; Loman 1982; Walsberg 1983; Reynolds and Székely 1997) and the related, formerly underappreciated importance of helpers for nuturing of related offspring (Brown 1978) have become recognized as critical. Also of broader recognition are the roles of mate guarding for evolution of parental care (van Rhijn 1991) and the prevention of independent breeding by young adults in early adulthood through cooperative breeding (Arnold and Owens 1999).

Brown (1987) found few useful phylogenetic patterns below familial level in the incidence of communal breeding in birds, but concluded that such systems are more frequent in taxa (including the Rallidae) at low latitudes, in hot climates, and possibly under regimes of unpredictable rainfall. Leisler et al. (2002) established a relationship between cooperative breeding and monogamy and poor habitats in acrocephaline warblers, in which productive habitats are conducive to polygyny, leading to a view of breeding systems as comparatively plastic traits. Among the Rallidae, variation in mating system and flight capacity is encountered among the basalmost nodes in the family (Livezey 1998; Appendix 6), an observation perhaps relevant to the cooperative polyandry of the related family Psophiidae (Sherman 1995a, b, 1996). However, the sister-family Heliornithidae (Livezey 1998) is characterized by monogamy (Percy 1963; Alvarez del Toro 1971). Variably pronounced communal breeding systems have been observed in at least some of the flighted populations of *Porphyrio*, whereas flightless *P. hochstetteri* evidently practices monogamy, perhaps of life-long duration (Jamieson and Craig 1987a, b; Craig and Jamieson 1990; Jamieson 1997; Birkhead 1998; Goldizen et al. 2000). In Pukekos, an advantage for females from multiple paternity is increased paternal care of the nest and brood. Such complex mating systems also carry implications for participating males (Ligon 1999), including both positive aspects (e.g., mutually advantageous alliances in paternal care) and negative implications (e.g., sperm competition); the latter may contribute to widespread territoriality among males (Birkhead 1998). Given that a low "cost" of sharing males by females with respect to parental care is an important component favoring avian polygyny (Searcy and Yasukawa 1995), it is not surprising that polygyny is rare among the Rallidae (P. B. Taylor 1996, 1998; Ligon 1999). Polygyny would seem to be even more unlikely among flightless rallids, in that the reduced mobility of males would place more stringent limits on the size of breeding territory that could be defended in most cases, thereby increasing the likelihood that polygyny would prove unfavorable to females simultaneously sharing mates.

Comparable variation in mating systems, in which variably frequent, sometimes serial polyandry has been documented, characterizes a number of other rallid genera having members favoring open-water habitats: *Porphyrio* (e.g., Wettin 1984), *Porzana* (Wintle and Taylor 1993; Bockheim and Mezzell 1999), *Tribonyx* (Maynard Smith and Ridpath 1972; Ridpath 1972a–c; Goldizen et al. 1993, 1998a, b, 2000), and *Gallinula* (Garnett 1978, 1980; McRae 1996a, b). Likewise, visual signals of reproductive and territorial status (e.g., size and seasonally varying coloration of cristae frontales) and age-related variance in reproduction (Gullion 1951; Crawford 1980) are notable in *Gallinula* (Eskell and Garnett 1979) and more-aquatic *Fulica* (Visser 1988). The similar frequencies of these departures from monogamy in rallids suggest a possible importance of habitat and flightlessness in the adoption of specialized breeding systems. With the exception of *Tribonyx mortierii,* flightless rails for which data are available show no such departures from the presumptive familial plesiomorphy of monogamy (Appendix 6; P. B. Taylor 1996, 1998). To the extent that polygamous matings or reproductive groups involve breeding among a restricted subset of individuals, such systems would not be favorable on genetic grounds in that small, insular demes would intensify the effects of inbreeding depression (MacArthur and Wilson 1967; M. Williamson 1981). Adler and Levins (1994) suggested that crowding and ultimately reduced reproductive effort and favored greater body size follow colonization in insular rodents. Both tendencies appear to apply to a number of flightless Rallidae (e.g., *Porphyrio hochstetteri, Gallirallus australis, Porzana atra,* and *Gallinula comeri* [Appendices 1 and 6]).

As reported for those rallids exhibiting variation in mating system (Appendix 6), the mating system of the paleognathous kiwis (Apterygidae: *Apteryx*)—which are similar to many rallids in their flightlessness, terrestrial feeding modes, and precocial development (Winkler and Walters 1983)—is essentially monogamy (Folch 1992; Taborsky and Taborsky 1999). The mating systems of other ratites range from monogamy to polyandry, some of which manifest considerable intraspecific variation principally related to climate (Folch 1992). Flightless neognaths of a number of higher taxa, most of which inhabit(ed) temperate latitudes and possess(ed) limited reproductive seasons, conserve(d) plesiomorphic conditions of monogamy (Livezey 1988, 1989a–c, 1990, 1992a, b). A notable exception evidently was *Pezophaps* of Rodriguez, in which observations of mating-related combat among males, possible leks, and carpometacarpal armature strongly indicate polygyny, in contrast with evident monogamy of the Dodo (Livezey

1993b). The plausibility of lek-breeding in the Solitaire is enhanced by the use of leks and polygamous mating of the Kakapo, which in combination with a highly skewed sex ratio, was to have unforeseen, potentially disastrous implications for the preservation of the species (Livezey 1992a).

Location and structure of nests.—Nests of terrestrial birds serve multiple purposes (Martin 1995, 1998; Soler et al. 1998), including enclosure of the clutch en mass for incubation, provision of cover from the elements for the incubating adult (Williams 1996b), and concealment of the eggs from predators or birds (often conspecific) that seek to lay eggs in the nests of others (i.e., nest parasites). In addition, predation on nests and eggs holds implications for sexual dichromatism, especially selecting for cryptic plumage patterns in incubating females where aerial predators are prevalent (Martin and Badyaev 1996). For similar reasons, crypsis of nests can be afforded by males feeding incubating females (Martin and Ghalambor 1999). The evolutionary ecology of nest parasitism (Yom-Tov 1980) and brood amalgamation (Weller 1959; Andersson 1984, 2001; Savard 1987; Eadie et al. 1988; Rohwer and Freeman 1989; Beauchamp 1997) has been the subject of comparatively intensive, continuing study. Most recently, models presented by Zink (2000) suggested that brood parasitism in birds was favored (relative to solitary or cooperative nesting) where relatedness among breeding adults was low and parasitic individuals could achieve high fecundity, whereas cooperative nesting was favored by high relatedness and low skewness in reproductive contributions of members of the population.

Behavior evidently contributory to concealment of nests was observed in nesting rallids (Arnold 1992a, b), as were relative rewards for defense of nests (Montgomerie and Weatherhead 1988). The construction of canopied nests among flighted and flightless rallids of diverse relationships is relatively widespread among species nesting on the ground or aquatic vegetation (Appendix 6; Hansell 2000). In addition to concealment from potential predators of incubating adults and eggs, the frequency of nest parasitism among rallids (Petrie 1986; Lyon 1991, 1993a, b; MacRae 1995, 1996b, 1997; Sorenson 1995; McRae and Burke 1996) alone may qualify this phenomenon as potentially critical on islands that host(ed) more than one species of rail.

Nesting habits of ratites reveal little about the physical accommodation of included clutches. For example, in struthionids and rheids, several females lay eggs in a common nest to be incubated by their shared mate, and the resultant volume necessitates a substantial nest bowl (Folch 1992). Among the Anseriformes, too, is a relative conservatism of nesting habits with the loss of flight, excepting those taxa for which nest sites are unknown (e.g., *Cnemiornis* and moa-nalos). Common to all others (e.g., *Tachyeres* spp., *Anas aucklandica,* and *Mergus australis*) is the predictable exclusion of nest sites that make terrestrial approach difficult or that are especially vulnerable to detection by aerial predators (e.g., skuas and gulls). A conservatism of at least aggregations of nests by the extinct *Chendytes* (Livezey 1993c)—possibly plesiomorphic for these apparent relatives of the colonial eiders (Anatidae: Somaterina)—may extend the conservatism of nesting habits with the evolution of flightlessness (Guthrie 1992). Similarly, nesting habits of flightless grebes reflect their limited locomotory capacity in the restriction of suitable nest sites to overwater platforms on the narrowly circumscribed, high-altitude lakes of each species (Livezey 1989d). Flightless seabirds of two orders, the Great Auk and Galapagos Cormorant, construct(ed) nests characteristic of respective confamilials, but limited this activity to shorelines approachable on foot from the sea (Livezey 1988, 1992a), similar in most respects to those of penguins (Livezey 1989c). Like penguins and flighted cormorants, the Galapagos Cormorant nests colonially, although these colonies are comparatively small and only loosely seasonal (Swarth 1934; Snow 1966; Harris 1973; Livezey 1992a). What is known of the nesting habits of the critically endangered Kakapo (Livezey 1992b) and the extinct raphids (Livezey 1993b) conforms with this pattern of limited change with flightlesseness, in which the terrestrial locations of nests that evolved or were conserved under insular conditions provided little protection upon the introduction of mammalian (including human) predators.

Sizes of clutches and eggs.—Investments related to reproduction account for a significant portion of the annual energetic budgets of most birds (Meijer and Drent 1999); may affect development of embryos via composition and size of eggs (Byerly 1932; Lack 1968a; Sotherland and Rahn 1987; Ricklefs and Starck 1998a); and are reflected in decreases in body mass, declines in reserves of protein and lipids, and ancillary changes in organs of adults before and during mating and nesting periods (Blem 1976, 1990; Jones and Ward 1976; Korschgen 1977; Fogden and Fogden 1979; Raveling 1979; Wypkema and Ankney 1979; J. R. King 1981; Robbins 1981; Houston et al. 1983; Ankney 1984; Gauthier et al. 1984; Austin and Fredrickson 1987). These physiological changes may be related to decreased immuno-competence (Deerenberg et al. 1997; Weber and Korpimäki 1998). However, suggestions of a disproportionate decrease in body mass in larger individuals during breeding evidently represent a statistical artifact (Gebhardt-Henrich 2000).

Fundamental, quantitative reproductive parameters in birds are sizes of clutches and dimensions of eggs, if for no other reason than ease of quantification. Despite the evident plesiomorphy of oviparity in Aves and Reptilia sensu stricto (Bellairs 1960), the implications of egg mass for flight led Blackburn and Evans (1986) to consider the evolutionary implications of oviparity in birds. These investigators concluded that the absence of viviparity among flightless birds (especially ratites), and the presence of viviparity in bats (Chiroptera), invalidated a traditional argument that the mechanical constraints of flight precluded intrauterine development of embryos in birds (Dorst 1974). Nonetheless, eggs are a critical currency of reproduction comprising several important parameters, some of which, notably clutch size, have been demonstrated to be heritable in natural populations (Lack 1947b, c; van Noordwijk et al. 1981; Gustafsson 1986; Findlay and Cooke 1987; Gibbs 1988; Godfray et al. 1991). Reiss (1985, 1989) found that larger species tend to invest relatively less in their offspring (including propagules) because of the comparatively high absolute energetic demands of maintenance of other essential bodily functions, but the complexity of relationships between body size and reproductive parameters (among other critical characteristics of life history) may defy simple interpretation (Brown et al. 1978).

Clutch size is an important and comparatively variable component of avian reproductive strategies (Cody 1966; Lack 1968b; Royama 1969; Klomp 1970; Yom-Tov and Hilborn 1981; Loman 1982; Cooke et al. 1990). An intuitive relationship between food supply and clutch size has been inferred in birds (Lack 1947b, 1948; Hussell 1972). Klomp (1970) argued that selection should favor large clutches, predicting the largest clutches to be the most successful, leading to a progressive increase in modal size. The latter prediction differs from the traditional expectation that the modal clutch size should approximate the observed mode (Lack 1954), a general correspondence that Högstedt (1980, 1981) reconciled with field data through an inferred dependence between optimal clutch size and quality of accessible habitat. Price and Liou (1989) confirmed the observation of Klomp (1970), but concluded that this reflected a correlation between condition of the female both to lay eggs and rear young. Reviews of avian clutch size from a perspective of life-historical trade-offs included those of Cody (1966), Loman (1982), and Charnov and Krebs (1974). Cooke et al. (1990) theorized that clutch size should conform comparatively directly with the predictions of fitness pioneered by Fisher (1930) and Wright (1949). Lepage et al. (1998) found enhanced fledging success in large clutches in a precocial species.

An important component of avian reproduction is the trade-off between number and size of offspring (Smith and Fretwell 1974; F. C. Rohwer 1989; Viñuela 1997; Sinervo 1999), and a general inverse relationship between these parameters has been confirmed for birds (Drent and Daan 1980; Rohwer 1988; Rohwer and Eisenhauer 1989; Roff 1992). Despite intraspecific variation in size and composition (Williams 1994), variable correlations have been inferred between egg size and body size (Huxley 1927), taxonomic group (Heinroth 1922; Amadon 1943b; Moreau 1944), relative size of yolk (Sotherland and Rahn 1987), and sexual dimorphism (Weatherhead and Teather 1994), including insular effects (Blondel et al. 2002) and recruitment (Lundberg and Väisänen 1979).

Among precocial birds, species laying large eggs also tend to produce hatchlings of larger size and greater energy stores, particularly if the enlarged eggs contain appreciably more yolk (Lack 1967, 1968a; Ar and Yom-Tov 1978; Carey et al. 1980). Such variations in physiological allocations apparently are controlled by endocrinological adjustments and may constrain tracking of environmental changes (especially in temperate zones) by avian populations (Jacobs and Wingfield 2000). Vleck and Bucher (1998) compiled mean compositions of eggs by avian order, listing those of gruiforms as averaging 32% yolk, 76% water, and having an energy density of 6 $kJ \cdot g^{-1}$. A direct relationship between egg size and survivorship of young has been documented in several species of waterfowl, an advantage effected in part by egg size and enhanced skills in foraging and locomotion upon hatching (Hepp et al. 1987; Dawson and Clark 1996; Erikstad et al. 1998; Anderson and Alisauskas 2001, 2002).

Large eggs also require longer incubation periods (Worth 1940; Rahn and Ar 1974; Drent 1975; Thomason et al. 1998), despite considerable variation in the relationship between egg size and incubation period at diverse taxonomic levels (Nice 1954; Boersma 1982; Ricklefs and Starck 1998a, b; Starck and Ricklefs 1998b, c), during which there can be higher risk of breakage and predation (Ar et al. 1979; Rahn and Paganelli 1989). Lima (1987) and Milonoff (1989) considered the risk of predation as critical in the evolution of clutch size and associated incubatory demands, and the former was extended to a suite of reproductive decisions by Lima and Dill (1990). Shine (1978, 1989b) regarded large size of propagules as indicative of the comparative security of eggs during development. This hypothesis of the "safe harbor" afforded eggs through incubation and brooding was judged by Ydenberg (1989) to be applicable to the Alcidae, but that this

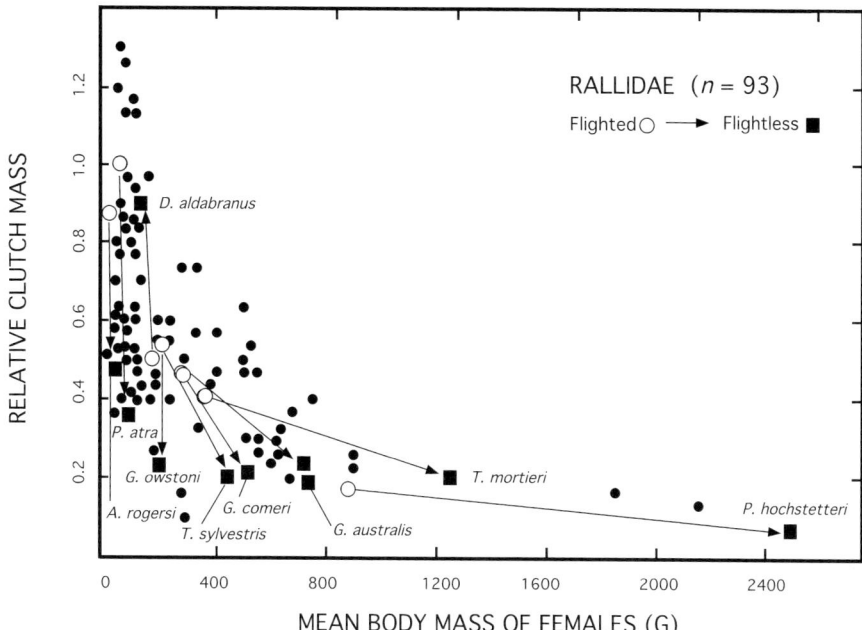

FIG. 125. Bivariate plot of relative clutch masses on body masses of females for 93 species of Rallidae (Appendix 7). Flightless species (squares) are related to flighted relatives (hollow circles) by inferred vectors of change, and additional flighted species are depicted (solid circles) to clarify general trends.

relative security is balanced against the comparatively higher developmental rates of juveniles attainable at sea. Precocial development such as characteristic of the Rallidae may suggest that the latter outweighs the former early in life in taxa so characterized.

General parameters of reproduction in the few comparatively typical (i.e., flighted) members of the Rallidae (Table 80; Appendix 6) are not exceptional for birds in general (Lack 1954; Amadon 1964; Alisankas and Ankney 1985); the complete absence of information for *Rallicula* and *Amaurornis ineptus* is noteworthy. Simple comparisons of clutch sizes (Appendix 7) indicate that most flightless lineages have undergone decreases in numbers of eggs laid, reductions in fecundity that not infrequently represented a halving of modal clutch size. A graphical analysis of clutch mass for members of the Rallidae, also including information on egg size with clutch size (Fig. 125), principally revealed that investment in clutches relative to body mass declined in flightless species having undergone increases in the latter. The few flightless species of rail for which reproductive data were available showed the opposite trend, in which reproductive investment essentially was conserved despite the decrease in body size (Fig. 125). One marginal exception to this pattern is the relatively large eggs laid by the flightless *Atlantisia rogersi* (Appendix 7; Rothschild 1928). These findings indicate that reproductive investments remained roughly constant in flightless lineages, in which changes in relative clutch mass largely reflect changes in body mass.

Other notable reproductive modalities include the apomorphic extremes of fecundity shown by a number of comparatively widespread, flighted, aquatic, and mid-sized species of the Northern Hemisphere (*Rallus aquaticus, R. limicola, R. elegans, R. longirostris, Porzana carolina, P. parva,* and *P. porzana*). That is, the rallids most familiar to European and North American ornithologists are characteristically familial extremes in fecundity and (in some cases) total clutch mass (Appendix 7). Also, the large eggs of American Coots (*Fulica americana*) contained proportionately more yolk than smaller eggs, a characteristic associated with higher mean lean dry masses of hatchlings (Alisauskas 1986), although clutches contain more eggs than normally can be reared to fledging (Forbes and Mock 2000) and vary considerably in the genus (Arnold 1991).

Other flightless birds vary greatly in the changes indicated for the key reproductive parameters of clutch and egg size, perhaps the most striking being the uniquely great egg size of *Apteryx* (Reid 1971a, b, 1972, 1977; Calder and Rowe 1977; Calder 1979, 1991), a proportionality that Reid and Williams (1975) inferred to reflect an apomorphic reversal in body size shown by the kiwis among ratites. Despite great absolute size of eggs of other ratites (e.g., *Struthio,* and the record-holding *Aepyornis*), neither relative egg size (given the great body mass of adults) nor clutch size is exceptional in other ratites (Schönwetter 1960a; Folch 1992).

With the exception of two species of flightless steamer-ducks, which are unremarkable in reproductive characteristics (Livezey and Humphrey 1985a, b), flightless neognaths other than rallids are insular. Corollaries of reproduction in diverse organisms in insular environments are numerous (Carlquist 1965, 1974; Abbott 1980), and examples involve virtually all strategies known for the taxonomic groups represented (Whittaker 1998). Insular waterfowl (notably *Anas*) are characterized by unusually small clutches of disproportionately large eggs (Schönwetter 1960b, 1961; Lack 1970b; Weller 1980; Rohwer 1988), consistent with a general inverse correlation between life span and simple clutch size among *K*-selected species (Haukioja and Hakala 1979), although decreases in reproductive investment relative to body mass are not sustained in most species (Lack 1967). However, the massive eggs and small clutches of the Auckland Islands Teal represent an extreme example of increased per capita reproductive investment (Weller 1975; Livezey 1990; Williams 1995), shifts that are indicated to a lesser degree in other, comparatively weakly modified *Anas* of oceanic islands (Livezey 1993a).

Changes in sizes of eggs or clutches of insular birds also were not demonstrated for the Great Auk (Livezey 1988), the Galapagos Cormorant (Livezey 1992a), or the Kakapo (Livezey 1992b). Information was not available for the extinct Auckland Islands Merganser (Livezey 1989b) and was only marginal for the raphids (Livezey 1993b). The Dodo laid clutches of single eggs (Livezey 1993b). Assuming that the raphids did not lay overlapping clutches (Burley 1980) and that eggs of columbiforms are calorically comparable (Ar and Yom-Tov 1978; Carey et al. 1980), extrapolations from relationships for members of the Columbiformes and other birds (Rahn et al. 1975; Saunders et al. 1984; Robertson 1988) indicate that the Dodo and Solitaire invested relatively little in their eggs.

Parental care and development of hatchlings.—A preliminary analysis of higher-order phylogenetic effects (conditioned on groups delimited by Sibley and Ahlquist [1990]) in the evolution of growth rates and mode of development indicated that body size was the predominant influence on ordinal differences in rate (Starck and Ricklefs 1998b); the mode may have been related to developmental timing and ultimate size of the nervous system (Ricklefs and Starck 1998b). An assumption that an observed growth schedule represents a global optimum for a given species is unrealistic, and instead reflects a minimum period attainable given the phylogenetically constrained and selectively maintained limits of metabolism (Kirkwood 1983) and provision of food (Ricklefs 1969, 1973, 1974; Case 1978b; Drent and Daan 1980). A strong correlation between body size and developmental time is both intuitive and well established, with development generally taking absolutely longer and transpiring at a slower rate in large species (Ricklefs 1973, 1983a, 1998a, b), despite variable heritability of body size (Kruuk et al. 2001). However, other factors have been implicated in the evolution of developmental rates, including the relative risks of predation for young at different developmental stages (Shine 1978, 1989b; Bosque and Bosque 1995), the relative size of adult females (Reiss 1987), comparatively variable levels of paternal care (Kendeigh 1952; Maynard Smith 1977; Ridley 1978; Silver et al. 1985; Clutton-Brock 1991; Ketterson and Nolan 1994; Soler et al. 1998), and sex-biased parental care (Lessells 1998). With respect to flightless rails, perhaps most promising are the competitive advantages of rapid development and large size in *K*-selected insular environments (Lack 1970a, b; M. Williamson 1981; Boyce 1984; Rohwer 1988).

Beyond the role of parents in development of young are several variables of potential importance for rails, including possible contributions of cooperative (often polyandrous) breeding and helping behavior among pairs and unmated adults (MacRoberts and MacRoberts 1976; Emlen 1978; Faaborg 1981; Koenig and Pitelka 1981; Gibbons 1986, 1987; Brown 1987; Koenig and Stacy 1990; Koenig et al. 1992; Cockburn 1998; Ward 2000). Helpers related to the breeding pair likely increase their probability of occupying the parental territory in subsequent years, thereby more than compensating for a delay in individual mating (Stacey and Ligon 1987, 1991). However, ecological constraints imposed by critical resources also are important (Houston 1987; Koenig and Stacey 1990), and arguments based on inclusive fitness have not been confirmed in some species (de Jong 1994), including *Porphyrio* (Craig and Jamieson 1990).

Amalgamation of unrelated broods—either through deposition of eggs in the nests of other, typically conspecific pairs (nest parasitism or prehatch amalgamation [Weller 1959; Yom-Tov 1980; Andersson 1984; Savard 1987; Eadie et al. 1988; Rohwer and Freeman 1989; Beauchamp 1997; Zink 2000]) or after departure from the nests (e.g., formation of crèches or posthatch amalgamation [Riedman 1982; Savard 1987; Eadie et al. 1988]) are not well understood from the standpoint of selective advantages; the former is considered among the *K*-selected traits for North American Anatidae by Eadie et al. (1988). In that both prehatch and posthatch amalgamation of broods has been observed in at least some rallids, especially comparatively aquatic species (Gibbons 1986; Lyon 1991, 1993a, b; McRae 1995, 1996a, b, 1997; Sorenson 1995; McRae and Burke 1996), the relative frequency of such ancillary activities in the reproduction of flightless rails is warranted. Communal nesting, involving nonreproductive helping by often related subordinate individuals, is frequent in continental *Porphyrio* and some members of the Fulicarina as well (Craig 1976, 1977, 1979, 1980a, b, 1984; Krekorian 1978; Craig and Jamieson 1985, 1988, 1990; Jamieson and Craig 1987a, b; Jamieson et al. 1994; Lambert et al. 1994; Jamieson 1997).

Although the apparently plesiomorphic condition of monogamy of the Rallidae may be primarily explanatory of biparental incubation, the need for frequent changeovers between incubating birds as a means to regulate temperatures of eggs (Grant 1982) may be a contributory factor for rails nesting on tropical islands (e.g., *Gallirallus wakensis* and *Porzana palmeri*). Based on analysis by Livezey (1998), the precocial development of the Rallidae is plesiomorphic for the Gruiformes, with the altricial condition of the Heliornithidae evidently being derived. This agrees with the broad polarities inferred by Starck and Ricklefs (1998a) for the order. Hatchlings of members of the Rallidae generally possess downy plumage, early motor and locomotor activity, and at an early age follow and are fed by their parents (P. B. Taylor 1996, 1998). Early attainment of locomotor capability distinguishes traditional altricial and precocial groups (Ricklefs and Starck 1998b; Starck and Ricklefs 1998a) and relative variation in lengths of developmental period indicate that this parameter plays a fundamental role in the increased body sizes of many flightless rails (Table 80). In other words, whereas dietary limitations may be responsible for the modest dwarfism or approximate stasis in adult body sizes of a minority of flightless rails, the variable increases in body mass shown by a majority of flightless rallids are produced simply by corresponding extensions of growth periods. During the protracted developmental periods of the larger flightless species in which the juveniles remain in the company of their parents or helpers, learning of complex or specialized techniques for foraging may be substantially improved (Heinsohn 1991).

The evolutionarily intriguing frequency of communal-nesting, pre- and post-hatch brood amalgamation in continental *Porphyrio, Porphyrula, Gallinula,* and *Fulica* (citations above)—typically denizens of open-water habitats—evidently has no strong association with insularity or with flightlessness per se. However, it is noteworthy, that examination of limited field data (P. B. Taylor 1996, 1998) indicates that these departures from plesiomorphic monogamy are either very infrequent or absent in flightless members of two of these genera (*Porphyrio hochstetteri* and *Gallinula comeri*). Exceptional in this regard are some of the variably constituted reproductive arrangements documented in *Tribonyx mortierii* of Tasmania (Goldizen et al. 1993, 2000; Gibbs et al. 1994). The implications of this apparent difference are challenging to infer in light of the multiplicity of other ecological differences confounding these pairwise comparisons (Ripley 1977; P. B. Taylor 1996, 1998).

In most ratites, among which the allocation of duties of incubation and broodcare vary considerably, large body size requires that developmental periods be protracted, and young sometimes remain with the attending adult(s) for a year or longer (Folch 1992). These generalities do not extend to the comparatively small, noctural *Apteryx,* in which monogamy is the rule, incubation is prolonged by the disadvantaged combination of large eggs and low body temperatures of adults, and in which the young attain independence (typically within the parental territory) several weeks after hatching (Calder 1979, 1991; Folch 1992). The relatively high vulnerability of immature individuals of the large, cursorial ratites (e.g., *Struthio*) to predation evidently is ameliorated to some extent by formation of crèches in several taxonomic families, over which adults genetically related to one or more of the constituent broods maintain vigilance for several months (Folch 1992).

With the exception of the Auckland Islands Teal, which is exceptional in the parental care contributed by males (Weller 1975; Livezey 1990), consistent differences between continental and insular *Anas* in parental care are not evident (Maynard Smith 1977; Weller 1980; Livezey 1991, 1993a). Predictably, there is a limit to total parental investment in dabbling ducks; this is reflected in an inverse relationship between number of offspring and per capita investment in offspring in insular *Anas* (Livezey 1990, 1991, 1993a). Unlike insular dabbling ducks (Livezey 1990, 1993a), meager evidence suggests that at least some populations of flightless steamer-ducks form crèches (Humphrey and Livezey 1985), evidently to permit coordinated vigilance by adults or the protection afforded to immature birds by foraging in groups during their protracted developmental periods.

Although the species nests in loose colonies, the Galapagos Cormorant does not form crèches (Livezey 1992a). *Pezophaps solitaria* may have formed crèches or social amalgamation of broods (Halliday 1978b; Livezey 1993b), and both raphids conformed to at least three of six characteristics shared by species showing alloparental care: production of single offspring, prolonged parental care, and limited lifetime reproductive output (Riedman 1982). In the comparatively solitary, polygynous Kakapo, in which the females mated to the same male locate nests in loose spatial association, young are raised with individual family units (Livezey 1992a).

Four Case-Histories Exemplifying Principles

Selection of Exemplars

A direct association of morphometric proportionalities indicative of flightlessness in the Rallidae (Figs. 32, 59, 60) with positions of species on the r–K continuum would provide a direct means with which to associate morphological shifts with ecological changes. However, the latter would require data not available for most modern flightless rallids and none of those taxa known only from subfossil skeletal remains, including detailed information on energy budgets of adults, relative alterations of reproductive parameters, and details of developmental rates (ontogenetic schedules) of the pectoral apparatus for precise calibration and qualitative diagnoses of heterochrony involved. The latter ontogenetic events are calibrated by developmental time as opposed to use of size as a surrogate estimate of age (Godfrey and Sutherland 1995).

Prolongation of developmental periods in rails (flighted and flightless) seems inextricably tied to increased adult body mass, further confounded by the effects of environmental stress, caloric throughput, and supplies of nutrients (Brody 1945; Ricklefs 1974, 1996; Boag 1987; Beintema and Visser 1989; Dykstra and Karasov 1993; Starck 1993; Thomas et al. 1993). Whether this relationship in rallids represents a departure from the generality described by Carrier and Auriemma (1992) for flighted birds of many avian families, in which time to fledging (defined therein as capacity for flight) evidently was delimited by attainment of definitive lengths of appendicular elements, is not clear. The vast majority of rallids (41 species in 15 genera) can be characterized as follows: young fully feathered at 4 weeks after hatch, acquisition of remiges completed at 5–6 weeks, and flight attained at 7–8 weeks, with "fledging" being set at roughly 5–7 weeks of age. However, those taxa having notably larger body masses for which developmental times are available (P. B. Taylor 1998; Eason and Williams 2001) demonstrate that greater body size requires protracted developmental periods, even when restricting comparisons to flighted taxa: *Porphyrio porphyrio*-group (mean adult mass, ~550–1,000 g; age at fledging, 10 weeks); *Aramides ypecaha* (~700 g; 8–9 weeks); *Eulabeornis castaneoventris* (~700 g; 8 weeks); *Gallinula chloropus* and *G. tenebrosa* (~300 and 500 g; 7–8 weeks); *Fulica atra* and *F. americana* (~500–850 g; 8–10 weeks); and *Fulica gigantea* (2,400 g; 16 weeks).

In lieu of such exhaustive comparisons, infeasible at present, an alternative is to perform comparatively detailed eco-ontogenetic contrasts of a minority of flightless rails and their flighted relatives for which comparative data are available (Table 84; Appendix 6). In addition to focusing on taxa for which desired data were maximal, contrasts will be made for a diversity of taxonomic groups within the Rallidae (four tribes or subtribes) and a range of body masses.

CASE HISTORIES FROM DIVERSE TAXA, LATITUDES, AND HABITATS

Swamphens (Tribe Porphyriornithini).—The flightless species involved in this comparison (*Porphyrio hochstetteri*) is the largest of the species to be considered in detail here, and examination of the data indicates that the species underwent a substantial increase in size concomitant with the loss of flight (Tables 71, 84; Appendix 1). In addition, other apomorphies of *P. hochstetteri* were increased sexual size dimorphism (Tables 24, 26, 55, 57), a shift to upland habitats and diet dominated by grasses (Gurr 1951; Carroll 1966, 1969; Fordham 1983; Norman and Mumford 1985; Clout and Hay 1989) with associated changes in oral anatomy (McCann 1964; Suttie and Fennessy 1992), a demographic shift to strongly prevalent (perhaps life-long) monogamy, and a halving of clutch size accompanied by only a modest increase in egg size. The comparatively few and massive eggs of *P. hochstetteri* require only several additional days to hatch, and a month passes before natal down begins to be replaced by feathers of the juvenal plumage, whereas time to fledging and independence of young are approximately three times as great as those of its smaller flighted congeners (Table 84).

Assistance by unmated birds of reproductive age, probably related to the mated pair and likely related to the modal 1-year delay in first breeding, is typical of the flightless species (Table 84), a behavioral pattern also practiced to varying degrees in the variably polygamous flighted members of *Porphyrio* and sister-genus *Porphyrula* (Wettin 1984; Hunter 1985, 1987a, b; Goldizen et al. 1993, 1998, 2000; Jamieson et al. 1994; Lambert et al. 1994; Jamieson 1997; Birkhead 1998). Definitive plumages are indistinguishable and (in good light) of comparable gaudiness to that of flighted *Porphyrio*; in shadow all swamphens can appear quite dark. On both ecological and demographic grounds, the Takahe typifies a shift toward K-selected attributes: larger body size, evidently increased longevity, and evidence that the species maintained stable populations at or near carrying capacity of a comparatively specialized habitat before human immigration (Table 84; Fig. 124).

Typical Rails (Subtribe Rallina).—Employing flighted *Gallirallus philippensis* as the primary standard for comparisons (Elliott 1987; Amerson et al. 1995; Robinson 1995), flightlessness in three of the best-studied typical rails (*Gallirallus*) of the Pacific region (*G. australis, G. owstoni,* and *Tricholimnas sylvestris*) presents for the most part a coherent suite of ecomorphological changes. In most

TABLE 84. Reproductive and ontogenetic parameters of selected flighted and flightless species of Rallidae, with inferred r–K shifts and underlying heterochrony. Flightless species are preceded by asterisks. Signed factor relates multiplicative factor of mean mass (g) of flightless species to that for flighted relative (Appendix 3). Egg masses (g) are given in means for one egg, followed parenthetically by percentage of mean body mass of adult female (Appendix 3). Age is defined by acquisition of remiges; capacity for flight is typically attained at least 1 week later in flighted species (Appendix 6). Heterochronic classes follow McNamara (1986), McKinney (1988a), and McKinney and McNamara (1991).

Taxon	Change in body mass	Modal clutch size	Mean egg mass	Incubation period (days)	Age at fledging (weeks)	Age at independence (months)	Age at first breeding (years)	"Helpers" involved	Shift in r–K continuum	Heterochrony–size class
Porphyrio melanotus	—	4	34 (4%)	23–27	8–10	1–2	1–2	Variable	—	—
*Porphyrio hochstetteri	+2.8	2	93 (10%)	29–31	9–20	6–8	2+	Typical	r → K	Paedomorphic giantism
Gallirallus philippensis	—	5	17 (8%)	18–25	6–8	2–3	1	No	—	—
*Gallirallus australis-group	+4.2	3	52 (7%)	26–28	12	4	1	No	Indeterminate	Paedomorphic giantism
*Gallirallus owstoni	+1.1	3	17 (8%)	19	12	4	1	No	Indeterminate	Paedomorphism
*Tricholimnas sylvestris	+2.5	3	32 (7%)	20–23	9–10	3–5	1±	No	Indeterminate	Paedomorphic giantism
Laterallus jamaicensis	—	6	5 (14%)	17–20	6	—	—	No	—	—
*Atlantisia rogersi	+1.1	2	9 (26%)	—	—	~12	1+	Variable	r → K	Paedomorphism
Porzana tabuensis	—	4	8 (21%)	19–22	6–9	4–5	1	No	—	—
*Porzana atra	+1.8	2	13 (18%)	21	6	3	1	No	Indeterminate	Paedomorphic giantism
*Porzana palmeri	+1.1	—	7 (16%)	—	—	—	—	No	Indeterminate	Paedomorphism
Tribonyx ventralis	—	6	24 (7%)	19–20	—	—	—	—	—	—
*Tribonyx mortierii	+3.2	6	45 (4%)	19–25	6–8	9–15*	1+	Typical	Indeterminate	Paedomorphic giantism
Gallinula chloropus-group	—	7	22 (7%)	17–22	6–7	2–3	1	Typical	—	—
*Gallinula comeri	+1.5	4	29 (6%)	21	—	—	—	Typical	Indeterminate	Paedomorphic giantism

* Estimate confounded by typical presence of "helpers" accompanying broods (Appendix 6).

dimensions of inferred change, the typical rails of the Northern Hemisphere and Neotropics (*Rallus*), distinguished ecologically by greater fecundity (excepting *R. caerulescens*) and (in many species) long-distance migratory habit, indicated similar changes by comparison with flightless *Gallirallus* (Walkinshaw 1937; Kozicky and Schmidt 1949; Meanley 1953, 1956, 1969; Meanley and Wetherbee 1962; Schmidt 1976; Kaufmann 1977; Johnson and Dinsmore 1985; Zembal and Fancher 1988; Conway et al. 1994). All species of *Gallirallus* and the closely related *Tricholimnas* and *Nesoclopeus* are characterized by monogamous mating systems, a plesiomorphy shared by related genera (Ripley 1977; P. B. Taylor 1996, 1998).

However, the three flightless species in this sample share modest increases in body size, a trend of wider phylogenetic prevalence (Fig. 12), and this increase in body size was accompanied by variably pronounced enhancements of sexual size dimorphism (Tables 24, 26, 55, 57), a 50% reduction in clutch size, and a modest increase in the size of the eggs that remain (Tables 71, 84; Appendices 1, 7). Of the three flightless species detailed here, both *Tricholimnas sylvestris* and the *Gallirallus australis*-group manifest (possibly synapomorphically) paedomorphic definitive plumages (Fig. 3; Ripley 1977). Whether the apparent crypsis provided by this comparatively dull plumage entailed selective advantages with respect to aerial predators (e.g., *Harpagornis moorei, Circus eylesi,* and *Ninox novaeseelandiae* with respect to adults and young of *G. australis* [cf. Holdaway 1989; Atkinson and Millener 1991]) is uncertain, but given evidence that extinct *Harpagornis* preyed on smaller species of moa (Dinornithiformes: Anomalopterygidae), predation on flightless rails by at least these eagles would appear likely (Worthy and Holdaway 1996). Also, the possibility exists that this external paedomorphosis is but an external corollary of a pervasive heterochrony primarily targeting the pectoral apparatus (Tables 71–75; Figs. 121, 123).

Incubation periods are extended in a manner broadly commensurate with the differences in body size, whereas all three flightless species are characterized by similar (roughly 50%) increases in ages at fledging and slightly less marked advancements (roughly 33%) in age at independence (Table 84). As for mating systems, no differences in age at first breeding or participation of parareproductive helpers distinguished flightless from flighted *Gallirallus* (Table 84; Appendix 6), including *G. owstoni* (Perez 1968). All three flightless species show dietary breadths at least as great as their flighted relatives, with the *Gallirallus australis*-group being notorious for its predation on the eggs of other birds (Ripley 1977). With the regrettable omission of the nanitic and potentially exceptional *G. wakensis* (Table 71), these summaries support a modest shift in the direction of *K*-selected attributes with respect to basic reproductive attributes, but aspects of mating systems and demographic parameters are equivocal in this respect.

Crakes (subtribe Crecina).—Data permit the inclusion of two sets of comparisons among the crakes—flighted *Laterallus jamaicensis* versus flightless *Atlantisia rogersi,* and flighted *Porzana tabuensis* versus flightless *P. atra* and *P. palmeri.* Change in body size was very similar in all three flightless species (Tables 71, 84), inferred to be slight increases relative to the close relatives but obscured by wider comparisons with flighted crakes (Inskipp and Round 1989; Woodall 1993; Bookhout 1995; Martinez et al. 1997; Bockheim and Mezzell 1999). The foregoing pattern of a substantial reduction in clutch size accompanied by modest increases in egg size applies to all of the flightless crakes for which data were available (Table 84; Appendix 7), whereas all of these taxa differ substantially in the high fecundity characteristic of migratory crakes of the Northern Hemisphere (Walkinshaw 1939, 1940; Meanley 1960, 1965; Kaufmann 1977; Rundle and Sayre 1983; Johnson and Dinsmore 1985; Flores and Eddleman 1995; Green and Rayment 1996).

Limitations of information regarding developmental rates notwithstanding, there are suggestions of modest protraction of developmental periods in *Atlantisia rogersi,* a weakly supported inference bolstered by the retention of an "immature" plumage by at least some individuals for 1 or 2 years after hatching, suggestive of delayed acquisition of adult plumage (Flood 1984; Studd and Robertson 1985; Fraser 1989; Hakkarainen et al. 1993). In contrast, if the marginal data available permit any inferences for *Porzana* (Table 84), flightless crakes appear to attain independence and sexual maturity in a slightly shorter period than their flighted congeners (Soper 1969; Onley 1982; Kaufmann and Lavers 1987). Virtual uniformity of mating system (limited variation in some other *Porzana* aside), reduction in clutch sizes, slight increases in sizes of eggs and adults, and heterogeneity of shifts in developmental periods render a single assessment regarding r–K shifts unrealistic. There are modest indications of shifts toward K-selected characteristics in *Atlantisia rogersi* (bolstered by reductions in metabolic rates [Ryan et al. 1989]), whereas flightless *Porzana* present a mosaic of r- and K-selected changes (Table 84).

Moorhens (Subtribe Fulicarina).—External similarities of moorhens belie an array of subtle social and demographic variations, but interspecific trends and generalities are adequate to sustain comparisons of two pairs of congeners (Appendices 6, 7)—flighted *Gallinula chloropus* versus flightless *G. comeri,* and flighted *Tribonyx ventralis* versus flightless *T. mortierii* (Table 84). Mating systems within *Tribonyx* in general and *T. mortierii* in particular are somewhat complex, with occurrence of polyandry, polygyny, and communal nesting reported (Goldizen et al. 1993, 2000; Gibbs et al. 1994) and complicated by significant patterns of dispersal (Goldizen et al. 2002). Likewise, enlarged breeding groups have been documented to occur in some *Gallinula* (Fredrickson 1971; Huxley and Wood 1976; Eden 1987; McRae 1996a), whereas mating systems of *G. comeri* are comparatively poorly known, although a role for "honest" signals of fitness has been documented (Fenoglio et al. 2002). In light of the variation in behaviors confirmed, the crucial comparisons that remain are as much the specification of relative frequencies of these behaviors with environmental conditions that influence these variations in flighted and flightless taxa. Participation by nonreproductive helpers, at least in some cases related to territoriality and aggression between parents and older offspring, also is not uncommon in both genera (Leonard et al. 1988, 1989; Gibbons 1989; Putland and Goldizen 2001). In both couplets of moorhens, flightlessness is associated with substantial increases in body mass (Tables 71, 84). *Tribonyx* has no reduction in clutch size, whereas flightlessness in *Gallinula* conformed with the general pattern of reduced fecundity. Both flightless moorhens lay larger eggs than the flighted species, despite the disparity in clutch sizes, differences with modest, associated increases in incubation (Table 84). Information on developmental periods is simply inadequate for detailed comparisons, and this deficiency and the variation in other critical aspects of breeding ecology permit no confident inferences regarding shifts on the r–K continuum related to flightlessness in moorhens. With respect to feeding ecology, *G. comeri* relied more heavily on upland habitats for foraging than most flighted *Gallinula*, whereas both species of *Tribonyx* exploit a mixture of wetland and upland habitats (Ripley 1977).

EMERGENT PATTERNS AND THE r–K CONTINUUM

In the cases detailed above (Table 84) and the majority of cases based on even minimal data (Table 71), variably pronounced giantism accompanied the loss of flight. In addition, a pronounced majority of cases indicates that several attributes conventionally considered to be K-selected—decreased clutch size, increased egg size, and protracted developmental periods (Pianka 1970, 1972; Fenchel 1974; Stearns 1976, 1977, 1989b, 1992; Western and Ssemakula 1982)—were characteristic of flightless taxa in comparison with flighted relatives. Exceptions to the modality of increased body size (e.g., *Dryolimnas aldabranus, Gallirallus wakensis,* and *Cabalus modestus* [Tables 71, 84]) and protracted developmental periods (e.g., some *Porzana* [Table 84]) exist, and variation precludes confident generalizations pertaining to mating systems and participation of helpers in nesting with respect to the r–K continuum. Despite the frequent suggestion that *Atlantisia rogersi* represents a dwarfed paedomorph, in part related to the status of the species as the smallest extant flightless bird (Ryan et al. 1989), the likelihood that the taxon is most closely related (Livezey 1998) to equally small flighted crakes (e.g., *Laterallus jamaicensis*) indicates that stasis in body size is the most-parsimonious interpretation for *A. rogersi* (Appendix 1).

Changes in (nonpectoral) size or qualitative maturity are primarily a function of variation in developmental period, whereas relative pectoral development is the result of paedomorphosis arising during these variable growth periods, the exact mechanisms of which (delayed offset or deceleration, or both) remain unclear and may vary among taxa. Predictably, those lineages showing the majority condition in which substantial pectoral paedomorphosis is combined with increased body size include some of the most extreme examples of relative pectoral reduction among the Rallidae (e.g., *Porphyrio hochstetteri, Diaphorapteryx hawkinsi,* and *Aphanapteryx bonasia*). Flighted *Porphyrula flavirostris,* which shows paedomorphosis of plumage pattern without pectoral consequences, also evidently underwent at least a 35% reduction in body mass (roughly 60% if compared to *P. martinica*).

The misfortune that all possibly flightless coots are extinct and will remain poorly known is ameliorated to some extent by the likely similarity of *Fulica gigantea* and *F. cornuta* to these species in locomotory ability, negligible seasonal movements, and perhaps basic ecological tendencies (Johnson 1964, 1965; MacFarlane 1975). In light of the pronounced territoriality of coots generally, in which pugnacious defense of nests, young, and feeding areas extends beyond confamilials (Gullion 1952, 1953b, 1954; Kiel 1955; Kornowski 1957; Navas 1960; Fredrickson 1970; Fjeldså 1973; Ryan and Dinsmore 1980; Gorenzel et al. 1982; Byrd et al. 1985; Cave et al. 1989; Impe and Lieckens 1993), the large body sizes of the three subfossil *Fulica* may represent analogs to the territoriality-related increases in body size (and negative allometry of wing size) in steamer-ducks (Livezey and Humphrey 1986). Although coots differ from *Tachyeres* in their largely vegetarian food habits (Jones 1940; Driver 1988), like steamer-ducks, the species of *Fulica* routinely dive during foraging (Jones 1940; Bakker and Fordham 1993), thereby enlarging the analogy to the advantages of increased body mass to counter buoyancy and increase locomotory efficiency (Livezey and Humphrey 1984b, 1986). Similarities between the large, possibly flightless *Fulica* and flightless steamer-ducks are striking and include increased body size; negatively allometric scaling of wing elements with body mass, resulting in a close approach to the threshold of flightlessness; only minimal reductions in flight-related skeletal features (e.g., carina sterni); negligible qualitative evidence of paedomorphosis; likely importance of protracted developmental periods to attain large adult body sizes; and likely advantages of increased body mass related to diving, territoriality, thermodynamic advantages in cool climates (at least in the *F. chathamensis*-group), and sexual selection (Livezey and Humphrey 1986). Similar comparisons of flightless rails with flightless members of other avian families are tempting (Table 79), perhaps the most appealing being those concerning the apparent progenesis of dwarfed, flightless lineages of the Rallidae such as *Dryolimnas aldabranus*, *Gallirallus wakensis*, and *Cabalus modestus* in comparison with that of the anatid *Anas aucklandica* (Livezey 1990).

As suggested by associations among variables in the correspondence analysis (Table 76; Fig. 116), there were relations between these morphological and developmental changes and the ecological circumstances under which these occurred. Although most of the species highlighted in the case studies (Table 84) occurred on islands of small size and at low or middle latitudes with no marked changes in habits, the two species showing comparatively pronounced shifts from r-selected to K-selected traits (*Porphyrio hochstetteri* and *Atlantisia rogersi*) also were lineages undergoing shifts to upland habitats (tussock grasslands) midlatitude islands. The latter shift in habitat also might apply to a third species detailed among the case studies (*Gallinula comeri*), one sympatric with *A. rogersi* in the Tristan da Cunha island group of the South Atlantic (Table 84).

Gould (1977:10) generalized that progenesis is r-selected, whereas neoteny is K-selected. If accurate for the Rallidae, then this generality suggests that the most likely heterochronic process underlying pectoral paedomorphosis in most flightless rails is neoteny. This generalization accurately associated the derived characteristics of the Dodo and Solitaire—substantial increases in body size, deposition of fat, developmental period, sexual dimorphism, tameness, and conspicuous changes in anatomical proportions related to flightlessness—with the K-selected selective regime of a comparatively predictable climate and moderately competitive ecological community of an isolated, oceanic island (Livezey 1993b, 1995a).

Stearns and Crandall (1981a) modeled the incidence of delayed maturity (one of several life-historical attributes of K-selected communities) based simply on selective advantages of increased fecundity and survivorship of young. However, such assessments are obscured by a general trend in which a majority of Southern-Hemispheric, largely nonmigratory, flighted species in a number of major taxonomic groups have fewer young in which there is greater parental investment, longer periods of parental care, and delayed ages at independence. The latter evidently is responsible for a lower mortality of young, especially from predation (Russell 2000). The lower mortality of young among Southern-Hemispheric taxa of birds evidently is related, at least in part, to stronger responses of parents to reduce risks (Ghalambor and Martin 2001).

Despite some strong indications that flightlessness in a number of rails is associated with *K*-selected attributes (Table 84), the correspondence was far from uniform, and some weakly substantiated shifts (e.g., shortened developmental periods in flightless *Porzana*) appear to be contradictory. Therefore, a view of rails as "generalized" (Olson 1973a; Ripley 1977) is countered by several specializations independently evolved by several flightless rails, for example, extreme modifications of the bill and changes in body size.

ADAPTIVE LANDSCAPES AND AVIAN FLIGHTLESSNESS

Variably intense natural selection has been documented in natural populations (Endler 1986; Mitton 1997). One of the most convincing examples of one aspect of relative fitness concerned weather-related natural mortality of House Sparrows collected by Bumpus (1899), and subsequently reanalyzed (Harris 1911; Calhoun 1947; Grant 1972a; Johnston et al. 1972; O'Donald 1973; Lowther 1977; Pugesek and Tomer 1996). A recent review revealed that gradients (magnitudes of slopes) in the literature were greatest among morphological traits and parameters bearing on reproductive success (Kingsolver et al. 2001), although compelling examples of natural selection in birds have been documented in additional taxa (e.g., G. Johnson 1987; Schluter 1988b). Adaptive optima for particular characters, although difficult to quantify empirically, have been assessed directly in natural populations of birds where two or more local optima are comparatively close, of relatively low dimensionality, or are related intuitively to diet (e.g., Smith and Temple 1982; Price et al. 1984; Grant 1985; Schluter et al. 1985; Smith 1990a–c, 1991).

Adaptive landscapes, two-dimensional depictions of the relative fitnesses of genotypes as surface contours, were conceptualized first by Wright (1932) and extended, among others, by Wright (1940, 1977, 1982), Lande (1979, 1986), Kirkpatrick (1982b), Barton and Rouhani (1987), and Barton and Turelli (1987). The graphical display of relative fitness in a continuous fashion for two dimensions allowed a natural depiction of the fitnesses of multiple optima and intervening forms that were crucial to a burgeoning literature related to evolutionary radiations, character divergence, disruptive selection, and peak shifts, which continues unabated to the present (e.g., Rouhani and Barton 1987; Grant and Grant 1993; Price et al. 1993; Wagner et al. 1994; Grant 1998b; Brock 2000; Schluter 2000a, b). Subsequent theoretical refinements, especially in the areas of epistasis and the multiplicative interactions among differentially selective genes, has revealed that traversing such landscapes can be exceedingly complex and include selective "shortcuts" between optima (Cheverud 2000; Phillips et al. 2000; Rice 2000), and that nonadditive epistasis can alter the covariance structure that specifies the topography of the landscape (Wolf et al. 2001).

Although most applications of adaptive landscapes and related concepts of local and global optima have been limited to a single set of conditions or a single selection regime, the method lends itself well to the display of change in selection related to shifts in time and space. In both evolutionary and geographical senses, rails colonizing oceanic islands arrive on new landscapes, which in turn imply new and novel evolutionary challenges to the founders. Therefore, in the context of flightless rails, a natural application is the hypothetical parsimonious passage from the selective optimum for a continental source (ancestral) population to that

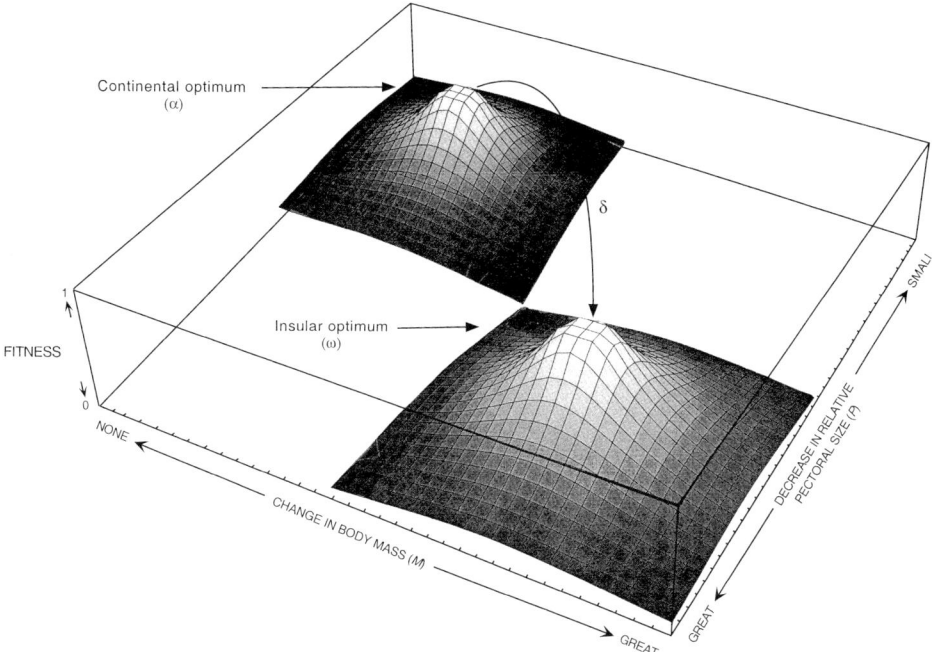

FIG. 126. Hypothetical adaptive landscapes, after Wright (1932, 1940, 1977, 1982), for continental (plesiomorphic) optimum (α) and insular (apomorphic) optimum (ω), depicting morphological shift (δ) required after colonization; the latter is simplistically reduced to absolute change in body mass (M), encompassing dwarfism, stasis, and giantism, and variably pronounced decrease in relative pectoral size (P). Example depicts two optima separated by deep trough of profoundly disadvantageous intermediate phenotypes.

of a (presumably) globally lower, but locally higher selective optimum for descendant (flightless) population on an oceanic island (Fig. 126), and such optima can assume diverse suites of changes (Figs. 117, 121).

This exercise underscores the importance of initial conditions (plesiomorphic position in this context) with respect to subsequent changes. Initial conditions also can influence the behavior of trajectories under certain selection regimes (e.g., Ruelle 1989). Also, polymorphism (see below) in either or both of the populations in question (e.g., Smith and Temple 1982; Smith 1990a–c, 1991; Smith and Girman 2000) would require multiple or variously conjoined optima within the corresponding landscapes. With improved phylogenetic resolution, the insular microdifferentiation of the *Gallirallus philippensis*-group (including *G. dieffenbachii* and *Cabalus modestus*) would provide for a particularly compelling use of hierarchical shifts in optima.

The utility of graphical landscapes was extended by Waddington (1957) to the context of epigenetics (Hall 1992). In these applications, the surface of the landscape reflects the relative ontogenetic ease of various developmental pathways, and a theoretical ontogeny traversing this landscape tends to follow the line(s) of least resistance (Merilä and Björklund 1999), that is, those representative of epigenetically favored trajectories (Rollo 1994; Schlichting and Pigliucci 1998). Topographies can accommodate a number of plausible ontogenetic circumstances, including multiple pathways, and provide for latitude of variation within each (Wagner and Misof 1993; Wagner et al. 1997). Such applications also are pertinent to the epigenetic determination of the essentially bifurcated ontogenetic pathways confronted by embryonic rails, in which the surface of the landscape (and the relative likelihoods of the flighted and flightless pathways) vary in large part with selection regimes of continental and insular environments.

BET-HEDGING: VAGRANCY AS MINORITY TACTIC

Seger and Brockmann (1987:182) posed the following query: "... it seems natural to ask whether plants or animals ever sacrifice some of their potential fitness in order to reduce their probabilities of complete failure. In other words, is there an evolutionary trade-off between *expected* fitness and the *variance* of fitness?" Such trade-offs are alleged to be examples of "bet-hedging" (Stearns and Crandall 1981b; Seger and Brockmann 1987), synonymously referred to as "risk-spreading" (den Boer 1968, 1981; Goodman 1984). Given more conventional mechanisms for the preservation of diversity in life histories (e.g., polymorphism), which in some instances offer an excess of plausible explanations for the adaptative nature of variation (Jones et al. 1977), it is not surprising that Seger and Brockmann (1987) concluded that bet-hedging is comparatively rare and is thought to be possible (sensu qualifying as an evolutionarily stable strategy [ESS]) only under narrow conditions.

Ricklefs (1983b) applied the expression bet-hedging to issues of avian reproduction. Inextricably involved with reproduction in many organisms is the issue of dispersal, especially of propagules or variably developed young (Johnson and Gaines 1990). The perennial debate concerning the possible selective advantages of variation in clutch sizes of some members of the Procellariidae, Sulidae, and Alcidae, in which the vast majority result in but one surviving hatchling, may represent a case of bet-hedging in that a minimal investment of one or two extra or "insurance" eggs (where relative egg size is small) is a hedge for the rare seasons in which resources permit the production of two chicks in a single reproductive cycle. However, augmentation of clutches to enhance the likelihood of successfully rearing a single chick in a normal year would not qualify (Lack 1954, 1968b; Dorward 1962; Ashmole 1963c; J. S. Nelson 1964; J. B. Nelson 1978; Cramp and Simmons 1977; Werschkul and Jackson 1979; Ricklefs 1982, 1993; D. J. Anderson 1989; Marchant and Higgins 1990; Warham 1990; Konarzewski et al. 1998). Perhaps of similar relevance to bet-hedging is the infanticide or aggressive neglect (e.g., "tousling") shown by some rallids (notably *Fulica*) toward the youngest members of their broods (or at least members of extended broods) in most years, especially in the event of low availability of food. Avoidance of investment in unrelated progeny in aggregations involving brood parasites is an alternative explanation (Horsfall 1984; Lyon 1993a, b). The phenomenon of brood reduction through neglect or outright aggression by siblings or parents pertains principally to species showing asynchronous hatching and altricial development, and is interpreted by some as a means to maximize survivorship of a subset of a clutch or brood (i.e., expected fitness) under stressful conditions (Lack 1947b, 1954; Ricklefs 1965, 1993; Nisbet and Cohen 1975; Stinson 1979; Richter 1982; Mock 1984, 1987; Shaw 1985; Mock and Parker 1986; Temme and Charnov 1987; D. J. Anderson 1989, 1990a, b; Pijanowski 1992).

Does the concept of bet-hedging hold potential for the hypothesis that rails maintain diverse modes of dispersal that contribute to the high frequency of long-distance vagrancy and insular endemism in the family? Despite circumstantial evidence of dispersal polymorphism in rails, the phenomenon has not been demonstrated in the family. Evidence that many migratory rallids approach or exceed carrying capacities on established breeding grounds, territoriality dominated by experienced adults that characterizes traditional wintering grounds (Matthysen 1993), and migrants (especially first-year birds) that engage in informed prospecting for new breeding sites (Reed et al. 1999) collectively suggest that the exceptionally high vagrancy and insular colonization by rails may represent more than mere accidents. An alternative hypothesis, clearly not amenable to testing at present, proposes that persistence of a proportion of offspring genetically predisposed to vagrancy—with a low probability of finding new habitable parcels, but a comparatively high probability of success if reached—may represent the parental investment in a long-odds strategy against the unpredictable success of hatchlings poised to follow traditional migratory routes or committed to often ephemeral, natal habitats (Remsen and Parker 1990). Even if such were demonstrated, whether such allocations represent bet-hedging sensu stricto (i.e., optimizing variance but not mean of fitness) would require sophisticated, long-term investigation.

A Metapopulational Approach to Avian Flightlessness

Metapopulations and Evolution of Life Histories

General tenets.—Recent years have seen a paradigmatic shift from some of the central issues of island biogeography (sensu MacArthur and Wilson 1967) and classic works on the evolution of life histories (Cole 1954; Maynard Smith 1962, 1966, 1970; Gadgil and Bossert 1970; Felsenstein 1979; Stearns 1982; Caswell 1983; Bruton 1989) to a perspective of metapopulational ecology (Haila and Järvinen 1982; Giles and Goudet 1997; Hanski and Simberloff 1997). Hanski and Simberloff (1997) chronicled this conceptual extension of the classic model proposed by MacArthur and Wilson (1963, 1967) to include quasi-insular subdivision of both continental and insular populations (Vandermeer and Carvajal 2001), and showed that the latter approach circumvented the treatment of entire species as uniformly interconnected demes (Hanski and Gilpin 1997; Hanski 1999). This change in paradigms also shifted attention from summary characteristics of communities (e.g., richness) to fates and attributes of the individual species that communities comprise, and provided a framework for analyses of ESSs and group dynamics that are free of traditional notions of group selection argued by a minority (e.g., Van Valen 1971; Hanski and Simberloff 1997) to be crucial for the evolution of dispersal.

Semantics of insular demography employed in metapopulational study underscore the conceptual relationship between the two paradigms (Harrison and Taylor 1997). For example, some of the more realistic models of metapopulation dynamics have been based on stochastic (Markovian) variations on incarnations of stepping-stone models of insularity (Giles and Goudet 1997; Olivieri and Gouyon 1997). Many of the genetic concepts pertaining to the founding and establishment of new, isolated populations apply to the constituent metapopulations of "species" as diagnosed by any of several viewpoints (Hedrick and Gilpin 1997). However, preliminary models suggest that influential parameters are sufficiently numerous and interactively complex that detailed predictions require information into both environmental and demographic stochasticity of a quality rarely achieved in studies in natural settings (Barton and Whitlock 1997; Foley 1997).

Applicability to the Rallidae.—Available information on extant rallids indicates that many members meet the suite of preconditions for metapopulational dynamics (Gyllenberg et al. 1997; Hanski 1997): patchiness of suitable habitat (Thomas 1965; Clinchy et al. 2002); vulnerability of even the largest breeding "populations" to extinction; patches of suitable habitat are of proximity to permit recolonization; and local "populations" are not completely synchronous in dynamic properties (Ripley 1977; P. B. Taylor 1996, 1998). Distributional patterns shown by the Rallidae span an extraordinary continuum encompassing single islands, large but contiguous areas of undetermined metapopulational dynamics, and distributions so highly fragmented as to invite formal recognition of numerous allospecies (Ripley 1977; P. B. Taylor 1996, 1997, 1998; Livezey 1998).

The marked distributional disjunctions of a number of continental species with flightless congeners (Ripley 1977; Remsen and Parker 1990; P. B. Taylor 1996, 1998) further distinguish these clades as candidates for qualification for metapopulational dynamics. These candidates include *Porphyrio porphyrio*-group (cf. flightless *P. mantelli* and *P. hochstetteri*), *Aramides cajanea*-group (cf. *Nesotrochis* spp.), *Ortygonax sanguinolentus* (cf. *Cyanolimnas cerverai*), *Rallus longirostris*-group (cf. *R. recessus*), *Gallirallus philippensis*-group (e.g., *G. dieffenbachii*, *G. owstoni*, and *G. wakensis*), *Habropteryx torquatus*-group (cf. *H. okinawae*), *Laterallus jamaicensis* and *Coturnicops noveboracensis* (cf. *Atlantisia rogersi* and "*Atlantisia*" *elpenor*), *Porzana fusca*, *P. pusilla*, *P. tabuensis* (cf. numerous flightless *Porzana*), *Amaurornis olivaceus*-group (cf. *A. ineptus*), *Gallinula chloropus*-group (cf. *G. comeri* and *Pareudiastes* spp.), and *Fulica atra*-group (cf. *F. chathamensis*, *F. prisca*, and *F. newtonii*). Several of these species (e.g., *R. longirostris*-group, *G. philippensis*-group, *L. jamaicensis*, *P. tabuensis*, and *G. chloropus*-group) are characterized both by fragmented distributions and variable migratory tendencies (Crawford et al. 1983; P. B. Taylor 1996, 1998; Olson 1997).

Several other rallids possess distributional subdivisions requisite for metapopulational structure (P. B. Taylor 1996, 1997, 1998), including *Canirallus oculeus* (several moderately large but well-separated distributional components in continental Africa); *Aramides calopterus* (small, distinct patches in central South America); *Amaurolimnas concolor* (disjunct populations through Central and South America); *Rallina fasciata* (fragmented Malaysian distribution); *Sarothrura lugens*, *S. affinis*, and *S. ayresi* (networks of isolated patches throughout Africa); *Pardirallus maculatus* and *P. nigricans* (fragmented resident populations throughout Central and South America); *Rallus limicola* (variably sized breeding areas in North, Central, and South America); *Rallus aquaticus*-group (fragmented breeding populations throughout Eurasia); *Gallirallus striatus* (multiple distinct populations throughout Indo-Malaysian region); *G. pectoralis* (populations scattered throughout New Guinea and eastern Australia); *Laterallus exilis* (several disjunct Neotropical natal areas); *Coturnicops notatus* (extremely scattered breeding patches throughout South America); *Micropygia schomburgkii* (three distantly separated, moderately large blocks in South America) and *M. flaviventer* (Caribbean and highly partitioned continental distributional range); *Porzana cinerea* (fragmented Indonesian distribution); *Porzana erythrops* and *P. albicollis* (patchy Neotropical range); *Amaurornis phoenicurus* (allopatric subspecies throughout Indo-Malaysia); and *Gallinula tenebrosa* (multiple allopatric resident populations throughout Australasia).

Also noteworthy is the limited or nonexistent vagrancy documented (P. B. Taylor 1996, 1997, 1998) for the tribe Sarothrurini (comprising *Rallina, Rallicula,* and *Sarothrura* [Livezey 1998]), one of few tribes of the Rallidae lacking flightless members (Table 2; Fig. 11). The fragmentation of distributions and variation in migratory habit typical of the Rallidae is reflected in the small minority of species that show a contiguous northern breeding distribution coupled with one or two, variably distinct, wintering ground(s), for example, *Porphyrula martinica, Crex crex, Porzana parva, P. porzana, P. carolina, P. marginalis, Gallicrex cinerea,* and *Fulica (a.) americana.* Taken as a whole, the distributional patterns, exceptional frequency of extralimital records, and variation in migratory habits of modern members of the Rallidae strongly suggest a dynamic evolutionary history including a subset of key taxa that through colonization by migratory strays, came to comprise multiple, largely independent, breeding populations. These then founded multiple sedentary allospecies, which if insular and well-isolated, typically evolved permanent flightlessness (Remsen and Parker 1990).

POLYMORPHISM OF FORM AND HABIT

> What evolutionary forces could mold a population's anatomical or psychological ability to disperse? Dispersal involves costs and potential benefits. The costs are the energy expenditure and risk of death in dispersing. The potential benefits are the possibility of finding more available resources than at the point of departure. The trade-off between these costs and benefits determines, and natural selection may produce, an optimal dispersal strategy. If, for instance, ... populations had for any reason lower long-term reproductive success on islands ..., and if individual[s] ... initially differed in their willingness to disperse across water, those that did have this willingness would leave fewer descendants, and the genetic basis for this willingness would eventually be selected out of the source population.—Diamond (1977b:252).

> Perhaps the flightlessness gene compensates for the water-crossing gene.—Ripley (1977:19)

Evolutionary origin and maintenance of polymorphisms.—Polymorphism, the coexistence of discrete alternative phenotypes within populations (Haldane and Jayakar 1963; Levins and MacArthur 1966; Gillespie 1973, 1974, 1975, 1981; Hartl and Cook 1973, 1976; Christiansen 1974, 1975; Maynard Smith and Hoekstra 1980; Karlin and Campbell 1981; Moran 1992), is considered widely to be related to heterogeneity of habitat (E. S. Brown 1951; Levins 1962, 1963, 1968, 1970; Maynard Smith 1970; Schaffer 1974; Giesel 1976; Hedrick et al. 1976). However, Hoekstra et al. (1985) inferred that models of polymorphism induced by heterogeneous environments are realistic only where the fitness of the heterozygote approaches those of the homozygotes favored under their respective optimal environments. Avian polymorphism includes plumage morphs (Huxley

1955; Butcher and Rohwer 1989) in geese (Anatidae), evidently maintained principally through sexual selection (Rockwell and Cooke 1977; Rockwell et al. 1985); herons (Ardeidae), apparently a reflection of differential selection for crypsis while hunting prey (Holyoak 1973; Murton 1971; Rohwer 1990); and hawks (Accipitridae) and jaegers (Stercorariidae), for which the "avoidance-image" hypothesis has been proposed (Brown and Amadon 1968; Paulson 1973; Arnason 1978; Furness and Furness 1980; Rohwer 1983b; Rohwer and Paulson 1987).

The role of polymorphisms in the evolution of life histories often distinguishes those treating "mating" and "trophic" polymorphisms (Roff 1996), although this distinction is overly simplistic in that a change in body size primarily related to sexual selection can involve a change in trophic status as well. Mating polymorphisms (i.e., those primarily related to mating success) include polymorphisms of body size, relative appendicular development, capacity for combat, locomotion, and territoriality, and have been documented for insects (Hamilton 1979; Fujisaki 1992; Roff and Fairbairn 1993), fishes (Nelson 1977), and behaviorally for birds (van Rhijn 1973, 1983). Trophic polymorphisms (i.e., those primarily related to acquisition of food) have been implicated in environmentally dynamic shifts in bill morphs in birds (Smith and Temple 1982; Smith 1987, 1990a, b, 1991, 1993, 1997; Smith and Girman 2000) and other tetrapods related to heterogeneous resources (Smith and Skúlason 1996). By contrast, the Cocos Finch (*Pinaroloxias inornata*) comprises distinct "ethomorphs" differing in foraging niches without significant differences in form (Werner and Sherry 1987).

Dispersal polymorphisms.—Of special interest in the context of avian flightlessness are "life-cycle" polymorphisms, which include those of wing development, flight capacity, migratory habit, and ability to escape predators, which are documented in greatest detail among insects (Roff 1984, 1990a, 1994b; Liebherr 1988; Denno et al. 1989; Dodson 1989; Fairbairn and Roff 1990; Masaki and Seno 1990; Roff and Fairbairn 1991; Fairbairn 1994) and some amphibians (Harris 1987; Jackson and Semlitsch 1993). An especially important class of polymorphism pertains to the maintenance of intraspecific variation in dispersal habit (e.g., frequency, conditions, and distances) and the evolutionary mechanisms by which it is maintained (Balkau and Feldman 1973; Felsenstein 1976; Comins et al. 1980; Comins 1982; Comins and Noble 1985; Frank 1986; Cohen and Motro 1989; Cohen and Levins 1991; Schoener 1991; Olivieri et al. 1995; Barton and Whitlock 1997; Hanski 1997; Harrison and Taylor 1997). Metapopulational analyses have provided insights into the possible origins of migratory behavior, and likewise have shed light on the circumstances surrounding the surrender of migratory habit (Olivieri and Gouyon 1997). Special attention has been given the relationship between the likelihood of migratory or dispersal traits, and interrelations of these with heterogeneity of habitats (Roff 1974, 1975; Denno et al. 1991; Denno 1994; Hawthorne 1997; Olivieri and Gouyon 1997), plasticity of development (Kaplan and Cooper 1984; Lively 1986a; Løvtrup 1989b), demographic structure (Lande and Orzack 1988; Kawecki 1993; Gyllenberg et al. 1997), impacts of genetic "turnover" within populational subunits (Gadgil 1971; Maruyama and Kimura 1980; Hedrick and Gilpin 1997), physiological parameters during ontogeny (Zera et al. 1994, 1997), relative prevalence of specializations for dispersal (Ellner and Shmida 1981; Mathias et al. 2001), implications for colonization (Ebenhard 1991; Gotelli 1991; Nichols and Hewitt 1994), and probability of extinction (Richter-Dyn and Goel 1972; Haila and Hanski 1993; Lande 1993; Foley 1994; Gomulkiewicz and Holt 1995; Foley 1997).

Olivieri and Gouyon (1997) listed several key findings concerning the polymorphism of "dispersal" conserved within complexes of metapopulations, including: decline in frequency of high-dispersal phenotypes with age of the population; higher incidence of low-dispersal phenotypes (including wingless forms of insects) in stable, isolated patches than in temporary habitats having high but ephemeral carrying capacities (Alerstam and Enckell 1979); elevated levels of evolutionarily stable rates of dispersal or migration among high-fecundity phenotypes; increased frequency of dispersal (sensu emigration) can be induced by conditions within the patch, including high population density and unpredictable levels of resources; and correlations among attributes of phenotypes rendering predictions complex and at times counterintuitive, for example, winglessness can be correlated with emergence time, or high rates of migration can be associated with increased selection for perennial habit or increased longevity, the latter in opposition to generalities of the "colonizer syndrome" as characterized by Baker and Stebbins (1965).

Although much of the fundamental work on dispersal polymorphism was based on insects and plants, the essential generalizations that have emerged extend to tetrapods under certain conditions (Hanski and Gilpin 1997). Accordingly, it is likely that migration-prone individuals represent nonrandom, genetically distinct subsamples of populations, and given the evidence for heritability and responsiveness to selection in such traits (Lindroth 1946; Dingle and Evans 1987; Wilson 1995; Dingle 1996), these individuals effect shifts in allelic frequences upon egress or ingress into conspecific populations (Hanski 1999). If emigration substantially exceeds immigration in frequency, as is probable in isolated (e.g., insular) populations, the associated gene flow would select for reduced rates of migration through the disproportionate loss of individuals so predisposed (Hanski 1999). In general, metapopulational models have shown that whereas migratory habit (or dispersal of any kind) is favored where resources show spatiotemporal heterogeneity, loss of migratory habit and (in extreme cases) flightlessness is favored where resources are more predictable, especially where immigration is diminished or negligible (Olivieri and Gouyon 1997; Hanski 1999). These predictions have received limited support from avian examples (e.g., Alerstam and Enckell 1979).

Although experimental investigation of the genetics of migration is in its infancy, migratory and nonmigratory populations within two Eurasian species of passerine birds—the Blackcap (*Sylvia atricapilla*) and the European Robin (*Erithacus rubecula*)—have been shown to have genetic bases (Berthold 1988, 1990, 1991, 1995). Furthermore, migratory and sedentary populations of the Blackcap differ in wing shape as well (Lockwood et al. 1998; Pérez-Tris and Tellería 2001). Additional genetic information for these and several other species (e.g., *Phoenicurus phoenicurus, P. ochruros, Sylvia borin,* and *Ficedula hypoleuca*) indicate that at least primary orientational direction exhibits patterns of inheritance consistent with polygenic determination and subject to comparatively rapid evolutionary changes (Berthold et al. 1990, 1992; Berthold 1996). Members of a number of avian families also show variation in migratory habit among breeding populations (Gauthreaux 1982; Chan 2001), including migratory, continental Water Pipits (*Anthus spinoletta*) and coastal, resident Rock Pipits (*A. petrosus*) of the British Isles; Winter Wrens (*Troglodytes troglodytes*), a circumpolar complex including large resident insular forms in the Pribilof and Aleutian islands; Sharp-tailed Sparrows (*Ammodramus caudacutus*-group), considered by some to comprise two allospecies (Greenlaw 1993; Rising and Avise 1993; American Ornithologists' Union 1998); Fox Sparrows (*Passerella iliaca*), which show an array of seasonal movements (Zink 1986); and crossbills (*Loxia* spp.), for which nomadism, morphological and genetic complexity, and far-flung populations characterize roughly two dozen circumpolar, Mediterranean, Philippine, Japanese, and Central American subspecies of Red Crossbills (*L. curvirostra*-group), and for which both migratory continental and resident Hispaniolan populations of White-winged Crossbills (*L. leucoptera*) are noteworthy (Benkman 1988a, 1989a, 1993; Groth 1993); and Song Sparrows (*Melospiza melodia*) and Savannah Sparrows (*Passerculus sandwichensis*), including significant variation and extreme forms in the Aleutian Islands (Johnson and Ohmart 1973; Aldrich 1984; Rising 2001).

Variation in flight capacity and migratory habit in Flying Steamer-Ducks (Humphrey and Livezey 1982; Livezey and Humphrey 1986) and short-term phenotypic responses in birds to changing climatic conditions (Gibbs and Grant 1987a, b; P. R. Grant and B. R. Grant 1987, 1998, 2000; Grant 1992; B. R. Grant and P. R. Grant 1993) enhance the possibility that some members of the Rallidae (e.g., *Gallirallus philippensis*-group and *Porzana tabuensis*) show metapopulational dynamics with respect to the maintenance of migratory habit and flight capacity. This likelihood is consistent with the positions of constituent populations of the *G. philippensis*-group and *Habropteryx torquatus*-group with respect to the morphological "threshold of flightlessness" (Fig. 27). Apparent variation in migratory habit (including nonmigratory populations or complex patterns at least as suggestive of malleability in this regard) occurs in several flighted species-complexes of Rallidae (*Porphyrio porphyrio*-group, *Rallus longirostris*-group, *Laterallus jamaicensis*-group, *Gallinula chloropus*-group), of which at least the *R. longirostris* complex also is alleged to include polymorphism of plumage (Ripley 1977; P. B. Taylor 1996, 1998; Olson 1997). Together with a propensity for migratory gregariousness, indicated by mortality of migrant flocks of rails (Tordoff and Mengel 1956; Pulich 1961; Stoddard 1962; Thompson and Ely 1989), the variable vagrancy of rails (e.g., Olson 1972; Crawford et al. 1983) qualifies them as potential insular colonists of the first order.

Insular colonization and interspecific variation in flight capacity in the Rallidae is at least consistent with polymorphism of dispersal habit. Many rallids combine migratory habit and exploitation of ephemeral, early-successional habitats such as meadows, marshes, and ecotones (Ripley 1977; P. B. Taylor 1996, 1998). This association with transient habitats is shared by some insects also showing dispersal polymorphism, including variation in flight capacity (E. S. Brown 1951; Southwood 1962; Roff 1994b; Denno et al. 1996), in some taxa conjoined with polymorphism in the flight apparatus (Vepsäläinen 1974; Järvinen and Vepsäläinen 1976; Harrison 1980; Roff 1986a–c; Kaitala 1988). Moreover, for migration to succeed at evolutionary timescales, it would have to be sufficiently plastic (in direction, distance, and duration of stays at either or both breeding and nonbreeding areas) to accommodate the inevitable climatic dynamics related to glaciations, tectonic movements, and catastrophic events (e.g., cyclonic storms [Wiedenfeld and Wiedenfeld 1993]).

PHENOTYPIC PLASTICITY AND EVOLUTIONARY CHANGE

Threshold traits.—Grant (1998d:312–315) posed four major questions concerning the evolution of vertebrates on islands: Why do some taxa undergo extensive radiations and others do not?; Does the temporal diversification of species within a radiation approach a uniform rate?; Is evolution on islands especially rapid?; and How are these adaptations to be interpreted? With respect to insular rails, answers to at least some of these seem to be close at hand. The foregoing discussions propose that a polymorphism of dispersal capacity characterizes at least a critical minority of the Rallidae, by which critical genetic diversity is maintained (at least in part) by strongly partitioned, metapopulational demographics. Furthermore, rails were shown to possess unusual capacities for dispersal and colonization, leading to an analogy of the Rallidae as including many "weedy" species. The likening of rails to invasive, adaptable, and phenotypically plastic plants may be especially fortuitous in that plasticity of habit and form in rails may be the capacity critical to the success of the Rallidae as colonists subsequently prone to flightlessness.

Variation not directly attributable to genotype has been demonstrated increasingly in diverse taxa (e.g., Blanckenhorn 1998; Losos et al. 2000). Variation of this kind elevated the concept of phenotypic plasticity from the status of nuisance variance to one of several key environmentally interactive, potentially adaptive, evolutionary factors influencing phenotype (de Jong 1995). Other factors involved in mapping genotypes into ranges of phenotypes include epigenetics, pleiotropy, epistasis, and constraint (Bull 1987; Scheiner 1993a, 1998; Gotthard and Nylin 1995; Hodin 2000; Pigliucci 2001). It is emphasized that a selective role for such plasticity does not invoke a neo-Lamarckian evolutionary mechanism (see below), but rather a means by which selectively conditioned ontogeny can accelerate advantageous, phenotypic response to novel circumstances, which in turn can contribute to directional shifts and (perhaps) play a role in incipient speciation (Pigliucci 2001).

Although many familiar examples of avian polymorphism conform with strictly Mendelian modes of inheritance, some cases are consistent with polygenic determination, including continuous variation that is constrained to two (rarely more) discrete states; permanent or delayed change in individuals (i.e., are distinct from seasonal polyphenism [Shapiro 1976]); involve a morphological component; and are subject to environmental induction. Such polymorphisms are likely to represent threshold traits (Roff 1996; Roff et al. 1997). Although some examples of threshold traits have shown Mendelian patterns of inheritance, many are polygenic and amenable to quantitative genetics (Roff 1986a–c, 1990b, 1994c, d; Falconer 1989; Roff and Shannon 1993), and examples include protective, trophic, life-cycle, or mating polymorphisms. The phenonenon provides for the maintenance of temporally varying morphs within populations and sexes (i.e., is not equivalent to variably expressed sexual dimorphism), and conditional inductability consistent with dynamic selective optima, the details of which often are mediated by ontogenetic mechanisms (Roff 1992, 1996). Importantly, threshold traits constrain continuous genetic variation into discrete phenotypes, principally through environmental induction.

Developmental reaction norms.—Recently, threshold traits were seen to be but one of a larger arsenal of means enabling environmental modification of phenotypes (Bradshaw 1973; Smith-Gill 1983; Via and Lande 1985; Stearns 1989c; Schlichting and Pigliucci 1998). An important basis for this recognition was the theory by Baldwin (1896), subsequently dubbed the "Baldwin effect" (Simpson 1953b). Phenotypic evolution currently emphasizes the integration of traits within phenotypes (Schlichting and Pigliucci 1995, 1998), quantitative expression of traits determined by multiple genes (Barton and Turelli 1989; Pigliucci and Schlichting 1997), potential advantages of developmental "instability" (Markow 1994), existence of plasticity (environmentally induced variation) in allometry among morphological parts (Schlichting and Pigliucci 1998), and influence of environmental cues (as constrained by ecology, mechanics, and development) on the mediation of genotypic information through ontogeny (West-Eberhard 1989; Bradshaw 1991; Raff et al. 1991; Scheiner 1993a; Nager et al. 2000). This perspective returned appropriate attention to the nongenetic component of phenotypes, a conceptualization of long, if underrated standing (van Tienderen and de Jong 1994; Schlichting and Pigliucci 1998).

Although most intensively studied in plants, where environmental influence on phenotype was amenable to readily quantified variates (e.g., Bradshaw 1965, 1973), eventually this complex of evolutionary mechanisms was extended to other taxa, including vertebrates (Blouin 1992; Van Buskirk and McCollum 1997; Losos et al. 2000). Geographic variation per se has received much ornithological attention (e.g., Gould and Johnston 1972; Niles 1973; Campbell and Saunders 1976; Hamilton and Johnston 1978; Johnson 1980; Handford 1983; Schnell et al. 1985; Rising 1987; Johnston 1992a, b; Engelmoer and Roselaar 1998), notably for recently introduced or colonist populations (Baker and Moeed 1979, 1980; Ross and Baker 1982), but the potential importance of environmental parameters on the ontogenetic bases of such patterns remains largely unexplored (but see James 1970, 1983). Although environmental influence on development—in utero, after egg-laying or parturition, and during postnatal development—has received greatest attention in nonavian reptiles and mammals, typically under the rubric of "maternal effects" (e.g., Janzen and Paukstis 1991; Janzen 1994, 1995; Rossiter 1996; Mousseau and Fox 1998; Robert and Thompson 2001), the possibility that environmental conditions may play a role in ontogeny is not without precedent in birds (James 1983; James and NeSmith 1989). Could developmental reaction norms (DRNs) provide an insight into the apparent facility with which the members of the Rallidae undergo suites of anatomical and physiological changes (including flightlessness) after the colonization of oceanic islands?

The combined roles of genotypic and environmental parameters acting during ontogeny to produce genetically delimited plasticity led to a conceptual formalization in terms of DRNs (de Jong 1990; Gomulkiewicz and Kirkpatrick 1992; van Tienderen and Koelewijn 1994; Schlichting and Pigliucci 1998). This view holds that many species confronted by spatiotemporal variation in resources and conditions possess selectively maintained capacity for mediation of the expression of key characters (Arnold and Peterson 2002; Day and Rowe 2002). Such variation in states comprises the reaction norm of the species (Hedrick et al. 1976;

Hedrick 1986; Lively 1986a; Via and Lande 1987; van Tienderen 1991; Brown and Pavlovic 1992; Zhirvotovsky et al. 1996), and examples so endowed include species possessed of high-dispersal modes or migratory habit (Fairbairn and Yadlowski 1997) or subject to peripheral isolation (Levin 1970).

Phenotypic plasticity can vary in magnitude, pattern, rate, ontogenetic interval of inductability, or reversibility (Gavrilets and Scheiner 1993; de Jong 1999; Pigliucci 2001; Day and Rowe 2002), and has been considered to be subject to selection and related to evolvability of systems (Via and Lande 1985; Via 1987; Alberch 1991; Scheiner 1993b, 1998; Wagner and Altenberg 1996). These dimensions raise the possibility that species for which such environmental cues are comparatively informative may possess "plasticity genes" that promote ontogenetic responsiveness (Padilla and Adolph 1996; Wagner 1996; Whitlock 1996; Pigliucci 1998, 2001; but see Via 1993a, b, 1994; Kaplan and Pigliucci 2001), and regulatory genes such as the *Hox* group may serve in this capacity (Raff et al. 1991; Schlichting and Pigliucci 1998).

Clearly, the magnitude of phenotypic plasticity shown by birds (in which growth is limited to a very short period) is far less than that displayed by plants (in which growth is protracted if not indefinite) or insects (in which environmental effects on ontogeny are better documented and generation times are relatively short). Even the limited variation in birds is unlikely to be entirely heritable and environmentally inducible, although most is likely to be subject to trade-offs between risks of adaptation to local conditions, specialization, and plasticity (Kisdi 2002; Sultan and Spencer 2002; Vázquez and Simberloff 2002). Nevertheless, several species or complexes thereof possess a combination of characters that are at least consistent with the conditions of DRNs, namely, complex morphological variation, especially in traits having trophic implications (e.g., bill shape); environmental diversity confronted either through migratory movements, ecotonal conditions, or ephemeral habitats; and limited or undetectable genetic differentiation among morphotypes. Environmental influence on the expression of genes has been documented in the Collared Flycatcher (Muscicapidae: *Ficedula albicollis*) in several morphometric traits (Merilä 1997; Merilä et al. 1994; Merilä and Björklund 1999). Notable additional candidates among birds are the Red Crossbill superspecies (Carduelinae: *Loxia curvirostra* and *L. scotica*) and the closely related Parrot Crossbill (*L. pytyopsittacus*), which manifest a range of continental and insular populations differing in vocalizations and bill type, the latter uniquely associated with foraging food resources (Parchman and Benkman 2002), but among which genetic differences are often negligible and in some cases not congruent with morphotypes (Lack 1944; Pulliainen 1972; Knox 1976, 1990; Massa 1987; Groth 1988, 1993; Benkman 1990; Benkman and Lindholm 1991; Piertney et al. 2001). Other taxa showing promise in this regard are the Black-bellied Seedcracker (Estrildidae: *Pyrenestes ostrinus*), which shows spatiotemporal variation in geographically restricted morphotypes, especially ecotones (T. B. Smith 1990a–c, 1991, 1993, 1997; Smith et al. 1997; Smith and Girman 2000); ecotypes of the Swamp Sparrow (*Melospiza georgiana*) manifesting morphological but no genetic differentiation (Greenberg et al. 1998); and the ground-finches of the Galápagos (Geospizinae: *Geospiza*), which have been shown to undergo extremely rapid phenotypic shifts (notably bill shape) with changes in temperature, humidity, food resources in response to El Niño events (Grant 1992).

TABLE 85. Environmental (direct) parameters (means and variances) qualifying as potential triggers for shifts in developmental reaction norms (DRNs) serving in colonization of oceanic islands. Direct changes comprise environmental and ecological triggers for DRNs or threshold traits (Pigliucci 2001) that are not conditional on evolutionary changes in parents in generations subsequent to initial colonization.

Parameter subject to change with insularity	Likely phenotypic attributes to be affected	Changes conducive to avian flightlessness	References
Intensity of predation	Body form, schedule of reproduction	None obvious	Lively 1986b, c; Bronmark and Miner 1992; Tollrian 1995; DeWitt 1998
Air temperature	Incubation, growth rate, allometry	Changes in definitive body mass (especially increase)	James 1983; James and NeSmith 1989
Air temperature	Sex of progeny	Males typically more massive, inclined to flightlessness	Janzen and Paukstis 1991; Jansen 1994, 1995; Robert and Thompson 2001
Relative humidity	Evaporation from eggs	Shorten incubation period, favor paedomorphosis	
Abundance of food	Egg size, growth rates	Modify size and shape	Rossiter 1996
Nutritional quality of food	Egg size, growth rates	Modify size and shape	
Seasonality	Schedule of reproduction	Timing of growth stage	
Diel photoperiodism	Timing of reproduction, growth	Timing of growth stage	Shimizu and Masaki 1993
Environmental stress	Body mass, reproduction	Muscular atrophy	Bradshaw and Harwick 1989; Ward 1994
Population density	Body mass, reproduction	Modify size and shape	Rossiter 1996

An extreme consistent with the cumulative results of reaction-norm shifts is the eventual loss of migratory habit through the evolution of flightlessness, an interpretation bolstered by the qualification of flightlessness as a threshold trait in insects on isolated oceanic islands (Dingle et al. 1980; Roff 1996) and the loss of migratory habit in some fishes (Roff 1988). Also, examination of preliminary data suggests a plausible relationship between the taxonomic prevalence of wing dimorphisms among continental insects and the frequencies of permanent flightlessness or winglessness in insular endemics (Shimizu and Masaki 1993). However, even if the greater selective risks of permanent traits in longer-lived organisms stemming from DRNs limit their frequency in birds, the phenomenon holds promise as an additional, environmentally reactive means for accelerating shifts among morphs within species or metapopulational subdivisions thereof. This likelihood also is fortified by the importance of reaction norms in groups prone to paedomorphosis (Denver 1997) or under stressful or novel environments (Bradshaw and Hardwick 1989; Ward 1994).

Environmentally plastic components of the phenotype that are related to colonizational success in rails deserve first consideration as potential DRNs critical to the evolution of flightlessness. Colonists of oceanic islands deriving from long-distance migrants would encounter a wide range of environments that might serve directly or indirectly as triggers of genetically bracketed shifts in ontogenetic trajectories in subsequent generations. Environmental heterogeneity for polymorphism and contributions by DRNs (Kawecki and Stearns 1993) hold promise for heretofore unexplored insights into changes in flightless rails and other insular tetrapods (Table 85).

The suite of phenotypic changes that characterizes flightless rails—including changes in morphology, migratory habit, ecology of feeding and reproduction, and metabolism—may derive from a highly developed, multifaceted ontogenetic plasticity that frees a given genotype to express different ecotypes under different environmental circumstances. The applicability of selection-based plasticity to the problem of paedomorphosis and flightlessness in rails seems especially promising in light of the progress already made with respect to the larger problem of reaction norms in body size and age at maturity (e.g., Lloyd 1987; Stearns and Koella 1986). This view of the shift from flighted, "normally" conformed, migratory species to insular, paedomorphic, sedentary endemic includes a suite of behavioral and morphological characteristics in one or more variably complex reaction norms, each triggered by changes in environmental conditions experienced by the progeny of the colonists. Effectively, changes in form and habitus are subsumed by a shift between continental and insular phenotypes. Even if this extreme, inclusive implementation of the reaction norm to insular endemism of rails is unrealistic to some degree, to the extent that environmental plasticity plays a role in these shifts (e.g., general body size, pectoral paedomorphosis, pelvic peramorphosis, or migratory habit), the efficacy of natural selection to reduce genetic variation in these lineages would be diminished accordingly (Schlichting and Pigliucci 1998).

Avian Flightlessness as Evolutionary Phenomenon

[They] had no call for practising or endeavoring to effect that hardest and most strenuous mode of locomotion to obtain sustenance or fulfill any of the conditions of preservation of the individual or of the species; they were never scared into the violent volant exercise.—Owen (1879, appendix 3:5; in reference to *Pezophaps*)

There is no doubt but that the avian [taxonomic] series is in general an ascending one.—Cope (1885:345)

It is therefore evident that the comparison of *Atlantisia* with other forms presents considerable difficulties. This very difficulty, taken with the fact that the almost wholly black coloration characteristic of the chicks of the Ralline family is apparently retained for a much longer time than is usual in immature examples, would seem to suggest that *Atlantisia* is a generalised and so presumably a near representative of some more primitive type.—Lowe (1928a:105)

Flight as Locomotory Blessing or Energetic Burden

Pennycuick (1987) listed four uses to which seabirds put flight: foraging, commuting to feeding areas, access to nest sites, and migration (Swingland and Greenwood 1983; Alerstam 1990). Flightlessness presumably influenced all of these activities in the Rallidae, through restriction of home ranges, augmented capacity for fasting associated with increased body size and redirection of energy formerly allocated to flight, improved diving capacity associated with a lifting of flight-imposed maximum for body mass (in *Fulica*), severe constraints on commuting distances, limitation of escape behaviors, strict dependence on availability of accessible nest sites, and appreciable limitations or impossibility of migration. Most of these dimensions of life history evidently also underwent similar changes in other flightless carinates (Humphrey and Livezey 1982; Livezey and Humphrey 1982, 1986; Livezey 1988, 1989a–e, 1990, 1992a, b, 1993a–c, 1995a), especially those exclusive of wing-propelled diving birds.

The letter by Darwin (1860) to Lyell (quoted above, p. 423) raised the philosophical merit of speculations concerning the absence of examples of flightlessness in insular bats. Lull (1935) deemed flightless bats to be evolutionarily proscribed because of the involvement of both pectoral and pelvic limbs in the chiropteran wing (see also Maina 1998), whereas the possession of dual locomotory modules in birds liberated of one of the limb pairs from flight function (Gatesy and Dial 1996a). These speculations prompt a consideration of those groups of birds for which flightlessness is unthinkable, examples of which would include widely foraging forms such terns and gulls or aerial foragers such as swifts and hummingbirds. A sacrifice of flight capacity appears to hold dire consequences for the majority of avian taxa endemic to oceanic islands (e.g., raptors, parrots, kingfishers, and passerines), whereas some groups that remained flighted (e.g., some ground-feeding *Geospiza* in which interisland dispersal is unimportant) conceivably could have survived the loss (cf. *Emberiza* from the Canary Islands [Rando et al. 1999]).

The capacity of rails to exploit diverse and innovative food resources comports with a view of rallids as presenting the "profile of an insular colonizer"—a coupling of a plesiomorphic capacity for long-distance movements with an apomorphic, contextually economical, (onto)genetic faculty for loss of flight. A return through the threshold of flightlessness, a verified example of reversal to flight capacity, has yet to be confirmed, and the data required to document such would appear to be unlikely to be recovered. The most enlightening substitutes for the latter may be extant species that closely approach the threshold—*Laterallus spilonotus* and *Fulica gigantea*—taxa that represent functional analogs to several taxa in the Anatidae (*Tachyeres patachonicus* and *Anas chlorotis* [Humphrey and Livezey 1982; Livezey and Humphrey 1986; Livezey 1990]).

Flightlessness as Degeneration Versus Developmental Economy

The functionally diminished pectoral apparatus of some flightless birds, notably ratites, prompted a view in which avian flightlessness assumed the role of evolutionary metaphor for "degeneration" (Lankester 1880; Duerden 1920). This antiquated perception in some respects has been reincarnated to be consistent in the hypothesis of the selectively neutral loss of "useless" structures through mutational stochasticity (Brace 1963; Kosswig 1963; Prout 1964; Peters and Peters 1968; Regal 1977; Peters 1988; Fong et al. 1995), invoked most frequently in the context of cavernicolous organisms (e.g., Eigenmann 1909; Culver 1982; Langecker et al. 1993; Culver et al. 1995) and subterranean (fossorial) mammals (Nevo 1999). These comparatively negative connotations were at least compatible with the early view of avian flightlessness as a defense against virtually hopeless attempts at dispersal from islands into the open sea (i.e., making the best of a bad situation), a notion deriving largely from studies of insular plants in which dispersal of progeny is limited to dissemination of propagules as seeds (Carlquist 1965, 1966, 1974).

An opposing view, in which "specializations" (such as appendicular modifications) are considered as plausible as thresholds of opportunities as "dead ends," was articulated by others (Amadon 1943c; Holmes 1977). Wake (1992) regarded the "regressive" evolution of sensory organs in caecilians as innovative, and the pectoral reduction in flightless birds on islands can be considered as a means for substantial "somatic savings" rendered permissible by a lack of terrestrial predators (Wiglesworth 1900; Lowe 1928a; Olson 1973a; Halliday 1978a; M. Williamson 1981; James and Olson 1983), an example of economic trade-offs among anatomical structures having selective values conditioned by environmental circumstance (Perrin and Sibly 1993). This interpretation is consistent with the "economy principle" of evolutionary maintenance (Curio 1969), and casts flightlessness as the selective redirection of energy from the pectoral apparatus to other demands, for example, caloric reserves (body fat), strengthened pelvic musculature, and augmented capacities for reproduction (Olson 1973a, b; Feduccia 1980; Livezey 1989b, d, 1990, 1992a, b). Emerson (1986) regarded pectoral paedomorphosis as functionally important in the saltatory capacity of frogs (Amphibia: Anura); others considered paedomorphosis in other amphibians as comparable to any other developmentally mediated change with respect to fitness (Alberch and Alberch 1981; Alberch and Gale 1983, 1985; Alberch 1987).

Viewed properly, avian flightlessness is a striking example of functional trade-offs, comparable to others not striking at the heart of the popular notion of things avian, for example, foraging behavior (e.g., Amadon 1950; Richards and Bock 1973; Pimm and Pimm 1982; James and Olson 1991), reproductive strategies (e.g., Lack 1968b; King 1973; Murton and Westwood 1977; Brown 1987; Koenig and Stacey 1990; Ligon 1999), and parameters of growth (e.g., Ricklefs 1968, 1973, 1983a; O'Connor 1984; Reiss 1989; Starck 1993; Starck and Ricklefs 1998a–c). The evolutionary trade-off represented by pectoral modifications in flightless, wing-propelled diving birds such as *Mancalla,* the Great Auk (*Pinguinus impennis*), and penguins is comparatively intuitive (Wiglesworth 1900; Storer 1960a, 1971a; Livezey 1988, 1989c; Raikow et al. 1988; Kooyman 1989; Cubo and Casinos 1997). As a result, the alcids and (especially) spheniscids are known at least as much for their rapid and energetically efficient submarine movement (Stonehouse 1967; Kooyman 1975; Hui 1983; Pennycuick 1987), and complementary structural and physiological changes (e.g., Andersen 1966; Walton et al. 1998), as for being aerially flightless. Similar considerations extend to other flightless diving birds—Galapagos Cormorant (Livezey 1992a), Auckland Islands Merganser (*Mergus australis* [Livezey 1989b]), and steamer-ducks (Livezey and Humphrey 1986).

The view of reductive evolution, exemplified by avian flightlessness, suggests that the morphological revolution of flightless rails is the facility with which members of the family ontogenetically implement a suite of phenotypic alternatives to the continental, flighted, often migratory norm to exploit insular environments. Relative magnitudes of this "reduction" differ significantly among species (Tables 71–75), contrary to the notion that the loss of flight in rallids is universally rapid, resulting in two distinct states (e.g., Olson 1973a; Feduccia 1980). In addition, speculations concerning the rapidity of the evolution of flightlessness based on the geological ages of occupied islands in hot-spot archipelagos may be misleading in that ancestral colonists may have originated on older islands in the archipelago (including islands no longer emergent) permitting substantially longer evolutionary intervals (Simkin 1984; Carson and Clague 1995).

PRECONDITIONS, CAUSES, CURRENCIES, AND CONVERGENCE OF CHANGE

At the outset, it is important to distinguish between *preconditions* (conditions *permitting* evolution of flightlessess) from *selective advantages* (selective differentials *promoting* flightlessness). Among the preconditions or primitive states, exclusive of taxonomic affiliation, that characterized the minority of insular avian endemics that eventually evolved flightless in situ were several ecological or functional attributes: foraging was principally or completely nonarboreal and nonaerial; nesting was terrestrial or semiterrestrial and for which access was not problematic in the absence of flight; and year-round survival was feasible in the absence of migration and repeated, short-term, long-distance movements from the island. Penguins and flightless alcids clearly represent exceptions to the last condition (Livezey 1988, 1989c), although it is unlikely that strictly insular taxa of penguins evolved flightlessness in situ. For example, the Galapagos Penguin (*Spheniscus mendiculus*), although accommodating life on a tropical archipelago (Boersma 1975, 1976), derives from a taxonomic order known to have been flightless from at least the early Cenozoic and principally distributed at the highest latitudes of Gondwanaland (Simpson 1946, 1975; Stonehouse 1975; Livezey 1989c; Williams 1995).

The analogy of insular colonization by wayward migrants as a permanent stopover provides a useful premise from which to reconstruct the evolution of avian flightlessness on oceanic islands. After initial colonization, isolated lineages possessed of the requisite life-historical qualities and (epi)genetic opportunities will undergo, to varying extents, changes in morphological and ecophysiological parameters during subsequent generations under the new selection regime, the more speculative issues of evolutionary rates and irreversibility aside (see above). Whether islands are considered environments permitting phenotypes otherwise selectively "forbidden" (sensu Vermeij 1987)—for example, by reduced risks of predation—or environments conducive to stresses and accelerated evolutionary change (sensu Parsons 1993b, 1996)—for example, through limitations of habitable areas and deme size—insular habitats clearly are central to flightlessness in rails and of greater evolutionary importance per unit area than are comparable continental areas. The widespread colonization of a wide diversity of insular communities has fostered in rails a rich diversity of form shown by flightless endemics—both in size and shape (Tables 7, 11, 13–23, 30–33, 40–42, 46–54, 68–75; Figs. 15, 20, 22–32, 42–44, 48–60, 111–115)—but the group remains less tractable with respect to geological, ecological, morphological, and phylogenetic diversity than the comparatively heavily studied cases such as the Galápagos (Darwin's) finches (Geospizinae [Lack 1947a; Bowman 1961, 1963; Grant 1986]) and Hawaiian honeycreepers (Drepanididae [Amadon 1950; Bock 1970; Dobzhansky 1977; Pimm and Pimm 1982; James and Olson 1991]).

Long-term "success" of flightless species in insular habitats is governed, as for any phylogenetic lineage, by the success of evolutionary accommodation to

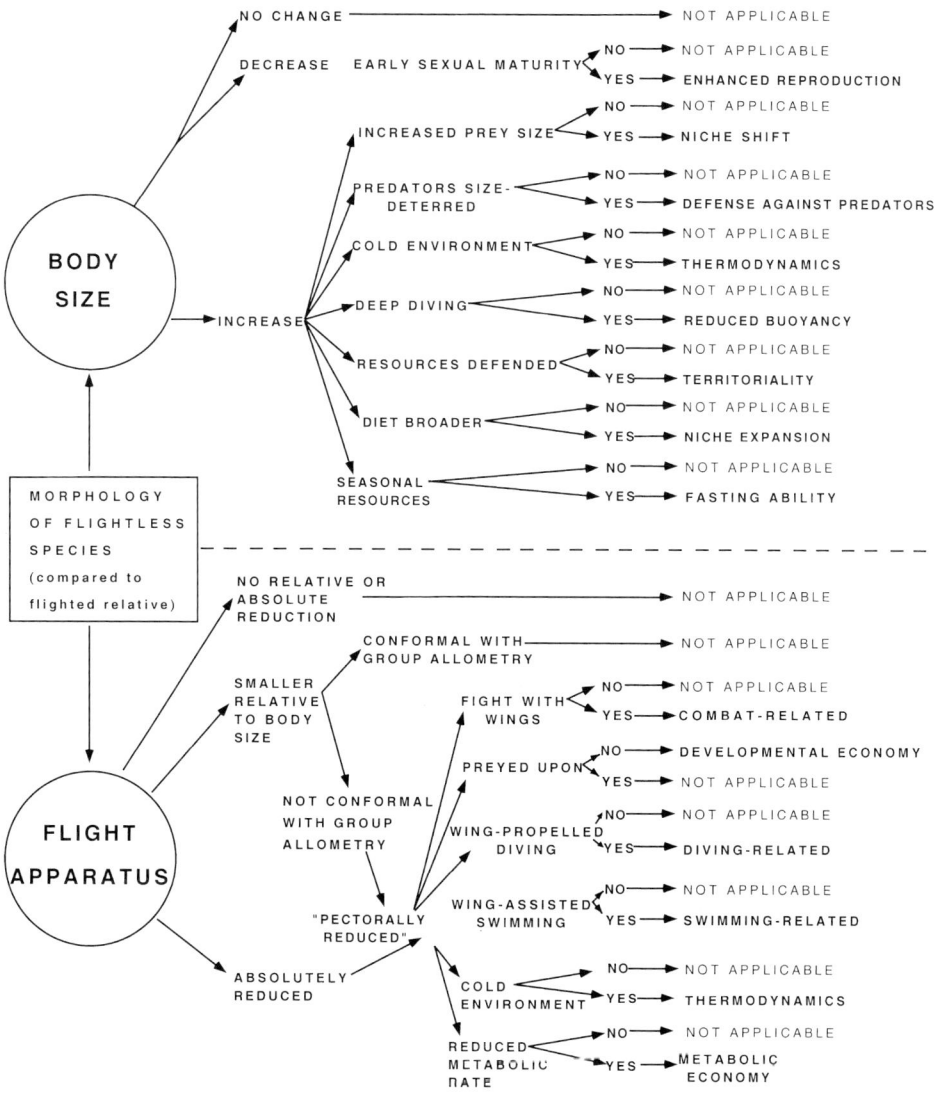

FIG. 127. Dichotomous decision-tree delineating selective advantage(s) of avian flightlessness based on deviations in body mass (among Rallidae, increases in body size typical, decreases or stasis rare), changes in pectoral apparatus (among Rallidae, variably profound reductions evident), and candidates for functional advantages associated with these in flightless birds are indicated at termina.

changing environments, hypothetically depicted as a spatiotemporal shift in selective optima before and after colonization (Fig. 126). The obvious dependence of insular flightless birds on isolation from terrestrial predators is important from a historical perspective, but provides few insights into the selection regimes that molded the diversity of flightless lineages. Exclusive of the catastrophic effects of human immigration, the changes in body form inferred for flightless birds can be subjected graphically to a simplistic, two-component "decision-tree" that provides a basic framework for delimiting viable candidates for potential selective advantages of flightlessness under pristine conditions (Fig. 127). Of these, advan-

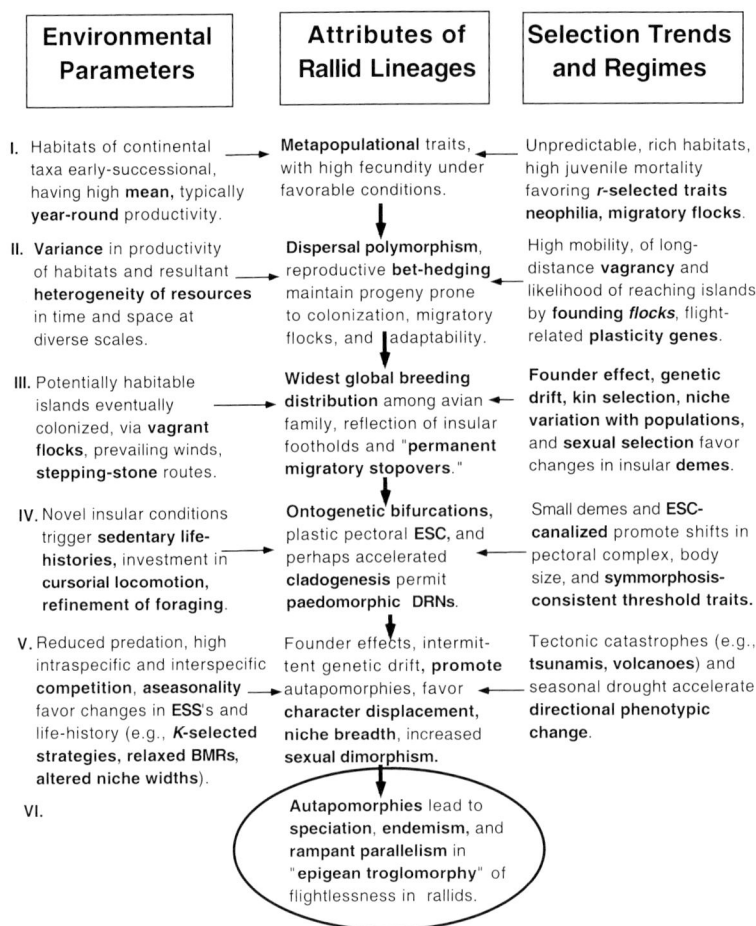

FIG. 128. Diagram of the environmental and selective parameters impinging on the attributes of Rallidae leading to the evolution of flightlessness. Acronyms included (see text for details) are BMR (basal metabolic rate), ESC (evolutionarily stable configuration), and ESS (evolutionarily stable strategy).

tages related to defense against predators (e.g., protection and concealment of nests and young) or combat with conspecifics are least likely to pertain to flightless rails; both were critical for morphological changes associated with flightlessness in steamer-ducks (Livezey and Humphrey 1986). Given the multitude of benefits that can accrue to the sacrifice of flight (Fig. 127), the suite of plausible mechanisms that perpetuate a capacity among rails to redirect the allocation of resources typically directed to flight toward other faculties of advantage under novel, insular conditions (Fig. 128), and the diversity of taxonomic groups in which flightless members have evolved (Table 4), it may be legitimate to seek the environmental circumstances and ontogenetic mechanisms responsible for the scarcity of avian flightlessness as opposed to the consistent view of this specialization as an evolutionary oddity or exception.

Macroevolutionary trends in phylogenetic lineages have received special attention since the paradigm proposed by Darwin (1859) assumed a central role in biology. Issues pertaining to the diversification and extinction of lineages, morphological evolution, and functional trends dominated subsequent literature (cf. Huxley 1943, 1953; Simpson 1944, 1953a; Bock 1970, 1979; Salthe 1975; Stanley 1979; Charlesworth and Lande 1982; Gould 1982; Levinton 1983, 1988; Benton 1987; Lemen and Freeman 1989; Baum and Larson 1991; Howarth 1991; Hansen and Martins 1996; Jablonski 1996, 2000). McShea (1998) listed eight large-scale trends or "live hypotheses" exemplified by modern organisms—entropy, energy intensiveness, evolutionary versatility, developmental depth, structural depth, adaptedness, size, and complexity. Some of these are comparatively esoteric, and several are validated by flightless rails beyond what is typical (if not unavoidable) for other organisms, especially tetrapods. Energy intensiveness, after Vermeij (1987), is expressed among rails in the morphological trade-offs involving locomotion favored under insular conditions. Evolutionary versatility is of special significance to flightless rails in that the ontogenetic plasticity that permits the sacrifice of the pectoral apparatus for other ecomorphological refinements is one aspect of this concept. Developmental depth, although inescapably germane to flightless rails in its emphasis on the hierarchical structure of ontogeny and the decrease in admissable variation that characterizes progressively earlier, generatively entrenched stages (Wimsatt 1986), merits special attention given the influence of differentially conservative heterochrony in the morphology of flightlessness. Adaptedness is problematic in currencies and timescales, in that flightless rails are successful in terms of taxonomic richness and distribution, but limited in ecological dominance by small occupied areas and recent extinctions. Finally, size and increases thereof (sensu Cope's rule) are clearly relevant to most flightless lineages of rails, although neither the selective advantage(s) of these convergent changes, rates of change, nor the sequence of events (e.g., whether increase in size preceded or led to flightlessness) is known in most instances.

This literature reveals body size to be focus of study in most macroevolutionary studies (e.g., Newell 1949; LaBarbera 1986; Bonner 1988; McKinney 1988b; Jablonski 1996; Kozlowski 1996). Apomophies of flightless rails, including increased body size (where inferred), are consistent in some respects with the traditional view that macroevolution is the cumulative outcome of microevolutionary processes, in that all degrees of intermediacy are evident in the family and within several genera (Figs. 32, 59, 60). This perspective was articulated clearly by (Bock 1970:712, emphasis in original): *"If the sequence of events in a major evolutionary change is outlined properly, then the known mechanisms of microevolution will provide a complete explanation for the macroevolutionary change."* It is likely that this correspondence can be approximated for the comparatively recent transitions reflected by flightless rails, in which a comparative wealth of intermediate forms share critical aspects with the Hawaiian honeycreepers examined by Bock (1970). Imperfections of the fossil record and the likelihood that resultant artifactual discontinuities that appear to contradict this view have been cited as potential flaws in prominent counter-reductionist hypotheses of this kind. An excellent example of these debates is the critique by Dennett (1995) of the reliance on fossil transitions in the context of speciation, the hypothesis of punctuated equilibria (Eldredge and Gould 1972; Gould and Eldredge 1977). These long-standing concerns have returned attention to the possibly misleading artifacts of paleontological lacunae in a series of evolutionary topics (e.g., Donovan 1989; Wheeler 1992; Kemp 1999).

Large body size is not the only apomorphy shared by many flightless carinates (Table 78). Large body size compromised the capacity for dispersal in these lineages—one of several, typically *K*-selected correlates of insularity shared by flightless rails and raphids, among others—but may have bestowed a number of correlated, selective advantages. Most notable among the latter is the growing substantiation of reduced BMR that attends the loss of flight capacity and its physiological demands, documented in *Atlantisia rogersi* and several other rallids and flightless carinates (Ryan et al. 1989; McNab 1994a, unpubl. data). Other advantages accruing with the loss of flight may include lower (resting) body temperature (Wetmore 1921; McNab 1983) and likely enhanced capacity for fasting (Livezey 1993b; McNab 2000). Perhaps in part a corollary of increased body size, sexual size dimorphism was exceptionally great in some flightless rallids (e.g., *Diaphorapteryx hawkinsi, Aphanapteryx bonasia,* and perhaps incipient in *Cabalus modestus*) and involved differences in body size and changes in proportions possibly related to foraging. Other apomorphies related to the loss of flight in at least some rails and members of other neognathous families include losses of remiges; peramorphosis of the skull, trunk, and pelvic limb; and comparatively great seasonal variation in deposition of body fat (Livezey and Humphrey 1982, 1986; Livezey 1993b, 1995a).

Appendicular reductions in other taxonomic groups of tetrapods, especially those entailing functional sacrifice, revealed that comparisons with flightless birds fail increasingly with detail (Table 78). Pectoral reduction in some reptiles (e.g., snakes), for which the evolutionary reduction and loss of limbs is comparatively well manifested in the fossil record (Caldwell and Lee 1997; Greene and Cundall 2000; Tchernov et al. 2000), was hindered by the plesiomorphic absence of both pectoral and pelvic appendages in entire higher taxonomic groups and the typical association of limblessness with corporal elongation and fossorial locomotion (Lande 1978; Carroll 1997), a deficiency overcome ultimately by classical comparative anatomy (Gressitt 1956; Gans 1975; Presch 1975) and increasingly explicit phylogenetic studies (Caputo et al. 1995; Lee 1998; Zaher and Rieppel 1999; Wiens and Slingluff 2001). Anatomical changes associated with avian flightlessness are minor, in part related to the plesiomorphy of reduction in the avian forelimb, exclusive of simple elongation in some segments (Chiappe 1995a; Padian and Chiappe 1998).

Extreme examples of convergence were provided by invertebrate taxa, primarily the Insecta (e.g., Jackson 1928; Holloway 1963; Hackman 1964, 1966; Otte 1979; Harrison 1980; Roff 1986a–c, 1990a, 1991, 1994a–c, 1995; Barbosa and Krischik 1989; Sattler 1991; Zera and Mole 1994; Finston and Peck 1995; Hunter 1995; Marden and Kramer 1995; Zera and Denno 1997). Flightlessness among insects was instrumental in the extension of concepts of metapopulations and threshold traits to the evolutionary realm of vertebrates, despite the nonhomology of the affected structures and issues of comparability of birds and insects, deriving from small body sizes, ectothermy, and short life spans of the invertebrates.

MODERN PERSPECTIVES: REGRESSIVE, PROGRESSIVE, AND NEUTRAL

The evolutionary innovation of the Rallidae embodied by the colonization of oceanic islands combines dispersal, perhaps largely expressed by immatures during inaugural migratory passages, married to an ontogenetic capacity to convert the "anatomical assets" of the flight apparatus into resources necessary for success if flocks stray to insular environments. The first component is a change in migratory habit having profound implications for energy budgets, annual schedules of activities, and physiological cycles. The second component in this "marriage of capabilities" turns on ontogenetic variation in morphology, in which an ontogenetic facility to engage heterochrony gives rise to a unique bifurcation of development, that is, a capacity to substantially alter investment in a flight apparatus that is consistent with an ESC (i.e., sensu Wagner and Schwenk [2000]) in most avian lineages. Conservative vestiges of avian developmental schedules preclude conversion of the entire pectoral limb to effect this trade-off, although moas (Dinornithiformes) most closely approached this level of completeness (Owen 1879; Forbes 1892f; Archey 1941; A. Anderson 1989). Nonetheless, the variable, lesser exchanges realized by flightless rails evidently were of sufficient advantage to effect a distributional revolution in the family.

The tendency of flightless rallids to shift toward larger size (Table 71; Appendix 1) and *K*-type reproductive parameters (Table 84) appears to have contributed to the vulnerability of flightless rails to anthropogenic agencies, in that large size (through increased selection as prey) and diminished reproductive potential (thereby reducing ability to recover) carry penalties for many avian taxa in the face of increased predation (Balouet and Olson 1987; Holdaway and Jacomb 2000). The observation that the largest, most reproductively conservative species of flightless rail were among the first to be extirpated (Table 2) lends credence to this hypothesis. This generalization fails for some insular rallids, as species suffered from assaults of variable intensities, design, and time frames, including unconscious destruction of many species through "siege" by commensals, "collateral damage" incurred through destruction of habitat (many species), as well as all-out "*Blitzkrieg*" (sensu Holdaway and Jacomb 2000) with focused intent (e.g., *Gallirallus wakensis* and *Cabalus modestus*). In addition, if a given locomotory capacity is accepted as an "escalative" stratagem, then flightlessness represents the exchange between two potential avenues for realization of proximal benefit; that is, flightlessness of rails may qualify as an example of "nonescalative" macroevolutionary change (sensu Vermeij 1987). Taken as a whole, these circumstances favored "opportunistic" (short-term) qualities as opposed to refinement of attributes of colonists for the long-term, including such unforeseen eventualities as human immigration (Berry 1992).

Paedomorphosis is potentially important for evolutionary innovation, a process by which "overspecialization" may be avoided and new potentials for cladogenesis attained (de Beer 1951; Hardy 1954; Godfrey 1985; Ayala 1988; Levinton 1988; Ruse 1992, 1996). Goldschmidt (1940), in reference to the flightless Galapagos Cormorant, noted such apomorphy as leading to an example of a "hopeful monster." Clearly favorable in original intent, this characterization was to foster escalating derision by a number of prominent evolutionary biologists; for example, Hull (1988) attributed to E. Mayr the derogation "hope*less* monsters" in reference to the hypothesis of Goldschmidt (1940). However, in a review of heterochrony as mechanism of evolutionary change, McKinney et al. (1990:279) extended this apellation to "More-than-hopeful Monsters" to renew recognition of the evolutionary potential of ontogenetic change. Within this perspective, such "monstrosities" often represent phenomena considered anomalies under conventional paradigms, natural "outliers" poised to foment scientific advancement (Alberch 1989; Lightman and Gingerich 1991). Also during recent years, circumstances and classes of "monstrosity" conducive to establishment of these divergent phenotypes have been hypothesized (Alberch 1982a; Arthur 1984; Werner and Gilliam 1984; Pimm 1986; McKinney 1988b).

Interpretation of avian flightlessness as degenerative or deviant in that it represents a departure from a natural or essentialist norm appears to have roots in an ideal of naturalism (Mayr 1982), and the pervasive, essentially religious of view of the earth in decay (Chambers 1844; Davies 1969; Bowler 1976, 1983; Levinton 1983). In time, such evolutionary outliers came to be viewed with restrained optimism. Grant (1998d), citing the swarms of flightless coleopteran (Roff 1994a, b; Finston and Peck 1995) and orthopteran (K. L. Shaw 1995) insects on islands, admitted that flightlessness is related to sedentariness, geographical isolation, and small demes. This led Grant (1998d) to acknowledge that flightlessness in these cases may have been key to phylogenetic diversification, but he questioned whether or not flightlessness qualified as a "key innovation."

Is it fair to single out extinct, flightless birds as "over-specialized?" Are other species, when denied essential requirements (e.g., suitable habitat) or presented with new, deadly threats (e.g., viruses), and driven to extinction, so judged? Insular endemics were numerous, but human-related change was so pervasive on oceanic islands as to render a repeatedly evolved, formerly successful specialization a literal "dead end." It may be more logical to consider the latter term as an evolutionary pathway (perhaps strongly favored in the short-term) having few options (e.g., selectively attainable bifurcations), deemed a posteriori as having a low probability of success.

Many flightless rallids are peramorphic in all but the pectoral appendage, but are probably most properly considered functionally paedomorphic (given the importance of pectoral paedomorphsis for flightlessness and extinction), and as such offer little support for the notion that paedomorphs tend to succeed as ancestors. Heterochronically generated evolutionary change, strongly evidenced in several flightless rails, may be critical in the origin of evolutionary novelties (Mayr 1960), the rate of such changes (Simpson 1944), and their phylogenetic patterns (Cracraft 1984, 1985, 1990; Müller and Wagner 1991). Furthermore, at least some of the flightless rails represent(ed) highly specialized members of insular environments that were characterized by extreme morphological and (probably) physiological changes. Before the arrival of humans, some may have been dominant consumers in their respective, if geographically limited, ecological communities (e.g., pristine densities documented for *Porzana palmeri*). However, in light of the difficulties of delimiting the anatomical changes enabling avian flight as a discrete innovation (Cracraft 1990), in many respects the characterization of flightlessness as innovation is similarly problematic.

The endangered status and extirpation of many birds (flighted and flightless), as well the formerly global distribution and diversity of flightless rails, present a confusing array of success in the sense of perseverance in the face of human devastation. Even if one adopts the extreme view that virtually all habitable oceanic islands at one time supported at least one endemic flightless rail (Steadman 1995), the global ecological impact of these small populations and the area occupied would have been vanishingly small. Even if limitations of timescales and evidence of intermediates are put aside, flightless birds do not permit an easy assessment in terms of "evolutionary progress" as conceptualized (or rejected) by most scholars (Simpson 1949, 1953a; Huxley 1953; Rensch 1959; Dobzhansky 1970; Kühne 1972; Ayala 1974; Fisher 1986; Benton 1987; Gould 1988b, c; Nitecki 1988; Rosenzweig and McCord 1991; Ruse 1996). That is, flightless lineages conform to a widespread view that a sustained trend is prerequisite for a consideration of "progressive" change, flightless lineages have descendants that have both gained or lost capacities or functions in the process, and the difficulty or impossibility of a relative valuation of these trade-offs precludes the prejudicially loaded attribution of "improvement," "increase," or "betterment" on the overall trend in most or all instances (including among the Rallidae).

Brock (2000) considered that anagenetic changes can be considered "progressive" in the evolutionary sense, a perspective that does not necessarily include assessments of adaptive properties normally implicit in the broader concept of "progress" as an emergent property of entire evolutionary systems. By this criterion, the multiple convergent transitions shown by flightless rails (and other neognaths) could be classified as progressive. Extinctions of insular avifaunas seem random or "wanton" with respect to the affected taxa or their habit. Well-understood exceptions to the ecoevolutionary indiscrimination typical of insular extirpations include the decreased extinction rates of high-elevational birds in Hawaii that reflect the failure of disease-carrying mosquitoes to live in mountainous habitats (Scott et al. 1986). However, to the extent that "chance" (i.e., changes in parameters not subject to predictability or selection) has played a role in the extinction of flightless birds, the insights to be gained from the standpoint of evolutionary "progress" or "success" are diminished (Eble 1999).

A related and more intensively contested issue surrounds the concept of "adaptation," principally in terms of the implications of the term with respect to selection inferred to have acted on historical sequences of events (Gould and Lewontin 1979; Gould and Vrba 1982) and the hierarchical level at which the underlying selection operates (Williams 1966, 1992; Lewontin 1970; Maynard Smith 1976; Alexander and Borgia 1978; Wade 1978; Wilson 1980, 1983, 1997; Dawkins 1982; Endler 1986). One of the counterproductive philosophical outcomes from the success of neo-Darwinism in the 20th century was the ascension of the concept of "adaptation" to a status tantamount to noumenon—an idea not amenable to empirical testing but considered essentially omnipotent. In the present work, selective advantage(s) of flightlessness are framed in terms of the individuals expressing the trait in question (i.e., individual selection, as extended via inclusive fitness to kin selection), as opposed to benefits gained by a collection of unrelated individuals (i.e., group selection [Wade 1978]). This perspective does not proscribe the collective effects of individual selection on the differential success of lineages. This emergent corollary at the level of species is sometimes referred to as "species selection" or "lineage sorting" (Brock 2000), and marks a conceptual distinction critical to the paradigm of metapopulations (Olivieri and Gouyon 1997; see above).

A focal point of debates regarding "adaptation" is the relative importance of original and current function (Gould and Lewontin 1979; Gould and Vrba 1982). Amundson and Lauder (1994) classified the philosophical interpretations of "function" in terms of the history of natural selection undergone by ancestral lineages, current causal properties, and the "selected effects" of a character. Amundson and Lauder (1994) also acknowledged the challenges and value of historical reconstructions of selection regimes and past function, and cited the issue of wing size and flightlessness of some insects in a comparison of these competing approaches. The current consensus minimally requires the phylogenetically based exclusion of plesiomorphy and demonstration of selective advantage of the feature in question for "adaptation" to be inferred (Eldredge and Cracraft 1980; West-Eberhard 1986; Taylor 1987; Lemen and Freeman 1989; Eldredge 1993; Leroi et al. 1994; Losos and Miles 1994; Vermeij 1994, 1996).

Accordingly, flightlessness is properly considered an indirect adaptation or "aptation," in the strict sense, in that no selective advantage (whether involved in its origin or simply acting post facto) strictly attends the loss of flight (Frazzetta 1975; Brandon 1978; Bock 1980; Clutton-Brock and Harvey 1984; Coddington 1988, 1994; Hailman 1988; Arnold 1989; Baum and Larson 1991; Reeve and Sherman 1993; Amundson 1996; Larson and Losos 1996). Instead, the economic implications of the sacrifice, that is, the gains realized by the redirection of resources formerly allocated to the flight apparatus and its employment, are the features of selective advantage. That selection toward optimization among essential dimensions of life histories, changes in body size, or refinements of metabolism to succeed on oceanic islands leads toward changes that reasonably would be considered "adaptive" seems inescapable. The critical point is not to mistake the aspects sacrificed in these shifts as the "adaptations" achieved. Unfortunately, the repeatedly affirmed evolutionary strategy involving the sacrifice of flight for other faculties, in a manner comparable to insular giantism of uncounted insular Mammalia, was rendered a Pyrrhic victory with the biotic calamity wrought during and since the age of human exploration.

FLIGHTLESS BIRDS AND CONSERVATION ON ISLANDS

OPPORTUNITIES LOST AND INVESTIGATIONAL PRIORITIES

> If there really are lost species, it can doubtless only be among the large animals which live on the dry parts of the earth; where man exercises absolute sway, and has compassed the destruction of all the individuals of some species which he has not wished to preserve or domesticate.—Lamarck (1809, vol. 1:76; translated by Elliot 1984:44)

Unfortunately, the optimism of Lamarck (1809) was to prove misplaced, and the admonition (quotation below) by Mayr (1967) came centuries too late for many taxa, avian and otherwise, and continues to be ignored for many more to the present. Islands unquestionably hold unique potential for evolutionary insights (Berry 1992), but as a result of the numerous extirpations of flightless rails and other insular endemics around the world, many important opportunities for ornithological research have been lost irrevocably. Oceanic islands simultaneously embody fragments of uniquely varying resource units, presenting natural settings for evolutionary "experiments." Perhaps to a degree attained by no other avian family, the family Rallidae has participated in these uncounted evolutionary trials, through which an unlikely proclivity for dispersal may have endowed virtually all habitable land fragments with one or more variably enduring colonizations

(Steadman 1995). But for the fortunate preservation of subfossil elements of some of these extraordinary species, any synthesis of flightless rails would be reduced to little more than a jeremiad of the destroyed faunas and lost evolutionary insights. These lost subjects of study include some of the most remarkable and derived flightless rails, for example, *Diaphorapteryx, Aphanapteryx, Erythromachus, Capellirallus,* and Hawaiian *Porzana* (a flightless analog of Darwin's finches). However, a number of key questions still can be addressed through study of the remaining minority of flightless rails, taxa that fortunately represent most of the clades that formerly included flightless members.

Among the general hypotheses having components uniquely accessible under insular conditions involve the degrees to which taxonomic diversity of avian communities may be structured by an avoidance of nest predators through increased diversity of nest sites (T. E. Martin 1981, 1987, 1988a–c, 1992, 1995, 1998; Martin and Clobert 1996); magnitude of sexual dichromatism may be related to intensity of nest predation (Martin and Badyaev 1996); "apparent competition" contributes to interspecific patterns of distribution (Holt 1977; Holt and Kotler 1987; Schmitt 1987; Holt et al. 1994); evolutionary changes in insular birds correspond in genetic and ontogenetic dimensions with those of troglomorphs in the subterranean isolation of caves (Culver et al. 1995); generalities of r- and K-selection characterize endemic species, subspecific populations, and recent immigrants (Murray 2001); and polymorphisms of dispersal both within and among species are maintained genetically and asynchronous temporal fluctuations in resources, induced as threshold traits, or evolved de novo (Johnson and Gaines 1990; Mathias et al. 2001), proposals that were made in preliminary form for rails by Ripley (1977) and Diamond (1977b).

Some investigators have judged that flightlessness evolves rapidly in rails and other insular birds (e.g., Olson 1973b), an intuitively appealing view epitomized by the suggestion by Olson (1999:13) that introduction of *Porzana pusilla* on Laysan Island would provide "an extremely interesting experiment to determine how rapidly such a species would adapt behaviorally and morphologically to insular conditions." It seems unlikely that such an experiment, which would require at least decades to document incipient trends, could reach a satisfactory conclusion in the face of human accidents and natural disasters. Nevertheless, the importance of experimental study of avian flightlessness is substantial and growing. Some such experiments may involve detailed study of insular radiations of focal species (e.g., *Porphyrio porphyrio*-group, *Gallirallus philippensis*-group, *Habropteryx torquatus*-group, *Porzana tabuensis,* and *Gallinula chloropus*-group) within a phylogenetic context. With respect to this enterprise, the midpoint of the 20th century saw the conclusion of the anecdotal phase of study, and the time has come to move from the descriptive to the experimental and theoretical frames. A number of field experiments are viable, and include assessments of differences between flighted and flightless congeners in metabolic rates, avoidance behavior, mating systems, parental care, ontogeny, diet, time–energy budgets (including molt), foraging behavior, and vigilance. Also, tests for interisland dispersal within the Galápagos by *Laterallus spilonotus* during periods of environmental stress (e.g., effects of El Niño events), as a possible source of episodic selection against complete loss of flight (possibly analogous to the situation in some marine populations of *Tachyeres patachonicus*), would provide a critical insight into the dynamics of change at the very threshold of flightlessness.

Of special relevance to the prevalence of flightlessness in rails and of particular theoretical interest in a broader context would be studies of possible environmental influence on avian ontogeny, initially by using logistically practical and empirically compelling examples from other families (e.g., *Loxia* and *Pyrenestes*), and followed (if confirmed) by parallel studies of key species of Rallidae. Most critical would be investigations of the most likely candidates for involvement in DRNs that are related to the morphological shifts associated with flightless in rails (Table 85), some of which could be designed under controlled conditions, thereby circumventing interruptions by external agencies. The seminal work by James (1983) provides the basic framework for a suite of variably sophisticated manipulations that would require samples of only the most common rallids (e.g., *Porphyrio porphyrio*-group, *Rallus aquaticus, Porzana carolina, Gallinula chloropus*-group, and *Fulica americana*-group) that would shed light on the possible importance of DRNs. In addition, a new standard of intensity and sophistication in the quantification of reproduction, selection intensity, and dispersal—including aspects germane to possible bet-hedging and dispersal polymorphism—would be particularly strategic for ornithology in general and the study of avian flightlessness in particular.

Many detailed phylogenetic overlays will be possible with refined reconstructions of phylogeny, especially for *Porzana* and *Gallirallus* (Livezey 1998), which will permit incorporation of morphometric trends in phylogenetic contexts (cf. Wake and Larson 1987). The most plausible candidates for such syntheses among flightless rails are *Porphyrio hochstetteri, Habroptila wallacii, Dryolimnas aldabranus, Gallirallus owstoni, G. australis, Tricholimnas sylvestris, Habropteryx okinawae, Atlantisia rogersi, Porzana atra, Amaurornis ineptus, Gallinula comeri,* and *Tribonyx mortierii*. Studies involving these species should be undertaken as soon as feasible, because some pose substantial logistical challenges (e.g., *H. wallacii* and *A. ineptus*) and most others are already threatened or endangered (e.g., *P. hochstetteri, G. owstoni,* and *H. okinawae*). Obviously, comparative assessments of metabolic rates, ontogeny, and molt would require comparatively frequent access to live birds (B. McNab, pers. comm.). The latter would be most feasible for those taxa currently having captive-rearing programs, which with foresight may provide the critically needed skeletal and fluid-preserved specimens (through unavoidable mortality of captives and released birds) and developmental series that are not currently available in adequate numbers for most members of the Rallidae (e.g., Eason and Williams 2001).

ANTHROPOGENIC IMPACTS AND CONSERVATIONAL IMPERATIVES

> Islands are an enormously important source of information and an unparalleled testing ground for various scientific theories. But this very importance imposes an obligation on us. Their biota is vulnerable and precious. We must protect it. We have an obligation to hand over these unique faunas and floras with a minimum of loss from generation to generation. What is once lost is lost forever because so much of the island biota is unique. Island faunas offer us a great deal scientifically and aesthetically. Let us do our share to live up to our obligations for their permanent preservation.—Mayr (1967:374)

Critical losses, historical and ongoing.—Extirpation of insular endemics, either directly by humans or indirectly through predation by commensals of humans or alteration of habitat (Martin and Colbert 1996), is an unfortunate and widespread legacy of recent centuries (Cassels 1984; Olson and James 1984; Simberloff 1986; Olson 1989; Steadman 1989a, b; Newton 1993; Carnutt and Pimm 2001), and human-caused mortality looms large among the agencies that limit populations of birds (Newton 1998). Risk of extinction of island birds can be exacerbated by a combination of the demographic characteristics of small populations and the destructive impacts of human colonists (Lande 1999), and it is becoming increasingly evident that comparatively specialized insular endemics are especially vulnerable to incursions by the latter (Freed 1999).

Vulnerability to extinction is especially intense for flightless rails simply on the grounds of minimal spatial requirements within limited land areas, small deme sizes, possibly lower densites of populations, and (for flighted species) the risks of diminished or terminated immigration (MacArthur and Wilson 1963, 1967; Diamond 1970b, 1984b; MacArthur et al. 1973; Simberloff 1974; M. Williamson 1981, 1989; Pimm et al. 1988; Tilman and Lehman 1997; Hanski 1999). These genetic characteristics, in combination with small population sizes, demographic fluctuations (e.g., negative impacts of tropical storms and tsunamis), destruction of habitat, illegal hunting, and introduced predators and disease continue to jeopardize most insular species of *Anas* and many rails (Diamond 1984a, b; Ralph and van Riper 1985; Simberloff 1986; Loope and Mueller-Dombois 1989). In combination with the failure of birds to meet the attributes discerned by Murphy et al. (1990) for achieving metapopulational characteristics for survival under fragmentation of requisite, specialized habitats—small body size (e.g., Insecta), high reproductive rates, and abbreviated generation intervals (Hanski 1999)—many insular birds (including large flightless rails) match the profile of species posing special challenges for conservation (Dennis et al. 1991; Calder 2000; Harte 2000). A recent comparison of success for introductions of birds and mammals to New Zealand indicates that the former possess a lower potential for introduction to new regions, itself a conservational tool of last resort (Forsyth and Duncan 2001).

Extinctions of many flightless birds (e.g., Dinornithiformes, *Apteryx* spp., *Raphus cucullatus, Pezophaps solitaria,* and *Gallirallus wakensis*) are directly attributable to the negative impacts of humans and associated introductions of predators and competitors (e.g., Strickland and Melville 1848; Owen 1866a, 1879; Diamond 1989b; Jolly 1989; Livezey 1993b; Holdaway and Jacomb 2000). Fortunately, threats to flightless rails by introgression (hybridization) with flighted relatives (Rhymer and Simberloff 1996), a means of genetic extinction frequent among small, recently isolated populations of insular *Anas* (Livezey 1991, 1993a), evidently are minimal or lacking because of a lack of heterosis in hybrids between flightless rallids and flighted immigrants (e.g., heterozygous advantages in viability, growth rates, or fecundity [Roff 1992]).

Insular futures both bleak and promising.—The importance of vocalizations of the rails as one of the earliest and most enduring familial characteristics leads naturally to a metaphor regarding likely futures. In countless and diverse marshes, meadows, and forests on islands worldwide, the matutinal or crepuscular choruses of flightless rails, testimony to the repeated evolutionary success of the sacrifice of flight, have been silenced forever. The scope of this phenomenon never will be known with precision, but the evidence at hand demonstrates that avian flightlessness on oceanic islands represents one of the most dramatic and redundantly manifested evolutionary compromises documented among vertebrates, a history wrought in defiance of recurrent, natural disasters of volcanoes, tsunamis, and earthquakes. This repeated pattern of radical functional change challenged such prevalent paradigms as scriptural decay of creation, essentialism, and typology, but affirmed the fundamental principles of natural selection and the interrelated roles of ontogeny, phylogeny, and ecological circumstance in the evolution of anatomical form and habitus. Surrender of the air under the special ecological circumstances of islands clearly conferred substantial advantages in reproduction, resilience to environmental variation, and proficiency in foraging and holding territories. In short, these specialists of isolation shed or eased the burden of costly contrivance in exchange for other advantages, only to be all but obliterated by the vast, unforeseen changes in habitat and adaptive landscapes wrought by human immigrants.

It may be less troubling to view the losses of the Dodo and Great Auk as failures of evolution, instead of acknowledging them as victims of unrestrained human destructiveness. Tragically, as it is in all species that ultimately succumb(ed) to extinction, the total population of any doomed species ultimately declined to a single bird. These terminal chapters of evolution, rarely recorded in nature (but see Cokinos 2000), document that each of the innumerable extinct rallids entailed the demise of an "omega" individual—for example, a young bird driven from a remnant of woodland by fire to a reception by club or noose, a lone female expiring from exposure on its charred nest in a devastated landscape, or an old survivor slipping into oblivion in the comparative peace of lonely senility. Appreciation of this sober reality is a critical dimension of the species-level implications of extinction, and a motivation to mobilize all possible resources to reduce this carnage.

Steadman (1995) judged that insular rallids, especially flightless endemics of oceanic islands, represent one of the most spectacular of avian radiations. The record of loss augurs poorly for those that remain, however, and the time for a substantive program of recovery and conservation of most of these remnant species lapsed decades ago. In a remarkably short period of time, most of the earth-bound swamphens, rails, crakes, moorhens, and coots that haunted innumerable islands were extinguished. Those few that remain face growing threats despite intensive programs of conservation, and it appears all but assured that wild populations of most flightless rails will not survive to be studied in natural settings by future generations of ornithologists.

ACKNOWLEDGMENTS

This study was supported by National Science Foundation grants BSR-8516623, BSR-9120545, BSR-9396249, and DEB-9815248; the M. Graham Netting Research Fund (CMNH); the National Museum of Natural History (USNM); and the American Museum of Natural History (AMNH). Access to specimens was permitted by the staff of the following institutions in the USA: Carnegie Museum of Natural History, Pittsburgh, PA (CM); Departments of Vertebrate Zoology and Paleobiology, National Museum of Natural History, Washington,

DC; National Museum of Natural History, Washington, DC; Department of Ornithology, American Museum of Natural History, New York, NY; Department of Ornithology, Academy of Natural Sciences, Philadelphia, PA; Delaware Museum of Natural History, Wilmington, DE (DEL); Division of Ornithology, Museum of Comparative Zoology, Harvard University, Cambridge, MA; Department of Ornithology, Peabody Museum of Natural History, Yale University, New Haven, CT (YPM); Florida State Museum of Natural History, University of Florida, Gainesville, FL; Department of Biology, University of Miami, Miami, FL; Museum of Natural Science, Louisiana State University, Baton Rouge, LA; Division of Birds, Museum of Zoology, University of Michigan, Ann Arbor, MI (UMMZ); Division of Birds, Field Museum of Natural History, Chicago, IL; Division of Ornithology, Natural History Museum, University of Kansas, Lawrence, KS (KUMNH); Stovall Museum of Science and History, University of Oklahoma, Norman, OK; Department of Ornithology, San Diego Natural History Museum, San Diego, CA; Department of Birds and Mammals, California Academy of Sciences, San Francisco, CA; Department of Ornithology, Los Angeles County Museum, Los Angeles, CA; Department of Biology, University of California–Long Beach, Long Beach, CA; Department of Biology, University of California–Los Angeles, Los Angeles, CA; Burke Museum, University of Washington, Seattle, WA; and Division of Vertebrate Zoology, Bernice P. Bishop Museum, Honolulu, HI.

The following foreign institutions provided specimens for study: Department of Ornithology, Royal Ontario Museum, Toronto, Ontario, Canada; Section de l'Ornithologie, Musée Canadien de la Nature, Ottawa, Ontario, Canada; Museum of Zoology, Department of Zoology, The Natural History Museum, Tring, Hertfordshire, United Kingdom (BMNH); Museum of Zoology, Cambridge University, Cambridge, United Kingdom; Department of Vertebrate Zoology, National Museums and Galleries on Merseyside, Liverpool Museum, Liverpool, United Kingdom; Department of Palaeontology, The Natural History Museum, London, United Kingdom; Zoological Collections, University Museum, Oxford University, Oxford, United Kingdom; Musée National d'Histoire Naturelle, Paris, France; Department des Sciences de la Terre, Universite Claude-Bernard, Lyon, France; Department of Ornithology, Musée Royal d'Afrique Central, Tervuren, Belgium; Zoologisk Museum, University of Copenhagen, Denmark; Institut Royal des Sciences Naturelles de Belgique, Bruxelles, Belgium; Rijksmuseum van Natuurlijke Histoire, Leiden, The Netherlands; Department of Ornithology, Transvaal Museum, Pretoria, South Africa; Cenozoic Palaeontology Department, South African Museum, Capetown, South Africa; Department of Ornithology, Natural History Museum of Zimbabwe, Bulawayo, Zimbabwe; Division of Wildlife Research, Commonwealth Science and Industry Research Organization (CSIRO), Canberra, New South Wales, Australia; Department of Ornithology, South Australian Museum, Adelaide, South Australia; Department of Ornithology, Museum of Victoria, New South Wales, Australia; Department of Ornithology, Department of Earth Sciences, Monash University, Clayton, Victoria, New South Wales, Australia; Department of Ornithology, Australian Museum, Sydney, New South Wales, Australia (SY); Queensland Museum, Fortitude Valley, Queensland, Australia; Queen Victoria Museum, Launceston, Tasmania, Australia; Tasmanian Museum and Art Gallery, Hobart, Tasmania, Australia; Department of Ornithology, National Museum of New Zealand, Wellington, New Zealand (NZNM); Department of Ornithology, Auckland Institute and Museum, Auckland, New Zealand; Department of Ornithology, Otago Museum, Dunedin, New Zealand; Canterbury Museum, University of Canterbury, Christchurch, New Zealand; and Waitomo Caves Museum, Waitomo Caves, New Zealand.

The following staff members at other institutions provided data on request: H. Hoerschelmann, Universität Hamburg, Zoologisches Institut und Museum, Hamburg, Germany; O. Frisch, Staatliches Naturhistorisches Museum, Braunschweig, Germany; D. Heinrich, Institut für Haustierkunde der Christian-Albrechts-Universität, Kiel, Germany; B. Stephan, Museum für Naturkunde, Humboldt-Universität zu Berlin, Berlin, Germany; S. Eck, Staatliches Museum für Tierkunde, Dresden, Germany; H. König, Museum Heineanum, Halberstadt, Germany; R. Winkler, Naturhistorisches Museum Basel, Basel, Switzerland; V. Loskot, Zoological Institute, Academy of Sciences, St. Petersburg, Russia; and B. Taylor, Department of Zoology and Entomology, University of Natal, Pietermaritzburg, South Africa. J. Cooper generously prepared X-ray photographs of skin specimens of *Gallirallus dieffenbachii* and *Cabalus modestus,* and fruitful discussions of the rallids of the Chatham Islands; B. K. McNab, J. K. Sailer, and D. W. Steadman (Florida Museum of Natural History, Gainesville, FL) shared of data for *Amaurornis ineptus*; J. P. Hume, for prepared the three color plates (Figs. 2–4) and provided data for *Aphanapteryx bonasia* at the Mauritius Institute.

Special thanks go to R. L. Zusi, P. C. Chu, J. C. Vanden Berge, and M. C. McKitrick for sharing their anatomical skills; P. R. Millener, G. R. Graves, and S. L. Olson for access to specimens, some recently collected; G. F. Barrowclough for permission to remove skeletal elements from study skins; R. Prys-Jones for permission to dissect anatomical specimens; D. Causey for assistance with correspondence analyses; B. DeWalt, P. Trunzo, and J. Wible for defrayment of publication charges; and R. L. Zusi, G. Mack, H. Levenson, B. Gillies, C. Fisher, C. Lefevre, E. Pasquet, R. C. Chandler, and W. E. Boles for their hospitality. T. Harper assisted with digital reproduction, J. P. Suhan performed digital SEM imaging, S. Townsend provided artistic assistance, and P. C. Rasmussen acted as trans-Atlantic courier of plates. D. A. Wiedenfeld, B. K. McNab, and two anonymous reviewers provided useful comments on the manuscript.

LITERATURE CITED

Abbott, I. 1974a. Numbers of plant, insect and land bird species on nineteen remote islands in the Southern Hemisphere. Biol. J. Linn. Soc. 6:143–152.

Abbott, I. 1974b. The avifauna of Kangaroo Island and causes of its impoverishment. Emu 7:124–134.

Abbott, I. 1974c. Morphological changes in isolated populations of some passerine bird species in Australia. Biol. J. Linn. Soc. 6:153–168.

Abbott, I. 1975. Coexistence of congeneric species in the avifaunas of Australian islands. Aust. J. Zool. 23:487–494.

Abbott, I. 1976a. Comparisons of habitat structure and plant, arthropod and bird diversity between mainland and island sites near Perth, Western Australia. Aust. J. Ecol. 1:275–280.

Abbott, I. 1976b. Is the avifauna of Kangaroo Island impoverished because of unsuitable habitat? Emu 76:43–44.

Abbott, I. 1977. The role of competition in determining differences between Victorian and Tasmanian passerine birds. Aust. J. Zool. 25:429–447.

Abbott, I. 1978a. Factors determining the number of land bird species on islands around southwestern Australia. Oecologia 33:221–233.

ABBOTT, I. 1978b. The significance of morphological variation in the finch species on Gough, Inaccessible and Nightingale islands, South Atlantic. J. Zool. Lond. 184:119–125.
ABBOTT, I. 1980. Theories dealing with the ecology of landbirds on islands. Pp. 329–371 *in* Advances in Ecological Research. Vol. 11 (A. MacFayden, Ed.). Academic Press, London, United Kingdom.
ABBOTT, I. 1981. The composition of landbird faunas of islands round southwestern Australia: Is there evidence for competitive exclusion? J. Biogeogr. 8:135–144.
ABBOTT, I., L. K. ABBOTT, AND P. R. GRANT. 1975. Seed selection and handling ability of four species of Darwin's finches. Condor 77:332–335.
ABBOTT, I., L. K. ABBOTT, AND P. R. GRANT. 1977. Comparative ecology of Galápagos ground finches (*Geospiza* Gould): evaluation of the importance of floristic diversity and interspecific competition. Ecol. Monogr. 47:151–184.
ABBOTT, I., AND P. R. GRANT. 1976. Nonequilibrial bird faunas on islands. Am. Nat. 110:507–528.
ABOUHEIF, E., AND D. J. FAIRBAIRN. 1997. A comparative analysis of allometry for sexual size dimorphism: assessing Rensch's rule. Am. Nat. 149:540–562.
ABOURACHID, A., AND S. RENOUS. 2000. Bipedal locomotion in ratites (Paleognatiform [sic]): examples of cursorial birds. Ibis 142:538–549.
ABS, M. 1983. Ontogeny and juvenile development. Pp. 3–18 *in* Physiology and Behaviour of the Pigeon (M. Abs, Ed.). Academic Press, London, United Kingdom.
ACKERLY, D. D. 2000. Taxon sampling, correlated evolution, and independent contrasts. Evolution 54:1480–1492.
ADLER, G. H. 1992. Endemism of birds of tropical Pacific islands. Evol. Ecol. 6:296–306.
ADLER, G. H. 1994. Avifaunal diversity and endemism on tropical Indian Ocean islands. J. Biogeogr. 21:85–95.
ADLER, G. H., AND R. LEVINS. 1994. The island syndrome in rodent populations. Q. Rev. Biol. 69: 473–490.
ADOBE SYSTEMS. 2000. Adope Photoshop 6.0. Adobe Systems Inc., San Jose, California.
AGASSIZ, L. 1849. Twelve Lectures on Comparative Embryology. Boston: H. Flanders.
AGASSIZ, L. 1857. Essay on classification. *In* Contributions to the Natural History of the United States. Vol. 1. Little, Brown, and Co., Boston, Massachusetts.
AGRESTI, A. 1984. Analysis of Ordinal Categorical Data. J. Wiley and Sons, New York, New York.
AIROLDI, J.-P., AND B. K. FLURY. 1988a. An application of common principal component analysis to cranial morphometry of *Microtus californicus* and *M. ochrogaster* (Mammalia, Rodentia). J. Zool. Lond. 216:21–36.
AIROLDI, J.-P., AND B. K. FLURY. 1988b. An application of common principal component analysis to cranial morphometry of *Microtus californicus* and *M. ochrogaster* (Mammalia, Rodentia): further remarks. J. Zool. Lond. 216:41–43.
ALATALO, R. V., AND L. GUSTAFSSON. 1988. Genetic component of morphological differentiation in Coal Tits under competitive release. Evolution 42:200–203.
ALATALO, R. V., J. HÖGLUND, AND A. LUNDBERG. 1988. Patterns of variation in tail ornament size in birds. Biol. J. Linn. Soc. 34:363–374.
ALATALO, R. V., AND J. MORENO. 1987. Body size, interspecific interactions, and use of foraging sites in tits (Paridae). Ecology 68:1773–1777.
ALBERCH, P. 1980. Ontogenesis and morphological diversification. Am. Zool. 20:653–667.
ALBERCH, P. 1982a. Developmental constraints in evolutionary processes. Pp. 313–332 *in* Evolution and Development (J. T. Bonner, Ed.). Springer-Verlag, Berlin, Germany.
ALBERCH, P. 1982b. The generative and regulatory roles of development in evolution. Pp. 19–36 *in* Environmental Adaptation and Evolution: A Theoretical and Empirical Approach (D. Mossakowski and G. Roth, Eds.). Gustav Fischer, Stuttgart, Germany.
ALBERCH, P. 1985. Problems with the interpretation of developmental sequences. Syst. Zool. 34: 46–58.
ALBERCH, P. 1987. Evolution of a developmental process: irreversibility and redundancy in amphibian metamorphosis. Pp. 23–46 *in* Development as an Evolutionary Process (R. A. Raff and E. C. Raff, Eds.). Alan R. Liss, New York, New York.
ALBERCH, P. 1989. The logic of monsters: evidence for internal constraint in development and evolution. Pp. 21–57 *in* Ontogenèse et Évolution (B. David, J.-L. Dommergues, J. Chaline, and B. Laurin, Coords.). Géobios Mém. Spéc. 12. Université Claude-Bernard, Lyon, France.

ALBERCH, P. 1991. From genes to phenotypes: dynamical systems and evolvability. Genetica 84: 5–11.
ALBERCH, P., AND J. ALBERCH. 1981. Heterochronic mechanisms of morphological diversification and evolutionary change in the Neotropical salamander, *Bolitoglossa occidentalis* (Amphibia: Plethodontidae). J. Morphol. 167:249–264.
ALBERCH, P., AND E. A. GALE. 1983. Size dependence during the development of the amphibian foot. Colchicine-induced digital loss and reduction. J. Embryol. Exp. Morphol. 76:177–197.
ALBERCH, P., AND E. A. GALE. 1985. A developmental analysis of an evolutionary trend: digital reduction in amphibians. Evolution 39:8–23.
ALBERCH, P., S. J. GOULD, G. F. OSTER, AND D. B. WAKE. 1979. Size and shape in ontogeny and phylogeny. Paleobiology 5:296–317.
ALBRECHT, G. H. 1978. Some comments on the use of ratios. Syst. Zool. 27:67–71.
ALBRECHT, G. H. 1979. The study of biological versus statistical variation in multivariate morphometrics: the descriptive use of multiple regression analysis. Syst. Zool. 28:338–344.
ALBRECHT, G. H. 1980. Multivariate analysis and the study of form, with special reference to canonical variate analysis. Am. Zool. 20:679–693.
ALCOVER, J. A., F. FLORIT, C. MOURER-CHAUVIRE, AND P. D. M. WEESIE. 1992. The avifaunas of the isolated Mediterranean islands during the middle and late Pleistocene. Pp. 273–283 *in* Papers in Avian Paleontology Honoring Pierce Brodkorb (K. E. Campbell, Jr., Ed.). Natural History Museum of Los Angeles County, Los Angeles, California.
ALCOVER, J. A., AND M. MCMINN. 1994. Predators of vertebrates on islands. BioScience 44:12–18.
ALDRICH, J. W. 1984. Ecogeographical variation in size and proportions of Song Sparrows (*Melospiza melodia*). Ornithol. Monogr. No. 35. American Ornithologists' Union, Washington, D.C.
ALERSTAM, T. 1979a. Wind as selective agent in bird migration. Ornis Scand. 10:76–93.
ALERSTAM, T. 1979b. Optimal use of wind by migrating birds: combined drift and overcompensation. J. Theor. Biol. 79:341–353.
ALERSTAM, T. 1990. Bird Migration. Cambridge University Press, Cambridge, United Kingdom.
ALERSTAM, T. 1991. Bird flight and optimal migration. Trends Ecol. Evol. 6:210–215.
ALERSTAM, T., AND P. H. ENCKELL. 1979. Unpredictable habitats and evolution of bird migration. Oikos 3:228–232.
ALERSTAM, T., AND A. HEDENSTRÖM. 1998. The development of bird migration theory. J. Avian Biol. 29:343–369.
ALERSTAM, T., AND G. HÖGSTEDT. 1980. Spring predictability and leap-frog migration. Ornis Scand. 11:196–200.
ALERSTAM, T., AND G. HÖGSTEDT. 1983. Regulation of reproductive success towards e^{-1} (= 37%) in animals with parental care. Oikos 40:140–145.
ALERSTAM, T., AND Å. LINDSTROM. 1990. Optimal bird migration: the relative importance of time, energy, and safety. Pp. 331–351 *in* Bird Migration: Physiology and Ecophysiology (E. Gwinner, Ed.). Springer-Verlag, Berlin, Germany.
ALEXANDER, R. D., AND G. BORGIA. 1978. Group selection, altruism and the levels of organization of life. Annu. Rev. Ecol. Syst. 9:449–475.
ALEXANDER, R. MCN. 1971. Size and Shape. Edward Arnold, London, United Kingdom.
ALEXANDER, R. MCN. 1977a. Terrestrial locomotion. Pp. 168–203 *in* Mechanics and Energetics of Animal Locomotion (R. McN. Alexander and G. Goldspink, Eds.). Chapman and Hall, London, United Kingdom.
ALEXANDER, R. MCN. 1977b. Flight. Pp. 249–278 *in* Mechanics and Energetics of Animal Locomotion (R. McN. Alexander and G. Goldspink, Eds.). Chapman and Hall, London, United Kingdom.
ALEXANDER, R. MCN. 1980. Size, shape, and structure for running and flight. Pp. 309–324 *in* A Companion to Animal Physiology (C. R. Taylor, K. Johansen, and L. Bolis, Eds.). Cambridge University Press, Cambridge, United Kingdom.
ALEXANDER, R. MCN. 1983a. Allometry of the leg bones of moas (Dinornithes) and other birds. J. Zool. Lond. 200:215–231.
ALEXANDER, R. MCN. 1983b. On the massive legs of a moa (*Pachyornis elephantopus*, Dinornithes). J. Zool. Lond. 201:363–376.
ALEXANDER, R. MCN. 1984. Optimum strengths for bones liable to fatigue and accidental fracture. J. Theor. Biol. 109:621–636.

ALEXANDER, R. McN. 1985a. Body support, scaling, and allometry. Pp. 26–37 *in* Functional Vertebrate Morphology (M. Hildebrand, D. M. Bramble, K. F. Liem, and D. B. Wake, Eds.). Belknap Press, Cambridge, United Kingdom.
ALEXANDER, R. McN. 1985b. The legs of ostriches (*Struthio*) and moas (*Pachyornis*). Acta Biotheor. 34:165–174.
ALEXANDER, R. McN. 1990a. Elastic mechanisms in the locomotion of vertebrates. Neth. J. Zool. 40: 93–105.
ALEXANDER, R. McN. 1990b. Size, speed and buoyancy adaptations in aquatic animals. Am. Zool. 30:189–196.
ALEXANDER, R. McN. 1996a. Optima for Animals. Rev. ed. Princeton University Press, Princeton, New Jersey.
ALEXANDER, R. McN. 1996b. Biophysical problems of small size in vertebrates. Symp. Zool. Soc. Lond. 69:3–14.
ALEXANDER, R. McN. 1998a. Symmorphosis and safety factors. Pp. 28–35 *in* Principles of Animal Design: The Optimization and Symmorphosis Debate (E. R. Weibel, C. R. Taylor, and L. Bolis, Eds.). Cambridge University Press, Cambridge, United Kingdom.
ALEXANDER, R. McN. 1998b. When is migration worthwhile for animals that walk, swim or fly? J. Avian Biol. 29:387–394.
ALEXANDER, R. McN., AND H. C. BENNETT-CLARK. 1977. Storage of elastic strain energy in muscle and other tissues. Nature 265:114–117.
ALEXANDER, R. McN., A. BRANDWOOD, J. D. CURREY, AND A. S. JAYES. 1984. Symmetry and precision of control of strength in limb bones of birds. J. Zool. Lond. 203:135–143.
ALI, S., AND S. D. RIPLEY. 1969. Handbook of the Birds of India and Pakistan. Vol. 2: Megapodes to Crab Plover. Oxford University Press, Bombay, India.
ALISAUSKAS, R. T. 1986. Variation in the composition of eggs and chicks of American Coots. Condor 88:84–90.
ALISAUSKAS, R. T. 1987. Morphometric correlates of age and breeding status in American Coots. Auk 104:640–646.
ALISAUSKAS, R. T., AND C. D. ANKNEY. 1985. Nutrient reserves and the energetics of reproduction in American Coots. Auk 102:133–144.
ALLEN, T. T. 1962. Myology of the Limpkin. Ph.D. dissertation, University of Florida, Gainesville, Florida.
ALTMAN, P. L., AND D. S. DITTMER. 1968. Metabolism. Federation of American Societies of Experimental Biology, Bethesda, Maryland.
ALVARENGA, H. M. F. 1995. A large and probably flightless anhinga from the Miocene of Chile. Cour. Forschungsinst. Senckenb. 181:149–162.
ALVARENGA, H. M. F., AND J. F. BONAPARTE. 1992. A new flightless landbird from the Cretaceous of Patagonia. Los Ang. Cty. Mus. Nat. Hist. (Sci. Ser.) 36:51–64.
ALVAREZ, F. 1993. Alertness signalling in two rail species. Anim. Behav. 46:1229–1231.
ALVAREZ, W. 1997. *T. rex* and the Crater of Doom. Princeton University Press, Princeton, New Jersey.
ALVAREZ DEL TORO, M. 1971. On the biology of the American Finfoot in southern Mexico. Living Bird 10:79–88.
AMADON, D. 1943a. Bird weights as an aid in taxonomy. Wilson Bull. 55:164–177.
AMADON, D. 1943b. Bird weights and egg weights. Auk 60:221–234.
AMADON, D. 1943c. Specialization and evolution. Am. Nat. 77:133–141.
AMADON, D. 1947. An estimated weight of the largest known bird. Condor 49:159–164.
AMADON, D. 1950. The Hawaiian honeycreepers (Drepaniidae). Bull. Am. Mus. Nat. Hist. 95:157–257.
AMADON, D. 1959. The significance of sexual differences in size among birds. Proc. Am. Philos. Soc. 103:531–536.
AMADON, D. 1964. The evolution of low reproductive rates in birds. Evolution 18:105–110.
AMADON, D. 1966a. Insular adaptive radiation among birds. Pp. 18–30 *in* The Galápagos: Proceedings of the Symposia of the Galápagos International Scientific Project (R. I. Bowman, Ed.). University of California Press, Berkeley, California.
AMADON, D. 1966b. Birds Around the World: A Geographical Look at Evolution and Birds. Natural History Press, Garden City, New York.
AMADON, D. 1977. Further comments on sexual size dimorphism in birds. Wilson Bull. 89:619–620.

AMADON, D., AND L. R. SHORT. 1976. Treatment of subspecies approaching species status. Syst. Zool. 25:161–167.
AMBROS, V. 1988. Genetic basis for heterochronic variation. Pp. 269–285 in Heterochrony in Evolution: A Multidisciplinary Approach (M. L. McKinney, Ed.). Plenum Press, New York, New York.
AMERICAN ORNITHOLOGISTS' UNION. 1983. Check-list of North American Birds. 6th ed. American Ornithologists' Union, Washington, D.C.
AMERICAN ORNITHOLOGISTS' UNION. 1998. Check-list of North American Birds. 7th ed. American Ornithologists' Union, Washington, D.C.
AMERSON, A. B., W. A. WHISTLER, AND T. D. SCHWANER. 1995. Breeding pattern in the Banded Rail (*Gallirallus philippensis*) in western Somoa. Notornis 42:46–48.
AMUNDSON, R. 1994. Two concepts of constraint: adaptationism and the challenge from developmental biology. Philos. Sci. 61:556–578.
AMUNDSON, R. 1996. Historical development of the concept of adaptation. Pp. 11–53 in Adaptation (M. R. Rose and G. V. Lauder, Eds.). Academic Press, San Diego, California.
AMUNDSON, R., AND G. V. LAUDER. 1994. Function without purpose: the uses of causal role function in evolutionary biology. Biol. Philos. 9:443–469.
ANDERBERG, M. R. 1973. Cluster Analysis for Application. Academic Press, New York, New York.
ANDERSEN, H. T. 1966. Physiological adaptations in diving vertebrates. Physiol. Rev. 46:212–243.
ANDERSON, A. 1989. Prodigious Birds: Moas and Moa-hunting in Prehistoric New Zealand. Cambridge University Press, Cambridge, United Kingdom.
ANDERSON, A. 1997. Prehistoric Polynesian impact on the New Zealand environment: Te Whenua Hou. Pp. 271–283 in Historical Ecology in the Pacific Islands (P. V. Kirch and T. L. Hunt, Eds.). Yale University Press, New Haven, Connecticut.
ANDERSON, D. J. 1989. The role of hatching asynchrony in siblicidal brood reduction of two booby species. Behav. Ecol. Sociobiol. 25:363–368.
ANDERSON, D. J. 1990a. Evolution of obligate siblicide in boobies. 1. A test of the insurance egg hypothesis. Am. Nat. 135:334–350.
ANDERSON, D. J. 1990b. Evolution of obligate siblicide in boobies. 2: Food limitation and parent–offspring conflict. Evolution 44:2069–2082.
ANDERSON, D. T. 1987. Developmental pathways and evolutionary rates. Pp. 143–155 in Rates of Evolution (K. S. W. Campbell and M. F. Day, Eds.). Allen and Unwin, London, United Kingdom.
ANDERSON, J. F., A. HALL-MARTIN, AND D. A. RUSSELL. 1985. Long-bone circumference and weight in mammals, birds and dinosaurs. J. Zool. Lond. 207:53–61.
ANDERSON, J. F., H. RAHN, AND H. D. PRANGE. 1979. Scaling of supportive tissue mass. Q. Rev. Biol. 54:139–148.
ANDERSON, R. P., AND C. O. HANDLEY, JR. 2002. Dwarfism in insular sloths: biogeography, selection, and evolutionary rate. Evolution 56:1045–1058.
ANDERSON, T. W. 1984. An Introduction to Multivariate Statistical Analysis. 2nd ed. J. Wiley, New York, New York.
ANDERSON, V. R., AND R. T. ALISAUKAS. 2001. Egg size, body size, locomotion, and feeding performance in captive King Eider ducklings. Condor 103:195–199.
ANDERSON, V. R., AND R. T. ALISAUKAS. 2002. Composition and growth of King Eider ducklings in relation to egg size. Auk 119:62–70.
ANDERSSON, M. 1982. Sexual selection, natural selection and quality advertisement. Biol. J. Linnean Soc. 17:375–393.
ANDERSSON, M. 1984. Brood parasitism within species. Pp. 195–228 in Producers and Scroungers: Strategies of Exploitation and Parasitism (C. J. Barnard, Ed.). Croom Helm, London, United Kingdom.
ANDERSSON, M. 1994. Sexual Selection. Princeton University Press, Princeton, New Jersey.
ANDERSSON, M. 1999. Phylogeny, behaviour, plumage evolution and neoteny in skuas Stercorariidae. J. Avian Biol. 30:205–215.
ANDERSSON, M. 2001. Relatedness and the evolution of conspecific brood parasitism. Am. Nat. 158:599–614.
ANDERSSON, M., AND R. Å. NORBERG. 1981. Evolution of reversed sexual size dimorphism and role

partitioning among predatory birds, with a size scaling of flight performance. Biol. J. Linn. Soc. 15:105–130.

ANDERSSON, S., AND M. ANDERSSON. 1994. Tail ornamentation, size dimorphism and wing length in the genus *Euplectes* (Ploceinae). Auk 111:80–86.

ANDORS, A. V. 1992. Reappraisal of the Eocene groundbird *Diatryma* (Aves: Anserimorphae). Pp. 109–125 *in* Papers in Avian Paleontology Honoring Pierce Brodkorb (K. E. Campbell, Jr., Ed.). Natural History Museum of Los Angeles County, Los Angeles, California.

ANDREEV, A. V. 1999. Energetics and survival of birds in extreme environments. Ostrich 70:13–22.

ANDREW, P., AND D. A. HOLMES. 1990. Sulawesi bird report. Kukila 5(1):4–26.

ANDREWS, C. W. 1896a. On the skull, sternum, and shoulder girdle of *Aepyornis*. Ibis 38:376–389.

ANDREWS, C. W. 1896b. On the extinct birds of the Chatham Islands. Part I. The osteology of *Diaphorapteryx hawkinsi*. Novit. Zool. 3:73–84.

ANDREWS, C. W. 1896c. On the extinct birds of the Chatham Islands. Part II. The osteology of *Paleolimnas chathamensis* and *Nesolimnas* (gen. nov.) *dieffenbachii*. Novit. Zool. 3:260–271.

ANDREWS, C. W. 1896d. Note on the skeleton of *Diaphorapteryx hawkinsi*, Forbes, a large extinct rail from the Chatham Islands. Geol. Mag. (New Ser., 4) 3:337–339.

ANDREWS, C. W. 1896e. Note on a nearly complete skeleton of *Aptornis defossor* (Owen). Geol. Mag. (New Ser.) 3:241–242.

ANDREWS, C. W. 1897. On some fossil remains of carinate birds from central Madagascar. Ibis 39: 343–359.

ANDREWS, C. W. 1899. On the extinct birds of Patagonia.—I. The skull and skeleton of *Phororhacos inflatus* Ameghino. Trans. Zool. Soc. Lond. 15:55–86.

ANDREWS, J. R. H. 1988. The Southern Ark: Zoological Discovery in New Zealand 1769–1900. University of Hawaii Press, Honolulu, Hawaii.

ANKNEY, C. D. 1984. Nutrient reserve dynamics of breeding and molting Brant. Auk 101:361–370.

AR, A., AND H. RAHN. 1980. Water in the avian egg: overall budget of incubation. Am. Zool. 20: 373–384.

AR, A., H. RAHN, AND C. V. PAGANELLI. 1979. The avian egg: mass and strength. Condor 81:331–337.

AR, A., AND Y. YOM-TOV. 1978. The evolution of parental care in birds. Evolution 32:655–669.

ARCHEY, G. 1941. The moa, a study of the Dinornithiformes. Bull. Auck. Inst. Mus. 1:1–145.

ARCHEY, G., AND C. LINDSAY. 1924. Notes on the birds of the Chatham Islands. Rec. Canterbury Mus. 2:187–201.

ARCHIBALD, G. W., AND C. D. MEINE. 1996. Family Gruidae (cranes). Pp. 60–89 *in* Handbook of the Birds of the World. Vol. 3: Hoatzin to Auks (J. del Hoyo, A. Elliott and J. Sargatal, Eds.). Lynx Edicions, Barcelona, Spain.

ARMSTRONG, J. T. 1965. Breeding home range in the Nighthawk and other birds: its evolutionary and ecological significance. Ecology 46:619–629.

ARNASON, E. 1978. Apostatic selection and kleptoparasitism in the Parasitic Jaeger. Auk 95:377–381.

ARNOLD, K. E., AND I. P. F. OWENS. 1999. Cooperative breeding in birds: the role of ecology. Behav. Ecol. 10:465–471.

ARNOLD, S. J. 1983. Sexual selection: the interface of theory and empiricism. Pp. 67–107 *in* Mate Choice (P. Bateson, Ed.). Cambridge University Press, Cambridge, United Kingdom.

ARNOLD, S. J., RAPPORTEUR. 1989. How do complex organisms evolve? [group report]. Pp. 403–433 *in* Complex Organismal Functions: Integration and Evolution in Vertebrates (D. B. Wake and G. Roth, Eds.). J. Wiley and Sons, Chichester, United Kingdom.

ARNOLD, S. J. 1992. Constraints on phenotypic evolution. Am. Nat. 140(Suppl.):85–107.

ARNOLD, S. J. 1994. Is there a unifying concept of sexual selection that applies to both plants and animals? Am. Nat. (Suppl.) 144:1–12.

ARNOLD, S. J., AND D. DUVALL. 1994. Animal mating systems: a synthesis based on selection theory. Am. Nat. 143:317–348.

ARNOLD, S. J., AND C. R. PETERSON. 2002. A model for optimal reaction norms: the case of the pregnant garter snake and her temperature-sensitive embryos. Am. Nat. 160:306–316.

ARNOLD, T. W. 1991. Intraclutch variation in egg size of American Coots. Condor 93:19–27.

ARNOLD, T. W. 1992a. Continuous laying by American Coots in response to partial clutch removal and total clutch loss. Auk 109:407–421.

ARNOLD, T. W. 1992b. The adaptive significance of eggshell removal by nesting birds: testing the egg-capping hypothesis. Condor 94:547–548.

ARNOLD, T. W., AND C. D. ANKNEY. 1997. The adaptive significance of nutrient reserves to breeding American Coots: a reassessment. Condor 99:91–103.

ARREDONDO, O. 1977. Distribución geográfica y descripción de algunos huesos de *Ornimegalonyx oteroi* Arredondo, 1958 (Strigiformes: Strigidae) del Pleistoceno Superior de Cuba. Soc. Cien. Nat. La Salle Mem. 35:133–190.

ARTHUR, W. 1982a. A developmental approach to the problem of variation in evolutionary rates. Biol. J. Linn. Soc. 18:243–261.

ARTHUR, W. 1982b. The evolutionary consequences of interspecific competition. Pp. 127–187 *in* Advances in Ecological Research. Vol. 12 (A. MacFayden and E. D. Ford, Eds.). Academic Press, London, United Kingdom.

ARTHUR, W. 1984. Mechanisms of Morphological Evolution: A Combined Genetic, Developmental and Ecological Approach. J. Wiley and Sons, Chichester, United Kingdom.

ARTHUR, W. 1988. A Theory of the Evolution of Development. J. Wiley and Sons, Chichester, United Kingdom.

ARTHUR, W. 1997. The Origin of Animal Body Plans: A Study in Evolutionary Developmental Biology. Cambridge University Press, Cambridge, United Kingdom.

ASCHOFF, J. 1981. Thermal conductance in mammals and birds: its dependence on body size and circadian pahse. Comp. Biochem. Physiol. 69:611–619.

ASCHOFF, J., AND H. POHL. 1970a. Der Ruheumasatz von Vögeln als Funktion der Tageszeit und der Körpergrösse. J. Ornithol. 3:38–48.

ASCHOFF, J., AND H. POHL. 1970b. Rhythmic variations in energy metabolism. Fed. Proc. 29:1541–1552.

ASHMOLE, N. P. 1963a. The extinct avifauna of St. Helena Island. Ibis 103b:390–408.

ASHMOLE, N. P. 1963b. Subfossil bird remains on Ascension Island. Ibis 103b(Suppl.):382–389.

ASHMOLE, N. P. 1963c. The regulation of numbers of tropical oceanic birds. Ibis 103b:458–473.

ASHMOLE, N. P., AND M. J. ASHMOLE. 1997. The land fauna of Ascension Island: new data from caves and lava flows, and a reconstruction of the prehistoric ecosystem. J. Biogeogr. 24:549–589.

ASQUITH, A. 1995. Evolution of *Sarona* (Heteroptera, Miridae): speciation on geographic and ecological islands. Pp. 90–120 *in* Hawaiian Biogeography: Evolution on a Hot Spot Archipelago (W. L. Wagner and W. A. Funk, Eds.). Smithsonian Institution Press, Washington, D.C.

ATCHLEY, W. R. 1978. Ratios, regression intercepts, and the scaling of data. Syst. Zool. 27:78–83.

ATCHLEY, W. R. 1984. Ontogeny, timing of development, and genetic variance–covariance structure. Am. Nat. 123:519–540.

ATCHLEY, W. R. 1987. Developmental quantitative genetics and the evolution of ontogenies. Evolution 41:316–330.

ATCHLEY, W. R. 1990. Heterochrony and morphological change: a quantitative genetic perspective. Semin. Dev. Biol. 1:289–297.

ATCHLEY, W. R., AND D. ANDERSON. 1978. Ratios and the statistical analysis of biological data. Syst. Zool. 27:71–78.

ATCHLEY, W. R., C. T. GASKINS, AND D. ANDERSON. 1976. Statistical properties of ratios. I. Empirical results. Syst. Zool. 25:137–148.

ATCHLEY, W. R., AND B. K. HALL. 1991. A model for the development and evolution of complex morphological structures. Biol. Rev. 66:101–157.

ATKINSON, I. A. E., AND B. D. BELL. 1973. Offshore and outlying islands. Pp. 372–392 *in* The Natural History of New Zealand: An Ecological Survey (G. R. Williams, Ed.). Reed Publications, Wellington, New Zealand.

ATKINSON, I. A. E., AND R. M. GREENWOOD. 1989. Relationships between moas and plants. N. Z. J. Ecol. 12(Suppl.):67–96.

ATKINSON, I. A. E., AND P. R. MILLENER. 1991. An ornithological glimpse into New Zealand's prehuman past. Pp. 127–192 *in* Acta XX Congressus Internationalis Ornithologici. Vol. 1 (B. D. Bell, R. O. Cossee, J. E. C. Flux, B. D. Heather, R. A. Hitchmough, C. J. R. Robertson, and M. J. Williams, Eds.). New Zealand Ornithological Congress Trust Board, Wellington, New Zealand.

ATWATER, B. F. 1987. Evidence for great Holocene earthquakes along the outer coast of Washington State. Science 236:942–944.

ATWATER, B. F. 1992. Geological evidence for earthquakes during the past 2000 years along the Copalis River, southern coastal Washington. J. Geophys. Res. (B2) 97:1901–1919.

ATWOOD, J. L. 1980. Social interactions in the Santa Cruz Island Scrub Jay. Condor 82:440–448.

AUBER, L. 1957. The distribution of structural colours and unusual pigments in the class Aves. Ibis 99:463–476.

AULIE, A. 1983. The fore-limb muscular system and flight. Pp. 117–129 in Physiology and Behaviour of the Pigeon (M. Abs, Ed.). Academic Press, London, United Kingdom.

AUSTIN, J. E., AND L. H. FREDRICKSON. 1987. Body and organ mass and body composition of post-breeding female Lesser Scaup. Auk 104:694–699.

AVISE, J. C., AND R. L. ZINK. 1988. Molecular genetic divergence between avian sibling species: King and Clapper Rails, Long-billed and Short-billed Dowitchers, Boat-tailed and Great-tailed Grackles, and Tufted and Black-crested Titmice. Auk 105:516–528.

AYALA, F. J. 1974. The concept of biological progress. Pp. 339–355 in Studies in the Philosophy of Biology: Reduction and Related Problems (F. J. Ayala and T. Dobzhansky, Eds.). University of California Press, Berkeley, California.

AYALA, F. J. 1988. Can "progress" be defined as a biological concept? Pp. 75–96 in Evolutionary Progress (M. H. Nitecki, Ed.). University of Chicago Press, Chicago, Illinois.

BADYAEV, A. V., AND G. E. HILL. 2000. The evolution of sexual dimorphism in the House Finch. I. Population divergence in morphological covariance structure. Evolution 54:1784–1794.

BADYAEV, A. V., G. E. HILL, P. O. DUNN, AND J. C. GLEN. 2001a. Plumage color as a composite trait: developmental and functional integration of sexual ornamentation. Am. Nat. 158:221–235.

BADYAEV, A. V., G. E. HILL, AND L. A. WHITTINGHAM. 2001b. The evolution of sexual dimorphism in the House Finch. IV. Population divergence in ontogeny. Evolution 55:2534–2549.

BADYAEV, A. V., AND T. E. MARTIN. 2000. Sexual dimorphism in relation to current selection in the House Finch. Evolution 54:987–997.

BADYAEV, A. V., AND A. QVARNSTRÖM. 2002. Putting sexual traits into the context of an organism: a life-history perspective in studies of sexual selection. Auk 119:301–310.

BADYAEV, A. V., L. A. WHITTINGHAM, AND G. E. HILL. 2001c. The evolution of sexual size dimorphism in the House Finch. III. Developmental basis. Evolution 55:176–189.

BAER, K. E. VON. 1868. Über Prof. Nic. Wagner's Entdeckung von larven, die sich fortpflanzen, Herrn Gerran's verwandte und ergänzende Beobachtung und über die pädogenesis überhaupt. Bull. Acad. Imp. Sci. St. Petersbourg 9:63–137.

BAILEY, A. M. 1956. Birds of Midway and Laysan islands. Denver Mus. Nat. Hist. Pict. 12:1–130.

BAILEY, R. C., AND J. BYRNES. 1990. A new, old method for assessing measurement error in both univariate and multivariate morphometric studies. Syst. Zool. 39:124–130.

BAIRD, R. F. 1984. The Pleistocene distribution of the Tasmanian Native-hen Gallinula mortierii mortierii. Emu 84:119–123.

BAIRD, R. F. 1985. Avian fossils from Quaternary deposits in 'Green Waterhole Cave', south eastern South Australia. Rec. Aust. Mus. 37:353–370.

BAIRD, R. F. 1986. Tasmanian Native-hen Gallinula mortierii: the first Late Pleistocene record from Queensland. Emu 86:121–122.

BAIRD, R. F. 1991a. Avian fossils from the Quaternary of Australia. Pp. 810–848 in Vertebrate Palaeontology of Australia (P. Vickers-Rich, J. M. Monaghan, R. F. Baird, and T. H. Rich, Eds.). Monash University Publications Committee, Melbourne, Australia.

BAIRD, R. F. 1991b. The dingo as a possible factor in the disappearance of Gallinula mortierii from the Australian mainland. Emu 91:121–122.

BAIRD, R. F. 1992. Fossil avian assemblage of pitfall origin from Holocene sediments in Amphitheatre Cave (G-2), south-western Victoria, Australia. Rec. Aust. Mus. 44:21–44.

BAKER, A. J. 1975. Morphological variation, hybridization and systematics of New Zealand oystercatchers (Charadriiformes: Haematopodidae). J. Zool. Lond. 175:357–390.

BAKER, A. J. 1980. Morphometric differentiation in New Zealand populations of the House Sparrow (Passer domesticus). Evolution 34:638–653.

BAKER, A. J. 1991. A review of New Zealand ornithology. Pp. 1–67 in Current Ornithology. Vol. 8 (D. M. Power, Ed.). Plenum Press, New York, New York.

BAKER, A. J., AND P. F. JENKINS. 1987. Founder effect and cultural evolution of songs in an isolated population of Chaffinches, Fringilla coelebs, in the Chatham Islands. Anim. Behav. 35:1793–1803.

BAKER, A. J., AND A. MOEED. 1979. Evolution in the introduced New Zealand populations of the Common Myna, *Acridotheres tristis* (Aves: Sturninae). Can. J. Zool. 57:570–584.

BAKER, A. J., AND A. MOEED. 1980. Morphometric variation in Indian samples of the Common Myna, *Acridotheres tristis* (Aves: Sturninae). Bijd. Dierkd. 50:351–363.

BAKER, H. G., AND G. L. STEBBINS, EDS. 1965. The Genetics of Colonizing Species. Academic Press, New York, New York.

BAKER, M. C. 1979. Morphological correlates of habitat selection in a community of shorebirds (Charadriiformes). Oikos 33:121–126.

BAKER, M. C., AND A. E. M. BAKER. 1973. Niche relationships among six species of shorebirds on their wintering and breeding ranges. Ecol. Monogr. 43:193–212.

BAKER, R. H. 1951. The avifauna of Micronesia, its origin, evolution, and distribution. Univ. Kans. Mus. Nat. Hist. Publ. 3:1–359.

BAKER, R. R., AND G. A. PARKER. 1979. The evolution of bird coloration. Philos. Trans. R. Soc. Lond. (Ser. B) 287:63–130.

BAKER-GABB, D. J. 1986. Ecological release and behavioral and ecological flexibility in Marsh Harriers on islands. Emu 86:71–81.

BAKKER, B. J., AND R. A. FORDHAM. 1993. Diving behavior of the Australian Coot in a New Zealand Lake. Notornis 40:131–136.

BALDWIN, J. M. 1896. A new factor in evolution. Am. Nat. 30:354–451.

BALDWIN, P. H. 1945. The fate of the Laysan Rail. Audubon Mag. 47:343–348.

BALDWIN, P. H. 1947. The life history of the Laysan Rail. Condor 49:14–21.

BALDWIN, P. H. 1953. Annual cycle, environment and evolution in the Hawaiian honeycreepers. Univ. Calif. Publ. Zool. 52:285–398.

BALDWIN, S. P., AND S. C. KENDEIGH. 1938. Variations in the weight of birds. Auk 55:416–467.

BALDWIN, S. P., H. C. OBERHOLSER, AND L. G. WORLEY. 1931. Measurements of birds. Cleve. Mus. Nat. Hist. Sci. Publ. 2:1–165.

BALKAU, B. J., AND M. W. FELDMAN. 1973. Selection for migration modification. Genetics 74:171–174.

BALL, E., AND J. GLUCKSMAN. 1975. Biological colonization of Motmot, a recently-created tropical island. Proc. R. Soc. Lond. (Ser. B Biol. Sci.) 190:421–442.

BALLANCE, A. 2001. Takahe: the bird that twice came back from the grave. Pp. 11–22 *in* The Takahe: Fifty Years of Conservation Management and Research. University of Otago Press, Dunedin, New Zealand.

BALMFORD, A., I. L. JONES, AND A. L. R. THOMAS. 1993. On avian asymmetry: evidence of natural selection for symmetrical tails and wings in birds. Proc. R. Soc. Lond. (Ser. B Biol. Sci.) 252: 245–251.

BALMFORD, A., I. L. JONES, AND A. L. R. THOMAS. 1994. How to compensate for costly sexually selected tails: the origin of sexually dimorphic wings in long-tailed birds. Evolution 48:1062–1070.

BALMFORD, A., L. R. THOMAS, AND I. L. JONES. 1993. Aerodynamics and the evolution of long tails in birds. Nature 361:628–630.

BALON, E. K. 1989. The epigenetic mechanisms of bifurcation and alternative life-history styles. Pp. 467–501 *in* Alternative Life-History Styles of Animals (M. N. Bruton, Ed.). Kluwer Academic Publications, Dordrecht, The Netherlands.

BALOUET, J. C. 1991. The fossil vertebrate record of New Caledonia. Pp. 1383–1409 *in* Vertebrate Palaeontology of Australia (P. Vickers-Rich, J. M. Monaghan, R. F. Baird, and T. H. Rich, Eds.). Monash University Publications Committee, Melbourne, Australia.

BALOUET, J. C., AND S. L. OLSON. 1987. A new extinct species of giant pigeon (Columbidae: *Ducula*) from archeological deposits on Wallis (Uvea) Island, South Pacific. Proc. Biol. Soc. Wash. 100: 769–775.

BALOUET, J. C., AND S. L. OLSON. 1989. Fossil birds from Late Quaternary deposits in New Caledonia. Smithson. Contrib. Zool. 469:1–38.

BALTZER, M. C. 1990. A report on the wetland avifauna of South Sulawesi. Kukila 5:27–55.

BANG, B. G. 1968. Olfaction in Rallidae (Gruiformes), a morphological study of thirteen species. J. Zool. Lond. 156:97–107.

BANG, B. G. 1971. Functional anatomy of the olfactory system in 23 orders of birds. Acta Anat. Suppl. 79:1–76.

BANGS, O. 1907. On the wood rails, genus *Aramides,* occurring north of Panama. Am. Nat. 41:177–187.
BANGS, O. 1930. Types of birds now in the Museum of Comparative Zoology. Bull. Mus. Comp. Zool. 70:147–426.
BANNASCH, R. 1986a. Morphologisch-funktionelle Untersuchung am Lokomotionsapparat der Pinguine als Grundlage für ein allgemeines Bewegungsmodell des "Unterwasserfluges." Tiel I. Gegenbaurs Morphol. Jahrb. 132:645–679.
BANNASCH, R. 1986b. Morphologisch-funktionelle Untersuchung am Lokomotionsapparat der Pinguine als Grundlage für ein allgemeines Bewegungsmodell des "Unterwasserfluges." Tiel I. Gegenbaurs Morphol. Jahrb. 132:757–817.
BANNASCH, R. 1987. Morphologisch-funktionelle Untersuchung am Lokomotionsapparat der Pinguine als Grundlage für ein allgemeines Bewegungsmodell des "Unterwasserfluges." Tiel III. Gegenbaurs Morphol. Jahrb. 133:39–59.
BARBAULT, R. 1988. Body size, ecological constraints, and the evolution of life-history strategies. Pp. 261–286 in Evolutionary Biology. Vol. 22 (M. K. Hecht, B. Wallace, and G. T. Prance, Eds.). Plenum Press, New York, New York.
BARBOSA, A., AND A. P. MØLLER. 1999. Aerodynamic costs of long tails in male Barn Swallows *Hirundo rustica* and the evolution of sexual size dimorphism. Behav. Ecol. 10:128–135.
BARBOSA, P., AND V. KRISCHIK. 1989. Life-history traits of forest-inhabiting flightless Lepidoptera. Am. Midl. Nat. 122:262–274.
BARBOUR, T. 1928. Notes on three Cuban birds. Auk 45:28–32.
BARNARD, C. J. 1984. Stasis: a coevolutionary model. J. Theor. Biol. 110:27–34.
BARNARD, P. 1991. Ornament and body size variation and their measurement in natural populations. Biol. J. Linn. Soc. 42:379–388.
BARNES, T. 1997. Finally found: the missing link between the egg and the adult stage of the Chestnut Rail—*Eulabeornis castaneoventris.* Pp. 219–221 in Zoos: Evolution or Extinction? (G. Hunt and L. Slater, Eds.). Healesville Santuary, Healesville, Australia.
BARNETT, V., AND T. LEWIS. 1984. Outliers in Statistical Data. 2nd ed. J. Wiley and Sons, Chichester, United Kingdom.
BARNIKOL, A. 1952. Korrelationen in der Ausgestaltung der Schädelform bei Vögeln. Gegenbaurs Morphol. Jahrb. 92:373–414.
BARNOSKY, A. D. 1989. The Late Pleistocene event as a paradigm for widespread mammal extinction. Pp. 235–254 in Mass Extinctions: Processes and Evidence (S. K. Donovan, Ed.). Columbia University Press, New York, New York.
BARR, T. G. 1968. Cave ecology and the evolution of troglobites. Pp. 25–102 in Evolutionary Biology. Vol. 2 (T. Dobzhansky, M. K. Hecht, and W. C. Steere, Eds.). Plenum Press, New York, New York.
BARROWCLOUGH, G. F. 1980. Gene flow, effective population sizes, and genetic variance components in birds. Evolution 34:789–798.
BART, J., R. A. STEHN, J. A. HERRICK, N. A. HEASLIP, T. A. BOOKHOUT, AND J. R. STENZEL. 1984. Survey methods for breeding Yellow Rails. J. Wildl. Manage. 48:1382–1386.
BARTHOLOMEW, D. J. 1987. Latent Variable Models and Factor Analysis. Oxford University Press, Oxford, United Kingdom.
BARTHOLOMEW, G. A., AND T. J. CADE. 1963. The water economy of land birds. Auk 80:504–539.
BARTON, N. H. 1989. Founder effect speciation. Pp. 229–256 in Speciation and Its Consequences (D. Otte and J. A. Endler, Eds.). Sinauer, Sunderland, Massachusetts.
BARTON, N. H. 1998. Natural selection and random genetic drift as causes of evolution on islands. Pp. 102–123 in Evolution on Islands (P. R. Grant, Ed.). Oxford University Press, Oxford, United Kingdom.
BARTON, N. H., AND B. CHARLESWORTH. 1984. Genetic revolutions, founder effects, and speciation. Annu. Rev. Ecol. Syst. 15:133–164.
BARTON, N. H., AND S. ROUHANI. 1987. The frequency of shifts between alternative equilibria. J. Theor. Biol. 125:397–418.
BARTON, N. H., AND M. TURELLI. 1987. Adaptive landscapes, genetic distance, and the evolution of quantitative characters. Genet. Res. 49:157–174.
BARTON, N. H., AND M. TURELLI. 1989. Evolutionary quantitative genetics: How little do we know? Annu. Rev. Genet. 23:337–370.

BARTON, N. H., AND M. C. WHITLOCK. 1997. The evolution of metapopulations. Pp. 183–214 *in* Metapopulation Biology: Ecology, Genetics, and Evolution (I. Hanski and M. E. Gilpin, Eds.). Academic Press, San Diego, California.

BASILEVSKY, A. 1994. Statistical Factor Analysis and Related Methods: Theory and Applications. J. Wiley and Sons, New York, New York.

BASSETT-HULL, A. F. 1910. The birds of Lord Howe and Norfolk islands. Proc. Linn. Soc. N. S. Wales 34:636–693.

BAUM, D. A., AND A. LARSON. 1991. Adaptation reviewed: a phylogenetic methodology for studying character macroevolution. Syst. Zool. 40:1–18.

BAUMEL, J. J. 1988. Functional morphology of the tail apparatus of the Pigeon (*Columba livia*). Adv. Anat. Embryol. Cell Biol. 110:1–115.

BAUMEL, J. J., AND R. J. RAIKOW. 1993. Arthrologia. Pp. 133–188 *in* Handbook of Avian Anatomy: Nomina Anatomica Avium. 2nd ed. (J. J. Baumel, A. S. King, J. E. Breazile, H. E. Evans, and J. C. Vanden Berge, Eds.). Publ. No. 23. Nuttall Ornithological Club, Cambridge, Massachusetts.

BAUMEL, J. J., J. A. WILSON, AND D. R. BERGREN. 1990. The ventilatory movements of the avian pelvis and tail: function of the muscles of the tail region of the Pigeon (*Columba livia*). J. Exp. Biol. 151:263–277.

BAUMEL, J. J., AND L. M. WITMER. 1993. Osteologia. Pp. 45–132 *in* Handbook of Avian Anatomy: Nomina Anatomica Avium. 2nd ed. (J. J. Baumel, A. S. King, J. E. Breazile, H. E. Evans, and J. C. Vanden Berge, Eds.). Publ. No. 23. Nuttall Ornithological Club, Cambridge, Massachusetts.

BEALE, G. 1985. A radiological study of the kiwi (*Apteryx australis mantelli*). J. R. Soc. N. Z. 15:187–200.

BEALE, G. 1991. The maturation of the skeleton of a kiwi (*Apteryx australis mantelli*)—a ten year radiological study. J. R. Soc. N. Z. 21:219–220.

BEAUCHAMP, A. J. 1986. A case of co-operative rearing in Wekas. Notornis 33:51–52.

BEAUCHAMP, A. J. 1987a. A population study of the Weka *Gallirallus australis* on Kapiti Island. Ph.D. dissertation, Victoria University, Wellington, New Zealand.

BEAUCHAMP, A. J. 1987b. The social structure of the Weka (*Gallirallus australis*) at Double Cove, Marlborough Sounds. Notornis 34:317–325.

BEAUCHAMP, A. J. 1988. Status of the Weka (*Gallirallus australis*) on Cape Brett, Bay of Islands. Notornis 35:282–284.

BEAUCHAMP, A. J. 1989. Panbiogeography and rails of the genus *Gallirallus*. N. Z. J. Zool. 16:763–772.

BEAUCHAMP, A. J. 1998a. The decline of North Island Weka (*Gallirallus australis greyi*) at Parekura Bay, Bay of Islands. Notornis 45:31–43.

BEAUCHAMP, A. J. 1998b. The aging of Weka (*Gallirallus australis*) using measurements, soft parts, plumage and wing spurs. Notornis 45:167–176.

BEAUCHAMP, A. J., AND R. CHAMBERS. 2000. Population density changes of adult North Island Weka (*Gallirallus australis greyi*) in the Mansion House Historic Reserve, Kawau Island, in 1992–1999. Notornis 47:82–89.

BEAUCHAMP, A. J., R. CHAMBERS, AND J. L. KENDRICK. 1993. North Island Weka on Rakitu Island. Notornis 40:309–312.

BEAUCHAMP, A. J., G. C. STAPLES, E. O. STAPLES, A. GRAEME, B. GRAEME, AND E. FOX. 2000. Failed establishment of North Island Weka (*Gallirallus australis greyi*) at Karangahake Gorge, North Island, New Zealand. Notornis 47:90–96.

BEAUCHAMP, A. J., AND T. H. WORTHY. 1988. Decline in the distribution of the Takahe *Porphyrio* (= *Notornis*) *mantelli*: a re-examination. J. R. Soc. N. Z. 18:103–118.

BEAUCHAMP, G. 1997. Determinants of intraspecific brood amalgamation in waterfowl. Auk 114:11–21.

BECKER, P. 1990. Kennzeichen und kleider der europäischen kleinen Rallen und Sumpfhühner *Rallus* und *Porzana*. Limicola 4:93–144.

BEDDARD, F. E. 1898. The Structure and Classification of Birds. Longmans, Green, and Co., London, England.

BEDNEKOFF, P. A., AND A. I. HOUSTON. 1994. Avian daily foraging patterns: effects of digestive constraints and variability. Evol. Ecol. 8:36–52.

BEEHLER, B. M., T. K. PRATT, AND D. A. ZIMMERMAN. 1986. Birds of New Guinea. Princeton University Press, Princeton, New Jersey.

BEHN, F., AND G. MILLIE. 1959. Beitrag zur Kenntnis des Rüsselbläßhuhn (*Fulica cornuta* Bonaparte). J. Ornithol. 100:119–131.

BEINTEMA, A. J. 1972. The history of the Island Hen (*Gallinula nesiotis*), the extinct flightless gallinule of Tristan da Cunha. Bull. Br. Ornithol. Club 92:106–113.

BEINTEMA, A. J., AND G. H. VISSER. 1989. The effect of weather on time budgets and development of chicks of meadow birds. Ardea 77:181–192.

BELL, B. D. 1991. Recent avifaunal changes and the history of ornithology in New Zealand. Pp. 195–230 *in* Acta XX Congressus Internationalis Ornithologici. Vol. 1 (B. D. Bell, R. O. Cossee, J. E. C. Flux, B. D. Heather, R. A. Hitchmough, C. J. R. Robertson, and M. J. Williams, Eds.). New Zealand Ornithological Congress Trust Board, Wellington, New Zealand.

BELLAIRS, R. 1960. Development of birds. Pp. 127–188 *in* Biology and Comparative Physiology of Birds. Vol. 1 (A. J. Marshall, Ed.). Academic Press, New York, New York.

BELLAMY, P. E., S. A. HINSLEY, AND I. NEWTON. 1996. Local extinctions and recolonisations of passerine bird populations in small woods. Oecologia 109:64–71.

BENHAM, W. B. 1898a. Notes on the fourth skin of *Notornis*. Trans. N. Z. Inst. 31:146–150.

BENHAM, W. B. 1898b. Notes on certain of the viscera of *Notornis*. Trans. N. Z. Inst. 31:151–156.

BENHAM, W. B. 1899. Notes on the internal anatomy of *Notornis*. Proc. Zool. Soc. Lond. 1899:88–96.

BENKMAN, C. W. 1987a. Food profitability and the foraging ecology of crossbills. Ecol. Monogr. 57:251–267.

BENKMAN, C. W. 1987b. Crossbill foraging behavior, bill structure, and patterns of food profitability. Wilson Bull. 99:351–368.

BENKMAN, C. W. 1988a. Seed handling efficiency, bill structure, and the cost of specialization for crossbills. Auk 105:715–719.

BENKMAN, C. W. 1988b. On the advantages of crossed mandibles: an experimental approach. Ibis 129:288–293.

BENKMAN, C. W. 1989a. On the evolution and ecology of island populations of crossbills. Evolution 43:1324–1330.

BENKMAN, C. W. 1989b. Intake rate maximization and the foraging behaviour of crossbills. Ornis Scand. 20:65–68.

BENKMAN, C. W. 1989c. Breeding opportunities, foraging rates, and parental care in White-winged Crossbills. Auk 106:483–485.

BENKMAN, C. W. 1990. Foraging rates and the timing of crossbill reproduction. Auk 107:376–386.

BENKMAN, C. W. 1991. Predation, seed size partitioning and the evolution of body size in seed-eating finches. Evol. Ecol. 5:118–127.

BENKMAN, C. W. 1993. Adaptation to single resources and the evolution of crossbill (*Loxia*) diversity. Ecol. Monogr. 63:305–325.

BENKMAN, C. W., W. C. HOLIMON, AND J. W. SMITH. 2001. The influence of a competitor on the geographic mosaic of coevolution between crossbills and lodgepole pine. Evolution 55:282–294.

BENKMAN, C. W., AND A. K. LINDHOLM. 1991. An experimental analysis of the advantages and evolution of a morphological novelty. Nature 349:519–520.

BENNETT, K. D. 1997. Evolution and Ecology: The Pace of Life. Cambridge University Press, Cambridge, United Kingdom.

BENNETT, M. B. 1996. Allometry of leg muscles of birds. J. Zool. Lond. 238:435–443.

BENNETT, P. M., AND P. H. HARVEY. 1985a. Relative brain size and ecology in birds. J. Zool. Lond. 207:151–169.

BENNETT, P. M., AND P. H. HARVEY. 1985b. Brain size, development and metabolism in birds and mammals. J. Zool. Lond. 207:491–509.

BENNETT, P. M., AND P. H. HARVEY. 1987. Active and resting metabolism in birds: allometry, physiology and ecology. J. Zool. Lond. 213:327–363.

BENSCH, S., AND B. NIELSEN. 1999. Autumn migration speed of juvenile Reed and Sedge Warblers in relation to date and fat loads. Condor 101:153–156.

BENSON, C. W., AND M. J. PENNY. 1971. The land birds of Aldabra. Philos. Trans. R. Soc. Lond. (Ser. B) 260:417–527.

BENTON, M. J. 1987. Progress and competition in macroevolution. Biol. Rev. 62:305–338.
BENTON, M. J. 1990. Evolution of large size. Pp. 147–152 *in* Palaeobiology: A Synthesis (D. E. G. Briggs and P. R. Crowther, Eds.). Blackwell Scientific Publications, Oxford, United Kingdom.
BENTON, M. J. 1991. Extinction, biotic replacements, and clade interactions. Pp. 89–102 *in* The Unity of Evolutionary Biology: Proceedings of the Fourth International Congress of Systematic and Evolutionary Biology. Vol. 1 (E. C. Dudley, Ed.). Dioscorides Press, Portland, Oregon.
BENTON, T. G., AND T. SPENCER. 1995. The birds of Henderson Island: ecological studies in a near pristine ecosystem. Biol. J. Linn. Soc. 56:147–148.
BERGER, A. J. 1956a. The appendicular myology of the Sandhill Crane, with comparative remarks on the Whooping Crane. Wilson Bull. 68:282–304.
BERGER, A. J. 1956b. The expansor secundariorum muscle, with special reference to passerine birds. J. Morphol. 99:137–168.
BERGER, A. J. 1966. The musculature. Pp. 301–349 *in* Biology and Comparative Physiology of Birds (A. J. Marshall, Ed.). Academic Press, New York, New York.
BERGER, M., AND J. S. HART. 1974. Physiology and energetics of flight. Pp. 415–477 *in* Avian Biology. Vol. 4 (D. S. Farner, J. R. King, and K. C. Parkes, Eds.). Academic Press, New York, New York.
BERGER, M., J. S. HART, AND O. Z. ROY. 1970. Respiration, oxygen consumption and heart rate in some birds during rest and flight. Z. Vgl. Physiol. 66:201–214.
BERGLUND, A., A. BISAZZA, AND A. PILASTRO. 1996. Armaments and ornaments: an evolutionary explanation of traits of dual utility. Biol. J. Linn. Soc. 58:385–399.
BERNSTEIN, M. H., S. P. THOMAS, AND K. SCHMIDT-NIELSEN. 1973. Power input during flight of the Fish Crow *Corvus ossifragus*. J. Exp. Biol. 58:401–410.
BERRIGAN, D., AND J. SEGER. 1998. Information and allometry. Evol. Ecol. 12:535–541.
BERRY, R. J. 1992. The significance of island biotas. Biol. J. Linn. Soc. 46:3–12.
BERRY, R. J. 1998. Evolution of small mammals. Pp. 35–50 *in* Evolution on Islands (P. R. Grant, Ed.). Oxford University Press, Oxford, United Kingdom.
BERTHOLD, P. 1988. Evolutionary aspects of migratory behavior in European warblers. J. Evol. Biol. 1:195–209.
BERTHOLD, P. 1990. Genetics of migration. Pp. 269–280 *in* Bird Migration: The Physiology and Ecophysiology (E. Gwinner, Ed.). Springer-Verlag, Heidelberg, Germany.
BERTHOLD, P. 1991. Genetic control of migratory behaviour in birds. Trends Ecol. Evol. 6:254–257.
BERTHOLD, P. 1993. Bird Migration: A General Survey. Oxford University Press, Oxford, United Kingdom.
BERTHOLD, P. 1995. Microevolution of migratory behaviour illustrated by the Blackcap *Sylvia atricapilla*: 1993 Witherby Lecture. Bird Study 42:89–100.
BERTHOLD, P. 1996. Control of Bird Migration. Chapman and Hall, London, United Kingdom.
BERTHOLD, P. 1999. A comprehensive theory for the evolution, control and adaptability of avian migration. Ostrich 70:1–11.
BERTHOLD, P., A. J. HELBIG, G. MOHR, AND U. QUERNER. 1992. Rapid microevolution of migratory behaviour in a wild bird species. Nature 360:668–669.
BERTHOLD, P., W. WILTSCHKO, H. MILTENBERGER, AND U. QUERNER. 1990. Genetic transmission of migratory behavior into a nonmigratory bird population. Experientia 46:107–108.
BERTRAM, B. C. R. 1996. Family Heliornithidae (finfoots). Pp. 210–217 *in* Handbook of the Birds of the World. Vol. 3: Hoatzin to Auks (J. del Hoyo, A. Elliott and J. Sargatal, Eds.). Lynx Edicions, Barcelona, Spain.
BIBBY, C. J. 1994a. Recent past and future extinctions in birds. Trans. R. Soc. Lond. 344:35–40.
BIBBY, C. J. 1994b. Recent past and future extinctions in birds. Pp. 98–110 *in* Extinction Rates (J. H. Lawton and R. M. May, Eds.). Oxford University Press, Oxford, United Kingdom.
BIEBACH, H. 1996. Energetics of winter and migratory fattening. Pp. 280–323 *in* Avian Energetics and Nutritional Ecology (C. Carey, Ed.). Chapman and Hall, New York, New York.
BIEWENER, A. A. 1982. Bone strength in small mammals and bipedal birds: Do safety factors change with body size? J. Exp. Biol. 98:289–301.
BIEWENER, A. A. 2000. Scaling of terrestrial support: differing solutions to mechanical constraints of size. Pp. 51–66 *in* Scaling in Biology (J. H. Brown and G. B. West, Eds.). Oxford University Press, Oxford, United Kingdom.

BIEWENER, A. A., K. P. DIAL, AND G. E. GOSLOW. 1992. Pectoralis muscle force and power output during flight in the Starling. J. Exp. Biol. 164:1–18.
BINFORD, L. C. 1989. A Distributional Survey of the Birds of the Mexican State of Oaxaca. Ornithol. Monogr. No. 43. American Ornithologists' Union, Washington, D.C.
BIRDLIFE INTERNATIONAL. 2000. Threatened Birds of the World. Lynx Editions and BirdLife International, Barcelona, Spain.
BIRKHEAD, T. R. 1998. Sperm competition in birds: mechanisms and function. Pp. 579–622 in Sperm Competition and Sexual Selection (T. R. Birkhead and A. P. Møller, Eds.). Academic Press, San Diego, California.
BIRKHEAD, T. R., AND A. P. MØLLER. 1992. Sperm Competition in Birds: Evolutionary Causes and Consequences. Academic Press, London, United Kingdom.
BISHOP, C. M. 1997. Heart mass and the maximum cardiac output of birds and mammals: implications for estimating aerobic power input of flying animals. Philos. Trans. R. Soc. Lond. (Ser. B) 352: 447–456.
BISHOP, C. M., AND P. J. BUTLER. 1995. Physiological modelling of oxygen consumption in birds during flight. J. Exp. Biol. 198:2153–2163.
BISHOP, C. M., P. J. BUTLER, A. J. EL-HAJ, S. EGGINTON, AND M. J. J. E. LOONEN. 1996. The morphological development of the locomotor and cardiac muscles of the migratory Barnacle Goose (*Branta leucopsis*). J. Zool. Lond. 239:1–15.
BISHOP, C. M., P. J. BUTLER, A. J. EL-HAJ, AND S. EGGINTON. 1998. Comparative development of captive and migratory populations of the Barnacle Goose. Physiol. Zool. 71:198–207.
BISHOP, K. D. 1983. Some notes on non-passerine birds of west New Britain. Emu 83:235–241.
BJÖRKLUND, M. 1991. Evolution, phylogeny, sexual dimorphism and mating system in the grackles (*Quiscalus* spp.; Icterinae). Evolution 45:608–621.
BJÖRKLUND, M. 1994. Allometric relations in three species of finches (Aves: Fringillidae). J. Zool. Lond. 233:657–668.
BJÖRKLUND, M. 1996. The importance of evolutionary constraints in ecological time scales. Evol. Ecol. 10:423–431.
BLABER, S. J. M. 1990. A checklist and notes on the current status of the birds of New Georgia, Western Province, Solomon Islands. Emu 90:205–214.
BLACKBURN, T. M., AND H. E. EVANS. 1986. Why are there no viviparous birds? Am. Nat. 128:165–190.
BLACKBURN, T. M., AND K. J. GASTON. 1996. Spatial patterns in the body sizes of bird species in the New World. Oikos 77:436–446.
BLACKMAN, T. M. 1945. War casualties among the birds. Nat. Hist. 54:298–299.
BLAKE, E. R. 1977. Manual of Neotropical Birds. Vol. 1: Spheniscidae (Penguins) to Laridae (Gulls and Allies). University of Chicago Press, Chicago, Illinois.
BLANCKENHORN, W. U. 1998. Adaptive phenotypic plasticity in growth, development, and body size in the yellow dung fly. Evolution 52:1394–1407.
BLANCKENHORN, W. U., R. F. PREZIOSI, AND D. J. FAIRBAIRN. 1995. Time and energy constraints and the evolution of sexual size dimorphism—to eat or to mate? Evol. Ecol. 9:369–381.
BLANDAMER, J. S., AND P. J. K. BURTON. 1979. Anatomical specimens of birds in the collections of the British Museum (Natural History). Bull. Br. Mus. Nat. Hist. (Zool. Ser.) 34:125–180.
BLAZQUEZ, M. C., R. RODRIGUEZ-ESTRELLA, AND M. DELIBES. 1997. Excape behaviour and predation risk of mainland and island spiny-tailed iguanas (*Ctenosaura hemilopha*). Ethology 103:990–998.
BLECHSCHMIDT, H. 1929. Messende Untersuchungen über die Fußanpassungen der Baum- und Laufvögel. Gegenbaurs Morphol. Jahrb. 61:517–547.
BLEDSOE, A. H. 1988. Status and hybridization of Clapper and King Rails in Connecticut. Conn. Warbler 8:61–65.
BLEM, C. R. 1975. Geographic variation in wing-loading of the House Sparrow. Wilson Bull. 87: 543–549.
BLEM, C. R. 1976. Patterns of lipid storage and utilization in birds. Am. Zool. 16:671–684.
BLEM, C. R. 1990. Avian energy storage. Pp. 59–113 in Current Ornithology. Vol. 7 (D. M. Power, Ed.). Plenum Press, New York, New York.
BLEM, C. R. 1999. Energy balance. Pp. 327–341 in Sturkie's Avian Physiology. 5th ed. (G. C. Whittow, Ed.). Academic Press, San Diego, California.

BLOCH, R., AND B. BRUDERER. 1982. The air speed of migrating birds and its relationship to the wind. Behav. Ecol. Sociobiol. 11:19–24.
BLONDEL, J. 1985a. Habitat selection in island versus mainland birds. Pp. 477–516 in Habitat Selection in Birds (M. L. Cody, Ed.). Academic Press, New York, New York.
BLONDEL, J. 1985b. Breeding strategies of the Blue Tit and Coal Tit (Parus) in mainland and island Mediterranean habitats: a comparison. J. Anim. Ecol. 54:531–556.
BLONDEL, J., D. CHESSEL, AND B. FROCHOT. 1988. Bird species impoverishment, niche expansion, and density inflation in Mediterranean island habitats. Ecology 69:1899–1917.
BLONDEL, J., P. PERRET, M.-C. ANSTETT, AND C. THÉBAUD. 2002. Evolution of sexual size dimorphism in birds. test of hypotheses using Blue Tits in contrasted Mediterranean habitats. J. Evol. Biol. 15:440–450.
BLOUIN, M. S. 1992. Comparing bivariate reaction norms among species: time and size at metamorphosis in three species of Hyla (Anura: Hylidae). Oecologia 90:288–293.
BLUEWEISS, L., H. FOX, V. KUDZMA, D. NAKASHIMA, R. PETERS, AND S. SAMS. 1978. Relationships between body size and some life history parameters. Oecologia 37:257–272.
BLUM, J. J. 1977. On the geometry of four-dimensions and the relationship between metabolism and body mass. J. Theor. Biol. 64:599–601.
BOAG, P. T. 1984. Growth and allometry of external morphology in Darwin's finches (Geospiza) on Isla Daphne Major, Galápagos. J. Zool. Lond. 204:413–441.
BOAG, P. T. 1987. Effects of nestling diet on growth and adult size of Zebra Finches (Peophila guttata). Auk 104:155–166.
BOAG, P. T. 1988. The genetics of island birds. Pp. 1550–1563 in Acta XIX Congress Internationalis Ornithologici. Vol. 2 (H. Ouellet, Ed.). University of Ottawa Press, Ottawa, Ontario, Canada.
BOAG, P. T., AND P. R. GRANT. 1984. The classical case of character release: Darwin's finches on Isla Daphne Major, Galápagos. Biol. J. Linn. Soc. 22:243–287.
BOCHENSKI, Z. 1989. Problems of skeletal proportions in fossil bird research. Pp. 445–450 in Trends in Vertebrate Morphology (H. Splechtna and H. Hilgers, Eds.). Gustav Fischer, Stuttgart, Germany.
BOCHENSKI, Z., AND Z. BOCHENSKI, JR. 1993. Correlation between the wing lengths of living birds and measurements of their bones. Belg. J. Zool. 122:123–132.
BOCK, W. J. 1963. The cranial evidence for ratite affinities. Pp. 39–54 in Proceedings XIII International Ornithological Congress (C. G. Sibley, Ed.). American Ornithologists' Union, Washington, D.C.
BOCK, W. J. 1964. Kinetics of the avian skull. J. Morphol. 114:1–42.
BOCK, W. J. 1966. An approach to the functional analysis of bill shape. Auk 83:10–51.
BOCK, W. J. 1970. Microevolutionary sequences as a fundamental consequence in macroevolutionary models. Evolution 24:704–722.
BOCK, W. J. 1976. Recent advances and the future of avian classification. Pp. 176–184 in Proceedings of the 16th International Ornithological Congress (H. J. Frith and J. H. Calaby, Eds.). Australian Academy of Science, Canberra, Australia.
BOCK, W. J. 1979. The synthetic explanation of macroevolutionary change—a reductionist approach. Pp. 20–69 in Models and Methodologies in Evolutionary Theory (J. H. Schwartz and H. B. Rollins, Eds.). Carnegie Mus. Nat. Hist. Bull. No. 13. Carnegie Museum of Natural History, Pittsburgh, Pennsylvania.
BOCK, W. J. 1980. The definition and recognition of biological adaptation. Am. Zool. 20:217–227.
BOCK, W. J. 1990. From biologische anatomie to ecomorphology. Neth. J. Zool. 40:254–277.
BOCK, W. J. 1992. Methodology in avian macrosystematics. Bull. Br. Ornithol. Club (Cent. Suppl.) 112A:53–72.
BOCK, W. J., AND P. BÜHLER. 1988. The evolution and biogeographic al history of the palaeognathous birds. Pp. 31–36 in Current Topics in Avian Biology (R. van den Elzen, K.-L. Schuchmann, and K. Schmidt-Koenig, Eds.). Deutsche Ornithologische-Gesellschaft, Bonn, Germany.
BOCK, W. J., AND J. FARRAND, JR. 1980. The number of species and genera of Recent birds: a contribution to comparative systematics. Am. Mus. Novit. 2703:1–29.
BOCK, W. J., AND C. R. SHEAR. 1972. A staining method for gross dissection of vertebrate muscles. Anat. Anz. 130:222–227.
BOCK, W. J., AND G. VON WAHLERT. 1965. Adaptation and the form–function complex. Evolution 19:269–299.

BOCKHEIM, G., AND S. MEZZELL. 1999. The Blake Crake *Amaurornis flavirostris* an effective cooperative breeder at Disney's Animal Kingdom, Florida, USA. Avic. Mag. 105:12–21.

BOERSMA, P. D. 1975. Adaptations of Galápagos Penguins for life in two environments. Pp. 101–114 *in* The Biology of Penguins (B. Stonehouse, Ed.). University Park Press, Baltimore, Maryland.

BOERSMA, P. D. 1976. An ecological and behavioral study of the Galápagos Penguin. Living Bird 15: 43–93.

BOERSMA, P. D. 1982. Why some birds take so long to hatch. Am. Nat. 120:733–750.

BÖKER, H. 1927. Die biologische Anatomie der Flugarten der Vögel und ihre Phylogenie. J. Ornithol. 75:304–371.

BOLNICK, D. I., R. SVANBÄCK, J. A. FORDYCE, L. H. YANG, J. M. DAVIS, C. D. HULSEY, AND M. L. FORISTER. 2003. The ecology of individuals: incidence and implications of individual specialization. Am. Nat. 161:1–28.

BOND, J. 1980. Birds of the West Indies. 4th ed. Houghton Mifflin, Boston, Massachusetts.

BONNER, J. T. 1952. Morphogenesis: An Essay on Development. Princeton University Press, Princeton, New Jersey.

BONNER, J. T. 1968. Size change in development and evolution. J. Paleontol. 42(Suppl.):1–15.

BONNER, J. T. 1974. On Development: The Biology of Form. Harvard University Press, Cambridge, Massachusetts.

BONNER, J. T. 1988. The Evolution of Complexity. Princeton University Press, Princeton, New Jersey.

BONNER, J. T., AND H. S. HORN. 1982. Selection for size, shape, and developmental timing. Pp. 259–276 *in* Evolution and Development (J. T. Bonner, Ed.). Springer-Verlag, Berlin, Germany.

BONNER, J. T., AND H. S. HORN. 2000. Allometry and natural selection. Pp. 25–35 *in* Scaling in Biology (J. H. Brown and G. B. West, Eds.). Oxford University Press, Oxford, United Kingdom.

BONSER, R. H. C., AND J. M. V. RAYNER. 1996. Measuring leg thrust forces in the Common Starling. J. Exp. Biol. 199:435–439.

BOOKHOUT, T. A. 1995. Yellow Rail (*Coturnicops noveboracensis*). Pp. 1–16 *in* The Birds of North America. No. 139 (A Poole and F. Gill, Eds.). Birds of North America, Inc., Philadelphia, Pennsylvania.

BOOKSTEIN, F. L. 1977. The study of shape transformation after D'Arcy Thompson. Math. Biosci. 34: 177–219.

BOOKSTEIN, F. L. 1978. The Measurement of Biological Shape and Shape Change. Springer-Verlag, Berlin, Germany.

BOOKSTEIN, F. L. 1980. When one form is between two others: an application of biorthogonal analysis. Am. Zool. 20:627–641.

BOOKSTEIN, F. L. 1984. Foundations of morphometrics. Annu. Rev. Ecol. Syst. 13:451–470.

BOOKSTEIN, F. L. 1989. "Size and shape": a comment on semantics. Syst. Zool. 38:173–180.

BOOKSTEIN, F. L. 1991. Morphometric Tools for Landmark Data: Geometry and Biology. Cambridge University Press, Cambridge, United Kingdom.

BOOKSTEIN, F. L., B. CHERNOFF, R. ELDER, J. HUMPHRIES, G. SMITH, AND R. STRAUSS. 1985. Morphometrics in Evolutionary Biology: The Geometry of Size and Shape Change, With Examples from Fishes. Acad. Nat. Sci. Phila. Spec. Publ. 15:1–277.

BOOKSTEIN, F. L., R. E. STRAUSS, J. M. HUMPHRIES, B. CHERNOFF, R. L. ELDER, AND G. R. SMITH. 1982. A comment upon the uses of Fourier methods in systematics. Syst. Zool. 31:76–84.

BORG, I., AND J. LINGOES. 1987. Multidimensional Similarity Structure Analysis. Springer-Verlag, New York, New York.

BOSQUE, C., AND M. T. BOSQUE. 1995. Nest predation as a selective factor in the evolution of developmental rates in altricial birds. Am. Nat. 145:234–260.

BOTKIN, D. B., AND R. S. MILLER. 1974. Mortality rates and survival of birds. Am. Nat. 108:181–192.

BOTTON, M. L., R. E. LOVELAND, AND T. R. JACOBSEN. 1994. Site selection by migratory shorebirds in Delaware Bay, and its relationship to beach characteristics and abundance of horseshoe crabs (*Limulus polyphemus*) eggs. Auk 111:605–616.

BOUBIER, M. 1934. Les oiseaux qui ne volent pas. Bull. Ornithol. Romand 1:77–89.

BOUCOT, A. J. 1975. Evolution and Extinction Rate Controls. Elsevier, Amsterdam, The Netherlands.

BOURNE, W. R. P., AND A. C. F. DAVID. 1981. Nineteenth century bird records from Tristan da Cunha. Bull. Br. Ornithol. Club 101:247–257.

BOURNE, W. R. P., AND A. C. F. DAVID. 1983. Henderson Island, central South Pacific, and its birds. Notornis 30:233–252.
BOURROUILH-LE, J. F. G., AND J. TALADIER. 2000. Sedimentation et fracturation de hautes énergie en milieu recifal: tsunamis, origins et cyclones et leurs éffects sur la sédimentologie et al géomorphologie d'un atoll: motu et hoa, à Rangiroa, Tuamotu Pacifique S E. Mar. Geol. 67:263–333.
BOWLER, P. J. 1976. Fossils and Progress: Paleontology and the Idea of Progressive Evolution in the Nineteenth Century. Science History Publications, New York, New York.
BOWLER, P. J. 1983. The Eclipse of Darwinism: Anti-Darwinian Evolution Theories in the Decades Around 1900. Johns Hopkins University Press, Baltimore, Maryland.
BOWMAN, R. I. 1961. Morphological differentiation and adaptation in the Galápagos finches. Univ. Calif. Publ. Zool. 59:1–302.
BOWMAN, R. I. 1963. Evolutionary patterns in Darwin's finches. Occas. Pap. Calif. Acad. Sci. 44:107–140.
BOWMAN, R. I., AND S. L. BILLEB. 1965. Blood-eating in a Galápagos finch. Living Bird 4:29–44.
BOWMAN, R. I., AND A. CARTER. 1971. Egg-pecking behavior in Galápagos mockingbirds. Living Bird 10:243–270.
BOYCE, M. S. 1984. Restitution of *r*- and *K*-selection as a model of density dependent natural selection. Annu. Rev. Ecol. Syst. 15:427–447.
BOYD, H. J., AND R. ALLEY. 1948. The function of the head-colouration of the nestling Coot and other nestling Rallidae. Ibis 90:582–593.
BRACE, C. L. 1963. Structural reduction in evolution. Am. Nat. 97:39–49.
BRACKENBURY, J. 1984. Physiological responses of birds to flight and running. Biol. Rev. 59:559–575.
BRADBURY, J. W. 1981. The evolution of leks. Pp. 138–169 *in* Natural Selection and Social Behavior (R. D. Alexander and D. W. Tinkle, Eds.). Chiron Press, New York, New York.
BRADBURY, J. W., AND N. B. DAVIES. 1987. Relative roles of intra- and intersexual selection. Pp. 143–163 *in* Sexual Selection: Testing the Alternatives (J. W. Bradbury and M. B. Andersson, Eds.). J. Wiley and Sons, Chichester, United Kingdom.
BRADBURY, J. W., AND R. M. GIBSON. 1983. Leks and mate choice. Pp. 109–138 *in* Mate Choice (P. Bateson, Ed.). Cambridge University Press, Cambridge, United Kingdom.
BRADSHAW, A. D. 1965. Evolutionary significance of phenotypic plasticity in plants. Adv. Genet. 13:115–155.
BRADSHAW, A. D. 1973. Environment and phenotypic plasticity. Brookhaven Symp. Biol. 25:74–94.
BRADSHAW, A. D. 1991. Genostasis and the limits to evolution. Philos. Trans. R. Soc. Lond. (Ser. B) 333:289–305.
BRADSHAW, A. D., AND K. HARDWICK. 1989. Evolution and stress—genotypic and phenotypic components. Biol. J. Linn. Soc. 37:137–155.
BRAITHWAITE, D. H. 1963. Another record of the Black-tailed Waterhen in New Zealand. Notornis 10:233.
BRAKEFIELD, P. M. 1989. The variance in genetic diversity among subpopulations is more sensitive to founder effects and bottlenecks than is the mean: a case study. Pp. 145–161 *in* Evolutionary Biology of Transient Unstable Populations (A. Fontdevila, Ed.). Springer-Verlag, Berlin, Germany.
BRAMLEY, G. N. 1996. A small predator removal experiment to protect North Island Weka (*Gallirallus australis greyi*) and the case for single-subject approaches in determining agents of decline. N. Z. J. Ecol. 20:37–43.
BRAMLEY, G. N. 2001. Dispersal by juvenile North Island Weka (*Gallirallus australis greyi*). Notornis 48:43–46.
BRAMLEY, G. N., AND C. J. VELTMAN. 2000. Call survey method for monitoring endangered North Island Weka (*Gallirallus australis greyi*). Notornis 47:154–159.
BRANDON, R. N. 1978. Adaptation and evolutionary theory. Stud. Hist. Philos. Sci. 9:181–206.
BRAZIL, M. 1985. Notes on the Okinawa Rail *Rallus okinawae*: observations at night and at dawn. Tori 33:125–127.
BRAZIL, M. 1991. The Birds of Japan. Christopher Helm, London, United Kingdom.
BRINCK, P. 1948. Coleoptera from Tristan da Cunha. Results of the Norwegian Scientific Expedition of Tristan da Cunha 1937–1938. No. 17. Komm. Hos. Jacob. Dybwad, Oslo, Norway.

BRITTON, P. L. 1970. Some non-passerine bird weights from East Africa. Bull. Br. Ornithol. Club 90: 142–144.
BROCK, J. P. 2000. The Evolution of Adaptive Systems. Academic Press, San Diego, California.
BRODIE, E. D., III. 2000. Why evolutionary genetics does not always add up. Pp. 3–19 in Epistasis and the Evolutionary Process (J. B. Wolf, E. D. Brodie III, and M. J. Wade, Eds.). Oxford University Press, Oxford, United Kingdom.
BRODKORB, P. 1963. Catalogue of fossil birds: part 1 (Archaeopterygiformes Through Ardeiformes). Bull. Fla. State Mus. (Biol. Sci.) 7:179–293.
BRODKORB, P. 1964. Catalogue of fossil birds: part 2 (Anseriformes through Galliformes). Bull. Fla. State Mus. (Biol. Sci.) 8:195–335.
BRODKORB, P. 1967. Catalogue of fossil birds: part 3 (Ralliformes, Ichthyornithiformes, Charadriiformes). Bull. Fla. State Mus. (Biol. Sci.) 11:99–220.
BRODKORB, P. 1971. Catalogue of fossil birds: part 4 (Columbiformes through Piciformes). Bull. Fla. State Mus. (Biol. Sci.) 15:163–266.
BRODY, S. 1945. Bioenergetics and Growth. Reinhold, Baltimore, Maryland.
BROEKHUYSEN, G. J., AND W. MACNAE. 1949. Observations on the birds of Tristan da Cunha Islands and Gough Islands in February and early March. Ardea 37:97–122.
BROM, T. G. 1986. Microscopic identification of feathers and feather fragments of Palearctic birds. Bijdr. Dierkd. 56:181–204.
BROM, T. G., AND T. G. PRINS. 1989. Microscopic investigation of feather remains from the head of the Oxford Dodo, *Raphus cucullatus*. J. Zool. Lond. 218:233–246.
BROMAN, I. 1941. Über die Entstehung und Bedeutung der Embryonaldunen. Gegenbaurs Morphol. Jahrb. 86:141–217.
BRONMARK, C., AND J. G. MINER. 1992. Predator-induced phenotypical change in body morphology in crucian carp. Science 258:1348–1350.
BROOK, B. W., L. LIM, R. HARDEN, AND R. FRANKHAM. 1997. How secure is the Lord Howe Island Woodhen? A population viability analysis using VORTEX. Pac. Conserv. Biol. 3:125–133.
BROOKE, R. K., AND T. W. WALLETT. 1976. Shark predation on seabirds in Natal waters. Ostrich 47: 126.
BROOKS, D. R., D. D. CUMMING, AND P. H. LEBLOND. 1988. Dollo's law and the second law of thermodynamics: analogy or extension? Pp. 189–224 in Entropy, Information, and Evolution (B. H. Weber, D. J. Depew, and J. D. Smith, Eds.). Massachusetts Institute of Technology Press, Cambridge, Massachusetts.
BROOKS, D. R., AND D. A. MCLENNAN. 1991. Phylogeny, Ecology, and Behavior: A Research Program in Comparative Biology. University of Chicago Press, Chicago, Illinois.
BROOKS, D. R., AND E. O. WILEY. 1988. Evolution as Entropy. 2nd ed. University of Chicago Press, Chicago, Illinois.
BROOM, R. 1907. On the early development of the appendicular skeleton of the Ostrich, with remarks on the origin of birds. Trans. S. Afr. Philos. Soc. 16:355–368.
BROTHERS, N. P., AND I. J. SKIRA. 1984. The Weka on Macquarie Island. Notornis 31:145–154.
BROWER, J. C., AND J. VEINUS. 1978. Multivariate analysis of allometry using point coordinates. J. Paleontol. 52:1037–1053.
BROWER, J. C., AND J. VEINUS. 1981. Allometry in pterosaurs. Paleontol. Contrib. Univ. Kans. 105: 1–32.
BROWN, D. A., K. S. W. CAMPBELL, AND K. A. W. CROOK. 1968. The Geological Evolution of Australia and New Zealand. Pergamon Press, Oxford, United Kingdom.
BROWN, E. S. 1951. The relation between migration rate and type of habitat in aquatic insects, with special reference to certain species of Corixidae. Proc. Zool. Soc. Lond. 121:382–394.
BROWN, J. H. 1989. Patterns, modes and extents of invasions by vertebrates. Pp. 85–109 in Biological Invasions: A Global Perspective (J. A. Drake, H. A. Mooney, F. di Castri, R. H. Groves, F. J. Kruger, M. Rejmanek, and M. Williamson, Eds.). J. Wiley and Sons, Chichester, United Kingdom.
BROWN, J. H., W. A. CALDER III, AND A. KODRIC-BROWN. 1978. Correlates and consequences of body size in nectar-feeding birds. Am. Zool. 18:687–700.
BROWN, J. H., AND A. C. GIBSON. 1983. Biogeography. C. V. Mosby, St. Louis, Missouri.
BROWN, J. H., AND A. KODRIC-BROWN. 1977. Turnover rates in insular biogeography: effect of immigration on extinction. Ecology 58:445–449.

Brown, J. H., P. A. Marquet, and M. L. Taper. 1993. Evolution of body size: consequences of an energetic definition of fitness. Am. Nat. 142:573–584.
Brown, J. H., and B. A. Maurer. 1986. Body size, ecological dominance and Cope's rule. Nature 324:248–250.
Brown, J. H., and G. B. West, Eds. 2000. Scaling in Biology. Oxford University Press, Oxford, United Kingdom.
Brown, J. H., G. B. West, and B. J. Enquist. 2000. Scaling in biology: patterns and processes, causes and consequences. Pp. 1–24 in Scaling in Biology (J. H. Brown and G. B. West, Eds.). Oxford University Press, Oxford, United Kingdom.
Brown, J. L. 1964. The evolution of diversity in avian territorial systems. Wilson Bull. 76:160–169.
Brown, J. L. 1969. Territorial behavior and population regulation in birds. Wilson Bull. 81:293–329.
Brown, J. L. 1978. Avian communal breeding systems. Annu. Rev. Ecol. Syst. 9:123–155.
Brown, J. L. 1987. Helping and Communal Breeding in Birds: Ecology and Evolution. Princeton University Press, Princeton, New Jersey.
Brown, J. S., and N. B. Pavlovic. 1992. Evolution in heterogeneous environments: effects of migration on habitat specialization. Evol. Ecol. 6:360–382.
Brown, L. H., and D. Amadon. 1968. Eagles, Hawks and Falcons of the World. 2 vols. McGraw-Hill Book Co., New York, New York.
Brown, L. H., E. K. Urban, and K. Newman, Eds. 1982. The Birds of Africa. Vol. 1. Academic Press, London, United Kingdom.
Brown, M. B., and A. B. Forsythe. 1974. Robust tests for the equality of variances. J. Am. Stat. Assoc. 69:364–367.
Brown, R. E., J. J. Baumel, and R. D. Klemm. 1994. Anatomy of the propatagium: the Great Horned Owl (*Bubo virginianus*). J. Morphol. 219:205–224.
Brown, R. E., J. J. Baumel, and R. D. Klemm. 1995. Mechanics of the avian propatagium: flexion–extension mechanism of the avian wing. J. Morphol. 225:91–106.
Brown, R. E., and D. K. Saunders. 1998. Regulated changes in body mass and muscle mass in molting Blue-winged Teal for an early return to flight. Can. J. Zool. 76:26–32.
Brown, R. H. J. 1948. The flight of birds: the flapping cycle of the pigeon. J. Exp. Biol. 25:322–333.
Brown, R. H. J. 1951. Flapping flight. Ibis 93:333–359.
Brown, R. H. J. 1953. The flight of birds. II. Wing function in relation to flight speed. J. Exp. Biol. 30:90–103.
Brown, R. H. J. 1961. Flight. Pp. 289–305 in Biology and Comparative Physiology of Birds. Vol. 2 (A. J. Marshall, Ed.). Academic Press, New York, New York.
Brown, W. L., Jr., and E. O. Wilson. 1956. Character displacement. Syst. Zool. 5:49–64.
Bruce, R. C. 1979. Evolution of paedomorphosis in salamanders in the genus *Gyrinophilus*. Evolution 33:998–1000.
Bruderer, B., and A. Boldt. 2001. Flight characteristics of birds: I. Radar measurements of speeds. Ibis 143:178–204.
Brühl, C. A. 1997. Flightless insects: a test case for historical relationships of African mountains. J. Biogeogr. 24:233–250.
Bruinzeel, L. W., T. Piersma, and M. Kersten. 2000. Low costs of terrestrial locomotion in waders. Ardea 87:199–205.
Bruner, P. L. 1972. The Birds of French Polynesia. Bernice P. Bishop Museum, Honolulu, Hawaii.
Bruton, M. N. 1989. The ecological significance of alternative life-history styles. Pp. 503–553 in Alternative Life-History Styles of Animals (M. N. Bruton, Ed.). Kluwer Academic Publications, Dordrecht, The Netherlands.
Bryan, D. C. 1996. Family Aramidae (Limpkin). Pp. 90–95 in Handbook of the Birds of the World. Vol. 3: Hoatzin to Auks (J. del Hoyo, A. Elliott and J. Sargatal, Eds.). Lynx Edicions, Barcelona, Spain.
Bryan, E. H., Jr., and J. C. Greenway, Jr. 1944. Contribution to the ornithology of the Hawaiian Islands. Bull. Mus. Comp. Zool. 94:79–142.
Bryan, W. A. 1915. Natural History of Hawaii. Hawaiian Gazette, Honolulu, Hawaii.
Bryant, D. M. 1988. Energy expenditure and body mass changes as measures of reproductive costs in birds. Funct. Ecol. 2:23–34.
Bryant, D. M. 1991. Constraints on energy expenditure by birds. Pp. 1989–2001 in Acta XX Con-

gressus Internationalis Ornithologici. Vol. 4 (B. D. Bell, R. O. Cossee, J. E. C. Flux, B. D. Heather, R. A. Hitchmough, C. J. R. Robertson, and M. J. Williams, Eds.). Ornithological Congress Trust Board, Christchurch, New Zealand.

BRYANT, D. M. 1997. Energy expenditure in wild birds. Proc. Nutr. Soc. 56:1025–1039.

BRYANT, D. M., AND P. TATNER. 1988. The costs of brood provisioning: effects of brood size and food supply. Pp. 364–379 in Acta XIX Congressus Internationalis Ornithologici. Vol. 1 (H. Ouellet, Ed.). University of Ottawa Press, Ottawa, Ontario, Canada.

BRYANT, E. A. 2001. Tsunami: The Underrated Hazard. Cambridge University Press, Cambridge, United Kingdom.

BRYANT, E. A., R. W. YOUNG, D. M. PRICE, D. J. WHEELER, AND M. I. PEASE. 1997. The impact of tsunami on the coastline of Jervis Bay, southeastern Australia. Phys. Geogr. 18:440–459.

BRYANT, E. H. 1986. On use of logarithms to accommodate scale. Syst. Zool. 35:552–559.

BRYANT, E. H. 1989. Multivariate morphometrics of bottlenecked populations. Pp. 19–31 in Evolutionary Biology of Transient Unstable Populations (A. Fontdevila, Ed.). Springer-Verlag, Berlin, Germany.

BUCHANAN, K. L., AND M. R. EVANS. 2000. The effect of tail streamer length on aerodynamic performance in the Barn Swallow. Behav. Ecol. 11:228–238.

BUFFON, G. L. L. 1770–1786. Historie Naturelle des Oiseaux. Imprimerie Royale, Paris, France.

BULL, J. J. 1987. Evolution of phenotypic variance. Evolution 41:303–315.

BULL, J. J., AND E. L. CHARNOV. 1985. On irreversible evolution. Evolution 39:1149–1155.

BULL, P. C., AND A. H. WHITAKER. 1975. The amphibians, reptiles, birds and mammals. Pp. 231–276 in Biogeography and Ecology in New Zealand (G. Kuschel, Ed.). Dr. W. Junk, The Hague, The Netherlands.

BULLER, W. L. 1868. Notes on Herr Finsch's review of Mr. Walter Buller's essay on New Zealand ornithology. Trans. N. Z. Inst. 1:49–57.

BULLER, W. L. 1873a. History of the Birds of New Zealand. 1st ed. J. Van Voorst, London, England.

BULLER, W. L. 1873b. Remarks on Captain Hutton's notes on certain species of New Zealand birds. Trans. N. Z. Inst. 6:123–126.

BULLER, W. L. 1873c. Notes on the ornithology of New Zealand. Trans. N. Z. Inst. 6:112–118.

BULLER, W. L. 1874a. [Untitled letter dated 11 November 1873.] Ibis 16:93–94.

BULLER, W. L. 1874b. On the ornithology of New Zealand. Trans. N. Z. Inst. 7:197–211.

BULLER, W. L. 1874c. Note on *Rallus modestus*. Trans. N. Z. Inst. 7:511.

BULLER, W. L. 1875. Remarks on Dr. Finsch's paper on New Zealand ornithology. Trans. N. Z. Inst. 8:194–196.

BULLER, W. L. 1876. On the alleged intercrossing of *Ocydromus earli* and the domestic fowl. Trans. N. Z. Inst. 9:341–342.

BULLER, W. L. 1877. On the species forming the genus *Ocydromus*, a peculiar group of brevi-pennate rails. Trans. N. Z. Inst. 10:213–216.

BULLER, W. L. 1878. Additions to list of species, and notices of rare occurrences, since the publication of "The Birds of New Zealand." Trans. N. Z. Inst. 11:361–366.

BULLER, W. L. 1881. On the *Notornis*. Trans. N. Z. Inst. 14:238–244.

BULLER, W. L. 1882. Manual of the Birds of New Zealand. Colonial Museum and Geological Survey Department, Wellington, New Zealand.

BULLER, W. L. 1888. History of the Birds of New Zealand. 2nd ed. Vols. 1, 2. Privately Published, London, England.

BULLER, W. L. 1891. Notes and observations on New Zealand birds. Trans. N. Z. Inst. 24:64–74.

BULLER, W. L. 1892a. Notes on New Zealand birds. Trans. N. Z. Inst. 25:53–80.

BULLER, W. L. 1892b. Note on the flightless rail of the Chatham Islands (*Cabalus modestus*). Trans. N. Z. Inst. 25:52–53.

BULLER, W. L. 1894. Illustrations of Darwinism; or, the avifauna of New Zealand considered in relation to the fundamental law of descent with modification. Trans. N. Z. Inst. 27:75–104.

BULLER, W. L. 1896. Notes on the ornithology of New Zealand. Trans. N. Z. Inst. 29:179–207.

BULLER, W. L. 1898. On the ornithology of New Zealand. Trans. N. Z. Inst. 31:1–37.

BULLER, W. L. 1905. Supplement to the Birds of New Zealand. Vol. 1. Privately Published, London, England.

BULMER, M. G. 1974. Density-dependent selection and character displacement. Am. Nat. 108:45–58.

BUMPUS, H. C. 1899. The elimination of the unfit as illustrated by the introduced sparrow, *Passer domesticus*. Mar. Biol. Lab. Lect. 1899:209–226.
BUNIN, J. S. 1995. Preliminary observations of behavioural interactions between Takahe (*Porphyrio mantelli*) and Pukeko (*P. porphyrio*) on Mana Island. Notornis 42:140–143.
BUNIN, J. S., AND I. G. JAMIESON. 1995. New approaches toward a better understanding of the decline of Takahe (*Porphyrio mantelli*) in New Zealand. Conserv. Biol. 9:100–106.
BUNIN, J. S., AND I. G. JAMIESON. 1996a. A cross-fostering experiment between the endangered Takahe (*Porphyrio mantelli*) and its closest relative, the Pukeko (*P. porphyrio*). N. Z. J. Ecol. 20:207–213.
BUNIN, J. S., AND I. G. JAMIESON. 1996b. Responses to a model predator of New Zealand's endangered Takahe and its closest relative, the Pukeko. Conserv. Biol. 10:1463–1466.
BUNIN, J. S., I. G. JAMIESON, AND D. EASON. 1997. Low reproductive success of the endangered Takahe *Porphyrio mantelli* on offshore island refuges in New Zealand. Ibis 139:144–151.
BURGER, J. 1973. Competition between American Coots and Franklin's Gulls for nest sites and egg predation by the coots. Wilson Bull. 85:449–451.
BURGER, J., AND J. R. TROUT. 1979. Additional data on body size as a difference related to niche. Condor 81:305–307.
BURLEY, N. 1980. Clutch overlap and clutch size: alternative and complementary reproductive tactics. Am. Nat. 115:223–246.
BURLEY, N. 1981. The evolution of sexual indistinguishability. Pp. 121–137 *in* Natural Selection and Social Behavior: Recent Research and New Theory (R. D. Alexander and D. W. Tinkle, Eds.). Chiron Press, New York, New York.
BURLEY, N. 1985. The organization of behavior and the evolution of sexually selected traits. Pp. 22–44 *in* Avian Monogamy (P. A. Gowarty and D. W. Mock, Eds.). Ornithol. Monogr. 37. American Ornithologists' Union, Washington, D.C.
BURNABY, P. 1966. Growth-invariant discriminant functions and generalized distances. Biometrics 22:96–110.
BURNEY, D. A., H. F. JAMES, F. V. GRADY, J.-G. RAFAMANTANANTSOA, RAMILISONINIA, H. T. WRIGHT, AND J. B. COWART. 1997. Environmental change, extinction and human activity: evidence from caves in NW Madagascar. J. Biogeogr. 24:755–767.
BURTON, P. J. K. 1974. Anatomy of head an neck in the Huia (*Heteralocha acutirostris*) with comparative notes on other Callaeidae. Bull. Br. Mus. (Nat. Hist.), Zool. 27:1–48.
BUSH, G. L. 1975. Modes of animal speciation. Annu. Rev. Ecol. Syst. 6:311–364.
BUSH, G. L. 1993. A reaffirmation of Santa Rosalia, or why are there so many kinds of *small* animals? Pp. 229–249 *in* Evolutionary Patterns and Processes (D. R. Lees and D. Edwards, Eds.). Linn. Soc. Symp. No. 14. Academic Press, London, United Kingdom.
BUSS, L. W. 1987. The Evolution of Individuality. Princeton University Press, Princeton, New Jersey.
BUTCHER, G. S., AND S. ROHWER. 1989. The evolution of conspicuous and distinctive coloration for communication in birds. Pp. 51–108 *in* Current Ornithology. vol. 6 (D. M. Power, Ed.). Plenum Press, New York, New York.
BUTENDIECK, E. 1980. Die Benennung des Skeletts beim Truthuhn (*Meleagris gallopavo*) unter Berücksichtigung der Nomina Anatomica Avium (1979). Ph.D. dissertation, Tierärztliche Hochschule Hannover, Hannover, Germany.
BUTENDIECK, E., AND H. WISSDORF. 1982. Beitrag zur Benennung der Knochen des Kopfes beim Truthuhn (*Meleagris gallopavo*) unter Berücksichtigung der Nomina Anatomica Avium (1979). Zool. Jahrb. Anat. 107:153–184.
BUTLER, P. J. 1991. Exercise in birds. J. Exp. Biol. 160:233–262.
BUTLER, P. J., AND C. M. BISHOP. 1999. Flight. Pp. 391–435 *in* Sturkie's Avian Physiology. 5th ed. (G. C. Whittow, Ed.). Academic Press, San Diego, California.
BUTLER, P. J., AND A. J. WOAKES. 1990. The physiology of bird flight. Pp. 300–318 *in* Bird Migration: Physiology and Ecophysiology (E. Gwinner, Ed.). Springer-Verlag, Berlin, Germany.
BUTLER, R. W., T. D. WILLIAMS, N. WARNOCK, AND M. A. BISHOP. 1997. Wind assistance: a requirement for migration of shorebirds? Auk 114:456–466.
BYERLY, T. C. 1932. Growth of the chick embryo in relation to its food supply. J. Exp. Biol. 9:15–44.
BYRD, G. V., R. A. COLEMAN, R. J. SHALLENBERGER, AND C. S. ARUME. 1985. Notes on the breeding biology of the Hawaiian race of the American Coot. Elepaio 45:57–63.

BYRD, G. V., AND G. F. ZEILLEMAKER. 1981. Ecology of nesting Hawaiian Common Gallinules at Hanalei, Hawaii. West. Birds 12:105–116.

CABANA, G., A. FREWIN, R. H. PETERS, AND L. RANDALL. 1982. The effect of sexual size dimorphism on variations in reproductive effort of birds and mammals. Am. Nat. 120:17–25.

CAIN, A. J., AND I. C. J. GALBRAITH. 1956. Field notes on birds of the eastern Solomon Islands. Ibis 98:100–134.

CAIN, B. W. 1970. Growth and plumage development of the Black-bellied Tree Duck, *Dendrocygna autumnalis* (Linneaus). Taius 3:25–48.

CALDER, W. A., III. 1968. Respiratory and heart rates of birds at rest. Condor 70:358–365.

CALDER, W. A., III. 1974. Consequences of body size for avian energetics. Pp. 86–144 in Avian Energetics (R. A. Paynter, Ed.). Publ. No. 15. Nuttal Ornithologists' Club, Cambridge, Massachusetts.

CALDER, W. A., III. 1978. The kiwi: a case of compensating divergences from allometric predictions. Pp. 239–242 in Respiratory Function in Birds, Adult and Embryonic (J. Piiper, Ed.). Springer-Verlag, Berlin, Germany.

CALDER, W. A., III. 1979. The kiwi and egg design: evolution as a package deal. BioScience 29:461–467.

CALDER, W. A., III. 1981. Scaling of physiological processes in homeothermic animals. Annu. Rev. Physiol. 43:301–322.

CALDER, W. A., III. 1982a. The relationship of the Gompertz constant and maximum potential lifespan to body mass. Exp. Gerontol. 17:383–385.

CALDER, W. A., III. 1982b. The pace of growth: an allometric approach to comparative embryonic and post-embryonic growth. J. Zool. Lond. 198:215–225.

CALDER, W. A., III. 1983a. Ecological scaling: mammals and birds. Annu. Rev. Ecol. Syst. 14:213–230.

CALDER, W. A., III. 1983b. Body size, mortality, and longevity. J. Theor. Biol. 102:135–144.

CALDER, W. A., III. 1984. Size, Function, and Life History. Harvard University Press, Cambridge, Massachusetts.

CALDER, W. A., III. 1991. The kiwi and its egg. Pp. 155–171 in Kiwis (E. Fuller, Ed.). Swan Hill Press, Shrewsburg, United Kingdom.

CALDER, W. A., III. 2000. Diversity and convergence: scaling for conservation. Pp. 297–323 in Scaling in Biology (J. H. Brown and G. B. West, Eds.). Oxford University Press, Oxford, United Kingdom.

CALDER, W. A., III, AND T. J. DAWSON. 1978. Resting metabolic rates of ratite birds: the kiwis and the Emu. Comp. Biochem. Physiol. (Ser. A) 60:479–481.

CALDER, W. A., III, AND J. R. KING. 1972. Body weight and the energetics of temperature regulation: a re-examination. J. Exp. Biol. 56:775–780.

CALDER, W. A., III, AND J. R. KING. 1974. Thermal and caloric relations of birds. Pp. 259–413 in Avian Biology. Vol. 4 (D. S. Farner and J. R. King, Eds.). Academic Press, New York, New York.

CALDER, W. A., III, AND B. ROWE. 1977. Body mass changes and energetics of the kiwi's egg cycle. Notornis 24:129–135.

CALDWELL, M. W., AND M. S. Y. LEE. 1997. A snake with legs from the marine Cretaceous of the Middle East. Nature 386:705–709.

CALHOUN, J. B. 1947. The role of temperature and natural selection in relation to the variations in the size of the English Sparrow in the United States. Am. Nat. 81:203–228.

CALMAESTRA, R. G., AND E. MORENO. 2001. A phylogenetically-based analysis on the relationship between wing morphology and migratory behaviour in passeriforms. Ardea 89:407–416.

CALOW, P. 1977. Ecology, evolution and energetics: a study in metabolic adaptation. Pp. 1–62 in Advances in Ecological Research. Vol. 10 (A. MacFadyen, Ed.). Academic Press, London, United Kingdom.

CALOW, P. 1979. The cost of reproduction—a physiological approach. Biol. Rev. 54:23–40.

CAMPBELL, K. E., JR., AND L. MARCUS. 1992. The relationship of hindlimb bone dimensions to body weight in birds. Pp. 395–412 in Papers in Avian Paleontology Honoring Pierce Brodkorb (K. E. Campbell, Jr., Ed.). Natural History Museum of Los Angeles County, Los Angeles, California.

CAMPBELL, K. E., JR., AND E. P. TONNI. 1983. Size and locomotion in teratorns (Aves: Teratornithidae). Auk 100:390–403.

CAMPBELL, N. A., AND W. R. ATCHLEY. 1981. The geometry of canonical variate analysis. Syst. Zool. 30:268–280.

CAMPBELL, N. A., AND D. A. SAUNDERS. 1976. Morphological variation in the White-tailed Black Cockatoo, *Calyptorhynchus baudinii*, in western Australia: 'a multivariate approach.' Aust. J. Zool. 24:589–595.

CANE, W. P. 1993. The ontogeny of postcranial integration in the Common Tern, *Sterna hirundo*. Evolution 47:1138–1151.

CAPDEVILA, J., AND J. C. IZPISÚA BELMONTE. 2000. Perspectives on the evolutionary origin of tetrapod limbs. Pp. 531–558 *in* The Character Concept in Evolutionary Biology (G. P. Wagner, Ed.). Academic Press, San Diego, California.

CAPUTO, V., B. LANZA, AND R. PALMIERI. 1995. Body elongation and limb reduction in the genus *Chalcides* Laurenti 1768 (Squamata: Scincidae): a comparative study. Trop. Zool. 8:95–152.

CAREY, C. 1983. Structure and function of avian eggs. Pp. 69–103 *in* Current Ornithology. Vol. 1 (R. F. Johnston, Ed.). Plenum Press, New York, New York.

CAREY, C., H. RAHN, AND P. PARISI. 1980. Calories, water, lipid and yolk in avian eggs. Condor 82: 335–343.

CARLQUIST, S. 1965. Island Life. Natural History Press, Garden City, New York.

CARLQUIST, S. 1966. The biota of long-distance dispersal. I. Principles of dispersal and evolution. Q. Rev. Biol. 41:247–270.

CARLQUIST, S. 1974. Island Biology. Columbia University Press, New York, New York.

CARLQUIST, S. 1980. Hawaii: A Natural History. Natural History Press, Garden City, New York.

CARMI, N., B. PINSHOW, P. PORTER, AND J. JAEGER. 1992. Water and energy limitations on flight duration in small migrating birds. Auk 109:268–276.

CARMICHAEL, D. 1818. Some account of the island of Tristan da Cunha and of its natural productions. Trans. Linn. Soc. Lond. 12:483–513.

CARNUTT, J., AND S. L. PIMM. 2001. How many bird species in Hawai'i and the central Pacific before first contact? Pp. 15–30 *in* Evolution, Ecology, Conservation, and Management of Hawaiian Birds: A Vanishing Avifauna (J. M. Scott, S. Conant, and C. van Riper III, Eds.). Stud. Avian Biol. No. 22. Cooper Ornithological Society, Los Angeles, California.

CARPENTER, R. E., AND M. A. STAFFORD. 1970. The secretory rates and the chemical stimulus for secretion of the nasal salt glands in the Rallidae. Condor 72:316–324.

CARR, G. S. 1980. Historic and prehistoric avian records from Easter Island. Pac. Sci. 34:19–20.

CARRASCAL, L. M., E. MORENO, AND A. VALIDA. 1994. Morphological evolution and changes in foraging behaviour of island and mainland populations of Blue Tit (*Parus caeruleus*)—a test of convergence and ecomorphological hypotheses. Evol. Ecol. 8:25–35.

CARRASCAL, L. M., J. L. TELLERÍA, AND A. VALIDO. 1992. Habitat distribution of Canary Chaffinches among islands: Competitive exclusion or species-specific habitat preferences? J. Biogeogr. 19: 383–390.

CARRIER, D., AND J. AURIEMMA. 1992. A developmental constraint on the fledging time of birds. Biol. J. Linn. Soc. 47:61–77.

CARRIER, D., AND L. R. LEON. 1990. Skeletal growth and function in the California Gull (*Larus californicus*). J. Zool. Lond. 222:375–389.

CARROLL, A. L. K. 1963a. Food habits of the North Island Weka. Notornis 10:281–284, 289–300.

CARROLL, A. L. K. 1963b. Breeding cycle of the North Island Weka. Notornis 10:300–302.

CARROLL, A. L. K. 1963c. Sexing of Wekas. Notornis 10:302–303.

CARROLL, A. L. K. 1966. Food habits of Pukeko (*Porphyrio melanotus*, Temminck). Notornis 13: 133–144.

CARROLL, A. L. K. 1969. The Pukeko (*Porphyrio melanotus*) in New Zealand. Notornis 16:101–120.

CARROLL, R. L. 1997. Patterns and Processes of Vertebrate Evolution. Cambridge University Press, Cambridge, United Kingdom.

CARSON, H. L., AND D. A. CLAGUE. 1995. Geology and biogeography of the Hawaiian Islands. Pp. 14–29 *in* Hawaiian Biogeography: Evolution in a Hotspot Archipelago (W. Wagner and V. Funk, Eds.). Smithsonian Institution Press, Washington, D.C.

CARSON, H. L., AND A. R. TEMPLETON. 1984. Genetic revolutions in relation to speciation phenomena: the founding of new populations. Annu. Rev. Ecol. Syst. 15:97–131.

CASE, T. J. 1978a. Optimal body size and an animal's diet. Acta Biotheor. 28:54–69.

CASE, T. J. 1978b. On the evolution and adaptive significance of postnatal growth rates in the terrestrial vertebrates. Q. Rev. Biol. 53:243–282.

CASE, T. J. 1978c. A general explanation for insular body size trends in terrestrial vertebrates. Ecology 59:1–18.

CASE, T. J. 1983. The reptiles: ecology. Pp. 159–209 in Island Biogeography in the Sea of Cortez (T. J. Case and M. L. Cody, Eds.). University California Press, Los Angeles, California.

CASE, T. J., J. FAABORG, AND R. SIDELL. 1983. The role of body size in the assembly of West Indian bird communities. Evolution 37:1062–1074.

CASSELS, R. 1984. The role of prehistoric man in the faunal extinctions of New Zealand and other Pacific islands. Pp. 741–767 in Quaternary Extinctions: A Prehistoric Revolution (P. S. Martin and R. G. Klein, Eds.). University of Arizona Press, Tucson, Arizona.

CASTRO, G. 1968. The role of competition in the evolution of bird migration. Evolution 22:180–192.

CASTRO, G. 1989. Energy costs and avian distributions: limitations or chance?—Comment. Ecology 70:1181–1182.

CASTRO, G., AND J. P. MYERS. 1988. A statistical method to estimate the cost of flight in birds. J. Field Ornithol. 59:369–380.

CASTRO, G., AND J. P. MYERS. 1993. Shorebird predation on eggs of horseshoe crabs during spring stopover on Delaware Bay. Auk. 110:927–930.

CASWELL, H. 1983. Phenotypic plasticity in life-history traits: demographic effects and evolutionary consequences. Am. Zool. 23:35–46.

CATELL, R. B. 1978. The Scientific Use of Factor Analysis in Behavioral and Life Sciences. Plenum Press, New York, New York.

CATRY, P., R. A. PHILLIPS, AND R. W. FURNESS. 1999. Evolution of reversed sexual dimorphism in skuas and jaegers. Auk 116:158–168.

CAUGHLEY, G. 1988. The colonisation of New Zealand by the Polynesians. J. R. Soc. N. Z. 18:245–270.

CAVE, A. J., J. VISSER, AND A. C. PERDECK. 1989. Size and quality of the Coot *Fulica atra* territory in relation to age of its tenants and neighbours. Ardea 77:87–98.

CHAI, P., J. S. C. CHEN, AND R. DUDLEY. 1997. Transient hovering performance of hummingbirds under conditions of maximal loading. J. Exp. Biol. 200:921–929.

CHAI, P., AND R. DUDLEY. 1999. Maximum flight performance of hummingbirds: capacities, constraints, and trade-offs. Am. Nat. 153:398–411.

CHAI, P., AND D. MILLARD. 1997. Flight and size constraints: hovering performance of large hummingbirds under maximal loading. J. Exp. Biol. 200:2757–2763.

CHAMBERS, R. 1844. Vestiges of the Natural History of Creation. 10th ed. John Churchill, London, England.

CHAN, K. 2001. Partial migration in Australian landbirds: a review. Emu 101:281–292.

CHANDLER, A. C. 1916. A study of the structure of feathers, with reference to their taxonomic significance. Univ. Calif. Publ. Zool. 13:243–446.

CHAPIN, J. P. 1939. The birds of the Belgian Congo. Part 2. Bull. Am. Mus. Nat. Hist. 75:1–632.

CHARLESWORTH, B. 1997. Is founder effect speciation defensible? Am. Nat. 149:600–603.

CHARLESWORTH, B., AND R. LANDE. 1982. Morphological stasis and developmental constraint: no problem for neo-Darwinism. Nature 296:610.

CHARLESWORTH, B., R. LANDE, AND M. SLATKIN. 1982. A neo-Darwinian commentary on macroevolution. Evolution 36:474–498.

CHARLESWORTH, B., AND D. B. SMITH. 1982. A computer model of speciation by founder effects. Genet. Res. 39:227–236.

CHARNOV, E. L. 1993. Life History Invariants: Some Explanations of Symmetry in Evolutionary Ecology. Oxford University Press, Oxford, United Kingdom.

CHARNOV, E. L., AND J. R. KREBS. 1974. On clutch-size and fitness. Ibis 116:217–219.

CHATTERJEE, S., A. HADI, AND B. PRICE. 1999. Regression Analysis by Example. 3rd ed. J. Wiley and Sons, New York, New York.

CHEKE, A. S. 1987. An ecological history of the Mascarene Islands, with particular reference to extinctions and introductions of land vertebrates. Pp. 5–89 in Studies of Mascarene Islands Birds (A. W. Diamond, Ed.). Cambridge University Press, Cambridge, United Kingdom.

CHEREL, Y., J.-B. CHARASSIN, AND Y. HANDRICH. 1993. Comparison of body reserve buildup in prefasting chicks and adults of King Penguins (*Aptenodytes patagonicus*). Physiol. Zool. 66:750–770.

CHEREL, T., J.-P. ROBIN, A. HEITZ, C. CLAGARI, AND Y. LEMAHO. 1992. Relationships between lipid availability and protein utilization during prolonged fasting in birds. J. Comp. Physiol. (Ser. B) 162:305–313.

CHEREL, T., J.-P. ROBIN, AND Y. LEMAHO. 1988. Physiology and biochemistry of long-term fasting in birds. Can. J. Zool. 66:159–166.

CHEVERUD, J. M. 1982. Relationships among ontogenetic, static, and evolutionary allometry. Am. J. Phys. Anthropol. 59:139–149.

CHEVERUD, J. M. 1984. Quantitative genetics and developmental constraints on evolution by selection. J. Theor. Biol. 110:155–171.

CHEVERUD, J. M. 2000. Detecting epistasis among quantitative trait loci. Pp. 58–81 *in* Epistasis and the Evolutionary Process (J. B. Wolf, E. D. Brodie III, and M. J. Wade, Eds.). Oxford University Press, Oxford, United Kingdom.

CHEVERUD, J. M., J. J. RUTLEDGE, AND W. R. ATCHLEY. 1983. Quantitative genetics of development: genetic correlations among age-specific trait values and the evolution of ontogeny. Evolution 37:895–905.

CHIAPPE, L. M. 1995a. The first 85 million years of avian evolution. Nature 378:349–355.

CHIAPPE, L. M. 1995b. The phylogenetic position of the Cretaceous birds of Argentina: Enantiornithes and *Patagopteryx deferrariisi*. Cour. Forschungsinst. Senckenb. 181:55–63.

CHIAPPE, L. M. 1996. Late Cretaceous birds of southern South America: anatomy and systematics of Enantiornithes and *Patagopteryx deferrariisi*. Muench. Geowiss. Abh. (Ser. A) 30:203–244.

CHIAPPE, L. M., M. A. NORELL, AND J. M. CLARK. 1998. The skull of a relative of the stem-group bird *Mononykus*. Nature 392:275–278.

CHOI, I.-H., AND R. E. RICKLEFS. 1997. Changes in protein and electrolyte concentrations in the pectoral and leg muscles during avian development. Auk 114:688–694.

CHOI, I. H., R. E. RICKLEFS, AND R. E. SHEA. 1993. Skeletal muscle growth, enzyme activities, and the development of thermogenesis: a comparison between altricial and precocial birds. Physiol. Zool. 66:455–473.

CHOWN, S. L., K. J. GASTON, AND P. H. WILLIAMS. 1998a. Global patterns in species richness of pelagic seabirds: the Procellariiformes. Ecography 21:342–350.

CHOWN, S. L., N. M. J. GREMMEN, AND K. J. GASTON. 1998b. Ecological biogeography of southern ocean islands: species–area relationships, human impacts, and conservation. Am. Nat. 152:562–575.

CHRISTIANSEN, F. B. 1974. Sufficient conditions for protected polymorphism in a subdivided population. Am. Nat. 108:257–166.

CHRISTIANSEN, F. B. 1975. Hard and soft selection in a subdivided population. Am. Nat. 109:11–19.

CHRISTIANSEN, K. A. 1961. Convergence and parallelism in cave Entomobryinae. Evolution 15:288–301.

CHRISTIANSEN, K. A. 1985. Regressive evolution in Collembola. Natl. Speleol. Soc. Bull. 47:89–100.

CHRISTIANSEN, K. A. 1992. Cave life in the light of modern evolutionary theory. Pp. 453–478 *in* The Natural History of Biospeleogy (A. Camacho, Ed.). Museo Nacional de Ciencias Naturales, Madrid, Spain.

CHRISTIANSEN, P. 1999. Scaling of the limb bones to body mass in terrestrial mammals. J. Morphol. 239:167–190.

CHRISTIE, D. M., R. A. DUNCAN, A. R. MCBIRNEY, M. A. RICHARDS, W. M. WHITE, K. S. HARPP, AND C. G. FOX. 1992. Drowned islands downstream from the Galápagos hotspot imply extended speciation times. Nature 355:246–248.

CHU, P. C. 1994. Historical examination of delayed plumage maturation in the shorebirds (Aves: Charadriiformes). Evolution 48:327–350.

CLANCEY, P. A. 1981. On birds from Gough Island, central South Atlantic. Durban Mus. Novit. 12:187–200.

CLANCEY, P. A. 1985. The Rare Birds of Southern Africa. Winchester Press, Johannesburg, South Africa.

CLAPPERTON, B. K. 1987. Individual recognition by voice in the Pukeko, *Porphyrio porphyrio melanotus* (Aves: Rallidae). N. Z. J. Zool. 14:11–18.

CLAPPERTON, B. K., AND P. F. JENKINS. 1984. Vocal repertoire of the Pukeko (Aves: Rallidae). N. Z. J. Zool. 11:71–84.

CLAPPERTON, B. K., AND P. F. JENKINS. 1987. Individuality in contact calls of the Pukeko (Aves: Rallidae). N. Z. J. Zool. 14:19–28.

CLARK, G. A., JR. 1979. Body weights of birds: a review. Condor 81:193–202.

CLARK, G. A., JR. 1993a. Anatomia topographica externa. Pp. 7–13 in Handbook of Avian Anatomy: Nomina Anatomica Avium. 2nd ed. (J. J. Baumel, A. S. King, J. E. Breazile, H. E. Evans, and J. C. Vanden Berge, Eds.). Publ. No. 23. Nuttall Ornithologists' Club, Cambridge, Massachusetts.

CLARK, G. A., JR. 1993b. Integumentum commune. Pp. 17–44 in Handbook of Avian Anatomy: Nomina Anatomica Avium. 2nd ed. (J. J. Baumel, A. S. King, J. E. Breazile, H. E. Evans, and J. C. Vanden Berge, Eds.). Publ. No. 23. Nuttall Ornithologists' Club, Cambridge, Massachusetts.

CLARK, J., AND R. M. ALEXANDER. 1975. Mechanics of running by quail (*Coturnix*). J. Zool. Lond. 176:87–113.

CLARK, K. E., AND L. J. NILES. 1993. Abundance and distribution of migrant shorebirds on Delaware Bay. Condor 95:694–705.

CLARK, R. J. 1971. Wing-loading—a plea for consistency in usage. Auk 88:927–928.

CLARKE, W. E. 1905. Ornithological results of the Scottish National Antarctic Expedition.—I. On the birds of Gough Island, South Atlantic Ocean. Ibis 47:247–268.

CLEVELAND, W. S. 1985. The Elements of Graphing Data. Wadsworth Advanced Books, Monterey, California.

CLINCHY, M., D. T. HAYDON, AND A. T. SMITH. 2002. Pattern does not equal process: What does patch occupancy really tell us about metapopulation dynamics? Am. Nat. 159:351–362.

CLOUT, M. N., AND J. L. CRAIG. 1995. The conservation of critically endangered flightless birds in New Zealand. Ibis 137(Suppl.):181–190.

CLOUT, M. N., AND J. R. HAY. 1989. The importance of birds as browsers, pollinators and seed dispersers in New Zealand forests. N. Z. J. Ecol. 12(Suppl.):27–33.

CLUTTON-BROCK, T. H. 1985. Size, sexual dimorphism, and polygyny in Primates. Pp. 51–60 in Size and Scaling in Primate Biology (W. L. Jungers, Ed.). Plenum Press, New York, New York.

CLUTTON-BROCK, T. H. 1991. The Evolution of Parental Care. Princeton University Press, Princeton, New Jersey.

CLUTTON-BROCK, T. H., AND P. H. HARVEY. 1979. Comparison and adaptation. Proc. R. Soc. Lond. (Ser. B Biol. Sci.) 205:547–565.

CLUTTON-BROCK, T. H., AND P. H. HARVEY. 1984. Comparative approaches to investigating adaptation. Pp. 7–29 in Behavioural Ecology: An Evolutionary Approach. 2nd ed. (J. R. Krebs and N. B. Davies, Eds.). Blackwell Scientific Publications, Oxford, United Kingdom.

COATES, B. J. 1985. The Birds of Papua New Guinea. Vol. 1. Non-passerines. Dove Publications, Alderley, Australia.

COCHRAN, W. G., AND G. M. COX. 1957. Experimental Designs. 2nd ed. Wiley, New York, New York.

COCK, A. G. 1966. Genetical aspects of metrical growth and form in animals. Q. Rev. Biol. 41:131–190.

COCKBURN, A. 1998. Evolution of helping behavior in cooperatively breeding birds. Annu. Rev. Ecol. Syst. 29:141–178.

CODDINGTON, J. A. 1988. Cladistic tests of adaptational hypotheses. Cladistics 4:1–22.

CODDINGTON, J. A. 1994. The roles of homology and convergence in studies of adaptation. Pp. 53–78 in Phylogenetics and Ecology (P. Eggleton and R. I. Vane-Wright, Eds.). Linn. Soc. Symp. 17. Academic Press, London, United Kingdom.

CODY, M. L. 1966. A general theory of clutch size. Evolution 20:174–184.

CODY, M. L. 1969. Convergent characteristics in sympatric species: a possible relation to interspecific competition and aggression. Condor 71:222–239.

CODY, M. L. 1973. Character convergence. Annu. Rev. Ecol. Syst. 4:189–211.

CODY, M. L. 1974. Competition and the Structure of Bird Communities. Princeton University Press, Princeton, New Jersey.

CODY, M. L., ED. 1985. Habitat Selection in Birds. Academic Press, Orlando, Florida.

CODY, M. L., AND J. M. DIAMOND, EDS. 1975. Ecology and Evolution of Communities. Belknap Press, Cambridge, Massachusetts.

COELHO, C. N. D., K. M. KRABBENHOFT, W. B. UPHLOT, J. F. FALLON, AND R. A. KOSHER. 1991. Altered expression of the chicken homeobox-containing genes *Ghox-7* and *Ghox-8* in the limb buds of limbless mutant chick embryos. Development 113:1487–1493.

COELHO, C. N. D., W. B. UPHLOT, AND R. A. KOSHER. 1992. Role of the chicken homeobox-containing genes *Ghox-4.6* and *Ghox-8* in the specification of positional identities during the development of normal and polydactylous limb buds. Development 115:629–637.

COHEN, D., AND S. A. LEVINS. 1991. Dispersal in patchy environments: the effects of temporal and spatial structure. Theor. Popul. Biol. 39:63–99.

COHEN, D., AND U. MOTRO. 1989. More on optimal rates of dispersal: taking into account the cost of the dispersal mechanism. Am. Nat. 134:659–663.

COHN, M. J., AND C. TICKLE. 1999. Developmental basis of limblessness and axial patterning in snakes. Nature 399:474–479.

COKINOS, C. 2000. Hope Is the Thing With Feathers: A Personal Chronicle of Vanished Birds. Warner Books, New York, New York.

COLBERT, E. H. 1955. Evolution of the Vertebrates: A History of the Backboned Animals Through Time. J. Wiley and Sons, New York, New York.

COLBERT, J., E. DANCHIN, A. A. DHONDT, AND J. D. NICHOLS, EDS. 2001. Dispersal. Oxford University Press. Oxford, United Kingdom.

COLE, B. J. 1981. Colonizing abilities, island size, and the number of species on archipelagoes. Am. Nat. 117:629–638.

COLE, L. C. 1954. The population consequences of life history phenomena. Q. Rev. Biol. 29:103–107.

COLEMAN, J. D., B. WARBURTON, AND W. Q. GREEN. 1983. Some population statistics and movements of the western Weka. Notornis 30:93–107.

COLLAR, N. J. 1993. The conservation status in 1982 of the Aldabra White-throated Rail *Dryolimnas cuvieri aldabranus*. Bird Conserv. Int. 3:299–305.

COLLAR, N. J., AND P. ANDREW. 1988. Birds to Watch: The ICBP World Check-list of Threatened Birds. Smithsonian Institution Press, Washington, D.C.

COLLAR, N. J., M. J. CROSBY, AND A. J. STATTERSFIELD. 1994. Birds to Watch 2: The World List of Threatened Birds. Birdlife Conserv. Ser. 4. Birdlife International, Cambridge, United Kingdom.

COLLAR, N. J., L. P. GONZAGA, N. KRABBE, A. MADROÑO, L. G. NARANJO, T. A. PARKER, AND D. C. WEGE. 1992. Threatened Birds of the Americas. International Council for Bird Preservation, Cambridge, United Kingdom.

COLLAR, N. J., AND S. N. STUART. 1985. Threatened Birds of Africa and Related Islands. International Council for Bird Preservation, Cambridge, United Kingdom.

COMINS, H. N. 1982. Evolutionarily stable strategies for localized dispersal in two dimensions. J. Theor. Biol. 94:579–606.

COMINS, H. N., W. D. HAMILTON, AND R. M. MAY. 1980. Evolutionary stable dispersal strategies. J. Theor. Biol. 82:205–230.

COMINS, H. N., AND I. R. NOBLE. 1985. Dispersal, variability, and transient niches: species coexistence in a uniformly variable environment. Am. Nat. 126:706–723.

CONANT, S. 1988. Geographic variation in the Laysan Finch (*Telyspiza cantans*). Evol. Ecol. 2:270–282.

CONNELL, J. H. 1980. Diversity and the coevolution of competitors, or the ghost of competition past. Oikos 35:131–138.

CONNOR, E. F., AND D. SIMBERLOFF. 1978. Species number and compositional similarity of the Galápagos flora and avifauna. Ecol. Monogr. 48:219–248.

CONNOR, E. F., AND D. SIMBERLOFF. 1979. The assembly of species communities: Chance or competition? Ecology 60:1132–1140.

CONNOR, E. F., AND D. SIMBERLOFF. 1984. Neutral models of species' co-occurence patterns. Pp. 316–331 *in* Ecological Communities: Conceptual Issues and the Evidence (D. R. Strong, Jr., D. Simberloff, L. G. Abele, and A. B. Thistle, Eds). Princeton University Press, Princeton, New Jersey.

CONOVER, H. B. 1934. A new species of rail from Paraguay. Auk 51:365–366.

CONOVER, M. R., J. G. REESE, AND A. D. BROWN. 2000. Costs and benefits of subadult plumage in Mute Swans: testing hypotheses for the evolution of delayed plumage maturation. Am. Nat. 156:193–200.

CONOVER, W. J. 1980. Practical Nonparametric Statistics. 2nd ed. J. Wiley and Sons, New York, New York.
CONTRERAS, J. R. 1979. Bird weights from northeastern Argentina. Bull. Br. Ornithol. Club 99: 21–24.
CONWAY, C. J., W. R. EDDLEMAN, AND S. H. ANDERSON. 1994. Nesting success and survival of Virginia Rails and Soras. Wilson Bull. 106:466–473.
COOKE, F., AND J. C. DAVIES. 1983. Assortative mating, mate choice and reproductive fitness in Snow Geese. Pp. 279–295 *in* Mate Choice (P. Bateson, Ed.). Cambridge University Press, Cambridge, United Kingdom.
COOKE, F., P. D. TAYLOR, C. M. FRANCIS, AND R. F. ROCKWELL. 1990. Directional selection and clutch size in birds. Am. Nat. 136:261–267.
COOMANS DE RUITER, L. 1947. Over de wederontdekking van *Aramidopsis plateni* (W. Blasius) in de Minahasa (Noord-Celebes) aldaar voorkomen van *Gymnocrex rosenbergii* (Schlegel) aldaar. Limosa 19:65–75.
COOPER, A., J. RYMER, H. F. JAMES, S. L. OLSON, C. E. MCINTOSH, M. D. SORENSON, AND R. C. FLEISCHER. 1996. Ancient DNA and island endemics. Nature 381:484.
COOPER, J. C. B. 1983. Factor analysis: an overview. Am. Stat. 37:141–147.
COOPER, R. A. 1989. New Zealand tectonostratigraphic terranes and panbiogeography. N. Z. J. Zool. 16:699–712.
COPE, E. D. 1868. On the origin of genera. Proc. Acad. Nat. Sci. Phila. 20:242–300.
COPE, E. D. 1872. On the Wyandotte Cave and its fauna. Am. Nat. 6:406–422.
COPE, E. D. 1885. On the evolution of the Vertebrata, progressive and retrogressive, 3 parts. Am. Nat. 19:140–148, 234–247, 341–353.
COPE, E. D. 1896. Primary Factors of Organic Evolution. Univ. Chicago Press, Chicago, Illinois.
CORBIN, K. W., B. C. LIVEZEY, AND P. S. HUMPHREY. 1988. Genetic differentiation among steamer-ducks (Anatidae: *Tachyeres*): an electrophoretic analysis. Condor 90:773–781.
CORDONNIER, P. 1985. Hand-rearing and growth of the Purple Gallinule *Porphyrio p. porphyrio* at the park of Villars-Les-Dombes, France. Avic. Mag. 89:205–210.
CORRUCCINI, R. S. 1972. Allometry correction in taximetrics. Syst. Zool. 21:375–383.
CORRUCCINI, R. S. 1977. Correlation properties of morphometric ratios. Syst. Zool. 26:211–214.
CORRUCCINI, R. S. 1987. Shape in morphometrics: comparative analyses. Am. J. Phys. Anthropol. 73: 289–303.
CORRUCCINI, R. S. 1988. Morphometric replicability using chords and cartesian coordinates of the same landmarks. J. Zool. Lond. 215:389–394.
COTGREAVE, P., AND P. H. HARVEY. 1992. Relationships between body size, abundance, and phylogeny in bird communities. Funct. Ecol. 8:219–228.
COTTAM, C., C. S. WILLIAMS, AND C. A. SOOTER. 1942. Flight and running speeds of birds. Wilson Bull. 54:121–131.
COTTON, P. A. 1996. Body size and the ecology of hummingbirds. Symp. Zool. Soc. Lond. 69:239–258.
COWLES, G. S. 1987. The fossil record. Pp. 90–100 *in* Studies of Mascarene Island Birds (A. W. Diamond, Ed.). Cambridge University Press, Cambridge, United Kingdom.
COX, G. W. 1985. The evolution of avian migration systems between temperate and tropical regions of the New World. Am. Nat. 126:451–474.
COX, G. W. 1990. Centres of speciation and ecological differentiation in the Galápagos land bird fauna. Evol. Ecol. 4:130–142.
COX, G. W., AND P. D. MOORE. 1993. Biogeography: An Ecological and Evolutionary Approach. 5th ed. Blackwell Scientific Publications, Oxford, United Kingdom.
COX, G. W., AND R. E. RICKLEFS. 1977. Species diversity, ecological release, and community structuring in Caribbean land bird faunas. Oikos 29:60–66.
COYNE, J. A., AND T. D. PRICE. 2000. Little evidence for sympatric speciation in island birds. Evolution 54:2166–2171.
CRACRAFT, J. 1971. The functional morphology of the hind limb of the domestic Pigeon, *Columba livia*. Bull. Am. Mus. Nat. Hist. 144:171–268.
CRACRAFT, J. 1972. The relationships of the higher taxa of birds: problems in phylogenetic reasoning. Condor 74:379–392.

CRACRAFT, J. 1973a. Systematics and evolution of the Gruiformes (class Aves): 3. Phylogeny of the suborder Grues. Bull. Am. Mus. Nat. Hist. 151:1–127.

CRACRAFT, J. 1973b. Continental drift, paleoclimatology, and the evolution and biogeography of birds. J. Zool. Lond. 169:455–545.

CRACRAFT, J. 1974a. Phylogeny and evolution of the ratite birds. Ibis 116:494–521.

CRACRAFT, J. 1974b. Evolution of the rails of the South Atlantic Islands (Aves: Rallidae), by S. L. Olson [review]. Wilson Bull. 86:484–485.

CRACRAFT, J. 1975. Mesozoic dispersal of terrestrial faunas around the southern end of the world. Mem. Mus. Natl. Hist. Nat. (Nouv. Ser. A Zool.) 88:29–54.

CRACRAFT, J. 1976a. Covariation patterns in the postcranial skeleton of moas (Aves, Dinornithidae): a factor analytic study. Paleobiology 2:166–173.

CRACRAFT, J. 1976b. The hindlimb of moas (Aves, Dinornithidae): a multivariate assessment of size and shape. J. Morphol. 150:495–526.

CRACRAFT, J. 1980. Biogeographic patterns of terrestrial vertebrates in the southwest Pacific. Palaeogeogr. Palaeoclimatol. Palaeoecol. 31:353–369.

CRACRAFT, J. 1983. Species concepts and speciation analysis. Pp. 159–187 in Current Ornithology. Vol. 1 (R. F. Johnston, Ed.). Plenum Press, New York, New York.

CRACRAFT, J. 1984. Conceptual and methodological aspects of the study of evolutionary rates, with some comments on bradytely in birds. Pp. 95–104 in Living Fossils (N. Eldredge and S. M. Stanley, Eds.). Springer-Verlag, New York, New York.

CRACRAFT, J. 1985. Biological diversification and its causes. Ann. Mo. Bot. Gard. 72:794–822.

CRACRAFT, J. 1986. The origin and early diversification of birds. Paleobiology 12:383–399.

CRACRAFT, J. 1987. DNA hybridization and avian phylogenetics. Pp. 47–96 in Evolutionary Biology. Vol. 21 (M. K. Hecht, B. Wallace, and G. T. Prance, Eds.). Plenum Press, New York, New York.

CRACRAFT, J. 1988a. Speciation and its ontology: the empirical consequences of alternative species concepts for understanding patterns and process of differentiation. Pp. 28–59 in Speciation and Its Consequences (D. Otte and J. A. Endler, Eds.). Sinauer Assoc., Sunderland, Massachusetts.

CRACRAFT, J. 1988b. The major clades of birds. Pp. 339–361 in The Phylogeny and Classification of the Tetrapods. Vol. 1 (M. J. Benton, Ed.). Clarendon Press, Oxford, United Kingdom.

CRACRAFT, J. 1990. The origin of evolutionary novelties: pattern and process at different hierarchical levels. Pp. 21–44 in Evolutionary Innovations (M. H. Nitecki, Ed.). University of Chicago Press, Chicago, Illinois.

CRACRAFT, J. 1992. Phylogeny and classification of birds: a study of molecular evolution [review of Sibley and Ahlquist 1990]. Mol. Biol. Evol. 9:182–186.

CRACRAFT, J., AND D. P. MINDELL. 1989. The early history of modern birds: a comparison of molecular and morphological evidence. Pp. 389–403 in The Hierarchy of Life (B. Fernholm, K. Bremer, and J. Jornvall, Eds.). Amsterdam: Exerptica Medica.

CRAIG, J. L. 1976. An interterritorial hierarchy: an advantage for a subordinate in a communal territory. Z. Tierpsychol. 42:200–205.

CRAIG, J. L. 1977. The behaviour of the Pukeko, Porphyrio porphyrio melanotus. N. Z. J. Zool. 4: 413–433.

CRAIG, J. L. 1979. Habitat variation in the social organization of a communal gallinule, the Pukeko, Porphyrio porphyrio melanotus. Behav. Ecol. Sociobiol. 5:331–358.

CRAIG, J. L. 1980a. Pair and group breeding behavior of a communal gallinule, the Pukeko, Porphyrio porphyrio melanotus. Anim. Behav. 28:593–603.

CRAIG, J. L. 1980b. Breeding success of a communal gallinule. Behav. Sociobiol. 6:289–295.

CRAIG, J. L. 1982. On the evidence for a "pursuit deterrent" function of alarm signals of swamphens. Am. Nat. 119:753–755.

CRAIG, J. L. 1984. Are communal Pukeko caught in the prisoner's dilemma? Behav. Ecol. Sociobiol. 14:147–150.

CRAIG, J. L. 2001. Takahe: breaking ground in species recovery. Pp. 114–122 in The Takahe: Fifty Years of Conservation Management and Research. University of Otago Press, Dunedin, New Zealand.

CRAIG, J. L., AND I. G. JAMIESON. 1985. The relationship between presumed gamete contribution and parental investment in communal breeder. Behav. Ecol. Sociobiol. 17:207–211.

CRAIG, J. L., AND I. G. JAMIESON. 1988. Incestuous mating in a communal bird: a family affair. Am. Nat. 131:58–70.

CRAIG, J. L., AND I. G. JAMIESON. 1990. Pukeko: different approaches and some different answers. Pp. 387–412 *in* Cooperative Breeding in Birds: Long-Term Studies of Ecology and Behavior (P. B. Stacey and W. D. Koenig, Eds.). Cambridge University Press, Cambridge, United Kingdom.

CRAIG, J. L., B. H. MCARDLE, AND P. D. WETTIN. 1980. Sex determination of the Pukeko or Purple Swamphen. Notornis 27:287–291.

CRAMP, S., AND K. E. L. SIMMONS. 1977. Handbook of the Birds of Europe, the Middle East and North Africa. Vol. 1: Ostrich to Ducks. Oxford University Press, Oxford, United Kingdom.

CRAMP, S., AND K. E. L. SIMMONS. 1980. Handbook of the Birds of Europe, the Middle East and North Africa. Vol. 2: Hawks to Bustards. Oxford University Press, Oxford, United Kingdom.

CRAW, R. C. 1988. Continuing the synthesis between panbiogeography, phylogenetic systematics and geology as illustrated by empirical studies on the biogeography of New Zealand and the Chatham Islands. Syst. Zool. 37:291–310.

CRAW, R. C., J. R. GREHAN, AND M. J. HEADS. 1999. Panbiogeography: Tracking the History of Life. Oxford University Press, Oxford, United Kingdom.

CRAWFORD, E. C., JR., AND R. C. LASIEWSKI. 1968. Oxygen consumption and respiratory evaporation of the Emu and Rhea. Condor 70:333–339.

CRAWFORD, R. D. 1978. Tarsal colour of American Coots in relation to age. Wilson Bull. 90:536–543.

CRAWFORD, R. D. 1980. Effects of age on reproduction in American Coots. J. Wildl. Manage. 44:183–189.

CRAWFORD, R. L., S. L. OLSON, AND W. KINGSLEY TAYLOR. 1983. Winter distribution of subspecies of Clapper Rails (*Rallus longirostris*) in Florida with evidence for long-distance and overland movements. Auk 100:198–200.

CRISTOL, D. A., M. B. BAKER, AND C. CARBONE. 1999. Differential migration revisited: latitudinal segregation by age and sex class. Pp. 33–88 *in* Current Ornithology. Vol. 15 (V. Nolan, Jr., Ed.). Plenum Press, New York, New York.

CRITCHLEY, F. 1985. Influence in principal components analysis. Biometrika 72:627–636.

CRNOKRAK, P., AND D. A. ROFF. 2002. Trade-offs to flight capacity in *Gryllus firmus*: the influence of whole organism respiration rate on fitness. J. Evol. Biol. 15:388–398.

CROW, J. F., AND M. KIMURA. 1970. An Introduction to Population Genetics Theory. Harper and Row, New York, New York.

CROWELL, K. L. 1962. Reduced interspecific competition among the birds of Bermuda. Ecology 43:75–88.

CUBO, J., AND W. ARTHUR. 2001. Patterns of correlated character evolution in flightless birds: a phylogenetic approach. Evol. Biol. 14:693–702.

CUBO, J., AND A. CASINOS. 1997. Flightlessness and long bone allometry in Palaeognathiformes and Sphenisciformes. Neth. J. Zool. 47:209–226.

CULVER, D. C. 1970. Analysis of simple cave communities. I. Caves as islands. Evolution 24:463–474.

CULVER, D. C. 1982. Cave Life: Evolution and Ecology. Harvard University Press, Cambridge, Massachusetts.

CULVER, D. C., AND J. R. HOLSINGER. 1992. How many species of troglobites are there? Nat. Speleol. Soc. Bull. 54:79–80.

CULVER, D. C., T. C. KANE, AND D. W. FONG. 1995. Adaptation and Natural Selection in Caves: The Evolution of *Gammarus minus*. Harvard University Press, Cambridge, Massachusetts.

CUMBERLAND, K. B. 1965. Man's role in modifying island environments in the southwest Pacific: with special reference to New Zealand. Pp. 187–205 *in* Man's Place in the Island Ecosystem (F. R. Fosberg, Ed.). Bishop Museum Press, Honolulu, Hawaii.

CUNNINGHAM, J. M. 1959. A colony of buff-colored Pukeko. Notornis 6:83–84.

CUNNINGHAM, R. O. 1871. On some points in the anatomy of the steamer-duck (*Micropterus cinereus*). Trans. Zool. Soc. Lond. 7:493–501.

CUPPY, W. 1951. How to Become Extinct. Garden City Books, New York, New York.

CURIO, E. 1969. Funktionsweise und stammesgeschichte des Flugfeinderkennens einiger Darwinfinken (*Geospizinae*). Z. Tierpsychol. 26:394–487.

CURIO, E. 1976. The Ethology of Predation. Springer-Verlag, Berlin, Germany.
CURREY, J. D. 1984. The Mechanical Adaptations of Bones. Princeton University Press, Princeton, New Jersey.
CURREY, J. D., AND R. MCN. ALEXANDER. 1985. The thickness of the walls of tubular bones. J. Zool. Lond. (Ser. A) 206:453–468.
CUVIER, G. 1802. [Lectures On Comparative Anatomy. Vol. 1: On the Organs of Motion.] Translated by J. MacCartney. T. N. Longmans and O. Rees, London, England.
DAAN, S., D. MASMAN, AND A. GROENEWOLD. 1990. Avian basal metabolic rates: their association with body composition and energy expenditure in nature. Am. J. Physiol. 259:333–340.
DABELOW, A. 1925. Die Schwimmanpassung der Vögel. Ein Beitrag zur biologischen Anatomie der Fortbewegung. Gegenbaurs Morphol. Jahrb. 54:288–321.
DAMUTH, J. 1993. Cope's rule, the island rule and the scaling of mammalian population density. Nature 365:748–750.
DAMUTH, J., AND B. J. MACFADDEN. 1990. Introduction: body size and its estimation. Pp. 1–10 in Body Size in Mammalian Paleobiology: Estimation and Biological Implications (J. Damuth and B. J. MacFaddem, Eds.). Cambridge University Press, Cambridge, United Kingdom.
DANIEL, C., AND F. S. WOOD. 1980. Fitting Equations to Data: Computer Analysis of Multifactor Data. 2nd ed. J. Wiley and Sons, New York, New York.
DARLINGTON, P. J., JR. 1943. Carabidae of mountains and islands: data on the evolution of isolated faunas, and on atrophy of wings. Ecol. Monogr. 13:36–61.
DARLINGTON, P. J., JR. 1957. Zoogeography: The Geographical Distribution of Animals. J. Wiley, New York, New York.
DARLINGTON, P. J., JR. 1965. Biogeography of the Southern End of the World. Harvard University Press, Cambridge, Massachusetts.
DARLINGTON, P. J., JR. 1970. Carabidae on tropical islands, especially the West Indies. Biotropica 2: 7–15.
DARROCH, J. N., AND J. E. MOSIMANN. 1985. Canonical and principal components of shape. Biometrika 72:241–252.
DARWIN, C. 1830. Journal of Researches into the Geology and Natural History of the Various Countries Visited by H. M. S. 'Beagle'. H. Colburn, London, England.
DARWIN, C. 1859. On the Origin of Species by Means of Natural Selection, or the Preservation of Favoured Races in the Struggle for Life. John Murray, London, England.
DARWIN, C. 1868. The Variation of Animals and Plants Under Domestication. Vol. 1. John Murray, London, England.
DARWIN, C. 1871. The Descent of Man, and Selection in Relation to Sex. John Murray, London, England.
DARWIN, F. 1887. The Life and Letters of Charles Darwin. 2 vols. D. Appleton, New York, New York.
DAVENPORT, L. J. 1979. Shag swallowed by monkfish. Bull. Br. Ornithol. Club 72:77–78.
DAVIES, G. L. 1969. The Earth in Decay, 1578–1878. Elsevier, New York, New York.
DAVIES, N. B. 1982. Ecological questions about territorial behaviour. Pp. 317–350 in Behavioural Ecology: An Evolutionary Approach (J. R. Krebs and N. B. Davies, Eds.). Blackwell Scientific Publications, Oxford, United Kingdom.
DAVIES, N. B. 1991. Mating systems. Pp. 263–299 in Behavioural Ecology. 3rd ed. (J. R. Krebs and N. B. Davies, Eds.). Blackwell Scientific Publications, Oxford, United Kingdom.
DA VINCI, L. 1505. Codice sul Volo degli Uccelli [Codex on the Flight of Birds]. Translated by A. Marinoni (1976). Harcourt Brace Jovanovich, New York, New York.
DAVIS, A. W. 1980. On the effects of moderate multivariate nonnormality on Wilks's likelihood ratio criterion. Biometrika 67:419–427.
DAVISON, G. W. H. 1985. Avian spurs. J. Zool. Lond. (Ser. A) 206:353–366.
DAVISON, M. L. 1983. Multidimensional Scaling. J. Wiley and Sons, New York, New York.
DAWKINS, R. 1982. The Extended Phenotype. W. H. Freeman, Oxford, United Kingdom.
DAWKINS, R., AND J. R. KREBS. 1979. Arms races between and within species. Proc. R. Soc. Lond. (Ser. B Biol. Sci.) 205:489–511.
DAWSON, A., F. J. MCNAUGHTON, A. R. GOLDSMITH, AND A. A. DEGEN. 1994. Ratite-like neoteny induced by neonatal thyroidectomy of European Starlings, *Sturnus vulgaris*. J. Zool. Lond. 232: 633–639.

DAWSON, A. G., AND S. SHI. 2000. Tsunami deposits. Pure Appl. Geophys. 157:875–897.
DAWSON, E. W. 1958. Re-discoveries of the New Zealand sub-fossil birds named by H. O. Forbes. Ibis 100:232–237.
DAWSON, R., AND R. CLARK. 1996. Effects of variation in egg size and hatching date on survival of Lesser Scaup (*Aythya affinis*) ducklings. Ibis 138:693–699.
DAWSON, W. R., AND J. W. HUDSON. 1970. Birds. Pp. 223–310 *in* Comparative Physiology of Thermoregulation. Vol. 1: Invertebrates and Nonmammalian Vertebrates (G. C. Whittow, Ed.). Academic Press, New York, New York.
DAWSON, W. R., AND R. L. MARSH. 1989. Metabolic acclimatization to cold and season in birds. Pp. 83–94 *in* Physiology of Cold Adaptation in Birds (C. Bech and R. E. Reinertsen, Eds.). Plenum Press, New York, New York.
DAWSON, W. R., R. L. MARSH, AND M. E. YACOE. 1983. Metabolic adjustments of small passerine birds for migration and cold. Am. J. Physiol. 245:755–767.
DAY, D. 1989. Vanished Species. 2nd ed. Gallery Books, New York, New York.
DAY, T., AND L. ROWE. 2002. Developmental thresholds and the evolution of reaction norms for age and size at life-history transitions. Am. Nat. 159:338–350.
DE BEER, G. 1937. The Development of the Vertebrate Skull. Clarendon Press, Oxford, United Kingdom.
DE BEER, G. 1940. Embryos and Ancestors. Oxford University Press, Oxford, United Kingdom.
DE BEER, G. 1951. Embryos and Ancestors. Rev. ed. Clarendon Press, Oxford, United Kingdom.
DE BEER, G. 1964. Phylogeny of the ratites. Pp. 681–685 *in* New Dictionary of Birds (A. Landsborough Thomson, Ed.). Nelson, London, United Kingdom.
DE BEER, G. 1975. The Evolution of Flying and Flightless Birds. Oxford University Press, Oxford, United Kingdom.
DEERENBERG, C., V. APANIUS, S. DAAN, AND N. BOS. 1997. Reproductive effort decreases antibody responsiveness. Proc. R. Soc. Lond. (Ser. B Biol. Sci.) 264:1021–1029.
DE HAAN, G. A. L. 1950. Notes on the invisible flightless rail of Halmahera (*Habroptila wallacii* Gray). Amsterdam Nat. 1:57–60.
DE JONG, G. 1989. Phenotypically plastic characters in isolated populations. Pp. 3–18 *in* Evolutionary Biology of Transient Unstable Populations (A. Fontdevila, Ed.). Springer-Verlag, Berlin, Germany.
DE JONG, G. 1990. Quantitative genetics of reaction norms. J. Evol. Biol. 3:447–468.
DE JONG, G. 1994. The fitness of fitness concepts and the description of natural selection. Q. Rev. Biol. 69:3–29.
DE JONG, G. 1995. Phenotypic plasticity as a product of selection in a variable environment. Am. Nat. 145:493–512.
DE JONG, G. 1999. Unpredictable selection in a structured population leads to local genetic differentiation in evolved reaction norms. J. Evol. Biol. 12:839–851.
DELACOUR, J. 1956. Waterfowl of the World. Vol. 2. Country Life, London, United Kingdom.
DEN BOER, P. J. 1968. Spreading of risk and stabilization of animal numbers. Acta Biotheor. 18:165–194.
DEN BOER, P. J. 1981. On the survival of populations in a heterogeneous and variable environment. Oecologia 50:39–53.
DENEBA SOFTWARE. 1998. Canvas, Version 6.0. Deneba Software, Miami, Florida.
DENNETT, D. C. 1995. Darwin's Dangerous Idea: Evolution and the Meanings of Life. Simon and Schuster, New York, New York.
DENNIS, B., P. L. MUNHOLLAND, AND J. M. SCOTT. 1991. Estimation of growth and extinction parameters for endangered species. Ecol. Monogr. 61:115–143.
DENNISON, M., AND A. J. BAKER. 1991. Morphometric variability in continental and Atlantic island population of Chaffinches (*Fringilla coelebs*). Evolution 45:29–39.
DENNO, R. F. 1994. The evolution of dispersal polymorphism in insects: the influence of habitat, host plants and mates. Res. Popul. Ecol. 36:127–135.
DENNO, R. F., K. L. OLMSTEAD, AND E. S. MCCLOUD. 1989. Reproductive cost of flight capacity: a comparison of life history traits in wing dimorphic planthoppers. Ecol. Entomol. 14:31–44.
DENNO, R. F., G. K. RODERICK, K. L. OLMSTEAD, AND H. G. DOBEL. 1991. Density-related migration in planthoppers (Homoptera, Delphacidae)—the role of habitat persistance. Am. Nat. 138:1513–1541.

DENNO, R. F., G. K. RODERICK, M. A. PETERSON, A. F. HUBERTY, H. G. DÖDEL, AND H. G. EUBANKS. 1996. Habitat persistence shapes the interspecific dispersal strategies of planthoppers. Ecol. Monogr. 66:389–408.

DENNY, M. W. 1990. Terrestrial *versus* aquatic biology: the medium and its message. Am. Zool. 30: 111–121.

DENVER, R. J. 1997. Proximate mechanisms of phenotypic plasticity in amphibian metamorphosis. Am. Zool. 37:172–184.

DE QUEIROZ, K., AND D. A. GOOD. 1997. Phenetic clustering in biology: a critique. Q. Rev. Biol. 72: 3–30.

DESALLE, R. 1995. Molecular approaches to biogeographic analysis of Hawaiian Drosophilidae. Pp. 72–89 *in* Hawaiian Biogeography: Evolution on a Hot Spot Archipelago (W. L. Wagner and W. A. Funk, Eds.). Smithsonian Institution Press, Washington, D.C.

DESMOND, A. 1982. Archetypes and Ancestors: Palaeontology in Victorian London 1850–1875. University of Chicago Press, Chicago, Illinois.

DEWAR, R. E. 1984. Extinctions in Madagascar: the loss of the subfossil fauna. Pp. 574–593 *in* Quaternary Extinctions: A Prehistoric Revolution (P. S. Martin and R. G. Klein, Eds.). University of Arizona Press, Tucson, Arizona.

DEWITT, T. J. 1998. Costs and limits of phenotypic plasticity: tests with predator-induced morphology and life history in a freshwater snail. J. Evol. Biol. 11:465–480.

DIAL, K. P. 1992. Avian forelimb muscles and nonsteady flight: Can birds fly without using the muscles in their wings? Auk 109:874–885.

DIAL, K. P., AND A. A. BIEWENER. 1993. Pectoralis muscle force and power output during different modes of flight in Pigeons (*Columba livia*). J. Exp. Biol. 176:31–54.

DIAL, K. P., A. A. BIEWENER, B. W. TOBALSKE, AND D. R. WARRICK. 1997. Mechanical power output of bird flight. Nature 390:67–70.

DIAL, K. P., G. E. GOSLOW, JR., AND F. A. JENKINS, JR. 1991. The functional anatomy of the shoulder in the European Starling (*Sturnus vulgaris*). J. Morphol. 207:327–344.

DIAL, K. P., S. R. KAPLAN, G. E. GOSLOW, JR., AND F. A. JENKINS, JR. 1987. Structure and neural control of the pectoralis in Pigeons: implications for flight mechanics. Anat. Rec. 218:529–536.

DIAL, K. P., AND J. M. MARZLUFF. 1988. Are the smallest organisms the most diverse? Ecology 69: 1620–1624.

DIAMOND, J. M. 1969. Preliminary results of an ornithological exploration of the North Coastal Range, New Guinea. Am. Mus. Novit. 2362:1–57.

DIAMOND, J. M. 1970a. Ecological consequences of island colonization by southwest Pacific birds, I. Types of niche shifts. Proc. Natl. Acad. Sci. USA 67:529–536.

DIAMOND, J. M. 1970b. Ecological consequences of island colonization by southwest Pacific birds, II. The effect of species diversity on total population density. Proc. Natl. Acad. Sci. USA 67: 1715–1721.

DIAMOND, J. M. 1972. Avifauna of the Eastern Highlands of New Guinea. Publ. 12. Nuttall Ornithologists' Club, Cambridge, Massachusetts.

DIAMOND, J. M. 1974. Colonization of exploded volcanic islands by birds: the supertramp strategy. Science 184:803–806.

DIAMOND, J. M. 1976. Dynamic Island Biogeography. Princeton University Press, Princeton, New Jersey.

DIAMOND, J. M. 1977a. Continental and insular speciation in Pacific land birds. Syst. Zool. 26:263–268.

DIAMOND, J. M. 1977b. Colonization cycles in man and beast. World Archaeol. 8:249–261.

DIAMOND, J. M. 1981. Flightlessness and fear of flying in island species. Nature 293:507–508.

DIAMOND, J. M. 1984a. Distributions of New Zealand birds on real and virtual islands. N. Z. J. Ecol. 7:37–55.

DIAMOND, J. M. 1984b. Historic extinctions: a rosetta stone for understanding prehistoric extinctions. Pp. 824–862 *in* Quaternary Extinctions: A Prehistoric Revolution (P. S. Martin and R. G. Klein, Eds.). University of Arizona Press, Tucson, Arizona.

DIAMOND, J. M. 1984c. "Normal" extinctions of isolated populations. Pp. 191–246 *in* Extinctions (M. H. Nitecki, Ed.). University of Chicago Press, Chicago, Illinois.

DIAMOND, J. M. 1985. Population processes in island birds: immigration, extinction and fluctuations.

Pp. 17–21 *in* Conservation of Island Birds (P. J. Moors, Ed.). ICBP Publ. No. 3. International Council Bird for Preservation, Cambridge, United Kingdom.

DIAMOND, J. M. 1987. Extant unless proven extinct? Or, extinct unless proven extant? Conserv. Biol. 1:77–79.

DIAMOND, J. M. 1989a. The present, past and future of human-caused extinctions. Philos. Trans. R. Soc. Lond. (Ser. B) 325:469–477.

DIAMOND, J. M. 1989b. Nine hundred kiwis and a dog. Nature 338:544.

DIAMOND, J. M. 1991. A new species of rail from the Solomon Islands and convergent evolution of insular flightlessness. Auk 108:461–470.

DIAMOND, J. M., AND M. E. GILPIN. 1982. Examination of the "null" model of Connor and Simberloff for species co-occurrences on islands. Oecologia 52:64–74.

DIAMOND, J. M., S. L. PIMM, M. E. GILPIN, AND M. LECROY. 1989. Rapid evolution of character displacement in myzomelid honeyeaters. Am. Nat. 134:675–708.

DIAMOND, J. M., AND C. R. VEITCH. 1981. Extinctions and introductions in the New Zealand avifauna: Cause and effect? Science 211:499–511.

DICKERMAN, R. W. 1968a. Notes on the Red Rail (*Laterallus ruber*). Wilson Bull. 80:94–99.

DICKERMAN, R. W. 1968b. Notes on the Ocellated Rail (*Micropygia schomburgkii*) with first record from Central America. Bull. Br. Ornithol. Club 88:25–30.

DICKERMAN, R. W., AND F. HAVERSCHMIDT. 1971. Further notes on the juvenal plumage of the Spotted Rail (*Rallus maculatus*). Wilson Bull. 83:444–446.

DICKERMAN, R. W., AND K. C. PARKES. 1969. Juvenal plumage of the Spotted Rail. Wilson Bull. 81:207–209.

DICKERMAN, R. W., AND D. W. WARNER. 1961. Distribution records from Tecolutla, Veracruz, with the first record of *Porzana flaviventer* for Mexico. Wilson Bull. 73:336–340.

DICKINSON, E. C. 1984. Notes on Philippine birds, I. The status of *Porzana paykullii* in the Philippines. Bull. Br. Ornithol. Club 104:71–72.

DICKISON, D. J. 1962. Flight of the Musk Duck (*Biziura lobata*). Aust. Bird Watcher 1:233–234.

DIEFFENBACH, E. 1843. Travels in New Zealand. Vol. 2. J. Murray, London, England.

DIEHL, S. R., AND G. L. BUSH. 1989. The role of habitat preference in adaptation and speciation. Pp. 345–365 *in* Speciation and Its Consequences (D. Otte and J. A. Endler, Eds.). Sinauer, Sunderland, Massachusetts.

DIETZ, M. W., AND R. E. RICKLEFS. 1997. Growth rate and maturation of skeletal muscles over a size range of galliform birds. Physiol. Zool. 70:502–510.

DILL, H. R., AND W. A. BRYAN. 1912. Report of an expedition to Laysan Island in 1911. U.S. Dep. Agric. (Biol. Surv.) Bull. 42:1–30.

DINGLE, H. 1996. Migration: The Biology of Life on the Move. Oxford University Press, Oxford, United Kingdom.

DINGLE, H., N. R. BLAKELY, AND E. R. MILLER. 1980. Variation in body size and flight performance in milkweed bugs (*Oncopeltus fasciatus*). Evolution 42:79–92.

DINGLE, H., AND K. E. EVANS. 1987. Responses in flight to selection on wing length in non-migratory milkweed bugs, *Oncopeltus fasciatus*. Entomol. Exp. Appl. 45:289–296.

DINIZ-FILHO, J. A. F. 2001. Phylogenetic autocorrelation under distinct evolutionary processes. Evolution 55:1104–1109.

DISNEY, H. J. DE S., AND C. N. SMITHERS. 1972. The distribution of terrestrial and freshwater birds on Lord Howe Island, in comparison with Norfolk Island. Aust. J. Zool. 17:1–11.

DIXON, W. J., CHIEF ED. 1992. BMDP Statistical Software Manual. 3 vols. University of California Press, Berkeley, California.

DJAWDAN, M., M. R. ROSE, AND T. J. BRADLEY. 1997. Does selection for stress resistance lower metabolic rate? Ecology 78:828–837.

DOBZHANSKY, T. 1965. Biological evolution in island populations. Pp. 65–74 *in* Man's Place in the Island Ecosystem (F. R. Fosberg, Ed.). Bishop Museum Press, Honolulu, Hawaii.

DOBZHANSKY, T. 1970. Genetics of the Evolutionary Process. Columbia University Press, New York, New York.

DOBZHANSKY, T. 1977. Evolution. Freeman, San Francisco, California.

DOD, A. S. 1980. First records of Spotted Rail *Pardirallus maculatus* on the island of Hispaniola. Auk 97:407.

DODSON, E. O. 1960. Evolution: Process and Product. Reinhold, New York, New York.

DODSON, S. 1989. Predator-induced reaction norms. BioScience 39:447–452.
DOHERTY, P. F., JR., J. B. WILLIAMS, AND T. C. GRUBB, JR. 2001. Field metabolism and water flux of Carolina Chickadees during breeding and nonbreeding seasons: a test of the "peak-demand" and "reallocation" hypotheses. Condor 103:370–375.
DOLE, S. B. 1869. A synopsis of the birds hitherto described from the Hawaiian Islands. Proc. Boston Soc. Nat. Hist. 12:294–309.
DOLE, S. B. 1879. Birds of the Hawaiian Islands. 2nd ed. T. G. Thrum, Honolulu, Hawaii.
DONHOFFER, SZ. 1986. Body size and metabolic rate: exponent and coefficient of the allometric equation: the role of units. J. Theor. Biol. 119:125–137.
DONOGHUE, M. J., AND D. D. ACKERLY. 1996. Phylogenetic uncertainties and sensitivity analyses in comparative biology. Philos. Trans. R. Soc. Lond. (Ser. B) 351:1241–1249.
DONOGHUE, M. J., J. A. DOYLE, J. GAUTHIER, A. G. KLUGE, AND T. ROWE. 1989. The importance of fossils in phylogeny reconstruction. Annu. Rev. Ecol. Syst. 20:431–460.
DONOVAN, S. K., ED. 1989. Mass Extinctions: Processes and Evidence. Columbia University Press, New York, New York.
DORST, J. 1962. The Migrations of Birds. Houghton Mifflin, Boston, Massachusetts.
DORST, J. 1974. The Life of Birds. 2 vols. Columbia University Press, New York, New York.
DORWARD, D. F. 1962. Comparative biology of the White Booby and the Brown Booby *Sula* spp. at Ascension. Ibis 103b:174–234.
DOVE, C. J. 2000. A descriptive and phylogenetic analysis of plumulaceous feather characters in Charadriiformes. Ornithol. Monogr. 51:1–163.
DOWNHOWER, J. F. 1976. Darwin's finches and the evolution of sexual dimorphism in body size. Nature 263:558–563.
DOWSETT, R. J., AND F. DOWSETT-LEMAIRE. 1980. The systematic status of some Zambian birds. Gerfaut 70:151–199.
DRAPER, N. R., AND H. SMITH. 1981. Applied Regression Analysis. 2nd ed. J. Wiley and Sons, New York, New York.
DRENT, R. H. 1975. Incubation. Pp. 333–420 *in* Avian Biology. Vol. 5 (D. S. Farner and J. R. King, Eds.). Academic Press, New York, New York.
DRENT, R. H., AND S. DAAN. 1980. The prudent parent: energetic adjustments in avian breeding. Ardea 68:225–252.
DRENT, R. H., AND B. STONEHOUSE. 1971. Thermoregulatory responses of the Peruvian Penguin, *Spheniscus humboldti*. Comp. Biochem. Physiol. (Ser. A) 40:689–710.
DRIEDZIC, W. R., K. L. CROWE, P. W. HICKLIN, AND D. H. SEPHTON. 1993. Adaptations in pectoralis muscle, heart mass, and energy metabolism during premigratory fattening in Semipalmated Sandpipers (*Calidris pusilla*). Can. J. Zool. 71:1602–1608.
DRIVER, E. A. 1988. Diet and behaviour of young American Coots. Wildfowl 39:34–42.
DROVETSKI, S. V. 1996. Influence of the trailing-edge notch on flight performance of galliforms. Auk 113:802–810.
DRUSCHEL, R. F., AND A. I. CAPLAN. 1991. Three-dimensional reconstruction and cross-sectional anatomy of the thigh musculature of the developing chick embryo (*Gallus gallus*). J. Morphol. 208: 293–309.
DUBOWY, P. J. 1988. Waterfowl communities and seasonal environments: temporal variability in interspecific competition. Ecology 69:1439–1453.
DUDLEY, R., AND G. J. VERMEIJ. 1992. Do the power requirements of flapping flight constrain folivory in flying animals? Funct. Ecol. 6:101–104.
DUERDEN, J. E. 1920. Methods of degeneration in the Ostrich. J. Genet. 9:131–193.
DUGATKIN, L. A. 1997. Cooperation Among Animals: An Evolutionary Perspective. Oxford University Press, Oxford, United Kingdom.
DULLEMEIJER, P. 1974. Concepts and Approaches in Animal Morphology. Van Gorcum, Assen, The Netherlands.
DULLEMEIJER, P. 1980. Functional morphology and evolution. Acta Biotheor. 29:151–250.
DULLEMEIJER, P. 1989. On the concept of integration in animal morphology. Pp. 3–18 *in* Trends in Vertebrate Morphology (H. Splechtna and H. Hilgers, Eds.). Gustav Fischer, Stuttgart, Germany.
DUNHAM, A. E., D. W. TINKLE, AND J. W. GIBBONS. 1978. Body size in island lizards: a cautionary tale. Ecology 59:1230–1238.
DUNNING, J. B., JR. 1993. CRC Handbook of Avian Body Masses. CRC Press, Boca Raton, Florida.

DYKSTRA, C. R., AND W. H. KARASOV. 1993. Nesting energetics of House Wrens (*Troglodytes aedon*) in relation to maximal rates of energy flow. Auk 110:481–491.

EADIE, J. MCA., F. P. KEHOE, AND T. D. NUDDS. 1988. Pre-hatch and post-hatch brood amalgamation in North American Anatidae: a review of hypotheses. Can. J. Zool. 66:1707–1721.

EARLS, K. D. 2000. Kinematics and mechanics of ground take-off in the Starling *Sturnis vulgaris* and the Quail *Coturnix coturnix*. J. Exp. Biol. 203:725–739.

EASON, D. K., C. C. MILLER, A. CREE, J. HALVERSON, AND D. M. LAMBERT. 2001. A comparison of five methods for assignment of sex in the Takahe (Aves: *Porphyrio mantelli*). J. Zool. Lond. 253:281–292.

EASON, D. K., AND M. WILLIAMS. 2001. Captive rearing: a management tool for the recovery of the endangered Takahe. Pp. 80–95 *in* The Takahe: Fifty Years of Conservation Management and Research. University of Otago Press, Dunedin, New Zealand.

EAST, R., AND G. R. WILLIAMS. 1984. Island biogeography and the conservation of New Zealand's indigeous forest-dwelling avifauna. N. Z. J. Ecol. 7:27–35.

EBENHARD, T. 1991. Colonization in metapopulations: a review of theory and observations. Biol. J. Linn. Soc. 42:105–121.

EBENMAN, B., AND S. G. NILSSON. 1982. Size patterns in Willow Warblers *Phylloscopus trochilus* on islands in a south Swedish lake and the nearby mainland. Ibis 123:528–534.

EBER, G. 1961. Vergleichende Untersuchungen am flugfähigen Teichhuhn *Gallinula chl. chloropus* und an der flugunfähigen Inselralle *Gallinula nesiotis*. Bonn. Zool. Beitr. 12:247–317.

EBERHARD, W. G. 1996. Female Control: Sexual Selection by Cryptic Female Choice. Princeton University Press, Princeton, New Jersey.

EBLE, G. J. 1999. On the dual nature of chance in evolutionary biology and paleobiology. Paleobiology 25:75–87.

ECONOMOS, A. C. 1979. On structural theories of basal metabolic rate. J. Theor. Biol. 80:445–450.

ECONOMOS, A. C. 1982. On the origin of biological similarity. J. Theor. Biol. 94:25–60.

EDDLEMAN, W. R., AND C. J. CONWAY. 1998. Clapper Rail. Part 340 *in* The Birds of North America (A. Poole and F. Gill, Eds.). American Ornithologists' Union, Washington, D.C.

EDDLEMAN, W. R., AND F. L. KNOPF. 1985. Determining age and sex of American Coots. J. Field Ornithol. 56:41–55.

EDDLEMAN, W. R., F. L. KNOPF, B. MEANLEY, F. A. REID, AND R. ZEMBAL. 1988. Conservation of North American rallids. Wilson Bull. 100:458–475.

EDE, D. 1991. Mutation and limb evolution. Pp. 365–371 *in* Developmental Patterning of the Vertebrate Limb (J. R. Hinchliffe, J. M. Hurle, and D. Summerbell, Eds.). Plenum Press, New York, New York.

EDEN, S. F. 1987. When do helpers help? Food availability and helping in the Moorhen, *Gallinula chloropus*. Behav. Ecol. Sociobiol. 21:191–195.

EDINGER, T. 1942. The pituitary body in giant animals, fossil and living: a survey and a suggestion. Q. Rev. Biol. 14:31–45.

EDWARDS, S. V. 1993. Mitochondrial gene genealogy and gene flow among island and mainland populations of a sedentary songbird, the Gray-crowned Babbler (*Pomatostomus temporalis*). Evolution 47:1118–1137.

EDWARDS, W. E. 1967. The Late-Pleistocene extinction and diminution in size of many mammalian species. Pp. 141–154 *in* Pleistocene Extinctions: The Search for a Cause (P. S. Martin and H. E. Wright, Eds.). Yale University Press, New Haven, Connecticut.

EHRLICH, R., R. B. PHARR, JR., AND N. HEALY-WILLIAMS. 1983. Comments on the validity of Fourier descriptors in systematics: a reply to Bookstein et al. Syst. Zool. 32:202–206.

EIGENMANN, C. 1909. Cave vertebrates of America, a study in degenerative evolution. Carnegie Inst. Wash. Publ. 104:1–241.

EIMER, T. 1888. Die Entstehund der Arten auf Grund von Vererbung erworbener Eigenshaften. Vol. 1. Gustav Fischer, Jena, Germany.

ELDREDGE, N. 1974. Character displacement in evolutionary time. Am. Zool. 14:1083–1097.

ELDREDGE, N. 1985. Unfinished Synthesis: Biological Hierarchies and Modern Evolutionary Thought. Oxford University Press, Oxford, United Kingdom.

ELDREDGE, N. 1993. History, function, and evolutionary biology. Pp. 33–50 *in* Evolutionary Biology. Vol. 27 (M. K. Hecht, R. J. MacIntyre, and M. T. Clegg, Eds.). Plenum Press, New York, New York.

ELDREDGE, N., AND J. CRACRAFT. 1980. Phylogenetic Patterns and the Evolutionary Process. Columbia University Press, New York, New York.

ELDREDGE, N., AND S. J. GOULD. 1972. Punctuated equlibria: an alternative to phyletic gradualism. Pp. 82–115 in Models in Paleobiology (T. J. M. Schopf, Ed.). Freeman, Cooper, and Co., San Francisco, California.

ELGAR, M. A., AND P. H. HARVEY. 1987. Basal metabolic rates in mammals: allometry, phylogeny and ecology. Funct. Ecol. 1:25–36.

ELKINS, N. 1988. Weather and Bird Behaviour. Poyser, Carlton, United Kingdom.

ELLINGTON, C. P. 1984. The aerodynamics of flapping animal flight. Am. Zool. 24:95–105.

ELLINGTON, C. P. 1991. Limitations on animal flight performance. J. Exp. Biol. 160:71–91.

ELLIOTT, C. C. H. 1969. Gough Island. Bokmakierie 21:17–19.

ELLIOTT, C. C. H. 1970. Additional note on the sea-birds of Gough Island. Ibis 112:112–114.

ELLIOTT, G. P. 1987. Habitat use by the Banded Rail. N. Z. J. Ecol. 10:109–115.

ELLIOTT, G., K. WALKER, AND R. BUCKINGHAM. 1991. The Auckland Island Rail. Notornis 38:199–209.

ELLIOTT, H. F. I. 1953. The fauna of Tristan da Cunha. Oryx 2:41–53.

ELLIOTT, H. F. I. 1957. A contribution to the ornithology of the Tristan da Cunha group. Ibis 99:545–586.

ELLNER, S., AND A. SHMIDA. 1981. Why are adaptations for long-range seed dispersal rare in desert plants? Oecologia 51:133–144.

ELMBERG, J., P. NUMMI, H. PÖYSÄ, AND K. SJÖBERG. 1994. Relationships between species number, lake size and resource diversity in assemblages of breeding waterfowl. J. Biogeogr. 21:75–84.

ELTON, C. 1946. Competition and the structure of ecological communities. J. Anim. Ecol. 15:54–68.

ELTON, C. 1958. The Ecology of Invasions by Animals and Plants. Methuen, London, United Kingdom.

ELY, C. A., AND R. C. CLAPP. 1973. The natural history of Laysan Island, Northwestern Hawaiian Islands. Atoll Res. Bull. 171:1–361.

EMERSON, S. B. 1978. Allometry and jumping in frogs: helping the twain to meet. Evolution 32:551–564.

EMERSON, S. B. 1984. Morphological variation in frog pectoral girdles: testing alternatives to a traditional adaptive explanation. Evolution 38:376–388.

EMERSON, S. B. 1986. Heterochrony and frogs: the relationship of a life history trait to morphological form. Am. Nat. 127:167–183.

EMERSON, S. B. 1988. Testing for historical patterns of change: a case study with frog pectoral girdles. Paleobiology 14:174–186.

EMERSON, S. B. 2000. Vertebrate secondary sexual characteristics—physiological mechanisms and evolutionary patterns. Am. Nat. 156:84–91.

EMERSON, S. B., AND D. M. BRAMBLE. 1993. Scaling, allometry, and skull design. Pp. 384–421 in The Skull. Vol. 3: Functional and Evolutionary Mechanisms (J. Hanken and B. K. Hall, Eds.). University of Chicago Press, Chicago, Illinois.

EMERSON, S. B., H. W. GREENE, AND E. L. CHARNOV. 1994. Allometric aspects of predator–prey interactions. Pp. 123–139 in Ecological Morphology: Integrative Organismal Biology (P. C. Wainwright and S. M. Reilly, Eds.). University of Chicago Press, Chicago, Illinois.

EMERSON, S. B., AND P. A. HASTINGS. 1998. Morphological correlations in evolution: consequences for phylogenetic analysis. Q. Rev. Biol. 73:141–162.

EMILIANI, C. 1982. Extinctive evolution. J. Theor. Biol. 97:13–33.

EMLEN, J. T. 1977. Land Bird Communities of Grand Bahama Island: The Structure and Dynamics of an Avifauna. Ornithol. Monogr. 24. American Ornithologists' Union, Washington, D.C.

EMLEN, S. T. 1978. The evolution of cooperative breeding in birds. Pp. 245–281 in Behavioural Ecology: An Evolutionary Approach (J. R. Krebs and N. B. Davies, Eds.). Blackwell Scientific Publications, Oxford, United Kingdom.

EMLEN, S. T., AND L. W. ORING. 1977. Ecology, sexual selection, and the evolution of mating systems. Science 197:215–223.

ENDLER, J. A. 1973. Gene flow and population differentiation. Science 179:243–250.

ENDLER, J. A. 1977. Geographic Variation, Speciation, and Clines. Princeton Monogr. Popul. Biol. 10. Princeton University Press, Princeton, New Jersey.

ENDLER, J. A. 1978. A predator's view of animal color patterns. Pp. 319–364 in Evolutionary Biology.

Vol. 11 (M. K. Hecht, W. C. Steere, and B. Wallace, Eds.). Plenum Press, New York, New York.

ENDLER, J. A. 1982. Problems in distinguishing historical from ecological factors in biogeography. Am. Zool. 22:441–452.

ENDLER, J. A. 1986. Natural Selection in the Wild. Princton University Press, Princeton, New Jersey.

ENDLER, J. A. 1992. Signals, signal conditions, and the direction of evolution. Am. Nat. 139(Suppl.): 125–153.

ENGBRING, J., AND H. D. PRATT. 1985. Endangered birds in Micronesia: their history, status and future prospects. Bird Conserv. 2:71–105.

ENGELMOER, M., AND C. S. ROSELAAR. 1998. Geographical Variation in Waders. Kluwer Academic Publications, Dordrecht, The Netherlands.

ENGELS, W. L. 1938a. Variations in bone length and limb proportions in the Coot (*Fulica americana*). J. Morphol. 62:599–607.

ENGELS, W. L. 1938b. Cursorial adaptations in birds. Limb proportions in the skeleton of *Geococcyx*. J. Morphol. 63:207–217.

ENGILIS, A., JR., L. W. ORING, E. CARRERA, J. W. NELSON, AND A. M. LOPEZ. 1998. Shorebird surveys in Ensenada Pabellones and Bahía Santa María, Sinaloa, Mexico: critical winter habitats for Pacific flyway shorebirds. Wilson Bull. 110:332–341.

ENGILIS, A., JR., AND T. K. PRATT. 1993. Status and population trends of Hawaii's native waterbirds, 1977–1987. Wilson Bull. 105:142–158.

EPTING, R. J. 1980. Functional dependence of the power for hovering or wing disc loading in hummingbirds. Physiol. Zool. 53:347–357.

EPTING, R. J., AND T. M. CASEY. 1973. Power output and wing disc loading in hovering hummingbirds. Am. Nat. 107:761–765.

ERDMANN, K. 1940. Zur Entwicklungsgeschichte der Knochen im Schädel des Huhnes bis zum Zeitpunkt des Ausschlüpfens aus dem Ei. Z. Morphol. Okol. Tiere 36:315–400.

ERIKSTAD, K., T. TVERAA, AND J. BUSTNES. 1998. Significance of intraclutch egg-size variation in Common Eider: the role of egg-size variation and quality of ducklings. J. Avian Biol. 29:3–9.

ERRITZOE, J. 1995. A small collection of birds from the Philippines with notes on body mass, distribution, and habitat. Nemouria 40:1–15.

ESHEL, I. 1977. On the founder effect and the evolution of altruistic traits: an ecogenetical approach. Theor. Popul. Biol. 11:410–424.

ESKELL, R., AND S. GARNETT. 1979. Notes on the colours of the legs, wings and flanks of the Dusky Moorhen *Gallinula tenebrosa*. Emu 79:143–146.

ESSEX, R. 1927. Studies in reptilian degeneration. Proc. Zool. Soc. Lond. 1927:879–945.

EVANS, M. I., A. F. A. HAWKINS, AND J. W. DUCKWORTH. 1996. Family Mesitornithidae (mesites). Pp. 34–43 in Handbook of the Birds of the World. Vol. 3: Hoatzin to Auks (J. del Hoyo, A. Elliott and J. Sargatal, Eds.). Lynx Edicions, Barcelona, Spain.

EVANS, M. R., T. L. F. MARTINS, AND M. HALEY. 1994. The asymmetrical cost of tail elongation in Red-billed Streamertails. Proc. R. Soc. Lond. (Ser. B Biol. Sci.) 256:97–103.

EVANS, M. R., M. ROSÉN, K. J. PARK, AND A. HEDENSTRÖM. 2002. How do birds' tails work? Deltawing theory fails to predict tail shape during flight. Proc. R. Soc. Lond. (Ser. B) 269:1053–1057.

EVANS, M. R., AND A. L. R. THOMAS. 1997. Testing the functional significance of tail streamers. Proc. R. Soc. Lond. (Ser. B Biol. Sci.) 264:211–217.

EVANS, P. R., AND N. C. DAVIDSON. 1990. Migration strategies and tactics of waders breeding in arctic and north temperate latitudes. Pp. 387–398 in Bird Migration: Physiology and Ecophysiology (E. Gwinner, Ed.) Springer Verlag, Berlin, Germany.

EVANS, P. R., N. C. DAVIDSON, J. D. UTTLEY, AND R. D. EVANS. 1992. Premigratory hypertrophy of flight muscles: an ultrastructural study. Ornis Scand. 23:238–243.

EVENS, J., AND G. W. PAGE. 1986. Predation on Black Rails during high tides in salt marshes. Condor 88:107–109.

FAABORG, J. 1977. Metabolic rates, resources, and the occurence of nonpasserines in terrestrial avian communities. Am. Nat. 111:903–916.

FAABORG, J. 1979. Qualitative patterns of avian extinction on Neotropical land-bridge islands: lessons for conservation. J. Appl. Ecol. 16:99–107.

FAABORG, J. 1980. Further observation on ecological release in Mona Island birds. Auk 97:624–627.

FAABORG, J. 1981. The characteristics and occurrence of cooperative polyandry. Ibis 123:477–484.
FAABORG, J. 1982. Trophic and size structure of West Indian bird communities. Proc. Natl. Acad. Sci. USA 79:1563–1567.
FAABORG, J., AND W. J. ARENDT. 1992. Rainfall correlates of bird populations fluctuations in a Puerto Rican dry forest: a 15-year study. Ornitol. Caribena 3:10–19.
FAABORG, J., AND W. J. ARENDT. 1995. Survival rates of Puerto Rican birds: Are islands really that different? Auk 112:503–507.
FAIRBAIRN, D. J. 1994. Wing dimorphism and the migratory syndrome: correlated traits for migratory tendency in wing dimorphic insects. Res. Popul. Ecol. 36:157–163.
FAIRBAIRN, D. J. 1997. Allometry for sexual size dimorphism: pattern and process in the coevolution of body size in males and females. Annu. Rev. Ecol. Syst. 28:659–687.
FAIRBAIRN, D. J., AND D. A. ROFF. 1990. Genetic correlations among traits determining migratory tendency in the sand cricket, *Gryllus firmus*. Evolution 44:1787–1795.
FAIRBAIRN, D. J., AND D. E. YADLOWSKI. 1997. Coevolution of traits determining migratory tendency: correlated response of a critical enzyme, juvenile hormone esterase, to selection on wing morphology. J. Evol. Biol. 10:495–513.
FAITH, D. P. 1989. Homoplasy as pattern: multivariate analysis of morphological convergence in Anseriformes. Cladistics 5:235–258.
FAIVRE, B., J. SECONDI, C. FERRY, L. CHASTRAGNAT, AND F. CÉZILLY. 1999. Morphological variation and the recent evolution of wing length in the Icterine Warbler: A case of unidirectional introgression? J. Avian Biol. 30:152–158.
FALCONER, D. S. 1989. Introduction to Quantitative Genetics. Longmans, London, United Kingdom.
FALLA, R. A. 1949. *Notornis* rediscovered. Emu 48:316–322.
FALLA, R. A. 1951. Nesting season of *Notornis*. Notornis 4:97–100.
FALLA, R. A. 1954. A new rail from cave deposits in the North Island of New Zealand. Rec. Auckl. Inst. Mus. 4:241–244.
FALLA, R. A. 1960. Notes on some bones collected by Dr. Watters and Mr. Lindsay at Chatham Islands. Notornis 8:226–227.
FALLA, R. A. 1964. Moa. Pp. 477–479 *in* New Dictionary of Birds (A. Landsborough Thomson, Ed.). Nelson, London, United Kingdom.
FALLA, R. A. 1967. An Auckland Island Rail. Notornis 14:107–113.
FALLA, R. A., AND J.-L. MOUGIN. 1979. Order Sphenisciformes. Pp. 121–134 *in* Check-list of Birds of the World. Vol. 1. 2nd ed. (E. Mayr and G. W. Cottrell, Eds.). Museum of Comparative Zoology, Harvard University, Cambridge, Massachusetts.
FARBER, P. L. 1982. The Emergence of Ornithology as a Scientific Discipline: 1760–1850. D. Reidel Publishing Co., Dordrecht, The Netherlands.
FARMER, A. H., AND A. H. PARENT. 1997. Effects of the landscape on shorebird movements at spring migration stopovers. Condor 99:698–707.
FARMER, A. H., AND J. A. WIENS. 1998. Optimal migration schedules depend on the landscape and the physical environment: a dynamic modeling view. J. Avian Biol. 29:405–415.
FEDAK, M. A., N. C. HEGLUND, AND C. R. TAYLOR. 1982. Energetics and mechanics of terrestrial locomotion. II. Kinetic energy changes of the limbs and body as a function of speed and body size in birds and mammals. J. Exp. Biol. 79:23–40.
FEDAK, M. A., B. PINSHOW, AND K. SCHMIDT-NIELSEN. 1974. Energy cost of bipedal running. Am. J. Physiol. 227:1038–1044.
FEDAK, M. A., AND H. J. SEEHERMAN. 1979. Reappraisal of energetics of locomotion shows identical cost in bipeds and quadrapeds including Ostrich and horse. Nature 282:713–716.
FEDER, M. E., A. F. BENNETT, W. W. BURGGREN, AND R. B. HUEY, EDS. 1987. New Directions in Ecological Physiology. Cambridge University Press, Cambridge, United Kingdom.
FEDUCCIA, A. 1980. The Age of Birds. Harvard University Press, Cambridge, Massachusetts.
FEDUCCIA, A. 1985. The morphological evidence for ratite monophyly: fact or fiction? Pp. 184–190 *in* Acta XVIII Congressus Internationalis Ornithologici (V. D. Ilyichev and V. M. Gavrilov, Eds.). Academy of Science, Moscow, USSR.
FEDUCCIA, A. 1995. Explosive evolution in Tertiary birds and mammals. Science 267:637–638.
FEDUCCIA, A. 1996. The Origin and Evolution of Birds. Yale University Press, New Haven, Connecticut.

FEDUCCIA, A., AND H. B. TORDOFF. 1979. Feathers of *Archeopteryx*: asymmetric vanes indicate aerodynamic function. Science 203:1021–1022.

FEINSINGER, P., AND S. B. CHAPLIN. 1975. On the relationship between wing disc loading and foraging strategy in hummingbirds. Am. Nat. 107:217–224.

FEINSINGER, P., R. K. COLWELL, J. TERBORGH, AND S. B. CHAPLIN. 1979. Elevation and the morphology, flight energetics, and foraging ecology of tropical hummingbirds. Am. Nat. 113:481–497.

FEINSINGER, P., AND L. A. SWARM. 1982. 'Ecological release', seasonal variation in food supply, and the hummingbird *Amazilia tobaci* on Trinidad and Tobago. Ecology 63:1574–1587.

FELDMAN, H. A., AND T. A. MCMAHON. 1983. The 3/4 mass exponent for energy metabolism is not a statistical artifact. Respir. Physiol. 52:149–163.

FELL, H. B. 1939. The origin and developmental mechanics of the avian sternum. Philos. Trans. R. Soc. Lond. (Ser. B) 229:407–464.

FELLOWS, M. 1995. The Art of Angels and Cherubs. Parragon Book Service, London, United Kingdom.

FELSENSTEIN, J. 1976. The theoretical population genetics of variable selection and migration. Annu. Rev. Genet. 10:253–280.

FELSENSTEIN, J. 1979. r- and K-selection in a completely chaotic population model. Am. Nat. 113:499–510.

FELTON, E. A., K. A. W. CROOK, AND B. H. KEATING. 2000. The Hulopoe Gravel, Lanai, Hawaii: new sedimentological data and their bearing on the "giant wave" (mega-tsunami) emplacement hypothesis. Pure Appl. Geophys. 157:1257–1284.

FENCHEL, T. 1974. Intrinsic rate of natural increase: the relationship with body size. Oecologia 14:317–326.

FENOGLIO, S., M. CUCCO, AND G. MALACARNE. 2002. Bill colour and body condition in the Moorhen *Gallinula chloropus*. Bird Study 49:89–92.

FIENBERG, S. E. 1980. The Analysis of Cross-classified Categorical Data. 2nd ed. Massachusetts Institute of Technology Press, Cambridge, Massachusetts.

FIGUEROLA, J., AND A. J. GREEN. 2000. The evolution of sexual dimorphism in relation to mating patterns, cavity nesting, insularity and sympatry in the Anseriformes. Funct. Ecol. 14:701–710.

FINDLAY, C. S., AND F. COOKE. 1987. Repeatability and heritability of clutch size in Lesser Snow Geese. Evolution 41:453.

FINK, W. L. 1982. The conceptual relationship between ontogeny and phylogeny. Paleobiology 8:254–264.

FINK, W. L. 1988. Phylogenetic analysis and the detection of ontogenetic patterns. Pp. 71–91 *in* Heterochrony in Evolution: A Multidisciplinary Approach (M. L. McKinney, Ed.). Plenum Press, New York, New York.

FINSCH, O. 1868. Notes on Mr. Walter Buller's "Essay on the ornithology of New Zealand." Trans. N. Z. Inst. 1:58–73.

FINSCH, O. 1874. Preliminary remarks on some New Zealand birds. Trans. N. Z. Inst. 7:226–236.

FINSCH, O. 1875. Further remarks on some New Zealand birds. Trans. N. Z. Inst. 8:200–204.

FINSCH, O. 1898. On the so-called "Sandwich Rail" in the Leiden Museum. Notes Leiden Mus. 20:77–80.

FINSTON, T. L., AND S. B. PECK. 1995. Population structure and gene flow in *Stomium*: a species swarm of flightless beetles of the Galápagos Islands. Heredity 75:390–397.

FISCHER, K., AND B. STEPHAN. 1971a. Ein flugunfähiger Kranich (*Grus cubensis* n. sp.) aus dem Pleistozän von Kuba—eine Osteologie der Familie der Kraniche (Gruidae). Wiss. Z. Humboldt-Univ. Berl. (Math.-Nat.) 20:541–592.

FISCHER, K., AND B. STEPHAN. 1971b. Weitere Vogelreste aus em Pleistozän der Pio-Domingo-Höhle in Kuba. Wiss. Z. Humboldt-Univ. Berl. (Math.-Nat.) 20:593–607.

FISHER, C. T. 1981. Specimens of extinct, endangered or rare birds in the Merseyside County Museums, Liverpool. Bull. Br. Ornithol. Club 101:276–285.

FISHER, D. C. 1986. Progress in organismal design. Pp. 99–117 *in* Patterns and Processes in the History of Life (D. M. Raup and D. Jablonski, Eds.). Springer-Verlag, Berlin, Germany.

FISHER, H. I. 1945. Flying ability and the anterior intermuscular line on the coracoid. Auk 62:125–129.

FISHER, H. I. 1955. Avian anatomy, 1925–1950, and some suggested problems. Pp. 57–104 *in* Recent Studies in Avian Biology (A. Wolfson, Ed.). University of Illinois Press, Urbana, Illinois.

FISHER, H. I., AND P. H. BALDWIN. 1946. War and the birds of Midway atoll. Condor 48:3–15.
FISHER, H. I., AND D. C. GOODMAN. 1955. The myology of the Whooping Crane, *Grus americana*. Ill. Biol. Monogr. 24:1–127.
FISHER, R. A. 1930. The Genetical Theory of Natural Selection. Clarendon Press, Oxford, United Kingdom.
FISHER, R. A. 1936. The use of multiple measurements in taxonomic problems. Ann. Eugen. 7:179–188.
FISHER, R. A. 1949. The Theory of Inbreeding. Oliver and Boyd, Edinburgh, United Kingdom.
FISHER, R. A. 1958. The Genetic Theory of Natural Selection. 2nd ed. Dover Publications, New York, New York.
FISHER, W. K. 1903. Notes on the birds peculiar to Laysan Island, Hawaiian Group. Auk 20:384–397.
FISHER, W. K. 1906. Birds of Laysan and the Leeward islands, Hawaiian group. Bull. U.S. Fish Comm. 23:769–807.
FITZGERALD, B. M., AND C. R. VEITCH. 1985. The cats of Herekopare Island, New Zealand; their history, ecology and affects on birdlife. N. Z. J. Zool. 12:319–330.
FITZPATRICK, S. 1997. Patterns of morphometric variation in birds' tails: length, shape and variability. Biol. J. Linn. Soc. 62:145–162.
FITZPATRICK, S. 1998. Intraspecific variation in wing length and male plumage coloration with migratory behaviour in continental and island populations. J. Avian Biol. 29:248–256.
FITZPATRICK, S. 1999. Tail length in birds in relation to tail shape, general flight ecology and sexual selection. J. Evol. Biol. 12:49–60.
FJELDSÅ, J. 1973. Territorial regulation of the progress of breeding in a population of Coots, *Fulica atra*. Dansk. Ornithol. Foren. Tidsskr. 67:115–127.
FJELDSÅ, J. 1977. Sex and age variation in wing-length in the Coot *Fulica atra*. Ardea 65:115–125.
FJELDSÅ, J. 1981. Biological notes on the Giant Coot *Fulica gigantea*. Ibis 123:423–437.
FJELDSÅ, J. 1983a. Systematic and biological notes on the Colombian Coot *Fulica americana colombiana* (Aves, Rallidae). Steenstrupia 9:209–215.
FJELDSÅ, J. 1983b. Ecological character displacement and character release in grebes Podicipedidae. Ibis 125:463–481.
FJELDSÅ, J. 1985. Origin, evolution, and status of the avifauna of Andean wetlands. Pp. 85–112 *in* Neotropical Ornithology (P. A. Buckley, M. S. Foster, E. S. Morton, R. S. Ridgely, and F. G. Buckley, Eds.). Ornithol. Monogr. 36. American Ornithologists' Union, Washington, D.C.
FJELDSÅ, J., AND N. KRABBE. 1990. Birds of the High Andes. Apollo Books, Svendborg, and Zoology Museum, Copenhagen, Denmark.
FJELDSÅ, J., E. LAMBIN, AND B. MERTENS. 1999. Correlation between endemism and local ecoclimatic stability documented by comparing Andean bird distributions and remotely sensed surface data. Ecography 22:63–78.
FLEAGLE, J. G. 1985. Size and adaptation in Primates. Pp. 1–19 *in* Size and Scaling in Primate Biology (W. L. Jungers, Ed.). Plenum Press, New York, New York.
FLEISCHER, R. C., AND R. F. JOHNSTON. 1982. Natural selection on body size and proportions in House Sparrows. Nature 298:747–749.
FLEMING, C. A. 1951. *Notornis* in February, 1950. Notornis 4:101–106.
FLEMING, C. A. 1960. History of the New Zealand land bird fauna. Notornis 9:270–274.
FLEMING, C. A. 1962. New Zealand biogeography. Tuatara 10:53–108.
FLEMING, C. A. 1975. The geological history of New Zealand and its biota. Pp. 1–86 *in* Biogeography and Ecology in New Zealand (G. Kuschel, Ed.). Dr. W. Junk, The Hague, The Netherlands.
FLEMING, C. A. 1979. The Geological History of New Zealand and Its Life. Oxford University Press, Oxford, United Kingdom.
FLEMING, J. H. 1939a. Birds of the Chatham Islands. Part I. Emu 38:381–413.
FLEMING, J. H. 1939b. Birds of the Chatham Islands. Part II. Emu 38:492–509.
FLESSA, K. W., RAPPORTEUR. 1986. Causes and consequences of extinction. Pp. 235–257 *in* Patterns and Processes in the History of Life (D. M. Raup and D. Jablonski, Eds.). Springer-Verlag, Berlin, Germany.
FLETCHER, J. A. 1909. Bird notes from Cleveland, Tasmania. Emu 8:210–214.
FLETCHER, J. A. 1912. Notes on the Native Hen (*Tribonyx mortierii*). Emu 11:250–252.

FLOOD, N. J. 1984. Adaptive significance of delayed plumage maturation in male Northern Orioles. Evolution 38:267–279.

FLORES, R. E., AND W. R. EDDLEMAN. 1995. California Black Rail use of habitat in southwestern Arizona. J. Wildl. Manage. 59:357–363.

FLOWER, W. H. 1885. An Introduction to the Osteology of the Mammalia. Macmillan, London, United Kingdom.

FLURY, B. 1988. Common Principal Components and Related Multivariate Models. J. Wiley and Sons, New York, New York.

FOGDEN, M. P. L. 1972. The seasonality and population dynamics of equatorial forest birds in Sarawak. Ibis 114:307–343.

FOGDEN, M. P. L., AND P. M. FOGDEN. 1979. The role of fat and protein reserves in the annual cycle of Grey-backed Camaroptera in Uganda (Aves: Sylviidae). J. Zool. Lond. 189:233–258.

FOLCH, A. 1992. Order Struthioniformes. Pp. 75–110 in Handbook of the Birds of the World. Vol. 1 (J. del Hoyo and J. Sargatal, Eds.). Lynx Edicions, Barcelona, Spain.

FOLEY, P. 1994. Predicting extinction times from environmental stochasticity and carrying capacity. Conserv. Biol. 8:124–137.

FOLEY, P. 1997. Extinction models for local populations. Pp. 215–246 in Metapopulation Biology: Ecology, Genetics, and Evolution (I. Hanski and M. E. Gilpin, Eds.). Academic Press, San Diego, California.

FONG, D. W., AND D. C. CULVER. 1985. A reconsideration of Ludwig's differential migration theory of regressive evolution. Natl. Speleol. Soc. Bull. 47:123–127.

FONG, D. W., T. C. KANE, AND D. C. CULVER. 1995. Vestigialization and loss of nonfunctional characters. Annu. Rev. Ecol. Syst. 26:249–268.

FOOTE, M. 1994. Morphological disparity in Ordovician–Devonian crinoids and the early saturation of morphological space. Paleobiology 20:320–344.

FORBES, H. O. 1891. Preliminary notice of additions to the extinct avifauna of New Zealand. Trans. N. Z. Inst. 24:185–189.

FORBES, H. O. 1892a. New extinct rail [telegram]. Nature 45:416.

FORBES, H. O. 1892b. New extinct rail. Ibis 34:473–474.

FORBES, H. O. 1892c. *Aphanapteryx* in the New Zealand region. Nature 45:580–581.

FORBES, H. O. 1892d. *Aphanapteryx* and other remains in the Chatham Islands. Nature 46:252–253.

FORBES, H. O. 1892e. [Untitled communication.] Bull. Br. Ornithol. Club 1:21–22.

FORBES, H. O. 1892f. Evidence of a wing in *Dinornis*. Nature 41:209.

FORBES, H. O. 1893a. [Untitled communication.] Ibis 35:253–254.

FORBES, H. O. 1893b. [Untitled communication.] Ibis 35:450–451.

FORBES, H. O. 1893c. [Untitled communication.] Bull. Br. Ornithol. Club 1:50–51.

FORBES, H. O. 1893d. A list of the birds inhabiting the Chatham Islands. Ibis 35:521–546.

FORBES, H. O. 1893e. [Untitled communication.] Ibis 35:253.

FORBES, H. O. 1893f. [Untitled communication.] Bull. Br. Ornithol. Club 1:20.

FORBES, H. O. 1893g. [Untitled communication.] Ibis 35:445.

FORBES, H. O. 1893h. [Untitled communication.] Bull. Br. Ornithol. Club 1:45–46.

FORBES, H. O. 1901. Notes on some rare birds in the Lord Derby Museum. Bull. Liverpool Mus. 3: 61–68.

FORBES, H. O. 1923. The ralline genus *Notornis*, Owen. Nature 112:762.

FORBES, S., AND D. W. MOCK. 2000. A tale of two strategies: life-history aspects of family strife. Condor 102:23–34.

FORD, H. H., AND D. C. PATON. 1975. Impoverishment of the avifauna of Kangaroo Island. Emu 75: 155–156.

FORDHAM, R. A. 1983. Seasonal dispersion and activity of the Pukeko *Porphyrio p. melanotus* (Rallidae) in swamp and pasture. N. Z. J. Ecol. 6:133–142.

FORDYCE, E. 1982. The fossil vertebrate record of New Zealand. Pp. 630–698 in The Fossil Vertebrate Record of Australasia (P. V. Rich and E. M. Thompson, Eds.). Monash University, Clayton, Australia.

FORSMAN, A. 1991. Adaptive variation in head size in *Vipera berus* L. populations. Biol. J. Linn. Soc. 43:281–296.

FORSYTH, D. M., AND R. P. DUNCAN. 2001. Propagule size and the relative success of exotic ungulate and bird introductions to New Zealand. Am. Nat. 157:583–595.

FOSBERG, F. R., M. H. SACHET, AND D. R. STODDART. 1983. Henderson Island (southeastern Polynesia): summary of current knowledge. Atoll Res. Bull. 272:1–47.

FOSTER, J. B. 1964. Evolution of mammals on islands. Nature 202:234–235.

FOSTER, M. S. 1987. Delayed maturation, neoteny, and social system differences in two manakins of the genus *Chiroxiphia*. Evolution 41:547–558.

FRANK, S. A. 1986. Dispersal polymorphisms in subdivided populations. J. Theor. Biol. 122:303–309.

FRANKLIN, A. B., D. A. CLARK, AND D. B. CLARK. 1979. Ecology and behavior of the Galápagos Rail. Wilson Bull. 91:202–221.

FRANKLIN, D. C., AND T. A. BARNES. 1998. The downy young and juvenile of the Chestnut Rail, with notes on development. Corella 22:64–66.

FRANKLIN, I. R. 1987. Population biology and evolutionary change. Pp. 156–174 *in* Rates of Evolution (K. S. W. Campbell and M. F. Day, Eds.). Allen and Unwin, London, United Kingdom.

FRASER, M. W. 1984. Foods of Subantarctic Skuas on Inaccessible Island. Ostrich 55:192–195.

FRASER, M. W. 1989. The Inaccessible Island Rail: smallest flightless bird in the world. Afr. Wildl. 43:14–19.

FRASER, M. W., W. R. J. DEAN, AND I. C. BEST. 1992. Observations on the Inaccessible Island Rail *Atlantisia rogersi*: the world's smallest flightless bird. Bull. Br. Ornithol. Club 112:12–22.

FRASIER, C. C. 1984. An explanation of the relationships between mass, metabolic rate and characteristic skeletal length for birds and mammals. J. Theor. Biol. 109:331–371.

FRAUENFELD, G. R. VON. 1868. Neu aufgegundende Abbildung des Dronte und eines Zweiten kurzflügeligen Vogels, wahrscheinlich des Poule Rouge au Bec de Becasse der Maskarenen in der Privatbibliothek S. M. des Verstorbenen Kaisers Franz. C. Überreuter'sche Buchdruckerei, Vienna, Austria.

FRAUENFELD, G. R. VON. 1869. Über den Artnamen von *Aphanapteryx*. Verh. Kaiserl.-Königl. Zool. Bot. Ges. Wien 19:761–764.

FRAZZETTA, T. H. 1975. Complex Adaptations in Evolving Populations. Sinauer Assoc., Sunderland, Massachusetts.

FREDRICKSON, L. H. 1970. Breeding biology of American Coots in Iowa. Wilson Bull. 82:445–457.

FREDRICKSON, L. H. 1971. Common Gallinule breeding biology and development. Auk 88:914–919.

FREED, L. A. 1999. Extinction and endangerment of Hawaiian honeycreepers: a comparative approach. Pp. 137–162 *in* Genetics and the Extinction of Species: DNA and the Conservation of Biodiversity (L. F. Landweber and A. P. Dobson, Eds.). Princeton University Press, Princeton, New Jersey.

FRENCH, T. W. 1981. Fish attack on Black Guillemot and Common Eider in Maine. Wilson Bull. 93:279–280.

FRETWELL, S. D. 1972. Populations in Seasonal Environments. Monogr. Popul. Biol. 5. Princeton University Press, Princeton, New Jersey.

FRETWELL, S. D. 1980. Evolution of migration in relation to factors regulating bird numbers. Pp. 517–527 *in* Migrant Birds in the Neotropics. Smithsonian Institution Press, Washington, D.C.

FRETWELL, S. D., AND H. L. LUCAS, JR. 1970. On territorial behaviour and other factors influencing habitat distribution in birds. Acta Biotheor. 19:16–36.

FRITH, C. B., AND D. W. FRITH. 1990. Nidification of the Chestnut Forest-Rail *Rallina rubra* (Rallidae) in Papua New Guinea and a review of *Rallina* nesting biology. Emu 90:254–259.

FRITTS, T. H., AND G. H. RODDA. 1998. The role of introduced species in the degradation of island ecosystems: a case history of Guam. Annu. Rev. Ecol. Syst. 29:113–140.

FROHAWK, F. W. 1892. Description of a new species of rail from Laysan Island (North Pacific). Ann. Mag. Nat. Hist. 9:247–249.

FROST, D. R., AND A. G. KLUGE. 1994. A reconsideration of epistemology in systematic biology, with special reference to species. Cladistics 10:259–294.

FRY, C. H. 1983. The jacanid radius and *Microparra*. A neotenic genus. Gerfaut 73:173–184.

FRY, C. H., I. J. FERGUSON-LEES, AND R. J. DOWSETT. 1972. Flight muscle hypertrophy and ecophysiological variation of Yellow Wagtail *Motacilla flava* races at Lake Chad. J. Zool. Lond. 167:293–306.

FUJIOKA, T. 1955. Time and order of ossification of centres in the chicken skeleton. Acta Anat. Nippon. 30:140–150.

FUJISAKI, K. 1992. A male fitness advantage to wing reduction in the oriental chinch bug, *Cavelerius saccharivorus* Okajima (Heteroptera: Lygaeidae). Res. Popul. Ecol. 34:173–183.

FULLAGAR, P. J. 1985. The Woodhens of Lord Howe Island. Avic. Mag. 91:15–30.
FULLAGAR, P. J., AND H. J. DE S. DISNEY. 1975. The birds of Lord Howe Island: a report on the rare and endangered species. Bull. Int. Counc. Bird Preserv. 12:187–202.
FULLAGAR, P. J., AND H. J. DE S. DISNEY. 1981. Discriminant functions for sexing Woodhens. Corella 5:106–108.
FULLAGAR, P. J., H. J. DE S. DISNEY, AND R. DE NAUROIS. 1982. Additional specimens of two rare rails and comments on the genus *Tricholimnas* of New Caledonia and Lord Howe Island. Emu 82:131–136.
FULLER, E. 1987. Extinct Birds. Viking/Reinhard, London, United Kingdom.
FULLER, E. 2001. Extinct Birds. 2nd ed. Cornell University Press, Ithaca, New York.
FUNK, V. A., AND D. R. BROOKS. 1990. Phylogenetic systematics as the basis of comparative biology. Smithsonian Contrib. Bot. 73:1–45.
FUNK, V.A., AND W. L. WAGNER. 1995. Biogeographic patterns in the Hawaiian Islands. Pp. 379–419 *in* Hawaiian Biogeography: Evolution on a Hot Spot Archipelago (W. L. Wagner and W. A. Funk, Eds.). Smithsonian Institution Press, Washington, D.C.
FÜRBRINGER, M. 1888. Untersuchungen zur Morphologie und Systematik der Vögel, zugleich ein Beitrag zur Anatomie der Stütz- und Bewegungsorgane. 2 vols. T. J. Van Holkema, Amsterdam, Holland.
FÜRBRINGER, M. 1902. Zur vergleichenden Anatomie des Brustscheulterapparates und der Schultermuskeln. V. Tiel. Vögel. Jena Z. Naturwiss. 36:289–736.
FURNESS, B. L., AND R. W. FURNESS. 1980. Apostatic selection and kleptoparasitism in the Parasitic Jaeger: a comment. Auk 97:832–836.
GADGIL, M. 1971. Dispersal: population consequences and evolution. Ecology 52:253–261.
GADGIL, M., AND W. BOSSERT. 1970. Life history consequences of natural selection. Am. Nat. 104:1–24.
GADOW, H. 1888. Remarks on the numbers and on the phylogenetic development of the remiges of birds. Proc. Zool. Soc. Lond. 1888:655–667.
GADOW, H. 1893. Vögel. II. Systematischer Theil. *In* Klassen und Ordnungen des Their-Reichs, Wissenschaftlich Dargestellt in Wort und Bild. Vol. 6. Number 4 (H. G. Bronn, Ed.). C. F. Winter, Leipzig, Germany.
GADOW, H. 1902. The wings and the skeleton of *Phalacrocorax harrisi*. Novit. Zool. 9:169–176.
GAERTNER, R. A. 1949. Development of the posterior trunk and tail of the chick embryo. J. Exp. Zool. 111:157–174.
GALILEI, G. 1637. [Dialogues Concerning the Two New Sciences. Proposition VIII.] Translated by H. Crew and A. de Salvio (1933). Macmillan, New York, New York.
GALLI, A. E., C. F. LECK, AND R. T. T. FORMAN. 1976. Avian distribution patterns in forest islands of different sizes in central New Jersey. Auk 93:356–364.
GANS, C. 1961. Studies on the amphisbaenids (Amphisbaenia: Reptilia). I. A taxonomic revision of the Trogonophinae and a functional interpretation of the amphisbaenian adaptive pattern. Bull. Am. Mus. Nat. 119:129–204.
GANS, C. 1973. Locomotion and burrowing in limbless vertebrates. Nature 242:414–415.
GANS, C. 1975. Tetrapod limblessness: evolution and functional corollaries. Am. Zool. 15:455–467.
GANS, C. 1979. Momentarily excessive construction as the basis for protoadaptation. Evolution 33:227–233.
GANS, C., AND W. J. BOCK. 1965. The functional significance of muscle architecture—a theoretical analysis. Rev. Anat. Embryol. Cell Biol. 38:115–142.
GARLAND, T., JR. 1983. The relation between maximal running speed and body mass in terrestrial mammals. J. Zool. Lond. 199:157–170.
GARNETT, S. T. 1978. The behaviour patterns of the Dusky Moorhen *Gallinula tenebrosa* Gould (Aves: Rallidae). Aust. Wildl. Res. 5:363–384.
GARNETT, S. T. 1980. The social organization of the Dusky Moorhen *Gallinula tenebrosa* Gould (Aves: Rallidae). Aust. Wildl. Res. 7:103–112.
GARNETT, S. T., ED. 1993. Threatened and Extinct Birds of Australia. 2nd ed. Rep. 82. Royal Australasian Ornithologists' Union, Victoria, Australia.
GARRIDO, O. H. 1985. Cuban endangered birds. Pp. 992–999 *in* Neotropical Ornithology (P. A. Buckley, M. S. Foster, E. S. Morton, R. S. Ridgely, and F. G. Buckley, Eds.). Ornithol. Monogr. 36. American Ornithologists' Union, Washington, D.C.

GARRIDO, O. H., AND A. KIRKCONNELL. 2000. Field Guide to the Birds of Cuba. Cornell University Press, Ithaca, New York.
GARSTANG, W. 1922. The theory of recapitulation: a critical re-statement of the biogenetic law. J. Linn. Soc. (Zool.) 35:81–101.
GASKELL, J. 2000. Who Killed the Great Auk? Oxford University Press, Oxford, United Kingdom.
GASTON, K. J., AND T. M. BLACKBURN. 1997a. Evolutionary age and risk of extinction in the global avifauna. Evol. Ecol. 11:557–565.
GASTON, K. J., AND T. M. BLACKBURN. 1997b. Age, area and avian diversification. Biol. J. Linn. Soc. 62:239–253.
GASTON, K. J., AND S. L. CHOWN. 1999. Geographic range size and speciation. Pp. 236–259 in Evolution of Biological Diversity (A. E. Magurran and R. M. May, Eds.). Oxford University Press, Oxford, United Kingdom.
GATESY, S. M. 1991. Hind limb scaling in birds and other theropods: implications for terrestrial locomotion. J. Morphol. 209:83–96.
GATESY, S. M. 1995. Functional evolution of the hindlimb and tail from basal theropods to birds. Pp. 219–234 in Functional Morphology in Vertebrate Paleontology (J. J. Thomason, Ed.). Cambridge University Press, Cambridge, United Kingdom.
GATESY, S. M., AND A. A. BIEWENER. 1991. Bipedal locomotion: effects of speed, size and limb posture in birds and humans. J. Zool. Lond. 224:127–147.
GATESY, S. M., AND K. P. DIAL. 1993. Tail muscle activity patterns in walking and flying Pigeons (*Columba livia*). J. Exp. Biol. 176:55–76.
GATESY, S. M., AND K. P. DIAL. 1996a. Locomotor modules and the evolution of avian flight. Evolution 50:331–340.
GATESY, S. M., AND K. P. DIAL. 1996b. From frond to fan: *Archaeopteryx* and the evolution of short-tailed birds. Evolution 50:2037–2048.
GAUNT, A. S., AND C. GANS. 1990. Architecture of chicken muscles: short-fibre patterns and their ontogeny. Proc. R. Soc. Lond. (Ser. B Biol. Sci.) 240:351–362.
GAUNT, A. S., R. S. HIKIDA, J. R. JEHL, JR., AND L. FENBERT. 1990. Rapid atrophy and hypertrophy of an avian flight muscle. Auk 107:649–659.
GAUTHIER, G., J. BEDARD, J. HUOT, AND Y. BEDARD. 1984. Spring accumulation of fat by Greater Snow Geese in two staging habitats. Condor 86:192–199.
GAUTHREAUX, S. A., JR. 1982. The ecology and evolution of avian migration systems. Pp. 93–167 in Avian Biology. Vol. 6 (D. S. Farner and J. R. King, Eds.). Academic Press, New York, New York.
GAVRILETS, S., AND A. HASTINGS. 1996. Founder effect speciation: a theoretical reassessement. Am. Nat. 147:466–491.
GAVRILETS, S., AND S. M. SCHEINER. 1993. The genetics of phenotypic plasticity. V. Evolution of reaction norm shape. J. Evol. Biol. 6:31–48.
GAYMER, R., R. A. A. BLACKMAN, P. G. DAWSON, M. PENNY, AND C. M. PENNY. 1969. The endemic birds of the Seychelles. Ibis 111:157–176.
GEBHARDT-HENRICH, S. G. 2000. When heavier birds lose more mass during breeding: statistical artifact or biologically meaningful? J. Avian Biol. 31:245–246.
GEBHARDT-HENRICH, S. G., AND H. RICHNER. 1998. Causes of growth variation and its consequences for fitness. Pp. 324–339 in Avian Growth and Development: Evolution Within the Altricial–Precocial Spectrum (J. M. Starck and R. E. Ricklefs, Eds.). Oxford University Press, Oxford, United Kingdom.
GEIST, V. 1987. Bergmann's rule is invalid. Can. J. Zool. 65:1035–1038.
GEORGE, J. C., AND A. J. BERGER. 1966. Avian Myology. Academic Press, New York, New York.
GEORGE, J. C., T. M. JOHN, AND K. J. MINHAS. 1987. Seasonal degradative, reparative and regenerative ultrastructural changes in the breast muscle of the migratory Canada Goose. Cytobios 52:109–126.
GEORGE, J. C., AND K. K. NAIR. 1952. The wing-spread and its significance in the flight of some common Indian birds. J. Univ. Bombay 20:1–5.
GEORGE, J. C., AND R. M. NAIR. 1957. Studies on the structure and physiology of the flight muscles of birds. 1. The variations in the structure of the *pectoralis major* muscle of a few representative types and their significance in the respective modes of flight. J. Anim. Morphol. Physiol. 4: 23–32.

GEORGE, J. C., AND R. M. NAIR. 1959a. Studies on the structure and physiology of the flight muscles of birds. 5. Some histological and cytochemical observations on the structure of the *pectoralis*. J. Anim. Morphol. Physiol. 6:16–23.

GEORGE, J. C., AND R. M. NAIR. 1959b. Studies on the structure and physiology of the flight muscles of birds. 6. Variation in the diameter of the fibres of the *pectoralis major* and its relation to the muscle size and mode of flight. J. Anim. Morphol. Physiol. 6:90–94.

GEORGE, J. C., AND R. M. NAIR. 1959c. Studies on the structure and physiology of the flight muscles of birds. 7. Structure of the *pectoralis major* muscle of the Pigeon in disuse atrophy. J. Anim. Morphol. Physiol. 6:95–102.

GERITZ, S. A. H., E. KISDI, G. MESZÉNA, AND J. A. J. METZ. 1998. Evolutionarily singular strategies and the adaptive growth and branching of the evolutionary tree. Evol. Ecol. 12:35–57.

GHALAMBOR, C. K., AND T. E. MARTIN. 2001. Fecundity–survival trade-offs and parental risk-taking in birds. Science 292:494–497.

GHISELIN, M. T. 1974. The Economy of Nature and the Evolution of Sex. University of California Press, Berkeley, California.

GHISELIN, M. T. 1980. The failure of morphology to assimilate Darwinism. Pp. 180–193 *in* The Evolutionary Synthesis: Perspectives on the Unification of Biology (E. Mayr and W. B. Provine, Eds.). Harvard University Press, Cambridge, Massachusetts.

GIBBONS, D. W. 1986. Brood parasitism and cooperative nesting in the Moorhen, *Gallinula chloropus*. Behav. Ecol. Sociobiol. 19:221–232.

GIBBONS, D. W. 1987. Juvenile helping in the Moorhen, *Gallinula chloropus*. Anim. Behav. 85:170–181.

GIBBONS, D. W. 1989. Seasonal reproductive success of the Moorhen *Gallinula chloropus*: the importance of male weight. Ibis 131:57–68.

GIBBS, D. 1996. Notes on Solomon Island birds. Bull. Br. Ornithol. Club 116:18–25.

GIBBS, H. L. 1988. Heritability and selection of clutch size in Darwin's Medium Ground Finches (*Geospiza fortis*). Evolution 42:750–762.

GIBBS, H. L., A. W. GOLDIZEN, C. BULLOUGH, AND A. R. GOLDIZEN. 1994. Parentage analysis of multimale social groups of Tasmanian Native-hens (*Tribonyx mortierii*): genetic evidence for monogamy and polyandry. Behav. Ecol. Sociobiol. 35:363–371.

GIBBS, H. L., AND P. R. GRANT. 1987a. Oscillating selection on Darwin's finches. Nature 327:511–513.

GIBBS, H. L., AND P. R. GRANT. 1987b. Ecological consequences of an exceptionally strong El Niño event on Darwin's finches. Ecology 68:1735–1746.

GIBSON, A. R., A. J. BAKER, AND A. MOEED. 1984. Morphometric variation in introduced populations of the Common Myna (*Acridotheres tristis*): an application of the jackknife to principal component analysis. Syst. Zool. 33:408–412.

GIESEL, J. T. 1976. Reproductive strategies as adaptations to life in temporally heterogeneous environments. Annu. Rev. Ecol. Syst. 7:57–79.

GIFFORD, E. W. 1913. Expedition of the California Academy of Sciences to the Galápagos Islands, 1905–1906. VIII: The birds of the Galápagos Islands, with observations on the birds of Cocos and Clipperton islands (Columbiformes to Pelecaniformes). Proc. Calif. Acad. Sci. (Ser. 4) 2:1–132.

GILBERT, F. S. 1980. The equilibrium theory of island biogeography: Fact or fiction? J. Biogeogr. 7:209–235.

GILES, B. E., AND J. GOUDET. 1997. A case study of genetic structure in a plant metapopulation. Pp. 429–454 *in* Metapopulation Biology: Ecology, Genetics, and Evolution (I. Hanski and M. E. Gilpin, Eds.). Academic Press, San Diego, California.

GILL, F. B. 1973. Intra-island variation in the Mascarene White-eye (*Zosterops borbonica*). Ornithol. Monogr. 12. American Ornithologists' Union, Washington, D.C.

GILLESPIE, J. H. 1973. Polymorphism in random environments. Theor. Popul. Biol. 4:193–195.

GILLESPIE, J. H. 1974. Polymorphism in patchy environments. Am. Nat. 108:145–151.

GILLESPIE, J. H. 1975. The role of migration in the genetic structure of populations in temporally and spatially varying environments. I. Conditions for polymorphism. Am. Nat. 109:127–135.

GILLESPIE, J. H. 1981. The role of migration in the genetic structure of populations in temporally and spatially varying environments. III. Migration modification. Am. Nat. 117:223–233.

GILLESPIE, J. H. 2001. Is the population size of a species relevant to its evolution? Evolution 55: 2161–2169.
GILLILAND, S. G., AND C. D. ANKNEY. 1992. Estimating age of young birds with a multivariate measure of body size. Auk 109:444–450.
GILPIN, M. E. 1980. The role of stepping-stone islands. Theor. Popul. Biol. 17:247–253.
GILPIN, M. E. 1987. Spatial structure and population vulnerability. Pp. 125–139 in Viable Populations for Conservation (M. E. Soulé, Ed.). Cambridge University Press, Cambridge, United Kingdom.
GILPIN, M. E., AND J. M. DIAMOND. 1976. Calculation of immigration and extinction curves from the species–area–distance relation. Proc. Natl. Acad. Sci. USA 73:4130–4134.
GILPIN, M. E., AND J. M. DIAMOND. 1981. Immigration and extinction probabilities for individual species: relation to incidence functions and species colonization curves. Proc. Natl. Acad. Sci. USA 78:392–396.
GILPIN, M. E., AND J. M. DIAMOND. 1982. Factors contributing to non-randomness in species co-occurrences on islands. Oecologia 52:75–84.
GILPIN, M. E., AND J. M. DIAMOND. 1984. Are species co-occurrences on islands non-random, and are null hypotheses useful in community ecology? Pp. 297–315 in Ecological Communities: Conceptual Issues and the Evidence (D. R. Strong, Jr., D. Simberloff, L. G. Abele, and A. B. Thistle, Eds.). Princeton University Press, Princeton, New Jersey.
GILPIN, M. E., AND M. E. SOULÉ. 1986. Minimum viable populations: processes of species extinction. Pp. 19–34 in Conservation Biology: The Science of Scarcity and Diversity (M. E. Soulé, Ed.). Sinauer Assoc., Sunderland, Massachusetts.
GILROY, D. 1975. The determination of the rate constants of island colonization. Ecology 56:915–923.
GITTINS, R. 1985. Canonical Analysis: A Review with Applications in Ecology. Springer-Verlag, Berlin, Germany.
GITTLEMAN, J. L., AND H.-K. LUH. 1992. On comparing comparative methods. Annu. Rev. Ecol. Syst. 23:383–404.
GIVNISH, T. J., K. J. SYTSMA, J. F. SMITH, AND W. J. HAHN. 1995. Molecular evolution, adaptive radiation, and geographic speciation in *Cyanea* (Campanulaceae, Lobelioideae). Pp. 288–337 in Hawaiian Biogeography: Evolution on a Hot Spot Archipelago (W. L. Wagner and W. A. Funk, Eds.). Smithsonian Institution Press, Washington, D.C.
GLADE, T., M. CROZIER, AND P. SMITH. 2000. Applying probability determination to refine landslide-triggering rainfall thresholds using an empirical "antecedent daily rainfall model." Pure Appl. Geophys. 157:1059–1079.
GLADKOV, N. A. 1937a. The importance of length of wing for the bird's flight. Arch. Mus. Zool. Univ. Moscow 4:35–47.
GLADKOV, N. A. 1937b. [The weight of pectoral muscles and wings of birds in gonnection [sic] with the character of their flight.] Zool. Zh. 16:677–687.
GLEGG, W. E. 1945. Fishes and other aquatic animals preying on birds. Ibis 87:422–433.
GLEGG, W. E. 1947. Fishes and other aquatic animals preying on birds: additional matter. Ibis 89: 433–435.
GLENNY, F. H. 1959. Specific and individual variation in reduction of the clavicles in parrots. Ohio J. Sci. 59:321–322.
GLENNY, F. H., AND H. FRIEDMANN. 1954. Reduction of the clavicles in the Mesoenatidae, with some remarks concerning the relationship of the clavicle to flight-function in birds. Ohio J. Sci. 54: 111–113.
GLUTZ VON BLOTZHEIM, U. N. 1958. Zur Morphologie und Ontogenese von Schultergürtel, Sternum und Becken von *Struthio, Rhea* und *Dromiceius*. Rev. Suisse Zool. 65:609–772.
GLUTZ VON BLOTZHEIM, U. N., K. M. BAUER, AND E. BEZZEL, EDS. 1973. Handbuch der Vögel Mitteleuropas. Vol. 5. Akad. Verlag, Frankfurt, Germany.
GMELIN, J. F. [K. LINNÉ]. 1789. Systema Naturae. 13th ed. Vol. 2. Emanuel Beer, Leipsig, Germany.
GNANADESIKAN, R. 1977. Methods for Statistical Data Analysis of Multivariate Observations. J. Wiley and Sons, New York, New York.
GODDARD, C., A. GRAY, H. GILHOOLEY, AND I. E. O'NEILL. 1993. The molecular biology and genetic control of growth in poultry. Pp. 165–184 in Manipulation of the Avian Genome (R. J. Etches and A. M. Verrinder Gibbins, Eds.). CRC Press, Boca Raton, Florida.

GODFRAY, H. C. J., L. PARTRIDGE, AND P. H. HARVEY. 1991. Clutch size. Annu. Rev. Ecol. Syst. 22: 409–429.

GODFREY, L. R. 1985. Darwinian, Spencerian, and modern perspectives on progress in biological evolution. Pp. 40–60 in What Darwin Began: Modern Darwinian and Non-Darwinian Perspectives on Evolution (L. R. Godfrey, Ed.). Allyn and Bacon, Boston, Massachusetts.

GODFREY, L. R., AND M. R. SUTHERLAND. 1995. Flawed inference: why size-based tests of heterochronic processes do not work. J. Theor. Biol. 172:43–61.

GOFF, D. J., AND C. J. TABIN. 1997. Analysis of *Hoxd-13* and *Hoxd-11* misexpression in chick limb buds reveals that *Hox* genes affect both bone condensation and growth. Development 124:627–636.

GOLD, A. 1973. Energy expenditure in animal locomotion. Science 181:275–276.

GOLDIZEN, A. W., J. C. BUCHAN, D. A. PUTLAND, A. R. GOLDIZEN, AND E. A. KREBS. 2000. Patterns of mate-sharing in a population of Tasmanian Native Hens *Gallinula mortierii*. Ibis 142:40–47.

GOLDIZEN, A. W., A. R. GOLDIZEN, AND T. DEVLIN. 1993. Unstable social structure associated with a population crash in the Tasmanian Native Hen *Tribonyx mortierii*. Anim. Behav. 46:1013–1016.

GOLDIZEN, A. W., D. A. PUTLAND, AND A. R. GOLDIZEN. 1998a. Variable mating patterns in Tasmanian Native Hens (*Gallinula mortierii*): correlates of reproductive success. J. Anim. Ecol. 67:307–317.

GOLDIZEN, A. W., A. R. GOLDIZEN, D. A. PUTLAND, D. M. LAMBERT, C. D. MILLAR, AND J. C. BUCHAN. 1998b. "Wife-sharing" in the Tasmanian Native Hen (*Gallinula mortierii*): Is it caused by a male-biased sex ratio? Auk 115:528–532.

GOLDIZEN, A. W., D. A. PUTLAND, AND K. A. ROBERTSON. 2002. Dispersal strategies in Tasmanian Native-hens. (*Gallinunla mortierii*). Behav. Ecol. 13:328–336.

GOLDSCHMIDT, R. 1940. The Material Basis of Evolution. Yale University Press, New Haven, Connecticut.

GOLDSPINK, G. 1977a. Energy cost of locomotion. Pp. 153–167 in Mechanics and Energetics of Animal Locomotion (R. McN. Alexander and G. Goldspink, Eds.). Chapman and Hall, London, United Kingdom.

GOLDSPINK, G. 1977b. Mechanics and energetics of muscle in animals of different sizes, with particular reference to the muscle fibre composition of vertebrate muscle. Pp. 37–55 in Scale Effects in Animal Locomotion (T. J. Pedley, Ed.). Academic Press, London, United Kingdom.

GOLDSPINK, G. 1981. The use of muscles during flying, swimming, and running from the point of view of energy saving. Symp. Zool. Soc. Lond. 48:219–238.

GOMULKIEWICZ, R., AND R. D. HOLT. 1995. When does evolution by natural selection prevent extinction? Evolution 49:201–207.

GOMULKIEWICZ, R., AND M. KIRKPATRICK. 1992. Quantitative genetics and the evolution of reaction norms. Evolution 46:390–411.

GONZÁLEZ, P. M., T. PIERSMA, AND Y. VERKUIL. 1996. Food, feeding, and refueling of Red Knots during northward migration at San Antonio Oeste, Rio Negro, Argentina. J. Field Ornithol. 67: 575–591.

GOODMAN, D. 1984. Risk spreading as an adaptive strategy in interoparous life histories. Theor. Popul. Biol. 25:1–20.

GOODMAN, S. M., AND P. C. GONZALES. 1989. Notes on Philippine birds, 12. Seven species new to Catanduanes Island. Bull. Br. Ornithol. Club 109:48–50.

GOODMAN, S. M., D. E. WILLARD, AND P. C. GONZALES. 1995. The birds of Sibuyan Island, Romblon Province, Philippines, with particular reference to elevational distribution and biogeographic affinities. Fieldiana, Zool. (New Ser.) 82:1–57.

GOODNIGHT, C. J. 1987. On the effect of founder events on the epistatic genetic variance. Evolution 41:80–91.

GOODNIGHT, C. J. 1988. Epistasis and the effect of founder events on the additive genetic variance. Evolution 42:441–454.

GOODNIGHT, C. J. 2000. Modeling gene interaction in structured populations. Pp. 129–145 in Epistasis and the Evolutionary Process (J. B. Wolf, E. D. Brodie III, and M. J. Wade, Eds.). Oxford University Press, Oxford, United Kingdom.

GOODRICH, E. S. 1930. Studies on the Structure and Development of Vertebrates. Macmillan, London, United Kingdom.

GOODWIN, B. C. 1982. Development and evolution. J. Theor. Biol. 97:43–55.
GOODWIN, B. C. 1984. Changing from an evolutionary to a generative paradigm in biology. Pp. 99–120 in Evolutionary Theory: Paths Into the Future (J. W. Pollard, Ed.). J. Wiley and Sons, Chichester, United Kingdom.
GOODWIN, D. 1974. Gruiformes. Pp. 62–66 in Birds of the Harold Hall Australian Expeditions: 1962–70 (B. P. Hall, Ed.). British Museum (Natural History), London, United Kingdom.
GORDON, M. S. 1977. Animal Physiology: Principles and Adaptations. 3rd ed. Macmillan, New York, New York.
GORENZEL, W. P., R. A. RYDER, AND C. E. BRAUN. 1982. Reproduction and nest site characteristics of American Coots at different altitudes in Colorado. Condor 84:59–65.
GORMAN, M. L. 1975. Habitats of the land-birds of Viti Levu, Fiji Islands. Ibis 117:152–161.
GOSLER, A. G., J. J. D. GREENWOOD, AND C. PERRINS. 1995. Predation risk and the cost of being fat. Nature 377:621–623.
GOSLOW, G. E., JR., RAPPORTEUR. 1989. How are locomotor systems integrated and how have evolutionary innovations been introduced? Pp. 205–218 in Complex Organismal Functions: Integration and Evolution in Vertebrates (D. B. Wake and G. Roth, Eds.). J. Wiley and Sons, Chichester, United Kingdom.
GOSLOW, G. E., JR., K. P. DIAL, AND F. A. JENKINS. 1989. The avian shoulder: an experimental approach. Am. Zool. 29:287–301.
GOTELLI, N. J. 1991. Metapopulation models: the rescue effect, the propagule rain, and the core-satellite hypothesis. Am. Nat. 138:768–776.
GOTELLI, N. J., AND L. G. ABELE. 1982. Statistical distributions of West Indian land bird families. J. Biogeogr. 9:421–435.
GOTELLI, N. J., AND G. R. GRAVES. 1996. Null Models in Ecology. Smithsonian Institution Press, Washington, D.C.
GÖTMARK, F. 1993. Conspicuous coloration in male birds is favoured by predation in some species and disfavoured in others. Proc. R. Soc. Lond. (Ser. B Biol. Sci.) 253:143–146.
GOTTHARD, K., AND S. NYLIN. 1995. Adaptive plasticity and plasticity as an adaptation: a selective review of plasticity in animal morphology and life history. Oikos 74:3–17.
GOULD, J. 1852a. Remarks on Notornis mantelli. Proc. Zool. Soc. Lond. 1850:212–214.
GOULD, J. 1852b. Remarks on Notornis mantellii [sic]. Trans. Zool. Soc. Lond. 4:73–74.
GOULD, S. J. 1966. Allometry and size in ontogeny and phylogeny. Biol. Rev. 41:587–640.
GOULD, S. J. 1968. Ontogeny and the explanation of form: an allometric analysis. J. Paleontol. 42(Suppl.):81–98.
GOULD, S. J. 1970a. Evolutionary paleontology and the science of form. Earth-Sci. Rev. 6:77–119.
GOULD, S. J. 1970b. Dollo on Dollo's law: irreversibility and the status of evolutionary laws. J. Hist. Biol. 3:189–212.
GOULD, S. J. 1971. Geometric similarity in allometric growth: a contribution to the problem of scaling in the evolution of size. Am. Nat. 105:113–136.
GOULD, S. J. 1972. Allometric fallacies and the evolution of Gryphaea: a new interpretation based on White's criterion of geometric similarity. Pp. 91–118 in Evolutionary Biology. Vol. 6 (T. Dobzhansky, M. K. Hecht, and W. C. Steere, Eds.). Appleton-Century-Crofts, New York, New York.
GOULD, S. J. 1974. The evolutionary significance of "bizarre" structures: antler size and skull size in the "Irish Elk," Megaloceres giganteus. Evolution 28:191–220.
GOULD, S. J. 1977. Ontogeny and Phylogeny. Belknap Press, Cambridge, Massachusetts.
GOULD, S. J. 1979. An allometric interpretation of species–area curves: the meaning of the coefficient. Am. Nat. 114:335–343.
GOULD, S. J. 1980a. The evolutionary biology of constraint. Daedalus 109:39–52.
GOULD, S. J. 1980b. Is a new and general theory of evolution emerging? Paleobiology 6:119–130.
GOULD, S. J. 1980c. Paleontology. Pp. 153–172 in The Evolutionary Synthesis: Perspectives on the Unification of Biology (E. Mayr and W. B. Provine, Eds.). Harvard University Press, Cambridge, Massachusetts.
GOULD, S. J. 1982. Change in developmental timing as a mechanism of macroevolution. Pp. 333–346 in Evolution and Development (J. T. Bonner, Ed.). Springer-Verlag, Berlin, Germany.
GOULD, S. J. 1984. Morphological channelling by structural constraint: convergence in styles of

dwarfing and gigantism. *Cerion*, with a description of two new fossil species and a report on the discovery of the largest *Cerion*. Paleobiology 10:172–194.

GOULD, S. J. 1988a. The uses of heterochrony. Pp. 1–13 *in* Heterochrony in Evolution (M. L. McKinney, Ed.). Plenum Press, New York, New York.

GOULD, S. J. 1988b. Trends as changes in variance: a new slant on progress and directionality in evolution. J. Paleontol. 62:319–329.

GOULD, S. J. 1988c. On replacing the idea of progress with an operational notion of directionality. Pp. 319–338 *in* Evolutionary Progress (M. H. Nitecki, Ed.). University of Chicago Press, Chicago, Illinois.

GOULD, S. J. 1990. Speciation and sorting as the source of evolutionary trends, or 'things are seldom what they seem.' Pp. 3–27 *in* Evolutionary Trends (K. J. McNamara, Ed.). University of Arizona Press, Tucson, Arizona.

GOULD, S. J. 1997. The exaptive excellence of spandrels as a term and prototype. Proc. Natl. Acad. Sci. USA 94:10750–10755.

GOULD, S. J., AND N. ELDREDGE. 1977. Punctuated equilibria: the tempo and mode of evolution reconsidered. Paleobiology 3:115–151.

GOULD, S. J., AND R. F. JOHNSTON. 1972. Geographic variation. Annu. Rev. Ecol. Syst. 3:457–498.

GOULD, S. J., AND R. C. LEWONTIN. 1979. The spandrels of San Marco and the Panglossian paradigm: a critque of the adaptionist programme. Proc. R. Soc. Lond. (Ser. B Biol. Sci.) 205:581–598.

GOULD, S. J., AND E. S. VRBA. 1982. Exaptation—a missing term in the science of form. Paleobiology 8:4–15.

GRAHAM, J. H., J. M. EMLEN, D. C. FREEMAN, L. J. LEAMY, AND J. A. KIESER. 1998. Directional asymmetry and the measurement of developmental instability. Biol. J. Linn. Soc. 64:1–16.

GRAJAL, A., S. D. STRAHL, R. PARRA, M. G. DOMINQUEZ, AND A. NEHER. 1989. Foregut fermentation in the Hoatzin, a Neotropical leaf-eating bird. Science 245:1236–1238.

GRANT, B. R. 1985. Selection on bill characters in a population of Darwin's finches: *Geospiza conirostris* on Isla Genovesa, Galápagos. Evolution 39:523–532.

GRANT, B. R., AND P. R. GRANT. 1979. Darwin's finches: population variation and sympatric speciation. Proc. Natl. Acad. Sci. USA 76:2359–2363.

GRANT, B. R., AND P. R. GRANT. 1982. Niche shifts and competition in Darwin's finches: *Geospiza conirostris* and congeners. Evolution 36:637–657.

GRANT, B. R., AND P. R. GRANT. 1983. Fission and fusion in a population of Darwin's finches: an example of the value of studying individuals in ecology. Oikos 41:530–547.

GRANT, B. R., AND P. R. GRANT. 1989. Evolutionary Dynamics of a Natural Population: The Large Cactus Finch of the Galápagos. University of Chicago Press, Chicago, Illinois.

GRANT, B. R., AND P. R. GRANT. 1993. Evolution of Darwin's finches caused by a rare climatic event. Proc. R. Soc. Lond. (Ser. B Biol. Sci.) 251:111–117.

GRANT, C. 1947. Frigate birds and the Laysan Rail. Condor 49:130.

GRANT, G. S. 1982. Avian incubation: egg temperature, nest humidity, and behavioral thermoregulation in a hot environment. Ornithol. Monogr. 30. American Ornithologists' Union, Washington, D.C.

GRANT, P. R. 1965a. The adaptive significance of some size trends in island birds. Evolution 19:355–367.

GRANT, P. R. 1965b. Plumage and the evolution of birds on islands. Syst. Zool. 14:47–52.

GRANT, P. R. 1965c. The fat condition of some island birds. Ibis 107:350–356.

GRANT, P. R. 1966a. Ecological compatibility of bird species on islands. Am. Nat. 100:451–462.

GRANT, P. R. 1966b. Further information on the relative length of the tarsus in land birds. Postilla 98:1–13.

GRANT, P. R. 1966c. The density of land birds on the Tres Marias Islands in Mexico. I. Numbers and biomass. Can. J. Zool. 44:391–400.

GRANT, P. R. 1966d. The density of land birds on the Tres Marias Islands in Mexico. II. Distribution of abundances in the community. Can. J. Zool. 44:1023–1030.

GRANT, P. R. 1967. Bill length variability in birds of the Tres Marias Islands, Mexico. Can. J. Zool. 45:805–815.

GRANT, P. R. 1968. Bill size, body size, and the ecological adaptations of birds species to competitive situations on islands. Syst. Zool. 17:319–333.

GRANT, P. R. 1969. Colonization of islands by ecologically dissimilar species of birds. Can. J. Zool. 47:41–43.
GRANT, P. R. 1971. Variation in the tarsus length of birds in island and mainland regions. Evolution 25:599–614.
GRANT, P. R. 1972a. Centripetal selection and the House Sparrow. Syst. Zool. 21:23–30.
GRANT, P. R. 1972b. Convergent and divergent character displacement. Biol. J. Linn. Soc. 4:39–68.
GRANT, P. R. 1972c. Bill dimensions of the three species of *Zosterops* on Norfolk Island. Syst. Zool. 21:289–291.
GRANT, P. R. 1975. The classical case of character displacement. Pp. 237–337 *in* Evolutionary Biology. Vol. 8 (T. D. Dobzhansky, M. K. Hecht, and W. C. Steere, Eds.). Plenum Press, New York, New York.
GRANT, P. R. 1976. Population variation on islands. Pp. 603–615 *in* Proceedings of the 16th International Ornithological Congress (H. J. Frith and J. H. Calaby, Eds.). Australian Academy of Science, Canberra, Australia.
GRANT, P. R. 1978. Recent evolution of *Zosterops lateralis* on Norfolk Island, Australia. Can. J. Zool. 56:1624–1626.
GRANT, P. R. 1979a. Ecological and morphological variation of Canary Island Blue Tits, *Parus caeruleus* (Aves: Paridae). Biol. J. Linn. Soc. 11:103–129.
GRANT, P. R. 1979b. Evolution of the Chaffinch, *Fringilla coelebs,* on the Atlantic Islands. Biol. J. Linn. Soc. 11:301–332.
GRANT, P. R. 1981a. Patterns of growth in Darwin's finches. Proc. R. Soc. Lond. (Ser. B Biol. Sci.) 212:403–432.
GRANT, P. R. 1981b. Speciation and the adaptive radiation of Darwin's finches. Am. Sci. 69:653–663.
GRANT, P. R. 1983a. The relative size of Darwin's finch eggs. Auk 100:228–229.
GRANT, P. R. 1983b. Inheritance of size and shape in a population of Darwin's finches, *Geospiza conirostris*. Proc. R. Soc. Lond. (Ser. B Biol. Sci.) 220:219–236.
GRANT, P. R. 1983c. The role of interspecific competition in the adaptive radiation of Darwin's finches. Pp. 187–199 *in* Patterns of Evolution in Galápagos Organisms (A. Levinton and R. I. Bowman, Eds.). Spec. Publ. 1. American Associaton for the Advancement of Science, Pacific Division, San Francisco, California.
GRANT, P. R. 1986. Ecology and Evolution of Darwin's Finches. Princeton University Press, Princeton, New Jersey.
GRANT, P. R. 1992. Systematics and micro-evolution. Bull. Br. Ornithol. Club (Cent. Suppl.) 112A: 97–106.
GRANT, P. R. 1994a. Ecological character displacement. Science 266:746–747.
GRANT, P. R. 1994b. Population variation and hybridization: comparison of finches from two archipelagos. Evol. Ecol. 8:598–617.
GRANT, P. R. 1998a. Patterns on islands and microevolution. Pp. 1–17 *in* Evolution on Islands (P. R. Grant, Ed.). Oxford University Press, Oxford, United Kingdom.
GRANT, P. R. 1998b. Radiations, communities, and biogeography. Pp. 196–209 *in* Evolution on Islands (P. R. Grant, Ed.). Oxford University Press, Oxford, United Kingdom.
GRANT, P. R. 1998c. Speciation. Pp. 82–101 *in* Evolution on Islands (P. R. Grant, Ed.). Oxford University Press, Oxford, United Kingdom.
GRANT, P. R. 1998d. Epilogue and questions. Pp. 305–319 *in* Evolution on Islands (P. R. Grant, Ed.). Oxford University Press, Oxford, United Kingdom.
GRANT, P. R., AND I. ABBOTT. 1980. Interspecific competition, island biogeography and null hypotheses. Evolution 34:332–341.
GRANT, P. R., AND B. R. GRANT. 1987. The extraordinary El Niño event of 1982–83: effects on Darwin's finches on Isla Genovesa, Galápagos. Oikos 49:55–66.
GRANT, P. R., AND B. R. GRANT. 1989. Sympatric speciation and Darwin's finches. Pp. 433–457 *in* Speciation and Its Consequences (D. Otte and J. A. Endler, Eds.). Sinauer, Sunderland, Massachusetts.
GRANT, P. R., AND B. R. GRANT. 1995. The founding of a new population of Darwin's finches. Evolution 49:229–240.
GRANT, P. R., AND B. R. GRANT. 1996. Speciation and hybridization in island birds. Philos. Trans. R. Soc. Lond. (Ser. B) 351:765–772.

GRANT, P. R., AND B. R. GRANT. 1998. Speciation and hybridization of birds on islands. Pp. 142–162 in Evolution on Islands (P. R. Grant, Ed.). Oxford University Press, Oxford, United Kingdom.

GRANT, P. R., AND B. R. GRANT. 2000. Quantitative variation in populations of Darwin's finches. Pp. 3–40 in Adaptive Genetic Variation in the Wild (T. A. Mousseau, B. Sinervo, and J. A. Endler, Eds.). Oxford University Press, Oxford, United Kingdom.

GRANT, P. R., B. R. GRANT, J. N. M. SMITH, I. J. ABBOTT, AND L. K. ABBOTT. 1976. Darwin's finches: population variation and natural selection. Proc. Natl. Acad. Sci. USA 73:257–261.

GRANT, P. R., AND D. SCHLUTER. 1984. Interspecific competition inferred from patterns of guild structure. Pp. 201–233 in Ecological Communities: Conceptual Issues and the Evidence (D. R. Strong, Jr., D. Simberloff, L. G. Abele, and A. B. Thistle, Eds.). Princeton University Press, Princeton, New Jersey.

GRASSÉ, P.-P. 1955. Ordre de Édentés. Pp. 1182–1266 in Traité de Zoologie. Vol. 17: Mammifères (P.-P. Grassé, Ed.). Masson et Cie, Paris, France.

GRASSÉ, P.-P. 1977. Evolution of Living Organisms: Evidence for a New Theory of Transformation. Academic Press, New York, New York.

GRAVES, G. R. 1992. The endemic land birds of Henderson Island, southeastern Polynesia: notes on natural history and conservation. Wilson Bull. 104:32–43.

GRAVES, G. R. 2001. Crayfish-processing behavior in the King Rail (*Rallus elegans*). J. Elisha Mitchell Sci. Soc. 117:71–72.

GRAVES, G. R., AND N. J. GOTELLI. 1983. Neotropical land-bridge islands: new approaches to null hypotheses in biogeography. Oikos 41:322–333.

GRAY, B. F. 1981. On the "surface law" and basal metabolic rate. J. Theor. Biol. 93:757–767.

GRAY, G. R. 1862. A list of the birds of New Zealand and the adjacent islands. Ibis 4:214–252.

GRAY, J. E. 1844. List of Specimens of Birds in the Collection of the British Museum, Part 3: Gallinæ, Grallæ, and Anseres. George Woodfall and Son, London, England.

GREEN, A. J. 2000. The scaling and selection of sexually dimorphic characters: an example using the Marbled Teal. J. Avian Biol. 31:345–350.

GREEN, A. J., J. FIGUEROLA, AND R. KING. 2001. Comparing interspecific and intraspecific allometry in the Anatidae. J. Ornithol. 142:321–334.

GREEN, M. 2000. Flight speeds and climb rates of Brent Geese: mass-dependent differences between spring and autumn migration. J. Avian Biol. 31:215–225.

GREEN, R. E., AND M. D. RAYMENT. 1996. Geographical variation in the abundance of the Corncrake *Crex crex* in Europe in relation to the intensity of agriculture. Bird Conserv. Int. 6:201–211.

GREENACRE, M. J. 1984. Theory and Applications of Correspondence Analysis. Academic Press, London, United Kingdom.

GREENBERG, R. S. 1983. The role of neophobia in determining foraging specialization of some migratory warblers. Am. Nat. 122:444–453.

GREENBERG, R. S. 1984a. Differences in feeding neophobia in tropical warblers, *Dendroica castanea* and *D. pensylvanica*. J. Comp. Psychol. 98:131–136.

GREENBERG, R. S. 1984b. Neophobia in the foraging site selection of a Neotropical migrant bird: an experimental study. Proc. Natl. Acad. Sci. USA 81:3778–3780.

GREENBERG, R. S. 1990. Ecological plasticity, neophobia, and resource use in birds. Pp. 431–437 in Avian Foraging: Theory, Methodology, and Applications (M. L. Morrison, C. J. Ralph, J. Verner, and J. R. Jehl, Jr., Eds.). Stud. Avian Biol. No. 13. Cooper Ornithological Society, Los Angeles, California.

GREENBERG, R. S., P. J. CORDERO, S. DROEGE, AND R. C. FLEISCHER. 1998. Morphological adaptation with no mitochondrial differentiation in the coastal plain Swamp Sparrow. Auk 115:706–712.

GREENBERG, R. S., AND C. METTKE-HOFMANN. 2001. Ecological aspects of neophobia and neophilia in birds. Pp. 119–178 in Current Ornithology. Vol. 16 (V. Nolan, Jr., and C. F. Thompson, Eds.). Kluwer Academic–Plenum Publishing, New York, New York.

GREENE, H. W., AND D. CUNDALL. 2000. Limbless tetrapods and snakes with legs. Science 287:1939–1941.

GREENEWALT, C. H. 1962. Dimensional relationships for flying animals. Smithson. Misc. Coll. 155:1–46.

GREENEWALT, C. H. 1975. The flight of birds. Trans. Am. Philos. Soc. (New Ser.) 65:1–67.

GREENEWALT, C. H. 1977. The energetics of locomotion—Is small size really disadvantageous? Proc. Am. Philos. Soc. 121:100–106.

GREENLAW, J. S. 1993. Behavioral and morphological diversification in Sharp-tailed Sparrows (*Ammodramus caudacutus*) of the Atlantic coast. Auk 110:286–303.
GREENSLADE, P. J. M. 1968. Island patterns in the Solomon Islands bird fauna. Evolution 22:751–761.
GREENWAY, J. C., JR. 1952. *Tricholimnas conditicius* is probably a synonym of *T. sylvestris* (Aves, Rallidae). Breviora 5:1–4.
GREENWAY, J. C., JR. 1967. Extinct and Vanishing Birds of the World. 2nd ed. Dover Publications, New York, New York.
GREENWAY, J. C., JR. 1968. Family Drepanididae, Hawaiian honeycreepers. Pp. 93–103 *in* Check-list of Birds of the World. Vol. 14 (R. A. Paynter, Jr., Ed.). Museum of Comparative Zoology, Harvard University, Cambridge, Massachusetts.
GREENWAY, J. C., JR. 1973. Type specimens of birds in the American Museum of Natural History, part 1. Bull. Am. Mus. Nat. Hist. 150:207–346.
GREENWOOD, P. H. 1989. Ontogeny and evolution: saltatory or otherwise? Pp. 245–259 *in* Alternative Life-History Styles of Animals (M. N. Bruton, Ed.). Kluwer Academic Publications, Dordrecht, The Netherlands.
GREENWOOD, P. J. 1980. Mating systems, philopatry and dispersal in birds and mammals. Anim. Behav. 28:1140–1162.
GREENWOOD, P. J., AND P. H. HARVEY. 1982. Natal and breeding dispersal in birds. Annu. Rev. Ecol. Syst. 13:1–21.
GREENWOOD, P. J., AND P. WHEELER. 1985. The evolution of sexual size dimorphism in birds and mammals: a 'hot-blooded hypothesis.' Pp. 287–299 *in* Evolution: Essays in Honour of John Maynard Smith (P. J. Greenwood, P. H. Harvey, and M. Slatkin, Eds.). Cambridge University Press, Cambridge, United Kingdom.
GREER, A. E. 1987. Limb reduction in the lizard genus *Lerista*. 1. Variation in the number of phalanges and presacral vertebrae. J. Herpetol. 21:267–276.
GREER, A. E. 1990. Limb reduction in the scincid genus *Lerista*. 2. Variation in the bone complements of the front and rear limbs and the number of presacral vertebrae. J. Herpetol. 24:142–150.
GREGORY, P. 1996. The New Guinea Flightless Rail (*Megacrex inepta*) in Gulf Province. Muruk 8: 38–39.
GREHAN, J. R., AND R. AINSWORTH. 1985. Orthogenesis and evolution. Syst. Zool. 34:174–192.
GREIG-SMITH, P. W., AND N. C. DAVIDSON. 1977. Weights of west African savanna birds. Bull. Br. Ornithol. Club 97:96–99.
GRESSITT, J. L. 1954. Insects of Micronesia: introduction. Bishop Mus. Occas. Pap. 1:1–257.
GRESSITT, J. L. 1956. Faunal distribution on Pacific islands. Syst. Zool. 5:11–32.
GRESSITT, J. L. 1961. Problems in the zoogeography of Pacific and Antarctic insects. Pac. Insects Monogr. 2:1–94.
GRESSITT, J. L. 1964. Insects of Campbell Island: summary. Pac. Insects Monogr. 7:531–600.
GRIFFITH, H. 1990. Miniaturization and elongation in *Eumeces* (Sauria: Scincidae). Copeia 1990:751–758.
GRIFFITHS, C. S. 1994. Monophyly of the Falconiformes based on syringeal morphology. Auk 111: 787–805.
GRINNELL, J. 1914. Barriers to distribution as regards birds and mammals. Am. Nat. 48:248–254.
GRINNELL, J. 1917. Field tests of theories concerning distributional control. Am. Nat. 51:115–128.
GROTH, J. G. 1988. Resolution of cryptic species in Appalachian Red Crossbills. Condor 90:745–760.
GROTH, J. G. 1993. Evolutionary differentiation in morphology, vocalizations, and allozymes among nomadic sibling species in the North American Red Crossbill (*Loxia curvirostra*) complex. Univ. Calif. Publ. Zool. 127:1–143.
GROTH, J. G., AND G. F. BARROWCLOUGH. 1999. Basal divergences in birds and the phylogenetic utility of the nuclear *RAG-1* gene. Mol. Phylogenet. Evol. 12:115–123.
GRUBB, B. 1983. Allometric relations of cardiovascular function in birds. Am. J. Physiol. 245:567–572.
GRUSON, E. S. 1972. Words for Birds: A Lexicon of North American Birds with Biographical Notes. Quadrangle Books, New York, New York.
GUILDAY, J. E. 1967. Differential extinction during Late-Pleistocene and Recent times. Pp. 121–140

in Pleistocene Extinctions: The Search for a Cause (P. S. Martin and H. E. Wright, Eds.). Yale University Press, New Haven, Connecticut.

GUILLEMETTE, M. 1994. Digestive-rate constraint in wintering Common Eiders (*Somateria mollissima*): implications for flying capabilities. Auk 111:900–909.

GUILLEMETTE, M. 2001. Foraging before spring migration and before breeding in Common Eiders: Does hyperphagia occur? Condor 103:633–638.

GULICK, A. 1932. Biological peculiarities of oceanic islands. Q. Rev. Biol. 7:405–427.

GULLION, G. W. 1951. The frontal shield of the American Coot. Wilson Bull. 63:157–166.

GULLION, G. W. 1952. The displays and calls of the American Coot. Wilson Bull. 64:83–97.

GULLION, G. W. 1953a. Observation on molting of the American Coot. Wilson Bull. 55:169–186.

GULLION, G. W. 1953b. Territorial behaviour in the American Coot. Condor 55:169–186.

GULLION, G. W. 1954. The reproductive cycle of American Coots in California. Auk 71:366–412.

GÜNTHER, A., AND E. NEWTON. 1879. The extinct birds of Rodriguez. Philos. Trans. R. Soc. Lond. 168:423–437.

GÜNTHER, B. 1975. Dimensional analysis and theory of biological similarity. Physiol. Rev. 55:659–699.

GÜNTHER, B., AND E. MORGADO. 1982. Theory of biological similarity revisited. J. Theor. Biol. 96:543–559.

GURR, L. 1951. Food of the chick of *Notornis hochstetteri*. Notornis 4:114.

GUSTAFSSON, L. 1986. Lifetime reproductive success and heritability: empirical support for Fisher's fundamental theorem. Am. Nat. 128:761–764.

GUTHRIE, D. A. 1992. A late Pleistocene avifauna from San Miguel Island, California. Pp. 319–327 *in* Papers in Avian Paleontology Honoring Pierce Brodkorb (K. E. Campbell, Jr., Ed.). Natural History Museum of Los Angeles County, Los Angeles, California.

GUTHRIE, R. D. 1984. Alaskan megabucks, megabulls, and megarams: the issue of Pleistocene gigantism. Pp. 482–510 *in* Contributions in Quaternary Vertebrate Paleontology: A Volume in Memorial to John E. Guilday (H. H. Genoways and M. R. Dawson, Eds.). Spec. Publ. No. 8. Carnegie Museum of Natural History, Pittsburgh, Pennsylvania.

GUYER, C., AND J. B. SLOWINSKI. 1993. Adaptive radiation and the topology of large phylogenies. Evolution 47:253–263.

GYLLENBERG, M., I. HANSKI, AND A. HASTINGS. 1997. Structured metapopulation models. Pp. 93–122 *in* Metapopulation Biology: Ecology, Genetics, and Evolution (I. Hanski and M. E. Gilpin, Eds.). Academic Press, San Diego, California.

HAARTMAN, L. VON. 1968. The evolution of resident versus migratory habit in birds. Some considerations. Ornis Fennica 45:1–7.

HACHISUKA, M. 1937. On the flightless heron of Rodriguez. Proc. Biol. Soc. Wash. 50:145–150.

HACHISUKA, M. 1953. The Dodo and Kindred Birds or the Extinct Birds of the Mascarene Islands. H. F. and G. Witherby, London, United Kingdom.

HACKETT, S. J., C. S. GRIFFITHS, J. M. BATES, AND N. KLEIN. 1995. One kilobase sequences demonstrate that the Hoatzin is not a chicken: a commentary on Avise et al. 1994. Mol. Phylogenet. Evol. 4:350–353.

HACKMAN, W. 1964. On reduction and loss of wings in Diptera. Notul. Entomol. 44:73–93.

HACKMAN, W. 1966. On wing reduction and loss of wings in Lepidoptera. Notul. Entomol. 46:1–16.

HADDEN, D. 2002. Woodford's Rail (*Nesoclopeus woodfordi*) on Bougainville Island, Papua New Guinea. Notornis 49:115–121.

HADDRATH, O., AND A. J. BAKER. 2001. Complete mitochondrial DNA genome sequences of extinct birds: ratite phylogenetics and the vicariance biobeography hypothesis. Proc. R. Soc. Lond. (Ser. B) 268:939–945.

HAECKEL, E. 1866. Generelle Morphologie der Organismen: Allgemeine Grundzüge der organischen Formen-Wissenschaft, mechanisch begründet durch die von Charles Darwin reformirte Descendez-Theorie. 2 vols. Georg Reimer, Berlin, Germany.

HAFNER, J. C., AND M. S. HAFNER. 1988. Heterochrony in rodents. Pp. 217–235 *in* Heterochrony in Evolution: A Multidisciplinary Approach (M. L. McKinney, Ed.). Plenum Press, New York, New York.

HAGEN, Y. 1952. Birds of Tristan da Cunha. Results of the Norwegian Scientific Expedition of Tristan da Cunha 1937–1938. No. 20. Kommisjon hos Jacob Dybwad, Oslo, Norway.

HAHN, P. 1963. Where Is That Vanished Bird? Royal Ontario Museum, University of Toronto, Toronto, Ontario, Canada.
HAIG, S. M., AND J. D. BALLOU. 1995. Genetic diversity in two avian species formerly endemic to Guam. Auk 112:445–455.
HAIG, S. M., J. D. BALLOU, AND N. J. CASNA. 1994. Identification of kin structure among Guam Rail founders: a comparison of pedigrees and DNA profiles. Mol. Ecol. 4:109–119.
HAIG, S. M., J. D. BALLOU, AND S. R. DERRICKSON. 1990. Management options for preserving genetic diversity: reintroduction of Guam Rails to the wild. Conserv. Biol. 4:290–300, 464.
HAILA, Y., AND I. K. HANSKI. 1993. Breeding birds on small British islands and extinction risks. Am. Nat. 142:1025–1029.
HAILA, Y., AND O. JÄRVINEN. 1982. The role of theoretical concepts in understanding the ecological theatre: a case study on island biogeography. Pp. 261–278 *in* Conceptual Issues in Ecology (E. Saarinen, Ed.). Reidel, Dordrecht, The Netherlands.
HAILA, Y., O. JÄRVINEN, AND S. KUUSELA. 1983. Colonization of islands by land birds: prevalence functions in a Finnish archipelago. J. Biogeogr. 10:499–531.
HAILA, Y., O. JARVINEN, AND R. A. VÄISÄNEN. 1979. Effect of mainland population changes on the terrestrial fauna of a northern island. Ornis Scand. 10:48–55.
HAILMAN, J. P. 1988. Operationalism, optimality and optimism: suitabilities versus adaptations of organisms. Pp. 85–116 *in* Evolutionary Processes and Metaphors (M.-W. Ho and S. W. Fox, Eds.). J. Wiley and Sons, Chicester, United Kingdom.
HAILS, C. J. 1979. A comparison of flight energetics in hirundines and other birds. Comp. Biochem. Physiol. (Ser. A) 63:581–585.
HAILS, C. J. 1983. The metabolic rate of tropical birds. Condor 85:61–65.
HAKKARAINEN, H., E. KORPIMÄKI, E. HUHTA, AND P. PALOKANGAS. 1993. Delayed maturation in plumage colour: evidence for the female-mimicry hypothesis in the Kestrel. Behav. Ecol. Sociobiol. 33:247–251.
HALDANE, J. B. S. 1928. On being the right size. Pp. 18–26 *in* Possible Worlds and Other Essays. Chatto and Windus, London, United Kingdom.
HALDANE, J. B. S. 1932. The time of action of genes, and its bearing on some evolutionary problems. Am. Nat. 66:5–24.
HALDANE, J. B. S., AND S. D. JAYAKAR. 1963. Polymorphism due to selection of varying direction. J. Genet. 58:237–242.
HALL, B. K. 1984a. Developmental processes underlying heterochrony as an evolutionary mechanism. Can. J. Zool. 62:1–7.
HALL, B. K. 1984b. Developmental mechanisms underlying the formation of atavisms. Biol. Rev. 59: 89–124.
HALL, B. K. 1990a. Heterochrony in vertebrate development. Semin. Dev. Biol. 1:237–243.
HALL, B. K. 1990b. Genetic and epigenetic contral of vertebrate embryonic development. Neth. J. Zool. 40:352–361.
HALL, B. K. 1992. Waddington's legacy in development and evolution. Am. Zool. 32:113–122.
HALL, B. K. 1996. *Baupläne,* phylotypic stages, and constraint: why there are so few types of animals. Pp. 215–261 *in* Evolutionary Biology. Vol. 29 (M. K. Hecht, R. J. MacIntyre, and M. T. Clegg, Eds.). Plenum Press, New York, New York.
HALL, B. K. 1998. Evolutionary Developmental Biology. 2nd ed. Kluwer Academic Publications, Dordrecht, The Netherlands.
HALL, B. K., AND T. MIYAKE. 1995. How do embryos measure time? Pp. 3–20 *in* Evolutionary Time and Heterochrony (K. J. McNamara, Ed.). J. Wiley, London, United Kingdom.
HALLAM, A., AND P. B. WIGNALL. 1997. Mass Extinctions and Their Aftermath. Oxford University Press, Oxford, United Kingdom.
HALLIDAY, T. R. 1978a. Vanishing Birds: Their Natural History and Conservation. Holt, Rinehart and Winston, New York, New York.
HALLIDAY, T. R. 1978b. Sexual selection and mate choice. Pp. 180–213 *in* Behavioural Ecology: An Evolutionary Approach (J. R. Krebs and N. B. Davies, Eds.). Blackwell Scientific Publications, Oxford, United Kingdom.
HALLIDAY, T. R. 1987. Physiological constraints on sexual selection. Pp. 247–264 *in* Sexual Selection: Testing the Alternatives (J. W. Bradbury and M. B. Andersson, Eds.). J. Wiley and Sons, Chichester, United Kingdom.

HAMBLER, C., J. NEWING, AND K. HAMBLER. 1993. Population monitoring for the flightless rail *Dryolimnas cuvieri aldabranus*. Bird Conserv. Int. 3:307–318.

HAMBURGER, V., AND H. L. HAMILTON. 1951. A series of normal stages in the development of the chick embryo. J. Morphol. 88:49–92.

HAMILTON, A. 1892. On the genus *Aptornis*, with more especial reference to *Aptornis defossor*, Owen. Trans. N. Z. Inst. 24:175–184.

HAMILTON, A. 1893. On the fissure and caves at Castle Rocks, Southland; with a description of the remains of the existing and extinct birds found in them. Trans. N. Z. Inst. 25:88–106.

HAMILTON, J. E. 1946. Seals preying on birds. Ibis 88:131–132.

HAMILTON, S., AND R. F. JOHNSTON. 1978. Evolution in the House Sparrow—VI. Variability and niche width. Auk 95:313–323.

HAMILTON, T. H. 1961. The adaptive significances of intraspecific trends of variation in wing length and body size among bird species. Evolution 15:180–195.

HAMILTON, T. H., AND N. E. ARMSTRONG. 1965. Environmental determination of insular variation in bird species abundance in the Gulf of Guinea. Nature 207:148–151.

HAMILTON, T. H., R. H. BARTH, JR., AND I. RUBINOFF. 1964. The environmental control of insular variation in bird species abundance. Proc. Natl. Acad. Sci. USA 52:132–140.

HAMILTON, T. H., AND I. RUBINOFF. 1963. Isolation, endemism, and multiplication of species in the Darwin finches. Evolution 17:388–403.

HAMILTON, T. H., AND I. RUBINOFF. 1964. On models predicting abundance of species and endemics for the Darwin finches in the Galápagos archipelago. Evolution 18:339–342.

HAMILTON, T. H., AND I. RUBINOFF. 1967. On predicting insular variation in endemism and sympatry for the Darwin finches in the Galápagos archipelago. Am. Nat. 101:161–171.

HAMILTON, T. H., I. RUBINOFF, R. H. BARTH, JR., AND G. L. BUSH. 1963. Species abundance: natural regulation of insular variation. Science 142:1575–1577.

HAMILTON, W. D. 1972. Altruism and related phenomena, mainly in social insects. Annu. Rev. Ecol. Syst. 3:193–232.

HAMILTON, W. D. 1979. Wingless and fighting males in fig wasps and other insects. Pp. 167–220 *in* Sexual Selection and Reproduction in Insects (M. S. Blum and N. A. Blum, Eds.). Academic Press, New York, New York.

HANDFORD, P. 1983. Continental patterns of morphological variation in a South American sparrow. Evolution 37:920–930.

HANDLY, J. W. 1895. Notes on some species of New Zealand birds. Trans. N. Z. Inst. 28:360–367.

HANKEN, J. 1984. Miniaturization and its effect on cranial morphology in plethodontid salamanders, genus *Thorius* (Amphibia: Plethodontidae): I. Osteological variation. Biol. J. Linn. Soc. 23: 55–75.

HANKEN, J. 1985. Morphological novelty in the limb skeleton accompanies miniaturization in salamanders. Science 229:871–873.

HANKEN, J., AND B. K. HALL. 1993. Mechanisms of skull diversity and evolution. Pp. 1–36 *in* The Skull. Vol. 3: Functional and Evolutionary Mechanisms (J. Hanken and B. K. Hall, Eds.). University of Chicago Press, Chicago, Illinois.

HANKEN, J., AND D. B. WAKE. 1993. Miniaturization of body size: organismal consequences and evolutionary significance. Annu. Rev. Ecol. Syst. 24:501–519.

HANSELL, M. 2000. Bird Nests and Construction Behaviour. Cambridge University Press, Cambridge, United Kingdom.

HANSEN, T. F., AND E. P. MARTINS. 1996. Translating between microevolutionary process and macroevolutionary patterns: the correlation structure of interspecific data. Evolution 50:1404–1417.

HANSKI, I. 1997. Metapopulation dynamics: from concepts and observations to predictive models. Pp. 69–91 *in* Metapopulation Biology: Ecology, Genetics, and Evolution (I. Hanski and M. E. Gilpin, Eds.). Academic Press, San Diego, California.

HANSKI, I. 1999. Metapopulation Ecology. Oxford University Press, Oxford, United Kingdom.

HANSKI, I. A., AND M. E. GILPIN, EDS. 1997. Metapopulation Biology: Ecology, Genetics, and Evolution. Academic Press, San Diego, California.

HANSKI, I., AND D. SIMBERLOFF. 1997. The metapopulation approach, its history, conceptual domain, and application to conservation. Pp. 5–26 *in* Metapopulation Biology: Ecology, Genetics, and Evolution (I. Hanski and M. E. Gilpin, Eds.). Academic Press, San Diego, California.

HARATO, T., AND K. OZAKI. 1993. Roosting behavior of the Okinawa Rail. J. Yamashina Inst. Ornithol. 25:40–53.
HARDY, A. C. 1954. Escape from specialization. Pp. 122–142 *in* Evolution as a Process (J. Huxley, A. C. Hardy, and E. B. Ford, Eds.). George Allen and Unwin, London, United Kingdom.
HARESTED, A. S., AND F. L. BUNNELL. 1979. Home range and body weight—a reevaluation. Ecology 60:389–402.
HÄRLID, A., AND U. ARNASON. 1998. Analyses of mitochondrial DNA nest ratite birds within the Neognathae: supporting a neotenous origin of ratite morphological characters. Proc. R. Soc. Lond. (Ser. B Biol. Sci.) 266:305–309.
HÄRLID, A., A. JANKE, AND U. ARNASON. 1997. The mtDNA sequence of the ostrich and the divergence between palaeognathous and neognathous birds. Mol. Biol. Evol. 14:754–761.
HÄRLID, A., A. JANKE, AND U. ARNASON. 1998. The complete mitochondrial genome of *Rhea americana* and early avian divergences. J. Mol. Evol. 46:669–679.
HARMAN, H. H. 1967. Modern Factor Analysis. University of Chicago Press, Chicago, Illinois.
HARPER, D. G. C. 1999. Feather mites, pectoral muscle condition, wing length and plumage coloration of passerines. Anim. Behav. 58:553–562.
HARRIS, C. L. 1981. Evolution: Genesis and Revelations. State University of New York Press, Albany, New York.
HARRIS, J. A. 1911. A neglected paper on natural selection in the English Sparrow. Am. Nat. 45: 314–318.
HARRIS, M. P. 1968. Egg-eating by Galápagos mockingbirds. Condor 70:269–270.
HARRIS, M. P. 1973. The Galápagos avifauna. Condor 75:265–278.
HARRIS, M. P. 1981. The waterbirds of Lake Junín, central Peru. Wildfowl 32:137–145.
HARRIS, R. N. 1987. Density-dependent paedomorphosis in the salamander *Notophthalamus viridescens dorsalis*. Ecology 68:705–712.
HARRISON, C. J. O., AND S. A. PARKER. 1967. The eggs of Woodford's Rail, Rouget's Rail, and the Malayan Banded Crake. Bull. Br. Ornithol. Club 87:14–16.
HARRISON, J. M. 1955. Fish and other aquatic fauna as predators of birds. Bull. Br. Ornithol. Club 75:110–113.
HARRISON, R. G. 1980. Dispersal polymorphisms in insects. Annu. Rev. Ecol. Syst. 11:95–118.
HARRISON, S., AND A. D. TAYLOR. 1997. Empirical evidence for metapopulation dynamics. Pp. 27–42 *in* Metapopulation Biology: Ecology, Genetics, and Evolution (I. Hanski and M. E. Gilpin, Eds.). Academic Press, San Diego, California.
HARSHMAN, J. 1994. Reweaving the tapestry: What can we learn from Sibley and Ahlquist (1990)? Auk 111:377–388.
HARTE, J. 2000. Scaling and self-similarity in species distributions: implications for extinction, species richness, abundance, and range. Pp. 325–342 *in* Scaling in Biology (J. H. Brown and G. B. West, Eds.). Oxford University Press, Oxford, United Kingdom.
HARTL, D. L., AND R. D. COOK. 1973. Balanced polymorphisms of quasineutral alleles. Theor. Popul. Biol. 4:163–172.
HARTL, D. L., AND R. D. COOK. 1976. Stochastic selection and the maintenance of genetic variation. Pp. 593–615 *in* Population Genetics and Ecology (S. Karlin and E. Nevo, Eds.). Academic Press, New York, New York.
HARTLAUB, G., AND O. FINSCH. 1871. On a collection of birds from Savai and Rarotonga islands in the Pacific. Proc. Zool. Soc. Lond. 1871:21–32.
HARTMAN, F. A. 1955. Heart weight in birds. Condor 57:221–238.
HARTMAN, F. A. 1961. Locomotor mechanisms of birds. Smithson. Misc. Coll. 143:1–91.
HARTT, L., AND J. W. HAEFNER. 1995. Inbreeding depression effects on extinction time in a predator–prey system. Evol. Ecol. 9:1–9.
HARVEY, P. H. 1982. On rethinking allometry. J. Theor. Biol. 95:37–41.
HARVEY, P. H., AND P. J. GREENWOOD. 1978. Anti-predator defence strategies: some evolutionary problems. Pp. 129–151 *in* Behavioural Ecology: An Evolutionary Approach (J. R. Krebs and N. B. Davies, Eds.). Blackwell Scientific Publications, Oxford, United Kingdom.
HARVEY, P. H., AND M. D. PAGEL. 1991. The Comparative Method in Evolutionary Biology. Oxford University Press, Oxford, United Kingdom.
HATCH, J. J. 1965. Only one species of Galápagos mockingbird feeds on eggs. Condor 67:354–355.

HAUKIOJA, E. 1971. Flightlessness in some moulting passerines in northern Europe. Ornis Fenn. 48: 102–116.
HAUKIOJA, E., AND T. HAKALA. 1979. On the relationship between avian clutch size and life span. Ornis Fenn. 56:45–55.
HAVERSCHMIDT, F. 1974. Notes on the Grey-breasted Crake *Laterallus exilis*. Bull. Br. Ornithol. Club 74:2–3.
HAWTHORNE, D. J. 1997. Ecological history and evolution in a novel environment: habitat heterogeneity and insect adaptation to a new host plant. Evolution 51:153–162.
HAY, R. 1986. Bird Conservation in the Pacific Islands. ICBP Study Rep. No. 7. International Council for Bird Preservation, Cambridge, United Kingdom.
HAYAMI, I. 1978. Notes on the rates and patterns of size change in evolution. Paleobiology 4:252–260.
HAYAMI, I., AND A. MATSUKUMA. 1970. Variation of bivariate characters from the standpoint of allometry. Paleontology 13:588–605.
HAZEVOET, C. J. 1996. Conservation and species lists: taxonomic neglect promotes the extinction of endemic birds, as exemplified by taxa from eastern Atlantic islands. Bird Conserv. Int. 6:181–196.
HEANEY, L. R. 1978. Island area and body size of insular mammals: evidence from the tri-colored squirrel (*Callosciurus prevosti*) of southeast Asia. Evolution 32:29–44.
HEATWOLE, H., AND R. LEVINS. 1972. Biogeography of the Puerto Rican bank: flotsam transport of terrestrial animals. Ecology 53:112–117.
HECTOR, J. 1873a. On *Cnemiornis calcitrans,* showing its affinity to the Natatores. Proc. Zool. Soc. Lond. 1873:763–771.
HECTOR, J. 1873b. On *Cnemiornis calcitrans,* Owen, showing its affinity to the lamellirostrate Natatores. Trans. N. Z. Inst. 6:76–84.
HEDENSTRÖM, A. 1992. Flight performance in relation to fuel loads in birds. J. Theor. Biol. 158:535–537.
HEDENSTRÖM, A. 1993. Migration by soaring or flapping flight in birds: the relative importance of energy cost and speed. Philos. Trans. R. Soc. Lond. (Ser. B) 342:353–361.
HEDENSTRÖM, A., AND T. ALERSTAM. 1992. Climbing performance of migratory birds as a basis for estimating limits for fuel-carrying capacity and muscle work. J. Exp. Biol. 164:19–38.
HEDENSTRÖM, A., AND T. ALERSTAM. 1995. Optimal flight speed of birds. Philos. Trans. R. Soc. Lond. (Ser. B) 348:471–487.
HEDENSTRÖM, A., AND T. ALERSTAM. 1997. Optimum fuel loads in migratory birds: distinguishing between time and energy minimization. J. Theor. Biol. 189:227–234.
HEDENSTRÖM, A., AND S. SUNADA. 1999. On the aerodynamics of moult gaps in birds. J. Exp. Biol. 202:67–76.
HEDRICK, P. W. 1986. Genetic polymorphism in heterogeneous environments: a decades later. Annu. Rev. Ecol. Syst. 17:535–566.
HEDRICK, P. W., AND M. E. GILPIN. 1997. Genetic effective size of a metapopulation. Pp. 165–181 *in* Metapopulation Biology: Ecology, Genetics, and Evolution (I. Hanski and M. E. Gilpin, Eds.). Academic Press, San Diego, California.
HEDRICK, P. W., M. E. GINEVAN, AND E. P. EWING. 1976. Genetic polymorphism in heterogeneous environments. Annu. Rev. Ecol. Syst. 7:1–32.
HEGLUND, N. C., G. A. CAVAGNA, AND C. R. TAYLOR. 1982a. Energetics and mechanics of terrestrial locomotion. III. Energy changes of the centre of mass as a function of speed and body size in birds and mammals. J. Exp. Biol. 79:41–56.
HEGLUND, N. C., M. A. FEDAK, C. R. TAYLOR, AND G. A. CAVAGNA. 1982b. Energetics and mechanics of terrestrial locomotion. IV. Total mechanical energy changes as a function of speed and body size in birds and mammals. J. Exp. Biol. 97:57–66.
HEGLUND, N. C., AND C. R. TAYLOR. 1988. Speed, stride frequency, and energy cost per stride: How do they change with body size and gait? J. Exp. Biol. 138:301–318.
HEINROTH, O. 1922. Die Beziehungen zwischen Vogelgewicht, Eigewicht, Gelegegewicht und Brutdauer. J. Ornithol. 70:172–285.
HEINROTH, O., AND M. HEINROTH. 1928. Die Vögel Mitteleuropas. Vol. 3. Hugo Bermuhler, Berlin, Germany.

HEINSOHN, R. G. 1991. Slow learning of foraging skills and extended parental care in cooperatively breeding White-winged Choughs. Am. Nat. 137:864–881.
HELLIWELL, D. R. 1976. The effects of size and isolation on the conservation value of wooded sites in Britain. J. Biogeogr. 3:407–416.
HELLMAYR, C. E., AND B. CONOVER. 1942. Catalog of birds of the Americas and adjacent islands, pt. 1 (1). Field Mus. Nat. Hist. Publ. (Zool. Ser.) 13:1–636.
HELM-BYCHOWSKI, K., AND J. CRACRAFT. 1993. Recovering phylogenetic signal from DNA sequences: relationships within the corvine assemblage (class Aves) as inferred from complete sequences of the mitochondrial DNA cytochrome-*b* gene. Mol. Biol. Evol. 10:1196–1214.
HEMMINGSEN, A. M. 1960. Energy metabolism as related to body size and respiratory surfaces, and its evolution. Rep. Steno Mem. Hosp. Nord. Insulin Lab. 9:6–110.
HENGEVELD, R. 1997. Impact of biogeography on a population-biological paradigm shift. J. Biogeogr. 24:541–547.
HENRY, R. 1899. On the probable origin of *Notornis mantelli*, and its extinction in New Zealand. Trans. N. Z. Inst. 32:53–54.
HENRY, R. 1903. Flightless birds of New Zealand; and other notes. J. House Rep. N. Z. 1(Append.): 118–123.
HENSHAW, H. W. 1902. Birds of the Hawaiian Islands, Being a Complete List of the Birds of the Hawaiian Possessions, With Notes on Their Habits. T. G. Thrum, Honolulu, Hawaii.
HEPP, G., D. STANGOHR, L. BAKER, AND R. KENNAMER. 1987. Factors affecting variation in the egg and duckling components of Wood Ducks. Auk 104:435–443.
HEPPNER, F. H. 1974. Avian flight formations. Bird-Banding 45:160–169.
HEPPNER, F. H., AND J. G. T. ANDERSON. 1985. Leg thrust important in flight take-off in the Pigeon. J. Exp. Biol. 114:285–288.
HERHOLDT, J. J. 1988. Bird weights from the Orange Free State (part 1: non-passerines). Safring News 17:3–14.
HERREID, C. F., II, AND B. KESSEL. 1967. Thermal conductance in birds and mammals. Comp. Biochem. Physiol. (Ser. A) 21:405–414.
HERREMANS, M. 1990. Trends in the evolution of insular land birds, exemplified by the Comoro, Seychelle and Mascarene islands. Pp. 249–260 *in* Vertebrates in the Tropics (G. Peters and R. Hutterer, Eds.). Alexander Koenig. Zoological Research Institute and Zoology Museum, Bonn, Germany.
HERTEL, F. 1995. Ecomorphological indicators of feeding behavior in Recent and fossil raptors. Auk 112:890–903.
HERTEL, F., AND L. T. BALLANCE. 1999. Wing morphology of seabirds from Johnston Atoll. Condor 101:549–556.
HESPENHEIDE, H. A. 1973. Ecological inferences from morphological data. Annu. Rev. Ecol. Syst. 4: 213–229.
HESSE, R. W., C. ALLEE, AND K. P. SCHMIDT. 1951. Ecological Animal Geography. Chapman and Hall, New York, New York.
HEWITT, O. H. 1950. The bullfrog as a predator on ducklings. J. Wildl. Manage. 14:244.
HIGGS, A. J. 1981. Island biogeography theory and nature reserve design. J. Biogeogr. 8:117–124.
HILDEBRAND, M. 1995. Analysis of Vertebrate Structure. 4th ed. J. Wiley and Sons, New York, New York.
HILL, G. E. 1988. The function of delayed plumage maturation in male Black-headed Grosbeaks. Auk 105:1–10.
HILL, G. E. 1996. Subadult plumage in the House Finch and tests of models for the evolution of delayed plumage maturation. Auk 113:858–874.
HILL, M. O. 1974. Correspondence analysis: a neglected multivariate method. J. R. Stat. Assoc. (Appl. Stat.) 23:340–354.
HILLIS, D. M., B. K. MABLE, AND C. MORITZ. 1996. Applications of molecular systematics: the state of the field and a look to the future. Pp. 515–543 *in* Molecular Systematics. 2nd ed. (D. M. Hillis, C. Moritz, and B. K. Mable, Eds.). Sinauer Assoc., Sunderland, Massachusetts.
HINCHLIFFE, J. R.. 1977. The chondrogenic pattern in chick limb morphogenesis: a problem of development *and* evolution. Pp. 293–309 *in* Vertebrate Limb and Somite Morphogenesis (D. A. Ede, J. R. Hinchliffe, and M. Balls, Eds.). Cambridge University Press, Cambridge, United Kingdom.

HINCHLIFFE, J. R. 1989. An evolutionary perspective of the developmental mechanisms underlying the patterning of the limb skeleton in birds and other tetrapods. Pp. 217–225 in Ontogenèse et Évolution (B. David, J.-L. Dommergues, J. Chaline, and B. Laurin, Coords.). Géobios Mém. Spéc. 12. Université Claude-Bernard, Lyon, France.

HINCHLIFFE, J. R., AND D. A. EDE. 1973. Cell death and the development of limb form and skeletal pattern in normal and *wingless* (*ws*) chick embryos. J. Embryol. Exp. Morphol. 30:753–772.

HINCHLIFFE, J. R., AND P. J. GRIFFITHS. 1983. The prechondrogenic patterns in tetrapod limb development and their phylogenetic significance. Pp. 99–121 in Development and Evolution (B. C. Goodwin, N. Holder, and C. C. Wylie, Eds.). Cambridge University Press, Cambridge, United Kingdom.

HINCHLIFFE, J. R., AND M. GUMPEL-PINOT. 1983. Experimental analysis of avian limb morphogenesis. Pp. 293–327 in Current Ornithology. Vol. 1 (R. F. Johnston, Ed.). Plenum Press, New York, New York.

HINDS, D. S., AND W. A. CALDER. 1971. Tracheal dead space in the respiration of birds. Evolution 25:429–440.

HINDWOOD, K. A. 1932. An historic diary. Emu 32:17–29.

HINDWOOD, K. A. 1940. Birds of Lord Howe Island. Emu 40:1–86.

HINDWOOD, K. A. 1964. Birds caught by octopuses. Emu 64:69–70.

HINDWOOD, K. A. 1965. John Hunter, a naturalist and artist of the first fleet. Emu 65:83–96.

HNATIUK, S. H. 1980. The numbers of land birds, waders, sea birds, land crustacea, and certain insects on the lagoon islands of Aldabra Atoll. Biol. J. Linn. Soc. 14:151–161.

HOBBS, J. N. 1956. Musk Duck flying. Emu 56:429–430.

HOBBS, J. N. 1967. Distraction displays by two species of crakes. Emu 66:299–300.

HOBBS, J. N. 1973. Reactions to predators by Black-tailed Native Hen. Aust. Bird Watcher 5:29–30.

HODIN, J. 2000. Plasticity and constraints in development and evolution. J. Exp. Zool. 288:1–20.

HOEKSTRA, R. F., R. BIJLSMA, AND A. J. DOLMAN. 1985. Polymorphism from environmental heterogeneity: models are only robust if the heterozygote is close in fitness to the favored homozygote in each environment. Genet. Res. 45:299–314.

HOFER, H. 1945. Untersuchungen über den Bau des Vogelschädels, besonders über den der Spechte und Steißhühner. Zool. Jahrb. Abt. Anat. 69:1–158.

HOFER, H. 1955. Neuere Untersuchungen zur Kopfmorphologie der Vögel. Pp. 104–137 in Acta XI Congressus Internationalis Ornithologici (A. Portmann and E. Sutter, Eds.). Birkhäuser Verlag, Basel, Switzerland.

HOFF, J. G. 1975. Clapper Rail feeding on water snake. Wilson Bull. 87:112.

HOFFMAN, A. A. 1985. Island biogeography and palaeobiology: in search for evolutionary equilibria. Biol. Rev. 60:455–471.

HOFSTETTER, R. 1958. Xenartha. Pp. 535–636 in Traité de Paléontologie. Vol. 2. No. 6: Mammifères Evolution (J. Piveteau, Ed.). Masson et Cie, Paris, France.

HOGG, D. A. 1980. A re-investigation of the centres of ossification in the avian skeleton at and after hatching. J. Anat. 130:725–743.

HOGG, D. A. 1982. Fusions occurring in the postcranial skeleton of the domestic fowl. J. Anat. 135:501–512.

HÖGLUND, J. 1989. Size and plumage dimorphism in lek-breeding birds: a comparative analysis. Am. Nat. 134:72–87.

HOGSTAD, O. 1976. Sexual dimorphism and divergence in winter foraging behaviour of Three-toed Woodpeckers (*Picoides tridactylus*). Ibis 118:41–50.

HOGSTAD, O. 1977. Seasonal change in intersexual niche differentiation of the Three-toed Woodpecker *Picoides tridactylus*. Ornis Scand. 8:101–111.

HOGSTAD, O. 1978. Sexual dimorphism in relation to winter foraging and territorial behaviour of the Three-toed Woodpecker *Picoides tridactylus* and three *Dendrocopos* species. Ibis 120:198–203.

HÖGSTEDT, G. 1980. Evolution of clutch size in birds: adaptive variation in relation to territory quality. Science 210:1148–1150.

HÖGSTEDT, G. 1981. Should there be a positive or negative correlation between survival of adults in a bird population and their clutch size? Am. Nat. 118:568–571.

HOLDAWAY, R. N. 1989. New Zealand's pre-human avifauna and its vulnerability. N. Z. J. Ecol. 12(Suppl.):11–25.

HOLDAWAY, R. N. 1990. Changes in the diversity of New Zealand forest birds. N. Z. J. Zool. 17: 309–321.
HOLDAWAY, R. N. 1999. Introduced predators and avifaunal extinction in New Zealand. Pp. 189–238 *in* Extinctions in Near Time: Causes, Contexts, and Consequences (R. D. E. MacPhee, Ed.). Kluwer Academic/Plenum Publishing, New York, New York.
HOLDAWAY, R. N., AND C. JACOMB. 2000. Rapid extinction of the moas (Aves: Dinornithiformes): model, test, and implication. Science 287:2250–2254.
HOLDER, N. 1983. Developmental constraints and the evolution of vertebrate digit patterns. J. Theor. Biol. 104:451–471.
HOLDGATE, M. W. 1965. The fauna of the Tristan da Cunha Islands. Part III *in* The Biological Report of the Royal Society Expedition to Tristan da Cunha, 1962. Philos. Trans. R. Soc. Lond. (Ser. B) 249:361–402.
HOLDGATE, M. W., AND N. M. WACE. 1959. The influence of man on the floras and faunas of southern islands. Polar Rec. 10:475–493.
HOLLAND, R. F., AND S. K. JAIN. 1981. Insular biogeography of vernal pools in the central valley of California. Am. Nat. 117:24–37.
HOLLOWAY, B. A. 1963. Wing development and evolution of New Zealand Lucanidae (Insecta: Coleoptera). Trans. R. Soc. N. Z. 3:99–116.
HOLMES, E. B. 1977. Is specialization a dead end? Am. Nat. 111:1021–1026.
HOLMES, R. T. 1986. Foraging patterns of forest birds: male–female differences. Wilson Bull. 98:196–213.
HOLMES, W. N., AND J. G. PHILLIPS. 1985. The avian salt gland. Biol. Rev. 60:213–256.
HOLMGREN, N. 1955. Studies on the phylogeny of birds. Acta Zool. 36:244–328.
HOLMGREN, N., H. ELLEGREN, AND J. PETTERSSON. 1993. The adaptation of moult pattern in migratory Dunlins *Calidris alpina*. Ornis Scand. 24:21–27.
HOLMGREN, N., AND A. HEDENSTRÖM. 1995. The scheduling of molt in migratory birds. Evol. Ecol. 9:354–368.
HOLMGREN, N., AND S. LUNDBERG. 1993. Despotic behaviour and the evolution of migration patterns in birds. Ornis Scand. 24:103–109.
HOLT, R. D. 1977. Predation, apparent competition, and the structure of prey communities. Theor. Popul. Biol. 12:197–229.
HOLT, R. D., J. GROVER, AND D. TILMAN. 1994. Simple rules for interspecific dominance in systems with exploitation and apparent competition. Am. Nat. 144:741–771.
HOLT, R. D., AND B. D. KOTLER. 1987. Short-term apparent competition. Am. Nat. 130:412–430.
HOLT, R. D., AND J. H. LAWTON. 1994. The ecological consequences of shared natural enemies. Annu. Rev. Ecol. Syst. 25:495–520.
HOLYOAK, D. T. 1970. Sex-differences in feeding behaviour and size in the Carrion Crow. Ibis 112: 397–400.
HOLYOAK, D. T. 1973. Significance of colour dimorphism in Polynesian populations of *Egretta sacra*. Ibis 115:419–420.
HOLYOAK, D. T., AND D. SAGER. 1970. Observations on captive Tasmanian Native Hens and their interactions with wild Moorhens. Avic. Mag. 76:56–57.
HOMMES, J. H. 1924. On the development of the clavicula and the sternum in birds and mammals. Tijdschr. Ned. Diergeneeskd. Ver. (Ser. 2) 19:10–51.
HOPKINS, J. W. 1966. Some considerations in multivariate allometry. Biometrics 22:747–760.
HORN, H. S., AND R. H. MACARTHUR. 1972. Competition among fugitive species in a harlequin environment. Ecology 53:749–752.
HORSFALL, J. A. 1984. Brood reduction and brood division in coots. Anim. Behav. 32:216–225.
HORTON, D. R. 1972. Speciation of birds in Australia, New Guinea and the south-western Pacific islands. Emu 72:91–109.
HOUCK, M. A., J. A. GAUTHIER, AND R. E. STRAUSS. 1990. Allometric scaling in the earliest fossil bird, *Archaeopteryx lithographica*. Science 247:195–198.
HOUDE, P. W. 1987. Critical evaluation of DNA hybridization studies in avian systematics. Auk 104: 17–32.
HOUDE, P. W. 1988. Paleognathous Birds from the Early Tertiary of the Northern Hemisphere. Publ. No. 22. Nuttall Ornithologists' Club, Cambridge, Massachusetts.

HOUDE, P. W. 1994. Evolution of the Heliornithidae: reciprocal illumination by morphology, biogeography and DNA hybridization (Aves: Gruiformes). Cladistics 10:1–19.

HOUDE, P. W., A. COOPER, E. LESLIE, A. E. STRAND, AND G. A. MONTAÑO. 1997. Phylogeny and evolution of 12S rDNA in Gruiformes (Aves). Pp. 121–158 in Avian Molecular Evolution and Systematics (D. P. Mindell, Ed.). Academic Press, San Diego, California.

HOUDE, P. W., AND S. L. OLSON. 1981. Paleognathous carinate birds from the early Tertiary of North America. Science 214:1236–1237.

HOULE, D. 1991. Genetic covariance of fitness correlates: what genetic correlations are made of and why it matters. Evolution 45:630–648.

HOULE, D., J. MEZEY, AND P. GALPERN. 2002. Interpretation of the results of common principal component analyses. Evolution 56:433–440.

HOUSTON, A. I. 1987. Optimal foraging by parent birds feeding dependent young. J. Theor. Biol. 124:251–274.

HOUSTON, A. I. 1998. Models of optimal avian migration: state, time and predation. J. Avian Biol. 29:395–404.

HOUSTON, A. I., AND J. M. MCNAMARA. 1990. The effect of environmental variability on growth. Oikos 59:15–20.

HOUSTON, A. I., AND J. M. MCNAMARA. 1993. A theoretical investigation of the fat reserves and mortality levels of small birds in winter. Ornis Scand. 24:205–219.

HOUSTON, A. I., J. M. MCNAMARA, AND J. M. C. HUTCHINSON. 1993. General results concerning the trade-off between gaining energy and avoiding predation. Philos. Trans. R. Soc. Lond. (Ser. B) 341:375–397.

HOUSTON, A. I., N. J. WELTON, AND J. M. MCNAMARA. 1997. Acquisition and maintenance costs in the long-term regulation of avian fat reserves. Oikos 78:331–340.

HOUSTON, D. C., P. J. JONES, AND R. M. SIBLY. 1983. The effect of female body condition on egg laying in Lesser Black-backed Gulls *Larus fuscus*. J. Zool. Lond. 200:509–520.

HOWARD, H. 1929. The avifauna of Emeryville Shellmound. Univ. Calif. Publ. Zool. 32:301–394.

HOWARTH, F. G. 1991. Hawaiian cave faunas: macroevolution on young islands. Pp. 285–295 in The Unity of Evolutionary Biology: Proceedings of the Fourth International Congress of Systematic and Evolutionary Biology. Vol. 1 (E. C. Dudley, Ed.). Dioscorides Press, Portland, Oregon.

HOWARTH, F. G. 1993. High-stress subterranean habitats and evolutionary change in cave-inhabiting arthropods. Am. Nat. 142(Suppl.):65–77.

HOWE, R. W. 1984. Local dynamics of bird assemblages in small forest habitat islands in Australia and North America. Ecology 65:1585–1601.

HOWELL, A. B. 1937. Morphogenesis of the shoulder architecture: Aves. Auk 54:363–375.

HOYT, D. F. 1979. Practical methods of estimating volume and fresh weight of bird eggs. Auk 96:73–77.

HUBBELL, T. H. 1968. The biology of islands. Proc. Natl. Acad. Sci. USA 60:22–32.

HUBERTY, C. J. 1994. Applied Discriminant Analysis. J. Wiley and Sons, New York, New York.

HUDSON, G. E. 1937. Studies on the muscles of the pelvic appendage in birds. Am. Midl. Nat. 18:1–108.

HUDSON, G. E., AND P. J. LANZILLOTTI. 1955. Gross anatomy of the wing muscles in the family Corvidae. Am. Midl. Nat. 53:1–44.

HUDSON, G. E., AND P. J. LANZILLOTTI. 1964. Muscles of the pectoral limb in galliform birds. Am. Midl. Nat. 71:1–113.

HUDSON, G. E., K. M. HOFF, J. VANDEN BERGE, AND E. C. TRIVETTE. 1969. A numerical study of the wing and leg muscles of Lari and Alcae. Ibis 111:459–524.

HUDSON, G. E., R. A. PARKER, J. VANDEN BERGE, AND P. J. LANZILLOTTI. 1966. A numerical analysis of the modifications of the appendicular muscles in various genera of gallinaceous birds. Am. Midl. Nat. 76:1–73.

HUDSON, G. E., D. O. SCHREIWEIS, S. Y. C. WANG, AND D. A. LANCASTER. 1972. A numerical study of the wing and leg muscles of tinamous (Tinamidae). Northwest Sci. 46:207–255.

HUGHES, A. L., AND M. K. HUGHES. 1986. Paternal investment and sexual size dimorphism in North American passerines. Oikos 46:171–175.

HUGHES, M. R. 1970. Relative kidney size in nonpasserine birds with functional salt glands. Condor 72:164–168.

HUI, C. A. 1983. Energetics of penguin swimming. Am. Zool. 23:956.

HULL, A. F. B. 1909. The birds of Lord Howe and Norfolk islands. Proc. Linn. Soc. N. S. Wales 34: 636–693.
HULL, D. L. 1980. Individuality and selection. Annu. Rev. Ecol. Syst. 11:311–332.
HULL, D. L. 1988. Science as a Process: An Evolutionary Account of the Social and Conceptual Development of Science. University of Chicago Press, Chicago, Illinois.
HUMPHREY, P. S., AND B. C. LIVEZEY. 1982. Flightlessness in Flying Steamer-Ducks. Auk 99:368–372.
HUMPHREY, P. S., AND B. C. LIVEZEY. 1985. Nest, eggs, and downy young of the White-headed Flightless Steamer-Duck. Pp.. 949–954 in Neotropical Ornithology (P. A. Buckley, M. S. Foster, E. S. Morton, R. S. Ridgely, and F. G. Buckley, Eds.). Ornithol. Monogr. 36. American Ornithologists' Union, Washington, D.C.
HUMPHREY, P. S., B. C. LIVEZEY, AND D. SIEGEL-CAUSEY. 1987. Tameness of birds of the Falkland Islands: an index and preliminary results. Bird Behav. 7:67–72.
HUMPHRIES, C. J., AND L. R. PARENTI. 1999. Cladistic Biogeography 2nd ed. Oxford University Press, Oxford, United Kingdom.
HUMPHRIES, J. M., F. L. BOOKSTEIN, B. CHERNOFF, G. R. SMITH, R. L. ELDER, AND S. G. POSS. 1981. Multivariate discrimination by shape in relation to size. Syst. Zool. 30:291–308.
HUNT, G. R. 1996. Family Rhynochetidae (Kagu). Pp. 218–225 in Handbook of the Birds of the World. Vol. 3: Hoatzin to Auks (J. del Hoyo, A. Elliott and J. Sargatal, Eds.). Lynx Edicions, Barcelona, Spain.
HUNTER, A. F. 1995. The ecology and evolution of reduced wings in forest macrolepidoptera. Evol. Ecol. 9:275–287.
HUNTER, L. A. 1985. The effects of helpers in cooperatively breeding Purple Gallinules. Behav. Ecol. Sociobiol. 18:147–153.
HUNTER, L. A. 1987a. Acquisition of territories by floaters in cooperatively breeding Purple Gallinules. Anim. Behav. 35:402–410.
HUNTER, L. A. 1987b. Cooperative breeding in Purple Gallinules: the role of helpers in feeding chicks. Behav. Ecol. Sociobiol. 20:171–177.
HÜPPOP, K. 1985. The role of metabolism in the evolution of cave animals. Natl. Speleol. Soc. Bull. 47:136–146.
HURTER, H. 1979. Nahrungsökologie des Bläßhuhns *Fulica atra* an den Überwinterungsgewässern im nördlichen Alpenvorland. Ornithol. Beob. 76:257–288.
HUSSELL, D. J. T. 1972. Factors affecting clutch size in arctic passerines. Ecol. Monogr. 42:317–364.
HUTTON, F. W. 1872a. Notes on some birds from the Chatham Islands, collected by H. H. Travers, Esq.; with descriptions of two new species. Ibis 14:243–250.
HUTTON, F. W. 1872b. Notes on some of the birds brought by Mr. Henry Travers from the Chatham Islands, with descriptions of the new species. Trans. N. Z. Inst. 5:222–226.
HUTTON, F. W. 1872c. On the geographical relations of the New Zealand fauna. Trans. N. Z. Inst. 5: 227–256.
HUTTON, F. W. 1873a. On *Rallus modestus* of New Zealand. Ibis 15:349–352.
HUTTON, F. W. 1873b. On a new genus of *Rallidæ*. Trans. N. Z. Inst. 6:108–110.
HUTTON, F. W. 1873c. Notes on the New Zealand wood-hens (*Ocydromus*). Trans. N. Z. Inst. 6:110–112.
HUTTON, F. W. 1879. On a new species of rail from Macquarie Island. Ibis 21:454–456.
HUTTON, I. 1991. Birds of Lord Howe Island. Privately Published, Coffs Harbour, Australia.
HUXLEY, C. R., AND R. WILKINSON. 1977. Vocalizations of the Aldabra White-throated Rail *Dryolimnas cuvieri aldabranus*. Proc. R. Soc. Lond. (Ser. B Biol. Sci.) 197:315–331.
HUXLEY, C. R., AND R. WILKINSON. 1979. Duetting and vocal recognition by Aldabra White-throated Rails *Dryolimnas cuvieri aldabranus*. Ibis 121:265–273.
HUXLEY, C. R., AND N. A. WOOD. 1976. Aspects of the breeding of the Moorhen in Britain. Bird Study 23:1–10.
HUXLEY, J. S. 1914. The courtship-habits of the Great Crested Grebe (*Podiceps cristatus*): with an addition to the theory of sexual selection. Proc. Zool. Soc. Lond. 1914:491–562.
HUXLEY, J. S. 1927. On the relation between egg-weight and body-weight in birds. J. Linn. Soc. (Zool.) 36:457–466.
HUXLEY, J. S. 1932. Problems of Relative Growth. Lincoln MacVeagh, New York, New York.
HUXLEY, J. S. 1938a. Threat and warning colouration in birds, with a general discussion of the

biological functions of colour. Pp. 430–455 *in* Proceedings of the Eighth International Ornithological Congress (F. C. R. Jourdain, Ed.). Oxford University Press, Oxford, United Kingdom.

HUXLEY, J. S. 1938b. Darwin's theory of sexual selection and the data subsumed by it, in the light of recent research. Am. Nat. 72:416–433.

HUXLEY, J. S. 1938c. The present standing of the theory of sexual selection. Pp. 11–41 *in* Evolution: Essays on Aspects of Evolutionary Biology (G. R. de Beer, Ed.). Oxford University Press, Oxford, United Kingdom.

HUXLEY, J. S. 1943. Evolution: The Modern Synthesis. Harper and Brothers, New York, New York.

HUXLEY, J. S. 1953. Evolution in Action. Harper, New York, New York.

HUXLEY, J. S. 1955. Morphism in birds. Pp. 303–328 *in* Acta XI Congressus Internationalis Ornithologici (A. Portmann and E. Sutter, Eds.). Birkhäuser Verlag, Basel, Switzerland.

HUXLEY, T. H. 1867. On the classification of birds; and on the taxonomic value of the modifications of certain of the cranial bones observable in that class. Proc. Zool. Soc. Lond. 1867:415–472.

IKENAGA, H. 1983. Appearance time and behavior of the Okinawa Rail at a water site in the late afternoon. Strix 2:11.

IKENAGA, H., AND T. GIMA. 1993. Vocal repertoire and duetting in the Okinawa Rail *Rallus okinawae*. J. Yamashina Inst. Ornithol. 25:28–39.

IMPE, J. VAN, AND H. LIECKENS. 1993. Aspects of the breeding biology of the Coot *Fulica atra* in an uncommon habitat. Oriolus 59:3–13.

INSKIPP, T., AND P. D. ROUND. 1989. A review of the Black-tailed Crake *Porzana bicolor*. Forktail 5: 3–15.

IREDALE, T. 1910. An additional note on the birds of Lord Howe and Norfolk islands. Proc. Linn. Soc. N. S. Wales 35:773–782.

JABLONSKI, D. 1991. Extinctions: a paleobiological perspective. Science 253:754–757.

JABLONSKI, D. 1996. Body size and macroevolution. Pp. 256–289 *in* Evolutionary Paleobiology (D. Jablonski, D. H. Erwin, and J. H. Lipps, Eds.). University of Chicago Press, Chicago, Illinois.

JABLONSKI, D. 2000. Micro- and macroevolution: scale and hierarchy in evolutionary biology and paleobiology. Paleobiology 26(Suppl.):15–52.

JABLONSKI, P. G., AND P. MATYJASIAK. 2002. Male wing-patch asymmetry and aggressive response to intruders in the Common Chaffinch (*Fringilla coelebs*). Auk 119:566–572.

JACKSON, D. A. 1993. Stopping rules in principal components analysis: a comparison of heuristical and statistical approaches. Ecology 74:2204–2214.

JACKSON, D. J. 1928. The inheritance of long and short wings in the weevil *Sitona hispidula,* with a discussion of wing reduction among beetles. Trans. R. Soc. Edinb. 55:665–735.

JACKSON, H. D. 1989. Weights of birds collected in the Mutare Municipal Area, Zimbabwe. Bull. Br. Ornithol. Club 109:100–106.

JACKSON, J. E. 1991. A User's Guide to Principal Components. J. Wiley and Sons, New York, New York.

JACKSON, M. E., AND R. D. SEMLITSCH. 1993. Paedomorphosis in the salamander *Ambystoma talpoideum:* effects of a fish predator. Ecology 74:342–350.

JACKSON, S., AND J. DIAMOND. 1996. Metabolic and digestive responses to artificial selection in chickens. Evolution 50:1638–1650.

JACOB, F. 1977. Evolution and tinkering. Science 196:1161–1166.

JACOB, F. 1982. The Possible and the Actual. Pantheon Books, New York, New York.

JACOBS, J. D., AND J. C. WINGFIELD. 2000. Endocrine control of life-cycle stages: A constraint on response to the environment? Condor 102:35–51.

JAMES, F. C. 1968. A more precise definition of Bergmann's rule. Am. Zool. 8:815–816.

JAMES, F. C. 1970. Geographic size variation in birds and its relationship to climate. Ecology 51:365–390.

JAMES, F. C. 1983. Environmental component of morphological differentiation in birds. Science 221: 184–186.

JAMES, F. C., AND W. J. BOECKLEN. 1984. Interspecific morphological relationships and the densities of birds. Pp. 458–477 *in* Ecological Communities: Conceptual Issues and the Evidence (D. R. Strong, Jr., D. Simberloff, L. G. Abele, and A. B. Thistle, Eds.). Princeton University Press, Princeton, New Jersey.

JAMES, F. C., AND C. E. MCCULLOCH. 1990. Multivariate analysis in ecology and systematics: panacea or Pandora's Box? Annu. Rev. Ecol. Syst. 21:129–166.

JAMES, F. C., AND C. NESMITH. 1989. Nongenetic effects in geographic differences among nestling populations of Red-winged Blackbirds. Pp. 1424–1433 *in* Acta XIX Congressus Internationalis Ornithologici. Vol. 2 (H. Ouellet, Ed.). University of Ottawa Press, Ottawa, Ontario, Canada.

JAMES, H. F. 1995. Prehistoric extinctions and ecological changes on oceanic islands. Pp. 87–102 *in* Islands: Biological Diversity and Ecosystem Function (P. M. Vitousek, L. L. Loope, and H. Adsersen, Eds.). Springer-Verlag, Berlin, Germany.

JAMES, H. F. 1997. The diet and ecology of Hawaii's extinct flightless waterfowl: evidence from coprolites. Biol. J. Linn. Soc. 62:279–297.

JAMES, H. F., AND S. L. OLSON. 1983. Flightless birds. Nat. Hist. 92:30–40.

JAMES, H. F., AND S. L. OLSON. 1991. Descriptions of Thirty-two New Species of Birds from the Hawaiian Islands. Part II. Passeriformes. Ornithol. Monogr. 46. American Ornithologists' Union, Washington, D.C.

JAMES, H. F., T. W. STAFFORD, S. L. OLSON, D. W. STEADMAN, P. S. MARTIN, AND P. MCCOY. 1987. Radiocarbon dates on bones of extinct birds from the Hawaiian Islands. Proc. Natl. Acad. Sci. 84:2350–2354.

JAMES, K. A. C., G. C. WAHORN, R. G. POWLESLAND, AND B. D. LLOYD. 1991. Supplementary feeding of Kakapo on Little Barrier Island. Proc. Nutr. Soc. N. Z. 16:93–102.

JAMIESON, I. G. 1997. Testing reproductive skew models in a communally breeding bird, the Pukeko, *Porphyrio porphyrio*. Proc. R. Soc. Lond. (Ser. B Biol. Sci.) 264:335–340.

JAMIESON, I. G., AND J. L. CRAIG. 1987a. Dominance and mating in a communal polygynandrous bird: cooperation or indifference towards mating competitors? Ethology 75:317–327.

JAMIESON, I. G., AND J. L. CRAIG. 1987b. Male–male and female–female courtship and copulation behaviour in a communally breeding bird. Anim. Behav. 35:1251–1252.

JAMIESON, I. G., J. S. QUINN, P. A. ROSE, AND B. N. WHITE. 1994. Shared paternity among non-relatives is a result of an egalitarian mating system in a communally breeding bird, the Pukeko. Proc. R. Soc. Lond. (Ser. B Biol. Sci.) 257:271–277.

JAMIESON, I. G., AND C. J. RYAN. 2001. Closure of the debate over the merits of introducing Fiordland Takahe to predator-free islands. Pp. 96–113 *in* The Takahe: Fifty Years of Conservation Management and Research. University of Otago Press, Dunedin, New Zealand.

JANSON, H. W. 1991. History of Art. 4th ed. H. N. Abrams, New York, New York.

JANZEN, D. H. 1967. Why mountain passes are higher in the tropics. Am. Nat. 101:233–249.

JANZEN, F. J. 1994. Climate change and temperature-dependent sex determination in reptiles. Proc. Natl. Acad. Sci. USA 91:7487–7490.

JANZEN, F. J. 1995. Experimental evidence for the evolutionary significance of temperature-dependent sex determination. Evolution 49:864–873.

JANZEN, F. J., AND G. L. PAUKSTIS. 1991. Environmental sex determination in reptiles: ecology, evolution, and experimental design. Q. Rev. Biol. 66:149–179.

JÄRVINEN, O., AND Y. HAILA. 1984. Assembly of land bird communities on northern islands: a quantitative analysis of insular impoverishment. Pp. 138–147 *in* Ecological Communities: Conceptual Issues and the Evidence (D. R. Strong, Jr., D. Simberloff, L. G. Abele, and A. B. Thistle, Eds.). Princeton University Press, Princeton, New Jersey.

JÄRVINEN, O., AND K. VEPSÄLÄINEN. 1976. Wing dimorphism as an adaptive strategy in water striders (*Gerris*). Hereditas 84:61–68.

JEFFRIES, M. J., AND J. H. LAWTON. 1984. Enemy-free space and the structure of ecological communities. Biol. J. Linn. Soc. 23:269–286.

JEHL, J. R., JR. 1970. Sexual selection for size differences in two species of sandpipers. Evolution 24:311–319.

JEHL, J. R., JR. 1990. Aspects of the molt migration. Pp. 102–113 *in* Bird Migration: Physiology and Ecophysiology (E. Gwinner, Ed.). Springer-Verlag, Berlin, Germany.

JEHL, J. R., JR. 1997. Fat loads and flightlessness in Wilson's Phalaropes. Condor 99:538–543.

JEHL, J. R., JR., J. FRANCINE, AND S. I. BOND. 1990. Growth patterns of two races of California Gull raised in a common environment. Condor 92:732–738.

JEHL, J. R., JR., AND B. G. MURRAY, JR. 1986. The evolution of normal and reverse sexual size dimorphism in shorebirds and other birds. Pp. 1–86 *in* Current Ornithology. Vol. 3 (R. F. Johnston, Ed.). Plenum Press, New York, New York.

JEIKOWSKI, H. 1971. Die Flügelbefiederung des Bleßhuhns (*Fulica atra* L.). J. Ornithol. 112:164–201.

JENKINS, F. A., JR., K. P. DIAL, AND G. E. GOSLOW, JR. 1988. A cineradiographic analysis of bird flight: the wishbone in Starlings is a spring. Science 241:1495–1498.
JENKINS, F. A., JR., AND D. M. WALSH. 1993. An early Jurassic caecilian with limbs. Nature 365:246–249.
JENKINS, J. M. 1979. Natural history of the Guam Rail. Condor 81:404–408.
JENKINS, J. M. 1983. The native forest birds of Guam. Ornithol. Monogr. No. 31. American Ornithologists' Union, Washington, D.C.
JENNI, D. A. 1974. Evolution of polyandry in birds. Am. Zool. 14:129–144.
JENNI, D. A., AND G. COLLIER. 1972. Polyandry in the American Jaçana (*Jacana spinosa*). Auk 89:743–765.
JENNI, L., AND S. JENNI-EIERMANN. 1998. Fuel supply and metabolic constraints in migrating birds. J. Avian Biol. 29:521–528.
JOHANSSON, C., E. T. LINDER, P. HARDIN, AND C. M. WHITE. 1998. Bill and body size in the Peregrine Falcon, north versus south: Is size adaptive? J. Biogeogr. 24:265–273.
JOHNSGARD, P. A. 1963. Behavioral isolating mechanisms in the family Anatidae. Pp. 531–543 *in* Proceedings of the 13th International Ornithological Congress. Vol. 1 (C. G. Sibley, Ed.). American Ornithologists' Union, Washington, D.C.
JOHNSGARD, P. A. 1983. The Grouse of the World. University of Nebraska Press, Lincoln, Nebraska.
JOHNSGARD, P. A. 1988. The Quails, Partridges, and Francolins of the World. Oxford University Press, Oxford, United Kingdom.
JOHNSGARD, P. A. 1991. Bustards, Hemipodes, and Sandgrouse: Birds of Dry Places. Oxford University Press, Oxford, United Kingdom.
JOHNSGARD, P. A. 1999. The Pheasants of the World. 2nd ed. Smithsonian Institution Press, Washington, D.C.
JOHNSON, A. W. 1964. The Giant Coot *Fulica gigantea* Eydoux and Souleyet. Bull. Br. Ornithol. Club 84:170–172.
JOHNSON, A. W. 1965. The Horned Coot, *Fulica cornuta* Bonaparte. Bull. Br. Ornithol. Club 84:84–88.
JOHNSON, C., AND C. L. MADER. 1994. Modeling the 105 ka Lanai Tsunami. Sci. Tsunami Hazards 11:33–38.
JOHNSON, D. R. 1986. The Genetics of the Skeleton: Animal Models of Skeletal Development. Clarendon Press, Oxford, United Kingdom.
JOHNSON, G. 1987. Selection against large size in the Sand Martin *Riparia riparia* during a dramatic population crash. Ibis 129:274–280.
JOHNSON, K. P., F. R. ADLER, AND J. L. CHERRY. 2000. Genetic and phylogenetic consequences of island biogeography. Evolution 54:387–396.
JOHNSON, M. E. 1987. Multivariate Statistical Simulation. J. Wiley and Sons, New York, New York.
JOHNSON, M. L., AND M. S. GAINES. 1990. Evolution of dispersal: theoretical models and empirical tests among birds and mammals. Annu. Rev. Ecol. Syst. 21:449–480.
JOHNSON, N. K. 1975. Controls of number of bird species on montane islands in the Great Basin. Evolution 29:545–567.
JOHNSON, N. K. 1980. Character variation and evolution of sibling species in the *Empidonax difficilis–falvescens* complex (Aves: Tyrannidae). Univ. Calif. Publ. Zool. 112:1–151.
JOHNSON, O. W., AND R. D. OHMART. 1973. Some features of water economy and kidney microstructure in the large-billed Savannah Sparrow (*Passerculus sandwichensis*). Physiol. Zool. 46:276–284.
JOHNSON, R. A, AND D. W. WICHERN. 1982. Applied Multivariate Statistical Analysis. Prentice-Hall, Englewood Cliffs, New Jersey.
JOHNSON, R. R., AND J. J. DINSMORE. 1985. Brood-rearing and postbreeding habitat use by Virginia Rails and Soras. Wilson Bull. 97:551–554.
JOHNSON, R. R., AND J. J. DINSMORE. 1986a. The use of tape-recorded calls to count Virginia Rails and Soras. Wilson Bull. 98:303–306.
JOHNSON, R. R., AND J. J. DINSMORE. 1986b. Habitat use by breeding Virginia Rails and Soras. J. Wildl. Manage. 50:387–392.
JOHNSON, T. H., AND A. J. STATTERFIELD. 1990. A global review of island endemic birds. Ibis 132:167–180.

JOHNSTON, D. W. 1975. Ecological analysis of the Cayman Island avifauna. Bull. Fla. State Mus. (Biol. Sci.) 19:235–300.
JOHNSTON, D. W. 1988. A morphological atlas of the avian uropygial gland. Br. Mus. (Nat. Hist.) Bull. (Zool.) 54:199–259.
JOHNSTON, R. F. 1969. Character variation and adaptation in European sparrows. Syst. Zool. 18:206–231.
JOHNSTON, R. F. 1973. Evolution in the House Sparrow, IV. Replicate studies in phenetic covariation. Syst. Zool. 22:219–226.
JOHNSTON, R. F. 1976a. Estimating variation in bony characters and a comment on the Kluge–Kerfoot effect. Univ. Kans. Mus. Nat. Hist. Occas. Pap. 53:1–8.
JOHNSTON, R. F. 1976b. Evolution in the House Sparrow, V. Covariation of skull and hindlimb sizes. Univ. Kans. Mus. Nat. Hist. Occas. Pap. 56:1–8.
JOHNSTON, R. F. 1992a. Evolution in the Rock Dove: skeletal morphology. Auk 109:530–542.
JOHNSTON, R. F. 1992b. Geographic size variation in Rock Pigeons, *Columba livia*. Boll. Zool. 59: 111–116.
JOHNSTON, R. F., AND R. C. FLEISCHER. 1981. Overwinter mortality and sexual size dimorphism in the House Sparrow. Auk 98:503–511.
JOHNSTON, R. F., D. M. NILES, AND S. A. ROHWER. 1972. Herman Bumpus and natural selection in the House Sparrow *Passer domesticus*. Evolution 26:20–31.
JOHNSTON, R. F., AND R. K. SELANDER. 1971. Evolution in the House Sparrow, II. Adaptive differentiation in North American populations. Evolution 25:1–28.
JOHNSTON, R. F., AND R. K. SELANDER. 1973. Evolution in the House Sparrow, III. Variation in size and sexual dimorphism in Europe and North and South America. Am. Nat. 107:373–390.
JOLICOEUR, P. 1963. The multivariate generalization of the allometry equation. Biometrics 19:497–499.
JOLICOEUR, P. 1990. Bivariate allometry: interval estimation of the slopes of the ordinary and standardized normal major axes and structural relationship. J. Theor. Biol. 14:275–285.
JOLLIE, M. T. 1957. The head skeleton of the chicken and remarks on the anatomy of this region in other birds. J. Morphol. 100:389–436.
JOLLIFFE, I. T. 1986. Principal Component Analysis. Springer-Verlag, New York, New York.
JOLLY, J. N. 1989. A field study of the breeding biology of the Little Spotted Kiwi (*Apteryx owenii*) with emphasis on the causes of nest failures. J. R. Soc. Lond. 19:433–448.
JONES, D. N., R. W. R. J. DEKKER, AND C. S. ROSELAAR. 1995. The Megapodes: *Megapodiidae*. Oxford University Press, Oxford, United Kingdom.
JONES, D. S., AND S. J. GOULD. 1999. Direct measurement of age in fossil *Gryphaea*: the solution to a classic problem in heterochrony. Paleobiology 25:158–187.
JONES, G. 1996. Does echolocation constrain the evolution of body size in bats? Symp. Zool. Soc. Lond. 69:111–128.
JONES, J. C. 1940. Food habits of the American Coot with notes on distribution. U.S. Dept. Inter. Wildl. Res. Bull. 2:1–52.
JONES, J. S., B. H. LEITH, AND P. RAWLINGS. 1977. Polymorphism in *Cepea*: a problem with too many solutions. Annu. Rev. Ecol. Syst. 8:109–143.
JONES, P., S. SCHUBEL, J. JOLLY, M. DE L. BROOKE, AND J. VICKERY. 1995. Behaviour, natural history, and annual cycle of the Henderson Island Rail *Porzana atra* (Aves: Rallidae). Biol. J. Linn. Soc. 56:167–183.
JONES, P. J., AND P. WARD. 1976. The level of reserve protein as the proximate factor controlling the timing of breeding and clutch-size in the Red-billed Quelea, *Quelea quelea*. Ibis 118:547–574.
JONES, R., D. C. CULVER, AND T. C. KANE. 1992. Are parallel morphologies of cave organisms the result of similar selection pressures? Evolution 46:353–365.
JORGENSEN, P. D., AND H. L. FERGUSON. 1982. Clapper Rail preys on Savannah Sparrow. Wilson Bull. 94:215.
JUANES, F. 1986. Population density and body size in birds. Am. Nat. 128:921–929.
JUBERTHRIE, C. 1989. Insularité et speciation souterraine. Mem. Biospeol. 16:3–14.
JULLIEN, M., AND J.-M. THIOLLAY. 1996. Effects of rain forest disturbance and fragmentation: comparative changes of the raptor community along natural and human-made gradients in French Guiana. J. Biogeogr. 23:7–25.
JUNGERS, W. L., ED. 1985. Size and Scaling in Primate Biology. Plenum Press, New York, New York.

JUVIK, J. O., AND A. P. AUSTRING. 1979. The Hawaiian avifauna: biogeographic theory in evolutionary time. J. Biogeogr. 6:205–224.
KADMON, R., AND H. R. PULLIAM. 1993. Island biogeography: effect of geographical isolation on species composition. Ecology 74:977–981.
KAELIN, J. 1941. Über den Coracoscapularwinkel und die Beziehungen der Rumpfform zum Lokomotionstypus der Vögeln. Rev. Suisse Zool. 48:553–557.
KAISER, H. E. 1974. Morphology of the Sirenia. S. Karger, New York, New York.
KAITALA, A. 1988. Wing muscle dimorphism: two reproductive pathways of water strider *Gerris thoracicus* in relation to habitat instability. Oikos 53:222–228.
KANE, T. C., AND R. C. RICHARDSON. 1985. Regressive evolution: an historical perspective. Natl. Speleol. Soc. Bull. 47:71–77.
KANESHIRO, K. Y. 1995. Evolution, speciation, and the genetic structure of island populations. Pp. 1–4 *in* Islands: Biological Diversity and Ecosystem Function (P. M. Vitousek, L. L. Loope, and H. Adsersen, Eds.). Springer-Verlag, Berlin, Germany.
KAPLAN, R. H., AND W. S. COOPER. 1984. The evolution of developmental plasticity in reproductive characteristics: an application of the 'adaptive coin-flipping' principle. Am. Nat. 123:393–410.
KAPLAN, J. M., AND M. PIGLIUCCI. 2001. Genes "for" phenotypes: a modern history view. Biol. Philos. 16:189–213.
KARASOV, W. H. 1990. Digestion in birds: chemical and physiological determinants and ecological implications. Pp. 337–352 *in* Avian Foraging: Theory, Methodology, and Applications (M. L. Morrison, C. J. Ralph, J. Verner, and J. R. Jehl, Jr., Eds.). Cooper Ornithological Society, Los Angeles, California.
KARASOV, W. H. 1996. Digestive plasticity in avian energetics and feeding ecology. Pp. 61–84 *in* Avian Energetics and Nutritional Ecology (C. Carey, Ed.). Chapman and Hall, New York, New York.
KARHU, S. 1973. On the development stages of chicks and adult Moorhens *Gallinula chloropus* at the end of a breeding season. Ornis Fenn. 50:1–17.
KARLIN, S., AND R. B. CAMPBELL. 1981. The existence of a protected polymorphism under conditions of soft as opposed to hard selection in a multideme population system. Am. Nat. 117:262–275.
KARR, J. R. 1982. Population variability and extinction in the avifauna of a tropical land-bridge island. Ecology 63:1975–1978.
KAUFFMAN, S. A. 1983. Developmental constraints: internal factors in evolution. Pp. 195–225 *in* Development and Evolution (B. C. Goodwin, N. Holder, and C. C. Wylie, Eds.). Cambridge University Press, Cambridge, United Kingdom.
KAUFMANN, G. W. 1977. Breeding requirements of the Virginia Rail and the Sora in captivity. Avic. Mag. 83:135–141.
KAUFMANN, G. W. 1983. Displays and vocalizations of the Sora and the Virginia Rail. Wilson Bull. 95:42–59.
KAUFMANN, G. W. 1987. Growth and development of Sora and Virginia Rail chicks. Wilson Bull. 99:432–440.
KAUFMANN, G. W. 1988. Development of Spotless Crake chicks. Notornis 35:324–327.
KAUFMANN, G. W. 1989. Breeding ecology of the Sora *Porzana carolina* and the Virginia Rail *Rallus limicola*. Can. Field-Nat. 103:270–282.
KAUFMANN, G. W., AND R. LAVERS. 1987. Observations of breeding behaviour of Spotless Crakes (*Porzana tabuensis*) and Marsh Crake (*P. pusilla*) at Pukepuke Lagoon. Notornis 34:193–205.
KAUFMANN, J. H. 1983. On the definitions and functions of dominance and territoriality. Biol. Rev. 58:1–20.
KAUFMANN, L., AND P. J. ROUSSEEUW. 1990. Finding Groups in Data: An Introduction to Cluster Analysis. J. Wiley and Sons, New York, New York.
KAWECKI, T. J. 1993. Age and size at maturity in a patchy environment: fitness maximization versus evolutionary stability. Oikos 66:309–317.
KAWECKI, T. J., AND S. C. STEARNS. 1993. The evolution of life histories in spatially heterogeneous environments: optimal reaction norms revisited. Evol. Ecol. 7:155–174.
KEAR, J. 1970. The adaptive radiation of parental care in waterfowl. Pp. 357–392 *in* Social Behaviour in Birds and Mammals: Essays on the Social Ethology of Animals and Man (J. H. Crook, Ed.). Academic Press, New York, New York.
KEAR, J., N. HILGARTH, AND S. ANDERS. 1980. Coots eating goose droppings. Br. Birds 73:410.

KEAST, A. 1968. Competitive interactions and the evolution of ecological niches as illustrated by the Australian honeyeater genus *Melithreptus* (Meliphagidae). Evolution 22:762–784.
KEAST, A. 1970. Adaptive evolution and shifts in niche occupation in island birds. Biotropica 2: 61–75.
KEAST, A. 1971. Adaptive evolution and shifts in niche occupation in island birds. Pp. 39–53 *in* Adaptive Aspects of Insular Evolution (W. L. Stern, Ed.). Washington State University Press, Pullman, Washington.
KEAST, A. 1976. Ecological opportunities and adaptive evolution on islands, with special reference to evolution in the isolated forest outliers of southern Australia. Pp. 571–584 *in* Proceedings of the 16th International Ornithological Congress (H. J. Frith and J. H. Calaby, Eds.). Australian Academy of Science, Canberra, Australia.
KEATING, B. H., AND W. J. MCGUIRE. 2000. Island edifice failures and associated tsunami hazards. Pure Appl. Geophys. 157:899–955.
KEEFER, D. K. 1984. Landslides caused by earthquakes. Geol. Soc. Am. Bull. 95:406–421.
KEITH, S. 1986. Rallidae: rails, flufftails, crakes, gallinules, moorhens and coots. Pp. 84–130 *in* The Birds of Africa. Vol. 2 (E. K. Urban, C. H. Fry, and S. Keith, Eds.). Academic Press, London, United Kingdom.
KEITH, S., C. W. BENSON, AND M. P. STUART IRWIN. 1970. The genus *Sarothrura* (Aves, Rallidae). Bull. Am. Mus. Nat. Hist. 143:1–84.
KELLER, L. F., AND P. ARCESE. 1998. No evidence for inbreeding avoidance in a natural population of Song Sparrows (*Melospiza melodia*). Am. Nat. 152:380–392.
KELLER, L. F., P. R. GRANT, B. R. GRANT, AND K. PETREN. 2002. Environmental conditions affect the magnitude of inbreeding depression in survival of Darwin's finches. Evolution 56:1229–1239.
KELLOGG, R. 1936. A review of the Archaeoceti. Carnegie Inst. Wash. Publ. 482:1–366.
KELLY, J. F., R. SMITH, D. M. FINCH, F. R. MOORE, AND W. YONG. 1999. Influence of summer biogeography on wood warbler stopover abundance. Condor 101:76–85.
KELLY, J. K. 2000. Epistasis, linkage, and balancing selection. Pp. 146–157 *in* Epistasis and the Evolutionary Process (J. B. Wolf, E. D. Brodie III, and M. J. Wade, Eds.). Oxford University Press, Oxford, United Kingdom.
KELSO, J. E. H. 1922. Birds using their wings as a means of propulsion under water. Auk 39:426–428.
KEMP, T. S. 1999. Fossils and Evolution. Oxford University Press, Oxford, United Kingdom.
KENDEIGH, S. C. 1952. Parental care and its evolution in birds. Ill. Biol. Monogr. 22:1–358.
KENDEIGH, S. C. 1969. Tolerance of cold and Bergmann's rule. Auk 86:13–25.
KENDEIGH, S. C. 1970. Energy requirements for existence in relation to size of bird. Condor 72: 60–65.
KENDEIGH, S. C. 1972. Energy control of size limits in birds. Am. Nat. 106:79–88.
KENDEIGH, S. C. 1976. Latitudinal trends in the metabolic adjustments of the House Sparrow. Ecology 57:509–519.
KENDEIGH, S. C., V. R. DOL'NIK, AND V. M. GAVRILOV. 1977. Avian energetics. Pp. 127–204 *in* Granivorous Birds in Ecosystems (J. Pinkowski and S. C. Kendeigh, Eds.). Cambridge University Press, Cambridge, United Kingdom.
KERSTEN, M., AND T. PIERSMA. 1987. High levels of energy expenditure in shorebirds: metabolic adaptations to an energetically expensive way of life. Ardea 75:175–187.
KETTERSON, E. D., AND V. NOLAN, JR. 1994. Male parental behavior in birds. Annu. Rev. Ecol. Syst. 25:601–628.
KIDWELL, J. F., AND H. B. CHASE. 1967. Fitting the allometric equation—a comparison of ten methods by computer simulation. Growth 31:165–179.
KIEL, W. H. 1955. Nesting studies of the Coot in southwestern Manitoba. J. Wildl. Manage. 19:189–198.
KIENY, M. 1977. Proximo-distal pattern formation in avian limb development. Pp. 87–103 *in* Vertebrate Limb and Somite Morphogenesis (D. A. Ede, J. R. Hinchliffe, and M. Balls, Eds.). Cambridge University Press, Cambridge, United Kingdom.
KILHAM, L. A. 1965. Differences in feeding behavior of male and female Hairy Woodpeckers. Wilson Bull. 77:134–145.
KILHAM, L. A. 1970. Feeding behavior of Downy Woodpeckers. I. Preference for paper birches and sexual differences. Auk 87:544–556.

KILHAM, L. A. 1979. Snake and pond snails as food of Grey-necked Wood-Rails. Condor 81:100–101.

KIMBALL, R. T., AND J. D. LIGON. 1999. Evolution of avian plumage dichromatism from a proximate perspective. Am. Nat. 154:182–193.

KING, A. S., AND D. Z. KING. 1979. Avian morphology: general principles. Pp. 1–38 in Form and Function in Birds. Vol. 1 (A. S. King and J. McLelland, Eds.). Academic Press, London, United Kingdom.

KING, A. S., M. K. VIDYADARAN, AND H. KASSIM. 1992. Quantitative pulmonary anatomy of a ground-dwelling bird, the White-breasted Water-hen (*Amaurornis phoenicurus*). J. Zool. Lond. 227:185–191.

KING, C. 1984. Immigrant Killers: Introduced Predators and the Conservation of Birds in New Zealand. Oxford University Press, Auckland, New Zealand.

KING, J. R. 1972. Adaptive periodic fat storage by birds. Pp. 200–217 in Proceedings of the XVth International Ornithological Congress (K. H. Voous, Ed.). E. J. Brill, Leiden, The Netherlands.

KING, J. R. 1973. Energetics of reproduction in birds. Pp. 78–120 in Breeding Biology of Birds (D. S. Farner, Ed.). National Academy of Science, Washington, D.C.

KING, J. R. 1981. Energetics of reproduction in birds. Pp. 312–317 in Acta XVII Congressus Internationalis Ornithologici. Vol. 1 (R. Nörhring, Ed.). Deutschen Ornithologische-Gesellschaft, Berlin, Germany.

KING, J. R., AND D. S. FARNER. 1961. Energy metabolism, thermoregulation and body temperature. Pp. 215–288 in Biology and Comparative Physiology of Birds. Vol. 2 (A. J. Marshall, Ed.). Academic Press, New York, New York.

KING, W. B. 1980. Ecological basis of extinction in birds. Pp. 905–911 in Acta Congressus Internationalis Ornithologici. Vol. 2 (R. Nöhring, Ed.). Verlag Deutschen Ornithologische-Gesellschaft, Berlin, Germany.

KING, W. B. 1981. Endangered Birds of the World: The ICBP Red Data Book. Rev. ed. Smithsonian Institution Press, Washington, D.C.

KINGSOLVER, J. G., H. E. HOEKSTRA, J. M. HOEKSTRA, D. BERRIGAN, S. N. VIGNIERI, C. E. HILL, A. HOANG, P. GILBERT, AND P. BEERLI. 2001. The strength of phenotypic selection in natural populations. Am. Nat. 157:245–261.

KINNEAR, P. K. 1977. Predation of seabirds by grey seals. Scott. Birds 9:342–347.

KINSKY, F. C., CONVENER. 1970. Annotated Checklist of the Birds of New Zealand, Including the Birds of the Ross Dependency. Ornithological Society of New Zealand, Wellington, New Zealand.

KIPP, F. A. 1959. Der Handflügel-Index als flugbiologisches Maß. Vogelwarte 20:77–88.

KIRCH, P. V. 1997. Introduction: the environmental history of oceanic islands. Pp. 1–21 in Historical Ecology in the Pacific Islands (P. V. Kirch and T. L. Hunt, Eds.). Yale University Press, New Haven, Connecticut.

KIRCH, P. V. 2000. On the Road of the Winds: An Archaeological History of the Pacific Islands Before European Contact. University of California Press, Berkeley, California.

KIRK, T. W. 1880. Notes on some additions to the collection of birds in the Colonial Museum. Trans. N. Z. Inst. 13:235–236.

KIRK, T. W. 1895. The displacement of species in New Zealand. Trans. N. Z. Inst. 28:1–27.

KIRKHAM, W. B., AND H. W. HAGGARD. 1915. A comparative study of the shoulder region of the normal and of a wingless fowl. Anat. Rec. 9:159–180.

KIRKPATRICK, M. 1982a. Sexual selection and the evolution of female choice. Evolution 36:1–12.

KIRKPATRICK, M. 1982b. Quantum evolution and punctuated equilibria in continuous genetic characters. Am. Nat. 119:833–848.

KIRKWOOD, J. K. 1983. A limit to metabolisable energy intake in mammals and birds. Comp. Biochem. Physiol. (Ser. A) 75:1–3.

KIRKWOOD, J. K., P. J. DUIGNAN, N. F. KEMBER, P. M. BENNETT, AND D. J. PRICE. 1989. The growth rate of the tarsometatarsus bone in birds. J. Zool. Lond. 217:403–416.

KISDI, É. 2002. Dispersal: risk spreading versus local adaptation. Am. Nat. 159:579–596.

KLASSEN, M. 1995. Water and energy limitations on flight range. Auk 112:260–262.

KLASSEN, M. 1996. Metabolic constraints on long-distance migration in birds. J. Exp. Biol. 199:57–64.

KLASSEN, M., AND Å. LINDSTRÖM. 1996. Departure fuel loads in time-minimizing migrating birds can be explained by the energy costs of being heavy. J. Theor. Biol. 183:29–34.

KLASSING, K. C. 1998. Comparative Avian Nutrition. CAB International, New York, New York.

KLEIBER, M. 1932. Body size and metabolism. Hilgardia 6:315–353.

KLEIBER, M. 1961. The Fire of Life: An Introduction to Animal Energetics. J. Wiley, New York, New York.

KLEIBER, M. 1972. Body size, conductance for animal heat flow and Newton's law of cooling. J. Theor. Biol. 37:139–150.

KLEIBER, M. 1975. Metabolic turnover rate: a physiological meaning of the metabolic rate per unit body weight. J. Theor. Biol. 53:199–204.

KLICKA, J., K. P. JOHNSON, AND S. M. LANYON. 2000. New World nine-primaried oscine relationships: constructing a mitochondrial DNA framework. Auk 117:321–336.

KLIMA, M. 1962. The morphogenesis of the avian sternum. Pr. Brnenské Základny Cesk. Akad. Ved 34:151–194.

KLINGENBERG, C. P. 1998. Heterochrony and allometry: the analysis of evolutionary change in ontogeny. Biol. Rev. 73:79–123.

KLINGENBERG, C. P., AND R. FROESE. 1991. A multivariate comparison of allometric growth patterns. Syst. Zool. 40:410–419.

KLOMP, H. 1970. The determination of clutch-size in birds: a review. Ardea 58:1–124.

KLUGE, A. G., AND W. C. KERFOOT. 1973. The predictability and regularity of character divergence. Am. Nat. 107:426–442.

KNAPP, G. 1999. Angels, Archangels and All the Company of Heaven. Prestel Verlag, Munich, Germany.

KNÖPFLI, W. 1918. Beiträge zur Morphologie und Entwicklungsgeschichte des Brustschulterskelettes bei den Vögeln. Jena Z. Naturwiss. 55:577–720.

KNOPPIEN, P. 1985. Rare male mating advantage: a review. Biol. Rev. 60:81–117.

KNOX, A. G. 1976. The taxonomic status of the Scottish Crossbill *Loxia* sp. Bull. Br. Ornithol. Club 90:15–19.

KNOX, A. G. 1990. The sympatric breeding of Common and Scottish Crossbills *Loxia curvirostra* and *L. scotia* and the evolution of crossbills. Ibis 132:454–466.

KNOX, A. G., AND M. P. WALTERS. 1994. Extinct and endangered birds in the collection of the Natural History Museum. Br. Ornithol. Club Occas. Publ. 1:1–292.

KOEHL, M. A. R. 2000. Cosequences of size change during ontogeny and evolution. Pp. 67–86 *in* Scaling in Biology (J. H. Brown and G. B. West, Eds.). Oxford University Press, Oxford, United Kingdom.

KOENIG, W. D., AND F. A. PITELKA. 1981. Ecological factors and kin selection in the evolution of cooperative breeding in birds. Pp. 261–280 *in* Natural Selection and Social Behavior: Recent Research and New Theory (R. D. Alexander and D. W. Tinkle, Eds.). Chiron Press, New York, New York.

KOENIG, W. D., F. A. PITELKA, W. J. CARMEN, R. L. MUMME, AND M. T. STANBACK. 1992. The evolution of delayed dispersal in cooperative breeders. Q. Rev. Biol. 67:111–150.

KOENIG, W. D., AND P. B. STACEY, EDS. 1990. Cooperative Breeding in Birds: Long-term Studies of Ecology and Behaviour. Cambridge University Press, Cambridge, United Kingdom.

KOKKO, H., AND P. LUNDBERG. 2001. Dispersal, migration, and offspring retention in saturated habitats. Am. Nat. 157:188–202.

KOKSHAYSKY, N. V. 1973. Functional aspects of some details of bird wing configuration. Syst. Zool. 22:442–450.

KOKSHAYSKY, N. V. 1977. Some scale dependent problems in aerial animal locomotion. Pp. 421–435 *in* Scale Effects in Animal Locomotion (T. J. Pedley, Ed.). Academic Press, New York, New York.

KONARZEWSKI, M., S. A. L. M. KOOIJMAN, AND R. E. RICKLEFS. 1998. Models for avian growth and development. Pp. 340–365 *in* Avian Growth and Development: Evolution Within the Altricial–Precocial Spectrum (J. M. Starck and R. E. Ricklefs, Eds.). Oxford University Press, Oxford, United Kingdom.

KOOYMAN, B. 1991. Implications of bone morphology for moa taxonomy and behavior. J. Morphol. 209:53–81.

KOOYMAN, G. L. 1975. Behaviour and physiology of diving. Pp. 115–137 *in* The Biology of Penguins (B. Stonehouse, Ed.). University Park Press, Baltimore, Maryland.

KOOYMAN, G. L. 1989. Diverse Divers: Physiology and Behavior. Springer-Verlag, Berlin, Germany.

KORNOWSKI, G. 1957. Beiträge zur Ethologie des Blässhuhns (*Fulica atra* L.). J. Ornithol. 98:318–355.

KORSCHGEN, C. E. 1977. Breeding stress of female Eiders in Maine. J. Wildl. Manage. 41:360–373.

KOSSWIG, C. 1963. Genetische Analyse konstruktiver und degenerativer Evolutionsprozesse. Z. Zool. Syst. Evolutionsforsch. 1:205–239.

KOZICKY, E. L., AND F. V. SCHMIDT. 1949. Nesting habits of the Clapper Rail in New Jersey. Auk 66: 355–364.

KOZLOWSKI, J. 1996. Optimal initial size and adult size of animals: consequences for macroevolution and community structure. Am. Nat. 147:101–114.

KOZLOWSKI, J. 2000. Does body size optimization alter the allometries for production and life history traits? Pp. 237–252 *in* Scaling in Biology (J. H. Brown and G. B. West, Eds.). Oxford University Press, Oxford, United Kingdom.

KOZLOWSKI, J., AND J. WEINER. 1997. Interspecific allometries are by-products of body size optimization. Am. Nat. 149:352–380.

KRAMER, G. 1946. Veranderungen von nachkommenzifer und nachtkommengrossse sowie der altersverteilung von inseleidechsen. Z. Naturforsch. 1:700–710.

KRAMER, G. 1953. Über Wachstum und Entwicklung der Vögel. J. Ornithol. 94:194–199.

KRAMER, G. 1960. Funktionsgerechte Allometrien. Pp. 426–436 *in* Proceedings of the XII International Ornithological Congress, Helsinki, 1958. Vol. 1. Tilgmannin Kirjapaino, Helsinki, Finland.

KRAMER, G., AND R. MERTENS. 1938. Rassenbildung bei west-istrianischen Inseleidechsen in abhängengheit von Isolierungsalter und Arealgröße. Arch. Nat. 7:189–234.

KRATTER, A. W., D. W. STEADMAN, C. F. SMITH, C. E. FILARDI, AND H. P. WEBB. 2001. Avifauna of a lowland forest site on Isabel, Solomon Islands. Auk 118:472–483.

KRAUS, C., O. BERNATH, K. ZBINDEN, AND G. PILLERI. 1975. Zum Verhalten und Vokalisation von *Tribonyx mortierii* du Bois, 1840 (Aves, Rallidae). Rev. Suisse Zool. 82:6–13.

KRAUSE, J., AND J.-G. J. GODIN. 1996. Influence of prey foraging posture on flight behavior and predation risk: predators take advantage of unwary prey. Behav. Ecol. 7:264–271.

KREKORIAN, C. O. 1978. Alloparental care in the Purple Gallinule. Condor 80:382–390.

KRUUK, L. E. B., J. MERILÄ, AND B. C. SHELDON. 2001. Phenotypic selection on a heritable size trait revisted. Am. Nat. 158:557–571.

KÜHNE, W. G. 1972. Progress in biological evolution. Pp. 281–284 *in* Studies in Vertebrate Evolution (K. A. Joysey and T. S. Kemp, Eds.). Winchester Press, New York, New York.

KUHRY, B., AND L. F. MARCUS. 1977. Bivariate linear models in biometry. Syst. Zool. 26:201–209.

KULCZYCKI, W. 1901. Zur Entwickelungsgeschichte des Schultergürtels bei den Vögeln, mit besonderer Berücksichtigung des Schlüsselbeines (*Gallus, Columba, Anas*). Anat. Anz. 19:577–590.

KULLBERG, C., T. FRANSSON, AND S. JAKOBSSON. 1996. Impaired predator evasion in fat Blackcaps (*Sylvia atricapilla*). Proc. R. Soc. Lond. (Ser. B Biol. Sci.) 263:1671–1675.

KULLBERG, C., N. B. METCALFE, AND D. C. HOUSTON. 2002. Impaired flight ability during incubation in the Pied Flycatcher. J. Avian Biol. 33:179–183.

KURODA, N. 1961. A note on the pectoral muscles of birds. Auk 78:261–263.

KURODA, N. 1993. Morpho-anatomy of the Okinawa Rail *Rallus okinawae*. J. Yamashina Inst. Ornithol. 25:12–27.

KURODA, N., T. MANO, AND K. OZAKI. 1984. The Rallidae, their insular distribution and conservation, with special note around discovery of *Rallus okinawae*. J. Yamashina Inst. Ornithol. (50 Yr. Retr.) 14:36–57.

KUZ'MINA, M. A. 1992. Tetraonidae and Phasianidae of the USSR: Ecology and Morphology. Smithsonian Institution Press, Washington, D.C.

LABARBERA, M. 1986. The evolution and ecology of body size. Pp. 69–98 *in* Patterns and Processes in the History of Life (D. M. Raup and D. Jablonski, Eds.). Springer-Verlag, Berlin, Germany.

LABARBERA, M. 1989. Analyzing body size as a factor in ecology and evolution. Annu. Rev. Ecol. Syst. 20:97–117.

LACHENBRUCH, P. A. 1975. Discriminant Analysis. Hafner, New York, New York.

LACHENBRUCH, P., AND R. M. MICKEY. 1968. Estimation of error rates in discriminant analysis. Technometrics 10:1–11.

LACK, D. 1942. Ecological features of the bird faunas of British small islands. J. Anim. Ecol. 11:9–36.

LACK, D. 1944. Correlation between beak and food in the crossbill (*L. curvirostra*). Ibis 86:552–553.

LACK, D. 1945. The Galápagos finches (Geospizinae): a study in variation. Occas. Pap. Calif. Acad. Sci. 21:1–158.

LACK, D. 1947a. Darwin's Finches. Cambridge University Press, Cambridge, United Kingdom.

LACK, D. 1947b. The significance of clutch-size. Ibis 89:302–352.

LACK, D. 1948. The significance of clutch-size. Part III.—Some interspecific comparisons. Ibis 90:25–45.

LACK, D. 1949. The significance of ecological isolation. Pp. 299–308 in Genetics, Paleontology, and Evolution (G. L. Jepsen, E. Mayr, and G. G. Simpson, Eds). Princeton University Press, Princeton, New Jersey.

LACK, D. 1954. The Natural Regulation of Animal Numbers. Oxford University Press, Oxford, United Kingdom.

LACK, D. 1967. The significance of clutch-size in waterfowl. Wildfowl 18:125–128.

LACK, D. 1968a. The proportion of yolk in the eggs of waterfowl. Wildfowl 19:67–69.

LACK, D. 1968b. Ecological Adaptations for Breeding in Birds. Methuen, London, United Kingdom.

LACK, D. 1968c. Bird migration and natural selection. Oikos 19:1–9.

LACK, D. 1969. The numbers of bird species on islands. Bird Study 16:193–209.

LACK, D. 1970a. Island birds. Biotropica 2:29–31.

LACK, D. 1970b. The endemic ducks of remote islands. Wildfowl 21:5–10.

LACK, D. 1971a. Ecological Isolation in Birds. Harvard University Press, Cambridge, United Kingdom.

LACK, D. 1971b. Island birds. Pp. 29–31 in Adaptive Aspects of Insular Evolution (W. L. Stern, Ed.). Washington State University Press, Pullman, Washington.

LACK, D. 1974. Evolution Illustrated by Waterfowl. Blackwell Scientific Publications, Oxford, United Kingdom.

LACK, D. 1975. Island Birds. University of California Press, Berkeley, California.

LACK, D. 1976. Island Biology: Illustrated by the Land Birds of Jamaica. University of California Press, Berkeley, California.

LAIRD, A. K. 1965. Dynamics of relative growth. Growth 29:249–263.

LAIRD, A. K. 1966a. Dynamics of embryonic growth. Growth 30:263–275.

LAIRD, A. K. 1966b. Postnatal growth of birds and mammals. Growth 30:349–363.

LAIRD, A. K., A. D. BARTON, AND S. A. TYLER. 1968. Growth and time: an interpretation of allometry. Growth 32:347–354.

LAMARCK, J. B. 1809. Philosophie Zoologique, ou Exposition des Considérations relatives à l'Historie Naturelle des Animaux. C. Martin's, Paris, France. Translated by H. Elliot (1984). University of Chicago Press, Chicago, Illinois.

LAMBERT, D. M., C. D. MILLER, K. JACK, S. ANDERSON, AND J. L. CRAIG. 1994. Single- and multilocus fingerprinting of communally breeding Pukeko: Do copulations of dominance ensure reproductive success? Proc. Natl. Acad. Sci. USA 91:9641–9645.

LAMBERT, F. R. 1989. Some field observations of the endemic Sulawesi rails. Kukila 4:34–36.

LAMBERT, F. R. 1998a. A new species of *Gymnocrex* from the Talaud Islands, Indonesia. Forktail 13:1–6.

LAMBERT, F. R. 1998b. A new species of *Amaurornis* from the Talaud Islands, Indonesia, and a review of taxonomy of bush hens occurring from the Philippines to Australia. Bull. Br. Ornithol. Club 118:67–82.

LAMBRECHT, K. 1931. Fortschritte der Palaeornithologie. Pp. 73–99 in Proceedings of the VIIth International Ornithological Congress (L. F. Beaufort, Ed.). C. de Boer, Jr., Den Helder, The Netherlands. Amsterdam, Holland.

LAMBRECHT, K. 1933. Handbuch der Palaeornithologie. Gebrüder Borntraeger, Berlin, Germany.

LANCASTER, F. M. 1968. Sex-linked winglessness in the fowl. Heredity 23:257–262.

LANDE, R. 1977. On comparing coefficients of variation. Syst. Zool. 26:214–217.

LANDE, R. 1978. Evolutionary mechanisms of limb loss in tetrapods. Evolution 32:73–92.

LANDE, R. 1979. Quantitative genetic analysis of multivariate evolution, applied to brain:body size allometry. Evolution 33:402–416.
LANDE, R. 1980a. Sexual dimorphism, sexual selection, and adaptation in polygenic characters. Evolution 34:292–307.
LANDE, R. 1980b. Genetic variation and phenotypic evolution during allopatric speciation. Am. Nat. 116:463–479.
LANDE, R. 1985a. Genetic and evolutionary aspects of allometry. Pp. 21–32 in Size and Scaling in Primate Biology (W. L. Jungers, Ed.). Plenum Press, New York, New York.
LANDE, R. 1985b. Expected time for random genetic drift of a population between stable phenotypic states. Proc. Natl. Acad. Sci. USA 82:7641–7645.
LANDE, R. 1986. The dynamics of peak shifts and the patterns of morphological evolution. Paleobiology 12:343–354.
LANDE, R. 1987a. Genetic correlations between the sexes in the evolution of sexual dimorphism and mating preferences. Pp. 83–94 in Sexual Selection: Testing the Alternatives (J. W. Bradbury and M. B. Andersson, Eds.). J. Wiley and Sons, Chichester, United Kingdom.
LANDE, R. 1987b. Extinction and thresholds in demographic models of territorial populations. Am. Nat. 130:624–635.
LANDE, R. 1992. Neutral theory of quantitative genetic variance in an island model with local extinction and colonization. Evolution 46:381–389.
LANDE, R. 1993. Risks of population extinction from demographic and environmental stochasticity and random catastrophes. Am. Nat. 142:911–927.
LANDE, R. 1999. Extinction risks from anthropogenic, ecological, and genetic factors. Pp. 1–22 in Genetics and the Extinction of Species: DNA and the Conservation of Biodiversity (L. F. Landweber and A. P. Dobson, Eds.). Princeton University Press, Princeton, New Jersey.
LANDE, R., AND S. ORZACK. 1988. Extinction dynamics of age-structured populations in a fluctuating environment. Proc. Natl. Acad. Sci. USA 85:7418–7421.
LANG, C. 1956. Das Cranium der Ratiten mit besonderer Berücksichtigung von *Struthio camelus*. Z. Wiss. Zool. 159:165–224.
LANGECKER, T. G., H. SCHMALE, AND H. WILKENS. 1993. Transcription of the opsin gene in degenerate eyes of cave-dwelling *Astayanax fasciatus* (Teleostei, Characidae) and of its conspecific epigean ancestor during early ontogeny. Cell Tissue Res. 273:183–192.
LANGRAND, O. 1990. Guide to the Birds of Madagascar. Yale University Press, New Haven, Connecticut.
LANGSTON, N. E., AND S. ROHWER. 1996. Molt–breeding tradeoffs in albatrosses: life history implications for big birds. Oikos 76:498–510.
LANKESTER, E. R. 1880. Degeneration: A Chapter in Darwinism. Macmillan, London, England.
LANSDOWN, A. B. G. 1970. A study of the normal development of the leg skeleton in the quail (*Coturnix coturnix japonica*). J. Anat. 106:147–160.
LANYON, L. E. 1981. Locomotor loading and functional adaptation in limb bones. Symp. Zool. Soc. Lond. 48:305–329.
LANYON, S. M. 1992. Phylogeny and classification of birds: a study in molecular evolution [review of Sibley and Ahlquist 1990]. Condor 94:304–307.
LARSON, S. 1957. The suborder Charadrii in arctic and boreal areas during the Tertiary and Pleistocene. Acta Vertebr. 1:1–84.
LARSSON, K., H. P. VAN DER JEUGD, I. T. VAN DER VEEN, AND P. FORSLUND. 1998. Body size declines despite positive directional selection on heritable size traits in a Barnacle Goose population. Evolution 52:1169–1184.
LASIEWSKI, R. C., AND W. A. CALDER, JR. 1971. A preliminary allometric analysis of respiratory variables in resting birds. Respir. Physiol. 11:152–166.
LASIEWSKI, R. C., AND W. R. DAWSON. 1967. A re-examination of the relation between standard metabolic rate and body weight in birds. Condor 69:13–23.
LATHAM, J. 1781–1802. A General Synopsis of Birds. 11 vols. Leigh and Sotheby, London, England.
LAUDER, G. V. 1981. Form and function: structural analysis in evolutionary morphology. Paleobiology 7:430–442.
LAUDER, G. V. 1982. Historical biology and the problem of design. J. Theor. Biol. 97:57–67.
LAUDER, G. V. 1990. Functional morphology and systematics: studying functional patterns in an historical context. Annu. Rev. Ecol. Syst. 21:317–340.

LAUDER, G. V. 1991. Biomechanics and evolution: integrating physical and historical biology in the study of complex systems. Pp. 1–19 in Biomechanics in Evolution (J. M. V. Rayner and R. J. Wootton, Eds.). Cambridge University Press, Cambridge, United Kingdom.

LAUDER, G. V. 1995. On the inference of function from structure. Pp. 1–18 in Functional Morphology in Vertebrate Paleontology (J. J. Thomason, Ed.). Cambridge University Press, Cambridge, United Kingdom.

LAURENTI, J. N. 1768. Specimen Medicum Exhibens Synopsin Reptilium Emendatum cum Experimentis circa Venena et Antidota Reptilium Austriacorum. Trattnern, Wein, Austria.

LAVERS, R., AND J. MILLS. 1984. Takahe. John McIndoe, Dunedin, New Zealand.

LAWLEY, D. N., AND A. E. MAXWELL. 1971. Factor Analysis as a Statistical Method. Butterworths, London, United Kingdom.

LAWTON, M. F., AND R. O. LAWTON. 1986. Heterochrony, deferred breeding, and avian sociality. Pp. 187–222 in Current Ornithology. Vol. 3 (R. F. Johnston, Ed.). Plenum Press, New York, New York.

LAYARD, E. L., AND E. C. L. LAYARD. 1882. Notes on the avifauna of New Caledonia. Ibis 24:493–550.

LEACH, E. P. 1943. Do fishes prey upon sea-birds? Ibis 85:220.

LEBART, L., A. MORINEAU, AND K. M. WARWICK. 1984. Multivariate Descriptive Statistical Analysis: Correspondence Analysis and Related Techniques for Large Matrices. J. Wiley and Sons, New York, New York.

LECK, C. F. 1979. Avian extinctions in an isolated tropical wet-forest preserve, Ecuador. Auk 96:343–352.

LE CORRE, V., AND A. KREMER. 1998. Cumulative effects of founding events during colonisation on genetic diversity and differentiation in an island and stepping-stone model. J. Evol. Biol. 11:495–512.

LEE, K., J. FEINSTEIN, AND J. CRACRAFT. 1997. The phylogeny of ratite birds: resolving conflicts between molecular and morphological data sets. Pp. 173–211 in Avian Molecular Evolution and Systematics (D. P. Mindell, Ed.). Academic Press, San Diego, California.

LEE, M. S. Y. 1998. Convergent evolution and character correlation in burrowing reptiles: towards a resolution of squamate relationships. Biol. J. Linn. Soc. 65:369–453.

LEE, W. G. 2001. Fifty years of Takahe conservation, research and management: What have we learnt? Pp. 49–60 in The Takahe: Fifty Years of Conservation Management and Research (W. G. Lee and I. G. Jamieson, Eds.). University of Otago Press, Dunedin, New Zealand.

LEGARE, M. L., AND W. R. EDDLEMAN. 2001. Home range size, nest-site selection and nesting success of Black Rails in Florida. J. Field Ornithol. 72:170–177.

LEGENDRE, R. 1941. Oiseaux peches par des poissons. Rev. Fr. Ornithol. 78:37–41.

LEGENDRE, S., J. CLOBERT, A. P. MØLLER, AND G. SORCI. 1999. Demographic stochasticity and social mating system in the process of extinction of small populations: the case of passerines introduced to New Zealand. Am. Nat. 153:449–463.

LEIGHTON, A. T., JR., P. B. SIEGEL, AND H. S. SIEGEL. 1966. Body weight and surface area of chickens (*Gallus domesticus*). Growth 30:229–238.

LEISLER, B., H. WINKLER, AND M. WINK. 2002. Evolution of breeding systems in acrocephaline warblers. Auk 119:379–390.

LEMEN, C. A., AND P. W. FREEMAN. 1989. Testing macroevolutionary hypotheses with cladistic analysis: evidence against rectangular evolution. Evolution 43:1538–1554.

LEONARD, M. L., A. G. HORN, AND S. F. EDEN. 1988. Parent–offspring aggression in Moorhens. Behav. Ecol. Sociobiol. 23:265–270.

LEONARD, M. L., A. G. HORN, AND S. F. EDEN. 1989. Does juvenile helping enhance breeder reproductive success? A removal experiment on Moorhens. Behav. Ecol. Sociobiol. 25:357–362.

LEPAGE, D., G. GAUTHIER, AND A. DESROCHERS. 1998. Larger clutch size increases fledging success and offspring quality in a precocial species. J. Anim. Ecol. 67:210–216.

LEROI, A. M., M. R. ROSE, AND G. V. LAUDER. 1994. What does the comparative method reveal about adaptation? Am. Nat. 143:381–402.

LESSELLS, C. M. 1998. A theoretical framework for sex-biased parental care. Anim. Behav. 56:395–407.

LESSON, R.-P. 1831. Traité d'Ornithologie. Vol. 1. Levrault, Paris, France.

LESTREL, P. E. 1997. Fourier Descriptors and Their Applications in Biology. Cambridge University Press, Cambridge, United Kingdom.
LEUTENEGGER, W., AND J. M. CHEVERUD. 1985. Sexual dimorphism in primates: the effects of size. Pp. 33–50 *in* Size and Scaling in Primate Biology (W. L. Jungers, Ed.). Plenum Press, New York, New York.
LÉVÊQUE, R., R. I. BOWMAN, AND S. L. BILLEB. 1966. Migrants in the Galápagos area. Condor 68: 81–101.
LEVEY, D. J., AND W. H. KARASOV. 1989. Digestive responses of temperate birds switched to fruit or insect diets. Auk 106:675–686.
LEVEY, D. J., AND C. MARTÍNEZ DEL RIO. 2001. It takes guts (and more) to eat fruit: lessons from avian nutritional ecology. Auk 118:819–831.
LEVIN, B. R., AND W. L. KILMER. 1974. Interdemic selection and the evolution of altruism: a computer simulation study. Evolution 28:527–545.
LEVIN, D. A. 1970. Developmental instability and evolution in peripheral isolates. Am. Nat. 104:343–353.
LEVINS, R. 1962. Theory of fitness in a heterogeneous environment. I. The fitness set and adaptive function. Am. Nat. 96:361–373.
LEVINS, R. 1963. Theory of fitness in a heterogeneous environment. II. Developmental flexibility and niche selection. Am. Nat. 97:75–90.
LEVINS, R. 1968. Evolution in Changing Environments. Princeton Monogr. Popul. Biol. No. 2. Princeton University Press, Princeton, New Jersey.
LEVINS, R. 1970. Extinction. Pp. 77–107 *in* Some Mathematical Problems in Biology: Lectures on Mathematics in the Life Sciences. Vol. 2 (M. Gerstenhaber, Ed.). American Mathematical Society, Providence, Rhode Island.
LEVINS, R., AND R. H. MACARTHUR. 1966. The maintenance of genetic polymorphism in a spatially heterogeneous environment: variations on a theme by Howard Levene. Am. Nat. 100:585–589.
LEVINTON, J. S. 1982a. Estimating stasis: Can a null hypothesis be too null? Paleobiology 8:307.
LEVINTON, J. S. 1982b. The body size–prey size hypothesis: the adequacy of body size as a vehicle for character displacement. Ecology 63:869–872.
LEVINTON, J. S. 1983. Stasis in progress: the empirical basis of macroevolution. Annu. Rev. Ecol. Syst. 14:103–137.
LEVINTON, J. S. 1986. Developmental constraints and evolutionary saltations: a discussion and critique. Pp. 253–288 *in* Genetics, Development, and Evolution (J. P. Gustafson, G. L. Stebbins, and F. J. Ayala, Eds.). Plenum Press, New York, New York.
LEVINTON, J. S. 1988. Genetics, Paleontology, and Macroevolution. Cambridge University Press, Cambridge, United Kingdom.
LEWONTIN, R. C. 1965. Selection for colonizing ability. Pp. 77–91 *in* The Genetics of Colonizing Species (H. G. Baker and G. L. Stebbins, Eds.). Academic Press, New York, New York.
LEWONTIN, R. C. 1966. On the measurement of relative variability. Syst. Zool. 15:141–143.
LEWONTIN, R. C. 1970. The units of selection. Annu. Rev. Ecol. Syst. 1:1–18.
LEWONTIN, R. C., AND K. KOJIMA. 1960. The evolutionary dynamics of complex polymorphisms. Evolution 14:458–472.
LIDAUER, R. M., H. PLENK, JR., AND F. GRUNDSCHOBER. 1985. Sternal histogenesis in blackbirds with respect to the use of wings. Pp. 85–88 *in* Functional Morphology in Vertebrates (H.-R. Duncker and G. Fleischer, Eds.). Gustav Fischer Verlag, New York, New York.
LIEBHERR, J. K. 1988. Gene flow in ground beetles (Coleoptera: Carabidae) of differing habitat preference and flight-wing development. Evolution 42:129–137.
LIECHTI, F. 1995. Modelling optimal heading and airspeed of migrating birds in relation to energy expenditure and wind influence. J. Avian Biol. 26:330–336.
LIECHTI, F., AND B. BRUDERER. 1998. The relevance of wind for optimal migration theory. J. Avian Biol. 29:561–568.
LIEM, K. F. 1990. Aquatic *versus* terrestrial feeding modes: possible impacts on the trophic ecology of vertebrates. Am. Zool. 30:209–221.
LIGHTHILL, M. J. 1974. Aerodynamic aspects of animal flight. Bull. Inst. Math. Appl. 10:369–393.
LIGHTHILL, M. J. 1977. Introduction to the scaling of aerial locomotion. Pp. 365–404 *in* Scale Effects in Animal Locomotion (T. J. Pedley, Ed.). Academic Press, New York, New York.
LIGHTMAN, A., AND O. GINGERICH. 1991. When do anomalies begin? Science 255:690–695.

LIGON, J. D. 1968. Sexual differences in foraging behavior in two species of *Dendrocopos* woodpeckers. Auk 85:203–215.
LIGON, J. D. 1993. The role of phylogenetic history in the evolution of contemporary avian mating and parental care systems. Pp. 1–46 *in* Current Ornithology. Vol. 10 (D. M. Power, Ed.). Plenum Press, New York, New York.
LIGON, J. D. 1999. The Evolution of Avian Breeding Systems. Oxford University Press, Oxford, United Kingdom.
LILLIE, F. R. 1942. On the development of feathers. Biol. Rev. 17:247–266.
LILLIE, F. R. 1952. Development of the Chick: An Introduction to Embryology. 3rd ed. (revised by H. L. Hamilton). Holt, Rinehart and Winston, New York, New York.
LIMA, S. L. 1986. Predation risk and unpredictable feeding conditions: determinants of body mass in birds. Ecology 67:377–385.
LIMA, S. L. 1987. Clutch size in birds: a predation perspective. Ecology 68:1062–1070.
LIMA, S. L. 1993. Ecological and evolutionary perspectives on escape from predatory attack: a survey of North American birds. Wilson Bull. 105:1–47.
LIMA, S. L. 1998. Stress and decision making under the risk of predation: recent developments from behavioral, reproductive, and ecological perspectives. Pp. 215–290 *in* Stress and Behavior. Advances in the Study of Behavior. Vol. 27 (Møller, A. P., M. Milinski, and P. J. B. Slater, Eds.). Academic Press, San Diego, California.
LIMA, S. L., AND L. M. DILL. 1990. Behavioral decisions made under risk of predation: a review and prospectus. Can. J. Zool. 68:619–640.
LIND, J., T. FRANSSON, S. JAKOBSSON, AND C. KULLBERG. 1999. Reduced take-off ability in Robins (*Erithacus rubecula*) due to migratory fuel load. Behav. Ecol. Sociobiol. 46:65–70.
LINDÉN, H., AND R. A. VÄISÄNEN. 1986. Growth and sexual dimorphism in the skull of the Capercaillie *Tetrao urogallus*: a multivariate study of geographical variation. Ornis Scand. 17:85–98.
LINDENFORS, P., AND B. S. TULLBERG. 1998. Phylogenetic analyses of primate size evolution: the consequences of sexual selection. Biol. J. Linn. Soc. 64:413–447.
LINDROTH, C. H. 1946. Inheritance of wing dimorphism in *Pterostichus anthracinus* Ill. Hereditas 32: 37–40.
LINDSAY, B. 1885. On the avian sternum. Proc. Zool. Soc. Lond. 1885:684–716.
LINDSTEDT, S. L., AND W. A. CALDER III. 1976. Body size and longevity in birds. Condor 78:91–145.
LINDSTEDT, S. L., AND W. A. CALDER III. 1981. Body size, physiological time, and longevity of homeothermic animals. Q. Rev. Biol. 56:1–16.
LINDSTRÖM, Å. 1991. Maximum fat deposition rates in migrating birds. Ornis Scand. 22:12–19.
LINDSTRÖM, Å., AND T. ALERSTAM. 1992. Optimal fat loads in migrating birds: a test of the time-minimization hypothesis. Am. Nat. 140:477–491.
LINDSTRÖM, A., AND T. PIERSMA. 1993. Mass changes in migratory birds: the evidence for fat and protein storage re-examined. Ibis 135:70–78.
LINNAEUS, C. 1758. Systema Naturae per Regna Tria Naturae. 10th ed. Vol. 1. Impensis Laurenti Salvii, Stockholm, Sweden.
LISTER, A. M. 1993. Mammoths in miniature. Nature 362:288–289.
LISTER, A. M. 1996. Dwarfing in island elephants and deer: processes in relation to time of isolation. Symp. Zool. Soc. Lond. 69:277–292.
LIVELY, C. M. 1986a. Canalization versus developmental conversion in a spatially variable environment. Am. Nat. 128:561–572.
LIVELY, C. M. 1986b. Competition, comparative life histories, and maintenance of shell dimorphism in a barnacle. Ecology 67:858–864.
LIVELY, C. M. 1986c. Predator-induced shell dimorphism in the acorn barnacle *Chthamalus anisopoma*. Evolution 40:232–242.
LIVEZEY, B. C. 1986a. A phylogenetic analysis of Recent anseriform genera using morphological characters. Auk 103:737–754.
LIVEZEY, B. C. 1986b. Geographic variation in skeletons of Flying Steamer-Ducks (*Tachyeres patachonicus*). J. Biogeogr. 13:511–525.
LIVEZEY, B. C. 1986c. Phylogeny and historical biogeography of steamer-ducks (Anatidae: *Tachyeres*). Syst. Zool. 35:458–469.
LIVEZEY, B. C. 1987. Personal perspectives and lack of data underlie different interpretations of interspecific aggression. Condor 89:444–445.

LIVEZEY, B. C. 1988. Morphometrics of flightlessness in the Alcidae. Auk 105:681–698.

LIVEZEY, B. C. 1989a. Phylogenetic relationships of several subfossil Anseriformes of New Zealand. Univ. Kans. Mus. Nat. Hist. Occas. Pap. 128:1–25.

LIVEZEY, B. C. 1989b. Phylogenetic relationships and incipient flightlessness in the extinct Auckland Islands Merganser. Wilson Bull. 101:410–435.

LIVEZEY, B. C. 1989c. Morphometric patterns in Recent and fossil penguins (Aves, Sphenisciformes). J. Zool. Lond. 219:269–307.

LIVEZEY, B. C. 1989d. Flightlessness in grebes (Aves, Podicipedidae): its independent evolution in three genera. Evolution 43:29–54.

LIVEZEY, B. C. 1989e. Feeding morphology, foraging behavior, and foods of steamer-ducks (Anatidae: *Tachyeres*). Univ. Kans. Mus. Nat. Hist. Occas. Pap. 126:1–41.

LIVEZEY, B. C. 1990. Evolutionary morphology of flightlessness in the Auckland Islands Teal. Condor 92:639–673.

LIVEZEY, B. C. 1991. A phylogenetic analysis and classification of Recent dabbling ducks (tribe Anatini) based on comparative morphology. Auk 108:471–508.

LIVEZEY, B. C. 1992a. Flightlessness in the Galápagos Cormorant (*Compsohalieus* [*Nannopterum*] *harrisi*): heterochrony, giantism, and specialization. Zool. J. Linn. Soc. 105:155–224.

LIVEZEY, B. C. 1992b. Morphological corollaries and ecological implications of flightlessness in the Kakapo (Psittaciformes: *Strigops habroptilus*). J. Morphol. 213:105–145.

LIVEZEY, B. C. 1993a. Comparative morphometrics of *Anas* ducks, with particular reference to the Hawaiian Duck *Anas wyvilliana*, Laysan Duck *A. laysanensis*, and Eaton's Pintail *A. eatoni*. Wildfowl 44:74–99.

LIVEZEY, B. C. 1993b. An ecomorphological review of the Dodo (*Raphus cucullatus*) and Solitaire (*Pezophaps solitaria*), flightless Columbiformes of the Mascarene Islands. J. Zool. Lond. 230: 247–292.

LIVEZEY, B. C. 1993c. Morphology of flightlessness in *Chendytes*, fossil seaducks (Anatidae: Mergini) of coastal California. J. Vertebr. Paleontol. 13:185–199.

LIVEZEY, B. C. 1994. The carpometacarpus of *Apterornis*. Notornis 41:51–60.

LIVEZEY, B. C. 1995a. Heterochrony and the evolution of avian flightlessness. Pp. 169–193 *in* Evolutionary Change and Heterochrony (K. J. McNamara, Ed.). J. Wiley, London, United Kingdom.

LIVEZEY, B. C. 1995b. Phylogeny and comparative ecology of stiff-tailed ducks (Anatidae: Oxyurini). Wilson Bull. 107:214–234.

LIVEZEY, B. C. 1995c. Phylogeny and evolutionary ecology of modern seaducks (Anatidae: Mergini). Condor 97:233–255.

LIVEZEY, B. C. 1995d. A phylogenetic analysis of the whistling and white-backed ducks (Anatidae: Dendrocygninae) using morphological characters. Ann. Carnegie Mus. 64:65–97.

LIVEZEY, B. C. 1996a. A phylogenetic analysis of geese and swans (Anseriformes: Anserinae), including selected fossil species. Syst. Biol. 45:415–450.

LIVEZEY, B. C. 1996b. A phylogenetic analysis of modern pochards (Anatidae: Aythyini). Auk 113: 74–93.

LIVEZEY, B. C. 1996c. A phylogenetic reassessment of the tadornine–anatine divergence (Aves: Anseriformes: Anatidae). Ann. Carnegie Mus. 65:27–88.

LIVEZEY, B. C. 1997a. A phylogenetic analysis of basal Anseriformes, the fossil *Presbyornis*, and the interordinal relationships of waterfowl. Zool. J. Linn. Soc. 121:361–428.

LIVEZEY, B. C. 1997b. A phylogenetic analysis of modern shelducks and sheldgeese (Anatidae, Tadornini). Ibis 139:51–66.

LIVEZEY, B. C. 1997c. A phylogenetic classification of waterfowl (Aves: Anseriformes), including selected fossil species. Ann. Carnegie Mus. 67:457–496.

LIVEZEY, B. C. 1998. A phylogenetic analysis of the Gruiformes (Aves) based on morphological characters, with an emphasis on the rails (Rallidae). Philos. Trans. R. Soc. (Ser. B) 353:2077–2151.

LIVEZEY, B. C., AND P. S. HUMPHREY. 1982. Escape behaviour of steamer ducks. Wildfowl 33:12–16.

LIVEZEY, B. C., AND P. S. HUMPHREY. 1983. Mechanics of steaming in steamer-ducks. Auk 100:485–488.

LIVEZEY, B. C., AND P. S. HUMPHREY. 1984a. Sexual dimorphism in continental steamer-ducks. Condor 86:368–377.

LIVEZEY, B. C., AND P. S. HUMPHREY. 1984b. Diving behaviour of steamer ducks *Tachyeres* spp. Ibis 126:257–260.

LIVEZEY, B. C., AND P. S. HUMPHREY. 1985a. Territoriality and interspecific aggression in steamer-ducks. Condor 87:154–157.

LIVEZEY, B. C., AND P. S. HUMPHREY. 1985b. [Commentary.] Condor 87:567–568.

LIVEZEY, B. C., AND P. S. HUMPHREY. 1986. Flightlessness in steamer-ducks (Anatidae: *Tachyeres*): its morphological bases and probable evolution. Evolution 40:540–558.

LIVEZEY, B. C., AND P. S. HUMPHREY. 1992. Taxonomy and identification of steamer-ducks (Anatidae: *Tachyeres*). Univ. Kans. Mus. Nat. Hist. Monogr. 8:1–125.

LIVEZEY, B. C., AND R. L. ZUSI. 2001. Higher-order phylogenetics of modern Aves based on comparative anatomy. Neth. J. Zool. 51:179–206.

LLOYD, B. D., AND R. G. POWLESLAND. 1994. The decline of Kakapo *Strigops habroptilus* and attempts at conservation by translocation. Biol. Conserv. 69:75–85.

LLOYD, D. G. 1987. Selection of offspring size at independence and other size-versus-number strategies. Am. Nat. 129:800–817.

LOCKWOOD, J. L., AND M. P. MOULTON. 1994. Ecomorphological pattern in Bermuda birds: the influence of competition and implications for nature preserves. Evol. Ecol. 8:53–60.

LOCKWOOD, R., J. P. SWADDLE, AND J. M. V. RAYNER. 1998. Avian wingtip shape reconsidered: wingtip shape indices and morphological adaptations to migration. J. Avian Biol. 29:273–292.

LÖFGREN, O. 1995. Niche expansion and increased maturation rate of *Clethrionomys glareolus* in the absence of competitors. J. Mammal. 76:1100–1112.

LOMAN, J. 1982. A modal of clutch size determination in birds. Oecologia 52:253–257.

LOMOLINO, M. V. 1984. Mammalian island biogeography: effects of area, isolation and vagility. Oecologia 61:376–382.

LOMOLINO, M. V. 1985. Body size of mammals on islands: the island rule reexamined. Am. Nat. 125:310–316.

LOMOLINO, M. V. 1996. Investigating causality of nestedness of insular communities: Selective immigrations or extinctions? J. Biogeogr. 23:699–703.

LOMON, J. 1982. A model of clutch size determination in birds. Oecologia 52:253–257.

LOOPE, L. L., AND D. MUELLER-DOMBOIS. 1989. Characteristics of invaded islands, with special reference to Hawaii. Pp. 257–280 *in* Biological Invasions: A Global Perspective (J. A. Drake, H. A. Mooney, F. di Castri, R. H. Groves, F. J. Kruger, M. Rejmánek, and M. Williamson, Eds.). J. Wiley and Sons, New York, New York.

LÓPEZ-FANJUL, C., A. FERNÁNDEZ, AND M. A. TORO. 2002. The effect epistasis on the excess of the additive and nonadditive variances after population bottlenecks. Evolution 56:865–876.

LOSOS, J. B., D. A. CREER, D. GLOSSIP, R. GOELLNER, A. HAMPTON, G. ROBERTS, N. HASKELL, P. TAYLOR, AND J. ETTLING. 2000. Evolution implications of phenotypic plasticity in the hindlimb of the lizard *Anolis sagrei*. Evolution 54:301–305.

LOSOS, J. B., AND D. B. MILES. 1994. Adaptation, constraint, and the comparative method: phylogenetic issues and methods. Pp. 60–98 *in* Ecological Morphology: Integrative Organismal Biology (P. C. Wainright and S. M. Reilly, Eds.). University of Chicago Press, Chicago, Illinois.

LOURIE-FRASER, G. 1982. Captive breeding of the Lord Howe Island Woodhen: an endangered rail. A. F. A. Watchbird 10:30–44.

LØVTRUP, S. 1989a. Recapitulation, epigenesis and heterochrony. Pp. 269–281 *in* Ontogenèse et Évolution (B. David, J.-L. Dommergues, J. Chaline, and B. Laurin, Coords.). Géobios Mém. Spéc. 12. Université Claude-Bernard, Lyon, France.

LØVTRUP, S. 1989b. On divergent and progressive evolution. Pp. 55–67 *in* Alternative Life-History Styles of Animals (M. N. Bruton, Ed.). Kluwer Academic Publications, Dordrecht, The Netherlands.

LØVTRUP, S., AND K. H. MILD. 1979. Limb bone robustness: Natural selection of allometric growth? Bull. Biol. Fr. Belg. 113:53–69.

LØVTRUP, S., F. RAHEMTULLA, AND N.-G. HÖGLUND. 1974. Fisher's axiom and the body size of animals. Zool. Scr. 3:53–58.

LOWE, P. R. 1926. On the callosities of the ostrich (and other Palæognathæ) in connection with the inheritance of acquired characters. Ibis 68:667–679.

LOWE, P. R. 1928a. A description of *Atlantisia rogersi*, the diminutive and flightless rail of Inaccessible Island (southern Atlantic), with some notes on flightless rails. Ibis 70:99–131.

Lowe, P. R. 1928b. Studies and observations bearing on the phylogeny of the Ostrich and its allies. Proc. Zool. Soc. Lond. 1928:185–247.

Lowe, P. R. 1930. On the relationships of the Æpyornithes to the other Struthiones as revealed by a study of the pelvis of *Mullerornis*. Ibis 72:470–490.

Lowe, P. R. 1933. On the primitive characters of the penguins, and their bearing on the phylogeny of birds. Proc. Zool. Soc. Lond. 1933:483–541.

Lowe, P. R. 1934. On the evidence for the existence of two species of steamer-duck (*Tachyeres*), and primary and secondary flightlessness in birds. Ibis 76:467–495.

Lowe, P. R. 1935. On the relationship of the Struthiones to the dinosaurs and to the rest of the avian class, with special reference to the position of *Archæopteryx*. Ibis 77:398–432.

Lowe, P. R. 1939. Some additional notes on Miocene penguins in relation to their origin and systematics. Ibis 81:281–296.

Lowe, P. R. 1942. Some additional anatomical factors bearing on the phylogeny of the Struthiones. Proc. Zool. Soc. Lond. 1942:1–20.

Lowe, P. R. 1944. An analysis of the characters of *Archæopteryx* and *Archæornis*. Were they reptiles or birds? Ibis 86:517–543.

Lowe, V. T. 1966. Notes on the Musk-Duck *Biziura lobata*. Emu 65:279–290.

Lowe, W. P. 1943. Do fishes prey upon sea-birds? Ibis 85:104.

Lowrey, T. K. 1995. Phylogeny, adaptive radiation, and biogeography of Hawaiian *Tetramolopium* (Asteraceae, Astereae). Pp. 195–220 *in* Hawaiian Biogeography: Evolution on a Hot Spot Archipelago (W. L. Wagner and W. A. Funk, Eds.). Smithsonian Institution Press, Washington, D.C.

Lowther, P. E. 1977. Selection intensity in North American House Sparrows (*Passer domesticus*). Evolution 31:649–656.

Lucas, F. A. 1916. Animals of the Past. 4th ed. American Museum of Natural History, New York, New York.

Lull, R. S. 1906. Volant adaptation in vertebrates. Am. Nat. 40:527–566.

Lull, R. S. 1935. Organic Evolution. Rev. ed. Macmillan, New York, New York.

Lumer, H. 1936. The relation between b and K in systems of relative growth functions of the form $y = bx^K$. Am. Nat. 70:188–191.

Lumer, H. 1939. The dimensions and interrelationship of the relative growth constants. Am. Nat. 73:339–346.

Lumer, H., B. G. Anderson, and A. H. Hersh. 1942. On the significance of the constant b in the law of allometry $y = bx^a$. Am. Nat. 76:364–375.

Lundberg, C., and R. A. Väisänen. 1979. Selective correlation of egg size with chick mortality in the Black-headed Gull (*Larus ridibundus*). Condor 81:146–156.

Lundberg, P. 1987. Partial bird migration and evolutionarily stable strategies. J. Theor. Biol. 125:351–360.

Lundberg, P. 1988. The evolution of partial migration in birds. Trends Ecol. Evol. 7:172–175.

Lundberg, S., and T. Alerstam. 1986. Bird migration patterns: conditions for stable geographical population segregation. J. Theor. Biol. 123:403–414.

Lundgren, B. O., and K.-H. Kiessling. 1985. Seasonal variation in catabolic enzyme activities in breast muscle of some migratory birds. Oecologia 66:468–471.

Lundgren, B. O., and K.-H. Kiessling. 1988. Comparative aspects of fibre types, areas, and capillary supply in the pectoralis muscle of some passerine birds with differing migratory behaviour. J. Comp. Physiol. (Ser. B) 158:165–173.

Lutz, H. 1942. Beitrag zur Stammesgeschichte der Ratiten: Vergleich zwischen Emu-Embryo und entsprechendem Carinatenstadium. Rev. Suisse Zool. 49:299–399.

Lydekker, R. 1891. Catalogue of the Fossil Birds in the British Museum (Natural History). British Museum (Natural History), London, England.

Lydekker, R. 1893. On the extinct giant birds of Argentina. Ibis 32:40–47.

Lynch, J. F., and N. K. Johnson. 1974. Turnover and equilibria in insular avifaunas, with special reference to the California Channel Islands. Condor 76:370–384.

Lynch, J. F., E. S. Morton, and M. E. Van der Voort. 1985. Habitat segregation between the sexes of over-wintering Hooded Warblers (*Wilsonia citrina*). Auk 102:714–721.

Lyon, B. E. 1991. Brood parasitism in American Coots: avoiding the constraints of parental care. Pp. 1023–1030 *in* Acta XX Congressus Internationalis Ornithologici. Vol. 2 (B. D. Bell, R. O.

Cossee, J. E. C. Flux, B. D. Heather, R. A. Hitchmough, C. J. R. Robertson, and M. J. Williams, Eds.). Ornithological Congress Trust Board, Christchurch, New Zealand.

Lyon, B. E. 1993a. Conspecific brood parasitism as a flexible reproductive tactic in American Coots. Anim. Behav. 46:911–928.

Lyon, B. E. 1993b. Tactics of parasitic American Coots: host choice and the pattern of egg dispersion among host nests. Behav. Ecol. Sociobiol. 33:87–100.

Lyon, B. E., and R. D. Montgomerie. 1986. Delayed plumage maturation in passerine birds: reliable signaling by subordinate males? Evolution 40:605–615.

MacArthur, R. H. 1958. Population ecology of some warblers of northeastern coniferous forests. Ecology 39:599–619.

MacArthur, R. H. 1959. On the breeding distribution pattern of North American migrant birds. Auk 76:318–325.

MacArthur, R. H. 1962. Some generalized theorems of natural selection. Proc. Natl. Acad. Sci. USA 48:1893–1897.

MacArthur, R. H. 1968. The theory of the niche. Pp. 159–176 in Population Biology and Evolution (R. C. Lewontin, Ed.). Syracuse University Press, Syracuse, New York.

MacArthur, R. H. 1969. Patterns of communities in the tropics. Biol. J. Linn. Soc. 1:19–30.

MacArthur, R. H. 1971. Patterns of terrestrial bird communities. Pp. 189–221 in Avian Biology. Vol. 1 (D. S. Farner and J. R. King, Eds.). Academic Press, New York, New York.

MacArthur, R. H. 1972a. Geographical Ecology: Patterns in the Distribution of Species. Princeton University Press, Princeton, New Jersey.

MacArthur, R. H. 1972b. Coexistence of species. Pp. 253–259 in Challenging Problems: Directions Toward Their Solution (J. A. Behnke, Ed.). Oxford University Press, New York, New York.

MacArthur, R. H., J. M. Diamond, and J. R. Karr. 1972. Density compensation in island faunas. Ecology 53:330–342.

MacArthur, R. H., and R. Levins. 1967. The limiting similarity, convergence, and divergence of coexisting species. Am. Nat. 101:377–385.

MacArthur, R. H., J. MacArthur, D. MacArthur, and A. MacArthur. 1973. The effect of island area on population densities. Ecology 54:657–658.

MacArthur, R. H., and E. O. Wilson. 1963. An equilibrium theory of insular biogeography. Evolution 17:373–387.

MacArthur, R. H., and E. O. Wilson. 1967. The Theory of Island Biogeography. Princeton University Press, Princeton, New Jersey.

MacBeth, N. 1980. Reflections on irreversibility. Syst. Zool. 29:402–404.

Mace, G. M., and P. H. Harvey. 1983. Energetic constraints on home-range size. Am. Nat. 121: 120–132.

MacFarlane, R. W. 1975. Notes on the Giant Coot (*Fulica gigantea*). Condor 77:324–327.

MacGillivray, W. D. K. 1926. Birds of the Capricorn Islands. Emu 25:229–238.

MacIntyre, R. J. 1982. Regulatory genes and adaptation. Evol. Biol. 15:247–285.

Macke, T. 1969. Die Entwicklung des Craniums von *Fulica atra* L. Gegenbaurs Morphol. Jahrb. 113:230–294.

Mackem, S., and K. Mahon. 1991. *Ghox 4.7*: a chick homeobox gene expressed primarily in limb buds with limb-type differences in expression. Development 112:791–806.

MacLeod, N. 1999. Generalizing and extending the eigenshape method of shape space visualization and analysis. Paleobiology 25:107–138.

MacMillan, B. W. H. 1990. Attempts to re-establish Wekas, Brown Kiwis and Red-crowned Parakeets in the Waitakere Ranges. Notornis 37:45–51.

MacPhee, R. D. E., Ed. 1999. Extinctions in Near Time: Causes, Contexts, and Consequences. Plenum Press, New York, New York.

MacRoberts, M. H., and B. R. MacRoberts. 1976. Social organization and behavior of the Acorn Woodpecker in central coastal California. Ornithol. Monogr. 21. American Ornithologists' Union, Washington, D.C.

Macworth-Praed, C. W., and C. H. B. Grant. 1957. African Handbook of Birds. Series One: Birds of Eastern and North Eastern Africa. Vol. 1. Longman, London, United Kingdom.

Macworth-Praed, C. W., and C. H. B. Grant. 1962. African Handbook of Birds. Series Two: Birds of the Southern Third of Africa. Vol. 1. Longman, London, United Kingdom.

MACWORTH-PRAED, C. W., AND C. H. B. GRANT. 1970. African Handbook of Birds. Series Three: Birds of West Central and Western Africa. Vol. 1. Longman, London, United Kingdom.

MADDEN, S. T., AND M. B. SCHMITT. 1976. Scavenging by some Rallidae in the Transvaal. Ostrich 47:68.

MADDISON, D. R., AND W. P. MADDISON. 2000. MacClade 4: Analysis of Phylogeny and Character Evolution. Sinauer Assoc., Sunderland, Massachusetts.

MADERSON, P. F. A., P. ALBERCH, B. C. GOODWIN, S. J. GOULD, A. HOFFMAN, J. D. MURRAY, D. M. RAUP, A. DE RIEGLÈS, G. P. WAGNER, AND D. B. WAKE. 1982. The role of development in macroevolutionary change. Pp. 279–312 in Evolution and Development (J. T. Bonner, Ed.). Springer-Verlag, Berlin, Germany.

MAGNAN, M. A. 1912. Modifications organiques conséctives chez les oiseaux à l'absence de vol. Bull. Mus. Hist. Nat. Paris 18:524–530.

MAGNAN, M. A. 1913a. Rapport de la surface alaire avec le poids du corps chez les oiseaux. Bull. Mus. Hist. Nat. Paris 19:45–52.

MAGNAN, M. A. 1913b. Variations de la surface alaire chez les oiseaux. Bull. Mus. Hist. Nat. Paris 19:119–125.

MAGNAN, M. A. 1922. Les caractéristiques des oiseaux suivant le mode de vol. Ann. Sci. Nat. (Ser. 10) 5:125–334.

MAGUIRE, B., JR. 1963. The passive dispersal of small aquatic organisms and their colonization of isolated bodies of water. Ecol. Monogr. 33:161–185.

MAIER, A. 1983. Differences in muscle spindle structure between pigeon muscles used in aerial and terrestrial locomotion. Am. J. Anat. 168:27–36.

MAILLARD, J. 1948. Recherches embyologiques sur *Catharacta skua* Brunn. Rev. Suisse Zool. 55 (Suppl.):1–114.

MAINA, J. N. 1998. The lungs of the flying vertebrates—birds and bats: Is their structure optimized for this elite mode of locomotion? Pp. 177–185 in Principles of Animal Design: The Optimization and Symmorphosis Debate (E. R. Weibel, C. R. Taylor, and L. Bolis, Eds.). Cambridge University Press, Cambridge, United Kingdom.

MAINA, J. N., A. S. KING, AND J. G. SETTLE. 1989. An allometric study of the pulmonary morphometric parameters in birds, with mammalian comparisons. Philos. Trans. R. Soc. Lond. (Ser. B) 326:1–57.

MAINA, J. N., AND J. G. SETTLE. 1982. Allometric comparisons of some morphometric parameters of avian and mammalian lungs. J. Physiol. Lond. 330:28P.

MAJERUS, M. E. N. 1998. Melanism: Evolution In Action. Oxford University Press, Oxford, United Kingdom.

MALLET, J. L. B., AND J. R. G. TURNER. 1998. Biotic drift or the shifting balance—Did forest islands drive the diversity of warming coloured butterflies? Pp. 262–280 in Evolution on Islands (P. R. Grant, Ed.). Oxford University Press, Oxford, United Kingdom.

MALOIY, G. M. O., R. MCN. ALEXANDER, R. NJAU, AND A. S. JAYES. 1979. Allometry of the legs of running birds. J. Zool. Lond. 187:161–167.

MALONEY, S. K., AND T. J. DAWSON 1993. Sexual dimorphism in basal metabolism and body temperature of a large bird, the Emu. Condor 95:1034–1037.

MALONEY, S. K., AND T. J. DAWSON. 1994. Thermoregulation in a large bird, the Emu (*Dromaius novaehollandiae*). J. Comp. Physiol. (Ser. B) 164:464–472.

MANASTER, B. J., AND S. MANASTER. 1975. Techniques for estimating allometric equations. J. Morphol. 147:299–307.

MANNE, L. L., S. L. PIMM, J. M. DIAMOND, AND T. M. REED. 1998. The form of the curves: a direct evaluation of MacArthur & Wilson's classic theory. J. Anim. Ecol. 67:784–794.

MANTELL, G. A. 1850. Notice of the discovery by Mr. Walter Mantell in the middle island of New Zealand, of a living specimen of the *Notornis,* a bird of the rail family, allied to *Brachypteryx,* and hitherto unknown to naturalists except in a fossil state. Proc. Zool. Soc. Lond. 1850:209–212.

MANTELL, G. A. 1852. Notice of the discovery by Mr. Walter Mantell in the Middle Island of New Zealand, of a living specimen of the *Notornis,* a bird of the rail family, allied to *Brachypteryx,* and hitherto unknown to naturalists except in a fossil state. Trans. Zool. Soc. Lond. 4:69–72.

MARCHANT, S., AND P. J. HIGGINS, EDS. 1990. Handbook of Australian, New Zealand and Antarctic Birds. Vol. 1: Ratites to Ducks. Oxford University Press, Melbourne, Australia.

MARCHANT, S., AND P. J. HIGGINS, EDS. 1993. Handbook of Australian, New Zealand and Antarctic Birds. Vol. 2: Raptors to Lapwings. Oxford University Press, Melbourne, Australia.

MARCUS, L. F. 1990. Traditional morphometrics. Pp. 77–122 in Proceedings of the Michigan Morphometrics Workshop (F. J. Rohlf and F. L. Bookstein, Eds.). Univ. Mich. Mus. Zool. Spec. Publ. No. 2. University of Michigan Press, Ann Arbor, Michigan.

MARCUS, L. F., M. CORTI, A. LOY, G. J. P. NAYLOR, AND D. E. SLICE, EDS. 1996. Advances in Morphometrics. Plenum Press, New York, New York.

MARDEN, J. H. 1987. Maximum lift production during take off in flying animals. J. Exp. Biol. 130: 235–258.

MARDEN, J. H. 1994. From damselflies to pterosaurs—how burst and sustainable flight performance scale with size. Am. J. Physiol. 266:R1077–1084.

MARDEN, J. H., AND M. G. KRAMER. 1995. Locomotor performance of insects with rudimentary wings. Nature 377:332–334.

MARDIA, K. V. 1975. Assessment of multinormality and the robustness of Hotelling's T^2 test. Appl. Stat. 24:163–171.

MARDIA, K. V., J. T. KENT, AND J. M. BIBBY. 1979. Multivariate Analysis. Academic Press, New York, New York.

MARKOW, T. A. 1994. Developmental Instability: Its Origins and Evolutionary Implications. Kluwer Academic Publications, Dordrecht, The Netherlands.

MARKS, J. S. 1993. Molt of Bristle-thighed Curlews in the northwestern Hawaiian Islands. Auk 110: 573–587.

MARKS, J. S., AND R. L. REDMOND. 1994. Migration of Bristle-thighed Curlews on Laysan Island: timing, behavior and estimated flight range. Condor 96:316–330.

MARKS, J. S., R. L. REDMOND, P. HENDRICKS, R. B. CLAPP, AND R. E. GILL, JR. 1990. Notes on longevity and flightlessness in Bristle-thighed Curlews. Auk 107:779–781.

MARPLES, B. J. 1930. The proportions of birds' wings and their changes during development. Proc. Zool. Soc. Lond. 1930:997–1008.

MARPLES, B. J. 1932. The structure and development of the nasal glands of birds. Proc. Zool. Soc. Lond. 2:829–844.

MARQUES, A. C., AND P. GNASPINI. 2001. The problem of characters susceptible to parallel evolution in phylogenetic reconstructions: suggestion of a practical method and its application to cave animals. Cladistics 17:371–381.

MARSH, O. C. 1880. Odontornithes: A Monograph on the Extinct Toothed Birds of North America. Report of the Geological Exploration of the 40th Parallel. U.S. Government Printing Office, Washington, D.C.

MARSH, R. L. 1981. Catabolic enzyme activities in relation to premigratory fattening and muscle hypertrophy in the Gray Catbird (*Dumetella carolinensis*). J. Comp. Physiol. 141:417–423.

MARSH, R. L. 1983. Adaptations of the Gray Catbird *Dumetella carolinensis* to long distance migration: energy stores and substrate concentrations in plasma. Auk 100:170–179.

MARSH, R. L. 1984. Adaptations of the Gray Catbird *Dumetella carolinensis* to long-distance migration: flight muscle hypertrophy associated with elevated body mass. Physiol. Zool. 57:105–117.

MARSH, R. L., AND W. R. DAWSON. 1989. Avian adjustments to cold. Pp. 205–253 in Advances in Comparative and Environmental Physiology. Vol. 4: Animal Adaptation to Cold (L. C. H. Wang, Ed.). Springer-Verlag, Berlin, Germany.

MARSH, R. L., AND R. W. STORER. 1981. Correlation of flight-muscle size and body mass in Cooper's Hawks: a natural analogue of power training. J. Exp. Biol. 91:363–368.

MARSHALL, A. J., AND T. H. HARRISON. 1941. The comparative economy of closely related birds on an island and a continent. Emu 40:310–318.

MARSHALL, H. D., AND A. J. BAKER. 1999. Colonization history of Atlantic island Common Chaffinches (*Fringilla coelebs*) revealed by mitochondrial DNA. Mol. Phylogenet. Evol. 11:201–212.

MARSHALL, L. G. 1988. Extinction. Pp. 219–254 in Analytical Biogeography (A. A. Myers and P. S. Giller, Eds.). Chapman and Hall, London, United Kingdom.

MARSHALL, L. G., AND R. S. CORRUCCINI. 1978. Variability, evolutionary rates, and allometry in dwarfing lineages. Paleobiology 4:101–119.

MARTIN, H. 1885. The protection of native birds. Trans. N. Z. Inst. 18:112–117.

MARTIN, J.-L. 1992. Niche expansion in an insular bird community: an autecological perspective. J. Biogeogr. 19:375–381.
MARTIN, J.-L., A. J. GASTON, AND S. HITIER. 1995. The effect of island size and isolation on old growth forest habitat and bird diversity in Gwaii Haanas (Queen Charlotte Islands, Canada). Oikos 72:115–131.
MARTIN, J.-L., AND J.-C. THIBAULT. 1996. Coexistence in Mediterranean warblers: Ecological differences or interspecific territoriality? J. Biogeogr. 23:169–178.
MARTIN, L. D. 1984. A new hesperornithid and the relationships of the Mesozoic birds. Trans. Kans. Acad. Sci. 87:141–150.
MARTIN, L. D. 1992. The status of the Late Pleistocene birds *Gastornis* and *Remiornis.* Pp. 97–108 *in* Papers in Avian Paleontology Honoring Pierce Brodkorb (K. E. Campbell, Ed.). Contrib. Sci. No. 330. Natural History Museum of Los Angeles County, Los Angeles, California.
MARTIN, L. D., AND J. TATE, JR. 1976. The skeleton of *Baptornis advenus* from the Cretaceous of Kansas. Smithson. Contrib. Paleobiol. 27:35–66.
MARTIN, P. S. 1967. Prehistoric overkill. Pp. 75–120 *in* Pleistocene Extinctions: The Search for a Cause (P. S. Martin and H. E. Wright, Eds.). Yale University Press, New Haven, Connecticut.
MARTIN, P. S. 1984. Catastrophic extinctions and Late Pleistocene Blitzkrieg: two radiocarbon tests. Pp. 153–189 *in* Extinctions (M. H. Nitecki, Ed.). University of Chicago Press, Chicago, Illinois.
MARTIN, P. S., AND D. W. STEADMAN. 1999. Prehistoric extinctions on islands and continents. Pp. 17–56 *in* Extinctions in Near Time: Causes, Contexts, and Consequences (R. D. E. MacPhee, Ed.). Kluwer Academic/Plenum Publishing, New York, New York.
MARTIN, R. D. 1981. Relative brain size and basal metabolic rate in terrestrial vertebrates. Nature 293:57–60.
MARTIN, R. D., AND A. D. BARBOUR. 1989. Aspects of line-fitting in bivariate allometric analyses. Folia Primatol. 53:65–81.
MARTIN, T. E. 1980. Diversity and abundance of spring migratory birds using habitat islands on the Great Plains. Condor 82:430–439.
MARTIN, T. E. 1981. Limitation in small habitat islands: chance or competition? Auk 98:715–734.
MARTIN, T. E. 1987. Food as a limit on breeding birds: a life-history perspective. Annu. Rev. Ecol. Syst. 18:453–487.
MARTIN, T. E. 1988a. Habitat and area effects on forest bird assemblages: Is nest predation an influence? Ecology 69:74–84.
MARTIN, T. E. 1988b. Processes organizing open-nesting bird assemblages: competition or nest predation? Evol. Ecol. 2:37–50.
MARTIN, T. E. 1988c. On the advantage of being different: nest predation and the coexistence of bird species. Proc. Natl. Acad. Sci. USA 85:2196–2199.
MARTIN, T. E. 1992. Interaction of nest predation and food limitation in reproductive strategies. Curr. Ornithol. 9:163–197.
MARTIN, T. E. 1995. Avian life history evolution in relation to nest sites, nest predation, and food. Ecol. Monogr. 65:101–127.
MARTIN, T. E. 1996. Life history evolution in tropical and south temperate birds: What do we really know? J. Avian Biol. 27:263–272.
MARTIN, T. E. 1998. Are microhabitat preferences of coexisting species under selection and adaptive? Ecology 79:656–670.
MARTIN, T. E., AND A. V. BADYAEV. 1996. Sexual dichromatism in birds: importance of nest predation and nest location for females versus males. Evolution 50:2454–2460.
MARTIN, T. E., AND J. COLBERT. 1996. Nest predation and avian life-history evolution in Europe versus North America: a possible role of humans? Am. Nat. 147:1028–1046.
MARTIN, T. E., AND C. K. GHALAMBOR. 1999. Males feeding females during incubation. I. Required by microclimate or constrained by nest predation? Am. Nat. 153:131–139.
MARTINEZ, J. 1987. Un nouveau cas probable d'endemisme insulaire: le canard de de l'ile Amsterdam. Pp. 211–218 *in* L'Evolution des Oiseaux d'Apres le Temoignage des Fossiles (C. Mourer-Chauviré, Coord.). Doc. Lab. Geol. Lyon 99:1–248.
MARTINEZ, M. M., M. S. BO, AND J. P. ISACCH. 1997. Habitat y abundancia de *Coturnicops notata* y *Porzana spiloptera* en Mar Chiquita, Prov. de Buenos Aires, Argentina. Hornero 14:274–277.
MARTINS, E. P. 1996. Conducting phylogenetic comparative studies when phylogeny is not known. Evolution 50:12–22.

MARUYAMA, T., AND M. KIMURA. 1980. Genetic variability and effective population size when local extinction and recolonization of subpopulations are frequent. Proc. Natl. Acad. Sci. USA 77: 6710–6714.
MARZLUFF, J. M., AND K. P. DIAL. 1991. Does social organization influence diversification? Am. Midl. Nat. 125:126–134.
MASAKI, S., AND E. SENO. 1990. Effect of selection on wing dimorphism in the ground cricket *Dianemobius fascipes* (Walker). Bol. Sanid. Veg. Plag. (Ser. Fuera) 20:381–393.
MASMAN, D., M. GORDIJN, S. DAAN, AND C. DIJKSTRA. 1986. Ecological energetics of the Kestrel: field estimates of energy intake throughout the year. Ardea 74:24–39.
MASMAN, D., AND M. KLAASSEN. 1987. Energy expenditure during free flight in trained and free-living Eurasian Kestrels (*Falco tinnunculus*). Auk 104:603–616.
MASSA, B. 1987. Variations in Mediterranean crossbills *Loxia cruvirostra*. Bull. Br. Ornithol. Club 107:118–129.
MASTERS, J. C., AND R. J. RAYNER. 1993. Competition and macroevolution: the ghost of competition yet to come? Biol. J. Linn. Soc. 49:87–98.
MATHEWS, G. M. 1911. The Birds of Australia. Vol. 1. Parts 4 and 5. H. F. and G. Witherby, London, United Kingdom.
MATHEWS, G. M. 1913. A List of the Birds of Australia. H. F. and G. Witherby, London, United Kingdom.
MATHEWS, G. M. 1928. The Birds of Norfolk and Lord Howe Islands and the Australasian Quadrant. H. F. and G. Witherby, London, United Kingdom.
MATHEWS, G. M. 1936. A Supplement to the Birds of Norfolk and Lord Howe Islands. H. F. and G. Witherby, London, United Kingdom.
MATHEWS, G. M., AND T. IREDALE. 1914. Description of a strange New Zealand wood-hen. Ibis 56: 293–297.
MATHIAS, A., E. KISDI, AND I. OLIVIERI. 2001. Divergent evolution of dispersal in a heterogeneous landscape. Evolution 55:246–259.
MATTFELDT, T., AND G. MALL. 1987. Statistical methods for growth allometric studies. Growth 51: 86–102.
MATTHYSEN, E. 1993. Nonbreeding social organization in migratory and resident birds. Pp. 93–141 in Current Ornithology. Vol. 11 (D. M. Power, Ed.). Plenum Press, New York, New York.
MAUNDER, J. E., AND W. THRELFALL. 1972. The breeding biology of the Black-legged Kittiwake in Newfoundland. Auk 89:789–816.
MAURER, B. A. 1996. Energetics of avian foraging. Pp. 250–279 in Avian Energetics and Nutritional Ecology (C. Carey, Ed.). Chapman and Hall, New York, New York.
MAURER, B. A., AND J. H. BROWN. 1988. Distribution of energy use and biomass among species of North American terrestrial birds. Ecology 69:1923–1932.
MAURER, B. A., J. H. BROWN, AND R. D. RUSLER. 1992. The micro and macro in body size evolution. Evolution 46:939–953.
MAXWELL, J. M. 2001. Fiordland Takahe: Population trends, dynamics and problems. Pp. 61–79 in The Takahe: Fifty Years of Conservation Management and Research. University of Otago Press, Dunedin, New Zealand.
MAXWELL, J. M., AND I. G. JAMIESON. 1997. Survival and recruitment of captive-reared and wild-reared Takahe in Fiordland, New Zealand. Conserv. Biol. 11:683–691.
MAYBURY, W. J., AND J. M. V. RAYNER. 2001. The avian tail reduces body parasitic drag by controlling flow separation and vortex shedding. Proc. R. Soc. Lond. (Ser. B Biol. Sci.) 268:1405–1410.
MAYBURY, W. J., J. M. V. RAYNER, AND L. B. COULDRICK. 2001. Lift generation by the avian tail. Proc. R. Soc. Lond. (Ser. B Biol. Sci.) 268:1443–1448.
MAYNARD SMITH, J. 1962. Disruptive selection, polymorphism and sympatric speciation. Nature 195: 60–62.
MAYNARD SMITH, J. 1966. Sympatric speciation. Am. Nat. 100:637–650.
MAYNARD SMITH, J. 1970. Genetic polymorphism in a varied environment. Am. Nat. 104:487–490.
MAYNARD SMITH, J. 1976. Group selection. Q. Rev. Biol. 51:277–283.
MAYNARD SMITH, J. 1977. Parental investment: a prospective analysis. Anim. Behav. 25:1–9.
MAYNARD SMITH, J. 1986. The Problems of Biology. Oxford University Press, Oxford, United Kingdom.

MAYNARD SMITH, J. 1989. The causes of extinction. Philos. Trans. R. Soc. Lond. (Ser. B) 325:241–252.
MAYNARD SMITH, J. 1998. Shaping Life: Genes, Embryos and Evolution. Yale University Press, New Haven, Connecticut.
MAYNARD SMITH, J., AND R. L. BROWN. 1986. Competition and body size. Theor. Popul. Biol. 30: 166–179.
MAYNARD SMITH, J., R. BURIAN, S. KAUFFMAN, P. ALBERCH, J. CAMPBELL, B. GOODWIN, R. LANDE, D. RAUP, AND L. WOLPERT. 1985. Developmental constraints and evolution. Q.. Rev. Biol. 60: 265–287.
MAYNARD SMITH, J., AND R. HOEKSTRA. 1980. Polymorphism in a varied environment: How robust are the models? Genet. Res. 35:45–57.
MAYNARD SMITH, J., AND M. G. RIDPATH. 1972. Wife sharing in the Tasmanian Native Hen, *Tribonyx mortierii*: A case of kin selection? Am. Nat. 106:447–452.
MAYO, O. 1983. Natural Selection and Its Constraints. Academic Press, London, United Kingdom.
MAYR, E. 1933. Birds collected during the Whitney South Sea Expedition. XXII. Am. Mus. Novit. 590:1–6.
MAYR, E. 1941. Taxonomic notes on the birds of Lord Howe Island. Emu 40:321–322.
MAYR, E. 1943. The zoogeographic position of the Hawaiian Islands. Condor 45:45–48.
MAYR, E. 1944. Wallace's Line in the light of recent zoogeographic studies. Q. Rev. Biol. 19:1–14.
MAYR, E. 1947. Ecological factors in speciation. Evolution 1:263–288.
MAYR, E. 1954. Change of genetic environment and evolution. Pp. 157–180 *in* Evolution as a Process (J. S. Huxley, A. C. Hardy, and E. B. Ford, Eds.). Allen and Unwin, London, United Kingdom.
MAYR, E. 1960. The emergence of evolutionary novelties. Pp. 349–380 *in* Evolution After Darwin. Vol. 1. The Evolution of Life: Its Origin, History and Future (S. Tax, Ed.). University of Chicago Press, Chicago, Illinois.
MAYR, E. 1963. Animal Species and Evolution. Belknap Press, Cambridge, Massachusetts.
MAYR, E. 1965. The nature of colonizations in birds. Pp. 29–43 *in* The Genetics of Colonizing Species (H. G. Baker and G. L. Stebbins, Eds.). Academic Press, New York, New York.
MAYR, E. 1967. The challenge of island faunas. Aust. Nat. Hist. 15:359–374.
MAYR, E. 1969. Principles of Systematic Zoology. McGraw-Hill, New York, New York.
MAYR, E. 1976. Evolution and the Diversity of Life. Belknap Press, Cambridge, Massachusetts.
MAYR, E. 1979. Order Struthioniformes. Pp. 3–11 *in* Check-list of Birds of the World. Vol. 1. 2nd ed. (E. Mayr and G. W. Cottrell, Eds.). Museum of Comparative Zoology, Harvard University, Cambridge, Massachusetts.
MAYR, E. 1980a. Prologue: some thoughts on the history of the evolutionary synthesis. Pp. 1–48 *in* The Evolutionary Synthesis: Perspectives on the Unification of Biology (E. Mayr and W. B. Provine, Eds.). Harvard University Press, Cambridge, Massachusetts.
MAYR, E. 1980b. Systematics. Pp. 123–136 *in* The Evolutionary Synthesis: Perspectives on the Unification of Biology (E. Mayr and W. B. Provine, Eds.). Harvard University Press, Cambridge, Massachusetts.
MAYR, E. 1982. The Growth of Biological Thought: Diversity, Evolution, and Inheritance. Belknap Press, Cambridge, Massachusetts.
MAYR, E. 1994. Recapitulation reinterpreted: the somatic program. Q. Rev. Biol. 69:223–232.
MAYR, E., AND D. AMADON. 1951. A classification of Recent birds. Am. Mus. Novit. 1496:1–42.
MAYR, E., AND J. M. DIAMOND. 1976. Birds on islands in the sky: origin of the montane avifauna of Northern Melanisia. Proc. Natl. Acad. Sci. USA 73:1765–1769.
MAYR, E., AND E. T. GILLIARD. 1954. Birds of central New Guinea. Bull. Am. Mus. Nat. Hist. 103: 315–374.
MCALLAN, I. A. W., AND M. D. BRUCE. 1988. The Birds of New South Wales. Bioconservation Research Group, Turramurra, Australia.
MCARDLE, B. H. 1988. The structural relationship: regression in biology. Can. J. Zool. 66:2329–2339.
MCCALL, R. A., S. NEE, AND P. H. HARVEY. 1998. The role of wing length in the evolution of avian flightlessness. Evol. Ecol. 12:569–580.
MCCANN, C. 1964. The tongues of the Pukeko and Takahe compared. Notornis 11:183–186.
MCCOY, E. D. 1982. The application of island-biogeographic theory to forest tracts: problems in the determination of turnover rates. Biol. Conserv. 22:217–227.

McCracken, K. G., D. C. Paton, and A. D. Afton. 2000. Sexual size dimorphism of the Musk Duck. Wilson Bull. 112:457–466.

McDonald, M. A., and M. H. Smith. 1994. Behavioral and morphological correlates of heterochrony in Hispaniolan Palm-Tanagers. Condor 96:433–446.

McDowall, R. M. 1969. Extinction and endemism in New Zealand land birds. Tuatara 17:1–12.

McDowell, S. 1948. The bony palate of birds. Part I, the Palaeognathae. Auk 65:520–549.

McFadgen, B. G. 1994. Archaeology and Holocene sand dune stratigraphy on Chatham Island. J. R. Soc. N. Z. 24:17–44.

McFarlane, R. W. 1975. Notes on the Giant Coot (*Fulica gigantea*). Condor 77:324–327.

McGillivray, W. B. 1985. Size, sexual dimorphism, and their measurement in Great Horned Owls in Alberta. Can. J. Zool. 63:2364–2372.

McGillivray, W. B., and R. F. Johnston. 1987. Differences in sexual size dimorphism and body proportions between adult and subadult House Sparrows in North America. Auk 104:681–687.

McGowan, C. 1979. The hind limb musculature of the Brown Kiwi *Apteryx australis mantelli*. J. Morphol. 160:33–74.

McGowan, C. 1982. The wing musculature of the Brown Kiwi *Apteryx australis mantelli* and its bearing on ratite affinities. J. Zool. Lond. 197:173–219.

McGowan, C. 1986. The wing musculature of the Weka (*Gallirallus australis*), a flightless rail endemic to New Zealand. J. Zool. Lond. (Ser. A) 210:305–346.

McGowan, C. 1989. Feather structure in flightless birds and its bearing on the question of the origin of feathers. J. Zool. Lond. 218:537–547.

McKean, J. L., and K. A. Hindwood. 1965. Additional notes on the birds of Lord Howe Island. Emu 64:79–97.

McKinney, M. L. 1986. Ecological causation of heterochrony: a test and implications for evolutionary theory. Paleobiology 12:282–289.

McKinney, M. L. 1988a. Classifying heterochrony: allometry, size, and time. Pp. 17–34 *in* Heterochrony in Evolution: A Multidisciplinary Approach (M. L. McKinney, Ed.). Plenum Press, New York, New York.

McKinney, M. L. 1988b. Heterochrony in evolution: an overview. Pp. 327–340 *in* Heterochrony in Evolution: An Interdisciplinary Approach (M. L. McKinney, Ed.). Plenum Press, New York, New York.

McKinney, M. L. 1990a. Trends in body-size evolution. Pp. 75–118 *in* Evolutionary Trends (K. J. McNamara, Ed.). University of Arizona Press, Tucson, Arizona.

McKinney, M. L. 1990b. Classifying and analyzing evolutionary trends. Pp. 28–58 *in* Evolutionary Trends (K. J. McNamara, Ed.). University of Arizona Press, Tucson, Arizona.

McKinney, M. L. 1999. Heterochrony: beyond words. Paleobiology 25:149–153.

McKinney, M. L., and J. L. Gittleman. 1995. Ontogeny and phylogeny: tinkering with covariation in life history, morphology and behaviour. Pp. 21–47 *in* Evolutionary Change and Heterochrony (K. J. McNamara, Ed.). J. Wiley, London, United Kingdom.

McKinney, M. L., and K. J. McNamara. 1991. Heterochrony: The Evolution of Ontogeny. Plenum Press, New York, New York.

McKinney, M. L., K. J. McNamara, and L. G. Zachos. 1990. Heterochronic hierarchies: application and theory in evolution. Hist. Biol. 3:269–287.

McKitrick, M. C. 1991a. Phylogenetic analysis of avian hindlimb musculature. Univ. Mich. Mus. Zool. Misc. Publ. 179:1–85.

McKitrick, M. C. 1991b. Forelimb myology of loons (Gaviiformes), with comments on the relationship of loons and tubenoses (Procellariiformes). Zool. J. Linn. Soc. 102:115–152.

McKitrick, M. C. 1993. Phylogenetic constraint in evolutionary theory: Has it any explanatory power? Annu. Rev. Ecol. Syst. 24:307–330.

McKnight, S. K., and G. Hepp. 1999. Molt chronology of American Coots in winter. Condor 101:893–897.

McLachlan, G. J. 1992. Discriminant Analysis and Statistical Pattern Recognition. J. Wiley and Sons, New York, New York.

McLachlan, G. J., and K. E. Basford. 1988. Mixture Models: Inference and Applications to Clustering. Marcel Dekker, New York, New York.

McLain, D. K., M. P. Moulton, and T. P. Redfern. 1995. Sexual selection and the risk of extinction of introduced birds on oceanic islands. Oikos 74:27–34.

McMahon, T. A. 1973. Size and shape in biology. Science 179:1201–1204.
McMahon, T. A. 1975a. Allometry and biomechanics: limb bones in adult ungulates. Am. Nat. 109: 547–563.
McMahon, T. A. 1975b. Using body size to understand the structural design of animals: quadrupedal locomotion. J. Appl. Physiol. 39:619–627.
McMahon, T. A. 1984. Muscles, Reflexes, and Locomotion. Princeton University Press, Princeton, New Jersey.
McNab, B. K. 1963. Bioenergetics and the determination of home range size. Am. Nat. 97:133–140.
McNab, B. K. 1966. An analysis of the body temperatures of birds. Condor 68:47–55.
McNab, B. K. 1970. Body weight and the energetics of temperature regulation. J. Exp. Biol. 53: 329–348.
McNab, B. K. 1971. On the ecological significance of Bergmann's rule. Ecology 52:845–854.
McNab, B. K. 1983. Energetics, body size, and the limits to endothermy. J. Zool. Lond. 199:1–29.
McNab, B. K. 1988. Food habits and the basal rate of metabolism in birds. Oecologia 77:343–349.
McNab, B. K. 1989. Body mass, food habits, and the use of torpor in birds. Pp. 283–291 in Physiology of Cold Adaptation in Birds (C. Bech and R. E. Reinertsen, Eds.). Plenum Press, New York, New York.
McNab, B. K. 1994a. Energy conservation and the evolution of flightlessness in birds. Am. Nat. 144: 628–642.
McNab, B. K. 1994b. Resource use and the survival of land and freshwater vertebrates on oceanic islands. Am. Nat. 144:643–660.
McNab, B. K. 1996. Metabolism and temperature regulation of kiwis (Apterygidae). Auk 113:687–692.
McNab, B. K. 1999. On the comparative ecological and evolutionary significance of total and mass-specific rates of metabolism. Physiol. Biochem. Zool. 72:642–644.
McNab, B. K. 2000. The influence of body mass, climate, and distribution on the energetics of South Pacific pigeons. Comp. Biochem. Physiol. (A) 127:309–329.
McNab, B. K. 2001. Functional adaptations to island life in the West Indies. Pp. 55–62 in Biogeography of the West Indies: Patterns and Perspectives. 2nd ed. (C. A. Woods and F. E. Sergile, Eds.). CRC Press, Boca Raton, Florida.
McNab, B. K., and F. J. Bonaccorso. 1995. The energetics of Australasian swifts, frogmouths, and nightjars. Pysiol. Zool. 68:245–261.
McNab, B. K, and C. A. Salisbury. 1995. Energetics of New Zealand's temperate parrots. N. Z. J. Zool. 22:339–349.
McNamara, J. M., and A. I. Houston. 1987. Starvation and predation as factors limiting population size. Ecology 68:1515–1519.
McNamara, J. M., and A. I. Houston. 1990. The value of fat reserves and the trade-off between starvation and predation. Acta Biotheor. 38:37–61.
McNamara, J. M., and A. I. Houston. 1994. The effect of a change in foraging options on intake rate and predation rate. Am. Nat. 144:978–1000.
McNamara, K. J. 1982. Heterochrony and phylogenetic trends. Paleobiology 8:130–142.
McNamara, K. J. 1986. A guide to the nomenclature of heterochrony. J. Paleontol. 60:4–13.
McNamara, K. J. 1990. The role of heterochrony in evolutionary trends. Pp. 59–74 in Evolutionary Trends (K. J. McNamara, Ed.). University of Arizona Press, Tucson, Arizona.
McNamara, K. J. 1995. Sexual dimorphism: the role of heterochrony. Pp. 65–89 in Evolutionary Change and Heterochrony (K. J. McNamara, Ed.). J. Wiley, Chichester, United Kingdom.
McNamara, K. J. 1997. Shapes of Time. Johns Hopkins University Press, Baltimore, Maryland.
McRae, S. B. 1995. Temporal variation in responses to intraspecific brood parasitism in the Moorhen, *Gallinula chloropus*. Anim. Behav. 49:1073–1088.
McRae, S. B. 1996a. Family values: costs and benefits of communal nesting in the Moorhen. Anim. Behav. 52:225–245.
McRae, S. B. 1996b. Brood parasitism in the Moorhen: brief encounters between parasites and hosts and the significance of an evening laying hour. J. Avian Biol. 27:311–320.
McRae, S. B. 1997. A rise in nest predation enhances the frequency of intraspecific brood parasitism in a Moorhen population. J. Anim. Ecol. 66:143–153.
McRae, S. B., and T. Burke. 1996. Intraspecific brood parasitism in the Moorhen: parentage and

parasite-host relationships determined by DNA fingerprinting. Behav. Ecol. Sociobiol. 38:115–129.
MCSHEA, D. W. 1998. Possible largest-scale trends in organismal evolution: eight "live hypotheses." Annu. Rev. Ecol. Syst. 29:293–318.
MEADE-WALDO, E. G. B. 1908. Abbott's Rail, *Rallus abbotti*. Avic. Mag. (New Ser.) 6:219–221.
MEANLEY, B. 1953. Nesting of the King Rail in the Arkansas ricefields. Auk 70:262–269.
MEANLEY, B. 1956. Food habits of the King Rail in the Arkansas rice fields. Auk 73:252–258.
MEANLEY, B. 1957. Notes on the courtship behavior of the King Rail. Auk 74:433–440.
MEANLEY, B. 1960. Fall food of the Sora Rail in the Arkansas rice fields. J. Wildl. Manage. 24:339.
MEANLEY, B. 1965. Early fall food and habitat of the Sora in the Patuxent River Marsh, Maryland. Chesapeake Sci. 6:226–237.
MEANLEY, B. 1969. Natural history of the King Rail. Bur. Sport Fish. Wildl. N. Am. Fauna 67:1–108.
MEANLEY, B., AND A. G. MEANLEY. 1958. Growth and development of the King Rail. Auk 75:381–386.
MEANLEY, B., AND D. K. WETHERBEE. 1962. Ecological notes on mixed populations of King Rails and Clapper Rails in Delaware Bay marshes. Auk 79:453–457.
MEDWAY, D. C. 1967. Avian remains from new caves in the Taumatamaire District. Notornis 14:158–160.
MEES, G. F. 1977. Enige gegevens over de uitgestorven val *Pareudiastes pacificus* Hartlaub & Finsch. Zool. Meded. 50:231–242.
MEES, G. F. 1982. Birds from the lowlands of southern New Guinea (Merauke and Koembe). Zool. Verh. 191:1–188.
MEFFERT, L. M. 2000. The evolutionary potential of morphology and mating behavior. Pp. 177–193 *in* Epistasis and the Evolutionary Process (J. B. Wolf, E. D. Brodie III, and M. J. Wade, Eds.). Oxford University Press, Oxford, United Kingdom.
MEIJER, T., AND R. DRENT. 1999. Re-examination of the capital and income dichotomy in breeding birds. Ibis 141:399–414.
MEINERTZHAGEN, R. 1912. On the birds of Mauritius. Ibis 54:82–108.
MEINERTZHAGEN, R. 1955. The speed and latitude of bird flight (with notes on other animals). Ibis 97:81–117.
MEISE, W. 1963. Verhalten der Straussartigen Vögel und Monophylie der Ratitae. Pp. 115–125 *in* Proceedings XIII International Ornithological Congress. Vol. 1 (C. G. Sibley, Ed.). Museum of Zoology, Louisiana State University, Baton Rouge, Louisiana.
MELLAND, E. 1889. Notes on a paper entitled "The Takahe in western Otago," by Mr. James Park, F. G. S. Trans. N. Z. Inst. 22:295–300.
MELTOFTE, H. 1996. Are African wintering waders really forced south by competition from northerly wintering conspecifics? Benefits and constraints of northern versus southern wintering and breeding in waders. Ardea 84:31–44.
MENDELSOHN, J. M., A. C. KEMP, H. C. BIGGS, R. BIGGS, AND C. J. BROWN. 1989. Wing areas, wing loadings and wing spans of 66 species of African raptors. Ostrich 60:35–42.
MEREDITH, C. W. 1991. Vertebrate fossil faunas from islands in Australasia and the southwest Pacific. Pp. 1345–1382 *in* Vertebrate Palaeontology of Australia (P. Vickers-Rich, J. M. Monaghan, R. F. Baird, and T. H. Rich, Eds.). Monash University Publications Committee, Melbourne, Australia.
MEREDITH, C. W., J. R. SPECHT, AND P. V. RICH. 1985. A minimum date for Polynesian visitation to Norfolk Island, southwest Pacific, from faunal evidence. Search 16:304–306.
MERILÄ, J. 1997. Expression of genetic variation in body size of the Collared Flycatcher under different environmental conditions. Evolution 51:526–536.
MERILÄ, J., AND M. BJÖRKLUND. 1999. Population divergence and morphometric integration in the Greenfinch (*Carduelis chloris*)—Evolution against the trajectory of least resistance? J. Evol. Biol. 12:103–112.
MERILÄ, J., M. BJÖRKLUND, AND L. GUSTAFSSON. 1994. Evolution of morphological differences with moderate genetic correlations among traits as exemplified by two flycatcher species (*Ficedula*; Muscicapidae). Biol. J. Linn. Soc. 52:19–30.
MERREM, B. 1813. Tentamen systematis naturalis avium. Abh. Königl. (Preussiche) Akad. Wiss. Berl. (Physikal.) 1812–1813 (1816):237–259.

MERTON, D. V. 1978. Controlling introduced predators and competitors on islands. Pp. 121–128 *in* Endangered Birds (S. A. Temple, Ed.). University of Wisconsin Press, Madison, Wisconsin.

METCALFE, N. B., AND R. W. FURNESS. 1984. Changing priorities: the effect of pre-migratory fattening on the trade-off between foraging and vigilance. Behav. Ecol. Sociobiol. 15:203–206.

METCALFE, N. B., AND S. E. URE. 1995. Diurnal variation in flight performance and hence potential predation risk in small birds. Proc. R. Soc. Lond. (Ser. B Biol. Sci.) 261:395–400.

METTKE, C. 1995. Explorationsverhalten von Papageien—Adaptation an die Umwelt? J. Ornithol. 136:468–471.

METTKE-HOFMANN, C. 1999. Niche expansion and exploratory behavior on islands: Are they linked? Pp. 878–885 *in* Proceedings of the 22nd International Ornithological Congress (N. Adams and R. Slowtow, Eds.). University of the Natal, Durban, Republic of South Africa.

MEUNIER, K. 1951. Korrelation und Umkonstuktion in den Größenbeziehungen zwischen Vogelflügel und Vogelkörper. Biol. Gen. 19:403–443.

MEUNIER, K. 1959a. Die Allometrie des Vogelflügels. Z. Wiss. Zool. 161:444–482.

MEUNIER, K. 1959b. Die Größenabhangigkeit der Körperform bei Vögeln. Z. Wiss. Zool. 162:328–355.

MEYER, A. B. 1883. Abbildungen von Vogel-Skeletten. Vol. 1. Parts 4 and 5. R. Friedlander, Dresden, Germany.

MEYER, A. B., AND L. W. WIGLESWORTH. 1898. The Birds of Celebes and the Neighboring Islands. Vol. 2. R. Friedlander and Sohn, Berlin, Germany.

MEYER DE SCHAUENSEE, R. 1966. The Species of Birds of South America and Their Distribution. Livingston, Narbeth, Pennsylvania.

MEYER DE SCHAUENSEE, R., AND W. H. PHELPS, JR. 1978. A Guide to the Birds of Venezuela. Princeton University Press, Princeton, New Jersey.

MIDDLETON, A. L. A., AND E. PRIGODA. 2001. What does 'fledging' mean? Ibis 143:296–298.

MIELKE, H. P. 1989. Patterns of Life: Biogeography of a Changing World. Unwin Hyman, Boston, Massachusetts.

MILBERG, P., AND T. TYRBERG. 1993. Naïve birds and noble savages—a review of man-caused prehistoric extinctions of island birds. Ecography 16:229–250.

MILLENER, P. R. 1980. The taxonomic status of extinct New Zealand coots, *Fulica chathamensis* subspp. (Aves: Rallidae). Notornis 27:363–367.

MILLENER, P. R. 1981. The subfossil distribution of extinct New Zealand coots *Fulica chathamensis* subspp. (Aves: Rallidae). Notornis 28:1 9.

MILLENER, P. R. 1988. Contributions to New Zealand's Late Quaternary avifauna. I: *Pachyplichas*, a new genus of wren (Aves Acanthisittidae), with two new species. J. R. Soc. N. Z. 18:383–406.

MILLENER, P. R. 1989. The only flightless passerine; the Stephens Island Wren (*Traversia lyalli*: Acanthisittidae). Notornis 36:280–284.

MILLENER, P. R. 1991. The Quaternary of New Zealand. Pp. 1317–1344 *in* Vertebrate Palaeontology of Australia (P. Vickers-Rich, J. M. Monaghan, R. F. Baird, and T. H. Rich, Eds.). Monash University Publications Committee, Melbourne, Australia.

MILLENER, P. R. 1999. The history of the Chatham Islands' bird fauna of the last 7000 years—a chronicle of change and extinction. Pp. 85–110 *in* Avian Paleontology at the Close of the 20th Century: Proceedings of the 4th International Meeting of the Society of Avian Paleontology and Evolution, Washington, D.C., 4–7 June 1996 (S. L. Olson, Ed.). Smithson. Contrib. Paleobiol. 89:1–344.

MILLENER, P. R., AND T. H. WORTHY. 1991. Contributions to New Zealand's Late Quaternary avifauna. II: *Dendroscansor decurvirostris*, a new genus and species of wren (Aves: Acanthisittidae). J. R. Soc. N. Z. 21:179–200.

MILLER, A. H. 1937. Structural modifications in the Hawaiian Goose (*Nesochen sandvicensis*). A study in adaptive evolution. Univ. Calif. Publ. Zool. 42:1–80.

MILLER, A. H. 1947. Panmixia and population size with reference to birds. Evolution 1:186–190.

MILLER, A. H. 1966. Animal evolution on islands. Pp. 10–17 *in* The Galápagos: Proceedings of the Symposia of the Galápagos International Scientific Project (R. I. Bowman, Ed.). University of California Press, Berkeley, California.

MILLER, B., AND T. KINGSTON. 1980. Lord Howe Island Woodhen. Parks Wildl. (Aug.) 1980:17–26.

MILLER, B., AND K. J. MULLETTE. 1985. Rehabilitation of an endangered Australian bird: the Lord Howe Island Woodhen *Tricholimnas sylvestris* (Sclater). Biol. Conserv. 34:55–95.

MILLER, C. M. 1944. Ecologic relationships and adaptations of the limbless lizards of the genus *Anniella*. Ecol. Monogr. 14:271–289.

MILLER, P. J. 1996. Miniature Vertebrates: Implications of Small Body Size. Oxford University Press, Oxford, United Kingdom.

MILLER, P. J., AND R. J. PIERCE. 1995. Distribution and decline of the North Island Brown Kiwi (*Apteryx australis mantelli*) in Northland. Notornis 42:203–211.

MILLER, R. G., JR. 1981. Simultaneous Statistical Inference. McGraw-Hill, New York, New York.

MILLER, R. S. 1967. Pattern and process in competition. Pp. 1–74 *in* Advances in Ecological Research. Vol. 4 (J. B. Cragg, Ed.). Academic Press, London, United Kingdom.

MILLER, R. S. 1969. Competition and species diversity. Pp. 63–70 *in* Diversity and Stability in Ecological Systems (G. M. Woodwell and H. H. Smith, Eds.). Symp. No. 22. Brookhaven National Laboratory, Upton, New York.

MILLER, W. DEW. 1924. Further notes on ptilosis. Am. Mus. Nat. Hist. Bull. 50:305–331.

MILLIKAN, G. C., AND R. I. BOWMAN. 1967. Observations on Galápagos tool-using finches in captivity. Living Bird 6:23–41.

MILLS, J. A. 1975. Population studies on Takahe, *Notornis mantelli* in Fiordland, New Zealand. Bull. Int. Counc. Bird Preserv. 12:140–146.

MILLS, J. A. 1978. Population studies of the Takahe. Pp. 52–68 *in* Proceedings of a Seminar on the Takahe and Its Habitat (J. P. Harty, Chairman). Fiordland National Park Board, Invercargill, New Zealand.

MILLS, J. A. 1985. Takahe. P. 170 *in* The Reader's Digest Complete Book of New Zealand Birds (C. J. R. Robertson, Ed.). Reader's Digest, Sidney, Australia.

MILLS, J. A., AND B. LAVERS. 1974. Preliminary results of research into the present status of Takahe (*Notornis mantelli*) in the Murchison Mountains. Notornis 21:312–317.

MILLS, J. A., B. LAVERS, AND W. G. LEE. 1984. The Takahe—a relict of the Pleistocene grassland avifauna of New Zealand. N. Z. J. Ecol. 7:57–70.

MILLS, J. A., B. LAVERS, AND W. G. LEE. 1988. The post-Pleistocene decline of the Takahe (*Notornis mantelli*): a reply. J. R. Soc. N. Z. 18:112–118.

MILLS, J. A., R. B. LAVERS, W. G. LEE, AND M. K. MARA. 1991. Food selection by Takahe *Notornis mantelli* in relation to chemical composition. Ornis Scand. 22:111–128.

MILLS, J. A., W. G. LEE, AND R. B. LAVERS. 1989. Experimental investigations of the effects of Takahe and deer grazing on *Chionochloa pallens* grassland, Fiordland, New Zealand. J. Appl. Ecol. 26:397–417.

MILLS, J. A., W. G. LEE, AND A. F. MARK. 1978. Takahe feeding studies—preferences and requirements. Pp. 74–94 *in* Proceedings of a Seminar on the Takahe and Its Habitat (J. P. Harty, Chairman). Fiordland National Park Board, Invercargill, New Zealand.

MILLS, J. A., W. G. LEE, A. F. MARK, AND R. B. LAVERS. 1980. Winter use by Takahe (*Notornis mantelli*) of the summer-green fern (*Hypolepis millefolium*) in relation to its annual cycle of carbohydrates and minerals. N. Z. J. Ecol. 3:131–137.

MILLS, J. A., AND A. F. MARK. 1977. Food preferences of Takahe in Fiordland National Park, New Zealand, and the effect of competition from introduced red deer. J. Anim. Ecol. 46:939–958.

MILNE-EDWARDS, A. 1867. Mémoire sure une espèce éteinte du genre *Fulica* qui habitait autrefois l'Ile Maurice. Ann. Sci. Nat. Zool. Paléontol. (Ser. 5) 8:190–220.

MILNE-EDWARDS, A. 1868. Observations sur les affinités zoologiques de l'*Aphanapteryx* espèce éteinte qui vivait encore à l'Ile Maurice an XVIIe siècle. Ann. Sci. Nat. Zool. Paleontol. (Ser. 5) 10: 325–348.

MILNE-EDWARDS, A. 1869. Researches into the zoological affinities of the bird recently described by Herr von Frauenfeld under the name of *Aphanapteryx imperialis*. Ibis 11:256–275.

MILNE-EDWARDS, A. 1874. Recherches sur la faune ancienne des Iles Mascareignes. Ann. Sci. Nat. Zool. Paleontol. (Ser. 5) 19:1–31.

MILNE-EDWARDS, A. 1875. Observations sur l'époque de la disparition de la faune ancienne de l'Ile Rodriques. C. R. Hebd. Seances Acad. Sci. 80:1212–1216.

MILNE-EDWARDS, A. 1896. Sur les ressemblances qui existent entre la faune des Iles Mascareignes et celle de certaines Iles de l'Océan Pacifique Austral. Ann. Sci. Nat. Zool. Paleontol. (Ser. 8) 2: 117–136.

MILONOFF, M. 1989. Can nest predation limit clutch size in precocial birds? Oikos 55:424–427.

MILONOFF, M., AND H. LINDEN. 1989. Sexual size dimorphism of body components in Capercaillie chicks. Ornis Scand. 20:29–35.

MINDELL, D. P., M. D. SORENSON, D. E. DIMCHEFF, M. HASEGAWA, J. C. AST, AND T. YURI. 1999. Interordinal relationships of birds and other reptiles based on whole mitochondrial genomes. Syst. Biol. 48:138–152.

MINDELL, D. P., M. D. SORENSON, C. J. HUDDLESTON, H. C. MIRANDA, JR., A. KNIGHT, S. J. SWACHUK, AND T. YURI. 1997. Phylogenetic relationships among and within select avian orders based on mitochondrial DNA. Pp. 213–247 in Avian Molecular Evolution and Systematics (D. P. Mindell, ed.). Academic Press, San Diego, California.

MISKELLY, C. M. 1981. Leg colour and dominance in Buff Wekas. Notornis 28:47–48.

MITTON, J. B. 1997. Selection in Natural Populations. Oxford University Press, Oxford, United Kingdom.

MLÍKOVSKY, J. 1982. Evolution of flightlessness in birds: an ecological approach. Pp. 693–730 in Evolution and Environment. Vol. 2 (V. J. A. Novák and J. Mlíkovsky, Eds.). Czechoslovakian Academyof Science, Praha, Czechoslovakia.

MOCK, D. W. 1984. Infanticide, siblicide, and vain nestling mortality. Pp. 3–30 in Infanticide: Comparative and Evolutionary Perspectives (G. Haufater and S. B. Hrdy, Eds.). Aldine, New York, New York.

MOCK, D. W. 1987. Siblicide, parent–offspring conflict, and unequal parental investment by egrets and herons. Behav. Ecol. Sociobiol. 20:247–256.

MOCK, D. W., AND G. A. PARKER. 1986. Advantages and disadvantages of egret and heron brood reduction. Evolution 40:459–470.

MØLLER, A. P. 1998. Developmental instability as a general measure of stress. Pp. 181–213 in Stress and Behavior. Advances in the Study of Behavior. Vol. 27 (Møller, A. P., M. Milinski, and P. J. B. Slater, Eds.). Academic Press, San Diego, California.

MØLLER. A. P., A. BARBOSA, J. J. CUERVO, F. DE LOPE, S. MERINO, AND N. SAINO. 1998. Sexual selection and tail streamers in the Barn Swallow. Proc. R. Soc. Lond. (Ser. B Biol. Sci.) 265:409–414.

MØLLER, A. P., AND T. R. BIRKHEAD. 1994. The evolution of plumage brightness in birds is related to extrapair paternity. Evolution 48:1089–1100.

MØLLER, A. P., F. DE LOPE, AND N. SAINO. 1995. Sexual selection in the Barn Swallow *Hirundo rustica*. 6. Aerodynamic adaptations. J. Evol. Biol. 8:671–687.

MØLLER, A. P., AND A. HEDENSTRÖM. 1999. Comparative evidence for costs of secondary sexual characters: adaptive vane emargination of ornamented feathers in birds. J. Evol. Biol. 12:296–305.

MØLLER, A. P., AND J. P. L. SWADDLE. 1997. Asymmetry, Developmental Stability, and Evolution. Oxford University Press, Oxford, United Kingdom.

MÖLLER, E. 1992. Feeding technique of a White-browed Crake *Porzana cinerea*. Forktail 7:154–155.

MÖNKKÖNEN, M. 1995. Do migrant birds have more pointed wings? A comparative study. Evol. Ecol. 9:520–528.

MONTAGNA, W. 1945. A re-investigation of the development of the wing of the fowl. J. Morphol. 76: 87–113.

MONTGOMERIE, R. D., AND B. E. LYON. 1986. Does longevity influence the evolution of delayed plumage maturation in passerine birds? Am. Nat. 128:930–936.

MONTGOMERIE, R. D., AND P. J. WEATHERHEAD. 1988. Risks and rewards of nest defence by parent birds. Q. Rev. Biol. 63:167–187.

MOOERS, A. Ø, AND P. COTGREAVE. 1994. Sibley & Alquist's tapestry dusted off. TREE 9:458–459.

MOOERS, A. Ø., AND A. P. MØLLER. 1996. Colonial breeding and speciation in birds. Evol. Biol. 10: 375–385.

MOOERS, A. Ø., S. M. VAMOSI, AND D. SCHLUTER. 1999. Using phyogenies to test macroevolutionary hypotheses of trait evolution in cranes (Gruinae). Am. Nat. 154:249–259.

MOORE, F. R., AND D. A. ABORN. 2000. Mechanisms of *en route* habitat selection: How do migrants make habitat decisions during stopover? Pp. 34–42 in Stopover Ecology of Nearctic–Neotropical Landbird Migrants: Habitat Relations and Conservation Implications (F. R. Moore, Ed.). Stud. Avian Biol. No. 20. Cooper Ornithological Society, Los Angeles, California.

MOORE, J. G., D. A. CLAGUE, R. T. HOLCOM, P. W. LIPMAN, W. R. NORMARK, AND M. E. TORRESAN. 1989. Prodigious submarine landslides on the Hawaiian Ridge. J. Geophys. Res. 94:17465–17484.

MOORE, J. G., AND G. W. MOORE. 1984. Deposit from a giant wave on the island of Lanai, Hawaii. Science 226:1312–1315.

MOORE, J. G., W. R. NORMARK, AND R. T. HOLCOMB. 1994a. Giant Hawaiian underwater landslides. Science 264:46–47.

MOORE, J. G., W. R. NORMARK, AND R. T. HOLCOMB. 1994b. Giant Hawaiian landslides. Annu. Rev. Earth Planet. Sci. 22:119–144.

MOORS, P. J., ED. 1982. Conservation of Island Birds. International Council for Bird Preservation, Cambridge, United Kingdom.

MORAN, M. A., AND J. GORNBEIN. 1988. Correspondence Analysis. Tech. Rep. 87. BMDP Statistical Software, Inc., Los Angeles, California.

MORAN, N. A. 1992. The evolutionary maintenance of alternative phenotypes. Am. Nat. 139:971–989.

MOREAU, R. E. 1944. Clutch size: a comparative study, with special reference to African birds. Ibis 86:286–347.

MOREAU, R. E. 1966. The Bird Faunas of Africa and Its Islands. Academic Press, New York, New York.

MORENO, J. 1989. Strategies of mass change in breeding birds. Biol. J. Linn. Soc. 37:297–310.

MORRIS, D. W., AND T. W. KNIGHT. 1996. Can consumer–resource dynamics explain patterns of guild assembly? Am. Nat. 147:558–575.

MORRIS, R. B. 1977. Observations of Takahe nesting behaviour at Mount Bruce. Notornis 24:54–58.

MORRIS, W., ED. 1973. The American Heritage Dictionary of the English Language. American Heritage Publishing Co., Boston, Massachusetts.

MORRISON, D. F. 1976. Multivariate Statistical Methods. 2nd ed. McGraw-Hill, New York, New York.

MORSE, D. H. 1971. The foraging of warblers isolated on small islands. Ecology 52:216–228.

MORSE, D. H. 1975. Ecological aspects of adaptive radiation in birds. Biol Rev. 50:167–214.

MORSE, D. H. 1977. The occupation of small islands by passerine birds. Condor 79:399–412.

MORSE, D. H. 1980. Behavioral Mechanisms in Ecology. Harvard University Press, Cambridge, Massachusetts.

MORTON, E. S. 1978. Avian arboreal folivores: Why not? Pp. 123–130 in The Ecology of Arboreal Folivores (G. Montgomery, Ed.). Smithsonian Institution Press, Washington, D.C.

MORTON, E. S. 1990. The evolution of habitat segregation by sex in the Hooded Warbler: experiments on proximate causation. Am. Nat. 135:319–333.

MOSIMANN, J. E. 1970. Size allometry: size and shape variables with characterizations of the lognormal and generalized gamma distributions. J. Am. Stat. Assoc. 65:930–945.

MOSIMANN, J. E. 1975a. Statistical problems of size and shape. I. Biological applications and basic theorems. Pp. 187–217 in Statistical Distributions in Scientific Work. Vol. 2 (G. P. Patil, S. Kotz, and K. Ord, Eds.). D. Reidel Publishing Co., Boston, Massachusetts.

MOSIMANN, J. E. 1975b. Statistical problems of size and shape. II. Characterizations of the lognormal, gamma and Dierichlet distributions. Pp. 219–239 in Statistical Distributions in Scientific Work. Vol. 2 (G. P. Patil, S. Kotz, and K. Ord, Eds.). D. Reidel Publishing Co., Boston, Massachusetts.

MOSIMANN, J. E., AND F. C. JAMES. 1979. New statistical methods for allometry with application to Florida Red-winged Blackbirds. Evolution 33:444–459.

MOTANI, R. 2001. Estimating body mass from silhouettes: testing the assumption of elliptical body cross-sections. Paleobiology 27:735–750.

MOTTA JUNIOR, J. C. 1991. [Predation of *Micropygia schomburgkii* (Aves: Rallidae) by *Chrysocyon brachyurus* (Mammalia: Canidae) in the Federal District, Brazil.] Ararajuba 2:87–88.

MOULTON, D. W., AND M. W. WELLER. 1984. Biology and conservation of the Laysan Duck (*Anas laysanensis*). Condor 86:105–117.

MOULTON, M. P., AND S. L. PIMM. 1983. The introduced Hawaiian avifauna: biogeographic evidence for competition. Am. Nat. 121:669–690.

MOULTON, M. P., AND S. L. PIMM. 1987. A morphological assortment in introduced Hawaiian passerines. Evol. Ecol. 1:113–124.

MOUNTAINSPRING, S., AND J. M. SCOTT. 1985. Interspecific competition among Hawaiian forest birds. Ecol. Monogr. 55:219–239.

MOURER-CHAUVIRÉ, C., R. BOUR, AND S. RIBES. 1995a. Was the solitaire of Réunion an ibis? Nature 373:568.

MOURER-CHAUVIRÉ, C., R. BOUR, AND S. RIBES. 1995b. Position systématique du Solitaire de la

Réunion: nouvelle interprétation basée sur les restes fossiles et les récits des anciens voyageurs. C. R. Acad. Sci. Paris (Ser. IIa) 320:1125–1131.

MOURER-CHAUVIRÉ, C., R. BOUR, S. RIBES, AND F. MONTOU. 1999. The avifauna of Réunion Island (Mascarene Islands) at the the time of the arrival of the first Europeans. Pp. 1–38 *in* Avian Paleontology at the Close of the 20th Century: Proceedings of the 4th International Meeting of the Society of Avian Paleontology and Evolution, Washington, D.C., 4–7 June 1996 (S. L. Olson, Ed.). Smithson. Contrib. Paleobiol. 89:1–344.

MOUSSEAU, T. A., AND C. W. FOX. 1998. Maternal Effects as Adaptations. Oxford University Press, Oxford, United Kingdom.

MUELLER, H. C., N. S. MUELLER, AND P. G. PARKER. 1981. Observation of a brood of Sharp-shinned Hawks in Ontario, with comments on the functions of sexual dimorphism. Wilson Bull. 93:85–92.

MÜLLENHOFF, K. 1885. Die Grösse der Flugflächen. Pfluegers Arch. Physiol. 35:407–453.

MÜLLER, B. B. 1991. Evolutionary transformation of limb pattern: heterochrony and secondary fusion. Pp. 395–405 *in* Developmental Patterning of the Vertebrate Limb (J. R. Hinchliffe, J. M. Hurle, and D. Summerbell, Eds.). Plenum Press, New York, New York.

MÜLLER, G. B. 1990. Developmental mechanisms at the origin of morphological novelty: a side-effect hypothesis. Pp. 99–130 *in* Evolutionary Innovations (M. H. Nitecki, Ed.). University of Chicago Press, Chicago, Illinois.

MÜLLER, G. B., AND J. STREICHER. 1989. Ontogeny of the syndesmosis tibiofibularis and the evolution of the bird hindlimb: a caenogenetic feature triggers phenotypic novelty. Anat. Embryol. 179:327–339.

MÜLLER, G. B., AND G. P. WAGNER. 1991. Novelty in evolution: restructuring the concept. Annu. Rev. Ecol. Syst. 22:229–256.

MÜLLER, H. J. 1939. Reversibility in evolution considered from the standpoint of genetics. Biol. Rev. 14:261–280.

MÜLLER, H. J. 1961. Die Morphologie und Entwicklung des Craniums von *Rhea americana* Linné. I. Das knorpelige Neurocranium. Z. Wiss. Zool. 165:221–319.

MÜLLER, H. J. 1963. Die Morphologie und Entwicklung des Craniums von *Rhea americana* Linné. II. Viszeralskelett, Mittelohr und Osteocranium. Z. Wiss. Zool. 168:35–118.

MÜLLER, H. J. 1964. Morphologische Untersuchungen am Vogelschädel in ihrer Bedeutung fur die Systematik. J. Ornithol. 105:67–77.

MÜLLER, W., AND G. PATONE. 1998. Air transmissivity of feathers. J. Exp. Biol. 201:2591–2599.

MULVIHILL, R. S., AND C. R. CHANDLER. 1991. A comparison of wing-shape between migratory and sedentary Dark-eyed Juncos (*Junco hyemalis*). Condor 93:172–175.

MUNRO, G. C. 1946. The war and Pacific birds. Nat. Mag. 39:125–127.

MUNRO, G. C. 1947. Notes on the Laysan Rail. Elepaio 8:24–25.

MURIE, J. 1888. [Letter included as footnote.] P. 123 *in* History of the Birds of New Zealand. 2nd ed. Vol. 2 (W. L. Buller). Privately Published, London, England.

MURIE, J. 1889. On the assumed hybridity between the common fowl and the Woodhen (*Ocydromus*). Trans. N. Z. Inst. 22:342–353.

MURPHY, D. D. K. E. FREAS, AND S. B. WEISS. 1990. An "environment-metapopulation" approach to population viability analysis for a threatened invertebrate. Conserv. Biol. 4:41–51.

MURPHY, E. C. 1985. Bergmann's rule, seasonality, and geographic variation in body size of House Sparrows. Evolution 39:1327–1334.

MURPHY, M. E. 1996. Energetics and nutrition of molt. Pp. 158–198 *in* Avian Energetics and Nutritional Ecology (C. Carey, Ed.). Plenum Press, New York, New York.

MURPHY, R. C. 1936. Oceanic Birds of South America. Vol. 2. American Museum of Natural History, New York, New York.

MURPHY, R. C. 1938. The need for insular exploration as illustrated by birds. Science 88:533–539.

MURRAY, B. G., JR. 2001. The evolution of passerine life histories on oceanic islands, and its implications for the dynamics of population decline and recovery. Pp. 281–290 *in* Evolution, Ecology, Conservation, and Management of Hawaiian Birds: A Vanishing Avifauna (J. M. Scott, S. Conant, and C. van Riper III, Eds.). Stud. Avian Biol. No. 22. Cooper Ornithological Society, Los Angeles, California.

MURRAY, P. D. F. 1963. Adventitious (secondary) cartilage in the chick embryo and the development of certain bones and articulations in the chick skull. Aust. J. Zool. 11:368–430.

MURRAY, P. D. F., AND D. B. DRACHMAN. 1969. The role of movement in the development of joints and related structures: the head and neck in the chick embryo. J. Embryol. Exp. Morphol. 22: 349–371.

MURRAY, P. R., AND D. MEGIRIAN. 1998. The skull of dromornithid birds: anatomical evidence for their relationship to Anseriformes. Rec. S. Aust. Mus. 31:51–97.

MURTON, R. K. 1971. Polymorphism in Ardeidae. Ibis 113:97–99.

MURTON, R. K., AND N. J. WESTWOOD. 1977. Avian Breeding Cycles. Clarendon Press, Oxford, United Kingdom.

MYLES, D. 1985. The Great Waves. Robert Hale Ltd., London, United Kingdom.

NAGAMURA, T., T. NISHIDA, AND S. NOMURA. 1974. The origin and insertion of thoracic-limb muscles in the Pigeon (*Columba libia* [sic]). Jpn. J. Vet. Sci. 36:145–162.

NAGAMURA, T., T. NISHIDA, AND S. NOMURA. 1976. The origin and insertion of pelvic limb muscles in the Pigeon (*Columba libia* [sic]). Jpn. J. Vet. Sci. 38:355–368.

NÄGELI, C. 1865. Enstehung und Begriff der Naturhistorischen Art. K. Bayr Acad., München, Germany.

NAGER, R. G., L. K. DELLER, AND A. J. VAN NOORDWIJK. 2000. Understanding natural selection on traits that are influenced by environmental conditions. Pp. 95–115 *in* Adaptive Genetic Variation in the Wild (T. A. Mousseau, B. Sinervo, and J. A. Endler, Eds.). Oxford University Press, New York, New York.

NAGY, K. A. 1987. Field metabolic rate and food requirement scaling in mammals and birds. Ecol. Monogr. 57:111–128.

NAGY, K. A., AND C. C. PETERSON. 1988. Scaling of water flux in animals. Univ. Calif. Publ. Zool. 120:1–172.

NAIR, K. K. 1952. The chemical composition of the pectoral muscles of some Indian birds and its bearing on their flight. J. Univ. Bombay (Ser. B) 21:90–98.

NAIR, K. K. 1954a. The bearing of the weight of the pectoral muscles on the flight of some common Indian birds. J. Anim. Morphol. Physiol. 1:71–76.

NAIR, K. K. 1954b. A comparison of the muscles in the forearm of a flapping and a soaring bird. J. Anim. Morphol. Physiol. 1:26–34.

NAUCK, E. T. 1930a. Beiträge zur Kenntnis des Skeletts der paarigen Gliedmaßen der Wirbeltiere. VII. Der Coracoscapularwinkel am Vogelschultergürtel. Gegenbaurs Morphol. Jahrb. 64:541–557.

NAUCK, E. T. 1930b. Die ontogenetischen Änderundgen des Coracoscapularwinkels beim Huhn. (Vorläufige Mitteilung). Anat. Anz. 68:416–418.

NAVAS, J. R. 1960. Comportamiento agresivo de "*Fulica armillata*" Vieillot (Aves, Rallidae). Rev. Mus. Argent. Cienc. Nat. Bernardino Rivadavia (Zool.) 6:103–129.

NEE, S., R. M. MAY, AND P. H. HARVEY. 1994. The reconstructed evolutionary process. Philos. Trans. R. Soc. Lond. (Ser. B) 344:305–311.

NEI, M., T. MARUYAMA, AND C. I. WU. 1983. Models of evolution of reproductive isolation. Genetics 103:557–579.

NELSON, C. E., B. A. MORGAN, A. C. BURKE, E. LAUFER, E. DIMAMBRO, L. C. MURTAUGH, E. GONZALES, L. TESSAROLLO, L. F. PARADA, AND C. TABIN. 1996. Analysis of *Hox* gene expression in chick limb bud. Development 122:1449–1466.

NELSON, J. B. 1978. The Sulidae: Gannets and Boobies. Oxford University Press, Oxford, United Kingdom.

NELSON, J. S. 1964. Factors influencing clutch-size and chick growth in the North Atlantic Gannet *Sula bassana*. Ibis 106:63–77.

NELSON, J. S. 1977. Evidence of a genetic basis for absence of the pelvic skeleton in brook stickleback, *Culaea inconstans,* and notes on the geographical distribution and the origin of the loss. J. Fish. Res. Board Can. 34:1314–1320.

NEMESCHKAL, H. L. 1999. Morphometric correlation patterns of adult birds (Fringillidae: Passeriformes and Columbiformes) mirror the expression of developmental control genes. Evolution 53: 899–918.

NENE, R. V., AND J. C. GEORGE. 1965. A histological study of some muscles of the avian pectoral appendage. Pavo 3:35–46.

NEVO, E. 1999. Mosaic Evolution of Subterranean Mammals: Regression, Progression and Global Convergence. Oxford University Press, Oxford, United Kingdom.

NEWELL, N. D. 1949. Phyletic size increase, an important trend illustrated by fossil invertebrates. Evolution 3:103–124.
NEWTON, A. 1896. A Dictionary of Birds. Part 4. Adam and Charles Black, London, England.
NEWTON, E., AND H. GADOW. 1893. On additional bones of the Dodo and other extinct birds of Mauritius obtained by Mr. Theodore Sauzier. Trans. Zool. Soc. Lond. 13:291–302.
NEWTON, I. 1993. Predation and limitation of bird numbers. Pp. 143–198 *in* Current Ornithology. Vol. 11 (D. M. Power, Ed.). Plenum Press, New York, New York.
NEWTON, I. 1998. Population Limitation in Birds. Academic Press, San Diego, California.
NEWTON, I., AND L. DALE. 1996. Relationship between migration and latitude among west European birds. J. Anim. Ecol. 65:137–146.
NICE, M. M. 1938. The biological significance of bird weights. Bird-Banding 9:1–11.
NICE, M. M. 1941. The role of territory in bird life. Am. Midl. Nat. 26:441–487.
NICE, M. M. 1954. Problems of incubation periods in North American birds. Condor 56:173–197.
NICE, M. M. 1962. Development of behavior in precocial birds. Trans. Linn. Soc. N. Y. 8:1–211.
NICHOLS, R. A., AND G. M. HEWITT. 1994. The genetic consequences of long-distance dispersal during colonization. Heredity 72:312–317.
NIKLAS, K. J. 1994. The scaling of plant and animal body mass, length, and diameter. Evolution 48: 44–54.
NILES, D. M. 1973. Adaptive variation in body size and skeletal proportions of Horned Larks of the southeastern United States. Evolution 27:405–426.
NILSSON, S. G., C. BJÖRKMAN, P. FORSLUND, AND J. HÖGLUND. 1985a. Nesting holes and food supply in relation to forest bird densities on islands and mainland. Oecologia 66:516–521.
NILSSON, S. G., C. BJÖRKMAN, P. FORSLUND, AND J. HÖGLUND. 1985b. Egg predation in forest bird communities on islands and mainland. Oecologia 66:511–515.
NILSSON, S. G., AND B. EBENMAN. 1981. Density changes and niche differences in island and mainland Willow Warblers *Phylloscopus trochilus* at a lake in southern Sweden. Ornis Scand. 12:62–67.
NILSSON, S. G., AND I. N. NILSSON. 1983. Are estimated species turnover rates on islands largely sampling error? Am. Nat. 121:595–597.
NISBET, I. C. T., AND M. E. COHEN. 1975. Asynchronous hatching in Common and Roseate Terns, *Sterna hirundo* and *S. dougallii*. Ibis 117:374–379.
NISBET, I. C. T., W. H. DRURY, JR., AND J. BAIRD. 1963. Weight-loss during migration, part I: deposition and consumption of fat by the Blackpoll Warbler *Dendroica striata*. Bird-Banding 34:107–159.
NITECKI, M. H. 1988. Discerning the criteria for concepts of progress. Pp. 3–24 *in* Evolutionary Progress (M. H. Nitecki, Ed.). University of Chicago Press, Chicago, Illinois.
NITZSCH, C. L. 1840. System der Pterylographie. Eduard Anton, Halle, Germany.
NOORDWIJK, A. J. VAN. 1984. Quantitative genetics in natural populations of birds illustrated with examples from the Great Tit, *Parus major*. Pp. 67–79 *in* Population Biology and Evolution (K. Wöhrmann and V. Loeschcke, Eds.). Springer, Berlin, Germany.
NOORDWIJK, A. J. VAN, AND W. SCHARLOO. 1981. Inbreeding in an island population of the Great Tit. Evolution 35:674–688.
NORBERG, R. Å. 1981. Temporary weight decrease in breeding birds may result in more fledged young. Am. Nat. 118:838–850.
NORBERG, R. Å. 1985. Function of vane asymmetry and shaft curvature in bird flight feathers; inferences on flight ability of *Archaeopteryx*. Pp. 303–318 *in* The Beginnings of Birds (M. K. Hecht, J. H. Ostrom, G. Viohl, and P. Wellnhofer, Eds.). Freunde des Jura-Mus. Eichstatt, Willibaldsburg, Germany.
NORBERG, R. Å. 1986. Treecreeper climbing: mechanics, energetics, and structural adaptations. Ornis Scand. 17:191–209.
NORBERG, R. Å. 1994. Swallow tail streamer is a mechanical device for self-deflection of tail leading-edge, enhancing aerodynamic efficiency and flight manoeuverability. Proc. R. Soc. Lond. (Ser. B, Biol. Sci.) 257:227–233.
NORBERG, R. Å., AND U. M. NORBERG. 1971. Take-off, landing, and flight speed during fishing flights of *Gavia stellata* (Pont.). Ornis Scand. 2:55–67.
NORBERG, U. M. 1979. Morphology of the wings, legs and tail of three coniferous forest tits, the Goldcrest, and the Treecreeper in relation to locomotor pattern and feeding station selection. Philos. Trans. R. Soc. Lond. (Ser. B, Biol. Sci.) 287:131–165.

NORBERG, U. M. 1981a. Flight, morphology, and the ecological niche in some bats and birds. Symp. Zool. Soc. Lond. 48:173–197.

NORBERG, U. M. 1981b. Allometry of bat wings and legs and comparison with bird wings. Philos. Trans. R. Soc. Lond. (Ser. B, Biol. Sci.) 292:359–398.

NORBERG, U. M. 1986. Evolutionary convergence in foraging niche and flight morphology in insectivorous aerial-hawking birds and bats. Ornis Scand. 17:253–260.

NORBERG, U. M. 1987. Wing form and flight mode in bats. Pp. 43–56 in Recent Advances in the Study of Bats (M. B. Fenton, P. A. Racey, and J. M. V. Rayner, Eds.). Cambridge University Press, Cambridge, United Kingdom.

NORBERG, U. M. 1990. Vertebrate Flight: Mechanics, Physiology, Morphology, Ecology and Evolution. Springer-Verlag, Berlin, Germany.

NORBERG, U. M. 1994. Wing design, flight performance, and habitat use in bats. Pp. 205–239 in Ecological Morphology: Integrative Organismal Biology (P. C. Wainwright and S. M. Reilly, Eds.). University of Chicago Press, Chicago, Illinois.

NORBERG, U. M. 1995a. How a long tail and changes in mass and wing shape affect the cost of flight in animals. Funct. Ecol. 9:48–54.

NORBERG, U. M. 1995b. Wing design and migratory flight. Isr. J. Zool. 41:297–305.

NORBERG, U. M. 1996. Energetics of flight. Pp. 199–249 in Avian Energetics and Nutritional Ecology (C. Carey, Ed.). Chapman and Hall, New York, New York.

NORBERG, U. M., AND J. M. V. RAYNER. 1987. Ecological morphology and flight in bats (Mammalia: Chiroptera): wing adaptations, flight performance, foraging strategy and echolocation. Philos. Trans. R. Soc. Lond. (Ser. B, Biol. Sci.) 316:335–427.

NORES, M. 1995. Insular biogeography of birds on mountain-tops in northwestern Argentina. J. Biogeogr. 22:61–70.

NORMAN, F. I., AND L. MUMFORD. 1985. Studies on the Purple Swamphen, *Porphyrio porphyrio*, in Victoria. Aust. Wildl. Res. 12:263–278.

NORTHCOTE, E. M. 1982. Size, form and habit of the extinct Maltese swan *Cygnus falconeri*. Ibis 124:148–158.

NORRIS, D. R., AND B. J. M. STUTCHBURY. 2002. Sexual differences in gap-crossing ability of a forest songbird in a fragmented landscape revealed through radiotracking. Auk 119:528–532.

NOTT, J. 1997. Extremely high-energy wave deposits inside the Great Barrier Reef, Australia: determining the cause—tsunami or tropical cyclone. Mar. Geol. 141:193–207.

NUDDS, T. D. 1982. Ecological separation of grebes and coots: Interference competition or microhabitat selection? Wilson Bull. 94:505–514.

NUDDS, T. D., K. F. ABRAHAM, C. D. ANKNEY, AND P. D. TEBBEL. 1981. Are size gaps in dabbling- and wading-bird arrays real? Am. Nat. 118:549–553.

NUDDS, T. D., AND R. M. KAMINSKI. 1984. Sexual size dimorphism in relation to resource partitioning in North American dabbling ducks. Can. J. Zool. 62:2009–2012.

NUDDS, T. D., AND R. G. WICKETT. 1994. Body size and seasonal coexistence of North American dabbling ducks. Can. J. Zool. 72:779–782.

NUNN, P. D. 1994. Oceanic Islands. Blackwell, Oxford, United Kingdom.

NUR, N. 1988. The cost of reproduction in birds: an examination of the evidence. Ardea 76:155–168.

NYLIN, S., AND N. WEDELL. 1994. Sexual size dimorphism and comparative methods. Pp. 253–280 in Phylogenetics and Ecology (P. Eggleton and R. I. Vane-Wright, Eds.). Linn. Soc. Symp. 17. Academic Press, London, United Kingdom.

OAKES, E. J. 1992. Lekking and the evolution of sexual dimorphism in birds: comparative approaches. Am. Nat. 140:665–684.

OBERHOLSER, H. C. 1937. A revision of the Clapper Rails (*Rallus longirostris* Boddaert). Proc. U.S. Natl. Mus. 84:313–354.

O'CONNOR, R. J. 1981. Comparisons between migrant and non-migrant birds in Britain. Pp. 167–195 in Animal Migration (D. J. Aidley, Ed.). Cambridge University Press, Cambridge, United Kingdom.

O'CONNOR, R. J. 1984. The Growth and Development of Birds. J. Wiley and Sons, Chichester, United Kingdom.

O'CONNOR, R. J. 1985. Behavioural regulation of bird populations: a review of habitat use in relation to migration and residency. Pp. 105–142 in Behavioural Ecology: Ecological Consequences of

Adaptative Behaviour (R. M. Sibly and R. H. Smith, Eds.). Blackwell Scientific Publications, Oxford, United Kingdom.

O'CONNOR, R. J. 1990. Some ecological aspects of migrants and residents. Pp. 175–182 in Bird Migration: Physiology and Ecology (E. Gwinner, Ed.). Springer-Verlag, Berlin, Germany.

O'DONALD, P. 1962. The theory of sexual selection. Heredity 17:541–552.

O'DONALD, P. 1973. A further analysis of Bumpus' data: the intensity of natural selection. Evolution 27:398–404.

O'DONALD, P. 1980. Sexual selection by female choice in a monogamous bird: Darwin's theory corroborated. Heredity 45:201–217.

O'DONALD, P. 1983. Sexual selection by female choice. Pp. 53–66 in Mate Choice (P. Bateson, Ed.). Cambridge University Press, Cambridge, United Kingdom.

O'DONALD, P., N. S. WEDD, AND J. W. F. DAVIS. 1974. Mating preferences and sexual selection in the Arctic Skua. Heredity 33:1–16.

ODUM, E. P., C. E. CONNELL, AND H. L. STODDARD, SR. 1961. Flight energy and estimated flight ranges of some migratory birds. Auk 78:515–527.

OKAMOTO, M. 1972. Four techniques of principal component analysis. J. Jpn. Stat. Soc. 2:63–69.

OKSANEN, L., S. D. FRETWELL, AND O. JÄRVINEN. 1979. Interspecific aggression and the limiting similarity of close competitors: the problem of size gaps in some community arrays. Am. Nat. 114:117–129.

OLIVER, H. C. 1968. Annotated index to some early New Zealand bird literature. Wildl. Publ. No. 106. Department of Internal Affairs, Wellington, New Zealand.

OLIVER, W. R. B. 1930. New Zealand Birds. Fine Arts, Wellington, New Zealand.

OLIVER, W. R. B. 1945. Avian evolution in New Zealand and Australia. Emu 45:119–152.

OLIVER, W. R. B. 1949. The moas of New Zealand and Australia. Dom. Mus. Bull. 15:1–206.

OLIVER, W. R. B. 1955. New Zealand Birds. 2nd ed. A. H. and A. W. Reed, Wellington, New Zealand.

OLIVIERI, I., AND P.-H. GOUYON. 1997. Evolution of migration rate and other traits: the metapopulation effect. Pp. 293–323 in Metapopulation Biology: Ecology, Genetics, and Evolution (I. Hanski and M. E. Gilpin, Eds.). Academic Press, San Diego, California.

OLIVIERI, I., Y. MICHALAKIS, AND P.-H. GOUYON. 1995. Metapopulation genetics and the evolution of dispersal. Am. Nat. 146:202–228.

OLSON, S. L. 1972. The American Purple Gallinule, *Porphyrula martinica,* on Ascension and St. Helena islands. Bull. Br. Ornithol. Club 92:92–93.

OLSON, S. L. 1973a. A classification of the Rallidae. Wilson Bull. 85:381–416.

OLSON, S. L. 1973b. Evolution of the rails of the South Atlantic islands (Aves: Rallidae). Smithson. Contrib. Zool. 152:1–53.

OLSON, S. L. 1974a. A new species of *Nesotrochis* from Hispaniola, with notes on other fossil rails from the West Indies (Aves: Rallidae). Proc. Biol. Soc. Wash. 87:439–450.

OLSON, S. L. 1974b. The Pleistocene rails of North America. Condor 76:169–175.

OLSON, S. L. 1975a. The south Pacific gallinules of the genus *Pareudiastes.* Wilson Bull. 87:1–5.

OLSON, S. L. 1975b. A review of the extinct rails of the New Zealand region (Aves: Rallidae). Natl. Mus. N. Z. Rec. 1:63–79.

OLSON, S. L. 1975c. Paleornithology of St. Helena Island, South Atlantic Ocean. Smithson. Contrib. Paleobiol. 23:1–49.

OLSON, S. L. 1975d. The fossil rails of C. W. De Vis, being mainly an extinct form of *Tribonyx mortierii* from Queensland. Emu 75:49–54.

OLSON, S. L. 1977. A synopsis of the fossil Rallidae. Pp. 339–373 in Rails of the World (S. D. Ripley, Ed.). David R. Godine, Boston, Massachusetts.

OLSON, S. L. 1978. A paleontological perspective of West Indian birds and mammals. Acad. Nat. Sci. Phila. Spec. Publ. 13:99–117.

OLSON, S. L. 1982. Natural history of vertebrates on the Brazilian islands of the mid South Atlantic. Natl. Geogr. Soc. Res. Rep. 13:481–492.

OLSON, S. L. 1983. Lessons from a flightless ibis. Nat. Hist. 92:40.

OLSON, S. L. 1985. The fossil record of birds. Pp. 79–238 in Avian Biology. Vol. 8 (D. S. Farner, J. R. King, and K. C. Parkes, Eds.). Academic Press, New York, New York.

OLSON, S. L. 1986. *Gallirallus sharpei* (Büttikofer), nov. comb. A valid species of rail (Rallidae) of unknown origin. Gerfaut 76:263–269.

OLSON, S. L. 1989. Extinction on islands: man as a catastrophe. Pp. 50–53 in Conservation for the

Twenty-first Century (D. Western and M. Pearl, Eds.). Oxford University Press, Oxford, United Kingdom.

OLSON, S. L. 1990. Osteology and systematics of the fernbirds (*Bowdleria*: Sylviidae). Notornis 37: 161–171.

OLSON, S. L. 1991. Patterns of avian diversity and radiation in the Pacific as seen through the fossil record. Pp. 314–318 *in* The Unity of Evolutionary Biology: Proceedings of the Fourth International Congress of Systematic and Evolutionary Biology. Vol. 1 (E. C. Dudley, Ed.). Dioscorides Press, Portland, Oregon.

OLSON, S. L. 1992. Requiescat for *Tricholimnas conditicius*, a rail that never was. Bull. Br. Ornithol. Club 112:174–179.

OLSON, S. L. 1994. The endemic vireo of Fernando de Noronha (*Vireo gracilirostris*). Wilson Bull. 106:1–17.

OLSON, S. L. 1996. History and ornithological journals of the *Tanager* Expedition of 1923 to the northwestern Hawaiian Islands, Johnston and Wake islands. Atoll Res. Bull. 433:1–210.

OLSON, S. L. 1997. Towards a less imperfect understanding of the systematics and biogeography of the Clapper and King Rail complex (*Rallus longirostris* and *R. elegans*). Pp. 93–111 *in* The Era of Allan R. Phillips: A Festschrift (R. W. Dickerman, Ed.). Nuttall Ornithologists' Club, Cambridge, Massachusetts.

OLSON, S. L. 1999. Laysan Rail (*Porzana palmeri*) and Hawaiian Rail (*Porzana sandwichensis*). Pp. 1–20 *in* The Birds of North America. No. 426 (A Poole and F. Gill, Eds.). Birds of North America, Inc., Philadelphia, Pennsylvania.

OLSON, S. L., J. C. BALOUET, AND C. T. FISHER. 1987. The owlet-nightjar of New Caledonia *Aegotheles savesi*, with comments on the systematics of the Aegothelidae. Gerfaut 77:341–352.

OLSON, S. L., AND Y. HASEGAWA. 1979. Fossil counterparts of giant penguins from the North Pacific. Science 206:688–689.

OLSON, S. L., AND Y. HASEGAWA. 1996. A new genus and two new species of gigantic Plotopteridae from Japan (Aves: Pelecaniformes). J. Vertebr. Paleontol. 16:742–751.

OLSON, S. L., AND W. B. HILGARTNER. 1982. Fossil and subfossil birds from the Bahamas. Pp. 22–60 *in* Fossil Vertebrates from the Bahamas (S. L. Olson, Ed.). Smithson. Contrib. Paleobiol. 48. Smithsonian Institution, Washington, D.C.

OLSON, S. L., AND H. F. JAMES. 1982a. Prodromus of the fossil avifauna of the Hawaiian Islands. Smithson. Contrib. Zool. 365:1–59.

OLSON, S. L., AND H. F. JAMES. 1982b. Fossil birds from the Hawaiian Islands: evidence for wholesale extinction by man before western contact. Science 217:633–635.

OLSON, S. L., AND H. F. JAMES. 1984. The role of Polynesians in the extinction of the avifauna of the Hawaiian Islands. Pp. 768–780 *in* Quaternary Extinctions: A Prehistoric Revolution (P. S. Martin and R. G. Klein, Eds.). University of Arizona Press, Tucson, Arizona.

OLSON, S. L., AND H. F. JAMES. 1991. Descriptions of Thirty-two New Species of Birds from the Hawaiian Islands. Part I. Non-passeriformes. Ornithol. Monogr. 45. American Ornithologists' Union, Washington, D.C.

OLSON, S. L., AND H. F. JAMES. 1994. A chronology of ornithological exploration in the Hawaiian Islands, from Cook to Perkins. Pp. 91–102 *in* A Century of Avifaunal Change in Western North America (J. R. Jehl, Jr., and N. K. Johnson, Eds). Stud. Avian Biol. No. 15. Cooper Ornithological Society, Los Angeles, California.

OLSON, S. L., AND P. JOUVENTIN. 1996. A new species of small flightless duck from Amsterdam Island, southern Indian Ocean (Anatidae: *Anas*). Condor 98:1–9.

OLSON, S. L., AND D. W. STEADMAN. 1977. A new genus of flightless ibis (Threskiornithidae) and other fossil birds from cave deposits in Jamaica. Proc. Biol. Soc. Wash. 90:447–457.

OLSON, S. L., AND D. W. STEADMAN. 1979. The humerus of *Xenicibis*, the extinct flightless ibis of Jamaica. Proc. Biol. Soc. Wash. 92:23–27.

OLSON, S. L., AND D. W. STEADMAN. 1987. Comments on the supposed suppression of *Rallus nigra* Miller 1784, N. N. (S.) 2276, and *Columba r. forsteri* Wagler, 1829. Z. N. (S.) 2277. Bull. Zool. Nomen. 44:126–127.

OLSON, S. L., AND A. WETMORE. 1976. Preliminary diagnoses of two extraordinary new genera of birds from Pleistocene deposits in the Hawaiian Islands. Proc. Biol. Soc. Wash. 89:247–258.

OLSON, S. L., AND D. B. WINGATE. 2000. Two new species of flightless rails (Aves: Rallidae) from the Middle Pleistocene "crane fauna" of Bermuda. Proc. Biol. Soc. Wash. 113:356–368.

OLSON, S. L., AND D. B. WINGATE. 2001. A new species of large flightless rail of the *Rallus longirostris/elegans* complex (Aves: Rallidae) from the late Pleistocene of Bermuda. Proc. Biol. Soc. Wash. 114:509–516.
ONLEY, D. 1982. The Spotless Crake (*Porzana tabuensis*) on Aorangi, Poor Knights Islands. Notornis 29:9–21.
OPDAM, P., D. VAN DORP, AND C. J. F. TER BRAAK. 1984. The effect of isolation on the number of woodland birds in small woods in the Netherlands. J. Biogeogr. 11:473–478.
ORIANS, G. H. 1969. On the evolution of mating systems in birds and mammals. Am. Nat. 103:589–603.
ORIANS, G. H. 1971. Ecological aspects of behavior. Pp. 513–546 *in* Avian Biology. Vol. 1 (D. S. Farner and J. R. King, Eds.). Academic Press, New York, New York.
ORIANS, G. H., AND M. F. WILSON. 1964. Interspecific territories of birds. Ecology 45:736–745.
ORING, L. W. 1964. Predation upon flightless ducks. Wilson Bull. 76:190.
ORING, L. W. 1982. Avian mating systems. Pp. 1–92 *in* Avian Biology. Vol. 6 (D. S. Farner and J. R. King, Eds.). Academic Press, New York, New York.
ORNAT, A. L., AND R. GREENBERG. 1990. Sexual segregation by habitat in migratory warblers in Quintana Roo, Mexico. Auk 107:539–543.
OSBORNE, D. R., AND S. R. BEISSINGER. 1979. The Paint-billed Crake in Guyana. Auk 96:425.
O'STEELE, S., A. J. CULLUM, AND A. F. BENNETT. 2002. Rapid evolution of escape ability in Trinidadian guppies (*Poecilia reticulata*). Evolution 56:776–784.
OTTE, D. 1979. Biogeographic patterns in flight capacity of Nearctic grasshoppers (Orthoptera: Acrididae). Entomol. News 90:153–158.
OUDHOF, G. 1975. Development and Growth of the Cranium: A Quantitative Experimental Study in the Chick Embryo. Nooy's Drukkerij, Purmerend, The Netherlands.
OUELLET, H. 1967. Dispersal of land birds on the islands of the Gulf of St. Lawrence, Canada. Can. J. Zool. 45:1149–1167.
OWEN, M., AND M. OGILVIE. 1979. Wing moult and weights of Barnacle Geese in Spitsbergen. Condor 81:42–52.
OWEN, R. 1841. On the anatomy of the southern Apteryx (*Apteryx australis,* Shaw). Trans. Zool. Soc. Lond. 2:257–302.
OWEN, R. 1848a. On the remains of the gigantic and presumed extinct wingless and terrestrial birds of New Zealand (*Dinornis* and *Palapteryx*), with indications of two other genera (*Notornis* and *Nestor*). Proc. Zool. Soc. Lond. 1848:1–11.
OWEN, R. 1848b. On *Dinornis* (part III.): containing a description of the skull and beak of that genus, and of the same characteristic parts of *Palapteryx,* and of two other genera of birds, *Notornis* and *Nestor*; forming part of an extensive series of ornithic remains discovered by Mr. Walter Mantell at Waingongoro, North Island of New Zealand. Trans. Zool. Soc. Lond. 3:345–378.
OWEN, R. 1851. On *Dinornis* (part IV.): containing the restoration of the feet of that genus and of *Palapteryx,* with a description of the sternum in *Palapteryx* and *Aptornis.* Trans. Zool. Soc. Lond. 4:1–20.
OWEN, R. 1856. On the affinities of the large extinct bird (*Gastornis parisiensis,* Hébert), indicated by a fossil femur and tibia discovered in the lowest Eocene formation near Paris. Proc. Geol. Soc. Lond. 1856:204–217.
OWEN, R. 1866a. On the osteology of the Dodo (*Didus ineptus,* Linn.). Trans. Zool. Soc. Lond. 6:49–86.
OWEN, R. 1866b. Comparative Anatomy and Physiology of Vertebrates. Vol. 2. Longmans and Green, London, England.
OWEN, R. 1870. On *Dinornis* (part XV.): containing a description of the skull, femur, tibia, fibula, and metatarsus of *Aptornis defossor,* Owen, from near Oamaru, Middle Island, New Zealand, with additional observations on *Aptornis otidiformis,* on *Notornis mantelli,* and on *Dinornis curtus.* Trans. Zool. Soc. Lond. 7:353–380.
OWEN, R. 1872. On *Dinornis* (part XVII.): containing a description of the sternum and pelvis, with an attempted restoration, of *Aptornis defossor,* Ow. Trans. Zool. Soc. Lond. 8:119–126.
OWEN, R. 1875. On *Dinornis* (part XX.): containing a restoration of the skeleton of *Cnemiornis calcitrans* Ow., with remarks on its affinities in the lamellirostral group. Trans. Zool. Soc. Lond. 9:253–292.
OWEN, R. 1879. Memoirs on the Extinct Wingless Birds of New Zealand; with an Appendix on Those

of England, Australia, Newfoundland, Mauritius, and Rodriguez. 2 vols. John Van Voorst, London, England.

OWEN, R. 1882. On the sternum of *Notornis* and on sternal characters. Proc. Zool. Soc. Lond. 1882: 689–697.

OXNARD, C. E. 1969a. The combined use of multivariate and clustering analyses in functional morphology. J. Biomech. 2:73–88.

OXNARD, C. E. 1969b. Mathematics, shape and function: a study in primate anatomy. Am. Sci. 57: 75–96.

OXNARD, C. E. 1978. One biologist's view of morphometrics. Annu. Rev. Ecol. Syst. 9:219–241.

PACKARD, A. S. 1888. The cave fauna of North America, with remarks on the anatomy of brain and the origin of the blind species. Mem. Natl. Acad. Sci. USA 4:1–156.

PACKARD, G. C. 1967. House Sparrows: evolution of populations from the Great Plains and Colorado Rockies. Syst. Zool. 16:73–89.

PACKARD, G. C., AND T. J. BOARDMAN. 1987. The misuse of ratios to scale physiological data that vary allometrically with body size. Pp. 216–239 *in* New Directions in Ecological Physiology (M. E. Feder, A. F. Bennett, W. W. Burggren, and R. B. Huey, Eds.). Cambridge University Press, Cambridge, United Kingdom.

PADIAN, K., AND L. M. CHIAPPE. 1998. The origin and early evolution of birds. Biol. Rev. 73:1–42.

PADILLA, D. K., AND S. C. ADOLPH. 1996. Plastic inducible morphologies are not always adaptive: the importance of time delays in a stochastic environment. Evol. Ecol. 10:105–117.

PAGEL, M. D., AND P. H. HARVEY. 1988. The taxon-level problem in the evolution of mammalian brain size: facts and artifacts. Am. Nat. 132:344–359.

PAGEL, M. D., AND P. H. HARVEY. 1989. Taxonomic differences in the scaling of brain on body weight among mammals. Science 244:1589–1593.

PAGEL, M. D., AND P. H. HARVEY. 1992. On solving the correct problem: wishing does not make it so. J. Theor. Biol. 156:425–430.

PAGEL, M., AND R. J. H. PAYNE. 1996. How migration affects estimation of the extinction threshold. Oikos 76:323–329.

PALADINO, F. V. 1985. Temperature effects on locomotion and activity bioenergetics of amphibians, reptiles, and birds. Am. Zool. 25:965–972.

PALMER, M. W. 1993. Putting things in even better order: the advantages of canonical correspondence analysis. Ecology 74:2215–2230.

PANEK, M., AND P. MAJEWSKI. 1990. Remex growth and body mass of Mallards during wing molt. Auk 107:255–259.

PARADIS, E., S. R. BAILLIE, W. J. SUTHERLAND, AND R. D. GREGORY. 1998. Patterns of natal and breeding dispersal in birds. J. Anim. Ecol. 67:518–536.

PARCHMAN, T. L., AND C. W. BENKMAN. 2002. Diversifying coevolution between crossbills and black spruce on Newfoundland. Evolution 56:1663–1672.

PARK, K. J., R. M. EVANS, AND K. L. BUCHANAN. 2000. Assessing the aerodynamic effects of tail elongations in the House Martin (*Delichon urbica*): implications for the initial selection pressures in the hirundines. Behav. Ecol. Sociobiol. 48:364–372.

PARK, J. 1888a. The Takahe (*Notornis mantelli*) in western Otago. Trans. N. Z. Inst. 21:226–230.

PARK, J. 1888b. On the Takahe (*Notornis mantelli*) in west Otago. Trans. N. Z. Inst. 21:503.

PARK, J. 1890. Takahe *versus* Kakapo. Trans. N. Z. Inst. 23:112–119.

PARKER, G. A. 1983. Mate quality and mating decisions. Pp. 141–166 *in* Mate Choice (P. Bateson, Ed.). Cambridge University Press, Cambridge, United Kingdom.

PARKER, T. J. 1885a. Notes on a skeleton of *Notornis,* recently acquired by the Otago University Museum. Trans. N. Z. Inst. 18:78–82.

PARKER, T. J. 1885b. On the skeleton of *Notornis mantelli*. Trans. N. Z. Inst. 18:245–258.

PARKER, T. J. 1892. Observations on the anatomy and development of *Apteryx*. Philos. Trans. R. Soc. Lond. 182:25–134.

PARKER, T. J. 1894. On the origin of the sternum. Trans. N. Z. Inst. 23:119–123.

PARKER, W. K. 1864. On the sternal apparatus of birds and other Vertebrata. Proc. Zool. Soc. Lond. 1864:339–341.

PARKER, W. K. 1866. On the structure and development of the skull in the ostrich tribe. Philos. Trans. R. Soc. Lond. (Part 1) 156:113–183.

PARKER, W. K. 1868. A Monograph on the Structure and Development of the Shoulder-Girdle and Sternum in the Vertebrata. Ray Society, London, England.

PARKER, W. K. 1869. On the osteology of the Kagu (*Rhinochetus jubatus*). Trans. Zool. Soc. Lond. 8:501–521.

PARKER, W. K. 1888. On the structure and development of the wing in the common fowl. Philos. Trans. R. Soc. Lond. (Ser. B) 179:385–398.

PARKES, K. C., D. P. KIBBE, AND E. L. ROTH. 1978. First records of the Spotted Rail (*Pardirallus maculatus*) for the United States, Chile, Bolivia and western Mexico. Am. Birds 32:295–299.

PARRISH, J. D. 1997. Patterns of frugivory and energetic condition in Nearctic landbirds during autumn migration. Condor 99:681–697.

PARRISH, J. D. 2000. Behavioral, energetic, and conservation implications of foraging plasticity during migration. Pp. 53–70 *in* Stopover Ecology of Nearctic–Neotropical Landbird Migrants: Habitat Relations and Conservation Implications (F. R. Moore, Ed.). Stud. Avian Biol. No. 20. Cooper Ornithological Society, Los Angeles, California.

PARROT, C. 1894. Über die Grössenverhältnisse des Herzens bei Vögeln. Zool. Jahrb. (Abt. Syst.) 7: 496–522.

PARROTT, G. C. 1970. Aerodynamics of gliding flight of a Black Vulture *Coragyps atratus*. J. Exp. Biol. 53:363–374.

PARSONS, P. A. 1982. Adaptive strategies of colonizing animal species. Biol. Rev. 57:117–148.

PARSONS, P. A. 1987. Evolutionary rates under environmental stress. Pp. 311–347 *in* Evolutionary Biology. Vol. 21 (M. K. Hecht, B. Wallace, and G. T. Prance, Eds.). Plenum Press, New York, New York.

PARSONS, P. A. 1990. Fluctuating asymmetry: an epigenetic measure of stress. Biol. Rev. 65:131–145.

PARSONS, P. A. 1992. Evolutionary adaptation and stress: the fitness gradient. Pp. 191–223 *in* Evolutionary Biology. Vol. 26 (M. K. Hecht, B. Wallace, and R. T. MacIntyre, Eds.). Plenum Press, New York, New York.

PARSONS, P. A. 1993a. Stress, extinctions and evolutionary change: from living organisms to fossils. Biol. Rev. 68:313–333.

PARSONS, P. A. 1993b. Stress, metabolic cost and evolutionary change: from living organisms to fossils. Pp. 140–156 *in* Evolutionary Patterns and Processes (D. R. Lees and D. Edwards, Eds.). Linn. Soc. Symp. No. 14. Academic Press, London, United Kingdom.

PARSONS, P. A. 1993c. The importance and consequences of stress in living and fossil populations: from life-history variation to evolutionary change. Am. Nat. 142 (Suppl.):5–20.

PARSONS, P. A. 1996. Stress, resources, energy balances, and evolutionary change. Pp. 39–72 *in* Evolutionary Biology. Vol. 29 (M. K. Hecht, R. J. MacIntyre, and M. T. Clegg, Eds.). Plenum Press, New York, New York.

PARTRIDGE, L. 1989. Lifetime reproductive success and life-history evolution. Pp. 421–440 *in* Lifetime Reproduction in Birds (I. Newton, Ed.). Academic Press, London, United Kingdom.

PARTRIDGE, L., AND J. A. ENDLER. 1987. Life history constraints on sexual selection. Pp. 265–277 *in* Sexual Selection: Testing the Alternatives (J. W. Bradbury and M. B. Andersson, Eds.). J. Wiley and Sons, Chichester, United Kingdom.

PASKOFF, R. 1991. Likely occurrence of a mega-tsunami near Coquimbo, Chile. Rev. Geol. Chile 18: 87–91.

PATON, P. W. C., E. J. MESSINA, AND C. R. GRIFFIN. 1994. A phylogenetic approach to reversed size dimorphism in diurnal raptors. Oikos 71:492–498.

PAULEY, G. 1990. Henderson Island: biogeography and evolution at the edge of the Pacific plate. Pp. 304–313 *in* The Unity of Evolutionary Biology: Proceedings of the Fourth International Congress of Systematic and Evolutionary Biology. Vol. 1 (E. C. Dudley, Ed.). Dioscorides Press, Portland, Oregon.

PAULLIN, D. C. 1987. Cannibalism in American Coots induced by severe spring weather and avian cholera. Condor 89:442–443.

PAULSON, D. R. 1973. Predator polymorphism and apostatic selection. Evolution 27:269–277.

PAYNE, R. B. 1984. Sexual selection, lek and arena behavior, and sexual dimorphism in birds. Ornithol. Monogr. 33:1–52.

PEAKER, M., AND J. L. LINZELL. 1975. Salt glands in birds and reptiles. Monogr. Physiol. Soc. No. 32. Cambridge University Press, Cambridge, United Kingdom.

PEASE, M. S. 1961. Wingless poultry. J. Heredity 52:109–110.

PELTONEN, A., AND I. HANSKI. 1991. Patterns of island occupancy explained by colonization and extinction rates in shrews. Ecology 72:1698–1708.
PENNY, M. J., AND A. W. DIAMOND. 1971. The White-throated Rail *Dryolimnas cuvieri* on Aldabra. Philos. Trans. R. Soc. Lond. (Ser. B) 260:529–548.
PENNYCUICK, C. J. 1960. Gliding flight in the Fulmar petrel. J. Exp. Biol. 37:330–338.
PENNYCUICK, C. J. 1967. The strength of the pigeon's wing bones in relation to their function. J. Exp. Biol. 46:219–233.
PENNYCUICK, C. J. 1968. Power requirements for horizontal flight in the Pigeon *Columba livia*. J. Exp. Biol. 49:527–555.
PENNYCUICK, C. J. 1969. The mechanics of bird migration. Ibis 111:525–556.
PENNYCUICK, C. J. 1971. Gliding flight of the White-backed Vulture *Gyps africanus*. J. Exp. Biol. 55: 13–38.
PENNYCUICK, C. J. 1972. Soaring behaviour and performance of some East African birds, observed from a motor-glider. Ibis 114:178–218.
PENNYCUICK, C. J. 1975. Mechanics of flight. Pp. 1–75 *in* Avian Biology. Vol. 5 (D. S. Farner and J. R. King, Eds.). Academic Press, New York, New York.
PENNYCUICK, C. J. 1978. Fifteen testable predictions about bird flight. Oikos 30:165–176.
PENNYCUICK, C. J. 1982. The flight of petrels and albatrosses (Procellariiformes), observed in South Georgia and its vicinity. Philos. Trans. R. Soc. Lond. (Ser. B) 300:75–106.
PENNYCUICK, C. J. 1986. Mechanical constraints on the evolution of flight. Pp. 83–88 *in* The Origin of Birds and the Evolution of Flight (K. Padian, Ed.). Mem. No. 8. California Academy of Sciences, San Francisco, California.
PENNYCUICK, C. J. 1987. Flight of seabirds. Pp. 43–62 *in* Seabirds: Feeding Ecology and Role in Marine Ecosystems (J. P. Croxall, Ed.). Cambridge University Press, Cambridge, United Kingdom.
PENNYCUICK, C. J. 1988. On the reconstruction of pterosaurs and their manner of flight, with notes on vortex wakes. Biol. Rev. 63:299–331.
PENNYCUICK, C. J. 1989. Bird Flight Performance: A Practical Calculation Manual. Oxford University Press, Oxford, United Kingdom.
PENNYCUICK, C. J. 1990. Predicting wingbeat frequency and wavelength of birds. J. Exp. Biol. 150: 171–185.
PENNYCUICK, C. J. 1996. Wingbeat frequency of birds in steady cruising flight: new data and improved predictions. J. Exp. Biol. 199:1613–1618.
PENNYCUICK, C. J. 1997. Actual and 'optimum' flight speeds: field data reassessed. J. Exp. Biol. 200: 2355–2361.
PENNYCUICK, C. J. 1998a. Field observations of thermals and thermal streets, and the theory of cross-country soaring flight. J. Avian Biol. 29:33–43.
PENNYCUICK, C. J. 1998b. Towards an optimal strategy for bird flight research. J. Avian Biol. 29: 449–457.
PENNYCUICK, C. J. 1998c. Computer simulation of fat and muscle burn in long distance bird migration. J. Theor. Biol. 191:47–61.
PENNYCUICK, C. J., M. R. FULLER, J. J. OAR, AND S. J. KIRKPATRICK. 1994. Falcon versus grouse: flight adaptations of a predator and prey. J. Avian Biol. 25:39–49.
PENNYCUICK, C. J., AND A. LOCK. 1976. Elastic energy storage in primary feather shafts. J. Exp. Biol. 64:677–689.
PENNYCUICK, C. J., H. H. OBRECHT III, AND M. R. FULLER. 1988. Empirical estimates of body drag of large waterfowl and raptors. J. Exp. Biol. 135:253–264.
PERCY, W. 1963. Further notes on the African Finfoot, *Podica senegalensis*. Bull. Br. Ornithol. Club 83:127–131.
PEREZ, G. S. A. 1968. Notes on the breeding season of the Guam Rails (*Rallus owstoni*). Micronesica 4:133–135.
PÉREZ-TRIS, J., AND J. L. TELLERÍA. 2001. Age-related variation in wing shape of migratory and sedentary Blackcaps *Sylvia atricapilla*. J. Avian Biol. 32:207–213.
PERKINS, R. C. L. 1903. Vertebrata. Pp. 365–466 *in* Fauna Hawaiiensis or the Zoology of the Sandwich (Hawaiian) Islands. Vol. 1. Part 4 (D. Sharp, Ed.). Cambridge University Press, Cambridge, England.
PERKINS, R. C. L. 1913. Introduction: being a review of the land-fauna of Hawaii. Pp. xv–ccxxviii

in Fauna Hawaiiensis or the Zoology of the Sandwich (Hawaiian) Isles. Vol. 1. Part 6 (D. Sharp, Ed.). Cambridge University Press, Cambridge, England.

PERLE, A., L. M. CHIAPPE, B. RINCHEN, J. M. CLARK, AND M. A. NORELL. 1994. Skeletal morphology of *Mononykus olecranus* (Theropoda: Avialae) from the Late Cretaceous of Mongolia. Am. Mus. Novit. 3105:1–29.

PERLE, A., M. A. NORELL, L. M. CHIAPPE, AND J. M. CLARK. 1993. Flightless bird from Cretaceous of Mongolia. Nature 362:623–626.

PERRIN, N. 1992. Optimal resource allocation and the marginal value of organs. Am. Nat. 139:1344–1369.

PERRIN, N., AND R. M. SIBLY. 1993. Dynamic models of energy allocation and investment. Annu. Rev. Ecol. Syst. 24:379–410.

PERRINS, C. M., AND T. R. BIRKHEAD. 1983. Avian Ecology. Blackie, Glasgow, United Kingdom.

PERSSON, L. 1985. Asymmetrical competition: Are larger animals competitively superior? Am. Nat. 126:261–266.

PETERS, D. S., AND W. F. GUTMANN. 1985. Constructional and functional preconditions for the transition to powered flight in vertebrates. Pp. 233–242 *in* The Beginnings of Birds (M. K. Hecht, J. H. Ostrom, G. Viohl, and P. Wellnhofer, Eds.). Jura-Museum Eichstätt, Willibaldsburg, Germany.

PETERS, J. L. 1931. Check-list of Birds of the World. Vol. 1. Harvard University Press, Cambridge, Massachusetts.

PETERS, J. L. 1934. Check-list of Birds of the World. Vol. 2. Harvard University Press, Cambridge, Massachusetts.

PETERS, J. L. 1937. Check-list of Birds of the World. Vol. 3. Harvard University Press, Cambridge, Massachusetts.

PETERS, N. 1988. Quantitative Aspekte der regressiven Evolution—dargestellt am Beispiel der Augenrückbildung bei Höhlenfischen. Z. Zool. Syst. Evolutionsforsch. 26:430–441.

PETERS, N., AND G. PETERS. 1968. Zur genetischen Interpretation morphologischer Gesetzmäßigkeiten der degenerativen Evolution. Z. Morphol. Tiere 62:211–244.

PETERS, R. H. 1983. The Ecological Implications of Body Size. Cambridge University Press, Cambridge, United Kingdom.

PETERS, R. H., AND K. WASSENBERG. 1983. The effect of body size on animal abundance. Oecologia 60:89–96.

PETIT, D. R. 2000. Habitat use by landbirds along Nearctic–Neotropical migration routes: implications for conservation of stopover habitats. Pp. 15–33 *in* Stopover Ecology of Nearctic–Neotropical Landbird Migrants: Habitat Relations and Conservation Implications (F. R. Moore, Ed.). Stud. Avian Biol. No. 20. Cooper Ornithological Society, Los Angeles, California.

PETRIE, M. 1983. Female Moorhens compete for small fat males. Science 220:413–415.

PETRIE, M. 1984. Territory size in the Moorhen (*Gallinula chloropus*): an outcome of RHP asymmetry between neighbors. Anim. Behav. 32:861–870.

PETRIE, M. 1986. Reproductive strategies of male and female Moorhens (*Gallinula chloropus*). Pp. 43–63 *in* Ecological Aspects of Social Evolution (D. I. Rubenstein and R. W. Wrangham, Eds.). Princeton University Press, Princeton, New Jersey.

PETRIE, M. 1988. Mating decisions by female Common Moorhens (*Gallinula chloropus*). Pp. 947–955 *in* Acta XIX Congressus Internationalis Ornithologici. Vol. 1 (H. Ouellet, Ed.). University of Ottawa Press, Ottawa, Ontario, Canada.

PETRIE, M. 1992. Are all secondary sexual display structures positively allometric and, if so, why? Anim. Behav. 43:173–175.

PFENNIG, D. W., AND P. J. MURPHY. 2002. Fluctuating competition and phenotypic plasticity mediate species divergence. Evolution 56:1217–1228.

PHILLIPPS, W. J. 1959. The last(?) occurrence of *Notornis* in the North Island. Notornis 8:93–94.

PHILLIPS, H. A. 1991. Eurasian Coot *Fulica atra* eating waterbird faeces. Aust. Bird Watcher 14:150.

PHILLIPS, P. C., S. P. OTTO, AND M. C. WHITLOCK. 2000. Beyond the average: the evolutionary importance of gene interactions and variability of epistatic effects. Pp. 20–38 *in* Epistasis and the Evolutionary Process (J. B. Wolf, E. D. Brodie III, and M. J. Wade, Eds.). Oxford University Press, Oxford, United Kingdom.

PIANKA, E. R. 1970. On *r*- and *K*-selection. Am. Nat. 104:592–597.

PIANKA, E. R. 1972. *r* and *K* selection or *b* and *d* selection? Am. Nat. 106:581–588.

PIANKA, E. R. 1976. Competition and niche theory. Pp. 114–141 *in* Theoretical Ecology: Principles and Applications (R. M. May, Ed.). W. B. Saunders, Philadelphia, Pennsylvania.

PIELOU, E. C. 1977. Mathematical Ecology. John Wiley and Sons, New York, New York.

PIENKOWSKI, M. W., AND P. R. EVANS. 1985. The role of migration in the population dynamics of birds. Pp. 331–352 *in* Behavioural Ecology: Ecological Consequences of Adaptive Behaviour (R. M. Sibly and R. H. Smith, Eds.). Blackwell Scientific Publications, Oxford, United Kingdom.

PIENKOWSKI, M. W., P. R. EVANS, AND D. J. TOWNSHEND. 1985. Leap-frog and other migration patterns of waders: a critique of the Alerstam and Högstedt hypothesis, and some alternatives. Ornis Scand. 16:61–70.

PIEPER, H. 1985. The fossil land birds of Madeira and Porto Santo. Bocagiana 88:1–6.

PIERSMA, T. 1987. Hop, skip or jump? Constraints on migration of arctic waders by feeding, fattening and flight speed. Limosa 60:185–194.

PIERSMA, T. 1988. Breast muscle atrophy and constraints on foraging during the flightless period of wing moulting Great Crested Grebes. Ardea 76:96–106.

PIERSMA, T. 1990. Pre-migratory "fattening" usually involves more than the deposition of fat alone. Ringing & Migr. 11:113–115.

PIERSMA, T. 1996a. Family Charadriidae (plovers): introductory narrative. Pp. 384–409 *in* Handbook of the Birds of the World. Vol. 3: Hoatzin to Auks (J. del Hoyo, A. Elliott and J. Sargatal, Eds.). Lynx Edicions, Barcelona, Spain.

PIERSMA, T. 1996b. Family Scolopacidae (sandpipers, snipes and phalaropes): introductory narrative. Pp. 444–487 *in* Handbook of the Birds of the World. Vol. 3: Hoatzin to Auks (J. del Hoyo, A. Elliott and J. Sargatal, Eds.). Lynx Edicions, Barcelona, Spain.

PIERSMA, T. 1997. Do global patterns of habitat use and migration strategies co-evolve with relative investments in immunocompetence due to spatial variation in parasite pressure? Oikos 80:623–631.

PIERSMA, T. 1998. Phenotypic flexibility during migration: optimization of organ size contingent on the risks and rewards of fueling and flight? J. Avian Biol. 29:511–520.

PIERSMA, T., AND N. C. DAVIDSON. 1991. Confusion of mass and size. Auk 108:441–442.

PIERSMA, T., AND R. E. GILL, JR. 1998. Guts don't fly: small digestive organs in obese Bar-tailed Godwits. Auk 115:196–203.

PIERSMA, T., A. KOOLHAAS, AND A. DEKINGA. 1993. Interactions between stomach structure and diet choice in shorebirds. Auk 110:552–564.

PIERSMA, T., AND Å. LINDSTRÖM. 1997. Rapid reversible changes in organ size as a component of adaptive behaviour. Trends Ecol. Evol. 12:134–138.

PIERTNEY, S. B., R. SUMMERS, AND M. MARQUISS. 2001. Microsatellite and mitochondrial DNA homogeneity among phenotypically diverse crossbill taxa in the UK. Proc. R. Soc. Lond. (Ser. B Biol. Sci.) 268:1511–1517.

PIGLIUCCI, M. 1998. Plasticity genes: What are they, and why should we care? Pp. 117–130 *in* The Co-Action Between Living Systems and the Planet (H. Greppin, R. D. Agosti, and C. Penel, Eds.). University of Geneva, Geneva, Switzerland.

PIGLIUCCI, M. 2001. Phenotypic Plasticity: Beyond Nature and Nurture. Johns Hopkins University Press, Baltimore, Maryland.

PIGLIUCCI, M., AND C. C. SCHLICHTING. 1997. On the limits of quantitative genetics for the study of phenotypic evolution. Acta Biotheor. 45:143–160.

PIJANOWSKI, B. C. 1992. A revision of Lack's brood reduction hypothesis. Am. Nat. 139:1270–1292.

PIMENTEL, R. A. 1979. Morphometrics: The Multivariate Analysis of Biological Data. Kendall-Hunt Publishing Company, Dubuque, Iowa.

PIMM, S. L. 1986. Filling niches carefully. Trends Ecol. Evol. 1:86–87.

PIMM, S. L. 1987. The snake that ate Guam. Trends Ecol. Evol. 2:293–295.

PIMM, S. L., J. M. DIAMOND, T. M. REDDS, G. J. RUSSEL, AND J. VERNER. 1993. Times to extinction for small populations of large birds. Proc. Natl. Acad. Sci. USA 90:10871–10875.

PIMM, S. L., H. L. JONES, AND J. DIAMOND. 1988. On the risk of extinction. Am. Nat. 132:757–785.

PIMM, S. L., M. P. MOULTON, AND L. J. JUSTICE. 1994a. Bird extinctions in the central Pacific. Philos. Trans. R. Soc. Lond. (Ser. B) 344:27–33.

PIMM, S. L., M. P. MOULTON, AND L. J. JUSTICE. 1994b. Bird extinctions in the central Pacific. Pp.

75–87 *in* Extinction Rates (J. H. Lawton and R. M. May, Eds.). Oxford University Press, Oxford, United Kingdom.
PIMM. S. L., AND J. W. PIMM. 1982. Resource use, competition, and resource availability in Hawaiian Honeycreepers. Ecology 63:1468–1480.
PIPER, W. H. 1997. Social dominance in birds: early findings and new horizons. Pp. 125–187 *in* Current Ornithology. Vol. 14 (V. Nolan, Jr., E. D. Ketterson, and C. F. Thompson, Eds.). Plenum Press, New York, New York.
PIVETEAU, J. 1945. Étude sur l'*Aphanapteryx* ouseau éteint de l'Ile Maurice. Ann. Paleontol. 31: 31–37.
PLATT, T., AND W. SILVERT. 1981. Ecology, physiology, allometry and dimensionality. J. Theor. Biol. 93:855–860.
POLO, V., AND L. M. CARRASCAL. 1999. Shaping the body mass distribution of Passeriformes: habitat use and body mass are evolutionarily and ecologically related. J. Anim. Ecol. 68:34–337.
POND, C. M. 1978. Morphological aspects and the ecological and mechanical consequences of fat deposition in wild vertebrates. Annu. Rev. Ecol. Syst. 9:519–570.
POOLE, E. L. 1938. Weights and wing areas in North American birds. Auk 55:511–517.
POPLIN, F., AND C. MOURER-CHAUVIRÉ. 1985. *Sylviornis neocaledoniae* (Aves, Galliformes, Megapodidae), oiseau géant éteint de l'Ile des Pins (Nouvelle-Calédonie). Géobios 18:73–97.
PORTER, A. H., AND N. A. JOHNSON. 2002. Speciation despite gene flow when developmental pathways evolve. Evolution 56:2103–2111.
PORTMANN, A. 1938. Beiträge zur Kenntnis der postembryonalen Entwicklung der Vogel. I. Vergleichende Untersuchungen über die Ontogenese der Hühner und Sperlingsvögel. Rev. Suisse Zool. 45:273–348.
POTTS, T. H. 1870. On the birds of New Zealand. Trans. N. Z. Inst. 3:59–109.
POULSON, T. L. 1963. Cave adaptation in amblyopsid fishes. Am. Midl. Nat. 70:257–290.
POULSON, T. L. 1985. Evolutionary reduction by neutral mutations: plausibility arguments and data from ambylopsid fishes and linyphiid spiders. Natl. Speleol. Soc. Bull. 47:109–117.
POWER, D. M. 1969. Evolutionary implications of wing and size variation in the Red-winged Blackbird in relation to geographic and climatic factors: a multiple regression analysis. Syst. Zool. 18:363–373.
POWER, D. M. 1970a. Geographic variation of Red-winged Blackbirds in central North America. Univ. Kans. Mus. Nat. Hist. Publ. 19:1–83.
POWER, D. M. 1970b. Geographic variation in the surface/volume ratio of the bill of Red-winged Blackbirds in relation to certain geographic and climatic factors. Condor 72:299–304.
POWER, D. M. 1971. Statistical analysis of character correlations in Brewer's Blackbirds. Syst. Zool. 20:186–203.
POWER, D. M. 1972. Numbers of bird species on the California islands. Evolution 26:451–463.
POWER, D. M. 1975. Similarity among avifaunas of the Galápagos Islands. Ecology 56:616–626.
POWER, D. M. 1976. Avifauna richness on the California Channel islands. Condor 78:394–398.
POWER, D. M. 1979. Evolution in peripheral isolated populations: *Carpodacus* finches on the California islands. Evolution 33:834–847.
POWLESLAND, R. G., AND B. D. LLOYD. 1994. Use of supplementary feeding to induce breeding in free-living Kakapo *Strigops habroptilus* in New Zealand. Biol. Conserv. 69:97–106.
POWLESLAND, R. G., B. D. LLOYD, H. A. BEST, AND D. V. MERTON. 1992. Breeding biology of the Kakapo *Strigops habroptilus* on Stewart Island, New Zealand. Ibis 134:361–373.
POWLESLAND, R. G., R. G. ROBERTS, B. D. LLOYD, AND D. V. MERTON. 1995. Number, fate and distribution of Kakapo (*Strigops habroptilus*) found on Steward Island, New Zealand. N. Z. J. Zool. 22:239–248.
PÖYSÄ, H. 1983. Morphology-mediated niche organization in a guild of dabbling ducks. Ornis Scand. 14:317–326.
PRACY, L. T. 1969. Weka liberations in the Palliser Bay region. Notornis 26:212–213.
PRAGER, E. J. 2000. Furious Earth: The Science and Nature of Earthquakes, Volcanoes, and Tsunamis. McGraw-Hill, New York, New York.
PRANCE, G. T. 1998. Islands in Amazonia. Pp. 241–261 *in* Evolution on Islands (P. R. Grant, Ed.). Oxford University Press, Oxford, United Kingdom.
PRANGE, H. D., J. F. ANDERSON, AND H. RAHN. 1979. Scaling of skeletal mass to body mass in birds and mammals. Am. Nat. 113:103–122.

PRATT, H. D. 1979. A Systematic Analysis of the Endemic Avifauna of the Hawaiian Islands. Ph.D. dissertation, Louisiana State University, Baton Rouge, Louisiana.

PRATT, H. D. 1987. Occurrence of the American Coot (*Fulica americana*) in the Hawaiian Islands, with comments on the taxonomy of the Hawaiian Coot. Elepaio 47:25–28.

PREGILL, G. K. 1986. Body size of insular lizards: a pattern of Holocene dwarfism. Evolution 40: 997–1008.

PREGILL, G. K., AND S. L. OLSON. 1981. Zoogeography of West Indian vertebrates in relation to Pleistocene climatic cycles. Annu. Rev. Ecol. Syst. 12:75–98.

PREIN, F. 1914. Die Entwickelung des vorderen Extremitätenskellettes beim Haushuhun. Anat. Hefte 51:643–690.

PRESCH, W. 1975. The evolution of limb reduction in the teiid lizard genus *Bachia*. Bull. S. Calif. Acad. Sci. 74:113–121.

PRESTON, F. W. 1951. Flight speed of the Common Loon (*Gavia immer*). Wilson Bull. 63:198.

PRESTON, F. W. 1953. The shapes of birds' eggs. Auk 70:160–182.

PRESTON, F. W. 1962a. The canonical distribution of commonness and rarity: part I. Ecology 43:185–215.

PRESTON, F. W. 1962b. The canonical distribution of commonness and rarity: part II. Ecology 43:410–432.

PRESTON, F. W. 1974. The volume of an egg. Auk 91:132–138.

PRICE, T. D. 1984a. The evolution of sexual size dimorphism in Darwin's finches. Am. Nat. 123:500–518.

PRICE, T. D. 1984b. Sexual selection on body size, territory and plumage variables in a population of Darwin's finches. Evolution 38:327–341.

PRICE, T. D., AND P. R. GRANT. 1984. Life history traits and natural selection for small body size in a population of Darwin's finches. Evolution 38:483–494.

PRICE, T. D., AND P. R. GRANT. 1985. The evolution of ontogeny in Darwin's finches: a quantitative genetic approach. Am. Nat. 125:169–188.

PRICE, T. D., P. R. GRANT, AND P. T. BOAG. 1984a. Genetic changes in the morphological differentiation of Darwin's ground finches. Pp. 49–66 *in* Population Biology and Evolution (K. Wöhrmann and V. Loeschcke, Eds.). Springer, Berlin, Germany.

PRICE, T. D., P. R. GRANT, H. L. GIBBS, AND P. T. BOAG. 1984b. Recurrent patterns of natural selection in a population of Darwin's finches. Nature 309:787–789.

PRICE, T. D., AND L. LIOU. 1989. Selection on clutch size in birds. Am. Nat. 134:950–959.

PRICE, T. D, M. TURELLI, AND M. SLATKIN. 1993. Peak shifts produced by correlated response to selection. Evolution 47:280–290.

PRIEDE, I. G. 1977. Natural selection for energetic efficiency and the relationship between activity level and mortality. Nature 267:610–611.

PROTHERO, J. 1986. Methodological aspects of scaling in biology. J. Theor. Biol. 118:259–286.

PROUT, T. 1964. Observations on structural reduction in evolution. Am. Nat. 98:239–249.

PROVINE, R. R. 1983. Development of wing-flapping and flight: a review. Bird Behav. 5:16–21.

PROVINE, R. R. 1984. Wing-flapping during development and evolution. Am. Sci. 72:448–455.

PROVINE, R. R. 1986. Behavioral neuroembryology: motor perspectives. Pp. 213–239 *in* Developmental Neuropsychobiology (W. T. Greenough and S. Suraska, Eds.). Academic Press, New York, New York.

PROVINE, W. B. 1989. Founder effects and genetic revolutions in microevolution and speciation: an historical perspective. Pp. 43–76 *in* Genetics, Speciation, and the Founder Principle (L. V. Giddings, K. Y. Kaneshiro, and W. W. Anderson, Eds.). Oxford University Press, Oxford, United Kingdom.

PUGESEK, B. H., AND A. TOMER. 1996. The Bumpus House Sparrow data: a reanalysis using structural equation models. Evol. Ecol. 10:387–404.

PULICH, W. M. 1961. A record of the Yellow Rail from Dallas County, Texas. Auk 78:639–670.

PULLIAINEN, E. 1972. Summer nutrition of crossbills (*Loxia pytyopsittacus, L. curvirostra* and *L. leucoptera*) in northeast Lapland. Ann. Zool. Fenn. 9:28–31.

PUTLAND, D. A., AND A. W. GOLDIZEN. 2001. Juvenile helping behaviour in the Dusky Moorhen, *Gallinula tenebrosa*. Emu 101:265–267.

PYCRAFT, W. P. 1899. Some facts concerning the so-called "aquintocubitalism" in the bird's wing. J. Linn. Soc. Lond. (Zool.) 27:236–256.

PYCRAFT, W. P. 1900. On the morphology and phylogeny of the Palæognathæ (Ratitæ and Crypturi) and Neognathæ (Carinatæ). Trans. Zool. Soc. Lond. 15:149–290.
PYCRAFT, W. P. 1910. A History of Birds. Methuen, London, England.
QUINN, J. F., AND A. E. DUNHAM. 1983. On hypothesis testing in ecology and evolution. Am. Nat. 122:602–617.
QUIRING, D. P. 1950. Functional Anatomy of the Vertebrates. McGraw-Hill, New York, New York.
RABOR, D. S. 1959. The impact of deforestation on birds of Cebu, Philippines, with new records for that island. Auk 76:37–43.
RADINSKY, L. B. 1985. Approaches in evolutionary morphology: a search for patterns. Annu. Rev. Ecol. Syst. 16:1–14.
RAFF, R. A. 1996. The Shape of Life: Genes, Development, and the Evolution of Animal Form. University of Chicago Press, Chicago, Illinois.
RAFF, R. A., B. A. PARR, A. L. PARKS, AND G. A. WRAY. 1990. Heterochrony and other mechanisms of radical evolutionary change in early development. Pp. 71–98 in Evolutionary Innovations (M. H. Nitecki, Ed.). University of Chicago Press, Chicago, Illinois.
RAFF, R. A., AND T. C. KAUFMAN. 1983. Embryos, Genes and Evolution: The Developmental Genetic Basis of Evolutionary Change. Macmillan, New York, New York.
RAFF, R. A., AND G. A. WRAY. 1989. Heterochrony: developmental mechanisms and evolutionary results. J. Evol. Biol. 2:409–434.
RAFF, R. A., G. A. WRAY, AND J. J. HENRY. 1991. Implications of radical evolutionary changes in early development for concepts of developmental constraint. Pp. 189–207 in New Perspectives on Evolution (L. Warren and H. Koprowksi, Eds.). Wiley-Liss, New York, New York.
RAFF, R. W., M. B. USHER, AND R. G. JEFFERSON. 1985. Birds on reserves: the influence of area and habitat on species richness. J. Appl. Ecol. 22:327–335.
RAHN, H. 1980. Comparison of embryonic development in birds and mammals: birth weight, time, and cost. Pp. 124–137 in A Companion to Animal Physiology (C. R. Taylor, K. Johansen, and L. Bolis, Eds.). Cambridge University Press, Cambridge, United Kingdom.
RAHN, H., AND A. AR. 1974. The avian egg: incubation time and water loss. Condor 76:147–152.
RAHN, H., AND C. V. PAGANELLI. 1989. Shell mass, thickness and density of avian eggs derived from the tables of Schönwetter. J. Ornithol. 130:59–68.
RAHN, H., C. V. PAGANELLI, AND A. AR. 1974. The avian egg: air-cell, gas tension, metabolism, and incubation time. Respir. Physiol. 22:297–309.
RAHN, H., C. V. PAGANELLI, AND A. AR. 1975. Relation of avian egg weight to body weight. Auk 92: 750–765.
RAIKOW, R. J. 1970. Evolution of diving adaptations in the stifftail ducks. Univ. Calif. Publ. Zool. 94:1–52.
RAIKOW, R. J. 1973. Locomotor mechanisms in North American ducks. Wilson Bull. 85:295–307.
RAIKOW, R. J. 1977. The origin and evolution of the Hawaiian honeycreepers (Drepanididae). Living Bird 15:95–117.
RAIKOW, R. J. 1985a. Locomotor system. Pp. 57–147 in Form and Function in Birds. Vol. 3 (A. S. King and J. McLellund, Eds.). Academic Press, London, United Kingdom.
RAIKOW, R. J. 1985b. Systematic and functional aspects of the locomotor system of the scurb-birds, Atrichornis, and lyrebirds, Menura (Passeriformes: Atrichornithidae and Menuridae). Rec. Aust. Mus. 37:211–228.
RAIKOW, R. J., L. BICANOVSKY, AND A. H. BLEDSOE. 1988. Forelimb joint mobility and the evolution of wing-propelled diving in birds. Auk 105:446–451.
RAKOTOMANANA, H. 1998. Negative relationship between relative tarsus and wing lengths in Malagasy rain forest birds. Jpn. J. Ornithol. 47:1–9.
RALLS, K. 1976. Extremes of sexual dimorphism in size in birds. Wilson Bull. 88:149–150.
RALPH, C. J., AND C. VAN RIPER III. 1985. Historical and current factors affecting Hawaiian native birds. Pp. 7–44 in Bird Conservation 2 (S. A. Temple, Ed.). University of Wisconsin Press, Madison, Wisconsin.
RAND, A. L. 1942. Results of the Archbold Expeditions. No. 42. Birds of the 1936–37 New Guinea Expedition. Am. Mus. Nat. Hist. Bull. 79:289–366.
RAND, A. L. 1954. On the spurs on birds' wings. Wilson Bull. 66:127–134.
RAND, A. L. 1955. The origin of the landbirds of Tristan da Cunha. Fieldiana (Zool.) 37:139–166.
RAND, A. L. 1961a. Wing length as an indicator of weight: a contribution. Bird-Banding 32:71–79.

RAND, A. L. 1961b. Some size gradients in North American birds. Wilson Bull. 73:46–56.

RAND, A. L., AND E. T. GILLIARD. 1967. Handbook of New Guinea Birds. Weidenfeld and Nicolson, London, United Kingdom.

RANDALL, B. M., R. M. RANDALL, AND L. J. V. COMPAGNO. 1988. Injuries to Jackass Penguins (*Spheniscus demersus*): evidence for shark involvement. J. Zool. Lond. 214:59–599.

RANDO, J. C., M. LOPEZ, AND B. SEGUÍ. 1999. A new species of extinct flightless passerine (Emberizidae: *Emberiza*) from the Canary Islands. Condor 101:1–13.

RASMUSSEN, J. F., C. RAHBEK, B. O. POULSEN, M. K. POULSEN, AND H. BLOCH. 1996. Distributional records and natural history notes on threatened and little known birds of southern Ecuador. Bull. Br. Ornithol. Club 116:26–46.

RASSMANN, K. 1997. Evolutionary age of the Galápagos iguanas predates the age of the present Galápagos Islands. Mol. Phylogenet. Evol. 7:158–172.

RAVELING, D. G. 1979. The annual cycle of body composition of Canada Geese with special reference to control of reproduction. Auk 96:234–252.

RAVELING, D. G., AND E. A. LEFEBVRE. 1967. Energy metabolism and theoretical flight range of birds. Bird-Banding 38:97–113.

RAWLES, M. E. 1960. The integumentary system. Pp. 189–240 *in* Biology and Comparative Physiology of Birds (A. J. Marshall, Ed.). Academic Press, New York, New York.

RAYNAUD, A. 1972. Morphogenese des membres rudimentaires chex les reptiles: un problème d'embryologie et d'évolution. Bull. Soc. Zool. Fr. 97:409–485.

RAYNAUD, A. 1985. Development of limbs and embryonic limb reduction. Pp. 59–148 *in* Biology of the Reptilia. Vol. 15 (C. Gans and F. Billett, Eds.). J. Wiley and Sons, New York, New York.

RAYNER, J. M. V. 1979a. A vortex theory of animal flight. Part 1. The vortex wake of a hovering animal. J. Fluid Mech. 91:697–730.

RAYNER, J. M. V. 1979b. A vortex theory of animal flight. Part 2. The forward flight of birds. J. Fluid Mech. 91:731–763.

RAYNER, J. M. V. 1979c. A new approach to animal flight mechanics. J. Exp. Biol. 80:17–54.

RAYNER, J. M. V. 1981. Flight adaptations in vertebrates. Symp. Zool. Soc. Lond. 48:137–172.

RAYNER, J. M. V. 1982. Avian flight energetics. Annu. Rev. Physiol. 44:109–119.

RAYNER, J. M. V. 1985a. Mechanical and ecological constraints on flight evolution. Pp. 279–288 *in* The Beginnings of Birds (M. K. Hecht, J. H. Ostrom, G. Viohl, and P. Wellnhofer, Eds.). Jura-Museum Eichstätt, Willibaldsburg, Germany.

RAYNER, J. M. V. 1985b. Bounding and undulating flight in birds. J. Theor. Biol. 117:47–77.

RAYNER, J. M. V. 1985c. Linear relations in biomechanics: the statistics of scaling functions. J. Zool. Lond. 206:415–439.

RAYNER, J. M. V. 1988a. Form and function in avian flight. Pp. 1–66 *in* Current Ornithology. Vol. 5 (R. F. Johnston, Ed.). Plenum Press, New York, New York.

RAYNER, J. M. V. 1988b. The evolution of vertebrate flight. Biol. J. Linn. Soc. 34:269–287.

RAYNER, J. M. V. 1990. The mechanics of flight and bird migration performance. Pp. 283–299 *in* Bird Migration: Physiology and Ecophysiology (E. Gwinner, Ed.). Springer-Verlag, Heidelberg, Germany.

RAYNER, J. M. V. 1993. On aerodynamics and the energetics of vertebrate flapping flight. Pp. 351–400 *in* Fluid Dynamics in Biology (A. Y Cheer and C. P. van Dam, Eds.). American Mathematical Society, Providence, Rhode Island.

RAYNER, J. M. V. 1995. Dynamics of vortex wakes of swimming and flying vertebrates. Symp. Soc. Exp. Biol. 69:131–155.

RAYNER, J. M. V. 1996. Biomechanical constraints on size in flying vertebrates. Symp. Zool. Soc. Lond. 69:83–110.

RECHER, H. F., AND S. S. CLARK. 1974. A biological survey of Lord Howe Island with recommendations for the conservation of the island's wildlife. Biol. Conserv. 6:263–273.

REDFERN, C. P. F. 1989. Rates of primary-feather growth in nestling birds. Ornis Scand. 20:59–64.

REECE, R. L., AND R. BUTLER. 1984. Some observations on the development of the long bones of ratite birds. Aust. Vet. J. 61:403–404.

REED, J. M., T. BOULINIER, E. DANCHIN, AND L. W. ORING. 1999. Informed dispersal: prospecting by birds for breeding sites. Pp. 189–259 *in* Current Ornithology. Vol. 15 (V. Nolan, Jr., Ed.). Plenum Press, New York, New York.

REED, T. M. 1983. The role of species–area relationships in reserve choice: a British example. Biol. Conserv. 25:263–271.
REED, T. M. 1984. The numbers of landbird species on the Isles of Scilly. Biol. J. Linn. Soc. 21:431–437.
REEVE, E. C. R., AND J. S. HUXLEY. 1945. Some problems in the study of allometric growth. Pp. 121–156 in Essays on Growth and Form Presented to D'Arcy Wentworth Thompson (W. E. LeGros Clark and P. B. Medawar, Eds.). Clarendon Press, Oxford, United Kingdom.
REEVE, H. K., AND P. W. SHERMAN. 1993. Adaptation and the goals of evolutionary research. Q. Rev. Biol. 68:1–32.
REGAL, P. J. 1977. Evolutionary loss of useless features: Is it molecular noise suppression? Am. Nat. 111:123–133.
REID, B. E. 1967. Some features of recent research on the Takahe (*Notornis mantelli*). Proc. N. Z. Ecol. Soc. 14:79–87.
REID, B. E. 1969. Survival status of the Takahe, *Notornis mantelli*, of New Zealand. Biol. Conserv. 1:237–240.
REID, B. E. 1971a. The weight of the kiwi and its egg. Notornis 18:245–249.
REID, B. E. 1971b. Composition of a kiwi egg. Notornis 18:250–252.
REID, B. E. 1972. North Island Brown Kiwi: *Apteryx australis mantelli*. Measurements and weights of a young chick. Notornis 19:261–266.
REID, B. E. 1974a. Sightings and records of the Takahe (*Notornis mantelli*) prior to its "official rediscovery" by Dr. G. B. Orbell in 1948. Notornis 21:277–295.
REID, B. E. 1974b. Faeces of Takahe (*Notornis mantelli*): a general discussion relating the quantity of faeces to the type of food and to the estimated energy requirements of the bird. Notornis 21:306–311.
REID, B. E. 1977. The energy value of the yolk reserve in a North Island Brown Kiwi chick (*Apteryx australis mantelli*). Notornis 24:194–195.
REID, B. E. 1978. The history of the Takahe (*Notornis mantelli*). A review of this species [sic] former distribution and its progressive decline. Pp. 14–50 in Proceedings of the Seminar on the Takahe and its Habitat: 1978 (J. P. Harty, Chairman). Fiordland Nature Board, Invercargill, New Zealand.
REID, B. E., AND D. J. STACK. 1974. An assessment of the number of Takahe in the 'special area' of the Murchison Mountains during the years 1963–1967. Notornis 21:296–305.
REID, B. E., AND G. R. WILLIAMS. 1975. The kiwi. Pp. 301–330 in Biogeography and Ecology in New Zealand (G. Kuschel, Ed.). Dr. W. Junk, The Hague, The Netherlands.
REILLY, S. M. 1997. An integrative approach to heterochrony: the distinction between interspecific and intraspecific phenomena. Biol. J. Linn. Soc. 60:119–143.
REINERTSEN, R. E. 1983. Nocturnal hypothermia and its energetic significance for small birds in the arctic and subarctic regions. A review. Polar Res. 1:269–284.
REINHART, R. H. 1959. A review of the Sirenia and Desmostylia. Univ. Calif. Publ. Geol. Sci. 36:1–146.
REISS, M. J. 1985. The allometry of reproduction: why larger species invest relatively less in their offspring. J. Theor. Biol. 113:529–544.
REISS, M. J. 1986. Sexual dimorphism in body size: Are larger species more dimorphic? J. Theor. Biol. 121:163–172.
REISS, M. J. 1987. The intraspecific relationship of parental investment to female body weight. Funct. Ecol. 1:105–107.
REISS, M. J. 1989. The Allometry of Growth and Reproduction. Cambridge University Press, Cambridge, United Kingdom.
REIST, J. D. 1985. An empirical evaluation of several univariate methods that adjust for size variation in morphometric data. Can. J. Zool. 63:1429–1439.
REIST, J. D. 1986. An empirical evaluation of coefficients used in residual and allometric adjustment of size covariation. Can. J. Zool. 64:1363–1368.
REMSEN, J. V., JR., AND T. A. PARKER III. 1990. Seasonal distribution of the Azure Gallinule (*Porphyrula flavirostris*), with comments of vagrancy in rails and gallinules. Wilson Bull. 102:380–399.
RENSCH, B. 1938. Einwirkung des Klimas bei der Ausprägung von Vögelrassen, mit besonderer Berücksichtigung der Flügelform und der Eizahl. Pp. 285–311 in Proceedings of the Eighth

International Ornithological Congress (F. C. R. Jourdain, Ed.). University Press, Oxford, United Kingdom.

RENSCH, B. 1948. Histological changes correlated with evolutionary changes of body size. Evolution 2:218–230.

RENSCH, B. 1950. Die Abhängigkeit der relativen Sexualdifferenze von der Körpergrösse. Bonn. Zool. Beitr. 1:58–69.

RENSCH, B. 1959. Evolution Above the Species Level. Columbia University Press, New York, New York.

RENSCH, B. 1960. The laws of evolution. Pp. 95–116 in Evolution After Darwin. Vol. 1. The Evolution of Life: Its Origin, History and Future (S. Tax, Ed.). University of Chicago Press, Chicago, Illinois.

REY, J. R., D. R. STRONG, JR., AND E. D. MCCOY. 1982. On overinterpretation of the species–area relationship. Am. Nat. 119:741–743.

REYMENT, R. 1962. Observations on homogeneity of covariance matrices in paleontologic biometry. Biometrics 18:1–11.

REYMENT, R. A. 1980. On the interpretation of the smallest principal component. Bull. Geol. Inst. Univ. Uppsala (New Ser.) 8:1–4.

REYMENT, R. A. 1983. Paleontological aspects of island biogeography: colonization and evolution of mammals on Mediterranean islands. Oikos 41:299–306.

REYMENT, R. A. 1991. Multidimensional Palaeobiology. Pergamon Press, Oxford, United Kingdom.

REYMENT, R. A., R. E. BLACKITH, AND N. A. CAMPBELL. 1984. Multivariate Morphometrics. 2nd ed. Academic Press, New York, New York.

REYMENT, R., AND K. G. JÖRESKOG. 1993. Applied Factor Analysis in the Natural Sciences. Cambridge University Press, Cambridge, United Kingdom.

REYNOLDS, J. D., AND T. SZÉKELY. 1997. The evolution of parental care in shorebirds: life histories, ecology, and sexual selection. Behav. Ecol. 8:126–134.

REYNOLDS, P. S., AND R. N. LEE III. 1996. Phylogenetic analysis of avian energetics: passerines and nonpasserines do not differ. Am. Nat. 147:735–759.

REYNOLDS, W. W. 1977. Skeleton weight allometry in aquatic and terrestrial vertebrates. Hydobiologia 56:35–37.

REZNICK, D. 1985. Costs of reproduction: an evaluation of the empirical evidence. Oikos 44:257–267.

RHYMER, J. M., AND D. SIMBERLOFF. 1996. Extinction by hybridization and introgression. Annu. Rev. Ecol. Syst. 27:83–109.

RICE, S. H. 1997. The analysis of ontogenetic trajectories: when a change in size or shape is not heterochrony. Proc. Natl. Acad. Sci. USA 94:907–912.

RICE, S. H. 2000. The evolution of developmental interactions: epistatsis, canalization, and integration. Pp. 82–98 in Epistasis and the Evolutionary Process (J. B. Wolf, E. D. Brodie III, and M. J. Wade, Eds.). Oxford University Press, Oxford, United Kingdom.

RICE, W. R. 1989. Analyzing tables of statistical tests. Evolution 43:223–225.

RICH, P. V. 1979. The Dromornithidae, An Extinct Family of Large Ground Birds Endemic to Australia. Bull. 184. Bureau of Natural Resources, Geology and Geophysics, Canberra, Australia.

RICH, P. V. 1981. Kukwiede's revenge: a view into New Caledonia's distant past. Hemisphere 26: 166–171.

RICH, P. V., A. R. MCEVEY, AND R. F. BAIRD. 1985. Osteological comparison of the scrub-birds, Atrichornis, and lyrebirds, Menura (Passeriformes: Atrichornithidae and Menuridae). Rec. Aust. Mus. 37:165–191.

RICH, P. V., AND R. J. SCARLETT. 1977. Another look at Megaegotheles, a large owlet-nightjar from New Zealand. Emu 77:1–8.

RICH, P. V., G. VAN TETS, K. ORTH, C. MEREDITH, AND P. DAVIDSON. 1983. Prehistory of the Norfolk Island biota. Pp. 7–29 in A Review of Norfolk Island Birds: Past and Present (R. Schodde, P. Fullager, and N. Hermes, Eds.). Aust. Natl. Parks Wildl. Serv. Spec. Publ. 8. Australian National Parks and Wildlife, Canberra, New South Wales, Australia.

RICHARDS, L. P., AND W. J. BOCK. 1973. Functional Anatomy and Adaptive Evolution of the Feeding Apparatus in the Hawaiian Honeycreeper Genus Loxops (Drepanididae). Ornithol. Monogr. 15. American Ornithologists' Union, Washington, D.C.

RICHARDS, O. W., AND A. J. KAVANAGH. 1945. The analysis of growing form. Pp. 188–230 in Essays on Growth and Form Presented to D'Arcy Wentworth Thompson (W. E. LeGros Clark and P. B. Medawar, Eds.). Clarendon Press, Oxford, United Kingdom.

RICHARDSON, M. E. 1984. Aspects of the ornithology of the Tristan da Cunha group and Gough Island. Cormorant 12:123–201.

RICHTER, W. 1982. Hatching asynchrony: the nest failure hypothesis and brood reduction. Am. Nat. 120:828–832.

RICHTER-DYN, N., AND N. S. GOEL. 1972. On the extinction of a colonizing species. Theor. Popul. Biol. 3:406–433.

RICKER, W. E. 1984. Computation and uses of central trend lines. Can. J. Zool. 62:1897–1905.

RICKLEFS, R. E. 1965. Brood reduction in the curve-billed Thrasher. Condor 67:505–510.

RICKLEFS, R. E. 1968. Patterns of growth in birds. Ibis 110:419–451.

RICKLEFS, R. E. 1969. Preliminary models for growth rates in altricial birds. Ecology 50:1031–1039.

RICKLEFS, R. E. 1970. Stage of taxon cycle and distribution of birds on Jamaica, Greater Antilles. Evolution 24:475–477.

RICKLEFS, R. E. 1972a. Taxon cycles in the West Indian avifauna. Am. Nat. 106:195–219.

RICKLEFS, R. E. 1972b. Stage of taxon cycle and distribution of birds on Jamaica, Greater Antilles. Evolution 24:475–477.

RICKLEFS, R. E. 1973. Patterns of growth in birds. II. Growth rate and mode of development. Ibis 115:177–201.

RICKLEFS, R. E. 1974. Energetics of reproduction in birds. Pp. 152–297 in Avian Energetics (R. A. Paynter, Ed.). Publ. No. 15. Nuttall Ornithologists' Club, Cambridge, Massachusetts.

RICKLEFS, R. E. 1979. Adaptation, constraint, and compromise in avian postnatal development. Biol. Rev. 54:269–290.

RICKLEFS, R. E. 1980. Geographical variation in clutch size among passerine birds: Ashmole's hypothesis. Auk 97:38–49.

RICKLEFS. R. E. 1982. Some considerations of sibling competition and avian growth rates. Auk 99: 141–147.

RICKLEFS, R. E. 1983a. Avian postnatal development. Pp. 1–83 in Avian Biology. Vol. 7 (D. S. Farner, J. R. King, and K. C. Parkes, Eds.). Academic Press, New York, New York.

RICKLEFS, R. E. 1983b. Comparative avian demography. Pp. 1–32 in Current Ornithology. Vol. 1 (R. F. Johnston, Ed.). Plenum Press, New York, New York.

RICKLEFS, R. E. 1984. The optimization of growth rate in altricial birds. Ecology 65:1602–1616.

RICKLEFS, R. E. 1993. Sibling competition, hatching asynchrony, incubation period and life span in altricial birds. Pp. 199–275 in Current Ornithology. Vol. 11 (D. M. Power, Ed.). Plenum Press, New York, New York.

RICKLEFS, R. E. 1996. Avian energetics, ecology, and evolution. Pp. 1–30 in Avian Energetics and Nutritional Ecology (C. Carey, Ed.). Chapman and Hall, New York, New York.

RICKLEFS, R. E. 1998. The concept of symmorphosis applied to growing birds. Pp. 56–62 in Principles of Animal Design: The Optimization and Symmorphosis Debate (E. R. Weibel, C. R. Taylor, and L. Bolis, Eds.). Cambridge University Press, Cambridge, United Kingdom.

RICKLEFS, R. E. 2000. Density dependence, evolutionary optimization, and the diversification of avian life histories. Condor 102:9–22.

RICKLEFS, R. E., AND E. BERMINGHAM. 1999. Taxon cycles in the Lesser Antillean avifauna. Ostrich 70:49–59.

RICKLEFS, R. E., AND G. W. COX. 1972. Taxon cycles in the West Indian avifauna. Am. Nat. 106: 195–219.

RICKLEFS, R. E., AND G. W. COX. 1978. Stage of taxon cycle, habitat distribution, and population density in the avifauna of the West Indies. Am. Nat. 112:875–895.

RICKLEFS, R. E., M. KONARZEWSKI, AND S. DAAN. 1996. The relationship between basal metabolic rate and daily energy expenditure in birds and mammals. Am. Nat. 147:1047–1071.

RICKLEFS, R. E., R. E. SHEA, AND I.-H. CHOI. 1994. Inverse relationship between functional maturity and exponential growth rate of avian skeletal muscle: a constraint on evolutionary response. Evolution 48:1080–1088.

RICKLEFS, R. E., AND J. M. STARCK. 1998a. Embryonic growth and development. Pp. 31–58 in Avian Growth and Development: Evolution Within the Altricial–Precocial Spectrum (J. M. Starck and R. E. Ricklefs, Eds.). Oxford University Press, Oxford, United Kingdom.

RICKLEFS, R. E., AND J. M. STARCK. 1998b. The evolution of the developmental mode in birds. Pp.

366–380 *in* Avian Growth and Development: Evolution Within the Altricial–Precocial Spectrum (J. M. Starck and R. E. Ricklefs, Eds.). Oxford University Press, Oxford, United Kingdom.

RICKLEFS, R. E., J. M. STARCK, AND M. KONARZEWSKI. 1998. Internal constraints on growth in birds. Pp. 266–287 *in* Avian Growth and Development: Evolution Within the Altricial–Precocial Spectrum (J. M. Starck and R. E. Ricklefs, Eds.). Oxford University Press, Oxford, United Kingdom.

RICKLEFS, R. E., AND T. WEBB. 1985. Water content, thermogenesis, and growth rate of skeletal muscles in the European Starling. Auk 102:369–376.

RIDGWAY, R., AND H. FRIEDMANN. 1941. The birds of North America. U.S. Natl. Mus. Bull. 50:1–254.

RIDLEY, M. 1978. Paternal care. Anim. Behav. 26:904–932.

RIDLEY, M. 1983. The Explanation of Organic Diversity: The Comparative Method and Adaptations for Mating. Oxford University Press, Oxford, United Kingdom.

RIDPATH, M. G. 1964. The Tasmanian Native Hen. Aust. Nat. Hist. 14:346–350.

RIDPATH, M. G. 1972a. The Tasmanian Native Hen, *Tribonyx mortierii*. I. Patterns of behaviour. Commonw. Sci. Ind. Res. Org. Wildl. Res. 17:1–52.

RIDPATH, M. G. 1972b. The Tasmanian Native Hen, *Tribonyx mortierii*. II. The individual, the group, and the population. Commonw. Sci. Ind. Res. Org. Wildl. Res. 17:53–90.

RIDPATH, M. G. 1972c. The Tasmanian Native Hen, *Tribonyx mortierii*. III. Ecology. Commonw. Sci. Ind. Res. Org. Wildl. Res. 17:91–117.

RIDPATH, M. G., AND G. K. MELDRUM. 1968a. Damage to pasture by the Tasmanian Native Hen *Tribonyx mortierii*. Commonw. Sci. Ind. Res. Org. Wildl. Res. 13:11–24.

RIDPATH, M. G., AND G. K. MELDRUM. 1968b. Damage to oat crops by the Tasmanian Native Hen *Tribonyx mortierii*. Commonw. Sci. Ind. Res. Org. Wildl. Res. 13:25–54.

RIDPATH, M. G., AND R. E. MOREAU. 1966. The birds of Tasmania: ecology and evolution. Ibis 108:348–393.

RIEBESELL, J. F. 1982. Arctic–alpine plants on mountaintops: agreement with island biogeography theory. Am. Nat. 119:657–674.

RIEDL, R. 1978. Order in Living Organisms: A Systems Analysis of Evolution. J. Wiley and Sons, New York, New York.

RIEDMAN, M. L. 1982. The evolution of alloparental care and adoption in mammals and birds. Q. Rev. Biol. 57:405–435.

RIEPPEL, O. C. 1988. Fundamentals of Comparative Biology. Birkhäuser Verlag, Basel, Germany.

RIEPPEL, O. 1996. Miniturization in tetrapods: consequences for skull morphology. Symp. Zool. Soc. Lond. 69:47–62.

RIJKE, A. M. 1967. The water repellency and feather structure of cormorants, Phalacrocoracidae. Ostrich 38:163–165.

RIJKE, A. M. 1968. The water repellency and feather structure of cormorants, Phalacrocoracidae. J. Exp. Biol. 48:185–189.

RIJKE, A. M. 1970a. Wettability and phylogenetic development of water repellency in water bird feathers. J. Exp. Biol. 52:469–479.

RIJKE, A. M. 1970b. The phylogenetic development of water repellency in water bird feathers. Ostrich 8 (Suppl.):67–76.

RIPLEY, S. D. 1954. Birds of Gough Island. Postilla 19:1–6.

RIPLEY, S. D. 1957a. Notes on the Horned Coot, *Fulica cornuta* Bonaparte. Postilla 30:1–8.

RIPLEY, S. D. 1957b. Additional notes on the Horned Coot. Postilla 32:1–2.

RIPLEY, S. D. 1964. A systematic and ecological study of birds of New Guinea. Peabody Mus. Nat. Hist. Bull. 19:1–85.

RIPLEY, S. D. 1977. Rails of the World. David R. Goodine, Boston, Massachusetts.

RIPLEY, S. D., AND B. M. BEEHLER. 1985. Rails of the world, a compilation of new information, 1975–1983 (Aves: Rallidae). Smithson. Contrib. Zool. 417:1–28.

RISING, J. D. 1987. Geographic variation of sexual dimorphism in size of Savannah Sparrows (*Passerculus sandwichensis*): a test of hypotheses. Evolution 41:514–524.

RISING, J. D. 1988. Geographic variation in sex ratios and body size of wintering flocks of Savannah Sparrows (*Passerculus sandwichensis*). Wilson Bull. 100:183–203.

RISING, J. D. 2001. Geographic variation in size and shape of Savannah Sparrows (*Passerculus sandwichensis*). Stud. Avian Biol. 23:1–65.

RISING, J. D., AND J. C. AVISE. 1993. Application of the genealogical-concordance principles to the

taxonomy and evolutionary history of the Sharp-tailed Sparrow (*Ammodramus caudacutus*). Auk 110:844–856.
RISING, J. D., AND K. M. SOMERS. 1989. The measurement of overall body size in birds. Auk 106: 666–674.
RISKA, B. 1986. Some models for development, growth, and morphometric correlation. Evolution 40: 1303–1311.
RISKA, B. 1989. Composite traits, selection response, and evolution. Evolution 43:1172–1191.
RISKA, B. 1991. Regression models in evolutionary allometry. Am. Nat. 138:283–299.
RITTER, M. W., AND J. A. SAVIDGE. 1999. A predictive model of wetland habitat use on Guam by endangered Mariana Common Moorhens. Condor 101:282–287.
RITTER, M. W., AND T. M. SWEET. 1993. Rapid colonization of a human-made wetland by Mariana Common Moorhen on Guam. Wilson Bull. 105:685–687.
ROBBINS, C. T. 1981. Estimation of the relative protein cost of reproduction in birds. Condor 83:177–179.
ROBERT, K. A., AND M. B. THOMPSON. 2001. Viviparous lizard selects sex of embryos. Nature 412: 698–699.
ROBERT, M., AND P. LAPORTE. 1997. Field techniques for studying breeding Yellow Rails. J. Field Ornithol. 68:56–63.
ROBERTSON, D. B. 1976. Weka liberation in Northland. Notornis 23:213–219.
ROBERTSON, H. A. 1988. Relationships between body weight, egg weight, and clutch size in pigeons and doves (Aves: Columbiformes). J. Zool. Lond. 215:217–229.
ROBERTSON, H. A., AND A. J. BEAUCHAMP. 1985. Weka. Pp. 168–169 *in* The Reader's Digest Complete Book of New Zealand Birds (C. J. R. Robertson, Ed.). Reader's Digest, Sidney, Australia.
ROBINSON, A. C. 1995. Breeding pattern in the Banded Rail (*Gallirallus philippensis*) in Western Samoa. Notornis 42:46–48.
ROCKWELL, R. F., AND F. COOKE. 1977. Gene flow and local adaptation in a colonially nesting dimorphic bird: the Lesser Snow Goose (*Anser caerulescens caerulescens*). Am. Nat. 111:91–97.
ROCKWELL, R. F., C. S. FINDLAY, F. COOKE, AND J. A. SMITH. 1985. Life history studies of the Lesser Snow Goose (*Anser caerulescens caerulescens*). IV. The selective value of plumage polymorphism: net viability, the timing of maturation, and breeding propensity. Evolution 39:178–189.
RODDA, P. 1994. Geology of Fiji. Pp. 131–151 *in* Geology and Submarine Resources of the Tonga–Lau–Fiji Region (A. J. Steveson, R. H. Herzer, and P. F. Balance, Eds.). SOPAC Tech. Bull. 8. SOPAC Secretariat, Suva, Fiji.
ROE, N. A., AND W. E. REES. 1979. Notes on the Puna avifauna of Azángaro Province, Department of Puna, southern Peru. Auk 96:475–482.
ROFF, D. A. 1974. The analysis of a population model demonstrating the importance of dispersal in a heterogeneous environment. Oecologia 15:259–275.
ROFF, D. A. 1975. Population stability and the evolution of dispersal in a heterogeneous environment. Oecologia 19:217–237.
ROFF, D. A. 1981. On being the right size. Am. Nat. 118:405–422.
ROFF, D. A. 1984. The cost of being able to fly: a study of wing polymorphism in two species of crickets. Oecologia 63:30–37.
ROFF, D. A. 1986a. The evolution of wing dimorphism in insects. Evolution 40:1009–1020.
ROFF, D. A. 1986b. The genetic basis of wing dimorphism in the sand cricket, *Gryllus firmus*, and its relevance to the evolution of wing dimorphisms in insects. Heredity 57:221–231.
ROFF, D. A. 1986c. The evolution of wing polymorphisms and its impact on life cycle adaptation in insects. Pp. 209–221 *in* The Evolution of Insect Life Cycles (F. Taylor and R. Karban, Eds.). Springer-Verlag, New York, New York.
ROFF, D. A. 1988. The evolution of migration and some life history parameters in marine fishes. Environ. Biol. Fishes 22:133–146.
ROFF, D. A. 1990a. The evolution of flightlessness in insects. Ecol. Monogr. 60:389–421.
ROFF, D. A. 1990b. Selection for changes in the incidence of wing dimorphism in *Gryllus firmus*. Heredity 65:163–168.
ROFF, D. A. 1991. Life history consequences of bioenergetic and biomechanical constraints on migration. Am. Zool. 31:205–215.
ROFF, D. A. 1992. The Evolution of Life Histories: Theory and Analysis. Chapman and Hall, New York, New York.

ROFF, D. A. 1994a. The evolution of flightlessness: Is history important? Evol. Ecol. 8:639–657.
ROFF, D. A. 1994b. Habitat persistence and the evolution of wing dimorphism in insects. Am. Nat. 144:772–798.
ROFF, D. A. 1994c. Evidence that the magnitude of the trade-off in a dichotomous trait is frequency-dependent. Evolution 48:1650–1656.
ROFF, D. A. 1994d. The evolution of dimorphic traits: effect of directional selection on heritability. Heredity 72:36–34.
ROFF, D. A. 1995. Antagonistic and reinforcing pleiotropy: a study of differences in development time in wing-dimorphic insects. J. Evol. Biol. 8:405–419.
ROFF, D. A. 1996. The evolution of threshold traits in animals. Q. Rev. Biol. 71:3–35.
ROFF, D. A. 2000. The evolution of the **G** matrix: Selection or drift? Heredity 84:135–142.
ROFF, D. A., AND D. J. FAIRBAIRN. 1991. Wing dimorphisms and the evolution of migratory polymorphisms among Insecta. Am. Zool. 31:243–251.
ROFF, D. A., AND D. J. FAIRBAIRN. 1993. The evolution of alternative morphologies: fitness and the wing morphology in male sand crickets. Evolution 47:1572–1584.
ROFF, D. A., AND P. SHANNON. 1993. Genetic and ontogenetic variation in behavior: its possible role in the maintenance of genetic variation in the wing dimorphism of *Gryllus firmus*. Heredity 71: 481–487.
ROFF, D. A., G. STIRLING, AND D. J. FAIRBAIRN. 1997. The evolution of threshold traits: a quantitative genetic analysis of the physiological and life-history correlates of wing dimorphism in the sand cricket. Evolution 51:1910–1919.
ROFSTAD, G. 1986. Growth and morphology of nestling Hooded Crows *Corvus corone cornix*, a sexually dimorphic bird species. J. Zool. Lond. (Ser. A) 208:299–323.
ROGERS, C. M., V. J. NOLAN, AND E. D. KETTERSON. 1994. Winter fattening in the Dark-eyed Junco: plasticity and possible interaction with migration trade-offs. Oecologia 97:526–532.
ROHLF, F. J. 1990. Fitting curves to outlines. Pp. 167–177 *in* Proceedings of the Michigan Morphometrics Workshop (F. J. Rohlf and F. L. Bookstein, Eds.). Univ. Mich. Mus. Zool. Spec. Publ. No. 2. University of Michigan, Ann Arbor, Michigan.
ROHLF, F. J. 2001. Comparative methods for the analysis of continuous variables: geometric interpretations. Evolution 55:2143–2160.
ROHLF, F. J., AND J. W. ARCHIE. 1984. A comparison of Fourier methods for the description of wing shape in mosquitoes (Diptera: Culicidae). Syst. Zool. 33:302–317.
ROHLF, F. J., AND F. L. BOOKSTEIN. 1987. A comment on shearing as a method for "size correction." Syst. Zool. 36:356–367.
ROHLF, F. J., A. J. GILMARTIN, AND G. HART. 1983. The Kluge–Kerfoot phenomenon—a statistical artifact. Evolution 37:180–202.
ROHLF, F. J., AND D. SLICE. 1990. Extensions of the Procrustes method for the optimal superimposition of landmarks. Syst. Zool. 39:40–59.
ROHWER, F. C. 1988. Inter- and intraspecific relationships between egg size and clutch size in waterfowl. Auk 105:161–176.
ROHWER, F. C. 1989. Egg mass and clutch size relationships in geese, eiders, and swans. Ornis Scand. 20:43–48.
ROHWER, F. C. AND D. I. EISENHAUER. 1989. Egg mass and clutch size relationships in geese, eiders, and swans. Ornis. Scand. 20:43–48.
ROHWER, F. C., AND S. FREEMAN. 1989. The distribution of conspecific nest parasitism in birds. Can. J. Zool. 67:239–253.
ROHWER, S. 1978. Passerine subadult plumages and the deceptive acquisition of resources: test of a critical assumption. Condor 80:173–179.
ROHWER, S. 1982. The evolution of reliable and unreliable badges of fighting ability. Am. Zool. 22: 531–546.
ROHWER, S. 1983a. Testing the female mimicry hypothesis of delayed plumage maturation: a comment on Procter-Gray and Holmes. Evolution 37:421–423.
ROHWER, S. 1983b. Formalizing the avoidance-image hypothesis: critique of an earlier prediction. Auk 100:971–974.
ROHWER, S. 1986. A previously unknown plumage of first-year Indigo Buntings and theories of delayed plumage maturation. Auk 103:281–292.

ROHWER, S. 1990. Foraging differences between white and dark morphos of the Pacific Reef Heron *Egretta sacra*. Ibis 132:21–36.
ROHWER, S., AND G. S. BUTCHER. 1988. Winter versus summer explanations of delayed plumage maturation in temperate passerine birds. Am. Nat. 131:556–572.
ROHWER, S., AND P. W. EWALD. 1981. The cost of dominance and advantage of subordination in a badge signaling system. Evolution 35:441–454.
ROHWER, S., S. D. FRETWELL, AND D. M. NILES. 1980. Delayed maturation in passerine plumages and the deceptive acquisition of resources. Am. Nat. 115:400–437.
ROHWER, S., AND D. R. PAULSON. 1987. The avoidance-image hypothesis and color polymorphism in *Buteo* hawks. Ornis Scand. 18:285–290.
ROLLO, C. D. 1994. Phenotypes: Their Epigenetics, Ecology and Evolution. Chapman and Hall, London, United Kingdom.
ROMANOFF, A. L. 1960. The Avian Embryo: Structural and Functional Development. Macmillan, New York, New York.
ROMER, A. S. 1968. Notes and Comments on Vertebrate Paleontology. University of Chicago Press, Chicago, Illinois.
ROMERO, A. 1985. Can evolution regress? Natl. Speleol. Soc. Bull. 47:86–88.
ROOT, T. 1988a. Environmental factors associated with avian distributional boundaries. J. Biogeogr. 15:489–505.
ROOT, T. 1988b. Energy constraints on avian distributions and abundances. Ecology 69:330–339.
ROSENZWEIG, M. L., AND R. D. MCCORD. 1991. Incumbent replacement: evidence for long-term evolutionary progress. Paleobiology 17:202–213.
ROSS, C. A. 1988. Weights of some New Caledonian birds. Bull. Br. Ornithol. Club 108:91–93.
ROSS, H. A. 1979. Multiple clutches and shorebird egg and body weight. Am. Nat. 111:917–938.
ROSS, H. A., AND A. J. BAKER. 1982. Variation in the size and shape of introduced Starlings, *Sturnus vulgaris* (Aves: Struninae), in New Zealand. Can. J. Zool. 60:3316–3325.
ROSSER, B. W. C. 1980. The wing musculature of the American Coot (*Fulica americana* Gmelin). Can. J. Zool. 58:1758–1773.
ROSSER, B. W. C., AND J. C. GEORGE. 1985. Effect of flightlessness during moult on the iron content in the pectoralis muscle of the Giant Canada Goose (*Branta canadensis maxima*). Can. J. Zool. 63:480–483.
ROSSER, B. W. C., AND J. C. GEORGE. 1986. The avian pectoralis: histochemical characterization and distribution of muscle fiber types. Can. J. Zool. 64:1174–1185.
ROSSER, B. W. C., AND J. C. GEORGE. 1987. Ultrastructural changes in the muscle fibers of the pectoralis of the Giant Canada Goose (*Branta canadensis maxima*) in disuse atrophy during molt. Cell Tissue Res. 247:689–696.
ROSSER, B. W. C., D. M. SECOY, AND P. W. RIEGERT. 1982. The leg muscles of the American Coot (*Fulica americana* Gmelin). Can. J. Zool. 60:1236–1256.
ROSSITER, M. C. 1996. Incidence and consequences of inherited environmental effects. Annu. Rev. Ecol. Syst. 27:451–476.
ROTH, V. L. 1981. Constancy in the size ratios of sympatric species. Am. Nat. 118:394–404.
ROTH, V. L. 1990. Insular dwarf elephants: a case study in body mass estimation and ecological inference. Pp. 151–179 *in* Body Size in Mammalian Paleobiology: Estimation and Biological Implications (J. Damuth and B. J. McFadden, Eds.). Cambridge University Press, Cambridge, United Kingdom.
ROTH, V. L. 1992. Inferences from allometry and fossils: dwarfing of elephants on islands. Pp. 259–288 *in* Oxford Surveys in Evolutionary Biology. Vol. 8 (D. Futuyma and J. Antonovics, Eds.). Oxford University Press, Oxford, United Kingdom.
ROTHSCHILD, W. 1893a. The Avifauna of Laysan and Neighboring Islands, Part 1. R. H. Porter, London, England.
ROTHSCHILD, W. 1893b. On a new rail from the Auckland Islands. Trans. N. Z. Inst. 26:180.
ROTHSCHILD, W. 1900. The Avifauna of Laysan and Neighboring Islands, Part 3. R. H. Porter, London, England.
ROTHSCHILD, W. 1903. [Untitled description of *Gallirallus wakensis*.] Bull. Br. Ornithol. Club 13:78.
ROTHSCHILD, W. 1907a. Extinct birds. Hutchinson, London, England.
ROTHSCHILD, W. 1907b. On extinct and vanishing birds. Pp. 191–217 *in* Proceedings of the Fourth

International Ornithological Congress, London, June 1905 (R. B. Sharpe, Ed.). Dulau, London, England.

ROTHSCHILD, W. 1928. [The eggs of *Atlantisia rogersi*.] Bull. Br. Ornithol. Club 48:121–124.

ROUGHGARDEN, J., AND S. PACALA. 1989. Taxon cycle among *Anolis* lizard populations: review of evidence. Pp. 403–432 *in* Speciation and Its Consequences (D. Otte and J. A. Endler, Eds.). Sinauer, Sunderland, Massachusetts.

ROUHANI, S., N. H. BARTON. 1987. The probability of peak shifts in a founder population. J. Theor. Biol. 126:51–62.

ROUSE, I. 1986. Migrations in Prehistory. Yale University Press, New Haven, Connecticut.

ROWLAND, B. 1978. Birds With Human Souls: A Guide To Bird Symbolism. University of Tennessee Press, Knoxville, Tennessee.

ROWLANDS, B. W., T. TRUEMAN, S. L. OLSON, M. N. MCCULLOCH, AND R. K. BROOKE. 1998. The Birds of St. Helena: An Annotated Checklist. British Ornithologists' Union, Natural History Museum, Tring, United Kingdom.

ROWLEY, I., E. RUSSELL, AND M. BROOKER. 1993. Inbreeding in birds. Pp. 304–328 *in* The Natural History of Inbreeding and Outbreeding: Theoretical and Empirical Perspectives (N. W. Thornhill, Ed.). University of Chicago Press, Chicago, Illinois.

ROYAMA, T. 1969. A model for global variation in clutch size in birds. Oikos 20:562–567.

RUBIN, C. T., AND L. E. LANYON. 1984. Dynamic strain similarity in vertebrates; an alternative to allometric limb bone scaling. J. Theor. Biol. 107:321–327.

RUELLE, D. 1989. Chaotic Evolution and Strange Attractors. Cambridge University Press, Cambridge, United Kingdom.

RUNDLE, W. D., AND M. W. SAYRE. 1983. Feeding ecology of migrant Soras in southeastern Missouri. J. Wildl. Manage. 47:1153–1159.

RÜPPELL, G. 1977. Bird Flight. Van Nostrand Reinhold, New York, New York.

RUSE, M. 1992. A threefold parallelism for our time? Progressive development in society, science, and the organic world. Pp. 149–178 *in* History and Evolution (M. W. Nitecki and D. V. Nitecki, Eds.). State University of New York Press, Albany, New York.

RUSE, M. 1996. Monad to Man: The Concept of Progress in Evolutionary Biology. Harvard University Press, Cambridge, Massachusetts.

RUSSELL, E. M. 2000. Avian life histories: Is extended parental care the southern secret? Emu 100:377–409.

RUSSELL, S. M. 1964. A distributional study of the birds of British Honduras. Ornithol. Monogr. No. 1. American Ornithologists' Union, Washington, D.C.

RUTSCHKE, E. 1960. Untersuchungen über Wasserfestigkeit und Struktur des Gefieders von Schwimmvögeln. Zool. Jahrb. Abt. Syst. 87:441–506.

RYAN, C. 1997. Observations on the breeding behaviour of the Takahe (*Porphyrio mantelli*) on Mana Island. Notornis 44:233–240.

RYAN, M. R., AND J. J. DINSMORE. 1980. The behavioral ecology of breeding American Coots in relation to age. Condor 82:320–327.

RYAN, P. G., C. L. MOLONEY, AND J. HUDON. 1994. Color variation and hybridization among *Neospiza* buntings on Inaccessible Island, Tristan da Cunha. Auk 111:314–327.

RYAN, P. G., B. P. WATKINS, AND W. R. SIEGFRIED. 1989. Morphometrics, metabolic rate and body temperature of the smallest flightless bird: the Inaccessible Island Rail. Condor 91:465–467.

RYAN, T. J., AND R. D. SEMLITSCH. 1998. Intraspecific heterochrony and life history evolution: decoupling somatic and sexual development in a facultatively paedomorphic salamander. Proc. Natl. Acad. Sci. USA 95:5643–5648.

SADLER, J. P. 1999. Biodiversity on oceanic islands: a palaeoecological assessment. J. Biogeogr. 26:75–87.

SAETHER, B.-E. 1989. Survival in relation to body weight in European birds. Ornis Scand. 20:13–21.

SAFINA, C. 1984. Selection for reduced male size in raptorial birds: the possible roles of female choice and mate guarding. Oikos 43:159–164.

SALMON, J. T. 1975. The influence of man on the biota. Pp. 643–661 *in* Biogeography and Ecology in New Zealand (G. Kuschel, Ed.). Dr. W. Junk, The Hague, The Netherlands.

SALOMONSEN, F. 1955. The evolutionary significance of bird-migration. Dan. Biol. Medd. 22:1–62.

SALOMONSEN, F. 1968. The moult migration. Wildfowl 19:5–24.

SALOMONSEN, F. 1976. The main problems concerning avian evolution on islands. Pp. 585–602 *in*

Proceedings of the 16th International Ornithological Congress (H. J. Frith and J. H. Calaby, Eds.). Australian Academy of Science, Canberra, Australia.

SALTHE, S. N. 1975. Problems of macroevolution (molecular evolution, phenotype definition, and canalization) as seen from a hierarchical viewpoint. Am. Zool. 15:275–314.

SALVADORI, T. 1893a. [Untitled communication.] Ibis 35:255.

SALVADORI, T. 1893b. [Untitled communication.] Bull. Br. Ornithol. Club 1:23.

SALVAN, J. 1970. Remarques sur l'evolution de l'avifauna Malgache depuis 1945. Alauda 38:191–203.

SALVIN, O. 1873. Note on the *Fulica alba* of White. Ibis 15:295.

SALZER, U. 1996. Vergleichende Untersuchungen zur Brutbiologie und Jugendentwicklung von Wachtelkönig Wasserralle (*Rallus aquaticus*). Unpublished thesis, University of Bremen, Bremen, Germany.

SAMPSON, A. F., AND A. F. SIEGEL. 1985. The measure of "size" independent of "shape" for multivariate lognormal populations. J. Am. Stat. Assoc. 80:910–914.

SANDERSON, M. J. 1993. Reversibility in evolution: a maximum likelihood approach to character gain/loss bins in phylogenies. Evolution 47:236–252.

SANFRIEL, U. N. 1975. On the significance of clutch size in nidifugous birds. Ecology 56:703–708.

SANGSTER, G. 1998. Purple Swamp-hen is a complex of species. Dutch Birding 20:13–22.

SARICH, V. M., C. W. SCHMID, AND J. MARKS. 1989. DNA hybridization as a guide to phylogenies: a critical analysis. Cladistics 5:3–32.

SATTLER, K. 1991. A review of wing reduction in Lepidoptera. Bull. Br. Mus. Nat. Hist. (Entomol.) 60:243–288.

SAUNDERS, D. A., G. T. SMITH, AND N. A. CAMPBELL. 1984. The relationship between body weight, egg weight, incubation period, nestling period and nest site in the Psittaciformes, Falconiformes, Strigiformes and Columbiformes. Aust. J. Zool. 32:57–65.

SAUNDERS, J. W., JR. 1948. The proximo-distal sequence of origin of the parts of the chick wing and the role of the ectoderm. J. Exp. Zool. 108:363–403.

SAVAGE, R. J. G. 1976. Review of early Sirenia. Syst. Zool. 25:344–351.

SAVARD, J.-P. L. 1987. Causes and functions of brood amalgamation in Barrow's Goldeneye and Bufflehead. Can. J. Zool. 65:1548–1553.

SAVIDGE, J. A. 1987. Extinction of an island forest avifauna by an introduced snake. Ecology 68:660–668.

SAVILE, D. B. O. 1957. Adaptive evolution in the avian wing. Evolution 11:212–224.

SAVILE, D. B. O. 1958. The loon wing. Evolution 12:263.

SAVILE, D. B. O. 1962. Gliding and flight in the vertebrates. Am. Zool. 2:161–166.

SAVOLAINEN, V., S. B. HEARD, M. P. POWELL, T. J. DAVIES, AND A. Ø MOOERS. 2002. Is cladogenesis heritable? Syst. Biol. 51:835–843.

SCARLETT, R. J. 1955a. A new rail from South Island swamps in New Zealand. Rec. Canterbury Mus. 6:265–266.

SCARLETT, R. J. 1955b. Further report on bird remains from Pyramid Valley. Rec. Canterbury Mus. 6:261–264.

SCARLETT, R. J. 1968. An owlet-nightjar from New Zealand. Notornis 15:254–266.

SCARLETT, R. J. 1970. The genus *Capellirallus*. Notornis 17:303–319.

SCARLETT, R. J. 1972. Bones for the New Zealand archaeologist. Canterbury Mus. Bull. 4:1–69.

SCARLETT, R. J. 1979. [Untitled letter dated 6 October 1978 regarding taxonomic status of "*Rallus dieffenbachii.*"] Notornis 26:99.

SCHAEFER, J. 1993. Translation of "Grundfragen und Paläontologie" [Basic Questions in Paleontology], by O. H. Schindewolf (1950). University of Chicago Press, Chicago, Illinois.

SCHAFFER, W. M. 1974. Optimal reproductive effort in fluctuating environments. Am. Nat. 108:783–790.

SCHARLOO, W. 1990. The effect of developmental constraints on selection response. Pp. 197–210 *in* Organizational Constraints on the Dynamics of Evolution (J. Maynard Smith and G. Vida, Eds.). Manchester University Press, Manchester, United Kingdom.

SCHARLOO, W. 1991. Canalization: genetic and developmental aspects. Annu. Rev. Ecol. Syst. 22:65–93.

SCHAUINSLAND, H. 1899. Drei Monate auf einer Korallen-Insel (Laysan). M. Nössler, Bremen, Germany.

SCHEFFER, V. B. 1942. Sea birds eaten by Alaska cod. Murrelet 23:17.
SCHEINER, S. M. 1993a. Genetics and evolution of phenotypic plasticity. Annu. Rev. Ecol. Syst. 24: 35–68.
SCHEINER, S. M. 1993b. Plasticity as a selectable trait: reply to Via. Am Nat. 142:371–373.
SCHEINER, S. M. 1998. The genetics of phenotypic plasticity. VII. Evolution in a spatially-structured environment. J. Evol. Biol. 11:303–320.
SCHEW, W. A., AND R. E. RICKLEFS. 1998. Developmental plasticity. Pp. 288–304 in Avian Growth and Development: Evolution Within the Altricial–Precocial Spectrum (J. M. Starck and R. E. Ricklefs, Eds.). Oxford University Press, Oxford, United Kingdom.
SCHINDEWOLF, O. H. 1950. Grundfragen und Paläontologie [Basic Questions in Paleontology]. E. Schweizerbart'sche Verlag. Stuttgart, Germany. Translated by J. Schaefer (1993) University of Chicago Press.
SCHINDLE, D. E. 1990. Unoccupied morphospace and the coiled geometry of gastropods: architectural constraint or geometric covariation? Pp. 270–304 in Causes of Evolution: A Paleontological Perspective (R. M. Ross and W. D. Allmon, Eds.). University of Chicago Press, Chicago, Illinois.
SCHINZ, H. R., AND R. ZANGERL. 1937a. Beiträge zur Osteogenese des Knoechensystems beim Haushuhn, bei der Haustabe und beim Haubenseißfuß. Denkschr. Schweiz. Naturforsch. Ges. 72: 117–164.
SCHINZ, H. R., AND R. ZANGERL. 1937b. Über die Osteogeneses des Skelettes beim Haushuhn, bie der Haustaube und beim Haubensteißfuß. Gegenbaurs Morphol. Jahrb. 80:620–628.
SCHLEGEL, H. 1866. On some extinct gigantic birds of the Mascarene Islands. Ibis 8:146–168.
SCHLICHTING, C. D., AND M. PIGLIUCCI. 1993. Control of phenotypic plasticity via regulatory genes. Am Nat. 142:366–370.
SCHLICHTING, C. D., AND M. PIGLIUCCI. 1995. Gene regulation, quantitative genetics and the evolution of reaction norms. Evol. Ecol. 9:154–168.
SCHLICHTING, C. D., AND M. PIGLIUCCI. 1998. Phenotypic Evolution: A Reaction Norm Perspective. Sinauer Assoc., Sunderland, Massachusetts.
SCHLUTER, D. 1988a. The evolution of finch communities on islands and continents: Kenya vs. Galápagos. Ecol. Monogr. 58:229–249.
SCHLUTER, D. 1988b. Estimating the form of natural selection on a quantitative trait. Evolution 42: 849–861.
SCHLUTER, D. 1994. Experimental evidence that competition promotes divergence in adaptive radiation. Science 266:798–801.
SCHLUTER, D. 1996a. Adaptive radiation along genetic lines of least resistance. Evolution 50:1766–1774.
SCHLUTER, D. 1996b. Ecological causes of adaptive radiation. Am. Nat. 148 (Suppl.):40–64.
SCHLUTER, D. 1998. Ecological speciation in postglacial fishes. Pp. 163–180 in Evolution on Islands (P. R. Grant, Ed.). Oxford University Press, Oxford, United Kingdom.
SCHLUTER, D. 2000a. Ecological character displacement in adaptive radiation. Am. Nat. 156 (Suppl.): 4–16.
SCHLUTER, D. 2000b. The Ecology of Adaptive Radiation. Oxford University Press, Oxford, United Kingdom.
SCHLUTER, D., AND P. R. GRANT. 1984a. Determinants of morphological patterns in communities of Darwin's finches. Am. Nat. 123:175–196.
SCHLUTER, D., AND P. R. GRANT. 1984b. Ecological correlates of morphological evolution in a Darwin's finch, *Geospiza difficilis*. Evolution 38:856–869.
SCHLUTER, D., T. D. PRICE, AND P. R. GRANT. 1985. Ecological character displacement in Darwin's finches. Science 227:1056–1059.
SCHLUTER, D., AND J. N. M. SMITH. 1986. Natural selection on beak and body size in the Song Sparrow. Evolution 40:221–231.
SCHMIDT, K. A., AND C. J. WHELAN. 1998. Predator-mediated interactions between and within guilds of nesting songbirds: experimental and observational evidence. Am. Nat. 152:393–402.
SCHMIDT, M. B. 1976. Observations on the Cape Rail in the southern Transvaal. Ostrich 47:16–26.
SCHMIDT-NIELSEN, K. 1972. Locomotion: energy cost of swimming, flying, and running. Science 177: 222–228.
SCHMIDT-NIELSEN, K. 1977. Problems of scaling: locomotion and physiological correlates. Pp. 1–21

in Scale Effects in Animal Locomotion (T. J. Pedley, Ed.). Academic Press, London, United Kingdom.
SCHMIDT-NIELSEN, K. 1984. Scaling: Why Is Animal Size So Important? Cambridge University Press, Cambridge, United Kingdom.
SCHMIDT-NIELSEN, K. 1997. Animal Physiology: Adaptation and Environment. 5th ed. Cambridge University Press, Cambridge, United Kingdom.
SCHMIDT-NIELSEN, K., AND Y. T. KIM. 1964. The effect of salt intake on the size and function of the salt gland of ducks. Auk 81:160–172.
SCHMITT, R. J. 1987. Indirect interactions between prey: apparent competition, predator aggregation, and habitat segregation. Ecology 68:1887–1897.
SCHNELL, G. D. 1970a. A phenetic study of the suborder Lari (Aves) I. Methods and results of principal components analyses. Syst. Zool. 19:35–57.
SCHNELL, G. D. 1970b. A phenetic study of the suborder Lari (Aves) II. Phenograms, discussion, and conclusions. Syst. Zool. 19:264–302.
SCHNELL, G. D., AND J. J. HELLACK. 1979. Bird flight speeds in nature: optimized or a compromise? Am. Nat. 113:53–66.
SCHNELL, G. D., G. L. WORTHEN, AND M. E. DOUGLAS. 1985. Morphometric assessment of sexual dimorphism in skeletal elements of California Gulls. Condor 87:484–493.
SCHODDE, R., AND R. DE NAUROIS. 1982. Patterns of variation and dispersal in the Buff-banded Rail (*Gallirallus philippensis*) in the southwest Pacific, with a description of a new subspecies. Notornis 29:131–142.
SCHODDE, R., P. FULLAGER, AND N. HERMES. 1983. A review of Norfolk Island birds: past and present. Aust. Natl. Parks Wildl. Serv. Spec. Publ. 8:1–119.
SCHOENER, T. W. 1965. The evolution of bill size differences among sympatric congeneric species of birds. Evolution 19:189–213.
SCHOENER, T. W. 1968. Sizes of feeding territories among birds. Ecology 49:123–141.
SCHOENER, T. W. 1974. Resource partitioning in ecological communities. Science 185:27–39.
SCHOENER, T. W. 1975. Presence and absence of habitat shift in some widespread lizard species. Ecol. Monogr. 45:233–258.
SCHOENER, T. W. 1984. Size differences among sympatric, bird-eating hawks: a worldwide survey. Pp. 254–281 *in* Ecological Communities: Conceptual Issues and the Evidence (D. R. Strong, Jr., D. Simberloff, L. G. Abele, and A. B. Thistle, Eds.). Princeton University Press, Princeton, New Jersey.
SCHOENER, T. W. 1991. Extinction and the nature of the metapopulation. Acta Oecol. 12:53–75.
SCHOENER, T. W., AND G. C. GORMAN. 1968. Some niche differences in three Lesser Antillean lizards of the genus *Anolis*. Ecology 49:819–830.
SCHOLANDER, P. F., R. HOCK, V. WALTERS, AND L. IRVING. 1950a. Adaptation to cold in arctic and tropical mammals and birds in relation to body temperature, insulation, and basal metabolic rate. Biol. Bull. 99:259–271.
SCHOLANDER, P. F., R. HOCK, V. WALTERS, F. JOHNSTON, AND L. IRVING. 1950b. Heat regulation in some arctic and tropical mammals and birds. Biol. Bull. 99:237–258.
SCHOLANDER, P. F., V. WALTERS, R. HOCK, AND L. IRVING. 1950c. Body insulation of some arctic and tropical mammals and birds. Biol. Bull. 99:225–236.
SCHÖNWETTER, M. 1960a. Handbuch der Oologie, Lieferung 1. Akademic-Verlag, Berlin, Germany.
SCHÖNWETTER, M. 1960b. Handbuch der Oologie, Lieferung 2. Akademic-Verlag, Berlin, Germany.
SCHÖNWETTER, M. 1961. Handbuch der Oologie, Lieferung 3. Akademic-Verlag, Berlin, Germany.
SCHÖNWETTER, M. 1962. Handbuch der Oologie, Lieferung 6. Akademic-Verlag, Berlin, Germany.
SCHORGER, A. W. 1947. The deep diving of the Loon and Old-squaw and its mechanism. Wilson Bull. 59:151–159.
SCHORGER, A. W. 1966. The Wild Turkey: Its History and Domestication. University of Oklahoma Press, Norman, Oklahoma.
SCHREIWEIS, D. O. 1982. A comparative study of the appendicular musculature of penguins (Aves: Sphenisciformes). Smithson. Contrib. Zool. 341:1–46.
SCHUBEL, S. E., AND D. W. STEADMAN. 1989. More bird bones from Polynesian archeological sites on Henderson Island, Pitcairn Group, South Pacific. Atoll Res. Bull. 325:1–14.
SCHÜLE, W. 1993. Mammals, vegetation and the initial human settlement of the Mediterranen islands: a palaeological approach. J. Biogeogr. 20:399–411.

SCHULENBERG, T. S., AND J. V. REMSEN, JR. 1982. Eleven bird species new to Bolivia. Bull. Br. Ornithol. Club 102:52–57.

SCHULTZ, B. B. 1985. Levene's test for relative variation. Syst. Zool. 34:449–456.

SCHUMACHER, G.-H., AND E. WOLFF. 1967a. Zur vergleichenden Osteogenese von *Gallus domesticus* L., *Larus ridibundus* L. und *Larus canus* L. I. Zeitliches Erscheinen der Ossifikationen bei *Gallus domesticus* L. Gegenbaurs Morphol. Jahrb. 110:359–373.

SCHUMACHER, G.-H., AND E. WOLFF. 1967b. Zur vergleichenden Osteogenese von *Gallus domesticus* L., *Larus ridibundus* L. und *Larus canus* L. II. Zeitliches Erscheinen der Ossifikationen bei *Larus ridibundus* L. und *Larus canus* L. Gegenbaurs Morphol. Jahrb. 110:620–635.

SCHUMMER, A., B. VOLLMERHAUS, F. SINOWATZ, J. FREWEIN, AND H. WAIBL. 1992. Anatomie der Vögel. Verlag Paul Parey, Berlin, Germany.

SCHWANER, T. D., AND S. D. SARRE. 1988. Body size of tiger snakes in southern Australia, with particular reference to *Notechis ater serventyi* (Elapidae) on Chappell Island. J. Herpetol. 22:24–33.

SCLATER, P. L. 1861. On the island-hen of Tristan d'Acunha. Proc. Zool. Soc. Lond. 1861:260–263.

SCLATER, P. L. 1868. Synopsis of the American rails (*Rallidæ*). Proc. Zool. Soc. Lond. 1868:442–470.

SCLATER, P. L. 1874. Exhibition of an egg of *Pareudiastes pacificus*. Proc. Zool. Soc. Lond. 1874:605–606.

SCOTT, D. M., AND C. D. ANKNEY. 1983. Do Darwin's finches lay small eggs? Auk 100:226–227.

SCOTT, J. M., S. MOUNTAINSPRING, F. L. RAMSEY, AND C. B. KEPLER. 1986. Forest Bird Communities of the Hawaiian Islands: Their Dynamics, Ecology, and Conservation. Stud. Avian Biol. No. 9. Cooper Ornithological Society, Los Angeles, California.

SEARCY, W. A. 1982. The evolutionary effects of mate selection. Annu. Rev. Ecol. Syst. 13:57–85.

SEARCY, W. A., AND K. YASUKAWA. 1995. Polygyny and Sexual Selection in Red-winged Blackbirds. Princeton University Press, Princeton, New Jersey.

SEDINGER, J. S. 1986. Growth and development of Canada Goose goslings. Condor 88:169–180.

SEGER, K., AND H. J. BROCKMAN. 1987. What is bet-hedging? Pp. 182–211 in Oxford Surveys in Evolutionary Biology. Vol. 4 (P. H. Harvey and L. Partridge, Eds.). Oxford University Press, Oxford, United Kingdom.

SEILACHER, A. 1975. Fabricational noise in adaptive morphology. Syst. Zool. 22:451–465.

SEIM, E., AND B.-E. SAETHER. 1983. On rethinking allometry: Which regression model to use? J. Theor. Biol. 104:161–168.

SELANDER, R. K. 1965. On mating systems and sexual selection. Am. Nat. 99:129–141.

SELANDER, R. K. 1966. Sexual dimorphism and differential niche utilization in birds. Condor 68:113–151.

SELANDER, R. K. 1972. Sexual selection and dimorphism in birds. Pp. 180–230 in Sexual Selection and the Descent of Man: 1871–1971 (B. Campbell, Ed.). Aldine Publishing, Chicago, Illinois.

SELANDER, R. K., AND R. F. JOHNSTON. 1967. Evolution in the House Sparrow. I. Intrapopulation variation in North America. Condor 69:217–258.

SELMI, S., AND T. BOULINIER. 2001. Ecological biogeography of southern ocean islands: the importance of considering spatial issues. Am. Nat. 158:426–437.

SÉLYS-LONGCHAMPS, E. DE. 1848. Résumé concernant les oiseaux brevipennes mentionnés dans l. ouvrage de M. Strickland sur le Dodo. Rev. Zool. 11:292–295.

SEMLITSCH, R. D. 1985. Reproductive strategy of a facultatively paedomorphic salamander *Ambystoma talpoideum*. Oecologia 65:305–313.

SEMLITSCH, R. D. 1987. Paedomorphosis in *Ambystoma talpoideum*: effects of density, food, and pond drying. Ecology 68:994–1102.

SEMLITSCH, R. D. 1990. Paedomorphosis in *Ambystoma talpoideum*: maintenance of population variation and alternative life-history pathways. Evolution 44:1604–1613.

SEMLITSCH, R. D., AND H. M. WILBUR. 1989. Artificial selection for paedomorphosis in the salamander *Ambystoma talpoideum*. Evolution 43:105–112.

SENAR, J. C., J. LLEONART, AND N. B. METCALFE. 1994. Wing-shape variation between resident and transient wintering Siskins *Carduelis spinus*. J. Avian Biol. 25:50–54.

SENAR, J. C., AND J. PASCUAL. 1997. Keel and tarsus length may provide a good predictor of avian body size. Ardea 85:269–274.

SEPKOSKI, J. J., JR., AND M. A. REX. 1974. Distribution of freshwater mussels: coastal rivers as biogeographic islands. Syst. Zool. 23:165–188.
SERLE, W. 1939. Observations on the breeding habits of some Nigerian Rallidæ. Oologists' Rec. 19: 61–70.
SERVENTY, D. L. 1960. Geographical distribution of living birds. Pp. 95–126 in Biology and Comparative Physiology of Birds. Vol. 1 (A. J. Marshall, Ed.). Academic Press, New York, New York.
SERVICE, P. M., AND M. R. ROSE. 1985. Genetic covariance among life-history components: the effect of novel environments. Evolution 39:943–945.
SHAFFER, M. 1987. Minimum viable populations: coping with uncertainty. Pp. 69–86 in Viable Populations for Conservation (M. E. Soulé Ed.). Cambridge University Press, Cambridge, United Kingdom.
SHAPIRO, A. M. 1976. Seasonal polyphenism. Pp. 259–333 in Evolutionary Biology. Vol. 9 (M. K. Hecht, W. C. Steere, and B. Wallace, Eds.). Plenum Press, New York, New York.
SHAPIRO, B., D. SIBTHORPE, A. RAMBAUT, J. AUSTIN, G. M. WRAGG, O. R. P. BININDA-EMONDS, P. L. M. LEE, AND A. COOPER. 2002. Flight of the Dodo. Science 295:1683.
SHAPIRO, M. D., AND T. F. CARL. 2001. Novel features of tetrapod limb development in two nontraditional model species: a skink and a direct-developing frog. Pp. 337–361 in Beyond Heterochrony: The Evolution of Development (M. L. Zelditch, Ed.). Wiley Interscience, New York, New York.
SHARLAND, M. 1945. Tasmanian Birds. Angus and Robertson, Sydney, Australia.
SHARLAND, M. 1973. Tasmanian Native Hen. Bird Watcher 1973:23–27.
SHARPE, R. B. 1875. Birds: Appendix. In The Zoology of the Voyage of H. M. S. Erebus & Terror. Vol. 1 (J. Richardson and J. E. Gray, Eds). E. W. Janson, London, England.
SHARPE, R. B. 1893a. [Untitled communication.] Bull. Br. Ornithol. Club 1:26–29.
SHARPE, R. B. 1893b. [Untitled communication.] Bull. Br. Ornithol. Club 1:29–30.
SHARPE, R. B. 1893c. [Untitled communication.] Ibis 35:445–446.
SHARPE, R. B. 1893d. [Untitled communication.] Bull. Br. Ornithol. Club 1:46.
SHARPE, R. B. 1893e. [Untitled communication.] Ibis 35:261–262.
SHARPE, R. B. 1894. Catalogue of the Fulicariæ (Rallidæ and Heliornithidæ) and Alectorides (Aramidæ, Eurypygidæ, Mesitidæ, Rhinochetidæ, Gruidæ, Psophiidæ, and Otididæ) in the collection of the British Museum. British Museum (Natural History), London, England.
SHAW, G. 1784. Cimelia Physica. B. and J. White, London, England.
SHAW, K. L. 1995. Biogeographic patterns of two independent Hawaiian cricket radiations (Laupala and Prognathogryllus). Pp. 39–56 in Hawaiian Biogeography: Evolution on a Hot Spot Archipelago (W. L. Wagner and W. A. Funk, Eds.). Smithsonian Institution Press, Washington, D.C.
SHAW, P. 1985. Brood reduction in the Blue-eyed Shag Phalacrocorax atriceps. Ibis 127:476–494.
SHEA, B. T. 1983. Paedomorphosis and neoteny in the pygmy chimpanzee. Science 222:521–522.
SHEA, B. T. 1985a. Ontogenetic allometry and scaling: a discusssion based on the growth and form of the skull in African apes. Pp. 175–205 in Size and Scaling in Primate Biology (W. L. Jungers, Ed.). Plenum Press, New York, New York.
SHEA, B. T. 1985b. Bivariate and multivariate growth allometry: statistical and biological considerations. J. Zool. Lond. 206:367–390.
SHEA, B. T. 1988. Heterochrony in Primates. Pp. 237–266 in Heterochrony in Evolution: A Multidisciplinary Approach (M. L. McKinney, Ed.). Plenum Press, New York, New York.
SHELDON, P. R. 1993. Making sense of microevolutionary patterns. Pp. 19–31 in Evolutionary Patterns and Processes (D. R. Lees and D. Edwards, Eds.). Linn. Soc. Symp. No. 14. Academic Press, London, United Kingdom.
SHERMAN, P. T. 1995a. Breeding biology of White-winged Trumpeters (Psophia leucoptera) in Peru. Auk 112:285–295.
SHERMAN, P. T. 1995b. Social organization of cooperatively polyandrous White-winged Trumpeters (Psophia leucoptera). Auk 112:296–309.
SHERMAN, P. T. 1996. Family Psophiidae (trumpeters). Pp. 96–107 in Handbook of the Birds of the World. Vol. 3: Hoatzin to Auks (J. del Hoyo, A. Elliott and J. Sargatal, Eds.). Lynx Edicions, Barcelona, Spain.
SHERRY, T. W. 1990. When are birds dietarily specialized? Distinguishing ecological from evolutionary approaches. Pp. 337–352 in Avian Foraging: Theory, Methodology, and Applications (M.

L. Morrison, C. J. Ralph, J. Verner, and J. R. Jehl, Jr., Eds.). Cooper Ornithological Society, Los Angeles, California.
SHIGESADA, N. 1997. Biological Invasions: Theory and Practice. Oxford University Press, Oxford, United Kingdom.
SHIMIZU, T., AND S. MASAKI. 1993. Genetic variability of the wing-form response to photoperiod in a subtropical population of the ground cricket, *Dianemobius fascipes*. Zool. Sci. 10:935–944.
SHINE, R. 1978. Propagule size and parental care: the "safe harbor" hypothesis. J. Theor. Biol. 75: 417–424.
SHINE, R. 1989a. Ecological causes for the evolution of sexual dimorphism: a review of the evidence. Q. Rev. Biol. 64:419–461.
SHINE, R. 1989b. Alternative models for the evolution of offspring size. Am. Nat. 134:311–317.
SHUBIN, N. H., AND P. ALBERCH. 1986. A morphogenetic approach to the origin and basic organization of the tetrapod limb. Pp. 319–387 in Evolutionary Biology. Vol. 20 (M. K. Hecht, B. Wallace, and G. T. Prance, Eds.). Plenum Press, New York, New York.
SIBLEY, C. G. 1957. The evolutionary and taxonomic significance of sexual dimorphism and hybridization in birds. Condor 59:166–191.
SIBLEY, C. G., AND J. E. AHLQUIST. 1990. Phylogeny and Classification of Birds: A Study in Molecular Evolution. Yale University Press, New Haven, Connecticut.
SIBLEY, C. G., J. E. AHLQUIST, AND P. BENEDICTUS. 1993. The phylogenetic relationships of the rails, based on DNA comparisons. J. Yamashina Inst. Ornithol. 25:1–11.
SIBLY, R. M., AND P. CALOW. 1986. Physiological Ecology of Animals: An Evolutionary Approach. Blackwell Scientific Publications, Oxford, United Kingdom.
SICK, H. 1937. Morphohlogisch-funktionelle Untersuchungen über die Feistruktur der Vogelfeder. J. Ornithol. 85:206–372.
SICK, H. 1979. Notes on some Brazilian birds. Bull. Br. Ornithol. Club 99:115–120.
SIEGEL-CAUSEY, D. 1990. Phylogenetic patterns of size and shape of the nasal gland depression in Phalacrocoracidae. Auk 107:110–118.
SIEGFRIED, W. R., AND P. G. H. FROST. 1973. Regular occurrence of *Porphyrula martinica* in South Africa. Bull. Br. Ornithol. Club 93:36–38.
SIEGLBAUER, F. 1911. Zur Entwicklung der Vogelextremität. Z. Wiss. Zool. 97:262–313.
SIGMUND, L. 1958. Die postembryonale Entwicklung der Wasserralle (*Rallus aquaticus*). Sylvia 15: 85–118.
SIGMUND, L. 1959. Mechanik und anatomische Grundlagen der Fortbewegung bei Wasserralle (*Rallus aquaticus* L.), Teichhuhn (*Gallinula chloropus* L.) und Bläßhuhn (*Fulica atra* L.). J. Ornithol. 100:3–24.
SIGURJÓNSDÓTTIR, H. 1981. The evolution of sexual size dimorphism in gamebirds, waterfowl and raptors. Ornis Scand. 12:249–260.
SILLEN, B., AND C. SOLBRECK. 1977. Effects of area and habitat diversity on bird species richness in lakes. Ornis Scand. 8:185–192.
SILLÉN-TULLBERG, B. 1993. The effect of biased inclusion of taxa on the correlation between discrete characters in phylogenetic trees. Evolution 47:1182–1191.
SILLÉN-TULLBERG, B., AND H. TEMRIN. 1994. On the use of discrete characters in phylogenetic trees with special reference to the evolution of avian mating systems. Pp. 311–322 in Phylogenetics and Ecology (P. Eggleton and R. I. Vane-Wright, Eds.). Linn. Soc. Symp. 17. Academic Press, London, United Kingdom.
SILVER, R., H. ANDREWS, AND G. F. BALL. 1985. Parental care in an ecological perspective: a quantitative analysis of avian subfamilies. Am. Zool. 25:823–840.
SIMBERLOFF, D. 1974. Equilibrium theory of island biogeography and ecology. Annu. Rev. Ecol. Syst. 5:161–182.
SIMBERLOFF, D. 1983a. Sizes of coexisting species. Pp. 404–430 in Coevolution (D. J. Futuyma and M. Slatkin, Eds.). Sinauer Assoc., Sunderland, Massachusetts.
SIMBERLOFF, D. 1983b. Competition theory, hypothesis testing, and other community ecological buzzwords. Am. Nat. 122:626–635.
SIMBERLOFF, D. 1984. Properties of coexisting bird species in two archipelagos. Pp. 234–253 in Ecological Communities: Conceptual Issues and the Evidence (D. R. Strong, D. Simberloff, L. G. Abele, and A. B. Thistle, Eds.). Princeton University Press, Princeton, New Jersey.

SIMBERLOFF, D. 1986. The proximate causes of extinction. Pp. 259–276 *in* Patterns and Processes in the History of Life (D. M. Raup and D. Jablonski, Eds.). Springer-Verlag, Berlin, Germany.

SIMBERLOFF, D., AND L. G. ABELE. 1982. Refuge design and island biogeographic theory: effects of fragmentation. Am. Nat. 120:41–50.

SIMBERLOFF, D., AND W. BOECKLEN. 1981. Santa Rosalia reconsidered: size ratios and competition. Evolution 35:1206–1228.

SIMBERLOFF, D., AND W. BOECKLEN. 1991. Patterns of extinction in the introduced Hawaiian avifauna: a reexamination of the role of competition. Am. Nat. 138:300–327.

SIMBERLOFF, D., AND J.-L. MARTIN. 1991. Nestedness of insular avifaunas: simple summary statistics masking complex species patterns. Ornis Fenn. 68:178–192.

SIMIC, V., AND V. ANDREJEVIC. 1963. Morphologie und Topographie der Brustmuskeln bei den Hausphasioniden und der Taube. Gegenbaurs Morphol. Jahrb. 104:546–560.

SIMIC, V., AND V. ANDREJEVIC. 1964. Morphologie und Topographie der Brustmuskeln bei den Hausschwinnmogeln. Gegenbaurs Morphol. Jahrb. 106:480–490.

SIMKIN, T. 1984. Geology of the Galápagos. Biol. J. Linn. Soc. 21:61–75.

SIMONETTA, A. M. 1960. On the mechanical implications of the avian skull and their bearing on the evolution and classification of birds. Q. Rev. Biol. 35:206–220.

SIMPSON, F. G. 1983. The flight mechanism of the Pigeon *Columba livia* during takeoff. J. Zool. Lond. 200:435–443.

SIMPSON, G. G. 1944. Tempo and Mode in Evolution. Columbia University Press, New York, New York.

SIMPSON, G. G. 1946. Fossil penguins. Bull. Am. Mus. Nat. Hist. 87:1–99.

SIMPSON, G. G. 1949. The Meaning of Evolution: A Study of the History of Life and of Its Significance for Man. Yale University Press, New Haven, Connecticut.

SIMPSON, G. G. 1953a. The Major Features of Evolution. Columbia University Press, New York, New York.

SIMPSON, G. G. 1953b. The Baldwin effect. Evolution 7:110–117.

SIMPSON, G. G. 1975. Fossil penguins. Pp. 19–56 *in* The Biology of Penguins (B. Stonehouse, Ed.). University Park Press, Baltimore, Maryland.

SINCLAIR, A. R. E. 1983. The function of distance movements in vertebrates. Pp. 240–258 *in* The Ecology of Animal Movement (I. R. Swingland and P. J. Greenwood, Eds.). Clarendon Press, Oxford, United Kingdom.

SINCLAIR, W. J., AND M. S. FARR. 1932. Aves of the Santa Cruz Beds. Reports of the Princeton University Expeditions to Patagonia, 1896–1899. Vol. 7: Palaeontology 4, Part 2. Princeton University Press, Princeton, New Jersey.

SINERVO, B. 1999. Mechanistic analysis of natural selection and a refinement of Lack's and Williams's principles. Am. Nat. 154 (Suppl.):26–42.

SKET, B. 1985. Why all cave animals don't look alike: a discussion on adaptive value of reduction processes. Natl. Speleol. Soc. Bull. 47:78–85.

SLATKIN, M. 1987. Quantitative genetics of heterochrony. Evolution 41:799–811.

SLATKIN, M. 1996. In defense of founder-flush theories of speciation. Am. Nat. 147:493–505.

SLIJPER, E. J. 1961. Locomotion and locomotory organs in whales and dolphins (Cetacea). Symp. Zool. Soc. Lond. 5:77–94.

SLIKAS, B., S. L. OLSON, AND R. C. FLEISCHER. 2002. Rapid, independent evolution of flightlessness in four species of Pacific Island rails (Rallidae): an analysis based on mitochondrial sequence data. J. Avian Biol. 33:5–14.

SMITH, A. B., AND C. PATTERSON. 1988. The influence of taxonomic method on the perception of patterns of evolution. Pp. 127–216 *in* Evolutionary Biology. Vol. 23 (M. K. Hecht and B. Wallace, Eds.). Plenum Press, New York, New York.

SMITH, C. C., AND S. D. FRETWELL. 1974. The optimal balance between size and number of offspring. Am. Nat. 108:499–506.

SMITH, G. R. 1990. Homology in morphometrics and phylogenetics. Pp. 325–338 *in* Proceedings of the Michigan Morphometrics Workshop (F. J. Rohlf and F. L. Bookstein, Eds.). Univ. Mich. Mus. Zool. Spec. Publ. 2:1–380.

SMITH, J. R., AND P. WESSEL. 2000. Isostatic consequences of giant landslides on the Hawaiian Ridge. Pure Appl. Geophys. 157:1097–1114.

SMITH, M. F., AND J. L. PATTON. 1988. Subspecies of pocket gophers: causal bases for geographic differentiation in *Thomomys bottae*. Syst. Zool. 37:163–178.

SMITH, R. J., R. Z. GERMAN, AND W. L. JUNGERS. 1986. Variability of biological similarity criteria. J. Theor. Biol. 118:287–293.

SMITH, T. B. 1987. Bill size polymorphism and intraspecific niche utilization in an African finch. Nature 329:717–719.

SMITH, T. B. 1990a. Natural selection on bill characters in the two bill morphs of the African finch *Pyrenestes ostrinus*. Evolution 44:832–842.

SMITH, T. B. 1990b. Resource use by bill morphs of an African finch: evidence for intraspecific competition. Ecology 71:1246–1257.

SMITH, T. B. 1990c. Patterns of morphological and geographic variation in trophic bill morphs of the African finch *Pyrenestes*. Biol. J. Linn. Soc. 41:381–414.

SMITH, T. B. 1991. Inter- and intra-specific diet overlap during lean times between *Quelea erythrops* and bill morphs of *Pyrenestes ostrinus*. Oikos 60:76–82.

SMITH, T. B. 1993. Disruptive selection and the genetic basis of bill size polymorphism in the African finch *Pyrenestes*. Nature 363:618–620.

SMITH, T. B. 1997. Adaptive significance of the mega-billed form in the polymorphic Black-bellied Seedcracker *Pyrenestes ostrinus*. Ibis 139:382–387.

SMITH, T. B., AND D. J. GIRMAN. 2000. Reaching new adaptive peaks: evolution of alternative bill forms in an African finch. Pp. 139–156 *in* Adaptive Genetic Variation in the Wild (T. A. Mousseau, B. Sinervo, and J. A. Endler, Eds.). Oxford University Press, Oxford, United Kingdom.

SMITH, T. B., AND S. SKÚLASON. 1996. Evolutionary significance of resource polymorphisms in fish, amphibians and birds. Annu. Rev. Ecol. Syst. 27:111–133.

SMITH, T. B., AND S. A. TEMPLE. 1982. Feeding habits and bill polymorphism in Hook-billed Kites. Auk 99:197–207.

SMITH, T. B., R. K. WAYNE, D. J. GIRMAN, AND M. W. BRUFORD. 1997. A role for ecotones in generating forest biodiversity. Science 276:1855–1857.

SMITH, W. W. 1893. Notes on certain species of New-Zealand birds. Ibis 35:509–521.

SMITHERS, C. N., AND H. J. DE S. DISNEY. 1969. The distribution of terrestrial and freshwater birds on Norfolk Island. Aust. J. Zool. 15:127–140.

SMITH-GILL, S. J. 1983. Developmental plasticity: developmental conversion versus phenotypic modulation. Am. Zool. 23:47–55.

SNEDECOR, G. W., AND W. G. COCHRAN. 1967. Statistical Methods. 6th ed. Iowa State University Press, Ames, Iowa.

SNEDECOR, G. W., AND W. G. COCHRAN. 1980. Statistical Methods. 7th ed. Iowa State University Press, Ames, Iowa.

SNOW, B. K. 1966. Observations on the behaviour and ecology of the Flightless Cormorant *Nannopterum harrisi*. Ibis 108:265–280.

SNOW, D. W. 1954. Trends in geographical variation in Palaearctic members of the genus *Parus*. Evolution 8:19–28.

SNOW, D. W. 1958. Climate and geographical variation in birds. New Biol. 25:64–84.

SNOW, D. W. 1987. Colonel Meinertzhagen's wing drawings. Bull. Br. Ornithol. Club 107:189–191.

SNYDER, F. F. R., AND J. W. WILEY. 1976. Sexual size dimorphism in hawks and owls of North America. Ornithol. Monogr. 20:1–96.

SOKAL, R. R. 1976. The Kluge–Kerfoot phenomenon re-examined. Am. Nat. 110:1077–1091.

SOKAL, R. R., AND C. A. BRAUMANN. 1980. Significance tests for coefficients of variation and variability profiles. Syst. Zool. 29:50–66.

SOLER, J. J., A. P. MØLLER, AND M. SOLER. 1998. Nest building, sexual selection and parental investment. Evol. Ecol. 12:427–441.

SOMERS, K. M. 1986. Multivariate allometry and removal of size with principal components analysis. Syst. Zool. 35:359–368.

SOMERS, K. M. 1989. Allometry, isometry and shape in principal components analysis. Syst. Zool. 38:169–173.

SONDAAR, P. Y. 1976. Insularity and its effect on mammal evolution. Pp. 671–707 *in* Major Patterns in Vertebrate Evolution (M. K. Hecht, P. C. Goody, and B. M. Hecht, Eds.). Plenum Press, New York, New York.

SOPER, M. F. 1969. Kermadec Island reports: the Spotless Crake (*Porzana tabuensis plumbea*). Notornis 16:219–220.
SORENSON, M. D. 1995. Evidence of conspecific nest parasitism and egg discrimination in the Sora. Condor 97:819–821.
SOTHERLAND, P. R., AND H. RAHN. 1987. On the composition of bird eggs. Condor 89:48–65.
SOULÉ, M. E. 1982a. Allomeric variation. I. The theory and some consequences. Am. Nat. 120:751–764.
SOULÉ, M. E. 1982b. Allomeric variation. II. Developmental instability of extreme phenotypes. Am. Nat. 120:751–764.
SOUTHEY, I. C. 1989. The biogeography of New Zealand's terrestrial vertebrates. N. Z. J. Zool. 16:651–663.
SOUTHWOOD, T. R. E. 1962. Migration of terrestrial arthropods in relation to habitat. Biol. Rev. 37:171–214.
SPEAKMAN, J. R. 1996. Energetics and the evolution of body size in small terrestrial mammals. Symp. Zool. Soc. Lond. 69:63–81.
SPEAR, L. B., AND D. G. AINLEY. 1997. Flight behaviour of seabirds in relation to wind direction and wing morphology. Ibis 139:221–233.
SPEARMAN, R. I. C., AND J. A. HARDY. 1985. Integument. Pp. 1–56 *in* Form and Function in Birds. Vol. 3 (A. S. King and J. McLelland, Eds.). Academic Press, London, United Kingdom.
SPEDDING, G. R. 1992. The aerodynamics of flight. Pp. 51–111 *in* Advances in Comparative and Environmental Physiology. Vol. 11 (R. McN. Alexander, Ed.). Springer-Verlag, Berlin, Germany.
SPEDDING, G. R., AND P. B. S. LISSAMAN. 1998. Technical aspects of microscale flight systems. J. Avian Biol. 29:458–468.
SPELLERBERG, I. F. 1975. The predators of penguins. Pp. 413–434 *in* The Biology of Penguins (B. Stonehouse, Ed.). University Park Press, Baltimore, Maryland.
SPRENT, P. 1972. The mathematics of size and shape. Biometrics 28:23–37.
SPRUGEL, D. G. 1983. Correcting for bias in log-transformed allometric equations. Ecology 64:209–210.
STACEY, P. B., AND J. D. LIGON. 1987. Territory quality and dispersal options in the Acorn Woodpecker, and a challenge to the habitat-saturation model of cooperative breeding. Am. Nat. 130:654–676.
STACEY, P. B., AND J. D. LIGON. 1991. The benefits-of-philopatry hypothesis for the evolution of cooperative breeding: variation in territory quality and group size effects. Am. Nat. 137:831–846.
STAHL, J.-C., J. L. MOUGIN, P. JOVENTIN, AND H. WEIMERSKIRCH. 1984. Le canard d'Eaton, *Anas eutoni drygalskii*, des Iles Crozat: systematique, comportement alimentaire et biologie de reproduction. Gerfaut 74:305–326.
STAMPS, J. A., AND M. BUECHNER. 1985. The territorial defense hypothesis and the ecology of insular vertebrates. Q. Rev. Biol. 60:155–181.
STAMPS, J. A., AND V. V. KRISHNAN. 2001. How territorial animals compete for divisible space: a learning-based model with unequal competitors. Am. Nat. 157:154–169.
STANLEY, S. M. 1973. An explanation for Cope's rule. Evolution 27:1–26.
STANLEY, S. M. 1979. Macroevolution: Pattern and Process. W. H. Freeman, San Francisco, California.
STANLEY, S. M. 1990. The general correlation between rate of speciation and rate of extinction: fortuitous causal linkages. Pp. 103–127 *in* Causes of Evolution: A Paleontological Perspective (R. M. Ross and W. D. Allmon, Eds.). University of Chicago Press, Chicago, Illinois.
STARCK, J. M. 1989. Zeitmaster der Ontogenesen bei nestfluchtenden und nesthockenden Vögeln. Cour. Forschungsinst. Senckenb. 114:1–319.
STARCK, J. M. 1993. Evolution of avian ontogenies. Pp. 275–366 *in* Current Ornithology. Vol. 10 (D. M. Power, Ed.). Plenum Press, New York, New York.
STARCK, J. M. 1998. Structural variants and invariants in avian embryonic and postnatal development. Pp. 59–88 *in* Avian Growth and Development: Evolution Within the Altricial–Precocial Spectrum (J. M. Starck and R. E. Ricklefs, Eds.). Oxford University Press, Oxford, United Kingdom.
STARCK, J. M., AND R. E. RICKLEFS. 1998a. Patterns of development: the altricial-precocial spectrum. Pp. 3–30 *in* Avian Growth and Development: Evolution Within the Altricial–Precocial Spectrum (J. M. Starck and R. E. Ricklefs, Eds.). Oxford University Press, Oxford, United Kingdom.

STARCK, J. M., AND R. E. RICKLEFS. 1998b. Variation, constraint, and phylogeny: comparative analysis of variation in growth. Pp. 247–265 in Avian Growth and Development: Evolution Within the Altricial–Precocial Spectrum (J. M. Starck and R. E. Ricklefs, Eds.). Oxford University Press, Oxford, United Kingdom.

STARCK, J. M., AND R. E. RICKLEFS. 1998c. Avian growth rate data set. Pp. 381–423 in Avian Growth and Development: Evolution Within the Altricial–Precocial Spectrum (J. M. Starck and R. E. Ricklefs, Eds.). Oxford University Press, Oxford, United Kingdom.

STARCK, J. M., AND R. E. RICKLEFS. 1998d. Data set of avian growth parameters. Pp. 381–415 in Avian Growth and Development: Evolution Within the Altricial–Precocial Spectrum (J. M. Starck and R. E. Ricklefs, Eds.). Oxford University Press, Oxford, United Kingdom.

STARCK, J. M., AND E. SUTTER. 2000. Patterns of growth and heterochrony in moundbuilders (Megapodiidae) and fowl (Phasianidae). J. Avian Biol. 31:527–547.

STARK, R. J., AND R. L. SEARLS. 1973. A description of chick wing bud development and a model of limb morphogenesis. Devel. Biol. 33:138–153.

STARKS, J. R., AND J. M. PETER. 1991. Dusky Moorhens *Gallinula tenebrosa* eating gull faeces. Aust. Bird Watcher 14:69.

STAUFFER, D. F., E. O. GARTON, AND R. K STEINHORST. 1985. A comparison of principal components from real and random data. Ecology 66:1693–1698.

ST. CLAIR, C. C., AND R. C. ST. CLAIR. 1992. Weka predation on eggs and chicks of Fiordland Crested Penguins. Notornis 39:60–63.

STEADMAN, D. W. 1985. Fossil birds from Mangaia, southern Cook Islands. Bull. Br. Ornithol. Club 105:58–66.

STEADMAN, D. W. 1986a. Two new species of rails (Aves: Rallidae) from Mangaia, southern Cook Islands. Pac. Sci. 40:27–41.

STEADMAN, D. W. 1986b. Fossil birds and biogeography in Polynesia. Pp. 1526–1534 in Acta XIX Congressus Internationalis Ornithologici. Vol. 2 (H. Ouellet, Ed.). National Museum of Natural Science, Ottawa, Ontario, Canada.

STEADMAN, D. W. 1988. A new species of *Porphyrio* (Aves: Rallidae) from archeological sites in the Marquesas Islands. Proc. Biol. Soc. Wash. 101:162–170.

STEADMAN, D. W. 1989a. Extinction of birds in eastern Polynesia: a review of the record, and comparisons with other Pacific island groups. J. Archaeol. Sci. 16:177–205.

STEADMAN, D. W. 1989b. New species and records of birds (Aves: Megapodiidae, Columbidae) from an archaeological site on Lifuka, Tonga. Proc. Biol. Soc. Wash. 102:537–552.

STEADMAN, D. W. 1992. Extinct and extirpated birds from Rota, Mariana Islands. Micronesica 25: 71–84.

STEADMAN, D. W. 1993a. Biogeography of Tongan birds before and after human impact. Proc. Natl. Acad. Sci. USA 90:818–822.

STEADMAN, D. W. 1993b. Birds from the To'aga Site, Ofu, American Samoa: prehistoric loss of seabirds and megapodes. Pp. 217–228 in The To'aga Site: Three Millenia of Polynesian Occupation in the Manu'a Islands, American Samoa (P. V. Kirch and T. L. Hunt, Eds.). University of California Press, Berkeley, California.

STEADMAN, D. W. 1993c. A chronostratigraphic analysis of landbird extinction on Tahuata, Marquesas Islands. J. Archeol. Soc. 23:81–94.

STEADMAN, D. W. 1995. Prehistoric extinctions of Pacific island birds: biodiversity meets zooarchaeology. Science 267:1123–1131.

STEADMAN, D. W. 1997a. Extinctions of Polynesian birds: reciprocal impacts of birds and people. Pp. 51–79 in Historical Ecology in the Pacific Islands (P. V. Kirch and T. L. Hunt, Eds.). Yale University Press, New Haven, Connecticut.

STEADMAN, D. W. 1997b. Human-caused extinctions of birds. Pp. 139–161 in Biodiversity II (M. L. Reaka-Kudia, D. E. Wilson, and E. O. Wilson, Eds.). Joseph Henry Press, Washington, D.C.

STEADMAN, D. W. 1997c. The historic biogeography and community ecology of Polynesian pigeons and doves. J. Biogeogr. 24:737–753.

STEADMAN, D. W. 1999. The prehistory of vertebrates, especially birds, on Tinian, Aguiguan, and Rota, northern Mariana Islands. Micronesica 31:319–345.

STEADMAN, D. W., AND M. INTOH. 1994. Biogeography and prehistoric exploitation of birds from Fais Island, Yap State, Federated States of Micronesia. Pac. Sci. 48:116–135.

STEADMAN, D. W., AND L. J. JUSTICE. 1998. Prehistoric exploitation of birds on Mangareva, Gambier Islands, French Polynesia. Man Cult. Oceania 14:81–98.

STEADMAN, D. W., AND P. V. KIRCH. 1990. Prehistoric extinction of birds on Mangaia, Cook Islands, Polynesia. Proc. Natl. Acad. Sci. USA 87:9605–9609.

STEADMAN, D. W., AND P. V. KIRCH. 1998. Biogeography and prehistoric exploitation of birds in the Mussau Islands, Bismarck Archipelago, Papua New Guinea. Emu 98:13–22.

STEADMAN, D. W., AND S. L. OLSON. 1985. Bird remains from an archaeological site on Henderson Island, South Pacific: man-caused extinctions on an "uninhabited" island. Proc. Natl. Acad. Sci. USA 82:6191–6195.

STEADMAN, D. W., AND D. S. PAHLAVAN. 1992. Extinction and biogeography of birds on Huahine, Society Islands, French Polynesia. Geoarchaeology 7:449–483.

STEADMAN, D. W., D. S. PAHLAVAN, AND P. V. KIRCH. 1990. Extinction, biogeography, and human exploitation of birds on Tikopia and Anuta, Polynesian outliers in the Solomon Islands. Bishop Mus. Occas. Pap. 30:118–153.

STEADMAN, D. W., AND B. ROLETT. 1996. A chronostratigraphic analysis of landbird extinction on Tahuata Marquesas Islands. J. Archaeol. Sci. 23:81–94.

STEADMAN, D. W., C. VARGAS, AND F. CRISTINO. 1994. Stratigraphy, chronology, and cultural context of early faunal assemblage from Easter Island. Asian Perspect. 33:79–96.

STEADMAN, D. W., J. P. WHITE, AND J. ALLEN. 1999. Prehistoric birds from New Ireland, Papua New Guinea: extinctions on a large Melanisian island. Proc. Natl. Acad. Sci. USA 96:2563–2568.

STEADMAN, D. W., T. H. WORTHY, A. J. ANDERSON, AND R. WALTER. 2000. New species and records of birds from prehistoric sites on Niue, Southwest Pacific. Wilson Bull. 112:165–186.

STEARNS, S. C. 1976. Life-history tactics: a review of the ideas. Q. Rev. Biol. 51:3–47.

STEARNS, S. C. 1977. The evolution of life-history tactics: a critique of the theory and a review of the data. Annu. Rev. Ecol. Syst. 8:147–171.

STEARNS, S. C. 1982. The role of development in the evolution of life histories. Pp. 237–258 in Evolution and Development (J. T. Bonner, Ed.). Springer-Verlag, Berlin, Germany.

STEARNS, S. C. 1986. Natural selection and fitness, adaptation and constraint. Pp. 23–44 in Patterns and Processes in the History of Life (R. M. Raup and D. Jablonski, Eds.). Springer-Verlag, Berlin, Germany.

STEARNS, S. C. 1989a. Comparative and experimental approaches to the evolutionary ecology of development. Pp. 349–355 in Ontogenèse et Évolution (B. David, J.-L. Dommergues, J. Chaline, and B. Laurin, Coords.). Géobios Mém. Spéc. 12. University Claude-Bernard, Lyon, France.

STEARNS, S. C. 1989b. Trade-offs in life-history evolution. Funct. Ecol. 3:259–268.

STEARNS, S. C. 1989c. The evolutionary significance of phenotypic plasticity. BioScience 39:436–445.

STEARNS, S. C. 1992. The Evolution of Life Histories. Oxford University Press, Oxford, United Kingdom.

STEARNS, S. C. 1993. The evolutionary links between fixed and variable traits. Acta Palaeontol. Pol. 38.1–17.

STEARNS, S. C., AND R. E. CRANDALL. 1981a. Quantitative predictions of delayed maturity. Evolution 35:455–463.

STEARNS, S. C., AND R. E. CRANDALL. 1981b. Bet-hedging and persistence as adaptations of colonizers. Pp. 371–383 in Evolution Today: Proceedings of the Second International Congress of Systematic and Evolutionary Biology (G. G. E. Scudder and J. L. Reveal, Eds.). Hunt Institute of Botanical Documentation, Carnegie Mellon University, Pittsburgh, Pennsylvania.

STEARNS, S. C., AND J. C. KOELLA. 1986. The evolution of phenotypic plasticity in life-history traits: predictions of reaction norms for age and size at maturity. Evolution 40:893–914.

STEINER, H. 1922. Die ontogenetische und phylogenetische Entwicklung des Vogelflügelskelettes. Acta Zool. 25:307–360.

STEINER, H. 1956. Die taxonomische und phylogenetische Bedeutung der Diastataxie des Vogelflügels. J. Ornithol. 97:1–20.

STEINMETZ, H. 1930. Die Embryonalentwicklung der Bläßhuhns (*Fulica atra*) unter besonderer Berücksichtigung der Allantois. Gegenbaurs Morphol. Jahrb. 64:275–338.

STEPHAN, B. 1992. Vorkommen und Ausbildung der Fingerkrallen bei rezenten Vögeln. J. Ornithol. 133:251–277.

STEPHENS, D. W., AND J. R. KREBS. 1986. Foraging Theory. Princeton University Press, Princeton, New Jersey.
STERN, D. L., AND P. R. GRANT. 1996. A phylogenetic reanalysis of allozyme variation among populations of Galápagos finches. Zool. J. Linnean Soc. 118:119–134.
STETTENHEIM, P. 1972. The integument of birds. Pp. 1–63 in Avian Biology. Vol. 2 (D. S. Farner, J. R. King, and K. C. Parkes, Eds.). Academic Press, New York, New York.
STEVENS, L. 1991. Genetics and Evolution of the Domestic Fowl. Cambridge University Press, Cambridge, United Kingdom.
STINSON, C. H. 1979. On the selective advantage of fratricide in raptors. Evolution 33:1219–1225.
STINSON, D. W., M. W. RITTER, AND J. D. REICHEL. 1991. The Mariana Common Moorhen: decline of an island endemic. Condor 93:38–43.
STODDARD, H. L., SR. 1962. Bird casualties at a Leon County, Florida TV tower, 1955–1961. Tall Timbers Res. Stn. Bull. 1:1–93.
STOKELY, P. S. 1947. Limblessness and correlated changes in the girdles of a comparative morphological series of lizards. Am. Midl. Nat. 38:725–754.
STOKES, T. 1979. On the possible existence of the New Caledonian Wood Rail *Tricholimnas lafresnayanus*. Bull. Br. Ornithol. Club 99:47–54.
STOLPE, M. 1932. Physiologisch-anatomische Untersuchungen über die hintere Extremität der Vögel. J. Ornithol. 80:160–247.
STOLPE, M., AND K. ZIMMER. 1939b. Der Vogelflug: Seine anatomisch-physiologischen und physikalisch-aerodynamischen Grundlagen. Akademie Verlag, Leipzig, Germany.
STONE, J. R. 1996. Computer-simulated shell size and shape variation in the Caribbean island snail genus *Cerion*: a test of geometrical constraints. Evolution 50:341–347.
STONE, J. R. 1997. The spirit of D'Arch Thompson swells in emprirical morphospace. Math. Biosci. 142:13–30.
STONE, J. R. 1998. Ontogenic tracks and evolutionary vestiges in morphospace. Biol. J. Linn. Soc. 64:223–238.
STONEHOUSE, B. 1967. The general biology and thermal balances of penguins. Pp. 131–196 in Advances in Ecological Research. Vol. 4 (J. B. Cragg, Ed.). Academic Press, London, United Kingdom.
STONEHOUSE, B. 1975. Introduction: the Spheniscidae. Pp. 1–15 in The Biology of Penguins (B. Stonehouse, Ed.). University Park Press, Baltimore, Maryland.
STONOR, C. R. 1942. Anatomical notes on the New Zealand Wattled Crow (*Callæas*), with especial reference to its powers of flight. Ibis 84:1–18.
STORER, R. W. 1945. Structural modifications in the hind limb in the Alcidae. Ibis 87:433–456.
STORER, R. W. 1960a. Evolution in the diving birds. Pp. 694–707 in Proceedings of the XII International Ornithological Congress, Helsinki, 1958. Vol. 2 (G. Bergman, K. O. Donner, and L. V. Hoartmann, Eds.). Tigmannin, Kirjapaino, Helsinki, Finland.
STORER, R. W. 1960b. The classification of birds. Pp. 57–93 in Biology and Comparative Physiology of Birds. Vol. 1 (A. J. Marshall, Ed.). Academic Press, New York, New York.
STORER, R. W. 1971a. Adaptive radiation of birds. Pp. 149–188 in Avian Biology. Vol. 1 (D. S. Farner and J. R. King, Eds.). Academic Press, New York, New York.
STORER, R. W. 1971b. Classification of birds. Pp. 1–19 in Avian Biology. Vol. 1 (D. S. Farner and J. R. Kings, Eds.). Academic Press, New York, New York.
STORER, R. W. 1981. The Rufous-faced Crake (*Laterallus xenopterus*) and its Paraguayan congeners. Wilson Bull. 93:137–144.
STORER, R. W. 1989. Notes on Paraguayan birds. Univ. Mich. Mus. Zool. Occas. Pap. 719:1–21.
STOWE, T. J., A. V. NEWTON, R. E. GREEN, AND E. MAYES. 1993. The decline of the Corncrake *Crex crex* in Britain and Ireland in relation to habitat. J. Appl. Ecol. 30:53–62.
STRANECK, R., B. C. LIVEZEY, AND P. S. HUMPHREY. 1983. Predation on steamer-ducks by killer whale. Condor 85:255–256.
STRAUSS, R. E., AND F. L. BOOKSTEIN. 1982. The truss: body form reconstructions in morphometrics. Syst. Zool. 31:113–135.
STREICHER, J., AND G. B. MÜLLER. 1992. Natural and experimental reduction of the avian fibula: developmental thresholds and evolutionary constraint. J. Morphol. 214:269–285.
STRESEMANN, E. 1931. Vorläufiges über die ornithologischer Ergebnisse der Expedition Heinrich 1930–1931. Ornithol. Monatsber. 39:167–171.

STRESEMANN, E. 1932. La structure des rémiges chez quelques rales physiologiquement aptères. Alauda 4:1–5.
STRESEMANN, E. 1941. Die Vögel von Celebes. J. Ornithol. 1:1–102.
STRESEMANN, E. 1953. Birds collected by Capt. Dugald Carmichael on Tristan da Cunha 1816–1817. Ibis 95:146–147.
STRESEMANN, E. 1958. Wie hat die Dronte (*Raphus cucullatus* L.) ausgesehen? J. Ornithol. 99:441–459.
STRESEMANN, E. 1963. Variations in the number of primaries. Condor 65:449–459.
STRESEMANN, E., AND V. STRESEMANN. 1966. Die Mauser der Vögel. J. Ornithol. (Sonderheft) 107:1–448.
STRICKLAND, H. E. 1859. On some bones of birds allied to the Dodo, in the collection of the Zoological Society of London. Trans. Zool. Soc. Lond. 4:187–196.
STRICKLAND, H. E., AND A. G. MELVILLE. 1848. The Dodo and Its Kindred; Or the History, Affinities and Osteology of the Dodo, Solitaire and Other Extinct Birds of the Islands Mauritius, Rodriquez, and Bourbon. Reeve, Benham, and Reeve, London, England.
STRONG, D. R. 1980. Null hypotheses in ecology. Synthése 43:271–286.
STRONG, D. R., D. SIMBERLOFF, L. G. ABELE, AND A. B. THISTLE, EDS. 1984. Ecological Communities: Conceptual Issues and the Evidence. Princeton University Press, Princeton, New Jersey.
STRONG, D. R., JR., L. A. SZYSKA, AND D. S. SIMBERLOFF. 1979. Tests of community-wide character displacement against null hypotheses. Evolution 33:897–913.
STROPHLET, J. J. 1946. Birds of Guam. Auk 65:534–540.
STUDD, M. V., AND R. J. ROBERTSON. 1985. Life span, competition, and delayed plumage maturation in male passerines: the breeding threshold hypothesis. Am. Nat. 126:101–105.
SUAREZ, R. K. 1996. Upper limits to mass-specific metabolic rates. Annu. Rev. Physiol. 58:583–605.
SULLIVAN, G. E. 1962. Anatomy and embryology of the wing musculature of the domestic fowl (*Gallus*). Aust. J. Zool. 10:458–518.
SULLIVAN, J. O. 1965. "Flightlessness" in the Dipper. Condor 67:537–538.
SULLIVAN, M. S., AND N. HILLGARTH. 1993. Mating system correlates of tarsal spurs in the Phasianidae. J. Zool. Lond. 231:203–214.
SULTAN, S. E., AND H. G. SPENCER. 2002. Metapopulation structure favors plasticity over local adaptation. Am. Nat. 160:271–283.
SUMMERS, R. W., D. C. JARDINE, AND M. MARQUISS, AND R. RAE. 2002. The distribution and habitats of crossbills *Loxia* spp. in Britain, with special reference to the Scottish Crossbill *Loxia scotica*. Ibis 144:393–410.
SUNDBERG, P. 1989. Shape and size-constrained principal component analysis. Syst. Zool. 38:166–168.
SUNDEVALL, C. J. 1886. On the wings of birds. Ibis 28:389–457.
SUTHERLAND, W. I. 1996. Predicting the consequences of habitat loss for migratory populations. Proc. R. Soc. Lond. (Ser. B Biol. Sci.) 263:1325–1327.
SUTHERLAND, W. J. 1998. Evidence for flexibility and constraint in migration systems. J. Avian Biol. 29.441–446.
SUTHERLAND, W. J., AND P. D. DOLMAN. 1994. Combining behavior and population dynamics with applications for predicting the consequences of habitat loss. Proc. R. Soc. Lond. (Ser. B Biol. Sci.) 255:133–138.
SUTTER, G. C., AND R. A. MACARTHUR. 1992. Development of thermoregulation in a precocial aquatic bird, the American Coot (*Fulica americana*). Comp. Biochem. Physiol. (Ser. A) 101:533–543.
SUTTIE, J. M., AND P. F. FENNESSY. 1992. Organ weight and weight relationships in Takahe and Pukeko. Notornis 39:47–53.
SWADDLE, J. P., AND M. S. WITTER. 1997. The effects of molt on the flight performance, body mass, and behavior of European Starlings (*Sturnus vulgaris*): an experimental approach. Can. J. Zool. 75:1135–1146.
SWARTH, H. S. 1934. The bird fauna of the Galápagos Islands in relation to species formation. Biol. Rev. 9:213–234.
SWEET, S. S. 1980. Allometric inference in morphology. Am. Zool. 20:643–652.
SWINGLAND, I. R. 1983. Intraspecific differences in movement. Pp. 102–115 *in* The Ecology of Animal Movement (I. R. Swingland and P. J. Greenwood, Eds.). Clarendon Press, Oxford, United Kingdom.

SWINGLAND, I. R., AND P. J. GREENWOOD, EDS. 1983. The Ecology of Animal Movement. Clarendon Press, Oxford, United Kingdom.
SWINTON, W. E. 1958. Fossil Birds. British Museum (Natural History), London, United Kingdom.
SWOFFORD, D. L., G. J. OLSEN, P. J. WADDELL, AND D. M. HILLIS. 1996. Phylogenetic inference. Pp. 407–514 in Molecular Systematics. 2nd ed. (D. M. Hillis, C. Moritz, and B. K. Mable, Eds.). Sinauer Assoc., Sunderland, Massachusetts.
SY, M. 1936. Funktionell-anatomische Untersuchungen am Vogelflügel. J. Ornithol. 84:199–296.
SZARO, R. C. 1985. Avian use of a desert riparian island and its adjacent scrub habitat. Condor 87: 511–519.
SZÉKELY, T., J. D. REYNOLDS, AND J. FIGUEROLA. 2000. Sexual size dimorphism in shorebirds, gulls, and alcids: the influence of sexual and natural selection. Evolution 54:1404–1413.
TABACHNICK, B. G., AND L. S. FIDELL. 1989. Using Multivariate Statistics. Harper and Row, New York, New York.
TABER, W. B., JR. 1932a. Curvature of wing and soaring flight. Wilson Bull. 44:19–22.
TABER, W. B., JR. 1932b. Curvature of wing and flapping flight. Wilson Bull. 44:75–78.
TABORSKY, B., AND M. TABORSKY. 1999. The mating system and stability of pairs in kiwi *Apteryx* spp. J. Avian Biol. 30:143–151.
TABORSKY, M. 1988. Kiwis and dog predation: observations in Waitangi State Forest. Notornis 35: 1–13.
TALLAMY, D. W. 1983. Equilibrium biogeography and its application to insect host–parasite systems. Am. Nat. 121:244–254.
TANNER, J. M. 1949. Fallacy of per-weight and per-surface area standards and their relation to spurious correlation. J. Appl. Physiol. 2:1–15.
TARR, C. L., AND R. C. FLEISCHER. 1995. Evolutionary relationships of the Hawaiian honeycreepers. Pp. 147–159 in Hawaiian Biogeography: Evolution on a Hot Spot Archipelago (W. L. Wagner and W. A. Funk, Eds.). Smithsonian Institution Press, Washington, D.C.
TATNER, P. 1984. Body component growth and composition of the Magpie *Pica pica*. J. Zool. Lond. 203:397–410.
TAVERNER, P. A. 1943. Do fishes prey upon sea-birds? Ibis 85:347.
TAYLOR, C. R. 1973. Energy cost of animal locomotion. Pp. 23–42 in Comparative Physiology: Locomotion, Respiration, Transport and Blood (L. Bolis, K. Schmidt-Nielsen, and S. H. P. Maddrell, Eds.). North-Holland Publishing, Amsterdam, The Netherlands.
TAYLOR, C. R. 1977. The energetics of terrestrial locomotion and body size in vertebrates. Pp. 127–141 in Scale Effects in Animal Locomotion (T. J. Pedley, Ed.). Academic Press, London, United Kingdom.
TAYLOR, C. R. 1980. Scaling limits of metabolism to body size: implications for animal design. Pp. 161–170 in A Companion to Animal Physiology (C. R. Taylor, K. Johansen, and L. Bolis, Eds.). Cambridge University Press, Cambridge, United Kingdom.
TAYLOR, C. R., R. DMI'EL, M. FEDAK, AND K. SCHMIDT-NIELSEN. 1971. Energetic cost of running and heat balance in a large bird, the Rhea. Am. J. Physiol. 221:597–601.
TAYLOR, C. R., N. C. HEGLUND, AND G. M. O. MALOIY. 1982. Energetics and mechanics of terrestrial locomotion. I. Metabolic energy consumption as a function of speed and body size in birds and mammals. J. Exp. Biol. 97:1–21.
TAYLOR, C. R., AND E. R. WEIBEL. 1981. Design of the mammalian respiratory system. I. Problem and strategy. Respir. Physiol. 44:1–10.
TAYLOR, P. B. 1996. Family Rallidae (rails, gallinules and coots). Pp. 108–209 in Handbook of the Birds of the World. Vol. 3: Hoatzin to Auks (J. del Hoyo, A. Elliott and J. Sargatal, Eds.). Lynx Edicions, Barcelona, Spain.
TAYLOR, P. B. 1997. Species accounts (Rallidae). Pp. 318–343 in The Atlas of Southern African Birds. Vol. 1 (J. A. Harrison, D. G. Allan, L. G. Underhill, C. J. Brown, A. J. Tree, V. Parker, and M. Herremans, Eds.). BirdLife South Africa, Johannesburg, Republic of South Africa.
TAYLOR, P. B. 1998. Rails: A Guide to the Rails, Crakes, Gallinules and Coots of the World. Yale University Press, New Haven, Connecticut.
TAYLOR, P. J. 1987. Historical versus selectionist explanations in evolutionary biology. Cladistics 3: 1–13.
TAYLOR, P. J. 1998. Regional patterns of small mammal abundance and community composition in protected areas in KwaZulu-Natal. Durban Mus. Novit. 23:42–51.

TAYLOR, R. J., AND P. J. REGAL. 1978. The peninsular effect on species diversity and the biogeography of Baja California. Am. Nat. 112:583–593.

TAYLOR, ST. C. S. 1968. Time taken to mature in relation to mature weight for sexes, strains and species of domesticated mammals and birds. Anim. Prod. 10:157–169.

TCHERNOV, E., O. RIEPPEL, H. ZAHER, M. J. POLCYN, AND L. L. JACOBS. 2000. A fossil snake with limbs. Science 287:2010–2013.

TEISSIER, G. 1948. La relation d'allométrie: sa signification statistique et biologique. Biometrics 4: 14–48.

TEISSIER, G. 1955. Allométrie de taille et variabilité chez *Maia squinado*. Arch. Zool. Exp. Gén. 92: 221–264.

TEIXEIRA, D. M., J. B. NACINOVIC, AND G. LUIGI. 1989. Notes on some birds of northeastern Brazil (4). Bull. Br. Ornithol. Club 109:152–157.

TEIXEIRA, D. M., J. B. NACINOVIC, AND M. S. TAVARES. 1986. Notes on some birds of northeastern Brazil. Bull. Br. Ornithol. Club 106:70–74.

TELLERÍA, J. L., J. PÉREZ-TRIS, AND R. CARBONELL. 2002. Seasonal changes in abundance and flight-related morphology reveal different migration patterns in Iberian forest passerines. Ardeola 48: 27–46.

TELLERIA, J. L., AND T. SANTOS. 1999. Distribution of birds in fragments of Mediterranean forests: the role of ecological densities. Ecography 22:13–19.

TEMME, D. H., AND E. L. CHARNOV. 1987. Brood size adjustment in birds: economical tracking in a temporally varying environment. J. Theor. Biol. 126:137–147.

TEMPLE, R. C., ED. 1919. The Travels of Peter Mundy in Europe and Asia, 1608–1667. Vol. 3. Part 2. Hakluyt Society, London, England.

TEMPLE, R. C., AND L. M. ANSTEY, EDS. 1936. The Travels of Peter Mundy in Europe and Asia 1608–1667. Vol. 5. Hakluyt Society, London, United Kingdom.

TEMPLE, S. A. 1981. Applied island biogeography and the conservation of endangered island birds in the Indian Ocean. Biol. Conserv. 20:147–161.

TEMPLE, S. A. 1985. Why endemic island birds are so vulnerable to extinction. Pp. 3–6 in Bird Conservation 2 (S. A. Temple, Ed.). University of Wisconsin Press, Madison, Wisconsin.

TEMPLE, S. A. 1986. The problem of avian extinctions. Pp. 453–485 in Current Ornithology. Vol. 3 (R. F. Johnston, Ed.). Plenum Press, New York, New York.

TEMPLETON, A. R. 1980. The theory of speciation *via* the founder principle. Genetics 94:1011–1038.

TEMPLETON, A. R. 1981. Mechanisms of speciation—a population genetic approach. Annu. Rev. Ecol. Syst. 12:23–48.

TEMPLIN, R. J. 1976. Insularity and its effect on mammal evolution. Pp. 411–421 in Major Patterns in Vertebrate Evolution (M. K. Hecht, P. C. Goody, and B. M. Hecht, Eds.). Plenum Press, New York, New York.

TEMRIN, H., AND B. SILLÉN-TULLBERG. 1994. The evolution of avian mating systems: a phylogenetic analysis of male and female polygamy and length of pair bond. Biol. J. Linn. Soc. 52:121–149.

TEMRIN, H., AND B. SILLÉN-TULLBERG. 1995. A phylogenetic analysis of the evolution of avian mating systems in relation to altricial and precocial young. Behav. Ecol. 6:296–307.

TENNYSON, A. J. D., AND P. R. MILLENER. 1994. Bird extinctions and fossil bones from Mangere Island, Chatham Islands. Notornis 41 (Suppl.):165–178.

TERBORGH, J. 1973. Chance, habitat and dispersal in the distribution of birds in the West Indies. Evolution 27:338–349.

TERBORGH, J. 1974. Preservation of natural diversity: the problem of extinction-prone species. BioScience 24:715–722.

TERBORGH, J., AND J. FAABORG. 1980. Saturation of bird communities in the West Indies. Am. Nat. 116:178–195.

TERBORGH, J., J. FAABORG, AND J. BROCKMANN. 1978. Island colonization by Lesser Antillean birds. Auk 95:59–72.

TERBORGH, J., AND B. WINTER. 1980. Some causes of extinction. Pp. 119–134 in Conservation Biology: An Evolutionary–Ecological Perspective (M. E. Soulé and B. Wilcox, Eds.). Sinauer Assoc., Sunderland, Massachusetts.

TERMAAT, B. M., AND J. P. RYDER. 1984. Differences in skeletal characters between the disjunct eastern and western populations of Ring-billed Gulls (*Larus delawarensis*). Can. J. Zool. 62:1067–1074.

THEWISSEN, J. G. M., AND F. E. FISH. 1997. Locomotor evolution in the earliest cetaceans: functional model, modern analogues, and paleontological evidence. Paleobiology 23:482–490.

THIEDE, U. 1982. "Yambaru kuina" (*Rallus okinawae*), eine neu entdeckte Rallenart in Japan. Vogelwelt 103:143–150.

THINES, G. 1969. L'evolution Régressive des Poissons Cavernicoles et Abyssaux. Masson, Paris, France.

THIOLLAY, J. M. 1994. Family Acciptridae (hawks and eagles). Pp. 52–105 *in* Handbook of the Birds of the World. Vol. 2: New World Vultures to Guineafowl (J. del Hoyo, A. Elliott and J. Sargatal, Eds.). Lynx Edicions, Barcelona, Spain.

THODAY, J. M. 1972. Review lecture: disruptive selection. Proc. R. Soc. Lond. (Ser. A) 182:109–143.

THOM, R. 1975. Structural Stability and Morphogenesis. Benjamin Cummings, Reading, Massachusetts.

THOM, R. 1983. Mathematical Models of Morphogenesis. Ellis Horwood, Chichester, United Kingdom.

THOMAS, A. L. R. 1993. On the aerodynamics of birds' tails. Philos. Trans. R. Soc. Lond. (Ser. B) 340:361–380.

THOMAS, A. L. R. 1996a. The flight of birds that have wings and tail: variable geometry expands the envelope of flight performance. J. Theor. Biol. 183:237–245.

THOMAS, A. L. R. 1996b. Why do birds have tails: the tail as a drag reducing flap, and trim control? J. Theor. Biol. 183:247–253.

THOMAS, A. L. R., AND A. BALMFORD. 1995. How natural selection shapes birds' tails. Am. Nat. 146:848–868.

THOMAS, A. L. R., AND A. HEDENSTRÖM. 1998. The optimum flight speeds of flying animals. J. Avian Biol. 29:469–477.

THOMAS, D. W., C. BOSQUE, AND A. ARENDS. 1993. Development of thermoregulation and the energetics of nestling Oilbirds (*Steatornis caripensis*). Physiol. Zool. 66:422–348.

THOMAS, R. D. K., AND W.-E. REIF. 1993. The skeleton space: a finite set of organic designs. Evolution 47:341–360.

THOMAS, W. L., JR. 1965. The variety of physical environments among Pacific islands. Pp. 7–37 *in* Man's Place in the Island Ecosystem (F. R. Fosberg, Ed.). Bishop Museum Press, Honolulu, Hawaii.

THOMASON, D. L., P. MONAGHAN, AND R. W. FURNESS. 1998. The demands of incubation and avian clutch size. Biol. Rev. 73:293–304.

THOMPSON, C. W. 1991. The sequence of molts and plumages in Painted Buntings and implications for theories of delayed plumage maturation. Condor 93:209–235.

THOMPSON, D'A. 1961. On Growth and Form, Abridged. Cambridge University Press, Cambridge, United Kingdom.

THOMPSON, J. N. 1999. Specific hypotheses on the geographic mosaic of coevolution. Am. Nat. 153 (Suppl.):1–14.

THOMPSON, M. C., AND C. ELY. 1989. Birds in Kansas. Vol. I. Museum of Natural History, University Kansas Press, Lawrence, Kansas.

THOMSON, D. L., P. MONAGHAN, AND R. W. FURNESS. 1998. The demands of incubation and avian clutch size. Biol. Rev. 73:293–304.

THOMSON, K. S. 1985. Essay review: the relationship between development and evolution. Pp. 220–233 *in* Oxford Series in Evolution Biology. Vol. 2 (R. Dawkins and M. Ridley, Eds.). Oxford University Press, Oxford, United Kingdom.

THORNTON, I. W. B. 1967. The measurement of isolation on archipelagos, and its relation to insular faunal size and endemism. Evolution 21:842–849.

THORPE, R. S. 1988. Multiple group principal component analysis and population differentiation. J. Zool. Lond. 216:37–40.

THORPE, R. S., H. BLACK, AND A. MALHOTRA. 1996. Matrix correspondence tests on the DNA phylogeny of the Tenerife lacertid elucidates both historical causes and morphological adaptation. Syst. Biol. 45:335–343.

THORPE, R. S., AND A. MALHOTRA. 1996. Molecular and morphological evolution within small islands. Philos. Trans. R. Soc. Lond. (Ser. B) 351:815–822.

THORPE, R. S., A. MALHOTRA. 1998. Molecular and morphological evolution within small islands. Pp. 67–82 *in* Evolution on Islands (P. R. Grant, Ed.). Oxford University Press, Oxford, United Kingdom.

THULBORN, R. A. 1984. The avian relationships of *Archaeopteryx,* and the origin of birds. Zool. J. Linnean Soc. 82:119–158.

TILMAN, D., AND C. L. LEHMAN. 1997. Habitat destruction and species extinctions. Pp. 223–249 *in* Spatial Ecology: The Role of Space in Population Dynamics and Interspecific Interactions (D. Tilman and P. Kareiva, Eds.). Monogr. Popul. Biol. 30. Princeton University Press, Princeton, New Jersey.

TINBERGEN, N. 1957. The functions of territory. Bird Study 4:14–27.

TIPTON, A. T. 1962. Myology of the Limpkin. Ph.D. dissertation, University of Florida, Gainesville, Florida.

TITTERINGTON, D. M. 1985. Statistical Analysis of Finite Mixture Distributions. J. Wiley and Sons, New York, New York.

TOBALSKE, B. W. 1996. Scaling of muscle composition, wing morphology, and intermittent flight behavior in woodpeckers. Auk 113:151–177.

TOBALSKE, B. W., AND K. P. DIAL. 1996. Flight kinematics of Black-billed Magpies and Pigeons over a wide range of speeds. J. Exp. Biol. 199:263–280.

TOBALSKE, B. W., AND K. P. DIAL. 2000. Effects of body size on take-off flight performance in the Phasianidae (Aves). J. Exp. Biol. 203:3319–3332.

TOLLRIAN, R. 1995. Predator-induced morphological defenses: costs, life history shifts, and maternal effects in *Daphnia pulex.* Ecology 76:1691–1705.

TORDOFF, H. B. 1952. Genera of birds bearing vestigial claws on the wings. Auk 69:200–201.

TORDOFF, H. B., AND R. M. MENGEL. 1956. Studies of birds killed in nocturnal migration. Univ. Kans. Mus. Nat. Hist. Publ. 10:1–44.

TORRE-BUENO, J. R., AND J. LAROCHELLE. 1978. The metabolic cost of flight in unrestrained birds. J. Exp. Biol. 78:223–229.

TOWNSEND, C. W. 1909. The use of wings and feet by diving birds. Auk 26:234–248.

TOWNSEND, C. W. 1924. Diving of grebes and loons. Auk 41:29–41.

TRACY, C. R., AND T. L. GEORGE. 1992. On the determinants of extinction. Am. Nat. 139:102–122.

TRAVERS, W. T. L. 1872. On the *Birds of the Chatham Islands,* by H. H. Travers, with introductory remarks on the avi-fauna and flora of the islands in their relation to those of New Zealand. Trans. N. Z. Inst. 5:212–222.

TRAVERS, W. T. L. 1882. Remarks upon the distribution within the New Zealand zoological sub-region of the birds of the orders *Accipitres, Passeres, Scansores, Columbæ, Gallinæ, Struthiones,* and *Grallæ.* Trans. N. Z. Inst. 15:178–187.

TRAVERS, W. T. L. 1883. Some remarks upon the distribution of the organic production of New Zealand. Trans. N. Z. Inst. 16:461–467.

TRAVIS, J. 1994. Evaluating the adaptive role of morphological plasticity. Pp. 99–122 *in* Ecological Morphology: Integrative Organismal Biology (P. C. Wainwright and S. M. Reilly, Eds.). University of Chicago Press, Chicago, Illinois.

TRAVIS, J., AND R. E. RICKLEFS. 1983. A morphological comparison of island and mainland assemblages of Neotropical birds. Oikos 41:434–441.

TREVELYAN, R., P. H. HARVEY, AND M. D. PAGEL. 1990. Metabolic rates and life histories in birds. Funct. Ecol. 4:135–141.

TREWICK, S. A. 1996a. Morphology and evolution of two takahe: flightless rails of New Zealand. J. Zool. 238:221–237.

TREWICK, S. A. 1996b. The diet of Kakapo (*Strigops habroptilus*), Takahe (*Porphyrio mantelli*) and Pukeko (*P. porphyrio melanotus*) studied by faecal analysis. Notornis 43:79–84.

TREWICK, S. A. 1997a. Flightlessness and phylogeny amongst endemic rails (Aves: Rallidae) of the New Zealand region. Philos. Trans. R. Soc. Lond. (Ser. B) 352:429–446.

TREWICK, S. A. 1997b. Sympatric flightless rails *Gallirallus dieffenbachii* and *G. modestus* on the Chatham Islands, New Zealand: morphometrics and alternative scenarios. J. R. Soc. N. Z. 27:451–464.

TREWICK, S. A. 1997c. On the skewed sex-ratio of the Kakapo *Strigops habroptilus*: sexual and natural selection in opposition? Ibis 139:652–663.

TREWICK, S. A., AND T. H. WORTHY. 2001. Origins and prehistoric ecology of the Takahe based on morphometric, molcular, and fossil data. Pp. 31–48 *in* The Takahe: Fifty Years of Conservation Management and Research. University of Otago Press, Dunedin, New Zealand.

TRIBUTSCH, H. 1984. How Life Learned to Live: Adaptation in Nature. MIT Press, Cambridge, Massachusetts.

TRIMBLE, S. A. 1976. Galápagos mockingbird pecks at sea lion mouth. Condor 78:567.

TRIVERS, R. L. 1971. The evolution of reciprocal altruism. Q. Rev. Biol. 46:35–57.

TRIVERS, R. L. 1972. Parental investment and sexual selection. Pp. 136–175 *in* Sexual Selection and the Descent of Man: 1871–1971 (B. G. Campbell, Ed.). Aldine Publishing, Chicago, Illinois.

TRUEB, L. 1973. Bones, frogs, and evolution. Pp. 65–132 *in* Evolutionary Biology of the Anurans: Contemporary Research on Major Problems (J. L. Vial, Ed.). University of Missouri Press, Columbia, Missouri.

TRUEB, L., AND P. ALBERCH. 1985. Miniaturization and the anuran skull: a case study of heterochrony. Pp. 113–121 *in* Functional Morphology in Vertebrates (H.-R. Duncker and G. Fleischer, Eds.). Gustav Fischer Verlag, New York, New York.

TRUEMAN, A. E. 1922. The use of *Gryphaea* in the correlation of the Lower Lias. Geol. Mag. 59: 256–268.

TUCKER, V. A. 1970. Energetic cost of locomotion in animals. Comp. Biochem. Physiol. (Ser. A) 34: 841–846.

TUCKER, V. A. 1971. Flight energetics in birds. Am. Zool. 11:115–124.

TUCKER, V. A. 1973a. Aerial and terrestrial locomotion: a companion of energetics. Pp. 63–76 *in* Comparative Physiology (L. Bolis, K. Schmidt-Nielsen, and S. H. P. Maddrell, Eds.). North-Holland Publishing, Amsterdam, The Netherlands.

TUCKER, V. A. 1973b. Bird metabolism during flight: evaluation of a theory. J. Exp. Biol. 58:689–709.

TUCKER, V. A. 1974. Energetics of natural avian flight. Pp. 298–334 *in* Avian Energetics (R. A. Paynter, Ed.). Publ. No. 15. Nuttall Ornithologists' Club, Cambridge, Massachusetts.

TUCKER, V. A. 1975. Flight energetics. Symp. Zool. Soc. Lond. 35:49–63.

TUCKER, V. A. 1977. Scaling and avian flight. Pp. 497–509 *in* Scale Effects in Animal Locomotion (T. J. Pedley, Ed.). Academic Press, New York, New York.

TUCKER, V. A. 1987. Gliding birds: the effect of variable wing span. J. Exp. Biol. 133:33–58.

TUCKER, V. A., AND G. C. PARROTT. 1970. Aerodynamics of gliding flight in a falcon and other birds. J. Exp. Biol. 52:345–367.

TUCKER, V. A., AND K. SCHMIDT-KOENIG. 1971. Flight speeds of birds in relation to energetics and wind direction. Auk 88:97–107.

TURBOTT, E. G. 1951a. Winter observations on *Notornis*. Notornis 4:107–113.

TURBOTT, E. G. 1951b. A selected bibliography on *Notornis*. Notornis 4:115–117.

TURBOTT, E. G., ED. 1967. Buller's Birds of New Zealand. Whitcombe and Tombs, Christchurch, New Zealand.

TURBOTT, E. G., CONVENER. 1990. Checklist of the Birds of New Zealand and the Ross Dependency, Antractica. 3rd ed. Ornithologists' Society of New Zealand, Wellington, New Zealand.

TURCEK, F. J. 1966. On plumage quantity in birds. Ekol. Pol. (Ser. A) 14:617–634.

TURNER, G. F. 1999. Explosive speciation of African ciclid fishes. Pp. 113–129 *in* Evolution of Biological Diversity (A. E. Magurran and R. M. May, Eds.). Oxford University Press, Oxford, United Kingdom.

UCHMANSKI, J. 1985. Differentiation and distribution of body weights of plants and animals. Philos. Trans. R. Soc. Lond. (Ser. B) 310:1–75.

ULFSTRAND, S. 1963. Ecological aspects of irruptive bird migration in northwestern Europe. Pp. 780–794 *in* Proceedings XIII International Ornithological Congress. Vol. 2 (C. G. Sibley, Ed.). American Ornithologists' Union, Washington, D.C.

UNWIN, D. M. 1988. Extinction and survival in birds. Pp. 295–318 *in* Extinction and Survival in the Fossil Record (G. P. Larwood, Ed.). Clarendon Press, Oxford, United Kingdom.

URBAN, E. K., C. H. FRY, AND S. KEITH, EDS. 1986. The Birds of Africa. Vol. 2. Academic Press, London, United Kingdom.

URGELES, R., M. CANALS, J. BRAZA, B. ALONSO, AND D. MASSON. 1997. The mosts recent megalandslides of the Canary Islands: El Golfo debris and Canary debris flow, west El Hierro Island. J. Geophys. Res. 102:20305–20323.

VAN BUSKIRK, J., AND S. A. MCCOLLUM. 1997. Natural selection for environmentally induced phenotypes in tadpoles. Evolution 51:1983–1192.
VANDEN BERGE, J. C. 1970. A comparative study of the appendicular musculature of the order Ciconiiformes. Am. Midl. Nat. 84:289–364.
VANDEN BERGE, J. C., AND G. A. ZWEERS. 1993. Myologia. Pp. 189–250 in Handbook of Avian Anatomy: Nomina Anatomica Avium. 2nd ed. (J. J. Baumel, A. S. King, J. E. Breazile, H. E. Evans, and J. C. Vanden Berge, Eds.). Publ. No. 23. Nuttall Ornithologists' Club, Cambridge, Massachusetts.
VAN DER KLAAUW, C. J. 1948. Size and position of the functional components of the skull. A contribution to the knowledge of the architecture of the skull, based on data in the literature. Arch. Neerl. Zool. 9 (Suppl.):1–176.
VAN DER KLAAUW, C. J. 1951. Size and position of the functional components of the skull. A contribution to the knowledge of the architecture of the skull, based on data in the literature (*continuation*). Arch. Neerl. Zool. 9 (Suppl.):177–368.
VAN DER KLAAUW, C. J. 1952. Size and position of the functional components of the skull. A contribution to the knowledge of the architecture of the skull, based on data in the literature (*conclusion*). Arch. Neerl. Zool. 9 (Suppl.):369–559.
VANDERMEER, J., AND R. CARVAJAL. 2001. Metapopulation dynamics and the quality of the matrix. Am. Nat. 158:211–220.
VAN NOORDWIJK, A. J., AND W. SCHARLOO. 1981. Inbreeding in an island population of the Great Tit. Evolution 35:674–688.
VAN NOORDWIJK, A. J., A. J. VAN BALEN, AND W. SCHARLOO. 1981. Genetic and environmental variation in the clutch size of the Great Tit (*Parus major*). Neth. J. Zool. 31:342–372.
VAN RHIJN, J. G. 1973. Behavioural dimorphism in male Ruffs, *Philomachus pugnax* (L.). Behaviour 47:153–229.
VAN RHIJN, J. G. 1983. On the maintenance and origin of alternative strategies in the Ruff *Philomachus pugnax*. Ibis 125:482–498.
VAN RHIJN, J. G. 1984. Phylogenetical constraints in the evolution of parental care strategies in birds. Neth. J. Zool. 34:103–122.
VAN RHIJN, J. G. 1990. Unidirectionality in the phylogeny of social organization, with special reference to birds. Behaviour 115:153–174.
VAN RHIJN, J. G. 1991. Mate guarding as a key factor in the evolution of parental care in birds. Anim. Behav. 41:963–970.
VAN RIPER, C., III, S. G. VAN RIPER, M. L. GOFF, AND M. LAIRD. 1986. The epizootiology and ecological signficance of malaria in Hawaiian land birds. Ecol. Monogr. 56:327–344.
VAN TIENDEREN, P. H. 1991. Evolution of generalists and specialists in spatially heterogeneous environments. Evolution 45:1317–1331.
VAN TIENDEREN, P. H., AND G. DE JONG. 1994. A general model of the relation between phenotypic selection and genetic response. J. Evol. Biol. 7:1–12.
VAN TIENDEREN, P. H., AND H. P. KOELEWIJN. 1994. Selection on reaction norms, genetic correlations and constraints. Genet. Res. 64:115–125.
VAN TUINEN, M., C. G. SIBLEY, AND S. B. HEDGES. 1998. Phylogeny and biogeography of ratite birds inferred from DNA sequences of the mitochondrial ribosomal genes. Mol. Biol. Evol. 15:370–376.
VAN TUINEN, M., C. G. SIBLEY, AND S. B. HEDGES. 2000. The early history of modern birds inferred from DNA sequences of nuclear and mitochondrial ribosomal genes. Mol. Biol. Evol. 17:451–457.
VAN TYNE, J., AND A. J. BERGER. 1976. Fundamentals of Ornithology. 2nd ed. John Wiley and Sons, New York, New York.
VAN VALEN, L. 1971. Group selection and the evolution of dispersal. Evolution 25:591–598.
VAN VALEN, L. 1973a. Body size and numbers of plants and animals. Evolution 27:27–35.
VAN VALEN, L. 1973b. A new evolutionary law. Evol. Theory 1:1–30.
VÁZQUEZ, D. P., AND D. SIMBERLOFF. 2002. Ecological specialization and susceptibility to disturbance: conjectures and refutations. Am. Nat. 159:606–623.
VAZQUEZ, R. J. 1992. Functional osteology of the avian wrist and the evolution of flapping flight. J. Morphol. 211:259–268.

VAZQUEZ, R. J. 1994. The automating skeletal and muscular mechanisms of the avian wing (Aves). Zoomorphology 114:59–71.
VEASEY, J. S., N. B. MECALFE, AND D. C. HOUSTON. 1998. A reassessment of the effect of body mass upon flight speed and predation risk in birds. Anim. Behav. 56:883–889.
VELTMAN, C. J., S. NEE, AND M. J. CRAWLEY. 1996. Correlates of introduction success in exotic New Zealand birds. Am. Nat. 147:542–557.
VEPSÄLÄINEN, K. 1974. Determination of wing lengths and diapause in water striders (*Gerris* Fabr., Heteroptera). Hereditas 77:163–176.
VERHEYEN, R. 1958. A propos de la mue des rémiges primaires. Gerfaut 48:101–114.
VERMEIJ, G. J. 1974. Adaptation, versatility and evolution. Syst. Zool. 22:466–477.
VERMEIJ, G. J. 1987. Evolution and Escalation: An Ecological History of Life. Princeton University Press, Princeton, New Jersey.
VERMEIJ, G. J. 1994. The evolutionary interaction among species: selection, escalation, and coevolution. Annu. Rev. Ecol. Syst. 25:219–236.
VERMEIJ, G. J. 1996. Adaptations of clades: resistance and response. Pp. 363–380 *in* Adaptation (M. R. Rose and G. V. Lauder, Eds.). Academic Press, San Diego, California.
VERNER, J. 1977. On the adaptive significance of territoriality. Am. Nat. 111:769–775.
VERNER, J., AND M. F. WILSON. 1969. Mating systems, sexual dimorphism, and the role of male North American passerine birds in the nesting cycle. Ornithol. Monogr. No. 9. American Ornithologists' Union, Washington, D.C.
VESTJENS, W. J. M. 1963. Remains of the extinct Banded Rail at Macquarie Island. Emu 62:249–250.
VIA, S. 1987. Genetic constraints on the evolution of phenotypic plasticity. Pp. 47–71 *in* Genetic Constraints on Adaptive Evolution (V. Loeschcke, Ed.). Springer-Verlag, Berlin, Germany.
VIA, S. 1993a. Adaptive phenotypic plasticity: Target or by-product of selection in a variable environment? Am. Nat. 142:352–365.
VIA, S. 1993b. Regulatory genes and reaction norms. Am. Nat. 142:374–378.
VIA, S. 1994. The evolution of phenotypic plasticity: What do we really know? Pp. 35–57 *in* Ecological Genetics (L. Real, Ed.). Princeton University Press, Princeton, New Jersey.
VIA, S., AND R. LANDE. 1985. Genotype–environment interaction and the evolution of phenotypic plasticity. Evolution 39:505–522.
VIA, S., AND R. LANDE. 1987. Evolution of genetic variability in a spatially heterogeneous environment: effects of genotype–environment interaction. Genet. Res. 49:147–156.
VIÑUELA, J. 1997. Adaptation vs. constraint: intraclutch egg-mass variation in birds. J. Anim. Ecol. 66:781–792.
VISSER, G. H. 1998. Development of temperature regulation. Pp. 117–156 *in* Avian Growth and Development: Evolution Within the Altricial–Precocial Spectrum (J. M. Starck and R. E. Ricklefs, Eds.). Oxford University Press, Oxford, United Kingdom.
VISSER, J. 1974. The post-embryonic development of the Coot *Fulica atra*. Ardea 62:172–189.
VISSER, J. 1976. An evaluation of factors affecting wing length and its variability in the Coot *Fulica atra*. Ardea 64:1–21.
VISSER, J. 1978. Fat and protein metabolism in the Coot, *Fulica atra*. Ardea 66:173–183.
VISSER, J. 1988. Seasonal changes in shield size in the Coot. Ardea 76:56–63.
VITOUSEK, P. M., H. ADSERSEN, AND L. L. LOOPE. 1995. Introduction—Why focus on islands? Pp. 1–4 *in* Islands: Biological Diversity and Ecosystem Function (P. M. Vitousek, L. L. Loope, and H. Adsersen, Eds.). Springer, Berlin, Germany.
VLECK, C. M., AND T. L. BUCHER. 1998. Energy metabolism, gas exchange, and ventilation. Pp. 89–116 *in* Avian Growth and Development: Evolution Within the Altricial–Precocial Spectrum (J. M. Starck and R. E. Ricklefs, Eds.). Oxford University Press, Oxford, United Kingdom.
VLECK, C. M., D. VLECK, AND D. F. HOYT. 1980a. Patterns of metabolism and growth in avian embryos. Am. Zool. 20:405–416.
VLECK, D., C. M. VLECK, AND D. F. HOYT. 1980b. Metabolism of avian embryos: ontogeny of oxygen consumption in the Rhea and Emu. Physiol. Zool. 53:125–135.
VOGEL, S. 1988. Life's Devices: The Physical World of Animals and Plants. Princeton University Press, Princeton, New Jersey.
VOGL, C., AND G. P. WAGNER. 1990. Interspecific variability in randomly evolving clades: models for

testing hypotheses on the relative evolutionary flexibility of quantitative traits. Syst. Zool. 39: 109–123.

VOISIN, J. F. 1979. Observations ornithologiques aux îles Tristan da Cunha et Gough. Alauda 47: 73–82.

VON BAER, K. E. 1828. Über Entwicklungsgeschichte der Thiere: Beobachtung und Reflexion. Gebrüder Bornträger, Königsberg, Germany.

VON HELMHOLTZ, H. 1874. Über ein Theorem, geometrisch ähnliche Bewegungen flüssiger Körper betreffend, nebst Anwendung auf das Problem, Luftballons zu Lenken. Monatsber. Dtsch. Acad. Wiss. Berl. 1873:501–514.

VOOUS, K. H. 1961. The generic distinction of the Gough Island Flightless Gallinule. Bijdr. Dierkd. 31:75–79.

VOOUS, K. H. 1962. Notes on a collection of birds from Tristan da Cunha and Gough Island. Beaufortia 9:105–114.

VRBA, E. S. 1989. Levels of selection and sorting with special reference to the species level. Pp. 111–168 in Oxford Surveys in Evolutionary Biology. Vol. 6 (P. H. Harvey and L. Partridge, Eds.). Oxford University Press, Oxford, United Kingdom.

VUILLEUMIER, F. 1970. Insular biogeography in continental regions. I. The northern Andes of South America. Am. Nat. 104:373–388.

VUILLEUMIER, F. 1973. Insular biogeography in continental regions. II. Cave faunas from Tessin, southern Switzerland. Syst. Zool. 22:64–76.

VUILLEUMIER, F. 1975. Zoogeography. Pp. 421–495 in Avian Biology. Vol. 5 (D. S. Farner, J. R. King, and K. C. Parkes, Eds.). Plenum Press, New York, New York.

VUILLEUMIER, F., AND M. GOCHFELD. 1976. Notes sur l'avifaune de Nouvelle-Calédonie. Alauda 63: 135–148.

VUILLEUMIER, F., M. LECROY, AND E. MAYR. 1992. New species of birds described from 1981 to 1990. Bull. Br. Ornithol. Club (Cent. Suppl.) 122A:267–309.

VUILLEUMIER, F., AND D. SIMBERLOFF. 1980. Ecology versus history as determinants of patchy and insular distributions in high Andean birds. Pp. 235–379 in Evolutionary Biology. Vol. 12 (M. K. Hecht, W. C. Steere, and B. Wallace, Eds.). Plenum Press, New York, New York.

WACE, N. M., AND M. W. HOLDGATE. 1976. Man and Nature in Tristan da Cunha. Monogr. No. 6. International Union for the Conservation of Nature (IUCN), Morges, Switzerland.

WADDINGTON, C. H. 1957. The Strategy of the Genes. Allen & Unwin, London, United Kingdom.

WADE, M. J. 1978. A critical review of the models of group selection. Q. Rev. Biol. 53:101–114.

WADE, M. J. 2000. Epistasis as a genetic constraint within populations and an accelerant of adaptive divergence among them. Pp. 213–231 in Epistasis and the Evolutionary Process (J. B. Wolf, E. D. Brodie III, and M. J. Wade, Eds.). Oxford University Press, Oxford, United Kingdom.

WAGNER, A. 1996. Does evolutionary plasticity evolve? Evolution 50:1008–1023.

WAGNER, A., G. P. WAGNER, AND P. SIMILION. 1994. Epistasis can facilitate the evolution of reproductive isolation by peak shifts: a two-locus two-allele model. Genetics 138:533–545.

WAGNER, G. P. 1988a. The significance of developmental constraints for phenotypic evolution by natural selection. Pp. 222–229 in Population Genetics and Evolution (G. de Jong, Ed.). Springer-Verlag, Berlin, Germany.

WAGNER, G. P. 1988b. The influence of variation and of developmental constraints on the rate of multivariate phenotypic evolution. J. Evol. Biol. 1:45–66.

WAGNER, G. P. 1997. A population genetic theory of canalization. Evolution 51:329–347.

WAGNER, G. P., AND L. ALTENBERG. 1996. Complex adaptations and the evolution of evolvability. Evolution 50:967–976.

WAGNER, G. P., G. BOOTH, AND H. BAGHERI-CHAICHIAN. 1997. A population genetic theory of canalization. Evolution 51:329–347.

WAGNER, G. P., AND B. Y. MISOF. 1993. How can a character be developmentally constrained despite variation in developmental pathways? J. Evol. Biol. 6:449–455.

WAGNER, G. P., AND K. SCHWENK. 2000. Evolutionarily stable configurations: functional integration and the evolution of phenotypic stability. Pp. 155–217 in Evolutionary Biology. Vol. 31 (M. K. Hecht, R. J. MacIntyre, and M. T. Clegg, Eds.). Plenum Press, New York, New York.

WAGNER, P. J. 2000. Exhaustion of morphologic character states among fossil taxa. Evolution 54: 365–386.

WAGNER, W. L., S. G. WELLER, AND A. K. SAKAI. 1995. Phylogeny and biogeography in *Schiedea* and *Alsinidendron* (Caryophyllaceae). Pp. 221–258 *in* Hawaiian Biogeography: Evolution on a Hot Spot Archipelago (W. L. Wagner and W. A. Funk, Eds.). Smithsonian Institution Press, Washington, D.C.

WAGSTAFFE, R. 1978. Type Specimens of Birds in the Merseyside County Museums. Merseyside County Museums, Liverpool, United Kingdom.

WAKE, D. B. 1991. Homoplasy: The result of natural selection, or evidence of design limitations? Am. Nat. 138:543–567.

WAKE, D. B., AND A. LARSON. 1987. Multidimensional analysis of an evolving lineage. Science 238: 42–48.

WAKE, D. B., G. ROTH, AND M. H. WAKE. 1983. On the problem of stasis in organismal evolution. J. Theor. Biol. 101:211–224.

WAKE, M. H. 1992. "Regressive" evolution of special sensory organs in caecilians (Amphibia: Gymnophiona). Zoomorphology 105:277–295.

WALKER, A. F. G. 1970. The moult migration of Yorkshire Canada Geese. Wildfowl 21:99–104.

WALKINSHAW, L. H. 1937. The Virginia Rail in Michigan. Auk 54:464–475.

WALKINSHAW, L. H. 1939. The Yellow Rail in Michigan. Auk 56:227–237.

WALKINSHAW, L. H. 1940. Summer life of the Sora Rail. Auk 57:153–168.

WALLACE, A. R. 1878. Tropical Nature and Other Essays. Macmillan, London, England.

WALLACE, A. R. 1880. Island Life: Or, the Phenomenon and Causes of Insular Faunas and Floras, Including a Revision and Attempted Solution of the Problem of Geological Climates. Macmillan, London, England.

WALLACE, R. A. 1974. Ecological and social implications of sexual dimorphism in five melanerpine woodpeckers. Condor 76:238–248.

WALLACE, R. A. 1978. Social behavior on islands. Pp. 167–204 *in* Perspectives in Ethology. Vol. 3: Social Behavior (P. P. G. Bateson and P. H. Klopfer, Eds.). Plenum Press, New York, New York.

WALSBERG, G. E. 1983. Avian ecological energetics. Pp. 161–220 *in* Avian Biology. Vol. 7 (D. S. Farner, J. R. King, and K. C. Parkes, Eds.). Academic Press, New York, New York.

WALSBERG, G. E., AND J. R. KING. 1978. The relationship of the external surface area of birds to skin surface area and body mass. J. Exp. Biol. 76:185–189.

WALTER, A., AND M. R. HUGHES. 1978. Total body water volume and turnover rate in the fresh and seawater adapted Glaucous-winged Gulls. Comp. Biochem. Physiol. (Ser. A) 61:233–237.

WALTERS, M. P. 1987. The provenance of the Gilbert rail *Tricholimnas conditicius* (Peters & Griscom). Bull. Br. Ornithol. Club 107:181–184.

WALTERS, M. P. 1988. Probable validity of *Rallus nigra* Miller, an extinct species from Tahiti. Notornis 35:265–269.

WALTON, P., G. D. RUXTON, AND P. MONAGHAN. 1998. Avian diving, respiratory physiology and the marginal value theorem. Anim. Behav. 56:165–174.

WARD, D. 2000. Do polyandrous shorebirds trade off egg size with egg number? J. Avian Biol. 31: 473–478.

WARD, P. J. 1994. Parent–offspring regression and extreme environments. Heredity 72:574–581.

WARHAM, J. 1977. Wing loadings, wing shapes and flight capabilities of Procellariiformes. N. Z. J. Zool. 4:73–83.

WARHAM, J. 1980. Recent trends in sub-Antarctic ornithology. Bull. Br. Ornithol. Club 100:96–102.

WARHAM, J. 1990. The Petrels: Their Ecology and Breeding Systems. Academic Press, London, United Kingdom.

WARHAM, J. 1996. The Behaviour, Population Biology and Physiology of the Petrels. Academic Press, London, United Kingdom.

WARNER, R. E. 1963. Recent history and ecology of the Laysan Duck. Condor 65:3–23.

WARNER, R. E. 1968. The role of introduced diseases in the extinction of the endemic Hawaiian avifauna. Condor 70:101–120.

WARREN, R. L. M., AND C. J. O. HARRISON. 1973. Type-Specimens of Birds in the British Museum (Natural History). Vol. 3: Systematic Index. British Museum (Natural History), London, United Kingdom.

WATERS, J. A. 1977. Quantification of shape by use of Fourier analyses: the Mississippian blastoid genus *Pentremites*. Paleobiology 3:28–299.

WATERS, N. F., AND J. H. BYWATERS. 1943. A lethal embryonic wing mutation in the domestic fowl. J. Hered. 34:213–217.

WATKINS, B. P., AND B. FURNESS. 1986. Population status, breeding and conservation of the Gough Moorhen. Ostrich 57:32–36.
WATLING, D. 1982. Birds of Fiji, Tonga and Samoa. Milwood Press, Wellington, New Zealand.
WATLING, D. 1983. Ornithological notes from Sulawesi. Emu 83:247–261.
WATSON, D. M., AND B. W. BENZ. 1999. The Paint-billed Crake breeding in Costa Rica. Wilson Bull. 111:422–424.
WATSON, G. E. 1962. Notes on the Spotted Rail in Cuba. Wilson Bull. 74:349–356.
WATSON, J. L. 2001. *Notornis rediviva.* Pp. 23–30 *in* The Takahe: Fifty Years of Conservation Management and Research. University of Otago Press, Dunedin, New Zealand.
WATTEL, J. 1973. Geographical Differentiation in the Genus *Accipiter.* Publ. No. 13. Nuttall Ornithologists' Club, Cambridge, Massachusetts.
WATTERSON, R. L. 1942. The morphogenesis of down feathers with special reference to the developmental history of melanophores. Physiol. Zool. 15:234–259.
WEATHERHEAD, P. J., AND R. G. CLARK. 1994. Natural selection and sexual size dimorphism in Red-winged Blackbirds. Evolution 48:1071–1079.
WEATHERHEAD, P. J., AND K. L. TEATHER. 1994. Sexual size dimorphism and egg-size allometry in birds. Evolution 48:671–678.
WEATHERS, W. W. 1979. Climatic adaptation in avian standard metabolic rate. Oecologia 42:81–89.
WEATHERS, W. W. 1981. Physiological thermoregulation in heat-stressed birds: consequences of body size. Physiol. Zool. 54:345–361.
WEBB, H. P. 1992. Field observations of the birds of Santa Isabel, Solomon Islands. Emu 92:52–57.
WEBB, M. 1957. The ontogeny of the cranial bones, cranial peripheral and cranial parasympathetic nerves, together with a study of the visceral muscles of *Struthio.* Acta Zool. 38:81–203.
WEBER, E. 1990. Zur Kraniogenese bei der Lachmöwe (*Larus ridibundus* L.), Zugleich ein Beitrag zur Rekonstruktion des Grundplans der Vögel. Gegenbaurs Morphol. Jahrb. 136:335–387.
WEBER, E., AND A. HESSE. 1995. The systematic position of *Aptornis,* a flightless bird from New Zealand. Cour. Forschungsinst. Senckenb. 181:293–301.
WEBER, M. 1928. Die Säugetiere. Gustav Fischer, Jena, Germany.
WEBER, T. P., T. ALERSTAM, AND A. HEDENSTRÖM. 1998a. Stopover decisions under wind influence. J. Avian Biol. 29:552–560.
WEBER, T. P., B. J. ENS, AND A. I. HOUSTON. 1998b. Optimal avian migration: a dynamic model of fuel stores and site use. Evol. Ecol. 12:377–401.
WEBER, T. P., AND A. I. HOUSTON. 1997. Flight costs, flight range and the stopover ecology of migrating birds. J. Anim. Ecol. 66:297–306.
WEBER, T. P., AND E. KORPIMÄKI. 1998. Resource levels, reproduction and resistance to haematozoan infections. Proc. R. Soc. Lond. (Ser. B Biol. Sci.) 265:1197–1201.
WEBER, T. P., AND T. PIERSMA. 1996. Basal metabolic rate and the mass of tissues differing in metabolic scope: migration-related covariation between individual Knots *Calidris canutus.* J. Avian Biol. 27:215–224.
WEBSTER, M. S. 1992. Sexual dimorphism, mating system and body size in New World blackbirds (Icterinae). Evolution 46:1621–1641.
WEGE, M. L., AND D. G. RAVELING. 1984. Flight speed and directional responses to wind by migrating Canada Geese. Auk 101:342–348.
WEIBEL, E. R. 2000. Symmorphosis: On Form and Function in Shaping Life. Harvard University Press, Cambridge, Massachusetts.
WEIBEL, E. F., C. R. TAYLOR, AND L. BOLIS. 1998. Principles of Animal Design: The Optimization and Symmorphosis Debate. Cambridge University Press, Cambridge, United Kingdom.
WEIMERSKIRCH, H., AND R. ZOTIER, AND P. JOUVENTIN. 1989. The avifauna of the Kerguelen Islands. Emu 89:15–29.
WEINSTEIN, G. N., C. ANDERSON, AND J. D. STEEVES. 1984. Functional characterization of limb muscles involved in locomotion of the Canada Goose, *Branta canadensis.* Can. J. Zool. 62:1596–1604.
WEISBERG, S. 1985. Applied Linear Regression. 2nd ed. J. Wiley and Sons, New York, New York.
WEISLER, M. I. 1995. Henderson Island prehistory: colonization and extinction on a remote Polynesian island. Biol. J. Linn. Soc. 56:377–404.
WELHAM, C. V. J. 1992. Flight speeds of migrating birds: a test of maximum range speed predictions from three aerodynamic equations. Behav. Ecol. 5:1–8.

WELLER, M. W. 1959. Parasitic egg laying in the Redhead (*Aythya americana*) and other North American Anatidae. Ecol. Monogr. 29:333–365.
WELLER, M. W. 1975. Ecological studies of the Auckland Islands Flightless Teal. Auk 92:280–297.
WELLER, M. W. 1980. The Island Waterfowl. Iowa State University Press, Ames, Iowa.
WELLER, M. W. 1999. Wetland Birds: Habitat Resources and Conservation Implications. Cambridge University Press, Cambridge, United Kingdom.
WELLS, D. J. 1993a. Muscle performance in hovering hummingbirds. J. Exp. Biol. 178:39–57.
WELLS, D. J. 1993b. Ecological correlates of hovering flight of hummingbirds. J. Exp. Biol. 178: 59–70.
WELSH, A. H., A. T. PETERSON, AND S. A. ALTMANN. 1988. The fallacy of averages. Am. Nat. 132: 277–288.
WERNER, E. E., AND J. F. GILLIAM. 1984. The ontogenetic niche and species interactions in size-structured populations. Annu. Rev. Ecol. Syst. 15:393–425.
WERNER, T. K., AND T. W. SHERRY. 1987. Behavioral feeding specialization in *Pinaroloxias inornata*, the Darwin's finch of Cocos Island, Costa Rica. Proc. Natl. Acad. Sci. USA 84:5506–5510.
WERSCHKUL, D. F., AND J. A. JACKSON. 1979. Sibling competition and avian growth rates. Ibis 121: 97–102.
WESOLOWSKI, T. 1994. On the origin of parental care and the early evolution of male and female parental roles in birds. Am. Nat. 143:39–58.
WEST, G. B., J. H. BROWN, AND B. J. ENQUIST. 1997. A general model for the origin of allometric scaling laws in biology. Science 276:122–126.
WEST, G. B., J. H. BROWN, AND B. J. ENQUIST. 2000. The origin of universal scaling laws in biology. Pp. 87–112 *in* Scaling in Biology (J. H. Brown and G. B. West, Eds.). Oxford University Press, Oxford, United Kingdom.
WEST-EBERHARD, M. J. 1975. The evolution of social behavior by kin selection. Q. Rev. Biol. 50: 1–33.
WEST-EBERHARD, M. J. 1983. Sexual selection, social competition, and speciation. Q. Rev. Biol. 58: 155–183.
WEST-EBERHARD, M. J. 1986. Alternative adaptations, speciation, and phylogeny (a review). Proc. Natl. Acad. Sci. USA 83:1388–1392.
WEST-EBERHARD, M. J. 1989. Phenotypic plasticity and the origins of diversity. Annu. Rev. Ecol. Syst. 20:249–278.
WESTERN, D., AND J. SSEMAKULA. 1982. Life history patterns in birds and mammals and their evolutionary interpretation. Oecologia 54:281–290.
WETMORE, A. 1918. Bones of birds collected by Theodoor de Booy from kitchen midden deposits in the islands of St. Thomas and St. Croix. U.S. Natl. Mus. Proc. 54:513–522.
WETMORE, A. 1921. A study of the body temperature of birds. Smithson. Misc. Coll. 72:1–52.
WETMORE, A. 1922a. Bird remains from the caves of Porto Rica. Am. Mus. Nat. Hist. Bull. 46: 297–333.
WETMORE, A. 1922b. The wing claw in swifts. Condor 22:197–199.
WETMORE, A. 1930. A systematic classification for the birds of the world. Proc. U.S. Natl. Mus. 76: 1–8.
WETMORE, A. 1937. Ancient records of birds from the islands of St. Croix with observations on extinct and living birds of Puerto Rico. J. Agric. Univ. P. R. 21:5–16.
WETMORE, A. 1938. Bird remains from the West Indies. Auk 55:51–55.
WETMORE, A. 1951. A revised classification for the birds of the world. Smithson. Misc. Coll. 117: 1–22.
WETMORE, A. 1956. A check-list of the fossil and prehistoric birds of North America and the West Indies. Smithson. Misc. Coll. 131:1–105.
WETMORE, A. 1960a. Pleistocene birds in Bermuda. Smithson. Misc. Coll. 140:1–11.
WETMORE, A. 1960b. A classification for the birds of the world. Smithson. Misc. Coll. 139:1–37.
WETMORE, A. 1963. An extinct rail from the island of St. Helena. Ibis 103:379–381.
WETTIN, P. 1984. Simultaneous polyandry in the Purple Swamphen. Emu 84:111–112.
WHEELER, W. C. 1992. Extinction, sampling, and molecular phylogenetics. Pp. 205–215 *in* Extinction and Phylogeny (M. J. Novacek and Q. D. Wheeler, Eds.). Columbia University Press, New York, New York.

WHITE, C. M. N., AND M. D. BRUCE. 1986. The Birds of Wallacea (Sulawesi, the Moluccas and the Lesser Sunda Islands, Indonesia). British Ornithologists' Union, London, United Kingdom.
WHITE, J. F., AND S. J. GOULD. 1965. Interpretation of the coefficient in the allometric equation. Am. Nat. 99:5–18.
WHITEHEAD, H., AND S. J. WALDE. 1993. Territoriality and the evolution of character displacement and sexual dimorphism. Ethol. Ecol. Evol. 5:303–318.
WHITEMAN, H. H. 1994. Evolution of facultative paedomorphosis in salamanders. Q. Rev. Biol. 69: 205–221.
WHITEMAN, H. H., S. A. WISSINGER, AND W. S. BROWN. 1996. Growth and foraging consequences of facultative paedomorphosis in the tiger salamander, *Ambystoma tigrinum nebulosum*. Evol. Ecol. 10:433–446.
WHITLOCK, M. C. 1995. Variance-induced peak shifts. Evolution 49:252–259.
WHITLOCK, M. C. 1996. The Red Queen beats the jack-of-all-trades: the limitations on the evolution of phenotypic plasticity and niche breadth. Am. Nat. 148 (Suppl.):65–77.
WHITLOCK, M. C. 1997. Founder effects and peak shifts without genetic drift: adaptive peak shifts occur easily when environments fluctuate slightly. Evolution 51:1044–1048.
WHITLOCK, M. C., P. C. PHILLIPS, AND K. FOWLER. 2002. Persistence of changes in the genetic covariance matrix after a bottleneck. Evolution 56:1968–1975.
WHITTAKER, R. J. 1998. Island Biogeography: Ecology, Evolution, and Conservation. Oxford University Press, Oxford, United Kingdom.
WHYTE, L. L. 1965. Internal Factors in Evolution. George Braziller, New York, New York.
WICKLER, W., AND U. SEIBT. 1983. Monogamy: an ambiguous concept. Pp. 33–50 *in* Mate Choice (P. Bateson, Ed.). Cambridge University Press, Cambridge, United Kingdom.
WIEDENFELD, D. A. 1985. Humerus length and foramen magnum area as indicators of size in birds. Anat. Anz. 158:193–198.
WIEDENFELD, D. A., AND M. G. WIEDENFELD. 1993. Large kill of Neotropical migrants by tornado and storm in Louisiana, April, 1993. J. Field Ornithol. 66:70–77.
WIENS, J. A. 1982. On size ratios and sequences in ecological communities: Are there no rules? Ann. Zool. Fenn. 19:297–308.
WIENS, J. A. 1984. On understanding a non-equilibrium world: myth and reality in community patterns and processes. Pp. 439–457 *in* Ecological Communities: Conceptual Issues and the Evidence (D. R. Strong, Jr., D. Simberloff, L. G. Abele, and A. B. Thistle, Eds.). Princeton University Press, Princeton, New Jersey.
WIENS, J. A. 1989. The Ecology of Bird Communities. 2 vols. Cambridge University Press, Cambridge, United Kingdom.
WIENS, J. A., AND J. T. ROTENBERRY. 1981. Morphological size ratios and competition in ecological communities. Am. Nat. 117:592–599.
WIENS, J. J., AND J. L. SLINGLUFF. 2001. How lizards turn into snakes: a phylogenetic analysis of body-form evolution in anguid lizards. Evolution 55:2303–2318.
WIGGINS, D. A., AND A. P. MØLLER. 1997. Island size, isolation, or interspecific competition? The breeding distribution of the *Parus* guild in the Danish archipelago. Oecologia 111:255–260.
WIGLESWORTH, J. 1900. Inaugural address on flightless birds. Trans. Liverpool Biol. Soc. 14:1–33.
WIJNANDTS, H. 1984. Ecological energetics of the Long-eared Owl (*Asio otus*). Ardea 72:1–92.
WILEY, R. H. 1991. Lekking in birds and mammals: behavioral and evolutionary issues. Pp. 201–291 *in* Advances in the Study of Behavior. Vol. 20 (P. J. B. Slater, J. S. Rosenblatt, C. Beer, and M. Milinski, Eds.). Academic Press, New York, New York.
WILKENS, H. 1985. The evolution of polygenic systems, studied on epigean and cave populations of *Astyanax fasciatus* (Characidae, Pisces). Natl. Speleol. Soc. Bull. 47:101–108.
WILKENS, H. 1986. The tempo of regressive evolution: studies of eye reduction in stygobiont fishes and decapod crustaceans of the Gulf Coast and West Atlantic region. Stygologia 2:130–143.
WILKENS, H. 1988. Evolution and genetics of epigean and cave *Astyanax fasciatus* (Characidae, Pisces). Support for the neutral mutation theory. Pp. 271–367 *in* Evolutionary Biology. Vol. 23 (M. K. Hecht and B. Wallace, Eds.). Plenum Press, New York, New York.
WILKENS, H. 1993. Neutrale Mutationen end evolutionäre Fortentwicklung. Z. Zool. Syst. Evolutionsforsch. 31:98–109.
WILKIE, D. R. 1977. Metabolism and body size. Pp. 23–36 *in* Scale Effects in Animal Locomotion (T. J. Pedley, Ed.). Academic Press, London, United Kingdom.

WILKINSON, R., AND C. R. HUXLEY. 1978. Vocalizations of chicks and juveniles and the development of adult calls in the Aldabra White-throated Rail *Dryolimnas cuvieri aldabranus* (Aves: Rallidae). J. Zool. Lond. 186:487–505.
WILLIAMS, D. M. 1983. Mate choice in Mallards. Pp. 297–309 *in* Mate Choice (P. Bateson, Ed.). Cambridge University Press, Cambridge, United Kingdom.
WILLIAMS, E. E. 1969. The ecology of colonization as seen in the zoogeography of anoline lizards on small islands. Q. Rev. Biol. 44:345–389.
WILLIAMS, E. E. 1972. The origin of faunas. Evolution of lizard congeners in a complex island fauna: a trial analysis. Pp. 47–88 *in* Evolutionary Biology. Vol. 6 (T. Dobzhansky, M. K. Hecht, and W. C. Steere, Eds.). Appleton-Century-Crofts, New York, New York.
WILLIAMS, G. C. 1966. Adaptation and Natural Selection. Princeton University Press, Princeton, New Jersey.
WILLIAMS, G. C. 1992. Natural Selection: Domains, Levels, and Challenges. Oxford University Press, Oxford, United Kingdom.
WILLIAMS, G. R. 1952. *Notornis* in March, 1951. A report of the sixth expedition. Notornis 4:202–208.
WILLIAMS, G. R. 1953. The dispersal from New Zealand and Australia of some introduced European passerines. Ibis 95:676–692.
WILLIAMS, G. R. 1957. Some preliminary data on the population dynamics of the Takahe (*Notornis mantelli* Owen). Notornis 7:165–171.
WILLIAMS, G. R. 1960. The Takahe (*Notornis mantelli* Owen, 1848): a general survey. Trans. R. Soc. N. Z. 83:235–258.
WILLIAMS, G. R. 1962. Extinction and the land and freshwater-inhabiting birds of New Zealand. Notornis 10:15–32.
WILLIAMS, G. R. 1964. Extinction and the Anatidae of New Zealand. Wildfowl 15:140–146.
WILLIAMS, G. R. 1973. Birds. Pp. 304–333 *in* The Natural History of New Zealand: An Ecological Survey (G. R. Williams, Ed.). A. H. and A. W. Reed, Wellington, New Zealand.
WILLIAMS, G. R. 1984. Has island biogeography theory any relevance to the design of biological reserves in New Zealand? J. R. Soc. N. Z. 14:7–10.
WILLIAMS, G. R., AND D. R. GIVEN. 1981. The Red Data Book of New Zealand. National Conservation Council, Wellington, New Zealand.
WILLIAMS, G. R., AND K. H. MIERS. 1958. A five-year banding study of the Takahe (*Notornis mantelli* Owen). Notornis 8:1–12.
WILLIAMS, J. B. 1980. Intersexual niche partitioning in Downy Woodpeckers. Wilson Bull. 92:439–451.
WILLIAMS, J. B. 1996a. A phylogenetic perspective of evaporative water loss in birds. Auk 113:457–472.
WILLIAMS, J. B. 1996b. Energetics of avian incubation. Pp. 375–416 *in* Avian Energetics and Nutritional Ecology (C. Carey, Ed.). Chapman and Hall, New York, New York.
WILLIAMS, M. 1995. Social structure, dispersion and breeding of the Auckland Island Teal. Notornis 42:219–262.
WILLIAMS, P. A., P. COOPER, P. NES, AND K. F. O'CONNOR. 1976. Chemical composition of tall-tussocks in relation to the diet of the Takahe (*Notornis mantelli* Owen), on the Murchison Mountains, Fiordland, New Zealand. N. Z. J. Bot. 14:55–61.
WILLIAMS, T. D. 1994. Intraspecific variation in egg size and egg composition in birds: effects on offspring fitness. Biol. Rev. 68:35–59.
WILLIAMS, T. D. 1995. The Penguins: *Spheniscidae*. Oxford University Press, Oxford, United Kingdom.
WILLIAMSON, K. 1958. Bergmann's rule and obligatory overseas migration. Br. Birds 51:209–232.
WILLIAMSON, M. 1981. Island Populations. Oxford University Press, Oxford, United Kingdom.
WILLIAMSON, M. 1989. Natural extinction on islands. Philos. Trans. R. Soc. Lond. (Ser. B) 325:457–468.
WILLIAMSON, P. G. 1981. Morphological stasis and developmental constraints: real problems for neo-Darwinism. Nature 294:214–215.
WILLIAMSON, P. G. 1987. Selection or constraint?: A proposal on the mechanism for stasis. Pp. 129–142 *in* Rates of Evolution (K. S. W. Campbell and M. F. Day, Eds.). Allen and Unwin, London, United Kingdom.

WILLIG, M. R., R. D. OWEN, AND R. L. COLBERT. 1986. Assessment of morphometric variation in natural populations: the inadequacy of the univariate approach. Syst. Zool. 35:195–203.

WILLIS, E. O. 1983. Tinamous, chickens, guans, rails and trumpeters as army ant followers. Rev. Bras. Biol. 43:9–22.

WILLIS, J. H. 1996. Measures of phenotypic selection are biased by partial inbreeding. Evolution 50: 1501–1511.

WILLIS, J. H., AND H. A. ORR. 1993. Increased heritable variation following population bottlenecks: the role of dominance. Evolution 47:949–956.

WILSON, A. E., AND M. K. SWALES. 1958. Flightless moorhens (*Porphyriornis c. comeri*) from Gough Island breed in captivity. Avic. Mag. 64:43–45.

WILSON, D. S. 1975. The adequacy of body size as a niche difference. Am. Nat. 109:769–784.

WILSON, D. S. 1980. The Natural Selection of Populations and Communities. Benjamin/Cummings, Menlo Park, California.

WILSON, D. S. 1983. The group selection controversy: history and current status. Annu. Rev. Ecol. Syst. 14:159–187.

WILSON, D. S. 1997. Altruism and organism: disentangling the themes of multilevel selection theory. Am. Nat. 150 (Suppl.):122–134.

WILSON, E. O. 1961. The nature of the taxon cycle in the Melanesian ant fauna. Am. Nat. 95:169–193.

WILSON, H. K. 1971. Ordinary Differential Equations: Introductory and Intermediate Courses Using Matrix Methods. Addison-Wesley, Reading, Massachusetts.

WILSON, K. 1995. Insect migration in heterogeneous environments. Pp. 243–263 *in* Insect Migration: Tracking Resources Through Space and Time (V. A. Drake and A. G. Gatehouse, Eds.). Cambridge University Press, Cambridge, United Kingdom.

WILSON, M. F. 1976. The breeding distribution of North American migrant birds: a critique of MacArthur (1959). Wilson Bull. 88:582–587.

WILSON, S. B., AND A. H. EVANS. 1890–1899. Aves Hawaiienses: The Birds of the Sandwich Islands. R. H. Porter, London, England.

WIMAN, C. 1935. Über Æpyornithes. Nova Acta R. Soc. Sci. Uppsala 9:1–57.

WIMSATT, W. C. 1986. Developmental constraints, generative entrenchment, and the innate-acquired distinction. Pp. 185–208 *in* Integrating Scientific Disciplines (W. Bechtell, Ed.). Martinus-Nijhoff, Dordrecht, The Netherlands.

WINDLEY, B. F. 1995. The Evolving Continents. 3rd ed. J. Wiley and Sons, Chichester, United Kingdom.

WING, L. W. 1956. Natural History of Birds: A Guide to Ornithology. Ronald Press, New York, New York.

WINKLER, D. W. 2000. The phylogenetic approach to avian life histories: an important complement to within-population studies. Condor 102:52–59.

WINKLER, D. W., AND J. R. WALTERS. 1983. The determination of clutch size in precocial birds. Pp. 33–68 *in* Current Ornithology Vol. 1 (R. F. Johnston, Ed.). Plenum Press, New York, New York.

WINTLE, C. C., AND P. B. TAYLOR. 1993. Sequential polyandry, behaviour and moult in captive Striped Crakes *Aenigmatolimnas marginalis*. Ostrich 64:115–122.

WISELY, B. 1956. Notes on the distribution of Takahe in the Murchison Ranges. Rec. Canterbury Mus. 7:1–9.

WITMER, L. M. 1995. The extant phylogenetic bracket and the importance of reconstructing soft tissues in fossils. Pp. 19–33 *in* Functional Morphology in Vertebrate Paleontology (J. J. Thomason, Ed.). Cambridge University Press, Cambridge, United Kingdom.

WITMER, L. M., AND K. D. ROSE. 1991. Biomechanics of the jaw apparatus of the gigantic Eocene bird *Diatryma*: implications for diet and mode of life. Paleobiology 17:95–120.

WITTE, F., C. D. N. BAREL, AND R. J. C. HOOGERHOUD. 1990. Phenotypic plasticity of anatomical structures and its ecomorphological significance. Neth. J. Zool. 40:278–298.

WITTER, M. S., AND I. C. CUTHILL. 1993. The ecological costs of avian fat storage. Philos. Trans. R. Soc. Lond. (Ser. B) 340:73–92.

WOAKES, A. J., AND P. J. BUTLER. 1986. Respiratory, circulatory and metabolic adjustments during swimming in the Tufted Duck, *Aythya fuligula*. J. Exp. Biol. 120:215–231.

WODZICKI, K. 1965. The status of some exotic vertebrates in the ecology of New Zealand. Pp. 425–

458 *in* The Genetics of Colonizing Species (H. G. Baker and G. L. Stebbins, Eds.). Academic Press, New York, New York.

WOINARSKI, J. C. Z., A. FISHER, K. BRENNAN, I. MORRIS, R. C. WILLAN, AND R. CHATTO. 1998. The Chestnut Rail *Eulabeornis castaneoventris* on the Wessel and English Company islands: notes on unusual habitats and use of anvils. Emu 98:74–78.

WOLF, J. B., E. D. BRODIE III, AND M. J. WADE, EDS. 2000. Epistasis and the evolutionary process. Oxford University Press, Oxford, United Kingdom.

WOLF, J. B., W. A. FRANKINO, A. F. AGRAWAL, E. D. BRODIE III, AND A. J. MOORE. 2001. Developmental interactions and the constituents of quantitative variation. Evolution 55:2332–245.

WOLF, L. L., AND F. R. HAINSWORTH. 1978. Energy: expenditures and intakes. Pp. 307–358 *in* Chemical Zoology. Vol. 10, Aves (M. Florkin and B. T. Scheer, Eds.). Academic Press, New York, New York.

WOLFRAM, S. 1996. The MATHEMATICA® Book. 3rd ed. Cambridge University Press, Cambridge, United Kingdom.

WOLPERT, L., R. BEDDINGTON, J. BROCKES, T. JESSELL, P. LAWRENCE, AND E. MEYEROWITZ. 1997. Principles of Development. Oxford University Press, Oxford, United Kingdom.

WOLTERS, H. E. 1975. Die Vogelarten der Erde, Lieferung 1. Paul Parey, Hamburg, Germany.

WOOD, C. J. 1973. The flight of albatrosses (a computer simulation). Ibis 111:244–256.

WOOD, D. S., AND G. D. SCHNELL. 1986. Revised World Inventory of Avian Skeletal Specimens, 1982. American Ornithologists' Union and Oklahoma Biological Survey, Norman, Oklahoma.

WOOD, D. S., R. L. ZUSI, AND M. A. JENKINSON. 1982. World Inventory of Avian Spirit Specimens, 1982. American Ornithologists' Union and Oklahoma Biological Survey, Norman, Oklahoma.

WOOD, N. A. 1974. The breeding biology and behaviour of the Moorhen. Brit. Birds 67:104–115, 135–138.

WOODALL, P. F. 1993. The distribution and abundance of the Australian Crake (*Porzana fluminea*) and Baillon's Crake (*Porzana pusilla*) and their association with lignum (*Muehlenbeckia cunningham*) at Lake Bindegolly, south-west Queensland. Queensl. Nat. 31:107–113.

WOODLAND, D. J., Z. JAAFAR, AND M.-L. KNIGHT. 1980. The "pursuit deterrent" function of alarm signals. Am. Nat. 115:748–753.

WOODREY, M. S., AND C. R. CHANDLER. 1997. Age-related timing of migration: geographic and interspecific patterns. Wilson Bull. 109:52–67.

WOODREY, M. S., AND F. R. MOORE. 1997. Age-related differences in the stopover of fall landbird migrants on the coast of Alabama. Auk 114:695–707.

WOOLFENDEN, G. E. 1967. Selection for a delayed simultaneous wing molt in loons (Gaviidae). Wilson Bull. 79:416–420.

WORCESTER, S. E. 1996. The scaling of the size and stiffness of primary flight feathers. J. Zool. Lond. 239:609–624.

WORTH, C. B. 1940. Egg volumes and incubation periods. Auk 57:44–60.

WORTHINGTON, D. J. 1998. Intra-island dispersal of the Mariana Common Moorhen: a recolonization by an endangered species. Wilson Bull. 110:414–417.

WORTHY, T. H. 1988a. Loss of flight ability in the extinct New Zealand duck *Euryanas finschi.* J. Zool. Lond. 215:619–628.

WORTHY, T. H. 1988b. An illustrated key to the main leg bones of moas (Aves: Dinornithiformes). Natl. Mus. N. Z. Misc. Ser. 17:1–37.

WORTHY, T. H. 1990. An analysis of the distribution and relative abundance of moa species (Aves: Dinornithiformes). N. Z. J. Zool. 17:213–241.

WORTHY, T. H. 1997a. Fossil deposits in the Hodges Creek Cave System, on the northern foothills of Mt. Arthur, Nelson, South Island, New Zealand. Notornis 44:111–124.

WORTHY, T. H. 1997b. A mid-Pleistocene rail from New Zealand. Alcheringa 21:71–78.

WORTHY, T. H. 1998a. The Quaternary fossil avifauna of Southland, South Island, New Zealand. J. R. Soc. N. Z. 28:537–589.

WORTHY, T. H. 1998b. Fossil avifaunas from Old Neck and Native Island, Steward Island—Polynesian middens or natural sites? Rec. Canterbury Mus. 12:49–82.

WORTHY, T. H. 1999a. The role of climate change versus human impacts—avian extinction on South Island, New Zealand. Pp. 111–124 *in* Avian Paleontology at the Close of the 20th Century: Proceedings of the 4th International Meeting of the Society of Avian Paleontology and Evo-

lution, Washington, D.C., 4–7 June 1996 (S. L. Olson, Ed.). Smithson. Contrib. Paleobiol. 89: 1–344.
WORTHY, T. H. 1999b. What was on the menu? Avian extinction in New Zealand. N. Z. J. Archaeol. 19:125–160.
WORTHY, T. H. 2001. A giant flightless pigeon *gen. et sp. nov.* and a new species of *Ducula* (Aves: Columbidae), from Quaternary deposits. J. R. Soc. N. Z. 31:763–794.
WORTHY, T. H., A. J. ANDERSON, AND R. E. MOLNAR. 1999. Megafaunal expression in a land without mammals—the first fossil faunas from terrestrial deposits in Fiji (Vertebrata: Amphibia, Reptilia, Aves). Senckenb. Biol. 79:237–242.
WORTHY, T. H., AND R. N. HOLDAWAY. 1994. Quaternary fossil faunas from caves in Takaka Valley and on Takaka Hill, northwest Nelson, South Island, New Zealand. J. R. Soc. N. Z. 24:297–391.
WORTHY, T. H., AND R. N. HOLDAWAY. 1996. Quaternary fossil faunas, overlapping taphonomies, and palaeofaunal reconstruction in North Canterbury, South Island, New Zealand. J. R. Soc. N. Z. 26:275–361.
WORTHY, T. H., R. N. HOLDAWAY, M. D. SORENSON, AND A. C. COOPER. 1997. Description of the first complete skeleton of the extinct New Zealand goose *Cnemiornis calcitrans* (Aves: Anatidae), and a reassessment of the relationships of *Cnemiornis*. J. Zool. Lond. 243:695–723.
WORTHY, T. H., AND S. L. OLSON. 2002. Relationships, adaptations, and habits of the extinct duck '*Euryanas finschi*'. Notornis 49:1–17.
WORTHY, T. H., R. WALTER, AND A. J. ANDERSON. 1998. Fossil and archaeological avifauna of Niue Island, Pacific Ocean. Notornis 45:177–190.
WRAGG, G. M. 1995. The fossil birds of Henderson Island, Pitcairn Group: natural turnover and human impact, a synopsis. Biol. J. Linn. Soc. 56:405–414.
WRAY, G. A. 1992. Rates of evolution in developmental processes. Am. Zool. 32:123–134.
WRAY, R. A., AND G. A. WRAY. 1989. Heterochrony: developmental mechanisms and evolutionary results. J. Evol. Biol. 2:409–434.
WRIGHT, S. 1932. The roles of mutation, inbreeding, crossbreeding and selection in evolution. Pp. 356–366 *in* Proceedings of the Sixth International Congress of Genetics. Vol. 1 (D. F. Jones, Ed.). Brooklyn Botanical Garden, Brooklyn, New York.
WRIGHT, S. 1940. Breeding structure of populations in relation to speciation. Am. Nat. 74:232–248.
WRIGHT, S. 1949. Adaptation and selection. Pp. 365–389 *in* Genetics, Paleontology and Evolution (G. L. Jepsen, G. G. Simpson, and E. Mayr, Eds.). Princeton University Press, Princeton, New Jersey.
WRIGHT, S. 1969. Evolution and Genetics of Populations. Vol. 2: The Theory of Gene Frequencies. University of Chicago Press, Chicago, Illinois.
WRIGHT, S. 1977. Evolution and the Genetics of Populations. Vol. 3: Experimental Results and Evolutionary Deductions. University of Chicago Press, Chicago, Illinois.
WRIGHT, S. 1982. Character change, speciation, and the higher taxa. Evolution 36:427–443.
WÜRDINGER, I. 1975. Vergleichend morphologische Untersuchungen zur Jugendentwicklung von *Anser*- und *Branta*-Arten. J. Ornithol. 116:65–86.
WYPKEMA, R. C. P., AND C. D. ANKNEY. 1979. Nutrient reserve dynamics of Lesser Snow Geese staging at James Bay, Ontario, Canada. Can. J. Zool. 57:213–219.
YAMASHINA, Y., AND T. MANO. 1981. A new species of rail from Okinawa Island. J. Yamashina Inst. Ornithol. 13:147–152.
YAPP, W. B. 1962a. The Life of Vertebrates. Clarendon, Oxford, United Kingdom.
YAPP, W. B. 1962b. Some physical limitations on migration. Ibis 104:86–89.
YARBROUGH, C. G. 1971. The influence of distribution and ecology on the thermoregulation of small birds. Comp. Biochem. Physiol. (Ser. A) 39:235–266.
YDENBERG, R. C. 1989. Growth-maturity trade-offs, parent–offspring conflict, and the evolution of juvenile life histories in the avian family, Alcidae. Ecology 70:1494–1506.
YEH, J. 2002. The effect of miniaturized body size on skeletal morphology in frogs. Evolution 56: 628–641.
YEZERINAC, S. M., S. C. LOUGHEED, AND P. HANDFORD. 1992. Measurement error and morphometric studies: statistical power and observer experience. Syst. Biol. 41:471–482.
YOM-TOV, Y. 1976. Sexual dimorphism and sex ratios in wild birds. Oikos 27:81–85.
YOM-TOV, Y. 1980. Intraspecific nest parasitism in birds. Biol. Rev. 55:93–108.

Yom-Tov, Y., and R. Hilborn. 1981. Energetic constraints on clutch size and time of breeding in temperate zone birds. Oecologia 48:234–243.

Yong, W., D. M. Finch, F. R. Moore, and F. Kelly. 1998. Stopover ecology and habitat use of migratory Wilson's Warblers. Auk 115:829–842.

Young, E. C. 1995. Conservation values, research and New Zealand's responsibilities for the southern ocean islands and Antarctica. Pac. Conserv. Biol. 2:99–112.

Young, R. W., and T. Bryant. 1992. Catastrophic wave erosion on the southeastern coast of Australia: Impact of the Lanai Tsunami ca. 105 ka? Geology 20:199–202.

Younker, J. L., and R. Ehrlich. 1977. Fourier biometrics: harmonic amplitudes as multivariate shape descriptors. Syst. Zool. 26:336–342.

Zack, S., and B. J. Stutchbury. 1992. Delayed breeding in avian social systems: the role of territory quality and "floater" tactics. Behaviour 123:194–219.

Zahavi, A. 1975. Mate selection—a selection for a handicap. J. Theor. Biol. 53:205–214.

Zahavi, A., and A. Zahavi. 1997. The Handicap Principle: A Missing Piece of Darwin's Puzzle. Oxford University Press, Oxford, United Kingdom.

Zaher, H., and O. Rieppel. 1999. The phylogenetic relationships of *Pachyrachis problematicus,* and the evolution of limblessness in snakes (Lepidosauria, Squamata). C. R. Acad. Sci. Paris 329: 831–837.

Zann, R. A., E. B. Male, and Darjono. 1990. Bird colonization of Anak Krakatau, an emergent volcanic island. Philos. Trans. R. Soc. Lond. (Ser. B) 328:95–121.

Zar, J. H. 1968a. Calculation and miscalculation of the allometric equation as a model in biological data. BioScience 18:1118–1121.

Zar, J. H. 1968b. Standard metabolism comparisons between orders of birds. Condor 70:278.

Zar, J. H. 1969. The use of the allometric model for avian standard metabolism–body weight relationships. Comp. Biochem. Physiol. (Ser. A) 29:227–234.

Zelditch, M. L., F. L. Bookstein, and B. L. Lundrigan. 1993. The ontogenetic complexity of developmental constraints. J. Evol. Biol. 6:621–641.

Zelditch, M. L., and W. L. Fink. 1996. Heterochrony and heterotopy: stability and innovation in the evolution of form. Paleobiology 22:241–254.

Zembal, R., and J. M. Fancher. 1988. Foraging behavior and foods of the Light-footed Clapper Rail. Condor 90:959–962.

Zeng, Z.-B. 1988. Long-term correlated response, interpopulation covariation, and interspecific allometry. Evolution 42:363–374.

Zera, A. J., and R. F. Denno. 1997. Physiology and ecology of dispersal polymorphism in insects. Annu. Rev. Entomol. 42:207–230.

Zera, A. J., and S. Mole. 1994. The physiological costs of flight capacity in wing-dimorphic crickets. Res. Popul. Ecol. 36:151–156.

Zera, A. J., S. Mole, and K. Rokke. 1994. Lipid, carbohydrate and nitrogen content of long- and short-winged *Gryllus firmus*: implications for the physiological cost of flight capability. J. Insect Physiol. 40:1037–1044.

Zera, A. J., J. Sall, and G. L. Cisper. 1997. Flight-muscle polymorphism in the cricket *Gryllus firmus*: muscle characteristics and their influence on the evolution of flightlessness. Physiol. Zool. 70:519–529.

Zeuthen, E. 1947. Body size and metabolic rate in the animal kingdom. C. R. Lab. Carlsberg (Ser. Chim.) 26:17–165.

Zeuthen, E. 1953. Oxygen uptake as related to body size in organisms. Q. Rev. Biol. 28:1–12.

Zhirvotovsky, L. A., M. W. Feldman, and A. Bergman. 1996. On the evolution of phenotypic plasticity in a spatially heterogeneous environment. Evolution 50:547–558.

Zimmerman, D. A., D. A. Turner, and D. J. Pearson. 1996. Birds of Kenya and Northern Tanzania. Princeton University Press, Princeton, New Jersey.

Zink, A. G. 2000. The evolution of intraspecific brood parasitism in birds and insects. Am. Nat. 155: 395–405.

Zink, R. M. 1983. Evolutionary and systematic significance of temporal variation in the Fox Sparrow. Syst. Zool. 32:223–238.

Zink, R. M. 1986. Patterns and evolutionary significance of geographic variation in the *schistacea* group of the Fox Sparrow (*Passerella iliaca*). Ornithol. Monogr. 40. American Ornithologists' Union, Washington, D.C.

ZINK, R. M., AND M. C. MCKITRICK. 1995. The debate over species concepts and its implications for ornithology. Auk 112:701–719.

ZINK, R. M., AND J. V. REMSEN, JR. 1986. Evolutionary processes and patterns of geographic variation in birds. Pp. 1–69 in Current Ornithology. Vol. 4 (R. F. Johnston, Ed.). Plenum Press, New York, New York.

ZINK, R. M., AND D. W. WINKLER. 1983. Genetic and morphological similarity of two California Gull populations with different life history traits. Biochem. Syst. Ecol. 11:397–403.

ZISWILER, V. 1967. Extinct and Vanishing Animals: A Biology of Extinction and Survival. Springer-Verlag, New York, New York.

ZUSI, R. L. 1985. Muscles of the neck, trunk and tail in the Noisy Scrub-bird, *Atrichornis clamosus*, and the Superb Lyrebird, *Menura novaehollandiae* (Passeriformes: Atrichornithidae and Menuridae). Rec. Aust. Mus. 37:229–242.

ZUSI, R. L. 1993. Patterns of diversity in the avian skull. Pp. 391–437 in The Skull. Vol. 2: Patterns of Structural and Systematic Diversity (J. Hanken and B. K. Hall, Eds.). University of Chicago Press, Chicago, Illinois.

ZUSI, R. L., AND G. D. BENTZ. 1978. The appendicular myology of the Labrador Duck (*Camptorhynchus labradorius*). Condor 80:407–418.

ZUSI, R. L., AND G. D. BENTZ. 1984. Myology of the Purple-throated Carib (*Eulampis jugularis*) and other hummingbirds (Aves Trochilidae). Smithson. Contrib. Zool. 385:1–70.

ZWARTS, L., B. J. ENS, M. KERSTEN, AND T. PIERSMA. 1990. Moult, mass and flight range of waders ready to take off for long-distance migrations. Ardea 78:339–364.

ZWEERS, G. A. 1979. Explanation of structure by optimization and systemization. Neth. J. Zool. 29: 418–440.

ZWILLING, E. 1974. Effects of contact between mutant (wingless) limb buds and those of genetically normal chick embryos: confirmation of a hypothesis. Dev. Biol. 39:37–48.

APPENDIX 1. Body masses and wing lengths of rails.

Measurements are body mass (g) and wing length (mm), by sex. Where only one summary is given for a species, data were insufficient for sex-specific estimates and data were pooled. Where two sample sizes are given, these correspond to estimates of mean and standard deviation, respectively; single estimates are means. Where the mean is based on an unspecified sample or approximated by the midpoint of a published range of measurements, the estimate is enclosed in square brackets. Values are mean ($\bar{x}$) ± one standard deviation (SD), followed by sample size (n) in parentheses.

Species	Body mass (g)		Wing length (mm)	
	Male	Female	Male	Female
Himantornis haematopus	533 ± 53 (3)	424 ± 34 (2)	220 (6)	212 (6)
Porphyrio bellus	1,024 ± 277 (2)	629 (1)	281 ± 10 (7)	268 ± 8 (5)
Porphyrio indicus	—	—	226 ± 10 (22)	224 ± 8 (15)
Porphyrio samoensis	612 ± 109 (3)		229 ± 9 (49)	222 ± 10 (57)
Porphyrio pulverulentus	802 ± 76 (3)	715 ± 10 (4)	245 ± 5 (3)	239 ± 4 (7)
Porphyrio porphyrio	869 ± 71 (37)	724 ± 79 (35)	265 ± 7. (13)	259 ± 6 (12)
Porphyrio poliocephalus	—	557 ± 16 (2)	254 ± 15 (27)	260 ± 7 (16)
Porphyrio madagascariensis	643 ± 66 (18, 7)	554 ± 120 (16, 8)	248 ± 13 (11)	244 ± 14 (17)
Porphyrio viridis	—	—	246 ± 8 (17)	240 ± 7 (8)
Porphyrio ellioti	800 ± 93 (9, 7)	679 ± 102 (9, 6)	233 ± 14 (22)	223 ± 15 (24)
Porphyrio albus	—	—	229 ± 1 (2)	
Porphyrio melanotus	1,088 ± 92 (331)	881 ± 93 (236)	284 ± 7 (114)	268 ± 8 (66)
Porphyrio melanopterus	—	—	245 ± 15 (13)	229 ± 13 (15)
Porphyrio hochstetteri	2,926 ± 278 (49, 31)	2,510 ± 244 (51, 33)	255 ± 6 (7)	232 ± 6 (6)
Porphyrula alleni	154 ± 23 (11, 7)	125 ± 18 (3)	154 ± 6 (23)	150 ± 4 (29)
Porphyrula martinica	249 ± 27 (31)	204 ± 37 (15, 7)	179 ± 4 (30, 11)	173 ± 4 (40, 13)
Porphyrula flavirostris	93 ± 8 (14, 5)	88 ± 22 (12, 4)	137 ± 4 (26)	132 ± 6 (26)
Gymnocrex rosenbergii	—	—	196 ± 8 (21)*	
Gymnocrex plumbeiventris†	—	—	189 ± 8 (26)*	
Habroptila wallacii	292 ± 33 (3)			
Eulabeornis castaneoventris	—	—	166 ± 6 (5)	169 ± 5 (10)
Aramides saracura	746 ± 103 (6)	628 (7)	219 ± 8 (17)	202 ± 11 (14)
Aramides calopterus	512 ± 40 (2)	390 ± 30 (2)	193 (7)	187 (7)
Aramides wolfi	367 (1)	345 (1)	166 (6)	165 (7)
Aramides specaha	454 (1)		176 (9)	172 (6)
Aramides mangle	737 ± 109 (3)	648 ± 129 (6)	226 ± 7 (16)	222 ± 6 (14)
Aramides axillaris	238 ± 11 (2)	228 ± 38 (6)		[164]
Aramides cajanea	453 ± 62 (10)	382 ± 55 (8)	169 (18)	164 (13)
Aramides plumbeicollis	[275]		183 ± 10 (18, 11)	172 ± 3 (12, 7)
Aramides mexicana	—	—	178 (5)	172 (8)
Aramides albiventris	466 (1)		178 (8)	175 (9)
Canirallus oculeus	278 (1)	278 ± 3 (3)	190 ± 7 (13, 3)	189 ± 8 (14, 8)
			175 ± 5 (25)	174 ± 7 (18)

APPENDIX 1. Continued.

Species	Body mass (g) Male	Body mass (g) Female	Wing length (mm) Male	Wing length (mm) Female
Canirallus kioloides	225 ± 75 (2)	276 ± 4 (2)	134 ± 4 (22)	132 ± 3 (18)
Anurolimnas castaneiceps	126 (1)	99 (1)	118 (20)	115 (22)
Amaurolimnas concolor	106 ± 15 (6)	96 ± 14 (5)	121 (17)	120 (10)
Rougetius rougetii	220 (1)	170 (1)	—	—
Rallina eurizonoides-group	118 (1)	95 ± 11 (4)	133 ± 3 (19)	131 ± 3 (18)
Rallina canningi	—	—	130 ± 5 (7)	130 ± 6 (9)
Rallina fasciata	135 (1)	—	160 ± 5 (8)	156 ± 5 (8)
Rallina tricolor-group	203 ± 34 (8)	78 (1)	127 ± 5 (10)	124 ± 4 (6)
Rallicula forbesi	90 ± 2 (2)	177 ± 31 (6)	145 ± 3 (15)	142 ± 4 (11)
Rallicula leucospila	120 ± 8 (2)	81 (1)	—	—
Rallicula rubra	82 ± 8 (5)	—	111 (8)	108 (10)
Rallicula mayri	28 ± 6 (5)	—	109 ± 2 (4)	[106]
Sarothrura pulchra	46 ± 2 (23, 14)	118 ± 6 (3)	98 ± 3 (6)	[93]
Sarothrura elegans	46 ± 4 (53)	44 ± 4 (12, 10)	114 (4)	[109]
Sarothrura rufa	39 ± 4 (14)	46 ± 4 (34)	81 (232)	79 (78)
Sarothrura ayresi	32 (1)	37 ± 5 (9)	89 ± 3 (115)	89 ± 3 (60)
Sarothrura watersi	27 (1)	—	77 ± 3 (68)	77 ± 3 (35)
Sarothrura lugens	—	—	76 (14)	77 (11)
Sarothrura boehmi	37 ± 4 (4)	—	74 ± 3 (7)	71 ± 2 (6)
Sarothrura insularis	—	33 ± 3 (2)	78 (31)	79 (15)
Sarothrura affinis-group	28 ± 2 (4)	30 (1)	86 ± 3 (33)	85 ± 2 (21)
Cyanolimnas cerverai	—	—	72 (20)	71 (11)
Pardirallus maculatus	189 ± 16 (7)	166 ± 24 (8)	74 ± 3 (13)	71 ± 3 (9)
Ortygonax sanguinolentus	149 ± 26 (4)	125 (1)	110 ± 1 (2)	102 ± 2 (7)
Ortygonax nigricans	179 ± 36 (3)	—	126 (14)	119 (9)
Dryolimnas cuvieri	259 ± 4 (2)	223 ± 11 (2)	140 ± 7 (36, 10)	131 ± 4 (30, 10)
Dryolimnas aldabranus	189 ± 18 (32)	176 ± 22 (22)	134 ± 7 (17, 13)	130 ± 3 (18, 6)
Dryolimnas abbotti	—	—	154 ± 6 (25)	148 ± 3 (31)
"Rallus" madagascariensis	148 (1)	105 (1)	117 ± 6 (15)	116 ± 5 (13)
Rallus caerulescens	180 ± 15 (66)	146 ± 12 (50)	136 ± 3 (5)	135 ± 1 (2)
Rallus aquaticus	128 ± 16 (355, 170)	102 ± 12 (302, 145)	112 ± 3 (10)	106 ± 3 (11)
Rallus indicus	—	—	122 ± 4 (54)	115 ± 4 (38)
Rallus wetmorei	—	—	125 ± 3 (126)	116 ± 2 (124)
Rallus antarcticus	—	—	129 (10)	125 (6)
Rallus semiplumbeus	—	60 (1)	135 (8)	123 (6)
Rallus limicola	91 ± 16 (27, 18)	75 ± 7 (11, 8)	96 ± 3 (3)	94 ± 1 (6)
Rallus longirostris-group	322 ± 21 (20, 13)	268 ± 45 (13, 6)	113 (10)	104 (10)
Rallus elegans-group	392 ± 46 (18, 9)	313 ± 68 (11, 2)	162 (29)	152 (24)
Gallirallus pectoralis	35 ± 10 (26, 18)	82 ± 14 (31, 22)	163 (18)	154 (14)
			103 ± 3 (27)	101 ± 2 (33)

APPENDIX 1. Continued.

Species	Body mass (g)		Wing length (mm)	
	Male	Female	Male	Female
Gallirallus muelleri	—	—	81 ± 4 (6)	106 ± 2 (4)
Gallirallus mirificus	93 ± 5 (4)	—	105 ± 2 (3)	119 ± 4 (12)
Gallirallus striatus	—	109 (1)	121 ± 4 (12)	
Gallirallus sharpei	116 ± 16 (4)	—		140 (1)
Gallirallus australis	1,050 ± 151 (95, 87)	730 ± 132 (38, 36)	184 ± 10 (126)	166 ± 13 (59)
Gallirallus greyi‡	912 (n_1)	699 (n_2)	179 ± 14 (18)	168 ± 11 (14)
Gallirallus philippensis†	193 ± 25 (23, 18)	171 ± 37 (17, 8)	146 ± 7 (30)	141 ± 5 (18)
Gallirallus macquariensis	—	—	[124]	
Gallirallus assimilis	171 ± 31 (7)	168 ± 37 (8)	140 ± 6 (48, 4)	134 ± 3 (35, 3)
Gallirallus christophori	—	—	147 ± 3 (5)	141 ± 3 (4)
Gallirallus sethsmithi	—	—	147 ± 8 (5)	139 ± 3 (6)
Gallirallus goodsoni	262 ± 46 (3)	229 ± 38 (4)	149 ± 7 (5)	143 ± 3 (5)
Gallirallus ecaudatus	—	—	139 ± 3 (4)	131 ± 3 (3)
Gallirallus dieffenbachii	—	—		122 (1)
Gallirallus owstoni	241 ± 42 (27, 18)	212 ± 57 (20, 18)	124 ± 4 (11)	116 ± 4 (17)
Gallirallus wakensis	—	—	91 ± 6 (29)	90 ± 5 (25)
Tricholimnas lafresnayanus	—	—	191 ± 5 (4)	180 ± 10 (7)
Tricholimnas sylvestris	536 ± 77 (49)	456 ± 69 (58)	136 ± 5 (46)	140 ± 5 (38)
Tricholimnas conditicius	—	—		132 (1)
Nesoclopeus poeciloperus-group	—	—	162 ± 13 (17)	161 ± 9 (13)
Aramidopsis plateni	—	—	152 ± 6 (6)	150 ± 4 (7)
Cabalus modestus	—	—	88 ± 4 (15)	79 ± 3 (13)
Habropteryx insignis	—	—	149 ± 4 (6)	139 ± 4 (11)
Habropteryx celebensis	—	—	151 ± 5 (16)	146 ± 5 (12)
Habropteryx torquatus	263 ± 18 (5)	228 ± 24 (4)	151 ± 6 (20)	144 ± 7 (23)
Habropteryx sulcirostris	—	—	157 ± 8 (7)	141 ± 4 (6)
Habropteryx okinawae	433 (1)		145 (1)	140 (1)
Atlantisia rogersi	42 ± 4 (6)	37 ± 3 (6)	56 ± 2 (24)	54 ± 1 (20)
Laterallus jamaicensis	35 ± 3 (33)	36 ± 5 (16)	74 ± 3 (60)	74 ± 3 (47)
Laterallus tuerosi	—	—	74 (2)	
Laterallus muivagans	—	—	[78]	
Laterallus spilonotus	42 ± 3 (~18)		68 ± 2 (39, 8)	67 ± 2 (38, 11)
Laterallus levraudi	56 ± 9 (10, 5)	53 ± 11 (13, 10)	81 (5)	82 (8)
Laterallus viridis	49 ± 8 (8)	44 ± 4 (7)	91 ± 4 (14)	92 (14)
Laterallus ruber	58 ± 2 (2)	51 ± 4 (3)	78 ± 2 (12)	76 (6)
Laterallus melanophaius	—	—	81 ± 3 (19)	80 (9)
Laterallus fasciatus	78 ± 7 (3)	65 ± 2 (2)	95 (5)	95 (8)

APPENDIX 1. Continued.

Species	Body mass (g)		Wing length (mm)	
	Male	Female	Male	Female
Laterallus leucopyrrhus	47 ± 6 (6)	42 ± 7 (3)	81 ± 2 (7, 5)	83 ± 1 (6, 5)
Laterallus albigularis-group	50 ± 6 (13)	45 ± 6 (12)	75 (20)	74 (16)
Laterallus xenopterus	57 ± 7 (5)			87 (5)
Laterallus exilis	32 (10)	35 (9)	73 (19)	73 (6)
Coturnicops noveboracensis	59 ± 4 (70)	52 ± 7 (6)	86 ± 2 (36, 14)	83 ± 1 (33, 16)
Coturnicops exquisitus	—	—	79 ± 2 (7)	
Coturnicops notatus	[30]		[72]	
Micropygia schomburgkii	30 (1)	24 (2)	75 (11)	75 (11)
Micropygia flaviventer	26 ± 2 (15, 5)	24 (12)	66 (30)	67 (27)
Crex crex	166 ± 20 (81, 33)	138 ± 16 (22, 13)	141 ± 4 (33)	136 ± 5 (16)
Crex egregia	124 ± 12 (10, 4)	93 (1)	123 (14)	124 (15)
Crex albicollis	113 ± 12 (6, 4)	95 ± 14 (4)	109 ± 4 (20, 18)	105 ± 5 (11, 13)
Porzana sandwichensis	—	—	68 ± 5 (7)	
Porzana fusca	66 ± 7 (5)	62 ± 5 (3)	102 ± 5 (15)	101 ± 4 (17)
Porzana paykullii	120 ± 8 (4)	102 (1)	129 ± 4 (13)	124 (9)
Porzana cinerea	72 ± 10 (16)	66 ± 13 (9)	95 ± 3 (93, 18)	93 ± 3 (85, 22)
Porzana marginalis	—	61 (1)	106 ± 2 (9)	107 ± 3 (11)
Porzana parva	51 ± 13 (20, 10)	49 ± 10 (21, 11)	117 ± 2 (70, 24)	111 ± 3 (49, 22)
Porzana palmeri	—	—	59 ± 2 (57)	58 ± 3 (30)
Porzana spiloptera	—	—	73 ± 3 (3)	76 (1)
Porzana porzana	81 (27)	94 (42)	122 ± 3 (46)	118 ± 3 (27)
Porzana carolina	85 ± 12 (23)	77 ± 10 (13)	113 ± 3 (18)	106 ± 3 (13)
Porzana pusilla	39 ± 9 (16, 11)	36 ± 9 (24, 27)	87 ± 2 (81)	86 ± 3 (70)
Porzana fluminea	65 ± 7 (18, 10)	59 ± 5 (9, 3)	101 ± 4 (34)	98 ± 2 (17)
Porzana tabuensis	45 ± 4 (25, 13)	39 ± 5 (18, 10)	86 ± 3 (27)	84 ± 3 (17)
Porzana monasa	—	—		
Porzana atra§	80 ± 5 (11)	74 ± 4 (11)	83 ± 4 (20, 9)	83 ± 2 (17, 5)
Porzana flavirostra	89 ± 10 (6)	78 ± 12 (7)	106 (37)	102 (31)
Porzana olivieri	—	—	103 ± 1 (4)	108 (3)
Porzana bicolor	61 ± 8 (2)	70 ± 12 (2)	116 ± 3 (14)	114 ± 4 (10)
Porzana erythrops			102 (14)	103 (16)
Porzana colombiana			103 (6)	102 (6)
Amaurornis olivaceus	303 ± 14 (2)	250 (1)	165 ± 7 (8)	154 ± 3 (6)
Amaurornis moluccanus†	204 ± 24 (9)	179 ± 27 (6)	138 ± 8 (40)	133 ± 4 (6+, 6)
Amaurornis ruficrissus	177 ± 26 (9)	150 ± 15 (4)	152 ± 4 (12)	144 ± 6 (6)
Amaurornis phoenicurus	224 ± 29 (5)	196 ± 22 (—, 6)	167 ± 8 (28)	157 ± 6 (31)
Amaurornis akool	[142]	[125]	129 ± 4 (21)	125 ± 4 (21)
Amaurornis isabellinus	—	—	171 ± 6 (27)	160 ± 6 (30)
Amaurornis ineptus	967 (3)	—	183 ± 7 (6)	177 ± 7 (17)

APPENDIX 1. Continued.

Species	Body mass (g)		Wing length (mm)	
	Male	Female	Male	Female
Gallicrex cinerea	512 ± 4 (9, 2)	340 (4)	207 ± 12 (20)	175 ± 7 (20)
Porphyriops melanops	202 ± 42 (2)	137 ± 12 (3)	130 (23)	124 (25)
Pareudiastes pacificus	—	—		
Pareudiastes silvestris	450 (1)	—	149 (1)	123 ± 8 (8)
Tribonyx ventralis	440 ± 62 (20, 12)	362 ± 14 (13, 8)	215 ± 8 (52)	195 ± 9 (34)
Tribonyx mortierii	1,334 ± 109 (152)	1,251 ± 103 (120)	195 ± 9 (31)	193 ± 9 (15)
Gallinula nesiotis	[400]	—	153 (1)	134 ± 5 (3)
Gallinula comeri	513 ± 14 (3)		148 ± 5 (16)	140 ± 3 (19)
Gallinula tenebrosa	570 ± 74 (5)	493 ± 121 (7)	209 ± 7 (19)	200 ± 7 (16)
Gallinula angulata	155 ± 9 (4)	109 ± 24 (3)	137 (9)	132 ± 2 (5)
Gallinula chloropus	340 (103)	265 (110)	171 ± 6 (43, 14)	165 ± 4 (33, 16)
Gallinula pyrrhorrhoa	—	—	160 ± 7 (9)	159 ± 7 (11)
Gallinula cachinnans	408 ± 14 (3)	325 ± 40 (3)	174 ± 6 (31, 14)	165 ± 6 (28, 16)
Gallinula galeata	497 ± 84 (3)	400 (1)	187 ± 8 (32, 6)	176 ± 9 (26, 5)
Gallinula sandvicensis	—	—	180 ± 4 (22)	170 ± 5 (7)
Fulica rufifrons	689 ± 86 (4)	539 ± 78 (4)	180 ± 5 (5)	169 ± 6 (10)
Fulica armillata	937 ± 213 (4)	888 (1)	208 ± 10 (8)	203 ± 10 (9)
Fulica leucoptera	584 ± 52 (5)	509 ± 75 (5)	192 (14)	187 ± 5 (11)
Fulica cornuta	2,100 (1)	1,857 ± 179 (3)	289 ± 10 (12)	280 ± 10 (7)
Fulica gigantea	2,566 ± 116 (3)	2,144 ± 458 (5)	263 ± 19 (20)	270 ± 13 (20)
Fulica ardesiaca	1,040 (1)	893 ± 105 (3)	233 ± 7 (9)	220 ± 10 (7)
Fulica alai	495 (1)	—	185 ± 8 (28)	174 ± 5 (23)
Fulica americana	686 ± 79 (125, 98)	535 ± 57 (119, 99)	191 ± 9 (38, 10)	181 ± 10 (31, 10)
Fulica caribaea	—	517 ± 97 (3)	189 ± 9 (10)	177 ± 10 (9)
Fulica atra, atra-group	905 (81)	758 (190)	219 ± 4 (21)	205 ± 4 (23)
Fulica atra, australis-group	552 ± 75 (10, 6)	502 ± 52 (8, 5)	186 ± 6 (10)	177 ± 4 (12)
Fulica cristata	841 ± 205 (15, 5)	683 ± 134 (15, 5)	227 ± 5 (15)	217 ± 5 (13)

* Provisionally partitioned sex-groups (Table 7) merged here in light of aberrant pattern of "reversed" sexual dimorphism in wing lengths.

† Three recently described rallids (*Gymnocrex talaudensis* [Lambert 1998a], *Gallirallus rovianae* [Diamond 1991], and *Amaurornis magnirostris* [Lambert 1998b]) are excluded, because neither body masses nor wing lengths were available for these species.

‡ Only total sample size for both sexes is given for body masses; that is, $n_1 + n_2 = 94$.

§ Includes data from specimens collected by G. Graves and data from live birds compiled by Jones et al. (1995).

APPENDIX 2. Statistical details and sources of data.

PRINCIPAL COMPONENTS ANALYSIS (PCA)

FUNDAMENTALS OF STANDARD R-MODE PCA

Assumptions and goals.—PCA is commonly used to display samples on synthetic, mutually orthogonal axes distilled from largely redundant suites of associated measurements. Less frequently, more explicit tests of hypotheses related to PCA are employed, all of which at least assume multivariate normality, including those for equality of all or a subset of the included eigenvalues (sphericity), equality of separate covariance matrices of included groups, proportionality of covariance matrices, eigenvalues being significantly greater than zero, and comparisons with hypothetical eigenvectors (Borg and Lingoes 1987; Flury 1988). Although many of the analytical contexts described herein meet the requisite assumptions, such tests were used sparingly. Eigenvalues reflect the variance associated with the corresponding eigenvectors, that is, the variance of the points projected onto the eigenvectors. Given the mutual orthogonality of eigenvectors, the total variance of the associated dispersion matrices are given by the following positive definite matrices:

total variance of $\boldsymbol{S} = Tr(\boldsymbol{S})$

$$= \sum_{i=1}^{p} \lambda_i \quad \text{(for orthogonalized covariance matrix } \boldsymbol{S}\text{)},$$

and

total variance of $\boldsymbol{R} = Tr(\boldsymbol{R})$

$$= \sum_{i=1}^{p} \lambda_i = p \quad \text{(for orthogonalized correlation matrix } \boldsymbol{R}\text{)}.$$

Accordingly, for the standardized eigenstructural dispersion matrix the sum of the eigenvalues of a dispersion matrix equals the total variance of the data set (under a given transformational scheme). Therefore, the proportion of the total variance summarized by the kth eigenvector is given by (expressed as a percentage and given in tables herein):

$$\frac{\lambda_k}{\sum_{i=1}^{p} \lambda_i} \times 100\%.$$

For the standardized dispersion matrix of full rank ($\boldsymbol{R}$), wherein all off-diagonal elements are unitary and all diagonal elements are zero, the sum of eigenvalues (λ_i) is:

$$\sum_{i=1}^{p} \lambda_i = p = \text{rank }(\boldsymbol{R}) = \text{rank }(\boldsymbol{S}).$$

Where moderate differences in scale of variables were considered informative, PCAs were based on covariance matrices ($\boldsymbol{S}$) of $\log_e$-transformed variables. Where disparities in magnitudes of measurements were considered uninformative or misleading (e.g., lengths of remiges) or for Q-mode applications (see below), correlation matrices ($\boldsymbol{R}$) were employed. Where p (number of variables) exceeded $n - 1$ (the number of specimens minus one) analyzed in a PCA, the dispersion (covariance or correlation) matrix was singular, necessitating a singular-value decomposition or pseudo-inverse (Gnanadesikan 1977; Anderson 1984; Dixon 1992). Colinearity among the p variables (reducing the rank through redundancy of two or more groups of the original variables) also may result in singularity where simple dimensionality might indicate otherwise. Where pseudo-inverses of dispersion matrices were used, tests of hypotheses regarding eigenstructure are not possible, and the components derived were used to derive a comparatively smaller number of synthetic axes.

Unless multivariate normality and other distributional assumptions are met (M. E. Johnson 1987), permitting formal tests for nonzero eigenvalues, the number of PCs considered to account for significant proportions of the total dispersion is generally somewhat subjective or decided by heuristic rules adopted a priori (Jackson 1993). The intention to restrict attention to c PCs ($c \leq p$ original variates) incorporating significant or interpretable variance (i.e., a empirical reduction in dimensionality) is formalized through partial PCs (Flury 1988), an extension not implemented here in that the hypothesis tests and distributional assumptions typical of the method were not applicable. The practice of using the residuals from the first principal component (PC-I; variance attributed to subsequent components) as "size-corrected" dimensions of shape in other multivariate assessments (e.g., Smith and Patton 1988) preserves the variance associated with the components of smallest eigenvalue as a component of the residuals to be modeled.

Interpretation of standard components.—In morphometric contexts, PC-I typically is dominated by strongly covarying variation in size of elements. It is customary to interpret such components as "general size," with the subsequent components (PC-II and beyond) to be mutually orthogonal, size-independent dimensions of "shape" (Reyment et al. 1984; Rising and Somers 1989; Marcus 1990). Such quantitative partitioning of "size" from "shape" can be comparatively straightforward (Bookstein 1989), especially where ontogenetic "size" in taxa showing protracted or indeterminate growth is of primary interest (Burnaby 1966; Mosimann and James 1979; Darroch and Mosimann 1985; Sampson and Siegel 1985; James and McCulloch 1990). However, such methods possess subtly different properties (Reist 1985, 1986; Rohlf and Bookstein 1987), "correct" for different metrics of "size" (e.g., within species or among species), and may create other interpretational difficulties (Packard and Boardman 1987). In the present study, a practical distinction between "size" and various dimensions of "shape" seemed justified and resilient with respect to a suite of approaches (Rising and Somers 1989), in part because "size" was not considered a nuisance variable but instead an important dimension of morphological change. That is, in the present study, definition of the first component as essentially indicative of "size" was straightforward, and the criticality of interpreting differences in "shape" in juxtaposition with "size" (instead of in isolation) was foremost. Accordingly, an element-wise "correction" for size through division by body mass and log-transformation, as advocated by Mosimann (1970, 1975a, b) and Mosimann and James (1979), was not pursued.

The correspondence among various metrics of multivariate "size" was performed based on quantitative comparisons of correlations among the PC-Is of dispersion matrices of external and skeletal variables partitioned into three primary components: total covariance (S_T), among-taxon covariance (S_A), and pooled within-taxon covariance (S_W), wherein $S_W \oplus S_A = S_T$. As adjunct to the quantification of "size" included in multivariate axes, the numerical relationship between mean body masses and mean scores of species on multivariate axes (both PCA and canonical analysis) was calculated and shown as a vector in multivariate plots. In such applications, the vector for "mass" indicates the approximate direction of variation in body masses in the multivariate plane. Specifically, in a bivariate plot of PC_i ($i = 1, 2$), the direction of the vector **Mass** is defined as the sum, weighted by the square roots of the proportion of the total dispersion ($\Sigma \lambda_i$) given by respective eigenvalues of (λ_l) of PC-I and PC-II, of the angles of direction (θ_l) of mass with each eigenvector (ξ_l). The latter are directly estimable as the arccosines of the correlation coefficients relating log-transformed mean body masses with the respective axes. Symbolically, the direction of vector **Mass** relative to PC-I (ξ_1) is specified by direction θ_{mass}:

$$\theta_{mass} = \lambda_1 / \left(\sum \lambda_i\right) \cdot \theta_1 + \lambda_2 / \left(\sum \lambda_i\right) \cdot (\theta_2 + \pi/2)$$

$$= \left[\lambda_1 / \left(\sum \lambda_i\right)\right]^{1/2} \cdot \{\arccos[r(mass, \xi_1)]\}$$

$$+ \left[\lambda_2 / \left(\sum \lambda_i\right)\right]^{1/2} \cdot \{\arccos[r(mass, \xi_2)] + \pi/2\},$$

given that, by orthogonality of eigenvectors ($\xi_i \perp \xi_j, \forall\, i, j = 1, \ldots, p;\, i \neq j$), PC-I is perpendicular to PC-II (i.e., angle of translation is $\pi/2$), and therefore the vector sum given above amounts to the hypotenuse of the right triangle of the two vectors of correlation. Vectors relating body masses to mean values on canonical variates were derived similarly.

ALTERNATIVE *R*-MODE PARTITIONING OF "SIZE" AND "SHAPE"

Other analytical perspectives on "size" and "shape" have been advanced (e.g., Corruccini 1987; Piersma and Davidson 1991). In some comparisons, with a strong association between extremes in size with marked, flightlessness-related apomorphy in shape (e.g., only the largest taxon shows significant pectoral reduction), special measures have been taken to preserve the approximate orthogonality between the components reflecting "size" (e.g., that paralleled by body mass) and "shape" (including pectoral reduction and other morphometric departures unrelated to body mass). One approach is to impose multivariate isometry on the first component (Somers 1986, 1989; Sundberg 1989), wherein the vector loadings for PC-I ($\mathbf{l}_1$) are set a priori to be:

$$l_1^T \equiv (l_{11}, \ldots, l_{1i}, \ldots, l_{1p}) = (p^{-1/2}, \ldots, p^{-1/2}), \qquad \forall i = 1, \ldots, p.$$

All subsequent components (which are mutually orthogonal but each oblique with respect to the first, isometric component) are derived from the remaining variance. Although this method offers the advantage of a simple, a priori definition of "size," it has been used only infrequently because of the limited applicability of this narrow notion of "size" and the typically nontrivial correlations of subsequent axes with this first dimension.

A more conservative method, intended specifically to partition "size" (represented by the first component) from flightlessness-related "shape" (reserved for one or more subsequent components), was accomplished for selected taxonomic groups by defining the first component (PC-I*) with the flightless species having zero weights; this avoids the tendency for outliers at either extreme of the first component to influence the total eigenstructure of a matrix (Critchley 1985). The vectors for flightless species–sex groups were projected onto this axis a posteriori. All subsequent components were defined by weighting all species–sex groups equally, and were based on residuals from PC-I* summarized by the corresponding partial covariance matrix (Cochran and Cox 1957; Snedecor and Cochran 1980). This method offers advantages for displaying multivariate variation related to avian flightlessness (Livezey 1988, 1992a, b). A more radical approach for the definition of multivariate skeletal "size," where some of the included taxa showed extreme reduction of pectoral elements, based the first modified component solely on the nonpectoral dimensions of flighted species (Livezey 1992a, 1993c). Derivation of PC-I based on a subset of the taxa used for subsequent components (PC-I*), however, typically results in partial colinearity (i.e., non-zero correlations) between PC-I* and the subsequent (mutually orthogonal) axes (PC-II*, and so on). The latter characteristic is shared by factors obtained by oblique rotations; however, most methods of factor rotation (whether orthogonal or oblique) optimize other qualities of axes (e.g., heterogeneity of loadings within factors) that are not pertinent here (Harman 1967; Reyment and Jöreskog 1993; Basilevsky 1994).

Symbolically, for a total of S_T vectors of p mean measurements comprising S_F flighted and S_N flightless species ($S_F \oplus S_N = S_T$), PC-I* is the first PC (corresponding to eigenvectors ξ_{1F}, with eigenvalue λ_{1F}, for D_F having rank $min[n_F - 1, p]$) for the S_F flighted species. Subsequent principal components (PC-II*, PC-III*, ..., PC-t*; t = rank of dispersion matrix D_T) are derived for residuals from PC-I* for all S_T species; this variation is defined by eigenvectors ξ_{1T} for D_T/ξ_{1F}, that is, the eigenspace orthogonal to that spanned by PC-I*, or (PC-I*)^C, with respective eigenvalues λ_{1T} (PC-I*)^C. Nonorthogonality between PC-I* and the set {PC-II*, PC-III*, ..., PC-t*}, stemming from the exclusion of the S_N flightless species from the definition of PC-I*, is proportional to the magnitude of noncolinearity of PC-I (all S_T species) and PC-I* (only S_F species), which in turn reflects the angles between ξ_{1F} and the set $\{\xi_{1T} (\text{PC-I*})^C\}$. Correlation coefficients for PC-I* and subsequent, mutually orthogonal components PC-II*, through PC-T* are given by the correlation matrix R*:

$$R^* = \begin{bmatrix} 1 & r_{21} & r_{31} & \cdots & r_{T1} \\ r_{12} & 1 & 0 & \cdots & 0 \\ r_{13} & 0 & 1 & \cdots & 0 \\ \vdots & \vdots & \vdots & \ddots & 0 \\ r_{1T} & 0 & 0 & 0 & 1 \end{bmatrix}.$$

Off-diagonal elements in either the first row or the first column are measures of variance shared between the oblique components. That is, elements of the matrix R* are:

$$r_{ij} = 1, \qquad \forall\, i = j; \qquad -1 < r_{i1} = r_{1j} < 1, \qquad \forall\, i, j; \quad \text{and}$$

$$r_{ij} = 0, \qquad \forall\, i, j \neq 1 \quad \text{and} \quad i \neq j.$$

In the present study, the diversity of size exhibited by both flighted and flightless species of Rallidae rendered such refinements of negligible advantage over standard PCAs in analyses of study skins and skeletons; therefore, only results of the traditional PCAs were presented for these data sets. However, the asymmetrical distribution of body sizes between flighted and flightless taxa dissected for myological analyses (with flightless *Porphyrio hochstetteri* being more than twice as massive as any flighted species sampled) prompted the use of the flighted-only PC-I* to define "size," with subsequent shape axes being based on residuals of all taxa from this narrowly defined first component as adjunct to traditional techniques in the myological comparisons. In addition, $\log_e$-transformed mean body mass was used to define "size" for the myological data set, with PCs of residuals from body mass ($PC_i | \log_e[mass]$; $i = 1, \ldots, min [n_T - 1, p]$), that is, the PCs of the partial covariance matrix for measurements corrected for mean body masses, defining a suite of mutually orthogonal "shape" axes. In the latter approach, all taxa were treated equally, regardless of flight capacity.

Q-Mode PCA

Applications of PCA described above are of the typical form in which relationships among n observations (taxa in the present context) are examined in a multivariate space spanned by p variables; this class of applications are examples of R-analysis (Reyment et al. 1984). An alternative approach, termed Q-analysis, implements an inverse assessment of p variates in a space defined by n observations, and is intended to reveal multivariate commonalities among variables based on the correlation structure manifested in the sample (Okamoto 1972; Jackson 1991). This approach is related to multidimensional scaling (Davison 1983) and especially to the variable-analytic aspect of correspondence analysis (Hill 1974; Greenacre 1984; Lebart et al. 1984; Palmer 1993). In this study, Q-analysis was used as a supplementary method of assessing differences between flighted and flightless rails in correlation structures in external and skeletal measurements, accomplished by PCAs of correlation matrices for species means of flighted and flightless rails separately. Where the covariance matrix was singular or the number of observations (n, here equal to the number of taxa having complete data) is less than the number of variables considered, a singular-value decomposition was required to calculate the latent roots and vectors. Correlation matrices in this context were of dimension $n \times n$ and rank was $min(n - 1, p)$, and where $p < n$ the matrix is by definition singular (Jackson 1991). In this study, n for external data was 68 and 28 for flighted and flightless species, respectively, and n for skeletal data was 39 and 23 for flighted and flighted species, respectively. The number of variables p was 6 and 36 for external and included skeletal dimensions, respectively.

Linear Models and Orthogonal Contrasts

General Linear Models and *ANOVA*

Most applications in this study were based on predefined groups subdivided by both species and sex. Relative contributions of interspecific differences, sexual dimorphism, and species–sex interactions in multivariate distances among groups were assessed by using stepwise multivariate analysis of variance (MANOVA) targeting these effects (on all canonical variates) and two-way ANOVA of scores of specimens on canonical variates. Methodological details of these techniques were described in previous papers (Livezey 1989b–d, 1990). For purposes of simplicity and following most other comparable investigations, MANOVAs were limited to linear models partitioned into relevant main, two-way, and (rarely) three-way interaction (random) effects for a given multivariate comparison of means across taxon–sex groups. That is, the additive model for a vector of overall means of p variables from population k could be partitioned into the following components of interest:

$\bar{x}_k = \mathbf{m}$ (overall mean) $+ \mathbf{g}_i$ (genus effect) $+ \mathbf{t}_j$ (species effect) $+ \mathbf{d}_l$ (sex effect)

$+ \mathbf{f}_m$ (flight effect) $+ \mathbf{g}_i\mathbf{t}_j$ (genus–species interaction effect)

$+ \mathbf{t}_j\mathbf{d}_l$ (taxon–sex interaction effect) $+ \mathbf{t}_j\mathbf{f}_m$ (species–flight interaction effect)

$+ \mathbf{d}_l\mathbf{f}_m$ (sex–flight interaction effect) $+ \cdots$.

Interaction effects of presumably lesser or negligible magnitude and only marginal interest (e.g., genus–sex interaction effects and three-way interactions), most of which circumvent an implicit hierarchy of effects consonant with phylogenetic structure, are not shown but are signified by the terminal ellipses. The vector of values for an individual multivariate observation (i.e., specimen s in population k) corresponds to the model for mean measurements (above) appended by independent random error term $\mathbf{e}_{ks} \sim N(0, \sigma_k^2)$.

Canonical Contrasts Between Groups

The linear effects of each MANOVA were specified by using vectors of integers defining orthogonal contrasts (c) of group means. Customarily required to comprise integers for which the sum is zero, contrasts used herein are as follows (s, taxa, two sexes; g, total taxon–sex groups having sample sizes greater than one; n_i, specimens in the ith group, with a total of n specimens summed across g groups):

(a) variance attributable to flightlessess across s taxa, that is, a contrast between flighted (males and females of taxa 1 through k) versus flightless (males and females of taxa $k + 1$ through s), given by the orthogonal contrast $c = [c_i]$, having length $\|c\|$:

$$c = [-1, -1, -1, -1, \ldots, -1, -1 \text{ (}k\text{th pair)}, 1, 1, 1, 1, \ldots, 1, 1 \text{ (}s\text{th pair)}],$$

$$\|c\| = [2(k + s)]^{1/2};$$

(b) variance attributable to sexual dimorphism across taxa, that is, males (first entry in each species couplet) versus females (second in each couplet) across species, given by the orthogonal contrast c:

$$c = [-1, 1, -1, 1, \ldots, -1, 1 \text{ (}k\text{th pair)}, -1, 1, -1, 1, \ldots, -1, 1 \text{ (}s\text{th pair)}],$$

$$\|c\| = [2(k + s)]^{1/2};$$

(c) variance attributable to differences among species (regardless of flight status), that is, simple species effects (s species sequenced by zero-sum integers regardless of flight status, and sexes within species having the same elements); where s is even, given by the contrast c:

$$c = [-(s - 1), -(s - 1), -(s - 3), -(s - 3), \ldots, -1, -1, 1, 1, \ldots,$$
$$(s - 3), (s - 3), (s - 1), (s - 1)];$$

and where s is odd, given by the contrast c:

$$c = [-1/2(s - 1), -1/2(s - 1), -1/2(s - 3), -1/2(s - 3), \ldots,$$
$$-1, -1, 0, 0, 1, 1, \ldots, 1/2(s - 3), 1/2(s - 3), 1/2(s - 1), 1/2(s - 1)];$$

(d) variance attributable to interspecific differences in sexual dimorphism (species–sex interaction effects), that is, the element-wise product of the contrast between the sexes within each of the s taxa (b) and the preceding orthogonal contrast for interspecific effects; for example, for the case where s is even, is given by the contrast c:

$$c = [(s - 1), -(s - 1), (s - 3), -(s - 3), \ldots, 1, -1, -1, 1, \ldots,$$
$$-(s - 3), (s - 3), -(s - 1), (s - 1)];$$

(e) variance attributable to differences between k flighted and $s - k$ flightless species in sexual dimorphism, that is, flightlessness–sex interactions, given by the orthogonal contrast c:

$$c = [1, -1, 1, -1, \ldots, 1, -1 \text{ (}k\text{th pair)}, -1, 1, -1, 1, \ldots, -1, 1 \text{ (}s\text{th pair)}],$$

$$\|c\| = [2(k + s)]^{1/2};$$

Those contrasts involving binary elements—that is, only zeros and ones (regardless of sign)—can be used to define mutually orthogonal interaction terms through element-wise multiplication. However, those combinations involving two or more taxon effects (i.e., vectors having elements with absolute values greater than one) generally are not orthogonal, that is, these share some components of variance with their constituent main effects. Such oblique contrasts are characterized by nonzero inner products, and the shared components of variance reflect the hierarchically nested nature of the effects under consideration (e.g., interspecific and intergeneric effects). In algebraic terms, for vectors of contrast $\mathbf{v}$ and $\mathbf{w}$ having dimension k:

$$\mathbf{v}^T \cdot \mathbf{w} = \mathbf{w}^T \cdot \mathbf{v} \equiv [w_1, \ldots, w_k]^T \cdot [v_1, \ldots, v_k] = \sum_{i=1}^{k} w_i v_i = 0 \Leftrightarrow \mathbf{v} \perp \mathbf{w}.$$

CATEGORICAL DATA FOR CORRESPONDENCE ANALYSIS

Categories for ecomorphological variables were partitioned as follows: change in body mass compared to flighted relative (decrease, −75–0%; small increase, 1–100%; large increase, 101–326%); decrease in wing length, measured as residual at given body mass from regression line for flighted rails (−75 to 51, −50 to 26, −25–0); Mahalanobis D from flighted relative in skin measurements (1–1.99, 2.00–4.10); Mahalanobis D from flighted relative in skeletal measurements (0–0.99, 1.00–1.99, 2.00–3.70); island area ($k = 0, 1, 2, 3, 4$, where k is exponent in expression, area (km^2) of island = $2 \cdot 10^k$); latitude (tropical, 0–23°C; warm temperate, 23–33°C; cool temperate, 34°C or more); minimal distance (km) to continent (small, 0–1,000; medium, 1,001–2,000; large, 2,001 or more).

Sources of Miscellaneous Information

Sources for Body Masses and Wing Areas

MacGillivray 1926; Heinroth and Heinroth 1928; Mayr 1933; Conover 1934; Chapin 1939; Stresemann 1941; Piveteau 1945; Sharland 1945, 1973; Strophlet 1946; Baldwin 1947; Baker 1951; Mayr and Gilliard 1954; Ripley 1954, 1957a, b; Macworth-Praed and Grant 1957, 1962, 1970; Behn and Millie 1959; Dickerman and Warner 1961; Watson 1962; Russell 1964; Moreau 1966; Falla 1967; Harrison and Parker 1967; Bang 1968, 1971; Dickerman 1968a, b; Ali and Ripley 1969; Diamond 1969, 1972; Britton 1970; Keith et al. 1970; Glutz von Blotzheim et al. 1973; Goodwin 1974; Haverschmidt 1974; McFarlane 1975; Vuilleumier and Gochfeld 1976; Blake 1977; Greig-Smith and Davidson 1977; Mees 1977, 1982; Wagstaffe 1978; Contreras 1979; Roe and Rees 1979; Sick 1979; Voisin 1979; Cramp and Simmons 1980; Dowsett and Dowsett-Lemaire 1980; Warham 1980; Fisher 1981; Fjeldså 1981, 1983a, b; M. P. Harris 1981; Storer 1981, 1989; Fullagar et al. 1982; Rosser et al. 1982; Schulenberg and Remsen 1982; Carey 1983; Watling 1983; Dickinson 1984; Byrd et al. 1985; Clancey 1985; Coates 1985; Robertson and Beauchamp 1985; Keith 1986; Teixeira et al. 1986, 1989; White and Bruce 1986; Fjeldså und Krabbe 1988; Herholdt 1988; McAllan and Bruce 1988; Ross 1988; Binford 1989; Goodman and Gonzales 1989; Jackson 1989; Frith and Frith 1990; Langrand 1990; Elliott et al. 1991; Webb 1992; Dunning 1993; Marchant and Higgins 1993; Erritzoe 1995; Gibbs 1996; Rasmussen et al. 1996; Zimmerman et al. 1996.

Body Masses and Wing Lengths of Non-Rallied Power Fliers

Schorger 1966; Blake 1977; Cramp and Simmons 1977, 1980; Brown et al. 1982; Johnsgard 1983, 1988, 1991, 1999; Urban et al. 1986; Marchant and Higgins 1990, 1993; Kuz'mina 1992; Jones et al. 1995.

APPENDIX 3. Anatomical terms and abbreviations.

The nomenclature follows Baumel (1993). Abbreviations closely approximate those advocated by Vanden Berge and Zweers (1993:244–247). Nomenclature of musculi alae is that of George and Berger (1966), except that the latter referred to digitalis alulae, digitalis majoris, and digitalis minoris as digits II, III, and IV, respectively.

Anatomical feature, by group	Abbreviation
Integumentum	
Partes pennae	
Folliculus	Fol.
Scapus pennae	Scap. pen.
Scapus pennae, calamus pennae	Calamus
Rhachis pennae	Rhac. pen.
Vexillum pennae	Vex. pen.
Vexillum pennae, vexilllum internum	Vex. int.
Vexillum pennae, vexillum externum	Vex. ext.
Vexillum pennae, pars plumacea	Vex. plum.
Vexillum pennae, pars pennacea	Vex. penn.
Barba pennae	Barb.
Barba pennae, ramus	Barb. ram.
Barba pennae, ramus, petiolus	Barb. pet.
Barba pennae, ramus, incisura rami	Barb. inc.
Barba pennae, ramus, crista dorsalis	Barb. cr. dors.
Barba pennae, ramus, crista ventralis	Barb. cr. vent.
Barba pennae, barbula rami	Barb. vex. barb.
Barba pennae, barbula rami, barbula proximalis	Barb. vex. barb. prox.
Barba pennae, barbula rami, barbula distalis	Barb. vex. barb. dist.
Barba pennae, barbula rami, basis barbulae	Barbl. basis
Barba pennae, barbula rami, basis barbulae, arcus dorsalis	Barbl. arc. dors.
Barba pennae, barbula rami, basis barbulae, dens ventralis	Barbl. dens vent.
Barba pennae, barbula rami, pennula	Barbl. penn.
Barba pennae, barbula rami, pennula, nodus	Barbl. penn. nod.
Barba pennae, barbula rami, pennula, internodus	Barbl. penn. inter.
Barba pennae, barbula rami, pennula, cilium dorsale	Barbl. penn. cil. dors.
Barba pennae, barbula rami, pennula, cilium ventrale	Barbl. penn. cil. vent.
Barba pennae, barbula rami, pennula, hamulus	Barbl. penn. ham.
Vexillum pennae, apex	Vex. pen. apex
Vexillum pennae, margo	Vex. pen. marg.
Vexillum pennae, margo, incisura vexilli	Vex. pen. inc. vex.
Pterylae et apteria	
Pteryla scapulohumeralis	Pter. scap.-hum.
Pteryla alae, remiges primarii	Rem. prim.

APPENDIX 3. Continued.

Anatomical feature, by group	Abbreviation
Pteryla alae, remiges secundarii	Rem. sec.
Pteryla alae, remiges secundarii, diastema remigum secundarium	Dias. rem.
Pteryla alae, remiges tertiarii	Rem. tert.
Pteryla alae, remex carpalis	Rem. carp.
Pteryla alae, remiges alulae	Rem. alu.
Pteryla alae, tectrices alae, tectrices marginales manus dorsales	Tect. marg. man. dors.
Pteryla alae, tectrices alae, tectrices marginales propatagii	Tect. marg. prop.
Pteryla alae, tectrices alae, tectrices dorsales, tectrices dorsales alulae	Tect. alu. dors.
Pteryla alae, tectrices alae, tectrices dorsales, tectrices dorsales propatagii	Tect. prop. dors.
Pteryla alae, tectrices alae, tectrices dorsales, tectrices dorsales primariae majores	Tect. prim. dors. maj.
Pteryla alae, tectrices alae, tectrices dorsales, tectrices dorsales primariae mediae	Tect. prim. dors. med.
Pteryla alae, tectrices alae, tectrices dorsales, tectrices carpalis dorsalis	Tect. carp. dors.
Pteryla alae, tectrices alae, tectrices dorsales, tectrices dorsales secundariae majores	Tect. sec. dors. maj.
Pteryla alae, tectrices alae, tectrices dorsales, tectrices dorsales secundariae mediae	Tect. sec. dors. med.
Pteryla alae, tectrices alae, tectrices dorsales, tectrices dorsales secundariae minores	Tect. sec. dors. min.
Apteria alae	Apt. alae
Pennae	
Patagia alae, propatagium	Prop.
Patagia alae, propatagium, patagium alulae	Pat. alu.
Patagia alae, metapatagium	Metapat.
Partes alae et caudae	
Brachium	Brach.
Cubitus	Cub.
Antebrachium	Antebrac.
Carpus	Carpus
Manus	Manus
Ungues digiti manus, unguis digiti alulae	Ung. dig. alu.
Alula	Alula
Osteologia	
Sternum	
Corpus sterni, facies muscularis sterni, linea intermuscularis	Lin. intermus. ster.
Corpus sterni, facies muscularis sterni, impressio m. pectoralis sternobrachialis	Impr. m. pect. sternobrach.
Corpus sterni, facies muscularis sterni, impressio m. supracoracoideus	Impr. m. supracor.
Corpus sterni, facies muscularis sterni, planum postcarinale	Plan. postcarn. ster.
Corpus sterni, facies visceralis sterni, pars cardiaca	Pars card. ster.

APPENDIX 3. Continued.

Anatomical feature, by group	Abbreviation
Corpus sterni, facies visceralis sterni, pars hepatica	Pars hep. ster.
Corpus sterni, margo costalis sterni, incisurae costalis	Inc. cost. ster.
Corpus sterni, margo costalis sterni, loculus costalis	Loc. cost. ster.
Corpus sterni, margo costalis sterni, pila costalis	Pila cost. ster.
Corpus sterni, margo cranialis sterni, sulcus articularis coracoideus	Sulc. art. cor.
Corpus sterni, margo cranialis sterni, labrum externum	Labr. ext. ster.
Corpus sterni, margo cranialis sterni, labrum externum, tuberculum labri externi	Tub. lab. ext. ster.
Corpus sterni, margo cranialis sterni, processus craniolateralis sterni	Proc. cran.-lat. ster.
Corpus sterni, margo cranialis sterni, impressio craniolateralis sterni	Impr. cran.-lat. ster.
Rostrum sterni, spina externa rostri	Spin. ext. rost.
Rostrum sterni, spatium intercoracoidale	Spat. intercor. ster.
Margo caudalis sterni, incisura medialis	Inc. med. ster.
Margo caudalis sterni, trabecula intermedia	Trab. inter. ster.
Margo caudalis sterni, trabecula mediana	Trab. med. ster.
Margo caudalis sterni, processus caudolateralis sterni	Proc. caudolat. ster.
Carina sterni, apex carinae	Apex car. ster.
Carina sterni, facies articularis furculae	Fac. art. fur. ster.
Carina sterni, facies lateralis carinae	Fac. lat. car. ster.
Carina sterni, margo cranialis carinae, pila carinae	Pil. car. ster.
Carina sterni, margo cranialis carinae, sulcus carinae	Sulc. car. ster.
Carina sterni, margo ventralis carinae	Marg. vent. car. ster.

Clavicula

Extremitas omalis claviculae [epicleideum]	Extr. om. clav.
Extremitas omalis claviculae, processus acrocoracoideus claviculae	Proc. acrocor. clav.
Extremitas omalis claviculae, processus acromialis claviculae	Proc. acrom. clav.
Extremitas sternalis claviculae, apophysis furculae [hypocleideum]	Apo. furc.
Extremitas sternalis claviculae, scapus [corpus] claviculae	Scap. clav.
Extremitas sternalis claviculae, processus interclavicularis	Proc. interclav.

Scapula

Extremitas cranialis [caput] scapulae, acromion	Acrom. scap.
Extremitas cranialis [caput] scapulae, facies articularis coracoidea	Fac. art. cor. scap.
Extremitas cranialis [caput] scapulae, tuberculum coracoideum	Tub. cor. scap.
Collum scapulae	Coll. scap.
Corpus scapulae, facies medialis [facies costalis]	Fac. med. scap.
Corpus scapulae, facies lateralis	Fac. lat. scap.
Corpus scapulae, margo dorsalis [margo vertebralis]	Marg. dors. scap.
Extremitas caudalis [spina] scapulae	Extr. caud. scap

APPENDIX 3. Continued.

Anatomical feature, by group	Abbreviation
Coracoideum	
Extremitas omalis coracoidei, processus acrocoracoideus, facies articularis clavicularis	Fac. art. clav. cor.
Extremitas omalis coracoidei, processus acrocoracoideus, impressio ligamentum acrocoracohumeralis	Impr. lig. acrocor. cor.
Extremitas omalis coracoidei, processus glenoidalis coracoidei, facies articularis humeralis	Fac. art. hum. cor.
Extremitas omalis coracoidei, processus procoracoideus, facies articularis scapularis	Fac. art. scap. cor.
Corpus coracoidei, facies dorsalis	Fac. dors. cor.
Corpus coracoidei, facies dorsalis, cotyla scapularis	Cot. scap. cor.
Corpus coracoidei, facies ventralis	Fac. vent. cor.
Corpus coracoidei, facies ventralis, margo lateralis	Marg. lat. cor.
Corpus coracoidei, facies ventralis, margo medialis	Marg. med. cor.
Corpus coracoidei, facies ventralis, foramen [incisura] nervi supracoracoidei	For. nerv. supracor.
Corpus coracoidei, facies ventralis, linea intermuscularis ventralis	Lin. intermus. vent. cor.
Extremitas sternalis coracoidei, crista articularis sternalis, facies externa	Crist. art. ster. ext. cor.
Extremitas sternalis coracoidei, crista articularis sternalis, crista intermedia	Crist. art. ster. inter. cor.
Extremitas sternalis coracoidei, crista articularis sternalis, facies interna	Crist. art. ster. int. cor.
Extremitas sternalis coracoidei, angulus medialis	Ang. med. cor.
Extremitas sternalis coracoidei, processus lateralis, angulus lateralis	Ang. lat. cor.
Extremitas sternalis coracoidei, processus lateralis, impressio m. sternocoracoidei	Impr. m. sternocor. cor.
Cavitas glenoidalis	Cav. glen. cor.
Humerus	
Extremitas proximalis humeri, caput humeri	Cap. hum.
Extremitas proximalis humeri, caput humeri, incisura capitis humeri	Inc. cap. hum.
Extremitas proximalis humeri, caput humeri, crista incisurae capitis	Crist. inc. cap. hum.
Extremitas proximalis humeri, caput humeri, tuberculum dorsale	Tub. dors. hum.
Extremitas proximalis humeri, crista deltopectoralis	Crist. delt. hum.
Extremitas proximalis humeri, crista deltopectoralis, angulus cristae	Ang. crist. hum.
Extremitas proximalis humeri, tuberculum ventrale	Tub. vent. hum.
Extremitas proximalis humeri, tuberculum ventrale, crista bicipitalis	Crist. bic. hum.
Extremitas proximalis humeri, fossa pneumotricipitalis, foramen pneumaticum	For. pneu. hum.
Extremitas proximalis humeri, fossa pneumotricipitalis, crus dorsale fossae	Crus dors. fos. hum.
Extremitas proximalis humeri, fossa pneumotricipitalis, crus ventrale fossae	Crus vent. fos. hum.
Extremitas proximalis humeri, planum intertuberculare, intumescentia humeri	Intum. hum.
Extremitas proximalis humeri, sulcus ligamentosus transversus	Sulc. trans. hum.
Extremitas proximalis humeri, impressio coracobrachialis	Impr. cor. hum.
Corpus humeri, facies cranialis	Fac. cran. hum.
Corpus humeri, margo caudalis	Marg. caud. hum.
Corpus humeri, margo dorsalis	Marg. dors. hum.
Corpus humeri, margo ventralis	Marg. vent. hum.

APPENDIX 3. Continued.

Anatomical feature, by group	Abbreviation
Corpus humeri, margo ventralis, linea m. latissimius dorsi	Lin. m. lat. hum.
Extremitas distalis humeri, condylus dorsalis humeri	Cond. dors. hum.
Extremitas distalis humeri, condylus ventralis humeri	Cond. vent. hum.
Extremitas distalis humeri, incisura intercondylaris	Inc. intercond. hum.
Extremitas distalis humeri, fossa m. brachialis	Fos. m. brach. hum.
Extremitas distalis humeri, epicondylus ventralis [entepicondylus]	Epicond. vent. hum.
Extremitas distalis humeri, processus flexorius	Proc. flex. hum.
Extremitas distalis humeri, tuberculum supracondylare ventrale	Tub. supr. vent. hum.
Extremitas distalis humeri, fossa olecrani	Fos. olec. hum.
Extremitas distalis humeri, fossa olecrani, sulcus scapulotricipitalis	Sulc. scap.-tric. hum.
Extremitas distalis humeri, fossa olecrani, sulcus humerotricipitalis	Sulc. hum.-tric. hum.

Ulna

Extremitas proximalis ulnae, processus cotylaris dorsalis, cotyla dorsalis	Cot. dors. uln.
Extremitas proximalis ulnae, processus cotylaris dorsalis, cotyla ventralis	Cot. vent. uln.
Extremitas proximalis ulnae, processus cotylaris dorsalis, crista intercotylaris	Crist. intercot. uln.
Extremitas proximalis ulnae, impressio m. brachialis	Impr. m. brach. uln.
Extremitas proximalis ulnae, incisura radialis	Inc. rad. uln.
Extremitas proximalis ulnae, olecranon	Olec. uln.
Extremitas proximalis ulnae, tuberculum b cipitale ulnae	Tub. bic. uln.
Corpus ulnae, facies caudodorsalis	Fac. caudodors. uln.
Corpus ulnae, facies caudoventralis	Fac. caudovent. uln.
Corpus ulnae, facies cranialis	Fac. cran. uln.
Corpus ulnae, margo caudalis	Marg. caud. uln.
Corpus ulnae, margo interosseus [cranialis]	Marg. inteross. uln.
Extremitas distalis ulnae, trochlea carpalis. condylus dorsalis ulnae	Cond. dors. uln.
Extremitas distalis ulnae, trochlea carpalis. condylus ventralis ulnae	Cond. vent. uln.
Extremitas distalis ulnae, trochlea carpalis. sulcus intercondylaris	Sulc. intercon. uln.
Extremitas distalis ulnae, tuberculum carpale	Tub. carp. uln.

Radius

Extremitas proximalis radii, caput radii	Cap. rad.
Extremitas proximalis radii, caput radii, cotyla humeralis	Cot. hum. rad.
Extremitas proximalis radii, facies articularis ulnaris	Fac. art. uln. rad.
Corpus radii, margo interosseus [caudalis]	Marg. inteross. rad.
Extremitas distalis radii, sulcus tendinosus	Sulc. tend. rad.
Extremitas distalis radii, tuberculum aponeurosis ventralis	Tub. apon. vent. rad.

Ossa carpi

Os carpi radiale	Os carpi rad.
Os carpi ulnare	Os carpi uln.

APPENDIX 3. Continued.

Anatomical feature, by group	Abbreviation
Carpometacarpus	
Extremitas proximalis carpometacarpi, os metacarpale alulare	Os met. alu.
Extremitas proximalis carpometacarpi, processus alularis, facies articularis alularis	Fac. art. alu. carp.
Extremitas proximalis carpometacarpi, processus alularis, processus extensorius	Proc. exten. carp.
Extremitas proximalis carpometacarpi, trochlea carpalis	Troch. carp.
Extremitas proximalis carpometacarpi, fossa infratrochlearis	Fos. infratroch. carp.
Extremitas proximalis carpometacarpi, fossa supratrochlearis	Fos. supratroch. carp.
Extremitas proximalis carpometacarpi, fovea carpalis caudalis	Fov. carp. caud.
Extremitas proximalis carpometacarpi, fovea carpalis cranialis	Fov. carp. cran.
Extremitas proximalis carpometacarpi, processus pisiformis	Proc. pis. carp.
Corpus carpometacarpi, facies dorsalis	Fac. dors. carp.
Corpus carpometacarpi, facies ventralis	Fac. vent. carp.
Corpus carpometacarpi, margo caudalis	Marg. caud. carp.
Corpus carpometacarpi, margo cranialis	Marg. cran. carp.
Corpus carpometacarpi, os metacarpale majus	Os metacarp. maj.
Corpus carpometacarpi, os metacarpale majus, sulcus tendinosus	Sulc. tend. metacarp.
Corpus carpometacarpi, os metacarpale majus, processus intermetacarpalis	Proc. intermetacarp.
Corpus carpometacarpi, os metacarpale minus	Os metacarp. min.
Corpus carpometacarpi, spatium intermetacarpale	Spat. intermet.
Extremitas distalis carpometacarpi, symphysis metacarpalis distalis	Symp. metacarp. dist.
Extremitas distalis carpometacarpi, symphysis metacarpalis distalis, sulcus interosseus	Sulc. inteross. metacarp.
Extremitas distalis carpometacarpi, facies articularis digitalis major	Fac. art. dig. maj. metacarp.
Extremitas distalis carpometacarpi, facies articularis digitalis minor	Fac. art. dig. min. metacarp.
Ossa digitorum manus	
Phalanx digiti alulae	Phal. dig. alu.
Phalanx proximalis digiti majoris	Phal. prox. dig. maj.
Phalanx distalis digiti majoris	Phal. dist. dig. maj.
Phalanx digiti minoris	Phal. dig. min.
Myologia	
Termini generales	
Fascia	Fas.
Fasciculus	Fasc.
Musculi pterylarum [subcutanei]	
M. cucullaris, m. cucullaris capitis, pars interscapularis	M. cuc. cap. interscap.
M. cucullaris, m. cucullaris capitis, pars propatagialis	M. cuc. cap. propat.
M. cucullaris, m. cucullaris capitis, pars clavicularis	M. cuc. cap. clavic.
M. latissimus dorsi, pars interscapularis	M. lat. dors. interscap.

APPENDIX 3. Continued.

Anatomical feature, by group	Abbreviation
M. latissimus dorsi, pars metapatagialis	M. lat. dors. metap.
M. latissimus dorsi, pars scapulohumeralis	M. lat. dors. scap.-hum.
M. serratus superficialis, pars scapulohumeralis	M. serr. sup. scap.-hum.
M. pectoralis, pars subcutanea thoracica	M. pect. subcut. thor.
M. pectoralis, pars subcutanea abdominalis	M. pect. subcut. abdom.
M. pectoralis, pars propatagialis	M. pect. propat.
Musculi trunci	
M. longus colli dorsalis, pars caudalis	M. long. col. dors. caud.
M. longissimus dorsi	M. long. dors.
Mm. intercostales externi	Mm. intercost. ext.
M. costosternalis	M. costostern.
M. sternocoracoideus	M. sternocor.
M. sternotrachealis	M. sternotrach.
M. obliquus externus abdominis	M. obl. ext. abdom.
Musculi alae	
M. rhomboideus superficialis	M. rhom. sup.
M. rhomboideus profundus	M. rhom. prof.
M. serratus superficialis, pars scapulohumeralis	M. serr. sup. scap.-hum.
M. serratus superficialis, pars cranialis	M. serr. sup. cran.
M. serratus superficialis, pars caudalis	M. serr. sup. caud.
M. serratus superficialis, pars metapatagialis	M. serr. sup. metap.
M. serratus profundus	M. serr. prof.
M. scapulohumeralis cranialis	M. scap.-hum. cran.
M. scapulohumeralis caudalis	M. scap.-hum. caud.
Mm. subcoracoscapulares, m. subscapularis, caput laterale	M. subscap. lat.
Mm. subcoracoscapulares, m. subscapularis, caput mediale	M. subscap. med.
Mm. subcoracoscapulares, m. subcoracoideus, caput dorsale	M. subcor. dors.
Mm. subcoracoscapulares, m. subcoracoideus, caput ventrale	M. subcor. vent.
M. coracobrachialis cranialis	M. cor.-bra. cran.
M. coracobrachialis caudalis	M. cor.-bra. caud.
M. pectoralis, pars sternobrachialis	M. pect. sternobra.
M. pectoralis, pars costobrachialis	M. pect. costobra.
M. supracoracoideus	M. supracor.
M. latissimus dorsi, pars scapulohumeralis	M. lat. dors. scap.-hum.
M. latissimus dorsi, pars cranialis	M. lat. dors. cran.
M. latissimus dorsi, pars caudalis	M. lat. dors. caud.
M. deltoideus, pars propatagialis	M. delt. propat.
M. deltoideus, pars propatagialis, caput craniale [pars longa]	M. delt. propat. cran.

APPENDIX 3. Continued.

Anatomical feature, by group	Abbreviation
M. deltoideus, pars propatagialis, caput caudale [pars brevis]	M. delt. propat. caud.
M. deltoideus, pars major	M. delt. maj.
M. deltoideus, pars major, caput craniale	M. delt. maj. cran.
M. deltoideus, pars major, caput caudale	M. delt. maj. caud.
M. deltoideus, pars minor, caput dorsale	M. delt. min. dors.
M. deltoideus, pars minor, caput ventrale	M. delt. min. vent.
Mm. triceps brachii, m. scapulotriceps	M. scap.-tric.
Mm. triceps brachii, m. humerotriceps	M. hum.-tric.
M. biceps brachii	M. bic. bra.
M. biceps brachii, caput coracoideum	M. bic. bra. cor.
M. biceps brachii, caput humerale	M. bic. bra. hum.
M. biceps brachii, pars propatagialis	M. bic. bra. propat.
M. expansor secundariorum	M. exp. sec.
M. expansor secundariorum, tendo proximalis	M. exp. sec. tend. prox.
M. expansor secundariorum, pars (tendo) distalis	M. exp. sec. tend. dist.
M. brachialis	M. brach.
M. pronator superficialis	M. pron. sup.
M. pronator profundus	M. pron. prof.
M. flexor carpi ulnaris	M. flex. carp. uln.
M. flexor carpi ulnaris, pars remigalis	M. flex. carp. uln. rem.
M. flexor digitorum superficialis	M. flex. dig. sup.
M. flexor digitorum profundus	M. flex. dig. prof.
M. extensor carpi radialis	M. ext. carp. rad.
M. extensor carpi radialis, caput dorsale	M. ext. carp. rad. dors.
M. extensor carpi radialis, caput ventrale	M. ext. carp. rad. vent.
M. extensor carpi ulnaris	M. ext. carp. uln.
M. extensor digitorum communis	M. ext. dig. comm.
M. extensor longus alulae	M. ext. long. alu.
M. extensor longus alulae, caput radiale	M. ext. long. alu. rad.
M. extensor longus alulae, caput ulnare	M. ext. long. alu. uln.
M. extensor longus digiti majoris (pars proximalis)	M. ext. long. dig. maj.
M. supinator	M. sup.
M. ectepicondylo-ulnaris	M. ect.-uln.
M. entepicondylo-ulnaris	M. ent.-uln.
M. ulnometacarpalis dorsalis	M. ulnomet. dors.
M. ulnometacarpalis dorsalis, caput dorsale	M. ulnomet. dors. dors.
M. ulnometacarpalis dorsalis, caput ventrale	M. ulnomet. dors. vent.

APPENDIX 3. Continued.

Anatomical feature, by group	Abbreviation
M. ulnometacarpalis ventralis	M. ulnomet. vent.
M. interosseus dorsalis	M. inteross. dors.
M. interosseus ventralis	M. inteross. vent.
M. extensor brevis alulae	M. ext. brev. alu.
M. abductor alulae, pars dorsalis	M. abd. alu. dors.
M. abductor alulae, pars ventralis	M. abd. alu. vent.
M. flexor alulae	M. flex. alu.
M. adductor alulae	M. add. alu.
M. abductor digiti majoris	M. abd. dig. maj.
M. flexor digiti minoris	M. flex. dig. min.
Musculi membri pelvici	
Mm. iliotibiales, m. iliotibialis cranialis	M. iliotib. cran.
Mm. iliotibiales, m. iliotibialis lateralis	M. iliotib. lat.
M. iliotrochanterici, m. iliotrochantericus caudalis	M. iliotroch. caud.
Arthrologia	
Membrana sternocoracoclavicularis	Memb. sternocor.-clav.
Membrana interossea antebrachii	Memb. int. antebr.
Retinaculum m. scapulotricipitis	Ret. m. scap.-tric.
Retinaculum m. extensoris metacarpi ulnaris	Ret. m. ext. met. uln.
Ligamentum propatagiale	Lig. propat.
Ligamentum propatagiale, pars elastica	Lig. propat. elas.
Ligamentum limitans cubiti	Lig. lim. cub.
Ligamentum humerocarpale	Lig. hum.-carp.
Ligamenta remigium primariorum et secondariorum, ligamentum elasticum interremigale major	Ligg. elas. interrem. maj.
Ligamenta remigium primariorum et secondariorum, ligamentum elasticum interremigale minor	Ligg. elas. interrem. min.
Ligamenta remigium primariorum et secondariorum, aponeurosis dorsalis antebrachii	Apon. dors. antebr.
Ligamenta remigium primariorum et secondariorum, aponeurosis ventralis antebrachii	Apon. vent. antebr.
Ligamenta remigium primariorum, retinaculum flexorium	Ret. flex.
Ligamenta remigium primariorum, aponeurosis ulnocarpo-remigalis	Apon. ulncarp.-rem.
Ligamenta remigium primariorum, retinaculum ulnocarpo-remigalis	Rett. ulncarp.-rem.
Ligamenta remigium primariorum, aponeurosis interphalango-remigalis	Apon. interphal.-rem.
Ligamenta remigium primariorum, ligamentum interphalango-remigale	Lig. interphal.-rem.
Ligamenta remigium primariorum, ligamena phalango-remigalia distalia	Ligg. phal.-rem. dist.
Ligamentum radio-radiocarpale dorsale	Lig. rad.-radiocarp. dors.
Articulatio metacarpophalangealis digiti majoris	Art. metacarp.-phal. dig. maj.
Articulatio interphalangealis digiti majoris	Art. interphal. dig. maj.

APPENDIX 4. Symbols used in mathematical expressions.

Symbol	Name	Meaning(s) in present study
α	alpha	Initial evolutionary (plesiomorphic) or ontogenetic state
β	beta	Intermediate evolutionary or ontogenetic state
γ	gamma	Relative onset or offset parameter
Δ, γ	delta	Change in evolutionary or ontogenetic state
ϵ	epsilon	Evolutionary transition or vector of transitions
ζ	zeta	Factor of change in ontogeny of body mass associated with flight
η	eta	Hypothetical avian colonist of island susceptible to flightlessness
θ	theta	Angle or degree of evolutionary or ontogenetic divergence in form
ι	iota	Interval of geological time after colonization
κ	kappa	Key evolutionary or ontogenetic event or innovation
μ	mu	Body mass of an evolutionary lineage at a given stage
π	pi	Vector comprising ontogenetic factors of change ζ and ψ
ρ	rho	Ontogenetic or evolutionary correlation matrix for morphology
τ	tau	Threshold of flightlessness
υ	upsilon	Morphological state of insular lineage
ϕ	phi	Flight status
ψ	psi	Factor of change in ontogeny of pectoral apparatus associated with flight
$\Omega; \omega$	omega	Terminal evolutionary (apomorphic) or ontogenetic state

APPENDIX 5. Solution of morphological trajectory.

Defining total body size as M and size of pectoral apparatus (e.g., wing area) as P, with associated first derivatives with respect to evolutionary time (t) being M' and P', one reasonable model of change is given by specifying change in M as being logistic with asymptote of optimal mass (μ), and change in P as doubly retarded by the logistic change in M and a logistic approach to threshold of flightlessness (τ) given by intraspecific allometry (Fig. 118).

Defining $M' = dM/dt$ and $P' = dP/dt$ with respect to optimal body size (μ) and functional threshold of flightlessness (τ) as $M' = \alpha(1 - M/\mu)$, and $P' = \beta(1 - M/\mu)[1 - M/\tau P)]$, wherein α and β are scaling parameters. Dividing these explicit differential equations yields:

$$\frac{P'}{M'} = \frac{\beta(1 - M/\mu)[1 - M/(\tau A)]}{\alpha(1 - M/\mu)} = \frac{\beta}{\alpha}\left[1 - \frac{M}{(\tau A)}\right],$$

a model implicit with respect to time (t). Substitution of scaling parameters $\gamma = \beta/\alpha$ and rearrangement of terms yields:

$$P \cdot \frac{dP}{dM} - \gamma P + \frac{\gamma}{\tau}M = 0,$$

which implies

$$\frac{dP}{dM} = \gamma\left[1 - \frac{M}{(\tau P)}\right],$$

a homogeneous differential equation. Substitution of variable $Y = P/M$, and rearrangement of terms yields:

$$\frac{dP}{dM} = \frac{dY}{dM} + Y.$$

Therefore, one can write:

$$M\frac{dY}{dM} + Y = \gamma\left[1 - \frac{1}{(\tau Y)}\right],$$

which implies

$$M\frac{dY}{dM} = \frac{\gamma Y - \gamma/\tau - Y^2}{Y}.$$

Separation of variables yields:

$$\frac{Y}{\gamma Y - \gamma/\tau - Y^2} dY = \frac{dM}{M},$$

which implies

$$\frac{(2Y - \gamma) + \gamma}{Y^2 - \gamma Y + \gamma/\tau} dY = \frac{-2}{M} dM.$$

Integration yields:

$$\ln|Y^2 - \gamma Y + \gamma/\tau| + \gamma \int \frac{1}{(Y - \gamma/2)^2 + \gamma(1/\tau - \gamma/4)} dY = -2 \ln|M| + C,$$

which implies

$$\ln[M^2(Y^2 - \gamma Y + \gamma\tau)] + \sqrt{\frac{\gamma}{(1/\tau - \gamma/4)}} \cdot \arctan\left[\frac{Y - \gamma/2}{\sqrt{\gamma(1/\tau - \gamma/4)}}\right] = C.$$

Of interest is the limit of this expression as mass (M) of zero is approached:

$$\lim_{M \to 0^+} \left\{ \ln\left[M^2\left(Y^2 - \gamma Y + \frac{\gamma}{\tau}\right)\right] + \sqrt{\frac{\gamma}{(1/\tau - \gamma/4)}} \cdot \arctan\left[\frac{Y - \gamma/2}{\sqrt{\gamma(1/\tau - \gamma/4)}}\right] - C \right\}$$

$$= \lim_{M \to 0^+} \left\{ \ln(P^2 - \gamma PM + \gamma M^2/\tau) + \sqrt{\frac{\gamma}{(\tau^{-1} - \gamma/4)}} \cdot \arctan\left[\frac{P/M - \gamma/2}{\gamma(\tau^{-1} - \gamma/4)}\right] - C \right\}$$

$$= \ln|P^2| + \sqrt{\frac{\gamma}{(1/\tau - \gamma/4)}} \cdot \arctan(``+\infty") - C = \ln|P^2| + \sqrt{\frac{\gamma}{(1/\tau - \gamma/4)}} \cdot \left(\frac{\pi}{2}\right) - C.$$

Solving for P yields:

$$\lim_{M \to 0^+} (P) = \left\{ \exp\left[C - \frac{\tau}{2}\sqrt{\frac{\gamma}{(1/\tau - \gamma/4)}}\right] \right\}^{1/2}.$$

The last expression and the limit for M were used recursively to obtain the points and estimates of change in the phase plane. The latter was estimable in many regions of interest, in that the system is continuous, defined, bounded, and meets the criteria of Lipschitz (Wilson 1971) for points near the line $P = M/\tau$ and values of $M < \mu$. That is, the last condition requires of the system f that for some constant k, $|f(M_1, P_1) - f(M_2, P_2)| \le k|(M_1, P_1) - (M_2, P_2)|$. However, at $M = \mu$, an entire line of critical points exists, rendering direct analysis impractical at or beyond this critical boundary.

APPENDIX 6. Generalizations of life histories of rails.

Principally based on summaries by Archibald and Meine (1996), Bryan (1996), Sherman (1996), Bertram (1996), and Taylor (1996). Additional sources were consulted for *Eulabeornis* (Barnes 1997; Franklin and Barnes 1998) and *Porphyrio* (Craig 1976, 1977, 1979, 1980a, b, 1982, 1984; Goldizen et al. 1993, 1998; Jamieson et al. 1994; Lambert et al. 1994).

Taxonomic group	Migratory status	Breeding habitat and seasonality	Primary diet	Age at maturity	Mating system	Typical nest site	Duration of pair bond	Parental investment	Changes associated with flightlessness
Psophiidae	Sedentary	Mature, moist tropical forest	Fruit, especially pulp	2 years	Cooperative polyandry	Elevated natural cavity, no nest construction	Evidently long-term, but complicated by extended group	Male cleans cavity, both sexes incubate and rear brood	—
Aramidae	Largely sedentary	Freshwater marshes	Molluscs, especially *Pomacea*	1 year	Monogamy	Variable in both site and construction	Short-term	Both sexes build nest, incubate, and rear brood	—
Gruidae: Balearicinae	Sedentary, but with irregular, local movements	Wetlands and grasslands; variable, rain-dependent	Varied, including plants, invertebrates, and vertebrates	2 (rarely) or 3 years	Monogamy	Elevated platform of grasses and sedges in freshwater	Long-term, probably lifelong	Both sexes construct nest, incubate, and rear brood	—
Gruidae: Gruinae	Most species migratory	Open, variably wet grassland, steppe, marsh, or muskeg	Varied, including plants, invertebrates, and vertebrates	2–5, mostly 3 or 4 years	Monogamy	Platform of grasses, sedges, leaves, peat, or mud	Long-term, probably lifelong	Both sexes construct nest, incubate, and rear brood	—
Heliornithidae	Sedentary	Freshwater marshes, at high water levels	Invertebrates	Not known	Monogamy	Loose, over-water platform of sticks	Probably long-term	Both sexes construct nest, incubate, and rear brood	—
Himantornis	Sedentary	Lowland rain forest	Invertebrates	Not known	Probably monogamy	Variably elevated, bowl constructed of brush	Probably long-term	Unknown	—

APPENDIX 6. Continued.

Taxonomic group	Migratory status	Breeding habitat and seasonality	Primary diet	Age at maturity	Mating system	Typical nest site	Duration of pair bond	Parental investment	Changes associated with flightlessness
Porphyrio	Sedentary with local movements	Sheltered wetlands; alpine grassland in *P. hochstetteri*	Seeds and shoots of aquatic plants; grasses in *P. hochstetteri*	1 or (more typically) 2 years	Monogamy in most; cooperative polygamy in some (e.g., *P. melanotus*)	Bulky, elevated or floating platform, often with loose canopy	Probably long-term in most; complicated in some by breeding groups	Both sexes construct nest(s), incubate, and rear brood(s)	Shift to dry alpine habitat, grass-based diet, and lifelong monogamy; protracted ontogeny
Porphyrula	Sedentary, except *P. martinica* in north	Freshwater marshes	Seeds, shoots of aquatic plants; invertebrates	1 year	Monogamy	Loosely built, over-water platform, sometimes with loose canopy	Probably long-term, with care afforded by older siblings in some *P. martinica*	Both sexes construct nest(s), incubate, and rear brood(s)	—
*Gymnocrex**	Largely sedentary	Variably moist forest; and swamps	Invertebrates, notably insects	Not known	Not known	Little, contradictory information	Not known	Not known	—
Habroptila	Probably sedentary	Dense swamps or marshes	Unknown	Not known	Not known	Not known	Not known	Not known	None documented
Eulabeornis	Probably sedentary	Mangroves, estuarian marshes	Invertebrates, notably crustaceans	Not known	Monogamy	Bulky, loose, elevated platform	Not known	Apparently biparental	—
Aramides	Sedentary	Swamps, variably moist forest, mangroves, marshes	Invertebrates, some species feeding heavily on crabs	Not known	Monogamy in few species for which data collected	Bowl of vegetation in tree or shrub, often near/over water	Evidently long-term, probably lifelong	Both sexes construct nest, incubate, and rear brood(s)	—
Canirallus	Sedentary	Rainforest	Invertebrates; some amphibians, reptiles	Not known	Not known	Moderately elevated, concealed bowl of plant material	Not known	Not known	—
Anurolimnas	Sedentary	Moist forest and secondary growth	Probably invertebrates	Not known	Not known	Not known	Not known	Not known	—

APPENDIX 6. Continued.

Taxonomic group	Migratory status	Breeding habitat and seasonality	Primary diet	Age at maturity	Mating system	Typical nest site	Duration of pair bond	Parental investment	Changes associated with flightlessness
Amaurolimnas	Sedentary	Moist forests, swamps, and streamside thickets	Invertebrates (e.g., annelids, molluscs, insects); frogs	Not known	Probably monogamy	Single nest a loose cup of leaves in hollow stump	Not known	Not known	—
Rougetius	Sedentary	High-altitude marshes, grasslands, and moors	Invertebrates (insects, crustaceans, and molluscs)	Not known	Probably monogamy	Pad of rushes near or over water	Not known; evidence of cooperative brood-rearing	Both sexes incubate and attend brood	—
Rallina	Most species at least partly migratory	Rainforest, swamps, and marshes	Invertebrates, including annelids, molluscs, and insects	Not known	Not known	Untidy, variably elevated, shallow bowl of vegetation	Not known	Both sexes incubate, and evidently both attend brood	—
Rallicula	Evidently sedentary	Montane forest	Evidently invertebrates	Not known	Not known; possibly polyandrous	Elevated, domed structure of moss, ferns, and leaves	Not known	Poorly known; both sexes of *R. rubra* incubate	—
Sarothrura	Most sedentary or with only altitudinal shifts; *S. boehmi* and *S. ayresi* migratory	Variable; lowland forest, swamps, brush, grasslands	Notably annelids, insects, gastropods	1 year	Monogamy	Concealed, shallow cups of grass, often with loose canopy	Probably long-term	Both sexes construct nest, incubate, and rear brood(s)	—
Cyanolimnas	Sedentary	Dense, brushy swamp	Not known	Not known	Not known	Overwater hummock of saw grass	Not known	Not known	None evident
Pardirallus	Sedentary, but exceptional vagrancy	Marshes, paddies, and swamps	Invertebrates	Not known	Monogamy	Slightly elevated platform or shallow bowl of grass	Not known	Not known	—

APPENDIX 6. Continued.

Taxonomic group	Migratory status	Breeding habitat and seasonality	Primary diet	Age at maturity	Mating system	Typical nest site	Duration of pair bond	Parental investment	Changes associated with flightlessness
Ortygonax	Possibly locally migratory	Marshes, densely vegetated ditches, and sparse swamps	Invertebrates	Not known	Probably monogamy	Poorly built bowl of grass on ground or low stump	Not known	Both sexes care for brood	—
Dryolimnas	Sedentary	Variable; forests, marshes, mangroves	Invertebrates	1 year	Monogamy	Variably robust, sessile bowl	Permanent	Both sexes care for brood	None evident
Rallus†	Only *R. limicola*, *R. aquaticus*-group, and *R. elegans* migratory	Marshes, moist meadows, mangroves, or swamps	Invertebrates, some small aquatic vertebrates and plants	1 year	Monogamy	Concealed platform or cup on ground or in vegetation, often with canopy	Seasonal, generally lasting only during period of territoriality	Both sexes build nest, incubate eggs, and care for brood	—
Gallirallus	Sedentary, except some *G. philippensis*-group	Marshes, forest, meadows, swamps, mangroves, or ponds	Invertebrates, some plants and small vertebrates	1 year, less in *R. owstoni*	Monogamy	Platform of vegetation on ground, in brush, under log, or in cavity, sometimes with canopy	Evidently permanent	Both sexes build nest, incubate eggs, and care for brood	Great variability in breeding season in *R. australis*-group, protracted ontogeny in *R. australis*, early maturity in *R. owstoni*
Tricholimnas	Sedentary	Dense, tropical forest	Invertebrates	1 year	Monogamy	Shallow, plant-lined depression	Permanent; older siblings assist with brood-rearing	Both sexes incubate and care for brood	All members flightless
Nesoclopeus	Sedentary	Lowland, typically moist forest	Invertebrates, small reptiles and frogs	Not known	Not known	Shallow bowl of plant material on ground	Not known	Not known	All members flightless
Aramidopsis	Sedentary	Dense bamboo at margins of forests	Invertebrates, principally crabs	Not known	Not known	Not known	Not known	Not known	All members flightless

APPENDIX 6. Continued.

Taxonomic group	Migratory status	Breeding habitat and seasonality	Primary diet	Age at maturity	Mating system	Typical nest site	Duration of pair bond	Parental investment	Changes associated with flightlessness
Cabalus‡	Sedentary	Forest	Invertebrates	Not known	Not known	Reportedly in preexisting holes in the ground	Not known	Not known	All members flightless
Habropteryx	Sedentary	Damp forest, grassland, marsh or swamp	Invertebrates, some small vertebrates, plants	Not known	Evidently monogamy	Little known; nest placed on ground	Probably permanent	Not known, but likely biparental throughout	None evident
Atlantisia	Sedentary	Tussock grass, sedge, ferns, thickets	Invertebrates, some seeds, fruits	1 or 2 years, based on subadult plumage	Monogamy	Ovate, exposed or covered bowl on ground, often with canopy, access ramp or tunnel	Permanent, but territories dynamic; older young may assist with siblings	Both sexes construct nest, incubate, and care for brood	Possibly increased group size and developmental periods
Laterallus	Sedentary, except *L. jamaicensis*	Meadows, marshes, and swamps, especially ecotonal	Invertebrates, principally insects; some plants	1 year likely	Evidently monogamy, but data few for many species	Dense, domed, elevated cup of vegetation with side entrance	Probably long-term, except perhaps in *L. jamaicensis*	Both sexes construct nest, incubate, and care for brood	—
Coturnicops	Migratory	Marsh, meadow, swamp, or savanna	Invertebrates and plant seeds	Not known	Little known; polygyny in some *C. novaboracensis*	Canopied cup of vegetation on ground	Seasonal, overlapping in some	Both sexes build nest; female alone incubates and broods	—
Micropygia§	Sedentary	Dense, variably moist grasslands	Invertebrates, principally insects	Not known	Monogamy	Dense, canopied, grass cup with side entrance	Permanent, possibly territorial year-round	Little known; female incubates	—
Crex	Migratory	Moist meadows and grasslands	Invertebrates, some seeds and small vertebrates	1 year	Monogamy typically, but serial polygyny not uncommon	Shallow, sessile or slightly elevated, sometimes canopied cup	Seasonal	Both sexes build nest; female alone incubates and cares for brood	—

APPENDIX 6. Continued.

Taxonomic group	Migratory status	Breeding habitat and seasonality	Primary diet	Age at maturity	Mating system	Typical nest site	Duration of pair bond	Parental investment	Changes associated with flightlessness
Porzana‖	Sedentary, except northern populations of *P. parva*, *P. pusilla*, *P. porzana*, *P. carolina*, *P. fusca*, *P. paykullii*, and southern *P. marginalis*	Freshwater wetlands, moist woodlands, peat bogs, interdune pans, irrigated croplands, mangroves, laggoons, forests	Invertebrates (principally insects); some small vertebrates and vegetation	1 year	Monogamous in species documented in wild; *P. flavirostra* and *P. marginalis* showing serial polyandry in captivity	Bulky bowl of vegetation, on ground, elevated, or floating, some concealed by loose canopy	Seasonal in migratory populations, more protracted or permanent in sedentary taxa	Where known, both sexes participate in nest building, incubation, and brood rearing	Habitats of *P. atra* and possibly *P. monasa* diverse, including dry forests and shorelines
Amaurornis¶	Sedentary, except some northern populations of *A. phoenicurus*	Moist meadows, brushy thickets, forest edges, mangroves, swamps	Invertebrates, some seeds and shoots	Not known	Monogamy, at least in *A. akool* and *A. moluccanus*	Slightly concave, moderately elevated, typically canopied platform	Probably permanent; at least seasonally territorial	Nest construction to attendance of broods probably biparental	Mangrove habitat atypical of genus; adept, swift climber; kicks defensively
Gallicrex	Migratory	Swamp, marsh, or moist fields	Seeds and shoots of plants; insects	Not known	Monogamy	Large, low, concave platform with canopy	Seasonal; frequently two broods per year	Not known	—
Porphyriops	Sedentary	Ponds, marshes, lagoons, meadows	Not known	Not known	Not known	Half-domed, concealed platform built on ground	Not known	Not known	—
Pareudiastes	Sedentary	Dense undergrowth of steeply sloping forest; possibly nocturnal	Invertebrates	Not known	Not known	Possibly burrows	Not known	Not known	Both members flightless (see text)

APPENDIX 6. Continued.

Taxonomic group	Migratory status	Breeding habitat and seasonality	Primary diet	Age at maturity	Mating system	Typical nest site	Duration of pair bond	Parental investment	Changes associated with flightlessness
Tribonyx	Sedentary, but some irregular movements, and *G. ventralis* subject to marked, seasonal irruptive movements	Variable, most types of freshwater or brackish wetlands	Invertebrates (many insects) and seeds of aquatic plants	1 year	*T. ventralis* evidently monogamous; *T. mortierii* forming trios including two male (rarely female) siblings	Variably elevated cup woven from vegetation, near water with ramp; strong coloniality in *T. ventralis*	At least seasonal, territorial during breeding; bonds probably long-term in *T. mortierii*	No data for *T. ventralis*; all duties shared by adult *T. mortierii*, often assisted by older offspring	*T. mortierii* distinctive in largely vegetable diet, breeding structure, and construction of multiple brood nests
Gallinula	Sedentary, except northern *G. chloropus*-group	Diversity of freshwater (rarely brackish) wetlands; *P. nesiotis*-group in tussock grass and dense brushland	Variable; both plants and invertebrates seasonally important	1 or 2 years	Monogamy in *G. chloropus*-group (polyandry rare), *G. angulata*, and *G. nesiotis*-group); promiscuity in *G. tenebrosa*	Variably elevated, cup-shaped platform of vegetation, typically with access ramp, sometimes with canopy	Variable duration of pair-bonds, all species territorial during breeding	Adults in pairs or groups share all duties, assisted by offspring of previous broods in at least *G. tenebrosa* and *G. chloropus*-group	*G. nesiotis*-group distinctive in drier habitats, eating carrion, protracted retention of juvenal plumage, and possibly permanent pair-bonds
Fulica	Sedentary, except northern *F. atra*-group, *F. americana*-group; irregular movements in some others	Comparatively open, freshwater (rarely brackish) pond, lakes, and marshes	Aquatic plants, often taken under water; *Phragmites* favored	1 year, perhaps older in larger species	Monogamy	Bulky, floating or ramped, often exposed platform of plants	At least seasonal, perhaps longer-term; marked territoriality during breeding	Both parents participate in all duties; sometimes older offspring assist with brood	Flightless members extinct

* Recently described *Gymnocrex talaudensis* (Lambert 1998a) is excluded because of a lack of reproductive data.
† Provisionally includes "*Rallus*" *madagascariensis*.
‡ Primarily taken from review by Greenway (1967).
§ Following Livezey (1998), includes *M. flaviventer*, included by many authorities within *Porzana*.
‖ Following the classification by Livezey (1998), includes *Porzana flavirostra*, *P. olivieri*, and *P. bicolor*, all of which were referred to *Amaurornis* by Taylor (1996), as well as *Neocrex*, *Poliolimnas*, and *Aenigmatolimnas*.
¶ Recently described *Amaurornis magnirostris* (Lambert 1998b) is excluded because of a lack of reproductive data. Following the classification by Livezey (1998), *Amaurornis* includes "*Megacrex*" *ineptus* as sole flightless member.

APPENDIX 7. Clutch sizes and dimensions of eggs of rails.

Modal clutch sizes, mean linear dimensions, and estimated egg masses of eggs. Modal clutch sizes are after Taylor (1996) and estimated egg masses are after Schönwetter (1962). Clutch mass is the product of mean egg mass and modal clutch size; relative egg and clutch masses are ratios of respective means divided by mean body mass for females (Appendix 1). Where tentative, entries are enclosed by square brackets. Excludes taxa for which egg measurements were not available; single relative egg mass of 27% attributed to *Rallicula rubra* by Taylor (1996) is excluded. Only mode is given if well documented; otherwise, range of clutch sizes is followed by provisional mode in parentheses, the latter being the median of the range, modified by available information (e.g., means).

Species	Clutch size (mode)*	Egg length × width (mm)	Mean egg mass Absolute (g)	Mean egg mass Relative (%)	Mean clutch mass Absolute (g)	Mean clutch mass Relative (%)
Himantornis haematopus	3 (3)	50.4 × 37.3	38.5	9.1	116	27
Porphyrio bellus	[2–6 (4)]	54.7 × 38.5	44.3	7.0	177	28
Porphyrio indicus	[2–6 (4)]	47.6 × 33.0	31.0	—	124	—
Porphyrio samoensis	[2–6 (4)]	49.4 × 35.4	34.1	5.6	136	22
Porphyrio porphyrio	[2–6 (4)]	54.6 × 37.3	41.5	5.7	166	23
Porphyrio poliocephalus	3–5 (4)	50.7 × 35.9	35.7	6.4	143	26
Porphyrio madagascariensis	3–7 (4)	53.7 × 37.2	40.6	7.3	162	29
Porphyrio ellioti	2–6 (4)	50.2 × 34.2	31.7	4.7	127	19
Porphyrio melanotus	[2–6 (4)]	49.4 × 35.5	34.1	3.9	136	16
Porphyrio melanopterus	2–6 (4)	48.5 × 34.8	32.3	—	129	—
Porphyrio hochstetteri	[2–6 (4)]	73.5 × 48.3	93.0	4.1	186	8
Porphyrula alleni	2 (2)	36.3 × 26.2	13.5	10.8	54	43
Porphyrula martinica	3–8 (4)	40.0 × 28.7	18.0	8.8	108	53
Porphyrula flavirostris	4–9 (6)	32.4 × 24.7	[11.0]	[12.5]	[44]	[50]
Gymnocrex plumbeiventris	4–5 (4)	40.4 × 30.7	20.9	[7.0]	—	—
Eulabeornis castaneoventris	4–5 (5)	51.7 × 35.9	36.8	5.9	184	29
Aramides saracura	4–5 (5)	50.8 × 36.2	36.4	9.3	182	47
Aramides ypecaha	4–5 (5)	57.0 × 38.0	45.0	6.9	225	35
Aramides mangle	—	51.0 × 35.2	34.6	21.1	—	—
Aramides axillaris	5 (5)	45.6 × 32.7	26.7	16.3	134	81
Aramides cajanea-group	3–7 (5)	49.3 × 34.9	33.0	8.6	165	43
Canirallus oculeus	2 (2)	43.3 × 31.8	23.8	8.6	48	17
Canirallus kioloides	2 (2)	36.0 × 26.6	13.8	5.0	28	10
Amaurolimnas concolor	4 (4)	33.0 × 26.5	13.0	13.5	52	54
Rougetius rougetii	4–5 (4)	35.3 × 24.7	11.8	6.9	47	28
Rallina eurizonoides-group	4–8 (6)	35.4 × 27.3	14.7	15.5	88	93
Rallina canningi	—	40.6 × 30.8	21.1	—	—	—
Rallina fasciata	3–6 (5)	30.6 × 23.7	9.4	12.1	47	60

APPENDIX 7. Continued.

Species	Clutch size (mode)*	Egg length × width (mm)	Mean egg mass Absolute (g)	Mean egg mass Relative (%)	Mean clutch mass Absolute (g)	Mean clutch mass Relative (%)
Rallina tricolor-group	5 (5)	36.9 × 27.4	15.2	8.6	76	43
Sarothrura pulchra	2 (2)	30.0 × 22.0	7.7	17.5	15	35
Sarothrura elegans	3–5 (4)	28.0 × 21.7	7.0	17.1	28	68
Sarothrura rufa	2–3 (3)	27.4 × 20.5	6.1	16.1	18	51
Sarothrura lugens	—	27.7 × 20.3	6.0	—	—	—
Sarothrura insularis	3–4 (3)	26.8 × 20.7	6.0	20.0	18	60
Pardirallus maculatus	2–7 (5)	37.9 × 28.2	16.7	10.1	84	50
Ortygonax sanguinolentus	4–6 (5)	42.2 × 30.9	22.4	9.6	112	48
Ortygonax nigricans	3 (3)	40.3 × 30.7	20.7	—	62	—
Dryolimnas cuvieri	3–6 (~4)	40.8 × 31.4	22.0	9.9	88	39
Dryolimnas aldabranus	3–6 (~4)	43.2 × 29.6	21.0	12.2	84	48
"*Rallus*" *madagascariensis*	—	42.7 × 33.1	25.5	24.3	—	—
Rallus caerulescens	2–6 (4)	36.2 × 27.3	14.8	10.1	59	41
Rallus aquaticus-group	6–11 (9)	36.0 × 25.7	12.7	12.5	114	112
Rallus limicola	8–9 (9)	31.3 × 23.4	9.4	12.5	85	113
Rallus longirostris-group	[3–14] (10)	42.4 × 29.6	19.8	7.4	198	74
Rallus elegans-group	10–12 (11)	43.0 × 30.2	20.8	6.6	229	73
Gallirallus pectoralis	3–8 (6)	34.6 × 25.9	13.0	15.9	78	95
Gallirallus striatus	5–9 (7)	34.9 × 26.7	13.7	12.6	96	88
Gallirallus australis	2–4 (3)	56.5 × 40.0	50.7	6.9	152	21
Gallirallus greyi	2–4 (3)	60.0 × 40.4	54.0	7.7	162	23
Gallirallus philippensis-group	4–8 (6)	37.6 × 28.2	16.6	8.8	100	53
Gallirallus owstoni	3–4 (3)	38.5 × 28.7	[17.0]	[7.5]	[510]	[22]
Tricholimnas lafresnayanus	—	55.9 × 33.4	34.3	—	—	—
Tricholimnas sylvestris	1–4 (3)	48.5 × 34.5	31.8	70	95	21
Nesoclopeus poecilopterus	4–6 (5)	49.8 × 36.6	35.4	—	177	—
Cabalus modestus	—	38.5 × 28.0	16.5	—	—	—
Habropteryx insignis	—	41.5 × 31.5	22.7	—	—	—
Habropteryx torquatus-group	3 (3)	39.2 × 28.7	17.9	7.9	54	24
Habropteryx okinawae	2–3 (3)	—	—	—	—	—
Atlantisia rogersi	2 (2)	33.4 × 23.0	9.4	25.4	19	51
Laterallus jamaicensis	6–7 (6)	25.5 × 19.4	5.1	15.0	31	90
Laterallus murivagans	—	26.9 × 20.3	6.0	—	—	—
Laterallus viridis	1–3 (2)	32.6 × 24.1	10.2	19.2	20	39
Laterallus melanophaius	2–3 (3)	29.9 × 23.3	8.8	17.3	26	52
Laterallus fasciatus	—	31.0 × 22.8	8.7	13.4	—	—

APPENDIX 7. Continued.

Species	Clutch size (mode)*	Egg length × width (mm)	Mean egg mass Absolute (g)	Mean egg mass Relative (%)	Mean clutch mass Absolute (g)	Mean clutch mass Relative (%)
Laterallus leucopyrrhus	2–3 (3)	33.0 × 24.8	11.0	26.2	33	79
Laterallus albigularis-group	2–5 (3)	31.2 × 23.8	9.5	21.1	29	63
Laterallus exilis	3 (3)	31.2 × 23.2	9.2	25.6	28	77
Coturnicops noveboracensis	8 (8)	28.3 × 20.7	6.5	14.1	52	112
Coturnicops exquisitus	—	29.1 × 21.0	7.0	—	—	—
Micropygia schomburgkii	—	32.6 × 23.7	10.0	—	—	—
Crex crex	2 (2)	31.9 × 22.8	8.3	34.6	17	69
Crex egregia	8–12 (10)	36.9 × 26.3	13.2	9.6	132	96
Crex albicollis	3–9 (6)	34.2 × 25.2	12.0	12.9	72	77
Porzana fusca	2–3 (3)	34.1 × 26.2	12.9	13.6	39	41
Porzana paykullii	3–9 (6)	31.8 × 23.0	9.2	14.8	55	89
Porzana cinerea	5–9 (7)	34.0 × 24.4	11.0	11.6	77	81
Porzana marginalis	4 (4)	29.6 × 22.4	8.1	12.3	32	49
Porzana parva	4–5 (5)	29.3 × 21.7	7.3	12.0	37	60
Porzana palmeri	7–9 (8)	30.4 × 21.7	7.9	16.1	63	129
Porzana porzana	—	28.6 × 21.3	7.0	—	—	—
Porzana carolina	8–12 (10)	33.5 × 24.3	10.9	11.6	109	116
Porzana pusilla	10–12 (11)	31.6 × 22.6	8.9	12.9	98	142
Porzana fluminea	4–11 (7)	27.9 × 20.2	6.2	17.2	43	121
Porzana tabuensis	3–7 (5)	30.9 × 22.9	8.9	15.1	45	75
Porzana atra†	3–4 (4)	30.1 × 22.4	8.4	21.5	34	86
Porzana flavirostra	2–3 (2)	35.1 × 25.2	[13.0]	[17.6]	[26]	[35]
Porzana bicolor	3 (3)	31.7 × 23.7	9.7	12.4	29	37
Porzana erythrops	5–8 (6)	33.9 × 26.1	12.6	—	76	—
Amaurornis olivaceus	3–7 (5)	30.0 × 21.8	8.4	12.0	42	60
Amaurornis moluccanus	—	42.2 × 31.5	23.2	9.1	—	—
Amaurornis ruficrissus	4–7 (6)	38.7 × 29.0	18.0	9.9	108	60
Amaurornis phoenicurus	—	39.2 × 28.0	16.9	9.3	—	—
Amaurornis akool	4–9 (6)	39.9 × 29.7	19.4	13.1	116	79
Amaurornis isabellinus	5–6 (6)	36.0 × 27.0	14.5	11.6	87	70
Amaurornis ineptus	2–5 (5)	40.6 × 31.5	22.0	—	—	—
Gallicrex cinerea	[3]	—	—	—	—	967
	3–6 (5)	42.3 × 31.1	22.5	6.6	113	33

APPENDIX 7. Continued.

Species	Clutch size (mode)*	Egg length × width (mm)	Mean egg mass Absolute (g)	Mean egg mass Relative (%)	Mean clutch mass Absolute (g)	Mean clutch mass Relative (%)
Porphyriops melanops	4–8 (6)	40.3 × 28.9	18.9	13.8	113	83
Pareudiastes pacificus	[2]	45.7 × 31.8	25.0	—	[50]	—
Tribonyx ventralis	5–7 (6)	45.0 × 31.4	24.0	6.6	144	40
Tribonyx mortierii	5–8 (6)	56.0 × 38.7	45.0	3.6	270	22
Gallinula nesiotis-group	2–5 (4)	49.5 × 33.0	28.5	12.4	114	50
Gallinula tenebrosa	8 (8)	47.1 × 33.3	28.8	5.8	230	47
Gallinula angulata	5 (5)	34.1 × 24.8	11.2	9.0	56	45
Gallinula chloropus	5–9 (6)	41.9 × 29.8	20.7	7.8	124	47
Gallinula pyrrhorrhoa	3–9 (6)	45.0 × 31.5	24.5	—	147	—
Gallinula cachinnans	7–9 (8)	43.9 × 30.8	22.8	7.0	182	56
Gallinula galeata	[7–9 (8)]	47.5 × 33.2	28.0	7.0	224	56
Gallinula sandvicensis	4–8 (6)	44.1 × 31.5	24.6	—	148	—
Fulica rufifrons	2–9 (6)	55.5 × 37.2	42.0	7.8	252	47
Fulica armillata	2–8 (5)	58.0 × 39.5	49.6	5.6	248	28
Fulica leucoptera	10–12 (11)	48.8 × 33.2	29.3	5.8	322	63
Fulica cornuta	3–5 (4)	66.2 × 47.1	[80.0]	[4.5]	320	18
Fulica gigantea	3–7 (4)	67.0 × 45.1	74.3	3.5	297	14
Fulica ardesiaca	4–5 (4)	59.2 × 40.1	51.9	5.8	208	23
Fulica alai	5 (5)	46.6 × 33.1	28.0	—	140	—
Fulica americana	6–15 (10)	48.5 × 33.0	28.9	5.4	289	54
Fulica caribaea	4–7 (5)	48.0 × 34.7	32.0	6.2	160	39
Fulica atra, atra-group	6–10 (8)	52.5 × 35.8	36.5	4.8	292	39
Fulica atra, australis-group	6–10 (8)	49.3 × 34.2	31.2	6.2	250	50
Fulica cristata	5–7 (6)	54.2 × 37.3	41.0	6.0	246	36

* Clutch sizes and measurements of eggs taken from Byrd and Zeillemaker (1981).
† Includes data from Jones et al. (1995).